COMBINATORICS

Second Edition

DISCRETE MATHEMATICS AND ITS APPLICATIONS

Series Editors

Miklos Bona
Patrice Ossona de Mendez
Douglas West

R. B. J. T. Allenby and Alan Slomson, How to Count: An Introduction to Combinatorics, Third Edition

Craig P. Bauer, Secret History: The Story of Cryptology

Jürgen Bierbrauer, Introduction to Coding Theory, Second Edition

Katalin Bimbó, Combinatory Logic: Pure, Applied and Typed

Katalin Bimbó, Proof Theory: Sequent Calculi and Related Formalisms

Donald Bindner and Martin Erickson, A Student's Guide to the Study, Practice, and Tools of Modern Mathematics

Francine Blanchet-Sadri, Algorithmic Combinatorics on Partial Words

Miklós Bóna, Combinatorics of Permutations, Second Edition

Miklós Bóna, Handbook of Enumerative Combinatorics

Miklós Bóna, Introduction to Enumerative and Analytic Combinatorics, Second Edition

Jason I. Brown, Discrete Structures and Their Interactions

Richard A. Brualdi and Dragoš Cvetković, A Combinatorial Approach to Matrix Theory and Its Applications

Kun-Mao Chao and Bang Ye Wu, Spanning Trees and Optimization Problems

Charalambos A. Charalambides, Enumerative Combinatorics

Gary Chartrand and Ping Zhang, Chromatic Graph Theory

Henri Cohen, Gerhard Frey, et al., Handbook of Elliptic and Hyperelliptic Curve Cryptography

Charles J. Colbourn and Jeffrey H. Dinitz, Handbook of Combinatorial Designs, Second Edition

Titles (continued)

Abhijit Das, Computational Number Theory

Matthias Dehmer and Frank Emmert-Streib, Quantitative Graph Theory: Mathematical Foundations and Applications

Martin Erickson, Pearls of Discrete Mathematics

Martin Erickson and Anthony Vazzana, Introduction to Number Theory

Steven Furino, Ying Miao, and Jianxing Yin, Frames and Resolvable Designs: Uses, Constructions, and Existence

Mark S. Gockenbach, Finite-Dimensional Linear Algebra

Randy Goldberg and Lance Riek, A Practical Handbook of Speech Coders

Jacob E. Goodman and Joseph O'Rourke, Handbook of Discrete and Computational Geometry, Second Edition

Jonathan L. Gross, Combinatorial Methods with Computer Applications

Jonathan L. Gross and Jay Yellen, Graph Theory and Its Applications, Second Edition

Jonathan L. Gross, Jay Yellen, and Ping Zhang Handbook of Graph Theory, Second Edition

David S. Gunderson, Handbook of Mathematical Induction: Theory and Applications

Richard Hammack, Wilfried Imrich, and Sandi Klavžar, Handbook of Product Graphs, Second Edition

Darrel R. Hankerson, Greg A. Harris, and Peter D. Johnson, Introduction to Information Theory and Data Compression, Second Edition

Darel W. Hardy, Fred Richman, and Carol L. Walker, Applied Algebra: Codes, Ciphers, and Discrete Algorithms, Second Edition

Daryl D. Harms, Miroslav Kraetzl, Charles J. Colbourn, and John S. Devitt, Network Reliability: Experiments with a Symbolic Algebra Environment

Silvia Heubach and Toufik Mansour, Combinatorics of Compositions and Words

Leslie Hogben, Handbook of Linear Algebra, Second Edition

Derek F. Holt with Bettina Eick and Eamonn A. O'Brien, Handbook of Computational Group Theory

David M. Jackson and Terry I. Visentin, An Atlas of Smaller Maps in Orientable and Nonorientable Surfaces

Richard E. Klima, Neil P. Sigmon, and Ernest L. Stitzinger, Applications of Abstract Algebra with Maple™ and MATLAB®, Second Edition

Richard E. Klima and Neil P. Sigmon, Cryptology: Classical and Modern with Maplets

Patrick Knupp and Kambiz Salari, Verification of Computer Codes in Computational Science and Engineering

William L. Kocay and Donald L. Kreher, Graphs, Algorithms, and Optimization, Second Edition

Donald L. Kreher and Douglas R. Stinson, Combinatorial Algorithms: Generation Enumeration and Search

Titles (continued)

Hang T. Lau, A Java Library of Graph Algorithms and Optimization

C. C. Lindner and C. A. Rodger, Design Theory, Second Edition

San Ling, Huaxiong Wang, and Chaoping Xing, Algebraic Curves in Cryptography

Nicholas A. Loehr, Bijective Combinatorics

Nicholas A. Loehr, Combinatorics, Second Edition

Toufik Mansour, Combinatorics of Set Partitions

Toufik Mansour and Matthias Schork, Commutation Relations, Normal Ordering, and Stirling Numbers

Alasdair McAndrew, Introduction to Cryptography with Open-Source Software

Pierre-Loïc Méliot, Representation Theory of Symmetric Groups

Elliott Mendelson, Introduction to Mathematical Logic, Fifth Edition

Alfred J. Menezes, Paul C. van Oorschot, and Scott A. Vanstone, Handbook of Applied Cryptography

Stig F. Mjølsnes, A Multidisciplinary Introduction to Information Security

Jason J. Molitierno, Applications of Combinatorial Matrix Theory to Laplacian Matrices of Graphs

Richard A. Mollin, Advanced Number Theory with Applications

Richard A. Mollin, Algebraic Number Theory, Second Edition

Richard A. Mollin, Codes: The Guide to Secrecy from Ancient to Modern Times

Richard A. Mollin, Fundamental Number Theory with Applications, Second Edition

Richard A. Mollin, An Introduction to Cryptography, Second Edition

Richard A. Mollin, Quadratics

Richard A. Mollin, RSA and Public-Key Cryptography

Carlos J. Moreno and Samuel S. Wagstaff, Jr., Sums of Squares of Integers

Gary L. Mullen and Daniel Panario, Handbook of Finite Fields

Goutam Paul and Subhamoy Maitra, RC4 Stream Cipher and Its Variants

Dingyi Pei, Authentication Codes and Combinatorial Designs

Kenneth H. Rosen, Handbook of Discrete and Combinatorial Mathematics

Yongtang Shi, Matthias Dehmer, Xueliang Li, and Ivan Gutman, Graph Polynomials

Douglas R. Shier and K.T. Wallenius, Applied Mathematical Modeling: A Multidisciplinary Approach

Alexander Stanoyevitch, Introduction to Cryptography with Mathematical Foundations and Computer Implementations

Jörn Steuding, Diophantine Analysis

Douglas R. Stinson, Cryptography: Theory and Practice, Third Edition

Titles (continued)

Roberto Tamassia, Handbook of Graph Drawing and Visualization

Roberto Togneri and Christopher J. deSilva, Fundamentals of Information Theory and Coding Design

W. D. Wallis, Introduction to Combinatorial Designs, Second Edition

W. D. Wallis and J. C. George, Introduction to Combinatorics, Second Edition

Jiacun Wang, Handbook of Finite State Based Models and Applications

Lawrence C. Washington, Elliptic Curves: Number Theory and Cryptography, Second Edition

DISCRETE MATHEMATICS AND ITS APPLICATIONS

COMBINATORICS

Second Edition

Nicholas A. Loehr

CRC Press is an imprint of the
Taylor & Francis Group, an **informa** business
A CHAPMAN & HALL BOOK

CRC Press
Taylor & Francis Group
6000 Broken Sound Parkway NW, Suite 300
Boca Raton, FL 33487-2742

First issued in paperback 2022

Version Date: 20170712

ISBN 13: 978-1-03-247671-1 (pbk)
ISBN 13: 978-1-4987-8025-4 (hbk)

DOI: 10.1201/9781315153360

Library of Congress Cataloging-in-Publication Data

Names: Loehr, Nicholas A. | Loehr, Nicholas A. Combinatorics.
Title: Combinatorics / Nicholas A. Loehr.
Description: Second edition. | Boca Raton : CRC Press, 2017. | Previous edition: Bijective combinatorics / Nicholas A. Loehr (Boca Raton, FL : Chapman & Hall/CRC, c2011).
Identifiers: LCCN 2017011283 | ISBN 9781498780254
Subjects: LCSH: Combinatorial analysis.
Classification: LCC QA164 .L64 2017 | DDC 511/.62--dc23
LC record available at https://lccn.loc.gov/2017011283

Visit the Taylor & Francis Web site at
http://www.taylorandfrancis.com

and the CRC Press Web site at
http://www.crcpress.com

Dedication

This book is dedicated to

my sister Heather (1973–2015),
my aunt Nanette (1963–2011),
and my grandmother Oliva (1926–1982).

Contents

Preface to the Second Edition

This book presents a general introduction to enumerative, bijective, and algebraic combinatorics. *Enumerative combinatorics* is the mathematical theory of counting. This branch of discrete mathematics has flourished in the last few decades due to its many applications to probability, computer science, engineering, physics, and other areas. *Bijective combinatorics* produces elegant solutions to counting problems by setting up one-to-one correspondences (bijections) between two sets of combinatorial objects. *Algebraic combinatorics* uses combinatorial methods to obtain information about algebraic structures such as permutations, polynomials, matrices, and groups. This relatively new subfield of combinatorics has had a profound influence on classical mathematical subjects such as representation theory and algebraic geometry.

Part I of the text covers fundamental counting tools including the Sum and Product Rules, binomial coefficients, recursions, bijective proofs of combinatorial identities, enumeration problems in graph theory, inclusion-exclusion formulas, generating functions, ranking algorithms, and successor algorithms. This part requires minimal mathematical prerequisites and could be used for a one-semester combinatorics course at the advanced undergraduate or beginning graduate level. This material will be interesting and useful for computer scientists, statisticians, engineers, and physicists, as well as mathematicians.

Part II of the text contains an introduction to algebraic combinatorics, discussing groups, group actions, permutation statistics, tableaux, symmetric polynomials, and formal power series. My presentation of symmetric polynomials is more combinatorial (and, I hope, more accessible) than the standard reference work [84]. In particular, a novel approach based on antisymmetric polynomials and abaci yields elementary combinatorial proofs of some advanced results such as the Pieri Rules and the Littlewood–Richardson Rule for multiplying Schur symmetric polynomials. Part II assumes a bit more mathematical sophistication on the reader's part (mainly some knowledge of linear algebra) and could be used for a one-semester course for graduate students in mathematics and related areas. Some relevant background material from abstract algebra and linear algebra is reviewed in an appendix. The final chapter consists of independent sections on optional topics that complement material in the main text. In many chapters, some of the harder material in later sections can be omitted without loss of continuity.

Compared to the first edition, this new edition has an earlier, expanded treatment of generating functions that focuses more on the combinatorics and applications of generating functions and less on the algebraic formalism of formal power series. In particular, we provide greater coverage of exponential generating functions and the use of generating functions to solve recursions, evaluate summations, and enumerate complex combinatorial structures. We cover successor algorithms in more detail in Chapter 6, providing automatic methods to create these algorithms directly from counting arguments based on the Sum and Product Rules. The final chapter contains some new material on quasisymmetric polynomials. Many chapters in Part I have been reorganized to start with elementary content most pertinent to solving applied problems, deferring formal proofs and advanced material until later. I hope this restructuring makes the second edition more readable and appealing than the first edition, without sacrificing mathematical rigor.

Each chapter ends with a summary, a set of exercises, and bibliographic notes. The book contains over 1200 exercises, ranging in difficulty from routine verifications to unsolved problems. Although we provide references to the literature for some of the major theorems and harder problems, no attempt has been made to pinpoint the original source for every result appearing in the text and exercises.

I am grateful to the editors, reviewers, and other staff at CRC Press for their help with the preparation of this second edition. Readers may communicate errors and other comments to the author by sending e-mail to `nloehr@vt.edu`.

Nicholas A. Loehr

Introduction

The goal of *enumerative combinatorics* is to count the number of objects in a given finite set. This may seem like a simple task, but the sets we want to count are often very large and complicated. Here are some examples of enumeration problems that can be solved using the techniques in this book. How many encryption keys are available using the 128-bit AES encryption algorithm? How many strands of DNA can be built using five copies of each of the nucleotides adenine, cytosine, guanine, and thymine? How many ways can we be dealt a full house in five-card poker? How many ways can we place five rooks on a chessboard with no two rooks in the same row or column? How many subsets of $\{1, 2, \ldots, n\}$ contain no two consecutive integers? How many ways can we write 100 as a sum of positive integers? How many connected graphs on n vertices have no cycles? How many integers between 1 and n are relatively prime to n? How many circular necklaces can be made with three rubies, two emeralds, and one diamond if all rotations of a given necklace are considered the same? How many ways can we tile a chessboard with dominos? The answers to such counting questions can help us solve a wide variety of problems in probability, cryptography, algorithm analysis, physics, abstract algebra, and other areas of mathematics.

Part I of this book develops the basic principles of counting, placing particular emphasis on the role of *bijections*. To give a *bijective proof* that a given set S has size n, one must construct an explicit one-to-one correspondence (bijection) from S onto the set $\{1, 2, \ldots, n\}$. More generally, one can prove that two sets A and B have the same size by exhibiting a bijection between A and B. For example, given a fixed positive integer n, let A be the set of all strings $w_1 w_2 \cdots w_{2n}$ consisting of n left parentheses and n right parentheses that are *balanced* (every left parenthesis can be matched to a right parenthesis later in the sequence). Let B be the set of all arrays

$$\begin{bmatrix} y_1 & y_2 & \cdots & y_n \\ z_1 & z_2 & \cdots & z_n \end{bmatrix}$$

such that every number in $\{1, 2, \ldots, 2n\}$ appears once in the array, $y_1 < y_2 < \cdots < y_n$, $z_1 < z_2 < \cdots < z_n$, and $y_i < z_i$ for every i. The sets A and B seem quite different at first glance. Yet, we can demonstrate that A and B have the same size using the following bijection. Given $w = w_1 w_2 \cdots w_{2n}$ in A, let $y_1, y_2, \ldots, y_n$ be the positions of the left parentheses in w (written in increasing order), and let $z_1, z_2, \ldots, z_n$ be the positions of the right parentheses in w (written in increasing order). For example, the string `(()())((()))()` in A maps to the array

$$\begin{bmatrix} 1 & 2 & 4 & 7 & 8 & 9 & 13 \\ 3 & 5 & 6 & 10 & 11 & 12 & 14 \end{bmatrix}.$$

One may check that the requirement $y_i < z_i$ for all i is equivalent to the fact that w is a *balanced* string of parentheses. The string w is uniquely determined by the array of y_i's and z_i's, and every such array arises from some string w in A. Thus we have defined the required one-to-one correspondence between A and B. We now know that the sets A and B have the same size, although we have not yet determined what that size is!

Bijective proofs, while elegant, can be very difficult to discover. For example, let C be the set of rearrangements of $1, 2, \ldots, n$ that have no decreasing subsequence of length three. It turns out that the sets B and C have the same size, so there must exist a bijection from

B to C. Can you find one? (Before spending too long on this question, you might want to read §12.13.)

Luckily, the field of enumerative combinatorics contains a whole arsenal of techniques to help us solve complicated counting problems. Besides bijections, some of these techniques include recursions, generating functions, group actions, inclusion-exclusion formulas, linear algebra, probabilistic methods, and symmetric polynomials. In the rest of this introduction, we describe several challenging enumeration problems that can be solved using these more advanced methods. These problems, and the combinatorial technology needed to solve them, will be discussed at greater length later in the text.

Standard Tableaux

Suppose we are given a diagram D consisting of a number of rows of boxes, left-justified, with each row no longer than the one above it. For example, consider this diagram:

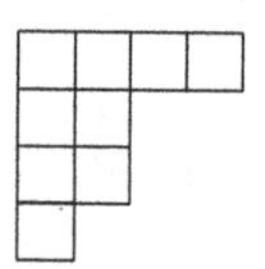

Let n be the total number of boxes in the diagram. A *standard tableau of shape D* is a filling of the boxes in D with the numbers $1, 2, \ldots, n$ (used once each) so that every row forms an increasing sequence (reading left to right), and every column forms an increasing sequence (reading top to bottom). For example, here are three standard tableaux of shape D, where D is the diagram pictured above:

1	2	3	4
5	6		
7	8		
9			

1	3	4	9
2	5		
6	7		
8			

1	2	3	7
4	8		
5	9		
6			

Question: *Given a diagram D of n cells, how many standard tableaux of shape D are there?*

There is a remarkable answer to this counting problem, known as the *Hook-Length Formula.* To state it, we need to define hooks and hook lengths. The *hook* of a box b in a diagram D consists of all boxes to the right of b in its row, all boxes below b in its column, and box b itself. The *hook length* of b, denoted $h(b)$, is the number of boxes in the hook of b. For example, if b is the first box in the second row of D, then the hook of b consists of the marked boxes in the following picture:

•	•		
•			
•			

So $h(b) = 4$. In the picture below, we have labeled each box in D with its hook length.

7	5	2	1
4	2		
3	1		
1			

Hook-Length Formula: *Given a diagram D of n cells, the number of standard tableaux of shape D is $n!$ divided by the product of the hook lengths of all the boxes in D.*

For the diagram D in our example, the formula says there are exactly

$$\frac{9!}{7\cdot 5\cdot 2\cdot 1\cdot 4\cdot 2\cdot 3\cdot 1\cdot 1} = 216$$

standard tableaux of shape D. Observe that the set B of $2\times n$ arrays (discussed above) can also be enumerated with the aid of the Hook-Length Formula. In this case, the diagram D consists of two rows of length n. The hook lengths for boxes in the top row are $n+1$, n, $n-1$, ..., 2, while the hook lengths in the bottom row are $n, n-1, \ldots, 1$. Since there are $2n$ boxes in D, the Hook-Length Formula asserts that

$$|B| = \frac{(2n)!}{(n+1)\cdot n\cdot (n-1)\cdot \ldots \cdot 2\cdot n\cdot (n-1)\cdot \ldots \cdot 1} = \frac{(2n)!}{(n+1)!n!}.$$

The fraction on the right side is an integer called the *nth Catalan number*. Since we previously displayed a bijection between B and A (the set of strings of balanced parentheses), we conclude that the size of A is also given by a Catalan number. As we will see, many different types of combinatorial structures are counted by the Catalan numbers.

How is the Hook-Length Formula proved? Many proofs of this formula have been found since it was originally discovered in 1954. There are algebraic proofs, probabilistic proofs, combinatorial proofs, and (relatively recently) bijective proofs of this formula. Here we discuss a *flawed* probabilistic argument that gives a little intuition for how the mysterious Hook-Length Formula arises. Suppose we choose a *random* filling F of the boxes of D with the integers $1, 2, \ldots, n$. What is the probability that this filling will actually be a standard tableau? We remark that the filling is standard if and only if for every box b in D, the entry in b is the smallest number in the hook of b. Since any of the boxes in the hook is equally likely to contain the smallest value, we see that the probability of this event is $1/h(b)$. Multiplying these probabilities together would give $1/\prod_{b\in D} h(b)$ as the probability that the random filling we chose is a standard tableau. Since the total number of possible fillings is $n!$ (by the Product Rule, discussed in Chapter 1), this leads us to the formula $n!/\prod_{b\in D} h(b)$ for the number of standard tableaux of shape D.

Unfortunately, the preceding argument contains a fatal error. The events "the entry in box b is the smallest in the hook of b," for various choices of b, are not necessarily *independent* (see §1.9). Thus we cannot find the probability that all these events occur by multiplying together the probabilities of each individual event. Nevertheless, remarkably, the final answer obtained by making this erroneous independence assumption turns out to be correct! The Hook-Length Formula can be justified by a more subtle probabilistic argument due to Greene, Nijenhuis, and Wilf. We describe this argument in §12.12.

Rook Placements

A *rook* is a chess piece that can travel any number of squares along its current row or column in a single move. We say that the rook *attacks* all the squares in its row and column. How many ways can we place eight rooks on an ordinary 8×8 chessboard so that no two rooks attack one another? The answer is $8! = 40,320$. More generally, we can show that there are $n!$ ways to place n non-attacking rooks on an $n\times n$ chessboard. To see this, first note that there must be exactly one rook in each of the n rows. The rook in the top row can occupy any of the n columns. The rook in the next row can occupy any of the $n-1$ columns not attacked by the first rook; then there are $n-2$ available columns for the next rook, and so on. By the Product Rule (discussed in Chapter 1), the total number of placements is therefore $n \times (n-1) \times (n-2) \times \cdots \times 1 = n!$.

Now consider an $(n+1) \times (n+1)$ chessboard with a *bishop* occupying the upper-left corner square. (A bishop is a chess piece that attacks all squares that can be reached from its current square by moving in a straight line northeast, northwest, southeast, or southwest along a diagonal of the chessboard.) **Question:** *How many ways can we place n rooks on this chessboard so that no two pieces attack one another?* An example of such a placement on a standard chessboard ($n+1=8$, so $n=7$) is shown below:

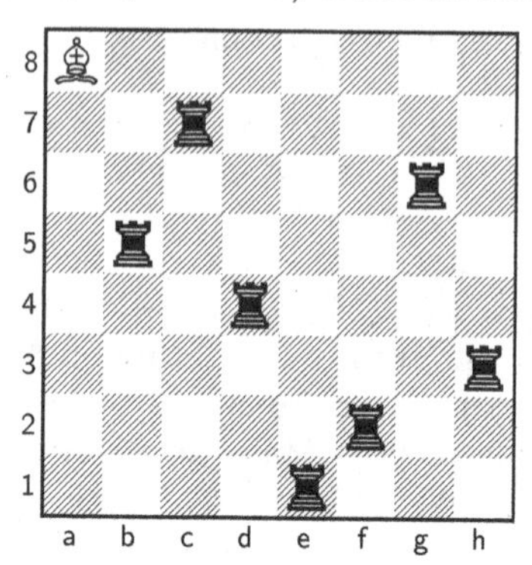

It turns out that *the number of non-attacking placements is the closest integer to $n!/e$*. Here, e is the famous constant $e = \sum_{k=0}^{\infty} 1/k! \approx 2.718281828$ that appears throughout the subject of calculus. When $n=7$, the number of placements is 1854 (note $7!/e = 1854.112\ldots$).

This answer follows from the Inclusion-Exclusion Formulas to be discussed in Chapter 4. We sketch the derivation now to indicate how the number e appears. First, there are $n!$ ways to place the n rooks on the board so that no two rooks attack each other, and no rook occupies the top row or the leftmost column (lest a rook attack the bishop). However, we have counted many configurations in which one or more rooks occupy the diagonal attacked by the bishop. To correct for this, we will subtract a term that accounts for configurations of this kind. We can build such a configuration by placing a rook in row i, column i, for some i between 2 and $n+1$, and then placing the remaining rooks in different rows and columns in $(n-1)!$ ways. So, presumably, we should subtract $n \times (n-1)! = n!$ from our original count of $n!$. But now our answer is zero! The trouble is that our subtracted term over-counts those configurations in which two or more rooks are attacked by the bishop. A naive count leads to the conclusion that there are $\frac{n(n-1)}{2}(n-2)! = n!/2!$ such configurations, but this figure over-counts configurations with three or more rooks on the main diagonal. Thus we are led to a formula (called an Inclusion-Exclusion Formula) in which we alternately add and subtract various terms to correct for all the over-counting. In the present situation, the final answer turns out to be

$$n! - n! + n!/2! - n!/3! + n!/4! - n!/5! + \cdots + (-1)^n n!/n! = n! \sum_{k=0}^{n} (-1)^k/k!.$$

Next, recall from calculus that $e^x = \sum_{k=0}^{\infty} x^k/k!$ for all real x. In particular, taking $x=-1$, we have

$$e^{-1} = \frac{1}{e} = 1 - 1 + 1/2! - 1/3! + 1/4! - 1/5! + \cdots = \sum_{k=0}^{\infty} (-1)^k/k!.$$

We see that the combinatorial formula stated above consists of the first $n+1$ terms in the infinite series for $n!/e$. It can be shown (see §4.5) that the "tail" of this series, namely $\sum_{k=n+1}^{\infty} (-1)^k n!/k!$, is always less than 0.5 in absolute value. Thus, rounding $n!/e$ to the nearest integer will produce the required answer.

Another interesting combinatorial problem arises by comparing non-attacking rook placements on two boards of different shapes. For instance, consider the two generalized chessboards shown here:

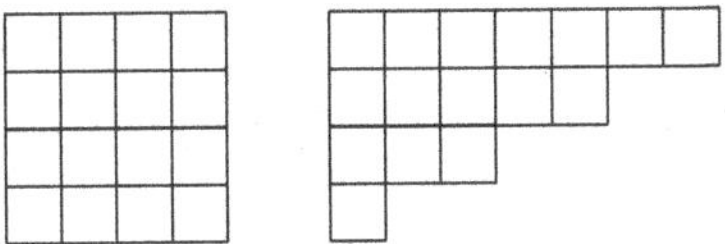

One can check that *for every $k \geq 1$, the number of ways to place k non-attacking rooks on the first board is the same as the number of ways to place k non-attacking rooks on the second board.* We say that two boards are *rook-equivalent* whenever this property holds. It turns out that an $n \times n$ board is always rook-equivalent to a board with successive row lengths $2n-1, 2n-3, \ldots, 5, 3, 1$. More generally, there is a simple criterion for deciding whether two boards "of partition shape" are rook-equivalent. We will present this criterion in §12.3.

Tilings

Now we turn to yet another problem involving chessboards. A *domino* is a rectangular object that can cover two horizontally or vertically adjacent squares on a chessboard. A *tiling* of a board is a covering of the board with dominos such that each square is covered by exactly one domino. For example, here is one possible tiling of a standard 8×8 chessboard:

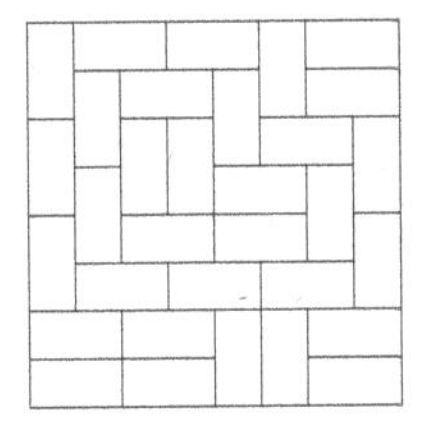

Question: *Given a board of dimensions $m \times n$, how many ways can we tile it with dominos?* This question may seem unfathomably difficult, so let us first consider the special case where $m = 2$. In this case, we are tiling a $2 \times n$ region with dominos. Let f_n be the number of such tilings, for $n = 0, 1, 2, \ldots$. One can see by drawing pictures that

$$f_0 = f_1 = 1, \; f_2 = 2, \; f_3 = 3, \; f_4 = 5, \; f_5 = 8, \; f_6 = 13, \ldots.$$

The reader may recognize these numbers as being the start of the famous *Fibonacci sequence.* This sequence is defined recursively by letting $F_0 = F_1 = 1$ and $F_n = F_{n-1} + F_{n-2}$ for all $n \geq 2$. Now, a routine counting argument can be used to prove that the tiling numbers f_n satisfy the same recursive formula $f_n = f_{n-1} + f_{n-2}$. To see this, note that a $2 \times n$ tiling either ends with one vertical domino or two stacked horizontal dominos. Removing this part of the tiling either leaves a $2 \times (n-1)$ tiling counted by f_{n-1} or a $2 \times (n-2)$ tiling counted by f_{n-2}. Since the sequences (f_n) and (F_n) satisfy the same recursion and initial conditions, we must have $f_n = F_n$ for all n.

Now, what about the original tiling problem? Since the area of a tiled board must be even, there are no tilings unless at least one of the dimensions of the board is even. For boards satisfying this condition, Kasteleyn, Temperley, and Fisher proved the following amazing result. *The number of domino tilings of an $m \times n$ chessboard (with m even) is* *exactly equal to*

$$2^{mn/2}\prod_{j=1}^{m/2}\prod_{k=1}^{n}\sqrt{\cos^2\left(\frac{j\pi}{m+1}\right)+\cos^2\left(\frac{k\pi}{n+1}\right)}.$$

The formula is especially striking since the individual factors in the product are transcendental numbers, yet the product of all these factors is a positive integer! When $m = n = 8$, the formula reveals that the number of domino tilings of a standard chessboard is 12,988,816. The proof of the formula involves *Pfaffians*, which are quantities analogous to determinants that arise in the study of skew-symmetric matrices. For details, see §12.15 and §12.16.

Notes

Different proofs of the Hook-Length Formula may be found in [37, 40, 55, 96, 103]. Treatments of various aspects of rook theory appear in [36, 48, 49, 69]. The Domino Tiling Formula was proved by Kasteleyn in [70] and discovered independently by Fisher and Temperley [31].

Part I

Counting

1

Basic Counting

This chapter develops the basic counting techniques that form the foundation of enumerative combinatorics. We apply these techniques to study fundamental combinatorial objects such as words, permutations, subsets, functions, and lattice paths. We also give some applications of combinatorics to probability theory.

1.1 The Product Rule

We begin with the Product Rule, which gives us a way to count objects that can be built up from smaller ingredients by making a sequence of choices.

1.1. The Product Rule. Suppose S is a set of objects such that every object in S can be constructed in exactly one way by making a sequence of k choices, where k is a fixed positive integer. Suppose the first choice can be made in n_1 ways; the second choice can be made in n_2 ways, no matter what the first choice was; and so on. In general, for $1 \leq i \leq k$, suppose the ith choice can be made in n_i ways, no matter what the previous choices were. Then the total number of objects in S is $n_1 n_2 \cdots n_k$, the product of the number of choices possible at each stage.

We prove the Product Rule later (see §1.15). When using this rule to count objects, the key question to ask is: how can I *build* the objects I want by making a sequence of choices? The following examples illustrate how this works.

1.2. Example: License Plates. A Virginia license plate consists of three uppercase letters followed by four digits. How many license plates are possible? To answer this with the Product Rule, we build a typical license plate by making a sequence of seven choices. First, choose the leftmost letter; there are $n_1 = 26$ ways to make this choice. Second, choose the next letter in any of $n_2 = 26$ ways. Third, choose the next letter in any of $n_3 = 26$ ways. Fourth, choose the first digit in any of $n_4 = 10$ ways. Continue similarly; we have $n_5 = n_6 = n_7 = 10$. By the Product Rule, the total number of license plates is

$$n_1 n_2 \cdots n_7 = 26^3 \cdot 10^4 = 175{,}760{,}000.$$

1.3. Example: Numbers. How many odd four-digit numbers have no repeated digits and do not contain the digit 8? For example, 3461 and 1705 are valid numbers, but 3189 and 1021 are not. We can build a number of the required type by choosing its digits one at a time in the following order. First, choose the *rightmost* digit. Since we want an odd number, there are $n_1 = 5$ possibilities for this digit (1 or 3 or 5 or 7 or 9). Second, choose the *leftmost* digit. Here there are $n_2 = 7$ possibilities: of the ten digits 0 through 9, we must avoid 0, 8, and the digit chosen in the first stage. Third, choose the second digit from the left. Now there are $n_3 = 7$ possibilities, since 0 is available, but we must avoid 8 and the two different digits chosen in the first two stages. Fourth, choose the third digit from the left; there are

$n_4 = 6$ possibilities (avoid 8 and the three different digits chosen previously). The object is now complete, so the Product Rule tells us there are

$$n_1n_2n_3n_4 = 5 \cdot 7 \cdot 7 \cdot 6 = 1470$$

numbers satisfying the given conditions.

In this example, we *cannot* arrive at the answer by choosing digits from left to right. Using this approach, we would find $n_1 = 8$ (avoid 0 and 8), $n_2 = 8$ (avoid 8 and the first digit), and $n_3 = 7$ (avoid 8 and the first two digits). But when we try to choose the last digit, what is n_4? The last digit needs to be odd and different from the three previously chosen digits. The number of available choices at the fourth stage depends on how many of the previous digits are odd. So n_4 depends on the particular choices made earlier, which violates the setup for the Product Rule. Thus, that rule does not apply to the construction method attempted here. In our solution above, we avoided this difficulty by choosing the rightmost digit first.

1.4. Example: Poker Hands. A *poker hand* is a set of five different cards from a 52-card deck. Each card in the deck has a suit (clubs, diamonds, hearts, or spades) and a value (2 through 10, jack, queen, king, or ace). For example, $H = \{4\heartsuit, 9\clubsuit, A\diamondsuit, 9\heartsuit, J\spadesuit\}$ is a poker hand. How many poker hands are there? It might seem we could build a poker hand by choosing the first card ($n_1 = 52$), then a second card different from the first ($n_2 = 51$), then the third card ($n_3 = 50$), then the fourth card ($n_4 = 49$), then the fifth card ($n_5 = 48$). Using the Product Rule would give $n_1n_2n_3n_4n_5 = 311,875,200$ possible hands. However, this argument is flawed because, by definition, a poker hand is an *unordered set* of cards. Thus we cannot speak of the "first" card in the hand. Another way to explain the error is to notice that the *same* hand can be constructed in many different ways by making different sequences of choices; but the Product Rule demands that each object to be counted must arise from *one and only one* sequence of choices. For example, the hand H shown above could be built by first choosing $4\heartsuit$, then $9\clubsuit$, then $A\diamondsuit$, then $9\heartsuit$, then $J\spadesuit$. But, we could build the *same* hand by first choosing $9\heartsuit$, then $J\spadesuit$, then $9\clubsuit$, then $4\heartsuit$, then $A\diamondsuit$. By the Product Rule, we see that there are $5 \cdot 4 \cdot 3 \cdot 2 \cdot 1 = 120$ different orderings of these five cards that would all produce the same (unordered) hand H. In fact, every poker hand (not just H) is constructed 120 times by the choice process described above. So we can obtain the correct count by dividing the initial answer by 120, getting $N = 2,598,960$ poker hands. The Subset Rule (see 1.27 below) generalizes this result.

1.5. Example: Straight Poker Hands. A *straight* poker hand consists of five cards with consecutive values, where the ace can be treated as either a low card or a high card. For example, $\{A\heartsuit, 2\clubsuit, 3\diamondsuit, 4\diamondsuit, 5\clubsuit\}$ and $\{K\clubsuit, 9\clubsuit, J\clubsuit, Q\clubsuit, 10\clubsuit\}$ are straight poker hands (note that the order of cards within the hand is irrelevant). How many straight poker hands are there? We can build all such hands by making the following choices:

1. Choose the lowest value in the hand. There are $n_1 = 10$ possible values here (ace, 2, ..., 10). Since the hand is a straight, all values in the hand are determined by this choice. If we choose 5 (say), then the partially constructed hand looks like {5–,6–,7–,8–,9–}.

2. Choose the suit for the lowest value in the hand; there are $n_2 = 4$ suits we could pick. In our sample object, suppose we pick spades; the object now looks like $\{5\spadesuit$,6–,7–,8–,9–}.

3. Choose the suit for the second lowest value in the hand; there are $n_3 = 4$ possibilities. Picking $\heartsuit$ for our sample object, we now have $\{5\spadesuit, 6\heartsuit$,7–,8–,9–}.

4. Choose the suit for the third lowest value in the hand, so $n_4 = 4$. Perhaps we pick hearts again; the partial hand is now $\{5\spadesuit, 6\heartsuit, 7\heartsuit$,8–,9–}.

5. Choose the suit for the fourth lowest value in the hand, so $n_5 = 4$. Say we pick spades; our object is now $\{5\spadesuit, 6\heartsuit, 7\heartsuit, 8\spadesuit, 9\text{–}\}$.

6. Choose the suit for the highest value in the hand, so $n_6 = 4$. If we pick diamonds, we obtain the complete object $\{5\spadesuit, 6\heartsuit, 7\heartsuit, 8\spadesuit, 9\diamondsuit\}$.

According to the Product Rule, there are $10 \cdot 4^5 = 10,240$ straight poker hands (which include straight flush hands). Now, by definition, the probability of a straight poker hand is the number of such hands divided by N, the total number of five-card poker hands from Example 1.4. This probability is approximately 0.00394. (We discuss probability in more detail in §1.7.)

In this example, we needed to think of a clever sequence of choices to build the straight poker hands that took advantage of the special structure of these hands. The more direct approach of choosing the cards in the hand one at a time might not work. For instance, if we choose an arbitrary card at stage 1 (rather than the low card), then the number of choices for cards in later stages depends on which choices were made earlier.

1.6. Example: Rook Placements. How many ways can we place four non-attacking rooks on the board shown below? (Recall that *non-attacking* means no two rooks are in the same row or column.)

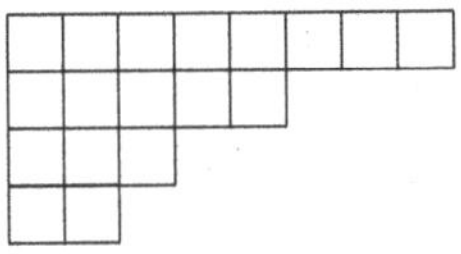

Evidently, each of the four rows must contain a rook. First, choose a square for the rook in the lowest row ($n_1 = 2$ ways). Second, choose a square for the rook in the next lowest row; there are $n_2 = 2$ possible squares, since we must avoid the column of the rook already placed. Third, choose a square for the rook in the next lowest row; there are $n_3 = 5 - 2 = 3$ choices here. Finally, place the rook in the highest row ($n_4 = 8 - 3 = 5$ ways). The Product Rule gives $n_1 n_2 n_3 n_4 = 2 \cdot 2 \cdot 3 \cdot 5 = 60$ rook placements. If we tried to place rooks starting in the top row and working down, the Product Rule could not be used since n_2 (for instance) would depend on what choice was made in the top row.

1.2 The Sum Rule

The next counting rule, called the Sum Rule, allows us to count a complicated set of objects by breaking the set into a collection of smaller, simpler subsets that can be counted separately.

1.7. The Sum Rule. Suppose S is a set of objects such that every object in S belongs to exactly one of k non-overlapping categories, where k is a fixed positive integer. Suppose the first category contains m_1 objects, the second category contains m_2 objects, and so on. Then the total number of objects in S is $m_1 + m_2 + \cdots + m_k$, the sum of the number of objects in each category.

When using the Sum Rule, we must create the appropriate categories (subsets of S) that can be counted by other means. Often we design the categories so that each category can be counted using the Product Rule. It is crucial that each object in the original set belong to *one and only one* of the smaller categories, and that all of the categories be subsets of the original collection S.

1.8. Example: Fraternity Names. The name of a fraternity is a sequence of two or three uppercase Greek letters. How many possible fraternity names are there? (There are 24 letters in the Greek alphabet.) To solve this, we divide the set S of all such names into two categories, where category 1 consists of the two-letter names, and category 2 consists of the three-letter names. We can build objects in category 1 by choosing the first letter in $n_1 = 24$ ways, and then choosing the second letter in $n_2 = 24$ ways. By the Product Rule, category 1 has size $n_1 n_2 = 24^2 = 576$. Similarly, category 2 has size $24^3 = 13824$. By the Sum Rule, the total number of names is $576 + 13824 = 14400$. Note that when using the Product Rule, the number of stages k in the choice sequence must be the same for all objects in the set being counted. For the original set S, we needed $k = 2$ stages for the two-letter names, and $k = 3$ stages for the three-letter names. This is why the Sum Rule was used in this example (but see Exercise 1-4).

Here and below, we write $|S|$ to denote the number of elements in a finite set S.

1.9. Example: Numbers. How many even four-digit numbers have no repeated digits and do not contain the digit 8? Write such a number as $abcd$ where a, b, c, d are digits between 0 and 9. If we try to imitate the argument used in Example 1.3, choosing d, a, b, and c in this order, we find that the number of choices for a depends on whether d is zero. This suggests dividing the set S of objects being counted into two categories, where category 1 consists of numbers in S ending in zero, and category 2 consists of numbers in S not ending in zero. To build an object in category 1, choose $d = 0$ ($n_1 = 1$ way), then choose a ($n_2 = 8$ ways, since we must avoid 0 and 8), then choose b ($n_3 = 7$ ways, since we must avoid a and d and 8), then choose c ($n_4 = 6$ ways). The Product Rule shows that there are $m_1 = 1 \cdot 8 \cdot 7 \cdot 6 = 336$ objects in category 1. To build an object in category 2, choose d ($n_1 = 3$ ways, since d can be 2 or 4 or 6), then choose a ($n_2 = 7$ ways, since we must avoid 0 and 8 and d), then choose b ($n_3 = 7$ ways, since we must avoid 8 and d and a), then choose c ($n_4 = 6$ ways). By the Product Rule, there are $m_2 = 3 \cdot 7 \cdot 7 \cdot 6 = 882$ objects in category 2. By the Sum Rule, S has size $336 + 882 = 1218$.

Here is another solution to the same problem. Let T be the set of all four-digit numbers with distinct digits unequal to 8. On one hand, an argument using the Product Rule shows that T has size $8 \cdot 8 \cdot 7 \cdot 6 = 2688$. On the other hand, we can write T as the union of two categories: S (the set of even numbers in T) and U (defined to be the set of odd numbers in T). In Example 1.3, we used the Product Rule to compute $|U| = 1470$. On the other hand, the Sum Rule tells us that $|T| = |S| + |U|$. So $|S| = |T| - |U| = 2688 - 1470 = 1218$.

1.10. Example: Rook Placements. How many ways can we place three non-attacking rooks on the board shown here?

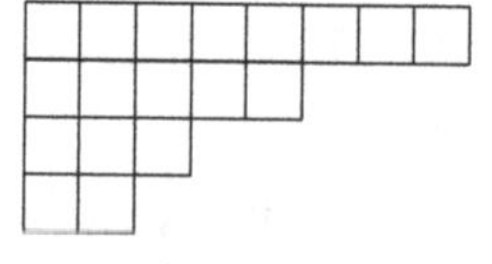

Let S be the set of all such rook placements. Since there are three rooks and four rows, every object in S has exactly one empty row. This suggests creating four categories S_1, S_2, S_3, and S_4, where S_i consists of those placements in S where the ith row from the top is empty. By the Sum Rule, $|S| = |S_1| + |S_2| + |S_3| + |S_4|$. Each set S_i can be counted using the Product Rule. For example, to build an object in S_3, place a rook in the lowest row ($n_1 = 2$ ways), then place a rook in the second row from the top ($n_2 = 5 - 1 = 4$ ways), then place a rook in the top row ($n_3 = 8 - 2 = 6$ ways). Thus, $|S_3| = 2 \cdot 4 \cdot 6 = 48$. Similarly, we find that $|S_1| = 2 \cdot 2 \cdot 3 = 12$, $|S_2| = 2 \cdot 2 \cdot 6 = 24$, and $|S_4| = 3 \cdot 4 \cdot 6 = 72$. So the total number of placements is $12 + 24 + 48 + 72 = 156$.

1.11. Example: Words. How many five-letter words (sequences of uppercase letters) have the property that any Q appearing in the word is immediately followed by U? To solve this problem with the Sum and Product Rules, we need to divide the given set of words into appropriate categories. One approach is to create templates showing exactly which positions in the word contain a Q. In the following templates, a star denotes a letter different from Q.

1. `*****` 2. `QU***` 3. `*QU**` 4. `**QU*`
5. `***QU` 6. `QUQU*` 7. `QU*QU` 8. `*QUQU`

For $1 \leq i \leq 8$, let S_i be the set of words matching the ith template. Since there are 25 choices for each starred position, the Product Rule gives $|S_1| = 25^5$, $|S_2| = |S_3| = |S_4| = |S_5| = 25^3$, and $|S_6| = |S_7| = |S_8| = 25$. Then the Sum Rule gives the answer $25^5 + 4 \cdot 25^3 + 3 \cdot 25 =$ 9,828,200 words.

Beginners sometimes get confused about whether to use the Sum Rule or the Product Rule. Remember that the Product Rule is used when we are *building* each object being counted via a *sequence of choices.* The Sum Rule is used when we are *subdividing* the objects being counted into *smaller, non-overlapping categories.*

1.3 Counting Words and Permutations

In the next few sections, we apply the Sum and Product Rules to count basic combinatorial structures such as words, permutations, subsets, etc. These objects often appear as building blocks used to solve more complicated counting problems.

1.12. Definition: Words. Let A be a finite set. A *word* in the alphabet A is a sequence $w = w_1 w_2 \cdots w_k$, where each $w_i \in A$ and $k \geq 0$. The *length* of $w = w_1 w_2 \cdots w_k$ is k. Two words $w = w_1 w_2 \cdots w_k$ and $z = z_1 z_2 \cdots z_m$ are *equal* iff[1] $k = m$ and $w_i = z_i$ for $1 \leq i \leq k$.

1.13. Example. Let $A = \{a, b, c, \ldots, z\}$ be the set of 26 lowercase letters in the English alphabet. Then *stop*, *opts*, and *stoops* are distinct words (of lengths 4, 4, and 6, respectively). If $A = \{0, 1\}$, the 8 words of length 3 in the alphabet A are

$$000, \quad 001, \quad 010, \quad 011, \quad 100, \quad 101, \quad 110, \quad 111.$$

There is exactly one word of length zero, called the *empty word.* It is sometimes denoted by the special symbols $\cdot$ or ϵ.

1.14. The Word Rule. If A is an n-letter alphabet and $k \geq 0$, then there are n^k words of length k over A.

Proof. For fixed $k > 0$, we can uniquely construct a typical word $w = w_1 w_2 \cdots w_k$ of length k by a sequence of k choices. First, choose $w_1 \in A$ to be any of the n letters in A. Second, choose $w_2 \in A$ in any of n ways. Continue similarly, choosing $w_i \in A$ in any of n ways for $1 \leq i \leq k$. By the Product Rule, the number of words is $n \times n \times \cdots \times n$ (k factors), which is n^k. Note that the empty word is the unique word of length 0 in the alphabet A, so our formula holds for $k = 0$ also. □

1.15. Definition: Permutations. Let A be an n-element set. A *permutation* of A is a word $w = w_1 w_2 \cdots w_n$ in which each letter of A appears exactly once. Permutations are also called *rearrangements* or *linear orderings* of A.

[1] Here and below, *iff* is an abbreviation for *if and only if.*

1.16. Example. The six permutations of $A = \{x, y, z\}$ are

$$xyz, \quad xzy, \quad yxz, \quad yzx, \quad zxy, \quad zyx.$$

Some linear orderings of $\{1, 2, 3, 4, 5\}$ are 35142, 54321, 24513, and 12345.

1.17. Definition: Factorials. For each integer $n \geq 1$, *n-factorial* is

$$n! = n \times (n-1) \cdot (n-2) \cdot \ldots \cdot 3 \cdot 2 \cdot 1,$$

which is the product of the first n positive integers. We also define $0! = 1$.

1.18. The Permutation Rule. There are $n!$ permutations of an n-letter alphabet A.

Proof. Build a typical permutation $w = w_1 w_2 \cdots w_n$ of A by making n choices. First, choose w_1 to be any of the n letters of A. Second, choose w_2 to be any of the $n-1$ letters of A different from w_1. Third, choose w_3 to be any of the $n-2$ letters of A different from w_1 and w_2. Proceed similarly; at the nth stage, choose w_n to be the unique letter of A that is different from $w_1, w_2, \ldots, w_{n-1}$. By the Product Rule, the number of permutations is $n \times (n-1) \times \cdots \times 1 = n!$. The result also holds when $n = 0$. □

Sometimes we want to count *partial permutations* in which not every letter of the alphabet gets used.

1.19. Definition: k-Permutations. Let A be an n-element set. A *k-permutation* of A is a word $w = w_1 w_2 \cdots w_k$ consisting of k *distinct* letters in A. An n-permutation of A is the same as a permutation of A.

1.20. Example. The twelve 2-permutations of $A = \{a, b, c, d\}$ are

$$ab, \quad ac, \quad ad, \quad ba, \quad bc, \quad bd, \quad ca, \quad cb, \quad cd, \quad da, \quad db, \quad dc.$$

1.21. The Partial Permutation Rule. Suppose A is an n-letter alphabet. For $0 \leq k \leq n$, the number of k-permutations of A is

$$n(n-1)(n-2)\cdots(n-k+1) = \frac{n!}{(n-k)!}.$$

For $k > n$, there are no k-permutations of A.

Proof. Build a typical k-permutation $w = w_1 w_2 \cdots w_k$ of A by making k choices. First, choose w_1 to be any of the n letters of A. Second, choose w_2 to be any of the $n-1$ letters of A different from w_1. Continue similarly. When we choose w_i (where $1 \leq i \leq k$), we have already used the $i-1$ distinct letters $w_1, w_2, \ldots, w_{i-1}$. Since A has n letters, there are $n-(i-1) = n-i+1$ choices available at stage i. In particular, for the kth and final choice, there are $n-k+1$ ways to choose w_k. By the Product Rule, the number of k-permutations is $\prod_{i=1}^{k}(n-(i-1)) = n(n-1)\cdots(n-k+1)$. Multiplying this expression by $(n-k)!/(n-k)!$, we obtain the product of the integers 1 through n in the numerator, which is $n!$. Thus the answer is also given by the formula $n!/(n-k)!$. □

1.22. Example. How many six-letter words: (a) begin with a consonant and end with a vowel; (b) have no repeated letters; (c) have no two *consecutive* letters equal; (d) have no two consecutive vowels? Assume we are using the alphabet $\{\mathrm{A}, \mathrm{B}, \ldots, \mathrm{Z}\}$ with vowels A, E, I, O, and U.

For (a), we build a word by choosing the consonant at the beginning ($n_1 = 21$ ways), then choosing the next four letters ($n_2 = 26^4$ ways, by the Word Rule), then choosing the vowel at the end ($n_3 = 5$ ways). By the Product Rule, the answer to (a) is $21 \cdot 26^4 \cdot 5$. For (b), we use the Partial Permutation Rule with $n = 26$ and $k = 6$ to get $26!/20!$ words. For (c), we choose the letters from left to right, getting $n_1 = 26$ and $n_2 = n_3 = \cdots = n_6 = 25$. Note that when we choose the ith letter with $i > 1$, we only need to avoid the one letter immediately preceding it in the word built so far. The answer to (c) is $26 \cdot 25^5$.

For (d), we use the Sum Rule, dividing the words we want into many smaller categories. We label each category with a template showing the possible positions of vowels and consonants in the words in that category. For example, the template CVCCVC labels the category of words with vowels in positions 2 and 5 and consonants elsewhere. By the Product Rule, this category has size $21^4 \cdot 5^2$ (and similarly for any other template with exactly two V's). There is one template CCCCCC with no vowels, which stands for a set of 21^6 words. There are six templates with one vowel, each of which stands for a set of $21^5 \cdot 5$ words. By explicitly listing them, we find there are ten templates with two vowels: VCVCCC, VCCVCC, VCCCVC, VCCCCV, CVCVCC, CVCCVC, CVCCCV, CCVCVC, CCVCCV, and CCCVCV. There are two templates with three vowels, namely VCVCVC and CVCVCV, each of which stands for a set of $21^3 \cdot 5^3$ words. Combining all these counts with the Sum Rule, the answer to (d) is

$$21^6 + 6 \cdot 21^5 \cdot 5 + 10 \cdot 21^4 \cdot 5^2 + 2 \cdot 21^3 \cdot 5^3 = 259,224,651.$$

1.23. Example. Three married couples and a single guy are standing in line at a bank. How many ways can the line be formed if: (a) the three women are at the front of the line; (b) men and women alternate in the line; (c) each husband stands adjacent to his wife in the line; (d) the single guy is not adjacent to a woman?

We solve these questions by combining the Sum, Product, and Permutation Rules. For (a), we build the line by first choosing a linear ordering of the three women ($n_1 = 3! = 6$ ways) and then choosing a permutation of the four men behind them ($n_2 = 4! = 24$ ways); the Product Rule gives $6 \cdot 24 = 144$ possible lines.

For (b), note that the men must occupy positions 1, 3, 5, and 7 in the line, and the women must occupy positions 2, 4, and 6. To build such a line, first choose a permutation of the men ($n_1 = 4! = 24$ ways) and then place the men (in this order) in the odd-numbered positions of the line. Next choose a permutation of the women ($n_2 = 3! = 6$ ways) and then place the women (in this order) in the even-numbered positions of the line. As in (a), the total count is $24 \cdot 6 = 144$.

For (c), construct the line by the following choices. First, choose a permutation of the four objects consisting of the three married couples and the single guy ($n_1 = 4! = 24$ ways). Second, for the married couple closest to the front of the line, choose whether the man or woman comes first ($n_2 = 2$ ways). Third, for the next married couple, choose who comes first ($n_3 = 2$ ways). Fourth, for the last married couple, choose who comes first ($n_4 = 2$ ways). The answer is $24 \cdot 2^3 = 192$. If all the married couples stood side by side (rather than adjacent in line), the answer would be 24.

For (d), we use the Sum Rule. Divide the lines being counted into three categories. Category 1 consists of lines with the single guy first; category 2 consists of lines with the single guy last; and category 3 consists of lines with the single guy somewhere in the middle. To build a line in category 1, put the single guy at the front (1 way); then choose a man to stand behind him (3 ways); then permute the remaining five people ($5! = 120$ ways). This gives $1 \cdot 3 \cdot 120 = 360$ objects in category 1. Similarly, category 2 has 360 objects. To build a line in category 3, put the single guy somewhere in the middle (5 ways); then choose a man to stand in front of him (3 ways); then choose a different man to stand behind him (2 ways); then permute the remaining four people and put them in the remaining positions in the line from front to back ($4! = 24$ ways). By the Product Rule, category 3 has $5 \cdot 3 \cdot 2 \cdot 24 = 720$ objects. The answer to (d) is $360 + 360 + 720 = 1440$.

1.4 Counting Subsets

In this section, we count subsets of a given finite set S. Recall that $T \subseteq S$ means that T is a subset of S, i.e., for all objects x, if $x \in T$ then $x \in S$. Sets S and T are *equal* iff $T \subseteq S$ and $S \subseteq T$, which means that S and T have exactly the same members.

1.24. Definition: Power Set. For any set S, the *power set* $\mathcal{P}(S)$ is the set of all subsets of S. Thus, $T \in \mathcal{P}(S)$ iff $T \subseteq S$.

1.25. Example. If $S = \{2, 5, 7\}$, then $\mathcal{P}(S)$ is the eight-element set

$$\{\emptyset, \{2\}, \{5\}, \{7\}, \{2,5\}, \{2,7\}, \{5,7\}, \{2,5,7\}\}.$$

1.26. The Power Set Rule. An n-element set has 2^n subsets. In other words, if $|S| = n$, then $|\mathcal{P}(S)| = 2^n$.

Proof. Suppose $S = \{x_1, \ldots, x_n\}$ is an n-element set. We can build a typical subset T of S by making a sequence of n choices. First, decide whether x_1 is or is not a member of T; there are $n_1 = 2$ possible choices here. Second, decide whether $x_2 \in T$ or $x_2 \notin T$; again there are two possibilities. Continue similarly; decide in the ith choice whether $x_i \in T$ or $x_i \notin T$ ($n_i = 2$ possible choices). This sequence of choices determines which x_j's belong to T. Since T is a subset of S, this information uniquely characterizes the set T. By the Product Rule, the number of subsets is $2 \times 2 \times \cdots \times 2$ (n factors), which is 2^n. □

To illustrate this proof, suppose $S = \{2, 5, 7\}$ and $x_1 = 2$, $x_2 = 5$, and $x_3 = 7$. When building a subset T, we might make these choices: 2 is in T; 5 is not in T; 7 is in T. This choice sequence creates the subset $T = \{2, 7\}$. Another choice sequence is: $2 \notin T$; $5 \notin T$; $7 \notin T$. This choice sequence constructs the empty set $\emptyset$.

Next we consider the enumeration of subsets of an n-element set having a fixed size k. For example, there are ten 3-element subsets of $\{a, b, c, d, e\}$:

$$\{a,b,c\}, \quad \{a,b,d\}, \quad \{a,b,e\}, \quad \{a,c,d\}, \quad \{a,c,e\},$$

$$\{a,d,e\}, \quad \{b,c,d\}, \quad \{b,c,e\}, \quad \{b,d,e\}, \quad \{c,d,e\}.$$

In this example, we present a given set by listing its members between curly braces. This notation forces us to list the members of each set in a particular order (alphabetical in this case). If we reorder the members of the list, the underlying set does not change. For example, the sets $\{a, c, d\}$ and $\{c, d, a\}$ and $\{d, c, a\}$ are all the same set. Similarly, listing an element more than once does not change the set: $\{a, c, d\}$ and $\{a, a, c, c, c, d\}$ and $\{d, d, a, c\}$ are all the same set. These assertions follow from the very definition of set equality: $A = B$ means that for every x, $x \in A$ iff $x \in B$. The sets mentioned earlier in this paragraph have members a and c and d and nothing else, so they are all equal. In contrast, the ordering and repetition of elements in a sequence (or word) definitely makes a difference. For instance, the words *cad*, *dac*, and *acadd* are unequal although they use the same three letters.

Suppose we try to count the k-element subsets of a given n-element set using the Product Rule. This rule requires us to construct objects by making an *ordered sequence* of choices. We might try to construct a subset by choosing its first element in n ways, then its second element in $n - 1$ ways, etc., which leads to the *incorrect* answer $n(n-1)\cdots(n-k+1)$. The trouble here is that there is no well-defined *first element* of a subset. In fact, our construction procedure generates each subset several times, once for each possible ordering of its members. There are $k!$ such orderings, so we obtain the correct answer by dividing the previous formula by $k!$ (cf. Example 1.4). The next proof gives an alternate version of this argument that avoids overcounting.

1.27. The Subset Rule. For $0 \leq k \leq n$, the number of k-element subsets of an n-element set is

$$\frac{n!}{k!(n-k)!}.$$

Proof. Fix n and k with $0 \leq k \leq n$. Let A be an n-element set, and let x denote the number of k-element subsets of A. Our goal is to show $x = n!/(k!(n-k)!)$. Let S be the set of all k-permutations of A. Recall that elements of S are *ordered* sequences $w_1 w_2 \cdots w_k$, where the w_i are distinct elements of A. We compute $|S|$ in two ways. On one hand, $|S| = n!/(n-k)!$ by the Partial Permutation Rule. On the other hand, we can build an object in S by first choosing an (unordered) k-element subset of A in any of x ways, and then choosing a linear ordering of these k objects in any of $k!$ ways. Every object in S can be constructed in exactly one way by this sequence of two choices. By the Product Rule, $|S| = x \cdot k!$. Comparing the two formulas for $|S|$ and solving for x, we get $x = n!/(k!(n-k)!)$ as needed. (This argument is an example of a *combinatorial proof*, in which we establish an algebraic formula by counting a given set of objects in two different ways. We study combinatorial proofs in more detail in Chapter 2.) □

The Subset Rule is used very frequently in counting arguments, so we introduce the following notation for the quotient of factorials appearing in this rule.

1.28. Definition: Binomial Coefficients. For $0 \leq k \leq n$, the *binomial coefficient* is

$$\binom{n}{k} = C(n,k) = \frac{n!}{k!(n-k)!}.$$

For $k < 0$ or $k > n$, we define $\binom{n}{k} = C(n,k) = 0$. Thus, for all $n \geq 0$ and all k, $\binom{n}{k}$ is the number of k-element subsets of an n-element set. In particular, $\binom{n}{k}$ is always an integer.

Here are some counting problems whose solutions use the Subset Rule.

1.29. Example: Flush Poker Hands. A five-card poker hand is a *flush* iff every card in the hand has the same suit. For example, $\{3\diamondsuit, 7\diamondsuit, 9\diamondsuit, J\diamondsuit, Q\diamondsuit\}$ is a flush. We can build a flush by first choosing the common suit for the five cards in $n_1 = 4$ ways, and then choosing a subset of five cards out of the set of 13 cards having this suit. By the Subset Rule, there are $n_2 = \binom{13}{5} = 1287$ ways to make this second choice. By the Product Rule, the number of flush poker hands is $4 \cdot 1287 = 5148$. To find the probability of a flush, we divide this number by the total number of poker hands. A poker hand is a 5-element subset of the set of 52 cards in a deck, so the Subset Rule tells us there are $\binom{52}{5}$ poker hands (cf. Example 1.4). The required probability is approximately 0.00198, about half the probability of getting a straight poker hand. For this reason, a flush beats a straight in poker.

1.30. Example: Full House Poker Hands. A five-card poker hand with three cards of one value and two cards of another value is called a *full house*. For example, $\{4\heartsuit, 4\clubsuit, 4\spadesuit, 7\spadesuit, 7\heartsuit\}$ is a full house. The best way to count full house poker hands (and hands with similar restrictions on the suits and values) is to construct the hand gradually by choosing suits and values rather than choosing individual cards. For instance, we can build a full house hand as follows.

1. Choose the value that will occur three times in the hand in $n_1 = 13$ ways. To build the sample hand above, we would choose the value 4 here, obtaining the partial hand {4–,4–,4–,?,?}.

2. Choose the value that will occur twice in the hand in $n_2 = 12$ ways (since we cannot reuse the value chosen first). In our example, we would choose the value 7 here, obtaining the partial hand {4–,4–,4–,7–,7–}.

3. Choose a subset of three suits (out of four) for the value that occurs three times. By the Subset Rule, this can be done in $\binom{4}{3} = 4$ ways. In our example, we choose the subset $\{\heartsuit, \clubsuit, \spadesuit\}$, leading to the partial hand $\{4\heartsuit, 4\clubsuit, 4\spadesuit, 7-, 7-\}$.

4. Choose a subset of two suits (out of four) for the value that occurs twice. By the Subset Rule, this can be done in $\binom{4}{2} = 6$ ways. In our example, we choose the subset $\{\spadesuit, \heartsuit\}$ for the 7's, giving us the completed hand displayed above.

The total number of full house hands is $13 \cdot 12 \cdot 4 \cdot 6 = 3744$. The probability of a full house is $3744/\binom{52}{5} \approx 0.00144$.

1.31. Example: Numbers. How many six-digit numbers contain exactly three copies of the digit 7? Since the first digit is special (it cannot be zero), we create two categories: let S_1 be the set of such numbers that start with 7, and let S_2 be the set of such numbers that do not start with 7. To build a number in S_1, first put 7 in the first position ($n_1 = 1$ way); then choose a subset of two of the remaining five positions for the other 7's ($n_2 = \binom{5}{2} = 10$ ways, by the Subset Rule); then fill the three unused positions from left to right with digits other than 7 ($n_3 = 9^3$ ways, by the Word Rule). The Product Rule gives $|S_1| = 1 \cdot 10 \cdot 9^3 = 7290$. To build a number in S_2, first choose a subset of three of the five non-initial positions to contain 7's ($n_1 = \binom{5}{3} = 10$ ways, by the Subset Rule); then choose a digit unequal to 0 or 7 for the first position ($n_2 = 8$ ways); then choose the two remaining digits ($n_3 = n_4 = 9$ ways). We see that $|S_2| = 10 \cdot 8 \cdot 9 \cdot 9 = 6480$, so the answer is $7290 + 6480 = 13770$. Another approach to this problem leads to the formula $\binom{6}{3}9^3 - \binom{5}{3}9^2 = 13770$; can you reconstruct the counting argument that leads to this formula?

1.5 Counting Anagrams

For our next counting problem, we enumerate words where each letter appears a specified number of times.

1.32. Definition: Anagrams. Suppose $a_1, \ldots, a_k$ are distinct letters from an alphabet A and $n_1, \ldots, n_k$ are nonnegative integers. Let $\mathcal{R}(a_1^{n_1} a_2^{n_2} \cdots a_k^{n_k})$ denote the set of all words $w = w_1 w_2 \cdots w_n$ that are formed by rearranging n_1 copies of a_1, n_2 copies of a_2, $\ldots$, and n_k copies of a_k (so that $n = n_1 + n_2 + \cdots + n_k$). Words in a given set $\mathcal{R}(a_1^{n_1} \cdots a_k^{n_k})$ are said to be *anagrams* or *rearrangements* of one another.

1.33. Example. The set of all anagrams of the word 00111 is

$$\mathcal{R}(0^2 1^3) = \{00111, 01011, 01101, 01110, 10011, 10101, 10110, 11001, 11010, 11100\}.$$

The set $\mathcal{R}(a^1 b^2 c^1 d^0)$ consists of the words

$$\{abbc, abcb, acbb, babc, bacb, bbac, bbca, bcab, bcba, cabb, cbab, cbba\}.$$

1.34. The Anagram Rule. Suppose $a_1, \ldots, a_k$ are distinct letters, $n_1, \ldots, n_k$ are nonnegative integers, and $n = n_1 + \cdots + n_k$. Then

$$|\mathcal{R}(a_1^{n_1} a_2^{n_2} \cdots a_k^{n_k})| = \frac{n!}{n_1! n_2! \cdots n_k!}.$$

We give two proofs of this result. *First Proof:* We give a combinatorial argument similar to our proof of the Subset Rule. Define a new alphabet A consisting of n *distinct letters* by attaching distinct numerical superscripts to each copy of the given letters $a_1, \ldots a_k$:

$$A = \{a_1^{(1)}, a_1^{(2)}, \ldots, a_1^{(n_1)}, a_2^{(1)}, \ldots, a_2^{(n_2)}, \ldots, a_k^{(1)}, \ldots, a_k^{(n_k)}\}.$$

Let $x = |\mathcal{R}(a_1^{n_1} a_2^{n_2} \cdots a_k^{n_k})|$. Let S be the set of all permutations of the n-letter alphabet A. We count $|S|$ in two ways. On one hand, we know $|S| = n!$ by the Permutation Rule. On the other hand, the following sequence of choices constructs each object in S exactly once. First, choose a word $v \in \mathcal{R}(a_1^{n_1} \cdots a_k^{n_k})$ in any of x ways. Second, choose a linear ordering of the superscripts 1 through n_1 and attach these superscripts (in the chosen order) to the n_1 copies of a_1 in v. By the Permutation Rule, this second choice can be made in $n_1!$ ways. Third, choose a linear ordering of 1 through n_2 and attach these superscripts to the n_2 copies of a_2 in v; there are $n_2!$ ways to make this choice. Continue similarly; at the last stage, we attach the superscripts 1 through n_k to the n_k copies of a_k in v in any of $n_k!$ ways. By the Product Rule, $|S| = x \cdot n_1! \cdot n_2! \cdot \ldots \cdot n_k!$. Since $|S| = n!$ also, solving for x gives the formula asserted in the Anagram Rule.

Second Proof. This proof makes repeated use of the Subset Rule followed by an algebraic manipulation of factorials. We construct a typical object $w = w_1 w_2 \cdots w_n \in \mathcal{R}(a_1^{n_1} \cdots a_k^{n_k})$ by making the following sequence of k choices. Intuitively, we are going to choose the positions of the a_1's, then the positions of the a_2's, etc. First, choose any n_1-element subset S_1 of $\{1, 2, \ldots, n\}$ in any of $\binom{n}{n_1}$ ways, and define $w_i = a_1$ for all $i \in S_1$. Second, choose any n_2-element subset S_2 of the $n - n_1$ unused positions in any of $\binom{n-n_1}{n_2}$ ways, and define $w_i = a_2$ for all $i \in S_2$. At the jth stage (where $1 \leq j \leq k$), we have already filled the positions in S_1 through S_{j-1}, so there are $n - n_1 - n_2 - \cdots - n_{j-1}$ remaining positions in the word. We choose any n_j-element subset S_j of these remaining positions in any of $\binom{n-n_1-\cdots-n_{j-1}}{n_j}$ ways, and define $w_i = a_j$ for all $i \in S_j$. By the Product Rule, the number of rearrangements is

$$\binom{n}{n_1}\binom{n-n_1}{n_2}\binom{n-n_1-n_2}{n_3}\cdots\binom{n-n_1-\cdots-n_{k-1}}{n_k} = \prod_{j=1}^{k} \frac{(n-n_1-\cdots-n_{j-1})!}{n_j!(n-n_1-\cdots-n_j)!}.$$

This is a telescoping product that simplifies to $n!/(n_1! n_2! \cdots n_k!)$. For instance, when $k = 4$, the product is

$$\frac{n!}{n_1!(n-n_1)!} \cdot \frac{(n-n_1)!}{n_2!(n-n_1-n_2)!} \cdot \frac{(n-n_1-n_2)!}{n_3!(n-n_1-n_2-n_3)!} \cdot \frac{(n-n_1-n_2-n_3)!}{n_4!(n-n_1-n_2-n_3-n_4)!},$$

which simplifies to $n!/(n_1! n_2! n_3! n_4!)$, using the fact that $(n - n_1 - n_2 - \cdots - n_k)! = 0! = 1$.

1.35. Example. We now illustrate the constructions in each of the two preceding proofs. For the first proof, suppose we are counting $\mathcal{R}(a^3 b^1 c^4)$. The alphabet A in the proof consists of eight distinct letters:

$$A = \{a^{(1)}, a^{(2)}, a^{(3)}, b^{(1)}, c^{(1)}, c^{(2)}, c^{(3)}, c^{(4)}\}.$$

Let us build a specific permutation of A using the second counting method. First, choose an element of $\mathcal{R}(a^3 b^1 c^4)$, say $v = baccaacc$. Second, choose a linear ordering of the superscripts $\{1, 2, 3\}$, say 312, and label the a's from left to right with these superscripts, obtaining $ba^{(3)}cca^{(1)}a^{(2)}cc$. Third, choose a linear ordering of $\{1\}$, namely 1, and label the b with this superscript, producing $b^{(1)}a^{(3)}cca^{(1)}a^{(2)}cc$. Finally, choose a linear ordering of $\{1, 2, 3, 4\}$, say 1243, and label the c's accordingly to get $b^{(1)}a^{(3)}c^{(1)}c^{(2)}a^{(1)}a^{(2)}c^{(4)}c^{(3)}$. We have now constructed a permutation of the alphabet A.

Next, let us see how to build the word *baccaacc* using the method of the second proof. Start with an empty 8-letter word, which we denote $--------$. We first choose the 3-element subset $\{2,5,6\}$ of $\{1,2,\ldots,8\}$ and put a's in those positions, obtaining the partial word $-a--aa--$. We then choose the 1-element subset $\{1\}$ of $\{1,3,4,7,8\}$ and put a b in that position, obtaining $ba--aa--$. Finally, we choose the 4-element subset $\{3,4,7,8\}$ of $\{3,4,7,8\}$ and put c's in those positions, obtaining the final word *baccaacc*.

1.36. Definition: Multinomial Coefficients. Suppose $n_1,\ldots,n_k$ are nonnegative integers and $n = n_1 + \cdots + n_k$. The *multinomial coefficient* is

$$\binom{n}{n_1,n_2,\ldots,n_k} = C(n;n_1,n_2,\ldots,n_k) = \frac{n!}{n_1!n_2!\cdots n_k!}.$$

This is the number of rearrangements of k letters where there are n_i copies of the ith letter. In particular, $\binom{n}{n_1,n_2,\ldots,n_k}$ is always an integer.

Binomial coefficients are a special case of multinomial coefficients. Indeed, it is immediate from the definitions that

$$\binom{a+b}{a} = \frac{(a+b)!}{a!b!} = \binom{a+b}{a,b}$$

for all integers $a,b \geq 0$.

1.37. Example. (a) Count the number of anagrams of MADAMIMADAM. (b) In how many anagrams do the four M's occur consecutively? (c) The word in (a) is a *palindrome* (a word that reads the same forward and backward). How many anagrams of this word are also palindromes?

We solve (a) using the Anagram Rule: we seek the size of $\mathcal{R}(\mathrm{A}^4\mathrm{D}^2\mathrm{I}^1\mathrm{M}^4)$, which is $\binom{4+2+1+4}{4,2,1,4} = 11!/(4!2!1!4!) = 34650$. Part (b) can also be solved with the Anagram Rule, if we think of the four consecutive M's as a single meta-letter $\boxed{\mathrm{MMMM}}$. Here the answer is

$$\left|\mathcal{R}\left(\mathrm{A}^4\mathrm{D}^2\mathrm{I}^1\boxed{\mathrm{MMMM}}^1\right)\right| = \binom{4+2+1+1}{4,2,1,1} = \frac{8!}{4!2!1!1!} = 840.$$

For (c), we can build the palindromic anagrams as follows. First, put the sole copy of I in the middle position (1 way). Second, choose an anagram of MADAM to place before the I in $\binom{5}{2,2,1} = 30$ ways. Third, complete the word by filling the last five positions with the reversal of the word in the first five positions (1 way). The Product Rule gives 30 as the answer.

1.6 Counting Rules for Set Operations

In set theory, one studies various operations for combining sets such as unions, intersections, set differences, and Cartesian products. We now examine some counting rules giving the sizes of the sets produced by these operations.

We begin with Cartesian products. Recall that $S \times T$ is the set of all ordered pairs (x,y) with $x \in S$ and $y \in T$. More generally, the Cartesian product of k sets $S_1, S_2, \ldots, S_k$, denoted $S_1 \times S_2 \times \cdots \times S_k$, is the set of all ordered k-tuples $(a_1, a_2, \ldots, a_k)$ such that $a_i \in S_i$ for $1 \leq i \leq k$. We write S^k for the Cartesian product of k copies of the set S.

1.38. The Cartesian Product Rule. Given k finite sets $S_1,\ldots,S_k$, $|S_1 \times S_2 \times \cdots \times S_k| = |S_1| \cdot |S_2| \cdot \cdots \cdot |S_k|$. In particular, $|S^k| = |S|^k$.

Proof. We build a typical object $(a_1, \ldots, a_k) \in S_1 \times \cdots \times S_k$ by making a sequence of k choices. First pick a_1 in $n_1 = |S_1|$ ways. Then pick a_2 in $n_2 = |S_2|$ ways. Continue similarly; for $1 \leq i \leq k$, pick a_i in $n_i = |S_i|$ ways. By the Product Rule, the product set has size $n_1 n_2 \cdots n_k$. □

Next we consider unions of sets. Recall that the *union* $A \cup B$ consists of all objects x such that $x \in A$ or $x \in B$. The *intersection* $A \cap B$ consists of all objects x such that $x \in A$ and $x \in B$. More generally, $S_1 \cup S_2 \cup \cdots \cup S_k$ consists of all x such that $x \in S_i$ for some i between 1 and k; and $S_1 \cap S_2 \cap \cdots \cap S_k$ consists of all x such that $x \in S_i$ for every i between 1 and k. Two sets A and B are *disjoint* iff $A \cap B = \emptyset$ (the empty set). We say that sets $S_1, \ldots, S_k$ are *pairwise disjoint* iff for all $i \neq j$, $S_i \cap S_j = \emptyset$. The next rule is nothing more than a restatement of the Sum Rule in the language of sets.

1.39. The Disjoint Union Rule. For all pairwise disjoint finite sets $S_1, \ldots, S_k$,

$$|S_1 \cup S_2 \cup \cdots \cup S_k| = |S_1| + |S_2| + \cdots + |S_k|.$$

For sets A and B, the *set difference* $A{-}B$ consists of all x such that $x \in A$ and $x \notin B$. We do not require that B be a subset of A here.

1.40. The Difference Rule. (a) If T is a subset of a finite set S, $|S{-}T| = |S| - |T|$.
(b) For all finite sets S and all sets U, $|S{-}U| = |S| - |S \cap U|$.

Proof. To prove (a), assume $T \subseteq S$. Then S is the union of the two disjoint sets T and $S{-}T$, so $|S| = |T| + |S{-}T|$ by the Disjoint Union Rule. Subtracting $|T|$ from both sides, we get (a). To prove (b), take $T = S \cap U$, which is a subset of S. Since $S{-}U = S{-}T$, we can use (a) to conclude that $|S{-}U| = |S{-}T| = |S| - |T| = |S| - |S \cap U|$. □

The next rule gives a formula for the size of the union of two arbitrary (not necessarily disjoint) finite sets.

1.41. The Union Rule for Two Sets. For all finite sets S and T, $|S \cup T| = |S| + |T| - |S \cap T|$.

Proof. The set $S \cup T$ is the union of the three disjoint sets $S{-}T$, $S \cap T$, and $T{-}S$. By the Disjoint Union Rule and the Difference Rule,

$$|S \cup T| = |S{-}T| + |S \cap T| + |T{-}S| = |S| - |S \cap T| + |S \cap T| + |T| - |T \cap S|,$$

which simplifies to $|S| + |T| - |S \cap T|$. □

By repeatedly using the Union Rule for Two Sets, we can deduce similar formulas for the union of three or more sets. For example, given any finite sets A, B, C, it can be shown (Exercise 1-39) that

$$|A \cup B \cup C| = |A| + |B| + |C| - |A \cap B| - |A \cap C| - |B \cap C| + |A \cap B \cap C|. \qquad (1.1)$$

The most general version of the Union Rule is called the Inclusion-Exclusion Formula; we discuss it in Chapter 4.

1.42. Example. (a) How many five-letter words in the alphabet {A,B,C,D,E,F,G} start with a vowel or end with a consonant? (b) How many five-letter words contain the letters C and F?

For (a), let S be the set of five-letter words starting with a vowel, and let T be the set of five-letter words ending with a consonant. We seek $|S \cup T|$. By the Word Rule and the

Product Rule, $|S| = 2 \cdot 7^4$, $|T| = 7^4 \cdot 5$, and $|S \cap T| = 2 \cdot 7^3 \cdot 5$. By the Union Rule for Two Sets, $|S \cup T| = |S| + |T| - |S \cap T| = 13377$.

To solve (b), we use some negative logic. Let X be the set of all five-letter words; let S be the set of words in X that do *not* contain C, and let T be the set of words in X that do *not* contain F. Each word in $S \cup T$ has no C *or* has no F; so each word in $Y = X-(S \cup T)$ has at least one C *and* at least one F. To find $|Y|$, we use the Union Rule and the Difference Rule. By the Word Rule, $|X| = 7^5$ and $|S| = 6^5 = |T|$. Words in $S \cap T$ are five-letter words using the alphabet {A,B,D,E,G}, hence $|S \cap T| = 5^5$. Now, $|S \cup T| = |S| + |T| - |S \cap T| = 2 \cdot 6^5 - 5^5$, so $|Y| = |X| - |S \cup T| = 7^5 - 2 \cdot 6^5 + 5^5 = 4380$.

1.7 Probability

The counting techniques discussed so far can be applied to solve many problems in probability theory. This section introduces some fundamental concepts of probability and considers several examples.

1.43. Definition: Sample Spaces and Events. A *sample space* is a set S whose members represent the possible outcomes of a random experiment. In this section, we only consider *finite* sample spaces. An *event* is a subset of the sample space.

Intuitively, an event consists of the set of outcomes of the random experiment that possess a particular property we are interested in. We imagine repeating the experiment many times. Each run of the experiment produces a single outcome $x \in S$; we say that an event $A \subseteq S$ has *occurred* on this run of the experiment iff x is in the subset A.

1.44. Example: Coin Tossing. Suppose the experiment consists of tossing a coin five times. We could take the sample space for this experiment to be $S = \{\mathrm{H}, \mathrm{T}\}^5$, the set of all 5-letter words using the letters H (for heads) and T (for tails). The element HHHTH $\in S$ represents the outcome where the fourth toss was tails and all other tosses were heads. The subset $A = \{w \in S : w_2 = \mathrm{H}\}$ is the event in which the second toss comes up heads. The subset $B = \{w \in S : w_1 \neq w_5\}$ is the event that the first toss is different from the last toss. The subset

$$C = \{w \in S : w_i = \mathrm{T} \text{ for an odd number of indices } i\}$$

is the event that there are an odd number of tails.

1.45. Example: Dice Rolling. Suppose the experiment consists of rolling a six-sided die three times. The sample space for this experiment is $S = \{1, 2, 3, 4, 5, 6\}^3$, the set of all 3-letter words over the alphabet $\{1, 2, \ldots, 6\}$. The subset $A = \{w \in S : w_1 + w_2 + w_3 \in \{7, 11\}\}$ is the event that the sum of the three numbers rolled is 7 or 11. The subset $B = \{w \in S : w_1 = w_2 = w_3\}$ is the event that all three numbers rolled are the same. The subset $C = \{w \in S : w \neq (4, 1, 3)\}$ is the event that we do not see the numbers 4, 1, 3 (in that order) in the dice rolls.

1.46. Example: Lotteries. Consider the following random experiment. We put 49 white balls (numbered 1 through 49) into a machine that mixes the balls and then outputs a sequence of six distinct balls, one at a time. We could take the sample space here to be the set S' of all 6-letter words w consisting of six distinct letters from $A = \{1, 2, \ldots, 49\}$. In lotteries, the order in which the balls are drawn usually does not matter, so it is more common to take the sample space to be the set S of all 6-element subsets of A. (We will see later that using S instead of S' does not affect the probabilities we are interested in.) Suppose a lottery player picks a (fixed and known) 6-element subset T_0 of A. For $0 \leq k \leq 6$, define events $B_k = \{T \in S : |T \cap T_0| = k\} \subseteq S$. Intuitively, the event B_k is the set of outcomes in which the player has matched exactly k of the winning lottery numbers.

1.47. Example: Special Events. For any sample space S, $\emptyset$ and S are events. The event $\emptyset$ contains no outcomes, and therefore never occurs. On the other hand, the event S contains all the outcomes, and therefore always occurs. If A and B are *events* (i.e., subsets of S), note that $A \cup B$, $A \cap B$, $S-A$, and $A-B$ are also *events*. Intuitively, $A \cup B$ is the event that either A occurs or B occurs (or both); $A \cap B$ is the event that both A and B occur; $S-A$ is the event that A does not occur; and $A-B$ is the event that A occurs but B does not occur.

Now we can formally define the concept of probability. Intuitively, for each event A, we want to define a number $P(A)$ that measures the probability or likelihood that A occurs. Numbers close to 1 represent more likely events, while numbers close to 0 represent less likely events. A probability-zero event is (virtually) impossible, while a probability-one event is (virtually) certain to occur. In general, if we perform the random experiment many times, $P(A)$ should approximately equal the ratio of the number of trials where A occurs to the total number of trials performed. These intuitive ideas are formalized in the following mathematical axioms for probability.

1.48. Definition: Probability. Assume S is a finite sample space. A *probability measure* for S is a function P assigning to each event $A \subseteq S$ a real number $P(A) \in [0,1]$ such that $P(\emptyset) = 0$, $P(S) = 1$, and for any two *disjoint* events A and B, $P(A \cup B) = P(A) + P(B)$.

By induction, it follows that P satisfies the *finite additivity* property

$$P(A_1 \cup A_2 \cup \cdots \cup A_n) = P(A_1) + P(A_2) + \cdots + P(A_n)$$

for all pairwise disjoint sets $A_1, A_2, \ldots, A_n \subseteq S$. Moreover, it can be shown (Exercise 1-51) that for all events A and B, $P(A-B) = P(A) - P(A \cap B)$ and $P(A \cup B) = P(A) + P(B) - P(A \cap B)$.

1.49. Example: Classical Probability Spaces. Suppose S is a finite sample space in which *all outcomes are equally likely.* Then we must have $P(\{x\}) = 1/|S|$ for each outcome $x \in S$. For any event $A \subseteq S$, finite additivity gives

$$P(A) = \frac{|A|}{|S|} = \frac{\text{number of outcomes in } A}{\text{total number of outcomes}}. \tag{1.2}$$

Thus the calculation of probabilities (in this classical setup) reduces to two counting problems: counting the number of elements in A and counting the number of elements in S. We can take Equation (1.2) as the *definition* of our probability measure P. Note that the axiom "$A \cap B = \emptyset$ implies $P(A \cup B) = P(A) + P(B)$" is then a consequence of the Sum Rule. Also note that this probability model will only be appropriate if all the possible outcomes of the underlying random experiment are equally likely to occur.

1.50. Example: Coin Tossing. Suppose we toss a fair coin five times. The sample space is $S = \{\mathrm{H}, \mathrm{T}\}^5$, so that $|S| = 2^5 = 32$. Consider the event $A = \{w \in S : w_2 = \mathrm{H}\}$ of getting a head on the second toss. By the Product Rule, $|A| = 2 \cdot 1 \cdot 2^3 = 16$, so $P(A) = 16/32 = 1/2$. Consider the event $B = \{w \in S : w_1 \neq w_5\}$ in which the first toss differs from the last toss. B is the disjoint union of $B_1 = \{w \in S : w_1 = \mathrm{H}, w_5 = \mathrm{T}\}$ and

$B_2 = \{w \in S : w_1 = \mathrm{T}, w_5 = \mathrm{H}\}$. The Product Rule shows that $|B_1| = |B_2| = 2^3 = 8$, so that $P(B) = (8+8)/32 = 1/2$. Finally, consider the event

$$C = \{w \in S : w_i = \mathrm{T} \text{ for an odd number of indices } i\}.$$

C is the disjoint union $C_1 \cup C_3 \cup C_5$, where (for $0 \leq k \leq 5$) C_k is the event of getting exactly k tails. Observe that $C_k = \mathcal{R}(\mathrm{T}^k\mathrm{H}^{5-k})$. So $|C_k| = \binom{5}{k,5-k} = \binom{5}{k}$ by the Anagram Rule, $P(C_k) = \binom{5}{k}/2^5$, and

$$P(C) = \frac{\binom{5}{1} + \binom{5}{3} + \binom{5}{5}}{2^5} = 16/32 = 1/2.$$

1.51. Example: Dice Rolling. Consider the experiment of rolling a six-sided die twice. The sample space is $S = \{1,2,3,4,5,6\}^2$, so that $|S| = 6^2 = 36$. Consider the event $A = \{x \in S : x_1 + x_2 \in \{7,11\}\}$ of rolling a sum of 7 or 11. By direct enumeration, we have

$$A = \{(1,6),(2,5),(3,4),(4,3),(5,2),(6,1),(5,6),(6,5)\}; \qquad |A| = 8.$$

Therefore, $P(A) = 8/36 = 2/9$. Consider the event $B = \{x \in S : x_1 \neq x_2\}$ of getting two different numbers on the two rolls. The Product Rule gives $|B| = 6 \cdot 5 = 30$, so $P(B) = 30/36 = 5/6$.

1.52. Example: Balls in Urns. Suppose an urn contains n_1 red balls, n_2 white balls, and n_3 blue balls. Let the random experiment consist of randomly drawing a k-element subset of balls from the urn. What is the probability of drawing k_1 red balls, k_2 white balls, and k_3 blue balls, where $k_1 + k_2 + k_3 = k$? We can take the sample space S to be all k-element subsets of the set

$$\{1,2,\ldots,n_1,n_1+1,\ldots,n_1+n_2,n_1+n_2+1,\ldots,n_1+n_2+n_3\}.$$

Here the first n_1 integers represent red balls, the next n_2 integers represent white balls, and the last n_3 integers represent blue balls. By the Subset Rule, $|S| = \binom{n_1+n_2+n_3}{k}$. Let A be the event where we draw k_1 red balls, k_2 white balls, and k_3 blue balls. To build a set $T \in A$, we choose a k_1-element subset of $\{1,2,\ldots,n_1\}$, then a k_2-element subset of $\{n_1+1,\ldots,n_1+n_2\}$, then a k_3-element subset of $\{n_1+n_2+1,\ldots,n_1+n_2+n_3\}$. By the Subset Rule and Product Rule, $|A| = \binom{n_1}{k_1}\binom{n_2}{k_2}\binom{n_3}{k_3}$. Therefore, the definition of the probability measure gives

$$P(A) = \binom{n_1}{k_1}\binom{n_2}{k_2}\binom{n_3}{k_3} \Big/ \binom{n_1+n_2+n_3}{k_1+k_2+k_3}.$$

This calculation can be generalized to the case where the urn has balls of more than three colors.

1.53. Example: General Probability Measures on a Finite Sample Space. We can extend the previous discussion to the case where not all outcomes of the random experiment are equally likely. Let S be a finite sample space and let p be a function assigning to each outcome $x \in S$ a real number $p(x) \in [0,1]$, such that $\sum_{x \in S} p(x) = 1$. Intuitively, $p(x)$ is the probability that the outcome x occurs. Now p is not a probability measure, since its domain is S (the set of *outcomes*) instead of $\mathcal{P}(S)$ (the set of *events*). We build a probability measure from p by defining $P(A) = \sum_{x \in A} p(x)$ for each event $A \subseteq S$. The axioms for a probability measure may be routinely verified.

1.54. Remark. In this section, we used counting techniques to solve basic probability questions. It is also possible to use probabilistic arguments to help solve counting problems. Examples of such arguments appear in §12.4 and §12.12.

1.8 Lotteries and Card Games

In this section, we give more examples of probability calculations by analyzing lotteries, bridge, and five-card poker.

1.55. Example: Lotteries. Consider the lottery described in Example 1.46. Here the sample space S consists of all 6-element subsets of $A = \{1, 2, \ldots, 49\}$, so $|S| = \binom{49}{6} = 13,983,816$. Suppose a lottery player picks a (fixed and known) 6-element subset T_0 of A. For $0 \leq k \leq 6$, define events $B_k = \{T \in S : |T \cap T_0| = k\}$. B_k occurs when the player matches exactly k of the winning numbers. We can build a typical object $T \in B_k$ by choosing k elements of T_0 in $\binom{6}{k}$ ways, and then choosing $6 - k$ elements of $A - T_0$ in $\binom{43}{6-k}$ ways. Hence,

$$P(B_k) = \binom{6}{k}\binom{43}{6-k} \bigg/ \binom{49}{6}.$$

We compute $P(B_3) \approx 0.01765$, $P(B_4) \approx 0.001$, $P(B_5) \approx 1.8 \times 10^{-5}$, and $P(B_6) = 1/|S| \approx 7.15 \times 10^{-8}$. We can view this example as the special case of Example 1.52 where the urn contains 6 balls of one color (representing the winning numbers) and 43 balls of another color (representing the other numbers).

In the lottery example, we could have taken the sample space to be the set S' of all *ordered* sequences of six distinct elements of $\{1, 2, \ldots, 49\}$. Let B'_k be the event that the player guesses exactly k numbers correctly (disregarding order, as before). Let P' be the probability measure on the sample space S'. It can be checked that $|S'| = \binom{49}{6} \cdot 6!$ and $|B'_k| = \binom{6}{k}\binom{43}{6-k} \cdot 6!$, so that

$$P'(B'_k) = \frac{\binom{6}{k}\binom{43}{6-k} 6!}{\binom{49}{6} 6!} = P(B_k).$$

This confirms our earlier remark that the two sample spaces S and S' give the same probabilities for events that do not depend on the order in which the balls are drawn.

1.56. Example: Powerball. A *powerball lottery* has two kinds of balls: white balls (numbered $1, \ldots, M$) and red balls (numbered $1, \ldots, R$). Each week, one red ball and a set of n distinct white balls are randomly chosen. Lottery players separately guess the numbers of the n white balls and the red ball, which is called the *powerball*. Players win prizes based on how many balls they guess correctly. Players always win a prize for matching the red ball, even if they incorrectly guess all the white balls.

To analyze this lottery, let the sample space be

$$S = \{(T, x) : T \text{ is an } n\text{-element subset of } \{1, 2, \ldots, M\} \text{ and } x \in \{1, 2, \ldots, R\}\}.$$

Let (T_0, x_0) be a fixed and known element of S representing a given player's lottery ticket. For $0 \leq k \leq n$, let A_k be the event $\{(T, x) \in S : |T \cap T_0| = k, x \neq x_0\}$ in which the player matches exactly k white balls but misses the powerball. Let B_k be the event $\{(T, x) \in S : |T \cap T_0| = k, x = x_0\}$ in which the player matches exactly k white balls and also matches the powerball. We have $|S| = \binom{M}{n} R$ by the Subset Rule and the Product Rule. To build a typical element in A_k, we first choose k elements of T_0, then choose $n - k$ elements of $\{1, 2, \ldots, M\} - T_0$, then choose $x \in \{1, 2, \ldots, R\} - \{x_0\}$. Thus, $|A_k| = \binom{n}{k}\binom{M-n}{n-k}(R - 1)$, so

$$P(A_k) = \frac{\binom{n}{k}\binom{M-n}{n-k}(R-1)}{\binom{M}{n} R}.$$

TABLE 1.1
Analysis of Powerball.

Matches	Probability	Prize Value
0 white, 1 red	0.0261 or 1 in 38	$4
1 white, 1 red	0.0109 or 1 in 92	$4
2 white, 1 red	0.00143 or 1 in 701	$7
3 white, 0 red	0.00172 or 1 in 580	$7
3 white, 1 red	0.000069 or 1 in 14,494	$100
4 white, 0 red	0.0000274 or 1 in 36,525	$100
4 white, 1 red	0.0000011 or 1 in 913,129	$50,000
5 white, 0 red	8.56×10^{-8} or 1 in 11.7 million	$1,000,000
5 white, 1 red	3.422×10^{-9} or 1 in 292 million	Jackpot

Note: The cost of a ticket is $2.

Similarly,

$$P(B_k) = \frac{\binom{n}{k}\binom{M-n}{n-k} \cdot 1}{\binom{M}{n}R}.$$

In one version of this lottery, we have $M = 69$, $R = 26$, and $n = 5$. The probabilities of certain events A_k and B_k are shown in Table 1.1 together with the associated prize amounts.

1.57. Example: Bridge. A bridge game has four players, called North, South, East, and West. All cards in the deck are dealt to the players so that each player receives a set of 13 cards. (a) How big is the sample space S? (b) Find the probability that North receives all four aces and South is void in clubs (i.e., has no clubs).

For (a), we can build an object in S by choosing a set of 13 cards out of 52 for North ($\binom{52}{13}$ ways), then choosing a set of 13 out of the remaining 39 cards for South ($\binom{39}{13}$ ways), then choosing a set of 13 of the remaining 26 cards for East ($\binom{26}{13}$ ways), then choosing 13 of the remaining 13 cards for West ($\binom{13}{13} = 1$ way). The resulting product of binomial coefficients simplifies to the multinomial coefficient $\binom{52}{13,13,13,13}$. This is not a coincidence: we can also think of an element of S as an anagram $w \in \mathcal{R}(\mathrm{N}^{13}\mathrm{S}^{13}\mathrm{E}^{13}\mathrm{W}^{13})$, as follows. Let $C_1C_2\cdots C_{52}$ be a fixed ordering of the 52 cards. Use the anagram $w = w_1 \cdots w_{52}$ to distribute cards to the bridge players by giving card C_i to North, South, East, or West when w_i is N, S, E, or W, respectively. This encoding of bridge hands via anagrams is an instance of the Bijection Rule, discussed in §1.11 below.

Let B be the event described in (b). We build an outcome in B as follows. First give North the four aces (one way). There are now 48 cards left, 12 clubs and 36 non-clubs. Choose a set of 13 of the 36 non-clubs for South ($\binom{36}{13}$ ways). Now give North 9 more of the remaining 35 cards ($\binom{35}{9}$ ways), then choose 13 of the remaining 26 cards for East ($\binom{26}{13}$ ways), then give the remaining 13 cards to West (1 way). By the Product Rule, $|B| = \binom{36}{13}\binom{35}{9}\binom{26}{13}$. Then

$$P(B) = \frac{\binom{36}{13}\binom{35}{9}}{\binom{52}{13}\binom{39}{13}} \approx 0.000032.$$

Next we compute more probabilities associated with five-card poker.

1.58. Example: Four-of-a-Kind Hands. A *four-of-a-kind* poker hand is a five-card hand such that some value in $\{2, 3, \ldots, 10, J, Q, K, A\}$ appears four times in the hand. For example, $\{3\heartsuit, 3\clubsuit, 5\clubsuit, 3\diamondsuit, 3\spadesuit\}$ is a four-of-a-kind hand. To find the probability of such a hand, we use the sample space S consisting of all five-element subsets of the 52-card deck.

We know $|S| = \binom{52}{5}$ by the Subset Rule. Define the event A to be the subset of S consisting of all four-of-a-kind hands. We build a typical hand in A by choosing the value to occur four times (there are $n_1 = 13$ possibilities), then putting all four cards of that value in the hand (in $n_2 = \binom{4}{4} = 1$ way), then selecting a fifth card ($n_3 = 48$ choices). By the Product Rule, $|A| = 13 \cdot 1 \cdot 48 = 624$. So $P(A) = |A|/|S| \approx 0.00024$.

1.59. Example: One-Pair Hands. A *one-pair* poker hand is a five-card hand containing four values, one of which occurs twice. For example, $\{5\heartsuit, 9\diamondsuit, 9\clubsuit, J\heartsuit, A\spadesuit\}$ is a one-pair hand. Let B be the event consisting of all one-pair hands. Build an object in B via these choices: first choose the value that appears twice ($n_1 = 13$ ways); then choose a subset of two cards out of the four cards with this value ($n_2 = \binom{4}{2} = 6$ ways); then choose a subset of three values from the twelve unused values ($n_3 = \binom{12}{3} = 220$ ways); then choose a suit for the lowest of these three values ($n_4 = 4$ ways); then choose a suit for the next lowest value ($n_5 = 4$ ways); then choose a suit for the last value ($n_6 = 4$ ways). The Product Rule gives $|B| = 1,098,240$, so $P(B) = |B|/|S| \approx 0.4226$. For instance, the sample hand above is built from the following choice sequence: choose value 9, then suits $\{\diamondsuit, \clubsuit\}$, then values $\{5, J, A\}$, then suit $\heartsuit$ for the 5, then suit $\heartsuit$ for the jack, then suit $\spadesuit$ for the ace. (Compare to Exercise 1-27.)

1.60. Example: Discarding Cards. Suppose we are dealt the five-card poker hand $H = \{3\heartsuit, 4\clubsuit, 5\clubsuit, 9\spadesuit, K\clubsuit\}$ from a 52-card deck. We now have the opportunity to discard k cards from our hand and receive k new cards. (a) If we discard the 9 and the king, what is the probability we will be dealt a straight hand? (b) If we discard the 3 and 9, what is the probability we will be dealt a flush?

In both parts, we can take the sample space to be the set S of all two-element subsets of the 47-element set consisting of the deck with the cards in H removed. Thus, $|S| = \binom{47}{2} = 1081$. For (a), we can build a straight starting from the partial hand $\{3\heartsuit, 4\clubsuit, 5\clubsuit\}$ as follows: choose the low value for the straight (there are three possibilities: ace, two, or three); choose the suit for the lowest value not in the original hand (four ways); choose the suit for the remaining value (four ways). The number of hands is $3 \cdot 4 \cdot 4 = 48$, and the probability is $48/1081 \approx 0.0444$.

For (b), we can build a flush starting from the partial hand $\{4\clubsuit, 5\clubsuit, K\clubsuit\}$ by choosing a subset of two of the remaining ten club cards in $\binom{10}{2} = 45$ ways. The probability here is $45/1081 \approx 0.0416$, just slightly less than the probability in (a).

1.9 Conditional Probability and Independence

Suppose that, in a certain random experiment, we are told that a particular event has occurred. Given this additional information, we can recompute the probability of other events occurring. This leads to the notion of conditional probability.

1.61. Definition: Conditional Probability. Suppose A and B are events in some sample space S such that $P(B) > 0$. The *conditional probability of A given B*, denoted $P(A|B)$, is defined by setting $P(A|B) = P(A \cap B)/P(B)$. In the case where S is a finite set of equally likely outcomes, we have $P(A|B) = |A \cap B|/|B|$.

To motivate this definition, suppose we know for certain that event B has occurred on some run of the experiment. Given this knowledge, another event A also occurred iff $A \cap B$ has occurred. This explains why $P(A \cap B)$ appears in the numerator. We divide by $P(B)$

to normalize the conditional probabilities, so that (for instance) $P(B|B) = 1$. The next example shows that the conditional probability of A given some other event can be greater than, less than, or equal to the original *unconditional* probability $P(A)$.

1.62. Example: Dice Rolling. Consider the experiment of rolling a fair die twice. What is the probability of getting a sum of 7 or 11, given that the second roll comes up 5? Here, the sample space is $S = \{1,2,3,4,5,6\}^2$. Let A be the event of getting a sum of 7 or 11, and let B be the event that the second die shows 5. We have $P(B) = 1/6$, and we saw earlier that $P(A) = 2/9$. Listing outcomes, we see that $A \cap B = \{(2,5),(6,5)\}$, so $P(A \cap B) = 2/36 = 1/18$. Therefore, the required conditional probability is

$$P(A|B) = \frac{P(A \cap B)}{P(B)} = \frac{1/18}{1/6} = 1/3 > 2/9 = P(A).$$

On the other hand, let C be the event that the second roll comes up 4. Here $A \cap C = \{(3,4)\}$, so

$$P(A|C) = \frac{1/36}{6/36} = 1/6 < 2/9 = P(A).$$

Next, let D be the event that the first roll is odd. Then $A \cap D = \{(1,6),(3,4),(5,2),(5,6)\}$, so

$$P(A|D) = \frac{4/36}{18/36} = 2/9 = P(A).$$

1.63. Example: Balls in Urns. Suppose an urn contains r red balls and b blue balls, where $r, b \geq 2$. Consider an experiment in which two balls are drawn from the urn in succession, without replacement. What is the probability that the first ball is red, given that the second ball is blue? We take the sample space to be the set S of all words w_1w_2, where $w_1 \neq w_2$ and

$$w_1, w_2 \in \{1, 2, \ldots, r, r+1, \ldots, r+b\}.$$

Here, the numbers 1 through r represent red balls, and the numbers $r+1$ through $r+b$ represent blue balls. The event of drawing a red ball first is the subset

$$A = \{w_1w_2 : 1 \leq w_1 \leq r\}.$$

The event of drawing a blue ball second is the subset

$$B = \{w_1w_2 : r+1 \leq w_2 \leq r+b\}.$$

By the Product Rule, $|S| = (r+b)(r+b-1)$, $|A| = r(r+b-1)$, $|B| = b(r+b-1)$, and $|A \cap B| = rb$. The conditional probability of A given B is

$$P(A|B) = P(A \cap B)/P(B) = r/(r+b-1).$$

In contrast, the unconditional probability of A is

$$P(A) = |A|/|S| = r/(r+b).$$

The conditional probability is slightly higher than the unconditional probability; intuitively, we are more likely to have gotten a red ball first if we know the second ball was not red. The probability that the second ball is blue, given that the first ball is red, is

$$P(B|A) = P(B \cap A)/P(A) = b/(r+b-1).$$

Note that $P(B|A) \neq P(A|B)$ (unless $r = b$).

1.64. Example: Card Hands. What is the probability that a 5-card poker hand is a full house, given that the hand is void in hearts (i.e., no card in the hand is a heart)? Let A be the event of getting a full house, and let B be the event of being void in hearts. We have $|B| = \binom{39}{5} = 575,757$ since we must choose a five-element subset of the $52 - 13 = 39$ non-heart cards. Next, we must compute $|A \cap B|$. To build a full house hand using no hearts, make the following choices: first, choose a value to occur three times (13 ways); second, choose the suits for this value (1 way, as hearts are forbidden); third, choose a value to occur twice (12 ways); fourth, choose the suits for this value ($\binom{3}{2} = 3$ ways). By the Product Rule, $|A \cap B| = 13 \cdot 1 \cdot 12 \cdot 3 = 468$. Accordingly, the probability we want is $P(A|B) = 468/575,757 \approx 0.000813$.

Next, what is the probability of getting a full house, given that the hand has at least two cards of the same value? Let C be the event that at least two cards in the hand have the same value; we seek $P(A|C) = P(A \cap C)/P(C) = |A \cap C|/|C|$. The numerator here can be computed quickly: since $A \subseteq C$, we have $A \cap C = A$ and hence $|A \cap C| = |A| = 3744$ (see Example 1.30). To compute the denominator, let us first enumerate $S{-}C$, where S is the full sample space of all five-card poker hands. Note that $S{-}C$ occurs iff all five cards in the hand have different values. Choose these values ($\binom{13}{5}$ ways), and then choose suits for each card (4 ways each). By the Product Rule, $|S{-}C| = \binom{13}{5}4^5 = 1,317,888$. By the Difference Rule, $|C| = |S| - |S{-}C| = 1,281,072$. The required conditional probability is $P(A|C) = |A \cap C|/|C| \approx 0.00292$.

In some situations, the knowledge that a particular event D occurs does not change the probability that another event A will occur. For instance, events D and A in Example 1.62 have this property because $P(A|D) = P(A)$. Writing out the definition of $P(A|D)$ and multiplying by $P(D)$, we see that the stated property is equivalent to $P(A \cap D) = P(A)P(D)$ (assuming $P(D) > 0$). This suggests the following definition, which is valid even when $P(D) = 0$.

1.65. Definition: Independence of Two Events. Two events A and D are called *independent* iff $P(A \cap D) = P(A)P(D)$.

Unlike the definition of conditional probability, this definition is symmetric in A and D. So, A and D are independent iff D and A are independent. As indicated above, when $P(D) > 0$, independence of A and D is equivalent to $P(A|D) = P(A)$. Similarly, when $P(A) > 0$, independence of A and D is equivalent to $P(D|A) = P(A)$. So, when considering two independent events of positive probability, knowledge that either event has occurred gives us no new information about the probability of the other event occurring.

1.66. Definition: Independence of a Collection of Events. Suppose $A_1, \ldots, A_n$ are events. This list of events is called *independent* iff for all choices of indices $1 \leq i_1 < i_2 < \cdots < i_k \leq n$, $P(A_{i_1} \cap A_{i_2} \cap \cdots \cap A_{i_k}) = P(A_{i_1})P(A_{i_2}) \ldots P(A_{i_k})$.

1.67. Example. Let $S = \{a, b, c, d\}$, and suppose each outcome in S occurs with probability $1/4$. Define events $B = \{a, b\}$, $C = \{a, c\}$, and $D = \{a, d\}$. One verifies immediately that B and C are independent; B and D are independent; and C and D are independent. However, the list of events B, C, D is *not* independent, because

$$P(B \cap C \cap D) = P(\{a\}) = 1/4 \neq 1/8 = P(B)P(C)P(D).$$

1.68. Example: Coin Tossing. Suppose we toss a fair coin five times. Take the sample space to be $S = \{\mathrm{H}, \mathrm{T}\}^5$. Let A be the event that the first and last toss agree; let B be the event that the third toss is tails; let C be the event that there are an odd number of heads. Routine counting arguments show that $|S| = 2^5 = 32$, $|A| = 2^4 = 16$, $|B| = 2^4 = 16$,

$|C| = \binom{5}{1}+\binom{5}{3}+\binom{5}{5} = 16$, $|A \cap B| = 2^3 = 8$, $|A \cap C| = 2(\binom{3}{1}+\binom{3}{3}) = 8$, $|B \cap C| = \binom{4}{1}+\binom{4}{3} = 8$, and $|A \cap B \cap C| = 4$. It follows that

$$P(A \cap B) = P(A)P(B); \quad P(A \cap C) = P(A)P(C); \quad P(B \cap C) = P(B)P(C);$$
$$\text{and } P(A \cap B \cap C) = P(A)P(B)P(C).$$

Thus, the list of events A, B, C is independent.

We often assume that unrelated physical events are independent (in the mathematical sense) to help us construct a probability model. The next examples illustrate this process.

1.69. Example. Urn 1 contains 3 red balls and 7 blue balls. Urn 2 contains 2 red balls and 8 blue balls. Urn 3 contains 4 red balls and 1 blue ball. If we randomly choose one ball from each urn, what is the probability that all three balls have the same color?

Let R be the event that all three balls are red, and let B be the event that all three balls are blue. Then $R = R_1 \cap R_2 \cap R_3$ where R_i is the event that the ball drawn from urn i is red. Since draws from different urns should not affect one another, it is reasonable to assume that the list R_1, R_2, R_3 is independent. So

$$P(R) = P(R_1 \cap R_2 \cap R_3) = P(R_1)P(R_2)P(R_3) = (3/10) \cdot (2/10) \cdot (4/5) = 0.048.$$

Similarly, $P(B) = (7/10) \cdot (8/10) \cdot (1/5) = 0.112$. The events R and B are disjoint, so the answer is $P(R \cup B) = P(R) + P(B) = 0.16$.

1.70. Example: Tossing an Unfair Coin. Consider a random experiment in which we toss an unbalanced coin n times in a row. Assume the coin comes up heads with probability q and tails with probability $1-q$, and successive coin tosses are unrelated to one another. Let the sample space be $S = \{\mathrm{H}, \mathrm{T}\}^n$. Since the coin is unfair, it is not appropriate to assume that every point of S occurs with equal probability. Given an outcome $w = w_1 w_2 \cdots w_n \in S$, what should the probability $p(w)$ be? Consider an example where $n = 5$ and $w = \mathrm{HHTHT}$. Define five events $B_1 = \{z \in S : z_1 = \mathrm{H}\}$, $B_2 = \{z \in S : z_2 = \mathrm{H}\}$, $B_3 = \{z \in S : z_3 = \mathrm{T}\}$, $B_4 = \{z \in S : z_4 = \mathrm{H}\}$, and $B_5 = \{z \in S : z_5 = \mathrm{T}\}$. Our physical assumptions suggest that $B_1, \ldots, B_5$ should be independent events because different tosses of the coin are unrelated. Moreover, $P(B_1) = P(B_2) = P(B_4) = q$, and $P(B_3) = P(B_5) = 1 - q$. Since $B_1 \cap B_2 \cap B_3 \cap B_4 \cap B_5 = \{w\}$, the definition of independence leads to

$$p(w) = P(B_1 \cap \cdots \cap B_5) = P(B_1)P(B_2) \cdots P(B_5) = qq(1-q)q(1-q) = q^3(1-q)^2.$$

Similar reasoning shows that if $w = w_1 w_2 \cdots w_n \in S$ is any outcome consisting of k heads and $n - k$ tails *arranged in one particular specified order*, then we should define $p(w) = q^k(1-q)^{n-k}$.

Next, define $P(A) = \sum_{w \in A} p(w)$ for every event $A \subseteq S$. For example, let A_k be the event that we get k heads and $n-k$ tails in *any* order. Note that $|A_k| = |\mathcal{R}(\mathrm{H}^k\mathrm{T}^{n-k})| = \binom{n}{k}$ by the Anagram Rule, and $p(w) = q^k(1-q)^{n-k}$ for each $w \in A_k$. It follows that

$$P(A_k) = \sum_{w \in A_k} p(w) = \binom{n}{k} q^k (1-q)^{n-k}.$$

We have not yet checked that $\sum_{w \in S} p(w) = 1$, which is needed to prove that we have defined a legitimate probability measure (see Example 1.53). This fact can be deduced from the Binomial Theorem (discussed in §2.3), as follows. Since S is the disjoint union of $A_0, A_1, \ldots, A_n$, we have

$$\sum_{w \in S} p(w) = \sum_{k=0}^{n} \sum_{w \in A_k} p(w) = \sum_{k=0}^{n} \binom{n}{k} q^k (1-q)^{n-k}.$$

By the Binomial Theorem 2.8, the right side is $(q + [1-q])^n = 1^n = 1$, as needed.

1.10 Counting Functions

This section counts various kinds of functions. We use the notation $f : X \to Y$ to mean that f is a *function* with *domain* X and *codomain* Y. This means that for each x in the set X, there exists exactly one y in the set Y with $y = f(x)$. We can think of X as the set of all inputs to f, and Y as the set of all potential outputs for f. The set of outputs that actually occur as function values is called the *image* of f, denoted $f[X] = \{f(x) : x \in X\}$.

When X and Y are finite sets, we can visualize f by an *arrow diagram*. We obtain this diagram by drawing a dot for each element of X and Y, and drawing an arrow from x to y whenever $y = f(x)$. The definition of a function requires that each input $x \in X$ have *exactly one* arrow emanating from it, and the arrow must point to an element of Y. On the other hand, a potential output $y \in Y$ may have zero, one, or more than one arrow hitting it. Figure 1.1 displays the arrow diagrams for four functions f, g, h, and p.

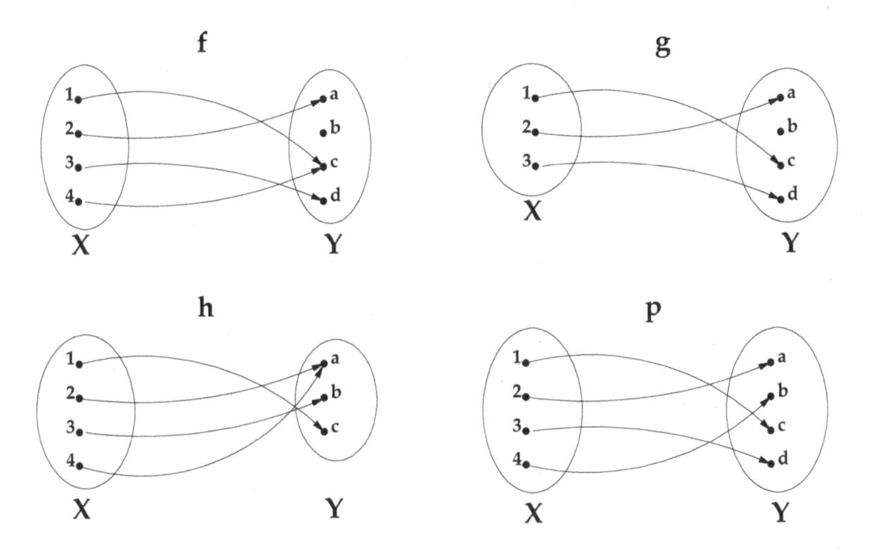

FIGURE 1.1
Arrow diagrams for four functions f, g, h, and p.

1.71. The Function Rule. Given a k-element domain X and an n-element codomain Y, there are n^k functions $f : X \to Y$.

Proof. Let $X = \{x_1, x_2, \ldots, x_k\}$ and $Y = \{y_1, y_2, \ldots, y_n\}$. To build a typical function $f : X \to Y$, we choose the function values one at a time. First, choose $f(x_1)$ to be any of the elements in Y; this choice can be made in $n_1 = n$ ways. Second, choose $f(x_2)$ to be any of the elements in Y; again there are $n_2 = n$ ways. Continue similarly; at the kth stage, we choose $f(x_k)$ to be any element of Y in $n_k = n$ ways. The Product Rule shows that the number of functions we can build is $n_1 n_2 \cdots n_k = n \cdot n \cdot \ldots \cdot n = n^k$. □

Next we study three special types of functions: injections, surjections, and bijections.

1.72. Definition: Injections. A function $f : X \to Y$ is an *injection* iff for all $u, v \in X$, $u \neq v$ implies $f(u) \neq f(v)$. Injective functions are also called *one-to-one* functions.

This definition says that different inputs must always map to different outputs when we apply an injective function. In the arrow diagram for an injective function, every $y \in Y$ has *at most one* arrow entering it. In Figure 1.1, g and p are one-to-one functions, but f and h are not.

1.73. The Injection Rule. Given a k-element domain X and an n-element codomain Y with $k \leq n$, there are $n!/(n-k)!$ injections $f : X \to Y$. If $k > n$, there are no one-to-one functions from X to Y.

Proof. First assume $k \leq n$. As above, we construct a typical injection f from $X = \{x_1, \ldots, x_k\}$ into Y by choosing the k function values $f(x_i)$, for $1 \leq i \leq k$. At the ith stage, we choose $f(x_i)$ to be an element of Y distinct from the elements $f(x_1), \ldots, f(x_{i-1})$ already chosen. Since the latter elements are pairwise distinct, we see that there are $n-(i-1) = n-i+1$ alternatives for $f(x_i)$, no matter what happened in the first $i-1$ choices. By the Product Rule, the number of injections is $n(n-1)\cdots(n-k+1) = n!/(n-k)!$.

On the other hand, suppose $k > n$. Try to build an injection f by choosing the values $f(x_1), f(x_2), \ldots$ as before. When we try to choose $f(x_{n+1})$, there are no elements of Y distinct from the previously chosen elements $f(x_1), \ldots, f(x_n)$. Since it is impossible to complete the construction of f, there are no injections from X to Y in this case. □

1.74. Definition: Surjections. A function $f : X \to Y$ is a *surjection* iff for every $y \in Y$ there exists $x \in X$ with $y = f(x)$. Surjective functions are also said to map X *onto* Y; by an abuse of grammar, one sometimes says "$f : X \to Y$ is an *onto* function."

In the arrow diagram for a surjective function, every $y \in Y$ has *at least one* arrow entering it. For example, the functions h and p in Figure 1.1 are surjective, but f and g are not. To count surjections in general, we need techniques not yet introduced; see §2.13. For now, we look at one special case that can be solved with rules already available. See also Exercise 1-65.

1.75. Example: Counting Surjections. How many functions $f : \{1,2,3,4,5,6\} \to \{a,b,c,d,e\}$ are surjective? Note that such a function is completely determined by the list of function values $(f(1), f(2), \ldots, f(6))$. We require that all elements of the codomain appear in this list, so exactly one such element must appear twice. Build the list by choosing which letter appears twice ($n_1 = 5$ ways); then choosing a subset of two positions out of six for that letter ($n_2 = \binom{6}{2} = 15$ ways), then filling the remaining positions from left to right with a permutation of the remaining letters ($n_3 = 4! = 24$ ways). The Product Rule gives 1800 surjections. For instance, if we choose d, then $\{2,4\}$, then *ebca*, we build the surjection given by $f(1) = e$, $f(2) = d$, $f(3) = b$, $f(4) = d$, $f(5) = c$, and $f(6) = a$. More generally, the same argument shows that there are $n\binom{n+1}{2}(n-1)! = (n+1)!n/2$ surjections mapping an $(n+1)$-element domain onto an n-element codomain.

1.76. Definition: Bijections. A function $f : X \to Y$ is a *bijection* iff f is both injective and surjective. This means that for every $y \in Y$, there exists a unique $x \in X$ with $y = f(x)$. Bijective functions are also called *one-to-one correspondences from* X *onto* Y.

In the arrow diagram for a bijective function, every $y \in Y$ has *exactly one* arrow entering it. For example, the function p in Figure 1.1 is a bijection, but the other three functions are not. The following theorem will help us count bijections.

1.77. Theorem. Let X and Y be **finite** sets with the *same* number of elements. For all functions $f : X \to Y$, the following conditions are equivalent: f is injective; f is surjective; f is bijective.

Proof. Suppose X and Y both have n elements, and write $X = \{x_1, \ldots, x_n\}$. First assume that $f : X \to Y$ is injective. Then the image $f[X] = \{f(x_1), \ldots, f(x_n)\}$ is a subset of Y consisting of n *distinct* elements. Since Y has n elements, this subset must be all of Y. This means that every $y \in Y$ has the form $f(x_i)$ for some $x_i \in X$, so that f is surjective.

Next assume $f : X \to Y$ is surjective. We prove that f is a bijection by contradiction. If f were not a bijection, then f must not be one-to-one, which means there exist i, j with $i \neq j$ and $f(x_i) = f(x_j)$. It follows that the set $f[X] = \{f(x_1), \ldots, f(x_n)\}$ contains fewer than n elements, since the displayed list of members of $f[X]$ contains at least one duplicate. Thus $f[X]$ is a proper subset of Y. Letting y be any element of $Y - f[X]$, we see that y does not have the form $f(x)$ for any $x \in X$. Therefore f is not surjective, which is a contradiction.

Finally, if f is bijective, then f is injective by definition. □

The previous result does *not* extend to infinite sets, as shown by the following examples. Let $\mathbb{Z}_{>0} = \{1, 2, 3, \ldots\}$ be the set of positive integers. The function $f : \mathbb{Z}_{>0} \to \mathbb{Z}_{>0}$ defined by $f(n) = n + 1$ for $n \in \mathbb{Z}_{>0}$ is injective but not surjective. The function $g : \mathbb{Z}_{>0} \to \mathbb{Z}_{>0}$ defined by $g(2k) = g(2k-1) = k$ for all $k \in \mathbb{Z}_{>0}$ is surjective but not injective. The function $\exp : \mathbb{R} \to \mathbb{R}$ defined by $\exp(x) = e^x$ is injective but not surjective. The function $h : \mathbb{R} \to \mathbb{R}$ defined by $h(x) = x(x-1)(x+1)$ is surjective but not injective.

1.78. The Bijection-Counting Rule. Suppose X is an n-element set and Y is an m-element set. If $n = m$, then there are $n!$ bijections $f : X \to Y$. If $n \neq m$, then there are no bijections from X to Y.

Proof. Suppose X has n elements and Y has m elements. If $n = m$, then bijections from X to Y are the same as injections from X to Y by the previous theorem, so the number of bijections is $n!$ by the Injection Rule.

If $n < m$, then the image of any $f : X \to Y$ has size at most n. So the image of f cannot be all of Y, which means there are no surjections (and hence no bijections) from X to Y. If $n > m$, then there must be a repeated value in the list $f(x_1), f(x_2), \ldots, f(x_n)$, since each entry in this list comes from the m-element set Y. This means there are no injections (and hence no bijections) from X to Y. □

1.79. Remark. Compare the Word Rule to the Function Rule, the Partial Permutation Rule to the Injection Rule, and the Permutation Rule to the Bijection-Counting Rule. You will notice that the same formulas appear in each pair of rules. This is not a coincidence. Indeed, we can formally define a word $w_1 w_2 \cdots w_k$ over an n-element alphabet A as the *function* $w : \{1, 2, \ldots, k\} \to A$ defined by $w(i) = w_i$. The number of such words (functions) is n^k. The word $w_1 w_2 \cdots w_k$ is a partial permutation of A iff the w_i's are all distinct iff w is an *injective* function. The word $w_1 w_2 \cdots w_k$ is a permutation of A iff w is a *bijective* function. Finally, note that w is surjective iff every letter in the alphabet A occurs among the letters $w_1, \ldots, w_k$.

1.11 Cardinality and the Bijection Rule

So far we have been using the notation $|S| = n$ informally as an abbreviation for the statement "the set S has n elements." The number $|S|$ is called the *cardinality of S*. We can give a more formal definition of this notation in terms of bijections.

1.80. Definition: Cardinality. For all sets S and all integers $n \in \mathbb{Z}_{\geq 0}$, $|S| = n$ means there exists a bijection $f : S \to \{1, 2, \ldots, n\}$. For all sets A and B (possibly infinite), $|A| = |B|$ means there exists a bijection $f : A \to B$.

If $f : X \to Y$ and $g : Y \to Z$ are functions, the *composition* of g and f is the function $g \circ f : X \to Z$ defined by $(g \circ f)(x) = g(f(x))$ for $x \in X$. We state the following theorem without proof.

1.81. Theorem: Properties of Bijections. Let X, Y, Z be any sets. (a) The identity map $\mathrm{id}_X : X \to X$, defined by $\mathrm{id}_X(x) = x$ for all $x \in X$, is a bijection. Hence, $|X| = |X|$. (b) A function $f : X \to Y$ is bijective iff there exists a function $g : Y \to X$ such that $g \circ f = \mathrm{id}_X$ and $f \circ g = \mathrm{id}_Y$. If such a g exists, it is unique; we call it the *two-sided inverse* of f and denote it by f^{-1}. This inverse is also a bijection, and $(f^{-1})^{-1} = f$. Hence, $|X| = |Y|$ implies $|Y| = |X|$. (c) The composition of two bijections is a bijection. Hence, if $|X| = |Y|$ and $|Y| = |Z|$ then $|X| = |Z|$.

It can also be shown that for all sets S and all $n, m \in \mathbb{Z}_{\geq 0}$, if $|S| = n$ and $|S| = m$ then $n = m$, so the cardinality of a finite set is well-defined. This fact is equivalent to the assertion that there are no bijections from an n-element set to an m-element set when $m \neq n$. We gave an intuitive explanation of this statement in our proof of the Bijection-Counting Rule, but to give a rigorous formal proof, a delicate induction argument is needed (we omit this). In any case, this fact justifies the following rule, which is the foundation of bijective combinatorics.

1.82. The Bijection Rule. If B is an n-element set and there is a bijection $f : A \to B$ or a bijection $g : B \to A$, then $|A| = n$.

Here are some initial examples to illustrate the Bijection Rule.

1.83. Example: Subsets. Let A be the set of all subsets of $\{1, 2, \ldots, n\}$. We apply the Bijection Rule to show that $|A| = 2^n$ (we obtained the same result by a different method in the proof of the Power Set Rule 1.26). To use the Bijection Rule, we need another set B whose size is already known to be 2^n. We take B to be $\{0, 1\}^n$, the set of all n-letter words in the alphabet $\{0, 1\}$, which has size 2^n by the Word Rule. Now we must define a bijection $f : A \to B$. An object S in the domain of f is a subset of $\{1, 2, \ldots, n\}$. We define $f(S) = w_1 w_2 \cdots w_n$, where $w_i = 1$ if $i \in S$, and $w_i = 0$ if $i \notin S$. For example, taking $n = 5$, we have $f(\{1, 3, 4\}) = 10110$, $f(\{3, 5\}) = 00101$, $f(\emptyset) = 00000$, and $f(\{1, 2, 3, 4, 5\}) = 11111$.

To see that f is a bijection, we produce a two-sided inverse $g : B \to A$ for f. Given a word $w = w_1 w_2 \cdots w_n \in B$, define $g(w) = \{i : w_i = 1\}$, which is the subset of positions in the word that contain a 1. For example, taking $n = 5$, we have $g(01110) = \{2, 3, 4\}$, $g(00010) = \{4\}$, and $g(10111) = \{1, 3, 4, 5\}$. One sees immediately that $g(f(S)) = S$ for all $S \in A$, and $f(g(w)) = w$ for all $w \in B$, so g is the inverse of f. Thus f and g are bijections, so $|A| = |B| = 2^n$ by the Bijection Rule.

1.84. Example: Increasing and Decreasing Words. A word $w = w_1 w_2 \cdots w_k$ in the alphabet $\{1, 2, \ldots, n\}$ is *strictly increasing* iff $w_1 < w_2 < \cdots < w_k$; the word w is *weakly increasing* iff $w_1 \leq w_2 \leq \cdots \leq w_k$. Strictly decreasing and weakly decreasing words are defined similarly. How many k-letter words in the alphabet $X = \{1, 2, \ldots, n\}$ are: (a) strictly increasing; (b) strictly decreasing; (c) weakly increasing?

Trying to build words by the Product Rule does not work here because the number of choices for a given position depends on what value was chosen for the preceding position. Instead, we use the Bijection Rule. Let A be the set of strictly increasing k-letter words in the alphabet X, and let B be the set of k-element subsets of X. We know $|B| = \binom{n}{k}$ by the Subset Rule. Define $f : A \to B$ by setting $f(w_1 w_2 \cdots w_k) = \{w_1, w_2, \ldots, w_k\}$. For example, when $k = 4$ and $n = 9$, $f(2358) = \{2, 3, 5, 8\}$. The inverse of f is the map $g : B \to A$

defined by letting $g(S)$ be the list of the k elements in the subset S written in increasing order. For example, $g(\{4,9,7,1\}) = 1479$ and $g(\{8,2,3,4\}) = 2348$. We have $g(f(w)) = w$ for all $w \in A$. Since presenting the elements of a set in a different order does not change the set, we also have $f(g(S)) = S$ for all $S \in B$. For example, $f(g(\{3,7,5,1\})) = f(1357) = \{1,3,5,7\} = \{3,7,5,1\}$. Thus g really is a two-sided inverse for f, proving that f and g are bijections. We conclude that $|A| = |B| = \binom{n}{k}$.

To solve (b), let A' be the set of strictly decreasing k-letter words in the alphabet X. The *reversal map* $r : A \to A'$ given by $r(w_1w_2w_3\cdots w_k) = w_k\cdots w_3w_2w_1$ is a one-to-one map from A onto A'. By the Bijection Rule, $|A'| = |A| = \binom{n}{k}$.

To solve (c), we need a more subtle bijection. Let C be the set of weakly increasing k-letter words in the alphabet X. The map f from part (a) can no longer be used, since we cannot guarantee that the output will always be a subset of size k, and collisions may occur. For example, the sequence 1133 would map to the subset $\{1,1,3,3\} = \{1,3\}$, and the sequence 1333 also maps to $\{1,3\}$. We get around this problem by the following clever device. Let D be the set of all k-letter strictly increasing words in the expanded alphabet $\{1,2,\ldots,n,n+1,\ldots,n+k-1\}$. By part (a) with n replaced by $n+k-1$, we know $|D| = \binom{n+k-1}{k}$. Define a map $h : C \to D$ by setting $h(w_1w_2\cdots w_k) = z_1z_2\cdots z_k$, where $z_i = w_i + i - 1$ for $1 \leq i \leq k$. In other words, h acts by adding 0 to w_1, 1 to w_2, 2 to w_3, and so on. For example, taking $k = 4$ and $n = 6$, $h(1133) = 1256$, $h(1333) = 1456$, and $h(2356) = 2479$. Since the input $w \in C$ satisfies $1 \leq w_1 \leq w_2 \leq w_3 \leq \cdots \leq w_k \leq n$, we see that $1 \leq w_1 + 0 < w_2 + 1 < w_3 + 2 \leq \cdots < w_k + k - 1 \leq n + k - 1$. This shows that the output word $h(w)$ really does belong to the codomain D. To see that h is a bijection, we produce the inverse map $h' : D \to C$. Given $u = u_1u_2\cdots u_k \in D$, define $h'(u) = v_1v_2\cdots v_k$, where $v_i = u_i - (i-1)$ for $1 \leq i \leq k$. For example, $h'(2458) = 2335$, $h'(1379) = 1256$, and $h'(6789) = 6666$. Since u satisfies $1 \leq u_1 < u_2 < \cdots < u_k \leq n+k-1$, it follows that v satisfies $1 \leq v_1 \leq v_2 \leq \cdots \leq v_k \leq n$, so h' really does map D into C. It is immediate that $h'(h(w)) = w$ for all $w \in C$ and $h(h'(u)) = u$ for all $u \in D$, so h is a bijection with inverse h'. By the Bijection Rule, $|C| = |D| = \binom{n+k-1}{k}$. Applying a bijection that reverses words, we can also conclude that there are $\binom{n+k-1}{k}$ weakly decreasing k-letter words in the alphabet X.

1.85. Remark: Proving Bijectivity. When using the Bijection Rule, we must check that a given formula or algorithm really does define a bijective function f from a set X to a set Y. This can be done in two ways. One way is to check that: (a) for each $x \in X$, there is exactly one associated output $f(x)$ (so that f is *well-defined* or *single-valued*); (b) for each $x \in X$, $f(x)$ does lie in the set Y (so that f maps *into* Y); (c) for all $u, v \in X$, if $f(u) = f(v)$ then $u = v$ (so that f is injective); and (d) for all $y \in Y$, there exists $x \in X$ with $y = f(x)$ (so that f is surjective).

The second way is to produce a two-sided inverse g for f. In this case, we must check that f is well-defined and maps into Y, that g is well-defined and maps into X, that $g(f(x)) = x$ for all $x \in X$, and that $f(g(y)) = y$ for all $y \in Y$. In simple situations, we usually check all these conditions by inspection without writing a detailed proof. However, for bijective proofs involving complex combinatorial objects, it is necessary to give these details. In particular, in part (c) of the example above, the most crucial point to check was that h and h' both mapped *into* their claimed codomains.

1.12 Counting Multisets and Compositions

Recall that the concepts of *order* and *repetition* play no role when deciding whether two *sets* are equal. For instance, $\{1,3,5\} = \{3,5,1\} = \{1,1,1,5,5,3,3\}$ since all these sets have the same members. We now introduce the concept of a *multiset*, in which order still does not matter, but repetitions of a given element are significant.

1.86. Definition: Multisets. A *multiset* is a pair $M = (S, m)$, where S is a set and $m : S \to \mathbb{Z}_{>0}$ is a function. For $x \in S$, the number $m(x)$ is called the *multiplicity* of x in M or the *number of copies of* x in M. The *number of elements of* M is $|M| = \sum_{x \in S} m(x)$.

We often display a multiset as a list $[x_1, x_2, \ldots, x_k]$, where each $x \in S$ occurs exactly $m(x)$ times in the list. However, we must remember that the order of the elements in this list does not matter when deciding equality of multisets. For example, $[1,1,2,3,3] \neq [1,1,1,2,3,3] = [3,2,3,1,1,1]$.

1.87. Notational Convention. We use *square brackets* for multisets (repetition matters, order does not). We use *curly braces* for sets (order and repetition do not matter). We use *round parentheses* or *word notation* for sequences (order and repetition do matter). For example, $[w, x, x, z, z, z]$ is a multiset, $\{w, x, z\}$ is a set, (x, z, w, x, w) is a sequence, and $xzwxw$ is a word.

1.88. The Multiset Rule. The number of k-element multisets using letters from an n-letter alphabet is

$$\binom{k+n-1}{k, n-1} = \frac{(k+n-1)!}{k!(n-1)!}.$$

We give two proofs of this result. *First Proof:* Let $X = (x_1, x_2, \ldots, x_n)$ be a fixed (ordered) n-letter alphabet, and let U be the set of all k-element multisets using letters from X. Introduce the symbols $\star$ (star) and $|$ (bar), and let $V = \mathcal{R}(\star^k\,|^{n-1})$ be the set of all rearrangements of k stars and $n-1$ bars. By the Anagram Rule, $|V| = \binom{k+n-1}{k, n-1}$. It therefore suffices to define a bijection $f : U \to V$.

Given a multiset $M = (S, m) \in U$ with $S \subseteq X$, extend m to a function defined on all of X by letting $m(x_i) = 0$ if $x_i \notin S$. Then define $f(M)$ to be the word consisting of $m(x_1)$ stars, then a bar, then $m(x_2)$ stars, then a bar, and so on; we end with $m(x_n)$ stars *not* followed by a bar. Using the notation $\star^j$ to denote a sequence of j stars, we can write

$$f(M) = \star^{m(x_1)}|\star^{m(x_2)}|\cdots|\star^{m(x_{n-1})}|\star^{m(x_n)}.$$

Since M is a multiset of size k, the word $f(M)$ contains k stars. Since there is no bar after the stars for x_n, the word $f(M)$ contains $n-1$ bars. Thus $f(M)$ does belong to the set V. For example, if $X = (w, x, y, z)$ and $k = 3$, we have

$$f([w,w,w]) = \star\star\star|||, \quad f([w,x,y]) = \star|\star|\star|, \quad f([x,x,x]) = |\star\star\star||, \quad f([x,z,z]) = |\star||\star\star.$$

To see that f is a bijection, we define a map $f' : V \to U$ by letting $f'(\star^{m_1}|\star^{m_2}|\cdots|\star^{m_n})$ be the unique multiset that has m_i copies of x_i for $1 \leq i \leq n$ (note each $m_i \geq 0$). Since $\sum_{i=1}^{n} m_i = k$, this is a k-element multiset using letters from X. For example, if $n = 6$, $k = 4$, and $X = (1,2,3,4,5,6)$, then $f'(||\star||\star|\star\star) = [3,5,6,6]$. It is routine to check that f' is the two-sided inverse of f.

Second Proof: By replacing x_i by i, we may assume without loss of generality that the alphabet X is $(1,2,\ldots,n)$. As above, let U be the set of all k-element multisets using letters from X. Let W be the set of weakly increasing sequences of elements of X having length k. We found $|W| = \binom{k+n-1}{k}$ in Example 1.84(c). By the Bijection Rule, it suffices to define a bijection $g : W \to U$. Given $w = (w_1 \le w_2 \le \cdots \le w_k)$ in W, let $g(w)$ be the multiset $[w_1, w_2, \ldots, w_k]$ in U. For example, taking $n = 6$ and $k = 4$, $g(1135) = [1,1,3,5]$. The inverse of g is the map $g' : U \to W$ sending a multiset $M \in U$ to the sequence of elements of M (repeated according to their multiplicity) written in weakly increasing order. For example, $g'([4,2,2,4]) = 2244$ and $g'([5,1,6,1]) = 1156$. We have $g(g'(M)) = M$ for all $M \in U$ since order does not matter in a multiset. Similarly, $g'(g(w)) = w$ for all $w \in W$, so g is a bijection with inverse g'.

We can also view the Multiset Rule as counting the number of solutions to a certain linear equation using variables ranging over $\mathbb{Z}_{\ge 0} = \{0,1,2,3,\ldots\}$.

1.89. The Integer Equation Rule. Let $n > 0, k \ge 0$ be fixed integers. (a) The number of sequences $(z_1, z_2, \ldots, z_n)$ with all $z_i \in \mathbb{Z}_{\ge 0}$ and $z_1 + z_2 + \cdots + z_n = k$ is $\binom{k+n-1}{k,n-1}$. (b) The number of sequences $(y_1, y_2, \ldots, y_n)$ with all $y_i \in \mathbb{Z}_{>0}$ and $y_1 + y_2 + \cdots + y_n = k$ is $\binom{k-1}{n-1}$.

Proof. (a) A particular solution $(z_1, z_2, \ldots, z_n)$ to the given equation corresponds to the k-element multiset M in the alphabet $\{1,2,\ldots,n\}$ where i occurs z_i times for $1 \le i \le n$. This correspondence is evidently a bijection, so part (a) follows from the Bijection Rule and the Multiset Rule.

(b) Note that $y_1 + y_2 + \cdots + y_n = k$ holds for given $y_i \in \mathbb{Z}_{>0}$ iff $(y_1 - 1) + (y_2 - 1) + \cdots + (y_n - 1) = k - n$ holds with each $y_i - 1 \in \mathbb{Z}_{\ge 0}$. Setting $z_1 = y_1 - 1, \ldots, z_n = y_n - 1$, we are reduced to counting solutions to $z_1 + \cdots + z_n = k - n$ where all $z_i \in \mathbb{Z}_{\ge 0}$. By (a) with k replaced by $k - n$, the number of solutions to this equation is $\binom{k-n+n-1}{k-n,n-1} = \binom{k-1}{n-1}$. □

A variation of the last problem is to count positive integer solutions to $y_1 + y_2 + \cdots + y_n = k$ where n is not fixed in advance. In this context, the following terminology is used.

1.90. Definition: Compositions. A *composition* of an integer $k > 0$ is a sequence $\alpha = (\alpha_1, \alpha_2, \ldots, \alpha_s)$ where each α_i is a positive integer and $\alpha_1 + \alpha_2 + \cdots + \alpha_s = k$. The *number of parts* of α is s. Let $\mathrm{Comp}(k)$ be the set of all compositions of k.

1.91. Example. The sequences $(1,3,1,3,3)$ and $(3,3,3,1,1)$ are two distinct compositions of 11 with five parts. The four compositions of 3 are (3), $(2,1)$, $(1,2)$, and $(1,1,1)$.

1.92. The Composition Rule. (a) For all $k > 0$, there are 2^{k-1} compositions of k. (b) For all $k, s > 0$, there are $\binom{k-1}{s-1}$ compositions of k with s parts.

Proof. For (a), we define a bijection $g : \mathrm{Comp}(k) \to \{0,1\}^{k-1}$. Given $\alpha = (\alpha_1, \alpha_2, \ldots, \alpha_s) \in \mathrm{Comp}(k)$, define

$$g(\alpha) = 0^{\alpha_1 - 1} 1 0^{\alpha_2 - 1} 1 \cdots 1 0^{\alpha_s - 1}.$$

Here, the notation 0^j denotes a sequence of j consecutive zeroes, and 0^0 denotes the empty word. For example, $g((3,1,3)) = 001100$. Since $\sum_{i=1}^{s} (\alpha_i - 1) = k - s$ and there are $s - 1$ ones, we see that $g(\alpha) \in \{0,1\}^{k-1}$. Now define $g' : \{0,1\}^{k-1} \to \mathrm{Comp}(k)$ as follows. We can uniquely write any word $w \in \{0,1\}^{k-1}$ in the form $w = 0^{b_1} 1 0^{b_2} 1 \cdots 1 0^{b_s}$ where $s \ge 1$, each $b_i \ge 0$, and $\sum_{i=1}^{s} b_i = (k-1) - (s-1) = k - s$ since there are $s - 1$ ones. Define $g'(w) = (b_1 + 1, b_2 + 1, \ldots, b_s + 1)$, which is a composition of k. For example, $g'(100100) = (1,3,3)$. It can be checked that g' is the two-sided inverse of g, so g is a bijection. By the Bijection Rule, $|\mathrm{Comp}(k)| = |\{0,1\}^{k-1}| = 2^{k-1}$.

Part (b) merely restates part (b) of the Integer Equation Rule. Alternatively, if we restrict the domain of g to the set of compositions of k with a given number s of parts, we get a bijection from this set onto the set of words $\mathcal{R}(0^{k-s} 1^{s-1})$, so part (b) also follows from the Anagram Rule. □

The bijections in the preceding proof are best understood pictorially. We represent an integer $i > 0$ as a sequence of i unit squares glued together. We visualize a composition $(\alpha_1, \ldots, \alpha_s)$ by drawing the squares for $\alpha_1, \ldots, \alpha_s$ in a single row, separated by gaps. For instance, the composition $(1, 3, 1, 3, 3)$ is represented by the picture

We now scan the picture from left to right and record what happens between each two successive boxes. If the two boxes in question are glued together, we record a 0; if there is a gap between the two boxes, we record a 1. The composition of 11 pictured above maps to the word $1001100100 \in \{0,1\}^{10}$. Going the other way, the word $0101000011 \in \{0,1\}^{10}$ leads first to the picture

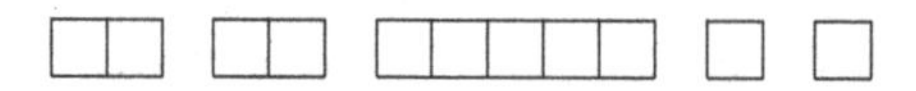

and then to the composition $(2, 2, 5, 1, 1)$. It can be checked that the pictorial operations just described correspond precisely to the maps g and g' in the proof above. When $n = 3$, we have:

$$g((3)) = 00; \quad g((2,1)) = 01; \quad g((1,2)) = 10; \quad g((1,1,1)) = 11.$$

1.13 Counting Balls in Boxes

Many counting questions can be reduced to the problem of counting the number of ways to distribute k balls into n boxes. There are several variations of this problem, depending on whether the balls are labeled or unlabeled, whether the boxes are labeled or unlabeled, and whether we place restrictions on the number of balls in each box. The two most common restrictions require that each box have at most one ball, or that each box have at least one ball. In this section, we study versions of the problem where the boxes are labeled and the balls may or may not be labeled. Versions involving unlabeled boxes will be solved later, when we study integer partitions and set partitions (see §2.11 through §2.13).

1.93. Counting Labeled Balls in Labeled Boxes. (a) The number of ways to put k balls labeled $1, 2, \ldots, k$ into n boxes labeled $1, 2, \ldots, n$ is n^k. (b) If each box can contain at most one ball, the number of distributions is $n!/(n-k)!$ for $k \leq n$ and 0 for $k > n$. (c) If each box must contain at least one ball, the number of distributions is $\sum_{j=0}^{n}(-1)^j\binom{n}{j}(n-j)^k$.

Proof. When the balls and boxes are both labeled, we can encode the distribution of balls into boxes by a function $f : \{1, 2, \ldots, k\} \to \{1, 2, \ldots, n\}$, by letting $y = f(x)$ iff the ball labeled x goes into the box labeled y. So (a) follows from the Function Rule. The requirement in (b) that each box (potential output y) contains at most one ball translates into the requirement that the corresponding function be injective. So (b) follows from the Injection Rule. Similarly, the requirement in (c) that each box be nonempty translates into the condition that the corresponding function be surjective. The formula in (c) is the number of surjections from a k-element set onto an n-element set; we will prove this formula in §4.3. □

1.94. Counting Unlabeled Balls in Labeled Boxes. (a) The number of ways to put k identical balls into n boxes labeled $1, 2, \ldots, n$ is $\binom{k+n-1}{k,n-1}$. (b) If each box can contain at most one ball, the number of distributions is $\binom{n}{k}$. (c) If each box must contain at least one ball, the number of distributions is $\binom{k-1}{n-1}$.

Proof. When the boxes are labeled but the balls are not, we can model a distribution of k balls into the n boxes by a *multiset* of size k using the alphabet $\{1, 2, \ldots, n\}$, where the multiset contains i copies of j iff the box labeled j contains i balls. So (a) follows from the Multiset Rule. The requirement in (b) that each box contains at most one ball means that the corresponding multiset contains at most one copy of each box label. In this case, the multiset is the same as an ordinary k-element subset of the n box labels. So (b) follows from the Subset Rule. The problem in (c) is the same as counting the number of solutions to $y_1 + y_2 + \cdots + y_n = k$, where each y_i is a *positive* integer giving the number of balls in box i. So (c) follows from part (b) of the Integer Equation Rule. □

1.95. Example. How many ways can we give twelve identical cookies and four different baseball cards to seven kids if each kid must receive at least one cookie and at most one baseball card? We solve this with the Product Rule, first choosing who gets the cookies and then choosing who gets the baseball cards. The first choice can be modeled as a placement of 12 unlabeled balls (the cookies) in 7 labeled boxes (the kids) where each box must contain at least one ball. As seen above, there are $\binom{12-1}{7-1} = 462$ ways to do this. The second choice can be modeled as a placement of 4 labeled balls (the baseball cards) in 7 labeled boxes (the kids) where each box contains at most one ball. There are $7!/(7-4)! = 840$ ways to do this. So the answer is $462 \cdot 840 = 388{,}080$.

1.96. Example. How many anagrams of `TETRAGRAMMATON` have no consecutive vowels? To begin the solution, let us first figure out how many valid patterns of consonants and vowels are possible. We can encode this pattern by a word in $\mathcal{R}(\mathrm{C}^9\mathrm{V}^5)$ with no two adjacent V's; for instance, the original word has vowel-consonant pattern CVCCVCCVCCVCVC. Each such word has the form

$$\mathrm{C}^{x_1}\mathrm{V}\mathrm{C}^{x_2}\mathrm{V}\mathrm{C}^{x_3}\mathrm{V}\mathrm{C}^{x_4}\mathrm{V}\mathrm{C}^{x_5}\mathrm{V}\mathrm{C}^{x_6},$$

where $x_1 + x_2 + \cdots + x_6 = 9$, x_1 and x_6 are nonnegative integers, and x_2, x_3, x_4, x_5 are strictly positive integers. Setting $y_1 = x_1$, $y_i = x_i - 1$ for $i = 2, 3, 4, 5$, and $y_6 = x_6$, the given equation in the x_i's is equivalent to the equation $y_1 + y_2 + \cdots + y_6 = 5$ with all $y_i \in \mathbb{Z}_{\geq 0}$. By the Integer Equation Rule, the number of solutions is $\binom{5+6-1}{5,6-1} = 252$. We could have also viewed this as counting the number of ways to put nine unlabeled balls (the consonants) into six labeled boxes (the spaces before and after the vowels) such that the four middle boxes must be nonempty.

We finish the problem with the Product Rule and Anagram Rule. Build an anagram by first choosing a valid vowel-consonant pattern (252 ways); then replacing the C's in the chosen pattern (from left to right) by an anagram in $\mathcal{R}(\mathrm{G}^1\mathrm{M}^2\mathrm{N}^1\mathrm{R}^2\mathrm{T}^3)$ in one of $\binom{9}{1,2,1,2,3} =$ 15,120 ways; then replacing the V's in the chosen pattern (from left to right) by an anagram in $\mathcal{R}(\mathrm{A}^3\mathrm{E}^1\mathrm{O}^1)$ in one of $\binom{5}{3,1,1} = 20$ ways. The final answer is 76,204,800.

1.14 Counting Lattice Paths

In this section, we give more illustrations of the Bijection Rule by counting combinatorial objects called lattice paths.

1.97. Definition: Lattice Paths. A *lattice path* in the plane is a sequence

$$P = ((x_0, y_0), (x_1, y_1), \ldots, (x_k, y_k)),$$

where the x_i's and y_i's are integers, and for each $i \geq 1$, either $(x_i, y_i) = (x_{i-1} + 1, y_{i-1})$ or $(x_i, y_i) = (x_{i-1}, y_{i-1} + 1)$. We say that P is a path *from* (x_0, y_0) *to* (x_k, y_k).

We often take (x_0, y_0) to be the origin $(0,0)$. The sequence P encodes a path consisting of line segments of length 1 from (x_{i-1}, y_{i-1}) to (x_i, y_i) for $1 \leq i \leq k$. For example, Figure 1.2 displays the ten lattice paths from $(0,0)$ to $(2,3)$.

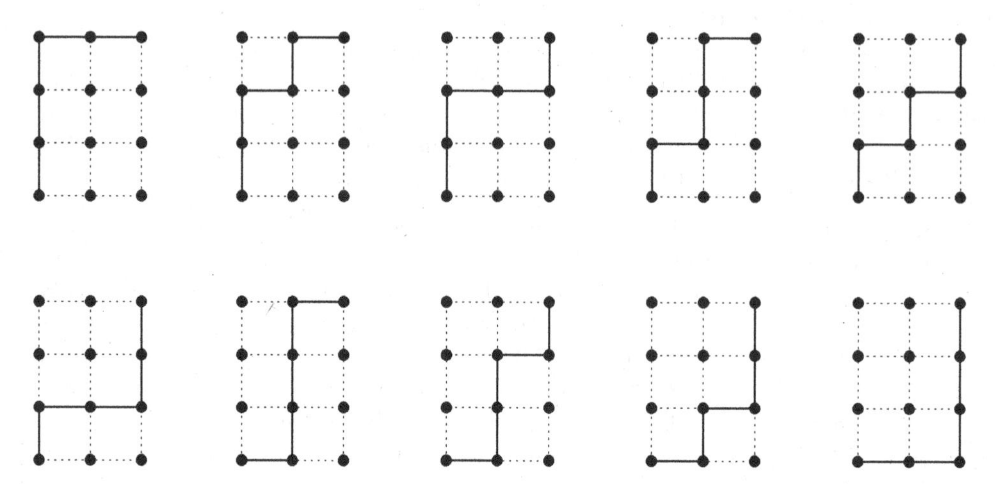

FIGURE 1.2
Lattice paths from $(0,0)$ to $(2,3)$.

1.98. The Lattice Path Rule. For all integers $a, b \geq 0$, there are $\binom{a+b}{a,b}$ lattice paths from $(0,0)$ to (a,b).

Proof. We can encode a lattice path P from $(0,0)$ to (a,b) as a word $w \in \mathcal{R}(\mathrm{E}^a\mathrm{N}^b)$ by setting $w_i = \mathrm{E}$ if $(x_i, y_i) = (x_{i-1}+1, y_{i-1})$ and $w_i = \mathrm{N}$ if $(x_i, y_i) = (x_{i-1}, y_{i-1}+1)$. Here, E stands for "east step," and N stands for "north step." Since the path ends at (a,b), w must have exactly a occurrences of E and exactly b occurrences of N. Thus we have a bijection between the given set of lattice paths and the set $\mathcal{R}(\mathrm{E}^a\mathrm{N}^b)$. Since $|\mathcal{R}(\mathrm{E}^a\mathrm{N}^b)| = \binom{a+b}{a,b}$, the result follows from the Bijection Rule. □

For example, the paths shown in Figure 1.2 are encoded by the words

NNNEE, NNENE, NNEEN, NENNE, NENEN,
NEENN, ENNNE, ENNEN, ENENN, EENNN.

More generally, we can consider lattice paths in $\mathbb{R}^d$. Such a path is a sequence of points $(v_0, v_1, \ldots, v_k)$ in $\mathbb{Z}^d$ such that for each i, $v_i = v_{i-1} + e_j$ for some standard basis vector $e_j = (0, \ldots, 1, \ldots, 0) \in \mathbb{R}^d$ (the 1 occurs in position j). We can encode a path P from $(0, \ldots, 0)$ to $(n_1, \ldots, n_d)$ by a word $w_1 w_2 \cdots w_n \in \mathcal{R}(e_1^{n_1} \cdots e_d^{n_d})$, where $n = n_1 + \cdots + n_d$ and $w_i = e_j$ iff $v_i = v_{i-1} + e_j$. By the Bijection Rule and the Anagram Rule, the number of such lattice paths is the multinomial coefficient $\binom{n_1+n_2+\cdots+n_d}{n_1,n_2,\ldots,n_d}$.

Henceforth, we usually will not distinguish between a lattice path (which is a sequence of lattice points) and the word that encodes the lattice path. We now turn to a more difficult enumeration problem involving lattice paths.

1.99. Definition: Dyck Paths. A *Dyck path of order n* is a lattice path from $(0,0)$ to (n,n) such that $y_i \geq x_i$ for all points (x_i, y_i) on the path. This requirement means that the path always stays weakly above the line $y = x$. For example, Figure 1.3 displays the five Dyck paths of order 3.

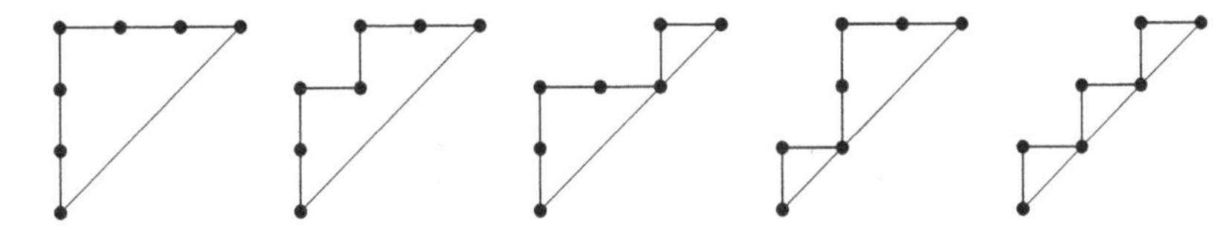

FIGURE 1.3
Dyck paths of order 3.

1.100. Definition: Catalan Numbers. For $n \geq 0$, the *nth Catalan number* is

$$C_n = \frac{1}{n+1}\binom{2n}{n,n} = \frac{1}{2n+1}\binom{2n+1}{n+1,n} = \frac{(2n)!}{n!(n+1)!} = \binom{2n}{n,n} - \binom{2n}{n+1,n-1}.$$

One may check that these expressions are all equal. For instance,

$$\binom{2n}{n,n} - \binom{2n}{n+1,n-1} = \frac{(2n)!}{n!n!} - \frac{(2n)!}{(n+1)!(n-1)!} = \frac{(2n)!}{n!n!}\left[1 - \frac{n}{n+1}\right] = \frac{1}{n+1}\binom{2n}{n,n}.$$

The first few Catalan numbers are

$$C_0 = 1, \quad C_1 = 1, \quad C_2 = 2, \quad C_3 = 5, \quad C_4 = 14, \quad C_5 = 42, \quad C_6 = 132, \quad C_7 = 429.$$

1.101. The Dyck Path Rule. For all $n \geq 0$, the number of Dyck paths of order n is the Catalan number $C_n = \binom{2n}{n,n} - \binom{2n}{n+1,n-1}$.

Proof. We present a remarkable bijective proof due to D. André [3]. Let A be the set of all lattice paths from $(0,0)$ to (n,n); let B be the set of all lattice paths from $(0,0)$ to $(n+1,n-1)$; let C be the set of all Dyck paths of order n; and let $D = A - C$ be the set of paths from $(0,0)$ to (n,n) that go strictly below the line $y = x$. Since $C = A - D$, the Difference Rule gives

$$|C| = |A| - |D|.$$

We already know that $|A| = \binom{2n}{n,n}$ and $|B| = \binom{2n}{n+1,n-1}$. To establish the required formula $|C| = C_n$, it therefore suffices to exhibit a bijection $r : D \to B$.

We define r as follows. Given a path $P \in D$, follow the path backwards from (n,n) until it goes below the diagonal $y = x$ for the first time. Let (x_i, y_i) be the first lattice point we encounter that is below $y = x$; this point must lie on the line $y = x - 1$. P is the concatenation of two lattice paths P_1 and P_2, where P_1 goes from $(0,0)$ to (x_i, y_i) and P_2 goes from (x_i, y_i) to (n,n). By choice of i, every lattice point of P_2 after (x_i, y_i) lies strictly above the line $y = x - 1$. Now, let P_2' be the path from (x_i, y_i) to $(n+1, n-1)$ obtained by reflecting P_2 in the line $y = x - 1$. Define $r(P)$ to be the concatenation of P_1 and P_2'. See Figure 1.4 for an example. Here, $(x_i, y_i) = (7,6)$, $P_1 =$ NEEENNENEEENN, $P_2 =$ NNNEENE, and $P_2' =$ EEENNEN. Note that $r(P)$ is a lattice path from $(0,0)$ to $(n+1, n-1)$, so $r(P) \in B$. Furthermore, (x_i, y_i) is the only lattice point of P_2' lying on the line $y = x - 1$.

The inverse map $r' : B \to D$ acts as follows. Given $Q \in B$, choose i maximal such that (x_i, y_i) is a point of Q on the line $y = x - 1$. Such an i must exist, since there is no way for a lattice path to reach $(n+1, n-1)$ from $(0,0)$ without passing through this line. Write $Q = Q_1Q_2$, where Q_1 goes from $(0,0)$ to (x_i, y_i) and Q_2 goes from (x_i, y_i) to $(n+1, n-1)$. Let Q_2' be the reflection of Q_2 in the line $y = x - 1$. Define $r'(Q) = Q_1Q_2'$, and note that this is a lattice path from $(0,0)$ to (n,n) which passes through (x_i, y_i), and hence

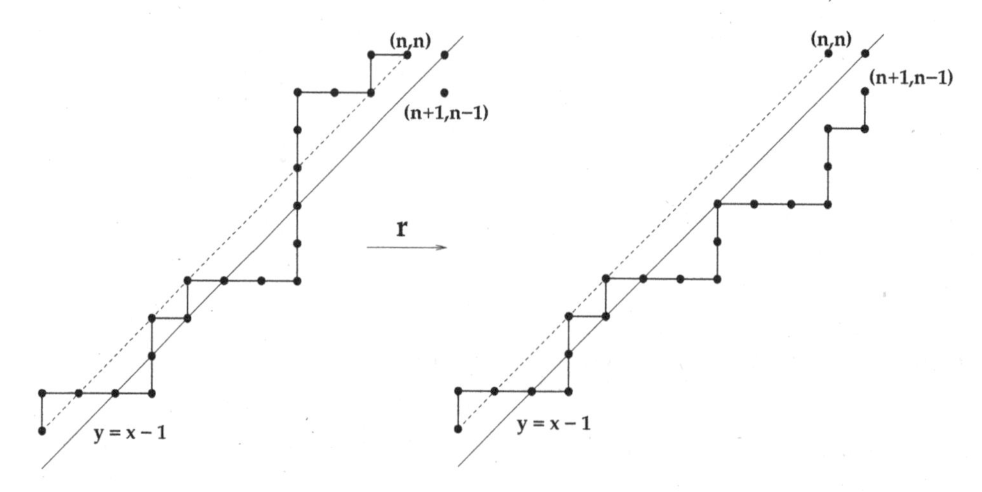

FIGURE 1.4
Example of the reflection map r.

lies in D. See Figure 1.5 for an example. Here, $(x_i, y_i) = (6, 5)$, $Q_1 =$ NNENEEEEENEN, $Q_2 =$ EEENENENENN, and $Q_2' =$ NNNENENENEE. From our observations about the point (x_i, y_i) in this paragraph and the last, it follows that r' is the two-sided inverse of r. □

The technique used in the preceding proof is called *André's Reflection Principle.* Another proof of the Dyck Path Rule, which leads directly to the formula $\frac{1}{2n+1}\binom{2n+1}{n+1,n}$, is given in §12.1. Yet another proof, which leads directly to the formula $\frac{1}{n+1}\binom{2n}{n,n}$, is given in §12.2.

1.15 Proofs of the Sum Rule and the Product Rule

Many people accept the Sum Rule and the Product Rule as being intuitively evident. Nevertheless, it is possible to give formal proofs of these results using induction and the definition of cardinality. We sketch these proofs now.

Step 1. Assume A and B are finite disjoint sets with $|A| = n$ and $|B| = m$; we prove $|A \cup B| = n + m$. The assumption $|A| = n$ means that there is a bijection $f : A \to \{1, 2, \ldots, n\}$. The assumption $|B| = m$ means that there is a bijection $g : B \to \{1, 2, \ldots, m\}$. Define a function $h : A \cup B \to \{1, 2, \ldots, n + m\}$ by setting

$$h(x) = \begin{cases} f(x) & \text{if } x \in A; \\ g(x) + n & \text{if } x \in B. \end{cases}$$

Since A and B have no common elements, h is a well-defined (single-valued) function. Moreover, since $f(x) \in \{1, 2, \ldots, n\}$ and $g(x) \in \{1, 2, \ldots, m\}$, h does map into the required codomain $\{1, 2, \ldots, n + m\}$. To see that h is a bijection, we display a two-sided inverse $h' : \{1, 2, \ldots, n + m\} \to A \cup B$. We define

$$h'(i) = \begin{cases} f^{-1}(i) & \text{if } 1 \le i \le n; \\ g^{-1}(i - n) & \text{if } n + 1 \le i \le n + m. \end{cases}$$

It is routine to verify that h' is single-valued, h' maps into the codomain $A \cup B$, and $h \circ h'$ and $h' \circ h$ are identity maps.

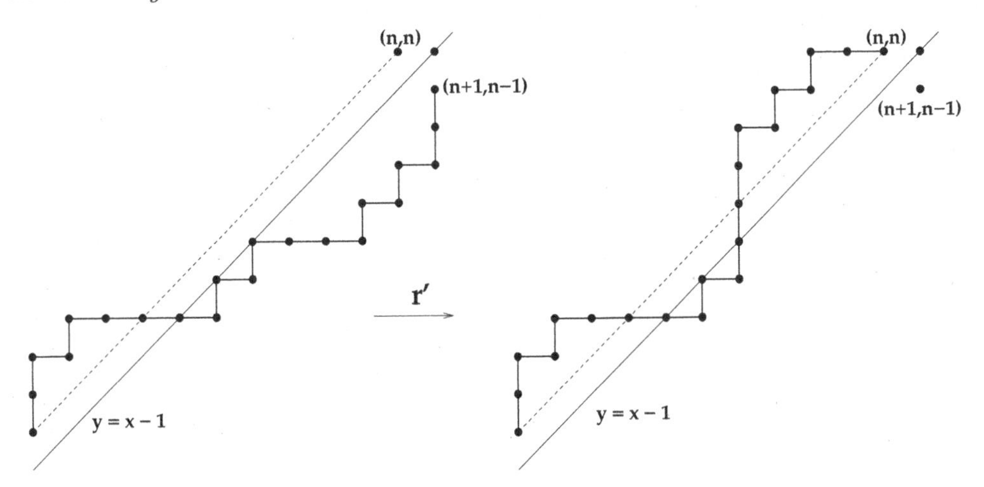

FIGURE 1.5
Example of the inverse reflection map.

Step 2. We prove the Sum Rule, as formulated in 1.39, by induction on $k \geq 1$. Assume $S_1, \ldots, S_k$ are finite, pairwise disjoint sets with $|S_i| = m_i$ for $i = 1, \ldots, k$; we want to prove $|S_1 \cup \cdots \cup S_k| = m_1 + \cdots + m_k$. This is immediate if $k = 1$; it follows from Step 1 if $k = 2$. For fixed $k > 2$, we may assume by induction that $|S_1 \cup \cdots \cup S_{k-1}| = m_1 + \cdots + m_{k-1}$ is already known. Now we apply Step 1 taking $A = S_1 \cup \cdots \cup S_{k-1}$ and $B = S_k$. Since S_k is disjoint from each of the sets $S_1, \ldots, S_{k-1}$, B is disjoint from A. Then Step 1 and the induction hypothesis give

$$|S_1 \cup \cdots \cup S_{k-1} \cup S_k| = |A \cup B| = |A| + |B| = (m_1 + \cdots + m_{k-1}) + m_k.$$

Step 3. Suppose $S_1, \ldots, S_k$ are any finite sets with $|S_i| = n_i$ for $i = 1, \ldots, k$; we prove $|S_1 \times S_2 \times \cdots \times S_k| = n_1 n_2 \cdots n_k$ by induction on $k \geq 1$. The result is immediate for $k = 1$. For fixed $k > 1$, we may assume by induction that the set $A = S_1 \times \cdots \times S_{k-1}$ has size $|A| = n_1 \cdots n_{k-1}$. Any set of the form $A \times \{c\}$ also has size $n_1 \cdots n_{k-1}$, since $f(x_1, \ldots, x_{k-1}) = (x_1, \ldots, x_{k-1}, c)$ defines a bijection from A onto this set. Now $S_1 \times \cdots \times S_{k-1} \times S_k = A \times S_k$ is the union of n_k pairwise disjoint sets of the form $A \times \{c\}$, as c ranges over the n_k elements in S_k. Each of these sets has size $n_1 \cdots n_{k-1}$, so Step 2 shows that

$$|S_1 \times \cdots \times S_k| = n_1 \cdots n_{k-1} + \cdots + n_1 \cdots n_{k-1} \ (n_k \text{ terms}) = n_1 \cdots n_{k-1} n_k.$$

Step 4. We prove the Product Rule 1.1 by formally restating the rule as follows. Suppose we build the objects in a set S by making a sequence of k choices, where there are n_i possible choices at stage i for $i = 1, 2, \ldots, k$. We model the choices available at stage i by the n_i-element set $S_i = \{1, 2, \ldots, n_i\}$, and we model the full choice sequence by an ordered k-tuple $(c_1, \ldots, c_k) \in S_1 \times \cdots \times S_k$. The hypothesis of the Product Rule amounts to the assumption that *there exists a bijection* $f : S_1 \times \cdots \times S_k \to S$. The bijection f tells us how to take a given list of choices $(c_1, \ldots, c_k)$ and build an object in S. (Note that *how* f *uses* the choice value c_i is allowed to depend on the previous choices $c_1, \ldots, c_{i-1}$; but the number of possibilities for c_i does not depend on previous choices. We usually specify f by an informal verbal description or an algorithm.) It follows from the Bijection Rule and Step 3 that $|S| = n_1 \cdots n_k$, as needed.

1.102. Remark. We can unravel the induction arguments above to obtain explicit bijective proofs of the general Sum Rule (for k sets) and the general Product Rule. These bijections will help us create algorithms (called ranking and unranking maps) for listing and storing collections of combinatorial objects. We discuss such algorithms in Chapter 6.

Summary

We end each chapter by summarizing the main definitions and results from the chapter.

- **Notation.**
  Factorials: $0! = 1$ and $n! = n \cdot (n-1) \cdot \ldots \cdot 3 \cdot 2 \cdot 1$.
  Binomial Coefficients: $\binom{n}{k} = \frac{n!}{k!(n-k)!}$ for $0 \leq k \leq n$; $\binom{n}{k} = 0$ otherwise.
  Multinomial Coefficients: $\binom{n}{n_1,\ldots,n_k} = \frac{n!}{n_1!n_2!\cdots n_k!}$, where $n = \sum_{i=1}^{k} n_i$ and $n_i \geq 0$.
  Anagram Sets: $\mathcal{R}(a_1^{n_1} \cdots a_k^{n_k})$ is the set of all words consisting of n_i copies of a_i.
  Types of Braces: $(a_1, \ldots, a_n)$ is a sequence (order and repetition matter);
  $\{a_1, \ldots, a_n\}$ is a set (order and repetition do not matter);
  $[a_1, \ldots, a_n]$ is a multiset (repetition matters, order does not).

- **Definitions of Combinatorial Objects.**
  Words: sequences $w_1 w_2 \cdots w_k$ (order matters, repeated letters can occur).
  Permutations of A: sequences $w_1 w_2 \cdots w_n$ using each letter in A exactly once.
  Partial Permutations: sequences $w_1 w_2 \cdots w_k$ (order matters, no repeated letters).
  Anagrams: words where each letter occurs a specified number of times.
  Sets: collections of objects where order and repetition do not matter.
  Multisets: collections of objects where repetition matters but order does not.
  Functions: $f : X \to Y$ means for all $x \in X$ there is a unique $y \in Y$ with $y = f(x)$.
  Injections: functions such that unequal inputs map to unequal outputs.
  Surjections: functions such that every y in the codomain is $f(x)$ for some x.
  Bijections: functions that are injective and surjective; functions with two-sided inverses.
  Compositions of k: sequences $\alpha = (\alpha_1, \ldots, \alpha_s)$ of positive integers summing to k.
  Lattice Paths: sequences of points in $\mathbb{Z}^2$ joined by north and east steps.
  Dyck Paths: lattice paths from $(0,0)$ to (n,n) not going below $y = x$.

- **Basic Counting Rules.**
  The Product Rule: If every object in S can be constructed in exactly one way by making a sequence of k choices, and there are n_i ways to make choice i no matter what choices were made earlier, then $|S| = n_1 n_2 \cdots n_k$.

  The Sum Rule: If every object in S belongs to exactly one of k non-overlapping categories, where category i contains m_i objects, then $|S| = m_1 + m_2 + \cdots + m_k$.

  The Word Rule: There are n^k word of length k using an n-letter alphabet.

  The Permutation Rule: There are $n!$ permutations of an n-letter alphabet.

  The Partial Permutation Rule: For k between 0 and n, the number of k-letter partial permutations of an n-letter alphabet is $n!/(n-k)!$.

  The Power Set Rule: An n-element set has 2^n subsets.

  The Subset Rule: The number of k-element subsets of an n-element set is $\binom{n}{k}$.

  The Anagram Rule: $|\mathcal{R}(a_1^{n_1} a_2^{n_2} \cdots a_k^{n_k})| = \binom{n_1+n_2+\cdots+n_k}{n_1,n_2,\ldots,n_k}$.

  The Union Rule for Two Sets: When S and T are finite, $|S \cup T| = |S| + |T| - |S \cap T|$.

The Disjoint Union Rule: If $|S_i| = m_i$ for $1 \leq i \leq k$ and $S_i \cap S_j = \emptyset$ for $i \neq j$, then $|S_1 \cup S_2 \cup \cdots \cup S_k| = m_1 + m_2 + \cdots + m_k$.

The Difference Rule: When S is finite, $|S - U| = |S| - |S \cap U|$.

The Cartesian Product Rule: If $|S_i| = n_i$ for $1 \leq i \leq k$, then $|S_1 \times S_2 \times \cdots \times S_k| = n_1 n_2 \cdots n_k$.

The Function Rule: When $|X| = k$ and $|Y| = n$, there are n^k functions $f : X \to Y$.

The Injection Rule: When $|X| = k$ and $|Y| = n \geq k$, there are $n!/(n-k)!$ injections $f : X \to Y$.

The Surjection Rule: When $|X| = k$ and $|Y| = n$, there are $\sum_{j=0}^{n} (-1)^j \binom{n}{j} (n-j)^k$ surjections $f : X \to Y$ (see §4.3 for the proof).

The Bijection-Counting Rule: If $|X| = |Y| = n$, there are $n!$ bijections $f : X \to Y$.

The Bijection Rule: If there exists a bijection $f : A \to B$ or $g : B \to A$, then $|A| = |B|$.

The Multiset Rule: The number of k-letter multisets using an n-letter alphabet is $\binom{k+n-1}{k,n-1}$.

The Integer Equation Rule: The number of nonnegative integer solutions to $z_1 + \cdots + z_n = k$ is $\binom{k+n-1}{k,n-1}$. The number of positive integer solutions to $y_1 + \cdots + y_n = k$ is $\binom{k-1}{n-1}$.

The Composition Rule: There are 2^{k-1} compositions of k. There are $\binom{k-1}{s-1}$ compositions of k with s parts.

Rules for Labeled Balls in Labeled Boxes: There are n^k ways to put k labeled balls in n labeled boxes. If at most one ball goes in each box, there are $n!/(n-k)!$ distributions for $k \leq n$. If at least one ball goes in each box, there are $\sum_{j=0}^{n} (-1)^j \binom{n}{j} (n-j)^k$ distributions.

Rules for Unlabeled Balls in Labeled Boxes: There are $\binom{k+n-1}{k,n-1}$ ways to put k identical balls in n labeled boxes. If at most one ball goes in each box, there are $\binom{n}{k}$ distributions. If at least one ball goes in each box, there are $\binom{k-1}{n-1}$ distributions.

The Lattice Path Rule: There are $\binom{a+b}{a,b}$ lattice paths from (x_0, y_0) to $(x_0 + a, y_0 + b)$.

The Dyck Path Rule: There are $C_n = \frac{1}{n+1}\binom{2n}{n}$ Dyck paths from $(0,0)$ to (n,n).

- **Probability Definitions.** A *sample space* is the set S of outcomes for some random experiment. An *event* is a subset of the sample space. When all outcomes in S are equally likely, the *probability* of an event A is $P(A) = |A|/|S|$. The *conditional probability of A given B* is $P(A|B) = P(A \cap B)/P(B)$, when $P(B) > 0$. Events A and B are *independent* iff $P(A \cap B) = P(A)P(B)$.

Exercises

1-1. (a) How many seven-digit phone numbers are possible? (b) How many such phone numbers have no repeated digit? (c) How many seven-digit phone numbers do not start with 0, 1, 911, 411, or 555?

1-2. A key for the *DES encryption system* is a word of length 56 in the alphabet $\{0, 1\}$. A key for a *permutation cipher* is a permutation of the 26-letter English alphabet. Which encryption system has more keys?

1-3. A key for the *AES encryption system* is a word of length 128 in the alphabet $\{0,1\}$. Suppose we try to decrypt an AES message by exhaustively trying every possible key. Assume six billion computers are running in parallel, where each computer can test one trillion keys per second. Estimate the number of years required for this attack to search the entire space of keys.

1-4. Solve Example 1.8 using only the Product Rule.

1-5. (a) How many four-digit numbers consist of distinct even digits? (b) How many five-digit numbers contain only odd digits with at least one repeated digit?

1-6. How many four-digit numbers start with an odd digit, are divisible by 5, and do not contain the digits 6 or 7?

1-7. How many four-digit numbers consist of four consecutive digits in some order (e.g., 3214 or 9678 or 1023)?

1-8. How many four-digit even numbers contain the digit 5 but not the digit 2?

1-9. Find the minimum k such that every printable character on a standard computer keyboard can be encoded by a distinct word in $\{0,1\}$ of length exactly k. Does the answer change if we allow nonempty words of length *at most* k?

1-10. (a) How many four-letter words w using an n-letter alphabet satisfy $w_i \neq w_{i+1}$ for $i = 1,2,3$? (b) How many of the words in (a) also satisfy $w_4 \neq w_1$?

1-11. Solve Example 1.11 again with the additional condition that all the letters in the word must be distinct.

1-12. How many n-letter words in the alphabet $\{A, B, \ldots, Z\}$ contain: (a) only vowels; (b) no vowels; (c) at least one vowel; (d) alternating vowels and consonants; (e) two vowels and $n-2$ consonants?

1-13. (a) How many license plates consist of three uppercase letters followed by three digits? (b) How many license plates consist of three letters and three digits in some order? (c) How many license plates consist of seven characters (letters or digits), where all the letters precede all the digits and there must be at least one letter and at least one digit?

1-14. (a) How many six-digit numbers contain the digit 8 but not 9? (b) How many six-digit numbers contain the digit 0 or 1?

1-15. A pizza shop offers ten toppings. How many pizzas can be ordered with: (a) three different toppings; (b) up to three different toppings; (c) three toppings, with repeats allowed; (d) four different toppings, but pepperoni and sausage cannot be ordered together?

1-16. (a) How many numbers between 1 and 1000 are divisible by 5 or 7? (b) How many such numbers are divisible by 5 or 7, but not both?

1-17. How many three-digit numbers: (a) do not contain the digits 5 or 7; (b) contain the digits 5 and 7; (c) contain the digits 5 or 7; (d) contain exactly one of the digits 5 or 7?

1-18. Count the number of five-digit integers x having each property. (a) All digits of x are distinct. (b) The digit 8 appears and x is divisible by 5. (c) x is odd with distinct digits. (d) x is even with distinct digits. (e) x starts with 7 and is even. (f) x starts with 7 or is even. (g) x contains the digit 3 but not the digits 4 or 5. (h) x contains consecutive digits 686. (i) Any digit 6 in x is immediately followed by the digit 7. (j) x contains the digit 2 and the digit 9. (k) x contains the digit 6 exactly three times. (l) x contains distinct digits in increasing order.

1-19. Let S be the set of eight-letter words in the alphabet $\{A, B, \ldots, Z\}$. (a) Find the size of S. (b) How many words in S have all letters distinct? (c) How many words in S have no two adjacent letters equal? (d) How many words in S begin with a vowel and end with a consonant? (e) How many words in S begin with a vowel or end with a consonant, but not both? (f) How many words in S are rearrangements of the word LOOPHOLE? (g) How many words in S have no two consecutive vowels and no two consecutive consonants? (h) How many words in S are palindromes? (i) How many words in S start with three vowels and end with five consonants? (j) How many words in S contain three vowels and five consonants in some order? (k) How many words in S contain four consecutive letters MATH (in this order) somewhere in the word?

1-20. A class consists of nine women (including Alice and Carla) and eleven men (including Bob and Dave). We want to form a committee of six members of the class. Count the number of committees satisfying each condition below. (a) Any six people can be on the committee. (b) The committee must have four men and two women. (c) The committee must have more men than women. (d) Alice and Carla must both be on the committee. (e) Bob or Dave must be on the committee. (f) Exactly two of Alice, Bob, Carla, and Dave must be on the committee.

1-21. Consider a business employing fifteen lawyers (including Jones), ten salesmen (including Smith), and eight clerks (including Reed). Find the number of committees satisfying each condition. (a) The size of the committee is five or six. (b) The committee consists of three lawyers, two salesmen, and one clerk. (c) The committee has six members, at least three of whom are lawyers. (d) The committee has seven members, including Reed and Jones but excluding Smith. (e) The committee has four lawyers (including Jones) and three salesmen (excluding Smith). (f) The committee has six members, including Jones or Smith. (g) The committee has six members, including at least one lawyer, one salesman, and one clerk.

1-22. How many ways can four husbands and wives be seated in a row of ten seats with each restriction below? (a) The two end seats must be empty; (b) there are no restrictions; (c) all the men sit to the left of all the women; (d) Bob does not sit immediately next to anyone; (e) men and women are seated alternately (ignoring empty seats); (f) each husband and wife sit immediately next to each other.

1-23. Suppose $|A| = m$, $|B| = n$, and $|A \cup B| = p$. Find $|(A \times B) - (B \times A)|$.

1-24. (a) How many anagrams of MISSISSIPPI are there? (b) How many of these anagrams begin and end with P? (c) In how many of these anagrams are the two P's adjacent? (d) In how many of these anagrams are no two I's adjacent?

1-25. (a) How many straight flush poker hands are there? (b) How many five-card poker hands are straights or flushes?

1-26. A *two-pair* poker hand is a hand with three different values, two of which occur twice; an example is $\{2\heartsuit, 2\spadesuit, 5\diamondsuit, 5\clubsuit, K\heartsuit\}$. Someone tries to count these hands by the following Product Rule argument: choose the value of the first pair ($n_1 = 13$ ways); choose two suits out of four for this pair ($n_2 = \binom{4}{2} = 6$ ways); choose the value of the second pair ($n_3 = 12$ ways); choose two suits out of four for this pair ($n_4 = \binom{4}{2} = 6$ ways); choose the value for the last card ($n_5 = 11$ ways); choose the suit for the last card ($n_6 = 4$ ways); the answer is $13 \cdot 6 \cdot 12 \cdot 6 \cdot 11 \cdot 4 = 247{,}104$. Explain exactly what is wrong with this counting argument, illustrating using the sample hand above. Then find the correct answer.

1-27. What is wrong with the following argument for counting one-pair poker hands? Choose any card ($n_1 = 52$ ways); then choose another card of the same value to make the pair ($n_2 = 3$ ways); then choose three more cards of different values ($n_3 = 48$ ways, $n_4 = 44$ ways, $n_5 = 40$ ways); the answer is 13,178,880. Also explain why this answer is an integer multiple of the true answer.

1-28. (a) How many five-card hands have at least one card of every suit? (b) How many six-card hands have at least one card of every suit?

1-29. For $1 \leq k \leq 5$, find the number of ways to place k non-attacking rooks on the board shown below.

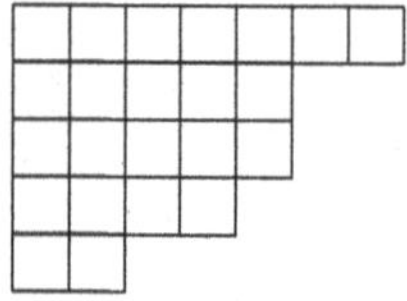

1-30. How many ways can we place k non-attacking rooks on an $n \times n$ chessboard?

1-31. (a) Verify the claim in the Introduction that for each $k \geq 1$, there are the same number of ways to place k non-attacking rooks on each of the two boards below.

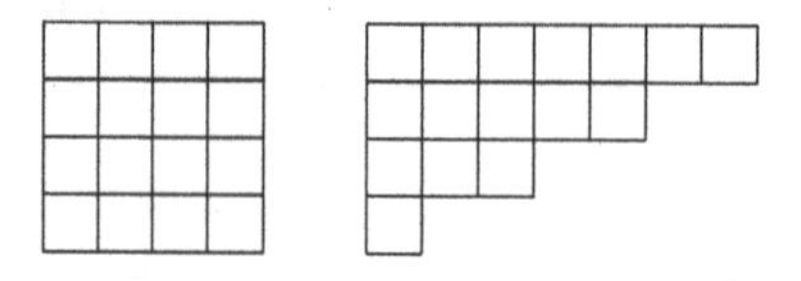

(b) Show that the board with row lengths $(6, 6, 2, 2)$ is also rook-equivalent to the two boards above.

1-32. Consider a board with left-justified rows of lengths $a_1 \geq a_2 \geq \cdots \geq a_m$. Find the number of ways to place k non-attacking rooks on this board for $k = 1$, $k = m$, and $k = m - 1$.

1-33. A DNA strand can be modeled as a word in the alphabet $\{A, C, G, T\}$. (a) How many DNA strands have length 12? (b) How many strands in (a) are palindromes? (c) How many strands of length 12 are there if we do not distinguish between a strand and its reversal? (d) Define the *complement* of a DNA strand to be the strand obtained by interchanging all A's and T's and interchanging all C's and G's. How many strands of length 12 are there if we do not distinguish between a strand and its complement?

1-34. Repeat the preceding question considering only strands consisting of four A's, four T's, two C's, and two G's.

1-35. (a) How many n-letter words using a k-letter alphabet are palindromes? (b) How many n-digit integers are palindromes? (c) How many words in $\mathcal{R}(a_1^{n_1} a_2^{n_2} \cdots a_k^{n_k})$ are palindromes?

1-36. (a) For a fixed $k < n$, count the number of permutations w of $\{1, 2, \ldots, n\}$ such that

$$w_1 < w_2 < \cdots < w_k > w_{k+1} < w_{k+2} < \cdots < w_n. \qquad (1.3)$$

(b) How many permutations satisfy (1.3) for some $k \in \{1, 2, \ldots, n-1\}$?

1-37. A *relation from* X *to* Y is any subset of $X \times Y$. Suppose X has n elements and Y has k elements. (a) How many relations from X to Y are there? (b) How many relations R satisfy the following property: for each $y \in Y$, there exists at most one $x \in X$ with $(x, y) \in R$?

1-38. For any sets S and T, define $S \Delta T = (S - T) \cup (T - S)$, so $x \in S \Delta T$ iff x belongs to exactly one of S and T. Prove: for all finte sets S and T, $|S \Delta T| = |S| + |T| - 2|S \cap T|$.

1-39. Use the Union Rule for Two Sets (see 1.41) to prove the Union Rule for Three Sets: for all finite sets A, B, C,

$$|A \cup B \cup C| = |A| + |B| + |C| - |A \cap B| - |A \cap C| - |B \cap C| + |A \cap B \cap C|.$$

1-40. How many positive integers less than 10,000 are not divisible by 7, 11, or 13? (Use the previous exercise.)

1-41. Two fair dice are rolled. Find the probability that: (a) the same number appears on both dice; (b) the sum of the numbers rolled is 8; (c) the sum of the numbers rolled is divisible by 3; (d) the two numbers rolled differ by 1.

1-42. In blackjack, you have been dealt two cards from a shuffled 52-card deck: 9♡ and 6♣. Find the probability that drawing one more card will cause the sum of the three card values to go over 21. (Here, an ace counts as 1 and other face cards count as 10.)

1-43. Find the probability that a random 5-letter word in $\{A, B, \ldots, Z\}$: (a) has no repeated letters; (b) contains no vowels; (c) is a palindrome.

1-44. A company employs ten men (one of whom is Bob) and eight women (one of whom is Alice). A four-person committee is randomly chosen. Find the probability that the committee: (a) consists of all men; (b) consists of two men and two women; (c) does not have both Alice and Bob as members.

1-45. A fair coin is tossed ten times. (a) Find the probability of getting exactly seven heads. (b) Find the probability of getting at least two heads. (c) Find the probability of getting exactly seven heads, given that the number of heads was prime.

1-46. A fair die is rolled ten times. What is the probability that, in these ten tosses, 1 comes up five times, 3 comes up two times, and 6 comes up three times?

1-47. We draw 10 balls (without replacement) from an urn containing 40 red, 30 blue, and 30 white balls. (a) What is the probability that no blue balls are drawn? (b) What is the probability of getting 4 red, 3 blue, and 3 white balls? (c) What is the probability that all 10 balls have the same color? (d) Answer the same questions assuming the balls are drawn with replacement.

1-48. A sequence of four cards is dealt from a 52-card deck (order matters). (a) What is the size of the sample space? (b) Find the probability of getting no clubs. (c) Find the probability of getting all four cards of the same value. (d) Find the probability that the first card is red and the last card is a spade. (e) Find the probability of getting one card from each suit. (f) Find the conditional probability that the last card is an ace given that the first two cards are not aces.

1-49. A sequence of four cards is dealt from a 52-card deck (order matters). (a) What is the size of the sample space? (b) Find the probability that your hand has at least one king but no diamonds. (c) Given that the hand contains 3♡ and 6♡, find the probability of a flush. (d) Given that the hand contains 3♡ and 6♡, find the probability of a straight. (e) Given that the hand contains 8♣ and 8♠, find the probability of a four-of-a-kind. (f) Given that the hand contains 8♣ and 8♠, find the probability of a full house.

1-50. A license plate consists of three uppercase letters followed by four digits. A random license plate is chosen. (a) Find the probability that the plate has three distinct letters and four distinct digits. (b) Find the probability that the plate has a Z but not an 8. (c) Find the probability that the plate has an 8 or does not have a Z. (d) Find the probability that the plate is a rearrangement of JZB-3553. (e) Find the probability that the plate starts with ACE and ends with four distinct digits in increasing order.

1-51. Let P be a probability measure. (a) Prove: for all events A and B, $P(A - B) = P(A) - P(A \cap B)$. (b) Prove: for all events A and B, $P(A \cup B) = P(A) + P(B) - P(A \cap B)$.

1-52. A 5-card poker hand is dealt from a 52-card deck. (a) What is the probability that the hand contains only red cards (i.e., hearts and diamonds)? (b) What is the probability that the hand contains exactly two 8's? (c) What is the probability that the hand contains only numerical cards (i.e., ace, jack, queen, and king may not appear)?

1-53. (a) Find the probability of getting a *three-of-a-kind* poker hand (this is a hand with three different values, one of which occurs three times; an example is $\{4\heartsuit, 4\spadesuit, 4\diamondsuit, 7\heartsuit, Q\spadesuit\}$). (b) Find the probability of getting a three-of-a-kind hand containing three fours.

1-54. A *bad* poker hand is a hand with five different non-consecutive values and at least two suits; an example is $\{2\heartsuit, 3\heartsuit, 4\heartsuit, 5\heartsuit, 7\diamondsuit\}$. (a) Use the Product Rule and Difference Rule to enumerate bad poker hands. (b) Check your answer by subtracting the number of *good* poker hands (found in the text and earlier exercises) from the total number of hands.

1-55. Consider the three urns from Example 1.69. An urn is selected at random, and then one ball is selected from that urn. What is the probability that: (a) the ball is blue, given that urn 2 was chosen; (b) the ball is blue; (c) urn 2 was chosen, given that the ball was blue?

1-56. A fair coin is tossed three times. (a) Describe the sample space. (b) Consider the following events. A: second toss is tails; B: second and third tosses disagree; C: all tosses are the same; D: the number of heads is even. Describe each event as a subset of the sample space. (c) Which pairs of events from $\{A, B, C, D\}$ are independent? (d) Is the list of events A, B, D independent? Explain.

1-57. Describe the sample space of the urn experiment in Example 1.69. Find the probability of each outcome in the sample space, and verify that these probabilities add to 1.

1-58. A 5-card poker hand is dealt from a 52-card deck. (a) What is the probability that the hand is a flush, given that the hand contains no clubs? (b) What is the probability that the hand contains at least one card from each of the four suits, given that the hand contains both a red and a black card? (c) What is the probability of getting a two-pair hand, given that at least two cards in the hand have the same value?

1-59. Let A, B, C be events in a probability space S. Assume A and C are independent, and B and C are independent. (a) Give an example where $A \cup B$ and C are not independent. (b) Prove that $A \cup B$ and C are independent if A and B are disjoint. (c) Must $A \cap B$ and C be independent? Explain.

1-60. (a) How many n-letter words using the alphabet $\{0, 1\}$ contain both the symbols 0 and 1? (b) How many n-letter multisets in the alphabet $\{0, 1\}$ contain both the symbols 0 and 1?

1-61. How many lattice paths from $(0, 0)$ to $(7, 5)$ pass through the point $(2, 3)$?

1-62. (a) How many five-digit numbers have strictly decreasing digits reading left to right? (b) How many five-digit numbers have weakly increasing digits reading left to right?

1-63. A *two-to-one function* is a function $f : X \to Y$ such that for every $y \in Y$, there exist *exactly two* elements $x_1, x_2 \in X$ with $f(x_1) = y = f(x_2)$. How many two-to-one functions are there from a $2n$-element set to an n-element set?

1-64. Suppose we try to adapt the argument before Rule 1.27 to count k-element multisets from an n-element alphabet, as follows. Choose the first element of the multiset in n ways, choose the second element in n ways, and so on; then divide by $k!$ since order does not matter in a multiset. The answer $n^k/k!$ cannot be correct because it is not even an integer in general. Explain exactly what is wrong with this counting argument.

1-65. Count the number of surjections from an n-element set to an m-element set for each of these special cases: (a) $m = 1$; (b) $m = 2$; (c) $n = m$; (d) $n < m$; (e) $n = m + 2$.

1-66. List all 5-letter words in $\{0, 1\}$ that do not contain two consecutive zeroes.

1-67. List all permutations $w_1w_2w_3w_4$ of $\{1, 2, 3, 4\}$ where $w_i \neq i$ for all i.

1-68. List all words in $\mathcal{R}(x^2y^3z^2)$ with no two consecutive y's.

1-69. List all bijections $f : \{1, 2, 3, 4\} \to \{1, 2, 3, 4\}$ such that $f \circ f$ is the identity map.

1-70. List all bijections $g : \{1, 2, 3, 4\} \to \{1, 2, 3, 4\}$ where $g^{-1} = g \circ g$.

1-71. List all surjections $g : \{1,2,3,4\} \to \{a,b\}$.

1-72. List all injections $h : \{a,b,c\} \to \{1,2,3,4,5,6,7\}$ such that the image of h does not contain two consecutive integers.

1-73. List all compositions of 4.

1-74. List all compositions of 7 with exactly three parts.

1-75. List all lattice paths from $(0,0)$ to $(4,2)$.

1-76. List all Dyck paths of order 4.

1-77. List all three-element multisets using the alphabet $\{a,b,c\}$.

1-78. List all subsets of the three-element set $\{\{1,2\},(3,4),[1,1]\}$.

1-79. Draw pictures of all compositions of 5. For each composition, determine the associated word in $\{0,1\}^4$ constructed in the proof of the Composition Rule.

1-80. How many lattice paths start at $(0,0)$ and end on the line $x+y=n$?

1-81. Let r be the bijection in the proof of the Dyck Path Rule. Compute

$$r(\text{NNEEEEENNNNEEEEENN}) \text{ and } r^{-1}(\text{NENEENNEEENEEENN}).$$

1-82. Draw all the non-Dyck lattice paths from $(0,0)$ to $(3,3)$ and compute their images under the reflection map r from the proof of the Dyck Path Rule.

1-83. Ten lollipops are to be distributed to four children. All lollipops of the same color are considered identical. How many distributions are possible if: (a) all lollipops are red; (b) all lollipops have different colors; (c) there are four red and six blue lollipops? (d) What are the answers if each child must receive at least one lollipop?

1-84. A *monomial* in N variables is a term of the form $x_1^{d_1}x_2^{d_2}\cdots x_N^{d_N}$, where each $d_i \geq 0$. The *degree* of this monomial is $d_1+d_2+\cdots+d_N$. How many monomials in N variables have degree (a) exactly d; (b) at most d?

1-85. How many multisets (of any size) can be formed from an n-letter alphabet if each letter can appear at most k times in the multiset?

1-86. Find a bijection on $\text{Comp}(n)$ that maps compositions with k parts to compositions with $n+1-k$ parts for all k.

1-87. (a) What is the probability that a random lattice path from $(0,0)$ to (n,n) is a Dyck path? (b) What is the probability that a random Dyck path of order n only touches $y=x$ at $(0,0)$ and (n,n)? (c) What is the probability that a lattice path from $(0,0)$ to (n,n) is a Dyck path, given that the path starts with a north step or ends with an east step?

1-88. How many 7-element subsets of $\{1,2,\ldots,20\}$ contain no two consecutive integers?

1-89. (a) How many points with positive integer coordinates lie on the plane $x+y+z=8$? (b) How many points with nonnegative integer coordinates lie on this plane?

1-90. Count the integer solutions to $x_1+x_2+\cdots+x_5=23$ if we require $x_i \geq i$ for each i.

1-91. (a) For fixed n and k, how many integer sequences $(x_1,x_2,\ldots,x_n)$ satisfy $x_1+x_2+\cdots+x_n=k$ with all x_i even and nonnegative? (b) How many sequences satisfy $x_1+\cdots+x_n=k$ with all x_i odd and positive?

1-92. How many solutions to $x_1+x_2+\cdots+x_7=k$ are there if we require all $x_i \in \mathbb{Z}_{\geq 0}$, x_1 is even, and $x_2 \in \{0,1\}$?

1-93. How many words $w_1\cdots w_k$ in the alphabet $\{1,2,\ldots,n\}$ satisfy $w_i \geq w_{i-1}+2$ for $i=2,3,\ldots,k$?

1-94. (a) How many ways can we put k unlabeled balls in n labeled boxes if not all balls need be used? (b) Repeat (a) assuming each box must contain at least one ball. (c) Repeat (a) using labeled balls.

1-95. Let $A = \{1, 2, 3, 4\}$, $B = \{u, v, w, x, y, z\}$, and define $f : A \to B$ by $f(1) = v$, $f(2) = x$, $f(3) = z$, $f(4) = u$. How many functions $g : B \to A$ satisfy $g \circ f = \text{id}_A$?

1-96. Counting Left Inverses. Given an injective function $f : X \to Y$ where $|X| = n$ and $|Y| = m$, count the functions $g : Y \to X$ such that $g \circ f = \text{id}_X$.

1-97. Let $A = \{1, 2, 3, 4, 5, 6, 7\}$, $B = \{x, y, z\}$, and define $f : A \to B$ by $f(1) = f(3) = f(4) = y$, $f(2) = f(5) = x$, $f(6) = f(7) = z$. How many functions $g : B \to A$ satisfy $f \circ g = \text{id}_B$?

1-98. Counting Right Inverses. Given a surjective function $f : X \to Y$ between finite sets, how many functions $g : Y \to X$ satisfy $f \circ g = \text{id}_Y$? (The answer depends on f, not just on the sizes of X and Y.)

1-99. Given a positive integer n, let the prime factorization of n be $n = p_1^{e_1} p_2^{e_2} \cdots p_k^{e_k}$, where each $e_j > 0$ and the p_j are distinct primes. How many positive divisors does n have? How many divisors does n have in $\mathbb{Z}$?

1-100. Let the prime factorization of $n!$ be $p_1^{e_1} p_2^{e_2} \cdots p_k^{e_k}$. Prove that $e_i = \sum_{k=1}^{\infty} \lfloor n/p_i^k \rfloor$. (The notation $\lfloor x \rfloor$ denotes the greatest integer not exceeding the real number x.) Hence determine the number of trailing zeroes in the decimal notation for 100!.

1-101. A password is a word using an alphabet containing 26 uppercase letters, 26 lowercase letters, 10 digits, and 30 special characters. (a) Compare the number of 5-character passwords with the number of 12-character passwords. (b) Find the least m for which there are more m-letter passwords using only uppercase letters than 7-letter passwords with no special characters. (c) How many 8-character passwords contain at least one digit and at least one special character? (d) How many 8-character passwords contain a letter, a digit, and a special character? (Use the Union Rule for Three Sets.)

1-102. (a) How many 10-letter words in the alphabet $\{A, B, \ldots, Z\}$ are such that every Q in the word is immediately followed by U? [Hint: Create categories based on the number of Q's in the word.] (b) What is the answer for n-letter words?

1-103. Consider an alphabet with A consonants and B vowels. (a) How many words in this alphabet consist of m consonants and n vowels with no two consecutive vowels? (b) Repeat (a) if all letters in the word must be distinct.

1-104. How many 15-digit numbers have the property that any two zero digits are separated by at least three nonzero digits?

1-105. (a) How many integers between 1 and 1,000,000 contain the digit 7? (b) If we write the integers from 1 to 1,000,000, how often will we write the digit 7? (c) What are the answers to (a) and (b) if 7 is replaced by 0?

1-106. For fixed $k, n, d \in \mathbb{Z}_{>0}$, how many k-element subsets S of $\{1, 2, \ldots, n\}$ are such that any two distinct elements of S differ by at least d?

1-107. How many positive integers x are such that x and $x + 3$ are both four-digit numbers with no repeated digits?

1-108. Properties of Injections. Prove the following statements about injective functions. (a) If $f : X \to Y$ and $g : Y \to Z$ are injective, then $g \circ f$ is injective. (b) If $g \circ f$ is injective, then f is injective but g may not be. (c) $f : X \to Y$ is injective iff for all sets W and all $g, h : W \to X$, $f \circ g = f \circ h$ implies $g = h$.

1-109. Properties of Surjections. Prove the following statements about surjective functions. (a) If $f : X \to Y$ and $g : Y \to Z$ are surjective, then $g \circ f$ is surjective. (b) If $g \circ f$ is surjective, then g is surjective but f may not be. (c) $f : X \to Y$ is surjective iff for all sets Z and all $g, h : Y \to Z$, $g \circ f = h \circ f$ implies $g = h$.

1-110. Sorting by Comparisons. Consider a game in which player 1 picks a permutation w of n letters, and player 2 must determine w by asking player 1 a sequence of yes or no questions. (Player 2 can choose later questions in the sequence based on the answers to earlier questions.) Let $K(n)$ be the minimum number such that, no matter what w player 1 chooses, player 2 can correctly identify w after at most $K(n)$ questions. (a) Prove that $(n/2)\log_2(n/2) \leq \lceil \log_2(n!) \rceil \leq K(n)$. (b) Prove that $K(n) = \lceil \log_2(n!) \rceil$ for $n \leq 5$. (c) Prove that (b) still holds if we restrict player 2 to ask only questions of the form "is $w_i < w_j$?" at each stage. (d) What can you conclude about the length of time needed to sort n distinct elements using an algorithm that makes decisions by comparing two data elements at a time?

1-111. (a) You are given 12 seemingly identical coins and a balance scale. One coin is counterfeit and is either lighter or heavier than the others. Describe a strategy that can be used to identify which coin is fake in only three weighings. (b) If there are 13 coins, can the fake coin always be found in three weighings? Justify your answer. (c) If there are N coins (one of which is fake), derive a lower bound for the number of weighings required to find the fake coin.

1-112. Suppose we randomly draw a 5-card hand from a 51-card deck where the queen of spades has been removed. Find the probability of the following hands: (a) four-of-a-kind; (b) three-of-a-kind; (c) full house; (d) straight; (e) flush; (f) two-pair; (g) one-pair.

1-113. Suppose we mix together two 52-card decks and randomly draw a 5-card hand. Find the probability of the following hands: (a) four-of-a-kind; (b) three-of-a-kind; (c) full house; (d) straight; (e) flush; (f) two-pair; (g) one-pair; (h) five-of-a-kind, defined to be a hand where all five values are the same (e.g., five kings).

1-114. Suppose we randomly draw a 5-card hand from a 48-card deck where the four jacks have been removed. Find the probability of the following hands: (a) four-of-a-kind; (b) three-of-a-kind; (c) full house; (d) straight; (e) flush; (f) two-pair; (g) one-pair; (h) a hand containing at least two diamonds.

1-115. Suppose we randomly draw a 5-card hand from a 49-card deck where $8\heartsuit$, $8\diamondsuit$, and $8\clubsuit$ have been removed. Find the probability of the following hands: (a) four-of-a-kind; (b) three-of-a-kind; (c) full house; (d) straight; (e) flush; (f) two-pair; (g) one-pair; (h) three-of-a-kind, given that $8\spadesuit$ is in the hand.

1-116. Suppose we randomly draw a 5-card hand from a 53-card deck containing a joker card. The joker can have any suit and any value to make the hand as good as possible. Find the probability of the following hands: (a) four-of-a-kind; (b) three-of-a-kind; (c) full house; (d) straight; (e) flush; (f) two-pair; (g) one-pair; (h) five-of-a-kind.

1-117. Let A be the event that a five-card poker hand contains the ace of spades. (a) Find the conditional probability of the various poker hands given A. Which of these events are independent of A? (b) With minimum additional calculation, answer (a) again taking A to be the event that your hand contains the six of spades.

1-118. Texas Hold 'em. In a popular version of poker, a player is dealt an *ordered* sequence $(C_1, C_2, \ldots, C_7)$ of seven distinct cards from a 52-card deck. The last five cards in this sequence are *community cards* shared with other players. In this exercise we concentrate on a single player, so we ignore this aspect of the game. The player uses these seven cards to form the best possible five-card poker hand. For example, if the seven-card sequence was $(4\heartsuit, 7\clubsuit, 3\diamondsuit, 9\clubsuit, 5\clubsuit, 6\clubsuit, Q\clubsuit)$, we would have a flush (the five club cards) since this beats the straight (3,4,5,6,7 of various suits). (a) Compute the size of the sample space. (b) What is the probability of getting four-of-a-kind? (c) What is the probability of getting a flush? (d) What is the probability of getting four-of-a-kind, given $C_1 = 3\heartsuit$ and $C_2 = 3\spadesuit$? (e) What is the probability of getting a flush, given $C_1 = 5\diamondsuit$ and $C_2 = 9\diamondsuit$?

1-119. You have been dealt the poker hand $\{A\clubsuit, A\diamondsuit, A\heartsuit, 3\spadesuit, 7\heartsuit\}$ from a 52-card deck. You can now discard k cards and receive k new cards. Which is more likely: getting a full house or four-of-a-kind by discarding the 3, or getting a full house or four-of-a-kind by discarding the 3 and the 7?

1-120. What is the probability that a random lattice path from $(0,0)$ to (n,n) has exactly one north step below $y=x$?

1-121. Define $f:\mathbb{Z}_{\geq 0}\times\mathbb{Z}_{\geq 0}\to\mathbb{Z}_{>0}$ by $f(a,b)=2^a(2b+1)$. Prove that f is a bijection.

1-122. Define $f:\mathbb{Z}_{\geq 0}\times\mathbb{Z}_{\geq 0}\to\mathbb{Z}_{\geq 0}$ by $f(a,b)=((a+b)^2+3a+b)/2$. Prove that f is a bijection.

1-123. Euler's ϕ Function. For each $n\geq 1$, let $\Phi(n)$ be the set of integers k between 1 and n such that $\gcd(k,n)=1$, and let $\phi(n)=|\Phi(n)|$. (a) Compute $\Phi(n)$ and $\phi(n)$ for $1\leq n\leq 12$. (b) Compute $\phi(p)$ for p prime. (c) Compute $\phi(p^e)$ for p prime and $e>1$. (The next exercise shows how to compute $\phi(n)$ for any n.)

1-124. Chinese Remainder Theorem. In this exercise, we write "a mod k" to denote the unique integer b in the range $\{1,2,\ldots,k\}$ such that k divides $(a-b)$. Suppose m and n are fixed positive integers. Define a map

$$f:\{1,2,\ldots,mn\}\to\{1,2,\ldots,m\}\times\{1,2,\ldots,n\}\text{ by setting }f(z)=(z\bmod m, z\bmod n).$$

(a) Show that $f(z)=f(w)$ iff $\operatorname{lcm}(m,n)$ divides $z-w$. (b) Show that f is injective iff $\gcd(m,n)=1$. (c) Deduce that f is a bijection iff $\gcd(m,n)=1$. (d) Prove that for $\gcd(m,n)=1$, f maps $\Phi(mn)$ bijectively onto $\Phi(m)\times\Phi(n)$, and hence $\phi(mn)=\phi(m)\phi(n)$. (See the previous exercise for the definition of Φ and ϕ.) (e) Suppose n has prime factorization $p_1^{e_1}\cdots p_k^{e_k}$. Prove that $\phi(n)=n\prod_{i=1}^{k}(1-1/p_i)$.

1-125. The Bijective Product Rule. For any positive integers m,n, define

$$g:\{0,1,\ldots,m-1\}\times\{0,1,\ldots,n-1\}\to\{0,1,\ldots,mn-1\}$$

by setting $g(i,j)=ni+j$. Carefully prove that g is a bijection.

1-126. Bijective Laws of Algebra. (a) For all sets X,Y,Z, prove that $X\cup Y=Y\cup X$, $(X\cup Y)\cup Z=X\cup(Y\cup Z)$, and $X\cup\emptyset=X=\emptyset\cup X$. (b) For all sets X,Y,Z, define bijections $f:X\times Y\to Y\times X$, $g:(X\times Y)\times Z\to X\times(Y\times Z)$, and (for Y,Z disjoint) $h:X\times(Y\cup Z)\to(X\times Y)\cup(X\times Z)$. (c) Use (a), (b), and counting rules to deduce the algebraic laws $x+y=y+x$, $(x+y)+z=x+(y+z)$, $x+0=x=0+x$, $xy=yx$, $(xy)z=x(yz)$, and $x(y+z)=xy+xz$, valid for all integers $x,y,z\geq 0$.

1-127. The Bijective Laws of Exponents. Let $\operatorname{Fun}(A,B)$ denote the set of all functions $f:A\to B$. (a) Given sets X,Y,Z with $Y\cap Z=\emptyset$, define a bijection from $\operatorname{Fun}(Y\cup Z,X)$ to $\operatorname{Fun}(Y,X)\times\operatorname{Fun}(Z,X)$. (b) If X,Y,Z are any sets, define a bijection from $\operatorname{Fun}(Z,(\operatorname{Fun}(Y,X)))$ to $\operatorname{Fun}(Y\times Z,X)$. (c) By specializing to finite sets, deduce the laws of exponents $x^{y+z}=x^yx^z$ and $(x^y)^z=x^{yz}$ for all integers $x,y,z\geq 0$.

1-128. Let X be any set (possibly infinite). Prove that every $f:X\to\mathcal{P}(X)$ is not surjective. (Given f, show that $S=\{x\in X: x\notin f(x)\}\in\mathcal{P}(X)$ is not in the image of f.) Conclude that $|X|\neq|\mathcal{P}(X)|$.

1-129. Let X be the set of infinite sequences $w=(w_k:k\geq 0)$ with each $w_k\in\{0,1\}$. Show there is no bijection $f:\mathbb{Z}_{\geq 0}\to X$. (Hint: See the previous exercise.)

1-130. A sample space S consists of 25 equally likely outcomes. Suppose we randomly choose an ordered pair (A,B) of events in S. (a) Find the probability that A and B are disjoint. (b) Find the probability that A and B are independent events.

Notes

Some general references on combinatorics include [1, 9, 12, 15, 19, 20, 24, 53, 107, 110, 121, 125, 128]. Detailed treatments of probability are given in texts such as [10, 27, 61, 88]. More information on set theory, including a discussion of cardinality for infinite sets, may be found in [59, 63, 90].

2

Combinatorial Identities and Recursions

Suppose we are proving an identity of the form $a = b$, where a and b are formulas that may involve factorials, binomial coefficients, powers, summations, etc. One way to prove such an identity is to give an *algebraic proof* using tools like the Binomial Theorem, proof by induction, or other algebraic techniques. This chapter introduces another powerful and elegant method of proving identities called a *combinatorial proof.* A combinatorial proof establishes the equality of two formulas by exhibiting a set of objects whose cardinality is given by both formulas. There are three main steps in a combinatorial proof of a formula $a = b$. First, define an appropriate set S of objects; second, give a counting argument showing that $|S| = a$; third, give a different counting argument showing that $|S| = b$. We illustrate this technique by giving combinatorial proofs of the Binomial Theorem, the Multinomial Theorem, the Geometric Series Formula, and other binomial coefficient identities.

The second part of the chapter studies combinatorial objects including set partitions, integer partitions, equivalence relations, surjections, ballot paths, pattern-avoiding permutations, and rook placements. To count these objects, we need the idea of *combinatorial recursions.* A recursion lets us compute a whole sequence of quantities by relating later values in the sequence to earlier values; the Fibonacci recursion $f_n = f_{n-1} + f_{n-2}$ is one famous example. We will see that recursions derived from counting arguments allow us to enumerate complicated combinatorial collections whose cardinalities may not be given by a simple closed formula. We also present a method for automatically finding exact solutions to certain types of recursions, including the Fibonacci recursion.

2.1 Initial Examples of Combinatorial Proofs

In this section, we give initial illustrations of the combinatorial proof technique by proving four binomial coefficient identities. Each proof follows the outline discussed in the introduction to this chapter: first we introduce a set of objects; then we count this set in two different ways, leading to the two formulas on each side of the identity to be proved. Our first sample identity shows how to evaluate a sum of binomial coefficients.

2.1. Theorem: Sums of Binomial Coefficients. For all $n \in \mathbb{Z}_{\geq 0}$,

$$\sum_{k=0}^{n} \binom{n}{k} = 2^n.$$

Proof. Fix $n \in \mathbb{Z}_{\geq 0}$. *Step 1.* We define S to be the set of all subsets of $\{1, 2, \ldots, n\}$. *Step 2.* By the Power Set Rule, we know that $|S| = 2^n$. (This rule motivated our choice of S in Step 1.) *Step 3.* Now we count S in a different way. For $0 \leq k \leq n$, let S_k be the set of all k-element subsets of $\{1, 2, \ldots, n\}$. On one hand, the Subset Rule tells us that $|S_k| = \binom{n}{k}$ for each k. On the other hand, S is the disjoint union of the sets $S_0, \ldots, S_n$. So the Sum

Rule tells us that $|S| = \sum_{k=0}^{n} |S_k| = \sum_{k=0}^{n} \binom{n}{k}$. Comparing Steps 2 and 3, we obtain the required formula. □

In our next example, we give an algebraic proof, a combinatorial proof, and a bijective proof of the identity in question.

2.2. Theorem: Symmetry of Binomial Coefficients. For all integers n, k with $0 \leq k \leq n$, we have

$$\binom{n}{k} = \binom{n}{n-k}.$$

Algebraic Proof. To prove the identity algebraically, we give a chain of known equalities leading from one side of the identity to the other. Specifically, by definition of binomial coefficients, we have

$$\binom{n}{n-k} = \frac{n!}{(n-k)!(n-(n-k))!} = \frac{n!}{(n-k)!k!} = \frac{n!}{k!(n-k)!} = \binom{n}{k}.$$

Combinatorial Proof. Step 1. We let $S = \mathcal{R}(0^k 1^{n-k})$ be the set of rearrangements of k zeroes and $n-k$ ones. *Step 2.* To build a word in S, choose a set of k positions out of n available positions to contain the zeroes, and fill the remaining positions with ones. By the Subset Rule, we see that $|S| = \binom{n}{k}$. *Step 3.* Alternatively, we can build a word in S by choosing a set of $n-k$ positions out of n positions to contain the ones, then filling the remaining positions with zeroes. By the Subset Rule, we see that $|S| = \binom{n}{n-k}$.

Bijective Proof. Here is a variation of the basic combinatorial proof template where the final step involves a bijection between two sets of objects. Let S be the set of all k-element subsets of $\{1, 2, \ldots, n\}$, and let T be the set of all $(n-k)$-element subsets of $\{1, 2, \ldots, n\}$. By the Subset Rule, we know $|S| = \binom{n}{k}$ and $|T| = \binom{n}{n-k}$. To prove $\binom{n}{k} = \binom{n}{n-k}$, it therefore suffices to define a bijection $f : S \to T$. For $A \in S$, define $f(A) = \{1, 2, \ldots, n\} - A$. Since A has size k, $f(A)$ has size $n-k$, so f does map S into the codomain T. Similarly, define $g : T \to S$ by $g(B) = \{1, 2, \ldots, n\} - B$ for $B \in T$. Since B has size $n-k$, $g(B)$ has size $n-(n-k) = k$ and therefore is in the codomain S. One sees immediately that $g \circ f = \text{id}_S$ and $f \circ g = \text{id}_T$, using the set-theoretic fact that $X - (X - A) = A$ whenever $A \subseteq X$. Thus, f and g are bijections.

Our next identity expresses the binomial coefficient $\binom{n}{k}$ as a sum of two other binomial coefficients. This identity is connected to *Pascal's Triangle*, which we discuss in §2.7.

2.3. Pascal's Binomial Coefficient Identity. For all $n \in \mathbb{Z}_{>0}$ and all $k \in \mathbb{Z}$,

$$\binom{n}{k} = \binom{n-1}{k} + \binom{n-1}{k-1}.$$

Algebraic Proof. If $k < 0$ or $k > n$, both sides of the identity are zero by definition. If $k = 0$, the left side is $\binom{n}{0} = 1$ and the right side is $\binom{n-1}{0} + 0 = 1$. If $k = n$, the left side is $\binom{n}{n} = 1$ and the right side is $0 + \binom{n-1}{n-1} = 1$. For the rest of the proof, assume $0 < k < n$. Expanding the right side of the identity, we find that

$$\binom{n-1}{k} + \binom{n-1}{k-1} = \frac{(n-1)!}{k!(n-1-k)!} + \frac{(n-1)!}{(k-1)!(n-k)!}.$$

To produce a common denominator, we multiply the first fraction by $(n-k)/(n-k)$ and the second fraction by k/k, obtaining

$$\frac{(n-1)!(n-k)}{k!(n-k)!} + \frac{(n-1)!k}{k!(n-k)!} = \frac{(n-1)!(n-k+k)}{k!(n-k)!} = \frac{n!}{k!(n-k)!} = \binom{n}{k}.$$

We have now transformed the right side of the identity into the left side.

Combinatorial Proof. As in the algebraic proof, it suffices to consider the case $0 < k < n$. *Step 1.* We define S to be the set of all k-element subsets of $\{1, 2, \ldots, n\}$. *Step 2.* By the Subset Rule, we know that $|S| = \binom{n}{k}$. *Step 3.* To count S in a new way, we write $S = S_1 \cup S_2$, where S_1 is the collection of subsets in S that *do not* contain n, and S_2 is the collection of subsets in S that *do* contain n. Symbolically, $S_1 = \{A \in S : n \notin A\}$ and $S_2 = \{B \in S : n \in B\}$. Since S_1 and S_2 are disjoint, the Sum Rule says that $|S| = |S_1| + |S_2|$. To count S_1, note that a k-element subset of $\{1, 2, \ldots, n\}$ that does not contain n is the same as a k-element subset of $\{1, 2, \ldots, n-1\}$. Thus, $|S_1| = \binom{n-1}{k}$ by the Subset Rule. We can build each object in S_2 by picking an arbitrary subset of $\{1, 2, \ldots, n-1\}$ of size $k-1$ (in $\binom{n-1}{k-1}$ ways), then adding n to this subset (in 1 way) to get a k-element subset of $\{1, 2, \ldots, n\}$ containing n. Thus, $|S_2| = \binom{n-1}{k-1}$, so $|S| = \binom{n-1}{k} + \binom{n-1}{k-1}$.

2.4. Theorem: Sums of Squared Binomial Coefficients. For all $n \geq 0$,

$$\sum_{k=0}^{n} \binom{n}{k}^2 = \binom{2n}{n}.$$

Proof. We give a combinatorial proof of the equivalent identity $\binom{2n}{n} = \sum_{k=0}^{n} \binom{n}{k}\binom{n}{n-k}$, in which we have replaced one copy of $\binom{n}{k}$ by $\binom{n}{n-k}$ (using Theorem 2.2). *Step 1.* Define S to be the set of all n-element subsets of $X = \{1, 2, \ldots, 2n\}$. *Step 2.* By the Subset Rule, we know that $|S| = \binom{2n}{n}$ (this fact motivated our choice of S in Step 1). *Step 3.* We count S in a new way, based on the degree of overlap of S with the subsets $X_1 = \{1, 2, \ldots, n\}$ and $X_2 = \{n+1, \ldots, 2n\}$. For $0 \leq k \leq n$, define

$$S_k = \{A \in S : |A \cap X_1| = k \text{ and } |A \cap X_2| = n - k\}.$$

S is the disjoint union of the S_k's, so that $|S| = \sum_{k=0}^{n} |S_k|$ by the Sum Rule. To compute $|S_k|$, we build a typical object $A \in S_k$ by making two choices. First, choose the k-element subset $A \cap X_1$ in any of $\binom{n}{k}$ ways. Second, choose the $(n-k)$-element subset $A \cap X_2$ in any of $\binom{n}{n-k}$ ways. We see that $|S_k| = \binom{n}{k}\binom{n}{n-k}$ by the Product Rule. Thus, $|S| = \sum_{k=0}^{n} \binom{n}{k}\binom{n}{n-k}$, completing the proof. □

2.2 The Geometric Series Formula

A *geometric series* is a sum of terms where each term arises from the preceding one by multiplying by a fixed *ratio* r. The following formulas show how to evaluate a finite or infinite geometric series.

2.5. The Geometric Series Formula. (a) For all $a \in \mathbb{R}$, $n \in \mathbb{Z}_{\geq 0}$, and $r \in \mathbb{R} - \{1\}$,

$$\sum_{k=0}^{n} ar^k = \frac{a(1 - r^{n+1})}{1 - r}.$$

(b) For all $a \in \mathbb{R}$ and all $r \in (-1, 1)$, $\sum_{k=0}^{\infty} ar^k = a/(1 - r)$.

Proof. To prove part (a) algebraically, fix a, n, and $r \neq 1$, and define

$$S = \sum_{k=0}^{n} ar^k = a + ar + ar^2 + ar^3 + \cdots + ar^{n-1} + ar^n.$$

Multiplying both sides of this equation by r, we get $rS = ar+ar^2+\cdots+ar^{n-1}+ar^n+ar^{n+1}$. Subtracting the new equation from the old one gives $(1-r)S = a - ar^{n+1} = a(1-r^{n+1})$, since all the middle terms cancel. Dividing by $1-r$ gives the formula in part (a). For part (b), note that $-1 < r < 1$ implies $\lim_{n\to\infty} r^n = 0$. Taking the limit of both sides of (a) as n goes to infinity, we obtain the formula in (b). □

When $r = 1$, we cannot divide by $1 - r$. In this case, however, it is immediate that $\sum_{k=0}^{n} ar^k = \sum_{k=0}^{n} a = a(n+1)$, since we are summing up $n+1$ copies of a. The Geometric Series Formula also holds for complex values of a and r (with the same proof), as long as we restrict to $r \neq 1$ in (a) and $|r| < 1$ in (b).

2.6. Example. What rational number is represented by the infinite repeating decimal $x = 4.1373737\cdots$? We have $x = 4.1+37(10^{-3}+10^{-5}+10^{-7}+\cdots) = 4.1+0.037\sum_{k=0}^{\infty} 10^{-2k}$. Applying the Geometric Series Formula with $r = 10^{-2} = 1/100$, the sum here is $1/(1-0.01) = 100/99$. Thus, $x = 41/10 + (37/1000)(100/99) = 41/10 + 37/990 = 2048/495$.

Now we give a combinatorial proof of the finite Geometric Series Formula. Here we take $a = 1$ and fix *positive integers* r and n. Multiplying both sides of the identity by $-(1-r)$, we are reduced to proving

$$(r-1)(1+r+r^2+\cdots+r^n) = r^{n+1} - 1. \tag{2.1}$$

Let S be the set of words in the alphabet $\{0, 1, \ldots, r-1\}$ of length $n+1$ excluding the word $00\cdots0$. On one hand, the Word Rule shows that $|S| = r^{n+1} - 1$. On the other hand, we can classify words in S based on the position of the first nonzero symbol. For $0 \leq j \leq n$, let S_j be the set of words w in S where $w_1 = \cdots = w_j = 0$ and $w_{j+1} \neq 0$. To build a word in S_j, put zeroes in the first j positions (1 way), then choose a nonzero symbol w_{j+1} ($r-1$ ways), then choose the remaining $n+1-(j+1) = n-j$ symbols arbitrarily (r ways each). By the Product Rule, $|S_j| = (r-1)r^{n-j}$. S is the disjoint union of $S_0, \ldots, S_n$. So the Sum Rule gives $|S| = \sum_{j=0}^{n} |S_j| = (r-1)r^n + (r-1)r^{n-1} + \cdots + (r-1)r^0$, completing the proof. When $r = 10$, this proof is classifying positive integers between 1 and $10^{n+1} - 1$ based on the number of digits in the decimal representation of the integer (excluding leading zeroes, as usual).

2.7. Remark. It may seem that the combinatorial proof yields a weaker result than the algebraic proof, since we had to assume that r was a positive integer (rather than a real number) in the combinatorial proof. But it turns out that if a *polynomial* identity of the form $p(r) = q(r)$ holds for infinitely many values of r (say for all positive integers), then this identity must automatically hold for all real numbers r. To see why, rewrite the given identity as $(p-q)(r) = 0$, where $p-q$ is a polynomial in the formal variable r. If $p-q$ were nonzero, say of degree d, then $p-q$ would have at most d real roots. Since we are assuming the equation holds for infinitely many values of r, we conclude that $p-q$ is the zero polynomial. Thus, $p(r) = q(r)$ must hold for every r. A similar result holds for polynomial identities involving multiple parameters.

2.3 The Binomial Theorem

The *Binomial Theorem* is a famous formula for expanding the nth power of a sum of two terms. Binomial coefficients are so named because of their appearance in this theorem.

2.8. The Binomial Theorem. For all $x, y \in \mathbb{R}$ and all $n \in \mathbb{Z}_{\geq 0}$,

$$(x+y)^n = \sum_{k=0}^{n} \binom{n}{k} x^k y^{n-k}.$$

For example,

$$(x+y)^4 = 1y^4 + 4xy^3 + 6x^2y^2 + 4x^3y + 1x^4.$$

We now give a combinatorial proof of the Binomial Theorem under the additional assumption that x and y are *positive integers.* (Since both sides of the formula are polynomials in x and y, we can deduce the general case using the result mentioned in Remark 2.7.) Let $A = \{V_1, \ldots, V_x, C_1, \ldots, C_y\}$ be an alphabet consisting of $x+y$ letters, where A consists of x vowels $V_1, \ldots, V_x$ and y consonants $C_1, \ldots, C_y$. Let S be the set of all n-letter words using the alphabet A. By the Word Rule, $|S| = (x+y)^n$. On the other hand, we can classify words in S based on how many vowels they contain. For $0 \leq k \leq n$, let S_k be the set of words in S that contain exactly k vowels (and hence $n-k$ consonants). To build a word in S_k, first choose a set of k positions out of n where the vowels will appear ($\binom{n}{k}$ ways, by the Subset Rule); then fill these positions with a sequence of vowels (x^k ways, by the Word Rule); then fill the remaining positions with a sequence of consonants (y^{n-k} ways, by the Word Rule). By the Product Rule, $|S_k| = \binom{n}{k} x^k y^{n-k}$. By the Sum Rule, $|S| = \sum_{k=0}^{n} |S_k| = \sum_{k=0}^{n} \binom{n}{k} x^k y^{n-k}$.

We now give an algebraic proof of the Binomial Theorem to illustrate the method of proof by induction. We hope the reader finds the combinatorial proof to be more illuminating and elegant than the computations that follow. Fix $x, y \in \mathbb{R}$. The base case for the induction proof occurs when $n = 0$. For this choice of n, the formula to be proved is $(x+y)^0 = \binom{0}{0} x^0 y^{0-0}$. Using the convention that $r^0 = 1$ for all real r (even $r = 0$) and noting that $\binom{0}{0} = 0!/(0!0!) = 1$, we see that both sides of the formula to be proved evaluate to 1.

For the induction step, fix $n \in \mathbb{Z}_{\geq 0}$. We may assume that the Binomial Theorem is already known for this fixed value of n, and we must then prove the corresponding formula with n replaced by $n+1$. Let us write the induction hypothesis in the form $(x+y)^n = \sum_{k \in \mathbb{Z}} \binom{n}{k} x^k y^{n-k}$, where we now sum over *all integers* k instead of just $k = 0, 1, \ldots, n$. This is permissible, since $\binom{n}{k} = 0$ for $k < 0$ or $k > n$. Using the distributive law, we can now compute

$$\begin{aligned} (x+y)^{n+1} &= (x+y)(x+y)^n = (x+y) \sum_{k \in \mathbb{Z}} \binom{n}{k} x^k y^{n-k} \\ &= \sum_{k \in \mathbb{Z}} \binom{n}{k} x^{k+1} y^{n-k} + \sum_{k \in \mathbb{Z}} \binom{n}{k} x^k y^{n+1-k}. \end{aligned}$$

In the first sum, replace the summation index k by $k-1$. This is allowed, since $k-1$ ranges over all integers as k ranges over all integers. We get

$$\sum_{k \in \mathbb{Z}} \binom{n}{k-1} x^k y^{n+1-k} + \sum_{k \in \mathbb{Z}} \binom{n}{k} x^k y^{n+1-k} = \sum_{k \in \mathbb{Z}} \left[\binom{n}{k-1} + \binom{n}{k} \right] x^k y^{n+1-k}.$$

Finally, use Pascal's Identity 2.3 (with n replaced by $n+1$) to replace the sum in square brackets by $\binom{n+1}{k}$. We have now shown $(x+y)^{n+1} = \sum_{k \in \mathbb{Z}} \binom{n+1}{k} x^k y^{n+1-k}$, which completes the induction step and the proof.

2.9. Example. What is the coefficient of z^7 in $(2z-5)^9$? We apply the Binomial Theorem taking $x = 2z$ and $y = -5$ and $n = 9$. We have

$$(2z-5)^9 = \sum_{k=0}^{9} \binom{9}{k} (2z)^k (-5)^{9-k}.$$

The only summand involving z^7 is the $k = 7$ summand. The corresponding coefficient is $\binom{9}{7} 2^7 (-5)^2 = 115,200$.

2.10. Remark. If r is any complex number and z is a complex number such that $|z| < 1$, there exists a power series expansion for $(1+z)^r$ called the *Extended Binomial Theorem.* This power series is discussed in §5.3.

2.4 The Multinomial Theorem

Just as binomial coefficients appear in the expansion of the nth power of a binomial, multinomial coefficients appear when we expand the nth power of a multinomial.

2.11. The Multinomial Theorem. For all $s, n \in \mathbb{Z}_{\geq 0}$ and all $z_1, \ldots, z_s \in \mathbb{R}$,

$$(z_1 + z_2 + \cdots + z_s)^n = \sum_{n_1+n_2+\cdots+n_s=n} \binom{n}{n_1, n_2, \ldots, n_s} z_1^{n_1} z_2^{n_2} \cdots z_s^{n_s}.$$

The summation here extends over all ordered sequences $(n_1, n_2, \ldots, n_s)$ of nonnegative integers that sum to n.

For example,

$$(x+y+z)^3 = x^3 + y^3 + z^3 + 3x^2y + 3x^2z + 3y^2z + 3xy^2 + 3xz^2 + 3yz^2 + 6xyz.$$

There is a combinatorial proof of the Multinomial Theorem similar to our proof of the Binomial Theorem. The idea is to count n-letter words using an alphabet with s types of letters, where there are z_1 letters of type 1, z_2 letters of type 2, and so on. Alternatively, the theorem can be proved algebraically using induction on n and the analogue of Pascal's Identity for multinomial coefficients. Yet another algebraic proof uses the Binomial Theorem and induction on s. We ask the reader to supply details of these proofs in the exercises. In the rest of this section, we show how the Multinomial Theorem may be deduced from a more general result called the Non-Commutative Multinomial Theorem.

Before stating this new result, let us consider an example. Suppose A, B, and C are $m \times m$ real matrices and we are trying to compute $(A+B+C)^n$. Recall that matrix multiplication is not commutative in general, but matrices do obey other laws such as associativity of addition and multiplication, commutativity of addition, and the distributive law. Using these facts, we first compute

$$\begin{aligned}
(A+B+C)^2 &= (A+B+C)(A+B+C) \\
&= A(A+B+C) + B(A+B+C) + C(A+B+C) \\
&= AA + AB + AC + BA + BB + BC + CA + CB + CC.
\end{aligned}$$

Here we have written AA instead of A^2, for reasons that will become clear in a moment.

Looking at the third power, we find

$$\begin{aligned}(A+B+C)^3 &= (A+B+C)(A+B+C)^2\\ &= (A+B+C)(AA+AB+AC+\cdots+CC)\\ &= A(AA+AB+AC+\cdots+CC)+B(AA+AB+AC+\cdots+CC)\\ &\quad +C(AA+AB+AC+\cdots+CC)\\ &= AAA+AAB+AAC+\cdots+ACC+BAA+BAB+BAC+\cdots\\ &\quad +BCC+CAA+CAB+CAC+\cdots+CCC.\end{aligned}$$

The final formula is the sum of all three-letter words in the alphabet $\{A,B,C\}$. If we continued on to $(A+B+C)^4=(A+B+C)(A+B+C)^3$, the distributive law would take the new factor $A+B+C$ and prepend each of the three letters A, B, C to each of the three-letter words formed at the previous stage. Thus, we see that $(A+B+C)^4$ is the sum of all four-letter words in $\{A,B,C\}$. Generalizing this argument, we are led to the following *Non-Commutative Multinomial Theorem.*

2.12. The Non-Commutative Multinomial Theorem. For all $s,n\in\mathbb{Z}_{\geq 0}$ and all $m\times m$ real matrices $Z_1,\ldots,Z_s$,

$$(Z_1+Z_2+\cdots+Z_s)^n=\sum_{w\in\{1,\ldots,s\}^n} Z_{w_1}Z_{w_2}\cdots Z_{w_n}.$$

Informally, the right side is the sum of all n-letter words in the alphabet $Z_1,\ldots,Z_s$.

Proof. We use induction on n. When $n=0$, both sides of the identity evaluate to I_m, the $m\times m$ identity matrix (provided we interpret an empty product as I_m). If the reader wishes to exclude the $n=0$ case, the $n=1$ case may also be verified directly by noting that both sides evaluate to $Z_1+\cdots+Z_s$. For the induction step, fix $n\geq 0$, assume the formula in the theorem holds for n, and prove the formula holds for $n+1$. We compute

$$\begin{aligned}(Z_1+Z_2+\cdots+Z_s)^{n+1} &= (Z_1+Z_2+\cdots+Z_s)^n(Z_1+\cdots+Z_s)\\ &= \left(\sum_{w\in\{1,\ldots,s\}^n} Z_{w_1}Z_{w_2}\cdots Z_{w_n}\right)\cdot\left(\sum_{j=1}^{s} Z_j\right)\\ &= \sum_{w\in\{1,\ldots,s\}^n}\sum_{j=1}^{s} Z_{w_1}Z_{w_2}\cdots Z_{w_n}Z_j\\ &= \sum_{v\in\{1,\ldots,s\}^{n+1}} Z_{v_1}Z_{v_2}\cdots Z_{v_{n+1}}.\end{aligned}$$

The first equality follows by definition of powers. The second equality holds by the induction hypothesis. The third equality uses the distributive law for matrices. The fourth equality arises by replacing the pair (w,j) by the word $v=w_1\cdots w_nj$. □

We still need to show how the original Multinomial Theorem follows from the non-commutative version. We begin with an example: given variables w,x,y,z representing real numbers, what is the coefficient of $w^3x^1y^2z^3$ in the expansion of $(w+x+y+z)^9$? Right now, we know that this expansion is the sum of all nine-letter words using the alphabet $\{w,x,y,z\}$. Words such as $wwwxyyzzz$, $wxyzwyzwz$, or $zzyywwwxz$ each contribute one copy of the monomial $w^3x^1y^2z^3$ to the full expansion. In general, any word containing exactly three w's, one x, two y's, and three z's (in some order) gives one copy of the monomial $w^3x^1y^2z^3$. Here we are using the fact that the real variables w, x, y, z commute with each other under multiplication. We see that the overall coefficient of $w^3x^1y^2z^3$ in the expansion is $|\mathcal{R}(w^3x^1y^2z^3)|$, the number of rearrangements of $wwwxyyzzz$. By the Anagram Rule, this coefficient is $\binom{9}{3,1,2,3} = 5040$.

The argument in this example extends to the general case. Given *commuting* variables $z_1, \ldots, z_s$, consider what happens to the right side of the multinomial formula

$$(z_1 + \cdots + z_s)^n = \sum_{w \in \{1,\ldots,s\}^n} z_{w_1} \cdots z_{w_s}$$

when we collect together all terms that yield a particular monomial $z_1^{n_1} z_2^{n_2} \cdots z_s^{n_s}$, where necessarily $n_1 + \cdots + n_s = n$. The number of times this monomial occurs in the original sum is the number of rearrangements of n_1 ones, n_2 twos, and so on. Thus, by the Anagram Rule, the coefficient of this monomial in the simplified sum is $\binom{n}{n_1, n_2, \ldots, n_s}$. This completes the proof of the commutative version of the Multinomial Theorem.

2.13. Remark. The Non-Commutative Multinomial Theorem is valid for elements $Z_1, \ldots, Z_s$ in any ring (see the Appendix for the definition of a ring). The commutative version of the formula holds whenever $Z_iZ_j = Z_jZ_i$ for all i, j. The Non-Commutative Multinomial Theorem is a special case of the Generalized Distributive Law, discussed in Exercise 2-16.

2.5 More Binomial Coefficient Identities

Lattice paths can often be used to give elegant, visually appealing combinatorial proofs of identities involving binomial coefficients. We give several examples of such proofs in this section. Our first result is a version of Pascal's Identity phrased in terms of multinomial coefficients.

2.14. Lattice Path Version of Pascal's Identity. For all $a, b \in \mathbb{Z}_{>0}$,

$$\binom{a+b}{a,b} = \binom{a+b-1}{a-1,b} + \binom{a+b-1}{a,b-1}.$$

Proof. Fix $a, b > 0$. Let S be the set of lattice paths from $(0,0)$ to (a,b). By the Lattice Path Rule, $|S| = \binom{a+b}{a,b}$. On the other hand, we can write $S = S_1 \cup S_2$, where S_1 is the set of paths in S ending with an east step, and S_2 is the set of paths in S ending with a north step. By the Sum Rule, $|S| = |S_1| + |S_2|$. We can build a typical path in S_1 by choosing any lattice path from $(0,0)$ to $(a-1,b)$ and then appending an east step; by the Lattice Path Rule, $|S_1| = \binom{a+b-1}{a-1,b}$. We can build a typical path in S_2 by choosing any lattice path from $(0,0)$ to $(a,b-1)$ and then appending a north step; by the Lattice Path Rule, $|S_2| = \binom{a+b-1}{a,b-1}$. Thus, $|S| = \binom{a+b-1}{a-1,b} + \binom{a+b-1}{a,b-1}$, as needed. We remark that the identity also holds if one (but not both) of a or b is zero, since both sides of the identity are 1 in that case. □

The key idea of the preceding proof was to classify a lattice path based on the last step in the path. We can prove more elaborate binomial coefficient identities by classifying lattice paths in more clever ways. For instance, let us reprove the identity for the sum of squares of binomial coefficients using lattice paths.

2.15. Lattice Path Proof of $\sum_{k=0}^{n}\binom{n}{k}^2=\binom{2n}{n}$. Let S be the set of all lattice paths from $(0,0)$ to (n,n). By the Lattice Path Rule, $|S|=\binom{2n}{n,n}=\binom{2n}{n}$. For $0\leq k\leq n$, let S_k be the set of all paths in S passing through the point $(k,n-k)$ on the line $x+y=n$. Every path in S must go through exactly one such point for some k between 0 and n, so S is the disjoint union of $S_0,S_1,\ldots,S_n$. See Figure 2.1. To build a path in S_k, first choose a path from $(0,0)$ to $(k,n-k)$ in any of $\binom{n}{k,n-k}=\binom{n}{k}$ ways. Second, choose a path from $(k,n-k)$ to (n,n). This is a path in a rectangle of width $n-k$ and height $n-(n-k)=k$, so there are $\binom{n}{n-k,k}=\binom{n}{k}$ ways to make this second choice. By the Sum Rule and Product Rule, we conclude that

$$|S|=\sum_{k=0}^{n}|S_k|=\sum_{k=0}^{n}\binom{n}{k}^2.$$

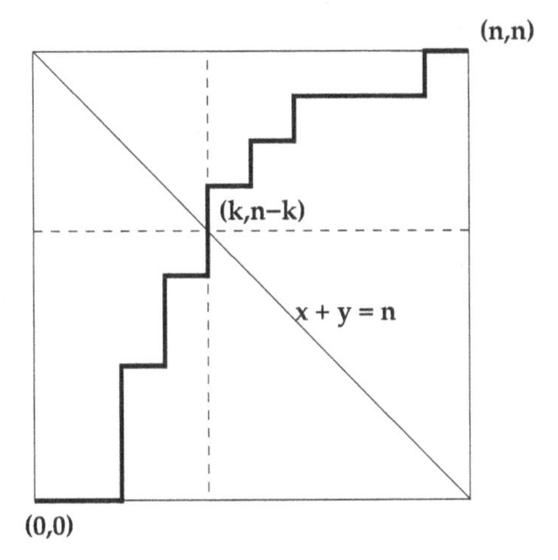

FIGURE 2.1
A combinatorial proof using lattice paths.

The next identity arises by classifying lattice paths based on the number of east steps following the final north step in the path.

2.16. Theorem. For all integers $a\geq 0$ and $b\geq 1$,

$$\binom{a+b}{a,b}=\sum_{k=0}^{a}\binom{k+b-1}{k,b-1}.$$

Proof. Let S be the set of all lattice paths from the origin to (a,b). We know that $|S|=\binom{a+b}{a,b}$. For $0\leq k\leq a$, let S_k be the set of paths in S such that the last north step of the path lies on the line $x=k$. See Figure 2.2. We can build a path in S_k by choosing any lattice path from the origin to $(k,b-1)$ in $\binom{k+b-1}{k,b-1}$ ways, and then appending one north step and $a-k$ east steps. Thus, the required identity follows from the Sum Rule. If we classify the paths by the final east step instead, we obtain the dual identity $\binom{a+b}{a,b}=\sum_{j=0}^{b}\binom{a-1+j}{a-1,j}$, valid for all $a\geq 1$ and $b\geq 0$. This identity also follows from the previous one by the known symmetry $\binom{a+b}{a,b}=\binom{b+a}{b,a}$. □

We can obtain a more general identity by classifying lattice paths based on the location of a particular north step (not necessarily the last one) in the path.

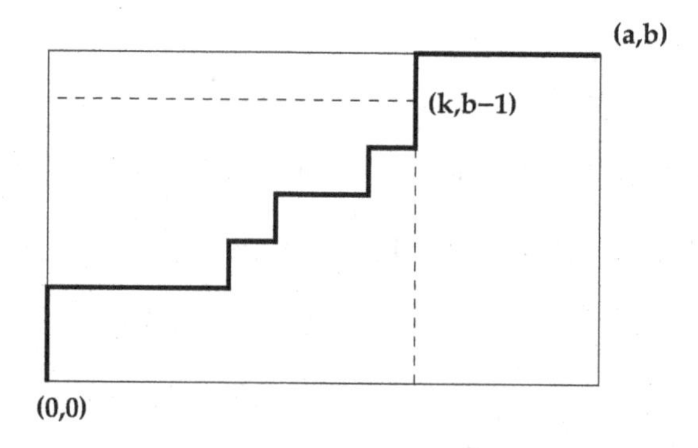

FIGURE 2.2
Another combinatorial proof using lattice paths.

2.17. The Chu–Vandermonde Identity. For all integers $a, b, c \geq 0$,

$$\binom{a+b+c+1}{a,\, b+c+1} = \sum_{k=0}^{a} \binom{k+b}{k,\, b} \binom{a-k+c}{a-k,\, c}.$$

Proof. Let S be the set of all lattice paths from the origin to $(a, b+c+1)$. By the Lattice Path Rule, $|S| = \binom{a+b+c+1}{a,b+c+1}$. For $0 \leq k \leq a$, let S_k be the set of paths in S that contain the north step from (k, b) to $(k, b+1)$. Since every path in S must cross the line $y = b + 1/2$ by taking a north step between the lines $x = 0$ and $x = a$, we see that S is the disjoint union of $S_0, S_1, \ldots, S_a$. See Figure 2.3. Now, we can build a path in S_k as follows. First, choose a lattice path from the origin to (k, b) in $\binom{k+b}{k,b}$ ways. Second, append a north step to this path. Third, choose a lattice path from $(k, b+1)$ to $(a, b+c+1)$. This is a path in a rectangle of width $a-k$ and height c, so there are $\binom{a-k+c}{a-k,c}$ ways to make this choice. Thus, $|S_k| = \binom{k+b}{k,b} \cdot 1 \cdot \binom{a-k+c}{a-k,c}$ by the Product Rule. The identity in the theorem now follows from the Sum Rule. □

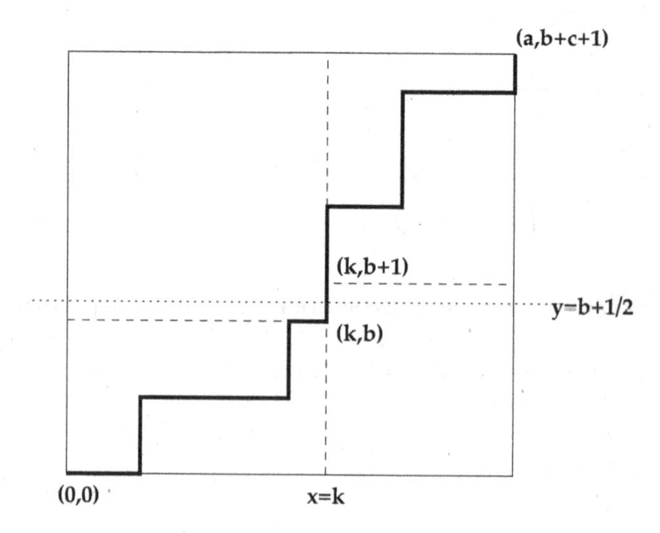

FIGURE 2.3
A third combinatorial proof using lattice paths.

2.6 Sums of Powers of Integers

One often needs to evaluate sums of the form $1^p + 2^p + \cdots + n^p$, where p is a fixed integer. The next result gives formulas for these sums when p is 1, 2, or 3.

2.18. Theorem: Sums of Powers. For all $n \in \mathbb{Z}_{>0}$,

$$\sum_{k=1}^{n} k = \frac{n(n+1)}{2}, \qquad \sum_{k=1}^{n} k^2 = \frac{n(n+1)(2n+1)}{6}, \qquad \sum_{k=1}^{n} k^3 = \frac{n^2(n+1)^2}{4}.$$

Each summation formula may be proved by induction on n (Exercise 2-26). A combinatorial proof of the sum-of-cubes formula is indicated in Exercise 2-27. Here we give an algebraic proof of these identities based on Theorem 2.16. Taking $b = p + 1$ and $a = n - p$ in that theorem, we see that

$$\binom{p}{p} + \binom{p+1}{p} + \binom{p+2}{p} + \cdots + \binom{n}{p} = \binom{n+1}{p+1} \quad \text{for } n \geq p \geq 0. \tag{2.2}$$

Letting $p = 1$ immediately yields $1 + 2 + 3 + \cdots + n = \binom{n+1}{2}$. Letting $p = 2$, we get $\sum_{k=1}^{n} \binom{k}{2} = \binom{n+1}{3}$. Since $\binom{k}{2} = (k^2 - k)/2$ for all k in the indicated range, we have

$$\frac{1}{2}\sum_{k=1}^{n} k^2 - \frac{1}{2}\sum_{k=1}^{n} k = \frac{(n+1)n(n-1)}{6}.$$

Using the previous formula for $\sum_{k=1}^{n} k$ and solving for the sum of squares, we obtain

$$\sum_{k=1}^{n} k^2 = \frac{n(n+1)}{2} + \frac{2n(n+1)(n-1)}{6} = \frac{n(n+1)(2n+1)}{6}.$$

Now letting $p = 3$ in (2.2), we get $\sum_{k=1}^{n} \binom{k}{3} = \binom{n+1}{4}$. Since $\binom{k}{3} = k(k-1)(k-2)/6 = (k^3 - 3k^2 + 2k)/6$, this becomes

$$\sum_{k=1}^{n} k^3 - 3\sum_{k=1}^{n} k^2 + 2\sum_{k=1}^{n} k = 6\frac{(n+1)n(n-1)(n-2)}{24}.$$

Inserting the previous formulas and doing some algebra, we eventually arrive at the formula for the sum of cubes. This method can be continued to evaluate the sum of fourth powers, fifth powers, and so on. The advantage of this technique compared to induction is that one does not need to know the final answer in advance.

2.7 Recursions

Suppose we are given some unknown quantities $a_0, a_1, \ldots, a_n, \ldots$. A *closed formula* for these quantities is an expression of the form $a_n = f(n)$, where the right side is some explicit formula involving the integer n but not involving any of the unknown quantities a_i. In contrast, a *recursive formula* for a_n is an expression of the form $a_n = f(n, a_0, a_1, \ldots, a_{n-1})$, where the right side is a formula that does involve one or more of the unknown quantities

a_i. A recursive formula is usually accompanied by one or more *initial conditions*, which are non-recursive expressions for a_0 and possibly other a_i's. Similar definitions apply to doubly indexed sequences $a_{n,k}$.

Now consider the problem of counting sets of combinatorial objects. Suppose we have several related families of objects, say $T_0, T_1, \ldots, T_n, \ldots$. We think of the index n as somehow measuring the size of the objects in T_n. Sometimes we can find a counting argument leading to a closed formula for $|T_n|$. In many cases, however, it is more natural to give a *recursive* description of T_n, which tells us how to construct a typical object in T_n by assembling smaller objects of the same kind from the sets $T_0, \ldots, T_{n-1}$. Such an argument leads to a recursive formula for $|T_n|$ in terms of one or more of the quantities $|T_0|, \ldots, |T_{n-1}|$. If we suspect that $|T_n|$ is also given by a certain closed formula, we can then prove this fact using induction. The following examples illustrate these ideas.

2.19. Example: Subset Recursion. For each integer $n \geq 0$, let T_n be the set of all subsets of $\{1, 2, \ldots, n\}$, and let $a_n = |T_n|$. We derive a recursive formula for a_n as follows. Suppose $n \geq 1$ and we are trying to build a typical subset $A \in T_n$. We can do this recursively by first choosing a subset $A' \subseteq \{1, 2, \ldots, n-1\}$ in any of $|T_{n-1}| = a_{n-1}$ ways, and then either adding or not adding the element n to this subset (two possibilities). By the Product Rule, we conclude that

$$a_n = a_{n-1} \cdot 2 \quad \text{for all } n \geq 1.$$

The initial condition is $a_0 = 1$, since $T_0 = \{\emptyset\}$.

Using the recursion and initial condition, we calculate:

$$(a_0, a_1, a_2, a_3, a_4, a_5, \ldots) = (1, 2, 4, 8, 16, 32, \ldots).$$

The pattern suggests that $a_n = 2^n$ for all $n \geq 0$. (This already follows from the Power Set Rule, but our goal is to reprove this fact using our recursion.) We prove that $a_n = 2^n$ by induction on n. In the base case ($n = 0$), we have $a_0 = 1 = 2^0$ by the initial condition. Assume that $n > 0$ and that $a_{n-1} = 2^{n-1}$ (this is the induction hypothesis). Using the recursion and the induction hypothesis, we see that

$$a_n = 2a_{n-1} = 2(2^{n-1}) = 2^n.$$

This completes the proof by induction.

2.20. Example: Fibonacci Words. Let W_n be the set of all words in $\{0, 1\}^n$ that do not have two consecutive zeroes, and let $f_n = |W_n|$. We now derive a recursion and initial conditions for the sequence of f_n's. To obtain initial conditions, note that $W_0 = \{\epsilon\}$ (where ϵ denotes the empty word), so $f_0 = 1$. Also $W_1 = \{0, 1\}$, so $f_1 = 2$. To obtain a recursion, fix $n \in \mathbb{Z}_{\geq 2}$. We use the Sum Rule to find a formula for $|W_n| = f_n$. The idea is to classify words in W_n based on their first symbol. Let $W'_n = \{w \in W_n : w_1 = 1\}$ and $W''_n = \{w \in W_n : w_1 = 0\}$. To build a word in W'_n, write a 1 followed by an arbitrary word of length $n-1$ with no consecutive zeroes. There are $|W_{n-1}| = f_{n-1}$ words of the latter type, so $|W'_n| = f_{n-1}$. On the other hand, to build a word in W''_n, first write a 0. Since consecutive zeroes are not allowed, the next symbol of the word must be 1. We can complete the word by choosing an arbitrary word of length $n-2$ with no consecutive zeroes. This choice can be made in f_{n-2} ways, so $|W''_n| = f_{n-2}$. The Sum Rule now shows that $|W_n| = |W'_n| + |W''_n|$, or

$$f_n = f_{n-1} + f_{n-2} \quad \text{for all } n \geq 2.$$

Using this recursion and the initial conditions, we compute

$$(f_0, f_1, f_2, f_3, f_4, f_5, \ldots) = (1, 2, 3, 5, 8, 13, 21, 34, 55, 89, 144, 233, \ldots).$$

The numbers f_n are called *Fibonacci numbers.* We will find an explicit closed formula for f_n in §2.16.

The next example involves a doubly indexed family of combinatorial objects. Our goal is to reprove the Subset Rule by developing a recursion for counting k-element subsets of $\{1, 2, \ldots, n\}$.

2.21. Example: Recursion for k-element Subsets. For all integers $n, k \geq 0$, let $C(n, k)$ be the number of k-element subsets of $\{1, 2, \ldots, n\}$. We will find a recursion and initial conditions for the numbers $C(n, k)$, then use this information to prove the earlier formula $C(n, k) = \binom{n}{k} = \frac{n!}{k!(n-k)!}$ by induction. The recursion is provided by Pascal's Identity 2.3. In the current notation, this identity says that

$$C(n, k) = C(n-1, k) + C(n-1, k-1) \quad \text{for } 0 < k < n.$$

Recall that the first term on the right counts k-element subsets of $\{1, 2, \ldots, n\}$ that do not contain n, whereas the second term counts those subsets that do contain n. Note that this reasoning establishes the recursion for $C(n, k)$ without advance knowledge of the Subset Rule. To get initial conditions, note that $C(n, k) = 0$ whenever $k > n$; $C(n, 0) = 1$ since the empty set is the unique zero-element subset of any set; and $C(n, n) = 1$ since $\{1, 2, \ldots, n\}$ is the only n-element subset of itself.

Now we use the recursion and initial conditions to prove $C(n, k) = \frac{n!}{k!(n-k)!}$ for $0 \leq k \leq n$. We proceed by induction on n. In the base case, $n = k = 0$, and we know $C(0, 0) = 1 = \frac{0!}{0!(0-0)!}$. Next, fix $n > 0$ and assume that for all j in the range $0 \leq j \leq n-1$,

$$C(n-1, j) = \frac{(n-1)!}{j!(n-1-j)!}.$$

Fix k with $0 \leq k \leq n$. If $k = 0$, the initial condition gives $C(n, k) = 1 = n!/(0!(n-0)!)$ as needed. Similarly, the result holds when $k = n$. If $0 < k < n$, we use the recursion and induction hypothesis (applied to $j = k-1$ and to $j = k$, which are integers between 0 and $n-1$) to compute

$$\begin{aligned}
C(n, k) &= C(n-1, k-1) + C(n-1, k) \\
&= \frac{(n-1)!}{(k-1)!((n-1)-(k-1))!} + \frac{(n-1)!}{k!((n-1)-k)!} \\
&= \frac{(n-1)!k}{k!(n-k)!} + \frac{(n-1)!(n-k)}{k!(n-k)!} \\
&= \frac{(n-1)!}{k!(n-k)!} \cdot [k + (n-k)] = \frac{n!}{k!(n-k)!}.
\end{aligned}$$

This completes the proof by induction. (Note that this calculation recreates the algebraic proof of Pascal's Identity.)

The reader may wonder what good it is to have a recursion for $C(n, k)$, since we already proved by other methods the explicit formula $C(n, k) = \frac{n!}{k!(n-k)!}$. There are several answers to this question. One answer is that the recursion for the $C(n, k)$'s gives us a fast method for calculating these quantities that is more efficient than computing with factorials. One popular way of displaying this calculation is called *Pascal's Triangle.* We build this triangle by writing the $n+1$ numbers $C(n, 0), C(n, 1), \ldots, C(n, n)$ in the nth row from the top. If we position the entries as shown in Figure 2.4, then each entry is the sum of the two entries directly above it. We compute $C(n, k)$ by calculating rows 0 through n of this triangle.

$$
\begin{array}{lccccccccccccccccc}
n=0: & & & & & & & & & 1 & & & & & & & & \\
n=1: & & & & & & & & 1 & & 1 & & & & & & & \\
n=2: & & & & & & & 1 & & 2 & & 1 & & & & & & \\
n=3: & & & & & & 1 & & 3 & & 3 & & 1 & & & & & \\
n=4: & & & & & 1 & & 4 & & 6 & & 4 & & 1 & & & & \\
n=5: & & & & 1 & & 5 & & 10 & & 10 & & 5 & & 1 & & & \\
n=6: & & & 1 & & 6 & & 15 & & 20 & & 15 & & 6 & & 1 & & \\
n=7: & & 1 & & 7 & & 21 & & 35 & & 35 & & 21 & & 7 & & 1 & \\
n=8: & 1 & & 8 & & 28 & & 56 & & 70 & & 56 & & 28 & & 8 & & 1
\end{array}
$$

FIGURE 2.4
Pascal's Triangle.

Note that computing $C(n,k)$ via Pascal's Recursion requires only addition operations. In contrast, calculation using the closed formula $\frac{n!}{k!(n-k)!}$ requires us to divide one large factorial by the product of two other factorials (although this work can be reduced somewhat by cancellation of common factors). For example, Pascal's Triangle quickly yields $C(8,4)=70$, while the closed formula gives $\binom{8}{4}=\frac{8!}{4!4!}=\frac{40{,}320}{24^2}=70$.

In bijective combinatorics, it turns out that the arithmetic operation of *division* is much harder to understand from a combinatorial standpoint than the operations of *addition* and *multiplication*. In particular, our original derivation of the formula $\binom{n}{k}=\frac{n!}{k!(n-k)!}$ was an indirect argument using the Product Rule, in which we divided by $k!$ at the end (see §1.4). For some applications (such as listing all k-element subsets of a given n-element set, discussed in Chapter 6), it is necessary to have a counting argument that does not rely on division.

A final reason for studying recursions for $C(n,k)$ is to emphasize that recursions are helpful and ubiquitous tools for analyzing combinatorial objects. Indeed, we will soon be considering combinatorial collections whose cardinalities may not be given by explicit closed formulas. Nevertheless, these cardinalities satisfy recursions that allow them to be computed quickly and efficiently.

2.8 Recursions for Multisets and Anagrams

This section develops recursions for counting multisets and anagrams.

2.22. Recursion for Multisets. In §1.12, we counted k-element multisets from an n-letter alphabet using bijective techniques. Now, we give a recursive analysis to reprove the enumeration results for multisets. For all integers $n,k\geq 0$, let $M(n,k)$ be the number of k-element multisets using letters from $\{1,2,\ldots,n\}$. The initial conditions are $M(n,0)=1$ for all $n\geq 0$ and $M(0,k)=0$ for all $k>0$. We now derive a recursion for $M(n,k)$ valid for $n>0$ and $k>0$. A typical multiset counted by $M(n,k)$ either does not contain n at all or contains one or more copies of n. In the former case, the multiset is a k-element multiset using letters from $\{1,2,\ldots,n-1\}$, and there are $M(n-1,k)$ such multisets. In the latter case, if we remove one copy of n from the multiset, we obtain an arbitrary $(k-1)$-element multiset using letters from $\{1,2,\ldots,n\}$. There are $M(n,k-1)$ such multisets. By the Sum

	$k=0$	$k=1$	$k=2$	$k=3$	$k=4$	$k=5$	$k=6$
$n=0:$	1	0	0	0	0	0	0
$n=1:$	1	1	1	1	1	1	1
$n=2:$	1	2	3	4	5	6	7
$n=3:$	1	3	6	10	15	21	28
$n=4:$	1	4	10	20	35	56	84
$n=5:$	1	5	15	35	70	126	210
$n=6:$	1	6	21	56	126	252	462

FIGURE 2.5
Table for computing $M(n,k)$ recursively.

Rule, we obtain the recursion

$$M(n,k) = M(n-1,k) + M(n,k-1) \quad \text{for } n>0 \text{ and } k>0.$$

It can now be proved by induction that for all $n \geq 0$ and all $k \geq 0$, $M(n,k) = \binom{k+n-1}{k,n-1}$. The proof is similar to the corresponding proof for $C(n,k)$, so we omit it.

We can use the recursion to compute values of $M(n,k)$. Here we use a left-justified table of entries in which the nth row contains the numbers $M(n,0), M(n,1), \ldots$. The values in the top row (where $n=0$) and in the left column (where $k=0$) are given by the initial conditions. Each remaining entry in the table is the sum of the number directly above it and the number directly to its left. See Figure 2.5. This table is a tilted version of Pascal's Triangle.

2.23. Recursion for Multinomial Coefficients. Let $n_1, \ldots, n_s$ be nonnegative integers that add to n. Let $\{a_1, \ldots, a_s\}$ be a given s-letter alphabet, and let $C(n; n_1, \ldots, n_s) = |\mathcal{R}(a_1^{n_1} \cdots a_s^{n_s})|$ be the number of n-letter words that are rearrangements of n_i copies of a_i for $1 \leq i \leq s$. We proved in §1.5 that $C(n; n_1, \ldots, n_s) = \binom{n}{n_1, \ldots, n_s} = \frac{n!}{n_1! n_2! \cdots n_s!}$. We now give a new proof of this result using recursions.

Assume first that every n_i is positive. For $1 \leq i \leq s$, let T_i be the set of words in $T = \mathcal{R}(a_1^{n_1} \cdots a_s^{n_s})$ that begin with the letter a_i. T is the disjoint union of the sets $T_1, \ldots, T_s$. To build a typical word $w \in T_i$, we start with the letter a_i and then append any element of $\mathcal{R}(a_1^{n_1} \cdots a_i^{n_i - 1} \cdots a_s^{n_s})$. There are $C(n-1; n_1, \ldots, n_i - 1, \ldots, n_s)$ ways to do this. Hence, by the Sum Rule,

$$C(n; n_1, \ldots, n_s) = \sum_{i=1}^{s} C(n-1; n_1, \ldots, n_i - 1, \ldots, n_s).$$

If we adopt the convention that $C(n; n_1, \ldots, n_s) = 0$ whenever any n_i is negative, then this recursion holds (with the same proof) for all choices of $n_i \geq 0$ and $n > 0$. The initial condition is $C(0; 0, 0, \ldots, 0) = 1$, since the empty word is the unique rearrangement of zero copies of the given letters.

Now let us prove that $C(n; n_1, \ldots, n_s) = n! / \prod_{k=1}^{s} n_k!$ by induction on n. In the base case, $n = n_1 = \cdots = n_s = 0$, and the required formula follows from the initial condition. For the induction step, assume that $n > 0$ and that

$$C(n-1; m_1, \ldots, m_s) = \frac{(n-1)!}{\prod_{k=1}^{s} m_k!}$$

whenever $m_1 + \cdots + m_s = n - 1$. Assume that we are given integers $n_1, \ldots, n_s \geq 0$ that sum to n. Using the recursion and induction hypothesis, we compute as follows:

$$\begin{aligned} C(n; n_1, \ldots, n_s) &= \sum_{k=1}^{s} C(n-1; n_1, \ldots, n_k - 1, \ldots, n_s) \\ &= \sum_{\substack{1 \leq k \leq s \\ n_k > 0}} \frac{(n-1)!}{(n_k - 1)! \prod_{j \neq k} n_j!} \\ &= \sum_{\substack{1 \leq k \leq s \\ n_k > 0}} \frac{(n-1)! n_k}{\prod_{j=1}^{s} n_j!} = \sum_{k=1}^{s} \frac{(n-1)! n_k}{\prod_{j=1}^{s} n_j!} \\ &= \frac{(n-1)!}{\prod_{j=1}^{s} n_j!} \left[\sum_{k=1}^{s} n_k \right] = \frac{n!}{\prod_{j=1}^{s} n_j!}. \end{aligned}$$

2.9 Recursions for Lattice Paths

Recursive techniques allow us to count many collections of lattice paths. We first consider the situation of lattice paths in a rectangle. Compare the next result to Identity 2.14.

2.24. Recursion for Paths in a Rectangle. For $a, b \geq 0$, let $L(a, b)$ be the number of lattice paths from the origin to (a, b). We have $L(a, 0) = L(0, b) = 1$ for all $a, b \geq 0$. If $a > 0$ and $b > 0$, note that any lattice path ending at (a, b) arrives there via an east step or a north step. We obtain lattice paths of the first kind by taking any lattice path ending at $(a-1, b)$ and appending an east step. We obtain lattice paths of the second kind by taking any lattice path ending at $(a, b-1)$ and appending a north step. Hence, by the Sum Rule,

$$L(a, b) = L(a-1, b) + L(a, b-1) \quad \text{for all } a, b > 0.$$

It can now be shown (by induction on $a + b$) that

$$L(a, b) = \binom{a+b}{a, b} = \frac{(a+b)!}{a!b!} \quad \text{for all } a, b \geq 0.$$

We can visually display and calculate the numbers $L(a, b)$ by labeling each lattice point (a, b) with the number $L(a, b)$. The initial conditions say that the lattice points on the axes are labeled 1. The recursion says that the label of some point (a, b) is the sum of the labels of the point $(a-1, b)$ to its immediate left and the point $(a, b-1)$ immediately below it. See Figure 2.6, in which we recognize yet another shifted version of Pascal's Triangle.

By modifying the boundary conditions, we can adapt the recursion in the previous example to count more complicated collections of lattice paths.

2.25. Recursion for Paths in a Triangle. For $b \geq a \geq 0$, let $T(a, b)$ be the number of lattice paths from the origin to (a, b) that always stay weakly above the line $y = x$. (In particular, $T(n, n)$ is the number of Dyck paths of order n.) By the same argument used above, we have

$$T(a, b) = T(a-1, b) + T(a, b-1) \quad \text{for } b > a > 0.$$

On the other hand, when $a = b > 0$, a lattice path can only reach $(a, b) = (a, a)$ by taking an east step, since the point $(a, b-1)$ lies below $y = x$. Thus,

$$T(a, a) = T(a-1, a) \quad \text{for } a > 0.$$

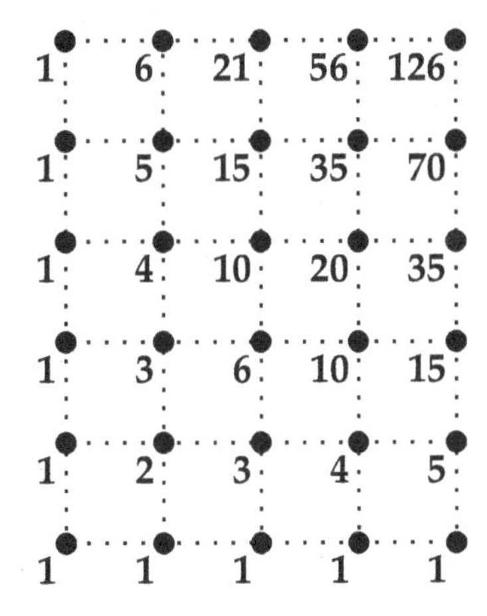

FIGURE 2.6
Recursive enumeration of lattice paths.

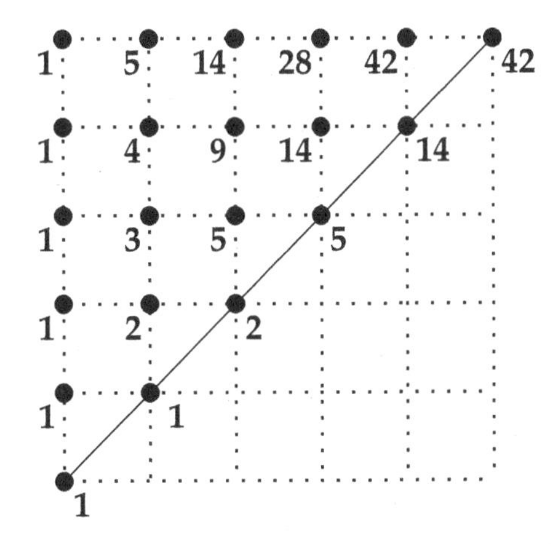

FIGURE 2.7
Recursive enumeration of lattice paths in a triangle.

The initial conditions are $T(0, b) = 1$ for all $b \geq 0$. Figure 2.7 shows how to compute the numbers $T(a, b)$ by drawing a picture. We see the Catalan numbers $1, 1, 2, 5, 14, \ldots$ appearing on the main diagonal.

It turns out that there is an explicit closed formula for the numbers $T(a, b)$, which are called *ballot numbers*.

2.26. Theorem: Ballot Numbers. For $b \geq a \geq 0$, the number of lattice paths from the origin to (a, b) that always stay weakly above the line $y = x$ is

$$\frac{b-a+1}{b+a+1}\binom{a+b+1}{a}.$$

In particular, the number of Dyck paths of order n is $C_n = \frac{1}{2n+1}\binom{2n+1}{n}$.

Proof. We show that

$$T(a,b) = \frac{b-a+1}{b+a+1}\binom{a+b+1}{a}$$

by induction on $a+b$. If $a+b=0$, so that $a=b=0$, then $T(0,0) = 1 = \frac{0-0+1}{0+0+1}\binom{0+0+1}{0}$. Now assume that $a+b>0$ and that $T(c,d) = \frac{d-c+1}{d+c+1}\binom{c+d+1}{c}$ whenever $d \geq c \geq 0$ and $c+d < a+b$. To prove the claimed formula for $T(a,b)$, we consider cases based on the recursions and initial conditions. First, if $a=0$ and $b \geq 0$, we have $T(a,b) = 1 = \frac{b-0+1}{b+0+1}\binom{0+b+1}{0}$. Second, if $a = b > 0$, we have

$$\begin{aligned} T(a,b) &= T(a,a) = T(a-1,a) = \frac{2}{2a}\binom{2a}{a-1} \\ &= \frac{(2a)!}{a!(a+1)!} = \frac{1}{2a+1}\binom{2a+1}{a} \\ &= \frac{b-a+1}{b+a+1}\binom{a+b+1}{a}. \end{aligned}$$

Third, if $b > a > 0$, we have

$$\begin{aligned} T(a,b) &= T(a-1,b) + T(a,b-1) = \frac{b-a+2}{a+b}\binom{a+b}{a-1} + \frac{b-a}{a+b}\binom{a+b}{a} \\ &= \frac{(b-a+2)(a+b-1)!}{(a-1)!(b+1)!} + \frac{(b-a)(a+b-1)!}{a!b!} \\ &= \left[\frac{a(b-a+2)}{a+b} + \frac{(b-a)(b+1)}{a+b}\right]\frac{(a+b)!}{a!(b+1)!} \\ &= \left[\frac{ab-a^2+2a+b^2-ab+b-a}{a+b}\right]\frac{(a+b+1)!}{(b+a+1)a!(b+1)!} \\ &= \left[\frac{(b-a+1)(a+b)}{a+b}\right]\frac{1}{b+a+1}\binom{a+b+1}{a} \\ &= \frac{b-a+1}{b+a+1}\binom{a+b+1}{a}. \quad \square \end{aligned}$$

The numbers $T(a,b)$ in the previous theorem are called ballot numbers for the following reason. Let $w \in \{\mathrm{N},\mathrm{E}\}^{a+b}$ be a lattice path counted by $T(a,b)$. Imagine that $a+b$ people are voting for two candidates (candidate N and candidate E) by casting an ordered sequence of $a+b$ ballots. The path w records this sequence of ballots as follows: $w_j = \mathrm{N}$ if the jth person votes for candidate N, and $w_j = \mathrm{E}$ if the jth person votes for candidate E. The condition that w stays weakly above $y=x$ means that candidate N always has at least as many votes as candidate E at each stage in the election process. The condition that w ends at (a,b) means that candidate N has b votes and candidate E has a votes at the end of the election.

Returning to lattice paths, suppose we replace the boundary line $y=x$ by the line $y=mx$ (where m is any positive integer). We can then derive the following more general result.

2.27. Theorem: m-Ballot Numbers. Let m be a fixed positive integer. For $b \geq ma \geq 0$, the number of lattice paths from the origin to (a,b) that always stay weakly above the line $y=mx$ is

$$\frac{b-ma+1}{b+a+1}\binom{a+b+1}{a}.$$

In particular, the number of such paths ending at (n, mn) is

$$\frac{1}{(m+1)n+1}\binom{(m+1)n+1}{n}.$$

Proof. Let $T_m(a, b)$ be the number of paths ending at (a, b) that never go below $y = mx$. Arguing as before, we have $T_m(0, b) = 1$ for all $b \geq 0$; $T_m(a, b) = T_m(a-1, b) + T_m(a, b-1)$ whenever $b > ma > 0$; and $T_m(a, ma) = T_m(a-1, ma)$ since the point $(a, ma-1)$ lies below the line $y = mx$. It can now be proved that

$$T_m(a, b) = \frac{b - ma + 1}{b + a + 1}\binom{a + b + 1}{a}$$

by induction on $a + b$. The proof is similar to the one given above, so we omit it. For a bijective proof of this theorem, see Exercise 12-3. □

When the slope m of the boundary line $y = mx$ is not an integer, we cannot use the formula in the preceding theorem. Nevertheless, the recursion (with appropriate initial conditions) can still be used to count lattice paths bounded below by this line. For example, Figure 2.8 illustrates the enumeration of lattice paths from $(0, 0)$ to $(6, 9)$ that always stay weakly above $y = (3/2)x$.

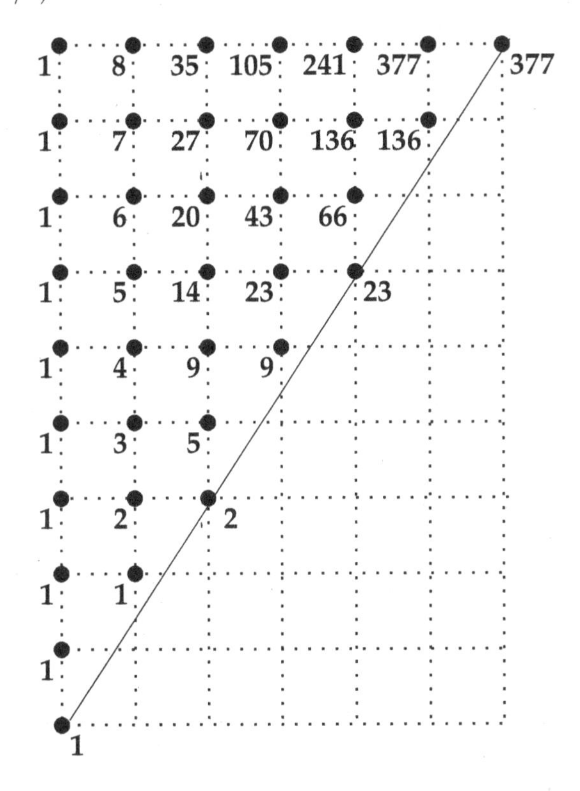

FIGURE 2.8
Recursive enumeration of lattice paths above $y = (3/2)x$.

2.10 Catalan Recursions

The recursions from the previous section provide one way of computing Catalan numbers, which are special cases of ballot numbers. This section discusses another recursion that involves only the Catalan numbers. This *convolution recursion* appears in many settings, thus leading to many different combinatorial interpretations for the Catalan numbers.

2.28. Theorem: Catalan Recursion. The Catalan numbers $C_n = \frac{1}{n+1}\binom{2n}{n}$ satisfy the recursion

$$C_n = \sum_{k=1}^{n} C_{k-1}C_{n-k} \qquad \text{for all } n > 0,$$

with initial condition $C_0 = 1$.

Proof. Recall from the Dyck Path Rule 1.101 that C_n is the number of Dyck paths of order n. There is one Dyck path of order 0, so $C_0 = 1$. Fix $n > 0$, and let A be the set of Dyck paths ending at (n, n). For $1 \leq k \leq n$, let A_k be the set of Dyck paths of order n that return to the diagonal line $y = x$ for the first time at the point (k, k). See Figure 2.9. Suppose w

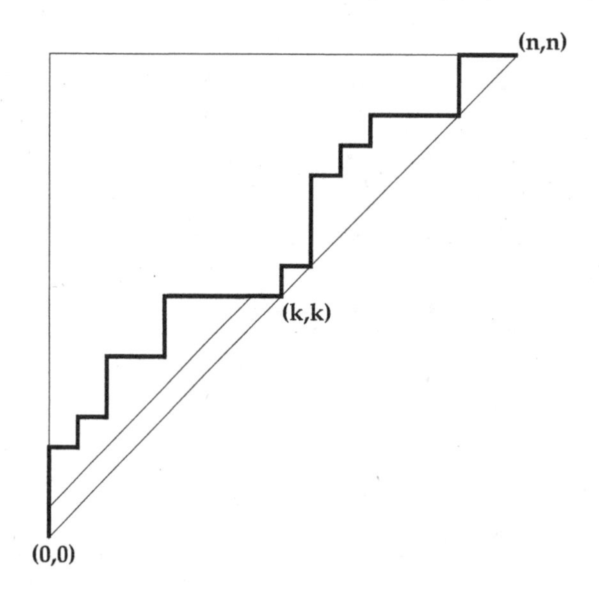

FIGURE 2.9
Proving the Catalan recursion by analyzing the first return to $y = x$.

is the word in $\{\mathrm{N}, \mathrm{E}\}^{2n}$ that encodes a path in A_k. Inspection of Figure 2.9 shows that we have the factorization $w = \mathrm{N}w_1\mathrm{E}w_2$, where w_1 encodes a Dyck path of order $k-1$ (starting at $(0, 1)$ in the figure) and w_2 encodes a Dyck path of order $n-k$ (starting at (k, k) in the figure). We can uniquely construct all paths in A_k by choosing w_1 and w_2 and then setting $w = \mathrm{N}w_1\mathrm{E}w_2$. There are C_{k-1} choices for w_1 and C_{n-k} choices for w_2. By the Product Rule and Sum Rule,

$$C_n = |A| = \sum_{k=1}^{n} |A_k| = \sum_{k=1}^{n} C_{k-1}C_{n-k}. \qquad \square$$

We now show that the Catalan recursion uniquely determines the Catalan numbers.

2.29. Proposition. Suppose $d_0, d_1, \ldots, d_n, \ldots$ is a sequence such that $d_0 = 1$ and $d_n = \sum_{k=1}^{n} d_{k-1}d_{n-k}$ for all $n > 0$. Then $d_n = C_n = \frac{1}{n+1}\binom{2n}{n}$ for all $n \geq 0$.

Proof. We argue by strong induction. For $n = 0$, we have $d_0 = 1 = C_0$. Assume that $n > 0$ and that $d_m = C_m$ for all $m < n$. Then $d_n = \sum_{k=1}^{n} d_{k-1}d_{n-k} = \sum_{k=1}^{n} C_{k-1}C_{n-k} = C_n$. □

We can now prove that various collections of objects are counted by the Catalan numbers. One proof method sets up a bijection between such objects and other objects (like Dyck paths) that are already known to be counted by Catalan numbers. A second proof method shows that the new collections of objects satisfy the Catalan recursion. We illustrate both methods in the examples below.

2.30. Example: Balanced Parentheses. For $n \geq 0$, let BP_n be the set of all words consisting of n left parentheses and n right parentheses, such that every left parenthesis can be matched with a right parenthesis later in the word. For example, BP_3 consists of the following five words:

((())) (())() ()(()) (()()) ()()()

We show that $|BP_n| = C_n$ for all n by exhibiting a bijection between BP_n and the set of Dyck paths of order n. Given $w \in BP_n$, replace each left parenthesis by N (which encodes a north step) and each right parenthesis by E (which encodes an east step). It can be checked that a string w of n left and n right parentheses is balanced iff for every $i \leq 2n$, the number of left parentheses in the prefix $w_1w_2\cdots w_i$ weakly exceeds the number of right parentheses in this prefix. Converting to north and east steps, this condition means that no lattice point on the path lies strictly below the line $y = x$. Thus we have mapped each $w \in BP_n$ to a Dyck path. This map is a bijection, so $|BP_n| = C_n$.

2.31. Example: Binary Trees. We recursively define the set of *binary trees with n nodes* as follows. The empty set is the unique binary tree with 0 nodes. If T_1 is a binary tree with h nodes and T_2 is a binary tree with k nodes, then the ordered triple $T = (\bullet, T_1, T_2)$ is a binary tree with $h + k + 1$ nodes. By definition, all binary trees arise by a finite number of applications of these rules. If $T = (\bullet, T_1, T_2)$ is a binary tree, we call T_1 the *left subtree* of T and T_2 the *right subtree* of T. Note that T_1 or T_2 (or both) may be empty. We can draw a picture of a nonempty binary tree T as follows. First, draw a *root node* of the binary tree at the top of the picture. If T_1 is nonempty, draw an edge leading down and left from the root node, and then draw the picture of T_1. If T_2 is nonempty, draw an edge leading down and right from the root node, and then draw the picture of T_2. For example, Figure 2.10 displays the five binary trees with three nodes. Figure 2.11 depicts a larger binary tree that is formally represented by the sequence

$$T = (\bullet, (\bullet, (\bullet, (\bullet, \emptyset, \emptyset), (\bullet, \emptyset, \emptyset)), (\bullet, \emptyset, \emptyset)), (\bullet, (\bullet, \emptyset, (\bullet, (\bullet, \emptyset, \emptyset), \emptyset)), \emptyset)).$$

Let BT_n denote the set of binary trees with n nodes. We show that $|BT_n| = C_n$ for all n by verifying that the sequence $(|BT_n| : n \geq 0)$ satisfies the Catalan recursion. First, $|BT_0| = 1$ by definition. Second, suppose $n \geq 1$. By the recursive definition of binary trees, we can uniquely construct a typical element of BT_n as follows. Fix k with $1 \leq k \leq n$. Choose a tree $T_1 \in BT_{k-1}$ with $k - 1$ nodes. Then choose a tree $T_2 \in BT_{n-k}$ with $n - k$ nodes. We assemble these trees (together with a new root node) to get a binary tree $T = (\bullet, T_1, T_2)$ with $(k - 1) + 1 + (n - k) = n$ nodes. By the Sum Rule and Product Rule, we have

$$|BT_n| = \sum_{k=1}^{n} |BT_{k-1}||BT_{n-k}|.$$

It follows from Proposition 2.29 that $|BT_n| = C_n$ for all $n \geq 0$.

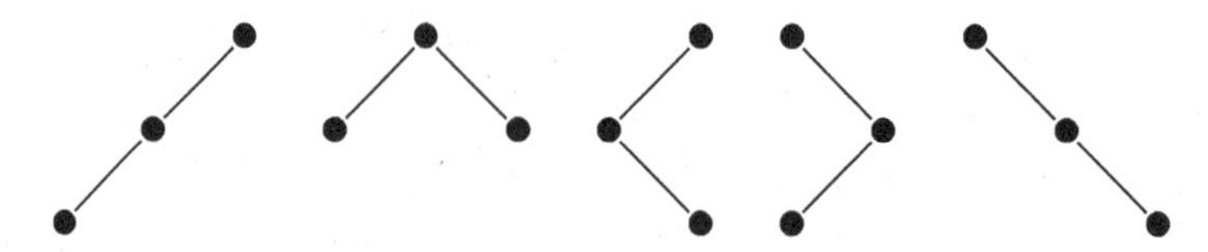

FIGURE 2.10
The five binary trees with three nodes.

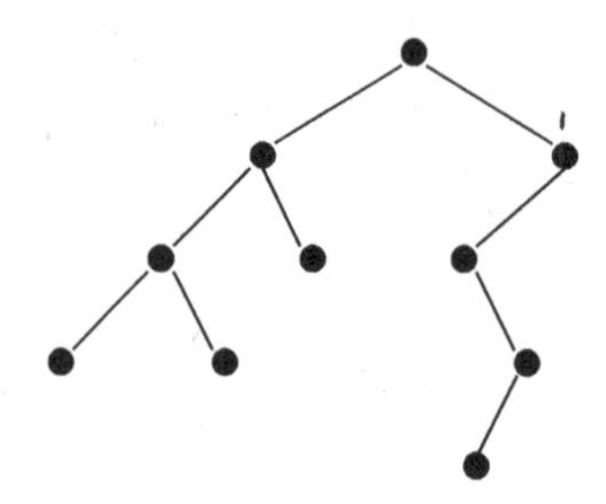

FIGURE 2.11
A binary tree with ten nodes.

2.32. Example: 231-avoiding permutations. Suppose $w = w_1w_2\cdots w_n$ is a permutation of n distinct integers. We say that w is *231-avoiding* iff there do not exist indices $i < k < p$ such that $w_p < w_i < w_k$. This means that no three-element subsequence $w_i \ldots w_k \ldots w_p$ in w has the property that w_p is the smallest number in $\{w_i, w_k, w_p\}$ and w_k is the largest number in $\{w_i, w_k, w_p\}$. For example, when $n = 4$, there are fourteen 231-avoiding permutations of $\{1, 2, 3, 4\}$:

$$1234,\ 1243,\ 1324,\ 1423,\ 1432,\ 2134,\ 2143,$$

$$3124,\ 3214,\ 4123,\ 4132,\ 4213,\ 4312,\ 4321.$$

The following ten permutations do contain occurrences of the pattern 231:

$$2314,\ 2341,\ 2431,\ 4231,\ 3421,\ 3412,\ 3142,\ 1342,\ 3241,\ 2413.$$

Let S_n^{231} be the set of 231-avoiding permutations of $\{1, 2, \ldots, n\}$. We prove that $|S_n^{231}| = C_n$ for all $n \geq 0$ by verifying the Catalan recursion. First, $|S_0^{231}| = 1 = C_0$ since the empty permutation is certainly 231-avoiding. Next, suppose $n > 0$. We construct a typical object $w \in S_n^{231}$ as follows. Consider cases based on the position of the letter n in w. Say $w_k = n$. For all $i < k$ and all $p > k$, we must have $w_i < w_p$; otherwise, the subsequence $w_i, w_k = n, w_p$ would be an occurrence of the forbidden 231 pattern. Assuming that $w_i < w_p$ whenever $i < k < p$, it can be checked that $w = w_1w_2\cdots w_n$ is 231-avoiding iff $w_1w_2\cdots w_{k-1}$ is 231-avoiding and $w_{k+1}\cdots w_n$ is 231-avoiding. Thus, for a fixed k, we can construct w by choosing an arbitrary 231-avoiding permutation w' of the $k-1$ letters $\{1, 2, \ldots, k-1\}$ in $|S_{k-1}^{231}|$ ways, then choosing an arbitrary 231-avoiding permutation w'' of the $n-k$ letters $\{k, \ldots, n-1\}$ in $|S_{n-k}^{231}|$ ways, and finally letting w be the concatenation of w', the letter n, and w''. By the Sum Rule and Product Rule, we have

$$|S_n^{231}| = \sum_{k=1}^{n} |S_{k-1}^{231}||S_{n-k}^{231}|.$$

By Proposition 2.29, $|S_n^{231}| = C_n$ for all $n \geq 0$.

2.33. Example: τ-avoiding permutations. Let $\tau : \{1, 2, \ldots, k\} \to \{1, 2, \ldots, k\}$ be a fixed permutation of k letters. A permutation w of $\{1, 2, \ldots, n\}$ is called *τ-avoiding* iff there do not exist indices $1 \leq i(1) < i(2) < \cdots < i(k) \leq n$ such that

$$w_{i(\tau^{-1}(1))} < w_{i(\tau^{-1}(2))} < \cdots < w_{i(\tau^{-1}(k))}.$$

This means that no subsequence of k entries of w consists of numbers in the same relative order as the numbers $\tau_1, \tau_2, \ldots, \tau_k$. For instance, $w = 15362784$ is not 2341-avoiding, since the subsequence 5684 matches the pattern 2341 (as does the subsequence 5674). On the other hand, w is 4321-avoiding, since there is no decreasing subsequence of w of length 4. Let S_n^τ denote the set of τ-avoiding permutations of $\{1, 2, \ldots, n\}$.

For general τ, the enumeration of τ-avoiding permutations is an extremely difficult problem that has received much attention in recent years. On the other hand, if τ is a permutation of $k = 3$ letters, then the number of τ-avoiding permutations of length n is always the Catalan number C_n, for all six possible choices of τ. We have already proved this in the last example for $\tau = 231$. The arguments in that example readily adapt to prove the Catalan recursion for $\tau = 132$, $\tau = 213$, and $\tau = 312$. However, more subtle arguments are needed to prove this result for $\tau = 123$ and $\tau = 321$ (see Theorem 12.84).

2.34. Remark: Catalan Bijections. Let $(A_n : n \geq 0)$ and $(B_n : n \geq 0)$ be two families of combinatorial objects such that $|A_n| = C_n = |B_n|$ for all n. Suppose that we have an explicit bijective proof that the numbers $|A_n|$ satisfy the Catalan recursion. This means that we can describe a bijection g_n from the set A_n onto the union of the disjoint sets $\{k\} \times A_{k-1} \times A_{n-k}$ for $k = 1, 2, \ldots, n$. (Such a bijection can often be constructed from a counting argument involving the Sum Rule and Product Rule.) Suppose we have similar bijections h_n for the sets B_n. We can combine these bijections to obtain recursively defined bijections $f_n : A_n \to B_n$. First, there is a unique bijection $f_0 : A_0 \to B_0$, since $|A_0| = 1 = |B_0|$. Second, fix $n > 0$, and assume that bijections $f_m : A_m \to B_m$ have already been defined for all $m < n$. Define $f_n : A_n \to B_n$ as follows. Given $x \in A_n$, suppose $g_n(x) = (k, y, z)$ where $1 \leq k \leq n$, $y \in A_{k-1}$, and $z \in A_{n-k}$. Set

$$f_n(x) = h_n^{-1}(k, f_{k-1}(y), f_{n-k}(z)).$$

The inverse map is defined analogously.

For example, let us recursively define a bijection ϕ from the set of binary trees to the set of Dyck paths such that trees with n nodes map to paths of order n. Linking together the first-return recursion for Dyck paths with the left/right-subtree recursion for binary trees as discussed in the previous paragraph, we obtain the rule

$$\phi(\emptyset) = \epsilon \text{ (the empty word)}; \qquad \phi((\bullet, T_1, T_2)) = \mathrm{N}\phi(T_1)\mathrm{E}\phi(T_2).$$

For example, the one-node tree $(\bullet, \emptyset, \emptyset)$ maps to the Dyck path $\mathrm{N}\epsilon\mathrm{E}\epsilon = \mathrm{NE}$. It then follows that

$$\phi((\bullet, (\bullet, \emptyset, \emptyset), \emptyset)) = \mathrm{N(NE)E}\epsilon = \mathrm{NNEE};$$
$$\phi((\bullet, \emptyset, (\bullet, \emptyset, \emptyset))) = \mathrm{N}\epsilon\mathrm{E(NE)} = \mathrm{NENE};$$
$$\phi((\bullet, (\bullet, \emptyset, \emptyset), (\bullet, \emptyset, \emptyset))) = \mathrm{N(NE)E(NE)} = \mathrm{NNEENE};$$

and so on. Figure 2.12 illustrates the recursive computation of $\phi(T)$ for the binary tree T shown in Figure 2.11.

As another example, let us recursively define a bijection ψ from the set of binary trees to the set of 231-avoiding permutations such that trees with n nodes map to permutations

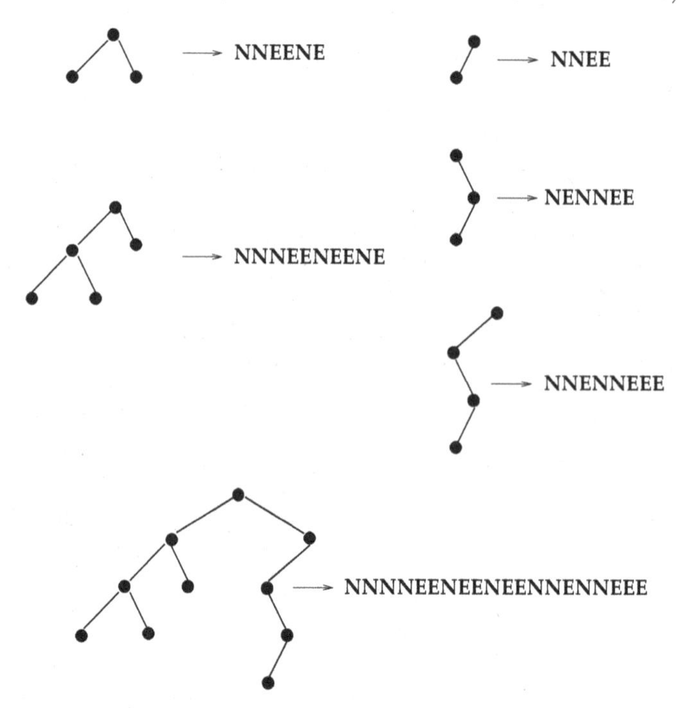

FIGURE 2.12
Mapping binary trees to Dyck paths.

of n letters. Linking together the two proofs of the Catalan recursion for binary trees and 231-avoiding permutations, we obtain the rule

$$\psi(\emptyset) = \epsilon, \qquad \psi((\bullet, T_1, T_2)) = \psi(T_1)\, n\, \psi'(T_2),$$

where $\psi'(T_2)$ is the permutation obtained by increasing each entry of $\psi(T_2)$ by $k-1 = |T_1|$. Figure 2.13 illustrates the recursive computation of $\psi(T)$ for the binary tree T shown in Figure 2.11.

2.11 Integer Partitions

This section introduces integer partitions, which are ways of writing a given positive integer as a sum of positive integers. Integer partitions are similar to compositions, but here the order of the summands does not matter. For example, $1+3+3$ and $3+1+3$ and $3+3+1$ are three different compositions of 7 that represent the same integer partition of 7. By convention, the summands in an integer partition are listed in weakly decreasing order, as in the following formal definition.

2.35. Definition: Integer Partitions. Let n be a nonnegative integer. An *integer partition* of n is a sequence $\mu = (\mu_1, \mu_2, \ldots, \mu_k)$ of positive integers such that $\mu_1 + \mu_2 + \cdots + \mu_k = n$ and $\mu_1 \geq \mu_2 \geq \cdots \geq \mu_k$. Each μ_i is called a *part* of the partition. Let $p(n)$ be the number of integer partitions of n, and let $p(n,k)$ be the number of integer partitions of n into exactly

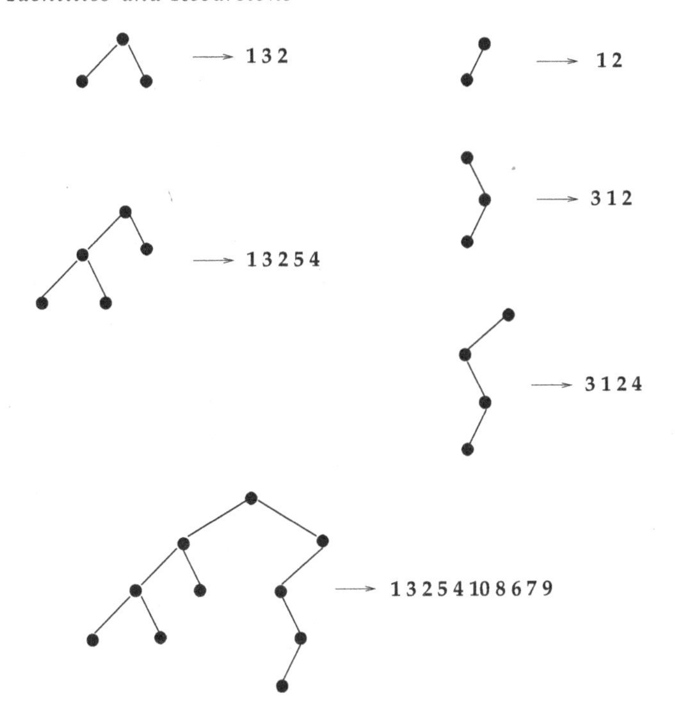

FIGURE 2.13
Mapping binary trees to 231-avoiding permutations.

k parts. If μ is a partition of n into k parts, we write $|\mu| = n$ and $\ell(\mu) = k$ and say that μ has *area* n and *length* k.

2.36. Example. The integer partitions of 5 are

$$(5),\ (4,1),\ (3,2),\ (3,1,1),\ (2,2,1),\ (2,1,1,1),\ (1,1,1,1,1).$$

Thus, $p(5) = 7$, $p(5,1) = 1$, $p(5,2) = 2$, $p(5,3) = 2$, $p(5,4) = 1$, and $p(5,5) = 1$. As another example, the empty sequence is the unique integer partition of 0, so $p(0) = 1 = p(0,0)$.

When presenting integer partitions, we often use the notation j^{a_j} to abbreviate a sequence of a_j copies of j. For example, the last three partitions of 5 in the preceding example could be written $(2^2, 1)$, $(2, 1^3)$, and (1^5). We now describe a way of visualizing integer partitions pictorially.

2.37. Definition: Diagram of a Partition. Let $\mu = (\mu_1, \mu_2, \ldots, \mu_k)$ be an integer partition of n. The *diagram* of μ is the set

$$\text{dg}(\mu) = \{(i,j) \in \mathbb{Z}_{>0} \times \mathbb{Z}_{>0} : 1 \leq i \leq k,\ 1 \leq j \leq \mu_i\}.$$

We can make a picture of $\text{dg}(\mu)$ by drawing an array of n boxes, with μ_i left-justified boxes in row i. For example, Figure 2.14 illustrates the diagrams for the seven integer partitions of 5. Note that $|\mu| = \mu_1 + \cdots + \mu_k = |\text{dg}(\mu)|$ is the total number of unit boxes in the diagram of μ, which is why we call $|\mu|$ the area of μ. The length $\ell(\mu)$ is the number of rows in the diagram of μ.

We know from the Composition Rule that there are 2^{n-1} compositions of n. There is no simple analogous formula for the partition number $p(n)$. Fortunately, the numbers $p(n,k)$ do satisfy the following recursion.

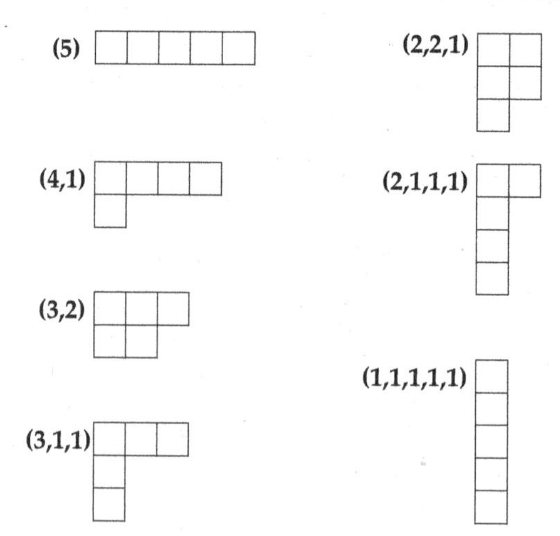

FIGURE 2.14
Partition diagrams.

2.38. Recursion for Integer Partitions. For all positive integers n and k,

$$p(n,k) = p(n-1,k-1) + p(n-k,k).$$

The initial conditions are $p(n,k) = 0$ for $k > n$ or $k < 0$, $p(n,0) = 0$ for $n > 0$, and $p(0,0) = 1$.

Proof. Fix $n, k \in \mathbb{Z}_{>0}$. Let A be the set of integer partitions of n into k parts, so $|A| = p(n,k)$ by definition. Write $A = B \cup C$, where $B = \{\mu \in A : \mu_k = 1\}$ and $C = \{\mu \in A : \mu_k > 1\}$. In terms of partition diagrams, B consists of those partitions in A with a single box in the lowest row, and C consists of those partitions in A with more than one box in the lowest row. We can build an object in B by choosing any partition ν of $n-1$ into $k-1$ parts (in any of $p(n-1,k-1)$ ways), then adding one new part of size 1 at the end. Thus, $|B| = p(n-1,k-1)$. Figure 2.15 illustrates the case $n = 13$, $k = 5$. We can build an object in C by choosing any partition ν of $n-k$ into k parts (in any of $p(n-k,k)$ ways), then adding 1 to each of the k parts of ν. This corresponds to adding a new column of k cells to the diagram of ν; see Figure 2.15. We conclude that $|C| = p(n-k,k)$. The recursion now follows from the Sum Rule. The initial conditions are readily verified. □

It is visually evident that if we interchange the rows and columns in a partition diagram, we obtain the diagram of another integer partition. This leads to the following definition.

2.39. Definition: Conjugate Partitions. Suppose μ is an integer partition of n. The *conjugate partition* of μ is the unique integer partition μ' of n satisfying

$$\text{dg}(\mu') = \{(j,i) : (i,j) \in \text{dg}(\mu)\}.$$

Figure 2.16 shows that the conjugate of $\mu = (7,4,3,1,1)$ is $\mu' = (5,3,3,2,1,1,1)$. Taking the conjugate twice restores the original partition: $\mu'' = \mu$ for all integer partitions μ.

Conjugation leads to the following new interpretation for the numbers $p(n,k)$.

2.40. Proposition. The number of integer partitions of n into k parts (namely, $p(n,k)$) equals the number of integer partitions of n with first part k.

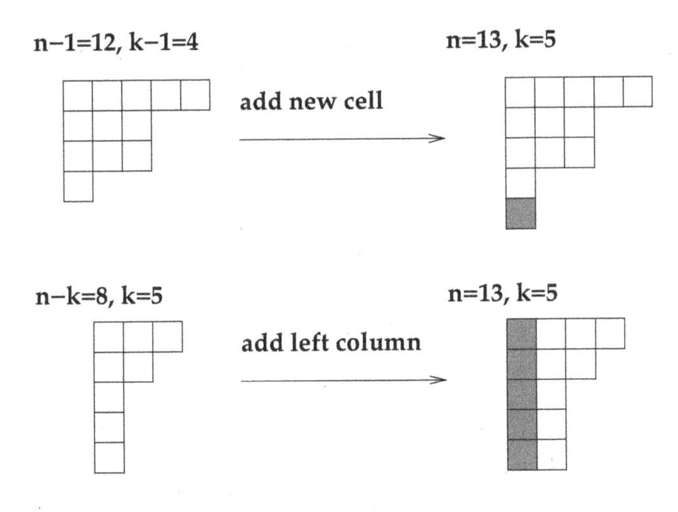

FIGURE 2.15
Proof of the recursion for $p(n,k)$.

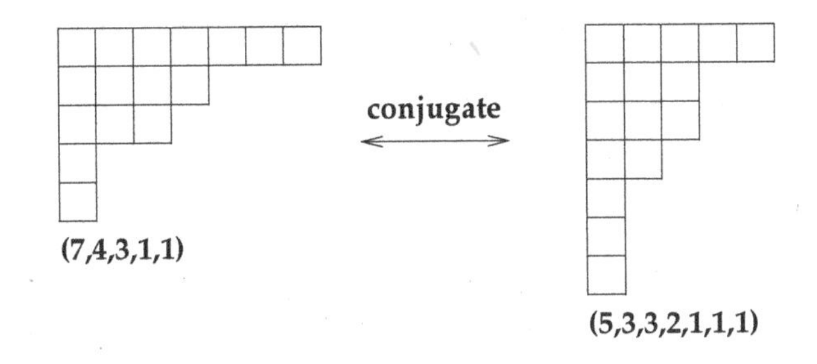

FIGURE 2.16
Conjugate of a partition.

Proof. Note that the length of a partition μ is the number of rows in $\text{dg}(\mu)$, whereas the first part of μ is the number of columns of $\text{dg}(\mu)$. Conjugation interchanges the rows and columns of a diagram. So, the map sending μ to μ' is a bijection from the set of integer partitions of n with k parts onto the set of integer partitions of n with first part k. □

Our next result counts integer partitions whose diagrams fit in a box with b rows and a columns.

2.41. Theorem: Partitions in a Box. The number of integer partitions μ such that $\text{dg}(\mu) \subseteq \{1,2,\ldots,b\} \times \{1,2,\ldots,a\}$ is

$$\binom{a+b}{a,b} = \frac{(a+b)!}{a!b!}.$$

Proof. We define a bijection between the set of integer partitions in the theorem statement and the set of all lattice paths from the origin to (a,b). We draw our partition diagrams in the box with corners $(0,0)$, $(a,0)$, $(0,b)$, and (a,b), as shown in Figure 2.17. Given a partition μ whose diagram fits in this box, the southeast boundary of $\text{dg}(\mu)$ is a lattice path from the origin to (a,b). We call this lattice path the *frontier* of μ (which depends on a and

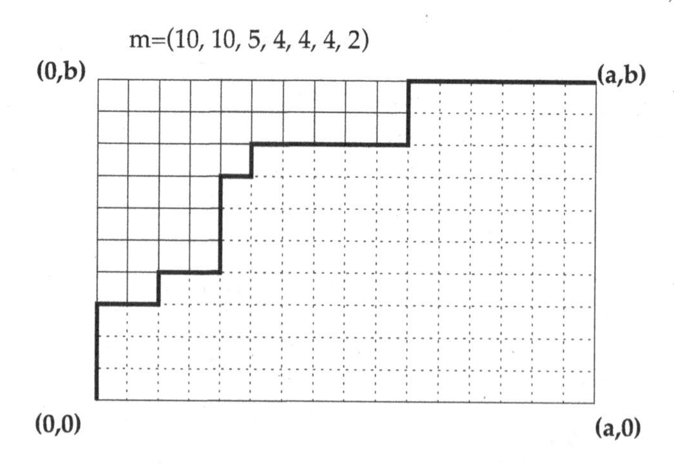

FIGURE 2.17
Counting partitions that fit in an $a \times b$ box.

b as well as μ). For example, if $a = 16$, $b = 10$, and $\mu = (10, 10, 5, 4, 4, 4, 2)$, we see from Figure 2.17 that the frontier of μ is

NNNEENEENNNNENEEEEENNEEEEEE.

Conversely, given any lattice path ending at (a, b), the set of lattice squares northwest of this path in the box uniquely determines the diagram of an integer partition. We already know that the number of lattice paths from the origin to (a, b) is $\binom{a+b}{a,b}$, so the theorem follows from the Bijection Rule. □

2.42. Remark: Euler's Partition Recursion. Our recursion for $p(n, k)$ gives a quick method for computing the quantities $p(n, k)$ and $p(n) = \sum_{k=1}^{n} p(n, k)$. The reader may wonder if the numbers $p(n)$ satisfy any recursion. In fact, Euler's study of the infinite product $\prod_{i=1}^{\infty}(1 - z^i)$ leads to the following recursion for $p(n)$ when $n > 0$:

$$\begin{aligned} p(n) &= \sum_{m=1}^{\infty} (-1)^{m-1}[p(n - m(3m-1)/2) + p(n - m(3m+1)/2)] \\ &= p(n-1) + p(n-2) - p(n-5) - p(n-7) + p(n-12) + p(n-15) \\ &\quad -p(n-22) - p(n-26) + p(n-35) + p(n-40) - p(n-51) - p(n-57) + \cdots . \end{aligned}$$

The initial conditions are $p(0) = 1$ and $p(j) = 0$ for all $j < 0$. It follows that, for each fixed n, the recursive expression for $p(n)$ is really a finite sum, since the terms become zero once the input to p becomes negative. For example, Figure 2.18 illustrates the calculation of $p(n)$ from Euler's recursion for $1 \leq n \leq 12$. We will prove Euler's recursion later (see Theorem 5.55).

2.12 Set Partitions

An *integer* partition decomposes a positive integer into a sum of smaller integers. In contrast, the *set* partitions introduced below decompose a given set into a union of disjoint subsets.

$$\begin{aligned}
p(1) &= p(0) = 1 \\
p(2) &= p(1) + p(0) = 1 + 1 = 2 \\
p(3) &= p(2) + p(1) = 2 + 1 = 3 \\
p(4) &= p(3) + p(2) = 3 + 2 = 5 \\
p(5) &= p(4) + p(3) - p(0) = 5 + 3 - 1 = 7 \\
p(6) &= p(5) + p(4) - p(1) = 7 + 5 - 1 = 11 \\
p(7) &= p(6) + p(5) - p(2) - p(0) = 11 + 7 - 2 - 1 = 15 \\
p(8) &= p(7) + p(6) - p(3) - p(1) = 15 + 11 - 3 - 1 = 22 \\
p(9) &= p(8) + p(7) - p(4) - p(2) = 22 + 15 - 5 - 2 = 30 \\
p(10) &= p(9) + p(8) - p(5) - p(3) = 30 + 22 - 7 - 3 = 42 \\
p(11) &= p(10) + p(9) - p(6) - p(4) = 42 + 30 - 11 - 5 = 56 \\
p(12) &= p(11) + p(10) - p(7) - p(5) + p(0) = 56 + 42 - 15 - 7 + 1 = 77
\end{aligned}$$

FIGURE 2.18
Calculating $p(n)$ using Euler's recursion.

2.43. Definition: Set Partitions. Let X be a set. A *set partition* of X is a collection P of nonempty, pairwise disjoint subsets of X whose union is X. Each element of P is called a *block* of the partition. The cardinality of P (which may be infinite) is called the *number of blocks* of the partition.

For example, if $X = \{1, 2, 3, 4, 5, 6, 7, 8\}$, then

$$P = \{\{3, 5, 8\}, \{1, 7\}, \{2\}, \{4, 6\}\}$$

is a set partition of X with four blocks. Note that the ordering of the blocks in this list, and the ordering of the elements within each block, is irrelevant when deciding the equality of two set partitions. For instance,

$$\{\{6, 4\}, \{1, 7\}, \{2\}, \{5, 8, 3\}\}$$

is the same set partition as P. We can visualize a set partition P by drawing the elements of X in a circle, and then drawing smaller circles enclosing the elements of each block of P. See Figure 2.19.

2.44. Definition: Stirling Numbers and Bell Numbers. Let $S(n, k)$ be the number of set partitions of $\{1, 2, \ldots, n\}$ into exactly k blocks. $S(n, k)$ is called a *Stirling number of the second kind.* Let $B(n)$ be the total number of set partitions of $\{1, 2, \ldots, n\}$. $B(n)$ is called a *Bell number.*

One can check that $S(n, k)$ is the number of partitions of *any* given n-element set into k blocks; similarly for $B(n)$. Unlike binomial coefficients, there are no simple closed expressions for Stirling numbers and Bell numbers (although there are summation formulas and generating functions for these quantities). However, the Stirling numbers satisfy a recursion that can be used to compute $S(n, k)$ and $B(n)$ rapidly.

2.45. Recursion for Stirling Numbers of the Second Kind. For all $n > 0$ and $k > 0$,

$$S(n, k) = S(n - 1, k - 1) + kS(n - 1, k).$$

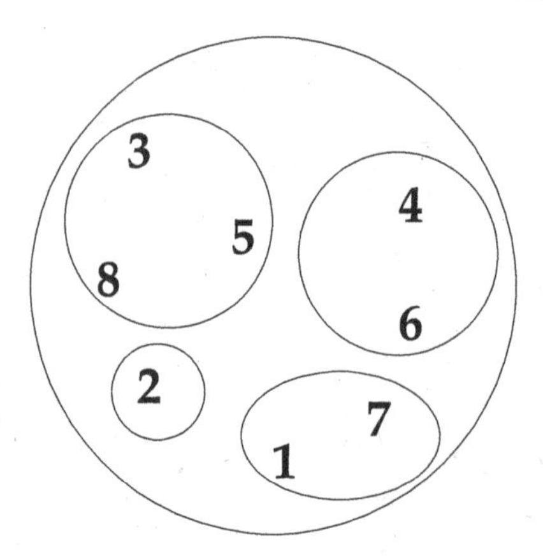

FIGURE 2.19
Diagram of the set partition $\{\{3,5,8\},\{1,7\},\{2\},\{4,6\}\}$.

The initial conditions are $S(0,0) = 1$, $S(n,0) = 0$ for $n > 0$, and $S(0,k) = 0$ for $k > 0$. Furthermore, $B(0) = 1$ and $B(n) = \sum_{k=1}^{n} S(n,k)$ for $n > 0$.

Proof. Fix $n, k > 0$. Let A be the set of set partitions of $\{1, 2, \ldots, n\}$ into exactly k blocks. Let $A' = \{P \in A : \{n\} \in P\}$ and $A'' = \{P \in A : \{n\} \notin P\}$. A is the disjoint union of A' and A'', where A' consists of those set partitions such that n is in a block by itself, and A'' consists of those set partitions such that n is in a block with some other elements. To build a typical partition $P \in A'$, we first choose an arbitrary set partition P_0 of $\{1, 2, \ldots, n-1\}$ into $k-1$ blocks in any of $S(n-1, k-1)$ ways. Then we add the block $\{n\}$ to P_0 to get P. To build a typical partition $P \in A''$, we first choose an arbitrary set partition P_1 of $\{1, 2, \ldots, n-1\}$ into k blocks in any of $S(n-1, k)$ ways. Then we choose one of these k blocks and add n as a new member of that block. By the Sum Rule and Product Rule,

$$S(n,k) = |A| = |A'| + |A''| = S(n-1, k-1) + kS(n-1, k).$$

The initial conditions are immediate from the definitions, noting that $P = \emptyset$ is the unique set partition of $X = \emptyset$. The formula for $B(n)$ follows from the Sum Rule. □

Figure 2.20 computes $S(n,k)$ and $B(n)$ for $n \leq 8$ using the recursion. The entry $S(n,k)$ in row n and column k is computed by taking the number immediately northwest and adding k times the number immediately above the given entry. The numbers $B(n)$ are found by adding the numbers in each row.

Next we study a recursion satisfied by the Bell numbers.

2.46. Recursion for Bell Numbers. For all $n > 0$,

$$B(n) = \sum_{k=0}^{n-1} \binom{n-1}{k} B(n-1-k).$$

The initial condition is $B(0) = 1$.

Proof. For $n > 0$, we construct a typical set partition P counted by $B(n)$ as follows. Let k be the number of elements in the block of P containing n, not including n itself; thus, $0 \leq k \leq n-1$. To build P, first choose k elements from $\{1, 2, \ldots, n-1\}$ that belong to the same block as n in any of $\binom{n-1}{k}$ ways. Then, choose an arbitrary set partition of the

	$k=0$	$k=1$	$k=2$	$k=3$	$k=4$	$k=5$	$k=6$	$k=7$	$k=8$	$B(n)$
$n=0:$	1	0	0	0	0	0	0	0	0	1
$n=1:$	0	1	0	0	0	0	0	0	0	1
$n=2:$	0	1	1	0	0	0	0	0	0	2
$n=3:$	0	1	3	1	0	0	0	0	0	5
$n=4:$	0	1	7	6	1	0	0	0	0	15
$n=5:$	0	1	15	25	10	1	0	0	0	52
$n=6:$	0	1	31	90	65	15	1	0	0	203
$n=7:$	0	1	63	301	350	140	21	1	0	877
$n=8:$	0	1	127	966	1701	1050	266	28	1	4140

FIGURE 2.20
Calculating $S(n,k)$ and $B(n)$ recursively.

$n-1-k$ elements that do not belong to the same block as n; this choice can be made in any of $B(n-1-k)$ ways. The recursion now follows from the Sum Rule and Product Rule. □

For example, assuming that $B(m)$ is already known for $m < 8$ (see Figure 2.20), we calculate

$$\begin{aligned} B(8) &= \binom{7}{0}B(7) + \binom{7}{1}B(6) + \binom{7}{2}B(5) + \cdots + \binom{7}{7}B(0) \\ &= 1\cdot 877 + 7\cdot 203 + 21\cdot 52 + 35\cdot 15 + 35\cdot 5 + 21\cdot 2 + 7\cdot 1 + 1\cdot 1 \\ &= 4140. \end{aligned}$$

2.13 Surjections, Balls in Boxes, and Equivalence Relations

This section shows how set partitions and Stirling numbers can be used to count surjections and equivalence relations. We also revisit the balls-in-boxes problem from §1.13. Recall that a function $f : X \to Y$ is surjective (or *onto*) iff for all $y \in Y$ there exists $x \in X$ with $f(x) = y$.

2.47. The Surjection Rule. The number of surjections from an n-element set onto a k-element set is $k!S(n,k)$, where $S(n,k)$ is a Stirling number of the second kind.

Proof. Without loss of generality, we may assume we are counting surjections with domain $\{1,2,\ldots,n\}$ and codomain $\{1,2,\ldots,k\}$. To build such a surjection f, first choose a set partition P of $\{1,2,\ldots,n\}$ into k blocks in any of $S(n,k)$ ways. Choose one of these blocks (in k ways), and let f map everything in this block to 1. Then choose a different block (in $k-1$ ways), and let f map everything in this block to 2. Continue similarly; at the last stage, there is 1 block left, and we let f map everything in this block to k. By the Product Rule, the number of surjections is

$$S(n,k)\cdot k\cdot (k-1)\cdot \ldots \cdot 1 = k!S(n,k).$$ □

2.48. Example. To illustrate this proof, suppose $n = 8$ and $k = 4$. In the first step, say we choose the set partition $P = \{\{1,4,7\},\{2\},\{3,8\},\{5,6\}\}$. In the next four steps, we choose a permutation (linear ordering) of the four blocks of P, say

$$\{2\},\{5,6\},\{3,8\},\{1,4,7\}.$$

Now we define the associated surjection f by setting

$$f(2) = 1,\ f(5) = f(6) = 2,\ f(3) = f(8) = 3,\ f(1) = f(4) = f(7) = 4.$$

There is also a recursion for surjections similar to the recursion for set partitions; see Exercise 2-52. Using integer partitions, set partitions, and surjections, we can now continue our discussion of the balls-in-boxes problem from §1.13.

2.49. Theorem: Balls in Nonempty Boxes. Consider distributions of a balls into b boxes where every box must be nonempty.
(a) For labeled balls and labeled boxes, there are $b!S(a,b)$ distributions.
(b) For unlabeled balls and labeled boxes, there are $\binom{a-1}{b-1}$ distributions.
(c) For labeled balls and unlabeled boxes, there are $S(a,b)$ distributions.
(d) For unlabeled balls and unlabeled boxes, there are $p(a,b)$ distributions.

Proof. (a) As in Rule 1.93, we can model the placement of balls in boxes by a function $f : \{1,2,\ldots,a\} \to \{1,2,\ldots,b\}$, where $f(x) = y$ means that ball x is placed in box y. The requirement that every box be nonempty translates to the requirement that f is surjective. So (a) follows from the Surjection Rule. Part (b) was proved in Rule 1.94.

(c) We can model a distribution of balls labeled $1,2,\ldots,a$ into b unlabeled boxes as a set partition of $\{1,2,\ldots,a\}$ with b blocks, where each block consists of the set of balls placed in the same box. For example, if $a = 8$ and $b = 4$, the set partition $\{\{3,5,8\},\{1,7\},\{2\},\{4,6\}\}$ models the distribution where one box contains balls 3, 5, and 8; another box contains balls 1 and 7; another box contains ball 2; and another box contains 4 and 6. Thus, the number of distributions in (c) is the Stirling number $S(a,b)$.

(d) We can model a distribution of a unlabeled balls into b unlabeled boxes as an integer partition of a into b parts, where each part counts the number of balls in one of the boxes. For example, if $a = 8$ and $b = 4$, the integer partition $(3,3,1,1)$ models the distribution where two boxes contain three balls and two boxes contain one ball. Thus, the number of distributions in (d) is the partition number $p(a,b)$. □

The rest of this section assumes the reader has some previous exposure to equivalence relations. We review the relevant definitions now.

2.50. Definition: Types of Relations. Let X be any set. A *relation* on X is any subset of $X \times X$. If R is a relation on X and $x,y \in X$, we may write xRy as an abbreviation for $(x,y) \in R$. We read this symbol as "x is related to y under R." A relation R on X is *reflexive on X* iff xRx for all $x \in X$. R is *irreflexive on X* iff xRx is false for all $x \in X$. R is *symmetric* iff for all $x,y \in X$, xRy implies yRx. R is *antisymmetric* iff for all $x,y \in X$, xRy and yRx imply $x = y$. R is *transitive* iff for all $x,y,z \in X$, xRy and yRz imply xRz. R is an *equivalence relation on X* iff R is symmetric, transitive, and reflexive on X. If R is an equivalence relation and $x_0 \in X$, the *equivalence class of x_0 relative to R* is the set $[x_0]_R = \{y \in X : x_0Ry\}$.

2.51. Theorem: Set Partitions and Equivalence Relations. Suppose X is a fixed set. Let $\mathcal{A}$ be the set of all set partitions of X, and let $\mathcal{B}$ be the set of all equivalence relations on X. There are canonical bijections $\phi : \mathcal{A} \to \mathcal{B}$ and $\phi' : \mathcal{B} \to \mathcal{A}$. If $P \in \mathcal{A}$, then the number of blocks of P equals the number of equivalence classes of $\phi(P)$. Hence, the Stirling number $S(n,k)$ is the number of equivalence relations on an n-element set having

k equivalence classes, and the Bell number $B(n)$ is the number of equivalence relations on an n-element set.

Proof. We only sketch the proof, leaving many details as exercises. Given a set partition $P \in \mathcal{A}$, define a relation $\phi(P)$ on X by

$$\phi(P) = \{(x, y) \in X \times X : \exists S \in P, x \in S \text{ and } y \in S\}.$$

In other words, x is related to y under $\phi(P)$ iff x and y belong to the same block of P. The reader may check that $\phi(P)$ is indeed an equivalence relation on X, i.e., that $\phi(P) \in \mathcal{B}$. Thus, ϕ is a well-defined function from $\mathcal{A}$ into $\mathcal{B}$.

Given an equivalence relation $R \in \mathcal{B}$, define

$$\phi'(R) = \{[x]_R : x \in X\}.$$

In other words, the blocks of $\phi'(R)$ are precisely the equivalence classes of R. The reader may check that $\phi'(R)$ is indeed a set partition of X, i.e., that $\phi'(R) \in \mathcal{A}$. Thus, ϕ' is a well-defined function from $\mathcal{B}$ into $\mathcal{A}$.

To complete the proof, one must check that ϕ and ϕ' are two-sided inverses of one another. In other words, for all $P \in \mathcal{A}$, $\phi'(\phi(P)) = P$; and for all $R \in \mathcal{B}$, $\phi(\phi'(R)) = R$. It follows that ϕ and ϕ' are bijections. □

2.14 Stirling Numbers and Rook Theory

Recall that the Stirling numbers $S(n, k)$ count set partitions of an n-element set into k blocks. This section gives another combinatorial interpretation of these Stirling numbers. We show that $S(n, k)$ counts certain placements of rooks on a triangular chessboard. A slight variation of this setup leads us to introduce the (signless) Stirling numbers of the first kind. The relationship between the two kinds of Stirling numbers will be studied in the following section.

2.52. Definition: Ferrers Boards and Rooks. A *Ferrers board* is the diagram of an integer partition, viewed as a collection of unit squares as in §2.11. A *rook* is a chess piece that can occupy any of the squares in a Ferrers board. In chess, a rook can move any number of squares horizontally or vertically from its current position in a single move. A rook located in row i and column j of a Ferrers board *attacks* all squares in row i and all squares in column j.

For example, in the Ferrers board shown below, the rook attacks all squares on the board marked with a dot (and its own square).

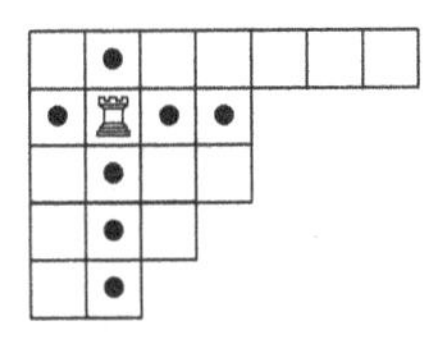

For each $n > 0$, let Δ_n denote the diagram of the partition $(n-1, n-2, \ldots, 3, 2, 1)$. Δ_n is a triangular Ferrers board with $n(n-1)/2$ total squares. For example, Δ_5 is shown below.

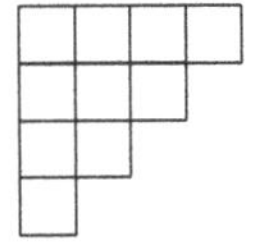

2.53. Definition: Non-attacking Rook Placements. A *placement* of k rooks on a given Ferrers board is a subset of k squares in the Ferrers board. These k squares represent the locations of k identical rooks on the board. A placement of rooks on a Ferrers board is called *non-attacking* iff no rook occupies a square attacked by another rook. Equivalently, all rooks in the placement occupy distinct rows and distinct columns of the board.

2.54. Example. The following diagram illustrates a non-attacking placement of three rooks on the Ferrers board corresponding to the partition $(7, 4, 4, 3, 2)$.

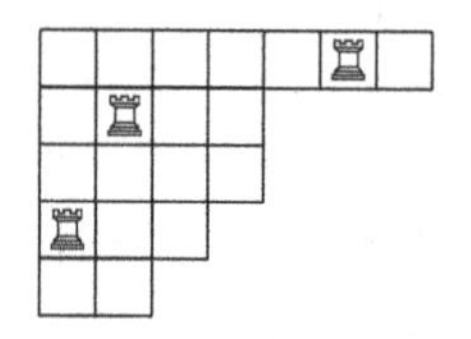

2.55. Rook Interpretation of Stirling Numbers of the Second Kind. For $n > 0$ and $0 \leq k \leq n$, let $S'(n, k)$ denote the number of non-attacking placements of $n - k$ rooks on the Ferrers board Δ_n. If $n > 1$ and $0 < k < n$, then

$$S'(n, k) = S'(n - 1, k - 1) + kS'(n - 1, k).$$

The initial conditions are $S'(n, 0) = 0$ for $n > 0$ and $S'(n, n) = 1$ for $n \geq 0$. Therefore, $S'(n, k) = S(n, k)$, a Stirling number of the second kind.

Proof. Fix $n > 1$ with $0 < k < n$. Let A, B, and C denote the set of placements counted by $S'(n, k)$, $S'(n-1, k-1)$, and $S'(n-1, k)$, respectively. Let A_0 consist of all rook placements in A with no rook in the top row, and let A_1 consist of all rook placements in A with a rook in the top row. A is the disjoint union of A_0 and A_1. Deleting the top row of the Ferrers board Δ_n produces the smaller Ferrers board Δ_{n-1}. It follows that deleting the (empty) top row of a rook placement in A_0 gives a bijection between A_0 and B (note that a placement in B involves $(n - 1) - (k - 1) = n - k$ rooks). On the other hand, we can build a typical rook placement in A_1 as follows. First, choose a placement of $n - k - 1$ non-attacking rooks from the set C, and use this rook placement to fill the bottom $n - 1$ rows of Δ_n. These rooks occupy $n-k-1$ distinct columns. This leaves $(n-1)-(n-k-1) = k$ columns in the top row in which we are allowed to place the final rook. By the Product Rule, $|A_1| = |C|k$. Using the Sum Rule, we conclude that

$$S'(n, k) = |A| = |A_0| + |A_1| = |B| + k|C| = S'(n - 1, k - 1) + kS'(n - 1, k).$$

For $n > 0$, we cannot place n non-attacking rooks on the Ferrers board Δ_n (which has only $n - 1$ columns), and hence $S'(n, 0) = 0$. On the other hand, for any $n \geq 0$ there is a unique placement of zero rooks on Δ_n. This placement is non-attacking (vacuously), and hence $S'(n, n) = 1$. Counting set partitions, we see that $S(n, 0) = 0$ for $n > 0$ and $S(n, n) = 1$ for $n \geq 0$. Since $S'(n, k)$ and $S(n, k)$ satisfy the same recursion and initial conditions, a routine induction argument (cf. Proposition 2.29) shows that $S'(n, k) = S(n, k)$ for all n and k. □

2.56. Remark. We have given combinatorial proofs that the numbers $S'(n, k)$ and $S(n, k)$ satisfy the same recursion. We can link together these proofs to get a recursively defined bijection between rook placements and set partitions, using the ideas in Remark 2.34. We can also directly define a bijection between rook placements and set partitions. We illustrate such a bijection through an example. Figure 2.21 displays a rook placement counted by $S'(8, 3)$. We write the numbers 1 through n below the last square in each column of the diagram, as shown in the figure. We view these numbers as labeling both the rows and

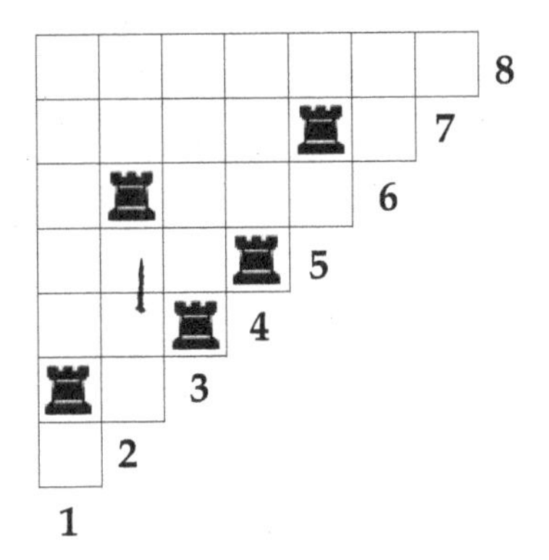

FIGURE 2.21
A rook placement counted by $S'(8,3)$.

columns of the diagram; note that the column labels increase from left to right, while row labels decrease from top to bottom. The bijection between non-attacking rook placements π and set partitions P acts as follows. For all $j < i \leq n$, there is a rook in row i and column j of π iff i and j are consecutive elements in the same block of P (writing elements of the block in increasing order). For example, the rook placement π in Figure 2.21 maps to the set partition

$$P = \{\{1,3,4,5,7\},\{2,6\},\{8\}\}.$$

The set partition $\{\{2\},\{1,5,8\},\{4,6,7\},\{3\}\}$ maps to the rook placement shown in Figure 2.22. It can be checked that a non-attacking placement of $n-k$ rooks on Δ_n corresponds to a set partition of n with exactly k blocks; furthermore, the rook placement associated to a given set partition is automatically non-attacking.

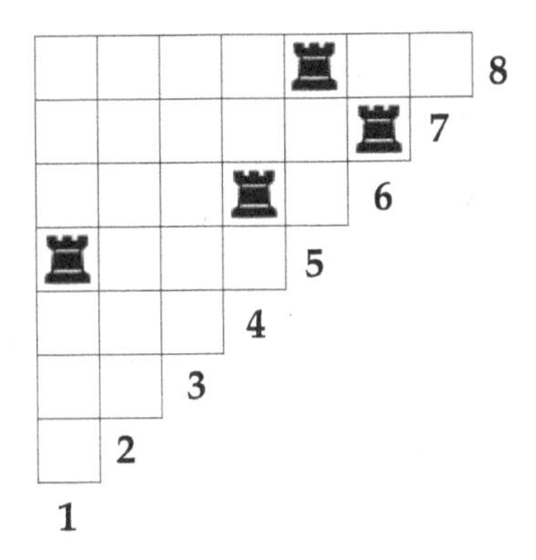

FIGURE 2.22
The rook placement associated to $\{\{2\},\{1,5,8\},\{4,6,7\},\{3\}\}$.

By modifying the way a rook can move, we obtain the Stirling numbers of the first kind described in the following definition.

2.57. Definition: File Rooks and Stirling Numbers of the First Kind. A *file rook* is a new chess piece that attacks only the squares in its column. For all $n > 0$ and $0 \leq k \leq n$, let $s'(n,k)$ denote the number of placements of $n-k$ non-attacking file rooks on the Ferrers board Δ_n. The numbers $s'(n,k)$ are called *signless Stirling numbers of the first kind.* The numbers $s(n,k) = (-1)^{n-k}s'(n,k)$ are called *(signed) Stirling numbers of the first kind.* By convention, we set $s(0,0) = 1 = s'(0,0)$.

Stirling numbers of the first kind also count permutations of n letters consisting of k disjoint cycles; we discuss this combinatorial interpretation in §3.6.

2.58. Recursion for Signless Stirling Numbers of the First Kind. For all $n > 1$ and $0 < k < n$,

$$s'(n,k) = s'(n-1,k-1) + (n-1)s'(n-1,k).$$

The initial conditions are $s'(n,0) = 0$ for $n > 0$ and $s'(n,n) = 1$ for $n \geq 0$.

Proof. Assume $n > 1$ and $0 < k < n$. Let A, B, and C denote the set of file rook placements counted by $s'(n,k)$, $s'(n-1,k-1)$, and $s'(n-1,k)$, respectively. Write A as the disjoint union of A_0 and A_1, where A_0 is the set of placements in A with no file rook in the leftmost column, and A_1 is the set of placements in A with a file rook in the leftmost column. We can get a bijection from A_0 to B by deleting the empty leftmost column of a placement in A_0. On the other hand, we can build a typical file rook placement in A_1 as follows. First, choose the position of the file rook in the leftmost column of Δ_n in $n-1$ ways. Second, choose any placement of $n-k-1$ non-attacking file rooks from the set C, and use this placement to fill the remaining $n-1$ columns of Δ_n. These file rooks do not attack the file rook in the first row. By the Sum Rule and Product Rule,

$$s'(n,k) = |A| = |A_0| + |A_1| = |B| + (n-1)|C| = s'(n-1,k-1) + (n-1)s'(n-1,k). \quad \square$$

We can use the recursion and initial conditions to compute the (signed or unsigned) Stirling numbers of the first kind. This is done in Figure 2.23; compare to the computation of Stirling numbers of the second kind in Figure 2.20.

	$k=0$	$k=1$	$k=2$	$k=3$	$k=4$	$k=5$	$k=6$	$k=7$
$n=0:$	1	0	0	0	0	0	0	0
$n=1:$	0	1	0	0	0	0	0	0
$n=2:$	0	−1	1	0	0	0	0	0
$n=3:$	0	2	−3	1	0	0	0	0
$n=4:$	0	−6	11	−6	1	0	0	0
$n=5:$	0	24	−50	35	−10	1	0	0
$n=6:$	0	−120	274	−225	85	−15	1	0
$n=7:$	0	720	−1764	1624	−735	175	−21	1

FIGURE 2.23
Signed Stirling numbers of the first kind.

There is a surprising relation between the two arrays of numbers in these figures, which explains the extra signs in the definition of $s(n,k)$. Specifically, for any fixed $n > 0$, consider the lower triangular $n \times n$ matrices $A = (s(i,j))_{1 \leq i,j \leq n}$ and $B = (S(i,j))_{1 \leq i,j \leq n}$. It turns out that A and B are inverse matrices. The reader may check this for small n using Figure 2.20 and Figure 2.23. We will prove this fact for all n in the next section.

2.15 Stirling Numbers and Polynomials

Our recursions for Stirling numbers (of both kinds) lead to algebraic proofs of certain polynomial identities. These identities express various polynomials as linear combinations of other polynomials, where the Stirling numbers are the coefficients involved in the linear combination. Thus, the Stirling numbers appear as entries in the transition matrices between some common bases for the vector space of one-variable polynomials. This linear-algebraic interpretation of Stirling numbers will be used to show the inverse relation between the two triangular matrices of Stirling numbers (cf. Figures 2.20 and 2.23).

2.59. Polynomial Identity for Stirling Numbers of the Second Kind. For all $n \geq 0$ and all real x,

$$x^n = \sum_{k=0}^{n} S(n,k)x(x-1)(x-2)\cdots(x-k+1). \tag{2.3}$$

We give an algebraic proof based on the Stirling recursion, as well as a combinatorial proof based on rook placements.

Algebraic Proof. Recall that $S(0,0) = 1$ and $S(n,k) = S(n-1,k-1) + kS(n-1,k)$ for $n \geq 1$ and $1 \leq k \leq n$. We prove (2.3) by induction on n. If $n = 0$, the right side is $S(0,0) = 1 = x^0$, so the identity holds. For the induction step, fix $n \geq 1$ and assume that $x^{n-1} = \sum_{k=0}^{n-1} S(n-1,k)x(x-1)\cdots(x-k+1)$. Multiplying both sides by $x = (x-k)+k$, we can write

$$\begin{aligned} x^n &= \sum_{k=0}^{n-1} S(n-1,k)x(x-1)\cdots(x-k+1)(x-k) \\ &\quad + \sum_{k=0}^{n-1} S(n-1,k)x(x-1)\cdots(x-k+1)k \\ &= \sum_{j=0}^{n-1} S(n-1,j)x(x-1)\cdots(x-j) + \sum_{k=0}^{n-1} kS(n-1,k)x(x-1)\cdots(x-k+1). \end{aligned}$$

In the first summation, replace j by $k-1$. The calculation continues:

$$\begin{aligned} &= \sum_{k=1}^{n} S(n-1,k-1)x(x-1)\cdots(x-k+1) + \sum_{k=0}^{n-1} kS(n-1,k)x(x-1)\cdots(x-k+1) \\ &= \sum_{k=0}^{n} S(n-1,k-1)x(x-1)\cdots(x-k+1) + \sum_{k=0}^{n} kS(n-1,k)x(x-1)\cdots(x-k+1) \\ &= \sum_{k=0}^{n} [S(n-1,k-1) + kS(n-1,k)]x(x-1)\cdots(x-k+1) \\ &= \sum_{k=0}^{n} S(n,k)x(x-1)\cdots(x-k+1), \end{aligned}$$

where the second equality uses the initial conditions $S(n-1,-1) = S(n-1,n) = 0$, and the final equality uses the recursion for $S(n,k)$.

Combinatorial Proof. Both sides of (2.3) are polynomials in x, so it suffices to verify the identity when x is a positive integer (see Remark 2.7). Fix $x, n \in \mathbb{Z}_{>0}$. Consider the *extended*

Ferrers board $\Delta_n(x) = (n, \ldots, n, n-1, n-2, n-3, \ldots, 2, 1)$ obtained by adding x new rows of length n above the board $\Delta_n = (n-1, n-2, \ldots, 2, 1)$. For example, $\Delta_5(3)$ is the board shown here.

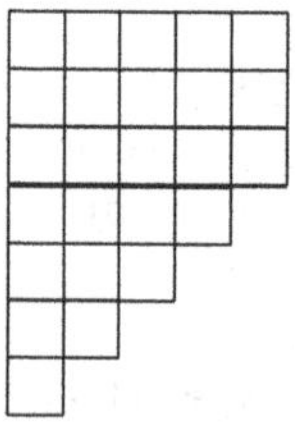

Let A be the set of placements of n non-attacking rooks on $\Delta_n(x)$. One way to build a typical rook placement in A is to place one rook in each column, working from right to left. The rook in the rightmost column can go in any of x squares. The rook in the next column can go in any of $(x+1)-1 = x$ squares, since one row is now attacked by the rook in the rightmost column. In general, the rook located $i \geq 0$ columns left of the rightmost column can go in any of $(x+i)-i = x$ squares, since i distinct squares in this column are already attacked by rooks placed previously when we choose the position of the rook in this column. The Product Rule therefore gives $|A| = x^n$.

On the other hand, for $0 \leq k \leq n$, let A_k be the set of rook placements in A in which there are exactly k rooks in the x new rows (and hence $n-k$ rooks on the board Δ_n). To build a typical rook placement in A_k, first place $n-k$ non-attacking rooks in Δ_n in any of $S(n,k)$ ways. There are now k unused columns of new squares, each of which consists of x squares. Visit these columns from left to right, placing one rook in each column. There are x choices for the first rook, then $x-1$ choices for the second rook (since the first rook's row must be avoided), then $x-2$ choices for the third rook, and so on. By the Product Rule, $|A_k| = S(n,k)x(x-1)(x-2)\cdots(x-k+1)$. Since A is the disjoint union of $A_0, A_1, \ldots, A_n$, the Sum Rule gives

$$|A| = \sum_{k=1}^{n} S(n,k)x(x-1)(x-2)\cdots(x-k+1).$$

The identity follows by comparing our two formulas for $|A|$.

2.60. Polynomial Identity for Signless Stirling Numbers of the First Kind. For all $n \geq 0$ and all real x,

$$x(x+1)(x+2)\cdots(x+n-1) = \sum_{k=0}^{n} s'(n,k)x^k. \tag{2.4}$$

Proof. We ask the reader to supply an algebraic proof in Exercise 2-73; we give a combinatorial proof here. Recall that $s'(n,k)$ counts placements of $n-k$ non-attacking file rooks on Δ_n. Fix an integer $x \geq 0$, and let A be the set of placements of n non-attacking file rooks on the extended Ferrers board $\Delta_n(x)$. Placing the file rooks on the board one column at a time, working from right to left, the Product Rule shows that

$$|A| = x(x+1)(x+2)\cdots(x+n-1).$$

On the other hand, we can write A as the disjoint union of sets A_k, where A_k consists of the file rook placements in A in which exactly k file rooks occupy the new squares above Δ_n. To build an object in A_k, first place $n-k$ non-attacking file rooks in Δ_n in any of $s'(n,k)$ ways. There are now k unused columns, each of which has x new squares, and k file rooks left to be placed. Visit the unused columns from left to right, and choose one of the x new squares in that column to be occupied by a file rook. By the Product Rule, $|A_k| = s'(n,k)x^k$. The Sum Rule now gives

$$|A| = \sum_{k=0}^{n} s'(n,k)x^k.$$

Equating the two expressions for $|A|$ completes the proof. □

The proof technique of using extended boards such as $\Delta_n(x)$ can be recycled to prove other results in rook theory, as we will see in §12.3.

2.61. Polynomial Identity for Signed Stirling Numbers of the First Kind. For all $n \geq 0$ and all real x,

$$x(x-1)(x-2)\cdots(x-n+1) = \sum_{k=0}^{n} s(n,k)x^k. \tag{2.5}$$

Proof. Replace x by $-x$ in (2.4) to obtain

$$(-x)(-x+1)(-x+2)\cdots(-x+n-1) = \sum_{k=0}^{n} s'(n,k)(-x)^k.$$

Factoring out -1's, we get

$$(-1)^n x(x-1)(x-2)\cdots(x-n+1) = \sum_{k=0}^{n}(-1)^k s'(n,k)x^k.$$

Moving the $(-1)^n$ to the right side and recalling that $s(n,k) = (-1)^{n+k}s'(n,k)$, the result follows. □

2.62. Summation Formula for Stirling Numbers of the First Kind. For all $n \geq 1$ and $1 \leq k \leq n$, we have

$$s'(n,k) = \sum_{1 \leq i_1 < i_2 < \cdots < i_{n-k} \leq n-1} i_1 i_2 \cdots i_{n-k}.$$

Proof. Recall that $s'(n,k)$ counts the number of placements of $n-k$ non-attacking file rooks on the triangular Ferrers board Δ_n. Since file rooks only attack cells in their columns, a placement is non-attacking iff all file rooks occupy distinct columns of Δ_n. Let us classify file rook placements based on which columns contain file rooks. Suppose the $n-k$ file rooks appear in the columns of lengths $i_1, i_2, \ldots, i_{n-k}$, where $1 \leq i_1 < i_2 < \cdots < i_{n-k} \leq n-1$. The Product Rule shows that the number of placements of file rooks in these rows is $i_1 i_2 \cdots i_{n-k}$. The formula in the theorem now follows from the Sum Rule. □

To understand the linear-algebraic significance of the preceding results, we need to introduce some bases for the vector space V of all polynomials in one variable with real coefficients. (See the Appendix for a review of the linear algebra concepts used here.)

2.63. Definition: Monomial, Falling Factorial, and Rising Factorial Bases. For any integer $n \geq 0$, define the *falling factorial* polynomials

$$(x){\downarrow_0} = 1, \quad (x){\downarrow_n} = x(x-1)(x-2)\cdots(x-n+1)$$

and the *rising factorial* polynomials

$$(x)\uparrow_0 = 1, \quad (x)\uparrow_n = x(x+1)(x+2)\cdots(x+n-1).$$

The *monomial basis* of V is the indexed set $M = \{x^n : n \geq 0\}$. The *falling factorial basis* of V is $F = \{(x)\downarrow_n : n \geq 0\}$. The *rising factorial basis* of V is $R = \{(x)\uparrow_n : n \geq 0\}$. Define $M_{\leq N} = \{x^n : 0 \leq n \leq N\}$, and define $F_{\leq N}$ and $R_{\leq N}$ similarly.

It can be checked that any indexed collection of polynomials $\{p_n(x) : n \geq 0\}$ such that $\deg(p_n) = n$ for all n is a basis of V. Since x^n, $(x)\downarrow_n$, and $(x)\uparrow_n$ all have degree n, it follows that M, F, and R are indeed bases of V. The three indexed collections $M_{\leq N}$, $F_{\leq N}$, and $R_{\leq N}$ are all bases of the subspace $V_{\leq N}$ consisting of polynomials in x of degree at most N.

We can now recast the preceding theorems in the language of linear algebra. Recall that if $B = (v_1, \ldots, v_n)$ and $C = (w_1, \ldots, w_n)$ are two ordered bases of a finite-dimensional vector space W, the *transition matrix from B to C* is the unique $n \times n$ matrix $A = (a_{ij})$ such that

$$v_j = \sum_{i=1}^{n} a_{ij} w_i \quad \text{for } 1 \leq j \leq n. \tag{2.6}$$

From linear algebra, we know that A is invertible, and A^{-1} is the transition matrix from C to B.

2.64. Theorem: Transition Matrices between Polynomial Bases. Fix $N \geq 0$.
(a) The matrix $\mathbf{S} = (S(n,k))_{0 \leq n,k \leq N}$ of Stirling numbers of the second kind is the transpose of the transition matrix from the basis $M_{\leq N}$ to the basis $F_{\leq N}$.
(b) The matrix $\mathbf{s}' = (s'(n,k))_{0 \leq n,k \leq N}$ of signless Stirling numbers of the first kind is the transpose of the transition matrix from $R_{\leq N}$ to $M_{\leq N}$.
(c) The matrix $\mathbf{s} = (s(n,k))_{0 \leq n,k \leq N}$ of signed Stirling numbers of the first kind is the transpose of the transition matrix from $F_{\leq N}$ to $M_{\leq N}$.
(d) The $(N+1) \times (N+1)$ matrices $\mathbf{S}$ and $\mathbf{s}$ are inverses of one another.

Proof. The first three statements follow from Equations (2.3), (2.4), (2.5), and the definition of transition matrices (2.6). The final statement is a special case of the fact that the transition matrix from B to C is the inverse of the transition matrix from C to B. □

Part (d) of the theorem says that we have matrix identities $\mathbf{S}\mathbf{s} = \mathbf{I} = \mathbf{s}\mathbf{S}$, where $\mathbf{I}$ is the $(N+1) \times (N+1)$ identity matrix. Writing what this means entry by entry, we obtain the formulas

$$\sum_k S(i,k)s(k,j) = \delta_{ij} = \sum_k s(i,k)S(k,j) \quad \text{for } i,j \geq 0,$$

where δ_{ij} is 1 if $i = j$ and 0 if $i \neq j$. A combinatorial proof of the second equality will be given later (see Example 4.25).

2.16 Solving Recursions with Constant Coefficients

In Example 2.20, we promised to find an explicit closed formula for the Fibonacci numbers. This section addresses the more general problem of finding exact solutions to certain types of recursions. We focus initially on solving recursions of the form

$$x_n = bx_{n-1} + cx_{n-2} \quad \text{for } n \geq 2,$$

where b and c are real constants, and $(x_n : n \geq 0)$ is an unknown real sequence satisfying the given recursion. We introduce the general method with a specific example.

2.65. Example. Let us seek solutions to the recursion $x_n = 2x_{n-1} + 8x_{n-2}$ (where $n \geq 2$). The key initial idea is to try sequences of the form $x_n = r^n$, where r is a fixed nonzero real number. Substituting into the recursion, we see that r must satisfy the equations $r^n = 2r^{n-1} + 8r^{n-2}$ for all $n \geq 2$. Dividing by r^{n-2}, we see that these equations are equivalent to the single condition $r^2 - 2r - 8 = 0$. Factoring this quadratic, we get $(r+2)(r-4) = 0$. So $r = -2$ or $r = 4$, yielding solutions $x_n = (-2)^n$ and $x_n = 4^n$ for the given recursion.

The next key observation is that *any linear combination of solutions to the recursion is also a solution.* More specifically, if sequences $(u_n : n \geq 0)$ and $(v_n : n \geq 0)$ solve the recursion and s, t are any scalars, then $x_n = su_n + tv_n$ also solves the recursion. To verify this, compute

$$\begin{aligned} 2x_{n-1} + 8x_{n-2} &= 2(su_{n-1} + tv_{n-1}) + 8(su_{n-2} + tv_{n-2}) \\ &= s(2u_{n-1} + 8u_{n-2}) + t(2v_{n-1} + 8v_{n-2}) \\ &= su_n + tv_n = x_n \qquad \text{for } n \geq 2. \end{aligned}$$

We now have an infinite family of solutions to the original recursion, namely $x_n = s \cdot (-2)^n + t \cdot 4^n$ for arbitrary scalars $s, t \in \mathbb{R}$. In fact, we will see momentarily that *all* real solutions of the recursion have this form.

To pick out one specific solution from our infinite family of solutions, we need two initial conditions. As an example, suppose we know $x_0 = 7$ and $x_1 = 10$. Using these values in the formula $x_n = s \cdot (-2)^n + t \cdot 4^n$, we find that $7 = s + t$ and $10 = -2s + 4t$. This is a system of two linear equations in the two unknowns s and t. Solving this system, we find that $s = 3$ and $t = 4$, so $x_n = 3(-2)^n + 4^{n+1}$ is a solution to the given recursion satisfying the given initial conditions.

More generally, suppose the initial conditions had been $x_0 = c_0$ and $x_1 = c_1$ for fixed constants c_0 and c_1. Here we would need to solve the system of linear equations $s + t = c_0$ and $-2s + 4t = c_1$. Since the coefficient matrix of this system is invertible, there is always a unique solution (s, t); specifically, Cramer's Rule gives $s = (4c_0 - c_1)/6$ and $t = (2c_0 + c_1)/6$. Thus the given recursion and initial conditions can be solved by exactly one sequence of the form $x_n = s \cdot (-2)^n + t \cdot 4^n$.

Now we can explain why *all* solutions to the recursion must have this form. Suppose $(z_n : n \geq 0)$ is any sequence satisfying the recursion. By the previous paragraph, there exist scalars s, t such that the sequence $x_n = s \cdot (-2)^n + t \cdot 4^n$ satisfies the recursion and initial conditions $x_0 = z_0$ and $x_1 = z_1$. We can now prove by strong induction that $z_n = x_n$ for all $n \geq 0$. This is true for $n = 0$ and $n = 1$ by the initial conditions. For fixed $n \geq 2$, we may assume by induction that $z_m = x_m$ for $0 \leq m < n$. Since both sequences satisfy the recursion, we have $z_n = 2z_{n-1} + 8z_{n-2} = 2x_{n-1} + 8x_{n-2} = x_n$, completing the induction.

The following result can be proved by generalizing the reasoning in the last example.

2.66. Theorem: Solutions to $x_n = ax_{n-1} + bx_{n-2}$. Suppose a, b are real constants such that $r^2 - ar - b = 0$ has two distinct real roots $r_1 \neq r_2$. Given any $c_0, c_1 \in \mathbb{R}$, there exist unique $s, t \in \mathbb{R}$ such that the recursion $x_n = ax_{n-1} + bx_{n-2}$ with initial conditions $x_0 = c_0$, $x_1 = c_1$ is solved by $x_n = sr_1^n + tr_2^n$. Every real solution to the recursion has this form.

The polynomial $r^2 - ar - b$ is called the *characteristic polynomial* of the recursion $x_n = ax_{n-1} + bx_{n-2}$.

2.67. Example: Fibonacci Recursion. We use the above method to solve $F_n = F_{n-1} + F_{n-2}$ with initial conditions $F_0 = 0$ and $F_1 = 1$. (The sequence f_n considered in Example 2.20 is a shift of this one, namely $f_n = F_{n+2}$.) Here $a = b = 1$, so we must find the roots of $r^2 - r - 1 = 0$. By the Quadratic Formula, the roots are $r_1 = (1+\sqrt{5})/2$ and $r_2 = (1-\sqrt{5})/2$. Thus, the general solution to the recursion is $sr_1^n + tr_2^n$. To find s and t, we use the initial conditions. We obtain the system of equations $s+t=0$, $r_1s+r_2t=1$. Solving leads to $s = 1/\sqrt{5}$ and $t = -1/\sqrt{5}$. Thus an exact formula for the Fibonacci numbers is

$$F_n = \frac{(1+\sqrt{5})^n - (1-\sqrt{5})^n}{2^n\sqrt{5}} \quad \text{for all } n \geq 0.$$

Note that $r_1 \approx 1.618$ and $r_2 \approx -0.618$, so $\lim_{n\to\infty} r_2^n = 0$. Thus, the subtracted term in the formula for F_n becomes negligible for large n, giving the approximation $F_n \approx r_1^n/\sqrt{5}$.

Everything said so far works equally well for complex sequences. In particular, the roots of the characteristic polynomial may be complex even if the recursion we are solving is real. Our solution method still applies as long as the two roots are distinct.

2.68. Example. Let us solve the recursion $x_n = 6x_{n-1} - 25x_{n-2}$ with initial conditions $x_0 = 0$, $x_1 = 1$. The characteristic polynomial is $r^2 - 6r + 25$, which has complex roots $r_1 = 3+4i$ and $r_2 = 3-4i$. The general (complex) solution of the recursion is $x = s(3+4i)^n + t(3-4i)^n$ where $s, t \in \mathbb{C}$. Using the initial conditions, we must have $s + t = 0$ and $s(3+4i)+t(3-4i) = 1$. Solving this system of linear equations in $\mathbb{C}$, we get $s = -i/8$ and $t = i/8$. Thus the particular solution to the given problem is $x_n = [-i(3+4i)^n + i(3-4i)^n]/8$. Although this formula involves complex numbers, each x_n is real (as can be seen by induction on n, using the recursion and initial conditions).

A special case occurs when the characteristic polynomial has a double root, i.e., $r^2 - ar - b$ factors as $(r - r_1)^2$. The next example shows what to do in this situation.

2.69. Example. Let us try to find all real solutions to the recursion $x_n = 6x_{n-1} - 9x_{n-2}$ for $n \geq 2$. Guessing that $x_n = r^n$ for all n, we must have $r^n = 6r^{n-1} - 9r^{n-2}$ for all $n \geq 2$, which leads (as before) to the characteristic equation $r^2 - 6r + 9 = 0$. In this case, $r_1 = 3$ is the only root of this equation, giving $x_n = 3^n$ as one solution to the recursion. More generally, for any real constant s, $x_n = s \cdot 3^n$ also solves the recursion, but we have not yet found all possible solutions. A new trick is needed here: we guess a solution might have the form $x_n = nr^n$ for some fixed r. Substituting into the recursion, we need $nr^n = 6(n-1)r^{n-1} - 9(n-2)r^{n-2}$ for all $n \geq 2$. Dividing by r^{n-2}, we need $nr^2 = 6(n-1)r - 9(n-2)$ for all $n \geq 2$. Rearranging this, we need $n(r^2 - 6r + 9) + 6(r - 3) = 0$ for all $n \geq 2$. This forces $r^2 - 6r + 9 = 0 = r - 3$, which holds precisely when $r = 3$. Thus, $x_n = n3^n$ is a new solution to the recursion. Taking linear combinations, we obtain the general solution $x_n = (s + tn)3^n$ for real constants s, t. If we also have initial conditions $x_0 = c_0$ and $x_1 = c_1$, we can find s and t by solving the linear system $c_0 = s$, $c_1 = 3(s + t)$. Evidently, the unique solution is $s = c_0$, $t = c_1/3 - c_0$. By the same argument used in our original example, it follows from this that *all* solutions of the recursion have the form $x_n = (s + tn)3^n$.

These ideas can be extended to solve recursions of the form $x_n = a_1x_{n-1} + a_2x_{n-2} + \cdots + a_dx_{n-d}$ for $n \geq d$, where $d \geq 1$ is fixed and $a_1, \ldots, a_d$ are given constants not depending on n. We only state the solution technique here, deferring the full proof until §11.5.

2.70. Method for Solving Recursions with Constant Coefficients. Let $a_1, \ldots, a_d$ be fixed real or complex numbers with $a_d \neq 0$. To solve the recursion

$$x_n = a_1x_{n-1} + a_2x_{n-2} + \cdots + a_dx_{n-d} \quad \text{for } n \geq d,$$

first find all complex roots of the *characteristic polynomial* $r^d - a_1r^{d-1} - a_2r^{d-2} - \cdots - a_d$. If

r is a root of this polynomial with multiplicity m, then each of the m sequences $(r^n : n \geq 0)$, $(nr^n : n \geq 0)$, $(n^2r^n : n \geq 0)$, ..., $(n^{m-1}r^n : n \geq 0)$ is a *basic solution* to the recursion. As r varies through all roots of the characteristic polynomial, we obtain d basic solutions. The general solution to the recursion is a linear combination (with complex scalars) of these d basic solutions.

To find the particular solution satisfying d given initial conditions, regard the d coefficients in the linear combination as unknowns, and substitute the general solution into the initial conditions to obtain a system of d linear equations in these unknowns. This system will have a unique solution giving the particular linear combination of basic solutions satisfying these initial conditions.

2.71. Example. Let us find the general solution of the recursion

$$x_n = 7x_{n-1} - 19x_{n-2} + 25x_{n-3} - 16x_{n-4} + 4x_{n-5} \quad \text{for } n \geq 5.$$

The characteristic polynomial $r^5 - 7r^4 + 19r^3 - 25r^2 + 16r - 4$ factors as $(r-1)^3(r-2)^2$. The root $r = 1$ gives three basic solutions $x_n = 1^n = 1$, $x_n = n1^n = n$, and $x_n = n^2 1^n = n^2$. The root $r = 2$ gives two more basic solutions $x_n = 2^n$ and $x_n = n2^n$. The general solution is

$$x_n = c_1 + c_2 n + c_3 n^2 + c_4 2^n + c_5 n 2^n \quad \text{for } n \geq 0.$$

To find the particular solution satisfying initial conditions $(x_0, x_1, x_2, x_3, x_4) = (0, 0, 1, 3, 1)$, we take $n = 0, 1, 2, 3, 4$ in the general solution to obtain the system of linear equations $c_1 + c_4 = 0$, $c_1 + c_2 + c_3 + 2c_4 + 2c_5 = 0$, $c_1 + 2c_2 + 4c_3 + 4c_4 + 8c_5 = 1$, $c_1 + 3c_2 + 9c_3 + 8c_4 + 24c_5 = 3$, $c_1 + 4c_2 + 16c_3 + 16c_4 + 64c_5 = 1$. After some linear algebra, we find $c_1 = -15$, $c_2 = -8$, $c_3 = -2$, $c_4 = 15$, and $c_5 = -5/2$. So the particular solution is

$$x_n = -15 - 8n - 2n^2 + 15 \cdot 2^n - (5/2)n2^n \quad \text{for } n \geq 0.$$

Once this solution is found, we can use strong induction to prove that it does indeed satisfy the given recursion.

Summary

- *Combinatorial Proofs.* To prove a formula of the form $a = b$ combinatorially, define an appropriate set S of objects, give a counting argument showing that $|S| = a$, then give a second counting argument showing that $|S| = b$.

- *Some Binomial Coefficient Identities.*
  (a) Symmetry of Binomial Coefficients: $\binom{n}{k} = \binom{n}{n-k}$.
  (b) Pascal's Identity: $\binom{n}{k} = \binom{n-1}{k} + \binom{n-1}{k-1}$; $\binom{a+b}{a,b} = \binom{a+b-1}{a-1,b} + \binom{a+b-1}{a,b-1}$.
  (c) Sum of Binomial Coefficients: $\sum_{k=0}^{n} \binom{n}{k} = 2^n$.
  (d) Sum of Squared Binomial Coefficients: $\sum_{k=0}^{n} \binom{n}{k}^2 = \binom{2n}{n}$.
  (e) Sum of Diagonal of Pascal's Triangle: $\sum_{k=0}^{a} \binom{k+b-1}{k,b-1} = \binom{a+b}{a,b}$.
  (f) The Chu–Vandermonde Identity: $\sum_{k=0}^{a} \binom{k+b}{k,b}\binom{a-k+c}{a-k,c} = \binom{a+b+c+1}{a,b+c+1}$.

- *Geometric Series.* For real or complex $r \neq 1$, $\sum_{k=0}^{n} ar^k = a(1 - r^{n+1})/(1 - r)$. When $|r| < 1$, $\sum_{k=0}^{\infty} ar^k = a/(1 - r)$.

- *The Binomial Theorem.* For all real x, y and $n \in \mathbb{Z}_{\geq 0}$, $(x + y)^n = \sum_{k=0}^{n} \binom{n}{k} x^k y^{n-k}$.

- *Multinomial Theorems.* When $z_1, \ldots, z_s$ commute,

$$(z_1 + z_2 + \cdots + z_s)^n = \sum_{n_1+n_2+\cdots+n_s=n} \binom{n}{n_1, n_2, \ldots, n_s} x_1^{n_1} x_2^{n_2} \cdots x_s^{n_s}.$$

When $Z_1, \ldots, Z_s$ do not necessarily commute,

$$(Z_1 + Z_2 + \cdots + Z_s)^n = \sum_{w \in \{1,\ldots,s\}^n} Z_{w_1} Z_{w_2} \cdots Z_{w_s}.$$

- *Sums of Powers of Integers.* For all integers $n \geq 0$,

$$\sum_{k=1}^{n} k = \frac{n(n+1)}{2}, \qquad \sum_{k=1}^{n} k^2 = \frac{n(n+1)(2n+1)}{6}, \qquad \sum_{k=1}^{n} k^3 = \frac{n^2(n+1)^2}{4}.$$

- *Recursions.* A collection of combinatorial objects can often be described *recursively*, by using smaller objects of the same kind to build larger objects. Induction arguments can be used to prove facts about recursively defined objects. The recursion and appropriate *initial conditions* uniquely determine the quantities under consideration. If two collections of objects satisfy the same recursion and initial conditions, one can link together two combinatorial proofs of the recursion to obtain recursively defined bijections between the two collections.

- *Fibonacci Recursion.* The Fibonacci numbers satisfy $F_n = F_{n-1} + F_{n-2}$ for $n \geq 2$, with initial conditions $F_0 = 0$ and $F_1 = 1$. (Other initial conditions are sometimes used.) For $n \geq 0$, F_{n+2} counts words in $\{0,1\}^n$ with no consecutive zeroes.

- *Subset Recursion.* Let $C(n,k)$ be the number of k-element subsets of an n-element set. Then $C(n,k) = C(n-1,k) + C(n-1,k-1)$ for $0 < k < n$, with initial conditions $C(n,0) = C(n,n) = 1$ for $n \geq 0$, and $C(n,k) = 0$ for $k < 0$ or $k > n$.

- *Multiset Recursion.* Let $M(n,k)$ be the number of k-element multisets using an n-letter alphabet. Then $M(n,k) = M(n-1,k) + M(n,k-1)$ for $n, k > 0$, with initial conditions $M(n,0) = 1$ for $n \geq 0$ and $M(0,k) = 0$ for $k > 0$.

- *Anagram Recursion.* Let $C(n; n_1, \ldots, n_s)$ be the number of rearrangements of n_1 copies of one letter, n_2 copies of another letter, and so on. Then $C(n; n_1, \ldots, n_s) = \sum_{i=1}^{s} C(n-1; n_1, \ldots, n_i - 1, \ldots, n_s)$ for $n > 0$, with initial conditions $C(n; n_1, \ldots, n_s) = 0$ if any $n_i < 0$, and $C(0; 0, \ldots, 0) = 1$.

- *Lattice Path Recursions.* For $m \in \mathbb{Z}_{>0}$, the m-ballot numbers $T_m(a,b)$ count lattice paths from $(0,0)$ to (a,b) that stay weakly above $y = mx$. These numbers satisfy $T_m(a,b) = T_m(a-1,b) + T_m(a,b-1)$ for $b > ma > 0$, $T_m(a,ma) = T_m(a-1,ma)$ for $a > 0$, and $T_m(0,b) = 1$ for $b \geq 0$. It follows that $T_m(a,b) = \frac{b-ma+1}{b+a+1}\binom{a+b+1}{a}$. Lattice paths in rectangles and other regions satisfy similar recursions, but the initial conditions change depending on the boundary of the region.

- *Catalan Recursion.* The Catalan numbers $C_n = \frac{1}{n+1}\binom{2n}{n}$ satisfy the recursion $C_n = \sum_{k=1}^{n} C_{k-1}C_{n-k}$ for $n > 0$, with initial condition $C_0 = 1$. If a sequence (d_n) satisfies this recursion and initial condition, then $d_n = C_n$ for all n.

- *Catalan Objects.* Examples of objects counted by Catalan numbers include Dyck paths, strings of balanced parentheses, binary trees, 231-avoiding permutations, and τ-avoiding permutations for any permutation τ of $\{1,2,3\}$.

- *Integer Partitions.* An integer partition of n into k parts is a weakly decreasing sequence of k positive integers whose sum is n. These partitions are counted by $p(n,k)$, which satisfies the recursion $p(n,k) = p(n-1,k-1) + p(n-k,k)$ for $n,k > 0$, with initial conditions $p(n,k) = 0$ for $k > n$ or $k < 0$, $p(n,0) = 0$ for $n > 0$, and $p(0,0) = 1$. Let $p(n)$ be the number of integer partitions of n; then $p(n) = \sum_{k=0}^{n} p(n,k)$ and $p(n) = \sum_{m=1}^{\infty}(-1)^{m-1}[p(n-m(3m-1)/2) + p(n-m(3m+1)/2)]$.

- *Partition Diagrams.* A partition $\mu = (\mu_1, \ldots, \mu_k)$ can be visualized as a diagram where there are μ_i left-justified squares in the ith row from the top. The conjugate partition μ' is found by interchanging rows and columns in the diagram of μ. By taking conjugates, $p(n,k)$ is the number of integer partitions of n with largest part k. The number of partitions whose diagrams fit in an $a \times b$ box is $\binom{a+b}{a,b}$.

- *Set Partitions.* A set partition of a set X is a set P of nonempty pairwise disjoint subsets of X (called the blocks of P) whose union is X. The Stirling number of the second kind, denoted $S(n,k)$, counts set partitions of an n-element set into k blocks. We have $S(n,k) = S(n-1,k-1) + kS(n-1,k)$ for $n,k > 0$, with initial conditions $S(0,0) = 1$, $S(n,0) = 0$ for $n > 0$, and $S(0,k) = 0$ for $k > 0$. The Bell number $B(n)$ counts all set partitions of an n-element set. We have $B(n) = \sum_{k=0}^{n-1}\binom{n-1}{k}B(n-1-k)$, with initial condition $B(0) = 1$.

- *Surjection Rule.* The number of surjective functions from an n-element set onto a k-element set is $k!S(n,k)$.

- *Balls in Nonempty Boxes.* When all boxes must be nonempty, there are:
  $b!S(a,b)$ ways to put a labeled balls into b labeled boxes;
  $\binom{a-1}{b-1}$ ways to put a unlabeled balls into b labeled boxes;
  $S(a,b)$ ways to put a labeled balls into b unlabeled boxes;
  $p(a,b)$ ways to put a unlabeled balls into b unlabeled boxes.

- *Equivalence Relations.* A relation R on a set X is an equivalence relation iff R is reflexive (aRa for all $a \in X$), symmetric (aRb implies bRa for all $a,b \in X$), and transitive (aRb and bRc imply aRc for all $a,b,c \in X$). The set of equivalence classes of an equivalence relation on X is a set partition of X; so there is a bijection from the set of equivalence relations on X to the set of set partitions of X. The Stirling number $S(n,k)$ is the number of equivalence relations on an n-element set with k equivalence classes. The Bell number $B(n)$ is the number of equivalence relations on an n-element set.

- *Rook Interpretations of Stirling Numbers.* The Stirling number of the second kind $S(n,k)$ is the number of placements of $n-k$ non-attacking rooks on the triangular board $\Delta_n = \mathrm{dg}(n-1, n-2, \ldots, 3, 2, 1)$. The signless Stirling number of the first kind $s'(n,k)$ is the number of placements of $n-k$ non-attacking file rooks on Δ_n (i.e., no two rooks appear in the same column). We have $s'(n,k) = s'(n-1,k-1) + (n-1)s'(n-1,k)$ for $0 < k < n$, with initial conditions $s'(n,0) = 0$ for $n > 0$ and $s'(n,n) = 1$ for $n \geq 0$. Moreover, $s'(n,k) = \sum_{1 \leq i_1 < i_2 < \cdots < i_{n-k} \leq n-1} i_1 i_2 \cdots i_{n-k}$.

- *Stirling Number Identities.* We have $x^n = \sum_{k=0}^{n} S(n,k)(x)\!\downarrow_k$, $(x)\!\uparrow_n = \sum_{k=0}^{n} s'(n,k)x^k$, and $(x)\!\downarrow_n = \sum_{k=0}^{n} s(n,k)x^k$, where $s(n,k) = (-1)^{n+k}s'(n,k)$ is the signed Stirling number of the first kind. So Stirling numbers of both kinds appear as entries in transition matrices between monomial bases, rising factorial bases, and falling factorial bases for vector spaces of polynomials. The matrices with entries $S(n,k)$ and $s(n,k)$ are inverses of each other.

- *Solving Recursions with Constant Coefficients.* To solve the recursion $x_n = a_1x_{n-1} + a_2x_{n-2} + \cdots + a_dx_{n-d}$, find the roots of the characteristic polynomial $r^d - a_1r^{d-1} - \cdots - a_d$. Each root r of multiplicity m yields m basic solutions $x_n = n^ir^n$ for $0 \leq i < m$. Every solution of the recursion is a linear combination of the d basic solutions. To find the coefficients in the linear combination, use initial conditions for $x_0, \ldots, x_{d-1}$ to set up a system of linear equations that can be solved for the coefficients.

Exercises

2-1. Expand each of the following powers: (a) $(x+y)^5$; (b) $(a-b)^6$; (c) $(1+r)^7$.

2-2. (a) Find the coefficient of x^6 in $(3x^2+5)^9$. (b) Find the coefficient of x^6 in $(2x^3+9)^5$.

2-3. Expand $(4x-3)^6$ into a sum of monomials.

2-4. Find the constant term in $(2x - x^{-1})^6$.

2-5. Find the coefficient of $a^4b^3cd^4$ in $(a+b+c+d)^{12}$, where a, b, c, d are real variables.

2-6. Find the coefficient of x^3 in $(x^2+x+1)^4$.

2-7. Suppose $z \in \mathbb{C}$ satisfies $z^7 = 1$ and $z \neq 1$. Find a cubic polynomial having $z + z^{-1}$ as a root.

2-8. Given real variables $z_1, \ldots, z_s$, write the expansions of $(z_1 + z_2 + \cdots + z_s)^2$ and $(z_1 + z_2 + \cdots + z_s)^3$ without using multinomial coefficients.

2-9. (a) Given arbitrary 3×3 matrices A, B, D, expand $(A+B+D)^3$. (b) How does the answer to (a) simplify if the three matrices commute? (c) How does the answer to (a) simplify if we only know that $AB = BA$?

2-10. (a) When we use the Non-Commutative Multinomial Theorem to expand $(z_1 + z_2 + \cdots + z_s)^n$, how many terms will there be?
(b) If all variables commute, how many terms will there be after collecting like terms? (For example, when $s = n = 3$ there are 10 terms.)

2-11. *Euler's Identity* states that $e^{it} = \cos t + i \sin t$ for all real t, where $i \in \mathbb{C}$ satisfies $i^2 = -1$. Use this and the Binomial Theorem to show that $\cos(3t) = 4\cos^3 t - 3\cos t$. Derive a similar identity for $\sin(3t)$.

2-12. Give a combinatorial proof of the Multinomial Theorem similar to the proof of the Binomial Theorem given in the text.

2-13. (a) Give an algebraic proof of this analogue of Pascal's Identity for multinomial coefficients:

$$\binom{n}{n_1, n_2, \ldots, n_s} = \binom{n-1}{n_1-1, n_2, \ldots, n_s} + \binom{n-1}{n_1, n_2-1, \ldots, n_s} + \cdots + \binom{n-1}{n_1, n_2, \ldots, n_s-1}.$$

(b) Give a combinatorial proof of this identity based on multidimensional lattice paths.

2-14. Give an algebraic proof of the Multinomial Theorem by induction on the power n, using the identity in the previous exercise.

2-15. Prove the Multinomial Theorem using the Binomial Theorem and induction on s. What identity relating binomial coefficients and multinomial coefficients do you need in your proof?

2-16. Generalized Distributive Law. Suppose R is a ring (possibly non-commutative),

s is a positive integer, $J_1, J_2, \ldots, J_s$ are finite pairwise disjoint index sets, and for each $j_k \in J_k$, a_{j_k} is a member of R. Prove:

$$\left(\sum_{j_1 \in J_1} a_{j_1}\right) \cdot \left(\sum_{j_2 \in J_2} a_{j_2}\right) \cdot \cdots \cdot \left(\sum_{j_s \in J_s} a_{j_s}\right) = \sum_{(j_1,\ldots,j_s) \in J_1 \times \cdots \times J_s} a_{j_1} a_{j_2} \cdots a_{j_s}.$$

Informally, this identity says that a product of sums expands into a sum of products of terms, where we build each term by choosing one factor from each sum and multiplying the chosen factors together. (The assumption that the index sets J_k are pairwise disjoint can always be achieved by replacing each J_k by $\{k\} \times J_k$.)

2-17. (a) Give an algebraic proof of the identity $\binom{ab}{b} = \sum_{r_1+r_2+\cdots+r_a=b} \binom{b}{r_1}\binom{b}{r_2}\cdots\binom{b}{r_a}$. (b) Give a combinatorial proof of this identity.

2-18. (a) Find $1+(2/5)+(2/5)^2+\cdots+(2/5)^8$. (b) Find $1-1/3+1/9-1/27+\cdots+1/6561$.

2-19. Evaluate each infinite series. (a) $1+(3/4)+(3/4)^2+(3/4)^3+\cdots$. (b) $1-(4/7)+(4/7)^2-(4/7)^3+\cdots$. (c) $1+r^3+r^6+r^9+\cdots$, where $|r|<1$. (d) $\sum_{k=2}^{\infty} e^{kx}$, where $x<0$.

2-20. Express each infinite repeating decimal as a rational number.
(a) $0.2222\cdots$ (b) $2.61616161\cdots$ (c) $0.123123123\cdots$ (d) $0.428571428571\cdots$.

2-21. (a) Prove algebraically that $\sum_{k=0}^{n}(-1)^k\binom{n}{k}=0$ for all $n>0$. (b) Rewrite the identity in (a) as $\sum_{k \text{ odd}}\binom{n}{k}=\sum_{k \text{ even}}\binom{n}{k}$, and give a bijective proof.

2-22. Prove $\sum_{k=0}^{n}\binom{n}{k}=2^n$ using lattice paths.

2-23. Prove $\binom{a+b}{a,b}=\binom{b+a}{b,a}$ using lattice paths.

2-24. Given $n \in \mathbb{Z}_{>0}$, evaluate $\sum_{0\le j<k\le n}\binom{n}{j}\binom{n}{k}$.

2-25. Given $m,n \in \mathbb{Z}_{>0}$, evaluate $\sum_{k_1+k_2+\cdots+k_m=n}(k_1!k_2!\cdots k_m!)^{-1}$.

2-26. Prove each identity by induction.

$$\text{(a)} \sum_{k=1}^{n} k = n(n+1)/2; \quad \text{(b)} \sum_{k=1}^{n} k^2 = n(n+1)(2n+1)/6; \quad \text{(c)} \sum_{k=1}^{n} k^3 = n^2(n+1)^2/4.$$

2-27. For fixed $n \ge 0$, let R be the rectangle with vertices $(0,0)$, $(n,0)$, $(0,n)$, and (n,n). Let S be the set of rectangles contained in R with sides parallel to the sides of R and with vertices at integer coordinates. By counting S in two ways, give a combinatorial proof of the formula for the sum of the first n cubes.

2-28. Use the technique of §2.6 to evaluate: (a) $\sum_{k=1}^{n} k^4$; (b) $\sum_{k=1}^{n} k^5$.

2-29. Use Pascal's Recursion to compute $\binom{9}{k}$ for $0 \le k \le 9$ and $\binom{10}{k}$ for $0 \le k \le 10$.

2-30. Compute the ballot numbers $T(a,7)$ for $0 \le a \le 7$ by drawing a picture.

2-31. For fixed $k \in \mathbb{Z}_{>0}$, let a_n be the number of n-letter words using the alphabet $\{0,1,\ldots,k\}$ that do not contain 00. Find a recursion and initial conditions for a_n. Then compute a_5.

2-32. How many words in $\{0,1,2\}^8$ do not contain three zeroes in a row?

2-33. How many lattice paths from $(1,1)$ to $(6,6)$ always stay weakly between the lines $y=2x/5$ and $y=5x/2$?

2-34. How many lattice paths go from $(1,1)$ to $(8,8)$ without ever passing through a point (p,q) such that p and q are both prime?

2-35. Find the number of lattice paths from $(0,0)$ to $(6,6)$ that stay weakly between the paths EENEENNEENNN and NNNENEENENEE and do not pass through $(1,2)$ or $(5,4)$.

2-36. Draw the diagrams of all integer partitions of $n = 6$ and $n = 7$. Indicate which partitions are conjugates of one another.

2-37. Show that the Catalan number C_n counts integer partitions μ such that $\mathrm{dg}(\mu) \subseteq \Delta_n$.

2-38. For each τ, list all permutations in S_4^τ: (a) $\tau = 123$; (b) $\tau = 132$; (c) $\tau = 213$; (d) $\tau = 312$; (e) $\tau = 321$.

2-39. Compute $p(8,3)$ by direct enumeration and by using a recursion.

2-40. Use Euler's recursion to compute $p(k)$ for $13 \leq k \leq 16$ (see Figure 2.18).

2-41. (a) List all the set partitions and rook placements counted by $S(5,2)$. (b) List all the set partitions and equivalence relations counted by $B(4)$. (c) Draw all the file rook placements counted by $s'(4,2)$.

2-42. Compute $S(9,k)$ for $0 \leq k \leq 9$ and $S(10,k)$ for $0 \leq k \leq 10$ (use Figure 2.20).

2-43. Compute the Bell number $B(k)$ for $9 \leq k \leq 12$ (use Figure 2.20).

2-44. Compute $s(8,k)$ for $0 \leq k \leq 8$ (use Figure 2.23).

2-45. Prove the identity $k\binom{n}{k} = n\binom{n-1}{k-1} = (n-k+1)\binom{n}{k-1}$ algebraically and combinatorially (where $1 \leq k \leq n$).

2-46. Prove the identity $\binom{n}{k}\binom{k}{s} = \binom{n}{s}\binom{n-s}{k-s}$ algebraically and combinatorially.

2-47. Prove that for all integers $n, k, i \geq 0$, $\binom{n-i}{k-i}\binom{n}{k+i}\binom{n+i}{k} = \binom{n-i}{k}\binom{n+i}{k+i}\binom{n}{k-i}$.

2-48. (a) Evaluate $\sum_{k=1}^{n} k(n+1-k)$. (b) Evaluate $\sum_{i=1}^{n}\sum_{j=i+1}^{n}\sum_{k=i}^{j} 1$.

2-49. Suppose X is an n-element set. Count the number of relations R on X satisfying each property: (a) no restrictions on R; (b) R is reflexive on X; (c) R is irreflexive; (d) R is symmetric; (e) R is irreflexive and symmetric; (f) R is antisymmetric.

2-50. Let X be a nine-element set and Y a four-element set. (a) Find the probability that a random function $f : X \to Y$ is surjective. (b) Find the probability that a random function $g : Y \to X$ is injective.

2-51. Let $p'(n,k)$ be the number of integer partitions of n with first part k. (a) Prove that $p'(n,k) = p'(n-1,k-1) + p'(n-k,k)$ for $n, k > 0$. What are the initial conditions? (b) Use (a) to prove $p'(n,k) = p(n,k)$ for all $n, k \geq 0$.

2-52. Let $\mathrm{Surj}(n,k)$ be the number of surjections from an n-element set onto a k-element set. (a) Find a recursion and initial conditions for $\mathrm{Surj}(n,k)$. (b) Use (a) to prove $\mathrm{Surj}(n,k) = k!S(n,k)$.

2-53. Verify equations (2.3), (2.4), and (2.5) by direct calculation for $n = 3$ and $n = 4$.

2-54. (a) How many ways can we put a labeled balls in b unlabeled boxes? (b) What is the answer if each box can contain at most one ball?

2-55. (a) How many ways can we put a unlabeled balls in b unlabeled boxes? (b) What is the answer if each box can contain at most one ball?

2-56. (a) Find the rook placement associated to the set partition

$$\{\{2,5\},\{1,4,7,10\},\{3\},\{6,8\},\{9\}\}$$

by the bijection in Remark 2.56. (b) Find the set partition associated to the following rook placement:

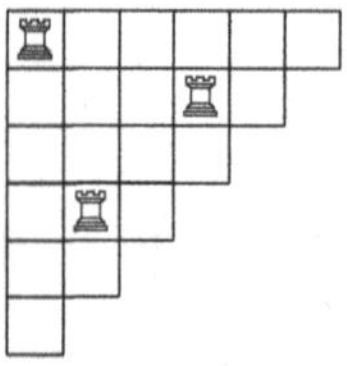

2-57. Let $f : \{1,2,\ldots,7\} \to \{1,2,3\}$ be the surjection given by $f(1) = 3$, $f(2) = 3$, $f(3) = 1$, $f(4) = 3$, $f(5) = 2$, $f(6) = 3$, $f(7) = 1$. In the proof of the Surjection Rule 2.47, what choice sequence can be used to construct f?

2-58. How many compositions of 20 only use parts of sizes 1, 3, or 5?

2-59. Use the recursion 2.22 for multisets to prove by induction that the number of k-element multisets using an n-element alphabet is $\frac{(k+n-1)!}{k!(n-1)!}$.

2-60. Given $a,b,c,n \in \mathbb{Z}_{>0}$ with $a+b+c=n$, prove combinatorially that $\binom{n}{a,b,c} = \sum_{k=0}^{c}\left[\binom{n-k-1}{a-1,b,c-k} + \binom{n-k-1}{a,b-1,c-k}\right]$.

2-61. Complete the proof of Theorem 2.27 by proving $T_m(a,b) = \frac{b-ma+1}{b+a+1}\binom{a+b+1}{a}$ by induction.

2-62. (a) Show that $|S_n^{132}| = C_n$. (b) Show that $|S_n^{213}| = C_n$. (c) Show that $|S_n^{312}| = C_n$.

2-63. Convert the binary tree in Figure 2.11 to a: (a) 132-avoiding permutation; (b) 213-avoiding permutation; (c) 312-avoiding permutation.

2-64. (a) Let G_n be the set of lists of integers $(g_0, g_1, \ldots, g_{n-1})$ where $g_0 = 0$, each $g_i \geq 0$, and $g_{i+1} \leq g_i + 1$ for all $i < n-1$. Prove that $|G_n| = C_n$. (b) For $m \in \mathbb{Z}_{>0}$, let $G_n^{(m)}$ be the set of lists of integers $(g_0, g_1, \ldots, g_{n-1})$ where $g_0 = 0$, each $g_i \geq 0$, and $g_{i+1} \leq g_i + m$ for all $i < n-1$. Prove that $|G_n^{(m)}| = T_m(n, mn)$, the number of lattice paths from $(0,0)$ to (n, mn) that never go below the line $y = mx$.

2-65. Consider the 231-avoiding permutation $w = 1\,5\,2\,4\,3\,11\,7\,6\,10\,8\,9$. Use recursive bijections based on the Catalan recursion to map w to objects of the following kinds: (a) a Dyck path; (b) a binary tree; (c) a 312-avoiding permutation; (d) an element of G_n (see the previous exercise).

2-66. Let π be the Dyck path NNENEENNNENNENNEEENENEEE. Use recursive bijections based on the Catalan recursion to map π to objects of the following kinds: (a) a binary tree; (b) a 231-avoiding permutation; (c) a 213-avoiding permutation.

2-67. Show that the number of possible rhyme schemes for an n-line poem using k different rhyme syllables is the Stirling number $S(n,k)$. (For example, ABABCDCDEFEFGG is a rhyme scheme with $n = 14$ and $k = 7$.)

2-68. Find explicit formulas for $S(n,k)$ when k is 1, 2, $n-1$, and n. Prove your formulas using counting arguments.

2-69. Give a combinatorial proof of the identity $kS(n,k) = \sum_{j=1}^{n}\binom{n}{j}S(n-j,k-1)$, where $1 \leq k \leq n$.

2-70. Prove $C_n = \sum_{k=0}^{\lfloor n/2 \rfloor} T(k, n-k)^2$ for $n \geq 1$.

2-71. Consider lattice paths that can take unit steps north (N), south (S), west (W), or east (E), with self-intersections allowed. How many such paths begin and end at $(0,0)$ and have 10 steps?

2-72. Use the recursions 2.45 and 2.55 and the ideas in Remark 2.34 to give a recursive definition of a bijection between rook placements counted by $S'(n,k)$ and set partitions counted by $S(n,k)$. Is this bijection the same as the bijection described in Remark 2.56?

2-73. Give an algebraic proof of (2.4) by induction on n.

2-74. Fix $n \in \mathbb{Z}_{>0}$, let μ be an integer partition of length $\ell(\mu) \leq n$, and set $\mu_k = 0$ for $\ell(\mu) < k \leq n$. Let $s'(\mu,k)$ be the number of placements of $n-k$ non-attacking file rooks on the board $\text{dg}(\mu')$. (a) Find a summation formula for $s'(\mu,k)$ analogous to 2.62. (b) Prove that

$$(x+\mu_1)(x+\mu_2)\cdots(x+\mu_n) = \sum_{k=0}^{n} s'(\mu,k)x^k.$$

(c) For $n = 7$ and $\mu = (8, 5, 3, 3, 1)$, find $s'(\mu, k)$ for $0 \le k \le 7$.

2-75. Recall the *Fibonacci numbers* satisfy $F_0 = 0$, $F_1 = 1$, and $F_n = F_{n-1} + F_{n-2}$ for all $n \ge 2$. (a) Show that F_{n+1} is the number of compositions of n in which every part has size 1 or 2. (b) Show that F_{n+2} is the number of subsets of $\{1, 2, \ldots, n\}$ that do not contain two consecutive integers.

2-76. (a) How many words in $\{0, 1\}^m$ do not contain consecutive zeroes and have exactly k ones? (b) Give a combinatorial proof that $F_{n+1} = \sum_{i \in \mathbb{Z}} \binom{n-i}{i}$.

2-77. (a) Show that the sequence $a_n = F_{2n}$ satisfies the recursion $a_n = 3a_{n-1} - a_{n-2}$ for $n \ge 2$. What are the initial conditions? (b) Show that a_{n+1} is the number of words in $\{\mathrm{A}, \mathrm{B}, \mathrm{C}\}^n$ in which A is never immediately followed by B.

2-78. For $n \ge 0$, let a_n be the number of words in $\{1, 2, \ldots, k\}^n$ in which 1 is never immediately followed by 2. Find and prove a recursion satisfied by the sequence $(a_n : n \ge 0)$.

2-79. Give algebraic or combinatorial proofs of the following formulas involving the Fibonacci numbers F_n.
(a) $\sum_{k=0}^{n} F_k = F_{n+2} - 1$
(b) $\sum_{k=0}^{n-1} F_{2k+1} = F_{2n}$
(c) $\sum_{k=0}^{n} F_{2k} = F_{2n+1} - 1$
(d) $\sum_{k=0}^{n} kF_k = nF_{n+2} - F_{n+3} + 2$
(e) $\sum_{k=0}^{n} F_k^2 = F_n F_{n+1}$
(f) $\sum_{k=0}^{a} \binom{a}{k} F_{b-k} F_n^k F_{n+1}^{a-k} = F_{an+b}$

2-80. Give a combinatorial proof of equation (2.3) by interpreting both sides as counting a certain collection of functions.

2-81. Let $C_{n,k}$ be the number of Dyck paths of order n that end with exactly k east steps. Prove the recursion

$$C_{n,k} = \sum_{r=1}^{n-k} \binom{k-1+r}{k-1, r} C_{n-k,r}.$$

2-82. Find the general solution to each recursion.
(a) $x_n = 8x_{n-1} - 15x_{n-2}$
(b) $x_n = 3x_{n-1} + 4x_{n-2}$
(c) $x_n = 7x_{n-1} - 6x_{n-2}$
(d) $x_n = -5x_{n-1} - 6x_{n-2}$

2-83. Find the general solution to each recursion.
(a) $x_n = 4x_{n-1} - x_{n-2}$
(b) $x_n = (7x_{n-1} - 2x_{n-2})/6$
(c) $x_n = -2x_{n-1} - x_{n-2}$
(d) $x_n = 4x_{n-1} - 29x_{n-2}$

2-84. Solve each recursion with initial conditions.
(a) $x_n = -x_{n-1} + 12x_{n-2}$, $x_0 = 4$, $x_1 = -1$.
(b) $x_n = 4x_{n-1} - 4x_{n-2}$, $x_0 = 2$, $x_1 = 3$.
(c) $x_n = 3x_{n-1} - x_{n-2}$, $x_0 = 5$, $x_1 = 0$.
(d) $x_n = 8x_{n-1} - 14x_{n-2}$, $x_0 = 0$, $x_1 = 1$.

2-85. Find the general solution to each recursion.
(a) $x_n = -2x_{n-1} + 11x_{n-2} + 12x_{n-3} - 36x_{n-4}$.
(b) $x_n = x_{n-1} + 5x_{n-2} + x_{n-3} + x_{n-4}$.
(c) $x_n = 6x_{n-1} - 3x_{n-2} - 14x_{n-3} - 6x_{n-4}$.
(d) $x_n = -4x_{n-1} - 6x_{n-2} - 4x_{n-3} - x_{n-4}$.

2-86. Prove Theorem 2.66.

2-87. Prove that Method 2.70 yields all solutions to the given recursion in the case where all roots of the characteristic polynomial are distinct. [*Hint:* You can use the fact that for distinct complex numbers $r_1, \ldots, r_d$, the $d \times d$ matrix with i,j-entry r_i^{d-j} has determinant $\prod_{a<b}(r_a - r_b) \neq 0$, hence this matrix is invertible. This is the Vandermonde Determinant Formula, which we prove in §12.11.]

2-88. Let p be prime. Prove that $\binom{p}{k}$ is divisible by p for $0 < k < p$. Can you find a combinatorial proof?

2-89. *Fermat's Little Theorem* states that $a^p \equiv a \pmod{p}$ for $a \in \mathbb{Z}_{>0}$ and p prime. Prove this by expanding $a^p = (1 + 1 + \cdots + 1)^p$ using the Multinomial Theorem (cf. the previous exercise).

2-90. Ordered Set Partitions. An *ordered set partition* of a set X is a sequence $P = (T_1, T_2, \ldots, T_k)$ of distinct sets such that $\{T_1, T_2, \ldots, T_k\}$ is a set partition of X. Let $B_o(n)$ be the number of ordered set partitions of an n-element set. (a) Show $B_o(n) = \sum_{k=1}^{n} k!S(n,k)$ for $n \geq 1$. (b) Find a recursion relating $B_o(n)$ to values of $B_o(m)$ for $m < n$. (c) Compute $B_o(n)$ for $0 \leq n \leq 5$.

2-91. (a) Let $B_1(n)$ be the number of set partitions of an n-element set such that no block of the partition has size 1. Find a recursion and initial conditions for $B_1(n)$, and use these to compute $B_1(n)$ for $1 \leq n \leq 6$. (b) Let $S_1(n,k)$ be the number of set partitions as in (a) with k blocks. Find a recursion and initial conditions for $S_1(n,k)$.

2-92. Let $p_d(n,k)$ be the set of integer partitions of n with first part k and all parts *distinct.* Find a recursion and initial conditions for $p_d(n,k)$.

2-93. Let $p_o(n,k)$ be the set of integer partitions of n with first part k and all parts *odd.* Find a recursion and initial conditions for $p_o(n,k)$.

2-94. Let $q(n,k)$ be the number of integer partitions μ of length k and area n such that $\mu' = \mu$ (such partitions are called *self-conjugate*). Find a recursion and initial conditions for $q(n,k)$.

2-95. Verify the statement made after Definition 2.63 that any indexed collection of polynomials $\{p_n(x) : n \geq 0\}$ such that $\deg(p_n) = n$ for all n is a basis for the real vector space of polynomials in one variable with real coefficients.

2-96. Lah Numbers. A *set partition with totally ordered blocks* is a set partition $\{B_1, B_2, \ldots, B_k\}$ together with a total ordering of the elements in each block B_j. For $1 \leq k \leq n$, let $L(n,k)$ be the number of set partitions of $[n]$ containing k totally ordered blocks. Prove that $L(n,k) = \binom{n-1}{k-1}\frac{n!}{k!}$.

2-97. Define $L(n,k)$ as in the previous exercise. Prove: for all $n \geq 1$,

$$(x)\!\uparrow_n = \sum_{k=1}^{n} (x)\!\downarrow_k L(n,k).$$

2-98. Use the previous exercise to find a summation formula expressing Lah numbers in terms of Stirling numbers of the first and second kind.

2-99. Complete the proof of Theorem 2.51 by verifying that: (a) $\phi(P) \in \mathcal{B}$ for all $P \in \mathcal{A}$; (b) $\phi'(R) \in \mathcal{A}$ for all $R \in \mathcal{B}$; (c) $\phi \circ \phi' = \mathrm{id}_{\mathcal{B}}$; (d) $\phi' \circ \phi = \mathrm{id}_{\mathcal{A}}$.

2-100. Consider a product $x_1 \times x_2 \times \cdots \times x_n$ where the binary operation $\times$ is not necessarily associative. Show that the number of ways to parenthesize this expression is the Catalan number C_{n-1}. For example, the five possible parenthesizations when $n = 4$ are

$$(((x_1 \times x_2) \times x_3) \times x_4),\ ((x_1 \times x_2) \times (x_3 \times x_4)),\ (x_1 \times ((x_2 \times x_3) \times x_4)),$$

$$(x_1 \times (x_2 \times (x_3 \times x_4))),\ ((x_1 \times (x_2 \times x_3)) \times x_4).$$

2-101. Let $f, g : \mathbb{R} \to \mathbb{R}$ have derivatives of all orders. Recall the Product Rule for Derivatives: $D(fg) = D(f)g + fD(g)$, where D denotes differentiation with respect to x. (a) Prove that the nth derivative of fg is given by

$$D^n(fg) = \sum_{k=0}^{n} \binom{n}{k} D^k(f) D^{n-k}(g).$$

(b) Find and prove a similar formula for $D^n(f_1 f_2 \cdots f_s)$, where $f_1, \ldots, f_s$ have derivatives of all orders.

2-102. Prove: for all $n \geq 0$, $\sum_{k=0}^{n} \binom{2k}{k}\binom{2n-2k}{n-k} = 4^n$.

2-103. Prove: for all $n \geq 0$, $\sum_{k \in \mathbb{Z}} \binom{n}{k}\binom{n-k}{k} 2^{n-2k} = (n+1)C_n$.

2-104. Prove: for all $n \geq 1$, $\sum_{k=1}^{n} k\binom{n}{k}^2 = n(2n-1)C_{n-1}$.

2-105. Prove: for all $n \geq 0$, $\sum_{k=0}^{n} \binom{n}{k} F_k = F_{2n}$.

2-106. Prove: for all $n \geq 1$, $\sum_{k \in \mathbb{Z}} \binom{n}{2k+1} 5^k = 2^{n-1} F_n$.

Notes

Gould's book [51] contains an extensive, systematic list of binomial coefficient identities. More recently, Wilf and Zeilberger developed an algorithm, called the WZ-method, that can automatically evaluate many hypergeometric summations (which include binomial coefficient identities) or prove that such a summation has no closed form. This method is described in [99]. For more information on hypergeometric series, see [75].

A wealth of information about integer partitions, including a discussion of the Hardy–Rademacher–Ramanujan summation formula for $p(n)$, may be found in [5]. There is a vast literature on pattern-avoiding permutations; for more information on this topic, consult [14].

A great many combinatorial interpretations have been discovered for the Catalan numbers C_n. A partial list appears in Exercise 6.19 of [121, Vol. 2]; this list continues in the "Catalan Addendum," available online at

`http://www-math.mit.edu/~rstan/ec/catadd.pdf`

3

Counting Problems in Graph Theory

Graph theory is a branch of discrete mathematics that studies networks composed of a number of sites (vertices) linked together by connecting arcs (edges). This chapter studies some enumeration problems that arise in graph theory. We begin by defining fundamental graph-theoretic concepts such as walks, paths, cycles, vertex degrees, connectivity, forests, and trees. This leads to a discussion of various counting problems involving different kinds of trees. Aided by ideas from matrix theory, we count walks in a graph, spanning trees of a graph, and Eulerian tours. We also investigate the chromatic polynomial of a graph; this polynomial counts the number of ways of coloring the vertices such that no two vertices joined by an edge receive the same color.

3.1 Graphs and Digraphs

Intuitively, a graph is a mathematical model for a network consisting of a collection of nodes and connections that link certain pairs of nodes. For example, the nodes could be cities and the connections could be roads between cities. The nodes could be computers and the connections could be network links between computers. The nodes could be species in an ecosystem and the connections could be predator-prey relationships between species. The nodes could be tasks and the connections could be dependencies among the tasks. There are many such applications that lead naturally to graph models. We now give the formal mathematical definitions underlying such models.

3.1. Definition: Graphs. A *graph* is an ordered triple $G = (V, E, \epsilon)$, where: $V = V(G)$ is a finite, nonempty set called the *vertex set* of G; $E = E(G)$ is a finite set called the *edge set* of G; and $\epsilon : E \to \mathcal{P}(V)$ is a function called the *endpoint function* such that, for all $e \in E$, $\epsilon(e)$ is either a one-element subset of V or a two-element subset of V. If $\epsilon(e) = \{v\}$, we call the edge e a *loop at vertex* v. If $\epsilon(e) = \{v, w\}$, we call v and w the *endpoints of* e and say that e is an edge *from* v *to* w. We also say that v and w are *adjacent in* G, v and w are *joined by* e, and e is *incident to* v and w.

We visualize a graph $G = (V, E, \epsilon)$ by drawing a collection of dots labeled by the elements $v \in V$. For each edge $e \in E$ with $\epsilon(e) = \{v, w\}$, we draw a line or curved arc labeled e between the two dots labeled v and w. Similarly, if $\epsilon(e) = \{v\}$, we draw a loop labeled e based at the dot labeled v.

3.2. Example. The left drawing in Figure 3.1 represents the graph defined formally by the ordered triple

$$G_1 = (\{1, 2, 3, 4, 5\}, \{a, b, c, d, e, f, g, h, i\}, \epsilon),$$

where ϵ acts as follows:

$$\begin{array}{lllll} \epsilon(a) = \{1, 4\}, & \epsilon(b) = \{4, 3\}, & \epsilon(c) = \{2, 3\}, & \epsilon(d) = \{1, 2\}, & \epsilon(e) = \{1, 2\}, \\ \epsilon(f) = \{3\}, & \epsilon(g) = \{2, 5\}, & \epsilon(h) = \{4, 5\}, & \epsilon(i) = \{4, 5\}. & \end{array}$$

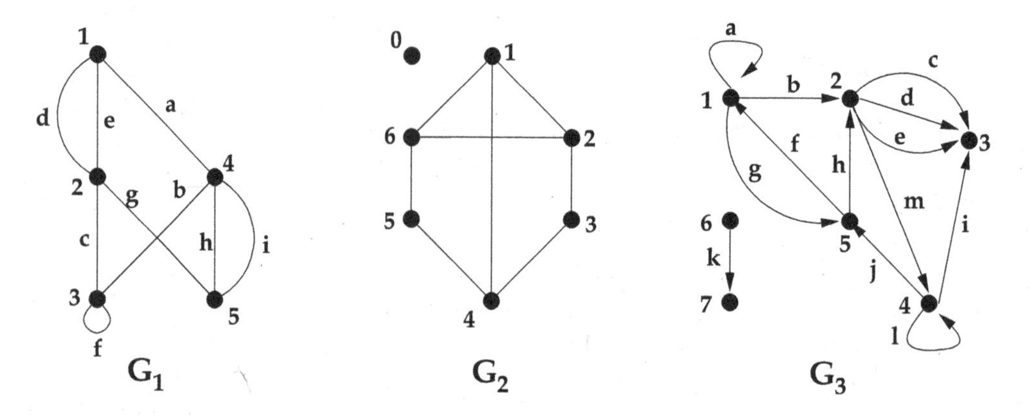

FIGURE 3.1
A graph, a simple graph, and a digraph.

Edge f is a loop at vertex 3; edges h and i both go between vertices 4 and 5; vertices 1 and 4 are adjacent, but vertices 2 and 4 are not.

In many applications, there are no loop edges, and there is never more than one edge between the same two vertices. This means that the endpoint function ϵ is a one-to-one map into the set of two-element subsets of V. So we can identify each edge e with its set of endpoints $\epsilon(e)$. This leads to the following simplified model in which edges are not explicitly named and there is no explicit endpoint function.

3.3. Definition: Simple Graphs. A *simple graph* is a pair $G = (V, E)$, where V is a finite nonempty set and E is a set of two-element subsets of V. We continue to use all the terminology introduced in Definition 3.1.

3.4. Example. The center drawing in Figure 3.1 depicts the simple graph G_2 with vertex set $V(G_2) = \{0, 1, 2, 3, 4, 5, 6\}$ and edge set

$$E(G_2) = \{\{1,2\}, \{2,3\}, \{3,4\}, \{4,5\}, \{5,6\}, \{1,6\}, \{2,6\}, \{1,4\}\}.$$

To model certain situations (such as predator-prey relationships, or one-way streets in a city), we need to introduce a *direction* on each edge. This leads to the notion of a digraph (directed graph).

3.5. Definition: Digraphs. A *digraph* is an ordered triple $D = (V, E, \epsilon)$, where V is a finite nonempty set of *vertices*, E is a finite set of *edges*, and $\epsilon : E \to V \times V$ is the *endpoint function*. If $\epsilon(e)$ is the *ordered pair* (v, w), we say that e is an edge *from v to w*.

In a digraph, an edge from v to w is not an edge from w to v when $v \neq w$, since $(v, w) \neq (w, v)$. On the other hand, in a graph, an edge from v to w is also an edge from w to v, since $\{v, w\} = \{w, v\}$.

3.6. Example. The right drawing in Figure 3.1 displays a digraph G_3. In this digraph, $\epsilon(j) = (4, 5)$, $\epsilon(a) = (1, 1)$, and so on. There are three edges from 2 to 3, but no edges from 3 to 2. There are edges in both directions between vertices 1 and 5.

As before, we can eliminate specific reference to the endpoint function of a digraph if there are no multiple edges with the same starting vertex and ending vertex.

3.7. Definition: Simple Digraphs. A *simple digraph* is an ordered pair $D = (V, E)$, where V is a finite, nonempty set and E is a subset of $V \times V$. Each ordered pair $(v, w) \in E$ represents an edge in D from v to w. Note that we do allow loops ($v = w$) in a simple digraph.

When investigating structural properties of graphs, the names of the vertices and edges are often irrelevant. The concept of graph isomorphism lets us identify graphs that are the same except for the names used for the vertices and edges.

3.8. Definition: Graph Isomorphism. Given two graphs $G = (V, E, \epsilon)$ and $H = (W, F, \eta)$, a *graph isomorphism* from G to H consists of two bijections $f : V \to W$ and $g : E \to F$ such that, for all $e \in E$, if $\epsilon(e) = \{v, w\}$ then $\eta(g(e)) = \{f(v), f(w)\}$ (we allow $v = w$ here). Digraph isomorphisms are defined similarly: $\epsilon(e) = (v, w)$ implies $\eta(g(e)) = (f(v), f(w))$. We say G and H are *isomorphic*, written $G \cong H$, iff there exists a graph isomorphism from G to H.

In the case of *simple* graphs $G = (V, E)$ and $H = (W, F)$, a graph isomorphism can be viewed as a bijection $f : V \to W$ that induces a bijection between the edge sets E and F. More specifically, this means that for all $v, w \in V$, $\{v, w\} \in E$ iff $\{f(v), f(w)\} \in F$.

3.9. Example. Let $G = (V, E) = (\{1, 2, 3, 4\}, \{\{1, 2\}, \{1, 3\}, \{1, 4\}\})$ and $H = (W, F) = (\{a, b, c, d\}, \{\{a, c\}, \{b, c\}, \{c, d\}\})$. The map $f : V \to W$ such that $f(1) = c$, $f(2) = a$, $f(3) = b$, and $f(4) = d$ is a graph isomorphism, as is readily verified, so $G \cong H$. In fact, for any bijection $g : V \to W$, g is a graph isomorphism iff $g(1) = c$. By the Product Rule, there are $3! = 6$ graph isomorphisms from G to H.

3.2 Walks and Matrices

We can travel through a graph by following several edges in succession. Formalizing this idea leads to the concept of a walk.

3.10. Definition: Walks, Paths, Cycles. Let $G = (V, E, \epsilon)$ be a graph or digraph. A *walk in* G is a sequence

$$W = (v_0, e_1, v_1, e_2, v_2, \ldots, v_{s-1}, e_s, v_s)$$

where $s \geq 0$, $v_i \in V$ for all i, $e_i \in E$ for all i, and e_i is an edge from v_{i-1} to v_i for $1 \leq i \leq s$. We say that W is a walk *of length* s *from* v_0 *to* v_s. The walk W is *closed* iff $v_0 = v_s$. The walk W is a *path* iff the vertices $v_0, v_1, \ldots, v_s$ are pairwise distinct (which forces the edges e_i to be distinct as well). The walk W is a *cycle* iff $s > 0$, $v_1, \ldots, v_s$ are distinct, $e_1, \ldots, e_s$ are distinct, and $v_0 = v_s$. A *k-cycle* is a cycle of length k. In the case of simple graphs and simple digraphs, the edges e_i are determined uniquely by their endpoints. So, in the simple case, we can regard a walk as a sequence of vertices $(v_0, v_1, \ldots, v_s)$ such that there is an edge from v_{i-1} to v_i in G for $1 \leq i \leq s$.

3.11. Remark. When considering cycles in a digraph, we often *identify* two cycles that are cyclic shifts of one another (unless we need to keep track of the starting vertex of the cycle). Similarly, we identify cycles in a graph that are cyclic shifts or reversals of one another.

3.12. Example. In the graph G_1 from Figure 3.1,

$$W_1 = (2, c, 3, f, 3, b, 4, h, 5, i, 4, i, 5)$$

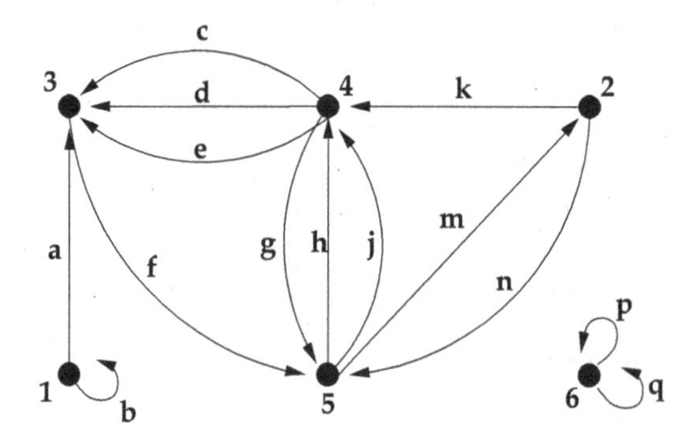

FIGURE 3.2
Digraph used to illustrate adjacency matrices.

is a walk of length 6 from vertex 2 to vertex 5. In the simple graph G_2 in the same figure, $W_2 = (1, 6, 2, 3, 4, 5)$ is a walk and a path of length 5, whereas $C = (6, 5, 4, 3, 2, 6)$ is a 5-cycle. We often identify C with the cycles $(5, 4, 3, 2, 6, 5)$, $(6, 2, 3, 4, 5, 6)$, etc. In the digraph G_3,

$$W_3 = (1, a, 1, g, 5, h, 2, m, 4, j, 5, h, 2, d, 3)$$

is a walk from vertex 1 to vertex 3; $(5, h, 2, m, 4, j, 5)$ is a 3-cycle; $(4, l, 4)$ is a 1-cycle; and $(5, f, 1, g, 5)$ is a 2-cycle. Observe that 1-cycles are the same as loop edges, and 2-cycles cannot exist in simple graphs. For any vertex v in a graph or digraph, (v) is a walk of length zero from v to v, which is a path but not a cycle.

We can now formulate our first counting problem: how many walks of a given length are there between two given vertices in a graph or digraph? We will develop an algebraic solution to this problem in which concatenation of walks is modeled by multiplication of certain matrices.

3.13. Definition: Adjacency Matrix. Let G be a graph or digraph with vertex set $X = \{x_i : 1 \leq i \leq n\}$. The *adjacency matrix of* G (relative to the given indexing of the vertices) is the $n \times n$ matrix A whose i, j-entry $A(i, j)$ is the number of edges in G from x_i to x_j.

3.14. Example. The adjacency matrix for the digraph G in Figure 3.2 is

$$A = \begin{bmatrix} 1 & 0 & 1 & 0 & 0 & 0 \\ 0 & 0 & 0 & 1 & 1 & 0 \\ 0 & 0 & 0 & 0 & 1 & 0 \\ 0 & 0 & 3 & 0 & 1 & 0 \\ 0 & 1 & 0 & 2 & 0 & 0 \\ 0 & 0 & 0 & 0 & 0 & 2 \end{bmatrix}.$$

3.15. Example. If G is a graph, edges from v to w are the same as edges from w to v. So, the adjacency matrix for G is a *symmetric* matrix ($A(i, j) = A(j, i)$ for all i, j). If G is a simple graph, the adjacency matrix consists of all 1's and 0's with zeroes on the main

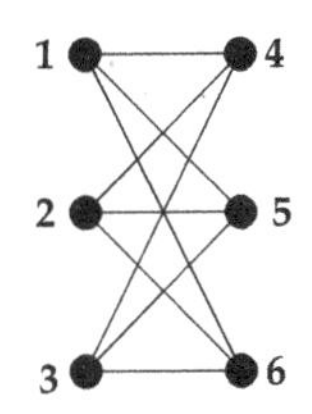

FIGURE 3.3
A simple graph.

diagonal. For example, the adjacency matrix of the simple graph in Figure 3.3 is

$$\begin{bmatrix} 0 & 0 & 0 & 1 & 1 & 1 \\ 0 & 0 & 0 & 1 & 1 & 1 \\ 0 & 0 & 0 & 1 & 1 & 1 \\ 1 & 1 & 1 & 0 & 0 & 0 \\ 1 & 1 & 1 & 0 & 0 & 0 \\ 1 & 1 & 1 & 0 & 0 & 0 \end{bmatrix}.$$

Recall from linear algebra the definition of the product of two matrices.

3.16. Definition: Matrix Multiplication. Suppose A is an $m \times n$ matrix and B is an $n \times p$ matrix. Then AB is the $m \times p$ matrix whose i,j-entry is

$$(AB)(i,j) = \sum_{k=1}^{n} A(i,k)B(k,j) \quad \text{for } 1 \leq i \leq m, 1 \leq j \leq p. \tag{3.1}$$

Matrix multiplication is associative, so we can write a product of three or more (compatible) matrices without any parentheses. The next theorem gives a formula for the general entry in such a product.

3.17. Theorem: Product of Several Matrices. Assume $A_1, \ldots, A_s$ are matrices such that A_i has dimensions $n_{i-1} \times n_i$. Then $A_1 A_2 \cdots A_s$ is the $n_0 \times n_s$ matrix whose k_0, k_s-entry is

$$(A_1 A_2 \cdots A_s)(k_0, k_s) = \sum_{k_1=1}^{n_1} \sum_{k_2=1}^{n_2} \cdots \sum_{k_{s-1}=1}^{n_{s-1}} A_1(k_0,k_1)A_2(k_1,k_2)A_3(k_2,k_3) \cdots A_s(k_{s-1},k_s) \tag{3.2}$$

for all k_0 and k_s such that $1 \leq k_0 \leq n_0$ and $1 \leq k_s \leq n_s$.

Proof. We use induction on s. The case $s = 1$ is immediate, and the case $s = 2$ is the definition of matrix multiplication (after a change in notation). Assume $s > 2$ and that (3.2) is known to hold for the product $B = A_1 A_2 \cdots A_{s-1}$. We can think of the given product $A_1 A_2 \cdots A_{s-1} A_s$ as the binary product BA_s. Therefore, using (3.1), the k_0, k_s-entry of $A_1 A_2 \cdots A_s$ is

$$\begin{aligned} (BA_s)(k_0,k_s) &= \sum_{k=1}^{n_{s-1}} B(k_0,k)A_s(k,k_s) \\ &= \sum_{k=1}^{n_{s-1}} \left(\sum_{k_1=1}^{n_1} \cdots \sum_{k_{s-2}=1}^{n_{s-2}} A_1(k_0,k_1)A_2(k_1,k_2) \cdots A_{s-1}(k_{s-2},k) \right) A_s(k,k_s) \\ &= \sum_{k_1=1}^{n_1} \cdots \sum_{k_{s-2}=1}^{n_{s-2}} \sum_{k=1}^{n_{s-1}} A_1(k_0,k_1)A_2(k_1,k_2) \cdots A_{s-1}(k_{s-2},k)A_s(k,k_s). \end{aligned}$$

Replacing k by k_{s-1} in the innermost summation, we obtain the result. □

Taking all A_i's in the theorem to be the same matrix A, we obtain the following formula for the entries of the powers of a given square matrix.

3.18. Corollary: Powers of a Matrix. Suppose A is an $n \times n$ matrix. For each integer $s > 0$ and all i, j in the range $1 \leq i, j \leq n$,

$$A^s(i,j) = \sum_{k_1=1}^{n} \cdots \sum_{k_{s-1}=1}^{n} A(i,k_1)A(k_1,k_2)\cdots A(k_{s-2},k_{s-1})A(k_{s-1},j). \tag{3.3}$$

The preceding formula may appear unwieldy, but it is precisely the tool we need to count walks in graphs.

3.19. The Walk Rule. Let G be a graph or digraph with vertex set $X = \{x_1, \ldots, x_n\}$, and let A be the adjacency matrix of G. For all i, j between 1 and n and all $s \geq 0$, the i,j-entry of A^s is the number of walks of length s in G from x_i to x_j.

Proof. The result holds for $s = 0$, since $A^0 = I_n$ (the $n \times n$ identity matrix) and there is exactly one walk of length zero from any vertex to itself. Now suppose $s > 0$. A walk of length s from x_i to x_j will visit $s - 1$ intermediate vertices (not necessarily distinct from each other or from x_i or x_j). Let $(x_i, x_{k_1}, \ldots, x_{k_{s-1}}, x_j)$ be the ordered list of vertices visited by the walk. To build such a walk, we choose any edge from x_i to x_{k_1} in $A(i, k_1)$ ways; then we choose any edge from x_{k_1} to x_{k_2} in $A(k_1, k_2)$ ways; and so on. By the Product Rule, the total number of walks associated to this vertex sequence is $A(i, k_1)A(k_1, k_2) \cdots A(k_{s-1}, j)$. This formula holds even if there are no walks with this vertex sequence, since some term in the product will be zero in this case. Applying the Sum Rule produces the right side of (3.3), and the result follows. □

3.20. Example. Consider again the adjacency matrix A of the digraph G in Figure 3.2. Some matrix computations show that

$$A^2 = \begin{bmatrix} 1 & 0 & 1 & 0 & 1 & 0 \\ 0 & 1 & 3 & 2 & 1 & 0 \\ 0 & 1 & 0 & 2 & 0 & 0 \\ 0 & 1 & 0 & 2 & 3 & 0 \\ 0 & 0 & 6 & 1 & 3 & 0 \\ 0 & 0 & 0 & 0 & 0 & 4 \end{bmatrix}, \quad A^3 = \begin{bmatrix} 1 & 1 & 1 & 2 & 1 & 0 \\ 0 & 1 & 6 & 3 & 6 & 0 \\ 0 & 0 & 6 & 1 & 3 & 0 \\ 0 & 3 & 6 & 7 & 3 & 0 \\ 0 & 3 & 3 & 6 & 7 & 0 \\ 0 & 0 & 0 & 0 & 0 & 8 \end{bmatrix}.$$

So, for example, there are six walks of length 2 from vertex 5 to vertex 3, and there are seven walks of length 3 that start and end at vertex 4. By looking at $A^9 = (A^3)^3$, we find that there are 1074 walks of length 9 from vertex 2 to vertex 4, but only 513 walks of length 9 from vertex 4 to vertex 2.

3.3 Directed Acyclic Graphs and Nilpotent Matrices

Next we consider the question of counting *all* walks (of any length) between two vertices in a digraph. The question is uninteresting for graphs, since the number of walks between two distinct vertices v, w in a graph is either zero or infinity. This follows since a walk is allowed to repeatedly traverse a particular edge along a path from v to w, which leads to arbitrarily long walks from v to w. Similarly, if G is a digraph that contains a cycle, we obtain arbitrarily long walks between two vertices on the cycle by going around the cycle again and again. To rule out these possibilities, we restrict attention to the following class of digraphs.

3.21. Definition: DAGs. A *DAG* is a digraph with no cycles. (The acronym DAG stands for "directed acyclic graph.")

To characterize adjacency matrices of DAGs, we need another concept from matrix theory.

3.22. Definition: Nilpotent Matrices. An $n \times n$ matrix A is called *nilpotent* iff $A^s = 0$ for some integer $s \geq 1$. The least such integer s is called the *index of nilpotence* of A.

Note that if $A^s = 0$ and $t \geq s$, then $A^t = 0$ also.

3.23. Example. The zero matrix is the unique $n \times n$ matrix with index of nilpotence 1. The square of the nonzero matrix $A = \begin{bmatrix} 0 & 1 \\ 0 & 0 \end{bmatrix}$ is zero, so A is nilpotent of index 2. Similarly, given any real x, y, z, the matrix

$$B = \begin{bmatrix} 0 & x & y \\ 0 & 0 & z \\ 0 & 0 & 0 \end{bmatrix}$$

satisfies

$$B^2 = \begin{bmatrix} 0 & 0 & xz \\ 0 & 0 & 0 \\ 0 & 0 & 0 \end{bmatrix} \quad \text{and} \quad B^3 = 0,$$

so we obtain examples of matrices that are nilpotent of index 3. The next result generalizes this example.

3.24. Theorem: Nilpotence of Strictly Triangular Matrices. Suppose A is an $n \times n$ *strictly upper-triangular* matrix, which means $A(i, j) = 0$ for all $i \geq j$. Then A is nilpotent of index at most n. A similar result holds for strictly lower-triangular matrices.

Proof. It suffices to show that A^n is the zero matrix. Fix k_0 and k_n between 1 and n. By (3.3), we have

$$A^n(k_0, k_n) = \sum_{k_1=1}^{n} \cdots \sum_{k_{n-1}=1}^{n} A(k_0, k_1)A(k_1, k_2) \cdots A(k_{n-1}, k_n).$$

We claim that each term in this sum is zero. Otherwise, there would exist $k_1, \ldots, k_{n-1}$ such that $A(k_{t-1}, k_t) \neq 0$ for each t between 1 and n. But since A is strictly upper-triangular, we would then have

$$k_0 < k_1 < k_2 < \cdots < k_n.$$

This cannot occur, since every k_t is an integer between 1 and n. □

The next theorem reveals the connection between nilpotent matrices and DAGs.

3.25. Theorem: DAGs and Nilpotent Matrices. Let G be a digraph with vertex set $X = \{x_1, \ldots, x_n\}$ and adjacency matrix A. G is a DAG iff A is nilpotent. When G is a DAG, there exists an ordering of the vertex set X for which A is a strictly lower-triangular matrix.

Proof. Assume first that G is not a DAG, so that G has at least one cycle. Let x_i be any fixed vertex involved in this cycle, and let $c \geq 1$ be the length of this cycle. By going around the cycle zero or more times, we obtain walks from x_i to x_i of lengths $0, c, 2c, 3c, \ldots$. By the Walk Rule, it follows that the (i,i)-entry of A^{kc} is at least 1, for all $k \geq 0$. This fact prevents any positive power of A from being the zero matrix, so A is not nilpotent.

Conversely, assume that A is not nilpotent. Then, in particular, $A^n \neq 0$, so there exist indices k_0, k_n with $A^n(k_0, k_n) \neq 0$. Using the Walk Rule again, we deduce that there is a walk in G visiting a sequence of vertices

$$(x_{k_0}, x_{k_1}, x_{k_2}, \ldots, x_{k_{n-1}}, x_{k_n}).$$

Since G has only n vertices, not all of the $n+1$ vertices listed here are distinct. If we choose i minimal and then $j > i$ minimal such that $x_{k_i} = x_{k_j}$, then there is a cycle in G visiting the vertices $(x_{k_i}, x_{k_{i+1}}, \ldots, x_{k_j})$. So G is not a DAG.

We prove the statement about lower-triangular matrices by induction on n. A DAG with only one vertex must have adjacency matrix $[0]$, so the result holds for $n = 1$. Suppose $n > 1$ and the result is known for DAGs with $n-1$ vertices. Create a walk $(v_0, e_1, v_1, e_2, v_2, \ldots)$ in G by starting at any vertex and repeatedly following any edge leading away from the current vertex. Since G has no cycles, the vertices v_i reached by this walk are pairwise distinct. Since there are only n available vertices, our walk must terminate at a vertex v_j with no outgoing edges. Let $x'_1 = v_j$. Deleting x'_1 and all edges with x'_1 as one endpoint produces an $(n-1)$-vertex digraph G' that is also a DAG, as one immediately verifies. By induction, there is an ordering $x'_2, \ldots, x'_n$ of the vertices of G' such that the associated adjacency matrix A' is strictly lower-triangular. Now, relative to the ordering $x'_1, x'_2, \ldots, x'_n$ of the vertices of G, the adjacency matrix of G has the form

$$\left[\begin{array}{c|c} 0 & 0 \cdots 0 \\ \hline * & \\ \vdots & A' \\ * & \end{array}\right],$$

where each $*$ denotes some nonnegative integer. This matrix is strictly lower-triangular. □

The next result leads to a formula for counting walks of any length in a DAG.

3.26. Theorem: Inverse of $I - A$ for Nilpotent A. Suppose A is a nilpotent $n \times n$ matrix with $A^s = 0$. Let I be the $n \times n$ identity matrix. Then $I - A$ is an invertible matrix with inverse

$$(I - A)^{-1} = I + A + A^2 + A^3 + \cdots + A^{s-1}.$$

Proof. Let $B = I + A + A^2 + \cdots + A^{s-1}$. By the distributive law for matrices,

$$(I - A)B = IB - AB = (I + A + A^2 + \cdots + A^{s-1}) - (A + A^2 + \cdots + A^{s-1} + A^s) = I - A^s.$$

Since $A^s = 0$, we see that $(I - A)B = I$. A similar calculation shows that $B(I - A) = I$. Therefore B is the two-sided matrix inverse of $I - A$. □

3.27. Remark. The previous result for matrices can be remembered by noting the analogy to the Geometric Series Formula for real or complex numbers: whenever $|r| < 1$,

$$(1 - r)^{-1} = \frac{1}{1 - r} = 1 + r + r^2 + \cdots + r^{s-1} + r^s + \cdots.$$

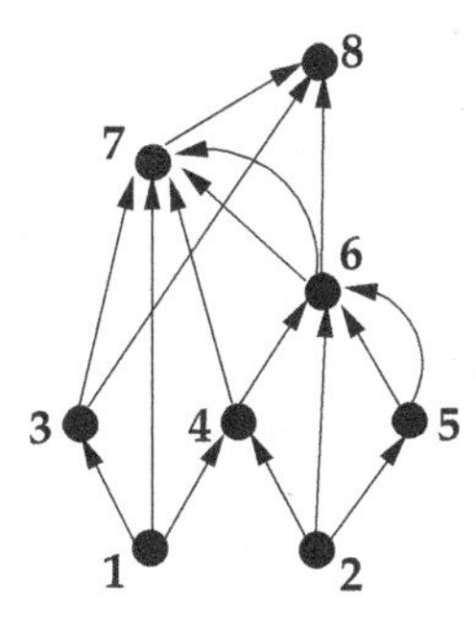

FIGURE 3.4
Example of a DAG.

3.28. The Path Rule for DAGs. Let G be a DAG with vertex set $\{x_1, \ldots, x_n\}$ and adjacency matrix A. For all i, j between 1 and n, the total number of paths from x_i to x_j in G (of any length) is the i, j-entry of $(I - A)^{-1}$.

Proof. By the Walk Rule, the number of walks of length $t \geq 0$ from x_i to x_j is $A^t(i, j)$. Because G is a DAG, we have $A^t = 0$ for all $t \geq n$. By the Sum Rule, the total number of walks from x_i to x_j is $\sum_{t=0}^{n-1} A^t(i, j)$. By Theorem 3.26, this number is precisely the i, j-entry of $(I - A)^{-1}$. Finally, one readily confirms that every walk in a DAG must in fact be a path. □

3.29. Example. Consider the DAG shown in Figure 3.4. Its adjacency matrix is

$$A = \begin{bmatrix} 0 & 0 & 1 & 1 & 0 & 0 & 1 & 0 \\ 0 & 0 & 0 & 1 & 1 & 1 & 0 & 0 \\ 0 & 0 & 0 & 0 & 0 & 0 & 1 & 1 \\ 0 & 0 & 0 & 0 & 0 & 1 & 1 & 0 \\ 0 & 0 & 0 & 0 & 0 & 2 & 0 & 0 \\ 0 & 0 & 0 & 0 & 0 & 0 & 2 & 1 \\ 0 & 0 & 0 & 0 & 0 & 0 & 0 & 1 \\ 0 & 0 & 0 & 0 & 0 & 0 & 0 & 0 \end{bmatrix}.$$

Using a computer algebra system, we compute

$$(I - A)^{-1} = \begin{bmatrix} 1 & 0 & 1 & 1 & 0 & 1 & 5 & 7 \\ 0 & 1 & 0 & 1 & 1 & 4 & 9 & 13 \\ 0 & 0 & 1 & 0 & 0 & 0 & 1 & 2 \\ 0 & 0 & 0 & 1 & 0 & 1 & 3 & 4 \\ 0 & 0 & 0 & 0 & 1 & 2 & 4 & 6 \\ 0 & 0 & 0 & 0 & 0 & 1 & 2 & 3 \\ 0 & 0 & 0 & 0 & 0 & 0 & 1 & 1 \\ 0 & 0 & 0 & 0 & 0 & 0 & 0 & 1 \end{bmatrix}.$$

So, for example, there are 13 paths from vertex 2 to vertex 8, and 4 paths from vertex 5 to vertex 7.

3.4 Vertex Degrees

The next definition introduces notation that records how many edges lead into or out of each vertex in a digraph.

3.30. Definition: Indegree and Outdegree. Let $G = (V, E, \epsilon)$ be a digraph. For each $v \in V$, the *outdegree* of v, denoted $\text{outdeg}_G(v)$, is the number of edges $e \in E$ leading away from v; the *indegree* of v, denoted $\text{indeg}_G(v)$, is the number of edges $e \in E$ leading to v. A *source* is a vertex of indegree zero. A *sink* is a vertex of outdegree zero.

3.31. Example. Let G_3 be the digraph on the right in Figure 3.1. We have

$$\begin{aligned}(\text{indeg}_{G_3}(1), \ldots, \text{indeg}_{G_3}(7)) &= (2, 2, 4, 2, 2, 0, 1); \\ (\text{outdeg}_{G_3}(1), \ldots, \text{outdeg}_{G_3}(7)) &= (3, 4, 0, 3, 2, 1, 0).\end{aligned}$$

Vertex 6 is the only source, whereas vertices 3 and 7 are sinks. A loop edge at v contributes 1 to both the indegree and outdegree of v. The sum of all the indegrees is 13, which is also the sum of all the outdegrees, and is also the number of edges in the digraph. This phenomenon is explained in the next theorem.

3.32. Theorem: Degree Sum Formula for Digraphs. In any digraph $G = (V, E, \epsilon)$,

$$\sum_{v \in V} \text{indeg}_G(v) = |E| = \sum_{v \in V} \text{outdeg}_G(v).$$

Proof. For each $v \in V$, let E_v be the set of edges $e \in E$ ending at v, meaning that $\epsilon(e) = (w, v)$ for some $w \in V$. By definition of indegree, $|E_v| = \text{indeg}_G(v)$ for each $v \in V$. The set E of all edges is the disjoint union of the sets E_v as v varies through V. By the Sum Rule, $|E| = \sum_{v \in V} |E_v| = \sum_{v \in V} \text{indeg}_G(v)$. The formula involving outdegree is proved similarly. □

Next we give analogous definitions and results for graphs.

3.33. Definition: Degree. Let $G = (V, E, \epsilon)$ be a graph. For each $v \in V$, the *degree of v in G*, denoted $\deg_G(v)$, is the number of edges in E having v as an endpoint, where any loop edge at v is counted twice. The *degree multiset of G*, denoted $\deg(G)$, is the multiset $[\deg_G(v) : v \in V]$. G is called *k-regular* iff every vertex in G has degree k. G is *regular* iff G is k-regular for some $k \geq 0$.

3.34. Example. For the graph G_1 in Figure 3.1, we have

$$(\deg_{G_1}(1), \ldots, \deg_{G_1}(5)) = (3, 4, 4, 4, 3); \qquad \deg(G_1) = [4, 4, 4, 3, 3].$$

The graph in Figure 3.3 is 3-regular. In both of these graphs, the sum of all vertex degrees is 18, which is twice the number of edges in the graph.

3.35. Theorem: Degree Sum Formula for Graphs. For any graph $G = (V, E, \epsilon)$,

$$\sum_{v \in V} \deg_G(v) = 2|E|.$$

Proof. First assume G has no loop edges. Let X be the set of pairs (v, e) such that $v \in V$, $e \in E$, and v is an endpoint of e. We count X in two ways. On one hand, X is the disjoint union of sets X_v, where X_v consists of pairs in X with first component v. The second component of such a pair can be any edge incident to v, so $|X_v| = \deg_G(v)$. By the Sum Rule, $|X| = \sum_{v\in V} |X_v| = \sum_{v\in V} \deg_G(v)$. On the other hand, we can build an object (v, e) in X by first choosing e (in any of $|E|$ ways) and then choosing one of the two endpoints v of e (2 ways, since G has no loop edges). So $|X| = 2|E|$ by the Product Rule, and the result holds in this case.

Next, if G has k loop edges, let G' be G with these loops deleted. Then

$$\sum_{v\in V} \deg_G(v) = \sum_{v\in V} \deg_{G'}(v) + 2k,$$

since each loop edge increases some vertex degree in the sum by 2. Using the result for loopless graphs,

$$\sum_{v\in V} \deg_G(v) = 2|E(G')| + 2k = 2|E(G)|,$$

since G has k more edges than G'. □

Vertices of low degree are given special names.

3.36. Definition: Isolated Vertices and Leaves. An *isolated vertex* in a graph is a vertex of degree 0. A *leaf* is a vertex of degree 1.

The following result will be used later in our analysis of trees.

3.37. The Two-Leaf Lemma. Suppose G is a graph. One of the following three alternatives must occur: (i) G has a cycle; (ii) G has no edges; or (iii) G has at least two leaves.

Proof. Suppose that G has no cycles and G has at least one edge; we prove that G has two leaves. Since G has no cycles, we can assume G is simple. Let $P = (v_0, v_1, \ldots, v_s)$ be a path of maximum length in G. Such a path exists, since G has only finitely many vertices and edges. Observe that $s > 0$ since G has an edge, and $v_0 \neq v_s$. Note that $\deg(v_s) \geq 1$ since $s > 0$. Assume v_s is not a leaf. Then there exists a vertex $w \neq v_{s-1}$ that is adjacent to v_s. Now, w is different from all v_j with $0 \leq j < s-1$, since otherwise $(v_j, v_{j+1}, \ldots, v_s, w = v_j)$ would be a cycle in G. But this means $(v_0, v_1, \ldots, v_s, w)$ is a path in G longer than P, contradicting maximality of P. So v_s must be a leaf. A similar argument shows that v_0 is also a leaf. □

3.5 Functional Digraphs

We can obtain structural information about functions $f : V \to V$ by viewing these functions as certain kinds of digraphs.

3.38. Definition: Functional Digraphs. A *functional digraph on* V is a simple digraph G with vertex set V such that $\text{outdeg}_G(v) = 1$ for all $v \in V$.

A function $f : V \to V$ can be thought of as a set E_f of ordered pairs such that for each $x \in V$, there exists exactly one $y \in V$ with $(x, y) \in E_f$, namely $y = f(x)$. Then (V, E_f) is a functional digraph on V. Conversely, a functional digraph $G = (V, E)$ determines a unique function $g : V \to V$ by letting $g(v)$ be the other endpoint of the unique edge in E departing from v. These comments establish a bijection between the set of functions on V and the set of functional digraphs on V.

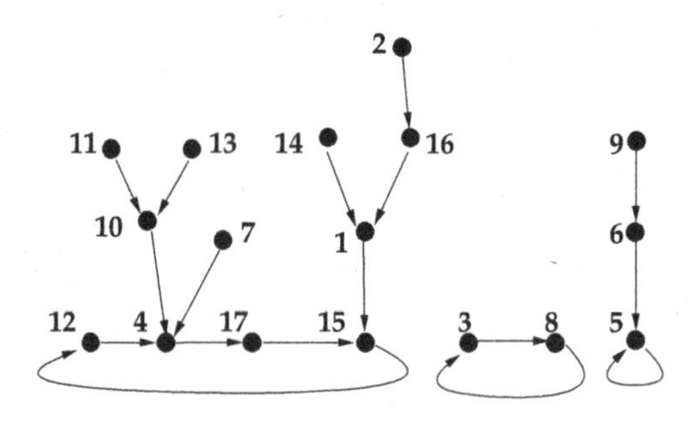

FIGURE 3.5
A functional digraph.

3.39. Example. Figure 3.5 displays the functional digraph associated to the following function:

$$\begin{array}{lllll} f(1) = 15; & f(2) = 16; & f(3) = 8; & f(4) = 17; & f(5) = 5; \\ f(6) = 5; & f(7) = 4; & f(8) = 3; & f(9) = 6; & f(10) = 4; \\ f(11) = 10; & f(12) = 4; & f(13) = 10; & f(14) = 1; & f(15) = 12; \\ f(16) = 1; & f(17) = 15. & & & \end{array}$$

Our next goal is to understand the structure of functional digraphs. Consider the digraph $G = (V, E)$ shown in Figure 3.5. Some of the vertices in this digraph are involved in cycles, which are drawn at the bottom of the figure. These cycles have length 1 or greater, and any two distinct cycles involve disjoint sets of vertices. The other vertices in the digraph all feed into these cycles at different points. We can form a set partition of the vertex set of the digraph by collecting together all vertices that feed into a particular vertex on a particular cycle. Each such collection can be viewed as a smaller digraph that has no cycles. We will show that these observations hold for all functional digraphs. To do this, we need a few more defintions.

3.40. Definition: Cyclic Vertices. Let G be a functional digraph on V. A vertex $v \in V$ is called *cyclic* iff v belongs to some cycle of G; otherwise, v is called *acyclic*.

3.41. Example. The cyclic elements for the functional digraph in Figure 3.5 are 3, 4, 5, 8, 12, 15, and 17.

Let $f : V \to V$ be the function associated to the functional digraph G. Then $v \in V$ is cyclic iff $f^i(v) = v$ for some $i \geq 1$, where f^i denotes the composition of f with itself i times. This fact follows since the only possible cycle involving v in G must look like $(v, f(v), f^2(v), f^3(v), \ldots)$.

3.42. Definition: Rooted Trees. A digraph G is called a *rooted tree with root v_0* iff G is a functional digraph and v_0 is the unique cyclic vertex of G.

3.43. Theorem: Structure of Functional Digraphs. Let G be a functional digraph on a nonempty set V with associated function $f : V \to V$. Let $C \subseteq V$ denote the set of cyclic vertices of G. C is nonempty, and each $v \in C$ belongs to exactly one cycle of G. Also, there exists a unique indexed set partition $\{S_v : v \in C\}$ of V such that the following hold for all $v \in C$: (i) $v \in S_v$; (ii) $x \in S_v$ and $x \neq v$ implies $f(x) \in S_v$; (iii) if $g : S_v \to S_v$ is defined by setting $g(x) = f(x)$ for $x \neq v$ and $g(v) = v$, then the functional digraph of g is a rooted tree with root v.

Proof. First, suppose $v \in C$. Since every vertex of G has exactly one outgoing edge, the only possible cycle involving v must be $(v, f(v), f^2(v), \ldots, f^i(v) = v)$. So each cyclic vertex (if any) belongs to a unique cycle of G. This implies that distinct cycles of G involve disjoint sets of vertices and edges.

Next we define a surjection $r : V \to C$. The existence of r will show that $C \neq \emptyset$, since $V \neq \emptyset$. Fix $u \in V$. By repeatedly following outgoing arrows, we obtain for each $k \geq 0$ a unique walk $(u = u_0, u_1, u_2, \ldots, u_k)$ in G of length k. Since V is finite, there must exist $i < j$ with $u_i = u_j$. Take i minimal and then j minimal with this property; then $(u_i, u_{i+1}, \ldots, u_j)$ is a cycle in G. We define $r(u) = u_i$, which is the first element on this cycle reached from u. It can be checked that $r(u) = u$ for all $u \in C$; this implies that r is surjective. On the other hand, if $u \notin C$, the definition of r shows that $r(u) = r(u_1) = r(f(u))$.

How shall we construct a set partition with the stated properties? For each $v \in C$, consider the *fiber* $S_v = r^{-1}(\{v\}) = \{w \in V : r(w) = v\}$; then $\{S_v : v \in C\}$ is a set partition of V indexed by C. The remarks at the end of the last paragraph show that this set partition satisfies (i) and (ii). To check (iii) for fixed $v \in C$, first note that the map g defined in (iii) does map S_v into S_v by (i) and (ii). Suppose $W = (w_0, w_1, \ldots, w_k)$ is a cycle in the functional digraph for g. Since $r(w_0) = v$, we will eventually reach v by following outgoing arrows starting at w_0. On the other hand, following these arrows keeps us on the cycle W, so some $w_i = v$. Since $g(v) = v$, the only possibility is that W is the 1-cycle (v). Thus (iii) holds for each $v \in C$.

To see that $\{S_v : v \in C\}$ is unique, let $\mathcal{P} = \{T_v : v \in C\}$ be another set partition with properties (i), (ii), and (iii). It is enough to show that $S_v \subseteq T_v$ for each $v \in C$. Fix $v \in C$ and $z \in S_v$. By (ii), every element in the sequence $(z, f(z), f^2(z), \ldots)$ belongs to the same set of $\mathcal{P}$, say T_w. Then $v = r(z) = f^i(z) \in T_w$, so (i) forces $w = v$. Thus $z \in T_v$ as needed. □

We can informally summarize the previous result by saying that *every functional digraph uniquely decomposes into disjoint rooted trees feeding into one or more disjoint cycles.* There are two extreme cases of this decomposition that are especially interesting — the case where there are no trees (i.e., $C = V$), and the case where the whole digraph is a rooted tree (i.e., $|C| = 1$). We study these types of functional digraphs in the next two sections.

3.6 Cycle Structure of Permutations

The functional digraph of a bijection (permutation) has special structure, as we see in the next example.

3.44. Example. Figure 3.6 displays the digraph associated to the following bijection:

$$h(1) = 7; \quad h(2) = 8; \quad h(3) = 4; \quad h(4) = 3; \quad h(5) = 10;$$
$$h(6) = 2; \quad h(7) = 5; \quad h(8) = 6; \quad h(9) = 9; \quad h(10) = 1.$$

We see that the digraph for h contains only cycles; there are no trees feeding into these cycles. To see why this happens, compare this digraph to the digraph for the non-bijective function f in Figure 3.5. The digraph for f has a rooted tree feeding into the cyclic vertex 15. Accordingly, f is not injective since $f(17) = 15 = f(1)$. Similarly, if we move backward through the trees in the digraph of f, we reach vertices with indegree zero (namely 2, 7, 9, 11, 13, and 14). The existence of such vertices shows that f is not surjective. Returning to

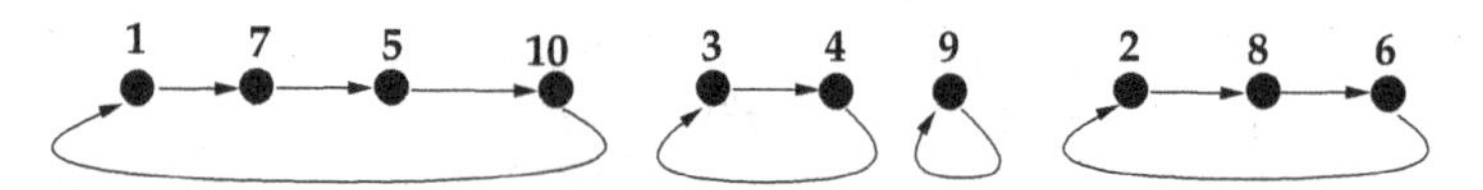

FIGURE 3.6
Digraph associated to a permutation.

the digraph of h, consider what happens if we reverse the direction of all the edges in the digraph. We obtain another functional digraph corresponding to the following function:

$$h'(1) = 10; \quad h'(2) = 6; \quad h'(3) = 4; \quad h'(4) = 3; \quad h'(5) = 7;$$
$$h'(6) = 8; \quad h'(7) = 1; \quad h'(8) = 2; \quad h'(9) = 9; \quad h'(10) = 5.$$

One sees immediately that h' is the two-sided inverse for h, so that h is bijective.

The next theorem explains the observations in the last example.

3.45. Theorem: Cycle Decomposition of Permutations. Let $f : V \to V$ be a function with functional digraph G. The map f is a bijection iff every $v \in V$ is a cyclic vertex in V. In this situation, G is a disjoint union of cycles.

Proof. Suppose $u \in V$ is a non-cyclic vertex. By Theorem 3.43, u belongs to a rooted tree S_v whose root v belongs to a cycle of G. Following edges outward from u will eventually lead to v; let y be the vertex in S_v just before v on this path. Let z be the vertex just before v in the unique cycle involving v. We have $y \neq z$, but $f(y) = v = f(z)$. Thus, f is not injective.

Conversely, suppose all vertices in V are cyclic. Then the digraph G is a disjoint union of directed cycles. So every $v \in V$ has indegree 1 as well as outdegree 1. Reversing the direction of every edge in G therefore produces another *functional* digraph G'. Let $f' : V \to V$ be the function associated to this new digraph. For all $a, b \in V$, we have $b = f(a)$ iff $(a, b) \in E(G)$ iff $(b, a) \in E(G')$ iff $a = f'(b)$. It follows that f' is the two-sided inverse for f, so that f and f' are both bijections. □

Recall that $S(n, k)$, the Stirling number of the second kind, is the number of set partitions of an n-element set into k blocks. Let $c(n, k)$ be the number of permutations of an n-element set whose functional digraph consists of k disjoint cycles. We will show that $c(n, k) = s'(n, k)$, the signless Stirling number of the first kind. Recall from Recursion 2.58 that the numbers $s'(n, k)$ satisfy $s'(n, k) = s'(n-1, k-1) + (n-1)s'(n-1, k)$ for $0 < k < n$, with initial conditions $s'(n, 0) = 0$ for $n > 0$ and $s'(n, n) = 1$ for $n \geq 0$.

3.46. Theorem: Recursion for $c(n, k)$. For $0 < k < n$, we have

$$c(n, k) = c(n-1, k-1) + (n-1)c(n-1, k).$$

The initial conditions are $c(n, 0) = 0$ for $n > 0$ and $c(n, n) = 1$ for $n \geq 0$. Therefore, $c(n, k) = s'(n, k)$ for $0 \leq k \leq n$.

Proof. The identity map is the unique permutation of an n-element set with n cycles (which must each have length 1), so $c(n, n) = 1$ for $n \geq 0$. The only permutation with zero cycles is the empty function on the empty set, so $c(n, 0) = 0$ for $n > 0$. Now suppose $0 < k < n$. Let A, B, C be the sets of permutations counted by $c(n, k)$, $c(n-1, k-1)$, and $c(n-1, k)$, respectively. Note that A is the disjoint union of the two sets

$$A_1 = \{f \in A : f(n) = n\} \text{ and } A_2 = \{f \in A : f(n) \neq n\}.$$

For each $f \in A_1$, we can restrict f to the domain $\{1, 2, \ldots, n-1\}$ to obtain a permutation of these $n-1$ elements. Since f has k cycles, one of which involves n alone, the restriction of f must have $k-1$ cycles. Since $f \in A_1$ is uniquely determined by its restriction to $\{1, 2, \ldots, n-1\}$, we have a bijection from A_1 onto B.

On the other hand, let us build a typical element $f \in A_2$ by making two choices. First, choose a permutation $g \in C$ in $c(n-1, k)$ ways. Second, choose an element $i \in \{1, 2, \ldots, n-1\}$ in $n-1$ ways. Let j be the unique number such that $g(j) = i$. Modify the digraph for g by removing the arrow from j to i and replacing it by an arrow from j to n and an arrow from n to i. Informally, we are splicing n into the cycle just before i. Let f be the permutation associated to the new digraph. Evidently, the splicing process does not change the number of cycles of g, and f satisfies $f(n) \neq n$. Thus, $f \in A_2$, and every element of A_2 arises uniquely by the choice process we have described. By the Sum Rule and Product Rule,

$$c(n, k) = |A| = |A_1| + |A_2| = c(n-1, k-1) + (n-1)c(n-1, k).$$

So $c(n, k)$ and $s'(n, k)$ satisfy the same recursion and initial conditions. A routine induction proof now shows that $c(n, k) = s'(n, k)$ for all integers $n, k \geq 0$. □

3.7 Counting Rooted Trees

Our goal in this section is to count rooted trees (see Definition 3.42) with a fixed root vertex.

3.47. The Rooted Tree Rule. For all $n > 1$, there are n^{n-2} rooted trees on the vertex set $\{1, 2, \ldots, n\}$ with root 1.

Proof. Let B be the set of rooted trees mentioned in the theorem. Let A be the set of all functions $f : \{1, 2, \ldots, n\} \to \{1, 2, \ldots, n\}$ such that $f(1) = 1$ and $f(n) = n$. The Product Rule shows that $|A| = n^{n-2}$. It therefore suffices to define maps $\phi : A \to B$ and $\phi' : B \to A$ that are mutual inverses. To define ϕ, fix $f \in A$. Let G_f be the functional digraph associated with f, which has directed edges $(i, f(i))$ for $1 \leq i \leq n$. By Theorem 3.43, we can decompose the vertex set $\{1, 2, \ldots, n\}$ of this digraph into some disjoint cycles $C_0, C_1, \ldots, C_k$ and (possibly) some trees feeding into these cycles. For $0 \leq i \leq k$, let ℓ_i be the largest vertex in cycle C_i, and write $C_i = (r_i, \ldots, \ell_i)$. We can choose the indexing of the cycles so that the numbers ℓ_i satisfy $\ell_0 > \ell_1 > \ell_2 > \cdots > \ell_k$. Since $f(1) = 1$ and $f(n) = n$, 1 and n belong to cycles of length 1, so that $\ell_0 = r_0 = n$, $C_0 = (n)$, $\ell_k = r_k = 1$, $C_k = (1)$, and $k > 0$. To obtain $\phi(f)$, we modify the digraph G_f by removing all edges of the form (ℓ_i, r_i) and adding new edges (ℓ_i, r_{i+1}), for $0 \leq i < k$. It can be checked that $\phi(f)$ is always a rooted tree with root 1.

3.48. Example. Suppose $n = 20$ and f is the function defined as follows:

$$\begin{array}{lllll} f(1) = 1; & f(2) = 19; & f(3) = 8; & f(4) = 17; & f(5) = 5; \\ f(6) = 5; & f(7) = 4; & f(8) = 3; & f(9) = 6; & f(10) = 1; \\ f(11) = 18; & f(12) = 4; & f(13) = 18; & f(14) = 20; & f(15) = 15; \\ f(16) = 1; & f(17) = 12; & f(18) = 4; & f(19) = 20; & f(20) = 20. \end{array}$$

We draw the digraph of f in such a way that all vertices involved in cycles occur in a horizontal row at the bottom of the figure, and the largest element in each cycle is the rightmost element of its cycle. We arrange these cycles so that these largest elements decrease from

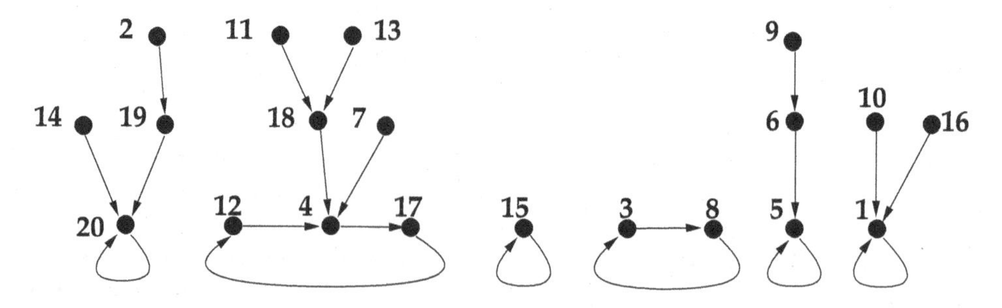

FIGURE 3.7
A functional digraph with cycles arranged in canonical order.

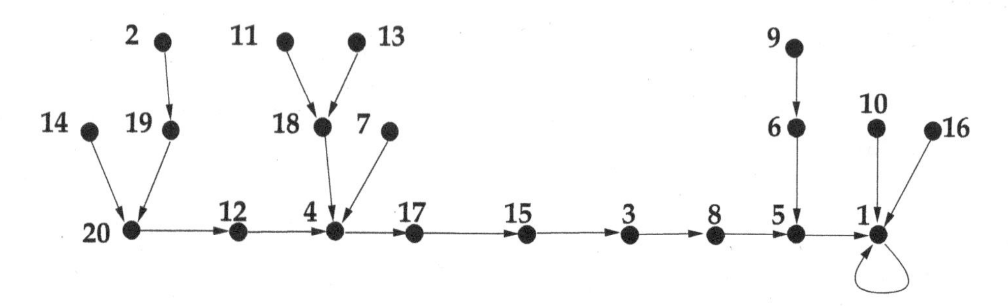

FIGURE 3.8
Conversion of the digraph to a rooted tree.

left to right; in particular, vertex n is always at the far left, and vertex 1 at the far right. See Figure 3.7. To compute $\phi(f)$, we cut the back-edges leading left from ℓ_i to r_i (which are loops if $\ell_i = r_i$) and add new edges leading right from ℓ_i to r_{i+1}. See Figure 3.8.

Continuing the proof, let us see why ϕ is invertible. Let T be a rooted tree on $\{1, 2, \ldots, n\}$ with root 1. Following outgoing edges from n must eventually lead to the unique cyclic vertex 1. Let $P = (v_0, v_1, \ldots, v_s)$ be the vertices encountered on the way from $v_0 = n$ to $v_s = 1$. We recursively recover the numbers $\ell_0, \ell_1, \ldots, \ell_k$ as follows. Let $\ell_0 = n$. Define ℓ_1 to be the largest number in P following ℓ_0. In general, after ℓ_{i-1} has been found, define ℓ_i to be the largest number in P following ℓ_{i-1}. After finitely many steps, we will get $\ell_k = 1$ for some k. Next, let $r_0 = n$, and for $i > 0$, let r_i be the vertex immediately following ℓ_{i-1} on the path P. Modify T by deleting the edges (ℓ_i, r_{i+1}) and adding edges of the form (ℓ_i, r_i), for $0 \leq i < k$. It can be verified that every vertex in the resulting digraph G' has outdegree exactly 1, and there are loop edges in G' at vertex 1 and vertex n. Thus, G' is a functional digraph that determines a function $f = \phi'(T) \in A$. It follows from the definition of ϕ' that ϕ' is the two-sided inverse of ϕ. □

3.49. Example. Suppose $n = 9$ and T is the rooted tree shown on the left in Figure 3.9. We first redraw the picture of T so that the vertices on the path P from n to 1 occur in a horizontal row at the bottom of the picture, with n on the left and 1 on the right. We recover ℓ_i and r_i by the procedure above, and then delete the appropriate edges of T and add appropriate back-edges to create cycles. The resulting functional digraph appears on

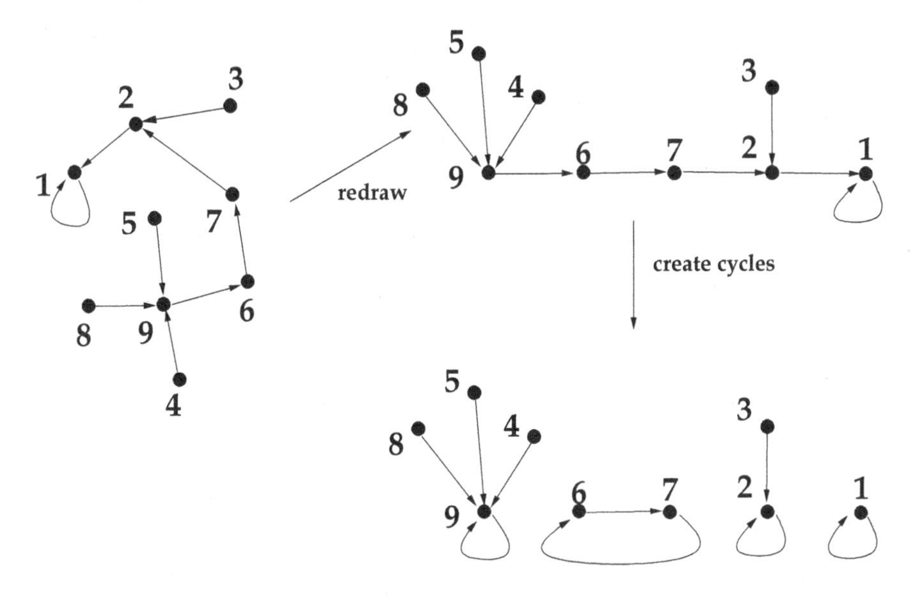

FIGURE 3.9
Conversion of a rooted tree to a functional digraph.

the bottom right in Figure 3.9. So $\phi'(T)$ is the function g defined as follows:

$$g(1) = 1; \quad g(2) = 2; \quad g(3) = 2; \quad g(4) = 9; \quad g(5) = 9;$$
$$g(6) = 7; \quad g(7) = 6; \quad g(8) = 9; \quad g(9) = 9.$$

3.8 Connectedness and Components

In many applications of graphs, we need to know whether every vertex is reachable from every other vertex.

3.50. Definition: Connectedness. Let $G = (V, E, \epsilon)$ be a graph or digraph. G is *connected* iff for all $u, v \in V$, there is a walk in G from u to v.

3.51. Example. The graph G_1 in Figure 3.1 is connected, but the simple graph G_2 and the digraph G_3 in that figure are not connected.

Connectedness can also be described using paths instead of walks.

3.52. Theorem: Walks and Paths. Let $G = (V, E, \epsilon)$ be a graph or digraph, and let $u, v \in V$. There is a walk in G from u to v iff there is a path in G from u to v.

Proof. Let $W = (v_0, e_1, v_1, \ldots, e_s, v_s)$ be a walk in G from u to v. We describe an algorithm to convert the walk W into a path from u to v. If all the vertices v_i are distinct, then the edges e_i must also be distinct, so W is already a path. Otherwise, choose i minimal such that v_i appears more than once in W, and then choose j maximal such that $v_i = v_j$. Then $W_1 = (v_0, e_1, v_1, \ldots, e_i, v_i, e_{j+1}, v_{j+1}, \ldots, e_s, v_s)$ is a walk from u to v of shorter length than

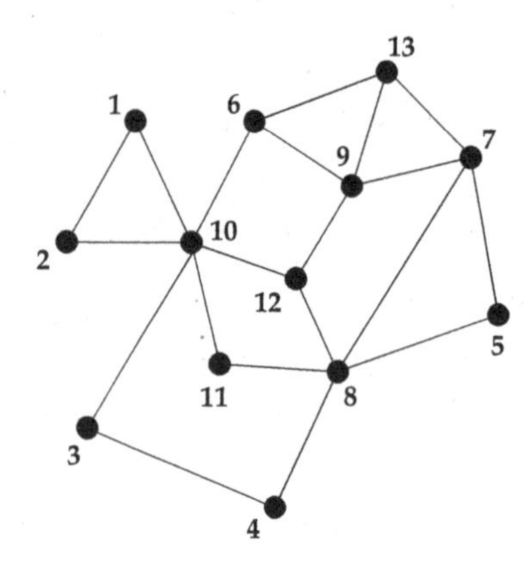

FIGURE 3.10
Converting a walk to a path.

W. If W_1 is a path, we are done. Otherwise, we repeat the argument to obtain a walk W_2 from u to v that is shorter than W_1. Since the lengths keep decreasing, this process must eventually terminate with a path W_k from u to v. (W_k has length zero if $u = v$.) The converse is immediate, since every path in G from u to v is a walk in G from u to v. □

3.53. Example. In the simple graph shown in Figure 3.10, consider the walk

$$W = (11, 10, 1, 2, 10, 3, 4, 8, 11, 8, 12, 10, 6, 9, 7, 13, 9, 12, 8, 5).$$

First, the repetition $v_0 = 11 = v_8$ leads to the walk

$$W_1 = (11, 8, 12, 10, 6, 9, 7, 13, 9, 12, 8, 5).$$

Eliminating the multiple visits to vertex 8 leads to the walk

$$W_2 = (11, 8, 5).$$

W_2 is a path from 11 to 5.

3.54. Corollary: Connectedness and Paths. A graph or digraph $G = (V, E, \epsilon)$ is connected iff for all $u, v \in V$, there is *at least one path* in G from u to v.

By looking at pictures of graphs, it becomes visually evident that any graph decomposes into a disjoint union of connected pieces, with no edge joining vertices in two separate pieces. These pieces are called the (connected) components of the graph. The situation for digraphs is more complicated, since there may exist directed edges between different components. To give a formal development of these ideas, we introduce the following equivalence relation.

3.55. Definition: Interconnection Relation. Let $G = (V, E, \epsilon)$ be a graph or digraph. Define a binary relation $\leftrightarrow_G$ on the vertex set V by setting $u \leftrightarrow_G v$ iff there exist walks in G from u to v and from v to u.

In the case of *graphs*, note that $u \leftrightarrow_G w$ iff there is a walk in G from u to w, since the reversal of such a walk is a walk in G from w to u. Now, for a graph or digraph G, let us verify that $\leftrightarrow_G$ is indeed an equivalence relation on V. First, for all $u \in V$, (u) is a walk of length 0 from u to u, so $u \leftrightarrow_G u$, and $\leftrightarrow_G$ is reflexive on V. Second, the symmetry of $\leftrightarrow_G$ is automatic from the way we defined $\leftrightarrow_G$: $u \leftrightarrow_G v$ implies $v \leftrightarrow_G u$ for all $u, v \in V$. Finally, to check transitivity, suppose $u, v, w \in V$ satisfy $u \leftrightarrow_G v$ and $v \leftrightarrow_G w$. Let W_1, W_2, W_3, and W_4 be walks in G from u to v, from v to w, from v to u, and from w to v, respectively. Then the concatenation of W_1 followed by W_2 is a walk in G from u to w, whereas the concatenation of W_4 followed by W_3 is a walk in G from w to u. Hence $u \leftrightarrow_G w$, as needed.

3.56. Definition: Components. Let $G = (V, E, \epsilon)$ be a graph or digraph. The *components* of G are the equivalence classes of the interconnection equivalence relation $\leftrightarrow_G$. Components are also called *connected components* or (in the case of digraphs) *strong components.*

Since $\leftrightarrow_G$ is an equivalence relation on V, the components of G form a set partition of the vertex set V. Given a component C of G, consider the graph or digraph (C, E', ϵ') obtained by retaining those edges in E with both endpoints in C and restricting ϵ to this set of edges. One may check that this graph or digraph is connected.

3.57. Example. The components of the graph G_2 in Figure 3.1 are $\{0\}$ and $\{1, 2, 3, 4, 5, 6\}$. The components of the digraph G_3 in that figure are $\{1, 2, 4, 5\}$, $\{3\}$, $\{6\}$, and $\{7\}$.

The next theorems describe how the addition or deletion of an edge affects the components of a graph.

3.58. Theorem: Edge Deletion and Components. Let $G = (V, E, \epsilon)$ be a graph with components $\{C_i : i \in I\}$. Let $e \in E$ be an edge with endpoints $v, w \in C_j$. Let $G' = (V, E', \epsilon')$ where $E' = E - \{e\}$ and ϵ' is the restriction of ϵ to E'. (a) If e appears in some cycle of G, then G and G' have the same components. (b) If e appears in no cycle of G, then G' has one more component than G. More precisely, the components of G' are the C_k with $k \neq j$, together with two disjoint sets A and B such that $A \cup B = C_j$, $v \in A$, and $w \in B$.

Proof. For (a), let $(v_0, e_1, v_1, e_2, \ldots, v_s)$ be a cycle of G containing e. Cyclically shifting and reversing the cycle if needed, we can assume $v_0 = v = v_s$, $e_1 = e$, and $v_1 = w$. Statement (a) will follow if we can show that the interconnection relations $\leftrightarrow_G$ and $\leftrightarrow_{G'}$ coincide. First, for all $y, z \in V$, $y \leftrightarrow_{G'} z$ implies $y \leftrightarrow_G z$ since every walk in the smaller graph G' is also a walk in G. On the other hand, does $y \leftrightarrow_G z$ imply $y \leftrightarrow_{G'} z$? We know there is a walk W from y to z in G. If W does not use the edge e, W is a walk from y to z in G'. Otherwise, we can modify W as follows. Every time W goes from $v = v_s = v_0$ to $w = v_1$ via e, replace this part of the walk by the sequence $(v_s, e_s, \ldots, e_2, v_1)$ obtained by taking a detour around the cycle. Make a similar modification each time W goes from w to v via e. This produces a walk in G' from y to z.

For (b), let us compute the equivalence classes of $\leftrightarrow_{G'}$. First, fix $z \in C_k$ where $k \neq j$. The set C_k consists of all vertices in V reachable from z by walks in G. It can be checked that none of these walks can use the edge e, so C_k is also the set of all vertices in V reachable from z by walks in G'. So C_k is the equivalence class of z relative to both $\leftrightarrow_G$ and $\leftrightarrow_{G'}$.

Next, let A and B be the equivalence classes of v and w (respectively) relative to $\leftrightarrow_{G'}$. By definition, A and B are two of the components of G' (possibly the same component). We now show that A and B are disjoint and that their union is C_j. If the equivalence classes A and B are not disjoint, then they must be equal. By Theorem 3.52, there must be a path $(v_0, e_1, v_1, \ldots, e_s, v_s)$ in G' from v to w. Appending e, v_0 to this path would produce a cycle in G involving the edge e, which is a contradiction. Thus A and B are disjoint.

We now show that $A \cup B \subseteq C_j$. If $z \in A$, then there is a walk in G' (and hence in G) from v to z. Since C_j is the equivalence class of v relative to $\leftrightarrow_G$, it follows that $z \in C_j$. Similarly, $z \in B$ implies $z \in C_j$ since C_j is also the equivalence class of w relative to $\leftrightarrow_G$. Next, we check that $C_j \subseteq A \cup B$. Let $z \in C_j$, and let $W = (w_0, e_1, w_1, \ldots, w_t)$ be a walk in G from v to z. If W does not use the edge e, then $z \in A$. If W does use e, then the portion of W following the last appearance of the edge e is a walk from either v or w to z in G';

thus $z \in A \cup B$. Since the union of A, B, and the C_k with $k \neq j$ is all of V, we have found all the components of G'. □

The previous result suggests the following terminology.

3.59. Definition: Cut-Edges. An edge e in a graph G is a *cut-edge* iff e does not appear in any cycle of G.

3.60. Theorem: Edge Addition and Components. Let $G = (V, E, \epsilon)$ be a graph with components $\{C_i : i \in I\}$. Let $G^+ = (V, E^+, \epsilon^+)$ be the graph obtained from G by adding a new edge e with endpoints $v \in C_j$ and $w \in C_k$. (a) If v and w are in the same component C_j of G, then e is involved in a cycle of G^+, and G and G^+ have the same components. (b) If v and w are in different components of G, then e is a cut-edge of G^+, and the components of G^+ are $C_j \cup C_k$ and the C_i with $i \neq j, k$.

This theorem follows readily from Theorem 3.58, so we omit the proof.

3.9 Forests

3.61. Definition: Forests. A *forest* is a graph with no cycles. Such a graph is also called *acyclic.*

A forest cannot have any loops or multiple edges between the same two vertices. So we can assume, with no loss of generality, that forests are *simple* graphs.

3.62. Example. Figure 3.11 displays a forest.

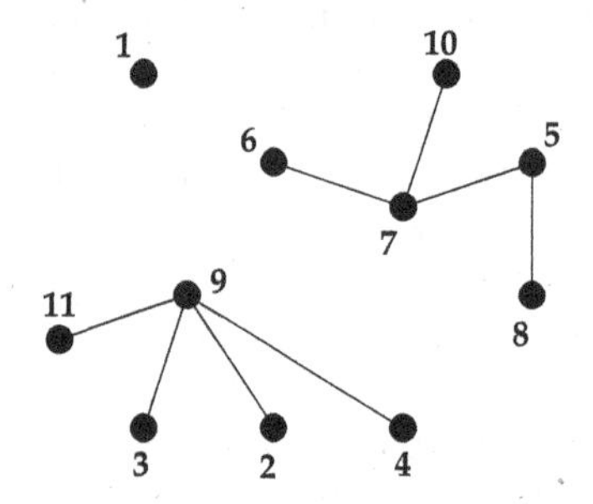

FIGURE 3.11
A forest.

Recall from Corollary 3.54 that a graph G is connected iff there exists *at least one* path between any two vertices of G. The next result gives an analogous characterization of forests.

3.63. Theorem: Forests and Paths. A graph G is acyclic iff G has no loops and for all u, v in $V(G)$, there is *at most one* path from u to v in G.

Proof. We prove the contrapositive in both directions. First suppose that G has a cycle $C = (v_0, e_1, v_1, \ldots, e_s, v_s)$. If $s = 1$, G has a loop. If $s > 1$, then $(v_1, e_2, \ldots, e_s, v_s)$ and (v_1, e_1, v_0) are two distinct paths in G from v_1 to v_0.

For the converse, we may assume G is simple since non-simple graphs must have cycles of length 1 or 2. The simple graph G has no loops, so we can assume that for some $u, v \in V(G)$, there exist two distinct paths P and Q from u to v in G. Among all such choices of u, v, P, and Q, choose one for which the path P has minimum length. Let P visit the sequence of vertices $(x_0, x_1, \ldots, x_s)$, and let Q visit the sequence of vertices $(y_0, y_1, \ldots, y_t)$, where $x_0 = u = y_0$, $x_s = v = y_t$, and s is minimal. We must prove G has a cycle.

First note that $s > 0$; otherwise $u = v$ and Q would not be a path. Second, we assert that $x_1 \neq y_1$; otherwise, we would have two distinct paths $P' = (x_1, \ldots, x_s)$ and $Q' = (y_1, \ldots, y_t)$ from x_1 to v with P' shorter than P, contradicting minimality of P. More generally, we claim that $x_i \neq y_j$ for all i, j satisfying $1 \leq i < s$ and $1 \leq j < t$. For if we had $x_i = y_j$ for some i, j in this range, then $P'' = (x_0, x_1, \ldots, x_i)$ and $Q'' = (y_0, y_1, \ldots, y_j)$ would be two paths from u to x_i with P'' shorter than P, and $P'' \neq Q''$ since $x_1 \neq y_1$. This again contradicts minimality of P. Since $x_s = v = y_t$ and $x_0 = u = y_0$, it now follows that

$$(x_0, x_1, \ldots, x_s, y_{t-1}, y_{t-2}, \ldots, y_1, y_0)$$

is a cycle in G. □

The following result gives a formula for the number of components in a forest.

3.64. Theorem: Components of a Forest. Let G be a forest with n vertices and k edges. The number of connected components of G is $n - k$.

Proof. We use induction on k. The result holds for $k = 0$, since G consists of n isolated vertices in this case. Assume that $k > 0$ and the result is already known for forests with n vertices and $k - 1$ edges. Given a forest G with n vertices and k edges, remove one edge e from G to get a new graph H. The graph H is acyclic and has n vertices and $k - 1$ edges. By induction, H has $n - (k - 1) = n - k + 1$ components. On the other hand, e must be a cut-edge since G has no cycles. It follows from Theorem 3.58 that H has one more component than G. Thus, G has $n - k$ components, as needed. □

3.10 Trees

3.65. Definition: Trees. A *tree* is a connected graph with no cycles.

For example, Figure 3.12 displays a tree.

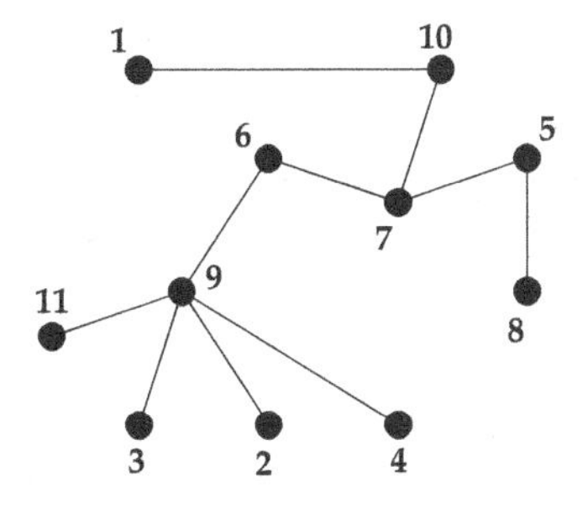

FIGURE 3.12
A tree.

Every component of a forest is a tree, so every forest is a disjoint union of trees. The next result is an immediate consequence of the Two-Leaf Lemma 3.37.

3.66. Theorem: Trees Have Leaves. If T is a tree with more than one vertex, then T has at least two leaves.

3.67. Definition: Pruning. Suppose $G = (V, E)$ is a simple graph, v_0 is a leaf in G, and e_0 is the unique edge incident to the vertex v_0. The *graph obtained by pruning* v_0 *from* G is the graph $(V - \{v_0\}, E - \{e_0\})$.

3.68. Pruning Lemma. If T is an n-vertex tree, v_0 is a leaf of T, and T' is obtained from T by pruning v_0, then T' is a tree with $n-1$ vertices.

Proof. First, T has no cycles, and the deletion of v_0 and the associated edge e_0 will not create any cycles. So T' is acyclic. Second, let $u, w \in V(T')$. There is a path from u to w in T. Since $u \neq v_0 \neq w$, this path will not use the edge e_0 or the vertex v_0. Thus there is a path from u to w in T', so T' is connected. □

Figure 3.13 illustrates the Pruning Lemma.

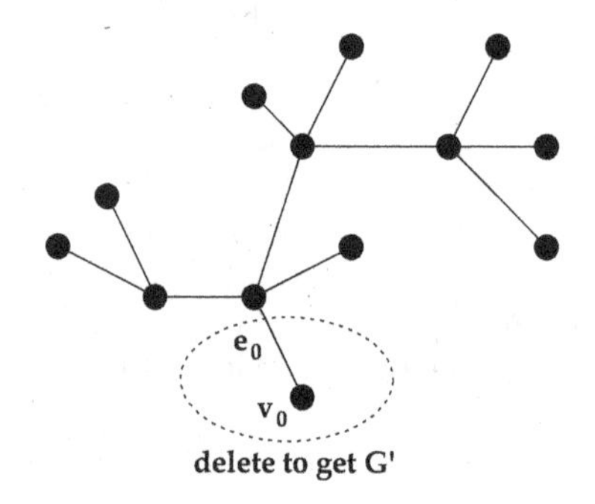

FIGURE 3.13
Pruning a leaf from a tree produces another tree.

To give an application of pruning, we now prove a fundamental relationship between the number of vertices and edges in a tree (this also follows from Theorem 3.64).

3.69. Theorem: Number of Edges in a Tree. If G is a tree with $n > 0$ vertices, then G has $n-1$ edges.

Proof. We use induction on n. If $n = 1$, then G must have 0 edges. Fix $n > 1$, and assume that the result holds for trees with $n-1$ vertices. Let T be a tree with n vertices. We know that T has at least one leaf; let v_0 be one such leaf. Let T' be the graph obtained by pruning v_0 from T. By the Pruning Lemma, T' is a tree with $n-1$ vertices. By induction, T' has $n-2$ edges. Hence, T has $n-1$ edges. □

3.70. Theorem: Characterizations of Trees. Let G be a graph with n vertices. The following conditions are logically equivalent:

(a) G is a tree (i.e., G is connected and acyclic).

(b) G is connected and has at most $n-1$ edges.

(c) G is acyclic and has at least $n-1$ edges.

(d) G has no loop edges, and for all $u, v \in V(G)$, there is a *unique* path in G from u to v.

Moreover, when these conditions hold, G has exactly $n-1$ edges.

Proof. First, (a) implies (b) and (a) implies (c) by Theorem 3.69. Second, (a) is equivalent to (d) by Corollary 3.54 and Theorem 3.63. Third, we prove (b) implies (a). Assume G is connected with $k \leq n-1$ edges. If G has a cycle, delete one edge on some cycle of G. The resulting graph is still connected (by Theorem 3.58) and has $k-1$ edges. Continue to delete edges in this way, one at a time, until there are no cycles. If we deleted i edges total, the resulting graph is a tree with $k-i \leq n-1-i$ edges and n vertices. By Theorem 3.69, we must have $i=0$ and $k=n-1$. So no edges were deleted, which means that the original graph G is in fact a tree.

Fourth, we prove (c) implies (a). Assume G is acyclic with $k \geq n-1$ edges. If G is not connected, add an edge joining two distinct components of G. The resulting graph is still acyclic (by Theorem 3.60) and has $k+1$ edges. Continue to add edges in this way, one at a time, until the graph becomes connected. If we added i edges total, the resulting graph is a tree with $k+i \geq n-1+i$ edges and n vertices. By Theorem 3.69, we must have $i=0$ and $k=n-1$. So no edges were added, which means that the original graph G is in fact a tree. □

3.11 Counting Trees

The next result, often attributed to Cayley, counts n-vertex trees.

3.71. The Tree Rule. For all $n \geq 1$, there are n^{n-2} trees with vertex set $\{1,2,\ldots,n\}$.

3.72. Example. Figure 3.14 displays all $4^{4-2}=16$ trees with vertex set $\{1,2,3,4\}$.

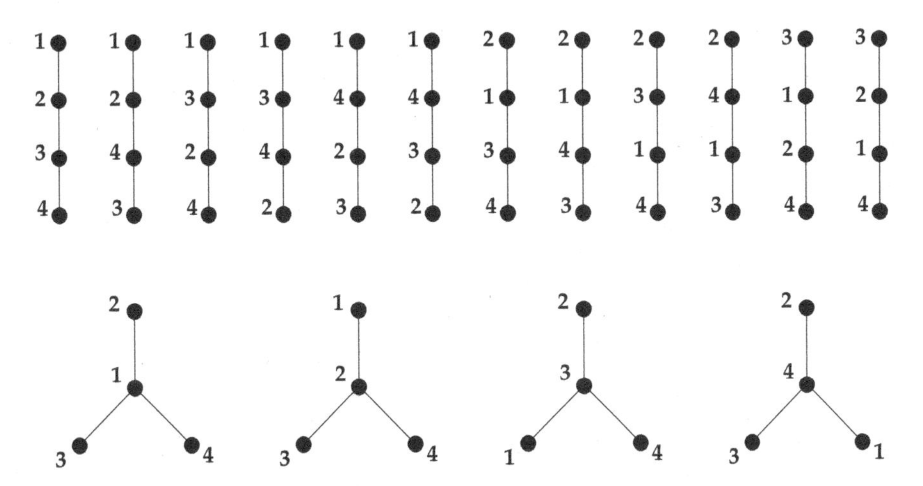

FIGURE 3.14
The 16 trees with vertex set $\{1,2,3,4\}$.

The Tree Rule 3.71 is an immediate consequence of the Rooted Tree Rule 3.47 and the following bijection.

3.73. Theorem: Trees and Rooted Trees. Let V be a finite set and $v_0 \in V$. There is a bijection from the set A of trees with vertex set V to the set B of rooted trees with vertex set V and root v_0.

Proof. We define maps $f : A \to B$ and $g : B \to A$ that are two-sided inverses. First, given $T = (V, E) \in A$, construct $f(T) = (V, E')$ as follows. For each $v \in V$ with $v \neq v_0$, there exists a unique path from v to v_0 in T. Letting $e = \{v, w\}$ be the first edge on this path, we add the directed edge (v, w) to E'. Also, we add the loop edge (v_0, v_0) to E'. Since T has no cycles, the only possible cycle in the resulting functional digraph $f(T)$ is the 1-cycle (v_0). It follows that $f(T)$ is a rooted tree on V with root v_0 (see Definition 3.42).

Next, given a rooted tree $S \in B$, define $g(S)$ by deleting the unique loop edge (v_0, v_0) and replacing every directed edge (v, w) by an undirected edge $\{v, w\}$. The resulting graph $g(S)$ has n vertices and $n-1$ edges. To see that $g(S)$ is connected, fix $y, z \in V$. Following outgoing edges from y (respectively z) in S produces a directed path from y (respectively z) to v_0 in S. In the undirected graph $g(S)$, we can concatenate the path from y to v_0 with the reverse of the path from z to v_0 to get a walk from y to z. It follows that $g(S)$ is a tree.

It is routine to check that $g \circ f = \mathrm{id}_A$, since f assigns a certain orientation to each edge of the original tree, and this orientation is then forgotten by g. It is somewhat less routine to verify that $f \circ g = \mathrm{id}_B$; we leave this as an exercise. The key point to be checked is that the edge orientations in $f(g(S))$ agree with the edge orientations in S, for each $S \in B$. □

A different bijective proof of the Tree Rule, based on parking functions, is presented in §12.5. We next prove a refinement of the Tree Rule that counts the number of trees such that each vertex has a specified degree. We give an algebraic proof first, and then convert this to a bijective proof in the next section.

3.74. Theorem: Counting Trees with Specified Degrees. Suppose $n \geq 2$ and $d_1, \ldots, d_n \geq 0$ are fixed integers. If $d_1 + \cdots + d_n = 2n - 2$, then there are

$$\binom{n-2}{d_1 - 1, d_2 - 1, \ldots, d_n - 1} = \frac{(n-2)!}{\prod_{j=1}^{n} (d_j - 1)!}$$

trees with vertex set $\{v_1, v_2, \ldots, v_n\}$ such that $\deg(v_j) = d_j$ for all j. If $d_1 + \cdots + d_n \neq 2n-2$, then there are no such trees.

Proof. The last statement holds because any tree T on n vertices has $n-1$ edges, and thus $\sum_{i=1}^{n} \deg(v_i) = 2(n-1)$. Assume henceforth that $d_1 + \cdots + d_n = 2n - 2$. We prove the result by induction on n. First consider the case $n = 2$. If $d_1 = d_2 = 1$, there is exactly one valid tree, and $\binom{n-2}{d_1-1, d_2-1} = 1$. For any other choice of d_1, d_2 adding to 2, there are no valid trees, and $\binom{n-2}{d_1-1, d_2-1} = 0$.

Now assume $n > 2$ and that the theorem is known to hold for trees with $n-1$ vertices. Let A be the set of trees T with $V(T) = \{v_1, \ldots, v_n\}$ and $\deg(v_j) = d_j$ for all j. If $d_j = 0$ for some j, then A is empty and the formula in the theorem is zero by convention. Now suppose $d_j > 0$ for all j. We must have $d_i = 1$ for some i, for otherwise $d_1 + \cdots + d_n \geq 2n > 2n - 2$. Fix an i with $d_i = 1$. Note that v_i is a leaf in T for every $T \in A$. Now, for each $k \neq i$ between 1 and n, define

$$A_k = \{T \in A : \{v_i, v_k\} \in E(T)\}.$$

A_k is the set of trees in A in which the leaf v_i is attached to the vertex v_k. A is the disjoint union of the sets A_k.

Fix $k \neq i$. Pruning the leaf v_i gives a bijection between A_k and the set B_k of all trees with vertex set $\{v_1, \ldots, v_{i-1}, v_{i+1}, \ldots, v_n\}$ such that $\deg(v_j) = d_j$ for $j \neq i, k$ and $\deg(v_k) = d_k - 1$. By induction hypothesis,

$$|B_k| = \frac{(n-3)!}{(d_k - 2)! \prod_{\substack{1 \leq j \leq n \\ j \neq i,k}} (d_j - 1)!}.$$

Therefore, by the Sum Rule and the Bijection Rule,

$$\begin{aligned}|A| &= \sum_{k\neq i}|A_k| = \sum_{k\neq i}|B_k| = \sum_{k\neq i}\frac{(n-3)!}{(d_k-2)!\prod_{j\neq k,i}(d_j-1)!}\\ &= \sum_{k\neq i}\frac{(n-3)!(d_k-1)}{\prod_{j\neq i}(d_j-1)!} = \sum_{k\neq i}\frac{(n-3)!(d_k-1)}{\prod_{j=1}^{n}(d_j-1)!}\\ &\qquad\text{(since } (d_i-1)! = 0! = 1)\\ &= \frac{(n-3)!}{\prod_{j=1}^{n}(d_j-1)!}\sum_{k\neq i}(d_k-1).\end{aligned}$$

Now, since $d_i = 1$, $\sum_{k\neq i}(d_k-1) = \sum_{k=1}^{n}(d_k-1) = (2n-2)-n = n-2$. Inserting this into the previous formula, we see that

$$|A| = \frac{(n-2)!}{\prod_{j=1}^{n}(d_j-1)!},$$

which completes the induction proof. □

3.75. Corollary: Second Proof of the Tree Rule. Let us sum the previous formula over all possible degree sequences $(d_1,\ldots,d_n)$. Making the change of variables $c_i = d_i - 1$ and invoking the Multinomial Theorem 2.11, we see that the total number of trees on this vertex set is

$$\begin{aligned}\sum_{\substack{d_1+\cdots+d_n=2n-2\\ d_i\geq 0}}\binom{n-2}{d_1-1,d_2-1,\ldots,d_n-1} &= \sum_{\substack{c_1+\cdots+c_n=n-2\\ c_i\geq 0}}\binom{n-2}{c_1,c_2,\ldots,c_n}1^{c_1}1^{c_2}\cdots 1^{c_n}\\ &= (1+1+\cdots+1)^{n-2} = n^{n-2}.\end{aligned}$$

3.12 Pruning Maps

We now develop a bijective proof of Theorem 3.74. Suppose $n \geq 2$ and $d_1,\ldots,d_n$ are positive integers that sum to $2n-2$. Let $V = \{v_1,\ldots,v_n\}$ be a vertex set consisting of n positive integers $v_1 < v_2 < \cdots < v_n$. Let A be the set of trees T with vertex set V such that $\deg(v_i) = d_i$ for all i. Let B be the set of words $\mathcal{R}(v_1^{d_1-1}v_2^{d_2-1}\cdots v_n^{d_n-1})$ (see Definition 1.32). Each word $w \in B$ has length $n-2$ and has $d_j - 1$ copies of v_j. To prove Theorem 3.74, it suffices to define a bijection $f : A \to B$.

Given a tree $T \in A$, we compute $f(T)$ by repeatedly pruning off the largest leaf of T, recording for each leaf the vertex adjacent to it in T. More formally, for i ranging from 1 to $n-1$, let x be the largest leaf of T; define w_i to be the unique neighbor of x in T; then modify T by pruning the leaf x. This process produces a word $w_1\cdots w_{n-1}$; we define $f(T) = w_1\cdots w_{n-2}$ (note that w_{n-1} is discarded).

3.76. Example. Let T be the tree shown in Figure 3.15. To compute $f(T)$, we prune leaves from T in the following order: $8,5,4,9,6,3,2,7$. Recording the neighbors of these leaves, we see that $w = f(T) = 9996727$. Observe that the algorithm computes $w_{n-1} = 1$, but this letter is not part of the output word w. Also observe that $w \in \mathcal{R}(1^0 2^1 3^0 4^0 5^0 6^1 7^2 8^0 9^3) = \mathcal{R}(1^{d_1-1}\cdots 9^{d_9-1})$.

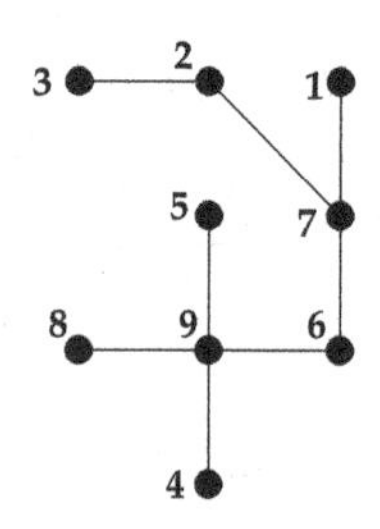

FIGURE 3.15
A tree with $(d_1, \ldots, d_9) = (1, 2, 1, 1, 1, 2, 3, 1, 4)$.

The observations in the last example hold in general. Given any $T \in A$, repeatedly pruning leaves from T will produce a sequence of smaller graphs, which are all trees by the Pruning Lemma 3.68. By Theorem 3.66, each such tree (except the last tree) has at least two leaves, so vertex v_1 will never be chosen for pruning. In particular, v_1 is always the last vertex left, so that w_{n-1} is always v_1. Furthermore, if v_j is any vertex different from v_1, then the number of occurrences of v_j in $w_1 w_2 \cdots w_{n-1}$ is exactly $d_j - 1$. For, every time a pruning operation removes an edge touching v_j, we set $w_i = v_j$ for some i, *except* when we are removing the last remaining edge touching v_j (which occurs when v_j has become the largest leaf and is being pruned). The same reasoning shows that v_1 (which never gets pruned) appears d_1 times in $w_1 \cdots w_{n-1}$. Since $w_{n-1} = v_1$, *every* vertex v_j occurs $d_j - 1$ times in the output word $w_1 \cdots w_{n-2}$.

To see that f is a bijection, we use induction on the number of vertices. The result holds when $n = 2$, since in this case, A consists of a single tree with two nodes, and B consists of a single word (the empty word). Now suppose $n > 2$ and the maps f (defined for trees with fewer than n vertices) are already known to be bijections. Given $w = w_1 \cdots w_{n-2} \in B$, we will show there exists exactly one $T \in A$ with $f(T) = w$. If such T exists, the leaves of T are precisely the vertices in $V(T)$ that do *not* appear in w. Thus, the first leaf that gets pruned when computing $f(T)$ must be the largest element z of $V(T) - \{w_1, \ldots, w_{n-2}\}$. By induction hypothesis, there exists exactly one tree T' on the vertex set $V(T) - \{z\}$ (with the appropriate vertex degrees) such that $f(T') = w_2 \cdots w_{n-2}$. This given, we will have $f(T) = w$ iff T is the tree obtained from T' by attaching a new leaf z as a neighbor of vertex w_1. One readily confirms that this graph *is* in A (i.e., the graph is a tree with the correct vertex degrees). This completes the induction proof. The proof also yields a recursive algorithm for computing $f^{-1}(w)$. The key point is to use the letters *not* seen in w (and its suffixes) to determine the identity of the leaf that was pruned at each stage.

3.77. Example. Given $w = 6799297$ and $V = \{1, 2, \ldots, 9\}$, let us compute the tree $f^{-1}(w)$ with vertex set V. The leaves of this tree must be $\{1, 3, 4, 5, 8\}$, which are the elements of V not seen in w. Leaf 8 was pruned first and was adjacent to vertex 6. So now we must compute the tree $f^{-1}(799297)$ with vertex set $V - \{8\}$. Here, leaf 6 was pruned first and was adjacent to vertex 7. Continuing in this way, we deduce that the leaves were pruned in the order $8, 6, 5, 4, 3, 2, 9, 7$; and the neighbors of these leaves (reading from w) were $6, 7, 9, 9, 2, 9, 7, 1$. Thus, $f^{-1}(w)$ is the tree shown in Figure 3.16.

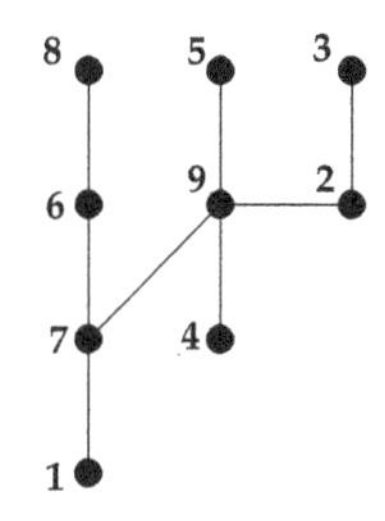

FIGURE 3.16
The tree computed by applying f^{-1} to $w = 6799297$.

3.13 Bipartite Graphs

Suppose m people are applying for n jobs, and we want to assign each person to a job that he or she is qualified to perform. We can model this situation with a graph in which there are two kinds of vertices: a set A of m vertices representing the people, and a disjoint set B of n vertices representing the jobs. For all $a \in A$ and $b \in B$, $\{a, b\}$ is an edge in the graph iff person a is qualified to perform job b. All edges in the graph join a vertex in A to a vertex in B. Graphs of this special type arise frequently; they are called bipartite graphs.

3.78. Definition: Bipartite Graphs. A graph G is *bipartite* iff there exist two sets A and B such that $A \cap B = \emptyset$, $A \cup B = V(G)$, and every edge of G has one endpoint in A and one endpoint in B. In this case, the sets A and B are called *partite sets* for G.

3.79. Example. The graph in Figure 3.3 is a bipartite graph with partite sets $A = \{1, 2, 3\}$ and $B = \{4, 5, 6\}$. This graph models the scenario where three people (labeled 1, 2, and 3) are all qualified to perform three jobs (labeled 4, 5, and 6). The graph G_1 in Figure 3.1 is not bipartite due to the loop edge f at vertex 3. However, deleting this edge gives a bipartite graph with partite sets $A = \{1, 3, 5\}$ and $B = \{2, 4\}$. The tree in Figure 3.12 is bipartite with partite sets $A = \{1, 7, 8, 9\}$ and $B = \{2, 3, 4, 5, 6, 10, 11\}$. These sets are also partite sets for the forest in Figure 3.11, but here we could use other partite sets such as $A' = \{1, 5, 6, 9, 10\}$ and $B' = \{2, 3, 4, 7, 8, 11\}$. Thus, the partite sets of a bipartite graph are not unique in general.

The next theorem gives a simple criterion for deciding whether a graph is bipartite.

3.80. Theorem: Bipartite Graphs. A graph G is bipartite iff G has no cycle of odd length iff G has no closed walk of odd length.

Proof. First assume G is a bipartite graph with partite sets A and B. Let $(v_0, e_1, v_1, e_2, v_2, \ldots, e_s, v_s)$ be any cycle in G; we must show that s is even. By switching A and B if needed, we may assume $v_0 \in A$. Since e_1 is an edge from v_0 to v_1, we must have $v_1 \in B$. Since e_2 is an edge from v_1 to v_2, we must have $v_2 \in A$. It follows by induction that for $0 \leq i \leq s$, $v_i \in A$ iff i is even, and $v_i \in B$ iff i is odd. Since $v_s = v_0$ is in A, we see that s must be even, as needed.

Second, we prove that if G has no cycle of odd length, then G has no closed walk of odd length. Using proof by contrapositive, assume there exists a closed walk of odd length in G; we prove there must exist a cycle of odd length in G. Choose a closed walk $(v_0, e_1, v_1, \ldots, e_s, v_s)$ of odd length with s as small as possible; we claim this walk must be a cycle. Otherwise, there would be an index i with $0 < i < s$ and $v_0 = v_i = v_s$. If i is odd, then $(v_0, e_1, v_1, \ldots, e_i, v_i)$ is a closed walk of odd length $i < s$, contradicting minimality of s. If i is even, then $(v_i, e_{i+1}, v_{i+1}, \ldots, e_s, v_s)$ is a closed walk of odd length $s - i < s$, contradicting minimality of s. Thus the original walk was already a cycle.

Third, we prove that if G has no closed walk of odd length, then G is bipartite. We begin with the special case where G is connected. Fix a vertex $v_0 \in V(G)$. Let A be the set of $v \in V(G)$ such that there is a walk of even length in G from v_0 to v; let B be the set of $w \in V(G)$ such that there is a walk of odd length in G from v_0 to w. Since G is connected, every vertex of G is reachable from v_0, so $V(G) = A \cup B$. We claim $A \cap B = \emptyset$. For if $v \in A \cap B$, there are walks W_1 and W_2 from v_0 to v where W_1 has even length and W_2 has odd length. The concatenation of W_1 with the reversal of W_2 would give a closed walk of odd length in G, contrary to our hypothesis on G. Next, we claim every edge of G must join a vertex in A to a vertex in B. For if $\{u, v\}$ is an edge where $u, v \in A$, then we get a closed walk of odd length by taking an even-length walk from v_0 to u, followed by the edge $\{u, v\}$, followed by an even-length walk from v to v_0. Similarly, we cannot have an edge $\{u, v\}$ with $u, v \in B$. We have now proved G is bipartite with partite sets A and B.

To treat the general case, let G have components $C_1, \ldots, C_m$. Since G has no closed walks of odd length, the same is true of each component. Thus, the argument in the last paragraph applies to each component, providing disjoint sets A_i, B_i such that $C_i = A_i \cup B_i$ and every edge with endpoints in C_i goes from a vertex in A_i to a vertex in B_i. It follows that $A = A_1 \cup \cdots \cup A_m$ and $B = B_1 \cup \cdots \cup B_m$ are partite sets for G, so G is bipartite. □

3.81. Example. The simple graph G_2 in Figure 3.1 is not bipartite because $(1, 2, 6, 1)$ is a cycle in G_2 of length 3.

3.14 Matchings and Vertex Covers

Consider again a job assignment graph where there is a set A of vertices representing people, a disjoint set B of vertices representing jobs, and an edge $\{a, b\}$ whenever person a is qualified for job b. We can model an assignment of people to jobs by a subset M of the edge set in which no two edges in M have a common endpoint. This condition means that each person can perform at most one of the jobs, and each job can be filled by at most one person. Ideally, we would like to choose M so that every person gets a job. If that is impossible, we would like to choose M as large as possible. This discussion motivates the following definition, which applies to any (not necessarily bipartite) graph.

3.82. Definition: Matchings. A *matching* of a graph G is a set M of edges of G such that no two edges in M share an endpoint. Let $m(G)$ be the maximum number of edges in any matching of G.

3.83. Example. The graph G in Figure 3.10 has a matching

$$M = \{\{1, 2\}, \{3, 10\}, \{4, 8\}, \{9, 12\}, \{6, 13\}, \{5, 7\}\}$$

of size 6. There can be no larger matching, since the graph has only 13 vertices. Thus, $m(G) = 6$.

3.84. Example. Consider the *path graph* P_n with vertex set $\{1, 2, \ldots, n\}$ and edge set $\{\{i, i+1\} : 1 \leq i < n\}$. Figure 3.17 shows the path graphs P_5 and P_6. We can write $n = 2k$ or $n = 2k+1$ for some integer $k \geq 0$. In either case, the set $M = \{\{1, 2\}, \{3, 4\}, \{5, 6\}, \ldots, \{2k-$

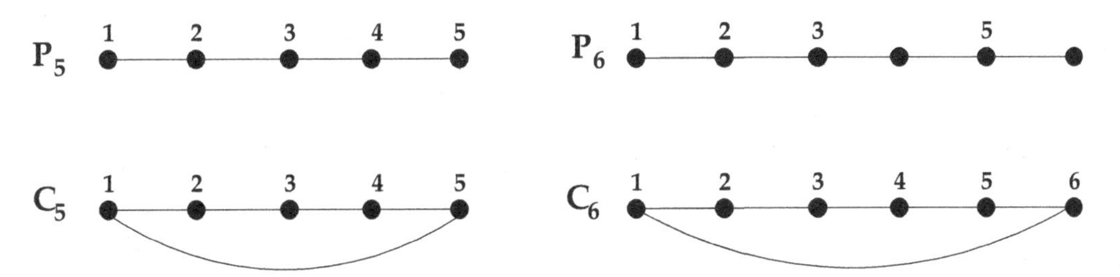

FIGURE 3.17
Path graphs and cycle graphs.

$1, 2k\}\}$ is a maximum matching of P_n, so $m(P_n) = k = \lfloor n/2 \rfloor$. Similarly, consider the *cycle graph* C_n obtained from P_n by adding the edge $\{1, n\}$. The matching M is also a maximum matching of C_n, so $m(C_n) = k = \lfloor n/2 \rfloor$.

The next definition introduces a new optimization problem that is closely related to the problem of finding a maximum matching in a graph.

3.85. Definition: Vertex Covers. A *vertex cover* of a graph G is a subset C of the vertex set of G such that every edge of G has at least one of its endpoints in C. Let $vc(G)$ be the minimum number of vertices in any vertex cover of G.

3.86. Example. The graph G in Figure 3.10 has vertex cover $C = \{1, 4, 7, 8, 9, 10, 13\}$. For $n = 2k$ or $n = 2k + 1$, the path graph P_n has vertex cover $D = \{2, 4, 6, \ldots, 2k\}$ of size $k = \lfloor n/2 \rfloor$. When $n = 2k$ is even, D is also a vertex cover of the cycle graph C_n. When $n = 2k + 1$ is odd, D is not a vertex cover of C_n, but $D \cup \{1\}$ is.

In the previous example, some additional argument is needed to determine whether we have found vertex covers of the minimum possible size. Similarly, in most cases, it is not immediately evident that a given matching of a graph has the maximum possible size. However, the next lemma shows that if we are able to build a matching M of G and a vertex cover C of G having the same size, then M must be a maximum matching and C must be a minimum vertex cover, so that $m(G) = |M| = |C| = vc(G)$.

3.87. Lemma: Matchings and Vertex Covers. Let G be a graph with matching M and vertex cover C. (a) $|M| \leq |C|$, and hence $m(G) \leq vc(G)$. (b) If $|M| = |C|$, then $m(G) = |M| = |C| = vc(G)$, and every vertex in C is an endpoint of some edge in M.

Proof. (a) We define a function $f : M \to C$ as follows. Given an edge $e \in M$, let v, w be the endpoints of this edge. Since C is a vertex cover, at least one of v or w must belong to C. Let $f(e) = v$ if $v \in C$, and $f(e) = w$ otherwise. We claim f is one-to-one. For suppose $e, e' \in M$ satisfy $f(e) = f(e')$. Then e and e' share a common endpoint $f(e) = f(e')$. Because M is a matching, this forces $e = e'$. So f is one-to-one, and hence $|M| \leq |C|$. Taking M to be a maximum matching and C to be a minimum vertex cover in this result, we deduce that $m(G) \leq vc(G)$.

(b) Suppose $|M| = |C|$. For any matching M' of G, part (a) shows that $|M'| \leq |C| = |M|$, so M is a maximum matching of G. For any vertex cover C' of G, part (a) shows that $|C'| \geq |M| = |C|$, so C is a minimum vertex cover of G. Thus, $m(G) = |M| = |C| = vc(G)$ in this situation. Moreover, because $|M| = |C|$, the injective map $f : M \to C$ constructed in part (a) must be bijective by Theorem 1.77. Surjectivity of f means that for any vertex $v \in C$, there is an edge $e \in M$ with $f(e) = v$. By definition of f, this means that each vertex in the minimum vertex cover C is an endpoint of some edge in the maximum matching M. □

3.88. Example. For the path graph P_n, we found a matching M and a vertex cover C of the same size. By the lemma, M is a maximum matching, C is a minimum vertex cover, and $m(P_n) = vc(P_n) = \lfloor n/2 \rfloor$. Similarly, for n even, we have $m(C_n) = vc(C_n) = n/2$. But, it can be checked that the odd cycle C_5 has $m(C_5) = 2$ and $vc(C_5) = 3$. So, for general graphs G, $m(G) < vc(G)$ can occur. We prove in the next section that for a *bipartite* graph G, $m(G)$ and $vc(G)$ are always equal.

3.15 Two Matching Theorems

3.89. The König–Egerváry Theorem. For all bipartite graphs G, $m(G) = vc(G)$.

Proof. Let G have vertex set V and edge set E. We proceed by strong induction on $|V|+|E|$. Fix a bipartite graph G with $N = |V| + |E|$, and assume that for any bipartite graph G' with $|V(G')| + |E(G')| < N$, $m(G') = vc(G')$. We will prove $m(G) = vc(G)$ by considering various cases. Case 1: G is isomorphic to a path graph P_n. The examples in the last section show that $m(G) = \lfloor n/2 \rfloor = vc(G)$ in this case. Case 2: G is isomorphic to a cycle graph C_n. Since bipartite graphs contain no odd cycles, n must be even, and we saw earlier that $m(G) = n/2 = vc(G)$ in this case.

Case 3: G is not simple. Now G has no loop edges (being bipartite), so there must exist multiple edges between some pair of vertices in G. Let G' be the bipartite graph obtained from G by deleting one of these multiple edges. By the induction hypothesis, $m(G') = vc(G')$. On the other hand, we see from the definitions that $m(G) = m(G')$ and $vc(G) = vc(G')$, so $m(G) = vc(G)$.

Case 4: G is not connected. Let $K_1, \ldots, K_m$ be the components of G. Form graphs $G_1, \ldots, G_m$ by letting G_i have vertex set K_i and edge set consisting of all edges in $E(G)$ with both endpoints in K_i. Since G is not connected, $m > 1$, so the induction hypothesis applies to the smaller graphs $G_1, \ldots, G_m$ (which are still bipartite). Let M_i be a maximum matching of G_i, and let C_i be a minimum vertex cover of G_i for each i; then $|M_i| = m(G_i) = vc(G_i) = |C_i|$ for all i. One immediately verifies that $M = \bigcup_{i=1}^{m} M_i$ is a matching of G of size $\sum_{i=1}^{m} |M_i|$, and $C = \bigcup_{i=1}^{m} C_i$ is a vertex cover of G of size $\sum_{i=1}^{m} |C_i|$. Since $|M| = |C|$, the Lemma on Matchings and Vertex Covers assures us that $m(G) = |M| = |C| = vc(G)$, as needed.

Case 5: G is simple and connected, but G is not isomorphic to a path P_n or a cycle C_n. It readily follows that there must exist a vertex u in G with $\deg(u) \geq 3$. Let v be a fixed vertex of G adjacent to u. Case 5a: Every maximum matching of G contains an edge with endpoint v. Let G' be the graph obtained from G by deleting vertex v and all edges having v as an endpoint. G' is still bipartite, and the induction hypothesis applies to show that $m(G') = vc(G')$. Let C' be a fixed minimum vertex cover of G'. Then $C = C' \cup \{v\}$ is a vertex cover of G, so $vc(G) \leq |C| = |C'| + 1 = vc(G') + 1 = m(G') + 1$. On the other hand, any maximum matching M' of G' is also a matching of G, but it cannot be a maximum matching of G by the assumption of Case 5a. So $m(G') < m(G)$, hence $vc(G) \leq m(G') + 1 \leq m(G)$. We already know $m(G) \leq vc(G)$, so $m(G) = vc(G)$ follows.

Case 5b: There exists a maximum matching M of G such that no edge of M has endpoint v. Recall $\deg(u) \geq 3$ and G is simple. Of the edges touching u, one leads to v, and at most one other edge appears in the matching M. Thus there must exist an edge $f \notin M$ such that u is an endpoint of f but v is not. Let G' be the graph obtained from G by deleting the edge

f. G' is still bipartite, and the induction hypothesis applies to show that $m(G') = vc(G')$. Because M did not use the edge f, M is a maximum matching of G' as well as G, so $m(G) = m(G')$. Let C' be a minimum vertex cover of G'. It suffices to show C' is also a vertex cover of G, for then G will have a matching and a vertex cover of the same size. Every edge of G except possibly f has an endpoint in C'. Also, since $|M| = |C'|$, the Lemma on Matchings and Vertex Covers (applied to the graph G') tells us that every vertex in C' is the endpoint of some edge in M. By the assumption in Case 5b, this forces $v \notin C'$. But the edge from u to v is in G', so $u \in C'$ by definition of a vertex cover. Since u is an endpoint of the edge f, we see that C' is a vertex cover of G, as needed. □

The next theorem will use the following notation. Given a set S of vertices in a graph G, a vertex v is a *neighbor* of S iff there exists $w \in S$ and an edge e in G with endpoints v and w. Let $N(S)$ be the set of all neighbors of S. Given a matching M of G and a set A of vertices of G, we say the matching *saturates* A iff every vertex in A is the endpoint of some edge in M.

3.90. Hall's Matching Theorem. Let G be a bipartite graph with partite sets A and B. There exists a matching of G saturating A iff for all $S \subseteq A$, $|S| \leq |N(S)|$.

Proof. First assume there is a matching M of G saturating A. Fix $S \subseteq A$. Define a function $f : S \to N(S)$ as follows. For each $v \in S$, there exists an edge e in M having v as one endpoint (since M saturates A), and this edge is unique (since M is a matching). Let $f(v)$ be the other endpoint of edge e, which lies in $N(S)$. Now f is one-to-one: if $f(v) = f(v')$ for some $v, v' \in S$, then the edge from v to $f(v)$ and the edge from v' to $f(v') = f(v)$ both appear in M, forcing $v = v'$ because M is a matching. So $|S| \leq |N(S)|$.

Conversely, assume $|S| \leq |N(S)|$ for all $S \subseteq A$. Let C be an arbitrary vertex cover of G. Let $A_1 = A \cap C$, $A_2 = A - C$, and $B_1 = B \cap C$. Every edge with an endpoint in A_2 must have its other endpoint in B_1, since C is a vertex cover of G. This means that $N(A_2) \subseteq B_1$. Taking $S = A_2$ in the assumption, we see that $|A_2| \leq |N(A_2)| \leq |B_1|$. Then $|C| = |A_1| + |B_1| \geq |A_1| + |A_2| = |A|$. On the other hand, the set A is a vertex cover of G since G is bipartite with partite sets A and B. Since $|A| \leq |C|$ holds for every vertex cover C, we see that A is a minimum vertex cover of G. By the previous theorem, $m(G) = vc(G) = |A|$. So there is a matching M of G consisting of $|A|$ edges. Each of these edges has one endpoint in A, and these endpoints are distinct because M is a matching. Thus, M is a matching that saturates A. □

3.16 Graph Coloring

This section introduces the graph coloring problem and some of its applications.

3.91. Definition: Colorings. Let $G = (V, E)$ be a simple graph, and let C be a finite set. A *coloring* of G using colors in C is a function $f : V \to C$. A coloring f of G is a *proper coloring* iff for every edge $\{u, v\} \in E$, $f(u) \neq f(v)$.

Intuitively, we are coloring each vertex of G using one of the available colors in the set C. For each $v \in V$, $f(v)$ is the color assigned to vertex v. A coloring is proper iff no two adjacent vertices in G are assigned the same color.

3.92. Definition: Chromatic Functions and Chromatic Numbers. Let G be a simple graph. For each positive integer x, let $\chi_G(x)$ be the number of proper colorings of G using colors in $\{1, 2, \ldots, x\}$. The function $\chi_G : \mathbb{Z}_{>0} \to \mathbb{Z}_{\geq 0}$ is called the *chromatic function* of G. The minimal x such that $\chi_G(x) > 0$ is called the *chromatic number* of G.

The chromatic number is the least number of colors required to obtain a proper coloring of G. The function χ_G is often called the *chromatic polynomial of* G because of Corollary 3.99 below.

3.93. Example. Suppose G is a simple graph with n vertices and no edges. By the Product Rule, $\chi_G(x) = x^n$ since we can assign any of the x colors to each vertex. The chromatic number for this graph is 1.

3.94. Example. At the other extreme, suppose G is a simple graph with n vertices such that there is an edge joining every pair of distinct vertices. Color the vertices one at a time. The first vertex can be colored in x ways. The second vertex must have a color different from the first, so there are $x-1$ choices. In general, the ith vertex must have a color distinct from all of its predecessors, so there are $x - (i-1)$ choices for the color of this vertex. The Product Rule gives $\chi_G(x) = x(x-1)(x-2)\cdots(x-n+1) = (x){\downarrow_n}$. The chromatic number for this graph is n. Recall from §2.15 that

$$(x){\downarrow_n} = \sum_{k=1}^{n} s(n,k)x^k,$$

so that the function χ_G in this example is a polynomial whose coefficients are the signed Stirling numbers of the first kind.

3.95. Example: Cycles. Consider the simple graph

$$G = (\{1,2,3,4\}, \{\{1,2\}, \{2,3\}, \{3,4\}, \{4,1\}\}).$$

G consists of four vertices joined in a 4-cycle. We might attempt to compute $\chi_G(x)$ via the Product Rule as follows. Color vertex 1 in x ways. Then color vertex 2 in $x-1$ ways, and color vertex 3 in $x-1$ ways. We run into trouble at vertex 4, because we do not know whether vertices 1 and 3 were assigned the same color. This example shows that we cannot always compute χ_G by the Product Rule alone. In this instance, we can classify proper colorings based on whether vertices 1 and 3 receive the same or different colors. If they receive the same color, the number of proper colorings is $x(x-1)(x-1)$ (color vertices 1 and 3 together, then color vertex 2 a different color, then color vertex 4 a different color from 1 and 3). If vertex 1 and 3 receive different colors, the number of proper colorings is $x(x-1)(x-2)(x-2)$ (color vertex 1, then vertex 3, then vertex 2, then vertex 4). Hence

$$\chi_G(x) = x(x-1)(x-1) + x(x-1)(x-2)(x-2) = x^4 - 4x^3 + 6x^2 - 3x.$$

The chromatic number for this graph is 2.

More generally, consider the cycle graph C_n consisting of n vertices joined in a cycle. It is routine to establish that the chromatic number of C_n is 1 for $n = 1$, is 2 for all even n, and is 3 for all odd $n > 1$. On the other hand, it is not immediately evident how to compute the chromatic function for C_n when $n > 4$. We will deduce a recursion for these functions shortly as a special case of a general recursion for computing chromatic functions.

Here is an application that can be analyzed using graph colorings and chromatic numbers. Suppose we are trying to schedule meetings for a number of committees. If two committees share a common member, they cannot meet at the same time. Consider the graph G whose vertices represent the various committees, and where there is an edge between two vertices iff the corresponding committees share a common member. Suppose there are x

available time slots in which meetings may be scheduled. A coloring of G with x colors represents a particular scheduling of committee meetings to time slots. The coloring is proper iff the schedule creates no time conflicts for any committee member. The chromatic number is the least number of time slots needed to avoid all conflicts, while $\chi_G(x)$ is the number of different conflict-free schedules using x (distinguishable) time slots.

3.96. Example. Six committees have members as specified in Table 3.1.

Committee	Members
A	Kemp, Oakley, Saunders
B	Gray, Saunders, Russell
C	Byrd, Oakley, Quinn
D	Byrd, Jenkins, Kemp
E	Adams, Jenkins, Wilson
F	Byrd, Gray, Russell

TABLE 3.1
Committee assignments in Example 3.96.

Figure 3.18 displays the graph G associated to this set of committees. To compute $\chi_G(x)$, consider cases based on whether vertices A and F receive the same color. If A and F are colored the same, the number of proper colorings is $x(x-1)(x-2)(x-1)(x-1)$ [color A and F, then C, D, B, and E]. If A and F receive different colors, the number of proper colorings is $x(x-1)(x-2)(x-3)(x-2)(x-1)$ [color A, F, C, D, B, E]. Thus,

$$\chi_G(x) = x(x-1)^2(x-2)(x-1+(x-2)(x-3)) = x^6 - 8x^5 + 26x^4 - 42x^3 + 33x^2 - 10x.$$

The chromatic number of G is 3.

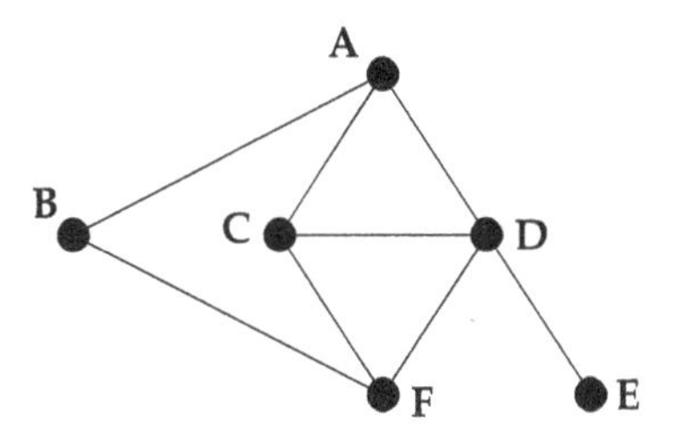

FIGURE 3.18
Conflict graph for six committees.

We are about to present a general recursion that can be used to compute the chromatic function of a simple graph. The recursion makes use of the following construction.

3.97. Definition: Collapsing an Edge. Let $G = (V, E)$ be a simple graph, and let $e_0 = \{v_0, w_0\}$ be a fixed edge of G. Let z_0 be a new vertex. We define a simple graph H called *the graph obtained from G by collapsing the edge e_0*. The vertex set of H is $(V - \{v_0, w_0\}) \cup \{z_0\}$. The edge set of H is

$$\begin{aligned}&\{\{x,y\} : x \neq v_0 \neq y \text{ and } x \neq w_0 \neq y \text{ and } \{x,y\} \in E\}\\ \cup&\{\{x,z_0\} : x \neq v_0 \text{ and } \{x,w_0\} \in E\}\\ \cup&\{\{x,z_0\} : x \neq w_0 \text{ and } \{x,v_0\} \in E\}.\end{aligned}$$

Pictorially, we construct H from G by shrinking the edge e_0 until the vertices v_0 and w_0 coincide. We replace these two overlapping vertices with a single new vertex z_0. All edges touching v_0 or w_0 (except the collapsed edge e_0) now touch z_0 instead. See Figure 3.19.

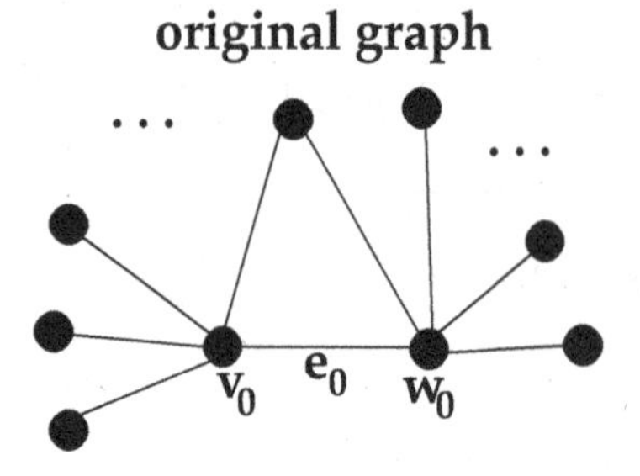

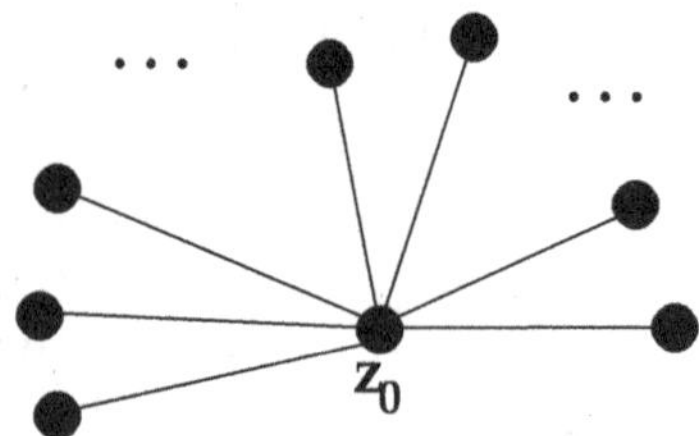

FIGURE 3.19
Collapsing an edge in a simple graph.

3.98. Theorem: Chromatic Recursion. Let $G = (V, E)$ be a simple graph. Fix any edge $e = \{v, w\} \in G$. Let $G' = (V, E - \{e\})$ be the simple graph obtained by deleting the edge e from G, and let G'' be the simple graph obtained from G by collapsing the edge e. Then

$$\chi_G(x) = \chi_{G'}(x) - \chi_{G''}(x).$$

Proof. Fix $x \in \mathbb{Z}_{>0}$, and let A, B, and C denote the set of proper colorings of G, G', and G'' (respectively) using x available colors. Write $B = B_1 \cup B_2$, where $B_1 = \{f \in B : f(v) = f(w)\}$ and $B_2 = \{f \in B : f(v) \neq f(w)\}$. Note that B_1 consists of the proper colorings of G' (if any) in which vertices v and w are assigned the same color. Let z be the new vertex in G'' that replaces v and w. Given a proper coloring $f \in B_1$, we define a corresponding coloring f'' of G'' by setting $f''(z) = f(v) = f(w)$ and $f''(u) = f(u)$ for all $u \in V$ different from v and w. Since f is proper, it follows from the definition of the edge set of G'' that f'' is a proper coloring as well. Thus we have a map $f \mapsto f''$ from B_1 to C. This map is invertible, since the color of z in a coloring of G'' determines the common color of v and w in a coloring of G' belonging to B_1. We conclude that $|B_1| = |C|$.

On the other hand, B_2 consists of the proper colorings of G' in which vertices v and w are assigned different colors. These are precisely the proper colorings of G (since G has an edge between v and w, and G is otherwise identical to G'). Thus, $B_2 = A$. It follows that

$$\chi_G(x) = |A| = |B_2| = |B| - |B_1| = |B| - |C| = \chi_{G'}(x) - \chi_{G''}(x). \qquad \square$$

3.99. Corollary: Polynomiality of Chromatic Functions. For any graph G, $\chi_G(x)$ is a polynomial in x with integer coefficients. (This justifies the terminology *chromatic polynomial.*)

Proof. We use induction on the number of edges in G. If G has k vertices and no edges, the Product Rule gives $\chi_G(x) = x^k$, which is a polynomial in x. Now assume G has $m > 0$ edges. Fix such an edge e, and define G' and G'' as in the preceding theorem. G' has one fewer edge than G. When passing from G to G'', we lose the edge e and possibly identify other edges in G (e.g., if both endpoints of e are adjacent to a third vertex). In any case, G'' has fewer edges than G. By induction on m, we may assume that both $\chi_{G'}(x)$ and $\chi_{G''}(x)$ are polynomials in x with integer coefficients. So $\chi_G(x) = \chi_{G'}(x) - \chi_{G''}(x)$ is also a polynomial with integer coefficients. $\square$

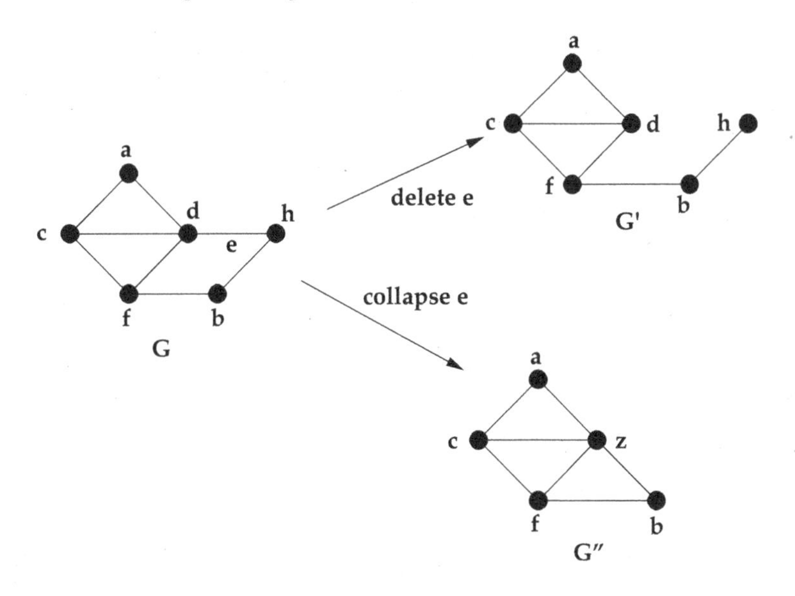

FIGURE 3.20
Using the chromatic recursion.

3.100. Remark. We can use the chromatic recursion to compute χ_G recursively for any graph G. The base case of the calculation is a graph with k vertices and no edges, which has chromatic polynomial x^k. If G has more than one edge, G' and G'' both have strictly fewer edges than G. Thus, the recursive calculation will terminate after finitely many steps. However, this is quite an inefficient method for computing χ_G if G has many vertices and edges. Thus, direct counting arguments using the Sum Rule and Product Rule may be preferable to repeatedly applying the chromatic recursion.

3.101. Example. Consider the graph G shown on the left in Figure 3.20. We compute $\chi_G(x)$ by applying the chromatic recursion to the edge $e = \{d, h\}$. The graphs G' and G'' obtained by deleting and collapsing this edge are shown on the right in Figure 3.20. Direct arguments using the Product Rule show that

$$\chi_{G'}(x) = x(x-1)(x-2)(x-2)(x-1)(x-1) \qquad \text{(color a, c, d, f, b, h);}$$

$$\chi_{G''}(x) = x(x-1)(x-2)(x-2)(x-2) \qquad \text{(color z, a, c, f, b).}$$

Therefore,

$$\chi_G(x) = x(x-1)(x-2)^2((x-1)^2 - (x-2)) = x^6 - 8x^5 + 26x^4 - 43x^3 + 36x^2 - 12x.$$

3.102. Chromatic Polynomials for Cycles. For each $n \geq 3$, let C_n denote a graph consisting of n vertices joined in a cycle. Let C_1 denote a one-vertex graph, and let C_2 denote a graph with two vertices joined by an edge. Finally, let $\chi_n(x) = \chi_{C_n}(x)$ be the chromatic polynomials for these graphs. We see directly that

$$\chi_1(x) = x, \qquad \chi_2(x) = x(x-1) = x^2 - x, \qquad \chi_3(x) = x(x-1)(x-2) = x^3 - 3x^2 + 2x.$$

Fix $n > 3$ and fix any edge e in C_n. Deleting this edge leaves a graph in which n vertices are joined in a line; the chromatic polynomial of such a graph is $x(x-1)^{n-1}$. On the other hand, collapsing the edge e in C_n produces a graph isomorphic to C_{n-1}. The chromatic recursion therefore gives

$$\chi_n(x) = x(x-1)^{n-1} - \chi_{n-1}(x).$$

Using this recursion to compute $\chi_n(x)$ for small n suggests the closed formula

$$\chi_n(x) = (x-1)^n + (-1)^n(x-1) \text{ for all } n \geq 2.$$

One may now prove this formula for $\chi_n(x)$ by induction, using the chromatic recursion.

3.17 Spanning Trees

This section introduces the notion of a spanning tree for a graph. A recursion resembling the Chromatic Recursion 3.98 will allow us to count the spanning trees for a given graph. This will lead to a remarkable formula, called the Matrix-Tree Theorem, that expresses the number of spanning trees as a certain determinant.

3.103. Definition: Subgraphs. Let $G = (V, E, \epsilon)$ and $H = (W, F, \eta)$ be graphs or digraphs. H is a *subgraph* of G iff $W \subseteq V$, $F \subseteq E$, and $\eta(f) = \epsilon(f)$ for all $f \in F$. H is an *induced subgraph* of G iff H is a subgraph such that F consists of *all* edges in E with both endpoints in W.

3.104. Definition: Spanning Trees. Given a graph $G = (V, E, \epsilon)$, a *spanning tree* for G is a subgraph H with vertex set V such that H is a tree. Let $\tau(G)$ be the number of spanning trees of G.

3.105. Example. Consider the graph G shown in Figure 3.21. This graph has 31 spanning

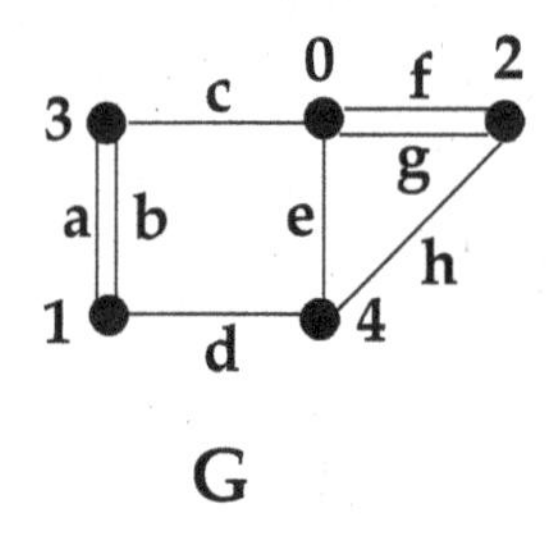

FIGURE 3.21
Graph used to illustrate spanning trees.

trees, which are specified by the following sets of edges:

$\{a,c,d,f\}$, $\{b,c,d,f\}$, $\{a,c,d,g\}$, $\{b,c,d,g\}$, $\{a,c,d,h\}$, $\{b,c,d,h\}$,
$\{c,d,e,f\}$, $\{c,d,e,g\}$, $\{c,d,e,h\}$, $\{a,c,e,f\}$, $\{a,d,e,f\}$, $\{b,c,e,f\}$,
$\{b,d,e,f\}$, $\{a,c,e,g\}$, $\{a,d,e,g\}$, $\{b,c,e,g\}$, $\{b,d,e,g\}$, $\{a,c,e,h\}$,
$\{a,d,e,h\}$, $\{b,c,e,h\}$, $\{b,d,e,h\}$, $\{a,c,f,h\}$, $\{a,d,f,h\}$, $\{b,c,f,h\}$,
$\{b,d,f,h\}$, $\{a,c,g,h\}$, $\{a,d,g,h\}$, $\{b,c,g,h\}$, $\{b,d,g,h\}$, $\{c,d,f,h\}$,
$\{c,d,g,h\}$.

We see that even a small graph can have many spanning trees. Thus we seek a systematic method for enumerating these trees.

We are going to derive a recursion involving the quantities $\tau(G)$. For this purpose, we need to adapt the ideas of *deleting an edge* and *collapsing an edge* (see Definition 3.97) from simple graphs to general graphs. Since loop edges are never involved in spanning trees, we only consider graphs without loops. Suppose we are given a graph $G = (V, E, \epsilon)$ and a fixed edge $z \in E$ with endpoints $u, v \in V$. To *delete* z from G, we replace E by $E' = E - \{z\}$ and replace ϵ by the restriction of ϵ to E'. To *collapse* the edge z, we act as follows: (i) delete z and any other edges linking u to v; (ii) replace V by $(V - \{u, v\}) \cup \{w\}$, where w is a new vertex; (iii) for each edge $y \in E$ that has exactly one endpoint in the set $\{u, v\}$, modify $\epsilon(y)$ by replacing this endpoint with the new vertex w.

3.106. Example. Let G be the graph shown in Figure 3.21. Figure 3.22 displays the graphs obtained from G by deleting edge f and collapsing edge f.

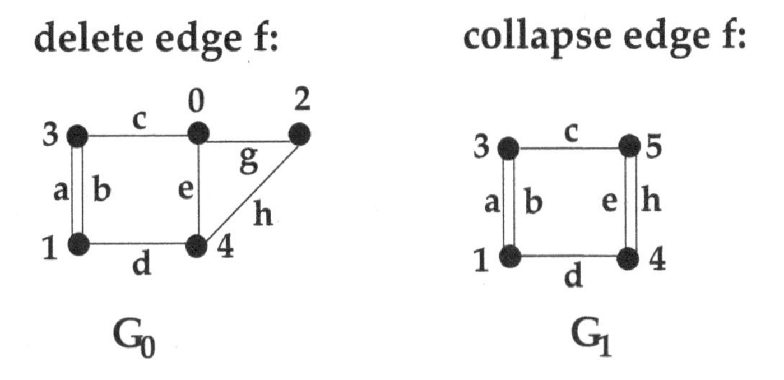

FIGURE 3.22
Effect of deleting or collapsing an edge.

3.107. Theorem: Spanning Tree Recursion. Let $G = (V, E, \epsilon)$ be a graph, and let $z \in E$ be a fixed edge. Let G_0 be the graph obtained from G by deleting z. Let G_1 be the graph obtained from G by collapsing z. Then

$$\tau(G) = \tau(G_0) + \tau(G_1).$$

The initial conditions are: $\tau(G) = 0$ if G is not connected, and $\tau(G) = 1$ if G is a tree with vertex set V.

Proof. For every graph K, let $Sp(K)$ be the set of all spanning trees of K, so $\tau(K) = |Sp(K)|$. Fix the graph G and the edge z. Let X be the set of trees in $Sp(G)$ that do not use the edge z, and let Y be the set of trees in $Sp(G)$ that do use the edge z. $Sp(G)$ is the disjoint union of X and Y, so $\tau(G) = |X| + |Y|$ by the Sum Rule. Now, it follows from the definition of edge deletion that the set X is precisely the set $Sp(G_0)$, so $|X| = \tau(G_0)$. To complete the proof, we need to show that $|Y| = \tau(G_1)$. It suffices to define a bijection $F : Y \to Sp(G_1)$.

Suppose $T \in Y$ is a spanning tree of G that uses the edge z with endpoints u, v. Define $F(T)$ to be the graph obtained from T by collapsing the edge z; this graph is a subgraph of G_1. Let n be the number of vertices of G; then T is a connected graph with $n-1$ edges, one of which is z. It is routine to check that $F(T)$ is still connected. Furthermore, since T is a tree, z is the only edge in T between u and v. It follows from the definition of collapsing that $F(T)$ has exactly $n-2$ edges. Since G_1 has $n-1$ vertices, it follows that $F(T)$ is a spanning tree of G_1. We see also that the edge set of $F(T)$ is precisely the edge set of T with z removed. So far, we have shown that F is a well-defined function mapping Y into $Sp(G_1)$.

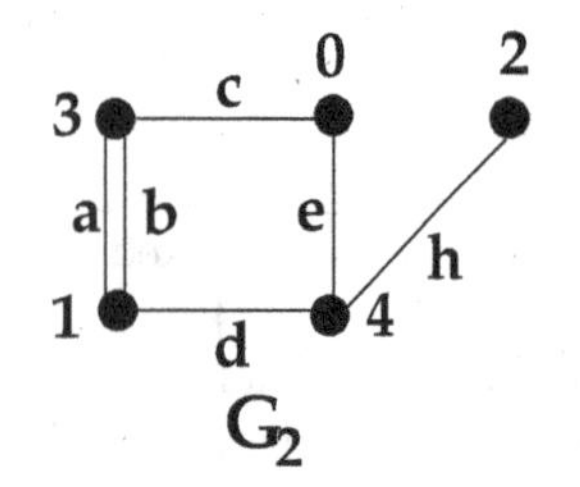

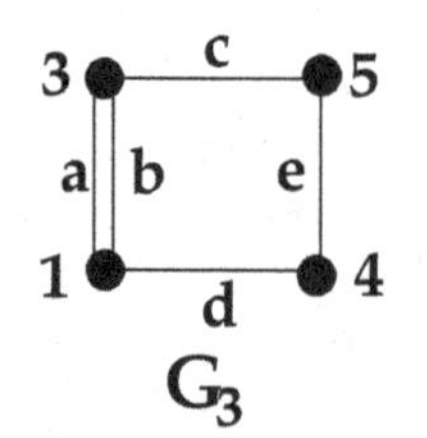

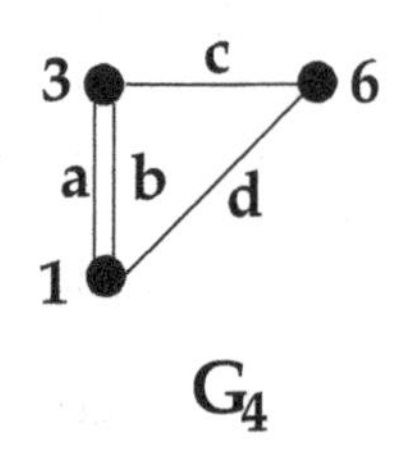

FIGURE 3.23
Auxiliary graphs used in the computation of $\tau(G)$.

Next we define a map $H : Sp(G_1) \to Y$ that is the two-sided inverse of F. Given $U \in Sp(G_1)$ with edge set $E(U)$, let $H(U)$ be the unique subgraph of G with vertex set V and edge set $E(U) \cup \{z\}$. We must check that $H(U)$ does lie in the claimed codomain Y. First, $H(U)$ is a subgraph of G with $n-1$ edges, one of which is the edge z. Furthermore, it can be checked that $H(U)$ is connected, since walks in U can be expanded using the edge z if needed to give walks in $H(U)$. Therefore, $H(U)$ is a spanning tree of G using z, and so $H(U) \in Y$. Since F removes z from the edge set while H adds it back, F and H are two-sided inverses of each other. Hence both are bijections, and the proof is complete. □

3.108. Example. We use the graphs in Figures 3.21 and 3.22 to illustrate the proof of the spanning tree recursion, taking $z = f$. The graph G_0 on the left of Figure 3.22 has 19 spanning trees; they are precisely the trees listed in Example 3.105 that do not use the edge f. Applying F to each of the remaining 12 spanning trees on the list produces the following subgraphs of G_1 (specified by their edge sets):

$$\begin{array}{llllll} \{a,c,d\}, & \{b,c,d\}, & \{c,d,e\}, & \{a,c,e\}, & \{a,d,e\}, & \{b,c,e\}, \\ \{b,d,e\}, & \{a,c,h\}, & \{a,d,h\}, & \{b,c,h\}, & \{b,d,h\}, & \{c,d,h\}. \end{array}$$

These are precisely the spanning trees of G_1.

Next, we illustrate the calculation of $\tau(G)$ using the recursion. We first delete and collapse edge f, producing the graphs G_0 and G_1 shown in Figure 3.22. We know that $\tau(G) = \tau(G_0) + \tau(G_1)$. Deletion of edge g from G_0 produces a new graph G_2 (Figure 3.23), while collapsing g in G_0 leads to another copy of G_1. So far, we have $\tau(G) = 2\tau(G_1) + \tau(G_2)$. Continuing to work on G_1, we see that deleting edge h leads to the graph G_3 in Figure 3.23, whereas collapsing edge h leads to the graph G_4 in that figure. On the other hand, deleting h from G_2 leaves a disconnected graph (which can be discarded), while collapsing h from G_2 produces another copy of G_3. Now we have $\tau(G) = 3\tau(G_3) + 2\tau(G_4)$. Deleting edge e from G_3 gives a graph that has two spanning trees (by inspection), while collapsing e in G_3 leads to G_4 again. So $\tau(G) = 3(2 + \tau(G_4)) + 2\tau(G_4) = 6 + 5\tau(G_4)$. Finally, deletion of d from G_4 leaves a graph with two spanning trees, while collapsing d produces a graph with three spanning trees. We conclude that $\tau(G_4) = 5$, so $\tau(G) = 6 + 25 = 31$, in agreement with the enumeration in Example 3.105.

Next we extend the preceding discussion to rooted spanning trees in digraphs.

3.109. Definition: Rooted Spanning Trees. Let $G = (V, E, \epsilon)$ be a digraph, and let $v_0 \in V$. A *spanning tree of G rooted at v_0* is a rooted tree T with root v_0 and vertex set V such that T (without the loop at v_0) is a subgraph of G. Let $\tau(G, v_0)$ be the number of spanning trees of G rooted at v_0.

The notions of edge deletion and contraction extend to digraphs. This leads to the following recursion for counting rooted spanning trees.

3.110. Theorem: Rooted Spanning Tree Recursion. Let v_0 be a fixed vertex in a digraph G, and let z be a fixed edge leading into v_0. Let G_1 be the digraph obtained from G by deleting z. Let G_2 be the digraph obtained from G by collapsing z, and let the new collapsed vertex in G_2 be v_0'. Then

$$\tau(G, v_0) = \tau(G_1, v_0) + \tau(G_2, v_0').$$

Proof. We modify the proof of Theorem 3.107. As before, the two terms on the right side count rooted spanning trees of G that do not contain z or do contain z. One may check that if T is a rooted spanning tree using the edge z, then the graph obtained from T by collapsing z is a rooted spanning tree of G_2 rooted at v_0'. Similarly, adding z to the edge set of a rooted spanning tree of G_2 rooted at v_0' produces a rooted spanning tree of G rooted at v_0. □

3.111. Remark. Our results for counting undirected spanning trees are special cases of the corresponding results for rooted spanning trees. Starting with any graph $G = (V, E, \epsilon)$, we consider the associated digraph obtained by replacing each $e \in E$ by two directed edges going in opposite directions. As in the proof of Theorem 3.73, there is a bijection between the set of rooted spanning trees of this digraph rooted at any given vertex $v_0 \in V$ and the set of spanning trees of G. In what follows, we shall only treat the case of digraphs.

3.18 The Matrix-Tree Theorem

There is a remarkable determinant formula for the number of rooted spanning trees of a digraph. The formula uses the following modified version of the adjacency matrix of the digraph.

3.112. Definition: Laplacian Matrix of a Digraph. Let G be a loopless digraph on the vertex set $V = \{v_0, v_1, \ldots, v_n\}$. The *Laplacian matrix* of G is the matrix $L = (L_{ij} : 0 \leq i, j \leq n)$ such that $L_{ii} = \text{outdeg}(v_i)$ and L_{ij} is the negative of the number of edges from v_i to v_j in G. We let L_0 be the $n \times n$ matrix obtained by erasing the row and column of L corresponding to v_0. The matrix $L_0 = L_0(G)$ is called the *truncated Laplacian matrix of G* (relative to v_0).

3.113. The Matrix-Tree Theorem. With the notation of the preceding definition, we have

$$\tau(G, v_0) = \det(L_0(G)).$$

We prove the theorem after considering two examples.

3.114. Example. Let G be the digraph associated to the undirected graph in Figure 3.21. In this case, L_{ii} is the degree of vertex i in the undirected graph, and L_{ij} is the negative of the number of undirected edges between i and j. So

$$L = \begin{bmatrix} 4 & 0 & -2 & -1 & -1 \\ 0 & 3 & 0 & -2 & -1 \\ -2 & 0 & 3 & 0 & -1 \\ -1 & -2 & 0 & 3 & 0 \\ -1 & -1 & -1 & 0 & 3 \end{bmatrix}.$$

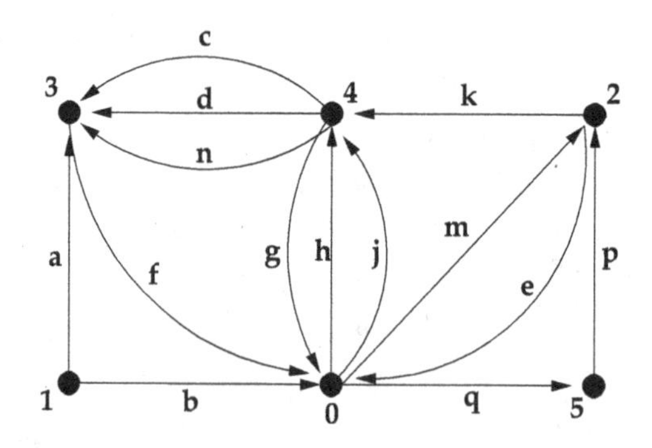

FIGURE 3.24
Digraph used to illustrate the Matrix-Tree Theorem.

Erasing the row and column corresponding to vertex 0 leaves

$$L_0 = \begin{bmatrix} 3 & 0 & -2 & -1 \\ 0 & 3 & 0 & -1 \\ -2 & 0 & 3 & 0 \\ -1 & -1 & 0 & 3 \end{bmatrix}.$$

We compute $\det(L_0) = 31$, which agrees with our earlier calculation of $\tau(G)$.

3.115. Example. Consider the digraph G shown in Figure 3.24. We compute

$$L = \begin{bmatrix} 4 & 0 & -1 & 0 & -2 & -1 \\ -1 & 2 & 0 & -1 & 0 & 0 \\ -1 & 0 & 2 & 0 & -1 & 0 \\ -1 & 0 & 0 & 1 & 0 & 0 \\ -1 & 0 & 0 & -3 & 4 & 0 \\ 0 & 0 & -1 & 0 & 0 & 1 \end{bmatrix}, \qquad L_0 = \begin{bmatrix} 2 & 0 & -1 & 0 & 0 \\ 0 & 2 & 0 & -1 & 0 \\ 0 & 0 & 1 & 0 & 0 \\ 0 & 0 & -3 & 4 & 0 \\ 0 & -1 & 0 & 0 & 1 \end{bmatrix},$$

and $\det(L_0) = 16$. So G has 16 spanning trees rooted at 0, as can be confirmed by direct enumeration. We use the matrix L_0 as a running example in the proof below.

3.116. Proof of the Matrix-Tree Theorem. Write $L_0 = L_0(G)$. First we prove that $\tau(G, v_0) = \det(L_0)$ in the case where $\mathrm{indeg}(v_0) = 0$. If v_0 is the only vertex of G, then $\tau(G, v_0) = 1$ and $\det(L_0) = 1$ by the convention that the determinant of a 0×0 matrix is 1. Otherwise, $\tau(G, v_0)$ is zero, and L_0 is a nonempty matrix. Using the condition $\mathrm{indeg}(v_0) = 0$ and the definition of L_0, we see that every row of L_0 sums to zero. Therefore, letting u be a column vector of n ones, we have $L_0 u = 0$, so that L_0 is singular and $\det(L_0) = 0$.

For the general case, we use induction on the number of edges in G. The case where G has no edges is covered by the previous paragraph. The only case left to consider occurs when $\mathrm{indeg}(v_0) > 0$. Let e be a fixed edge in G that leads from some v_i to v_0. Let G_1 be the graph obtained from G by deleting e, and let G_2 be the graph obtained from G by collapsing e. Both graphs have fewer edges than G, so the induction hypothesis tells us that

$$\tau(G_1, v_0) = \det(L_0(G_1)) \text{ and } \tau(G_2, v_0') = \det(L_0(G_2)), \tag{3.4}$$

where v_0' is the new vertex created after collapsing e. Using Theorem 3.110, we conclude that

$$\tau(G, v_0) = \det(L_0(G_1)) + \det(L_0(G_2)). \tag{3.5}$$

Next, we evaluate the determinant $\det(L_0(G))$. We use the fact that the determinant of a matrix is a linear function of each row of the matrix. More precisely, for a fixed matrix A and row index i, let $A[y]$ denote the matrix A with the ith row replaced by the row vector y; then $\det(A[y+z]) = \det(A[y]) + \det(A[z])$ for all y, z. This linearity property can be proved directly from the definition of the determinant (see Definition 12.40 and Theorem 12.48 for details). To apply this result, write the ith row of $L_0 = L_0(G)$ in the form $y + z$, where $z = (0, 0, \ldots, 1, 0, \ldots, 0)$ has a 1 in position i. Then

$$\det(L_0(G)) = \det(L_0[y]) + \det(L_0[z]). \tag{3.6}$$

For example, if G is the digraph in Figure 3.24 and e is the edge from 2 to 0 (so $i = 2$), then $y = (0, 1, 0, -1, 0)$, $z = (0, 1, 0, 0, 0)$,

$$L_0[y] = \begin{bmatrix} 2 & 0 & -1 & 0 & 0 \\ 0 & 1 & 0 & -1 & 0 \\ 0 & 0 & 1 & 0 & 0 \\ 0 & 0 & -3 & 4 & 0 \\ 0 & -1 & 0 & 0 & 1 \end{bmatrix}, \qquad L_0[z] = \begin{bmatrix} 2 & 0 & -1 & 0 & 0 \\ 0 & 1 & 0 & 0 & 0 \\ 0 & 0 & 1 & 0 & 0 \\ 0 & 0 & -3 & 4 & 0 \\ 0 & -1 & 0 & 0 & 1 \end{bmatrix}.$$

Comparing equations (3.5) and (3.6), we see that it suffices to prove $\det(L_0(G_1)) = \det(L_0[y])$ and $\det(L_0(G_2)) = \det(L_0[z])$.

How does the removal of e from G affect $L(G)$? Answer: The i,i-entry drops by 1, while the $i,0$-entry increases by 1. Since the zeroth column is ignored in the truncated Laplacian, we see that we can obtain $L_0(G_1)$ from $L_0(G)$ by decrementing the i,i-entry by 1. In other words, $L_0(G_1) = L_0[y]$, and hence $\det(L_0(G_1)) = \det(L_0[y])$.

Next, let us calculate $\det(L_0[z])$ by expanding the determinant along row i. The only nonzero entry in this row is the 1 in the diagonal position, so $\det(L_0[z]) = (-1)^{i+i}\det(M) = \det(M)$, where M is the matrix obtained from $L_0[z]$ (or equivalently, from L_0) by erasing row i and column i. In our running example,

$$M = \begin{bmatrix} 2 & -1 & 0 & 0 \\ 0 & 1 & 0 & 0 \\ 0 & -3 & 4 & 0 \\ 0 & 0 & 0 & 1 \end{bmatrix}.$$

We claim that $M = L_0(G_2)$, which will complete the proof. Consider the k,j-entry of M, where $k, j \in \{0, 1, \ldots, n\} - \{0, i\}$. If $k = j$, this entry is $\text{outdeg}_G(v_j)$, which equals $\text{outdeg}_{G_2}(v_j)$ because v_j is not v_0, v_i, or v_0'. For the same reason, if $k \neq j$, the k,j-entry of M is the negative of the number of edges from v_k to v_j, which is the same in G and G_2.

3.19 Eulerian Tours

3.117. Definition: Eulerian Tours. Let $G = (V, E, \epsilon)$ be a digraph. An *Eulerian tour* in G is a walk $W = (v_0, e_1, v_1, e_2, v_2, \ldots, e_n, v_n)$ such that W visits every vertex in V, and W uses every edge in E exactly once. Such a tour is called *closed* iff $v_n = v_0$.

3.118. Example. For the digraph G shown in Figure 3.25, one closed Eulerian tour of G is

$$W_1 = (0, m, 2, l, 5, e, 1, a, 3, c, 4, b, 3, d, 5, f, 4, g, 5, k, 0, i, 4, h, 5, j, 0).$$

To specify the tour, it suffices to list only the edges in the tour. For instance, here is the edge sequence of another closed Eulerian tour of G:

$$W_2 = (i, g, e, a, d, f, b, c, h, j, m, l, k).$$

3.119. Example. The digraph G shown in Figure 3.2 does not have any closed Eulerian tours, since there is no way to reach vertex 6 from the other vertices. Even if we delete vertex 6 from the graph, there are still no closed Eulerian tours. The reason is that no tour can use both edges leaving vertex 2, since only one edge enters vertex 2.

The previous example indicates two necessary conditions for a digraph to have a closed Eulerian tour: the digraph must be connected, and also *balanced* in the sense that $\text{indeg}(v) = \text{outdeg}(v)$ for every vertex v. We now show that these necessary conditions are also sufficient to guarantee the existence of a closed Eulerian tour.

3.120. Theorem: Existence of Closed Eulerian Tours. A digraph $G = (V, E, \epsilon)$ has a closed Eulerian tour iff G is connected and balanced.

Proof. First suppose G has a closed Eulerian tour W starting at v_0. Since W visits every vertex, we can obtain a walk from any vertex to any other vertex by following certain edges of W. So G is connected. Next, let v be any vertex of G. The walk W arrives at v via an incoming edge exactly as often as the walk leaves v via an outgoing edge; this is true even if $v = v_0$. Since the walk uses every edge exactly once, it follows that $\text{indeg}(v) = \text{outdeg}(v)$.

Conversely, assume that G is connected and balanced. Let $W = (v_0, e_1, v_1, \ldots, e_n, v_n)$ be a walk of maximum length in G that never repeats an edge. We claim that $v_n = v_0$. Otherwise, the walk W would enter vertex v_n one more time than it leaves v_n. Since $\text{indeg}(v_n) = \text{outdeg}(v_n)$, there must be an outgoing edge from v_n that has not been used by W. So we could use this edge to extend W, contradicting maximality of W. Next, we claim that W uses *every* edge of G. If not, let e be an edge not used by W. Since G is connected, we can find such an edge that is incident to one of the vertices v_i visited by W. Since $v_n = v_0$, we can cyclically shift the walk W to get a new walk $W' = (v_i, e_{i+1}, v_{i+1}, \ldots, e_n, v_n = v_0, e_1, \ldots, e_i, v_i)$ that starts and ends at v_i. By adding the edge e to the beginning or end of this walk (depending on its direction), we could again produce a longer walk than W with no repeated edges, violating maximality. Finally, W must visit every vertex of G, since W uses every edge of G and (unless G has one vertex and no edges) every vertex has an edge leaving it. □

Our goal in the rest of this section is to count the number of closed Eulerian tours in G

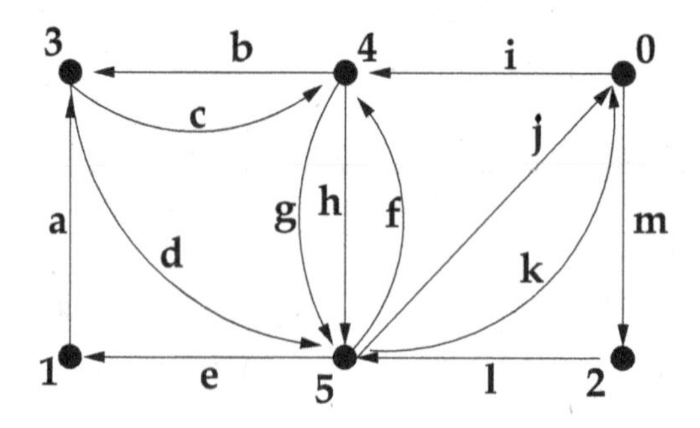

FIGURE 3.25
Digraph used to illustrate Eulerian tours.

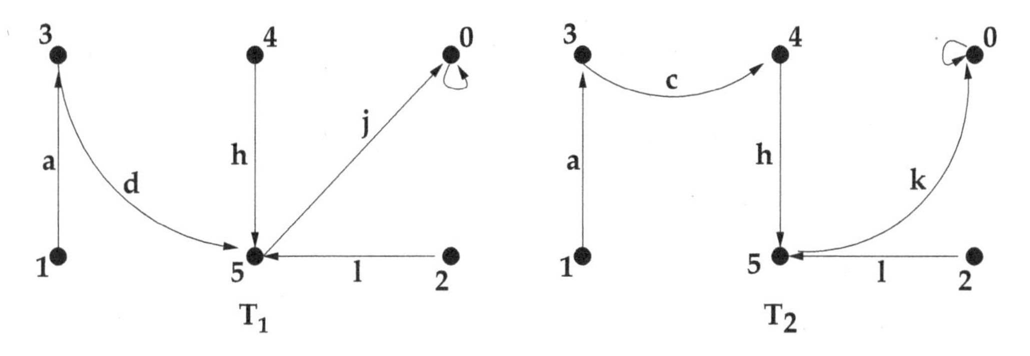

FIGURE 3.26
Rooted spanning trees associated to Eulerian tours.

starting at a given vertex v_0. Recall that $\tau(G, v_0)$ is the number of rooted spanning trees of G rooted at v_0.

3.121. The Eulerian Tour Rule. Let $G = (V, E, \epsilon)$ be a connected, balanced digraph. For each $v_0 \in V$, the number of closed Eulerian tours of G starting at v_0 is

$$\tau(G, v_0) \cdot \text{outdeg}(v_0)! \cdot \prod_{v \neq v_0} (\text{outdeg}(v) - 1)!. \tag{3.7}$$

Let $\{v_0, v_1, \ldots, v_n\}$ be the vertex set of G. Let X be the set of all closed Eulerian tours of G starting at v_0. Let $\text{SpTr}(G, v_0)$ be the set of spanning trees of G rooted at v_0. Let Y be the set of all tuples $(T, w_0, w_1, w_2, \ldots, w_n)$ satisfying these conditions: $T \in \text{SpTr}(G, v_0)$; w_0 is a permutation of all the edges leaving v_0; and, for $1 \leq i \leq n$, w_i is a permutation of those edges leaving v_i other than the unique outgoing edge from v_i that belongs to T (see Definition 3.42). By the Product Rule, the cardinality of Y is given by the right side of (3.7). So it suffices to define a bijection $f : X \to Y$.

Given an Eulerian tour $W \in X$, define $f(W) = (T, w_0, \ldots, w_n)$ as follows. For each i between 0 and n, let w'_i be the permutation of *all* edges leading out of v_i, taken in the order in which they occur in the walk W. Call w'_i the *departure word* of vertex v_i. Next, set $w_0 = w'_0$ and for $i > 0$, let w_i be the word w'_i with the last symbol erased. Finally, let T be the subgraph of G whose edges are given by the last symbols of $w'_1, \ldots, w'_n$, augmented by a loop edge at v_0. It is not immediately evident that $T \in \text{SpTr}(G, v_0)$; we prove this shortly.

Next we define a map $g : Y \to X$ that is the two-sided inverse of f. Fix $(T, w_0, \ldots, w_n) \in Y$. For every $i > 0$, form w'_i by appending the unique edge of T leaving v_i to the end of the word w_i; let $w'_0 = w_0$. Starting at v_0, we use the words w'_i to build a walk through G, one edge at a time, as follows. If we are currently at some vertex v_i, use the next unread symbol in w'_i to determine which edge to follow out of v_i. Repeat this process until the walk reaches a vertex in which all the outgoing edges have already been used. The resulting walk W is $g(T, w_0, \ldots, w_n)$. The edges occurring in W are pairwise distinct, but it is not immediately evident that W must use *all* edges of G; we prove this shortly.

Once we check that f and g map into their stated codomains, the definitions just given show that $f \circ g$ and $g \circ f$ are both identity maps. Before proving that f maps into Y and g maps into X, we consider an example.

3.122. Example. We continue the analysis of Eulerian tours in the digraph G from Example 3.118. The walk W_1 in that example has departure words $w'_0 = mi$, $w'_1 = a$, $w'_2 = l$,

$w_3' = cd$, $w_4' = bgh$, and $w_5' = efkj$. Therefore,

$$f(W_1) = (T_1, mi, \cdot, \cdot, c, bg, efk),$$

where $\cdot$ denotes an empty word and T_1 is the graph shown on the left in Figure 3.26. Similarly, for W_2 we compute $w_0' = im$, $w_1' = a$, $w_2' = l$, $w_3' = dc$, $w_4' = gbh$, $w_5' = efjk$, and

$$f(W_2) = (T_2, im, \cdot, \cdot, d, gb, efj).$$

We now calculate $g(T_1, im, \cdot, \cdot, c, bg, fke)$. First, we use the edges of T_1 to recreate the departure words $w_0' = im$, $w_1' = a$, $w_2' = l$, $w_3' = cd$, $w_4' = bgh$, and $w_5' = fkej$. We then use these words to guide our tour through the graph. We begin with $0, i, 4$, since i is the first letter of w_0'. Consulting w_4' next, we follow edge b to vertex 3, then edge c to vertex 4, then edge g to vertex 5, and so on. We obtain the tour

$$W_3 = (0, i, 4, b, 3, c, 4, g, 5, f, 4, h, 5, k, 0, m, 2, l, 5, e, 1, a, 3, d, 5, j, 0).$$

Similarly, we compute

$$g(T_2, mi, \cdot, \cdot, d, bg, jfe) = (m, l, j, i, b, d, f, g, e, a, c, h, k).$$

To complete the proof of Rule 3.121, we must prove two things. First, to show that $f(W) \in Y$ for all $W \in X$, we must show that the digraph T obtained from the last letters of the departure words w_i' (for $i > 0$) is a rooted spanning tree of G rooted at v_0. Since W visits every vertex of G, the definition of T shows that $\text{outdeg}_T(v_i) = 1$ for all $i \geq 0$. We need only show that T has no cycles other than the loop at v_0 (see Definition 3.42). We can view the tour W as a certain permutation of all the edges in G. Let us show that if e, h are two non-loop edges in T with $\epsilon(e) = (x, y)$ and $\epsilon(h) = (y, z)$, then e must precede h in the permutation W. Note that y cannot be v_0, since the only outgoing edge from v_0 in T is a loop edge. Thus, when the tour W uses the edge e to enter y, the following edge in the tour exists and is an outgoing edge from y. Since h is, by definition, the *last* such edge used by the tour, e must precede h in the tour. Now suppose $(z_0, e_1, z_1, \ldots, e_n, z_n)$ is a cycle in T that is not the 1-cycle at v_0. Using the previous remark repeatedly, we see that e_i precedes e_{i+1} in W for all i, and also e_n precedes e_1 in W. These statements imply that e_1 precedes itself in W, which is impossible. We conclude that $f(W) \in Y$.

Second, we must show that g maps Y into X. Fix $(T, w_0, \ldots, w_n) \in Y$ and $W = g(T, w_0, \ldots, w_n)$, and let w_i' be the departure word constructed from T and w_i. We know from the definition of g that W is a walk in G starting at v_0 that never repeats an edge. We must show that W ends at v_0 and uses every edge in G. Suppose, at some stage in the construction of W, that W has just reached v_i for some $i > 0$. Then W has entered v_i one more time than it has left v_i. Since G is balanced, there must exist an unused outgoing edge from v_i. This edge corresponds to an unused letter in w_i'. So W does not end at v_i. The only possibility is that W ends at the starting vertex v_0.

To prove that W uses every edge of G, we claim that it is enough to prove that W uses every non-loop edge of T. To establish the claim, consider a vertex $v \neq v_0$ of G. If W uses the unique outgoing edge from v that is part of T, then W must have previously used all other outgoing edges from v, by definition of W. Since W ends at v_0, W certainly uses all outgoing edges from v_0. All edges are accounted for in this way, proving the claim.

Finally, to get a contradiction, assume that some edge e in T from x to y is not used by W. Since T is a rooted tree rooted at v_0, we can choose such an e so that the distance from y to v_0 through edges in T is minimal. If $y \neq v_0$, minimality implies that the unique edge leading out of y in T *does* belong to W. Then, as noted in the last paragraph, every outgoing edge from y in G is used in W. Since G is balanced, every incoming edge into y in

G must also appear in W, contradicting the assumption that e is not used by W. On the other hand, if $y = v_0$, we see similarly that W uses every outgoing edge from y in G and hence every incoming edge to y in G. Again, this contradicts the assumption that e is not in W. This completes the proof of the Eulerian Tour Rule.

Summary

Table 3.2 contains brief definitions of the terminology from graph theory used in this chapter.

- *Facts about Matrix Multiplication.* If $A_1, \ldots, A_s$ are matrices such that A_t is $n_{t-1} \times n_t$, then the i, j-entry of the product $A_1 A_2 \cdots A_s$ is

$$\sum_{k_1=1}^{n_1} \sum_{k_2=1}^{n_2} \cdots \sum_{k_{s-1}=1}^{n_{s-1}} A_1(i, k_1) A_2(k_1, k_2) A_3(k_2, k_3) \cdots A_s(k_{s-1}, j).$$

  If $A^s = 0$ (i.e., A is nilpotent), then $I - A$ is invertible, and

$$(I - A)^{-1} = I + A + A^2 + A^3 + \cdots + A^{s-1}.$$

  This formula applies (with $s = n$) when A is a strictly upper or lower triangular $n \times n$ matrix.

- *Adjacency Matrices and the Walk Rule.* Given a graph or digraph G with vertex set $\{v_1, \ldots, v_n\}$, the adjacency matrix of G is the matrix A such that $A(i, j)$ is the number of edges from v_i to v_j in G. For all $s \geq 0$, $A^s(i, j)$ is the number of walks in G of length s from v_i to v_j. G is a DAG iff $A^n = 0$, in which case A will be strictly lower-triangular under an appropriate ordering of the vertices. When G is a DAG, $(I - A)^{-1}(i, j)$ is the total number of paths (or walks) from v_i to v_j.

- *Degree Sum Formulas.* For a graph $G = (V, E, \epsilon)$, $\sum_{v \in V} \deg_G(v) = 2|E|$. For a digraph $G = (V, E, \epsilon)$, $\sum_{v \in V} \text{indeg}_G(v) = |E| = \sum_{v \in V} \text{outdeg}_G(v)$.

- *Functional Digraphs.* For a finite set X, every function $f : X \to X$ has an associated functional digraph with vertex set X and edge set $\{(x, f(x)) : x \in X\}$. Every functional digraph decomposes uniquely into one or more disjoint cycles together with disjoint rooted trees rooted at the vertices on these cycles. For each vertex x_0 in a functional digraph, there exist unique walks of each length k starting at x_0, which are found by repeatedly following the unique outgoing edge from the current vertex. Such walks eventually reach a cycle in the functional digraph.

- *Cycle Structure of Permutations.* For a finite set X, a map $f : X \to X$ is a bijection iff the functional digraph of f is a disjoint union of directed cycles. The signless Stirling number of the first kind, $s'(n, k)$, counts the number of bijections f on an n-element set such that the functional digraph of f has k cycles. We have

$$s'(n, k) = s'(n-1, k-1) + (n-1)s'(n-1, k) \quad \text{for } 0 < k < n.$$

- *Connectedness and Components.* The vertex set of any graph or digraph G is the disjoint union of connected components. Two vertices belong to the same component iff each vertex is reachable from the other by a walk. G is connected iff there is only one

TABLE 3.2
Terminology used in graph theory.

Term	Brief Definition
graph	(V, E, ϵ) where $\epsilon(e) = \{v, w\}$ means edge e has endpoints v, w
digraph	(V, E, ϵ) where $\epsilon(e) = (v, w)$ means edge e goes from v to w
simple graph	graph with no loops or multiple edges
simple digraph	digraph with no multiple edges
$G \cong H$	G becomes H under some renaming of vertices and edges
walk	$(v_0, e_1, v_1, \ldots, e_s, v_s)$ where each e_i is an edge from v_{i-1} to v_i
closed walk	walk starts and ends at same vertex
path	walk visiting distinct vertices
cycle	closed walk visiting distinct vertices and edges, except at end
DAG	digraph with no cycles
$\text{indeg}_G(v)$	number of edges leading to v in digraph G
$\text{outdeg}_G(v)$	number of edges leading from v in digraph G
$\deg_G(v)$	number of edges incident to v in graph G (loops count as 2)
isolated vertex	vertex of degree zero
leaf	vertex of degree one
functional digraph	simple digraph with $\text{outdeg}(v) = 1$ for all vertices v
cyclic vertex	vertex in functional digraph that belongs to a cycle
rooted tree	functional digraph with a unique cyclic vertex (the root)
G is connected	for all $u, v \in V(G)$, there is a walk in G from u to v
cut-edge of G	edge belonging to no cycle of the graph G
forest	graph with no cycles
acyclic graph	graph with no cycles
tree	connected graph with no cycles
bipartite graph	all edges in graph go from $A \subseteq V(G)$ to $B \subseteq V(G)$ with $A \cap B = \emptyset$
matching of G	set M of edges in G where no two edges in M share an endpoint
$m(G)$	size of a maximum matching of G
vertex cover of G	set C of vertices where every edge of G has an endpoint in C
$vc(G)$	size of a minimum vertex cover of G
$N(S)$	set of vertices reachable from vertices in S by following one edge
proper coloring	map $f : V(G) \to C$ assigning unequal colors to adjacent vertices
$\chi_G(x)$	number of proper colorings of G using x available colors
chromatic number	least x with $\chi_G(x) > 0$
subgraph of G	graph G' with $V(G') \subseteq V(G)$, $E(G') \subseteq E(G)$ (same endpoints)
induced subgraph	subgraph G' where all edges in G with ends in $V(G')$ are kept
spanning tree of G	subgraph of G that is a tree using all vertices
$\tau(G)$	number of spanning trees of G
rooted spanning tree	rooted tree using all vertices of a digraph
$\tau(G, v_0)$	number of rooted spanning trees of G with root v_0
Eulerian tour	walk visiting each vertex that uses every edge once

component iff for all $u, v \in V(G)$ there exists at least one *path* from u to v in G. Deleting a cut-edge splits a component of G in two, whereas deleting a non-cut-edge has no effect on components.

- *Forests.* A graph G is a forest (acyclic) iff G has no loops and for each $u, v \in V(G)$, there is at most one path from u to v. A forest with n vertices and k edges has $n-k$ components.

- *Trees.* The following conditions on an n-vertex simple graph G are equivalent and characterize trees: (a) G is connected with no cycles; (b) G is connected with at most $n-1$ edges; (c) G is acyclic with at least $n-1$ edges; (d) for all $u, v \in V(G)$, there exists a unique path in G from u to v. An n-vertex tree has $n-1$ edges and (for $n > 1$) at least two leaves. Pruning any leaf from a tree produces another tree with one less vertex and one less edge.

- *Tree Counting Rules.* There are n^{n-2} trees with vertex set $\{1, 2, \ldots, n\}$. There are n^{n-2} rooted trees on this vertex set rooted at 1. For $d_1 + \cdots + d_n = 2(n-1)$, there are $\binom{n-2}{d_1-1,\ldots,d_n-1}$ trees on this vertex set with $\deg(j) = d_j$ for all j. Bijective proofs of these facts use the following ideas:

  - Functions on $\{1, 2, \ldots, n\}$ fixing 1 and n correspond to rooted trees by arranging the cycles of the functional digraph in a certain order, breaking back edges, and linking the cycles to get a tree (see Figures 3.7 and 3.8).
  - Trees correspond to rooted trees by directing each edge of the tree toward the root vertex.
  - Trees with $\deg(j) = d_j$ correspond to words in $\mathcal{R}(1^{d_1-1} \cdots n^{d_n-1})$ by repeatedly pruning the largest leaf and appending the leaf's neighbor to the end of the word.

- *Matchings and Vertex Covers.* For any matching M and vertex cover C of a graph G, $|M| \leq |C|$. If equality holds, then M must be a maximum matching of G and C must be a minimum vertex cover of G. We have $m(G) \leq vc(G)$ for all graphs G, but equality does not always hold.

- *Bipartite Graphs.* A graph G is bipartite iff G has no cycles of odd length. For bipartite graphs G, $m(G) = vc(G)$. If G has partite sets A and B, there exists a matching of G saturating A iff for all $S \subseteq A$, $|S| \leq |N(S)|$.

- *Chromatic Polynomials.* For any edge e in a simple graph G, the chromatic function of G satisfies the recursion $\chi_G = \chi_{G-\{e\}} - \chi_{G_e}$, where $G-\{e\}$ is G with e deleted, and G_e is G with e collapsed. It follows that $\chi_G(x)$ is a polynomial function of x. The signed Stirling numbers of the first kind, $s(n, k)$, are the coefficients in the chromatic polynomial for an n-vertex graph with an edge between each pair of vertices.

- *Recursion for Spanning Trees.* For any edge e in a graph G, the number $\tau(G)$ of spanning trees of G satisfies the recursion $\tau(G) = \tau(G-\{e\}) + \tau(G_e)$, where $G-\{e\}$ is G with e deleted, and G_e is G with e collapsed. A similar recursion holds for rooted spanning trees of a digraph.

- *The Matrix-Tree Theorem.* Given a digraph G and $v_0 \in V(G)$, let $L_{ii} = \text{outdeg}_G(v_i)$, let $-L_{ij}$ be the number of edges from i to j in G, and let L_0 be the matrix obtained from $[L_{ij}]$ by erasing the row and column indexed by v_0. Then $\det(L_0)$ is $\tau(G, v_0)$, the number of rooted spanning trees of G with root v_0.

- *Eulerian Tours.* A digraph G has a closed Eulerian tour iff G is connected and balanced (indegree equals outdegree at every vertex). In this case, the number of such tours starting at v_0 is

$$\tau(G, v_0) \cdot \text{outdeg}_G(v_0)! \cdot \prod_{v \neq v_0} (\text{outdeg}_G(v) - 1)!.$$

The proof associates to each tour a rooted spanning tree built from the last departure edge from each vertex, together with (truncated) departure words for each vertex giving the order in which the tour used the other outgoing edges.

Exercises

3-1. Draw pictures of the following simple graphs.
(a) $C = (\{1,2,3,4\}, \{\{1,2\},\{1,3\},\{1,4\}\})$ (the *claw graph*)
(b) $P = (\{1,2,3,4\}, \{\{1,2\},\{1,3\},\{1,4\},\{2,3\}\})$ (the *paw graph*)
(c) $K = (\{1,2,3,4\}, \{\{1,2\},\{1,3\},\{1,4\},\{2,3\},\{2,4\}\})$ (the *kite graph*)
(d) $B = (\{1,2,3,4,5\}, \{\{1,2\},\{2,3\},\{1,3\},\{1,4\},\{2,5\}\})$ (the *bull graph*)
(e) $K_n = (\{1,2,\ldots,n\}, \{\{i,j\} : 1 \leq i < j \leq n\})$ (the *complete graph* on n vertices)

3-2. Let V be an n-element set. (a) How many simple graphs have vertex set V? (b) How many simple digraphs have vertex set V?

3-3. Let V and E be sets with $|V| = n$ and $|E| = m$. (a) How many digraphs have vertex set V and edge set E? (b) How many graphs have vertex set V and edge set E?

3-4. Let V be an n-element set. Define a bijection between the set of simple graphs with vertex set V and the set of symmetric, irreflexive binary relations on V. Conclude that simple graphs can be viewed as certain kinds of simple digraphs.

3-5. Let G, H, and K be graphs. (a) Prove $G \cong G$. (b) Prove $G \cong H$ implies $H \cong G$. (c) Prove $G \cong H$ and $H \cong K$ imply $G \cong K$. Thus, graph isomorphism is an equivalence relation on any given set of graphs.

3-6. Find all isomorphism classes of simple graphs with at most four vertices.

3-7. Find the adjacency matrices for the graphs in Exercise 3-1.

3-8. Let G be the simple graph in Figure 3.10. For $1 \leq k \leq 8$, find the number of walks of length k in G from vertex 1 to vertex 10.

3-9. Let G be the graph in Figure 3.21. Find the number of walks in G of length 5 between each pair of vertices.

3-10. Let G be the digraph in Figure 3.24. Find the number of closed walks in G of length 10 that begin at vertex 0.

3-11. Let G be a graph with adjacency matrix A. (a) Find a formula for the number of paths in G of length 2 from v_i to v_j. (b) Find a formula for the number of paths in G of length 3 from v_i to v_j.

3-12. Consider the DAG G shown here.

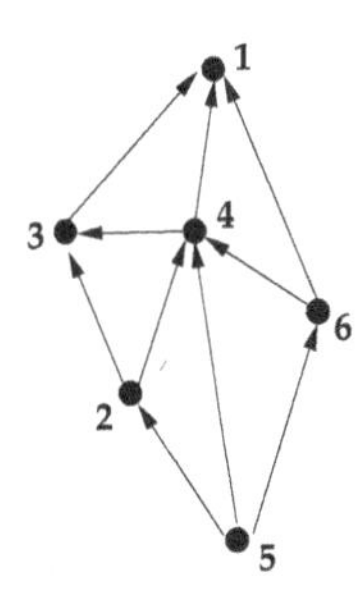

(a) Find all total orderings of the vertices for which the adjacency matrix of G is strictly lower-triangular. (b) How many paths in G go from vertex 5 to vertex 1?

3-13. A *strict partial order* on a set X is an irreflexive, transitive binary relation on X. Given a strict partial order R on a finite set X, show that the simple digraph (X, R) is a DAG.

3-14. For each of the following sets X and strict partial orders R, draw the associated DAG and calculate the number of paths from the smallest element to the largest element of the partially ordered set.

(a) $X = \{1, 2, 3, 4, 5\}$ under the ordering $1 < 2 < 3 < 4 < 5$.
(b) X is the set of subsets of $\{1, 2, 3\}$, and $(S, T) \in R$ iff $S \subsetneq T$.
(c) X is the set of positive divisors of 60, and $(a, b) \in R$ iff $a < b$ and a divides b.

3-15. Let $X = \{1, 2, \ldots, n\}$ ordered by $1 < 2 < \cdots < n$. In the associated DAG, how many paths go from 1 to n? Can you find a combinatorial (not algebraic) proof of your answer?

3-16. Let X be the set of subsets of $\{1, 2, \ldots, n\}$ ordered by strict set inclusion. In the associated DAG, how many paths go from $\emptyset$ to $\{1, 2, \ldots, n\}$?

3-17. Given a digraph G, construct a simple digraph H as follows. The vertices of H are the strong components of G. Given $C, D \in V(H)$ with $C \neq D$, there is an edge from C to D in H iff there exists $c \in C$ and $d \in D$ such that there is an edge from c to d in G. (a) Prove that H is a DAG. (b) Conclude that some strong component C of G has no incoming edges from outside C, and some strong component D has no outgoing edges. (c) Draw the DAGs associated to the digraph G_3 in Figure 3.1 and the functional digraph in Figure 3.5.

3-18. (a) Find the degree multiset for the graph in Figure 3.10, and verify Theorem 3.35 in this case. (b) Compute the indegrees and outdegrees at each vertex of the digraph in Figure 3.24, and verify Theorem 3.32 in this case.

3-19. Find necessary and sufficient conditions for a multiset $[d_1, d_2, \ldots, d_n]$ to be the degree multiset of a graph G.

3-20. Consider the cycle graph C_n defined in Example 3.84. (a) What is $\deg(C_n)$? (b) Show that any connected graph with the degree multiset in (a) must be isomorphic to C_n. (c) How many graphs with vertex set $\{1, 2, \ldots, n\}$ are isomorphic to C_n? (d) How many isomorphism classes of graphs have the same degree multiset as C_n? (e) How many isomorphism classes of *simple* graphs have the same degree multiset as C_n?

3-21. Consider the path graph P_n defined in Example 3.84. (a) What is $\deg(P_n)$? (b) Show that any connected graph with the degree multiset in (a) must be isomorphic to P_n. (c) How many graphs with vertex set $\{1, 2, \ldots, n\}$ are isomorphic to P_n? (d) How many isomorphism classes of graphs have the same degree multiset as P_n?

3-22. Find two simple graphs G and H with the smallest possible number of vertices, such that $\deg(G) = \deg(H)$ but $G \not\cong H$.

3-23. Prove or disprove: there exists a simple graph G with more than one vertex such that the degree multiset $\deg(G)$ contains no repetitions.

3-24. Prove or disprove: there exists a graph G with no loops and more than one vertex such that the degree multiset $\deg(G)$ contains no repetitions.

3-25. Given a graph $G = (V, E, \epsilon)$, we can encode the endpoint function ϵ by a $|V| \times |E|$ matrix M, with rows indexed by V and columns indexed by E, such that $M(v, e)$ is 2 if e is a loop edge at v, 1 if e is a non-loop edge incident to v, and 0 otherwise. M is called the *incidence matrix* of G. Prove the Degree Sum Formula 3.35 by computing the sum of all entries of M in two ways.

3-26. Draw the functional digraphs associated to each of the following functions $f : X \to X$. For each digraph, find the set C of cyclic vertices and the set partition $\{S_v : v \in C\}$ described in Theorem 3.43. (a) $X = \{1, 2, 3, 4\}$, f is the identity map on X; (b) $X = \{0, 1, \ldots, 6\}$, $f(x) = (x^2+1) \bmod 7$; (c) $X = \{0, 1, \ldots, 12\}$, $f(x) = (x^2+1) \bmod 13$; (d) $X = \{0, 1, \ldots, 10\}$, $f(x) = 3x \bmod 11$; (e) $X = \{0, 1, \ldots, 11\}$, $f(x) = 4x \bmod 12$.

3-27. Let $X = \{0, 1, 2, \ldots, 9\}$. (a) Define $f : X \to X$ by setting $f(x) = (3x + 7) \bmod 10$. Draw the functional digraphs for f, f^{-1} and $f \circ f$. What is the smallest integer $k > 0$ such that $f \circ f \circ \cdots \circ f$ (k factors) is the identity map on X? (b) Define $g : X \to X$ by setting $g(x) = (2x + 3) \bmod 10$. Draw the functional digraphs for g and $g \circ g$.

3-28. Let X be a finite set, let $x_0 \in X$, and let $f : X \to X$ be any function. Recursively define $x_{m+1} = f(x_m)$ for all $m \geq 0$. Show that there exists $i > 0$ with $x_i = x_{2i}$.

3-29. Pollard-rho Factoring Algorithm. Suppose $N > 1$ is an integer. Let $X = \{0, 1, \ldots, N-1\}$, and define $f : X \to X$ by $f(x) = (x^2 + 1) \bmod N$. (a) Show that the following algorithm always terminates and returns a divisor of N greater than 1. (Use the previous exercise.)

Step 1. Set $u = f(0)$, $v = f(f(0))$, and $d = \gcd(v - u, N)$.
Step 2. While $d = 1$: set $u = f(u)$, $v = f(f(v))$, and $d = \gcd(v - u, N)$.
Step 3. Return d.

(b) Trace the steps taken by this algorithm to factor $N = 77$ and $N = 527$.

3-30. Suppose X is a finite set of size k and $f : X \to X$ is a random function (which means that for all $x, y \in X$, $P(f(x) = y) = 1/k$, and these events are independent for different choices of x). Let $x_0 \in X$, define $x_{m+1} = f(x_m)$ for all $m \geq 0$, and let S be the least index such that $x_S = x_t$ for some $t < S$. (a) For each $s \geq 0$, find the exact probability that $S > s$. (b) Argue informally that the expected value of S is at most $2\sqrt{k}$. (c) Use (b) to argue informally that the expected number of gcd computations needed by the Pollard-rho factoring algorithm to find a divisor of a composite number N is bounded above by $2N^{1/4}$.

3-31. Let V be an n-element set, and let $v_0 \notin V$. A function $f : V \to V$ is called *acyclic* iff all cycles in the functional digraph of f have length 1. Count these functions by setting up a bijection between the set of acyclic functions on V and the set of rooted trees on $V \cup \{v_0\}$ with root v_0.

3-32. How many bijections f on an 8-element set are such that the functional digraph of f has: (a) five cycles; (b) three cycles; (c) one cycle?

3-33. Let X be an n-element set. Let Y be the set of all functional digraphs for bijections $f : X \to X$. How many equivalence classes does Y have under the equivalence relation of graph isomorphism?

3-34. How many functional digraphs with vertex set $\{1, 2, \ldots, n\}$ have a_1 cycles of length 1, a_2 cycles of length 2, etc., where $\sum_i i a_i = n$?

3-35. Referring to the proof of the Rooted Tree Rule, draw pictures of the set A of functions, the set B of trees, and the bijection $\phi : A \to B$ when $n = 4$.

3-36. Compute the rooted tree associated to the function below by the map ϕ in the proof of the Rooted Tree Rule.

$$\begin{array}{lllll} f(1)=1; & f(2)=19; & f(3)=8; & f(4)=30; & f(5)=5; \\ f(6)=15; & f(7)=8; & f(8)=9; & f(9)=26; & f(10)=23; \\ f(11)=21; & f(12)=30; & f(13)=27; & f(14)=13; & f(15)=28; \\ f(16)=16; & f(17)=13; & f(18)=23; & f(19)=25; & f(20)=11; \\ f(21)=5; & f(22)=19; & f(23)=25; & f(24)=30; & f(25)=18; \\ f(26)=9; & f(27)=16; & f(28)=15; & f(29)=7; & f(30)=30. \end{array}$$

3-37. Compute the function associated to the rooted tree with edge set

$$\{(1,1),(2,12),(3,1),(4,3),(5,10),(6,17),(7,15),(8,7),(9,3),\\ (10,3),(11,12),(12,1),(13,4),(14,10),(15,1),(16,4),(17,4)\}$$

by the map ϕ^{-1} in the proof of the Rooted Tree Rule.

3-38. Formulate a theorem for rooted trees similar to Theorem 3.74, and prove it by analyzing the bijection in the Rooted Tree Rule.

3-39. Let G be the digraph in Figure 3.2. Use the algorithm in Theorem 3.52 to convert the walk

$$W=(1,b,1,b,1,a,3,f,5,m,2,n,5,h,4,c,3,f,5,j,4,g,5,m,2,k,4)$$

to a path in G from 1 to 4.

3-40. What are the strong components of a functional digraph?

3-41. Show that a connected graph G with n vertices has n edges iff G has exactly one cycle.

3-42. Prove that a graph G is not connected iff there exists an ordering of the vertices of G for which the adjacency matrix of G is block-diagonal with at least two diagonal blocks.

3-43. Prove Theorem 3.60.

3-44. How many connected simple graphs have vertex set $\{1,2,3,4\}$?

3-45. Prove that every forest is bipartite.

3-46. How many connected simple graphs on the vertex set $\{1,2,3,4,5\}$ have exactly five edges?

3-47. Prove that a graph G with no odd-length cycles is bipartite by induction on the number of edges in G.

3-48. How many bipartite simple graphs have partite sets $A=\{1,2,\dots,m\}$ and $B=\{m+1,\dots,m+n\}$?

3-49. Suppose G is a bipartite graph with c components. Count the number of decompositions of $V(G)$ into an ordered pair of partite sets (A,B).

3-50. Suppose G is a k-regular graph with n vertices. (a) How many edges are in G? (b) If $k>0$ and G is bipartite with partite sets A and B, prove that $|A|=|B|$.

3-51. Fix $k\geq 2$. Prove or disprove: there exists a k-regular bipartite graph G such that G has a cut-edge.

3-52. Suppose G is a graph, and G' is obtained from G by deleting some edges of G. What is the relation between $m(G)$ and $m(G')$? What is the relation between $vc(G)$ and $vc(G')$?

3-53. For each $k\geq 0$, give a specific example of a graph G such that $vc(G)-m(G)=k$.

3-54. Find a maximum matching and a minimum vertex cover for the graph shown here.

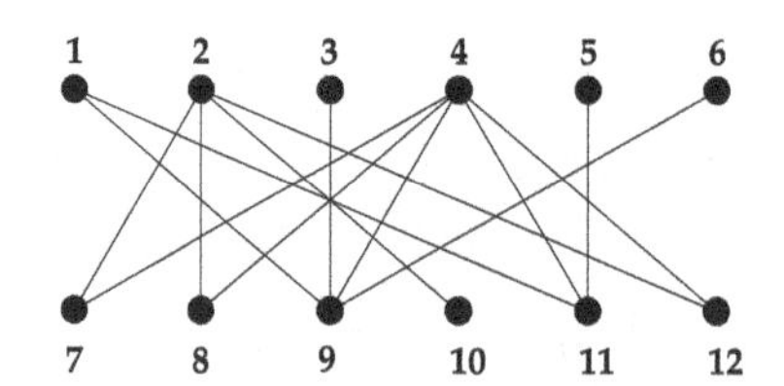

3-55. For $m, n \geq 1$, the *grid graph* $G_{m,n}$ has vertex set $\{1, 2, \ldots, m\} \times \{1, 2, \ldots, n\}$ with edges from (i, j) to $(i + 1, j)$ for $1 \leq i < m$, $1 \leq j \leq n$, and edges from (i, j) to $(i, j + 1)$ for $1 \leq i \leq m$, $1 \leq j < n$. For example, Figure 1.2 displays several copies of the graph $G_{3,4}$. Find a maximum matching and a minimum vertex cover for each graph $G_{m,n}$, and hence determine $m(G_{m,n})$ and $vc(G_{m,n})$.

3-56. Suppose X is a finite set, and $P = \{S_1, \ldots, S_k\}$ is a collection of subsets of X. A *system of distinct representatives* for P is a list $a_1, \ldots, a_k$ of k distinct elements of X such that $a_i \in S_i$ for all i. Prove that P has a system of distinct representatives iff for all $I \subseteq \{1, 2, \ldots, k\}$, $|I| \leq |\bigcup_{i \in I} S_i|$.

3-57. An *independent set* of a graph G is a subset I of the vertex set of G such that no two vertices in I are adjacent. Let $i(G)$ be the size of a maximum independent set of G. An *edge cover* of G is a subset C of the edge set of G such that every vertex of G is the endpoint of some edge in C. Let $ec(G)$ be the size of a minimum edge cover of G. Prove an analogue of the Lemma on Matchings and Vertex Covers for the numbers $i(G)$ and $ec(G)$.

3-58. Show that I is an independent set of a graph G iff $V(G) - I$ is a vertex cover of G. Conclude that $i(G) + vc(G) = |V(G)|$.

3-59. Let G be a graph with no isolated vertex. (a) Given any maximum matching M of G, use M to construct an edge cover of G of size $|V(G)| - |M|$. (b) Given any minimum edge cover C of G, use C to construct a matching of G of size $|V(G)| - |C|$. (c) Conclude that $m(G) + ec(G) = |V(G)|$.

3-60. Use the preceding exercises to prove that for a bipartite graph G with no vertex of degree zero, $i(G) = ec(G)$.

3-61. Use Hall's Matching Theorem to prove the König-Egerváry Theorem.

3-62. Prove that an n-vertex graph G in which every vertex has degree at least $(n - 1)/2$ must be connected.

3-63. Let G be a forest with n vertices and k connected components. Compute $\sum_{v \in V(G)} \deg_G(v)$ in terms of n and k.

3-64. The *arboricity* of a simple graph G, denoted $\text{arb}(G)$, is the least n such that there exist n forests F_i with $V(G) = \bigcup_{i=1}^{n} V(F_i)$ and $E(G) = \bigcup_{i=1}^{n} E(F_i)$. Prove that

$$\text{arb}(G) \geq \max_H \left\lceil \frac{|E(H)|}{|V(H)| - 1} \right\rceil,$$

where H ranges over all induced subgraphs of G with more than one vertex. (It can be shown that equality holds [94].)

3-65. Show that any tree not isomorphic to a path graph P_n must have at least three leaves.

3-66. Let T be a tree. Show that $\deg_T(v)$ is odd for all $v \in V(T)$ iff for all $e \in E(T)$, both connected components of $(V(T), E(T) - \{e\})$ have an odd number of vertices.

3-67. Helly Property of Trees. Suppose T, T_1, ..., T_k are trees, each T_i is a subgraph of T, and $V(T_i) \cap V(T_j) \neq \emptyset$ for all $i, j \leq k$. Show that $\bigcap_{i=1}^{k} V(T_i) \neq \emptyset$.

3-68. Let G be a tree with leaves $\{v_1, \ldots, v_m\}$. Let H be a tree with leaves $\{w_1, \ldots, w_m\}$.

Suppose that, for each i and j, the length of the unique path in G from v_i to v_j equals the length of the unique path in H from w_i to w_j. Prove $G \cong H$.

3-69. For $1 \leq n \leq 7$, count the number of isomorphism classes of trees with n vertices.

3-70. (a) How many isomorphism classes of n-vertex trees have exactly three leaves? (b) How many trees with vertex set $\{1, 2, \ldots, n\}$ have exactly three leaves?

3-71. How many trees with vertex set $\{1, 2, \ldots, n\}$ have exactly k leaves?

3-72. Let K_n be the complete graph on n vertices (see Exercise 3-1). (a) Give a bijective or probabilistic proof that every edge of K_n appears in the same number of spanning trees of K_n. (b) Use Cayley's Theorem to count the spanning trees of K_n that do not use the edge $\{1, 2\}$.

3-73. Use Theorem 3.74 to find the number of trees T with $V(T) = \{1, 2, \ldots, 8\}$ and $\deg(T) = [3, 3, 3, 1, 1, 1, 1, 1]$.

3-74. Let t_n be the number of trees on a given n-element vertex set. Without using Cayley's Theorem, prove the recursion

$$t_n = \sum_{k=1}^{n-1} k \binom{n-2}{k-1} t_k t_{n-k}.$$

3-75. (a) Use the pruning bijection to find the word associated to the tree

$$T = (\{0, 1, \ldots, 8\}, \{\{1, 5\}, \{2, 8\}, \{3, 7\}, \{7, 0\}, \{6, 2\}, \{4, 7\}, \{5, 4\}, \{2, 4\}\}).$$

(b) Use the inverse of the pruning bijection to find the tree with vertex set $\{0, 1, \ldots, 8\}$ associated to the word 1355173.

3-76. Use the inverse of the pruning bijection to find all trees with vertex set $\{1, 2, \ldots, 7\}$ associated to the words in $\mathcal{R}(11334)$.

3-77. Let G be the graph with vertex set $\{\pm 1, \pm 2, \ldots, \pm n\}$ and with an edge between i and $-j$ for all $i, j \in \{1, 2, \ldots, n\}$. (a) Show that any spanning tree in G has at least one positive leaf and at least one negative leaf. (b) Develop an analogue of the pruning map that sets up a bijection between the set of spanning trees of G and pairs of words (u, v), where $u \in \{1, \ldots, n\}^{n-1}$ and $v \in \{-1, \ldots, -n\}^{n-1}$. Conclude that G has n^{2n-2} spanning trees.

3-78. Let $\chi_n(x)$ be the chromatic polynomial for the graph C_n consisting of n vertices joined in a cycle. Prove that

$$\chi_n(x) = (x-1)^n + (-1)^n (x-1) \quad \text{for all } n \geq 2.$$

3-79. Find the chromatic polynomials for the graphs in Exercise 3-1.

3-80. Find the chromatic polynomial and chromatic number for the graph G_2 in Figure 3.1.

3-81. Find two non-isomorphic simple graphs with the same chromatic polynomial.

3-82. A certain department needs to schedule meetings for a number of committees, whose members are listed in the following table.

Committee	Members
Advisory	Driscoll, Loomis, Lasker
Alumni	Sheffield, Loomis
Colloquium	Johnston, Tchaikovsky, Zorn
Computer	Loomis, Clark, Spade
Graduate	Kennedy, Loomis, Trotter
Merit	Lee, Rotman, Fowler, Sheffield
Personnel	Lasker, Schreier, Tchaikovsky, Trotter
Undergraduate	Jensen, Lasker, Schreier, Trotter, Perkins

(a) What is the minimum number of time slots needed so that all committees can meet with no time conflicts? (b) How many non-conflicting schedules are possible if there are six (distinguishable) time slots available? (c) Repeat (a) and (b), assuming that Zorn becomes a member of the Merit Committee (and remains a member of the Colloquium Committee).

3-83. Let K_n be the complete graph on n vertices (see Exercise 3-1). (a) How many subgraphs does K_n have? (b) How many induced subgraphs does K_n have?

3-84. Prove that a graph G has at least one spanning tree iff G is connected.

3-85. Fill in the details of the proof of Theorem 3.110.

3-86. Use the Spanning Tree Recursion 3.107 to find $\tau(G_1)$ for the graph G_1 in Figure 3.1.

3-87. Let T_1 and T_2 be spanning trees of a graph G. (a) If $e_1 \in E(T_1)-E(T_2)$, prove there exists $e_2 \in E(T_2)-E(T_1)$ such that

$$T_3 = (V(G), (E(T_1)-\{e_1\}) \cup \{e_2\})$$

is a spanning tree of G. (b) If $e_1 \in E(T_1)-E(T_2)$, prove there exists $e_2 \in E(T_2)-E(T_1)$ such that

$$T_4 = (V(G), (E(T_2) \cup \{e_1\})-\{e_2\})$$

is a spanning tree of G.

3-88. Fix $k \geq 3$. For each $n \geq 1$, let G_n be a graph obtained by gluing together n regular k-sided polygons in a row along shared edges. The figure below illustrates the case $k = 6$, $n = 5$.

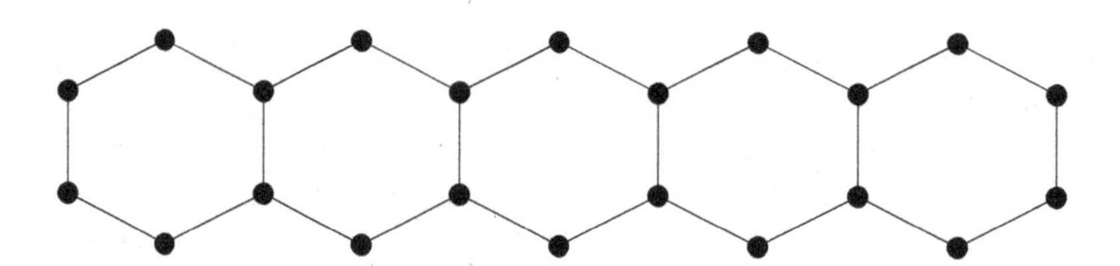

Let G_0 consist of a single edge. Prove the recursion

$$\tau(G_n) = k\tau(G_{n-1}) - \tau(G_{n-2}) \quad \text{for all } n \geq 2.$$

What are the initial conditions?

3-89. Find $m(G)$ and $vc(G)$ for the graph G displayed in the previous exercise.

3-90. Given a simple graph G, let $G-v$ be the induced subgraph with vertex set $V(G)-\{v\}$. Assume $|V(G)| = n \geq 3$. (a) Prove that $|E(G)| = (n-2)^{-1} \sum_{v \in V(G)} |E(G-v)|$. (b) Prove that, for $v_0 \in V(G)$, $\deg_G(v_0) = (n-2)^{-1} \sum_{v \in V(G)} |E(G-v)| - |E(G-v_0)|$.

3-91. For each graph in Exercise 3-1, count the number of spanning trees by direct enumeration, and again by the matrix-tree theorem.

3-92. Confirm by direct enumeration that the digraph in Figure 3.24 has 16 spanning trees rooted at 0.

3-93. Let G be the graph with vertex set $\{0,1\}^3$ such that there is an edge between $v, w \in V(G)$ iff the words v and w differ in exactly one position. Find the number of spanning trees of G.

3-94. Let I be the $m \times m$ identity matrix, let J be the $m \times m$ matrix all of whose entries are 1, and let t, u be scalars. Show that $\det(tI - uJ) = t^m - mt^{m-1}u$.

3-95. Deduce Cayley's Theorem 3.71 from the Matrix-Tree Theorem 3.113.

3-96. Let A and B be disjoint sets of size m and n, respectively. Let G be the simple graph with vertex set $A \cup B$ and edge set $\{\{a, b\} : a \in A, b \in B\}$. Show that $\tau(G) = m^{n-1}n^{m-1}$.

3-97. How many closed Eulerian tours starting at vertex 5 does the digraph in Figure 3.25 have?

3-98. Find necessary and sufficient conditions for a graph to have a (not necessarily closed) Eulerian tour.

3-99. Consider a *digraph with indistinguishable edges* consisting of a vertex set V and a *multiset* of directed edges $(u, v) \in V \times V$. Formulate the notion of a closed Eulerian tour for such a digraph, and prove an analogue of Theorem 3.121.

3-100. de Bruijn Sequences. Let $A = \{x_1, \ldots, x_n\}$ be an n-letter alphabet. For each $k \geq 2$, show that there exists a word $w = w_0 w_1 \cdots w_{n^k-1}$ such that the n^k words

$$w_i w_{i+1} \cdots w_{i+k-1} \quad \text{(where } 0 \leq i < n^k \text{ and subscripts are reduced mod } n^k\text{)}$$

consist of all possible k-letter words over A.

3-101. The *Petersen graph* is the graph G with vertex set consisting of all two-element subsets of $\{1, 2, 3, 4, 5\}$, and with edge set $\{\{A, B\} : A \cap B = \emptyset\}$. (a) Compute the number of vertices and edges in G. (b) Show that G is isomorphic to each of the graphs shown here.

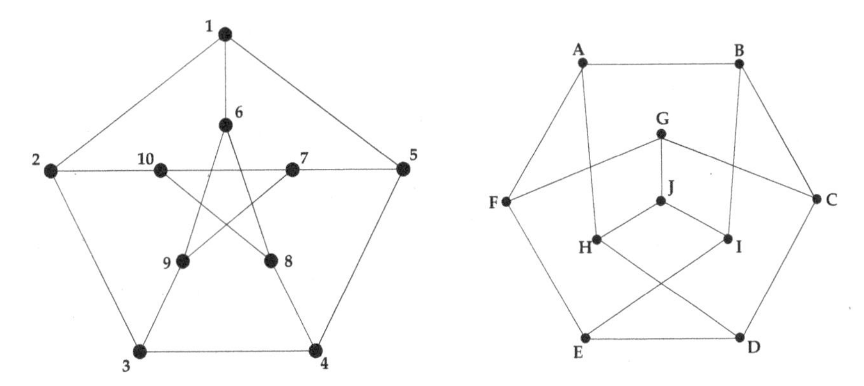

(c) Show that G is 3-regular. (d) Is G bipartite? (e) Show that any two non-adjacent vertices in G have exactly one common neighbor.

3-102. Find (with proof) all k such that the Petersen graph has a cycle of length k.

3-103. Given any edge e in the Petersen graph G, count the number of cycles of length 5 in G that contain e. Use this to count the total number of cycles of length 5 in G.

3-104. (a) Prove that the Petersen graph G has exactly ten cycles of length 6. (b) How many claws (see Exercise 3-1) appear as induced subgraphs of G?

3-105. How many spanning trees does the Petersen graph have?

Notes

Our coverage of graph theory in this chapter has been limited to a few enumerative topics. Systematic expositions of graph theory may be found in [13, 16, 17, 25, 52, 60, 130, 136]; the text by West is especially recommended. Roberts [109] gives a treatment of graph theory that emphasizes applications.

The bijection used to enumerate rooted trees in the Rooted Tree Rule 3.47 is due to Eğecioğlu and Remmel [28]. The original proof of Cayley's Theorem appears in [21]. The pruning bijection described in §3.12 is due to Prüfer [100]; the image of a tree under this map is often called the *Prüfer code* of the tree. For more on the enumeration of trees, see [91].

Our discussion of the König-Egerváry Theorem is based on Rizzi's proof [108]. The matrix-tree theorem for undirected graphs is often attributed to Kirchhoff [71]; Tutte extended the theorem to digraphs [126]. The enumeration of Eulerian tours in the Eulerian Tour Rule 3.121 was proved by van Aardenne-Ehrenfest and de Bruijn [127].

4

Inclusion-Exclusion, Involutions, and Möbius Inversion

This chapter studies combinatorial techniques that are related to the arithmetic operation of subtraction: inclusion-exclusion formulas, involutions, and Möbius inversion. The Inclusion-Exclusion Formula extends the Disjoint Union Rule to a rule for computing $|S_1 \cup S_2 \cup \cdots \cup S_n|$ in the case where the sets S_i need not be pairwise disjoint. Involutions allow us to give bijective proofs of identities involving both positive and negative terms, including the Inclusion-Exclusion Formula. The chapter concludes by discussing a generalization of inclusion-exclusion called the Möbius Inversion Formula for Posets, which has many applications in number theory and algebra as well as combinatorics.

4.1 The Inclusion-Exclusion Formula

Recall the Disjoint Union Rule: if $S_1, S_2, \ldots, S_n$ are *pairwise disjoint* finite sets, then

$$|S_1 \cup S_2 \cup \cdots \cup S_n| = |S_1| + |S_2| + \cdots + |S_n|.$$

Can we find a formula for $|S_1 \cup \cdots \cup S_n|$ in the case where the given sets S_i are not necessarily disjoint? The answer is provided by the Inclusion-Exclusion Formula, which we discuss now.

We have already seen the smallest case of the Inclusion-Exclusion Formula. Specifically, if S and T are any two finite sets, the Union Rule for Two Sets states that

$$|S \cup T| = |S| + |T| - |S \cap T|.$$

Intuitively, the sum $|S| + |T|$ overestimates the cardinality of $|S \cup T|$ because elements of $|S \cap T|$ are included twice in this sum. To correct this, we exclude one copy of each of the elements in $S \cap T$ by subtracting $|S \cap T|$.

Now consider three finite sets S, T, and U. The sum $|S| + |T| + |U|$ overcounts the size of $|S \cup T \cup U|$ since elements in the overlaps between these sets are counted twice (or three times, in the case of elements $z \in S \cap T \cap U$). We may try to account for this by subtracting $|S \cap T| + |S \cap U| + |T \cap U|$ from $|S| + |T| + |U|$. If x belongs to S and U but not T (say), this subtraction will cause x to be counted only once in the overall expression. A similar comment applies to elements in $(S \cap T) - U$ and $(T \cap U) - S$. However, an element $z \in S \cap T \cap U$ is counted three times in $|S| + |T| + |U|$ and subtracted three times in $|S \cap T| + |S \cap U| + |T \cap U|$. So we must include such elements once again by adding the term $|S \cap T \cap U|$. In summary, we have given an informal argument suggesting that the formula

$$|S \cup T \cup U| = |S| + |T| + |U| - |S \cap T| - |S \cap U| - |T \cap U| + |S \cap T \cap U|$$

should be true.

Generalizing the pattern in the preceding example, we arrive at the following formula.

4.1. The Inclusion-Exclusion Formula (General Union Rule). Suppose $n > 0$ and $S_1, \ldots, S_n$ are any finite sets. Then

$$|S_1 \cup S_2 \cup \cdots \cup S_n| = \sum_{k=1}^{n} (-1)^{k-1} \sum_{1 \leq i_1 < i_2 < \cdots < i_k \leq n} |S_{i_1} \cap S_{i_2} \cap \cdots \cap S_{i_k}|. \tag{4.1}$$

4.2. Example. If $n = 4$, the Inclusion-Exclusion Formula says that $|S_1 \cup S_2 \cup S_3 \cup S_4|$ equals

$$\begin{aligned}
&|S_1| + |S_2| + |S_3| + |S_4| \\
&-|S_1 \cap S_2| - |S_1 \cap S_3| - |S_1 \cap S_4| - |S_2 \cap S_3| - |S_2 \cap S_4| - |S_3 \cap S_4| \\
&+|S_1 \cap S_2 \cap S_3| + |S_1 \cap S_2 \cap S_4| + |S_1 \cap S_3 \cap S_4| + |S_2 \cap S_3 \cap S_4| \\
&-|S_1 \cap S_2 \cap S_3 \cap S_4|.
\end{aligned}$$

4.3. Remark. By setting $I = \{i_1, i_2, \ldots, i_k\}$, the Inclusion-Exclusion Formula can also be written

$$|S_1 \cup \cdots \cup S_n| = \sum_{\emptyset \neq I \subseteq \{1,2,\ldots,n\}} (-1)^{|I|-1} \left| \bigcap_{i \in I} S_i \right|.$$

We now give a proof of the Inclusion-Exclusion Formula using induction. This proof may be omitted without loss of continuity. A more combinatorial proof of the formula will be given later (§4.7) after we discuss involutions.

4.4. Proof of Inclusion-Exclusion by Induction. We prove that (4.1) holds for all $n > 0$ and all finite sets $S_1, \ldots, S_n$ by induction on n. The formula reduces to $|S_1| = |S_1|$ for $n = 1$, which is true. For $n = 2$, the formula becomes

$$|S_1 \cup S_2| = |S_1| + |S_2| - |S_1 \cap S_2|,$$

and this is the Union Rule for Two Sets proved previously (see 1.41). Now assume $n > 2$ and that formula (4.1) is already known to hold for any union of $n-1$ finite sets. Let $S_1, \ldots, S_n$ be fixed finite sets. The union of n sets $S_1 \cup \cdots \cup S_n$ can be regarded as the union of the two sets $S = S_1 \cup S_2 \cup \cdots \cup S_{n-1}$ and $T = S_n$. Hence, by the Union Rule for Two Sets,

$$|S_1 \cup \cdots \cup S_n| = |S_1 \cup \cdots \cup S_{n-1}| + |S_n| - |(S_1 \cup \cdots \cup S_{n-1}) \cap S_n|.$$

Since the set operations $\cap$ and $\cup$ obey the distributive law, we can write the subtracted term as

$$|(S_1 \cap S_n) \cup (S_2 \cap S_n) \cup \cdots \cup (S_{n-1} \cap S_n)|,$$

which is the union of the $n-1$ finite sets $S_i \cap S_n$ for i in the range $1 \leq i \leq n-1$. So we can apply the induction hypothesis to this term, and to the first term $|S_1 \cup \cdots \cup S_{n-1}|$. We obtain

$$\begin{aligned}
|S_1 \cup \cdots \cup S_n| &= \sum_{k=1}^{n-1} (-1)^{k-1} \sum_{1 \leq i_1 < \cdots < i_k \leq n-1} |S_{i_1} \cap \cdots \cap S_{i_k}| \\
&\quad + |S_n| - \sum_{j=1}^{n-1} (-1)^{j-1} \sum_{1 \leq i_1 < \cdots < i_j \leq n-1} |(S_{i_1} \cap S_n) \cap \cdots \cap (S_{i_j} \cap S_n)|.
\end{aligned}$$

We modify the second line of this formula as follows. First, observe that

$$\bigcap_{r=1}^{j} (S_{i_r} \cap S_n) = S_{i_1} \cap S_{i_2} \cap \cdots \cap S_{i_j} \cap S_n.$$

Next, change the summation index by setting $k = j + 1$ and defining $i_k = n$. The full formula now reads

$$\begin{aligned}|S_1 \cup \cdots \cup S_n| &= \sum_{k=1}^{n-1}(-1)^{k-1} \sum_{1 \leq i_1 < \cdots < i_k < n} |S_{i_1} \cap \cdots \cap S_{i_k}| \\ &\quad + |S_n| + \sum_{k=2}^{n}(-1)^{k-1} \sum_{1 \leq i_1 < \cdots < i_{k-1} < i_k = n} |S_{i_1} \cap \cdots \cap S_{i_k}|.\end{aligned}$$

We can absorb $|S_n|$ into the sum on the second line by allowing k to range from 1 to n there. Also, letting k range from 1 to n in the first summation does not introduce any new terms. After making these adjustments, the only difference between the formulas on the first and second lines is that $i_k < n$ in the first line while $i_k = n$ in the second line. We can now combine the two summations to obtain

$$|S_1 \cup \cdots \cup S_n| = \sum_{k=1}^{n}(-1)^{k-1} \sum_{1 \leq i_1 < \cdots < i_k \leq n} |S_{i_1} \cap \cdots \cap S_{i_k}|, \tag{4.2}$$

which is the required formula (4.1). This completes the induction proof.

4.5. The Union-Avoiding Rule. Suppose $S_1, \ldots, S_n$ are subsets of a finite set X. The number of elements $x \in X$ that belong to *none* of the S_i is

$$|X - (S_1 \cup \cdots \cup S_n)| = |X| + \sum_{k=1}^{n}(-1)^k \sum_{1 \leq i_1 < i_2 < \cdots < i_k \leq n} |S_{i_1} \cap S_{i_2} \cap \cdots \cap S_{i_k}|.$$

This equation follows by applying the original Inclusion-Exclusion Formula and the Difference Rule 1.40.

Intuitively, the preceding formula is applicable when we are trying to count objects in X that must simultaneously avoid a number of specified *bad* properties. Each set S_i consists of those objects in X that have the ith bad property (and possibly other bad properties too). We give examples of such counting problems in the next section.

In many applications of inclusion-exclusion, the sizes of the various intersections $|S_{i_1} \cap S_{i_2} \cap \cdots \cap S_{i_k}|$ (for fixed k) are all the same. When this happens, we have the following simplified version of the Inclusion-Exclusion Formula.

4.6. Simplified Version of the Inclusion-Exclusion Formula. Let $S_1, \ldots, S_n$ be finite sets. Suppose that for all $k \geq 1$, the intersection of any k distinct sets among the S_j's always has cardinality $N(k)$. In other words, $|S_{i_1} \cap S_{i_2} \cap \cdots \cap S_{i_k}| = N(k)$ for all choices of indices $i_1 < i_2 < \cdots < i_k$. Then

$$|S_1 \cup \cdots \cup S_n| = \sum_{k=1}^{n}(-1)^{k-1}\binom{n}{k}N(k).$$

If all S_j are subsets of a given finite set X, we also have

$$|X - (S_1 \cup \cdots \cup S_n)| = |X| + \sum_{k=1}^{n}(-1)^k\binom{n}{k}N(k).$$

These equations follow by substituting $N(k)$ for each summand $|S_{i_1} \cap \cdots \cap S_{i_k}|$ in the previous inclusion-exclusion formulas and noting (by the Subset Rule) that there are $\binom{n}{k}$ such summands.

4.2 Examples of the Inclusion-Exclusion Formula

We can use inclusion-exclusion formulas to enumerate complicated collections of objects that would be very difficult to count using only the rules in Chapter 1. We begin with some problems illustrating the Union-Avoiding Rule. The key to using this rule is to choose the sets $S_1, \ldots, S_n$ so that each set S_i consists of those objects in some big set X that have a certain bad property. We want to count the objects in X that avoid all of the bad properties. The set of all these good objects is $X-(S_1 \cup \cdots \cup S_n)$, and the cardinality of this set is given by the Union-Avoiding Rule. For this approach to succeed, we must be able to count the sets $S_{i_1} \cap S_{i_2} \cap \cdots \cap S_{i_k}$ using other counting rules.

4.7. Example: Bridge Hands. A *bridge hand* is a 13-element subset of a 52-card deck. A *face card* is a jack, queen, king, or ace. How many bridge hands have at least one of each kind of face card? To answer this question, let X be the set of all bridge hands; we know $|X| = \binom{52}{13}$ by the Subset Rule. In this problem, a hand is *bad* iff it lacks a particular kind of face card. So we introduce the set S_1 of all 13-card hands that do not contain a jack. Similarly, let S_2 be the set of hands in X containing no queen; let S_3 be the set of hands in X containing no king; and let S_4 be the set of hands in X containing no ace. Then the good hands we are trying to count are precisely the members of the set $X-(S_1 \cup S_2 \cup S_3 \cup S_4)$.

To use the Union-Avoiding Rule, we must now compute the sizes of the various intersections $S_{i_1} \cap \cdots \cap S_{i_k}$. Note that $|S_1| = \binom{48}{13}$ since we can build a hand in S_1 by choosing 13 cards out of the 48 non-jacks in the deck. Similarly, $|S_2| = |S_3| = |S_4| = \binom{48}{13}$. Next, $|S_1 \cap S_3| = \binom{44}{13}$ since we can build hands in $S_1 \cap S_3$ by choosing 13 cards out of the 44 cards in the deck that are neither jacks nor kings. The same formula holds for all other intersections of the form $S_{i_1} \cap S_{i_2}$. Similarly, each intersection $S_{i_1} \cap S_{i_2} \cap S_{i_3}$ has size $\binom{40}{13}$, while $|S_1 \cap S_2 \cap S_3 \cap S_4| = \binom{36}{13}$. Observe that the simplified version of the Union-Avoiding Rule can be used here, with $N(k) = \binom{52-4k}{13}$ for $k = 1, 2, 3, 4$. Thus, the answer to the original question is

$$\binom{52}{13} - 4\binom{48}{13} + 6\binom{44}{13} - 4\binom{40}{13} + \binom{36}{13} = 128{,}971{,}619{,}088.$$

Next, how many 13-card bridge hands have at least one jack, at least one queen, and at least one king, but do not contain any ace cards or spade cards? The last condition can be dealt with as follows: throw out the $13+4-1 = 16$ aces and spades at the outset, leaving $52-16 = 36$ cards. An inclusion-exclusion argument like the one in the last paragraph now leads to the answer

$$\binom{36}{13} - 3\binom{33}{13} + 3\binom{30}{13} - \binom{27}{13} = 930{,}511{,}530.$$

4.8. Example: Words. How many words in $X = \mathcal{R}(1^2 2^2 3^2 \cdots n^2)$ never have two adjacent letters that are equal? Note first that $|X| = \binom{2n}{2,2,\ldots,2} = (2n)!/2^n$ by the Anagram Rule. Next we must define sets $S_1, \ldots, S_n$ to model the bad properties that we need to avoid. For $1 \leq i \leq n$, let S_i be the set of words in X in which the two copies of letter i are adjacent to each other. Our goal is to count the words in $X-(S_1 \cup \cdots \cup S_n)$. To do so, fix $i_1 < i_2 < \cdots < i_k$ and consider the intersection $S_{i_1} \cap \cdots \cap S_{i_k}$. Given a word w in this intersection, form a new word by replacing the two consecutive copies of i_j by a single copy of i_j, for $1 \leq j \leq k$. This operation defines a bijection from $S_{i_1} \cap \cdots \cap S_{i_k}$ onto the set $\mathcal{R}(1^{a_1} 2^{a_2} \cdots n^{a_n})$, where $a_i = 1$ if $i = i_j$ for some j, and $a_i = 2$ otherwise. (The inverse bijection replaces each i_j by two consecutive copies of i_j.) By the Bijection Rule and the Anagram Rule, we conclude that

$$|S_{i_1} \cap \cdots \cap S_{i_k}| = \binom{1k + 2(n-k)}{\underbrace{1,\ldots,1}_{k},\underbrace{2,\ldots,2}_{n-k}} = (2n-k)!/2^{n-k}.$$

This expression depends only on k, not on the indices $i_1, \ldots, i_k$. Also, when $k = 0$, this expression reduces to $|X|$. Using the Simplified Inclusion-Exclusion Formula 4.6, we conclude that

$$|X-(S_1 \cup \cdots \cup S_n)| = \sum_{k=0}^{n} (-1)^k \binom{n}{k} \frac{(2n-k)!}{2^{n-k}}.$$

4.9. Example: Integer Equations. For given n, m, and b, how many integer sequences $(z_1, z_2, \ldots, z_n)$ solve the equation $z_1 + z_2 + \cdots + z_n = m$ and also satisfy $0 \leq z_i \leq b$ for all i? To answer this, let X be the set of all solutions to this equation with each $z_i \in \mathbb{Z}_{\geq 0}$. For $1 \leq i \leq n$, let S_i be the set of solutions where $z_i > b$. By the Integer Equation Rule, $|X| = \binom{m+n-1}{m,n-1}$. Next, fix indices $i_1, \ldots, i_k$ with $1 \leq i_1 < \cdots < i_k \leq n$. By subtracting $b+1$ from each z_{i_j}, we obtain a bijection from the set of sequences in $S_{i_1} \cap \cdots \cap S_{i_k}$ onto the set of sequences $(y_1, \ldots, y_n)$ satisfying $y_i \in \mathbb{Z}_{\geq 0}$ for all i and $y_1 + \cdots + y_n = m - k(b+1)$. By the Bijection Rule and the Integer Equation Rule, we conclude that $|S_{i_1} \cap \cdots \cap S_{i_k}| = \binom{m-k(b+1)+n-1}{m-k(b+1),n-1}$. This quantity depends on k but not on $i_1, \ldots, i_k$. Using the Simplified Inclusion-Exclusion Formula 4.6, we obtain the answer

$$|X-(S_1 \cup \cdots \cup S_n)| = \sum_{k=0}^{n} (-1)^k \binom{n}{k} \binom{m - k(b+1) + n - 1}{n-1}.$$

For instance, if $n = 5$, $m = 20$, and $b = 6$, the number of solutions is $\binom{24}{4} - 5\binom{17}{4} + 10\binom{10}{4} = 826$.

4.3 Surjections and Stirling Numbers

We now use inclusion-exclusion to prove the Surjection Rule, which was stated without proof in Chapter 1.

4.10. The Surjection Rule. For $m \geq n \geq 1$, the number of surjections from an m-element set onto an n-element set is $\sum_{k=0}^{n} (-1)^k \binom{n}{k} (n-k)^m$.

Proof. Fix $m \geq n \geq 1$; we count the surjections with domain $A = \{a_1, a_2, \ldots, a_m\}$ and codomain $B = \{b_1, b_2, \ldots, b_n\}$. Let X be the set of *all* functions $f : A \to B$. By the Function Rule, $|X| = n^m$. For $1 \leq i \leq n$, let S_i consist of all functions $f \in X$ such that b_i is *not* in the image of f. A function $f \in X$ is a surjection iff f belongs to none of the S_i. Thus, we must compute $|X-(S_1 \cup \cdots \cup S_n)|$. Consider one of the intersections $S_{i_1} \cap \cdots \cap S_{i_k}$, where $1 \leq i_1 < i_2 < \cdots < i_k \leq n$. By shrinking the codomain, each function f belonging to this intersection corresponds bijectively to an arbitrary function mapping A into the $(n-k)$-element codomain $B-\{b_{i_1}, b_{i_2}, \ldots, b_{i_k}\}$. By the Function Rule, the number of such functions is $(n-k)^m$, a value which depends on k but not on $i_1, \ldots, i_k$. Using the Simplified Inclusion-Exclusion Formula 4.6, we find that

$$|X-(S_1 \cup \cdots \cup S_n)| = n^m + \sum_{k=1}^{n} (-1)^k \binom{n}{k} (n-k)^m.$$

We can absorb n^m into the sum by letting k start at 0, which gives the formula in the statement of the Surjection Rule. □

In §2.13, we proved the following alternate version of the Surjection Rule: the number of surjections from an m-element set onto an n-element set is $S(m,n)n!$, where $S(m,n)$ is a Stirling number of the second kind. Comparing this to the rule just proved, we obtain the following summation formula for Stirling numbers.

4.11. Corollary: Summation Formula for Stirling Numbers of the Second Kind. For $m \geq n \geq 1$,

$$S(m,n) = \frac{1}{n!}\sum_{k=0}^{n}(-1)^k\binom{n}{k}(n-k)^m = \sum_{k=0}^{n}(-1)^k\frac{(n-k)^m}{k!(n-k)!}.$$

4.4 Euler's ϕ Function

Our next illustration of inclusion-exclusion comes from number theory. Recall that for any positive integers a and b, $\gcd(a,b)$ denotes the greatest common divisor of a and b, which is the largest integer dividing both a and b.

4.12. Definition: Euler's ϕ Function. For each integer $m \geq 1$, let $\phi(m)$ be the number of integers $x \in \{1,2,\ldots,m\}$ such that $\gcd(x,m) = 1$.

For example, if $m = 12$, then the relevant integers x are 1, 5, 7, and 11, so $\phi(12) = 4$. The function ϕ is prominent in algebra and number theory and has applications to modern cryptography.

4.13. Theorem: Product Formula for $\phi(m)$. Suppose an integer $m > 1$ has prime factorization $m = p_1^{e_1}p_2^{e_2}\cdots p_n^{e_n}$, where the p_i are distinct primes and each $e_i \geq 1$. Then

$$\phi(m) = \prod_{i=1}^{n} p_i^{e_i-1}(p_i - 1) = m\prod_{i=1}^{n}(1 - 1/p_i).$$

Proof. Let $X = \{1,2,\ldots,m\}$; for $1 \leq i \leq n$, let S_i be the set of $x \in X$ such that p_i is a factor of x. By the Fundamental Theorem of Arithmetic, $x \in X$ is not relatively prime to m iff x and m have a common factor greater than 1 iff x and m have a common *prime* factor. It follows that

$$\phi(m) = |X - (S_1 \cup S_2 \cup \cdots \cup S_n)|,$$

so we can compute $\phi(m)$ using the Union-Avoiding Rule. The relevant inclusion-exclusion formula can be written

$$|X - (S_1 \cup \cdots \cup S_n)| = \sum_{I \subseteq \{1,2,\ldots,n\}} (-1)^{|I|}\left|\bigcap_{i \in I} S_i\right|,$$

using the convention that $\bigcap_{i \in \emptyset} S_i$ is the set X. Fix a subset $I = \{i_1 < \cdots < i_k\} \subseteq \{1,2,\ldots,n\}$, and consider the intersection $S_{i_1} \cap \cdots \cap S_{i_k}$. An integer $x \in X$ lies in this intersection iff p_{i_j} divides x for $1 \leq j \leq k$ iff the product $q = p_{i_1}p_{i_2}\cdots p_{i_k}$ divides x iff x is a multiple of q. Now, the number of multiples of q between 1 and m is $m/q = m/\prod_{i \in I} p_i$. If

$I = \emptyset$ and the empty product is interpreted as 1, this expression becomes $m = |X|$. Inserting these expressions into the formula above, we find that

$$\phi(m) = |X - (S_1 \cup \cdots \cup S_n)| = m \sum_{I \subseteq \{1,2,\ldots,n\}} \frac{(-1)^{|I|}}{\prod_{i \in I} p_i}.$$

Comparing to the theorem statement, we see that it suffices to prove that

$$\prod_{i=1}^{n}\left(1 - \frac{1}{p_i}\right) = \sum_{I \subseteq \{1,2,\ldots,n\}} \frac{(-1)^{|I|}}{\prod_{i \in I} p_i}. \tag{4.3}$$

This identity can be deduced from the Generalized Distributive Law (Exercise 2-16), or as a special case of Exercise 4-68. Here we give a proof by induction on n. When $n = 1$, both sides of the formula are $1 - 1/p_1$. Fix $n > 1$, and assume $\prod_{i=1}^{n-1}(1 - p_i^{-1}) = \sum_{I \subseteq \{1,\ldots,n-1\}} (-1)^{|I|}/\prod_{i \in I} p_i$ is already known. Multiplying both sides by $1 - 1/p_n$, we get

$$\prod_{i=1}^{n}\left(1 - \frac{1}{p_i}\right) = (1 - 1/p_n) \sum_{I \subseteq \{1,2,\ldots,n-1\}} \frac{(-1)^{|I|}}{\prod_{i \in I} p_i}.$$

Using the distributive law, the right side becomes

$$1 \cdot \sum_{I \subseteq \{1,\ldots,n-1\}} \frac{(-1)^{|I|}}{\prod_{i \in I} p_i} - (1/p_n) \cdot \sum_{I \subseteq \{1,\ldots,n-1\}} \frac{(-1)^{|I|}}{\prod_{i \in I} p_i}.$$

By moving $-1/p_n$ inside the second sum and replacing the summation index I by $I \cup \{n\}$, the second sum becomes

$$\sum_{I \subseteq \{1,\ldots,n\}:\, n \in I} \frac{(-1)^{|I|}}{\prod_{i \in I} p_i}.$$

On the other hand, we can think of the first sum as ranging over all $I \subseteq \{1, \ldots, n\}$ such that $n \notin I$. The two sums can now be combined into a single sum over all subsets I of $\{1, \ldots, n\}$, which completes the induction step. □

4.14. Remark. Here is a sketch of an alternative derivation of the formula for $\phi(m)$ that avoids inclusion-exclusion, but assumes some results from algebra and number theory. For any commutative ring R, we let $R^\times$ be the set of *units* in R, i.e., the set of $x \in R$ such that there exists $y \in R$ with $xy = yx = 1_R$. The following facts must now be verified. First, if R and T are isomorphic rings, then $|R^\times| = |T^\times|$. Second, given a product ring $R \times S$, we have $(R \times S)^\times = R^\times \times S^\times$ and hence (by the Product Rule) $|(R \times S)^\times| = |R^\times| \cdot |S^\times|$. Third, $\gcd(x, n) = 1$ iff there exist integers y, z with $xy + nz = 1$ iff x has a multiplicative inverse in the ring of integers modulo n. So $\phi(n) = |(\mathbb{Z}/n\mathbb{Z})^\times|$. Fourth, by the Chinese Remainder Theorem, the rings $\mathbb{Z}/mn\mathbb{Z}$ and $\mathbb{Z}/m\mathbb{Z} \times \mathbb{Z}/n\mathbb{Z}$ are isomorphic whenever $\gcd(m, n) = 1$. Combining these four facts, we see that $\gcd(m, n) = 1$ implies

$$\phi(mn) = |(\mathbb{Z}/mn\mathbb{Z})^\times| = |(\mathbb{Z}/m\mathbb{Z} \times \mathbb{Z}/n\mathbb{Z})^\times| = |(\mathbb{Z}/m\mathbb{Z})^\times| \cdot |(\mathbb{Z}/n\mathbb{Z})^\times| = \phi(m)\phi(n).$$

Iteration of this result gives

$$\phi(p_1^{e_1} \cdots p_n^{e_n}) = \prod_{i=1}^{n} \phi(p_i^{e_i})$$

whenever $p_1, \ldots, p_n$ are distinct primes. Thus, it suffices to evaluate ϕ at prime powers. A direct counting argument using the Difference Rule and the definition of ϕ shows that $\phi(p^e) = p^e - p^{e-1} = p^{e-1}(p - 1)$ when p is prime and $e \geq 1$. So we obtain the first formula for $\phi(n)$ given in Theorem 4.13.

4.5 Derangements

The Inclusion-Exclusion Formula allows us to enumerate a special class of permutations called derangements. Intuitively, a derangement of $1, 2, \ldots, n$ is a rearrangement of these n symbols such that no symbol remains in its original position. The formal definition is as follows.

4.15. Definition: Derangements. A *derangement* of a set S is a bijection $f : S \to S$ such that $f(x) \neq x$ for all $x \in S$. For $n \geq 0$, let D_n be the set of derangements of $\{1, 2, \ldots, n\}$, and let $d_n = |D_n|$.

We have $d_0 = 1$ (since the empty function with domain and codomain $\emptyset$ satisfies the definition of derangement), while $d_1 = 0$. To give more examples of derangements, let us identify an element $f \in D_n$ with the word $f(1)f(2)\cdots f(n)$. Then $d_2 = 1$ since 21 is the unique derangement of $\{1, 2\}$. The derangements of $\{1, 2, 3\}$ are 312 and 231, so that $d_3 = 2$. The permutation 5317426 is a derangement of $\{1, 2, 3, 4, 5, 6, 7\}$.

4.16. Summation Formula for Derangements. For $n \geq 1$, the number of derangements of an n-element set is

$$d_n = n! \sum_{k=0}^{n} \frac{(-1)^k}{k!}.$$

Consequently, for all $n \geq 1$, d_n is the closest integer to $n!/e$.

Proof. Let X be the set of all permutations of $\{1, 2, \ldots, n\}$; by the Permutation Rule, $|X| = n!$. For $1 \leq i \leq n$, let $S_i = \{f \in X : f(i) = i\}$. The set D_n consists of precisely those elements in X that belong to none of the S_i, so $D_n = X - (S_1 \cup \cdots \cup S_n)$. To apply the Union-Avoiding Rule, we must determine the size of each intersection $S_{i_1} \cap S_{i_2} \cap \cdots \cap S_{i_k}$, where $1 \leq i_1 < i_2 < \cdots < i_k \leq n$. A permutation $f \in X$ belongs to this intersection iff f fixes $i_1, \ldots, i_k$ and permutes the remaining $n - k$ symbols among themselves. The number of such permutations is $(n - k)!$. This number depends only on k and not on the indices $i_1, \ldots, i_k$. Applying the Simplified Union-Avoiding Rule 4.6, we obtain

$$d_n = n! + \sum_{k=1}^{n} (-1)^k \binom{n}{k} (n-k)! = n! + \sum_{k=1}^{n} (-1)^k \frac{n!}{k!} = n! \sum_{k=0}^{n} \frac{(-1)^k}{k!}.$$

To relate this formula to the expression $n!/e$, recall from calculus that for all real x,

$$e^x = \sum_{k=0}^{\infty} \frac{x^k}{k!}.$$

Taking $x = -1$, we see that

$$1/e = \sum_{k=0}^{\infty} \frac{(-1)^k}{k!}.$$

Multiplying by $n!$ and comparing to our formula for d_n, we see that

$$n!/e - d_n = n! \sum_{k=n+1}^{\infty} \frac{(-1)^k}{k!}.$$

It now suffices to show that the right side of this formula is less than 1/2 in absolute value. Factoring out $\frac{1}{(n+1)!}$ from each term in the series, we obtain

$$|n!/e - d_n| = \frac{1}{n+1}\left|1 - \frac{1}{n+2} + \frac{1}{(n+2)(n+3)} - \frac{1}{(n+2)(n+3)(n+4)} + \cdots\right|.$$

The series within the absolute values on the right side is an alternating series that converges to a sum strictly less than 1. Since $n \geq 1$, it follows that

$$|n!/e - d_n| < \frac{1}{n+1} \cdot 1 \leq 1/2.$$

□

The following table lists the first few values of d_n.

n	0	1	2	3	4	5	6	7	8	9
d_n	1	0	1	2	9	44	265	1854	14,833	133,496

Like any permutation, a derangement has a functional digraph consisting of a disjoint union of one or more cycles. A permutation is a derangement iff there are no 1-cycles in its functional digraph. This observation leads to the following recursion for derangements.

4.17. First Recursion for Derangements. We have $d_0 = 1$, $d_1 = 0$, and

$$d_n = (n-1)d_{n-1} + (n-1)d_{n-2} \quad \text{for all } n \geq 2.$$

Proof. Fix $n \geq 2$. Write the set of derangements D_n as the disjoint union of sets A and B, where A consists of those derangements in which n is involved in a cycle of length 2, and B consists of the derangements where n is in a cycle of length greater than 2. To build an object in A, choose the partner of n in its 2-cycle ($n-1$ ways), and then choose a derangement of the remaining objects (D_{n-2} ways). To build an object in B, choose a derangement of the first $n-1$ objects (D_{n-1} ways), consider the functional digraph of this derangement, and splice n into a cycle just before any of the $n-1$ available elements. Since all original cycles have length at least 2, this construction will ensure that n appears in a cycle of length at least 3. The recursion now follows from the Product Rule and the Sum Rule. □

This recursion expresses d_n in terms of the previous two derangement numbers. Our next result shows how to transform this formula into another recursion giving d_n as a function of d_{n-1}.

4.18. Second Recursion for Derangements. We have $d_0 = 1$ and

$$d_n = nd_{n-1} + (-1)^n \quad \text{for all } n \geq 1.$$

Proof. We use induction on n. If $n = 1$, then

$$d_n = d_1 = 0 = 1 \cdot 1 + (-1)^1 = nd_{n-1} + (-1)^n.$$

Now assume $n > 1$ and that $d_{n-1} = (n-1)d_{n-2} + (-1)^{n-1}$. We can use this assumption to eliminate $(n-1)d_{n-2}$ in the derangement recursion proved earlier. We thereby obtain

$$d_n = (n-1)d_{n-1} + (n-1)d_{n-2} = (n-1)d_{n-1} + (d_{n-1} - (-1)^{n-1}) = nd_{n-1} + (-1)^n.$$

This completes the induction step. □

4.6 Involutions

In Chapter 2, we saw how bijections could be used to prove combinatorial identities. Some identities involve a mixture of positive and negative terms. One can use special bijections called *involutions* to give combinatorial proofs of such identities. We introduce this idea with the following binomial coefficient identity.

4.19. Theorem. For all $n \geq 1$, $\sum_{k=0}^{n}(-1)^k\binom{n}{k} = 0$.

Proof. The result can be proved algebraically by using the Binomial Theorem 2.8 to expand the left side of $(-1+1)^n = 0$. To prove the identity combinatorially, let X be the set of all subsets of $\{1, 2, \ldots, n\}$. For each $S \in X$, we define the *sign* of S to be $\operatorname{sgn}(S) = (-1)^{|S|}$. Since there are $\binom{n}{k}$ subsets S of size k, and $\operatorname{sgn}(S) = (-1)^k$ for all such subsets, we see that

$$\sum_{S\in X}\operatorname{sgn}(S) = \sum_{k=0}^{n}(-1)^k\binom{n}{k}.$$

Thus we have found a combinatorial model for the left side of the identity to be proved, where the model involves *signed objects.*

To continue, we define a function $I : X \to X$ as follows. Given $S \in X$, let $I(S) = S\cup\{1\}$ if $1 \notin S$, and let $I(S) = S-\{1\}$ if $1 \in S$. For example, $I(\{2,4\}) = \{1,2,4\}$ and $I(\{1,3\}) = \{3\}$. Observe that $I(I(S)) = S$ for all $S \in X$; in other words, $I \circ I = \mathrm{id}_X$. Thus, I is a bijection that is equal to its own inverse. Furthermore, since $|I(S)| = |S| \pm 1$, $\operatorname{sgn}(I(S)) = -\operatorname{sgn}(S)$ for all $S \in X$. It follows that I pairs each positive object in X with a negative object in X. Consequently, the number of positive objects in X equals the number of negative objects in X, and so $\sum_{S\in X}\operatorname{sgn}(S) = 0$. □

The general setup for involution proofs is described as follows.

4.20. Definition: Involutions. An *involution* on a set X is a function $I : X \to X$ such that $I \circ I = \mathrm{id}_X$. Equivalently, I is a bijection on X and $I = I^{-1}$. Given an involution I, the *fixed point set* of I is the set $\mathrm{Fix}(I) = \{x \in X : I(x) = x\}$, which may be empty. If $\operatorname{sgn} : X \to \{+1, -1\}$ is a function that attaches a sign to every object in X, we say that I is a *sign-reversing* involution iff for all $x \in X-\mathrm{Fix}(I)$, $\operatorname{sgn}(I(x)) = -\operatorname{sgn}(x)$.

4.21. Involution Theorem. Given a finite set X of signed objects and a sign-reversing involution I on X,

$$\sum_{x\in X}\operatorname{sgn}(X) = \sum_{x\in\mathrm{Fix}(I)}\operatorname{sgn}(X).$$

Proof. Define

$$X^+ = \{x \in X-\mathrm{Fix}(I) : \operatorname{sgn}(x) = +1\} \text{ and } X^- = \{x \in X-\mathrm{Fix}(I) : \operatorname{sgn}(x) = -1\}.$$

By definition, I restricts to X^+ and X^- to give functions $I^+ : X^+ \to X^-$ and $I^- : X^- \to X^+$ that are mutually inverse bijections. Therefore, $|X^+| = |X^-|$ and

$$\begin{aligned}\sum_{x\in X}\operatorname{sgn}(X) &= \sum_{x\in X^+}\operatorname{sgn}(x) + \sum_{x\in X^-}\operatorname{sgn}(x) + \sum_{x\in\mathrm{Fix}(I)}\operatorname{sgn}(x)\\ &= |X^+| - |X^-| + \sum_{x\in\mathrm{Fix}(I)}\operatorname{sgn}(x) = \sum_{x\in\mathrm{Fix}(I)}\operatorname{sgn}(x). \quad \square\end{aligned}$$

As a first illustration of the Involution Theorem, we prove a variation of Theorem 4.19.

4.22. Theorem. For all $n \geq 1$,

$$\sum_{k=0}^{n}(-1)^k\binom{2n}{k} = (-1)^n\binom{2n-1}{n}.$$

Proof. Let X be the set of all subsets of $\{1, 2, \ldots, 2n\}$ of size at most n, and let the sign of a subset T be $(-1)^{|T|}$. The left side of the identity to be proved is $\sum_{T\in X}\text{sgn}(T)$. Next, define an involution I on X as follows. If $T \in X$ and $1 \in T$, let $I(T) = T-\{1\}$. If $T \in X$ and $1 \notin T$ and $|T| < n$, let $I(T) = T \cup \{1\}$. Finally, if $T \in X$ and $1 \notin T$ and $|T| = n$, let $I(T) = T$. Note that I is a sign-reversing involution. The fixed points of I are the n-element subsets of $\{1, 2, \ldots, 2n\}$ not containing 1. There are $\binom{2n-1}{n}$ such subsets, and each of them has sign $(-1)^n$. So $\sum_{T\in\text{Fix}(I)}\text{sgn}(T)$ is the right side of the identity to be proved. We complete the proof by invoking the Involution Theorem. □

Before giving more examples of involutions, we introduce a convenient notational device.

4.23. Definition: The Truth Function χ. For any logical statement P, define $\chi(P) = 1$ if P is true, and $\chi(P) = 0$ if P is false.

4.24. Theorem. For all integers $n \geq 0$,

$$\sum_{k=0}^{n}(-1)^k\binom{n}{k}^2 = (-1)^{n/2}\binom{n}{n/2}\chi(n \text{ is even.})$$

(The right side is zero when n is odd, and the right side is $(-1)^k\binom{2k}{k}$ when $n = 2k$ is even.)

Proof. Let X be the set of all pairs (S, T), where S and T are subsets of $\{1, 2, \ldots, n\}$ of the same size. Define $\text{sgn}(S, T) = (-1)^{|S|}$. Then the left side of the identity to be proved is $\sum_{(S,T)\in X}\text{sgn}(S, T)$. We define an involution I on X as follows. Given $(S, T) \in X$, let i be the least integer in $\{1, 2, \ldots, n\}$ (if there is one) such that either $i \notin S$ and $i \notin T$, or $i \in S$ and $i \in T$. In the former case, let $I(S, T) = (S \cup \{i\}, T \cup \{i\})$; in the latter case, let $I(S, T) = (S-\{i\}, T-\{i\})$; if no such i exists, let $I(S, T) = (S, T)$. For example, taking $n = 6$, we find that $I(\{1, 3, 5\}, \{3, 5, 6\}) = (\{1, 2, 3, 5\}, \{2, 3, 5, 6\})$, $I(\{1, 2, 4\}, \{2, 3, 4\}) = (\{1, 4\}, \{3, 4\})$, and $(\{3, 5, 6\}, \{1, 2, 4\})$ is in $\text{Fix}(I)$.

It is routine to check that I is a sign-reversing involution; in particular, the designated integer i in the definition of $I(S, T)$ is the same as the i used to calculate $I(I(S, T))$, so $I(I(S, T)) = (S, T)$. By the Involution Theorem,

$$\sum_{k=0}^{n}(-1)^k\binom{n}{k}^2 = \sum_{(S,T)\in\text{Fix}(I)}(-1)^{|S|}.$$

Note that $(S, T) \in \text{Fix}(I)$ iff for every $i \in \{1, 2, \ldots, n\}$, i lies in exactly one of the two sets S or T. Since S and T must have the same size, (S, T) is a fixed point of I iff n is even and $|S| = |T| = n/2$ and $S = \{1, 2, \ldots, n\}-T$. If n is odd, the fixed point set is empty, so the given signed sum of squared binomial coefficients is zero. If n is even, we can construct an arbitrary element of $\text{Fix}(I)$ by choosing any subset S of size $n/2$ and letting T be the complementary subset $\{1, 2, \ldots, n\}-S$. Since there are $\binom{n}{n/2}$ choices for S, each with sign $(-1)^{n/2}$, the formula in the theorem is proved. □

4.25. Example: Stirling Numbers. Recall from §3.6 that $s(n,k) = (-1)^{n-k}c(n,k)$, where $c(n,k)$ is the number of permutations of an n-element set whose functional digraph consists of k cycles. We will show that for all $n \geq 1$,

$$\sum_{k=1}^{n} s(n,k) = \chi(n=1).$$

Both sides are 1 when $n = 1$, so assume $n > 1$. Let X be the set of all permutations of $\{1,2,\ldots,n\}$. If $w \in X$ is a permutation with k cycles, define $\mathrm{sgn}(w) = (-1)^k$. Now $\sum_{w\in X} \mathrm{sgn}(w) = (-1)^n \sum_{k=1}^{n} s(n,k)$, so it suffices to define a sign-reversing involution I on X with no fixed points. Given $w \in X$, the numbers 1 and 2 either appear in the same cycle of w or in different cycles. If 1 and 2 are in the same cycle, let the elements on this cycle (starting at 1) be

$$(1, x_1, x_2, \ldots, x_i, 2, y_1, y_2, \ldots, y_j),$$

where $i, j \geq 0$. Define $I(w)$ by replacing this cycle by the two cycles

$$(1, x_1, x_2, \ldots, x_i)(2, y_1, y_2, \ldots, y_j)$$

and leaving all other cycles the same. Similarly, if 1 and 2 are in different cycles of w, write these cycles as

$$(1, x_1, x_2, \ldots, x_i)(2, y_1, y_2, \ldots, y_j)$$

and define $I(w)$ by replacing these two cycles by the single cycle

$$(1, x_1, x_2, \ldots, x_i, 2, y_1, y_2, \ldots, y_j).$$

It is immediate that $I \circ I = \mathrm{id}_X$, I is sign-reversing, and I has no fixed points.

We can modify the preceding involution to obtain a combinatorial proof of the identity

$$\sum_{k\geq 0} s(i,k)S(k,j) = \chi(i=j),$$

which we proved algebraically in Theorem 2.64(d). If $i < j$, then for every k, either $s(i,k) = 0$ or $S(k,j) = 0$. So both sides of the identity are zero in this case. If $i = j$, the left side reduces to $s(i,i)S(i,i) = 1 = \chi(i = j)$. If $j = 0$, the identity is true. So we may assume i and j are fixed numbers such that $i > j > 0$. Let X be the set of pairs (w,U), where w is a permutation of $\{1,2,\ldots,i\}$ (viewed as a functional digraph) and U is a set partition of the set of cycles in w into j blocks. If w has k cycles, let $\mathrm{sgn}(w,U) = (-1)^k$. Then

$$\sum_{(w,U)\in X} \mathrm{sgn}(w,U) = (-1)^i \sum_{k=j}^{i} s(i,k)S(k,j),$$

and $\chi(i=j) = 0$. So it suffices to define a sign-reversing involution I on X with no fixed points. Given $(w,U) \in X$, there must exist a block of U such that the cycles in this block collectively involve more than one point in $\{1,2,\ldots,i\}$. This follows from the fact that i (the number of points) exceeds j (the number of blocks). Among all such blocks in U, choose the block that contains the smallest possible element in $\{1,2,\ldots,i\}$. Let this smallest element be a, and let the second-smallest element in this block be b. To calculate $I(w,U)$, modify the cycles in this block of U as we did above, with a and b playing the roles of 1 and 2. More specifically, a cycle of the form

$$(a, x_1, \ldots, x_r, b, y_1, \ldots, y_s)$$

gets replaced (within its block) by

$$(a, x_1, \ldots, x_r)(b, y_1, \ldots, y_s)$$

and vice versa. It is routine to check that I is a sign-reversing involution on X with no fixed points. For example, suppose $i = 10$, $j = 3$, w has cycles $(1), (3,5), (2,6,9), (4,8), (7), (10)$, and

$$U = \{\{(1)\}, \{(3,5), (10)\}, \{(2,6,9), (4,8), (7)\}\}.$$

Here the block of U modified by the involution is $\{(2,6,9), (4,8), (7)\}$, $a = 2$, and $b = 4$. We compute $I(w, U)$ by replacing the cycles $(2,6,9)$ and $(4,8)$ in w by the single cycle $(2,6,9,4,8)$ and letting the new set partition be

$$U' = \{\{(1)\}, \{(3,5), (10)\}, \{(2,6,9,4,8), (7)\}\}.$$

Note that the original object has sign $(-1)^6 = +1$, whereas $I(w, U)$ has sign $(-1)^5 = -1$.

4.7 Involutions Related to Inclusion-Exclusion

This section gives more examples of involution-based proofs. We begin by reproving the original Inclusion-Exclusion Formula with an involution.

4.26. Involution Proof of the Inclusion-Exclusion Formula. Our goal is to prove formula (4.1). Moving all terms in this formula to the left side, we can rewrite the goal as

$$\sum_{k=0}^{n}(-1)^k \sum_{1\leq i_1<\cdots<i_k\leq n} |S_{i_1} \cap \cdots \cap S_{i_k}| = 0, \tag{4.4}$$

where the summand corresponding to $k = 0$ is defined to be $|S_1 \cup \cdots \cup S_n|$. We will prove this formula by introducing an involution on a certain set of signed objects.

Let X be the set of all sequences $(x; i_1, i_2, \ldots, i_k)$ such that $1 \leq i_1 < i_2 < \cdots < i_k \leq n$, $0 \leq k \leq n$, and $x \in S_{i_1} \cap \cdots \cap S_{i_k}$. (If $k = 0$, then the object looks like $(x;)$, and the last condition is interpreted to mean $x \in S_1 \cup \cdots \cup S_n$.) Define $\operatorname{sgn}(x; i_1, i_2, \ldots, i_k) = (-1)^k$. It follows from the Sum Rule that $\sum_{z\in X} \operatorname{sgn}(z)$ is the left side of (4.4). So it suffices to define a sign-reversing involution on X with no fixed points.

Given $z = (x; i_1, \ldots, i_k) \in X$, we must have $x \in S_1 \cup \cdots \cup S_n$ no matter what the value of k is. Let i be the minimum index in $\{1, 2, \ldots, n\}$ such that $x \in S_i$. By definition of X, we either have $k = 0$ or $i < i_1$ or $i = i_1$. If $k = 0$ or $i < i_1$, define $I(z) = (x; i, i_1, i_2, \ldots, i_k)$. If instead $i = i_1$, define $I(z) = (x; i_2, \ldots, i_k)$. It is immediate that $I(I(z)) = z$ and $\operatorname{sgn}(I(z)) = -\operatorname{sgn}(z)$ for all $z \in X$.

The preceding proof is very ingenious because it establishes a rather complicated formula by a remarkably simple bookkeeping bijection. On the other hand, we would also like to have a combinatorial proof of inclusion-exclusion that is tied more closely to the intuitive "including and excluding" arguments we used originally to guess the formula for $|S \cup T \cup U|$. For such a proof, see Exercise 4-69.

The identities in the next lemma are needed in the next section to prove generalized versions of the Inclusion-Exclusion Formula. We prove each identity using an involution.

4.27. Lemma. For all integers $p, j \geq 0$,

(a) $\displaystyle\sum_{k=j}^{p}(-1)^{k-j}\binom{k}{j}\binom{p}{k} = \chi(p = j)$. (b) $\displaystyle\sum_{k=j}^{p}(-1)^{k-j}\binom{k-1}{j-1}\binom{p}{k} = \chi(p \geq j > 0)$.

Proof. (a) If $p < j$, both sides of (a) are zero. If $p = j$, both sides of (a) are 1. So assume $p > j$. Let X be the set of words in $\{0, 1, 2\}^p$ containing exactly j zeroes. For a word $w \in X$ containing s ones, let $\operatorname{sgn}(w) = (-1)^s$. We can evaluate $\sum_{w \in X} \operatorname{sgn}(w)$ by letting k be the number of letters in w equal to 0 or 1. We must have $j \leq k \leq p$. For a fixed k in this range, we can build a word w with j zeroes and $k - j$ ones by picking k positions out of p available positions where the 0's and 1's will go, then picking j of these k positions to contain the zeroes. The sign of the resulting object is $(-1)^{k-j}$. By the Sum Rule, we conclude that the left side of (a) is equal to $\sum_{w \in X} \operatorname{sgn}(w)$.

To complete the proof of (a), we define an involution $I : X \to X$ with no fixed points. Given $w \in X$, choose the minimal i with $w_i \neq 0$. Such an i must exist, since $j < p$. If $w_i = 1$, replace this symbol by a 2. If $w_i = 2$, replace this symbol by a 1. This produces a new word $I(w) \in X$ with the opposite sign as w. For example, if $j = 3$ and $p = 7$, $I(0120110) = 0220110$, where $\operatorname{sgn}(0120110) = (-1)^3$ and $\operatorname{sgn}(0220110) = (-1)^2$. Evidently $I(I(w)) = w$ for all $w \in X$, so I has no fixed points.

(b) If $j = 0$ or $p < j$ or $p = j > 0$, the identity is true. So assume $p > j > 0$. Define X as in the proof of (a), and let Y be the subset of X consisting of those words w where there is no 1 to the left of the first 0. To build a word in Y with j zeroes, $k - j$ ones, and $p - k$ twos (for fixed k in the range $j \leq k \leq p$), first choose k positions out of p for the zeroes and ones; then put a zero in the first of these k positions; then choose $j - 1$ of the remaining $k - 1$ positions to contain the remaining zeroes. The sign of each such object is $(-1)^{k-j}$, so the Sum Rule shows that the left side of (b) is equal to $\sum_{w \in Y} \operatorname{sgn}(w)$.

We complete the proof of (b) by defining an involution $I : Y \to Y$ with exactly one fixed point. For most words $w \in Y$, we form $I(w)$ by looking for the *rightmost* position i with $w_i \neq 0$, and toggling the symbol in this position between 1 and 2. For example, $I(2010120) = 2010110$. The action of I produces a new object in the set Y with the opposite sign as w, except in the case where all nonzero letters in w precede all the zeroes in w. In this case, since $w \in Y$, w must be the word consisting of $p - j$ twos followed by j zeroes. This word is the unique fixed point of I, and its sign is $+1$. □

4.8 Generalized Inclusion-Exclusion Formulas

Let $S_1, S_2, \ldots, S_n$ be subsets of a finite set X. The inclusion-exclusion formulas presented earlier allow us to count the number of objects that belong to *at least one* of the sets S_i, as well as the number of objects in X belonging to *none* of the sets S_i. Our next goal is to generalize these results to inclusion-exclusion formulas that count the objects in *at least* j of the sets S_i, or *exactly* j of the sets S_i, where j is a fixed integer between 1 and n.

4.28. Rule for Counting Objects with Exactly j Properties. Let $S_1, S_2, \ldots, S_n$ be distinct subsets of a finite set X. For fixed $j \in \{1, 2, \ldots, n\}$, the number of objects belonging to exactly j of the sets S_i is

$$\sum_{k=j}^{n} (-1)^{k-j} \binom{k}{j} \sum_{1 \leq i_1 < i_2 < \cdots < i_k \leq n} |S_{i_1} \cap S_{i_2} \cap \cdots \cap S_{i_k}|. \tag{4.5}$$

Proof. We start by rewriting (4.5) in more concise notation. Let $[n]$ denote the set $\{1, 2, \ldots, n\}$. We encode the sequence of subscripts $i_1 < i_2 < \cdots < i_k$ as a k-element subset I of $[n]$. Then (4.5) can be written

$$\sum_{k=j}^{n}(-1)^{k-j}\binom{k}{j}\sum_{\substack{I\subseteq[n]\\|I|=k}}\left|\bigcap_{i\in I}S_i\right|.$$

Next, for any set $S\subseteq X$, observe that $|S|=\sum_{x\in X}\chi(x\in S)$. Since $x\in\bigcap_{i\in I}S_i$ iff $x\in S_i$ for every $i\in I$, the previous formula now becomes

$$\sum_{k=j}^{n}(-1)^{k-j}\binom{k}{j}\sum_{\substack{I\subseteq[n]\\|I|=k}}\sum_{x\in X}\chi(\forall i\in I,x\in S_i).$$

Using the distributive law and commutative law for sums of finitely many terms, we can rewrite this as

$$\sum_{x\in X}\sum_{k=j}^{n}(-1)^{k-j}\binom{k}{j}\left[\sum_{\substack{I\subseteq[n]\\|I|=k}}\chi(\forall i\in I,x\in S_i)\right]. \tag{4.6}$$

For each $x\in X$, define $P(x)=\{i\in[n]:x\in S_i\}$, so $|P(x)|$ tells us how many sets S_i contain x. To complete the proof, it suffices to show that the summand in (4.6) indexed by x evaluates to $\chi(|P(x)|=j)$. Fix $x\in X$, and let $p=|P(x)|$. As k and I vary within the summand indexed by this x, we obtain a nonzero contribution to this summand iff $j\leq k\leq p$ and $I\subseteq P(x)$. So we want to show that

$$\sum_{k=j}^{p}(-1)^{k-j}\binom{k}{j}\left[\sum_{\substack{I\subseteq P(x)\\|I|=k}}1\right]=\chi(p=j).$$

The inner sum is the number of k-element subsets of the p-element set $P(x)$, which is $\binom{p}{k}$ by the Subset Rule. So we are reduced to showing that $\sum_{k=j}^{p}(-1)^{k-j}\binom{k}{j}\binom{p}{k}=\chi(p=j)$, which is the identity we proved in Lemma 4.27(a). □

4.29. Rule for Counting Objects with at Least j Properties. Let $S_1,S_2,\ldots,S_n$ be distinct subsets of a finite set X. For fixed $j\in\{1,2,\ldots,n\}$, the number of objects belonging to at least j of the sets S_i is

$$\sum_{k=j}^{n}(-1)^{k-j}\binom{k-1}{j-1}\sum_{1\leq i_1<i_2<\cdots<i_k\leq n}|S_{i_1}\cap S_{i_2}\cap\cdots\cap S_{i_k}|. \tag{4.7}$$

The proof is nearly identical to the one just given, so we leave it as an exercise.

4.30. Example. How many permutations of $[n]=\{1,2,\ldots,n\}$ have exactly j fixed points? To solve this, let X be the set of all permutations of $[n]$. For $1\leq i\leq n$, let $S_i=\{f\in X:f(i)=i\}$. We saw in an earlier example that for any $i_1<i_2<\cdots<i_k$, $|S_{i_1}\cap S_{i_2}\cap\cdots\cap S_{i_k}|=(n-k)!$. Since there are $\binom{n}{k}$ choices for $\{i_1,i_2,\ldots,i_k\}$, formula (4.5) gives

$$\sum_{k=j}^{n}(-1)^{k-j}\binom{k}{j}\binom{n}{k}(n-k)!=\frac{n!}{j!}\sum_{k=j}^{n}\frac{(-1)^{k-j}}{(k-j)!}=\frac{n!}{j!}\sum_{m=0}^{n-j}\frac{(-1)^m}{m!}$$

as the answer. Alternatively, we can solve this problem using derangements. We build a typical permutation of $[n]$ with j fixed points by first choosing which of the j inputs will be fixed (there are $\binom{n}{j}$ possibilities), then choosing a derangement of the remaining $n-j$ values (there are d_{n-j} possibilities). By the Product Rule, the answer is $\binom{n}{j}d_{n-j}$. Replacing d_{n-j} by the summation given in 4.16 shows that our two answers agree. If we instead ask for the number of permutations of $[n]$ with at least j fixed points, (4.7) gives the answer

$$\sum_{k=j}^{n}(-1)^{k-j}\binom{k-1}{j-1}\binom{n}{k}(n-k)! = \frac{n!}{(j-1)!}\sum_{k=j}^{n}\frac{(-1)^{k-j}}{k(k-j)!} = \frac{n!}{(j-1)!}\sum_{m=0}^{n-j}\frac{(-1)^m}{m!(m+j)}.$$

4.9 Möbius Inversion in Number Theory

We conclude this chapter with an introduction to the theory of Möbius inversion, which generalizes the inclusion-exclusion techniques studied so far. We begin in this section by describing the number-theoretic Möbius function and the corresponding Möbius Inversion Formula. Later sections discuss the generalization of the Möbius function and inversion formula to posets.

4.31. Definition: Number-Theoretic Möbius Function. Suppose $m \geq 1$ is an integer with prime factorization $m = p_1^{e_1}p_2^{e_2}\cdots p_n^{e_n}$, where $n \geq 0$, $e_i > 0$, and the p_i's are distint primes. (We take $n = 0$ when $m = 1$.) The *Möbius function* $\mu : \mathbb{Z}_{>0} \to \{-1, 0, 1\}$ is defined by $\mu(m) = 0$ if $e_i > 1$ for some i, and $\mu(m) = (-1)^n$ if $e_i = 1$ for all i.

In other words, $\mu(m)$ is zero if m is divisible by the square of a prime; $\mu(m) = +1$ if m is the product of an even number of distinct primes; and $\mu(m) = -1$ if m is the product of an odd number of distinct primes. For example,

$$\mu(1) = 1,\ \mu(7) = -1,\ \mu(10) = 1,\ \mu(12) = 0,\ \mu(30) = -1.$$

The following lemma is the key to proving the Möbius Inversion Formula. Here and below, we use the notation $d|m$ to mean that d divides m, i.e., $m = cd$ for some integer c. The symbol $\sum_{d|m}$ indicates a sum ranging over all *positive* divisors d of m. The reader should take care not to confuse the *statement* $d|m$ with the *rational number* d/m.

4.32. Lemma. For all integers $m \geq 1$, $\sum_{d|m}\mu(d) = \chi(m = 1)$.

Proof. When $m = 1$, we have $\sum_{d|1}\mu(d) = \mu(1) = 1 = \chi(m = 1)$. Suppose next that $m > 1$ and m has prime factorization $p_1^{e_1}\cdots p_n^{e_n}$. Instead of summing $\mu(d)$ over *all* divisors d of m, we may equally well sum over just the *square-free* divisors d of m, which give the only nonzero contributions to the sum. Examining prime factorizations, we see that there are 2^n such square-free divisors, which have the form $\prod_{i\in T}p_i$ as T ranges over all subsets of $[n] = \{1, 2, \ldots, n\}$. Therefore,

$$\sum_{d|m}\mu(d) = \sum_{T\subseteq[n]}\mu\left(\prod_{i\in T}p_i\right) = \sum_{T\subseteq[n]}(-1)^{|T|}.$$

Collecting together summands indexed by subsets T of the same size k, we conclude that

$$\sum_{d|m}\mu(d) = \sum_{k=0}^{n}\sum_{\substack{T\subseteq[n]\\|T|=k}}(-1)^{|T|} = \sum_{k=0}^{n}\binom{n}{k}(-1)^k = 0,$$

where the last step follows from Theorem 4.19. □

4.33. Number-Theoretic Möbius Inversion Formula. Suppose $f, g : \mathbb{Z}_{>0} \to \mathbb{R}$ are functions such that for all positive integers m,

$$f(m) = \sum_{d|m} g(d).$$

Then for all $m \geq 1$,

$$g(m) = \sum_{d|m} f(m/d)\mu(d) = \sum_{d|m} f(d)\mu(m/d).$$

Proof. We use the definition of f to expand the first claimed formula for $g(m)$:

$$\sum_{d|m} f(m/d)\mu(d) = \sum_{d|m} \left(\sum_{c|(m/d)} g(c) \right) \mu(d) = \sum_{(c,d)\in S} g(c)\mu(d),$$

where $S = \{(c,d) \in \mathbb{Z}_{>0} \times \mathbb{Z}_{>0} : d|m \text{ and } c|(m/d)\}$. It follows routinely from the definition of divisibility that

$$S = \{(c,d) : d|m \text{ and } cd|m\} = \{(c,d) : c|m \text{ and } cd|m\} = \{(c,d) : c|m \text{ and } d|(m/c)\}.$$

Therefore, the calculation continues as follows:

$$\begin{aligned}\sum_{(c,d)\in S} g(c)\mu(d) &= \sum_{c|m}\sum_{d|(m/c)} g(c)\mu(d) = \sum_{c|m} g(c) \left(\sum_{d|(m/c)} \mu(d) \right) \\ &= \sum_{c|m} g(c)\chi(m/c = 1) = g(m).\end{aligned}$$

The next-to-last step used Lemma 4.32 to simplify the inner sum. We conclude that

$$g(m) = \sum_{d|m} f(m/d)\mu(d) = \sum_{d|m} f(d)\mu(m/d),$$

where the final equality results by replacing the summation variable d by m/d. This is permissible, since m/d ranges over all positive divisors of m as d ranges over all positive divisors of m. □

To give examples of the Möbius Inversion Formula, we first introduce some functions that are studied in number theory.

4.34. Definition: Number-Theoretic Functions τ, σ, and σ_2. For each positive integer m, define

$$\tau(m) = \sum_{d|m} 1; \qquad \sigma(m) = \sum_{d|m} d; \qquad \sigma_2(m) = \sum_{d|m} d^2.$$

Thus, $\tau(m)$ is the number of positive divisors of m; $\sigma(m)$ is the sum of these divisors; and $\sigma_2(m)$ is the sum of the squares of these divisors.

4.35. Example. Taking m to be 1, 4, 7, 12, and 30, we calculate:

$$\begin{array}{lllll}\tau(1) = 1, & \tau(4) = 3, & \tau(7) = 2, & \tau(12) = 6, & \tau(30) = 8; \\ \sigma(1) = 1, & \sigma(4) = 7, & \sigma(7) = 8, & \sigma(12) = 28, & \sigma(30) = 72; \\ \sigma_2(1) = 1, & \sigma_2(4) = 21, & \sigma_2(7) = 50, & \sigma_2(12) = 210, & \sigma_2(30) = 1300.\end{array}$$

If m has prime factorization $p_1^{e_1} \cdots p_n^{e_n}$, then the divisors of m have the form $p_1^{f_1} \cdots p_n^{f_n}$ where $0 \leq f_i \leq e_i$ for all i. The Product Rule therefore gives $\tau(m) = \prod_{i=1}^{n}(e_i + 1)$ (build a divisor by choosing $f_1, \ldots, f_n$). Using the Generalized Distributive Law (Exercise 2-16) and the Geometric Series Formula, it can also be checked that

$$\sigma(m) = \prod_{i=1}^{n} \left(\sum_{f_i=0}^{e_i} p_i^{f_i} \right) = \prod_{i=1}^{n} \frac{p_i^{e_i+1} - 1}{p_i - 1}.$$

Applying the Möbius Inversion Formula to the definitions of τ, σ, and σ_2, we obtain the following identities.

4.36. Theorem. For all integers $m \geq 1$, we have

$$1 = \sum_{d|m} \tau(m/d)\mu(d); \qquad m = \sum_{d|m} \sigma(m/d)\mu(d); \qquad m^2 = \sum_{d|m} \sigma_2(m/d)\mu(d).$$

The next result uses Möbius inversion to deduce information about Euler's ϕ function.

4.37. Theorem: ϕ and μ. For all integers $m \geq 1$,

$$m = \sum_{d|m} \phi(d) \qquad \text{and so} \qquad \phi(m) = \sum_{d|m} \mu(d)(m/d).$$

Proof. To prove the first formula, fix $m \geq 1$. For each divisor d of m, let

$$S_d = \{x \in \mathbb{Z}_{>0} : 1 \leq x \leq m \text{ and } \gcd(x, m) = d\}.$$

It is immediate that the m-element set $\{1, 2, \ldots, m\}$ is the disjoint union of the sets S_d as d ranges over the positive divisors of m. Whenever d divides m, we have $\gcd(x, m) = d$ iff d divides x and $\gcd(x/d, m/d) = 1$. It follows that division by d gives a bijection from the set S_d onto the set of numbers counted by $\phi(m/d)$. Therefore, $|S_d| = \phi(m/d)$. By the Sum Rule,

$$m = \sum_{d|m} |S_d| = \sum_{d|m} \phi(m/d) = \sum_{d|m} \phi(d),$$

where the last equality follows by replacing the summation variable d by m/d. Applying Möbius inversion (with $f(m) = m$ and $g(m) = \phi(m)$), we obtain the second formula in the theorem. □

Some applications of these results to field theory are presented in §12.6.

4.10 Partially Ordered Sets

The Inclusion-Exclusion Formula 4.1 and the Möbius Inversion Formula 4.33 are special cases of the general Möbius Inversion Formula for partially ordered sets (posets). Before discussing this, we review some definitions and examples concerning posets.

Recall from Definition 2.50 the definition of a relation and the notions of reflexive, irreflexive, symmetric, antisymmetric, and transitive relations. Given a relation R on a finite set X, the pair (X, R) is a digraph G with vertex set X and directed edge set R. Reflexivity means that *every* vertex of G has a loop edge; irreflexivity means that *no* vertex of G has a loop edge. Symmetry means that the reversal of every edge is also an edge (so we can think of G as undirected); antisymmetry means that it is never true that a non-loop edge and its reversal are both in G. Finally, transitivity means that whenever there is a walk (x, y, z) of length 2 in G, the edge (x, z) is also present in G. More generally, we see by induction that when R is transitive, there exists a walk from x to z in G of positive length iff the edge (x, z) is present in G.

4.38. Poset Definitions. A *partial order relation* on X is a relation that is antisymmetric, transitive, and reflexive on X. A *strict order relation* on X is a relation that is transitive and irreflexive on X. A *partially ordered set (poset)* is a pair $(X, \leq)$ where $\leq$ is a partial order relation on X. A *totally ordered set* is a poset $(X, \leq)$ such that for all $x, y \in X$, either $x \leq y$ or $y \leq x$.

4.39. Example. Let $X = \{1, 2, \ldots, n\}$ and take $\leq$ to be the usual ordering of integers. Then $(X, \leq)$ is an n-element totally ordered poset. More generally, for any $S \subseteq \mathbb{R}$, $(S, \leq)$ is a totally ordered poset.

4.40. Example: Boolean Posets. Let S be any set, and let $X = \mathcal{P}(S)$ be the set of all subsets of S. Then $(X, \subseteq)$ is a poset, where $A \subseteq B$ means that A is a subset of B. In particular, $(\mathcal{P}(\{1, 2, \ldots, n\}), \subseteq)$ is a poset of size 2^n. This poset is not totally ordered when $n > 1$, since the statements $\{1\} \subseteq \{2\}$ and $\{2\} \subseteq \{1\}$ are both false.

4.41. Example: Divisibility Posets. Consider the divisibility relation $|$ on $\mathbb{Z}_{>0}$ defined by $a|b$ iff $b = ac$ for some $c \in \mathbb{Z}_{>0}$. Then $(\mathbb{Z}_{>0}, |)$ is an infinite poset. Given a fixed positive integer n, let X be the set of all divisors of n. Restricting $|$ to X gives a finite poset $(X, |)$. This poset is a totally ordered set iff n is a prime power.

The next result shows that partial order relations and strict order relations are essentially equivalent concepts.

4.42. Theorem: Partial Orders and Strict Orders. Let X be a set, let P be the set of all partial order relations on X, and let S be the set of all strict order relations on X. There are canonical bijections $f : P \to S$ and $g : S \to P$.

Proof. Let $\Delta = \{(x, x) : x \in X\}$ be the diagonal of $X \times X$. Define $f : P \to S$ by setting $f(R) = R - \Delta$ for each partial ordering R on X. Define $g : S \to P$ by setting $g(T) = T \cup \Delta$ for each strict ordering T on X. Viewing relations as digraphs as explained above, f removes self-loops from all vertices, and g restores the self-loops. One may now check that f does map P into S, g does map S into P, and $f \circ g$ and $g \circ f$ are both identity maps. □

4.11 Möbius Inversion for Posets

This section introduces Möbius functions for posets and proves a generalization of the Möbius Inversion Formula.

4.43. Definition: Matrix of a Relation. Let $X = \{x_1, x_2, \ldots, x_n\}$ be a finite set, and let R be a relation on X. Define the *matrix of* R to be the $n \times n$ matrix $A = A(R)$ with i, j-entry $A_{ij} = \chi(x_i R x_j)$.

Note that $A(R)$ is the adjacency matrix of the digraph (X, R).

4.44. Theorem. Let $\leq$ be a partial ordering of $X = \{x_1, \ldots, x_n\}$, and let $<$ be the associated strict ordering of X (see Theorem 4.42). Consider the matrices $Z = A(\leq)$ and $N = A(<)$. Then $Z = I + N$; N is nilpotent; Z is invertible; and

$$Z^{-1} = I - N + N^2 - N^3 + \cdots + (-1)^{n-1}N^{n-1}. \tag{4.8}$$

Proof. The matrix identity $Z = I + N$ holds since $(X, \leq)$ is obtained from $(X, <)$ by adding loop edges at each $x \in X$. Next, we claim that the digraph $(X, <)$ is acyclic. For if $(z_1, z_2, \ldots, z_k, z_1)$ were a directed cycle in this digraph, we must have $z_1 < z_2 < \cdots < z_k < z_1$. Then transitivity gives $z_1 < z_1$, which contradicts irreflexivity. By Theorem 3.25, N is nilpotent. The statements about the inverse of Z now follow from Theorem 3.26, taking A there to be $-N$. □

4.45. Definition: Möbius Function of a Finite Poset. Keeping the notation of the preceding theorem, define $\mu : X \times X \to \mathbb{Z}$ by setting $\mu(x_i, x_j)$ to be the i,j-entry of Z^{-1}. The function μ is called the *Möbius function of the poset* $(X, \leq)$. When discussing several posets at once, we sometimes write μ_X or $\mu_{\leq}$ or $\mu_{(X,\leq)}$ to denote the Möbius function of $(X, \leq)$.

4.46. Example. Let $X = \{1, 2, 3, 4\}$ with the total ordering $1 < 2 < 3 < 4$. For this poset, we have

$$Z = A(\leq) = \begin{bmatrix} 1 & 1 & 1 & 1 \\ 0 & 1 & 1 & 1 \\ 0 & 0 & 1 & 1 \\ 0 & 0 & 0 & 1 \end{bmatrix}, \qquad N = A(<) = \begin{bmatrix} 0 & 1 & 1 & 1 \\ 0 & 0 & 1 & 1 \\ 0 & 0 & 0 & 1 \\ 0 & 0 & 0 & 0 \end{bmatrix}.$$

The powers of N are

$$N^2 = \begin{bmatrix} 0 & 0 & 1 & 2 \\ 0 & 0 & 0 & 1 \\ 0 & 0 & 0 & 0 \\ 0 & 0 & 0 & 0 \end{bmatrix}, \qquad N^3 = \begin{bmatrix} 0 & 0 & 0 & 1 \\ 0 & 0 & 0 & 0 \\ 0 & 0 & 0 & 0 \\ 0 & 0 & 0 & 0 \end{bmatrix}, \qquad N^4 = 0.$$

The inverse of Z is

$$Z^{-1} = I - N + N^2 - N^3 = \begin{bmatrix} 1 & -1 & 0 & 0 \\ 0 & 1 & -1 & 0 \\ 0 & 0 & 1 & -1 \\ 0 & 0 & 0 & 1 \end{bmatrix}.$$

So for $i, j \in X$, $\mu(i, j) = 1$ if $j = i$, $\mu(i, j) = -1$ if $j = i + 1$, and $\mu(i, j) = 0$ otherwise.

4.47. Example: Möbius Function of a Totally Ordered Poset. The preceding example generalizes as follows. Let $X = \{1, 2, \ldots, n\}$ with the ordering $1 < 2 < \cdots < n$. We have $Z_{i,j} = 1$ for $i \leq j$ and $Z_{i,j} = 0$ for $i > j$. For all $i, j \in X$, let $M_{i,j} = 1$ if $j = i$, $M_{i,j} = -1$ if $j = i + 1$, and $M_{i,j} = 0$ otherwise. A routine matrix calculation shows that $ZM = MZ = I$. So for this poset,

$$\mu(i, i) = 1, \qquad \mu(i, i + 1) = -1, \qquad \mu(i, j) = 0 \text{ for } j \neq i, i + 1.$$

4.48. Example: Möbius Function of a Boolean Poset. Consider the poset $(X, \subseteq)$, where X consists of all subsets of $[n] = \{1, 2, \ldots, n\}$. In this example, we will index the rows and columns of matrices by subsets of $[n]$. For $S, T \subseteq [n]$, the S, T-entry of Z is 1 if $S \subseteq T$, and 0 otherwise. We claim that the inverse matrix $M = Z^{-1}$ has S, T-entry

$\mu(S,T) = (-1)^{|T-S|}$ if $S \subseteq T$, and zero otherwise. To verify this, let us show that $ZM = I$. The S,T-entry of ZM is

$$(ZM)(S,T) = \sum_{U\subseteq[n]} Z(S,U)M(U,T) = \sum_{U:S\subseteq U\subseteq T} (-1)^{|T-U|}.$$

If $S = T$, this sum is 1; while if $S \not\subseteq T$, this sum is 0. Now consider the case where $S \subsetneq T$. Let S have a elements and T have $a+b$ elements, where $b > 0$. For $0 \le c \le b$, the number of sets U with $S \subseteq U \subseteq T$ and $|T-U| = c$ is $\binom{b}{c}$, since we can build such a subset U by choosing a subset of c elements from the b-element set $T-S$ and removing these elements from T to get U. Grouping terms in the sum for $(ZM)(S,T)$ based on the size of $|T-U|$, we see from Theorem 4.19 that

$$(ZM)(S,T) = \sum_{c=0}^{b}(-1)^c\binom{b}{c} = 0.$$

So the Möbius function for this poset is

$$\mu(S,T) = (-1)^{|T-S|}\chi(S \subseteq T) \qquad \text{for all } S,T \subseteq [n].$$

An alternate proof of this formula will be given in Example 4.61 below.

4.49. Example: Möbius Function of a Divisibility Poset. Let n be a fixed positive integer, let X be the set of positive divisors of n, and consider the divisibility poset $(X, |)$. There is a close relation between the number-theoretic Möbius function μ and the Möbius function μ_X for this poset. More precisely, we claim that

$$\mu(d) = \mu_X(1,d) \qquad \text{for all } d \text{ dividing } n.$$

To verify this, let us work with matrices whose rows and columns are indexed by the positive divisors of n, considered in increasing order. As above, let Z be the matrix such that $Z_{d,e} = \chi(d|e)$; let M be the inverse matrix, which is uniquely determined by Z; and let $\mathbf{v}$ be the row vector $(\mu(d) : d|n)$. The identity $\sum_{d|m}\mu(d) = \chi(m = 1)$, which is valid for all m dividing n, can now be rewritten as the vector identity $\mathbf{v}Z = (1,0,\ldots,0)$. This shows that $\mathbf{v}$ must be the first row of M. More generally, we show in Example 4.62 that $\mu_X(d,e) = \mu(e/d)$ whenever $d|e$ and $e|n$, whereas $\mu_X(d,e) = 0$ if d does not divide e.

The next definition will be used to give a combinatorial interpretation for the values of the Möbius function.

4.50. Definition: Chains in a Poset. Let $(X, \le)$ be a poset. A *chain of length* k in X is a sequence $C = (z_0, z_1, \ldots, z_k)$ of elements of X such that $z_0 < z_1 < \cdots < z_k$. We say that C is a chain *from* z_0 *to* z_k and write $\text{len}(C) = k$. The *sign* of the chain C is $\text{sgn}(C) = (-1)^k$.

4.51. Theorem: Möbius Functions and Signed Chains. Let $(X, \le)$ be a finite poset. Given $y, z \in X$, let S be the set of all chains in X from y to z. Then

$$\mu_{(X,\le)}(y,z) = \sum_{C\in S}\text{sgn}(C).$$

In particular, if $y \not\le z$, then $\mu_{(X,\le)}(y,z) = 0$.

Proof. We know from (4.8) that

$$\mu_{(X,\le)}(y,z) = \sum_{k\ge0}(-1)^k N^k(y,z),$$

where N is the adjacency matrix of the digraph $G = (X, <)$. A chain of length k from y to z is the same as a walk (or path) of length k from y to z in G. By the Walk Rule 3.19, the number of such walks is $N^k(y,z)$, and each walk has sign $(-1)^k$. The theorem now follows from the Sum Rule. □

4.52. Theorem: Möbius Inversion Formula for Posets. Let $(X, \leq)$ be a finite poset with Möbius function μ. For any functions $f, g : X \to \mathbb{R}$,

$$\left(\forall x \in X, g(x) = \sum_{y \leq x} f(y)\right) \quad \text{iff} \quad \left(\forall x \in X, f(x) = \sum_{y \leq x} g(y)\mu(y,x)\right).$$

Proof. Let $X = \{x_1, \ldots, x_n\}$, and define $Z = A(\leq)$ and $M = Z^{-1}$ as in Theorem 4.44. Also define row vectors $F = [f(x_1) \ \ldots \ f(x_n)]$ and $G = [g(x_1) \ \ldots \ g(x_n)]$. The first formula in the theorem is equivalent to the matrix identity $G = FZ$, since $G_j = g(x_j)$ and

$$(FZ)_j = \sum_{k=1}^{n} F_k Z_{k,j} = \sum_{k=1}^{n} f(x_k)\chi(x_k \leq x_j) = \sum_{y \leq x_j} f(y).$$

Similarly, keeping in mind that $\mu(y,x) \neq 0$ implies $y \leq x$, the second formula in the theorem is equivalent to the matrix identity $F = GM$. Since M and Z are inverse matrices, $G = FZ$ is equivalent to $GM = F$. □

4.53. Example. In the special case where $X = \{1, 2, \ldots, n\}$ with the ordering $1 < 2 < \cdots < n$, Theorem 4.52 reduces to the following statement: given $f_1, \ldots, f_n \in \mathbb{R}$ and $g_1, \ldots, g_n \in \mathbb{R}$, we have ($g_i = f_1 + f_2 + \cdots + f_i$ for all i) iff ($f_1 = g_1$ and $f_i = g_i - g_{i-1}$ for $1 < i \leq n$).

4.54. Example. In the special case where X is the set of positive divisors of n ordered by divisibility, Theorem 4.52 reduces to the number-theoretic Möbius Inversion Formula 4.33, using the fact that $\mu_X(d,e) = \mu(e/d)$ when $d|e$, and $\mu_X(d,e) = 0$ otherwise.

4.55. Example. In the special case where $X = \mathcal{P}([n])$ ordered by containment of subsets, Theorem 4.52 reduces to the following statement:

$$\left(\forall T \subseteq [n], g(T) = \sum_{S \subseteq T} f(S)\right) \quad \text{iff} \quad \left(\forall T \subseteq [n], f(T) = \sum_{S \subseteq T} (-1)^{|T-S|} g(S)\right).$$

If instead we use the opposite poset $(X, \supseteq)$, we obtain:

$$\left(\forall T \subseteq [n], g(T) = \sum_{S \supseteq T} f(S)\right) \quad \text{iff} \quad \left(\forall T \subseteq [n], f(T) = \sum_{S \supseteq T} (-1)^{|S-T|} g(S)\right).$$

We now use this result to rederive a version of the original Inclusion-Exclusion Formula. Let $Z_1, \ldots, Z_n$ be given subsets of a finite set Z. For $S \subseteq [n]$, let $f(S)$ be the number of objects $z \in Z$ such that $z \in Z_i$ if and only if $i \in S$. For $S \subseteq [n]$, let $g(S)$ be the number of objects $z \in Z$ such that $z \in Z_i$ if $i \in S$. Regarding Z_i as the set of objects in Z with a certain property i, we can say that $f(S)$ counts objects that have *exactly* the properties in S, whereas $g(S)$ counts the objects that have *at least* the properties in S. It follows from this that $g(T) = \sum_{S \supseteq T} f(S)$ for all T, so Theorem 4.52 tells us that $f(T) = \sum_{S \supseteq T} (-1)^{|S-T|} g(S)$ for all T. Now, $f(\emptyset) = |Z - (Z_1 \cup \cdots \cup Z_n)|$ and $g(\{i_1, \ldots, i_k\}) = |Z_{i_1} \cap \cdots \cap Z_{i_k}|$. The Union-Avoiding Rule 4.5 follows from these observations.

4.12 Product Posets

This section introduces a product construction for posets that leads to alternative derivations of the Möbius functions for Boolean posets and divisibility posets.

4.56. Definition: Product Posets. Let $(X_1, \leq_1), \ldots, (X_n, \leq_n)$ be posets. Recall that the Cartesian product $X = X_1 \times \cdots \times X_n$ consists of all n-tuples $x = (x_1, \ldots, x_n)$ with $x_i \in X_i$ for $1 \leq i \leq n$. For $x = (x_i)$ and $y = (y_i)$ in X, define $x \leq y$ iff $x_i \leq_i y_i$ for $1 \leq i \leq n$. The poset $(X, \leq)$ is called the *product* of the posets $(X_i, \leq_i)$.

One immediately verifies that the relation $\leq$ in the preceding definition really is a partial ordering on X.

4.57. Example. Let $X_1 = X_2 = \{1, 2\}$ with the ordering $1 < 2$. Both X_1 and X_2 are totally ordered posets, but $X = X_1 \times X_2$ is not totally ordered. For example, $(1, 2)$ and $(2, 1)$ are two incomparable elements of X.

4.58. Theorem: Möbius Function for a Product Poset. Let $(X, \leq)$ be the product of posets $(X_i, \leq_i)$ for $1 \leq i \leq k$. Given $x = (x_i)$ and $y = (y_i)$ in X, we have

$$\mu_{(X,\leq)}(x, y) = \prod_{i=1}^{k} \mu_{(X_i,\leq_i)}(x_i, y_i).$$

Proof. For brevity, write $\mu = \mu_{(X,\leq)}$ and $\mu_i = \mu_{(X_i,\leq_i)}$. By induction, we can reduce to the case $k = 2$. We have the matrices

$$\begin{array}{ll} Z_1 = [\chi(u_1 \leq_1 v_1) : u_1, v_1 \in X_1], & M_1 = [\mu_1(u_1, v_1) : u_1, v_1 \in X_1], \\ Z_2 = [\chi(u_2 \leq_2 v_2) : u_2, v_2 \in X_2], & M_2 = [\mu_2(u_2, v_2) : u_2, v_2 \in X_2], \\ Z = [\chi(u \leq v) : u, v \in X], & M = [\mu(u, v) : u, v \in X], \end{array}$$

which satisfy $Z_1 M_1 = I$, $Z_2 M_2 = I$ and $ZM = I$. Define a matrix M', with rows and columns indexed by elements of X, such that for $u = (u_1, u_2)$ and $v = (v_1, v_2)$ in X, the u, v-entry of M' is $\mu_1(u_1, v_1)\mu_2(u_2, v_2)$. Note that the u, v-entry of Z is $\chi((u_1, u_2) \leq (v_1, v_2)) = \chi(u_1 \leq_1 v_1)\chi(u_2 \leq_2 v_2)$. The following computation verifies that $ZM' = I$, and hence $M' = M$ since the inverse of Z is unique. For all $u = (u_1, u_2)$ and $w = (w_1, w_2)$ in X,

$$\begin{aligned} (ZM')(u, w) &= \sum_{v \in X} Z(u, v) M'(v, w) \\ &= \sum_{v_1 \in X_1} \sum_{v_2 \in X_2} \chi(u_1 \leq_1 v_1)\chi(u_2 \leq_2 v_2)\mu_1(v_1, w_1)\mu_2(v_2, w_2) \\ &= \left(\sum_{v_1 \in X_1} \chi(u_1 \leq_1 v_1)\mu_1(v_1, w_1) \right) \cdot \left(\sum_{v_2 \in X_2} \chi(u_2 \leq_2 v_2)\mu_2(v_2, w_2) \right) \\ &= (Z_1 M_1)(u_1, w_1) \cdot (Z_2 M_2)(u_2, w_2) \\ &= \chi(u_1 = w_1)\chi(u_2 = w_2) = \chi(u = w). \quad \square \end{aligned}$$

4.59. Definition: Poset Isomorphisms. Given posets $(X, \leq)$ and $(X', \leq')$, a *poset isomorphism* is a bijection $f : X \to X'$ such that for all $u, v \in X$, $u \leq v$ iff $f(u) \leq' f(v)$.

4.60. Theorem. If $f : X \to X'$ is a poset isomorphism between $(X, \leq)$ and $(X', \leq')$, then for all $u, v \in X$,

$$\mu_{(X',\leq')}(f(u), f(v)) = \mu_{(X,\leq)}(u, v).$$

Proof. This follows from Theorem 4.51. The chains of a given length from u to v in $(X, \leq)$ correspond bijectively to the chains of that length from $f(u)$ to $f(v)$ in $(X', \leq')$; the bijection applies f to each element in the chain. □

4.61. Example: Möbius Function of a Boolean Poset. Consider once more the divisibility poset $X = (\mathcal{P}([n]), \subseteq)$. For $1 \leq i \leq n$, take $Y_i = \{0, 1\}$ with the ordering $0 < 1$, and let $Y = Y_1 \times \cdots \times Y_n$ be the product poset. There is a bijection f from $\mathcal{P}([n])$ to $\{0, 1\}^n$ that sends a subset S to the word $f(S) = w = w_1 w_2 \cdots w_n$ with $w_i = 1$ for $i \in S$ and $w_i = 0$ for $i \notin S$. One readily sees that f is a poset isomorphism, so $\mu_X(S, T) = \mu_Y(f(S), f(T))$. Writing $f(T) = z = z_1 z_2 \cdots z_n$, Theorem 4.58 shows that $\mu_Y(w, z) = \prod_{i=1}^n \mu_{Y_i}(w_i, z_i)$. As in Example 4.47, we see that

$$\mu_{Y_i}(0, 0) = \mu_{Y_i}(1, 1) = 1; \quad \mu_{Y_i}(0, 1) = -1; \quad \mu_{Y_i}(1, 0) = 0.$$

So $w \not\leq z$ implies $\mu_Y(w, z) = 0$. If $w \leq z$ and z has k more 1's than w does, we see that $\mu_Y(w, z) = (-1)^k$. Translating back to subsets via f^{-1}, this says that $\mu_X(S, T) = 0$ when $S \not\subseteq T$, and $\mu_X(S, T) = (-1)^{|T-S|}$ when $S \subseteq T$.

4.62. Example: Möbius Function of a Divisibility Poset. Let n be a fixed positive integer with prime factorization $n = p_1^{n_1} \cdots p_k^{n_k}$, and consider the divisibility poset $(X, |)$, where $X = \{d \in \mathbb{Z}_{>0} : d|n\}$. For $1 \leq i \leq k$, let $Y_i = \{0, 1, \ldots, n_i\}$ with the ordering $0 < 1 < \cdots < n_i$, and take Y to be the product poset $Y_1 \times \cdots \times Y_k$. Any $d \in X$ has prime factorization $d = p_1^{d_1} \cdots p_k^{d_k}$ where $0 \leq d_i \leq n_i$ for $1 \leq i \leq k$. The map sending d to $(d_1, \ldots, d_k)$ is readily seen to be a poset isomorphism from X to Y. So

$$\mu_X(d, e) = \mu_Y((d_1, \ldots, d_k), (e_1, \ldots, e_k)) = \prod_{i=1}^{k} \mu_{Y_i}(d_i, e_i).$$

As in Example 4.47, we see that $\mu_{Y_i}(d_i, e_i) = \chi(e_i = d_i) - \chi(e_i = d_i + 1)$. It follows that $\mu_X(d, e) = 0$ unless e is obtained from d by multiplying by a set of s *distinct* prime factors chosen from $\{p_1, \ldots, p_k\}$, in which case $\mu_X(d, e) = (-1)^s$. It is now routine to check that whenever $d|e$, $\mu_X(d, e) = \mu(e/d)$, where μ is the number-theoretic Möbius function.

Summary

1. **Inclusion-Exclusion Formulas.** Let $S_1, \ldots, S_n$ be subsets of a finite set X.
   - *General Union Rule:*

$$|S_1 \cup S_2 \cup \cdots \cup S_n| = \sum_{k=1}^{n} (-1)^{k-1} \sum_{1 \leq i_1 < i_2 < \cdots < i_k \leq n} |S_{i_1} \cap S_{i_2} \cap \cdots \cap S_{i_k}|.$$

   - *Union-Avoiding Rule:*

$$|X - (S_1 \cup \cdots \cup S_n)| = |X| + \sum_{k=1}^{n} (-1)^k \sum_{1 \leq i_1 < i_2 < \cdots < i_k \leq n} |S_{i_1} \cap S_{i_2} \cap \cdots \cap S_{i_k}|.$$

   - *Number of Objects in Exactly j of the Sets S_i:*

$$\sum_{k=j}^{n} (-1)^{k-j} \binom{k}{j} \sum_{1 \leq i_1 < i_2 < \cdots < i_k \leq n} |S_{i_1} \cap S_{i_2} \cap \cdots \cap S_{i_k}|.$$

• *Number of Objects in at Least j of the Sets S_i:*

$$\sum_{k=j}^{n}(-1)^{k-j}\binom{k-1}{j-1}\sum_{1\le i_1<i_2<\cdots<i_k\le n}|S_{i_1}\cap S_{i_2}\cap\cdots\cap S_{i_k}|.$$

• *Simplified Versions:* Often $|S_{i_1}\cap\cdots\cap S_{i_k}|$ is some value $N(k)$ depending only on k, not on $i_1,\ldots,i_k$. In this case, each sum over $i_1,\ldots,i_k$ in the preceding formulas can be replaced by $\binom{n}{k}N(k)$.

2. **Applications of Inclusion-Exclusion.**
• *Surjections and Stirling Numbers.* For $m\ge n\ge 1$, the number of surjections from an m-element set onto an n-element set is $\sum_{k=0}^{n}(-1)^k\binom{n}{k}(n-k)^m$. A summation formula for the Stirling number of the second kind is

$$S(m,n)=\sum_{k=0}^{n}(-1)^k\frac{(n-k)^m}{k!(n-k)!}.$$

• *Euler's ϕ Function.* For $m\ge 1$, $\phi(m)$ is the number of integers x with $1\le x\le m$ and $\gcd(x,m)=1$. We have $\phi(m)=m\prod_{p|m}(1-p^{-1})$, where the product ranges over all prime divisors p of m. For $m=p^e$ with p prime, $\phi(p^e)=p^e-p^{e-1}$. If $\gcd(m,n)=1$, then $\phi(mn)=\phi(m)\phi(n)$. For $m\ge 1$, $\sum_{d|m}\phi(d)=m$.

• *Derangements.* A *derangement of S* is a bijection $f:S\to S$ with $f(x)\ne x$ for all $x\in S$. Let d_n be the number of derangements of an n-element set. Then $d_n=n!\sum_{k=0}^{n}(-1)^k/k!$, which is the closest integer to $n!/e$. Moreover, the numbers d_n satisfy the recursions

$$d_n=(n-1)d_{n-1}+(n-1)d_{n-2}\quad\text{for all } n\ge 2;$$

$$d_n=nd_{n-1}+(-1)^n\quad\text{for all } n\ge 1.$$

3. **Truth Function χ.** For any statement P, $\chi(P)=1$ if P is true, and $\chi(P)=0$ if P is false.

4. **Involutions.** An involution is a function $I:X\to X$ with $I\circ I=\mathrm{id}_X$. The fixed point set of I is $\mathrm{Fix}(I)=\{x\in X: I(x)=x\}$. When X consists of signed objects, I is sign-reversing iff $\mathrm{sgn}(I(x))=-\mathrm{sgn}(x)$ for all $x\in X-\mathrm{Fix}(I)$. For a sign-reversing involution I with domain X,

$$\sum_{x\in X}\mathrm{sgn}(x)=\sum_{x\in\mathrm{Fix}(I)}\mathrm{sgn}(x).$$

Involutions provide combinatorial proofs of identities that involve signed terms.

5. **Möbius Functions and Posets.**
• *Number-theoretic Möbius Function.* Define $\mu:\mathbb{Z}_{>0}\to\{-1,0,1\}$ by $\mu(n)=(-1)^s$ if n is the product of $s\ge 0$ distinct primes, and $\mu(n)=0$ otherwise. Then $\sum_{d|m}\mu(d)=\chi(m=1)$. Given functions f and g such that $f(m)=\sum_{d|m}g(d)$ for all $m\ge 1$, the number-theoretic Möbius Inversion Formula states that

$$g(m)=\sum_{d|m}f(m/d)\mu(d)=\sum_{d|m}f(d)\mu(m/d)\quad\text{for all } m\ge 1.$$

It follows that $\phi(m) = \sum_{d|m} \mu(d)m/d$.

• *Posets.* A *partial ordering* of X is a relation $\leq$ on X that is reflexive, antisymmetric, and transitive; the pair $(X, \leq)$ is called a *poset.* A *strict ordering* of X is a relation $<$ on X that is irreflexive and transitive. There is a bijection mapping partial orders on X to strict orders on X defined by removing the diagonal $\{(x,x) : x \in X\}$. A *chain of length* k in a poset $(X, \leq)$ is a sequence $(z_0, z_1, \ldots, z_k)$ with $z_0 < z_1 < \cdots < z_k$. Such a chain *goes from* z_0 *to* z_k and has *sign* $(-1)^k$.

• *Möbius Functions for Posets.* Given a poset $(X, \leq)$ with $X = \{x_1, \ldots, x_n\}$, define $n \times n$ matrices Z, N, and M by $Z_{ij} = \chi(x_i \leq x_j)$, $N_{ij} = \chi(x_i < x_j)$, and $M_{ij} =$ the signed sum of all chains in the poset from x_i to x_j. Then $Z = I + N$; N is nilpotent; and M is the matrix inverse of Z. We write $\mu(x_i, x_j) = M_{ij}$ and call μ the *Möbius function* of the poset $(X, \leq)$. Suppose f and g are functions with domain X. The *Möbius Inversion Formula for Posets* states that

$$g(x) = \sum_{y \leq x} f(y) \text{ for all } x \in X \quad \text{iff} \quad f(x) = \sum_{y \leq x} g(y)\mu(y,x) \text{ for all } x \in X.$$

• *Product Posets.* Given posets $(X_i, \leq_i)$ for $1 \leq i \leq n$, the product set $X = X_1 \times \cdots \times X_n$ becomes a poset by defining $(x_1, \ldots, x_n) \leq (y_1, \ldots, y_n)$ iff $x_i \leq_i y_i$ for all i between 1 and n. The Möbius function for the product poset satisfies

$$\mu_X((x_1, \ldots, x_n), (y_1, \ldots, y_n)) = \prod_{i=1}^{n} \mu_{X_i}(x_i, y_i).$$

• *Examples of Möbius Functions.* The poset $X = \{1, 2, \ldots, n\}$ with the total ordering $1 < 2 < \cdots < n$ has Möbius function

$$\mu_X(i,i) = 1, \quad \mu_X(i, i+1) = -1, \quad \mu_X(i,j) = 0 \text{ for } j \neq i, i+1.$$

The Boolean poset $(\mathcal{P}(X), \subseteq)$ of subsets of $\{1, 2, \ldots, n\}$ ordered by inclusion has Möbius function

$$\mu(S,T) = (-1)^{|T-S|}\chi(S \subseteq T) \qquad \text{for } S, T \subseteq [n].$$

If n has prime factorization $p_1^{n_1} \cdots p_k^{n_k}$, then the poset of positive divisors of n under the divisibility ordering has Möbius function

$$\mu(d,e) = \begin{cases} (-1)^s & \text{if } e/d \text{ is a product of } s \text{ distinct primes;} \\ 0 & \text{otherwise.} \end{cases}$$

These results follow since the Boolean poset is isomorphic to the product of n copies of the totally ordered set $\{0, 1\}$, whereas the divisibility poset is isomorphic to the product poset $\{0, 1, \ldots, n_1\} \times \cdots \times \{0, 1, \ldots, n_k\}$.

Exercises

4-1. Given that $|S| = 15$, $|T| = 13$, $|U| = 12$, $|S \cap T| = 6$, $|S \cap U| = 3$, $|T \cap U| = 4$, and $|S \cap T \cap U| = 1$, find: (a) $|S \cup T|$; (b) $|S \cup T \cup U|$; (c) the number of objects in exactly one of the sets S, T, U.

4-2. Given $S, T, U \subseteq X$ with $|X| = 35$, $|S| = 12$, $|T| = 14$, $|U| = 15$, $|S \cap T| = 5 = |S \cap U|$, $|T \cap U| = 6$, and $|(S \cup T) \cap U| = 9$, find: (a) $|S \cap T \cap U|$; (b) $|X - (S \cup T \cup U)|$; (c) the number of objects in exactly two of the sets S, T, U.

4-3. List all the derangements of $\{1, 2, 3, 4\}$.

4-4. Compute d_{10} in four ways: (a) by rounding $10!/e$ to the nearest integer; (b) by using the summation formula in 4.16; (c) by using the recursion in 4.17; (d) by using the recursion in 4.18.

4-5. Compute $\phi(n)$, $\mu(n)$, $\tau(n)$, and $\sigma(n)$ for the following choices of n: (a) 6; (b) 11; (c) 28; (d) 60; (e) 1001; (f) 121.

4-6. Verify Theorem 4.37 by direct calculation for (a) $m = 24$; (b) $m = 30$.

4-7. Given n married couples, how many ways can the n men and n women be paired so that no pair consists of a man and his wife?

4-8. How many five-card poker hands have at least one card of every suit?

4-9. How many five-card poker hands have at least one face card, at least one diamond, and do not contain both a 2 and a 3?

4-10. How many ten-digit numbers contain at least one 4, one 5, and one 7?

4-11. How many bridge hands are void in clubs and have at least one card of value p for each prime $p < 10$?

4-12. A keyboard has 26 lowercase letters, 26 uppercase letters, 10 digits, and 32 special characters. How many n-letter passwords contain at least one character from each category?

4-13. How many ways can we place m labeled balls into n labeled boxes such that at least one box is empty?

4-14. How many ways can we place m identical balls into n labeled boxes such that at least one box is empty? Solve this problem in two ways, and thereby deduce a binomial coefficient identity.

4-15. (a) How many n-letter words using the alphabet $\{a_1, \ldots, a_k\}$ (where $k \geq 3$) contain at least one copy of a_1, a_2, and a_3? (b) Repeat (a) assuming all letters in the word must be distinct.

4-16. How many surjections $f : \{1, 2, \ldots, m\} \to \{1, 2, \ldots, n\}$ have the property that $f(x) = 1$ for exactly one x?

4-17. How many ways can we put thirty identical balls into eight labeled boxes if the first four boxes must each contain at most five balls and the last four boxes must each contain at least two balls?

4-18. For even $n \geq 2$, determine the number of integers $x \leq n$ with $\gcd(x, n) = 2$.

4-19. For $k \geq 0$ and $m \geq 1$, let $\sigma_k(m) = \sum_{d|m} d^k$. (a) Find a formula for $\sigma_k(m)$ in terms of the prime factorization of m. (b) Find a formula for m^k involving σ_k and μ.

4-20. Use Theorem 4.13 to show that $\phi(mn) = \phi(m)\phi(n)$ iff $\gcd(m, n) = 1$.

4-21. Explicitly compute how the first involution discussed in Example 4.25 matches up the 24 objects counted by $\sum_{k=1}^{4} s(4, k)$ into pairs of objects of opposite sign.

4-22. Suppose w has cycles (1), (2), (3, 8, 7), (5, 6, 9), (4), and

$$U = \{\{(1)\}, \{(2)\}, \{(4), (5, 6, 9)\}, \{(3, 8, 7)\}\}.$$

Compute $I(w, U)$, where I is the involution defined at the end of Example 4.25.

4-23. Consider the derangement $w = 436215 \in D_6$. Find the six derangements in D_7 and the seven derangements in D_8 that can be built from w by the construction in the proof of Theorem 4.17.

4-24. Use the recursion for derangements in 4.18 to give a proof by induction of the summation formula for derangements in 4.16.

4-25. Give the details of the proof of Theorem 4.42.

4-26. (a) Give an algebraic proof that $\sum_{k=0}^{n} \binom{n}{k} 2^k (-1)^{n-k} = 1$ for $n \geq 0$. (b) Prove the identity in (a) using an involution.

4-27. For integers $a \geq b > 0$, evaluate $\sum_{k=0}^{n} \binom{n}{k} a^{n-k} (-b)^k$ by using an involution.

4-28. Let $S \subseteq T$ be given finite sets. (a) Use an involution to prove $\sum_{U:\ S \subseteq U \subseteq T} (-1)^{|T-U|} = \chi(S = T)$ (cf. Example 4.48). (b) In a similar manner, evaluate $\sum_{U:\ S \subseteq U \subseteq T} (-1)^{|U-S|}$.

4-29. Given $d, e \in \mathbb{Z}_{>0}$ with $d|e$, use an involution to prove $\sum_{k:\ d|k|e} \mu(e/k) = \chi(d = e)$. Interpret this result in terms of the Möbius function of a poset.

4-30. Count the $n \times n$ matrices A with entries in $\{0, 1, 2\}$ such that: (a) no row of A contains all zeroes; (b) every column of A contains at least one zero; (c) there is no index j with $A(i, j) > 0$ and $A(j, i) > 0$ for all i.

4-31. An *arrowless vertex* in a simple digraph D is a vertex with indegree and outdegree zero. How many simple digraphs with vertex set $\{1, 2, \ldots, n\}$ have no arrowless vertices?

4-32. An *isolated vertex* in a simple digraph D is a vertex v such that there is no edge (u, v) or (v, u) in D with $u \neq v$. How many simple digraphs with vertex set $\{1, 2, \ldots, n\}$ have no isolated vertices?

4-33. How many simple graphs with vertex set $\{1, 2, \ldots, n\}$ have no isolated vertices?

4-34. (a) How many anagrams in $\mathcal{R}(1^3 2^3 \cdots n^3)$ never have three equal letters in a row? (b) How many anagrams in $\mathcal{R}(1^k 2^k \cdots n^k)$ never have k equal letters in a row?

4-35. (a) Count the permutations w of $\{1, 2, \ldots, n\}$ such that $w_{i+1} \neq w_i + 1$ for all i in the range $1 \leq i < n$. (b) Express your answer to (a) in terms of the derangement numbers d_k.

4-36. Given sequences $0 < a_1 \leq a_2 \leq \cdots \leq a_k < A$ and $0 < b_1 \leq b_2 \leq \cdots \leq b_k < B$, use inclusion-exclusion to derive a formula for the number of lattice paths from $(0, 0)$ to (A, B) that avoid all of the points (a_i, b_i) for $1 \leq i \leq k$.

4-37. Recursion for Möbius Functions. (a) Show that the Möbius function of a poset $(X, \leq)$ can be computed recursively via $\mu(x, z) = -\sum_{y:\ x \leq y < z} \mu(x, y)$ for $x < z$, with initial conditions $\mu(x, x) = 1$ and $\mu(x, z) = 0$ whenever $x \not\leq z$. (b) Show that the Möbius function also satisfies the recursion $\mu(x, z) = -\sum_{y:\ x < y \leq z} \mu(y, z)$ for $x < z$.

4-38. Poset Associated to a DAG. Suppose $G = (X, R)$ is a DAG. Prove that there exists a unique smallest irreflexive, transitive relation $<$ that contains R. The corresponding poset $(X, \leq)$ is called the *poset associated to the DAG G*.

4-39. Let $(X, \leq)$ be the poset associated to the DAG

$$(\{a, b, c, d, e\}, \{(a, b), (b, e), (a, c), (c, e), (a, d), (d, e)\}).$$

Compute the Möbius function μ_X in two ways, by: (a) inverting the matrix Z; (b) enumerating signed chains in $(X, \leq)$.

4-40. Let $(X, \leq)$ be the poset associated to the DAG

$$(\{a, b, c, d, e, f\}, \{(a, b), (a, c), (b, d), (b, e), (c, d), (d, f), (e, f)\}).$$

Compute the Möbius function μ_X in two ways, by: (a) inverting the matrix Z; (b) enumerating signed chains in $(X, \leq)$.

4-41. A *subposet* of a poset $(X, \leq)$ is a poset $(Y, \leq')$, where Y is a subset of X, and for $a, b \in Y$, $a \leq' b$ iff $a \leq b$. An *interval* in X is a subposet of the form $[x, z] = \{y \in X : x \leq y \leq z\}$. Show that for all $a, b, c, d \in X$, if the intervals $[a, b]$ and $[c, d]$ are isomorphic posets, then $\mu_X(a, b) = \mu_X(c, d)$.

4-42. Assume that X_1 and X_2 are finite disjoint sets. The *disjoint union* of the posets $(X_1, \leq_1)$ and $(X_2, \leq_2)$ is $(X, \leq)$ where $X = X_1 \cup X_2$ and for $a, b \in X$, $a \leq b$ iff $a, b \in X_1$ and $a \leq_1 b$, or $a, b \in X_2$ and $a \leq_2 b$. Determine μ_X in terms of μ_{X_1} and μ_{X_2}.

4-43. Given a poset $(X, \leq)$, define a new poset $(Y, \leq')$ by setting $Y = X \cup \{0\}$ (where 0 is a new symbol not in X), and letting $\leq'$ be the extension of $\leq$ such that $0 \leq' y$ for all $y \in Y$. Informally, $(Y, \leq')$ is obtained from $(X, \leq)$ by adjoining a new least element. Determine μ_Y in terms of μ_X.

4-44. Given posets $(X_1, \leq_1)$ and $(X_2, \leq_2)$ where X_1 and X_2 are finite disjoint sets, define a new poset $(X, \leq)$ by setting $X = X_1 \cup X_2$ and, for $a, b \in X$, $a \leq b$ iff $a, b \in X_1$ and $a \leq_1 b$, or $a, b \in X_2$ and $a \leq_2 b$, or $a \in X_1$ and $b \in X_2$. Informally, $(X, \leq)$ is obtained from X_1 and X_2 by making everything in X_1 less than everything in X_2. Determine μ_X in terms of μ_{X_1} and μ_{X_2}.

4-45. Prove Theorem 4.58 by counting signed chains in the product poset X.

4-46. Given events $S_1, \ldots, S_n$ in a sample space X, find and prove an inclusion-exclusion formula for $P(S_1 \cup \cdots \cup S_n)$.

4-47. Let $S_1, \ldots, S_n$ be independent events in a sample space X (see Definition 1.66). Prove that for $1 \leq i \leq n$, the events $S_1, S_2, \ldots, X - S_i, \ldots, S_n$ are independent.

4-48. Let $S_1, \ldots, S_n$ be independent events in a sample space X, with $P(S_i) = p_i$ for each i. Find the probability that none of the events S_i occurs first by using inclusion-exclusion, and then by iterating the previous exercise. Show algebraically that the two answers agree.

4-49. Use an involution to prove that for all $i, n \in \mathbb{Z}_{\geq 0}$, $\sum_{k=0}^{i} (-1)^k \binom{n}{k} \binom{n-k}{i-k} = \chi(i = 0)$.

4-50. Use an involution to prove that for $0 \leq k \leq n$, $\sum_{i=k}^{n} (-1)^{i-k} \binom{n}{i} \binom{i}{k} 2^{n-i} = \binom{n}{k}$.

4-51. Prove that for all $n, j > 0$, $n^j = \sum_{k=0}^{j} (-1)^{j-k} k! S(j, k) \binom{n+k-1}{k}$.

4-52. For $n > 0$, evaluate $\sum_{k=0}^{n-1} (-1)^k \binom{n}{k} (n-k)^n$.

4-53. Use an involution to prove the following identity satisfied by Catalan numbers: $C_n = \sum_{1 \leq k \leq (n+1)/2} (-1)^{k-1} C_{n-k} \binom{n+1-k}{k}$.

4-54. Let A be an $n \times n$ matrix with $A(i, j) = \binom{i-1}{j-1}$ for $1 \leq i, j \leq n$. Find and prove a formula for A^{-1}.

4-55. How many bijections $f : \{1, 2, \ldots, n\} \to \{1, 2, \ldots, n\}$ are such that the functional digraph of f contains no cycle of length k?

4-56. How many anagrams in $\mathcal{R}(a^3 b^3 c^3 d^3)$ never have two consecutive equal letters?

4-57. Prove or disprove: for every integer $y \geq 1$, there exist only finitely many integers $x \geq 1$ with $\phi(x) = y$.

4-58. How many compositions of n have k parts each of size at most m?

4-59. Call a function $f : X \to Y$ *doubly surjective* iff for all $y \in Y$, there exist at least two $x \in X$ with $f(x) = y$. Count the number of doubly surjective functions from an m-element set to an n-element set, where $m \geq 2n$. What is the answer when $m = 11$ and $n = 4$?

4-60. How many integers between 1 and 2311 are divisible by exactly two of the primes in $\{2, 3, 5, 7\}$?

4-61. Let (F_n) be the Fibonacci sequence ($F_0 = 0$, $F_1 = 1$, $F_n = F_{n-1} + F_{n-2}$ for $n \geq 2$). Find a formula for $\sum_{k=0}^{n} (-1)^k F_k$ and prove it algebraically or by using an involution.

4-62. Find and prove a formula for $\sum_{k=0}^{n}(-1)^k F_k F_{n-k}$.

4-63. Prove $F_{n-1}F_{n+1} - F_n^2 = (-1)^n$ using an involution on domino tilings.

4-64. Prove: for $i \leq n \leq i+m$, $F_m F_n - F_{m+i}F_{n-i} = (-1)^{n-i}F_{m+i-n}F_i$.

4-65. For each integer $x \geq 1$, evaluate $\sum_{k=1}^{x} \mu(k)\lfloor x/k \rfloor$.

4-66. Let $\mathrm{Surj}(n,k)$ be the number of surjections from an n-element set onto a k-element set. For $n > 0$, evaluate $\sum_{k=1}^{n}(-1)^k \mathrm{Surj}(n,k)$.

4-67. For $n > 0$, evaluate $\sum_{k=1}^{n-1}(-1)^k (k-1)! S(n,k)$.

4-68. Given elements a_k, b_k in $\mathbb{R}$ (or in any commutative ring), prove:

$$\prod_{k=1}^{n}(a_k + b_k) = \sum_{I \subseteq \{1,2,\ldots,n\}} \prod_{\substack{i=1 \\ i \in I}}^{n} a_i \prod_{\substack{j=1 \\ j \notin I}}^{n} b_j.$$

4-69. Counting Proof of Inclusion-Exclusion. Let $S_1, \ldots, S_n$ be given subsets of a finite set X. Let A be a matrix with rows indexed by elements $x \in X$ and columns indexed by nonempty subsets I of $[n] = \{1,2,\ldots,n\}$. Define the entry of A in row x and column I to be $(-1)^{|I|-1}\chi(x \in \bigcap_{i \in I} S_i)$. Prove the Inclusion-Exclusion Formula for $|\bigcup_{i=1}^{n} S_i|$ by computing the sum of all entries of A in two ways. Discuss how this proof keeps track of the number of times each $x \in X$ is included and excluded in the Inclusion-Exclusion Formula.

4-70. Let I_n be the number of involutions on an n-element set. (a) Find I_n for $1 \leq n \leq 5$. (b) Find a recursion for computing I_n from I_{n-1} and I_{n-2}. (c) Find a summation formula for I_n.

4-71. How many 10-letter words using the alphabet $\{\mathrm{A}, \mathrm{B}, \ldots, \mathrm{Z}\}$ contain exactly three of the five vowels A, E, I, O, U? (Vowels that are used may appear more than once.)

4-72. How many 13-card bridge hands contain exactly 10 of the 13 possible card values (2 through ace)?

4-73. How many numbers between 1 and 1,000,000 are divisible by at least three of the primes in the set $\{2,3,5,7,11,13\}$?

4-74. How many words in $\mathcal{R}(1^2 2^2 \cdots n^2)$ have at least k pairs of adjacent letters that are equal?

4-75. How many functions from an m-element set to an n-element set have image of size k?

4-76. Prove the generalized inclusion-exclusion formula (4.7) using Lemma 4.27(b).

4-77. Prove: for $j \leq k$, $\sum_{p=j}^{k}(-1)^p \binom{k}{p} = (-1)^j \binom{k-1}{j-1}$.

4-78. Prove (4.7) by summing appropriate instances of the formula (4.5).

4-79. Generalize the involution-based proof given in 4.26 to prove (4.5).

4-80. Combinatorial Interpretation for Coefficients of Chromatic Polynomials. Given a simple graph G with vertex set V and edge set E, a *vertex-spanning subgraph* of G is a graph H with vertex set V and edge set $E' \subseteq E$. Let $n(G,e,c)$ be the number of vertex-spanning subgraphs of G with e edges and c connected components. Prove the following formula for the chromatic polynomial of G:

$$\chi_G(x) = \sum_{e,c \geq 0} (-1)^e n(G,e,c) x^c.$$

4-81. Use the previous exercise to compute the chromatic polynomial of the graph with vertex set $\{1,2,3,4\}$ and edge set $\{\{1,2\},\{1,3\},\{1,4\},\{2,3\}\}$.

4-82. (a) What is the chromatic polynomial for the 4-cycle C_4? (b) For each coefficient of this chromatic polynomial, draw the vertex-spanning subgraphs of C_4 counted by that coefficient.

4-83. Show that if G is a simple graph with c connected components, then the chromatic polynomial $\chi_G(x)$ must be divisible by x^c.

4-84. For a poset $(X, \leq)$ with $X = \{x_1, \ldots, x_n\}$, define $n \times n$ matrices Z and M by setting $Z_{i,j} = \chi(x_i \leq x_j)$ and letting $M_{i,j}$ be the sum of the signs of all chains in X from x_i to x_j. Use an involution to prove the matrix identity $ZM = I_n$.

4-85. Use an involution to prove the Surjection Rule 4.10.

4-86. Use an involution to prove the Summation Formula for Derangements 4.16.

4-87. Consider an $n \times n$ lower-triangular matrix A such that $A(n,k)$ is the number of Dyck paths ending with exactly k east steps, for $1 \leq k \leq n$. Find a combinatorial description of A^{-1}, and prove that this is the inverse of A using an involution.

4-88. Garsia–Milne Involution Principle. Suppose I and J are involutions defined on finite signed sets X and Y, respectively. Suppose $f : X \to Y$ is a *sign-preserving* bijection, i.e., $\mathrm{sgn}(f(x)) = \mathrm{sgn}(x)$ for all $x \in X$. Suppose also that every object in $\mathrm{Fix}(I)$ and $\mathrm{Fix}(J)$ has positive sign. Construct an explicit bijection $g : \mathrm{Fix}(I) \to \mathrm{Fix}(J)$.

4-89. Bijective Subtraction. Suppose A, B, and C are finite, pairwise disjoint sets and $f : A \cup B \to A \cup C$ is a given bijection. Construct an explicit bijection $g : B \to C$.

4-90. Bijective Division by Two. Suppose A and B are finite sets. Given a bijection $f : \{0,1\} \times A \to \{0,1\} \times B$, can you use f to construct an explicit bijection $g : A \to B$?

4-91. In §4.6 we proved combinatorially that $\sum_k s(i,k)S(k,j) = \chi(i = j)$. Can you find a combinatorial proof that $\sum_k S(i,k)s(k,j) = \chi(i = j)$? (Compare to Theorem 2.64(d).)

4-92. Find a bijective proof of the derangement recursion $d_n = nd_{n-1} + (-1)^n$. (For a solution, see [105].)

4-93. Let X_n be the set of set partitions of $\{1, 2, \ldots, n\}$. Define the *refinement ordering* on X_n by setting, for $P, Q \in X_n$, $P \preceq Q$ iff every block $S \in P$ is contained in some block $T \in Q$. (a) Show that $(X_n, \preceq)$ is a poset. (b) Compute the Möbius function of this poset for $1 \leq n \leq 4$. (c) Show that any interval $[P, Q] = \{R \in X_n : P \preceq R \preceq Q\}$ in X_n is isomorphic to a poset $(X_k, \preceq)$ for some k. (d) Compute μ_{X_n} for all n.

4-94. Prove: for $n \geq 1$, $\sum_{k=1}^{n} (-1)^{k-1} \binom{n}{k} F_k = F_n$, where F_n denotes a Fibonacci number.

4-95. Prove: for $0 \leq m \leq n$, $\sum_{k=0}^{n} (-1)^{k-1} \binom{n}{k} F_{k+m} = (-1)^m F_{n-m}$.

4-96. Prove: for $n \geq 1$, $\sum_{k=1}^{2n} (-1)^k \binom{2n}{k} 2^{k-1} F_k = 0$.

4-97. Prove: for $n \geq 1$, $\sum_{k=1}^{n} (-1)^k \binom{n}{k} F_{2k} = (-1)^n F_n$.

4-98. Suppose $n \geq 2$, $0 < m < n$, and n is even. Prove $\sum_{k=1}^{n} (-1)^k \binom{2n}{n+k} k^m = 0$.

4-99. Prove: for $m, n \geq 0$, $\sum_{k=0}^{2n} (-1)^k \binom{m}{k} \binom{m}{2n-k} = (-1)^n \binom{m}{n}$.

4-100. Prove: for $n \geq 0$, $\sum_{k=0}^{\lfloor n/2 \rfloor} (-1)^k \binom{n}{k} \binom{2n-2k-1}{n-1} = 1$.

4-101. Prove: for $n \geq 0$, $\sum_{k=0}^{\lfloor n/2 \rfloor} (-1)^k \binom{n}{k} \binom{2n-2k}{n} = 2^n$.

4-102. Prove: for $n \geq 0$, $\sum_{k=0}^{2n} (-1)^k \binom{3n+1}{k} \binom{3n-k}{n}^3 = 1$.

4-103. Prove: for $n \geq 0$, $\sum_{k=0}^{2n} (-1)^k \binom{3n}{k} \binom{3n-k}{n}^3 = \binom{3n}{n,n,n}$.

Notes

A thorough treatment of posets from the combinatorial viewpoint appears in Chapter 3 of [121]. See [113] for one of the seminal papers on Möbius inversion in combinatorics. A classic text on posets is the book by Birkhoff [11]. The Garsia-Milne Involution Principle (Exercise 4-88) was introduced in [43, 44]. For applications and extensions of this principle, the reader may consult [50, 68, 82, 83, 103, 104, 133]. An application of Bijective Subtraction (Exercise 4-89) is presented in [80].

5

Generating Functions

This chapter introduces generating functions, which are powerful tools for solving many combinatorial problems. Intuitively, a generating function is an infinite series $\sum_{n=0}^{\infty} a_n z^n$ whose coefficients a_n count a family of combinatorial objects. We can obtain a great deal of information about the numbers a_n by performing algebraic or analytic operations on the associated generating functions. For example, generating functions can be used to find closed formulas for recursively defined sequences a_n. They also give an automatic method for evaluating a summation $\sum_{k=0}^{n} a_k$. Generating functions often deliver quick answers to otherwise intractable counting questions.

This chapter focuses on manipulating generating functions to obtain explicit solutions to concrete problems. Later, in Chapter 11, we look more closely at more theoretical aspects of generating functions. Some concepts used informally in this chapter will be justified rigorously at that time.

5.1 What is a Generating Function?

Generating functions can be defined in three different ways: combinatorially, analytically, or algebraically. According to the combinatorial viewpoint, a generating function is something that lets us count a set of *weighted* objects that might be infinite. More precisely, we define a *weighted set* to be a set S together with a function $\text{wt} : S \to \mathbb{Z}_{\geq 0}$. For each object $u \in S$, $\text{wt}(u)$ is a nonnegative integer called the *weight* of u. As part of the definition, we require that for each $n \geq 0$, the set $A_n = \{u \in S : \text{wt}(u) = n\}$ is *finite*. The *generating function* of the weighted set S is defined to be the *formal sum*

$$\text{GF}(S; z) = \sum_{u \in S} z^{\text{wt}(u)},$$

where z is a *formal variable* used to keep track of the weights of the objects in S. In the sum appearing here, a given power z^n will appear once for each object in S having weight n. Collecting together all of these powers, the generating function for S can be rewritten

$$\text{GF}(S; z) = \sum_{n=0}^{\infty} a_n z^n, \text{ where } a_n = |A_n| = |\{u \in S : \text{wt}(u) = n\}|. \tag{5.1}$$

Momentarily, we will explain precisely what is meant by this sum when S is infinite.

5.1. Example. Let S be the set of all finite sequences of 0's and 1's. For a sequence $u \in S$, let $\text{wt}(u)$ be the length of u. For example, $\text{wt}(10110) = 5$. In this case, the set A_n of objects in S having weight n is $\{0, 1\}^n$. By the Word Rule, $|A_n| = 2^n$. Therefore, the generating function for the weighted set S is

$$\text{GF}(S; z) = \sum_{n=0}^{\infty} 2^n z^n.$$

Next we discuss the analytic definition of generating functions. According to this viewpoint, a *generating function* is an actual function $G : D \to \mathbb{C}$ given by a power series $G(z) = \sum_{n=0}^{\infty} a_n z^n$, where $a_0, a_1, \ldots, a_n, \ldots$ is a fixed sequence of complex numbers, D is a neighborhood of 0 in $\mathbb{C}$, and z is a complex variable. A key technical point is that the infinite series of complex numbers defining $G(z)$ is required to *converge* for each z in the domain D. This means that for each $z \in D$, there exists a complex number $G(z)$ with

$$G(z) = \lim_{N\to\infty} \sum_{n=0}^{N} a_n z^n.$$

Each limit must be a complex number, not $\pm\infty$. By the theory of power series (reviewed in more detail below), for each sequence (a_n) there exists $R \in [0, \infty]$, called the *radius of convergence* of the power series, such that $\sum_{n=0}^{\infty} a_n z^n$ converges for all z with $|z| < R$, and $\sum_{n=0}^{\infty} a_n z^n$ does not converge for all z with $|z| > R$. When $R > 0$, all the coefficients a_n are uniquely determined by the associated function $G(z) = \sum_{n=0}^{\infty} a_n z^n$ via Taylor's formula $a_n = G^{(n)}(0)/n!$. When $R = 0$, the series $\sum_{n=0}^{\infty} a_n z^n$ converges only at $z = 0$, so the resulting function of z has domain $\{0\}$. This domain is not a neighborhood of 0, so we do not get an analytic generating function in this case. The function's only value is $G(0) = a_0$, and we learn nothing from this function about the coefficients a_n with $n > 0$.

5.2. Example. Our previous example led us to the generating function $G(z) = \sum_{n=0}^{\infty} 2^n z^n$. Viewing G analytically as a function of the complex variable z, we can use the Geometric Series Formula 2.2 to write

$$G(z) = \sum_{n=0}^{\infty} (2z)^n = \frac{1}{1-2z}.$$

The series converges to the indicated sum for all $z \in \mathbb{C}$ satisfying $|2z| < 1$, or $|z| < 1/2$. The radius of convergence of this power series is $1/2$.

One powerful feature of the analytic view of generating functions is that it allows us to obtain combinatorial information about weighted sets by manipulating functions of z (using techniques such as geometric series, partial fraction decompositions, the Quadratic Formula, and the Extended Binomial Theorem discussed below). The next example shows how the Binomial Theorem can be used to simplify a generating function.

5.3. Example. Fix a positive integer m, and let T be the set of all subsets of $\{1, 2, \ldots, m\}$. For an object $B \in T$, let $\mathrm{wt}(B) = |B|$ be the number of elements in the subset B. By the Subset Rule, we know there are $\binom{m}{n}$ objects in T having weight n. So the generating function for T is

$$H(z) = \mathrm{GF}(T; z) = \sum_{B\in T} z^{\mathrm{wt}(B)} = \sum_{n=0}^{m} \binom{m}{n} z^n.$$

The function H is given as a power series with only finitely many nonzero coefficients. In other words, H is a *polynomial* in z. Thus, the power series has infinite radius of convergence, and $H : \mathbb{C} \to \mathbb{C}$ is a polynomial function defined on all of $\mathbb{C}$. This happens whenever we use generating functions to enumerate a *finite* weighted set T, since $\{u \in T : \mathrm{wt}(u) = n\}$ has size 0 for all n exceeding the maximum weight of any object in T. For the particular polynomial H considered here, the Binomial Theorem shows that the series defining H can be simplified to $H(z) = (1 + z)^m$.

5.4. Example. For each $n \geq 0$, let S_n be the set of permutations of $\{1, 2, \ldots, n\}$. Let $S = \bigcup_{n=0}^{\infty} S_n$, and let the weight of a word $u \in S$ be the length of that word. By the

Permutation Rule, we have $n! = |S_n| = |\{u \in S : \text{wt}(u) = n\}|$, so the generating function for S is

$$\text{GF}(S; z) = \sum_{u \in S} z^{\text{wt}(u)} = \sum_{n=0}^{\infty} n! z^n.$$

Unfortunately, this power series has radius of convergence zero (as can be checked using the Ratio Test, which we review below). So we cannot obtain combinatorial information about factorials using analytic power series.

The last example reveals the main drawback of using analytically defined generating functions: when the combinatorial coefficients $a_n = |\{u \in S : \text{wt}(u) = n\}|$ grow too quickly, the associated power series only converges at $z = 0$. This problem can be addressed in two ways. One way is to use a modified version of the combinatorial generating function, called an *exponential generating function*, defined by $\text{EGF}(S; z) = \sum_{n=0}^{\infty} (a_n/n!) z^n$. In the previous example, we find that $\text{EGF}(S; z) = \sum_{n=0}^{\infty} z^n = 1/(1 - z)$, which converges for all z with $|z| < 1$. Generating functions as defined in (5.1) are sometimes called *ordinary generating functions* (OGFs) to distinguish them from exponential generating functions (EGFs). We discuss EGFs later in this chapter after developing the theory of OGFs.

However, even using EGFs, there are examples of weighted sets where the associated power series have radius of convergence equal to zero. The second way around this problem is to use the algebraic definition of generating functions. Algebraically, a *generating function* is nothing more than an infinite sequence $A = (a_n : n \in \mathbb{Z}_{\geq 0})$, where each a_n is a complex number (or more generally, an element of some commutative ring R). The power series notation $A = A(z) = \sum_{n=0}^{\infty} a_n z^n$ is still used as a way to present the sequence (a_n), and we still call a_n "the coefficient of z^n in $A(z)$." The crucial point here is that z is no longer a complex variable, and $\sum_{n=0}^{\infty} a_n z^n$ no longer represents an infinite series of complex numbers. The symbol z is nothing more than a *formal variable*, a placeholder that lets us write symbolic expressions displaying all the coefficients a_n. To emphasize this viewpoint, we call $A(z)$ a *formal power series* in z with coefficients in R. When working with a formal power series, the issue of the convergence of the series at particular values of z does not arise, since the series is not even a function of z. Returning to some previous examples, $\sum_{n=0}^{\infty} 2^n z^n$ is just notation for the sequence $(1, 2, 4, 8, \ldots, 2^n, \ldots)$; $\sum_{n=0}^{m} \binom{m}{n} z^n$ is notation for $(1, m, \binom{m}{2}, \ldots, \binom{m}{m}, 0, 0, \ldots)$, and $\sum_{n=0}^{\infty} n! z^n$ denotes the sequence $(1, 1, 2, 6, 24, \ldots, n!, \ldots)$.

Why use formal power series notation to discuss sequences? The answer is that analytic operations on actual power series (such as addition, multiplication, and differentiation) suggest algebraic definitions of analogous formal operations on formal power series. For example, the product of two formal power series (discussed below) has a very strange looking definition using sequence notation. Using formal power series notation instead, the definition is seen to be a natural extension of the familiar rule for multiplying two polynomials.

Our goal in this chapter is to apply generating functions to gain combinatorial information about weighted sets. In order to extract this information, we need to have some facility manipulating generating functions either analytically or algebraically. It turns out that most of these manipulations can be justified formally, using only the algebraic properties of operations on sequences. But the analytic versions of these manipulations are probably more familiar to most readers based on previous exposure to calculus. So we focus initially on computations with analytic power series, referring to other texts for proofs of convergence tests and other needed facts from analysis. This approach lets us get to combinatorial calculations more rapidly, without getting bogged down in a mass of technical details. Later, we revisit various manipulations on generating functions, showing how these calculations can be justified algebraically at the level of formal power series.

5.2 Convergence of Power Series

The next three sections review some facts from calculus about analytic power series. We begin with some tests for the convergence of series of complex numbers. Given a series $\sum_{n=0}^{\infty} c_n$ with all $c_n \in \mathbb{C}$, we say the series *converges absolutely* iff the real series $\sum_{n=0}^{\infty} |c_n|$ converges. If a series converges absolutely, then the series converges, but the converse does not always hold. For example, the alternating harmonic series $\sum_{n=1}^{\infty}(-1)^n/n$ converges but does not converge absolutely, since the harmonic series $\sum_{n=1}^{\infty} 1/n$ diverges. The Ratio Test and the Root Test can be used to detect the absolute convergence of a given series.

5.5. The Ratio Test. Suppose $\sum_{n=0}^{\infty} c_n$ is a series of complex numbers such that $L = \lim_{n\to\infty} |c_{n+1}/c_n|$ exists in $[0,\infty]$. If $L < 1$, then the series converges absolutely. If $L > 1$, then the series diverges.

5.6. The Root Test. Suppose $\sum_{n=0}^{\infty} c_n$ is a series of complex numbers such that $L = \lim_{n\to\infty} \sqrt[n]{|c_n|}$ exists in $[0,\infty]$. If $L < 1$, then the series converges absolutely. If $L > 1$, then the series diverges.

We have already mentioned that each complex power series has a radius of convergence $R \in [0,\infty]$ such that the series converges at all points inside the circle $\{z \in \mathbb{C} : |z| = R\}$ and diverges at all points outside this circle. The next theorem formally states this fact and gives formulas for R based on the Ratio Test and the Root Test. These formulas use the conventions $1/0 = \infty$ and $1/\infty = 0$.

5.7. Theorem: Radius of Convergence of Power Series. (a) For every sequence of complex numbers $(a_n : n \geq 0)$, there exists $R \in [0,\infty]$ such that for all $z \in \mathbb{C}$,

$$|z| < R \Rightarrow \sum_{n=0}^{\infty} a_n z^n \text{ converges absolutely;} \qquad |z| > R \Rightarrow \sum_{n=0}^{\infty} a_n z^n \text{ diverges.}$$

(b) The radius of convergence R is given by

$$R = \lim_{n\to\infty} \left|\frac{a_n}{a_{n+1}}\right| \quad \text{and} \quad R = \frac{1}{\lim_{n\to\infty} \sqrt[n]{|a_n|}},$$

when these limits exist in $[0,\infty]$.

Proof. We prove this theorem in the case where one of the limits mentioned in (b) exists, by applying the Ratio Test or the Root Test. Fix $z \in \mathbb{C}$, and let $c_n = a_n z^n$ for $n \geq 0$. On one hand, if the limit $R_1 = \lim_{n\to\infty} |a_n/a_{n+1}|$ exists in $[0,\infty]$, then

$$\lim_{n\to\infty} \left|\frac{c_{n+1}}{c_n}\right| = \lim_{n\to\infty} \left|\frac{a_{n+1}z^{n+1}}{a_n z^n}\right| = |z|/R_1.$$

By the Ratio Test, if $|z| < R_1$, then $|z|/R_1 < 1$ and the power series converges absolutely; if $|z| > R_1$, then $|z|/R_1 > 1$ and the power series diverges. Thus (a) holds with $R = R_1$. On the other hand, if the limit $R_2 = 1/\lim_{n\to\infty} \sqrt[n]{|a_n|}$ exists in $[0,\infty]$, then

$$\lim_{n\to\infty} \sqrt[n]{|c_n|} = \lim_{n\to\infty} \sqrt[n]{|a_n z^n|} = |z|/R_2.$$

Now the Root Test shows that the power series converges absolutely for $|z| < R_2$ and diverges for $|z| > R_2$, so that (a) holds with $R = R_2$. □

In fact, it can be shown that the Root Test is always strong enough to prove existence of the radius of convergence, if we replace the second limit in (b) by $R = 1/\limsup_{n\to\infty} \sqrt[n]{|a_n|}$. For a proof, see Theorem 3.39 in [115]. However, it is often easier to find the radius of convergence using the Ratio Test. We also remark that none of the theorems mentioned so far gives any information about the convergence or divergence of the power series on the *circle of convergence* $\{z \in \mathbb{C} : |z| = R\}$.

5.8. Example. Find the radius of convergence of these power series: (a) $\sum_{n=0}^{\infty} b^n z^n$, where $b \in \mathbb{C}$ is fixed; (b) $\sum_{n=0}^{\infty} n^n z^n$; (c) $\sum_{n=0}^{\infty}(2^n/n!)z^n$; (d) $\sum_{k=0}^{\infty}((-1)^k/k!)z^{2k+1}$. We can solve (a) using either formula for R in Theorem 5.7. The formula based on the Ratio Test gives $R = \lim_{n\to\infty} |b^n/b^{n+1}| = 1/|b|$. The formula based on the Root Test gives $R = 1/\lim_{n\to\infty} \sqrt[n]{|b^n|} = 1/|b|$. In contrast, for (b) the Root Test formula gives $R = 1/\lim_{n\to\infty} \sqrt[n]{|n^n|} = 1/\lim_{n\to\infty} n = 0$, so this power series converges only at $z = 0$. For (c), we take ratios:

$$R = \lim_{n\to\infty} \left| \frac{2^n/n!}{2^{n+1}/(n+1)!} \right| = \lim_{n\to\infty} \frac{n+1}{2} = \infty.$$

For (d), the ratio-based formula for R in Theorem 5.7 technically does not apply, since $a_n = 0$ for all even n. But the original version of the Ratio Test can be used on the series of nonzero terms $c_k = (-1)^k z^{2k+1}/k!$ indexed by k. For any $z \in \mathbb{C}$, we find that

$$\lim_{k\to\infty} \left| \frac{c_{k+1}}{c_k} \right| = \lim_{k\to\infty} \left| \frac{(-1)^{k+1} z^{2k+3}/(k+1)!}{(-1)^k z^{2k+1}/k!} \right| = \lim_{k\to\infty} \frac{|z|^2}{k+1} = 0 < 1.$$

So this series converges for every $z \in \mathbb{C}$, and the radius of convergence is $R = \infty$.

5.3 Examples of Analytic Power Series

We continue to review facts from analysis about complex power series. The next theorem is a fundamental result of complex analysis that explains the close connection between power series and analytic functions of a complex variable. For $0 < R \leq \infty$, we use the notation $D(0; R) = \{z \in \mathbb{C} : |z| < R\}$ for the open disk with center 0 and radius R in the complex plane. A function $G : D(0; R) \to \mathbb{C}$ is called *analytic* iff G has a complex derivative at each point of $D(0; R)$. This means that $G'(z) = \lim_{h\to 0, h\in\mathbb{C}} \frac{G(z+h) - G(z)}{h}$ exists for all $z \in D(0; R)$. The notation $G^{(n)}(z_0)$ denotes the nth complex derivative of G at z_0, when this exists.

5.9. Theorem: Power Series and Analytic Functions. (a) Suppose $\sum_{n=0}^{\infty} a_n z^n$ is a complex power series with radius of convergence $R > 0$. Then the function $G : D(0; R) \to \mathbb{C}$ defined by $G(z) = \sum_{n=0}^{\infty} a_n z^n$ for $|z| < R$ has complex derivatives of all orders, and $a_n = G^{(n)}(0)/n!$ for all $n \geq 0$.

(b) Suppose $R > 0$ and $G : D(0; R) \to \mathbb{C}$ is an analytic function. Then G has derivatives of all orders on $D(0; R)$, $G(z) = \sum_{n=0}^{\infty} \frac{G^{(n)}(0)}{n!} z^n$ for all $z \in D(0; R)$, and the radius of convergence of this power series is at least R.

See Chapter 5 of [23] for a proof of this theorem. In part (a), the formula $a_n = G^{(n)}(0)/n!$ shows that the coefficients a_n in the given power series are uniquely determined from the analytic function G defined by that power series, provided the radius of convergence is positive. Part (b) lets us manufacture many examples of convergent power series by starting with known analytic functions.

5.10. Example. The complex exponential function $G(z) = e^z$ is defined and analytic on the entire complex plane. Like the real exponential function, G satisfies $G'(z) = e^z = G(z)$, so $G^{(n)}(z) = e^z$ and $G^{(n)}(0) = 1$ for all $n \geq 0$. By part (b) of the previous theorem, we have the power series expansion

$$e^z = \sum_{n=0}^{\infty} \frac{z^n}{n!} = 1 + z + z^2/2! + z^3/3! + z^4/4! + \cdots,$$

which is valid for all complex z (i.e., $R = \infty$). Similarly, by taking derivatives repeatedly, we obtain the following expansions:

$$\begin{aligned}
\sin z &= \sum_{k=0}^{\infty} \frac{(-1)^k}{(2k+1)!} z^{2k+1} = z - z^3/3! + z^5/5! - z^7/7! + \cdots && \text{for all } z \in \mathbb{C};\\
\cos z &= \sum_{k=0}^{\infty} \frac{(-1)^k}{(2k)!} z^{2k} = 1 - z^2/2! + z^4/4! - z^6/6! + \cdots && \text{for all } z \in \mathbb{C};\\
\sinh z &= \sum_{k=0}^{\infty} \frac{z^{2k+1}}{(2k+1)!} = z + z^3/3! + z^5/5! + z^7/7! + \cdots && \text{for all } z \in \mathbb{C};\\
\cosh z &= \sum_{k=0}^{\infty} \frac{z^{2k}}{(2k)!} = 1 + z^2/2! + z^4/4! + z^6/6! + \cdots && \text{for all } z \in \mathbb{C}.
\end{aligned}$$

Here the *hyperbolic trigonometric functions* are defined by $\sinh z = (e^z - e^{-z})/2$ and $\cosh z = (e^z + e^{-z})/2$.

5.11. Example. In complex analysis, there is a version of the natural logarithm function, denoted Log, which is defined and analytic on the domain $\mathbb{C} - \{z \in \mathbb{R} : z \leq 0\}$. The derivative of $\operatorname{Log} z$ is $1/z$ on this domain. It follows that $G(z) = \operatorname{Log}(1+z)$ is defined and analytic on the disk $D(0;1)$, with $G'(z) = (1+z)^{-1}$, $G''(z) = -(1+z)^{-2}$, $G'''(z) = 2(1+z)^{-3}$, and in general $G^{(n)}(z) = (-1)^{n-1}(n-1)!(1+z)^{-n}$ for all $n \geq 1$. Setting $z = 0$ gives $G(0) = 0$ and $G^{(n)}(0) = (-1)^{n-1}(n-1)!$ for $n > 0$. By Theorem 5.9 on Power Series and Analytic Functions, we obtain the power series expansion

$$\operatorname{Log}(1+z) = \sum_{n=1}^{\infty} \frac{(-1)^{n-1}}{n} z^n = z - z^2/2 + z^3/3 - \cdots \qquad \text{for } |z| < 1.$$

Noting that $|z| < 1$ iff $|-z| < 1$, we can replace z by $-z$ to obtain

$$-\operatorname{Log}(1-z) = \sum_{n=1}^{\infty} \frac{z^n}{n} = z + z^2/2 + z^3/3 + \cdots \qquad \text{for } |z| < 1.$$

For nonzero $b \in \mathbb{C}$, we have already seen (§2.2) the Geometric Series expansion

$$(1-bz)^{-1} = \frac{1}{1-bz} = \sum_{n=0}^{\infty} b^n z^n = 1 + bz + b^2 z^2 + b^3 z^3 + \cdots \qquad \text{for } |z| < 1/|b|.$$

For a positive integer m, we have also seen (§2.3) the Binomial Theorem

$$(1+z)^m = \sum_{n=0}^{m} \binom{m}{n} z^m \qquad \text{for } z \in \mathbb{C}.$$

We will make constant use of these expansions as well as a generalization called the *Extended Binomial Theorem*, which we discuss now.

Given any complex constant r, there is an analytic function $G(z) = (1+z)^r$ with domain $D(0;1)$ given by $G(z) = e^{r \operatorname{Log}(1+z)}$ for $|z| < 1$. When $r \in \mathbb{Z}$, it can be shown that this definition coincides with the algebraic definition of integer powers; specifically, for $r \in \mathbb{Z}_{>0}$, $G(z)$ is the product of r copies of $(1+z)$, and when $r \in \mathbb{Z}_{<0}$, $G(z)$ is the product of $|r|$ copies of the multiplicative inverse of $1+z$. Using properties of complex exponentials and logarithms, it can also be shown that $(1+z)^{r+s} = (1+z)^r(1+z)^s$ for all $r, s \in \mathbb{C}$, and $G'(z) = r(1+z)^{r-1}$. More generally, for any $n \geq 0$, $G^{(n)}(z) = (r)\!\downarrow_n (1+z)^{r-n}$ where $(r)\!\downarrow_0= 1$ and $(r)\!\downarrow_n= r(r-1)(r-2)\cdots(r-n+1)$ for $n > 0$. Applying Theorem 5.9 on Power Series and Analytic Functions, we obtain the following expansion.

5.12. The Extended Binomial Theorem. Given any $r \in \mathbb{C}$,

$$(1+z)^r = \sum_{n=0}^{\infty} \frac{(r)\!\downarrow_n}{n!} z^n = 1 + rz + \frac{r(r-1)}{2} z^2 + \frac{r(r-1)(r-2)}{3!} z^3 + \cdots \qquad \text{for } |z| < 1.$$

By taking r to be a positive integer m, we recover the original Binomial Theorem, since $(m)\!\downarrow_n= 0$ for all $n > m$, whereas $(m)\!\downarrow_n /n! = \binom{m}{n}$ for $0 \leq n \leq m$. Taking $r = -1$ and replacing z by $-bz$, we recover the Geometric Series formula, since $(-1)\!\downarrow_n /n! = (-1)^n$ and $|-bz| < 1$ iff $|z| < 1/|b|$.

Another frequently needed special case of the Extended Binomial Theorem is obtained by taking r to be a negative integer $-m$ and replacing z by $-z$. Noting that

$$(-m)\!\downarrow_n (-z)^n/n! = m(m+1)\cdots(m+n-1)z^n/n! = \binom{n+m-1}{n, m-1} z^n,$$

we deduce the following expansion.

5.13. The Negative Binomial Theorem. Given any integer $m > 0$,

$$(1-z)^{-m} = \sum_{n=0}^{\infty} (-1)^n \frac{(-m)\!\downarrow_n}{n!} z^n = \sum_{n=0}^{\infty} \binom{n+m-1}{n, m-1} z^n \qquad \text{for } |z| < 1.$$

For example, given any $z \in D(0;1)$,

$$\begin{aligned}
(1-z)^{-1} &= \sum_{n=0}^{\infty} \binom{n+0}{0} z^n = \sum_{n=0}^{\infty} z^n = 1 + z + z^2 + z^3 + z^4 + \cdots; \\
(1-z)^{-2} &= \sum_{n=0}^{\infty} \binom{n+1}{1} z^n = \sum_{n=0}^{\infty} (n+1) z^n = 1 + 2z + 3z^2 + 4z^3 + 5z^4 + \cdots; \\
(1-z)^{-3} &= \sum_{n=0}^{\infty} \binom{n+2}{2} z^n = 1 + 3z + 6z^2 + 10z^3 + 15z^4 + \cdots; \\
(1-z)^{-4} &= \sum_{n=0}^{\infty} \binom{n+3}{3} z^n = 1 + 4z + 10z^2 + 20z^3 + 35z^4 + \cdots.
\end{aligned}$$

Replacing z by bz, where $b \in \mathbb{C}$ is nonzero, we also see:

the coefficient of z^n *in the power series for* $(1-bz)^{-m}$ *is* $b^n \binom{n+m-1}{n, m-1}$.

5.4 Operations on Power Series

Given analytic functions F and G defined on some disk $D(0;R)$, we can form new analytic functions by performing operations such as addition, multiplication, and differentiation. The next theorem gives formulas for the coefficients in the power series representations of these new functions.

5.14. Theorem: Operations on Analytic Power Series. Suppose $F(z)=\sum_{n=0}^{\infty} a_n z^n$ and $G(z)=\sum_{n=0}^{\infty} b_n z^n$ are analytic functions on $D(0;R)$, where $R>0$. The following rules hold for all $z \in D(0;R)$.

(a) *Equality Rule.* $F=G$ on $D(0;R)$ iff $a_n=b_n$ for all $n \geq 0$.

(b) *Sum Rule.* $F(z)+G(z)=\sum_{n=0}^{\infty}(a_n+b_n)z^n$.

(c) *Scalar Multiple Rule.* For $c \in \mathbb{C}$, $cF(z)=\sum_{n=0}^{\infty}(ca_n)z^n$.

(d) *Power-Shifting Rule.* For $k \in \mathbb{Z}_{\geq 0}$, $z^k F(z)=\sum_{n=k}^{\infty} a_{n-k} z^n$.

(e) *Product Rule.* $F(z)G(z)=\sum_{n=0}^{\infty}\left(\sum_{k=0}^{n} a_k b_{n-k}\right) z^n$.

(f) *Derivative Rule.* $F'(z)=\sum_{n=1}^{\infty} n a_n z^{n-1}=\sum_{n=0}^{\infty}(n+1)a_{n+1}z^n$.

(g) *Shifted Derivative Rule.* $zF'(z)=\sum_{n=0}^{\infty} n a_n z^n$.

Proof. Each part is proved by combining known derivative rules with Theorem 5.9 on Power Series and Analytic Functions. We prove (b), (e), and (f) as illustrations. For (b), the Sum Rule for Derivatives shows that $S(z)=F(z)+G(z)$ is analytic on $D(0;R)$ with $S'(z)=F'(z)+G'(z)$. More generally, $S^{(n)}(z)=F^{(n)}(z)+G^{(n)}(z)$ for all $n \geq 0$. We know there is a power series expansion $S(z)=\sum_{n=0}^{\infty} c_n z^n$ valid for $z \in D(0;R)$, and moreover

$$c_n=\frac{S^{(n)}(0)}{n!}=\frac{F^{(n)}(0)}{n!}+\frac{G^{(n)}(0)}{n!}=a_n+b_n \quad \text{for all } n \geq 0.$$

For (e), define $P(z)=F(z)G(z)$. Repeatedly using the Product Rule for Derivatives, we find that $P'(z)=F'(z)G(z)+F(z)G'(z)$, hence $P''=F''G+2F'G'+FG''$, then $P'''=F'''G+3F''G'+3F'G''+FG'''$, and so on. By induction on n, it can be proved that

$$P^{(n)}(z)=\sum_{k=0}^{n}\binom{n}{k}F^{(k)}(z)G^{(n-k)}(z) \tag{5.2}$$

(note the analogy to the Binomial Theorem). By Theorem 5.9(b), there is a power series expansion $P(z)=\sum_{n=0}^{\infty} d_n z^n$ valid for $z \in D(0;R)$, where

$$d_n=\frac{P^{(n)}(0)}{n!}=\sum_{k=0}^{n}\frac{F^{(k)}(0)}{k!}\frac{G^{(n-k)}(0)}{(n-k)!}=\sum_{k=0}^{n} a_k b_{n-k} \quad \text{for all } n \geq 0.$$

We remark that if F and G are finite series (i.e., polynomials in z), then (e) can be proved algebraically using the Generalized Distributive Law.

Finally, we prove (f), which says that the power series representation of an analytic function can be differentiated term by term. We know F' is analytic on $D(0;R)$ since F has derivatives of all orders. So there is an expansion $F'(z) = \sum_{n=0}^{\infty} e_n z^n$ for $z \in D(0;R)$, where

$$e_n = \frac{(F')^{(n)}(0)}{n!} = \frac{(n+1)F^{(n+1)}(0)}{(n+1)!} = (n+1)a_{n+1} \quad \text{for all } n \geq 0. \qquad \square$$

The preceding theorem motivates the following definitions of algebraic operations on *formal* power series. Recall that a formal power series $A = A(z) = \sum_{n=0}^{\infty} a_n z^n$ is defined to be the infinite sequence of coefficients $(a_n : n \geq 0)$. The coefficients a_n can belong to any fixed commutative ring R with identity 1_R, although we usually take $R = \mathbb{C}$.

5.15. Definitions: Operations on Formal Power Series. Let $A(z) = \sum_{n=0}^{\infty} a_n z^n$ and $B(z) = \sum_{n=0}^{\infty} b_n z^n$ be formal power series.

(a) *Equality.* $A(z) = B(z)$ iff $a_n = b_n$ for all $n \geq 0$.

(b) *Addition.* $A(z) + B(z) = \displaystyle\sum_{n=0}^{\infty} (a_n + b_n) z^n$.

(c) *Scalar Multiples.* For $c \in R$, $cA(z) = \displaystyle\sum_{n=0}^{\infty} (ca_n) z^n$.

(d) *Power-Shifting Rule.* For $k \in \mathbb{Z}_{\geq 0}$, $z^k A(z) = \displaystyle\sum_{n=k}^{\infty} a_{n-k} z^n$.

(e) *Multiplication.* $A(z)B(z) = \displaystyle\sum_{n=0}^{\infty} \left(\sum_{k=0}^{n} a_k b_{n-k} \right) z^n$.

(f) *Formal Differentiation.* $A'(z) = \displaystyle\sum_{n=1}^{\infty} n a_n z^{n-1} = \sum_{n=0}^{\infty} (n+1) a_{n+1} z^n$.

Now that operations on formal power series have been defined, it is necessary to *prove* algebraic identities and derivative rules similar to rules already known for analytic functions. For example, there are formal versions of the associative laws, the commutative laws, the distributive laws, the Sum Rule for Derivatives, the Product Rule for Derivatives, and Taylor's Formula. We leave most of these verifications as exercises, but we prove the distributive law to illustrate the method.

Fix formal power series $A = \sum_{n=0}^{\infty} a_n z^n$, $B = \sum_{n=0}^{\infty} b_n z^n$, and $C = \sum_{n=0}^{\infty} c_n z^n$; we prove $A(B+C) = AB + AC$. On the left side, $B + C = \sum_{n=0}^{\infty} (b_n + c_n) z^n$, so the coefficient of z^n in $A(B+C)$ is $\sum_{k=0}^{n} a_k (b_{n-k} + c_{n-k}) = \sum_{k=0}^{n} (a_k b_{n-k} + a_k c_{n-k})$. On the right side, $AB = \sum_{n=0}^{\infty} (\sum_{k=0}^{n} a_k b_{n-k}) z^n$ and $AC = \sum_{n=0}^{\infty} (\sum_{k=0}^{n} a_k c_{n-k}) z^n$. So the coefficient of z^n in $AB + AC$ is $\sum_{k=0}^{n} a_k b_{n-k} + \sum_{k=0}^{n} a_k c_{n-k}$. This equals the coefficient of z^n in $A(B+C)$ for each $n \geq 0$. So $A(B+C) = AB + AC$ follows from the definition of equality of formal power series.

5.5 Solving Recursions with Generating Functions

We have now developed enough computational machinery to describe our first application of generating functions: solving recursions. In §2.16, we briefly discussed a recipe to solve homogeneous recursions with constant coefficients. Generating functions provide a powerful general method for solving these recursions and many others. The following examples illustrate the technique.

5.16. Example. Define a sequence $(a_n : n \geq 0)$ by $a_0 = 2$ and $a_n = 3a_{n-1} - 1$ for all $n \geq 1$. What is a closed formula for a_n? Our approach is to introduce a generating function $F(z) = \sum_{n=0}^{\infty} a_n z^n$ whose coefficients encode all of the unknown quantities a_n. The next step is to use the recursion for a_n to deduce an algebraic equation satisfied by this generating function. Since the recursion is only valid for $n \geq 1$, we first subtract the constant term $a_0 = 2$ from $F(z)$. We then compute

$$F(z) - 2 = F(z) - a_0 = \sum_{n=1}^{\infty} a_n z^n = \sum_{n=1}^{\infty} (3a_{n-1} - 1)z^n = 3z \sum_{n=1}^{\infty} a_{n-1} z^{n-1} - z \sum_{n=1}^{\infty} z^{n-1}.$$

We can simplify these sums by changing the summation variable to $m = n - 1$. This leads to

$$F(z) - 2 = 3z \sum_{m=0}^{\infty} a_m z^m - z \sum_{m=0}^{\infty} z^m = 3zF(z) - \frac{z}{1-z}.$$

Using algebra to rearrange this equation, we get $(1 - 3z)F(z) = 2 - z/(1 - z)$, or

$$F(z) = \frac{2 - 3z}{(1-z)(1-3z)}.$$

This equation gives a *generating function solution* to the original recursion. To finish, we need only determine the coefficient of z^n in this generating function. This can be achieved using the *partial fraction decomposition* learned in calculus. (See §11.4 for a detailed development of this technique.) We write

$$F(z) = \frac{2 - 3z}{(1-z)(1-3z)} = \frac{B}{1-z} + \frac{C}{1-3z} \tag{5.3}$$

for some unknown constants B and C. Clearing denominators gives

$$2 - 3z = B(1 - 3z) + C(1 - z).$$

Setting $z = 1$ gives $-1 = -2B$, so $B = 1/2$. Setting $z = 1/3$ gives $1 = (2/3)C$, so $C = 3/2$. Putting these values back into (5.3) and using the Geometric Series Formula, we get

$$\sum_{n=0}^{\infty} a_n z^n = F(z) = B \sum_{n=0}^{\infty} z^n + C \sum_{n=0}^{\infty} 3^n z^n = \sum_{n=0}^{\infty} (1/2 + 3^{n+1}/2) z^n.$$

We conclude that $a_n = (3^{n+1} + 1)/2$ for all $n \geq 0$.

Before leaving this example, we must consider a subtle logical point. At the very end of the example, we found a_n by invoking the Equality Rule in part (a) of Theorem 5.14. But this rule can be used only if we already know that the analytic power series $\sum_{n=0}^{\infty} a_n z^n$ has a positive radius of convergence. One way to resolve this problem is to prove by induction that the claimed solution $a_n = (3^{n+1} + 1)/2$ really does satisfy the recursion and initial condition.

Another approach is to note that $G(z) = B(1 - z)^{-1} + C(1 - 3z)^{-1}$ converges for all $z \in D(0; 1/3)$. Reversing the algebraic steps that led to (5.3), we see that G solves the generating function identity $G(z) - 2 = 3zG(z) - z/(1 - z)$. In turn, it can be checked that this identity implies that the coefficients a'_n of $G(z)$ satisfy the original recursion and initial condition. A quick induction on n shows that there is exactly one solution to that recursion and initial condition. So the coefficients of G, namely $a'_n = (3^{n+1}+1)/2$, do solve the original problem.

By using *formal* power series, we can avoid the awkwardness of doing an extra induction argument or checking the reversibility of intricate generating function manipulations. In this approach, we start by defining $F = (a_n : n \geq 0)$ to be the unique sequence satisfying the given recursion and initial conditions. Then we perform all the algebraic steps above (always operating on formal power series) to see that F is the sequence $(\frac{3^{n+1}+1}{2} : n \geq 0)$. By definition of equality of formal power series, $a_n = (3^{n+1}+1)/2$ follows. The one catch here is that we must already know that formal power series obey all the laws of algebra including rules for manipulating fractions. This raises the question of what is meant when we *divide* one formal power series by another. In the present case, it suffices to know that the formal geometric series $\sum_{n=0}^{\infty} b^n z^n$ is the multiplicative inverse of the formal power series $1 - bz$ (taking $b = 1$ and $b = 3$). We examine multiplicative inverses of formal power series more closely later (see §11.3). For now, let us look at more examples of solving recursions.

5.17. Example. Define $a_0 = 2$, $a_1 = 1$, and $a_n = a_{n-2} + 2n$ for $n \geq 2$. What is a closed formula for a_n? Proceeding as before, we define a generating function $F(z) = \sum_{n=0}^{\infty} a_n z^n$ whose coefficients are the unknown values a_n. To use the recursion, we first subtract the first two terms $a_0 + a_1 z = 2 + z$ from $F(z)$:

$$F(z) - 2 - z = \sum_{n=2}^{\infty} a_n z^n = \sum_{n=2}^{\infty} (a_{n-2} + 2n) z^n = z^2 \sum_{n=2}^{\infty} a_{n-2} z^{n-2} + 2z \sum_{n=2}^{\infty} n z^{n-1}.$$

Introducing the new summation variable $m = n - 2$, the first term on the right side becomes $z^2 \sum_{m=0}^{\infty} a_m z^m = z^2 F(z)$. The second term looks like a derivative, suggesting the calculation:

$$\sum_{n=2}^{\infty} n z^{n-1} = \frac{d}{dz}\left[\sum_{n=2}^{\infty} z^n\right] = \frac{d}{dz}[(1-z)^{-1} - z - 1] = (1-z)^{-2} - 1.$$

So now we have

$$F(z) - 2 - z = z^2 F(z) + 2z(1-z)^{-2} - 2z.$$

Solving for $F(z)$ gives

$$F(z) = \frac{2z(1-z)^{-2} - z + 2}{1 - z^2} = \frac{-z^3 + 4z^2 - 3z + 2}{(1-z)^3(1+z)}.$$

In this case, the partial fraction decomposition looks like

$$F(z) = \frac{-z^3 + 4z^2 - 3z + 2}{(1-z)^3(1+z)} = \frac{A}{1-z} + \frac{B}{(1-z)^2} + \frac{C}{(1-z)^3} + \frac{D}{1+z}, \text{ or}$$

$$-z^3 + 4z^2 - 3z + 2 = A(1-z)^2(1+z) + B(1-z)(1+z) + C(1+z) + D(1-z)^3. \quad (5.4)$$

Setting $z = 1$ gives $2C = 2$, so $C = 1$. Setting $z = -1$ gives $8D = 10$, so $D = 5/4$. Plugging these values in and doing a little more algebra, we find that $A = 1/4$ and $B = -1/2$. Now, by the Negative Binomial Theorem,

$$\sum_{n=0}^{\infty} a_n z^n = F(z) = A \sum_{n=0}^{\infty} z^n + B \sum_{n=0}^{\infty} (n+1) z^n + C \sum_{n=0}^{\infty} \binom{n+2}{2} z^n + D \sum_{n=0}^{\infty} (-1)^n z^n.$$

Equating coefficients of z^n, we finally reach the solution

$$a_n = (1/4) - (1/2)(n+1) + \binom{n+2}{2} + (5/4)(-1)^n = n^2/2 + n + 3/4 + (5/4)(-1)^n.$$

It can be checked by induction that this solution does satisfy the original recursion and initial conditions. Alternatively, we can reverse all the steps, noting that the tentative solution $F(z)$ has radius of convergence 1. Even better, we can solve the whole problem using formal power series, assuming that the necessary facts about formal multiplicative inverses are available.

5.18. Example. Let us solve the recursion $a_n = 4a_{n-1} - 4a_{n-2} + n^2 2^n$ for $n \geq 2$, with initial conditions $a_0 = a_1 = 1$. Let (a_n) be the unique solution to this problem, and define $F(z) = \sum_{n=0}^{\infty} a_n z^n$. We compute

$$\begin{aligned} F(z) - 1 - z &= \sum_{n=2}^{\infty} a_n z^n = \sum_{n=2}^{\infty} (4a_{n-1} - 4a_{n-2} + n^2 2^n) z^n \\ &= 4z \sum_{n=2}^{\infty} a_{n-1} z^{n-1} - 4z^2 \sum_{n=2}^{\infty} a_{n-2} z^{n-2} + \sum_{n=2}^{\infty} n^2 2^n z^n. \end{aligned}$$

The first sum is $\sum_{m=1}^{\infty} a_m z^m = F(z) - 1$, and the second sum is $\sum_{m=0}^{\infty} a_m z^m = F(z)$.

But what is the third sum? This sum resembles the geometric series $G(z) = \sum_{n=0}^{\infty} 2^n z^n = (1-2z)^{-1}$, but each summand has an extra n^2 and the sum starts at $n = 2$ not $n = 0$. We can make this n^2 appear by repeated use of rule (g) from Theorem 5.14. Using rule (g) once, we see that $zG'(z) = \sum_{n=1}^{\infty} n 2^n z^n$. Taking the derivative of $(1-2z)^{-1}$ and then multiplying by z, we deduce that $H(z) = \sum_{n=0}^{\infty} n 2^n z^n = 2z(1-2z)^{-2}$. Now use rule (g) again on H, taking another derivative and multiplying by z. We get

$$\sum_{n=0}^{\infty} n^2 2^n z^n = 2z(1-2z)^{-2} + 8z^2(1-2z)^{-3}. \tag{5.5}$$

The third sum above starts at $n = 2$, not $n = 0$, so we must subtract $0 + 2z$ from this expression.

Returning to the equation for $F(z)$, we now have

$$F(z) - 1 - z = 4z(F(z) - 1) - 4z^2 F(z) + 2z(1-2z)^{-2} + 8z^2(1-2z)^{-3} - 2z.$$

This rearranges to

$$(1 - 4z + 4z^2)F(z) = 1 - 5z + 2z(1-2z)^{-2} + 8z^2(1-2z)^{-3}.$$

Since $1 - 4z + 4z^2 = (1-2z)^2$, we finally obtain

$$F(z) = (1-2z)^{-2} - 5z(1-2z)^{-2} + 2z(1-2z)^{-4} + 8z^2(1-2z)^{-5}.$$

We can use the Negative Binomial Theorem and the Power-Shifting Rule to get the series expansion for each term on the right. For instance, $2z(1-2z)^{-4} = \sum_{k=0}^{\infty} 2z\binom{k+3}{3}(2z)^k = \sum_{n=1}^{\infty} \binom{n+2}{3} 2^n z^n$ (letting $n = k+1$). The coefficient of z^n in $F(z)$ is

$$\begin{aligned} a_n &= \binom{n+1}{1} 2^n - 5\binom{n}{1} 2^{n-1} + \binom{n+2}{3} 2^n + 8\binom{n+2}{4} 2^{n-2} \\ &= \frac{n^4 + 4n^3 + 5n^2 - 16n + 12}{12} \cdot 2^n. \end{aligned}$$

For a general theorem regarding generating function solutions to recursions, see §11.5.

5.6 Evaluating Summations with Generating Functions

This section discusses another application of generating functions: evaluating summations. Suppose we are given a sequence $(a_n : n \geq 0)$ and are asked to find the sums $s_n = \sum_{k=0}^{n} a_k$ for each $n \geq 0$. The product formula for generating functions (Theorem 5.14(e)) allows us to find the generating function $S(z)$ for the sequence of sums $(s_n : n \geq 0)$, as follows. Define $F(z) = \sum_{n=0}^{\infty} a_n z^n$ and $G(z) = \sum_{n=0}^{\infty} 1z^n = (1-z)^{-1}$. So $G(z) = \sum_{n=0}^{\infty} b_n z^n$ where $b_n = 1$ for all n. Now, the rule for multiplying power series gives

$$\frac{F(z)}{1-z} = F(z)G(z) = \sum_{n=0}^{\infty}\left(\sum_{k=0}^{n} a_k b_{n-k}\right) z^n = \sum_{n=0}^{\infty}\left(\sum_{k=0}^{n} a_k\right) z^n = \sum_{n=0}^{\infty} s_n z^n.$$

Thus, $\sum_{k=0}^{n} a_k$ *is the coefficient of* z^n *in the power series* $S(z) = F(z)/(1-z)$. We can use partial fraction decompositions and similar techniques to find these coefficients, as seen in the following examples.

5.19. Example. Let us use generating functions to evaluate the finite geometric series $\sum_{k=0}^{n} 3^k$. Here $a_n = 3^n$ and $F(z) = \sum_{n=0}^{\infty} 3^n z^n = (1-3z)^{-1}$, so the generating function for the sums is $S(z) = (1-3z)^{-1}(1-z)^{-1}$. The partial fraction decomposition of $S(z)$ is

$$S(z) = \frac{1}{(1-3z)(1-z)} = \frac{3/2}{1-3z} + \frac{-1/2}{1-z}.$$

Thus the coefficient of z^n in $S(z)$ is

$$\sum_{k=0}^{n} 3^k = (3/2)3^n - (1/2)1^n = \frac{3^{n+1}-1}{2}.$$

Of course, this agrees with the formula for a finite geometric series derived in §2.2.

5.20. Example. Next, we compute $s_n = \sum_{k=0}^{n} k^3$ using generating functions. Here $a_n = n^3$ and $F(z) = \sum_{n=0}^{\infty} n^3 z^n$. We can evaluate this series using Theorem 5.14(g). Starting with $(1-z)^{-1} = \sum_{n=0}^{\infty} z^n$, we take a derivative and multiply by z three times in a row. We get:

$$\begin{aligned}
z(1-z)^{-2} &= \sum_{n=0}^{\infty} nz^n;\\
z(1-z)^{-2} + 2z^2(1-z)^{-3} &= \sum_{n=0}^{\infty} n^2 z^n;\\
z(1-z)^{-2} + 6z^2(1-z)^{-3} + 6z^3(1-z)^{-4} &= \sum_{n=0}^{\infty} n^3 z^n.
\end{aligned}$$

Multiplying by $(1-z)^{-1}$ gives the generating function for s_n:

$$z(1-z)^{-3} + 6z^2(1-z)^{-4} + 6z^3(1-z)^{-5} = \sum_{n=0}^{\infty} s_n z^n.$$

Using the Negative Binomial Theorem and the Power-Shifting Rule, we obtain

$$s_n = \binom{n-1+2}{2} + 6\binom{n-2+3}{3} + 6\binom{n-3+4}{4} = \frac{n^4+2n^3+n^2}{4} = \left[\frac{n(n+1)}{2}\right]^2.$$

Compare this derivation to the method used in §2.6.

5.21. Example. Next we find $s_n = \sum_{k=0}^{n} k^2 2^k$. In an earlier example (see (5.5)), we found

$$F(z) = \sum_{n=0}^{\infty} n^2 2^n z^n = 2z(1-2z)^{-2} + 8z^2(1-2z)^{-3}.$$

Therefore,

$$F(z)(1-z)^{-1} = \sum_{n=0}^{\infty} s_n z^n = 2z(1-2z)^{-2}(1-z)^{-1} + 8z^2(1-2z)^{-3}(1-z)^{-1}.$$

The partial fraction expansion (found with a computer algebra system) is

$$F(z)(1-z)^{-1} = \frac{-6}{1-z} + \frac{12}{1-2z} + \frac{-10}{(1-2z)^2} + \frac{4}{(1-2z)^3}.$$

Extracting the coefficient of z^n, we get

$$s_n = \sum_{k=0}^{n} k^2 2^k = -6 + 12 \cdot 2^n - 10\binom{n+1}{1} 2^n + 4\binom{n+2}{2} 2^n = (n^2 - 2n + 3)2^{n+1} - 6.$$

5.7 Generating Function for Derangements

Recall that a *derangement* of length n is a permutation $w = w_1 w_2 \cdots w_n$ with $w_i \neq i$ for $1 \leq i \leq n$. Let d_n be the number of derangements of length n. This section analyzes the exponential generating function $D(z) = \sum_{n=0}^{\infty} \frac{d_n}{n!} z^n$ for derangements.

In §4.5, we saw that the derangement numbers satisfy the recursion

$$d_n = n d_{n-1} + (-1)^n \qquad \text{for all } n \geq 1,$$

with initial condition $d_0 = 1$. Let us see what happens when we try to solve this recursion by the generating function method.

We start, as before, by subtracting off the initial coefficient of $D(z)$ and then applying the recursion:

$$D(z) - 1 = \sum_{n=1}^{\infty} \frac{d_n}{n!} z^n = \sum_{n=1}^{\infty} \frac{n d_{n-1} + (-1)^n}{n!} z^n = z \sum_{n=1}^{\infty} \frac{d_{n-1}}{(n-1)!} z^{n-1} + \sum_{n=1}^{\infty} \frac{(-z)^n}{n!}.$$

Letting $m = n - 1$, we see that the first sum here is $D(z)$. The second sum is the power series for e^{-z} with the $n = 0$ term deleted. Therefore, $D(z) - 1 = zD(z) + e^{-z} - 1$. So $(1-z)D(z) = e^{-z}$, and we conclude that

$$D(z) = \sum_{n=0}^{\infty} \frac{d_n}{n!} z^n = \frac{e^{-z}}{1-z}.$$

Let $F(z) = e^{-z} = \sum_{n=0}^{\infty} \frac{(-1)^n}{n!} z^n$. Recalling the technique of §5.6, we see that the power series $D(z)$ is precisely the ordinary generating function for the sequence of partial sums of the terms $a_n = (-1)^n/n!$. In other words,

$$\frac{d_n}{n!} = \sum_{k=0}^{n} \frac{(-1)^k}{k!} \qquad \text{for all } n \geq 0.$$

Thus, generating functions have led us to the Summation Formula for Derangements in §4.5.

5.8 Counting Rules for Weighted Sets

Now that we have some facility with algebraic manipulations of generating functions, it is time to reveal the combinatorial significance of the basic operations on generating functions. In Chapter 1, we studied three fundamental counting rules: the Sum Rule, the Product Rule, and the Bijection Rule. We now state versions of these rules that can be used to find ordinary generating functions for weighted sets. Recall that a *weighted set* consists of a set S and a weight function $\mathrm{wt}_S : S \to \mathbb{Z}_{\geq 0}$ such that $A_n = \{u \in S : \mathrm{wt}_S(u) = n\}$ is finite for all $n \geq 0$. The ordinary generating function (OGF) of this weighted set is $\mathrm{GF}(S; z) = \sum_{u \in S} z^{\mathrm{wt}(u)} = \sum_{n=0}^{\infty} |A_n| z^n$.

5.22. The Sum Rule for Weighted Sets. Suppose S is a weighted set that is the disjoint union of k weighted sets $S_1, S_2, \ldots, S_k$. Assume $\mathrm{wt}_{S_i}(u) = \mathrm{wt}_S(u)$ whenever $1 \leq i \leq k$ and $u \in S_i$. Then

$$\mathrm{GF}(S; z) = \mathrm{GF}(S_1; z) + \mathrm{GF}(S_2; z) + \cdots + \mathrm{GF}(S_k; z).$$

5.23. The Product Rule for Weighted Sets. Suppose k is a fixed positive integer and $S_1, \ldots, S_k$ are weighted sets. Suppose S is a weighted set such that every $u \in S$ can be constructed in exactly one way by choosing $u_1 \in S_1$, choosing $u_2 \in S_2$, and so on, and then assembling the chosen objects $u_1, \ldots, u_k$ in a prescribed manner. Assume that whenever u is constructed from $u_1, u_2, \ldots, u_k$, the *weight-additivity condition* $\mathrm{wt}_S(u) = \mathrm{wt}_{S_1}(u_1) + \mathrm{wt}_{S_2}(u_2) + \cdots + \mathrm{wt}_{S_k}(u_k)$ holds. Then

$$\mathrm{GF}(S; z) = \mathrm{GF}(S_1; z) \cdot \mathrm{GF}(S_2; z) \cdot \ldots \cdot \mathrm{GF}(S_k; z).$$

5.24. The Bijection Rule for Weighted Sets. Suppose S and T are weighted sets and $f : S \to T$ is a *weight-preserving* bijection, meaning that $\mathrm{wt}_T(f(u)) = \mathrm{wt}_S(u)$ for all $u \in S$. Then $\mathrm{GF}(S; z) = \mathrm{GF}(T; z)$.

We prove these rules in the rest of this section; these proofs may be omitted without loss of continuity. Simpler proofs of the special case of *finite* weighted sets are given in §8.2.

Step 1. We prove the Sum Rule for Weighted Sets when $k = 2$. Suppose S is the disjoint union of weighted sets A and B. For each $n \geq 0$, let $A_n = \{x \in A : \mathrm{wt}_A(x) = n\}$, $a_n = |A_n|$, $B_n = \{y \in B : \mathrm{wt}_B(y) = n\}$, and $b_n = |B_n|$. By definition, $\mathrm{GF}(A; z) = \sum_{n=0}^{\infty} a_n z^n$ and $\mathrm{GF}(B; z) = \sum_{n=0}^{\infty} b_n z^n$. Now let $C_n = \{u \in S : \mathrm{wt}_S(u) = n\}$ and $c_n = |C_n|$ for $n \geq 0$. Since S is the disjoint union of A and B, and since the weight function for S agrees with the weight functions for A and B, C_n is the disjoint union of A_n and B_n. So the Sum Rule for unweighted sets tells us that $c_n = a_n + b_n$ for all $n \geq 0$. By the formula for the sum of two power series,

$$\mathrm{GF}(S; z) = \sum_{n=0}^{\infty} c_n z^n = \sum_{n=0}^{\infty} (a_n + b_n) z^n = \sum_{n=0}^{\infty} a_n z^n + \sum_{n=0}^{\infty} b_n z^n = \mathrm{GF}(A; z) + \mathrm{GF}(B; z).$$

The general case of the Sum Rule follows from Step 1 by induction on k.

Step 2. We prove the Product Rule for Weighted Sets when $k = 2$. Suppose every object u in the weighted set S can be constructed uniquely by choosing $x \in A$, choosing $y \in B$, and assembling x and y in some manner to produce u. Also assume $\mathrm{wt}_S(u) = \mathrm{wt}_A(x) + \mathrm{wt}_B(y)$ when u is constructed from x and y. Define A_n, B_n, C_n, a_n, b_n, and c_n as in Step 1. Fix $n \in \mathbb{Z}_{\geq 0}$; we use the Sum Rule and Product Rule for unweighted sets to build objects in C_n. The key observation is that every $u \in C_n$ must be constructed from a pair $(x, y) \in A \times B$ where $\mathrm{wt}_A(x) = k$ for some $k \in \{0, 1, \ldots, n\}$ and $\mathrm{wt}_B(y) = \mathrm{wt}_S(u) - \mathrm{wt}_A(x) = n - k$. For fixed k between 0 and n, we can choose x from the set A_k in a_k ways, then choose y from the set B_{n-k} in b_{n-k} ways. The Product Rule gives $a_k b_{n-k}$ pairs (x, y) for this value of k. The Sum Rule now shows that $c_n = |C_n| = \sum_{k=0}^{n} a_k b_{n-k}$. By the formula for the product of two power series,

$$\begin{aligned} \text{GF}(S; z) &= \sum_{n=0}^{\infty} c_n z^n = \sum_{n=0}^{\infty} \left(\sum_{k=0}^{n} a_k b_{n-k} \right) z^n \\ &= \left(\sum_{n=0}^{\infty} a_n z^n \right) \cdot \left(\sum_{n=0}^{\infty} b_n z^n \right) = \text{GF}(A; z) \cdot \text{GF}(B; z). \end{aligned}$$

The general case of the Product Rule follows from Step 2 by induction on k.

Step 3. We prove the Bijection Rule for Weighted Sets. Let $f : S \to T$ be a weight-preserving bijection with two-sided inverse $g : T \to S$. One checks that g is also a weight-preserving bijection. For each $n \geq 0$, f and g restrict to bijections $f_n : S_n \to T_n$ and $g_n : T_n \to S_n$, where $S_n = \{u \in S : \text{wt}_S(u) = n\}$ and $T_n = \{v \in T : \text{wt}_T(v) = n\}$. By the Bijection Rule for unweighted sets, $|S_n| = |T_n|$ for all $n \geq 0$. Now, by the criterion for equality of power series,

$$\text{GF}(S; z) = \sum_{n=0}^{\infty} |S_n| z^n = \sum_{n=0}^{\infty} |T_n| z^n = \text{GF}(T; z).$$

5.9 Examples of the Product Rule for Weighted Sets

This section gives examples of counting problems that can be solved using the Product Rule for Weighted Sets.

5.25. Example. Let us count nonnegative integer solutions to the equation $x_1 + x_2 + x_3 + x_4 = n$ subject to the restrictions $0 \leq x_1 \leq 3$, $x_3 \geq 2$, and x_4 is even. Define S to be the set of all 4-tuples (x_1, x_2, x_3, x_4) satisfying the given restrictions, and define $\text{wt}(x_1, x_2, x_3, x_4) = x_1 + x_2 + x_3 + x_4$. We seek the coefficient of z^n in the generating function $\text{GF}(S; z)$.

To find this generating function, note that each object in S arises by choosing x_1 from the set $S_1 = \{0, 1, 2, 3\}$, choosing x_2 from the set $S_2 = \{0, 1, 2, \ldots\}$, choosing x_3 from the set $S_3 = \{2, 3, 4, \ldots\}$, and choosing x_4 from the set $S_4 = \{0, 2, 4, 6, \ldots\}$. Define $\text{wt}(a) = a$ for each integer a in any of the sets S_i. Then the weight-additivity condition

$$\text{wt}_S(x_1, x_2, x_3, x_4) = \text{wt}_{S_1}(x_1) + \text{wt}_{S_2}(x_2) + \text{wt}_{S_3}(x_3) + \text{wt}_{S_4}(x_4)$$

is satisfied. By the Product Rule for Weighted Sets, $\text{GF}(S; z) = \prod_{i=1}^{4} \text{GF}(S_i; z)$.

To find the generating function for S_i, we use the definition $\text{GF}(S_i; z) = \sum_{a \in S_i} z^{\text{wt}(a)}$. Making repeated use of the Geometric Series Formula, we compute:

$$\begin{aligned} \text{GF}(S_1; z) &= z^0 + z^1 + z^2 + z^3; \\ \text{GF}(S_2; z) &= z^0 + z^1 + z^2 + \cdots = (1 - z)^{-1}; \\ \text{GF}(S_3; z) &= z^2 + z^3 + z^4 + \cdots = z^2 \sum_{k=0}^{\infty} z^k = z^2 (1 - z)^{-1}; \\ \text{GF}(S_4; z) &= 1 + z^2 + z^4 + z^6 + \cdots = \sum_{k=0}^{\infty} z^{2k} = (1 - z^2)^{-1} = (1 - z)^{-1}(1 + z)^{-1}. \end{aligned}$$

Therefore,

$$\mathrm{GF}(S;z) = (1+z+z^2+z^3)z^2(1-z)^{-3}(1+z)^{-1} = (z^2+z^4)(1-z)^{-3}.$$

Using a computer algebra system, we find the partial fraction decomposition of this generating function to be

$$\mathrm{GF}(S;z) = -3 - z + 2(1-z)^{-3} - 6(1-z)^{-2} + 7(1-z)^{-1}.$$

Extracting the coefficient of z^n, the number of solutions for $n \geq 2$ is

$$2\binom{n+2}{2} - 6\binom{n+1}{1} + 7 = n^2 - 3n + 3.$$

When $n = 0$ or $n = 1$, there are no solutions since we required $x_3 \geq 2$.

5.26. Example. Suppose there are k varieties of donuts. How many bags of n donuts contain an odd number of donuts of each variety? We can model this counting problem using the integer equation $x_1 + x_2 + \cdots + x_k = n$, where each x_i must be an odd positive integer. Define S be the set of k-tuples $(x_1, \ldots, x_k)$ where every x_i is odd and positive, and let $\mathrm{wt}(x_1, \ldots, x_k) = x_1 + \cdots + x_k$. Let $T = \{1, 3, 5, \ldots\}$ be the set of odd positive integers, with $\mathrm{wt}(a) = a$ for $a \in T$. We can build each object in S by choosing x_1 from T, then x_2 from T, and so on. By the Product Rule for Weighted Sets, $\mathrm{GF}(S;z) = \prod_{i=1}^{k} \mathrm{GF}(T;z) = \mathrm{GF}(T;z)^k$. By the Geometric Series Formula,

$$\mathrm{GF}(T;z) = z^1 + z^3 + z^5 + \cdots = z\sum_{k=0}^{\infty} z^{2k} = z(1-z^2)^{-1}.$$

Therefore, $\mathrm{GF}(S;z) = z^k(1-z^2)^{-k}$. By the Negative Binomial Theorem (with z replaced by z^2),

$$z^k(1-z^2)^{-k} = z^k \sum_{j=0}^{\infty} \binom{j+k-1}{j, k-1} z^{2j}.$$

The total power of z is n when $k + 2j = n$, or equivalently $j = (n-k)/2$. So the answer to the counting question is $\binom{(n-k)/2+k-1}{k-1}$ when $n-k$ is even and nonnegative, and zero otherwise.

On the other hand, we can factor the generating function for S, obtaining $\mathrm{GF}(S;z) = z^k(1-z)^{-k}(1+z)^{-k}$. Using the Negative Binomial Theorem twice and the definition of the product of generating functions,

$$\begin{aligned}
\mathrm{GF}(S;z) &= z^k \left[\sum_{n=0}^{\infty} \binom{n+k-1}{k-1} z^n\right] \cdot \left[\sum_{n=0}^{\infty} (-1)^n \binom{n+k-1}{k-1} z^n\right] \\
&= \sum_{m=0}^{\infty} \left(\sum_{j=0}^{m} (-1)^{m-j} \binom{j+k-1}{k-1}\binom{m-j+k-1}{k-1}\right) z^{m+k}.
\end{aligned}$$

Taking $m = n - k$ gives the power z^n. Extracting this coefficient and comparing to our earlier answer, we deduce the combinatorial identity

$$\sum_{j=0}^{n-k} (-1)^{n-k-j} \binom{j+k-1}{k-1}\binom{n-j-1}{k-1} = \begin{cases} \binom{(n-k)/2+k-1}{k-1} & \text{if } (n-k)/2 \in \mathbb{Z}_{\geq 0}; \\ 0 & \text{otherwise.} \end{cases}$$

5.27. Example. How many ways can we make a multiset of 35 marbles using red, white, blue, green, and yellow marbles, if each color of marble must be used between 4 and 9 times? We build such a multiset by choosing the number of red marbles, the number of white marbles, and so on. The generating function for each individual choice is

$$F(z) = z^4 + z^5 + z^6 + z^7 + z^8 + z^9 = z^4 \sum_{k=0}^{5} z^k = \frac{z^4(1-z^6)}{1-z}.$$

By the Product Rule for Weighted Sets, the generating function for all multisets of marbles satisfying the color restrictions (weighted by the size of the multiset) is

$$F(z)^5 = z^{20}(1-z^6)^5(1-z)^{-5}.$$

We need the coefficient of z^{35} here, which equals the coefficient of z^{15} in $(1-z^6)^5(1-z)^{-5}$. Using the Binomial Theorem to expand $(1-z^6)^5$ and the Negative Binomial Theorem to expand $(1-z)^{-5}$, we get

$$(1 - 5z^6 + 10z^{12} - 10z^{18} + 5z^{24} - z^{30}) \cdot \left[1 + \binom{5}{4}z + \binom{6}{4}z^2 + \cdots + \binom{n+4}{4}z^n + \cdots\right].$$

When we multiply this out, the terms involving z^0, z^6, and z^{12} in the first factor can be matched with the terms involving z^{15}, z^9, and z^3 (respectively) in the second factor to get a total power of z^{15}. No other pair of terms can be multiplied to give this power of z. Thus, the coefficient we need is

$$1\binom{19}{4} - 5\binom{13}{4} + 10\binom{7}{4} = 651.$$

5.28. Example. Define the weight of a multiset to be the number of objects in the multiset (counting repetitions). Let us find the generating function for multisets using the alphabet $\{1, 2, 3\}$ where the number of 1's is unrestricted, the number of 2's is a multiple of 4, and the number of 3's cannot be a multiple of 3. We can build such a multiset by choosing how many 1's it contains, how many 2's it contains, and how many 3's it contains. The generating function for the first choice is $1 + z + z^2 + \cdots = (1-z)^{-1}$. The generating function for the second choice is $1 + z^4 + z^8 + \cdots = (1-z^4)^{-1}$. The generating function for the third choice is

$$z^1 + z^2 + z^4 + z^5 + z^7 + z^8 + \cdots = \sum_{n=0}^{\infty} z^n - \sum_{n=0}^{\infty} z^{3n} = (1-z)^{-1} - (1-z^3)^{-1}.$$

By the Product Rule for Weighted Sets, the required generating function is

$$(1-z)^{-1}(1-z^4)^{-1}[(1-z)^{-1} - (1-z^3)^{-1}] = \frac{z}{(1-z)^3(1+z^2)(1+z+z^2)}.$$

If we redefine the weight of a multiset to be the sum of all its members, the generating function would be

$$(1-z)^{-1}(1-z^8)^{-1}[(1-z^3)^{-1} - (1-z^9)^{-1}].$$

5.10 Generating Functions for Trees

This section illustrates the Sum Rule and Product Rule for Weighted Sets by deriving the generating functions for various kinds of trees.

5.29. Example: Binary Trees. Let S be the set of all binary trees, weighted by the number of vertices. By the definition of binary trees in Example 2.31, every tree $t \in S$ is either empty or is an ordered triple $(\bullet, t_1, t_2)$, where t_1 and t_2 are binary trees. Let S_0 be the one-element set consisting of the empty binary tree, let $S^+ = S - S_0$ be the set of nonempty binary trees, and let $N = \{\bullet\}$ be a one-element set such that $\text{wt}(\bullet) = 1$. By definition of the generating function for a weighted set, we have $\text{GF}(S_0; z) = z^0 = 1$ and $\text{GF}(N; z) = z^1 = z$. By the Sum Rule for Weighted Sets, we conclude that

$$\text{GF}(S; z) = \text{GF}(S_0; z) + \text{GF}(S^+; z) = 1 + \text{GF}(S^+; z).$$

By the recursive definition of nonempty binary trees, we can uniquely construct every tree $t \in S^+$ by (i) choosing the root node $\bullet$ from the set N; (ii) choosing the left subtree t_1 from the set S; (iii) choosing the right subtree t_2 from the set S; and then assembling these choices to form the tree $t = (\bullet, t_1, t_2)$. Observe that the weight-additivity condition $\text{wt}_{S^+}(t) = \text{wt}_N(\bullet) + \text{wt}_S(t_1) + \text{wt}_S(t_2)$ holds, since the weight of each tree is the number of vertices in the tree. It follows from the Product Rule for Weighted Sets that

$$\text{GF}(S^+; z) = \text{GF}(N; z)\,\text{GF}(S; z)\,\text{GF}(S; z) = z\,\text{GF}(S; z)^2.$$

Letting $F(z)$ denote the unknown generating function $\text{GF}(S; z)$, we conclude that F satisfies the equation $F(z) = 1 + zF(z)^2$, or equivalently $zF^2 - F + 1 = 0$. Furthermore, $F(0) = 1$ since there is exactly one binary tree with zero vertices.

We can formally solve this equation for F using the Quadratic Formula: $ax^2 + bx + c = 0$ has roots $(-b \pm \sqrt{b^2 - 4ac})/2a$. Taking $x = F$, $a = z$, $b = -1$, and $c = 1$ leads to

$$F(z) = \frac{1 \pm \sqrt{1 - 4z}}{2z}.$$

We use the Extended Binomial Theorem to get the power series expansion of $\sqrt{1 - 4z}$:

$$(1 - 4z)^{1/2} = \sum_{n=0}^{\infty} \frac{(1/2)\!\downarrow_n}{n!}(-4)^n z^n.$$

Expanding the definition of the falling factorial, the coefficient of z^n in $(1 - 4z)^{1/2}$ is

$$\frac{(1/2)(-1/2)(-3/2)\cdots(-(2n-3)/2)(-1)^n 2^n 2^n}{n!} = -\frac{1 \cdot 1 \cdot 3 \cdot 5 \cdot \ldots \cdot (2n-3) 2^n}{n!}.$$

To continue simplifying, multiply the numerator and denominator by $n!$, and notice that $2^n n! = 2 \cdot 4 \cdot 6 \cdot \ldots \cdot 2n$. The new numerator is the product of all integers from 1 to $2n$ except $2n - 1$. Multiplying numerator and denominator by $2n - 1$, we get

$$(1 - 4z)^{1/2} = \sum_{n=0}^{\infty} \frac{-1}{(2n-1)} \frac{(2n)!}{n!n!} z^n = 1 - 2z - 2z^2 - 4z^3 - 10^4 - 28z^5 - \cdots.$$

Returning to the expression $F(z) = (1 \pm \sqrt{1 - 4z})/(2z)$, do we choose the plus or minus sign? If we choose the plus sign, the resulting expression contains a term z^{-1}, which becomes unbounded as z approaches zero. If we choose the minus sign, this term goes away, and moreover the constant coefficient is $+1$ as needed. So we use the minus sign. Deleting the constant term of $(1 - 4z)^{1/2}$, negating, and dividing by $2z$, we find that

$$\begin{aligned} F(z) &= \sum_{n=1}^{\infty} \frac{(2n)!}{2(2n-1)n!n!} z^{n-1} = \sum_{m=0}^{\infty} \frac{(2m+2)!}{2(2m+1)(m+1)!(m+1)!} z^m \\ &= \sum_{m=0}^{\infty} \frac{(2m+2)(2m+1)!}{(2m+1)(2m+2)m!(m+1)!} z^m = \sum_{m=0}^{\infty} \frac{1}{2m+1}\binom{2m+1}{m, m+1} z^m. \end{aligned}$$

Taking the coefficient of z^m gives the number of binary trees with m vertices, which is the Catalan number C_m. A more combinatorial approach to this result was given in §2.10.

In the previous example, we can retroactively justify the formal application of the Quadratic Formula by confirming that $F(z) = (1 - (1 - 4z)^{1/2})/(2z)$ really does satisfy the functional equation $F(z) = 1 + zF(z)^2$ and initial condition $F(0) = 1$. In particular, the power series we found for F converges for all $z \in \mathbb{C}$ with $|z| < 1/4$. Alternatively, all of these algebraic computations can be justified at the level of formal power series, without worrying about convergence (see Chapter 11).

5.30. Example: Full Binary Trees. A binary tree is called *full* iff for every vertex in the tree, the left subtree and right subtree for that vertex are either both empty or both nonempty. A vertex whose left and right subtrees are both empty is called a *leaf* of the tree. Let S be the set of nonempty full binary trees, weighted by the number of leaves. S is the disjoint union of $S_1 = \{(\bullet, \emptyset, \emptyset)\}$ and $S_{\geq 2} = S - S_1$. We can build a tree t in $S_{\geq 2}$ by choosing any t_1 in S as the nonempty left subtree of the root, and then choosing any t_2 in S as the nonempty right subtree of the root. Note that $\text{wt}(t) = \text{wt}(t_1) + \text{wt}(t_2)$ since the weight is the number of leaves. So, by the Product Rule for Weighted Sets, $\text{GF}(S_{\geq 2}; z) = \text{GF}(S; z)^2$. We see directly that $\text{GF}(S_1; z) = z$. Letting $G(z) = \text{GF}(S; z)$, the Sum Rule for Weighted Sets now gives the relation $G(z) = z + G(z)^2$. Moreover, $G(0) = 0$ since there are no nonempty full binary trees with zero leaves. Solving the quadratic equation $G^2 - G + z = 0$ by calculations analogous to those in Example 5.29, we find that

$$G(z) = \frac{1 - \sqrt{1 - 4z}}{2} = zF(z),$$

where $F(z)$ is the generating function considered in the previous example. So for all $n \geq 1$, the coefficient of z^n in G equals the coefficient of z^{n-1} in F, namely the Catalan number C_{n-1}. We conclude that *there are C_{n-1} full binary trees with n leaves* for each $n \geq 1$.

The next example illustrates a generalization of the Sum Rule for Weighted Sets, in which a weighted set S is written as a disjoint union of *infinitely many* subsets S_k. By analogy with the finite version of the Sum Rule, we are led to the formula $\text{GF}(S; z) = \sum_{k=0}^{\infty} \text{GF}(S_k; z)$, but what is the meaning of the infinite sum of power series appearing on the right side? For now, we work with this infinite sum informally and see where the calculation leads us. A rigorous discussion of infinite sums and products of formal power series is given later (see §11.1).

5.31. Example: Ordered Trees. We recursively define *ordered trees* as follows. First, the list $t = (0)$ is an ordered tree with $\text{wt}(t) = 1$. Second, for every integer $k \geq 1$ and every list $t_1, t_2, \ldots, t_k$ of k ordered trees, the $(k+1)$-tuple $t = (k, t_1, t_2, \ldots, t_k)$ is an ordered tree with $\text{wt}(t) = 1 + \text{wt}(t_1) + \cdots + \text{wt}(t_k)$. All ordered trees arise by applying these two rules a finite number of times. The first rule can be considered a degenerate version of the second rule in which $k = 0$. Informally, the list $(k, t_1, \ldots, t_k)$ represents a tree whose root has k subtrees t_i, which appear in order from left to right, and where each subtree is also an ordered tree. The weight of a tree is the number of vertices in the tree.

Let S be the weighted set of all ordered trees. We now find the generating function $G(z) = \text{GF}(S; z)$. First, we write S as the disjoint union of sets S_k, for $k \in \mathbb{Z}_{\geq 0}$, where S_k consists of all trees $t \in S$ such that the root node has k subtrees. By the generalized Sum Rule for Weighted Sets,

$$G(z) = \text{GF}(S; z) = \sum_{k=0}^{\infty} \text{GF}(S_k; z).$$

For fixed $k \geq 0$, we can build each tree $t = (k, t_1, t_2, \ldots, t_k) \in S_k$ by choosing each entry in the sequence. Since the root vertex contributes 1 to the weight of t, the Product Rule for Weighted Sets shows that $\text{GF}(S_k; z) = zG(z)^k$. Substitution into the previous formula gives the functional equation

$$G(z) = \sum_{k=0}^{\infty} zG(z)^k = \frac{z}{1 - G(z)};$$

the last step is a formal version of the Geometric Series Formula (see §11.3 for a rigorous justification). This equation simplifies to the quadratic $G^2 - G + z = 0$, which is the same equation that occurred in the previous example. So we conclude, as before, that

$$G(z) = \frac{1 - \sqrt{1 - 4z}}{2} = \sum_{n=1}^{\infty} C_{n-1} z^n,$$

where C_{n-1} is a Catalan number. Thus, *there are C_{n-1} ordered trees with n vertices* for all $n \geq 1$.

5.11 Tree Bijections

The generating function calculations in the previous section prove the following (perhaps unexpected) enumeration result: *the set of binary trees with n vertices, the set of full binary trees with $n + 1$ leaves, and the set of ordered trees with $n + 1$ vertices all have cardinality C_n.* Although this is a purely combinatorial statement, the generating function method arrives at this result in a very mysterious and indirect fashion, which requires algebraic manipulations of quadratic expressions and the power series expansion of $\sqrt{1 - 4z}$.

Having found this result with the aid of generating functions, we may ask for a *bijective proof* in which the three sets of trees are linked by explicitly defined bijections. Some methods for building such bijections from recursions were studied in Chapter 2. Here we are seeking weight-preserving bijections on infinite sets, which can be defined as follows.

Let S denote the set of all binary trees weighted by number of vertices, and let T be the set of all nonempty full binary trees weighted by number of leaves. We define a bijection $f : S \to T$ recursively by setting $f(\emptyset) = (\bullet, \emptyset, \emptyset)$ and

$$f(\bullet, t_1, t_2) = (\bullet, f(t_1), f(t_2)).$$

See Figure 5.1. We claim that f preserves weights in the sense that $\text{wt}_T(f(t)) = \text{wt}_S(t) + 1$ for all $t \in S$ (we add 1 on the right side since $\text{GF}(T; z) = z\,\text{GF}(S; z)$). We verify this claim by induction. For the base case, note that the zero-vertex tree $\emptyset$ is mapped to the one-leaf tree $(\bullet, \emptyset, \emptyset)$. For the induction step, suppose t_1 and t_2 are trees in S with a vertices and b vertices, respectively. By induction, $f(t_1)$ and $f(t_2)$ are nonempty full binary trees with $a + 1$ leaves and $b + 1$ leaves, respectively. It follows that f sends the tree $t = (\bullet, t_1, t_2)$ with $a + b + 1$ vertices to a full binary tree with $(a + 1) + (b + 1) = (a + b + 1) + 1$ leaves, as needed. The inverse of f has an especially simple pictorial description: just erase all the leaves! Since f^{-1} respects weights, we see that a nonempty full binary tree always has one more leaf vertex than internal (non-leaf) vertex.

Next, let U be the set of all ordered trees, weighted by number of vertices. We define a weight-preserving bijection $g : T \to U$. First, define $g(\bullet, \emptyset, \emptyset) = (0)$. Second, given $t = (\bullet, t_1, t_2) \in T$ with t_1 and t_2 nonempty, define $g(t)$ by starting with the ordered tree $g(t_1)$ and appending the ordered tree $g(t_2)$ as the new rightmost subtree of the root node of $g(t_1)$.

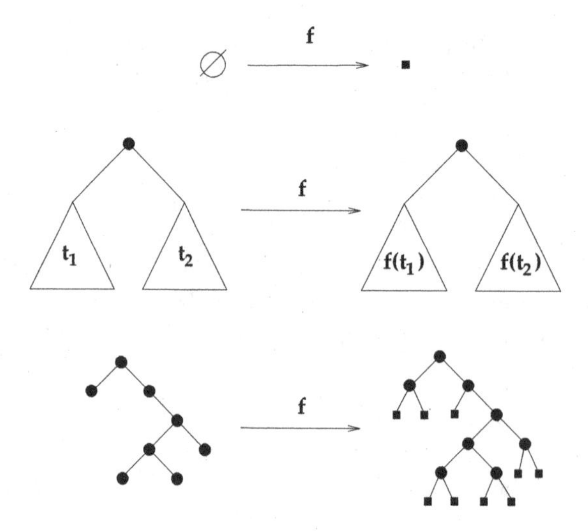

FIGURE 5.1
Bijection between binary trees and full binary trees.

See Figure 5.2. More formally, if $g(t_1) = (k, u_1, \ldots, u_k)$, let $g(t) = (k + 1, u_1, \ldots, u_k, g(t_2))$. One may check by induction that the number of vertices in $g(t)$ equals the number of leaves in t, as required.

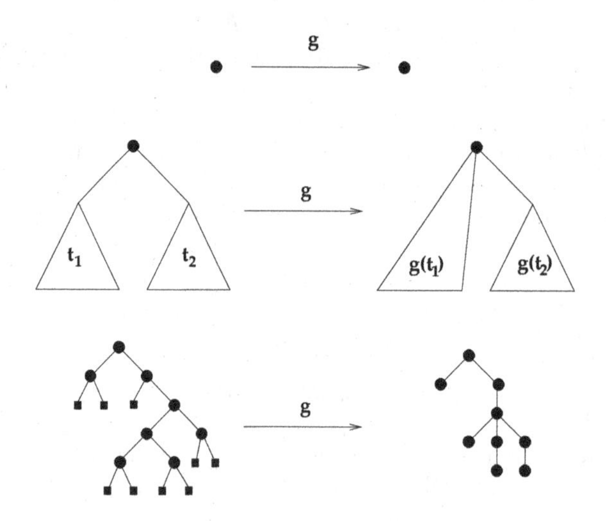

FIGURE 5.2
Bijection between full binary trees and ordered trees.

5.32. Remark. These examples show that generating functions are powerful algebraic tools for deriving enumeration results. However, once such results are found, we often seek direct combinatorial proofs that do not rely on generating functions. In particular, bijective proofs are more informative (and often more elegant) than algebraic proofs in the sense that they give us an explicit pairing between the objects in two sets.

5.12 Exponential Generating Functions

This section introduces exponential generating functions (EGFs) for weighted sets. We then develop a version of the Product Rule that explains the combinatorial meaning of the product of EGFs.

5.33. Definition: Exponential Generating Functions. Given a set S with weight function $\mathrm{wt} : S \to \mathbb{Z}_{\geq 0}$, the *exponential generating function* of the weighted set S is

$$\mathrm{EGF}(S;z) = \sum_{u\in S} \frac{z^{\mathrm{wt}(u)}}{\mathrm{wt}(u)!} = \sum_{n=0}^{\infty} \frac{a_n}{n!} z^n, \text{ where } a_n = |\{u \in S : \mathrm{wt}(u) = n\}|.$$

5.34. Example. Suppose $S = \mathbb{Z}_{\geq 0}$ and $\mathrm{wt}(k) = k$ for $k \in S$. Then $a_n = 1$ for all $n \in \mathbb{Z}_{\geq 0}$, so $\mathrm{EGF}(S;z) = \sum_{n=0}^{\infty} z^n/n! = e^z$. This is why the term "exponential generating function" is used. In contrast, $\mathrm{GF}(S;z) = \sum_{n=0}^{\infty} z^n = (1-z)^{-1}$.

Our goal is to find a version of the Product Rule applicable to EGFs. To prepare for this, we first recall the rule for multiplying two formal power series (Definition 5.15(e)). This rule can be written

$$\left(\sum_{n=0}^{\infty} a_n z^n\right) \cdot \left(\sum_{n=0}^{\infty} b_n z^n\right) = \sum_{n=0}^{\infty} \left(\sum_{\substack{(i,j)\in\mathbb{Z}_{\geq 0}^2: \\ i+j=n}} a_i b_j \right) z^n. \tag{5.6}$$

More generally, consider a product of k formal power series $F_1(z), \ldots, F_k(z)$. Write $F_j(z) = \sum_{n=0}^{\infty} f_{j,n} z^n$ for $1 \leq j \leq k$. The following formula can be proved using (5.6) and induction on k.

5.35. Theorem: Products of OGFs. For all $k \in \mathbb{Z}_{>0}$ and all formal power series $F_1, F_2, \ldots, F_k$ with $F_j = \sum_{m=0}^{\infty} f_{j,m} z^m$,

$$F_1(z)F_2(z)\cdots F_k(z) = \sum_{n=0}^{\infty} \left(\sum_{\substack{(i_1,i_2,\ldots,i_k)\in\mathbb{Z}_{\geq 0}^k: \\ i_1+i_2+\cdots+i_k=n}} f_{1,i_1} f_{2,i_2} \cdots f_{k,i_k} \right) z^n. \tag{5.7}$$

Intuitively, this formula arises by expanding the product of k sums on the left side using the Distributive Law (see Exercise 2-16). This expansion produces a sum of various terms, where each term is a product of k monomials $f_{1,i_1} z^{i_1}$, $f_{2,i_2} z^{i_2}$, ..., $f_{k,i_k} z^{i_k}$. This product will contribute to the coefficient of z^n iff $i_1 + i_2 + \cdots + i_k = n$; the contribution to that coefficient will be $f_{1,i_1} f_{2,i_2} \cdots f_{k,i_k}$. Adding over all possible choices of $(i_1, i_2, \ldots, i_k)$ gives the right side of (5.7).

Let us see how the product formula changes when we use EGFs. Now set $G_j(z) = \sum_{n=0}^{\infty} (g_{j,n}/n!) z^n$ for $1 \leq j \leq k$. Replacing $f_{j,i}$ by $g_{j,i}/i!$ in (5.7), we get

$$G_1(z)G_2(z)\cdots G_k(z) = \sum_{n=0}^{\infty} \left(\sum_{\substack{(i_1,i_2,\ldots,i_k)\in\mathbb{Z}_{\geq 0}^k: \\ i_1+i_2+\cdots+i_k=n}} \frac{g_{1,i_1} g_{2,i_2} \cdots g_{k,i_k}}{i_1! i_2! \cdots i_k!} \right) z^n. \tag{5.8}$$

If we multiply and divide the coefficient of z^n by $n!$, a multinomial coefficient appears. Thus we can write

$$G_1(z)G_2(z)\cdots G_k(z) = \sum_{n=0}^{\infty}\left(\sum_{\substack{(i_1,i_2,\ldots,i_k)\in\mathbb{Z}_{\geq 0}^k:\\ i_1+i_2+\cdots+i_k=n}}\binom{n}{i_1,i_2,\ldots,i_k}g_{1,i_1}g_{2,i_2}\cdots g_{k,i_k}\right)\frac{z^n}{n!}. \quad (5.9)$$

We can reverse engineer a combinatorial interpretation for the coefficient of $z^n/n!$ using the Anagram Rule, the Product Rule, and the Sum Rule. This leads to the following Product Rule for EGFs. (Compare to the version of the Product Rule for OGFs, given in 5.23.)

5.36. The EGF Product Rule. Suppose k is a fixed positive integer and $S_1,\ldots,S_k$ are weighted sets. Suppose S is a weighted set such that every $u \in S$ can be constructed in exactly one way as follows. For $1 \leq j \leq k$, choose $u_j \in S_j$ and let $i_j = \mathrm{wt}_{S_j}(u_j)$. Then choose an anagram $w \in \mathcal{R}(1^{i_1}2^{i_2}\cdots k^{i_k})$. Finally, assemble the chosen objects $u_1,\ldots,u_k,w$ in a prescribed manner. Assume that whenever u is constructed from $u_1,u_2,\ldots,u_k,w$, the *weight-additivity condition* $\mathrm{wt}_S(u) = i_1+\cdots+i_k = \mathrm{wt}_{S_1}(u_1)+\mathrm{wt}_{S_2}(u_2)+\cdots+\mathrm{wt}_{S_k}(u_k)$ holds. Then

$$\mathrm{EGF}(S;z) = \mathrm{EGF}(S_1;z)\cdot\mathrm{EGF}(S_2;z)\cdot\ldots\cdot\mathrm{EGF}(S_k;z).$$

The following examples illustrate the use of this rule.

5.37. Example. Fix a positive integer k, and let S be the set of all words using the alphabet $\{1,2,\ldots,k\}$, weighted by length. On one hand, the definition of EGFs and the Word Rule show that

$$\mathrm{EGF}(S;z) = \sum_{n=0}^{\infty}\frac{k^n}{n!}z^n = e^{kz}.$$

On the other hand, we can derive this result from the EGF Product Rule as follows. Take all of the sets $S_1,S_2,\ldots,S_k$ to be the weighted set $\mathbb{Z}_{\geq 0}$ from Example 5.34. We can build an arbitrary word u in S as follows. Choose $i_1 \in S_1$, which represents the number of times 1 will appear in u. Choose $i_2 \in S_2$, which is the number of 2's in u. Choose $i_3,\ldots,i_k$ similarly. Finally, choose an anagram $w \in \mathcal{R}(1^{i_1}\cdots k^{i_k})$, and let $u = w$. The weight-additivity condition holds, since the length of u is $i_1+i_2+\cdots+i_k$. By the EGF Product Rule and Example 5.34,

$$\mathrm{EGF}(S;z) = [\mathrm{EGF}(\mathbb{Z}_{\geq 0};z)]^k = (e^z)^k = e^{kz}.$$

We did not really need the EGF Product Rule in the last example, since we already knew how many words of length n there are. But by modifying the sets S_j in the previous example, we can use EGFs to solve far more complicated counting problems.

5.38. Example. How many 15-letter words using the alphabet $\{1,2,3,4\}$ have an odd number of 1's, an even number of 2's, and at most three 4's? To solve this, we find the EGF for the set S of all words (of any length) satisfying the given restrictions, and extract the coefficient of z^{15}. For this problem, take S_1 to be the set of odd positive integers, S_2 to be the set of even nonnegative integers, $S_3 = \mathbb{Z}_{\geq 0}$, and $S_4 = \{0,1,2,3\}$. In each case, let $\mathrm{wt}_{S_j}(i) = i$ for $i \in S_j$. By definition, the EGF for S_1 is

$$\mathrm{EGF}(S_1;z) = \sum_{n \text{ odd}} z^n/n! = z + z^3/3! + z^5/5! + z^7/7! + \cdots = \sinh z.$$

Similarly, the EGF for S_2 is $\cosh z$, the EGF for S_3 is e^z, and the EGF for S_4 is $1+z+z^2/2!+z^3/3!$. We build $u \in S$ by first choosing $i_1 \in S_1$, $i_2 \in S_2$, $i_3 \in S_3$, and $i_4 \in S_4$, where

i_j is the number of j's that will appear in u. Then we choose $w \in \mathcal{R}(1^{i_1}2^{i_2}3^{i_3}4^{i_4})$ and set $u = w$. Since the length of u is $i_1+i_2+i_3+i_4$, the EGF Product Rule applies. We conclude that

$$\mathrm{EGF}(S;z) = (\sinh z)(\cosh z)e^z(1+z+z^2/2+z^3/6).$$

Using a computer algebra system, we find the coefficient of z^{15} in this power series and multiply by 15! to obtain the answer 123,825,662.

5.39. Example: Surjections. Fix a positive integer k. Let us find the EGF for the set S of all surjective functions mapping some set $\{1,2,\ldots,n\}$ onto $\{1,2,\ldots,k\}$, where $n \geq k$. Take the weight of a function f in S to be the size n of the domain of f. On one hand, by the Surjection Rule, we know that the number of objects of weight n in S is $S(n,k)k!$, where $S(n,k)$ is a Stirling number of the second kind. So by definition, the EGF of S is $\mathrm{EGF}(S;z) = \sum_{n=k}^{\infty} S(n,k)k!\frac{z^n}{n!}$. On the other hand, we can build objects in S by the EGF Product Rule, as follows. Take all of the sets $S_1,\ldots,S_k$ to be $\mathbb{Z}_{>0}$ weighted by $\mathrm{wt}(i) = i$. Then $\mathrm{EGF}(S_j;z) = \sum_{n=1}^{\infty} z^n/n! = e^z - 1$ for $1 \leq j \leq k$. Build $f \in S$ as follows. Choose positive integers $i_1 \in S_1,\ldots,i_k \in S_k$, and then choose an anagram $w \in \mathcal{R}(1^{i_1}\cdots k^{i_k})$. Let $n = i_1+\cdots+i_k$, and define $f(r) = w_r$ for $1 \leq r \leq n$. This function f is surjective since every i_j is strictly positive. The EGF Product Rule applies to show $\mathrm{EGF}(S;z) = (e^z-1)^k$. Comparing to the previous expression for this EGF, we deduce the following generating function identity for Stirling numbers of the second kind:

$$\sum_{n=k}^{\infty} S(n,k)\frac{z^n}{n!} = \frac{(e^z-1)^k}{k!}.$$

We can also interpret this example as finding the EGF for words w in the alphabet $\{1,2,\ldots,k\}$ that use every letter at least once.

In all the examples so far, every set S_j had at most one element of any given weight n, and the final object constructed by the EGF Product Rule was the anagram w. The next example illustrates a more complex situation where objects chosen from the sets S_j are combined with w to build a more elaborate structure.

5.40. Example: Digraphs with k Cycles. Fix a positive integer k. Let S be the set of functional digraphs G such that the vertex set of G is $\{1,2,\ldots,n\}$ for some $n \geq k$; G consists of k disjoint directed cycles; and each cycle in G has been given a distinct *label* chosen from $\{1,2,\ldots,k\}$. Let the weight of G be n, the number of vertices in G. Recalling the definition of $c(n,k)$ from §3.6, we see that the EGF for S is $\mathrm{EGF}(S;z) = \sum_{n=k}^{\infty} c(n,k)k!z^n/n!$. We obtain another formula for this EGF using the EGF Product Rule. For $1 \leq j \leq k$, let S_j be the set of permutations of $\{1,2,\ldots,n\}$ (for some $n \geq 1$) where the permutation's digraph is a single directed cycle. There are $(n-1)!$ such permutations of $\{1,2,\ldots,n\}$, and therefore

$$\mathrm{EGF}(S_j;z) = \sum_{n=1}^{\infty}(n-1)!\frac{z^n}{n!} = \sum_{n=1}^{\infty}\frac{z^n}{n} = -\operatorname{Log}(1-z).$$

We build an object in S as follows. Choose cycles $g_1 \in S_1$, ..., $g_k \in S_k$ of respective lengths $i_1,\ldots,i_k$, choose $w \in \mathcal{R}(1^{i_1}\cdots k^{i_k})$, and let $n = i_1+\cdots+i_k$. Suppose the i_1 copies of 1 in w occur in positions $p(1) < p(2) < \ldots < p(i_1)$. Replace the symbols $1,2,\ldots,i_1$ in the cycle g_1 by the symbols $p(1),p(2),\ldots,p(i_1)$ (respectively) to get a cycle of length i_1 involving the symbols $p(1),\ldots,p(i_1)$. Attach the label 1 to this cycle. Renumber the entries in the other cycles $g_2,\ldots,g_k$ similarly. In general, if the i_j copies of j in w occur in positions $p(1) < p(2) < \cdots < p(i_j)$, then we replace the symbols $1,2,\ldots,i_j$ in cycle g_j by the symbols $p(1),p(2),\ldots,p(i_j)$ in this order, and we attach the label j to this cycle. The union of all the renumbered cycles is a functional digraph on $\{1,2,\ldots,n\}$ consisting of k cycles labeled $1,2,\ldots,k$. Finally, by the EGF Product Rule, $\mathrm{EGF}(S;z)=\prod_{j=1}^{k}\mathrm{EGF}(S_j;z)$, so

$$\sum_{n=k}^{\infty} c(n,k)\frac{z^n}{n!} = \frac{[-\mathrm{Log}(1-z)]^k}{k!}. \tag{5.10}$$

5.13 Stirling Numbers of the First Kind

So far, we have considered generating functions in a single variable z. It is sometimes necessary to use generating functions involving two or more variables. For example, suppose S is a set of objects and wt_1 and wt_2 are two weight functions for S. The ordinary generating function for S relative to these weights is

$$\mathrm{GF}(S;t,z)=\sum_{u\in S} t^{\mathrm{wt}_1(u)} z^{\mathrm{wt}_2(u)} = \sum_{n=0}^{\infty}\sum_{k=0}^{\infty} a_{k,n} t^k z^n,$$

where $a_{k,n} = |\{u\in S : \mathrm{wt}_1(u)=k \text{ and } \mathrm{wt}_2(u)=n\}|$. We can think of $\mathrm{GF}(S;t,z)$ as a formal power series in z, where each coefficient is a formal power series in t. We could also consider the exponential generating function

$$\mathrm{EGF}(S;t,z)=\sum_{n=0}^{\infty}\sum_{k=0}^{\infty} a_{k,n}\frac{t^k}{k!}\cdot\frac{z^n}{n!},$$

or a mixed version where only one variable is divided by a factorial.

In this section and the next, we illustrate manipulations of two-variable generating functions by developing generating functions for the Stirling numbers of the first and second kind. Recall from §3.6 that the signless Stirling number of the first kind, denoted $c(n,k)$, counts the number of permutations of n objects whose functional digraphs consist of k disjoint cycles. These numbers are determined by the recursion

$$c(n,k)=c(n-1,k-1)+(n-1)c(n-1,k) \qquad \text{for } 0<k<n,$$

with initial conditions $c(n,0)=0$ for $n>0$, $c(n,n)=1$ for $n\geq 0$, and $c(n,k)=0$ for $k<0$ or $k>n$.

We define the mixed generating function for the numbers $c(n,k)$ by

$$F(t,z)=\sum_{n=0}^{\infty}\sum_{k=0}^{n}\frac{c(n,k)}{n!}t^k z^n.$$

Observe that the coefficient of $t^k z^n$ in F, namely $c(n,k)/n!$, is the probability that a randomly chosen permutation of n objects will have k cycles. We will use the recursion for $c(n,k)$ to show that F satisfies the partial differential equation $(1-z)\partial F/\partial z = tF$, where $\partial F/\partial z$ denotes the formal partial derivative of F with respect to the variable z. Using the recursion, compute

$$\begin{aligned}\frac{\partial F}{\partial z} &= \sum_{n=0}^{\infty}\sum_{k=0}^{n}\frac{nc(n,k)}{n!}t^k z^{n-1} = \sum_{n=0}^{\infty}\sum_{k=0}^{n}\frac{n[c(n-1,k-1)+(n-1)c(n-1,k)]}{n!}t^k z^{n-1}\\ &= \sum_{n=0}^{\infty}\sum_{k=0}^{n}\frac{c(n-1,k-1)}{(n-1)!}t^k z^{n-1} + \sum_{n=0}^{\infty}\sum_{k=0}^{n}\frac{c(n-1,k)}{(n-2)!}t^k z^{n-1}.\end{aligned}$$

In the first summation, let $m = n - 1$ and $j = k - 1$. After discarding summands equal to zero, we see that

$$\sum_{n=0}^{\infty}\sum_{k=0}^{n}\frac{c(n-1,k-1)}{(n-1)!}t^k z^{n-1} = t\sum_{m=0}^{\infty}\sum_{j=0}^{m}\frac{c(m,j)}{m!}t^j z^m = tF.$$

On the other hand, letting $m = n - 1$ in the second summation shows that

$$\sum_{n=0}^{\infty}\sum_{k=0}^{n}\frac{c(n-1,k)}{(n-2)!}t^k z^{n-1} = z\sum_{m=0}^{\infty}\sum_{k=0}^{m}c(m,k)t^k\frac{z^{m-1}}{(m-1)!} = z\frac{\partial F}{\partial z},$$

since $\frac{\partial}{\partial z}(z^m/m!) = z^{m-1}/(m-1)!$. So we indeed have $\partial F/\partial z = tF + z\partial F/\partial z$, as claimed.

We now know a formal differential equation satisfied by $F(t,z)$, together with the initial condition $F(t,0) = 1$. To find an explicit formula for F, we solve for F by techniques that may be familiar from calculus. However, one must remember that all our computations need to be justifiable at the level of formal power series. We defer these technical matters for the time being (see Chapter 11). To begin solving, divide both sides of the differential equation $(1-z)\partial F/\partial z = tF$ by $(1-z)F$, obtaining

$$\frac{\partial F/\partial z}{F} = \frac{t}{1-z}.$$

We recognize the left side as the partial derivative of $\operatorname{Log}[F(t,z)]$ with respect to z. On the other hand, the right side is the partial derivative of $\operatorname{Log}[(1-z)^{-t}]$ with respect to z. Taking the antiderivative of both sides with respect to z, we therefore get

$$\operatorname{Log}[F(t,z)] = \operatorname{Log}[(1-z)^{-t}] + C(t).$$

By letting $z = 0$ on both sides, we see that the integration constant $C(t)$ is zero. Finally, we exponentiate both sides to arrive at the generating function

$$F(t,z) = (1-z)^{-t}. \tag{5.11}$$

Having discovered this formula for F, we can now give an independent verification of its correctness by invoking our earlier results on Stirling numbers and generalized powers:

$$\begin{aligned}
(1-z)^{-t} &= \sum_{n=0}^{\infty}\frac{(-t)\!\downarrow_n}{n!}(-z)^n \quad \text{by the Extended Binomial Theorem}\\
&= \sum_{n=0}^{\infty}\frac{(t)\!\uparrow_n}{n!}z^n \quad \text{by Definition 2.63}\\
&= \sum_{n=0}^{\infty}\left(\sum_{k=0}^{n}c(n,k)t^k\right)\frac{z^n}{n!} \quad \text{by Theorem 2.60}\\
&= F(t,z).
\end{aligned}$$

5.14 Stirling Numbers of the Second Kind

In this section, we derive a generating function for Stirling numbers of the second kind. Recall from §2.12 that $S(n,k)$ is the number of set partitions of an n-element set into k

nonempty blocks. Define the two-variable generating function

$$G(t,z)=\sum_{n=0}^{\infty}\sum_{k=0}^{n}\frac{S(n,k)}{n!}t^k z^n.$$

The following recursion will help us find a differential equation satisfied by G.

5.41. Theorem: Recursion for Stirling Numbers. For all $n \geq 0$ and $0 \leq k \leq n+1$,

$$S(n+1,k)=\sum_{i=0}^{n}\binom{n}{i}S(n-i,k-1).$$

The initial conditions are $S(0,0)=1$ and $S(n,k)=0$ whenever $k<0$ or $k>n$.

Proof. Consider set partitions of $\{1,2,\ldots,n{+}1\}$ into k blocks such that the block containing $n+1$ has i other elements in it (where $0 \leq i \leq n$). To build such a set partition, choose the i elements that go in the block with $n+1$ in $\binom{n}{i}$ ways, and then choose a set partition of the remaining $n-i$ elements into $k-1$ blocks. The recursion now follows from the Sum Rule and Product Rule. (Compare to the proof of Theorem 2.46.) □

5.42. Theorem: Differential Equation for G. The generating function $G(t,z) = \sum_{n=0}^{\infty}\sum_{k=0}^{n} S(n,k)t^k z^n/n!$ satisfies $G(t,0)=1$ and $\partial G/\partial z = te^z G$.

Proof. The partial derivative of G with respect to z is

$$\frac{\partial G}{\partial z}=\sum_{m=0}^{\infty}\sum_{k=0}^{m}\frac{S(m,k)}{m!}t^k m z^{m-1}=\sum_{n=0}^{\infty}\sum_{k=0}^{n+1}\frac{S(n+1,k)}{n!}t^k z^n,$$

where we have set $n=m-1$. Using Theorem 5.41 transforms this expression into

$$\sum_{n=0}^{\infty}\sum_{k=0}^{n+1}\sum_{i=0}^{n}\frac{1}{n!}\binom{n}{i}S(n-i,k-1)t^k z^n=\sum_{n=0}^{\infty}\sum_{k=0}^{n+1}\sum_{i=0}^{n}\frac{S(n-i,k-1)t^k z^{n-i}}{(n-i)!}\cdot\frac{z^i}{i!}.$$

Setting $j=k-1$, the formula becomes

$$t\sum_{n=0}^{\infty}\sum_{j=0}^{n}\sum_{i=0}^{n}\frac{S(n-i,j)t^j z^{n-i}}{(n-i)!}\cdot\frac{z^i}{i!}=t\sum_{n=0}^{\infty}\sum_{i=0}^{n}\sum_{j=0}^{n-i}\frac{S(n-i,j)t^j z^{n-i}}{(n-i)!}\cdot\frac{z^i}{i!}.$$

Finally, by the definition of the product of formal power series in z, the last expression equals

$$t\left(\sum_{i=0}^{\infty}\frac{z^i}{i!}\right)\cdot\left(\sum_{m=0}^{\infty}\left[\sum_{j=0}^{m}\frac{S(m,j)}{m!}t^j\right]z^m\right)=te^z G.$$ □

5.43. Theorem: Generating Function for Stirling Numbers of the Second Kind.

$$\sum_{n=0}^{\infty}\sum_{k=0}^{n}\frac{S(n,k)}{n!}t^k z^n=e^{t(e^z-1)}.$$

Proof. We solve the differential equation $\partial G/\partial z = te^z G$ with initial condition $G(t,0) = 1$. Dividing both sides by G and taking the antiderivative with respect to z, we get $\operatorname{Log}(G(t,z)) = te^z + C(t)$. Setting $z=0$ gives $0=t+C(t)$, so the constant of integration is $C(t)=-t$. Exponentiating both sides, we find that

$$G(t,z)=e^{te^z-t}=e^{t(e^z-1)}.$$ □

5.44. Theorem: Generating Function for $S(n,k)$ for fixed k. For all $k \geq 0$,

$$\sum_{n=k}^{\infty} S(n,k)\frac{z^n}{n!} = \frac{1}{k!}(e^z - 1)^k.$$

Proof. Using $e^u = \sum_{k=0}^{\infty} u^k/k!$ with $u = t(e^z - 1)$, we have

$$e^{t(e^z-1)} = \sum_{k=0}^{\infty} t^k \frac{(e^z-1)^k}{k!}.$$

Extracting the coefficient of t^k and using Theorem 5.43, we obtain the required formula. (Compare to Example 5.39, which derives the same formula using the EGF Product Rule.) □

5.15 Generating Functions for Integer Partitions

This section uses generating functions to prove some fundamental results involving integer partitions. Recall from §2.11 that an *integer partition* is a weakly decreasing sequence of positive integers. The *area* of a partition $\mu = (\mu_1, \ldots, \mu_k)$ is $|\mu| = \mu_1 + \cdots + \mu_k$, the sum of the parts of μ. Let Par denote the set of all integer partitions, weighted by area.

To get a generating function for Par, we need an extension of the Product Rule for Weighted Sets involving infinite products. We introduce this extension informally here, delaying a rigorous discussion until §11.2. Intuitively, in this version of the Product Rule, we build objects in a given set S by making a potentially infinite sequence of choices. However, we require that any particular object in the set can be completely built after making only finitely many choices, where the number of choices needed depends on the object. If the weight-additivity condition holds, then $\text{GF}(S; z)$ is the infinite product of the generating functions for each separate choice in the construction sequence. The following theorem gives a concrete illustration of this setup.

5.45. Theorem: Partition Generating Function.

$$\text{GF}(\text{Par}; z) = \sum_{\mu \in \text{Par}} z^{|\mu|} = \prod_{i=1}^{\infty} \frac{1}{1 - z^i}.$$

Proof. We use the infinite version of the Product Rule for Weighted Sets. We build a typical partition $\mu \in \text{Par}$ by making an infinite sequence of choices, as follows. First, choose how many parts of size 1 will occur in μ. The set of possible choices here is $\mathbb{Z}_{\geq 0} = \{0, 1, 2, 3, \ldots\}$. For this first choice, we let $\text{wt}(k) = k$ for $k \in \mathbb{Z}_{\geq 0}$. The generating function for this choice is $1+z+z^2+z^3+\cdots = (1-z)^{-1}$. Second, choose how many parts of size 2 will occur in μ. Again the set of possibilities is $\{0, 1, 2, 3, \ldots\}$, but now we use the weight $\text{wt}(k) = 2k$. The reason is that including k parts of size 2 in the partition μ will contribute $2k$ to $|\mu|$. So the generating function for the second choice is $1 + z^2 + z^4 + z^6 + \cdots = (1 - z^2)^{-1}$. Proceed similarly, choosing for every $i \geq 1$ how many parts of size i will occur in μ. Since choosing k parts of size i increases $|\mu|$ by ki, the generating function for choice i is $\sum_{k=0}^{\infty}(z^i)^k = (1 - z^i)^{-1}$. Multiplying the generating functions for all the choices gives the infinite product in the theorem. □

It may be helpful to consider a finite version of the previous proof. For fixed $N \in \mathbb{Z}_{\geq 0}$, let $\mathrm{Par}^{\leq N}$ be the set of all integer partitions μ such that $\mu_i \leq N$ for all i, or equivalently $\mu_1 \leq N$. Using the finite version of the Product Rule for Weighted Sets, the construction in the preceding proof shows that

$$\mathrm{GF}(\mathrm{Par}^{\leq N}; z) = \sum_{\mu \in \mathrm{Par}^{\leq N}} z^{|\mu|} = \prod_{i=1}^{N} \frac{1}{1-z^i}.$$

Informally, when we take the limit as N goes to infinity, the sets $\mathrm{Par}^{\leq N}$ get larger and larger and approach the full set Par consisting of all integer partitions. More precisely, $\bigcup_{N=1}^{\infty} \mathrm{Par}^{\leq N} = \mathrm{Par}$. So it is plausible that taking the limit of the generating functions of the sets $\mathrm{Par}^{\leq N}$ should give the generating function for the full set Par:

$$\lim_{N\to\infty} \mathrm{GF}(\mathrm{Par}^{\leq N}; z) = \mathrm{GF}(\mathrm{Par}; z), \text{ or } \lim_{N\to\infty} \prod_{i=1}^{N} \frac{1}{1-z^i} = \prod_{i=1}^{\infty} \frac{1}{1-z^i}.$$

The only difficulty is that we have not yet precisely defined the limit of a sequence of (formal) power series. We return to this technical point in §11.1.

We can add another variable to the partition generating function to keep track of additional information. Recall that, for $\mu \in \mathrm{Par}$, μ_1 is the length of the first (longest) part of μ, and $\ell(\mu)$ is the number of nonzero parts of μ.

5.46. Theorem: Enumerating Integer Partitions by Area and Length.

$$\sum_{\mu \in \mathrm{Par}} t^{\ell(\mu)} z^{|\mu|} = \prod_{i=1}^{\infty} \frac{1}{1-tz^i} = \sum_{\mu \in \mathrm{Par}} t^{\mu_1} z^{|\mu|}.$$

Proof. To prove the first equality, we modify the proof of Theorem 5.45 to keep track of the t-weight. At stage i, suppose we choose k copies of the part i for inclusion in μ. This will increase $\ell(\mu)$ by k and increase $|\mu|$ by ki. So the generating function for the choice made at stage i is

$$\sum_{k=0}^{\infty} t^k z^{ki} = \sum_{k=0}^{\infty} (tz^i)^k = \frac{1}{1-tz^i}.$$

The result now follows from the infinite version of the Product Rule for Weighted Sets. To prove the second equality, observe that conjugation is a bijection on Par that preserves area and satisfies $(\mu')_1 = \ell(\mu)$. So the result follows from the Bijection Rule for Weighted Sets. □

We can use variations of the preceding proofs to derive generating functions for various subcollections of integer partitions.

5.47. Theorem: Partitions with Odd Parts. Let OddPar be the set of integer partitions all of whose parts are odd. Then

$$\sum_{\mu \in \mathrm{OddPar}} z^{|\mu|} = \prod_{k=1}^{\infty} \frac{1}{1-z^{2k-1}}.$$

Proof. Repeat the proof of Theorem 5.45, but now make choices only for the odd part lengths 1, 3, 5, 7, etc. □

5.48. Theorem: Partitions with Distinct Parts. Let DisPar be the set of integer partitions all of whose parts are distinct. Then

$$\sum_{\mu \in \text{DisPar}} z^{|\mu|} = \prod_{i=1}^{\infty}(1+z^i).$$

Proof. We build a partition $\mu \in$ DisPar via the following choice sequence. For each part length $i \geq 1$, either choose to not use that part in μ or to include that part in μ (note that the part is only allowed to occur once). The generating function for this choice is $1+z^i$. The result now follows from the infinite version of the Product Rule for Weighted Sets. □

By comparing the generating functions in the last two theorems, we are led to the following unexpected result.

5.49. Theorem: OddPar vs. DisPar.

$$\sum_{\mu \in \text{OddPar}} z^{|\mu|} = \sum_{\nu \in \text{DisPar}} z^{|\nu|}.$$

Proof. We formally manipulate the generating functions as follows:

$$\begin{aligned}
\sum_{\mu \in \text{OddPar}} z^{|\mu|} &= \prod_{k=1}^{\infty} \frac{1}{1-z^{2k-1}} \\
&= \prod_{k=1}^{\infty} \frac{1}{1-z^{2k-1}} \prod_{j=1}^{\infty} \frac{1}{1-z^{2j}} \prod_{j=1}^{\infty}(1-z^{2j}) \\
&= \prod_{i=1}^{\infty} \frac{1}{1-z^i} \prod_{j=1}^{\infty}[(1-z^j)(1+z^j)] \\
&= \prod_{j=1}^{\infty} \frac{1}{1-z^j} \prod_{j=1}^{\infty}(1-z^j) \prod_{j=1}^{\infty}(1+z^j) \\
&= \prod_{j=1}^{\infty}(1+z^j) = \sum_{\nu \in \text{DisPar}} z^{|\nu|}. \quad \square
\end{aligned}$$

This "proof" glosses over many technical details about infinite products of formal power series. For example, when going from the first line to the second line, it is permissible to multiply by 1 in the form $\prod_{j=1}^{N}(1-z^{2j})^{-1}\prod_{j=1}^{N}(1-z^{2j})$. (This can be justified rigorously by induction on N.) But how do we know that this manipulation is allowed for $N=\infty$?

Similarly, to reach the third line, we need to know that

$$\prod_{i \geq 1, i \text{ odd}} \frac{1}{1-z^i} \prod_{i \geq 1, i \text{ even}} \frac{1}{1-z^i} = \prod_{i \geq 1} \frac{1}{1-z^i}.$$

Regarding the fourth line, how do we know that $\prod_{j=1}^{\infty}[(1-z^j)(1+z^j)]$ can be split into $\prod_{j=1}^{\infty}(1-z^j)\prod_{j=1}^{\infty}(1+z^j)$? And on the final line, we again need to know an infinite version of the cancellation property $(1-z^j)^{-1}(1-z^j)=1$.

Each of these steps can indeed by justified rigorously, but we defer a careful discussion until Chapter 11. The main point now is that we must not blindly assume that algebraic manipulations valid for finite sums and products will automatically carry over to infinite sums and products (although they often do).

5.16 Partition Bijections

We saw above that the generating function for partitions with odd parts (weighted by area) coincides with the generating function for partitions with distinct parts. We gave an *algebraic proof* of this result based on manipulation of infinite products of formal power series. However, from a combinatorial standpoint, we would like to have a *bijective proof* that $\mathrm{GF}(\mathrm{OddPar}; z) = \mathrm{GF}(\mathrm{DisPar}; z)$. By the Bijection Rule for Weighted Sets, it is enough to construct an area-preserving bijection $F : \mathrm{OddPar} \to \mathrm{DisPar}$. Two such bijections are presented in this section.

5.50. Sylvester's Bijection. Define $F : \mathrm{OddPar} \to \mathrm{DisPar}$ as follows. Given $\mu \in \mathrm{OddPar}$, draw a *centered* version of the Ferrers diagram of μ in which the middle boxes of the parts of μ are all drawn in the same column; an example is shown in Figure 5.3. Note that each part of μ does have a middle box, because the part is odd. Label the columns in the centered diagram of μ as $-k, \ldots, -2, -1, 0, 1, 2, \ldots, k$ from left to right, so the center column is column 0. Label the rows $1, 2, 3, \ldots$ from top to bottom. We define $\nu = F(\mu)$ by dissecting the centered diagram of μ into a sequence of disjoint L-shaped pieces (described below), and letting the parts of ν be the number of cells in each piece. The first L-shaped piece consists of all cells in column 0 together with all cells to the right of column 0 in row 1. The second L-shaped piece consists of all cells in column -1 together with all cells left of column -1 in row 1. The third piece consists of the unused cells in column 1 (so row 1 is excluded) together with all cells right of column 1 in row 2. The fourth piece consists of the unused cells in column -2 together with all cells left of column -2 in row 2. We proceed similarly, working outward in both directions from the center column, cutting off L-shaped pieces that alternately move up and right, then up and left (see Figure 5.3).

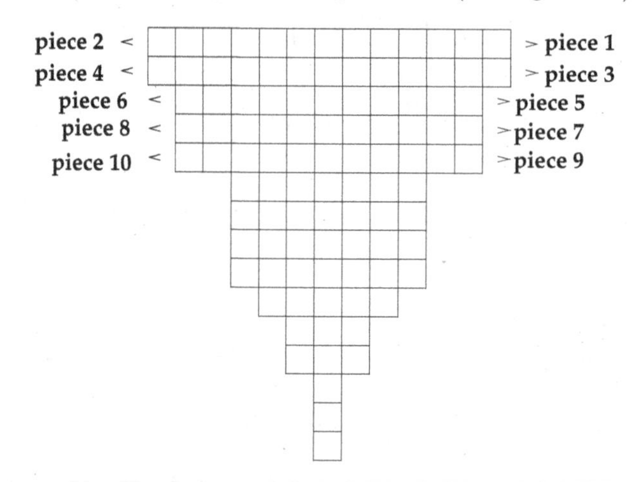

FIGURE 5.3
Sylvester's partition bijection.

We see from the geometric construction that the size of each L-shaped piece is strictly less than the size of the preceding piece. It follows that $F(\mu) = \nu = (\nu_1 > \nu_2 > \cdots)$ is indeed an element of DisPar. Furthermore, since $|\mu|$ is the sum of the sizes of all the L-shaped pieces, the map $F : \mathrm{OddPar} \to \mathrm{DisPar}$ is area-preserving. We must also check that

F is a *bijection* by constructing a map $G : \text{DisPar} \to \text{OddPar}$ that is the two-sided inverse of F.

To see how to define G, let us examine more closely the dimensions of the L-shaped pieces that appear in the definition of $F(\mu)$. Note that each L-shaped piece consists of a corner square, a *vertical portion* of zero or more squares below the corner, and a *horizontal portion* of zero or more squares to the left or right of the corner. Let y_0 be the number of cells in column 0 of the centered diagram of μ (so $y_0 = \ell(\mu)$). For all $i \geq 1$, let x_i be the number of cells in the horizontal portion of the $(2i-1)$th L-shaped piece for μ. For all $i \geq 0$, let y_i be the number of cells in the vertical portion of the $2i$th L-shaped piece for μ. For example, in Figure 5.3 we have $(y_0, y_1, y_2, \ldots) = (15, 11, 8, 6, 1, 0, 0, \ldots)$ and $(x_1, x_2, \ldots) = (6, 5, 3, 2, 1, 0, 0, \ldots)$. Note that for all $i \geq 1$, $y_{i-1} > y_i$ whenever $y_{i-1} > 0$, and $x_i > x_{i+1}$ whenever $x_i > 0$. Moreover, by the symmetry of the centered diagram of μ and the definition of F, we see that

$$\begin{aligned} \nu_1 &= y_0 + x_1, & \nu_2 &= x_1 + y_1, \\ \nu_3 &= y_1 + x_2, & \nu_4 &= x_2 + y_2, \\ \nu_5 &= y_2 + x_3, & \nu_6 &= x_3 + y_3, \end{aligned}$$

and, in general,

$$\nu_{2i-1} = y_{i-1} + x_i \quad \text{for all } i \geq 1; \qquad \nu_{2i} = x_i + y_i \quad \text{for all } i \geq 1. \tag{5.12}$$

To compute $G(\nu)$ for $\nu \in \text{DisPar}$, we need to solve the preceding system of equations for x_i and y_i, given the part lengths ν_j. Noting that ν_k, x_k, and y_k must all be zero for large enough indices k, we can solve for each variable by taking the alternating sum of all the given equations from some point forward. This forces us to define

$$\begin{aligned} y_i &= \nu_{2i+1} - \nu_{2i+2} + \nu_{2i+3} - \nu_{2i+4} + \cdots \quad \text{for all } i \geq 0; \\ x_i &= \nu_{2i} - \nu_{2i+1} + \nu_{2i+2} - \nu_{2i+3} + \cdots \quad \text{for all } i \geq 1. \end{aligned}$$

It is readily verified that these choices of x_i and y_i do indeed satisfy the equations $\nu_{2i-1} = y_{i-1} + x_i$ and $\nu_{2i} = x_i + y_i$. Furthermore, because the nonzero parts of ν are distinct, the required inequalities ($y_{i-1} > y_i$ whenever $y_{i-1} > 0$, and $x_i > x_{i+1}$ whenever $x_i > 0$) also hold. Now that we know the exact shape of each L-shaped piece, we can fit the pieces together to recover the centered diagram of $\mu = G(\nu) \in \text{OddPar}$. For example, given $\nu = (9, 8, 5, 3, 1, 0, 0, \ldots)$, we compute

$$\begin{aligned} y_0 &= 9 - 8 + 5 - 3 + 1 = 4 \\ x_1 &= 8 - 5 + 3 - 1 = 5 \\ y_1 &= 5 - 3 + 1 = 3 \\ x_2 &= 3 - 1 = 2 \\ y_2 &= 1. \end{aligned}$$

Using this data to reconstitute the centered diagram, we find that $G(\nu) = (11, 7, 5, 3)$. To finish, note that bijectivity of F follows from the fact that, for each $\nu \in \text{DisPar}$, the system of equations in (5.12) has exactly one solution for the unknowns x_i and y_i.

5.51. Glaisher's Bijection. We define a map $H : \text{DisPar} \to \text{OddPar}$ as follows. Each integer $k \geq 1$ can be written uniquely in the form $k = 2^e c$, where $e \geq 0$ and c is odd. Given $\nu \in \text{DisPar}$, we replace each part k in ν by 2^e copies of the part c (where $k = 2^e c$, as above). Sorting the resulting odd numbers into decreasing order gives us an element $H(\nu)$ in OddPar such that $|H(\nu)| = |\nu|$. For example,

$$\begin{aligned} H(15,12,10,8,6,3,1) &= \text{sort}(15,3,3,3,3,5,5,1,1,1,1,1,1,1,1,3,3,3,1) \\ &= (15,5,5,3,3,3,3,3,3,3,1,1,1,1,1,1,1,1,1). \end{aligned}$$

The inverse map K : OddPar $\to$ DisPar is defined as follows. Consider a partition $\mu \in$ OddPar. For each odd number c that appears as a part of μ, let $n = n(c) \geq 1$ be the number of times c occurs in μ. We can write n uniquely as a sum of distinct powers of 2 (this is the base-2 expansion of the integer n). Say $n = 2^{d_1} + 2^{d_2} + \cdots + 2^{d_s}$. We replace the n copies of c in μ by parts of size $2^{d_1}c$, $2^{d_2}c$, ..., $2^{d_s}c$. These parts are distinct from one another (since $d_1, \ldots, d_s$ are distinct), and they are also distinct from the parts obtained in the same way from other odd values of c appearing as parts of μ. Sorting the parts thus gives a partition $K(\mu) \in$ DisPar. For example,

$$K(7,7,7,7,7,3,3,3,3,3,3,1,1,1) = \text{sort}(28,7,12,6,2,1) = (28,12,7,6,2,1).$$

It is readily verified that $H \circ K$ and $K \circ H$ are identity maps.

Glaisher's bijection generalizes to prove the following theorem.

5.52. Theorem: Glaisher's Partition Identity. For all $d \geq 2$ and $N \geq 0$, the number of partitions of N where no part repeats d or more times equals the number of partitions of N with no part divisible by d.

Proof. For fixed d, let A be the set of partitions where no part repeats d or more times, and let B be the set of partitions with no part divisible by d. It suffices to describe weight-preserving maps $H : A \to B$ and $K : B \to A$ such that $H \circ K$ and $K \circ H$ are identity maps. We define K by analogy with what we did above for the case $d = 2$. Fix $\mu \in B$. For each c that appears as a part of μ, let $n = n(c)$ be the number of times the part c occurs in μ. Write n in base d as

$$n = \sum_{k=0}^{s} a_k d^k \quad \text{where } 0 \leq a_k < d \text{ for all } k,$$

and $n, a_0, \ldots, a_s$ all depend on c. To construct $K(\mu)$, we replace the n copies of c in μ by a_0 copies of d^0c, a_1 copies of d^1c, ..., a_k copies of d^kc, ..., and a_s copies of d^sc. One checks that the resulting partition does lie in the codomain A, using the fact that no part c of μ is divisible by d.

To compute $H(\nu)$ for $\nu \in A$, note that each part m in ν can be written uniquely in the form $m = d^kc$ for some $k \geq 0$ and some $c = c(m)$ not divisible by d. Adding up all such parts of ν that have the same value of c produces an expression of the form $\sum_{k\geq 0} a_k d^k c$, where $0 \leq a_k < d$ by definition of A. To get $H(\nu)$, we replace all these parts by $\sum_{k\geq 0} a_k d^k$ copies of the part c, for every possible c not divisible by d. Comparing the descriptions of H and K, it follows that these two maps are inverses. □

5.53. Remark: Rogers–Ramanujan Identities. A vast multitude of partition identities have been discovered, which are similar in character to the one we just proved. Two especially famous examples are the *Rogers–Ramanujan Identities*. The first such identity says that, for all N, the number of partitions of N into parts congruent to 1 or 4 modulo 5 equals the number of partitions of N into distinct parts $\nu_1 > \nu_2 > \cdots > \nu_k > 0$ such that $\nu_i - \nu_{i+1} \geq 2$ for i in the range $1 \leq i < k$. The second identity says that, for all N, the number of partitions of N into parts congruent to 2 or 3 modulo 5 equals the number of partitions of N into distinct parts $\nu_1 > \nu_2 > \cdots > \nu_k > 0 = \nu_{k+1}$ such that $\nu_i - \nu_{i+1} \geq 2$ for i in the range $1 \leq i \leq k$. One can seek algebraic and bijective proofs for these and other identities. Proofs of both types are known for the Rogers–Ramanujan Identities, but the bijective proofs are all quite complicated.

5.17 Euler's Pentagonal Number Theorem

We have seen that $\prod_{i=1}^{\infty}(1+z^i)$ is the generating function for partitions with distinct parts, whereas $\prod_{i=1}^{\infty}(1-z^i)^{-1}$ is the generating function for all integer partitions. This section investigates the infinite product $\prod_{i=1}^{\infty}(1-z^i)$, which is the multiplicative inverse for the partition generating function. The next theorem gives the power series expansion for this infinite product, which has a surprisingly high fraction of coefficients equal to zero. This expansion leads to a remarkable recursion for $p(n)$, the number of partitions of n.

5.54. Euler's Pentagonal Number Theorem.

$$\begin{aligned}\prod_{i=1}^{\infty}(1-z^i) &= 1+\sum_{n=1}^{\infty}(-1)^n[z^{n(3n-1)/2}+z^{n(3n+1)/2}] \\ &= 1-z-z^2+z^5+z^7-z^{12}-z^{15}+z^{22}+z^{26}-z^{35}-z^{40}+\cdots.\end{aligned}$$

Proof. Consider the set DisPar of integer partitions with distinct parts, weighted by area. For $\mu \in \text{DisPar}$, define the *sign* of μ to be $(-1)^{\ell(\mu)}$. By modifying the proof of Theorem 5.48 to include these signs, we obtain

$$\prod_{i=1}^{\infty}(1-z^i) = \sum_{\mu\in\text{DisPar}}(-1)^{\ell(\mu)}z^{|\mu|}.$$

We now define an ingenious weight-preserving, sign-reversing involution I on DisPar (due to Franklin). Given a partition $\mu = (\mu_1 > \mu_2 > \cdots > \mu_s) \in \text{DisPar}$, let $a \geq 1$ be the largest index such that the part sizes $\mu_1, \mu_2, \ldots, \mu_a$ are consecutive integers, and let $b = \mu_s$ be the smallest part of μ. Figure 5.4 shows how a and b can be found by visual inspection of the Ferrers diagram of μ. For most partitions μ, we define I as follows. If $a < b$, let $I(\mu)$ be the partition obtained by decreasing the first a parts of μ by 1 and adding a new part of size a to the end of μ. If $a \geq b$, let $I(\mu)$ be the partition obtained by removing the last part of μ (of size b) and increasing the first b parts of μ by 1 each. See the examples in Figure 5.4. I is weight-preserving and sign-reversing, since $I(\mu)$ has one more part or one fewer part than μ. It is also routine to check that $I(I(\mu)) = \mu$. Thus we can cancel out all the pairs of objects $\{\mu, I(\mu)\}$.

It may seem at first glance that we have canceled *all* the objects in DisPar! However, there are some choices of μ where the definition of $I(\mu)$ in the previous paragraph fails to produce a partition with distinct parts. Consider what happens in the overlapping case $a = \ell(\mu)$. If $b = a+1$ in this situation, the prescription for creating $I(\mu)$ leads to a partition whose smallest two parts both equal a. On the other hand, if $b = a$, the definition of $I(\mu)$ fails because there are not enough parts left to increment by 1 after dropping the smallest part of μ. In all other cases, the definition of I works even when $a = \ell(\mu)$. We see now that there are two classes of partitions that cannot be canceled by I (see Figure 5.5). First, there are partitions of the form $(2n, 2n-1, \ldots, n+1)$, which have length n and area $n(3n+1)/2$, for all $n \geq 1$. Second, there are partitions of the form $(2n-1, 2n-2, \ldots, n)$, which have

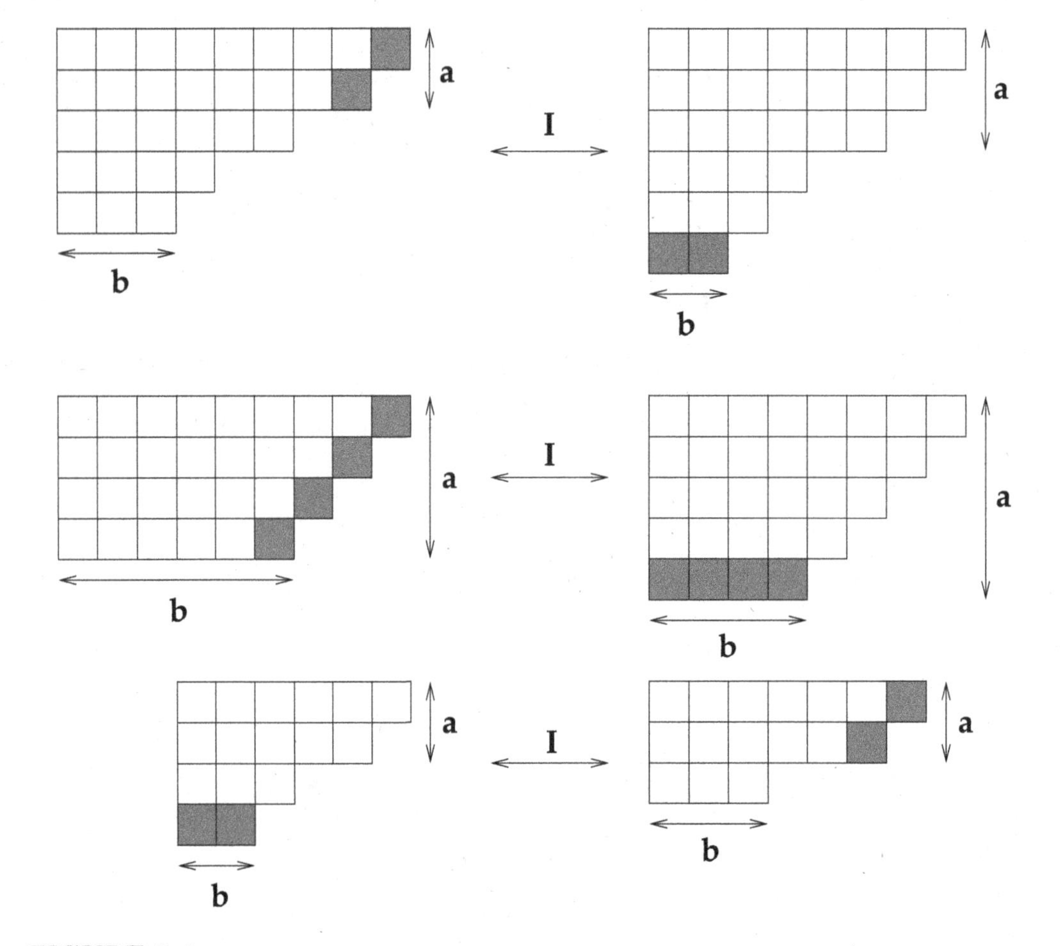

FIGURE 5.4
Franklin's partition involution.

length n and area $n(3n-1)/2$, for all $n \geq 1$. Furthermore, the empty partition is not canceled by I. Adding up these signed, weighted objects gives the right side of the equation in the theorem. □

We can now deduce Euler's recursion for counting integer partitions, which we stated without proof in §2.11.

5.55. Theorem: Partition Recursion. For every $n \in \mathbb{Z}$, let $p(n)$ be the number of integer partitions of n. The numbers $p(n)$ satisfy the recursion

$$\begin{aligned} p(n) &= p(n-1)+p(n-2)-p(n-5)-p(n-7)+p(n-12)+p(n-15)-\cdots \\ &= \sum_{k=1}^{\infty}(-1)^{k-1}[p(n-k(3k-1)/2)+p(n-k(3k+1)/2)] \end{aligned} \tag{5.13}$$

for all $n \geq 1$. The initial conditions are $p(0)=1$ and $p(n)=0$ for all $n<0$.

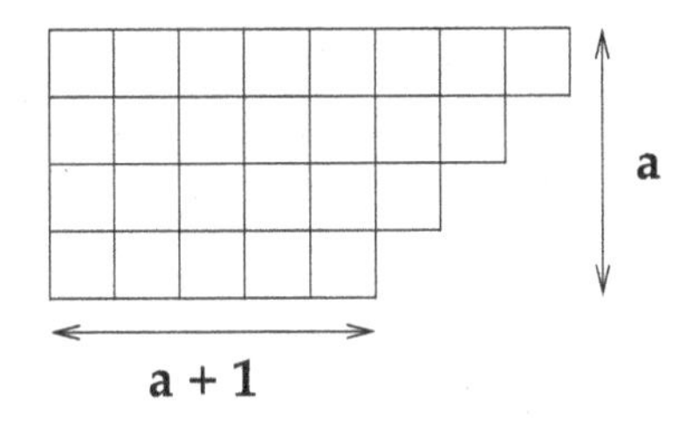

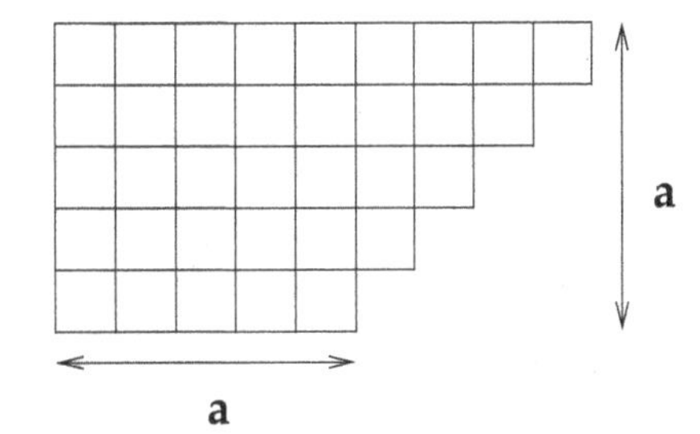

FIGURE 5.5
Fixed points of Franklin's involution.

Proof. We have proved the identities

$$\prod_{i=1}^{\infty} \frac{1}{1-z^i} = \sum_{\mu \in \mathrm{Par}} z^{|\mu|} = \sum_{n=0}^{\infty} p(n) z^n;$$
$$\prod_{i=1}^{\infty} (1-z^i) = 1 + \sum_{k=1}^{\infty} (-1)^k [z^{k(3k-1)/2} + z^{k(3k+1)/2}].$$

The product of the left sides of these two identities is 1, so the product of the right sides is also 1. Thus, for each $n \geq 1$, the coefficient of z^n in the product

$$\left(\sum_{n=0}^{\infty} p(n) z^n\right) \cdot \left(1 + \sum_{k=1}^{\infty} (-1)^k [z^{k(3k-1)/2} + z^{k(3k+1)/2}]\right)$$

is zero. This coefficient also equals $p(n) - p(n-1) - p(n-2) + p(n-5) + p(n-7) - \cdots$. Solving for $p(n)$ yields the recursion in the theorem. □

Summary

- *Definitions of Generating Functions.* Given a set S with a weight function $\mathrm{wt} : S \to \mathbb{Z}_{\geq 0}$ such that $a_n = |\{u \in S : \mathrm{wt}(u) = n\}|$ is finite for all $n \geq 0$, the ordinary generating function (OGF) for S is $\mathrm{GF}(S; z) = \sum_{n=0}^{\infty} a_n z^n$. The exponential generating function (EGF) for S is $\mathrm{EGF}(S; z) = \sum_{n=0}^{\infty} a_n z^n / n!$. Many generating functions can be viewed *analytically*, as analytic functions of a complex variable z specified by a power series that converges in a neighborhood of zero. All generating functions can be viewed *algebraically*, by defining the formal power series $\sum_{n=0}^{\infty} a_n z^n$ to serve as notation for the infinite sequence $(a_n : n \geq 0)$.

- *Convergence of Power Series.* Given a complex sequence (a_n), there exists a radius of convergence $R \in [0, \infty]$ such that for all $z \in \mathbb{C}$ with $|z| < R$, $\sum_{n=0}^{\infty} a_n z^n$ converges absolutely, and for all $z \in \mathbb{C}$ with $|z| > R$, this series diverges. The radius of convergence can be computed as

$$R = \lim_{n \to \infty} \left| \frac{a_n}{a_{n+1}} \right| \quad \text{and} \quad R = \frac{1}{\lim_{n \to \infty} \sqrt[n]{|a_n|}},$$

when these limits exist in $[0, \infty]$.

- *Power Series and Analytic Functions.* A power series $\sum_{n=0}^{\infty} a_n z^n$ with radius of convergence $R > 0$ defines an analytic function G on the open disk $D(0;R)$, and $a_n = G^{(n)}(0)/n!$. Conversely, any analytic function G defined on $D(0;R)$ has the power series expansion $G(z) = \sum_{n=0}^{\infty} G^{(n)}(0)z^n/n!$, valid for all z in this disk.

- *Examples of Power Series.* Some frequently used power series include:
  - Exponential Function: $e^z = \sum_{n=0}^{\infty} z^n/n!$ (valid for all $z \in \mathbb{C}$);
  - Logarithm Function: $\text{Log}(1+z) = \sum_{n=1}^{\infty}(-1)^{n-1}z^n/n$ (valid for $|z| < 1$);
  - Geometric Series: $(1-bz)^{-1} = \sum_{n=0}^{\infty} b^n z^n$ (valid for $|z| < 1/|b|$);
  - Negative Binomial Expansion: $(1-bz)^{-m} = \sum_{n=0}^{\infty} \binom{n+m-1}{n,m-1} b^n z^n$ (where $m \in \mathbb{Z}_{>0}$ and $|z| < 1/|b|$);
  - General Binomial Expansion: $(1+z)^r = \sum_{n=0}^{\infty}(r)\downarrow_n z^n/n!$ (valid for $|z| < 1$).

- *Operations on Formal Power Series.* Given $F = \sum_{n=0}^{\infty} a_n z^n$ and $G = \sum_{n=0}^{\infty} b_n z^n$, $F = G$ iff $a_n = b_n$ for all $n \geq 0$. Algebraic operations on formal power series are defined as follows:

$$F + G = \sum_{n=0}^{\infty}(a_n + b_n)z^n, \quad cF = \sum_{n=0}^{\infty}(ca_n)z^n,$$
$$z^k F = \sum_{n=k}^{\infty} a_{n-k}z^n, \qquad FG = \sum_{n=0}^{\infty}\left(\sum_{k=0}^{n} a_k b_{n-k}\right) z^n,$$
$$F' = \sum_{n=0}^{\infty}(n+1)a_{n+1}z^n, \qquad zF' = \sum_{n=0}^{\infty} na_n z^n.$$

  Similar formulas hold for analytic power series.

- *Solving Recursions via Generating Functions.* When a sequence (a_n) is defined by a recursion and initial conditions, the generating function $F(z) = \sum_{n=0}^{\infty} a_n z^n$ satisfies an algebraic equation determined by the recursion. To find this equation, subtract the terms of F coming from the initial conditions, apply the recursion, and simplify. One can often get an explicit expression for a_n by finding the partial fraction decomposition of F and using the Negative Binomial Theorem.

- *Evaluating Summations via Generating Functions.* Given $F(z) = \sum_{n=0}^{\infty} a_n z^n$, the coefficient of z^n in the power series expansion of $F(z)/(1-z)$ is $\sum_{k=0}^{n} a_k$. This coefficient can often be found using partial fraction decompositions and the Negative Binomial Theorem.

- *Generating Function for Derangements.* Let d_n be the number of derangements of $\{1, 2, \ldots, n\}$ (permutations with no fixed points). The EGF for these numbers is $\sum_{n=0}^{\infty} d_n z^n/n! = e^{-z}/(1-z)$.

- *Sum Rule for Weighted Sets.* If the weighted set S is the disjoint union of weighted sets $S_1, \ldots, S_k$ and $\text{wt}_{S_i}(u) = \text{wt}_S(u)$ for all $u \in S_i$, then

$$\text{GF}(S; z) = \text{GF}(S_1; z) + \text{GF}(S_2; z) + \cdots + \text{GF}(S_k; z).$$

- *Product Rule for Weighted Sets.* Given fixed $k \in \mathbb{Z}_{>0}$ and weighted sets $S, S_1, \ldots, S_k$, suppose every $u \in S$ can be constructed in exactly one way by choosing $u_1 \in S_1$, $u_2 \in S_2$, $\ldots$, $u_k \in S_k$, and assembling these choices in some prescribed manner. If $\text{wt}(u) = \sum_{j=1}^{k} \text{wt}(u_j)$ for all $u \in S$, then

$$\text{GF}(S; z) = \text{GF}(S_1; z) \cdot \text{GF}(S_2; z) \cdot \ldots \cdot \text{GF}(S_k; z).$$

- *Bijection Rule for Weighted Sets.* If $f : S \to T$ is a weight-preserving bijection between two weighted sets, then $\mathrm{GF}(S;z) = \mathrm{GF}(T;z)$.

- *Generating Functions for Trees.* The formal power series
$$\frac{1-\sqrt{1-4z}}{2} = \sum_{n=1}^{\infty} C_{n-1} z^n = \sum_{n=1}^{\infty} \frac{1}{2n-1}\binom{2n-1}{n-1,n} z^n$$
is the generating function for the following sets of weighted trees: (a) binary trees, weighted by number of vertices plus 1; (b) nonempty full binary trees, weighted by number of leaves; (c) ordered trees, weighted by number of vertices.

- *EGF Product Rule.* Given fixed $k \in \mathbb{Z}_{>0}$ and weighted sets $S, S_1, \ldots, S_k$, suppose every $u \in S$ can be constructed in exactly one way as follows. For $1 \le j \le k$, choose $u_j \in S_j$ of weight i_j; choose an anagram $w \in \mathcal{R}(1^{i_1}2^{i_2}\cdots k^{i_k})$; and assemble the chosen objects $u_1, \ldots, u_k, w$ in a prescribed manner. If $\mathrm{wt}(u) = i_1 + \cdots + i_k = \sum_{j=1}^{k} \mathrm{wt}(u_j)$ for all $u \in S$, then
$$\mathrm{EGF}(S;z) = \mathrm{EGF}(S_1;z) \cdot \mathrm{EGF}(S_2;z) \cdot \ldots \cdot \mathrm{EGF}(S_k;z).$$
This rule is useful for counting words with restrictions on the number of times each letter can be used.

- *Generating Functions for Stirling Numbers.* The recursions for Stirling numbers lead to formal differential equations for the associated generating functions. Solving these gives the identities
$$\sum_{n=0}^{\infty}\sum_{k=0}^{n} \frac{c(n,k)}{n!} t^k z^n = (1-z)^{-t}; \qquad \sum_{n=0}^{\infty}\sum_{k=0}^{n} \frac{S(n,k)}{n!} t^k z^n = e^{t(e^z-1)}.$$
Hence, $\sum_{n=k}^{\infty} S(n,k) z^n/n! = (e^z-1)^k/k!$ for all $k \ge 0$.

- *Partition Generating Functions.* Some generating function identities for integer partitions include:
$$\sum_{\mu \in \mathrm{Par}} t^{\ell(\mu)} z^{|\mu|} = \prod_{i=1}^{\infty} \frac{1}{1-tz^i} = \sum_{\mu \in \mathrm{Par}} t^{\mu_1} z^{|\mu|};$$
$$\sum_{\mu \in \mathrm{OddPar}} z^{|\mu|} = \prod_{k=1}^{\infty} \frac{1}{1-z^{2k-1}} = \prod_{i=1}^{\infty}(1+z^i) = \sum_{\mu \in \mathrm{DisPar}} z^{|\mu|}.$$
Sylvester's bijection dissects the centered Ferrers diagram of $\mu \in \mathrm{OddPar}$ into L-shaped pieces that give a partition in DisPar. Glaisher's bijection replaces each part $k = 2^e c$ in a partition $\nu \in \mathrm{DisPar}$ (where $e \ge 0$ and c is odd) by 2^e copies of c, giving a partition in OddPar.

- *Pentagonal Number Theorem.* Franklin proved
$$\prod_{i=1}^{\infty}(1-z^i) = 1 + \sum_{n=1}^{\infty}(-1)^n [z^{n(3n-1)/2} + z^{n(3n+1)/2}]$$
by an involution on signed partitions with distinct parts. The map moves boxes between the staircase at the top of the partition and the bottom row; this move cancels all partitions except the ones counted by the right side. Since $\prod_{i\ge 1}(1-z^i)$ is the inverse of the partition generating function, we deduce the partition recursion
$$p(n) = \sum_{k=1}^{\infty}(-1)^{k-1}[p(n-k(3k-1)/2) + p(n-k(3k+1)/2)].$$

Exercises

5-1. Let $S = \{\text{nitwit}, \text{blubber}, \text{oddment}, \text{tweak}\}$. (a) Find $\text{GF}(S;z)$ if the weight of a word is its length. (b) Find $\text{GF}(S;z)$ if the weight of a word is the number of distinct letters in it. (c) Find $\text{GF}(S;z)$ if the weight of a word is the number of distinct vowels in it.

5-2. Fix $q \in \mathbb{Z}_{>0}$, let S be the set of all words using the alphabet $\{1, 2, \ldots, q\}$, and let $\text{wt}(u)$ be the length of u for $u \in S$. (a) Find $\text{GF}(S;z)$. (b) Thinking of $\text{GF}(S;z)$ as an analytic power series, evaluate the series and state the radius of convergence.

5-3. Let $S = \bigcup_{n=0}^{\infty}\{1, 2, \ldots, n\}^n$, and let $\text{wt}(u)$ be the length of u for $u \in S$. (a) Find $F(z) = \text{GF}(S;z)$. What is the radius of convergence of F? (b) Viewing F as a formal power series, find $F'(z)$.

5-4. Fix $m > 0$, let T be the set of nonempty subsets of $\{1, 2, \ldots, m\}$, and let $\text{wt}(B)$ be the least element of B for $B \in T$. Find $\text{GF}(T;z)$ and then simplify the resulting series.

5-5. Let S be the set of five-card poker hands. Define the weight of a hand to be the number of aces in the hand. Find $\text{GF}(S;z)$.

5-6. For fixed $k \in \mathbb{Z}_{>0}$, describe a weighted set S of combinatorial objects such that $GF(S;z) = \sum_{n=0}^{\infty} n^k z^n$.

5-7. For fixed $m \in \mathbb{Z}_{>0}$, describe a weighted set S of combinatorial objects such that $\text{GF}(S;z) = (1-z)^{-m}$.

5-8. Let S be the set of all finite sequences of 0's and 1's, where the weight of a sequence is its length. Find $\text{EGF}(S;z)$, evaluate the resulting series, and state the radius of convergence.

5-9. (a) Let $S = \mathbb{Z}_{\geq 0}$, and let $\text{wt}(n) = n$ for $n \in S$. Find $\text{GF}(S;z)$ and $\text{EGF}(S;z)$. For each series, simplify fully and state the radius of convergence. (b) Repeat (a) using $S = \mathbb{Z}_{\geq 2}$. (c) Repeat (a) using $\text{wt}(n) = 3n$.

5-10. Suppose S is a weighted set with weight function $\text{wt} : S \to \mathbb{Z}_{\geq 0}$ and generating function $G(z) = \text{GF}(S;z)$. How does the generating function change if we use the new weight $\text{wt}'(u) = a \cdot \text{wt}(u) + b$, where $a, b \in \mathbb{Z}_{\geq 0}$ are fixed?

5-11. Find a weighted set S such that $\text{EGF}(S;z)$ has radius of convergence equal to zero.

5-12. Use the Ratio Test or Root Test to determine whether each series converges.
(a) $\sum_{n=0}^{\infty} 2^n n^3/n!$
(b) $\sum_{n=0}^{\infty}(3n)^n/n^{4n}$
(c) $\sum_{n=0}^{\infty} n^5/7^n$
(d) $\sum_{n=0}^{\infty}(1+1/n)^{2n^2}$

5-13. Find the radius of convergence of each complex power series.
(a) $\sum_{n=0}^{\infty} n^3 z^n$
(b) $\sum_{n=0}^{\infty} z^n/(n^2 4^n)$
(c) $\sum_{n=0}^{\infty} n^n z^n/n!$
(d) $\sum_{n=0}^{\infty}(z/n)^n$
(e) $\sum_{n=0}^{\infty} z^{3n}/(3n)!$

5-14. Let $a_n = \prod_{i=1}^{n}(2i-1)/(2i)$ for $n \geq 1$. Find the radius of convergence of $\sum_{n=1}^{\infty} a_n z^n$.

5-15. Find the radius of convergence of $J(z) = \sum_{k=0}^{\infty} \frac{(-1)^k z^{2k+1}}{k!(k+1)!2^{2k+1}}$.

5-16. For fixed $k \in \mathbb{Z}_{>0}$ and $b \in \mathbb{C}_{\neq 0}$, find the radius of convergence of each complex power series.
(a) $\sum_{n=0}^{\infty} n^k b^n z^n$
(b) $\sum_{n=0}^{\infty} n^k b^n z^n/n!$

(c) $\sum_{n=0}^{\infty} n^n b^n z^n/n!$
(d) $\sum_{n=0}^{\infty}(n!)^k z^n/(kn)!$

5-17. In Example 5.10, verify the power series representations for $\sin z$, $\cos z$, $\sinh z$, and $\cosh z$ by taking derivatives.

5-18. (a) Use power series to prove $e^{iz} = \cos z + i\sin z$ for all $z \in \mathbb{C}$. (b) Use power series to prove $\cos z = (e^{iz} + e^{-iz})/2$ for all $z \in \mathbb{C}$. (c) Find and prove a formula for $\sin z$ similar to the one in (b).

5-19. Define formal versions of e^z, $\sin z$, $\cos z$, $\sinh z$, $\cosh z$, and $\text{Log}(1+z)$ by viewing the series expansions in §5.3 as formal power series. Compute the formal derivatives of these six formal power series.

5-20. Use the Extended Binomial Theorem to write the first seven terms of each power series: (a) $(1-z)^{-5}$; (b) $(1+z)^{5/2}$; (c) $(1-8z)^{1/2}$; (d) $(1+z^2)^{2/3}$.

5-21. Expand each function as a power series and determine the radius of convergence: (a) $z(1-3z)^{-7}$; (b) $\sqrt{1+z}$; (c) $1/\sqrt{1-z^2}$; (d) $(1+5z)^{1/3}$.

5-22. Prove $(1+z)^{-1/2} = \sum_{n=0}^{\infty}\binom{2n}{n}(-z/4)^n$ for $|z| < 1$.

5-23. Find the coefficient of z^{10} in each power series: (a) $(1-z)^{-3}$; (b) $(1+2z)^{-5}$; (c) $(1-z^2)^8$; (d) $1/(1-7z+10z^2)$.

5-24. Suppose F and G are analytic on $D(0;R)$, and define $P(z) = F(z)G(z)$ for $z \in D(0;R)$. Use the Product Rule for Derivatives and induction to prove that for all $n \in \mathbb{Z}_{\geq 0}$,

$$P^{(n)}(z) = \sum_{k=0}^{n}\binom{n}{k}F^{(k)}(z)G^{(n-k)}(z).$$

5-25. Prove parts (a), (c), (d), and (g) of Theorem 5.14.

5-26. Suppose $F(z) = \sum_{n=0}^{\infty} a_n z^n$ for $z \in D(0;R)$. Find power series representations for all antiderivatives of F, which are functions $G : D(0;R) \to \mathbb{C}$ satisfying $G' = F$.

5-27. (a) Given $F(z) = \sum_{n=0}^{\infty} a_n z^n$, what is the coefficient of z^n in $F^{(k)}(z)$? (b) Let zD denote the operation of taking a derivative with respect to z and then multiplying by z. What is the coefficient of z^n in $(zD)^k(F)$?

5-28. (a) What is the sequence of coefficients corresponding to the formal power series z^k? (b) Verify that the product of the formal power series z^k and z^m is z^{k+m}. (c) Verify that parts (c) and (d) of Definition 5.15 are special cases of part (e) of that definition.

5-29. Suppose F and G are formal power series, and c is a scalar. Prove the following rules for formal derivatives. (a) The Sum Rule: $(F+G)' = F' + G'$. (b) The Scalar Rule: $(cF)' = c(F')$. (c) The Product Rule: $(F \cdot G)' = F' \cdot G + F \cdot G'$. (d) The Power Rule: For $n \in \mathbb{Z}_{>0}$, $(F^n)' = nF^{n-1} \cdot F'$.

5-30. Consider the formal power series $F(z) = 1 - bz$ and $G(z) = \sum_{n=0}^{\infty} b^n z^n$. Use the definition of multiplication of formal power series to confirm that $F(z)G(z) = 1$. This justifies the notation $G(z) = 1/(1-bz)$.

5-31. Solve each recursion using generating functions. (a) $a_n = 2a_{n-1}$ for $n \geq 1$ and $a_0 = 1$. (b) $a_n = 2a_{n-1}$ for $n \geq 1$ and $a_0 = 3$. (c) $a_n = 2a_{n-1} + 1$ for $n \geq 1$ and $a_0 = 0$.

5-32. (a) Given $a_n = 3a_{n-1}$ for $n \geq 1$ and $a_0 = 2$, solve for a_n. (b) Given $a_n = 3a_{n-1} + 3n$ for $n \geq 1$ and $a_0 = 2$, solve for a_n. (c) Given $a_n = 3a_{n-1} + 3^n$ for $n \geq 1$ and $a_0 = 2$, solve for a_n.

5-33. Solve the recursion $a_n = 6a_{n-1} - 8a_{n-2} + g(n)$ for $n \geq 2$ with initial conditions $a_0 = 0$, $a_1 = 2$ for the following choices of $g(n)$: (a) $g(n) = 0$; (b) $g(n) = 1$; (c) $g(n) = 2^n$; (d) $g(n) = n4^n$.

5-34. Define $a_0 = 1$, $a_1 = 2$, and $a_n = a_{n-1} + 6a_{n-2}$ for $n \geq 2$. Use generating functions to solve for a_n.

5-35. Define $a_0 = 3$, $a_1 = 0$, and $a_n = 2a_{n-1} - a_{n-2} - n$ for $n \geq 2$. Use generating functions to solve for a_n.

5-36. Define $a_0 = 0$, $a_1 = 1$, and $a_n = 5a_{n-1} - 6a_{n-2} + 2^n$ for $n \geq 2$. Use generating functions to solve for a_n.

5-37. Define $a_0 = b$, $a_1 = c$, and $a_n = -6a_{n-1} - 5a_{n-2} + (-1)^n$ for $n \geq 2$. Use generating functions to solve for a_n.

5-38. Define *Fibonacci numbers* by $f_0 = 0$, $f_1 = 1$, and $f_n = f_{n-1} + f_{n-2}$ for all $n \geq 2$. Use generating functions to derive a closed formula for f_n.

5-39. Define *Lucas numbers* by setting $L_0 = 1$, $L_1 = 3$, and $L_n = L_{n-1} + L_{n-2}$ for all $n \geq 2$. Use generating functions to find a closed formula for L_n.

5-40. Solve the recursion $a_n = -3a_{n-1} + 2a_{n-2} + 6a_{n-3} - a_{n-4} - 3a_{n-5}$ for $n \geq 5$ with initial conditions $a_k = k$ for $0 \leq k < 5$.

5-41. Repeat the previous exercise with initial conditions $a_k = 3$ for $0 \leq k < 5$.

5-42. Use generating functions to solve the recursion $a_n = ba_{n-1} + c$ and initial condition $a_0 = d$, where $c \neq 0$ and $b \neq 1$.

5-43. Use generating functions to solve the recursion $a_n = ba_{n-1} + nb^n$ and initial condition $a_0 = c$, where $b \neq 0$.

5-44. Use generating functions to prove Theorem 2.66.

5-45. Use generating functions to compute the following sums. (a) $\sum_{k=0}^{n} k^2$; (b) $\sum_{k=0}^{n} k5^k$; (c) $\sum_{k=0}^{n} \binom{k+3}{3}$.

5-46. For fixed $r \neq 1$, compute $\sum_{k=0}^{n}(k+1)r^k$ with generating functions.

5-47. Evaluate $\sum_{k=0}^{n} k^4$.

5-48. Prove $\sum_{i+j=n} C_{2i}C_{2j} = 4^n C_n$, where C_n denotes a Catalan number.

5-49. Let $F(z) = \sum_{n=0}^{\infty} a_n z^n$. Find the generating function for each sequence (s_n).
(a) $s_n = \sum_{k=0}^{n} a_k (-1)^{n-k}$
(b) $s_n = \sum_{k=0}^{n} a_k/(n-k)!$
(c) $s_n = \sum_{k=0}^{n}(n-k)^2 a_k$
(d) $s_n = \sum_{i+j+k=n} a_i a_j a_k$

5-50. Find $\sum_{k=0}^{n}(-1)^k k^2$.

5-51. Use generating functions to find $\sum_{k=0}^{n} 3^k/(k!(n-k)!)$.

5-52. Suppose $b_0 = 1$ and $b_n = b_0 + b_1 + \cdots + b_{n-1} + 1$ for all $n \geq 1$. Find $\sum_{n=0}^{\infty} b_n z^n$.

5-53. Suppose $(c_n : n \in \mathbb{Z})$ satisfies $c_0 = 0$, $c_1 = 1$, and $c_n = (c_{n-1} + c_{n+1})/L$ for all $n \in \mathbb{Z}$, where $L \in \mathbb{R}_{>0}$ is a constant. Find an explicit formula for c_n.

5-54. Sums of Powers of Integers. Recall the notation $(a){\uparrow_n} = a(a+1)(a+2)\cdots(a+n-1)$. Prove: for all $k, n \in \mathbb{Z}_{>0}$,

$$1^k + 2^k + \cdots + n^k = \sum_{j=0}^{k} \frac{S(k+1, j+1)}{j+1}(-1)^{k+j}(n+1){\uparrow_{j+1}}.$$

5-55. Suppose $f : S \to T$ is a bijection such that for some $k \in \mathbb{Z}_{>0}$ and all $u \in S$, $\mathrm{wt}_T(f(u)) = \mathrm{wt}_S(u) + k$. How are $\mathrm{GF}(S; z)$ and $\mathrm{GF}(T; z)$ related?

5-56. Fix a positive integer m, and let S be the set of all subsets of $\{1, 2, \ldots, m\}$. (a) Define

$\text{wt}(B) = |B|$ for $B \in S$. Use the Product Rule for Weighted Sets to show that $\text{GF}(S; z) = (1 + z)^m$ (cf. Example 5.3). (b) Now define $\text{wt}(B) = \sum_{i \in B} i$ for $B \in S$. Find $\text{GF}(S; z)$.

5-57. (a) Let S be the set of m-letter words using the alphabet $\{0, 1, \ldots, k-1\}$. Let the weight of a word be the sum of all letters in the word. Find $\text{GF}(S; z)$. (b) Repeat (a) using the alphabet $\{1, 2, \ldots, k\}$.

5-58. Find a formula for $\sum_{n=0}^{\infty} a_n z^n$, where a_n is the number of positive integer solutions to $x_1 + x_2 + x_3 + x_4 + x_5 = n$ such that x_1, x_3, and x_5 are odd, $x_2 \geq 4$, and $x_3 \leq 9$.

5-59. An urn contains r red balls, b blue balls, and w white balls. Let S be the set of all multisets of balls that can be drawn from this urn, weighted by the number of balls in the multiset. Find $\text{GF}(S; z)$.

5-60. Suppose we distribute 30 identical lollipops to 10 children so that each child receives between 2 and 4 lollipops. How many ways can this be done?

5-61. A candy store has 6 chocolate truffles, 8 coconut creams, 5 caramel nougats, and an unlimited supply of cashew clusters. (a) How many ways can a box of ten pieces of candy be chosen from this inventory? (b) Repeat (a) assuming there must be at least one of each type of candy.

5-62. Let a_n be the number of ways to pay someone n cents using pennies, nickels, dimes, and quarters. Find a formula for $\sum_{n=0}^{\infty} a_n z^n$.

5-63. Suppose we have three types of stamps that cost r cents, s cents, and t cents (respectively). Find the generating function for a_n, the number of ways to pay n cents postage using a multiset of these stamps.

5-64. State and prove versions of the Sum Rule and Bijection Rule for EGFs.

5-65. Find an EGF for words in the alphabet $\{1, 2, \ldots, k\}$ where every symbol in the alphabet is used at least twice.

5-66. Find an EGF for words in the alphabet $\{1, 2, \ldots, k\}$ where every symbol is used between three and eight times.

5-67. Find an EGF for words in the alphabet $\{a, b, \ldots, z\}$ where each vowel can be used at most twice.

5-68. How many n-digit strings using the symbols $\{0, 1, 2, 3\}$ have an odd number of 0's and an odd number of 1's?

5-69. Count the n-digit strings using the symbols $\{0, 1, 2, 3\}$ in which the total number of 0's and 1's is odd.

5-70. An *ordered set partition* of X is a list $(B_1, \ldots, B_k)$ of distinct sets such that $\{B_1, \ldots, B_k\}$ is a set partition of X. Fix $k > 0$, and let S be the set of ordered set partitions of $\{1, 2, \ldots, n\}$ (for varying $n \geq k$) into exactly k blocks where every block has odd size. Find the EGF for S.

5-71. Repeat the previous exercise assuming the size of each block is even (and positive).

5-72. Let S be the set of k-element subsets of $\mathbb{Z}_{\geq 0}$, weighted by the largest element in S. (a) Find $\text{GF}(S; z)$. (b) What happens if we weight a subset by its smallest element?

5-73. Fix $k \in \mathbb{Z}_{>0}$. Use the Sum and Product Rules for Weighted Sets to find the generating function for the set of all compositions with k parts, weighted by the sum of the parts.

5-74. Compute the images of these partitions under Sylvester's Bijection 5.50:
(a) $(15, 5^2, 3^7, 1^9)$; (b) $(7^5, 3^6, 1^3)$; (c) $(11, 7, 5, 3)$; (d) (9^8); (e) $(2n-1, 2n-3, \ldots, 5, 3, 1)$. (The notation a^b denotes b parts equal to a.)

5-75. Compute the images of these partitions under the inverse of Sylvester's Bijection.
(a) $(15, 12, 10, 8, 6, 3, 1)$

(b) $(28, 12, 7, 6, 2, 1)$
(c) $(11, 7, 5, 3)$
(d) $(21, 17, 16, 13, 11, 9, 8, 3, 2, 1)$
(e) $(n, n-1, \ldots, 3, 2, 1)$

5-76. Compute the images of these partitions under Glaisher's Bijection 5.51:
(a) $(9, 8, 5, 3, 1)$; (b) $(28, 12, 7, 6, 2, 1)$; (c) $(11, 7, 5, 3)$; (d) $(21, 17, 16, 13, 11, 9, 8, 3, 2, 1)$.

5-77. Compute the images of these partitions under the inverse of Glaisher's Bijection:
(a) $(15, 5^2, 3^7, 1^9)$; (b) $(13^2, 11^3, 7^4, 5, 3^2, 1^3)$ (c) $(11, 7, 5, 3)$; (d) (9^8); (e) (1^n).

5-78. Which partitions map to themselves under Glaisher's Bijection? Which partitions map to themselves under the generalized bijection in Theorem 5.52?

5-79. Let H and K be the bijections in the proof of Theorem 5.52. (a) Find $H((25, 17, 17, 10, 9, 6, 6, 5, 2, 2))$ for $d = 3, 4, 5$. (b) Find $K((8^{10}, 7^7, 2^{20}, 1^{30}))$ for $d = 3, 5, 6$.

5-80. Calculate the image of each partition under Franklin's Involution (§5.17):
(a) $(17, 16, 15, 14, 13, 10, 8, 7, 4)$; (b) $(17, 16, 15, 14, 13, 10, 8)$; (c) $(n, n-1, \ldots, 3, 2, 1)$;
(d) (n).

5-81. Find the generating function for the set of all integer partitions that satisfy each restriction below (weighted by area): (a) all parts are divisible by 3; (b) all parts are distinct and even; (c) odd parts appear at most twice; (d) each part is congruent to 1 or 4 mod 7; (e) for each $i > 0$, there are at most i parts of size i.

5-82. Give combinatorial interpretations for the coefficients in the following formal power series: (a) $\prod_{i\geq 1}(1-z^{5i})^{-1}$; (b) $\prod_{i\geq 0}(1+z^{6i+1})(1+z^{6i+5})$; (c) $\prod_{i\geq 2}(1-z^{2i}+z^{4i})/(1-z^i)$.

5-83. (a) Show that the first Rogers–Ramanujan Identity (see Remark 5.53) can be written $\prod_{n=0}^{\infty} \frac{1}{(1-z^{5n+1})(1-z^{5n+4})} = 1 + \sum_{k=1}^{\infty} \frac{z^{k^2}}{(1-z)(1-z^2)\cdots(1-z^k)}$. (b) Find a similar formulation of the second Rogers–Ramanujan Identity. (c) Verify the Rogers–Ramanujan Identities for partitions of $n = 12$ by explicitly listing all the partitions satisfying the relevant restrictions.

5-84. Prove by induction that a nonempty full binary tree with a leaves has $a-1$ non-leaf vertices.

5-85. Let f be the bijection in Figure 5.1. Compute $f(T)$, where T is the binary tree in Figure 2.11.

5-86. Let g be the bijection shown in Figure 5.2. Verify that the number of vertices in $g(t)$ equals the number of leaves in t, for each full binary tree t.

5-87. (a) Describe the inverse of the bijection g shown in Figure 5.2. (b) Calculate the image under g^{-1} of the ordered tree shown here.

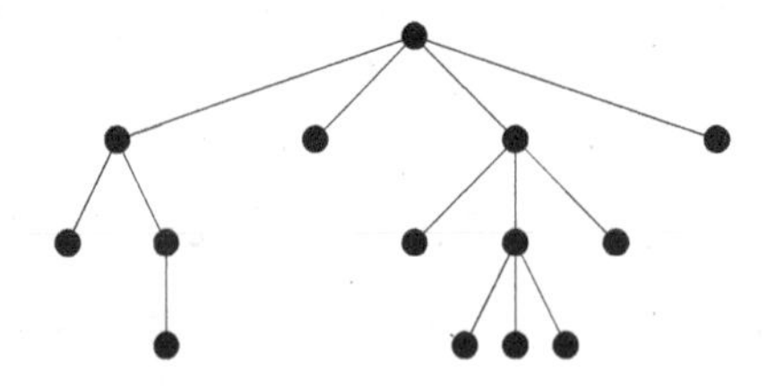

5-88. List all full binary trees with five leaves, and compute the image of each tree under the map g in Figure 5.2.

5-89. Give an algebraic proof of Theorem 5.52 using formal power series.

5-90. (a) Carefully verify that the maps H and K in 5.51 are two-sided inverses. (b) Repeat part (a) for the maps H and K in Theorem 5.52.

5-91. (a) Verify that the partition $(2n, 2n-1, \ldots, n+1)$ (one of the fixed points of Franklin's

Involution) has area $n(3n+1)/2$. (b) Verify that the partition $(2n-1, 2n-2, \ldots, n)$ has area $n(3n-1)/2$.

5-92. (a) Find the generating function for the set of all Dyck paths, where the weight of a path ending at (n,n) is n. (b) A *marked* Dyck path is a Dyck path in which one step (north or east) has been circled. Find the generating function for marked Dyck paths.

5-93. Recall that $\sum_{n=0}^{\infty}\sum_{k=0}^{n}\frac{S(n,k)}{n!}t^k z^n = e^{t(e^z-1)}$. Use partial differentiation of this generating function to find generating functions for: (a) the set of set partitions where one block in the partition has been circled; (b) the set of set partitions where one element of one block in the partition has been circled.

5-94. Let S be the set of paths that start at $(0,0)$ and take horizontal steps (right 1, up 0), vertical steps (right 0, up 1), and diagonal steps (right 1, up 1). By considering the final step of a path, find an equation satisfied by $G(z) = \mathrm{GF}(S; z)$ and solve for $G(z)$, taking the weight of a path ending at (c,d) to be: (a) the number of steps in the path; (b) $c+d$; (c) c.

5-95. For fixed $k \geq 1$, find the generating function for integer partitions with: (a) k nonzero parts; (b) k nonzero distinct parts. (c) Deduce summation formulas for the infinite products $\prod_{i=1}^{\infty}(1-z^i)^{-1}$ and $\prod_{i=1}^{\infty}(1+z^i)$.

5-96. A *ternary tree* is either $\emptyset$ or a 4-tuple $(\bullet, t_1, t_2, t_3)$, where each t_i is a ternary tree. Find an equation satisfied by the generating function for the set of ternary trees, weighted by number of vertices.

5-97. Let S be the set of ordered trees where every vertex has at most two children, weighted by the number of vertices. (a) Use the Sum and Product Rules to find an equation satisfied by $\mathrm{GF}(S; z)$. (b) Solve this equation for $\mathrm{GF}(S; z)$. (c) How many trees in S have seven vertices?

5-98. Prove that the number of integer partitions of n in which no even part appears more than once equals the number of partitions of n in which no part appears four or more times.

5-99. Prove that the number of integer partitions of n that have no part equal to 1 and no parts that differ by 1 equals the number of partitions of n in which no part appears exactly once.

5-100. (a) Find an infinite product that is the generating function for integer partitions with odd, distinct parts. (b) Show that the generating function for self-conjugate partitions (i.e., partitions such that $\lambda' = \lambda$) is $1 + \sum_{k=1}^{\infty} z^{k^2}/((1-z^2)(1-z^4)\cdots(1-z^{2k}))$. (c) Find an area-preserving bijection between the sets of partitions in (a) and (b), and deduce an associated formal power series identity.

5-101. Evaluate $\sum_{k=1}^{\infty} z^k(1-z)^{-k}$.

5-102. How many integer partitions of n have the form $((i+1)^k, i^j)$ for some $i, j, k > 0$?

5-103. Dobinski's Formula. Prove that the Bell numbers (see Definition 2.44) satisfy $B(n) = e^{-1}\sum_{k=0}^{\infty}(k^n/k!)$ for $n \geq 0$.

5-104. (a) Use an involution on the set $\mathrm{Par} \times \mathrm{DisPar}$ to give a combinatorial proof of the identity $\prod_{n=1}^{\infty}\frac{1}{1-z^n}\prod_{n=1}^{\infty}(1-z^n) = 1$. (b) More generally, for $S \subseteq \mathbb{Z}_{>0}$, prove combinatorially that $\prod_{n\in S}\frac{1}{1-z^n}\prod_{n\in S}(1-z^n) = 1$.

5-105. Let $d(n,k)$ be the number of derangements in S_n with k cycles. Find a formula for $\sum_{n=0}^{\infty}\sum_{k=0}^{n}\frac{d(n,k)}{n!}t^k z^n$.

5-106. Let $S_1(n,k)$ be the number of set partitions of $\{1,2,\ldots,n\}$ into k blocks where no block consists of a single element. Find a formula for $\sum_{n=0}^{\infty}\sum_{k=0}^{n}\frac{S_1(n,k)}{n!}t^k z^n$.

5-107. Let $\mathrm{Par}(n)$ be the set of integer partitions of n. Show that, for all $n \geq 0$,

$$(-1)^n|\,\mathrm{OddPar}\cap\mathrm{DisPar}\cap\mathrm{Par}(n)| = |\{\mu\in\mathrm{Par}(n) : \ell(\mu)\text{ is even}\}| - |\{\mu\in\mathrm{Par}(n) : \ell(\mu)\text{ is odd}\}|.$$

5-108. Involution Proof of Euler's Partition Recursion. (a) For fixed $n \geq 1$, prove $\sum_{j\in\mathbb{Z}}(-1)^j p(n-(3j^2+j)/2) = 0$ by verifying that the following map I is a sign-reversing involution with no fixed points. The domain of I is the set of pairs (j,λ) with $j \in \mathbb{Z}$ and $\lambda \in \mathrm{Par}(n-(3j^2+j)/2)$. To define $I(j,\lambda)$, consider two cases. If $\ell(\lambda)+3j \geq \lambda_1$, set $I(j,\lambda) = (j-1,\mu)$ where μ is formed by preceding the first part of λ by $\ell(\lambda)+3j$ and then decrementing all parts by 1. If $\ell(\lambda)+3j < \lambda_1$, set $I(j,\lambda) = (j+1,\nu)$ where ν is formed by deleting the first part of λ, incrementing the remaining nonzero parts of λ by 1, and appending an additional $\lambda_1 - 3j - \ell(\lambda) - 1$ parts equal to 1. (b) Given $n = 21$, $j = 1$, $\lambda = (5,5,4,3,2)$, compute $I(j,\lambda)$ and verify that $I(I(j,\lambda)) = (j,\lambda)$.

5-109. We say that an integer partition λ *extends* a partition μ iff for all k, k occurs in λ at least as often as k occurs in μ; otherwise, λ *avoids* μ. Suppose $\{\mu^i : i \geq 1\}$ and $\{\nu^i : i \geq 1\}$ are two sequences of distinct, nonzero partitions such that for all finite $S \subseteq \mathbb{Z}_{>0}$, $\sum_{i\in S}|\mu^i| = \sum_{i\in S}|\nu^i|$. (a) Prove that for every n, the number of partitions of n that avoid every μ^i equals the number of partitions of n that avoid every ν^i. (b) Show how Theorem 5.52 can be deduced from (a). (c) Use (a) to prove that the number of partitions of n into parts congruent to 1 or 5 mod 6 equals the number of partitions of n into distinct parts not divisible by 3.

5-110. Prove that absolute convergence of a complex series implies convergence of that series.

5-111. Prove that if $\sum_{n=0}^{\infty} c_n$ exists in $\mathbb{C}$, then $\lim_{n\to\infty} c_n = 0$. Is the converse true?

5-112. Comparison Test for Nonnegative Series. Suppose $a_n, b_n \in \mathbb{R}_{\geq 0}$ satisfy $a_n \leq b_n$ for all $n \geq n_0$. Prove: if $\sum_{n=0}^{\infty} b_n$ converges, then $\sum_{n=0}^{\infty} a_n$ converges. [*Hint:* Use the Monotone Convergence Theorem, which states that a weakly increasing sequence of real numbers that is bounded above converges to a real limit.]

5-113. Prove the Ratio Test. [*Hint:* If $L < 1$, compare $|c_n|$ to an appropriate geometric series.]

5-114. Prove the Root Test. [*Hint:* If $L < 1$, compare $|c_n|$ to an appropriate geometric series.]

5-115. (a) Show that $\sum_{n=0}^{\infty} z^{3n}/(3n)! = (1/3)e^z + (2/3)\cos(z\sqrt{3}/2)e^{-z/2}$. (b) Find similar formulas for $\sum_{n=0}^{\infty} z^{3n+1}/(3n+1)!$ and $\sum_{n=0}^{\infty} z^{3n+2}/(3n+2)!$.

5-116. Use complex power series to prove: for all $z, w \in \mathbb{C}$, $e^{z+w} = e^w e^z$. [*Hint:* Hold $w \in \mathbb{C}$ fixed and view both sides of the identity as analytic functions of z. Show that the power series expansions of these two functions have the same coefficient of z^n for every n.]

5-117. For $r \in \mathbb{C}$, let $G_r(z) = (1+z)^r = e^{r\,\mathrm{Log}(1+z)}$ for $z \in D(0;1)$. (a) Prove by induction on $n \in \mathbb{Z}_{>0}$ that $G_n(z)$ equals the product of n copies of $1+z$. (b) Prove that for all $n \in \mathbb{Z}_{>0}$, $G_{-n}(z)$ equals the product of n copies of $1/(1+z)$.

5-118. Define $G_r(z)$ as in the previous exercise. (a) Prove: for all $r, s \in \mathbb{C}$ and $z \in D(0;1)$, $G_{r+s}(z) = G_r(z)G_s(z)$. (b) Prove $G_r'(z) = r(1+z)^{r-1}$. (c) Prove $G_r^{(n)}(z) = (r){\downarrow_n}\,(1+z)^{r-n}$ for $n \in \mathbb{Z}_{>0}$.

5-119. For r in any commutative ring R containing $\mathbb{Q}$, define a formal power series version of $(1+z)^r$ by setting

$$(1+z)^r = \sum_{n=0}^{\infty} \frac{(r){\downarrow_n}}{n!} z^n.$$

(a) Verify that the formal derivative of $(1+z)^r$ is $r(1+z)^{r-1}$. (b) Compute the higher formal derivatives of $(1+z)^r$.

5-120. Prove the formal power series identity $(1+z)^{r+s} = (1+z)^r(1+z)^s$. [*Hint:* First find a combinatorial proof of the identity $(x+y)\downarrow_n = \sum_{k=0}^{n} \binom{n}{k} (x)\downarrow_k (y)\downarrow_{n-k}$, where $n \in \mathbb{Z}_{\geq 0}$ and x, y are formal variables.]

5-121. (a) Prove: for all $r \in \mathbb{Z}_{>0}$, the formal power series $(1+z)^r$ equals the formal product of r copies of $(1+z)$. (b) Prove: for all $r \in \mathbb{Z}_{>0}$, the formal power series $(1+z)^{-r}$ equals the formal product of r copies of $1/(1+z) = \sum_{n=0}^{\infty}(-1)^n z^n$.

5-122. For any commutative ring R, let $R[[z]]$ denote the set of formal power series with coefficients in R. Show that $R[[z]]$ is a commutative ring using the operations in Definition 5.15.

5-123. Let F and G be formal power series. Prove that the formal derivative operation satisfies the Sum Rule $(F+G)' = F' + G'$ and the Product Rule $(F \cdot G)' = F' \cdot G + F \cdot G'$.

5-124. Given a formal power series $F = (a_n : n \geq 0)$, let $F(0) = a_0$ be the constant coefficient of F. Prove: for all $n \geq 0$, $n!a_n = F^{(n)}(0)$.

5-125. Recursion for Divide-and-Conquer Algorithms. Many algorithms use a divide-and-conquer approach in which a problem of size n is divided into a subproblems of size n/b, and the solutions to these subproblems are then combined in time cn^k to give the solution to the original problem. Letting $T(n)$ be the time needed to solve a problem of size n, $T(n)$ satisfies the recursion $T(n) = aT(n/b) + cn^k$ and initial condition $T(1) = d$ (where $a, b, c, d > 0$ and $k \geq 0$ are given constants). Assume for simplicity that n ranges over powers of b. (a) Find a recursion and initial condition satisfied by $S(m) = T(b^m)$, where m ranges over $\mathbb{Z}_{\geq 0}$. (b) Use generating functions to solve the recursion in (a). Deduce that, for some constant C and large enough n,

$$T(n) \leq \begin{cases} Cn^k & \text{if } a < b^k \text{ (combining time dominates);} \\ Cn^k \log_2 n & \text{if } a = b^k \text{ (dividing and combining times balance);} \\ Cn^{\log_b a} & \text{if } a > b^k \text{ (time to solve subproblems dominates).} \end{cases}$$

5-126. Merge Sort. Suppose we need to sort a given sequence of integers $x_1, \ldots, x_n$ into increasing order. Consider the following recursive method: if $n = 1$, the sequence is already sorted. For $n > 1$, divide the list into two halves, sort each half recursively, and merge the resulting sorted lists. Let $T(n)$ be the time needed to sort n objects using this algorithm. Find a recursion satisfied by $T(n)$, and use the previous exercise to show that $T(n) \leq Cn \log_2 n$ for some constant C. (You may assume n ranges over powers of 2.)

5-127. Fast Binary Multiplication. (a) Given $x = ak + b$ and $y = ck + d$ (where $a, b, c, d, k \in \mathbb{Z}_{\geq 0}$), verify that $xy = (ak+b)(ck+d) = ack^2 + bd + ((a+b)(c+d) - ac - bd)k$. Take $k = 2^n$ in this identity to show that one can multiply two $2n$-bit numbers by recursively computing three products of n-bit numbers and doing several binary additions. (b) Find a recursion describing the number of bit operations needed to multiply two n-bit numbers by the recursive method suggested in (a). (c) Solve the recursion in (b) to determine the time complexity of this recursive algorithm (you may assume n is a power of 2).

Notes

For more applications of generating functions to combinatorial problems, see [9, 121, 132]. Two older references are [85, 107]. For an introduction to the vast subject of partition identities, the reader may consult [5, 97]. Sylvester's bijection appears in [123], Glaisher's bijection in [47], and Franklin's involution in [39]. The Rogers–Ramanujan identities are discussed in [101, 112]; Garsia and Milne gave the first bijective proof of these identities [43, 44].

6

Ranking, Unranking, and Successor Algorithms

In computer science, one often needs algorithms to loop through a set of combinatorial objects, generate a random object from a set, or store information about such objects in a linear array. This chapter studies techniques for creating such algorithms. We begin by discussing *ranking* and *unranking* algorithms, which implement explicit bijections mapping combinatorial objects to integers and vice versa. Later we examine *successor* algorithms, which traverse all the objects in a given set in a particular order by going from the current object to the next object in a systematic way. We develop automatic methods for translating counting arguments based on the Sum Rule and Product Rule into ranking algorithms and successor algorithms. Combinatorial recursions derived using these rules can be treated similarly, leading to recursively defined ranking algorithms and successor algorithms. Before beginning, we introduce the following notation that will be used constantly.

6.1. Definition: The Set $[\![n]\!]$. For each positive integer n, let $[\![n]\!]$ denote the n-element set $\{0, 1, \ldots, n-1\}$.

6.1 Introduction to Ranking and Successor Algorithms

This section introduces the general notions of ranking, unranking, and successor algorithms. Suppose S is a finite set consisting of n objects. A *ranking algorithm* for S is a specific procedure implementing a particular bijection $\mathrm{rk} : S \to [\![n]\!]$. This procedure takes an object $x \in S$ as input, performs some computation on x, and outputs an integer $\mathrm{rk}(x)$ called the *rank* of x. An *unranking algorithm* for S is a specific procedure implementing a bijection $\mathrm{unrk} : [\![n]\!] \to S$. Here the input is an integer j between 0 and $n-1$, and the output $\mathrm{unrk}(j)$ is the object in S whose rank is j. There can be many different ranking and unranking algorithms for a given set S, since there are many bijections between S and the set $[\![n]\!]$.

We can use ranking and unranking algorithms to solve the problems mentioned in the chapter opening, as follows. To loop through all objects in the n-element set S, loop through all integers j in the range 0 to $n-1$. For each such j, use an unranking algorithm to compute the object $\mathrm{unrk}(j) \in S$, and perform whatever additional processing is needed for this object. To select a random object from S, first select a random integer $j \in [\![n]\!]$ (using any standard algorithm for generating pseudorandom numbers), then return the object $\mathrm{unrk}(j)$. To store information about the objects in S in an n-element linear array, use a ranking algorithm to find the position $\mathrm{rk}(x)$ in the array where information about the object $x \in S$ should be stored.

Successor algorithms provide another, potentially more efficient approach to the task of looping through all the objects in S. We begin with a particular total ordering of the n objects in S, say $x_0 < x_1 < \cdots < x_{n-1}$. A successor algorithm for S (relative to this ordering) consists of three subroutines called `first`, `last`, and `next`. The `first` subroutine returns the object x_0. The `last` subroutine returns the object x_{n-1}. The `next` subroutine takes as input an object $x_i \in S$ where $0 \leq i < n-1$, and returns the object x_{i+1} immediately following x_i in the given total ordering of S. We can loop through all objects in S using the following pseudocode.

```
x=first(S);
process the object x;
while (x is not last(S)) do
{ x=next(x,S);
  process the object x;
}
```

Suppose we have built a ranking algorithm and an unranking algorithm for S, which implement a bijection $\text{rk} : S \to [\![n]\!]$ and its inverse $\text{unrk} : [\![n]\!] \to S$. We have an associated total ordering of S defined by setting $\text{unrk}(0) < \text{unrk}(1) < \cdots < \text{unrk}(n-1)$. We can use these algorithms to create a successor algorithm for S (relative to this ordering) as follows. Define $\texttt{first}(S) = \text{unrk}(0)$; define $\texttt{last}(S) = \text{unrk}(n-1)$; and define $\texttt{next}(x, S) = \text{unrk}(\text{rk}(x) + 1)$. Thus the problem of finding successor algorithms can be viewed as a special case of the problem of finding ranking and unranking algorithms. However, there may be more efficient ways to implement the `next` subroutine besides ranking x, adding 1, and unranking the new value. In the second half of this chapter, we develop techniques for building successor algorithms that never explicitly compute the ranks of the objects involved.

6.2 The Bijective Sum Rule

We begin our study of ranking and unranking by revisiting the fundamental counting rules from Chapter 1. Our first rule lets us assemble ranking (or unranking) maps for two disjoint finite sets to obtain a ranking (or unranking) map for the union of these sets.

6.2. The Bijective Sum Rule for Two Sets. Let S and T be disjoint finite sets. Given bijections $f : S \to [\![n]\!]$ and $g : T \to [\![m]\!]$, there is a bijection $h : S \cup T \to [\![n+m]\!]$ defined by

$$h(x) = \begin{cases} f(x) & \text{if } x \in S; \\ g(x) + n & \text{if } x \in T. \end{cases}$$

The inverse of h is given by

$$h^{-1}(j) = \begin{cases} f^{-1}(j) & \text{if } 0 \leq j < n; \\ g^{-1}(j-n) & \text{if } n \leq j < n+m. \end{cases}$$

We sometimes use the notation $h = f + g$ and $h^{-1} = f^{-1} + g^{-1}$.

We omit the detailed verification of this rule, except to remark that the disjointness of S and T is critical when showing that h is a well-defined function and that the claimed inverse of h is injective.

Observe that the *order* in which we combine the bijections makes a difference: although $S \cup T = T \cup S$ and $n + m = m + n$, the bijection $f + g : S \cup T \to [\![n+m]\!]$ is not the same as the bijection $g + f : T \cup S \to [\![m+n]\!]$. Intuitively, the ranking bijection $f + g$ assigns earlier ranks to elements of S (using f to determine these ranks) and assigns later ranks to elements of T (using g); $g + f$ does the opposite. Similarly, the unranking map $f^{-1} + g^{-1}$

generates a list in which elements of S occur first, followed by elements of T; $g^{-1} + f^{-1}$ lists elements of T first, then S.

Iterating the Bijective Sum Rule for Two Sets leads to the following general version of this rule.

6.3. The Bijective Sum Rule. Assume $S_1, \ldots, S_k$ are pairwise disjoint finite sets with $|S_i| = n_i$, and suppose we are given bijections $f_i : S_i \to [\![n_i]\!]$ for $1 \leq i \leq k$. Let $S = S_1 \cup \cdots \cup S_k$ and $n = |S| = n_1 + \cdots + n_k$. There is a bijection $f : S \to [\![n]\!]$ defined by

$$f(x) = f_i(x) + \sum_{j<i} n_j \quad \text{for } x \in S_i.$$

For $j \in [\![n]\!]$, we may compute $f^{-1}(j)$ as follows. There exists a unique i such that $1 \leq i \leq k$ and $n_1 + \cdots + n_{i-1} \leq j < n_1 + \cdots + n_i$; we have

$$f^{-1}(j) = f_i^{-1}(j - [n_1 + \cdots + n_{i-1}]).$$

We introduce notation $f = \sum_{i=1}^{k} f_i$ and $f^{-1} = \sum_{i=1}^{k} f_i^{-1}$.

As before, we omit the formal proof of the Bijective Sum Rule. One can give a direct proof that the maps in question are bijections, or use an induction argument involving the Bijective Sum Rule for Two Sets to show that $\sum_{i=1}^{k} f_i = \sum_{i=1}^{k-1} f_i + f_k$. Compare to §1.15.

Intuitively, $f_1 + \cdots + f_k$ is the ranking map that assigns elements of S_1 to positions 0 through $n_1 - 1$ using f_1, assigns elements of S_2 to positions n_1 through $n_1 + n_2 - 1$ using f_2, etc. The unranking map $f_1^{-1} + \cdots + f_k^{-1}$ generates a listing of S in which objects in S_1 occur first, then objects in S_2, and so on.

For programming purposes, it is helpful to have a pseudocode version of the Bijective Sum Rule. Here and below, we use notation such as `rank(x,S_i)` to denote the ranking function for the set S_i applied to the input object x. The pseudocode appears in Figure 6.1.

6.3 The Bijective Product Rule for Two Sets

Our next goal is to develop a bijective version of the Product Rule that can be used to build ranking and unranking maps. We approach the full rule gradually through a sequence of lemmas. The reader may already know the following theorem about integer division: for all integers a and m with $m > 0$, there exist unique integers q and r with $a = qm + r$ and $0 \leq r < m$. We call q and r the *quotient* and *remainder* when a is divided by m, writing $q = a \operatorname{div} m$ and $r = a \bmod m$. Our first lemma is a bijective reformulation of integer division that is a key ingredient in the creation of ranking and unranking algorithms.

6.4. Lemma. For all $n, m \in \mathbb{Z}_{>0}$, there is a bijection $p_{n,m} : [\![n]\!] \times [\![m]\!] \to [\![nm]\!]$ defined by

$$p_{n,m}(i, j) = im + j \quad \text{for all } i \in [\![n]\!] \text{ and } j \in [\![m]\!].$$

For all $a \in [\![nm]\!]$,

$$p_{n,m}^{-1}(a) = (a \operatorname{div} m, a \bmod m).$$

Proof. If one already knows the Integer Division Theorem quoted above, then the surjectivity and injectivity of $p_{n,m}$ quickly follow from the existence and uniqueness of the quotient and remainder when a is divided by m. Here we give a different proof in which we appeal to

```
Assumptions: S is the disjoint union of finite sets S_1,...,S_k; for all i,
 we already know |S_i| and ranking and unranking functions for S_i.

define procedure rank(x,S):
{ value=0;
  for i=1 to k do
  { if (x is in S_i) then return value+rank(x,S_i);
    else value=value+|S_i|;
  }
  return "error: x is not in S";
}

define procedure unrank(j,S):
{ for i=1 to k do
  { if (0<=j<|S_i|) then return unrank(j,S_i);
    else j=j-|S_i|;
  }
  return "error: j is outside the valid range for S";
}
```

FIGURE 6.1
Pseudocode for the Bijective Sum Rule.

the Bijective Sum Rule to construct the bijection $p_{n,m}$ automatically. For each $i \in [\![n]\!]$, define a set $S_i = \{i\} \times [\![m]\!]$. Also define a ranking bijection $f_i : S_i \to [\![m]\!]$ by letting $f_i(i,j) = j$ for all $j \in [\![m]\!]$. Since $S \times T$ is the union of the pairwise disjoint sets $S_0, \ldots, S_{n-1}$, the Bijective Sum Rule assembles these bijections to produce a bijection $f = \sum_{i=0}^{n-1} f_i$ from $[\![n]\!] \times [\![m]\!]$ onto $[\![nm]\!]$. To finish, we need only check that $p_{n,m}$ (as defined in the lemma statement) equals the function f. Given $(i,j) \in [\![n]\!] \times [\![m]\!]$, we calculate $f(i,j)$ by noting that $(i,j) \in S_i$ and $|S_k| = m$ for all $k < i$, so

$$f(i,j) = f_i(i,j) + \sum_{0 \leq k < i} |S_k| = j + im = p_{n,m}(i,j).$$

The Bijective Sum Rule guarantees the invertibility of f. Writing what this means in more detail, we see that any $a \in [\![nm]\!]$ can be written in exactly one way in the form $a = im + j$, where $i \in [\![n]\!]$ and $j \in [\![m]\!]$. This is exactly the existence and uniqueness assertion of the Integer Division Theorem (for the specified range of a's). So the claimed formula for $p_{n,m}^{-1}$ is well-defined and correct. □

Notice that the previous proof implicitly contains an algorithm for computing the quotient and remainder when a is divided by m. Indeed, because $p_{n,m}^{-1}$ is the unranking map $\sum_{i=0}^{n-1} f_i^{-1}$, we can use the algorithm in Figure 6.1 to find $p_{n,m}^{-1}(a) = (a \operatorname{div} m, a \bmod m)$. Since each $|S_i| = m$ in this case, the unranking algorithm proceeds by repeatedly subtracting m from a until a value less than m is reached. That value is the remainder $a \bmod m$, whereas the number of copies of m that we subtracted is the quotient $a \operatorname{div} m$. This is none other than the *iterated subtraction* algorithm for performing integer division. We hasten to point out that other, much more efficient division algorithms exist (such as the *long division* technique learned in grade school).

Here is another technical point. Although multiplication of numbers is commutative, the maps $p_{n,m}$ and $p_{m,n}$ are not equal when $n \neq m$, since their domains are not the same. Consider the bijection $s : [\![n]\!] \times [\![m]\!] \to [\![m]\!] \times [\![n]\!]$ given by $s(i,j) = (j,i)$. The maps $p_{n,m}$ and $p_{m,n} \circ s$ have the same domain but are still unequal, because $p_{n,m}(i,j) = im + j$, whereas $p_{m,n}(s(i,j)) = jn + i$. We could have built the map $p_{m,n} \circ s$ from the Bijective Sum Rule by writing $[\![n]\!] \times [\![m]\!]$ as the disjoint union of the slices $[\![n]\!] \times \{j\}$, for $j \in [\![m]\!]$.

To use the maps $p_{n,m}$ to create ranking algorithms for sets of combinatorial objects, we need the following version of the Bijection Rule.

6.5. The Bijection Rule for Ranking and Unranking Maps. Suppose $F : X \to Y$ is a bijection, and we know algorithms to compute F and F^{-1}. If $\text{rk} : Y \to [\![n]\!]$ is a ranking map for Y, then $\text{rk} \circ F$ is a ranking map for X. If $\text{unrk} : [\![n]\!] \to Y$ is an unranking map for Y, then $F^{-1} \circ \text{unrk}$ is an unranking map for X.

Proof. The assertions follow immediately from the facts that the composition of bijections is a bijection, and the inverse of a bijection is a bijection. □

We apply this rule in two stages to reach the Bijective Product Rule for Two Sets.

6.6. The Bijective Cartesian Product Rule for Two Sets. Let S and T be finite sets. Given ranking bijections $f : S \to [\![n]\!]$ and $g : T \to [\![m]\!]$, there is a bijection $p : S \times T \to [\![nm]\!]$ defined by

$$p(x,y) = f(x)m + g(y) \qquad \text{for all } x \in S \text{ and } y \in T.$$

For all $a \in [\![nm]\!]$,

$$p^{-1}(a) = (f^{-1}(a \operatorname{div} m), g^{-1}(a \bmod m)).$$

Proof. Define $F : S \times T \to [\![n]\!] \times [\![m]\!]$ by $F(x,y) = (f(x), g(y))$ for $x \in S$ and $y \in T$. One readily checks that F is a bijection with inverse $F^{-1}(i,j) = (f^{-1}(i), g^{-1}(j))$ for $i \in [\![n]\!]$ and $j \in [\![m]\!]$. We also have the bijection $p_{n,m} : [\![n]\!] \times [\![m]\!] \to [\![nm]\!]$ from the earlier lemma. Composing these bijections and using the formulas for $p_{n,m}$ and $p_{n,m}^{-1}$, we obtain the formulas for p and p^{-1} stated above. □

More generally, if $f : S \to A$ and $g : T \to B$ are bijections, the map $F : S \times T \to A \times B$ sending $(x,y) \in S \times T$ to $(f(x), g(y))$ is a bijection; we denote the map F by $f \times g$.

6.7. The Bijective Product Rule for Two Sets. Suppose S and T are finite sets such that $|S| = n$, $|T| = m$, and we know ranking and unranking algorithms for S and T. Suppose X is a set such that every object $x \in X$ can be constructed in exactly one way by first choosing an object $s \in S$, then choosing an object $t \in T$, and finally assembling these objects via some bijection $F : S \times T \to X$. There is a ranking bijection $\text{rk}_X : X \to [\![nm]\!]$ given by

$$\text{rk}_X(x) = \text{rk}_S(s)m + \text{rk}_T(t) \text{ where } F(s,t) = x.$$

The corresponding unranking bijection is given by

$$\text{unrk}_X(a) = F(\text{unrk}_S(a \operatorname{div} m), \text{unrk}_T(a \bmod m)) \text{ for } a \in [\![nm]\!].$$

Proof. This follows from the Bijection Rule by composing the previously constructed ranking and unranking maps for $S \times T$ with the bijections F^{-1} and F, respectively. □

Figure 6.2 gives pseudocode for the Bijective Product Rule for Two Sets.

```
Assumptions: S is an n-element set, T is an m-element set;
 we know a bijection F:S x T -> X and its inverse G:X -> S x T;
 we know ranking and unranking functions for S and T.

define procedure rank(x,X):
{ compute (s,t)=G(x);
  return rank(s,S)*|T|+rank(t,T);
}

define procedure unrank(a,X):
{ q=a div |T|;
  r=a mod |T|;
  s=unrank(q,S);
  t=unrank(r,T);
  return F(s,t);
}
```

FIGURE 6.2
Pseudocode for the Bijective Product Rule for Two Sets.

6.4 The Bijective Product Rule

This section extends the product rules from the previous section to products of k sets, for any fixed positive integer k. The key arithmetical ingredient is the following generalization of the product map $p_{n,m}$.

6.8. Lemma. For all $n_1, n_2, \ldots, n_k \in \mathbb{Z}_{>0}$, there is a bijection

$$p_{n_1,n_2,\ldots,n_k} : [\![n_1]\!] \times [\![n_2]\!] \times \cdots \times [\![n_k]\!] \to [\![n_1 n_2 \cdots n_k]\!]$$

given by

$$p_{n_1,\ldots,n_k}(c_1, c_2, \ldots, c_k) = c_1 n_2 n_3 \cdots n_k + c_2 n_3 \cdots n_k + \cdots + c_{k-1} n_k + c_k = \sum_{i=1}^{k} \left(c_i \prod_{j=i+1}^{k} n_j \right).$$

For all $a \in [\![n_1 n_2 \cdots n_k]\!]$, the following algorithm computes $p^{-1}_{n_1,\ldots,n_k}(a)$. For i looping from k down to 1, let $r_i = a \bmod n_i$, and then replace a by $a \operatorname{div} n_i$; then $p^{-1}_{n_1,\ldots,n_k}(a) = (r_1, r_2, \ldots, r_k)$.

Proof. We use induction on k to show that $p_{n_1,\ldots,n_k}$ is a bijection. When $k = 1$, p_{n_1} is the identity map on $[\![n_1]\!]$. When $k = 2$, the result follows from Lemma 6.4. Now assume $k > 2$ and the result is already known for products of $k - 1$ sets. Writing $n = n_1 n_2 \cdots n_k$ and $n' = n_2 n_3 \cdots n_k$, observe that

$$\begin{aligned} p_{n_1,\ldots,n_k}(c_1, \ldots, c_k) &= c_1(n_2 \cdots n_k) + \sum_{i=2}^{k} c_i \prod_{j=i+1}^{k} n_j \\ &= c_1 n' + p_{n_2,\ldots,n_k}(c_2, \ldots, c_k) \\ &= p_{n_1,n'}(c_1, p_{n_2,\ldots,n_k}(c_2, \ldots, c_k)). \end{aligned}$$

This means that $p_{n_1,\ldots,n_k}$ is the composition of the two bijections

$$\mathrm{id}_{[\![n_1]\!]} \times p_{n_2,\ldots,n_k} : [\![n_1]\!] \times ([\![n_2]\!] \times \cdots \times [\![n_k]\!]) \to [\![n_1]\!] \times [\![n']\!] \text{ and } p_{n_1,n'} : [\![n_1]\!] \times [\![n']\!] \to [\![n]\!], \tag{6.1}$$

so $p_{n_1,\ldots,n_k}$ is a bijection.

We verify the algorithm for $p^{-1}_{n_1,\ldots,n_k}$ by showing that applying this algorithm to $a = p_{n_1,\ldots,n_k}(c_1,\ldots,c_k)$ does recover $(c_1,\ldots,c_k)$. Again we use induction on k. The case $k=1$ is immediate. Fix $k>1$, and assume we have already verified the correctness of the algorithm for $p^{-1}_{n_1,\ldots,n_{k-1}}$. Observe that

$$p_{n_1,\ldots,n_k}(c_1,\ldots,c_k) = \sum_{i=1}^{k} c_i \prod_{i<j\leq k} n_j = n_k \left(\sum_{i=1}^{k-1} c_i \prod_{i<j<k} n_j \right) + c_k.$$

Since $0 \leq c_k < n_k$, this expression shows that the quotient and remainder when we divide a by n_k are

$$a \operatorname{div} n_k = \sum_{i=1}^{k-1} c_i \prod_{i<j\leq k-1} n_j = p_{n_1,\ldots,n_{k-1}}(c_1,\ldots,c_{k-1})$$

and $a \bmod n_k = c_k$, respectively. So the first division step successfully recovers c_k and replaces a by $p_{n_1,\ldots,n_{k-1}}(c_1,\ldots,c_{k-1})$. By induction hypothesis, the remaining divisions compute

$$p^{-1}_{n_1,\ldots,n_{k-1}}(p_{n_1,\ldots,n_{k-1}}(c_1,\ldots,c_{k-1})) = (c_1,\ldots,c_{k-1}).$$

So the algorithm for inverting $p_{n_1\ldots,n_k}$ is correct. □

6.9. Example. Suppose $(n_1,n_2,n_3,n_4,n_5) = (4,6,5,4,2)$, so $n = n_1n_2n_3n_4n_5 = 960$. Then

$$p(3,1,0,2,1) = 3\cdot(6\cdot5\cdot4\cdot2) + 1\cdot(5\cdot4\cdot2) + 0\cdot(4\cdot2) + 2\cdot(2) + 1 = 765.$$

To compute $p^{-1}(222)$, first divide 222 by $n_5 = 2$ to get $q_5 = 111$ and $r_5 = 0$. Then divide 111 by $n_4 = 4$ to get $q_4 = 27$ and $r_4 = 3$. Then divide 27 by $n_3 = 5$ to get $q_3 = 5$ and $r_3 = 2$. Then divide 5 by $n_2 = 6$ to get $q_2 = 0$ and $r_2 = 5$. Finally, divide 0 by $n_1 = 4$ to get $q_1 = 0$ and $r_1 = 0$. We conclude that $p^{-1}(222) = (0,5,2,3,0)$.

6.10. Remark. Here is another algorithm for computing $p^{-1}_{n_1,\ldots,n_k}(a) = (c_1,\ldots,c_k)$, based on (6.1), that recovers the sequence $(c_1,\ldots,c_k)$ from left to right. First, divide a by $n_2n_3\cdots n_k$ to obtain a quotient q_1 and a remainder r_1. Set $c_1 = q_1$, and recover $(c_2,\ldots,c_k)$ by recursively computing $p^{-1}_{n_2,\ldots,n_k}(r_1)$ in the same fashion.

6.11. Remark. If $n_1 = \cdots = n_k = b$ and $p^{-1}_{b,b,\ldots,b}(a) = (c_1,c_2,\ldots,c_k)$, then we have

$$a = p_{b,b,\ldots,b}(c_1,\ldots,c_k) = c_1b^{k-1} + c_2b^{k-2} + \cdots + c_{k-1}b + c_k.$$

Thus, the algorithm for $p^{-1}_{b,\ldots,b}$ computes the digits in the *base b representation* of a. The fact that $p_{b,\ldots,b}$ is a bijection means that every integer $a \in [\![b^k]\!]$ has a unique such expansion consisting of k *digits* c_i coming from the range $[\![b]\!] = \{0,1,\ldots,b-1\}$. Taking $b = 10$, we recover the familiar decimal representation of nonnegative integers. The maps $p^{-1}_{n_1,\ldots,n_k}$ provide generalizations of these representations in which the allowable digits in position i come from the set $[\![n_i]\!]$, and the *place value* of position i is $n_{i+1}\cdots n_k$.

6.12. The Bijective Cartesian Product Rule. Suppose $S_1, \ldots, S_k$ are finite sets, $|S_i| = n_i$, and $f_i : S_i \to [\![n_i]\!]$ is a ranking bijection for $1 \leq i \leq k$. There is a bijection $h : S_1 \times \cdots \times S_k \to [\![n_1 \cdots n_k]\!]$ defined by

$$h(x_1, \ldots, x_k) = p_{n_1,\ldots,n_k}(f_1(x_1), \ldots, f_k(x_k)) \quad \text{for all } x_i \in S_i.$$

For all $a \in [\![n_1 \cdots n_k]\!]$,

$$h^{-1}(a) = (f_1^{-1}(c_1), \ldots, f_k^{-1}(c_k)), \text{ where } (c_1, \ldots, c_k) = p^{-1}_{n_1,\ldots,n_k}(a).$$

Proof. This follows from Lemma 6.8 and the Bijection Rule for Ranking Maps, since the map $f_1 \times \cdots \times f_k : S_1 \times \cdots \times S_k \to [\![n_1]\!] \times \cdots \times [\![n_k]\!]$ sending $(x_1, \ldots, x_k)$ to $(f_1(x_1), \ldots, f_k(x_k))$ is a bijection. □

Finally, we can state the most general version of the Product Rule for ranking and unranking maps.

6.13. The Bijective Product Rule. Suppose $S_1, \ldots, S_k$ are finite sets with $|S_i| = n_i$, and we know ranking and unranking algorithms for each S_i. Suppose X is a set such that every object $x \in X$ can be constructed uniquely by choosing $x_i \in S_i$ for $1 \leq i \leq k$ and assembling these choices via some bijection $F : S_1 \times \cdots \times S_k \to X$. There is a ranking bijection $\mathrm{rk}_X : X \to [\![n_1 \cdots n_k]\!]$ given by

$$\mathrm{rk}_X(x) = p_{n_1,\ldots,n_k}(\mathrm{rk}_{S_1}(x_1), \ldots, \mathrm{rk}_{S_k}(x_k)) \text{ where } F(x_1, \ldots, x_k) = x.$$

The corresponding unranking bijection acts on $a \in [\![n_1 \cdots n_k]\!]$ via

$$\mathrm{unrk}_X(a) = F(\mathrm{unrk}_{S_1}(c_1), \ldots, \mathrm{unrk}_{S_k}(c_k)), \text{ where } (c_1, \ldots, c_k) = p^{-1}_{n_1,\ldots,n_k}(a).$$

A pseudocode implementation of this rule appears in Figure 6.3.

6.5 Ranking Words

In the next several sections, we give sample applications of the Bijective Sum Rule and Bijective Product Rule by constructing ranking and unranking maps for various sets of combinatorial objects.

6.14. Example: Four-Letter Words. Let S be the set of all four-letter words using the 26-letter alphabet $A = \{\mathrm{a}, \mathrm{b}, \ldots, \mathrm{z}\}$. We can think of S as the Cartesian product $A^4 = A \times A \times A \times A$ of four copies of A. Using the Bijective Cartesian Product Rule, we obtain a ranking map for S from a ranking map for A. The standard alphabetical ordering of letters induces a ranking bijection $\mathrm{rk}_A : A \to [\![26]\!]$ given by $\mathrm{rk}_A(\mathrm{a}) = 0$, $\mathrm{rk}_A(\mathrm{b}) = 1$, ..., $\mathrm{rk}_A(\mathrm{z}) = 25$. The corresponding ranking bijection for S is the map $\mathrm{rk}_S : S \to [\![26^4]\!]$ given by

$$\mathrm{rk}_S(w_1w_2w_3w_4) = p_{26,26,26,26}(\mathrm{rk}_A(w_1), \mathrm{rk}_A(w_2), \mathrm{rk}_A(w_3), \mathrm{rk}_A(w_4)).$$

For example,

$$\mathrm{rk}_S(\mathrm{goop}) = p_{26,26,26,26}(6, 14, 14, 15) = 6 \cdot 26^3 + 14 \cdot 26^2 + 14 \cdot 26^1 + 15 = 115,299;$$

$$\mathrm{rk}_S(\mathrm{pogo}) = p_{26,26,26,26}(15, 14, 6, 14) = 15 \cdot 26^3 + 14 \cdot 26^2 + 6 \cdot 26^1 + 14 = 273,274.$$

```
Assumptions: We know a bijection F:S_1 x ... x S_k -> X and its inverse
 G:X -> S_1 x ... x S_k; for all i, we know |S_i| and
 ranking and unranking functions for the finite set S_i.

define procedure rank(x,X):
{ compute (x_1,...,x_k)=G(x);
  prod=1;
  value=0;
  for i=k downto 1 do
  { value=value + rank(x_i,S_i)*prod;
    prod=prod*|S_i|;
  }
  return value;
}

define procedure unrank(a,X):
{ for i=k downto 1 do
  { q=a div |S_i|;
    r=a mod |S_i|;
    x_i=unrank(r,S_i);
    a=q;
  }
  return F(x_1,...,x_k);
}
```

FIGURE 6.3
Pseudocode for the Bijective Product Rule.

To compute the associated unranking map $\text{unrk}_S : [\![26^4]\!] \to S$ on $a \in [\![26^4]\!]$, first find $(c_1, c_2, c_3, c_4) = p^{-1}_{26,26,26,26}(a)$. Note that $c_1, \ldots, c_4$ are the digits in the base 26 representation of a. Return the answer $\text{unrk}_S(a) = \text{unrk}_A(c_1)\,\text{unrk}_A(c_2)\,\text{unrk}_A(c_3)\,\text{unrk}_A(c_4)$. For example, to unrank 200,000, we first calculate $p^{-1}_{26,26,26,26}(200,000) = (11, 9, 22, 8)$. Then

$$\text{unrk}_S(200,000) = \text{rk}_A^{-1}(11)\,\text{rk}_A^{-1}(9)\,\text{rk}_A^{-1}(22)\,\text{rk}_A^{-1}(8) = \text{ljwi}.$$

In general, if A is an m-letter alphabet with ranking map $\text{rk}_A : A \to [\![m]\!]$, then a ranking map for the set of k-letter words A^k is given by

$$\text{rk}(w_1w_2\cdots w_k) = p_{m,m,\ldots,m}(\text{rk}_A(w_1), \ldots, \text{rk}_A(w_k)) = \sum_{i=1}^{k} \text{rk}_A(w_i)m^{k-i}.$$

To unrank an integer $a \in [\![m^k]\!]$, find the base m representation $(c_1, \ldots, c_k) = p^{-1}_{m,\ldots,m}(a)$ and then replace each digit c_i by the letter $\text{rk}_A^{-1}(c_i)$.

6.15. Example: Three-Letter Words. Consider $A = \{\text{a}, \text{b}, \text{c}\}$ and $S = A^3$. Define $\text{rk}_A : A \to [\![3]\!]$ by $\text{rk}_A(\text{a}) = 0$, $\text{rk}_A(\text{b}) = 1$, and $\text{rk}_A(\text{c}) = 2$. For $w = w_1w_2w_2 \in S$, we have

$$\text{rk}_S(w) = 9\,\text{rk}_A(w_1) + 3\,\text{rk}_A(w_2) + \text{rk}_A(w_3),$$

where $\text{rk}_A(w_i)$ is the ith digit from the left in the base 3 expansion of $\text{rk}_S(w)$. Table 6.1

displays the 27 words in S and their corresponding ranks (given first in base 3 and then in base 10). Note that words are ranked in the same order that they would appear in a dictionary.

aaa ↔ 000=0	baa ↔ 100=9	caa ↔ 200=18
aab ↔ 001=1	bab ↔ 101=10	cab ↔ 201=19
aac ↔ 002=2	bac ↔ 102=11	cac ↔ 202=20
aba ↔ 010=3	bba ↔ 110=12	cba ↔ 210=21
abb ↔ 011=4	bbb ↔ 111=13	cbb ↔ 211=22
abc ↔ 012=5	bbc ↔ 112=14	cbc ↔ 212=23
aca ↔ 020=6	bca ↔ 120=15	cca ↔ 220=24
acb ↔ 021=7	bcb ↔ 121=16	ccb ↔ 221=25
acc ↔ 022=8	bcc ↔ 122=17	ccc ↔ 222=26

TABLE 6.1
Ranking of three-letter words.

More generally, a ranking map built using the Bijective Cartesian Product Rule will induce a certain *lexicographic ordering* of the set $S = S_1 \times \cdots \times S_k$. Informally, the ranking map rk_S treats the first component of the k-tuple as most significant and the last component as the least significant. If we unrank $0, 1, \ldots, |S| - 1$ in this order, we obtain a list that begins with all the k-tuples that have $\mathrm{rk}_{S_1}^{-1}(0)$ in the first component. Next we get all the k-tuples that have $\mathrm{rk}_{S_1}^{-1}(1)$ in the first component, and so on. Each sublist is also arranged lexicographically based on the values in the remaining components of the k-tuples.

We can describe the total ordering $\leq_{\mathrm{lex}}$ induced by rk_S more formally as follows. Each factor S_i has a total ordering $\leq_i$ determined by the ranking map rk_{S_i} (namely, for all $x_i, y_i \in S_i$, $x_i \leq_i y_i$ iff $\mathrm{rk}_{S_i}(x_i) \leq \mathrm{rk}_{S_i}(y_i)$). It can be checked that for all $x = (x_1, \ldots, x_k)$ and $y = (y_1, \ldots, y_k)$ in S, $x \leq_{\mathrm{lex}} y$ iff $x = y$ or $x_i \leq_i y_i$ for the least index i such that $x_i \neq y_i$.

6.16. Example: Words with Restrictions. Let S be the set of four-letter words $w_1w_2w_3w_4$ that begin and end with consonants and have a vowel in the second position. Choosing letters from left to right and using the Product Rule, we see that $|S| = 21 \cdot 5 \cdot 26 \cdot 21 = 57{,}330$. Let C, V, and A denote the set of consonants, vowels, and all letters, respectively. Using alphabetical order, we get ranking bijections $\mathrm{rk}_C : C \to [\![21]\!]$, $\mathrm{rk}_V : V \to [\![5]\!]$, and $\mathrm{rk}_A : A \to [\![26]\!]$. For example,

$$\mathrm{rk}_V(\mathrm{a}) = 0,\ \mathrm{rk}_V(\mathrm{e}) = 1,\ \mathrm{rk}_V(\mathrm{i}) = 2,\ \mathrm{rk}_V(\mathrm{o}) = 3,\ \mathrm{rk}_V(\mathrm{u}) = 4.$$

Since $S = C \times V \times A \times C$, the Bijective Cartesian Product Rule provides a ranking map $\mathrm{rk}_S : S \to [\![57,330]\!]$ given by

$$\mathrm{rk}_S(w_1w_2w_3w_4) = p_{21,5,26,21}(\mathrm{rk}_C(w_1), \mathrm{rk}_V(w_2), \mathrm{rk}_A(w_3), \mathrm{rk}_C(w_4)).$$

For example,

$$\mathrm{rk}_S(\mathrm{host}) = p_{21,5,26,21}(5, 3, 18, 15) = 5 \cdot (5 \cdot 26 \cdot 21) + 3 \cdot (26 \cdot 21) + 18 \cdot (21) + 15 = 15{,}681.$$

We unrank by applying $p_{21,5,26,21}^{-1}$ and then decoding to letters. For example, repeated division shows that

$$p_{21,5,26,21}^{-1}(44001) = (16, 0, 15, 6),$$

and therefore $u(44001) = \text{vapj}$. This unranking method generates the words in S in alphabetical order.

6.17. Example: License Plates. A California license plate consists of a digit, followed by three letters, followed by three digits. We can use the preceding ideas to rank and unrank license plates. For instance,

$$r(\text{3PZY292}) = p_{10,26,26,26,10,10,10}(3, 15, 25, 24, 2, 9, 2) = 63,542,292.$$

6.6 Ranking Permutations

In the examples considered so far, the choices made at each stage of the Product Rule did not depend on what choices were made in previous stages. This section studies the more complicated case where the available choices do depend on what happened earlier. We illustrate this situation by solving the ranking and unranking problems for permutations.

Suppose A is an n-letter alphabet. Recall that a k-permutation of A is a word $w = w_1 w_2 \cdots w_k$, where the w_i's are *distinct* elements of A. Let S be the set of all k-permutations of A. Using the ordinary Product Rule, we build elements w in S by choosing $w_1 \in A$ in n ways, then choosing $w_2 \in A - \{w_1\}$ in $n-1$ ways, and so on. At the ith stage (where $1 \leq i \leq k$), we choose $w_i \in A - \{w_1, w_2, \ldots, w_{i-1}\}$ in $n - (i-1)$ ways. Thus, $|S| = n(n-1)\cdots(n-k+1)$. Notice that the set of choices available at the ith stage depends on the choices made earlier, but the *cardinality* of this set is independent of previous choices. This last fact is a key hypothesis of the Product Rule.

Let us rephrase the preceding counting argument to obtain a bijection between S and the product set $[\![n]\!] \times [\![n-1]\!] \times \cdots \times [\![n-k+1]\!]$. Fix a total ordering $(x_0, x_1, \ldots, x_{n-1})$ of the letters in A; equivalently, fix an unranking bijection $\text{unrk}_A : [\![n]\!] \to A$ and let $x_i = \text{unrk}_A(i)$. Suppose $w = w_1 w_2 \cdots w_k \in S$. We must map w to a k-tuple $(j_1, j_2, \ldots, j_k)$, where $0 \leq j_i < n - (i-1)$. To compute j_1, locate w_1 in the sequence $x = (x_0, x_1, \ldots, x_{n-1})$, let j_1 be the number of letters preceding w_1 in the sequence, and then erase w_1 from the sequence to get a new sequence x'. To compute j_2, find w_2 in the sequence x', let j_2 be the number of letters preceding it, and then erase w_2 to get a new sequence x''. Continue similarly to generate $j_3, \ldots, j_k$. This process is reversible, as demonstrated in the next example, so we have defined the required bijection. Combining this bijection (and its inverse) with the maps $p_{n,n-1,\ldots,n-k+1}$ and $p^{-1}_{n,n-1,\ldots,n-k+1}$, we obtain ranking and unranking maps for S. It can be verified that these maps correspond to the alphabetical order of permutations determined by the given total ordering of the alphabet A.

6.18. Example. Let $n = 8$, $k = 5$, and $A = (\text{a,b,c,d,e,f,g,h})$ with the standard alphabetical ordering. Let $w = \text{cfbgd} \in S$. We compute $(j_1, \ldots, j_5)$ as follows:

2 letters precede c in (a,b,c,d,e,f,g,h), so	$j_1 = 2;$
4 letters precede f in (a,b,d,e,f,g,h), so	$j_2 = 4;$
1 letter precedes b in (a,b,d,e,g,h), so	$j_3 = 1;$
3 letters precede g in (a,d,e,g,h), so	$j_4 = 3;$
1 letter precedes d in (a,d,e,h), so	$j_5 = 1.$

Thus, cfbgd maps to $(j_1, \ldots, j_5) = (2, 4, 1, 3, 1)$. The rank of the word cfbgd is therefore

$$p_{8,7,6,5,4}(2, 4, 1, 3, 1) = 2 \cdot (7 \cdot 6 \cdot 5 \cdot 4) + 4 \cdot (6 \cdot 5 \cdot 4) + 1 \cdot (5 \cdot 4) + 3 \cdot (4) + 1 = 2193.$$

Next, let us unrank the integer 982. First, repeated division gives

$$p_{8,7,6,5,4}^{-1}(982) = (1,1,1,0,2).$$

Since $j_1 = 1$, the first letter of the word must be b. Removing b from the alphabet gives (a,c,d,e,f,g,h). Since $j_2 = 1$, the second letter of the word is c. Removing c from the previous list gives (a,d,e,f,g,h). Continuing in this way, we see that 982 unranks to give the word bcdag.

6.19. Example. Let S be the set of permutations of $\{1,2,3,4,5,6\}$. Using the procedure above to rank the permutation 462153, we first compute $(j_1,\ldots,j_6) = (3,4,1,0,1,0)$ and then calculate $p_{6,5,4,3,2,1}(3,4,1,0,1,0) = 463$. To unrank the integer 397, first calculate $p_{6,5,4,3,2,1}^{-1}(397) = (3,1,2,0,1,0)$. Then use these position numbers to recover the permutation 425163.

6.7 Ranking Subsets

According to the Subset Rule in §1.4, the number of k-element subsets of an n-element set is $\binom{n}{k} = \frac{n!}{k!(n-k)!}$. This result was obtained *indirectly*, by enumerating k-permutations of an n-element set in two ways and then dividing the resulting equation by $k!$. In general, the operation of division presents serious problems when attempting to construct bijections. Therefore, we adopt a different approach to the problem of ranking and unranking subsets. Instead of using the Bijective Product Rule, we apply the Bijective Sum Rule to the recursion characterizing binomial coefficients. This leads to recursive algorithms for ranking and unranking subsets.

Write $C(n,k)$ for the number of k-element subsets of an n-element set. In Example 2.21, we saw that these numbers satisfy the recursion

$$C(n,k) = C(n-1,k) + C(n-1,k-1) \quad \text{for } 0 < k < n, \tag{6.2}$$

with initial conditions $C(n,0) = C(n,n) = 1$. This recursion came from a combinatorial argument involving the Sum Rule. Using the Bijective Sum Rule instead leads to recursively defined bijections for ranking and unranking. For each alphabet A, introduce the temporary notation $S_k(A)$ to denote the set of all k-element subsets of A. We assume that all alphabets to be considered have been given some fixed total ordering that allows us to rank and unrank individual letters of the alphabet. Suppose $A = \{x_0 < x_1 < \cdots < x_{n-1}\}$ is such an alphabet with n letters. We can write $S_k(A)$ as the disjoint union of sets T and U, where T consists of all subsets that do not contain x_{n-1} and U consists of all subsets that do contain x_{n-1}. Note that $T = S_k(A-\{x_{n-1}\})$, and there is a bijection from U onto $S_{k-1}(A-\{x_{n-1}\})$ that acts by deleting x_{n-1} from a subset belonging to U. We can use recursion to obtain ranking and unranking maps for $S_k(A-\{x_{n-1}\})$ and $S_{k-1}(A-\{x_{n-1}\})$, as the members of these sets are subsets drawn from smaller alphabets. Then we combine these maps using the Bijective Sum Rule to get ranking and unranking maps for $S_k(A)$.

By expanding the definitions, we arrive at the following recursive ranking algorithm for mapping a subset $B \in S_k(A)$ to an integer:

- If $k = 0$ (so $B = \emptyset$), then return the answer 0.
- If $k > 0$ and the last letter x in A does not belong to B, then return the ranking of B relative to the set $S_k(A-\{x\})$, which we compute recursively using this very algorithm.

- If $k > 0$ and the last letter x in A does belong to B, let i be the rank of $B' = B - \{x\}$ relative to the set $S_{k-1}(A - \{x\})$ (computed recursively), and return the answer $C(n-1,k) + i$. Note that $C(n-1,k)$ can be computed using the recursion (6.2) for binomial coefficients.

The inverse map is the following recursive unranking algorithm that maps an integer $m \in [\![C(n,k)]\!]$ to a subset $B \in S_k(A)$:

- If $k = 0$ (so m must be zero), then return $\emptyset$.
- If $k > 0$ and $0 \leq m < C(n-1,k)$, then return the result of unranking m relative to the set $S_k(A - \{x\})$, where x is the last letter of A.
- If $k > 0$ and $C(n-1,k) \leq m < C(n,k)$, then let B' be the subset obtained by unranking $m - C(n-1,k)$ relative to the set $S_{k-1}(A - \{x\})$, and return $B' \cup \{x\}$.

6.20. Example. Let $A = \{a,b,c,d,e,f,g,h\}$ ordered alphabetically, and let us rank the subset $B = \{c,d,f,g\} \in S_4(A)$. Since the last letter of A (namely h) is not in B, we recursively proceed to rank B relative to the 7-letter alphabet $A_1 = \{a,b,c,d,e,f,g\}$. The new last letter g does belong to B, so we must add $C(7-1,4) = 15$ to the rank of $B_1 = \{c,d,f\}$ as a three-element subset of $\{a,b,c,d,e,f\}$. The last letter f belongs to B_1, so we must add $C(6-1,3) = 10$ to the rank of $B_2 = \{c,d\}$ as a two-element subset of $\{a,b,c,d,e\}$. Since e is not in B_2, this rank is the same as the rank of B_2 relative to $\{a,b,c,d\}$, which is $C(4-1,2) = 3$ plus the rank of $\{c\}$ as a one-element subset of $\{a,b,c\}$. In turn, this rank is $C(3-1,1) = 2$ plus the rank of $\emptyset$ as a zero-element subset of $\{a,b\}$. The rank of the empty set is 0 by the base case of the algorithm. Adding up the contributions, we see that the rank of B is

$$\binom{3-1}{1} + \binom{4-1}{2} + \binom{6-1}{3} + \binom{7-1}{4} = 30.$$

Generalizing the pattern in the previous example, we can convert the recursive algorithm to the following summation formula for the rank of a subset.

6.21. Theorem: Sum Formula for Ranking Subsets. If $A = \{x_0 < x_1 < \cdots < x_{n-1}\}$ and $B = \{x_{i_1}, x_{i_2}, \ldots, x_{i_k}\}$ where $i_1 < i_2 < \cdots < i_k$, then the rank of B as a member of $S_k(A)$ is $\sum_{j=1}^{k} \binom{i_j}{j}$.

This formula can be proved by induction on k.

6.22. Example. Now we illustrate the recursive unranking algorithm. Let us unrank the integer 53 to obtain an object $B \in S_4(A)$, where $A = \{a,b,c,d,e,f,g,h\}$. Here $n = 8$ and $k = 4$. Since $C(7,4) = 35 \leq 53$, we know that $h \in B$. We proceed by unranking $53 - 35 = 18$ to get a three-element subset of $\{a,b,c,d,e,f,g\}$. This time $C(6,3) = 20 > 18$, so g is not in the subset. We proceed to unrank 18 to get a three-element subset of $\{a,b,c,d,e,f\}$. Now $C(5,3) = 10 \leq 18$, so f does belong to B. We continue, unranking $18 - 10 = 8$ to get a two-element subset of $\{a,b,c,d,e\}$. Since $C(4,2) = 6 \leq 8$, $e \in B$ and we continue by unranking 2 to get a one-element subset of $\{a,b,c,d\}$. We have $C(3,1) = 3 > 2$, so $d \notin B$. But at the next stage $C(2,1) = 2 \leq 2$, so $c \in B$. We conclude, finally, that $B = \{c,e,f,h\}$.

As before, we can describe this algorithm iteratively instead of recursively.

6.23. Unranking Algorithm for Subsets. Suppose $A = \{x_0 < x_1 < \cdots < x_{n-1}\}$ and we are unranking an integer m to get a k-element subset B of A. Repeatedly perform the following steps until k becomes zero: let i be the largest integer such that $\binom{i}{k} \leq m$; declare that $x_i \in B$; replace m by $m - \binom{i}{k}$ and decrement k by 1.

We close with a remark about the ordering of subsets associated to the ranking and unranking algorithms described above. Let x be the last letter of A. If we unrank the integers $0, 1, 2, \ldots$ in this order to obtain a listing of $S_k(A)$, we will obtain all k-element subsets of A not containing x first, and all k-element subsets of A containing x second. Each of these sublists is internally ordered in the same way according to the next-to-last letter of A, and so on recursively. In contrast, if we had applied the Bijective Sum Rule using the recursion

$$C(n,k) = C(n-1,k-1) + C(n-1,k)$$

(in which the order of the summands is swapped), then the ordering rules at each level of this hierarchy would be reversed. Similarly, the reader can construct variant ranking algorithms in which the first letter of the alphabet is considered most significant, etc. Some of these variants are explored in the exercises.

6.8 Ranking Anagrams

Next we study the problem of ranking and unranking anagrams. Recall that $\mathcal{R}(a_1^{n_1} \cdots a_k^{n_k})$ is the set of all words of length $n = n_1 + \cdots + n_k$ consisting of n_i copies of a_i for $1 \leq i \leq k$. By the Anagram Rule, we know these sets are counted by the multinomial coefficients:

$$|\mathcal{R}(a_1^{n_1} \cdots a_k^{n_k})| = \binom{n}{n_1, n_2, \ldots, n_k} = \frac{n!}{n_1! n_2! \cdots n_k!}.$$

There are at least three ways of deriving this formula. One way counts permutations of n distinct letters in two ways, and solves for the number of anagrams by division. This method is not easily converted into a ranking algorithm. A second way uses the Product Rule, choosing the positions for the n_1 copies of a_1, then the positions for the n_2 copies of a_2, and so on. Combining the Bijective Product Rule with the ranking algorithm for subsets presented earlier, this method does lead to a ranking algorithm for anagrams. A third way to count anagrams involves finding recursions satisfied by multinomial coefficients (§2.8). This is the approach we pursue here.

Let $C(n; n_1, \ldots, n_k)$ be the number of rearrangements of n letters, where there are n_i letters of type i. Classifying words by their first letter leads to the recursion

$$C(n; n_1, \ldots, n_k) = \sum_{i=1}^{k} C(n-1; n_1, \ldots, n_i - 1, \ldots, n_k).$$

Applying the Bijective Sum Rule to this recursion, we are led to recursive ranking and unranking algorithms for anagrams.

Here are the details of the algorithms. We recursively define ranking maps

$$r_{n_1,\ldots,n_k} : \mathcal{R}(a_1^{n_1} \cdots a_k^{n_k}) \to [\![M]\!],$$

where $M = (n_1 + \cdots + n_k)!/(n_1! \cdots n_k!)$. If any n_i is negative, $r_{n_1,\ldots,n_k}$ is the empty function. If all n_i's are zero, $r_{n_1,\ldots,n_k}$ is the function sending the empty word to 0. To compute $r_{n_1,\ldots,n_k}(w)$ in the remaining case, suppose a_i is the first letter of w. Write $w = a_i w'$. Return the answer

$$r_{n_1,\ldots,n_k}(w) = \sum_{j<i} C(n-1; n_1, \ldots, n_j - 1, \ldots, n_k) + r_{n_1,\ldots,n_i-1,\ldots,n_k}(w'),$$

where the rank of w' is computed recursively by the same algorithm.

Next, we define the corresponding unranking maps

$$u_{n_1,\ldots,n_k} : [\![M]\!] \to \mathcal{R}(a_1^{n_1} \cdots a_k^{n_k}).$$

Use the only possible maps if some $n_i < 0$ or if all $n_i = 0$. Otherwise, to unrank $s \in [\![M]\!]$, first find the maximal index i such that $n_i > 0$ and $\sum_{j<i} C(n-1; n_1, \ldots, n_j - 1, \ldots, n_k) \leq s$; let s' be the difference between s and this sum. Recursively compute the word

$$w' = u_{n_1,\ldots,n_i-1,\ldots,n_k}(s'),$$

and return the answer $w = a_i w'$. This unranking algorithm induces a listing of the anagrams in $\mathcal{R}(a_1^{n_1} \cdots a_k^{n_k})$ in alphabetical order relative to the alphabet ordering $a_1 < a_2 < \cdots < a_k$.

6.24. Example. Let us compute the rank of the word $w = \text{abbcacb}$ in $\mathcal{R}(\text{a}^2\text{b}^3\text{c}^2)$; here $n = 7$, $n_1 = 2$, $n_2 = 3$, and $n_3 = 2$. Erasing the first letter a, we see that the rank of w equals zero plus the rank of $w_1 = \text{bbcacb}$; now $n = 6$, $n_1 = 1$, $n_2 = 3$, and $n_3 = 2$. Erasing b, we must now add $\binom{5}{0,3,2} = 10$ to the rank of $w_2 = \text{bcacb}$; now $n = 5$, $n_1 = 1$, $n_2 = 2$, and $n_3 = 2$. Erasing the next b, we must add $\binom{4}{0,2,2} = 6$ to the rank of $w_3 = \text{cacb}$; now $n = 4$, $n_1 = 1$, $n_2 = 1$, and $n_3 = 2$. Erasing c, we must add $\binom{3}{0,1,2} + \binom{3}{1,0,2} = 6$ to the rank of $w_4 = \text{acb}$; now $n = 3$, $n_1 = 1$, $n_2 = 1$, and $n_3 = 1$. Continuing in this way, we see that the rank of acb is 1. Thus, the rank of the original word is $10 + 6 + 6 + 1 = 23$.

Next, let us unrank 91 to obtain a word w in $\mathcal{R}(\text{a}^2\text{b}^3\text{c}^2)$. To determine the first letter of w, note that $0 \leq 91$, $\binom{6}{1,3,2} = 60 \leq 91$, but $\binom{6}{1,3,2} + \binom{6}{2,2,2} = 150 > 91$. Thus, the first letter is b, and we continue by unranking $91 - 60 = 31$ to obtain a word in $\mathcal{R}(\text{a}^2\text{b}^2\text{c}^2)$. This time, we have $0 \leq 31$, $\binom{5}{1,2,2} = 30 \leq 31$, but $\binom{5}{1,2,2} + \binom{5}{2,1,2} = 60 > 31$. So the second letter is b, and we continue by unranking $31 - 30 = 1$ to obtain a word in $\mathcal{R}(\text{a}^2\text{b}^1\text{c}^2)$. It is routine to check that the next two letters are both a, and we continue by unranking 1 to obtain a word in $\mathcal{R}(\text{a}^0\text{b}^1\text{c}^2)$. The word we get is cbc, so the unranking map sends 91 to the word $w = \text{bbaacbc}$.

6.9 Ranking Integer Partitions

In this section, we devise ranking and unranking algorithms for integer partitions by applying the Bijective Sum Rule to the recursion 2.38. Let $P(n, k)$ be the set of integer partitions of n with largest part k, and let $p(n, k) = |P(n, k)|$. We have seen that these numbers satisfy the recursion

$$p(n, k) = p(n - k, k) + p(n - 1, k - 1) \quad \text{for } 0 < k < n. \tag{6.3}$$

The first term on the right counts elements of $P(n, k)$ in which the largest part occurs at least twice (deleting the first copy of this part gives a bijection onto $P(n - k, k)$). The second term on the right counts elements of $P(n, k)$ in which the largest part occurs exactly once (reducing this part by 1 gives a bijection onto $P(n - 1, k - 1)$). Combining these bijections with the Bijective Sum Rule, we obtain recursively determined ranking maps $r_{n,k} : P(n, k) \to [\![p(n, k)]\!]$. To find $r_{n,k}(\mu)$, consider three cases. If μ has only one part (which happens when $k = n$), return 0. If $k = \mu_1 = \mu_2$, return $r_{n-k,k}((\mu_2, \mu_3, \ldots))$. If $k = \mu_1 > \mu_2$, return $p(n-k, k) + r_{n-1,k-1}((\mu_1 - 1, \mu_2, \ldots))$. The unranking maps $u_{n,k} : [\![p(n, k)]\!] \to P(n, k)$ operate as follows. To compute $u(m)$ where $0 \leq m < p(n, k)$, consider three cases. If $k = n$ (so $m = 0$), return $\mu = (n)$. If $0 \leq m < p(n - k, k)$, recursively compute $\nu = u_{n-k,k}(m)$

and return the answer $\mu = (k, \nu_1, \nu_2, \ldots)$. If $p(n-k,k) \leq m < p(n,k)$, recursively compute $\nu = u_{n-1,k-1}(m - p(n-k,k))$ and return the answer $\mu = (\nu_1 + 1, \nu_2, \nu_3, \ldots)$.

6.25. Example. Let us compute $r_{8,3}(\mu)$, where $\mu = (3,3,1,1)$. Since $\mu_1 = \mu_2$, the rank is $r_{5,3}(\nu)$, where $\nu = (3,1,1)$. Next, since $\nu_1 \neq \nu_2$, we have

$$r_{5,3}(3,1,1) = p(2,3) + r_{4,2}(2,1,1) = r_{4,2}(2,1,1).$$

The first two parts of the new partition are again different, so

$$r_{4,2}(2,1,1) = p(2,2) + r_{3,1}(1,1,1) = 1 + r_{3,1}(1,1,1).$$

After several more steps, we find that $r_{3,1}(1,1,1) = 0$, so $r_{8,3}(\mu) = 1$. Thus μ is the second partition in the listing of $P(8,3)$ implied by the ranking algorithm; the first partition in this list, which has rank 0, is $(3,3,2)$.

Next, let us compute $\mu = u_{10,4}(6)$. First, $p(6,4) = 2 \leq 6$, so μ is obtained by adding 1 to the first part of $\nu = u_{9,3}(4)$. Second, $p(6,3) = 3 \leq 4$, so ν is obtained by adding 1 to the first part of $\rho = u_{8,2}(1)$. Third, $p(6,2) = 3 > 1$, so ρ is obtained by adding a new first part of length 2 to $\xi = u_{6,2}(1)$. Fourth, $p(4,2) = 2 > 1$, so ξ is obtained by adding a new first part of length 2 to $\zeta = u_{4,2}(1)$. Fifth, $p(2,2) = 1 \leq 1$, so ζ is obtained by adding 1 to the first part of $\omega = u_{3,1}(0)$. We must have $\omega = (1,1,1)$, this being the unique element of $P(3,1)$. Working our way back up the chain, we successively find that

$$\zeta = (2,1,1), \quad \xi = (2,2,1,1), \quad \rho = (2,2,2,1,1), \quad \nu = (3,2,2,1,1),$$

and finally $\mu = u_{10,4}(6) = (4,2,2,1,1)$.

Now that we have algorithms to rank and unrank the sets $P(n,k)$, we can apply the Bijective Sum Rule to the identity

$$p(n) = p(n,n) + p(n,n-1) + \cdots + p(n,1)$$

to rank and unrank the set $P(n)$ of all integer partitions of n.

6.26. Example. Let us enumerate all the integer partitions of 6. We obtain this list of partitions by concatenating the lists associated to the sets

$$P(6,6),\ P(6,5),\ \cdots,\ P(6,1),$$

written in this order. In turn, each of these lists can be constructed by applying the unranking maps $u_{6,k}$ to the integers $0,1,2,\ldots,p(6,k)-1$. The reader can verify that this procedure leads to the following list:

$$(6),\ (5,1),\ (4,2),\ (4,1,1),\ (3,3),\ (3,2,1),\ (3,1,1,1),$$

$$(2,2,2),\ (2,2,1,1),\ (2,1,1,1,1),\ (1,1,1,1,1,1).$$

It can also be checked that the list obtained in this way presents the integer partitions of n in decreasing lexicographic order.

6.10 Ranking Set Partitions

Next, we consider the ranking and unranking of set partitions (which are counted by Stirling numbers of the second kind and Bell numbers). The recursion for Stirling numbers involves both addition and multiplication, so our recursive algorithms use both the Bijective Sum Rule and the Bijective Product Rule.

Let $SP(n,k)$ be the set of all set partitions of $\{1,2,\ldots,n\}$ into exactly k blocks, and let $S(n,k) = |SP(n,k)|$ be the associated Stirling number of the second kind. Recall from Recursion 2.45 that

$$S(n,k) = S(n-1,k-1) + kS(n-1,k) \quad \text{for } 0 < k < n.$$

The first term counts set partitions in $SP(n,k)$ such that n is in a block by itself; removal of this block gives a bijection onto $SP(n-1,k-1)$. The second term counts set partitions π in $SP(n,k)$ such that n belongs to a block with other elements. Starting with any set partition π' in $SP(n-1,k)$, we can build such a set partition $\pi \in SP(n,k)$ by adding n to any of the k nonempty blocks of π'. We index the blocks of π' using $0,1,\ldots,k-1$ by arranging the minimum elements of these blocks in increasing order. For example, if $\pi' = \{\{6,3,5\},\{2\},\{1,7\},\{8,4\}\}$, then block 0 of π' is $\{1,7\}$, block 1 is $\{2\}$, block 2 is $\{3,5,6\}$, and block 3 is $\{4,8\}$.

The ranking maps $r_{n,k} : SP(n,k) \to [\![S(n,k)]\!]$ are defined recursively as follows. Use the only possible maps if $k \leq 0$ or $k \geq n$. For $0 < k < n$, compute $r_{n,k}(\pi)$ as follows. If $\{n\} \in \pi$, return the answer $r_{n-1,k-1}(\pi - \{\{n\}\})$. Otherwise, let π' be obtained from π by deleting n from whatever block contains it, and let i be the index of the block of π that contains n. Return $S(n-1,k-1) + p_{k,S(n-1,k)}(i, r_{n-1,k}(\pi'))$.

We define the unranking maps $u_{n,k} : [\![S(n,k)]\!] \to SP(n,k)$ as follows. Assume $0 < k < n$ and we are computing $u_{n,k}(m)$. If $0 \leq m < S(n-1,k-1)$, then return $u_{n-1,k-1}(m) \cup \{\{n\}\}$. If $S(n-1,k-1) \leq m < S(n,k)$, first compute $(i,j) = p^{-1}_{k,S(n-1,k)}(m - S(n-1,k-1))$. Next, calculate the partition $\pi' = u_{n-1,k}(j)$ by unranking j recursively, and finally compute π by adding n to the ith block of π'.

6.27. Example. Let us compute the rank of $\pi = \{\{1,7\},\{2,4,5\},\{3,8\},\{6\}\}$ relative to the set $SP(8,4)$. In the first stage of the recursion, removal of the largest element 8 from block 2 leaves the set partition $\pi' = \{\{1,7\},\{2,4,5\},\{3\},\{6\}\}$. Therefore,

$$r_{8,4}(\pi) = S(7,3) + 2S(7,4) + r_{7,4}(\pi') = 301 + 2 \cdot 350 + r_{7,4}(\pi').$$

(See Figure 2.20 for a table of Stirling numbers, which are calculated using the recursion for $S(n,k)$.) In the second stage, removing 7 from block 0 leaves the set partition $\pi'' = \{\{1\},\{2,4,5\},\{3\},\{6\}\}$. Hence,

$$r_{7,4}(\pi') = S(6,3) + 0S(6,4) + r_{6,4}(\pi'') = 90 + r_{6,4}(\pi'').$$

In the third stage, removing the block $\{6\}$ leaves the set partition $\pi^{(3)} = \{\{1\},\{2,4,5\},\{3\}\}$, and

$$r_{6,4}(\pi'') = r_{5,3}(\pi^{(3)}).$$

In the fourth stage, removing 5 from block 1 leaves the set partition $\pi^{(4)} = \{\{1\},\{2,4\},\{3\}\}$, and

$$r_{5,3}(\pi^{(3)}) = S(4,2) + 1S(4,3) + r_{4,3}(\pi^{(4)}) = 7 + 6 + r_{4,3}(\pi^{(4)}).$$

In the fifth stage, removing 4 from block 1 leaves the set partition $\pi^{(5)} = \{\{1\},\{2\},\{3\}\}$, and

$$r_{4,3}(\pi^{(4)}) = S(3,2) + 1S(3,3) + r_{3,3}(\pi^{(5)}) = 3 + 1 + r_{3,3}(\pi^{(5)}).$$

Now $r_{3,3}(\pi^{(5)})$ is 0, since $|SP(3,3)| = 1$. We deduce in sequence

$$r_{4,3}(\pi^{(4)}) = 4,\ r_{6,4}(\pi'') = r_{5,3}(\pi^{(3)}) = 17,\ r_{7,4}(\pi') = 107,\ r_{8,4}(\pi) = 1108.$$

Next, let us compute $u_{7,3}(111)$. The input 111 weakly exceeds $S(6,2) = 31$, so we must first compute $p_{3,90}^{-1}(111 - 31) = (0, 80)$. This means that 7 goes in block 0 of $u_{6,3}(80)$. Now $80 \geq S(5,2) = 15$, so we compute $p_{3,25}^{-1}(80 - 15) = (2, 15)$. This means that 6 goes in block 2 of $u_{5,3}(15)$. Now $15 \geq S(4,2) = 7$, so we compute $p_{3,6}^{-1}(15 - 7) = (1, 2)$. This means that 5 goes in block 1 of $u_{4,3}(2)$. Now $2 < S(3,2) = 3$, so 4 is in a block by itself in $u_{4,3}(2)$. To find the remaining blocks, we compute $u_{3,2}(2)$. Now $2 \geq S(2,1) = 1$, so we compute $p_{2,1}^{-1}(2 - 1) = (1, 0)$. This means that 3 goes in block 1 of $u_{2,2}(0)$. Evidently, $u_{2,2}(0) = \{\{1\}, \{2\}\}$. Using the preceding information to insert elements $3, 4, \ldots, 7$, we conclude that

$$u_{7,3}(111) = \{\{1, 7\}, \{2, 3, 5\}, \{4, 6\}\}.$$

The ranking procedure given here lists the objects in $SP(n, k)$ in the following order. Set partitions with n in its own block appear first. Next come the set partitions with n in block 0 (i.e., n is in the same block as 1); then come the set partitions with n in block 1, etc. By applying the sum and product bijections in different orders, one can obtain different listings of the elements of $SP(n, k)$.

Let $SP(n)$ be the set of all set partitions of n, so $|SP(n)|$ is the nth Bell number. The preceding results lead to ranking and unranking algorithms for this collection, by applying the Bijective Sum Rule to the disjoint union

$$SP(n) = SP(n, 1) \cup SP(n, 2) \cup \cdots \cup SP(n, n).$$

Another approach to ranking $SP(n)$ is to use Recursion 2.46 for Bell numbers; see the exercises for details.

6.11 Ranking Trees

By the Rooted Tree Rule in §3.7, there are n^{n-2} rooted trees on the vertex set $\{1, 2, \ldots, n\}$ rooted at vertex 1. Let B be the set of such trees; we seek ranking and unranking algorithms for B. One way to obtain these algorithms is to use the bijective proof of the Rooted Tree Rule. In that proof, we described a bijection $\phi' : B \to A$, where A is the set of all functions $f : \{1, 2, \ldots, n\} \to \{1, 2, \ldots, n\}$ such that $f(1) = 1$ and $f(n) = n$. Let C be the set of words of length $n - 2$ in the alphabet $\{0, 1, \ldots, n - 1\}$. The map $\psi : A \to C$ such that $\psi(f) = w_1 \cdots w_{n-2}$ with $w_i = f(i+1) - 1$, is a bijection. The map $p_{n,n,\ldots,n}$ gives a bijection from $C = [\![n]\!]^{n-2}$ to $[\![n^{n-2}]\!]$. Composing all these bijections, we get the required ranking algorithm. Inverting the bijections gives the associated unranking algorithm.

6.28. Example. Consider the rooted tree T shown in Figure.3.9. In Example 3.49, we computed $\phi'(T)$ to be the function g such that

$$(g(1), g(2), \ldots, g(9)) = (1, 2, 2, 9, 9, 7, 6, 9, 9).$$

Hence, $\psi(g)$ is the word 1188658. Applying the map $p_{9,9,\ldots,9}$ (or equivalently, interpreting the given word as a number written in base 9), we find that $\text{rk}(T) = 649,349$.

This application shows how valuable a bijective proof of a counting result can be. If we have a bijection from a complicated set of objects to a simpler set of objects (such as functions or words), we can compose the bijection with standard ranking maps to obtain ranking and unranking algorithms for the complicated objects. In contrast, if a counting result is obtained by an algebraic manipulation involving division or generating functions, it may not be so straightforward to extract an effective ranking mechanism.

6.12 The Successor Sum Rule

We turn now from ranking algorithms to successor algorithms. Recall the setup from §6.1: given a totally ordered finite set S, our goal is to find algorithms `first`, `last`, and `next` such that `first`(S) is the least element of S, `last`(S) is the greatest element of S, and `next`(x, S) returns the immediate successor of x. Our strategy is to develop versions of the Bijection Rule, the Sum Rule, and the Product Rule that automatically create these subroutines.

6.29. Example: Successor Algorithm for $[\![n]\!]$. Consider the set $[\![n]\!] = \{0, 1, 2, \ldots, n-1\}$ with the standard ordering $0 < 1 < \cdots < n-1$. We define `first`($[\![n]\!]$) $= 0$, `last`($[\![n]\!]$) $= n-1$, and `next`($x, [\![n]\!]$) $= x + 1$ for $0 \leq x < n - 1$.

6.30. The Bijection Rule for Successor Algorithms. Suppose $F : X \to Y$ is a bijection with inverse G, and we already know successor algorithms for the finite set Y. There are successor algorithms for X defined by `first`(X) $= G($`first`$(Y))$, `last`(X) $=$ $G($`last`$(Y))$, and `next`(x, X) $= G($`next`$(F(x), Y))$ for $x \in X$ with $x \neq$ `last`(X).

6.31. The Successor Sum Rule. Assume the nonempty set S is the union of pairwise disjoint finite sets $S_1, \ldots, S_k$, and we already know successor algorithms for each S_i. We can then define a successor algorithm for the set S using the pseudocode in Figure 6.4.

```
Assumptions: S is the disjoint union of finite sets S_1,...,S_k;
 S is nonempty; we already know successor subroutines for each S_i.

define procedure first(S):
{ i=1; while (S_i is empty) do { i=i+1; }
  return first(S_i);
}

define procedure last(S):
{ i=k; while (S_i is empty) do { i=i-1; }
  return last(S_i);
}

define procedure next(x,S):  %% assumes x is not last(S)
{ find the unique i with x in S_i;
  if (x==last(S_i)) then
  { j=i+1; while (S_j is empty) do { j=j+1; }
    return first(S_j);
  }
  else return next(x,S_i);
}
```

FIGURE 6.4
Pseudocode for the Successor Sum Rule.

Figure 6.5 provides visual intuition for what the Successor Sum Rule is doing in the case where every S_i is nonempty. We are totally ordering the set $S = S_1 \cup S_2 \cup \cdots \cup S_k$ by putting all the elements of S_1 first, then all the elements of S_2, and so on. The first element

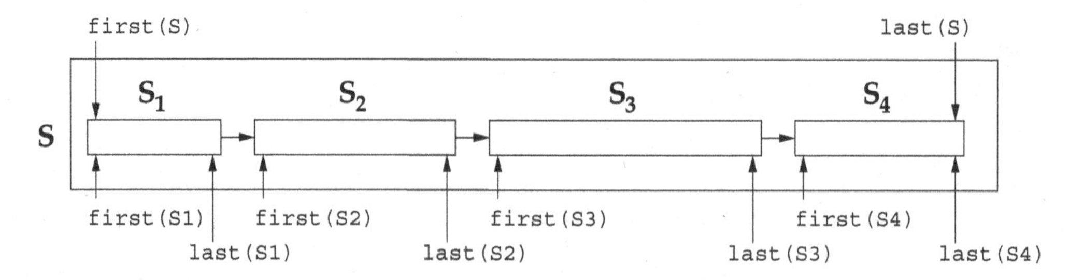

FIGURE 6.5
Schematic diagram for the Successor Sum Rule.

of S is the first element of S_1, and the last element of S is the last element of S_k. Given $x \in S$, we compute $\texttt{next}(x, S)$ as follows. If x is the last element in S_i where $i < k$, the successor of x in S is the first element in S_{i+1}. If $x \in S_i$ is not last in S_i, the successor of x in S is the same as the successor of x in S_i.

The pseudocode in the general case is a bit more complex since the subroutines must skip over any sets S_j that happen to be empty. We see in the next section that the Successor Sum Rule often leads to recursively defined successor algorithms. Although all the sets S_j at the top level of the recursion are typically nonempty, it is quite possible that empty sets will be encountered in one of the recursive calls.

6.13 Successor Algorithms for Anagrams

We now illustrate the rules in the previous section by converting a combinatorial recursion for multinomial coefficients into a successor algorithm for sets of anagrams. This algorithm specializes to give successor algorithms for permutations and subsets of a fixed size.

6.32. Successor Algorithm for Anagrams. Let $\{a_1 < a_2 < \cdots < a_k\}$ be a fixed ordered alphabet. We simultaneously construct successor algorithms for all the anagram sets $\mathcal{R}(a_1^{n_1} a_2^{n_2} \cdots a_k^{n_k})$ consisting of rearrangemenets of n_i copies of a_i. We assume that $n_i \geq 0$ for $1 \leq i \leq k$, so that each set of anagrams is nonempty. For a fixed sequence $(n_1, \ldots, n_k)$, note that $S = \mathcal{R}(a_1^{n_1} a_2^{n_2} \cdots a_k^{n_k})$ is the disjoint union of sets $S_1, S_2, \ldots, S_k$, where S_i is the set of all words in S (if there are any) starting with letter a_i. When S_i is nonempty (which happens iff $n_i > 0$), deleting the first letter a_i of a word in S_i gives a bijection from S_i onto the anagram set $\mathcal{R}(a_1^{n_1} \cdots a_i^{n_i - 1} \cdots a_k^{n_k})$. We can assume by induction on $n = n_1 + \cdots + n_k$ that successor algorithms are already available for each of the latter sets. Using the successor rules, we are led to the recursive algorithm shown in Figure 6.6. It can be checked that the `first` subroutine returns the word $a_1^{n_1} a_2^{n_2} \cdots a_k^{n_k}$, whereas the `last` subroutine returns the reversal of this word. It can also be proved by induction that this successor algorithm generates the words in each anagram class in alphabetical order.

As an example, let us find the successor of $x = \text{bdcaccb}$ in the anagram set $\mathcal{R}(\text{a}^1\text{b}^2\text{c}^3\text{d}^1)$. In the first call to `next`, we have $i = 2$, $a_i = \text{b}$, and $y = \text{dcaccb}$. This word is not last in its anagram class, so we continue by finding $\texttt{next}(y)$. Repeated recursive calls lead us to consider the words caccb, then accb, then ccb. But ccb is the last word in $\mathcal{R}(\text{a}^0\text{b}^1\text{c}^2\text{d}^0)$. To continue calculating `next`(accb), we therefore look for the next available letter after a, which is b. We concatenate b and $\texttt{first}(\mathcal{R}(\text{a}^1\text{b}^0\text{c}^2\text{d}^0)) = \text{acc}$ to obtain `next`(accb)=bacc. Going

```
Assumptions: n[1]...n[k] is an array where n[i] is the number of copies
 of letter a_i in the anagram set under consideration; all n[i] >= 0.

define procedure first(n[1]...n[k]):
{ if (k==0) then return the empty word;
  i=1; while (n[i]==0) do { i=i+1; }
  n[i]=n[i]-1;
  return a_i.first(n[1]...n[k]);  %% here . denotes concatenation of words
}

define procedure last(n[1]...n[k]):
{ if (k==0) then return the empty word;
  i=k; while (n[i]==0) do { i=i-1; }
  n[i]=n[i]-1;
  return a_i.last(n[1]...n[k]);  %% here . denotes concatenation of words
}

define procedure next(x,n[1]...n[k]):  %% assumes x is not last(S), so k>0
{ let x=a_i.y;
  let m[i]=n[i]-1, m[s]=n[s] for all s unequal to i;
  if (y==last(m[1]...m[k])) then
  { j=i+1; while (n[j]==0) do { j=j+1; }
    let p[j]=n[j]-1, p[s]=n[s] for all s unequal to j;
    return a_j.first(p[1]...p[k]);
  }
  else return a_i.next(y,m[1]...m[k]);
}
```

FIGURE 6.6
Pseudocode for the Anagram Successor Algorithm.

back up the chain of recursive calls, we then find `next`(caccb)=cbacc, `next`(dcaccb)=dcbacc, and finally `next`(bdcaccb)=bdcbacc.

If we apply the `next` function to the new word bdcbacc, we strip off initial letters one at a time until we reach the suffix cc, which is the last word in its anagram class. So, to compute `next`(acc), we must find the next *available* letter after a, namely c, and append to this letter the word `first`($\mathcal{R}(\mathrm{a}^1\mathrm{b}^0\mathrm{c}^1\mathrm{d}^0)$) = ac. Thus, `next`(acc)=cac, and working back up the recursive calls leads to a final answer of bdcbcac.

The pattern in these examples applies in general, leading to the following non-recursive description of the `next` subroutine for anagrams. To compute the next word after $w = w_1w_2\cdots w_n$, find the largest position i such that letter w_i precedes letter w_{i+1} in the given ordering of the alphabet. Modify the suffix $w_i\cdots w_n$ by replacing w_i with the next larger letter appearing in this suffix, and then sorting the remaining letters of this suffix into weakly increasing order. For example, `next`(cbcbdca)=cbccabd.

6.33. Successor Algorithm for Permutations. Permutations of the alphabet $\{a_1, a_2, \ldots, a_k\}$ are the same thing as anagrams in $\mathcal{R}(a_1^1a_2^1\cdots a_k^1)$. Thus the successor algorithm for anagrams specializes to give a successor algorithm for permutations. We can use the simplified (non-recursive) versions of the `first`, `last`, and `next` subroutines derived above. The permutations of $\{1,2,3,4\}$ are generated in the following order:

$$1234,\ 1243,\ 1324,\ 1342,\ 1423,\ 1432,\ 2134,\ 2143,\ \ldots,\ 4321.$$

As another example, repeatedly applying `next` starting with the permutation 72648531, we obtain 72651348, then 72651384, then 72651438, then 72651483, and so on.

6.34. Successor Algorithm for k-Element Subsets. Suppose $Z = \{z_1, \ldots, z_n\}$ is a given set of size n, and $P_k(Z)$ is the set of all k-element subsets of Z. To obtain a successor algorithm for $P_k(Z)$, we define a bijection $F : P_k(Z) \to \mathcal{R}(1^k 0^{n-k})$ and then apply the successor algorithm for anagrams. The bijection F maps a subset S of Z to the word $F(S) = w_1 \cdots w_n \in \mathcal{R}(1^k 0^{n-k})$, where $w_i = 1$ if $z_i \in S$ and $w_i = 0$ if $z_i \notin S$. As an example, let $n = 5$, $k = 2$, and $Z = \{1,2,3,4,5\}$. Using the alphabet ordering $1 < 0$ on $\mathcal{R}(1^2 0^3)$, the successor algorithm generates the words in this anagram class in this order:

$$11000,\ 10100,\ 10010,\ 10001,\ 01100,\ 01010,\ 01001,\ 00110,\ 00101,\ 00011.$$

Applying F^{-1}, the associated subsets are:

$$\{1,2\},\ \{1,3\},\ \{1,4\},\ \{1,5\},\ \{2,3\},\ \{2,4\},\ \{2,5\},\ \{3,4\},\ \{3,5\},\ \{4,5\}.$$

In general, the method used here lists k-element subsets of $\{1,2,\ldots,n\}$ in lexicographic order. Using the alphabet ordering $0 < 1$ would have produced the reversal of the list displayed above.

In contrast, the ranking method discussed in §6.7 lists the subsets according to a different ordering, in which all subsets not containing n are listed first, followed by all subsets that do contain n, and so on recursively. The latter method produces the following list of subsets:

$$\{1,2\},\ \{1,3\},\ \{2,3\},\ \{1,4\},\ \{2,4\},\ \{3,4\},\ \{1,5\},\ \{2,5\},\ \{3,5\},\ \{4,5\}.$$

6.14 The Successor Product Rule

This section develops a version of the Product Rule for successor algorithms. We begin with a rule for the Cartesian product of two sets.

6.35. The Successor Product Rule for Two Sets. Suppose S and T are finite, nonempty sets, and we know successor algorithms for S and T. The following subroutines define a successor algorithm for $S \times T$:

$$\begin{aligned}
\texttt{first}(S\times T) &= (\texttt{first}(S),\texttt{first}(T));\\
\texttt{last}(S\times T) &= (\texttt{last}(S),\texttt{last}(T));\\
\texttt{next}((a,b),S\times T) &= \begin{cases} (a,\texttt{next}(b,T)) & \text{if } b \neq \texttt{last}(T);\\ (\texttt{next}(a,S),\texttt{first}(T)) & \text{if } b = \texttt{last}(T).\end{cases}
\end{aligned}$$

Intuitively, to find the successor of (a, b), we hold the first coordinate a fixed and replace b by its successor in T. This works unless b is the last element of T, in which case we replace a by its successor in S and replace b by the first element of T. As in the case of ranking algorithms, the successor algorithm for $S \times T$ can be derived as a special case of the Successor Sum Rule. To do so, write $S = \{s_1, \ldots, s_m\}$ (using the ordering determined by the given successor algorithm for S) and view $S \times T$ as the union of the pairwise disjoint sets $S_i = \{s_i\} \times T$. The pseudocode in Figure 6.4 specializes to the formulas given in Rule 6.35.

Similarly, by iterating the Successor Product Rule for Two Sets, we are led to the Successor Product Rule for a Cartesian product of k nonempty sets. Combining this rule with the Bijection Rule for Successor Algorithms, we obtain the general version of the Successor Product Rule given by the pseudocode in Figure 6.7.

```
Assumptions: We know a bijection F:S_1 x ... x S_k -> X
 and its inverse G:X -> S_1 x ... x S_k; we already know
 successor algorithms for each nonempty finite set S_i.

define procedure first(S):
{ for i=1 to k do
  { x_i = first(S_i); }
  return F(x_1,...,x_k);
}

define procedure last(S):
{ for i=1 to k do
  { x_i = last(S_i); }
  return F(x_1,...,x_k);
}

define procedure next(x,S):  %% assumes x is not last(S)
{ compute (x_1,...,x_k) = G(x);
  i=k; while (x_i == last(S_i)) do
  { i = i-1; }
  x_i = next(x_i,S_i);
  for j=i+1 to k do
  { x_j = first(S_j); }
  return F(x_1,...,x_k);
}
```

FIGURE 6.7
Pseudocode for the Successor Product Rule.

6.36. Example. If we apply the Successor Product Rule to the product set $[\![b]\!]^k$, the resulting successor algorithm implements counting in base b. For example, taking $b = 3$ and $k = 4$, we obtain the following sequence of words:

$$0000,\ 0001,\ 0002,\ 0010,\ 0011,\ 0012,\ 0020,\ 0021,\ 0022,\ 0100,\ \ldots,\ 2222.$$

6.37. Example. Consider license plates consisting of three letters followed by four digits. We compute `next`(WYZ-9999)=WZA-0000 and `next`(ZZZ-9899)=ZZZ-9900.

6.38. Example. For the set S of four-letter words that begin and end with consonants and have a vowel in the second position, we find that `next`(duzz)=faab and `next`(satz)=saub.

6.15 Successor Algorithms for Set Partitions

In this section, we build successor algorithms for the sets $SP(n,k)$ consisting of all set partitions of $\{1,2,\ldots,n\}$ having k blocks. We do this by applying the Sum Rule, Product Rule, and Bijection Rule for Successor Algorithms to the recursion for Stirling numbers of the second kind.

Recall that $S(n,k) = |SP(n,k)|$ satisfies initial conditions $S(n,n) = 1$ for all $n \geq 0$, $S(n,1) = 1$ for all $n \geq 1$, and $S(n,k) = 0$ when $k = 0 < n$ or $k > n$. For $1 < k < n$, these numbers satisfy the recursion

$$S(n,k) = S(n-1,k-1) + kS(n-1,k),$$

where both terms on the right side are nonzero. The first term counts set partitions in $SP(n,k)$ where n appears in a block by itself, and the second term counts set partitions where n appears in a block with other elements. As in §6.10, we number the blocks of a set partition π from 0 to $k-1$, ordering the blocks by increasing minimum element.

We now describe the `first`, `last`, and `next` subroutines for the sets $SP(n,k)$. Whenever $S(n,k) = 1$, the subroutines `first` and `last` return the unique set partition in $SP(n,k)$. This set partition is $\{\{1,2,\ldots,n\}\}$ when $k = 1$, and it is $\{\{1\},\{2\},\ldots,\{n\}\}$ when $k = n$. For $1 < k < n$, the first object in $SP(n,k)$ is defined recursively to be the first object in $SP(n-1,k-1)$ with the block $\{n\}$ adjoined. The last object in $SP(n,k)$ is defined to be the last object in $SP(n-1,k)$ with n inserted into the block with the largest minimum element.

Now suppose $\pi \in SP(n,k)$ is not the last element in this set. We define $\texttt{next}(\pi)$ recursively via several cases. Case 1: $\{n\}$ is one of the blocks of π; let π' be π with this block removed. If π' is not last in $SP(n-1,k-1)$, then $\texttt{next}(\pi)$ is $\texttt{next}(\pi')$ with $\{n\}$ adjoined. If π' is last in $SP(n-1,k-1)$, then $\texttt{next}(\pi)$ is the first object in $SP(n-1,k)$ with n inserted into the lowest-numbered block. Case 2: n appears with other elements in the jth block of π; let π' be π with n deleted from its block. If π' is not last in $SP(n-1,k)$, then $\texttt{next}(\pi)$ is $\texttt{next}(\pi')$ with n inserted into the jth block. If π' is last in $SP(n-1,k)$, then $\texttt{next}(\pi)$ is the first object of $SP(n-1,k)$ with n inserted into block $j+1$. This completes the recursive definition of the `next` subroutine.

We can unravel the recursive calls to obtain more explicit descriptions of the successor subroutines. We find that the first and last objects in $SP(n,k)$ are the set partitions

$$\texttt{first}(SP(n,k)) = \{\{1,2,\ldots,n-k+1\},\{n-k+2\},\ldots,\{n-1\},\{n\}\};$$

$$\texttt{last}(SP(n,k)) = \{\{1\},\{2\},\ldots,\{k-1\},\{k,k+1,\ldots,n-1,n\}\}.$$

For $\pi \in SP(n,k)$, we can compute $\texttt{next}(\pi)$ as follows. Write out all the blocks of π and start erasing the symbols n, $n-1$, $n-2$, one at a time, until first encountering a set partition that is the last object in its class. As each symbol is erased, remember whether it was in a block by itself or the number of the block it was in. Suppose the final erased symbol is m, leaving the set partition π' that is last in $SP(m-1,r)$. Case 1: If m was in a block by itself just before being erased, replace π' by the first set partition in $SP(m-1,r+1)$, and insert m in the lowest-numbered block of this new set partition. Case 2: If m was in the jth block of π' just before being erased, replace π' by the first set partition in $SP(m-1,r)$, and insert m in block $j+1$ of this new set partition. In both cases, continue by restoring the symbols $m+1$ through n, one at a time, giving them the same block status they had originally. For example, if $m+1$ was originally in a block by itself when it got erased, put it back into the set partition in a block by itself. If instead $m+1$ was erased from the sth block, put it back into the sth block of the current set partition.

Here are some examples of the algorithm. Given $\pi = \{\{1\},\{2,8\},\{3,4,5,6\},\{7,9\}\}$, we delete 9 from block 3, then delete 8 from block 1, then delete 7 from a block by itself. We now have the set partition $\pi' = \{\{1\},\{2\},\{3,4,5,6\}\}$, which is the last object in $SP(6,3)$. Following Case 1 with $m = 7$, we replace π' with the first object in $SP(6,4)$, namely $\{\{1,2,3\},\{4\},\{5\},\{6\}\}$. Now we add 7 to block 0, add 8 to block 1, and add 9 to block 3, obtaining the final output $\texttt{next}(\pi) = \{\{1,2,3,7\},\{4,8\},\{5\},\{6,9\}\}$. To find the next object after this one, we delete 9 from block 3, 8 from block 1, 7 from block 0, 6 from a block by itself, 5 from a block by itself, and 4 from a block by itself, finally obtaining $\{\{1,2,3\}\}$ which is last in its class. Applying Case 1 with $m = 4$, we pass to the first object in $SP(3,2)$, namely $\{\{1,2\},\{3\}\}$, and put 4 in block 0. Restoring the remaining elements eventually produces the set partition $\{\{1,2,4,7\},\{3,8\},\{5\},\{6,9\}\}$.

Now suppose $\pi = \{\{1\},\{2,7\},\{3,4,5,6\},\{8,9\}\}$. To compute the next object, delete 9 from block 3, then delete 8 from a block by itself, then delete 7 from block 1, producing the set partition $\pi' = \{\{1\},\{2\},\{3,4,5,6\}\}$ that is the last object in $SP(6,3)$. Following Case 2 with $m = 7$, we replace π' by the first object in $SP(6,3)$, namely $\{\{1,2,3,4\},\{5\},\{6\}\}$. Now we add 7 to block 2, then add 8 in a block by itself, then add 9 to block 3, obtaining $\texttt{next}(\pi) = \{\{1,2,3,4\},\{5\},\{6,7\},\{8,9\}\}$. The successor of this object is found to be $\{\{1,2,3,5\},\{4\},\{6,7\},\{8,9\}\}$. To find the successor of $\{\{1,8\},\{2,7\},\{3,6\},\{4,5\}\}$, remove 8 from block 0, remove 7 from block 1, and remove 6 from block 2, producing the object $\{\{1\},\{2\},\{3\},\{4,5\}\}$ that is last in $SP(5,4)$. The first object in $SP(5,4)$ is $\{\{1,2\},\{3\},\{4\},\{5\}\}$. Now we add 6 to block 3, add 7 to block 1, and add 8 to block 0, obtaining the answer $\{\{1,2,8\},\{3,7\},\{4\},\{5,6\}\}$.

6.16 Successor Algorithms for Dyck Paths

We now use the First-Return Recursion for Catalan Numbers from §2.10 to develop a successor algorithm for Dyck paths. Let $DP(n)$ denote the set of Dyck paths of order n; recall that such a path is a sequence of n north steps and n east steps that never go below the line $y = x$. The Catalan numbers $C_n = |DP(n)|$ satisfy the recursion

$$C_n = C_0C_{n-1} + C_1C_{n-2} + \cdots + C_{k-1}C_{n-k} + \cdots + C_{n-1}C_0 \quad \text{for all } n > 0, \tag{6.4}$$

with initial condition $C_0 = 1$. The term $C_{k-1}C_{n-k}$ in this recursion counts Dyck paths of the form $\pi = \mathrm{N}\pi_1\mathrm{E}\pi_2$, where π_1 is a Dyck path of order $k-1$ (shifted to start at $(0,1)$) and π_2 is a Dyck path of order $n-k$ (shifted to start at (k,k)); the index k records where the full path π first returns to the diagonal line $y = x$.

Applying the Successor Sum Rule and the Successor Product Rule to the recursion above, we are led to the following successor subroutines. Define $\texttt{first}(C_0)$ and $\texttt{last}(C_0)$ to be the unique path of length zero. For $n > 0$, recursively define

$$\texttt{first}(C_n) = \mathrm{N}\,\texttt{first}(C_0)\,\mathrm{E}\,\texttt{first}(C_{n-1}); \quad \texttt{last}(C_n) = \mathrm{N}\,\texttt{last}(C_{n-1})\,\mathrm{E}\,\texttt{last}(C_0).$$

Expanding these recursive definitions, we find that

$$\texttt{first}(C_n) = \mathrm{NENENE}\cdots\mathrm{NE} = (\mathrm{NE})^n; \quad \texttt{last}(C_n) = \mathrm{NN}\cdots\mathrm{NEE}\cdots\mathrm{E} = \mathrm{N}^n\mathrm{E}^n.$$

To compute $\texttt{next}(\pi)$ for $\pi \in C_n$, first find the *first-return factorization* $\pi = \mathrm{N}\pi_1\mathrm{E}\pi_2$ where $\pi_1 \in DP(k-1)$ and $\pi_2 \in DP(n-k)$. We use three cases to define $\pi^* = \texttt{next}(\pi, DP(n))$.

1. If $\pi_2 \neq \texttt{last}(DP(n-k))$, let $\pi^* = \mathrm{N}\pi_1\mathrm{E}\,\texttt{next}(\pi_2, DP(n-k))$.
2. If $\pi_2 = \texttt{last}(DP(n-k))$ and $\pi_1 \neq \texttt{last}(DP(k-1))$, let

$$\pi^* = \mathrm{N}\,\texttt{next}(\pi_1, DP(k-1))\,\mathrm{E}\,\texttt{first}(DP(n-k)).$$

3. If $\pi_2 = \texttt{last}(DP(n-k))$ and $\pi_1 = \texttt{last}(DP(k-1))$, let

$$\pi^* = \mathrm{N}\,\texttt{first}(DP(k))\,\mathrm{E}\,\texttt{first}(DP(n-k-1)).$$

6.39. Example. Let us compute $\texttt{next}(\text{N NENNEE E NNNEEE}, DP(7))$. The given input π factors as $\pi = \mathrm{N}\pi_1\mathrm{E}\pi_2$ where $\pi_1 = \text{NENNEE}$ and $\pi_2 = \text{NNNEEE}$. Here, $k = 4$, $k-1 = 3$, and $n-k = 3$. Since $\pi_2 = \texttt{last}(DP(3))$ but $\pi_1 \neq \texttt{last}(DP(3))$, we follow Case 2. We must now recursively compute $\pi_1' = \texttt{next}(\pi_1, DP(3))$. Here, π_1 factors as $\pi_1 = \mathrm{N}\pi_3\mathrm{E}\pi_4$, where π_3 is the empty word and $\pi_4 = \text{NNEE}$. This time, π_4 is the last Dyck path of order 2 and π_3 is the last Dyck path of order 0. Following Case 3, we have

$$\pi_1' = \mathrm{N}\,\texttt{first}(DP(1))\,\mathrm{E}\,\texttt{first}(DP(1)) = \text{N NE E NE}.$$

Using this result in the original calculation, we obtain

$$\texttt{next}(\pi) = \mathrm{N}\ \pi_1'\ \mathrm{E}\ \texttt{first}(DP(3)) = \text{N NNEENE E NENENE}.$$

Summary

- *Definitions.* For each positive integer n, let $[\![n]\!] = \{0, 1, 2, \ldots, n-1\}$. Given an n-element set S, a *ranking map* for S is a bijection $\mathrm{rk}_S : S \to [\![n]\!]$ given by a specific algorithm. An *unranking map* for S is a bijection $\mathrm{unrk}_S : [\![n]\!] \to S$ given by a specific algorithm. Given a total ordering $s_0 < s_1 < \cdots < s_{n-1}$ of S, a *successor algorithm* relative to this ordering consists of subroutines $\texttt{first}$, $\texttt{last}$, and $\texttt{next}$, where $\texttt{first}(S) = s_0$, $\texttt{last}(S) = s_{n-1}$, and $\texttt{next}(s_i, S) = s_{i+1}$ for $0 \leq i < n-1$.

- *The Bijective Sum Rule.* Suppose a set S is the union of pairwise disjoint finite sets $S_1, \ldots, S_k$. Given ranking maps $f_i : S_i \to [\![n_i]\!]$, there is a ranking map $f : S \to [\![n_1 + \cdots + n_k]\!]$ given by

$$f(x) = \sum_{j<i} n_j + f_i(x) \quad \text{for } x \in S_i.$$

  The map f is denoted $\sum_{i=1}^k f_i$ and depends on the given ordering of the sets S_i. To compute $f^{-1}(a)$, find the unique index i such that $\sum_{j<i} n_j \leq a < \sum_{j\leq i} n_j$, and let $f^{-1}(a) = f_i^{-1}(a - \sum_{j<i} n_j)$. See the pseudocode in Figure 6.1.

- *The Bijection Rule for Ranking Maps.* If $F : X \to Y$ is a bijection and $\mathrm{rk} : Y \to [\![n]\!]$ is a ranking map for Y, then $\mathrm{rk} \circ F$ is a ranking map for X. If $\mathrm{unrk} : [\![n]\!] \to Y$ is an unranking map for Y, then $F^{-1} \circ \mathrm{unrk}$ is an unranking map for X.

- *Product Maps.* Given positive integers $n_1, n_2, \ldots, n_k$, there is a bijection

$$p_{n_1,n_2,\ldots,n_k} : [\![n_1]\!] \times [\![n_2]\!] \times \cdots \times [\![n_k]\!] \to [\![n_1 n_2 \cdots n_k]\!] \quad \text{given by}$$

$p_{n_1,n_2\ldots,n_k}(c_1,c_2,\ldots,c_k) = c_1n_2\cdots n_k + c_2n_3\cdots n_k + \cdots + c_{k-1}n_k + c_k$ for $0 \le c_i < n_i$. The following algorithm computes $p^{-1}_{n_1,\ldots,n_k}(a)$: for i looping from k down to 1, let $r_i = a \bmod n_i$, then replace a by $a \operatorname{div} n_i$; the output is $(r_1,\ldots,r_k)$. If $n_i = b$ for every i, then $p^{-1}_{b,\ldots,b}(a)$ is the base b representation of a.

- *The Bijective Product Rule.* Given sets $S_1,\ldots,S_k$ with $|S_i| = n_i$ and a bijection $F : S_1 \times \cdots \times S_k \to X$, suppose we know ranking maps for each S_i. A ranking map for X is given by

$$\mathrm{rk}_X(x) = p_{n_1,\ldots,n_k}(\mathrm{rk}_{S_1}(x_1),\ldots,\mathrm{rk}_{S_k}(x_k)) \text{ where } (x_1,\ldots,x_k) = F^{-1}(x).$$

An unranking map for X is given by

$$\mathrm{unrk}_X(a) = F(\mathrm{unrk}_{S_1}(c_1),\ldots,\mathrm{unrk}_{S_k}(c_k)) \text{ where } (c_1,\ldots,c_k) = p^{-1}_{n_1,\ldots,n_k}(a).$$

This rule is implemented by the pseudocode in Figure 6.3. For the special case $k = 2$, see Figure 6.2.

- *Ranking Partial Permutations.* The following rules rank k-permutations $w = w_1w_2\cdots w_k$ of an ordered alphabet $A = \{x_0 < x_1 < \cdots < x_{n-1}\}$. To find $\mathrm{rk}(w)$, first compute $(j_1,\ldots,j_k)$ by letting j_i be the number of letters preceding w_i in A that are different from $w_1,\ldots,w_{i-1}$. Then calculate $\mathrm{rk}(w) = p_{n,n-1,\ldots,n-k+1}(j_1,\ldots,j_k)$. To unrank a, find $p^{-1}_{n,n-1,\ldots,n-k+1}(a) = (j_1,\ldots,j_k)$, and then recover $w_1,\ldots,w_k$ from left to right by letting w_i be the (j_i+1)th smallest letter in $A - \{w_1,\ldots,w_{i-1}\}$.

- *Ranking Subsets.* Suppose we are ranking k-element subsets of an ordered alphabet $A = \{x_0 < x_1 < \cdots < x_{n-1}\}$. If $B = \{x_{i_1} < x_{i_2} < \cdots < x_{i_k}\}$, then $\mathrm{rk}(B) = \sum_{j=1}^{k}\binom{i_j}{j}$. To unrank a given integer a, recover $i_k,\ldots,i_1$ (in this order) by choosing the largest possible value that will not cause the partial sum of binomial coefficients $\binom{i_j}{j}$ to exceed a. This ranking method leads to a listing of k-element subsets in which all subsets containing x_{n-1} appear after all subsets not containing x_{n-1}; each sublist is ordered in the same way relative to x_{n-2}, and so on.

- *Ranking Anagrams.* The following recursive formula can be used to rank words $w \in \mathcal{R}(a_1^{n_1}\cdots a_k^{n_k})$ in alphabetical order: if $w = a_iw'$, then

$$\mathrm{rk}(w) = \sum_{j<i}\frac{(n_1+\cdots+n_k-1)!}{n_1!\cdots(n_j-1)!\cdots n_k!} + \mathrm{rk}(w').$$

To unrank a given integer b, choose i as large as possible so that the sum in the previous formula does not exceed b; subtract the sum for this choice of i from b; unrank the result recursively; and prepend the letter a_i to obtain the final answer.

- *The Successor Sum Rule.* Assume S is the union of nonempty, pairwise disjoint finite sets $S_1,\ldots,S_k$. Given successor algorithms for each S_i, we define $\texttt{first}(S) = \texttt{first}(S_1)$, $\texttt{last}(S) = \texttt{last}(S_k)$, $\texttt{next}(x,S) = \texttt{next}(x,S_i)$ if $x \in S_i$ is not last in S_i, and $\texttt{next}(x,S) = \texttt{first}(S_{i+1})$ if $x \in S_i$ is last in S_i. When some sets S_j may be empty, we modify these subroutines by skipping over any empty sets; see the pseudocode in Figure 6.4.

- *Successor Algorithm for Anagrams.* Given the ordered alphabet $\{a_1 < a_2 < \cdots < a_k\}$ and integers $n_i \ge 0$, the first anagram in $\mathcal{R}(a_1^{n_1}\cdots a_k^{n_k})$ is $a_1^{n_1}\cdots a_k^{n_k}$; the last anagram is the reversal of this word. To obtain the next anagram after $w = w_1\cdots w_n$, choose the maximal i with $w_i < w_{i+1}$; modify $w_i\cdots w_n$ by replacing w_i with the next larger letter in this suffix, then sorting the remaining letters of the suffix into weakly increasing order. See the pseudocode in Figure 6.6. This method can also be used to find successors of permutations and k-element subsets (viewed as binary words).

- *The Successor Product Rule.* Assume we know successor algorithms for the finite nonempty sets $S_1, \ldots, S_k$ and a bijection $F : S_1 \times \cdots \times S_k \to X$. Define

$$\texttt{first}(X) = F(\texttt{first}(S_1), \ldots, \texttt{first}(S_k)) \text{ and } \texttt{last}(X) = F(\texttt{last}(S_1), \ldots, \texttt{last}(S_k)).$$

  To compute $\texttt{next}(x, X)$, first compute $(x_1, \ldots, x_k) = F^{-1}(x)$. Find the maximal index i with $x_i \neq \texttt{last}(S_i)$, replace x_i by $\texttt{next}(x_i, S_i)$, replace x_j by $\texttt{first}(S_j)$ for all $j > i$, and return the answer $F(x_1, \ldots, x_k)$. See the pseudocode in Figure 6.7.

Exercises

6-1. Suppose $f : \{a, b, c\} \to [\![3]\!]$ and $g : \{d, e\} \to [\![2]\!]$ are defined by $f(a) = 1$, $f(b) = 2$, $f(c) = 0$, $g(d) = 1$, $g(e) = 0$. Compute the bijections $f + g$ and $g + f$.

6-2. (a) Use the maps f and g in the previous exercise and the Bijective Cartesian Product Rule to find a bijection from $\{a, b, c\} \times \{d, e\}$ onto $[\![6]\!]$. (b) Use f and g to build a bijection from $\{d, e\} \times \{a, b, c\}$ onto $[\![6]\!]$.

6-3. Let $S = \{w < z < y < x\}$ and $T = \{c < a < b\}$. (a) Describe the `first`, `last`, and `next` functions constructed by the Successor Sum Rule for the set $S \cup T$. (b) Repeat (a) for the set $T \cup S$. (c) Find the successor subroutines constructed by the Successor Product Rule for the set $S \times T$. (d) Repeat (c) for the set $T \times S$.

6-4. Compute: (a) $p_{7,5}(4, 3)$; (b) $p_{7,5}(3, 4)$; (c) $p_{5,7}(4, 3)$; (d) $p_{5,7}(3, 4)$; (e) $p_{7,5}^{-1}(22)$; (f) $p_{5,7}^{-1}(22)$.

6-5. Find: (a) $p_{2,2,2,2,2}(0, 1, 1, 0, 1)$; (b) $p_{2,2,2,2,2}^{-1}(29)$; (c) $p_{7,7,7}(3, 0, 6)$; (d) $p_{7,7,7}^{-1}(306)$; (e) $p_{10,10,10}^{-1}(306)$.

6-6. Find: (a) $p_{5,4,3,2,1}(3, 3, 0, 1, 0)$; (b) $p_{5,4,3,2,1}^{-1}(111)$; (c) $p_{3,6,2,6}(2, 5, 0, 4)$; (d) $p_{3,6,2,6}^{-1}(150)$; (e) $p_{6,2,6,3}^{-1}(150)$; (f) $p_{6,6,3,2}^{-1}(150)$.

6-7. Consider the product set $X = [\![3]\!] \times [\![4]\!]$. (a) View X as the disjoint union of the sets $X_i = \{i\} \times [\![4]\!]$, for $i = 0, 1, 2$. Let $f_i : X_i \to [\![4]\!]$ be the bijection $f_i(i, y) = y$. Compute the bijections $f_0 + f_1 + f_2$ and $f_2 + f_1 + f_0$, which map X to $[\![12]\!]$. (b) View X as the disjoint union of the sets $X^{(j)} = [\![3]\!] \times \{j\}$, for $j = 0, 1, 2, 3$. Let $g_j : X^{(j)} \to [\![3]\!]$ be the bijection $g_j(x, j) = x$. Compute the bijection $g_0 + g_1 + g_2 + g_3 : X \to [\![12]\!]$. (c) Compute the bijection $p_{3,4} : X \to [\![12]\!]$. Is this one of the maps found in (a) or (b)? (d) Let $t : X \to [\![4]\!] \times [\![3]\!]$ be the bijection $t(i, j) = (j, i)$. Compute the bijection $p_{4,3} \circ t : X \to [\![12]\!]$. Is this one of the maps found in (a) or (b)?

6-8. Rank the following four-letter words: (a) alto; (b) zone; (c) rank; (d) four; (e) word.

6-9. Unrank the following numbers in $[\![26^4]\!]$ to obtain four-letter words: (a) $115,287$; (b) $396,588$; (c) $392,581$; (d) $338,902$; (e) $275,497$.

6-10. (a) Rank the six-letter word "unrank." (b) Unrank 199,247,301 to get a 6-letter word. (c) What happens if we unrank 199,247,301 to get a k-letter word where $k > 6$?

6-11. A fraternity name consists of either two or three capital Greek letters. There are 24 letters in the Greek alphabet, ordered as follows:

ΑΒΓΔΕΖΗΘΙΚΛΜΝΞΟΠΡΣΤΥΦΧΨΩ.

Assume an ordering of fraternity names consisting of all two-letter names in alphabetical order, followed by all three-letter names in alphabetical order. Compute the rank of: (a) ΦΒΚ; (b) ΔΔ; (c) ΔΔΔ; (d) ΑΧΩ. Now, unrank: (e) 144; (f) 1440; (g) 13931.

6-12. Repeat (a)–(g) of the previous exercise, assuming the names are ordered so that all three-letter names precede all two-letter names, with names of each length in alphabetical order.

6-13. Repeat (a)–(g) of the previous exercise, assuming the names are ordered in alphabetical order (so that, for example, ΔΔ is immediately preceded by ΔΓΩ and immediately followed by ΔΔA).

6-14. Consider the set of four-digit even numbers (no leading zeroes allowed) that do not contain the digit 6. (a) Use the Product Rule to count this set. (b) Find a ranking bijection that will list these numbers in increasing numerical order. (c) Use (b) to rank 1234, 2500, and 9708. (d) Now unrank 1234, 2501, and 666.

6-15. Consider five-letter palindromes, ranked in alphabetical order. (a) Rank the palindromes LEVEL and MADAM. (b) Unrank 1581 and 12,662. (c) Find the first and last palindromes in the ranking that are real English words.

6-16. A Virginia license plate consists of three uppercase letters followed by four digits. For arcane bureaucratic reasons, license plate 0 is ZZZ-9999, followed by ZZZ-9998, ..., ZZZ-0000, ZZY-9999, etc. Use this system to rank the license plates: (a) ZCF-2073; (b) JXB-2007; (c) ABC-1234. Now, unrank: (d) 7,777,777; (e) 123,456,789.

6-17. Repeat the previous exercise assuming a new ordering, honoring the 400th anniversary of Jamestown, where license plate 0 is JAM-1607, and license plates count forward in lexicographic order (wrapping around from ZZZ-9999 to AAA-0000).

6-18. Let $A = \{\text{a,b,c,d,e,f}\}$. (a) Compute the ranks of bfdc and fdac among all 4-permutations of A. (b) Unrank 232 to get a 4-permutation of A.

6-19. Let $A = \{\text{a,b,c,d,e,f,g}\}$. (a) Compute the rank of ecagdb among all permutations of A. (b) Unrank 583 to get a permutation of A.

6-20. (a) Compute the rank of 42153 among all permutations of $\{1, 2, \ldots, 5\}$. (b) Unrank 46 to obtain a permutation of $\{1, 2, \ldots, 5\}$.

6-21. (a) Compute the rank of 36281745 among all permutations of $\{1, 2, \ldots, 8\}$. (b) Unrank 23,419 to obtain a permutation of $\{1, 2, \ldots, 8\}$.

6-22. Let $A = \{\text{a,b,c,d,e,f,g,h}\}$. (a) Use the ranking formula for 4-element subsets of A to rank the subsets {a,c,e,g}, {b,c,d,h}, and {d,e,f,h}. (b) Unrank 30, 40, and 50 to obtain 4-element subsets of A.

6-23. (a) Devise a ranking algorithm for k-element subsets of an n-element alphabet based on the recursion $C(n,k) = C(n-1,k-1) + C(n-1,k)$, which differs from the recursion in §6.7 due to the reversal of the order of terms on the right side. (b) Describe informally the order in which the ranking algorithm in (a) will produce the k-element subsets. (c) Answer the ranking and unranking questions in the previous exercise using this new ranking algorithm.

6-24. (a) Find the ranks of bbccacba and cabcabbc in the set $\mathcal{R}(\text{a}^2\text{b}^3\text{c}^3)$, ordered alphabetically. (b) Unrank 206 and 497 to get anagrams in $\mathcal{R}(\text{a}^2\text{b}^3\text{c}^3)$.

6-25. (a) Compute the rank of MISSISSIPPI among the set of all anagrams in $\mathcal{R}(\text{I}^4\text{MP}^2\text{S}^4)$ (listed alphabetically). (b) Which anagram in this set has rank 33,333?

6-26. (a) Use the ranking maps in §6.9 to rank the integer partitions $(3,3,3)$, $(5,2,2)$, and $(4,3,2,1)$. (b) Which integer partition in $P(12,3)$ has rank 6? (c) Which integer partition in $P(15,4)$ has rank 22? (d) Which integer partition in $P(20,6)$ has rank 47?

6-27. Use the ranking algorithm from §6.9 to list all integer partitions of 8 into four parts.

6-28. Follow the method used in Example 6.26 to enumerate all integer partitions of 7.

6-29. Use the ranking algorithm from §6.10 to list all set partitions of $\{1, 2, 3, 4, 5\}$ with three blocks.

6-30. (a) Use the algorithms in §6.10 to rank the following set partitions relative to the set $SP(n,k)$: $\{\{1,3\},\{2,4,5\}\}$; $\{\{1,5,7\},\{2\},\{3,4,8\},\{6\}\}$. (b) Unrank 247 to obtain a set partition in $SP(7,4)$. (c) Unrank 1492 to obtain a set partition in $SP(8,4)$.

6-31. (a) Use the method of §6.11 to find the rank of the rooted tree

$$T = \{(1,1),(2,1),(3,1),(4,1),(10,1),(7,1),(6,7),(5,6),(8,5),(11,6),(9,11)\}.$$

(b) Unrank 1,609,765 to obtain a rooted tree on nine vertices rooted at vertex 1.

6-32. Use the algorithms in §6.13 to find the successor of each word in the appropriate set of anagrams: (a) ccbabdc; (b) abcddcba; (c) 01101011; (d) 33212312; (e) UKULELE.

6-33. Find the next ten anagrams following 112021220 in $\mathcal{R}(0^2 1^3 2^4)$.

6-34. Find the successor of each permutation: (a) 3641275; (b) 6754321; (c) 135798642; (d) 123567984.

6-35. Find the next ten permutations following 416573829.

6-36. Find the successor of each subset in the set of all 4-element subsets of {a,b,c,d,e,f,g,h}: (a) {b,c,d,e}; (b) {a,c,f,h}; (c) {d,e,g,h}; (d) {a,b,c,h}.

6-37. In the set of all 6-element subsets of $\{0,1,\ldots,9\}$, find the next seven subsets following $\{2,3,4,6,7,8\}$.

6-38. Find the successor of each set partition in $SP(8,4)$: (a) $\{\{1,7\},\{2\},\{3,5,6\},\{4,8\}\}$; (b) $\{\{1,2\},\{3,4\},\{5,6\},\{7,8\}\}$; (c) $\{\{1,4,5\},\{6,8\},\{2,3\},\{7\}\}$; (d) $\{\{1\},\{2,3,6,7\},\{4,5\},\{8\}\}$.

6-39. Find the next ten set partitions following $\{\{1,4,6\},\{2,5,9\},\{3,7,8\}\}$ in $SP(9,3)$.

6-40. Use the first-return recursion (6.4) to create ranking and unranking algorithms for Dyck paths.

6-41. Use the unranking algorithm in the previous exercise to list all Dyck paths of order (a) 4; (b) 5.

6-42. Rank the following Dyck paths.
(a) NNENEENNNEEE
(b) NNNEEENNENEE
(c) NNNENEENNEENNENEEE
(d) NENENENENNNNNEEEEE

6-43. (a) Unrank 52 to get a Dyck path of order 6. (b) Unrank 335 to get a Dyck path of order 7. (c) Unrank 1000 to get a Dyck path of order 8.

6-44. Find the successor of each Dyck path in $DP(6)$.
(a) NNENEENNNEEE
(b) NNNEEENNENEE
(c) NNENEENENENE
(d) NENENENNENEE

6-45. Find the next ten Dyck paths following NENNENNEENEENENNEENE in $DP(10)$.

6-46. Fix $m,n \in \mathbb{Z}_{>0}$. Write ranking and unranking algorithms for lattice paths from $(0,0)$ to (mn,n) that never go below the line $x = my$ based on Recursion 2.25.

6-47. Write a successor algorithm for lattice paths from $(0,0)$ to (mn,n) that never go below the line $x = my$ based on Recursion 2.25.

6-48. Use Recursion 2.22 to create ranking and unranking algorithms for the set of k-element multisets using an n-element ordered alphabet.

6-49. Use the previous exercise to find the rank of each multiset using the alphabet {a,b,c,d,e,f}. (a) [b,b,c,d,d,d]; (b) [a,a,d,f,f]; (c) [a,c,d,f]; (d) [c,c,c,c,c].

6-50. Unrank the following integers to get 6-element multisets using the alphabet {a,b,c,d,e}. (a) 132; (b) 31; (c) 207; (d) 99.

6-51. Find a successor algorithm for the set of k-element multisets using an n-element ordered alphabet.

6-52. Use the previous exercise to find the successor of each multiset using the alphabet $\{1, 2, 3, 4, 5, 6\}$: (a) $[1, 3, 4, 6]$; (b) $[5, 5, 6, 6]$; (c) $[4, 4, 4, 4]$; (d) $[2, 5, 5, 5]$.

6-53. Repeat the previous five exercises, but now use ranking and successor algorithms based on one of the bijections in §1.12.

6-54. Write a successor algorithm for the set $P(n, k)$ of integer partitions of n with first part k, based on the recursion (6.3).

6-55. Use the previous exercise to find the successors of the following integer partitions: (a) $(9, 3)$; (b) $(7, 4, 2, 1)$; (c) $(3, 3, 1, 1, 1)$.

6-56. (a) Write a successor algorithm for the set $P'(n, k)$ of all integer partitions with exactly k parts. (b) Repeat the previous exercise using this successor algorithm.

6-57. Let $CP(n, k)$ be the set of permutations of $\{1, 2, \ldots, n\}$ with k cycles. Create ranking and unranking algorithms for these sets based on Recursion 3.46.

6-58. Use the previous exercise to rank the following permutations in the sets $CP(n, k)$: (a) 35412; (b) 231564798; (c) 23451.

6-59. Unrank the following integers to obtain permutations in $CP(7, 3)$: (a) 377; (b) 901; (c) 1616.

6-60. Write a successor algorithm for permutations in $CP(n, k)$ based on Recursion 3.46.

6-61. Use the previous exercise to find the successor of each permutation in $CP(n, k)$: (a) 35412; (b) 231564798; (c) 23451.

6-62. Create ranking and unranking algorithms for derangements based on Recursion 4.17.

6-63. Use the previous exercise to rank the following derangements: (a) 43512; (b) 3527614; (c) 789123654.

6-64. Unrank the following integers to obtain derangements in D_7: (a) 35; (b) 419; (c) 1776.

6-65. Write a successor algorithm for derangements based on Recursion 4.17.

6-66. Use the previous exercise to find the successor of each derangement: (a) 25431; (b) 7543216; (c) 214365.

6-67. Write ranking and unranking algorithms for the set $SP(n)$ of all set partitions of an n-element set based on Recursion 2.46 for Bell numbers.

6-68. Use the previous exercise to rank the following set partitions in $SP(n)$.
(a) $\{\{1, 2, 4\}, \{3, 5, 6\}, \{7, 8\}\}$
(b) $\{\{1, 7\}, \{2, 4, 5\}, \{3, 8\}, \{6\}\}$
(c) $\{\{1, 8\}, \{2, 7\}, \{3, 6\}, \{4, 5\}\}$

6-69. Unrank the following integers to get set partitions of $\{1, 2, \ldots, 8\}$: (a) 1394; (b) 2758; (c) 4026.

6-70. Write a successor algorithm for the set $SP(n)$ of all set partitions of $\{1, 2, \ldots, n\}$ based on Recursion 2.46.

6-71. Use the previous exercise to find the successor of each set partition.
(a) $\{\{1, 7\}, \{2, 4, 5\}, \{3, 8\}, \{6\}\}$
(b) $\{\{1, 8\}, \{2, 7\}, \{3, 6\}, \{4, 5\}\}$
(c) $\{\{1\}, \{2, 3, 4\}, \{5, 6, 7, 8\}\}$

6-72. Develop ranking and unranking algorithms for 231-avoiding permutations. Find the rank of 1 5 2 4 3 11 7 6 10 8 9. Unrank 231 to get a 231-avoiding permutation of length 7.

6-73. Create ranking and unranking algorithms for the set of subsets of $\{1, 2, \ldots, n\}$ that do not contain two consecutive integers.

6-74. Use the previous exercise to rank the following subsets (take $n = 10$): (a) $\{1, 3, 9\}$; (b) $\{2, 5, 7, 10\}$; (c) $\{2, 4, 6, 9\}$.

6-75. Unrank the following integers to get subsets of $\{1, 2, \ldots, 10\}$ with no two consecutive integers: (a) 1; (b) 42; (c) 130.

6-76. Write a successor algorithm for the set of subsets of $\{1, 2, \ldots, n\}$ that do not contain two consecutive integers.

6-77. Use the previous exercise to find the successor of each subset: (a) $\{1, 3, 5\}$; (b) $\{2, 5, 7, 10\}$; (c) $\{4, 6, 8, 10\}$.

6-78. Create algorithms to rank and unrank integer partitions of n into k *distinct* parts. Use your algorithms to rank the partition $(10, 7, 6, 3, 1)$ and unrank 10 to get a partition of 20 into three distinct parts.

6-79. We can view a deck of playing cards as the set $D = S \times V$, where $S = \{\clubsuit, \diamondsuit, \heartsuit, \spadesuit\}$ is the set of suits and $V = \{A, 2, 3, 4, 5, 6, 7, 8, 9, 10, J, Q, K\}$ is the set of values. (a) Describe a ranking algorithm for D. Find the rank of $6\diamondsuit$, $2\clubsuit$, and $K\spadesuit$. (b) Describe an unranking algorithm for D. Unrank 47, 13, and 31. (c) Describe a successor algorithm for D. Find the successor of $10\diamondsuit$, $K\heartsuit$, and $A\spadesuit$.

6-80. Repeat the previous exercise, now viewing the deck of cards as the set $D = V \times S$.

6-81. (a) Rank the hand $\{5\clubsuit, 7\diamondsuit, 8\spadesuit, 10\heartsuit, J\heartsuit\}$ among all possible poker hands. (b) Unrank 1,159,403 to get one of the $\binom{52}{5}$ possible poker hands. In this and later exercises, use the ordering of cards in the deck

$$A\clubsuit < 2\clubsuit < \cdots < K\clubsuit < A\diamondsuit < \cdots < A\heartsuit < \cdots < A\spadesuit < \cdots < K\spadesuit.$$

6-82. Create ranking and unranking maps for the set of flush poker hands by applying the Bijective Product Rule. (Include straight flushes.)

6-83. Use the previous exercise to rank these flush poker hands: (a) $\{3\heartsuit, 7\heartsuit, 10\heartsuit, J\heartsuit, K\heartsuit\}$; (b) $\{A\spadesuit, 2\spadesuit, 4\spadesuit, 5\spadesuit, 6\spadesuit\}$; (c)$\{4\diamondsuit, 5\diamondsuit, 8\diamondsuit, Q\diamondsuit, K\diamondsuit\}$.

6-84. Unrank these integers to get flush poker hands: (a) 4716; (b) 2724; (c) 295.

6-85. Create a successor algorithm for the set of flush poker hands.

6-86. Use the previous exercise to find the successor of each flush poker hand.
(a) $\{3\heartsuit, 7\heartsuit, 10\heartsuit, J\heartsuit, K\heartsuit\}$
(b) $\{A\spadesuit, 2\spadesuit, 4\spadesuit, 5\spadesuit, 6\spadesuit\}$
(c) $\{9\diamondsuit, 10\diamondsuit, J\diamondsuit, Q\diamondsuit, K\diamondsuit\}$.

6-87. Create ranking and unranking maps for the set of straight poker hands (including flushes).

6-88. Use the previous exercise to rank these straight poker hands: (a) $\{3\clubsuit, 4\heartsuit, 5\diamondsuit, 6\spadesuit, 7\clubsuit\}$; (b) $\{A\spadesuit, 2\heartsuit, 3\diamondsuit, 4\clubsuit, 5\heartsuit\}$; (c) $\{J\heartsuit, Q\heartsuit, K\clubsuit, 10\diamondsuit, A\spadesuit\}$.

6-89. Unrank these integers to get straight poker hands: (a) 1574; (b) 8877; (c) 4900.

6-90. Create a successor algorithm for the set of straight poker hands.

6-91. Use the previous exercise to find the successor of each straight poker hand.
(a) $\{3\clubsuit, 4\heartsuit, 5\diamondsuit, 6\spadesuit, 7\clubsuit\}$
(b) $\{A\spadesuit, 2\spadesuit, 3\spadesuit, 4\spadesuit, 5\spadesuit\}$
(c) $\{J\heartsuit, Q\heartsuit, K\heartsuit, 10\heartsuit, A\heartsuit\}$

6-92. Create ranking and unranking maps for the set of four-of-a-kind poker hands.

6-93. Use the previous exercise to rank these four-of-a-kind hands: (a) $\{3\clubsuit, 8\heartsuit, 8\diamondsuit, 8\spadesuit, 8\clubsuit\}$; (b) $\{5\spadesuit, 8\heartsuit, 5\diamondsuit, 5\clubsuit, 5\heartsuit\}$; (c) $\{A\heartsuit, Q\heartsuit, Q\clubsuit, Q\diamondsuit, Q\spadesuit\}$.

6-94. Unrank these integers to get four-of-a-kind hands: (a) 600; (b) 264; (c) 117.

6-95. Create a successor algorithm for the set of four-of-a-kind poker hands.

6-96. Use the previous exercise to find the successor of each four-of-a-kind hand.
(a) $\{3\clubsuit, 8\heartsuit, 8\diamondsuit, 8\spadesuit, 8\clubsuit\}$
(b) $\{K\spadesuit, 4\heartsuit, K\diamondsuit, K\clubsuit, K\heartsuit\}$
(c) $\{K\spadesuit, 10\heartsuit, 10\clubsuit, 10\diamondsuit, 10\spadesuit\}$

6-97. Create ranking and unranking maps for the set of full house poker hands.

6-98. Use the previous exercise to rank these full house hands: (a) $\{J\clubsuit, J\diamondsuit, J\spadesuit, 9\clubsuit, 9\heartsuit\}$. (b) $\{3\clubsuit, 3\heartsuit, 3\diamondsuit, 9\spadesuit, 9\diamondsuit\}$. (c) $\{A\spadesuit, A\heartsuit, A\diamondsuit, K\heartsuit, K\clubsuit\}$.

6-99. Unrank these integers to get full house hands: (a) 515; (b) 3082; (c) 483.

6-100. Create a successor algorithm for the set of full house poker hands.

6-101. Use the previous exercise to find the successor of each full house hand.
(a) $\{K\spadesuit, K\heartsuit, 3\diamondsuit, 3\spadesuit, 3\clubsuit\}$
(b) $\{5\spadesuit, 5\heartsuit, K\diamondsuit, K\spadesuit, K\heartsuit\}$
(c) $\{J\spadesuit, J\heartsuit, 10\heartsuit, 10\diamondsuit, 10\spadesuit\}$

6-102. Create ranking and unranking maps for the set of three-of-a-kind poker hands.

6-103. Use the previous exercise to rank these three-of-a-kind hands:
(a) $\{J\clubsuit, J\diamondsuit, J\spadesuit, 9\clubsuit, 5\heartsuit\}$. (b) $\{3\clubsuit, 3\heartsuit, 3\diamondsuit, A\spadesuit, 2\diamondsuit\}$. (c) $\{A\spadesuit, A\heartsuit, A\diamondsuit, Q\heartsuit, K\clubsuit\}$.

6-104. Unrank these integers to get three-of-a-kind hands: (a) 21,751; (b) 8; (c) 50,004.

6-105. Create a successor algorithm for the set of three-of-a-kind poker hands.

6-106. Use the previous exercise to find the successor of each three-of-a-kind hand.
(a) $\{J\clubsuit, J\diamondsuit, J\spadesuit, 9\clubsuit, 5\heartsuit\}$
(b) $\{K\diamondsuit, K\heartsuit, K\spadesuit, 5\spadesuit, 6\heartsuit\}$
(c) $\{3\spadesuit, 3\heartsuit, 3\diamondsuit, Q\spadesuit, K\spadesuit\}$

6-107. Create ranking and unranking maps for the set of two-pair poker hands.

6-108. Use the previous exercise to rank these two-pair hands: (a) $\{2\spadesuit, 2\clubsuit, 9\clubsuit, 9\heartsuit, K\diamondsuit\}$; (b) $\{A\diamondsuit, A\heartsuit, 7\spadesuit, Q\diamondsuit, Q\heartsuit\}$; (c) $\{K\heartsuit, K\clubsuit, Q\clubsuit, Q\diamondsuit, J\clubsuit\}$.

6-109. Unrank these integers to get two-pair hands: (a) 71,031; (b) 99,482; (c) 1417.

6-110. Create a successor algorithm for the set of two-pair poker hands.

6-111. Use the previous exercise to find the successor of each two-pair hand.
(a) $\{4\clubsuit, 4\heartsuit, 8\diamondsuit, 8\spadesuit, K\spadesuit\}$
(b) $\{K\spadesuit, Q\heartsuit, K\heartsuit, Q\clubsuit, J\heartsuit\}$
(c) $\{A\spadesuit, A\heartsuit, 9\clubsuit, 10\diamondsuit, 9\spadesuit\}$

6-112. Fix $s, t \in \mathbb{Z}_{>0}$. (a) Use the Bijective Sum Rule to construct a bijection F from $[\![s]\!] \times [\![t]\!]$ to $[\![st]\!]$ by writing the domain of F as the disjoint union of sets $S_i = [\![s]\!] \times \{j\}$ for $j \in [\![t]\!]$. (b) How is the map F related to the bijection $p_{t,s}$?

6-113. Fix $s, t \in \mathbb{Z}_{>0}$. Describe the successor algorithm for the product set $[\![s]\!] \times [\![t]\!]$ obtained from the Successor Sum Rule by writing this product set as the disjoint union of sets $S_i = [\![s]\!] \times \{j\}$ for $j \in [\![t]\!]$.

6-114. Give careful proofs of the Bijective Sum Rules 6.2 and 6.3.

6-115. Integer Division Theorem. Prove that for all $a \in \mathbb{Z}$ and all nonzero $b \in \mathbb{Z}$, there exist unique $q, r \in \mathbb{Z}$ with $a = bq + r$ and $0 \le r < |b|$. Describe an algorithm for computing q and r given a and b.

6-116. Prove that for all $a \in \mathbb{Z}$ and all nonzero $b \in \mathbb{Z}$, there exist unique $q, r \in \mathbb{Z}$ with $a = bq + r$ and $-|b|/2 < r \leq |b|/2$. Describe an algorithm for computing q and r given a and b.

6-117. Prove that the algorithm in Remark 6.10 correctly computes $p^{-1}_{n_1,\ldots,n_k}(a)$.

6-118. Given $a, b \in \mathbb{Z}$ with $b > 0$, recall that $a \bmod b$ is the unique remainder $r \in [\![b]\!]$ such that $a = bq + r$ for some $q \in \mathbb{Z}$. (a) Given $s, t > 0$, consider the map $f : [\![st]\!] \to [\![s]\!] \times [\![t]\!]$ given by $f(x) = (x \bmod s, x \bmod t)$ for $x \in [\![st]\!]$. Prove that f is a bijection iff $\gcd(s,t) = 1$. (b) Generalize (a) to maps from $[\![s_1 s_2 \cdots s_k]\!]$ to $[\![s_1]\!] \times [\![s_2]\!] \times \cdots \times [\![s_k]\!]$. (c) Can you find an algorithm for inverting the maps in (a) and (b) when they are bijections?

6-119. Fix $k \in \mathbb{Z}_{>0}$. Prove that every $m \in \mathbb{Z}_{\geq 0}$ can be written in exactly one way in the form $m = \sum_{j=1}^{k} \binom{i_j}{j}$, where $0 \leq i_1 < i_2 < \cdots < i_k$.

6-120. Fix $k \in \mathbb{Z}_{>0}$. Use the previous exercise to find an explicit formula for a bijection $f : \mathbb{Z}^k_{\geq 0} \to \mathbb{Z}_{\geq 0}$.

6-121. Suppose we rewrite the recursion for Stirling numbers in the form

$$S(n,k) = S(n-1,k)k + S(n-1,k-1) \quad \text{for all } n,k > 0.$$

(a) Use the Bijective Product Rule and Sum Rule (taking terms in the order written here) to devise ranking and unranking algorithms for set partitions in $SP(n,k)$. (b) Rank the partition $\pi = \{\{1,7\},\{2,4,5\},\{3,8\},\{6\}\}$ and unrank 111 to obtain an element of $SP(7,3)$ (cf. Example 6.27). (c) Repeat Exercise 6-30 using the new ranking algorithm.

6-122. (a) Use the pruning bijections in §3.12 to develop ranking and unranking algorithms for the set of trees with vertex set $\{v_1, \ldots, v_n\}$ such that $\deg(v_i) = d_i$ for all i (where $d_1, \ldots, d_n$ are positive integers summing to $2n-2$). (b) Given $(d_1, \ldots, d_9) = (1,2,1,1,1,2,3,1,4)$, find the rank of the tree shown in Figure 3.15. (c) Unrank 129 to obtain a tree with the degrees d_i from part (b).

6-123. Describe a successor algorithm for ranking rooted trees with vertex set $\{1, 2, \ldots, n\}$ rooted at vertex 1. Compute the successor of the tree shown in Figure 3.9.

6-124. Given a totally ordered finite set $S = \{s_0 < s_1 < \cdots < s_{n-1}\}$, a *predecessor algorithm* for S consists of subroutines `first`, `last`, and `prev`, where `first`$(S) = s_0$, `last`$(S) = s_{n-1}$, and `prev`$(s_i, S) = s_{i-1}$ for $0 < i \leq n-1$. Formulate versions of the Bijection Rule, Sum Rule, and Product Rule for creating predecessor algorithms.

6-125. (a) Write an algorithm for finding the predecessor of a word $w \in \mathcal{R}(a_1^{n_1} \cdots a_k^{n_k})$, listing anagrams alphabetically. (b) Find the ten anagrams preceding cbbadcbabcb.

6-126. (a) Write an algorithm for finding the predecessor of a Dyck path based on the first-return recursion (6.4). (b) Find the ten Dyck paths preceding NNNENEENNEENNENEEE.

6-127. Devise a ranking algorithm for four-letter words in which Q is always followed by U (so Q cannot be the last letter). Use your algorithm to rank AQUA and QUIT and to unrank 1000. Can you find an algorithm that generates these words in alphabetical order? Can you generalize to n-letter words?

6-128. Devise a ranking algorithm for five-letter words that never have two consecutive vowels. Use your algorithm to rank BILBO and THIRD and to unrank 9999. Can you find an algorithm that generates these words in alphabetical order? Can you generalize to n-letter words?

Notes

Our presentation of ranking and unranking maps emphasizes the automatic construction of bijections via the Bijective Sum Rule and the Bijective Product Rule. For a somewhat different approach based on a multigraph model, see the papers [134, 135]. Other discussions of ranking and related problems can be found in the texts [9, 95, 122]. An encyclopedic treatment of algorithms for generating combinatorial objects may be found in Knuth's comprehensive treatise [73, §7.2]. The exposition of successor algorithms in this chapter is based on [81].

Part II

Algebraic Combinatorics

7

Groups, Permutations, and Group Actions

This chapter contains an introduction to some aspects of group theory that are directly related to combinatorial problems. The first part of the chapter defines the initial concepts of group theory and derives some fundamental properties of permutations and symmetric groups. The second part of the chapter discusses *group actions*, which have many applications to algebra and combinatorics. In particular, group actions can be used to solve counting problems in which symmetry must be taken into account. For example, how many ways can we color a 5×5 chessboard with seven colors if all rotations and reflections of a given colored board are considered the same? The theory of group actions provides systematic methods for solving problems like this one.

7.1 Definition and Examples of Groups

A group is an abstract structure in which any two elements can be combined using an operation (analogous to addition or multiplication of numbers) obeying the algebraic axioms in the following definition.

7.1. Definition: Groups. A *group* is a set G with a binary operation $\star : G \times G \to G$ satisfying these axioms:

$$\begin{array}{ll} \forall x, y, z \in G, x \star (y \star z) = (x \star y) \star z & \text{(associativity);} \\ \exists e \in G, \forall x \in G, x \star e = x = e \star x & \text{(identity);} \\ \forall x \in G, \exists y \in G, x \star y = e = y \star x & \text{(inverses).} \end{array}$$

The requirement that $\star$ map $G \times G$ *into* G is often stated explicitly as the following axiom:

$$\forall x, y \in G, x \star y \in G \qquad \text{(closure).}$$

A group G is called *Abelian* or *commutative* iff G satisfies this additional axiom:

$$\forall x, y \in G, x \star y = y \star x \qquad \text{(commutativity).}$$

7.2. Example: Additive Groups. The set $\mathbb{Z}$ of all integers, with addition as the operation, is a commutative group. The identity element is $e = 0$, and the (additive) inverse of $x \in \mathbb{Z}$ is $-x \in \mathbb{Z}$. Similarly, $\mathbb{Q}$ and $\mathbb{R}$ and $\mathbb{C}$ are all commutative groups under addition. The set $\mathbb{Z}_{>0}$ is not a group under addition because there is no identity element *in the set* $\mathbb{Z}_{>0}$. The set $\mathbb{Z}_{\geq 0}$ is not a group under addition because 1 is in $\mathbb{Z}_{\geq 0}$, but 1 has no additive inverse *in the set* $\mathbb{Z}_{\geq 0}$. The three-element set $S = \{-1, 0, 1\}$ is not a group under addition because closure fails: $1 \in S$ and $1 \in S$, but $1 + 1 = 2 \notin S$.

7.3. Example: Multiplicative Groups. The set $\mathbb{Q}_{>0}$ of strictly positive rational numbers is a commutative group under multiplication. The identity element is $e = 1$, and the inverse of $a/b \in \mathbb{Q}_{>0}$ is $b/a \in \mathbb{Q}_{>0}$. Similarly, the set $\mathbb{R}_{>0}$ of strictly positive real numbers is a group under multiplication. The set $\mathbb{Q}$ is not a group under multiplication because 0 has no inverse. On the other hand, $\mathbb{Q}-\{0\}$, $\mathbb{R}-\{0\}$, and $\mathbb{C}-\{0\}$ are groups under multiplication. So is the two-element set $\{-1,1\} \subseteq \mathbb{Q}$ and the four-element set $\{1,i,-1,-i\} \subseteq \mathbb{C}$.

7.4. Example: Symmetric Groups. Let X be any set, and let $\mathrm{Sym}(X)$ be the set of all bijections $f : X \to X$. For $f,g \in \mathrm{Sym}(X)$, define $f \circ g : X \to X$ to be the composite function that sends $x \in X$ to $f(g(x))$. Then $f \circ g \in \mathrm{Sym}(X)$ since the composition of bijections is a bijection, so the closure axiom holds. Given $f,g,h \in \mathrm{Sym}(X)$, note that both of the functions $(f \circ g) \circ h : X \to X$ and $f \circ (g \circ h) : X \to X$ send $x \in X$ to $f(g(h(x)))$. So these functions are equal, proving the associativity axiom. Next, take e to be the bijection $\mathrm{id}_X : X \to X$, which is defined by $\mathrm{id}_X(x) = x$ for all $x \in X$. One immediately checks that $f \circ \mathrm{id}_X = f = \mathrm{id}_X \circ f$ for all $f \in \mathrm{Sym}(X)$, so the identity axiom holds. Finally, given a bijection $f \in \mathrm{Sym}(X)$, there exists an inverse function $f^{-1} : X \to X$ that is also a bijection, and which satisfies $f \circ f^{-1} = \mathrm{id}_X = f^{-1} \circ f$. So the inverse axiom holds. This completes the verification that $\mathrm{Sym}(X)$ is a group. This group is called the *symmetric group on X*, and elements of $\mathrm{Sym}(X)$ are called *permutations of X*. Symmetric groups play a central role in group theory and are closely related to group actions. In the special case when $X = \{1,2,\ldots,n\}$, we write S_n to denote the group $\mathrm{Sym}(X)$.

Most of the groups $\mathrm{Sym}(X)$ are *not* commutative. For instance, define $f,g \in S_3$ by

$$f(1)=2,\ f(2)=1,\ f(3)=3; \quad g(1)=3,\ g(2)=2,\ g(3)=1.$$

We see that $(f \circ g)(1) = f(g(1)) = 3$, whereas $(g \circ f)(1) = g(f(1)) = 2$. So $f \circ g \neq g \circ f$, and the commutativity axiom fails.

7.5. Example: Integers Modulo n. Let n be a fixed positive integer. Let $\mathbb{Z}_n$ be the set $[\![n]\!] = \{0,1,2,\ldots,n-1\}$. We define a binary operation on $\mathbb{Z}_n$ by setting, for all $x,y \in \mathbb{Z}_n$,

$$x \oplus y = \begin{cases} x+y & \text{if } x+y<n; \\ x+y-n & \text{if } x+y \geq n. \end{cases}$$

Closure follows from this definition, once we note that $0 \leq x+y \leq 2n-2$ for all $x,y \in \mathbb{Z}_n$. The identity element is 0. The inverse of 0 is 0, while for $x > 0$ in $\mathbb{Z}_n$, the inverse of x is $n-x \in \mathbb{Z}_n$. Associativity follows from the relations

$$(x \oplus y) \oplus z = \left\{ \begin{array}{ll} x+y+z & \text{if } x+y+z<n; \\ x+y+z-n & \text{if } n \leq x+y+z<2n; \\ x+y+z-2n & \text{if } 2n \leq x+y+z<3n; \end{array} \right\} = x \oplus (y \oplus z), \tag{7.1}$$

which can be established by a tedious case analysis. Commutativity of $\oplus$ follows from the definition and the commutativity of ordinary integer addition. We conclude that $\mathbb{Z}_n$ is an additive commutative group containing n elements. In particular, for every positive integer n, there exists a group of cardinality n.

7.6. Definition: Multiplication Tables. Given a finite group $G = \{x_1,\ldots,x_n\}$ with operation $\star$, a *multiplication table* for G is an $n \times n$ table, with rows and columns labeled by the group elements $x_i \in G$, such that the element in row i and column j is $x_i \star x_j$. When the operation is written additively, we refer to this table as the *addition table* for G. It is customary, but not mandatory, to take x_1 to be the identity element of G.

7.7. Example. The multiplication table for $\{1,i,-1,-i\} \subseteq \mathbb{C}$ and the addition table for

$\mathbb{Z}_4$ are shown here:

$\times$	1	i	-1	$-i$
1	1	i	-1	$-i$
i	i	-1	$-i$	1
-1	-1	$-i$	1	i
$-i$	$-i$	1	i	-1

$\oplus$	0	1	2	3
0	0	1	2	3
1	1	2	3	0
2	2	3	0	1
3	3	0	1	2

The reader may notice a relationship between the two tables: each row within the table is obtained from the preceding one by a cyclic shift one step to the left. Using terminology to be discussed later, this happens because each of the two groups under consideration is cyclic of size 4.

One can define a group operation by specifying its multiplication table. For example, here is the table for another group of size 4, which turns out not to be cyclic:

$\star$	a	b	c	d
a	a	b	c	d
b	b	a	d	c
c	c	d	a	b
d	d	c	b	a

The identity and inverse axioms can be checked from inspection of the table; we see that a is the identity, and every element is equal to its own inverse. There is no quick way to verify associativity by visual inspection of the multiplication table, but this axiom can be checked exhaustively using the table entries.

All of the groups in this example are commutative. This can be seen from the multiplication tables by noting the symmetry about the main diagonal line: the entry $x_i \star x_j$ in row i and column j always equals the entry $x_j \star x_i$ in row j and column i.

7.2 Basic Properties of Groups

We now collect some facts about groups that follow from the defining axioms.

First, *the identity element e in a group G is unique.* For, suppose $e' \in G$ also satisfies the identity axiom. On one hand, $e \star e' = e$ since e' is an identity element. On the other hand, $e \star e' = e'$ since e is an identity element. So $e = e'$. We use the symbol e_G to denote the unique identity element of an abstract group G. When the operation is addition or multiplication, we write 0_G or 1_G instead, dropping the G if it is understood from context.

Similarly, *the inverse of an element x in a group G is unique.* For suppose $y, y' \in G$ both satisfy the condition in the inverse axiom. Using the associativity axiom and the identity axiom,

$$y = y \star e = y \star (x \star y') = (y \star x) \star y' = e \star y' = y'.$$

We denote the unique inverse of x in G by the symbol x^{-1}. When the operation is written additively, the symbol $-x$ is used.

A product such as $x \star y$ is often written xy, except in the additive case. The associativity axiom can be used to show that any parenthesization of a product $x_1 x_2 \cdots x_n$ gives the same answer, so it is permissible to omit parentheses in products like these.

7.8. Theorem: Cancellation Laws and Inverse Rules. For all a, x, y in a group G: (a) if $ax = ay$ then $x = y$ (left cancellation); (b) if $xa = ya$ then $x = y$ (right cancellation); (c) $(x^{-1})^{-1} = x$ (double inverse rule); (d) $(xy)^{-1} = y^{-1}x^{-1}$ (inverse rule for products).

Proof. For (a), fix $a, x, y \in G$ and assume $ax = ay$; we must prove $x = y$. Multiply both sides of $ax = ay$ on the left by a^{-1} to get $a^{-1}(ax) = a^{-1}(ay)$. Then the associativity axiom gives $(a^{-1}a)x = (a^{-1}a)y$; the inverse axiom gives $ex = ey$; and the identity axiom gives $x = y$. The right cancellation law (b) is proved similarly. For (c), note that

$$(x^{-1})^{-1}x^{-1} = e = xx^{-1}$$

by the definition of the inverse of x^{-1} and the inverse of x. Right cancellation of x^{-1} yields $(x^{-1})^{-1} = x$. Similarly, routine calculations using the group axioms show that

$$(xy)^{-1}(xy) = e = (y^{-1}x^{-1})(xy),$$

so right cancellation of xy gives the inverse rule for products. □

7.9. Definition: Exponent Notation. Let G be a group written multiplicatively. Given $x \in G$, recursively define $x^0 = 1 = e_G$ and $x^{n+1} = x^n \star x$ for all $n \geq 0$. Define negative powers of x by $x^{-n} = (x^{-1})^n$ for all $n > 0$, where the x^{-1} on the right side denotes the inverse of x in G.

Informally, for positive n, x^n is the product of n copies of x. For negative n, x^n is the product of $|n|$ copies of the inverse of x. Note in particular that $x^1 = x$, and the negative power x^{-1} (as just defined) does reduce to the inverse of x. When G is written additively, we write nx instead of x^n; this denotes the sum of n copies of x for $n > 0$, or the sum of $|n|$ copies of $-x$ for $n < 0$.

7.10. Theorem: Laws of Exponents. Suppose G is a group, $x \in G$, and $m, n \in \mathbb{Z}$. In multiplicative notation, $x^{m+n} = x^m x^n$ and $x^{mn} = (x^n)^m$. If $x, y \in G$ satisfy $xy = yx$, then $(xy)^n = x^n y^n$. In additive notation, these results read: $(m+n)x = mx+nx$; $(mn)x = m(nx)$; and if $x + y = y + x$, then $n(x + y) = nx + ny$.

We omit the proof; the main idea is to use induction to establish the results for $m, n \geq 0$, and then use case analyses to handle the situations where m or n is negative.

7.3 Notation for Permutations

Permutations and symmetric groups appear frequently in the theory of groups and group actions. So it will be helpful to develop some notation for describing and visualizing permutations.

7.11. Definition: Two-Line Form of a Function. Let X be a finite set, and let $x_1, \ldots, x_n$ be a list of all the distinct elements of X in some fixed order. The *two-line form* of a function $f : X \to X$ relative to this ordering is the array

$$f = \begin{bmatrix} x_1 & x_2 & \cdots & x_n \\ f(x_1) & f(x_2) & \cdots & f(x_n) \end{bmatrix}.$$

If $X = \{1, 2, \ldots, n\}$, we usually display the elements of X on the top line in the order $1, 2, \ldots, n$.

7.12. Example. The notation $f = \begin{bmatrix} a & b & c & d & e \\ b & c & e & a & b \end{bmatrix}$ defines a function on the set $X =$

$\{a, b, c, d, e\}$ such that $f(a) = b$, $f(b) = c$, $f(c) = e$, $f(d) = a$, and $f(e) = b$. This function is not a permutation, since b occurs twice in the bottom row, and d never occurs.

The notation $g = \begin{bmatrix} 1 & 2 & 3 & 4 & 5 \\ 2 & 4 & 5 & 1 & 3 \end{bmatrix}$ defines an element of S_5 such that $g(1) = 2$, $g(2) = 4$, $g(3) = 5$, $g(4) = 1$, and $g(5) = 3$. Observe that the inverse of g sends 2 to 1, 4 to 2, and so on. So, we obtain one possible two-line form of g^{-1} by interchanging the rows in the two-line form of g:

$$g^{-1} = \begin{bmatrix} 2 & 4 & 5 & 1 & 3 \\ 1 & 2 & 3 & 4 & 5 \end{bmatrix}.$$

It is customary to write the numbers in the top line in increasing order. This can be accomplished by sorting the columns of the previous array:

$$g^{-1} = \begin{bmatrix} 1 & 2 & 3 & 4 & 5 \\ 4 & 1 & 5 & 2 & 3 \end{bmatrix}.$$

Recall that the group operation in $\mathrm{Sym}(X)$ is composition. We can compute the composition of two functions written in two-line form by tracing the effect of the composite function on each element. For instance,

$$\begin{bmatrix} a & b & c & d \\ b & d & a & c \end{bmatrix} \circ \begin{bmatrix} a & b & c & d \\ a & c & d & b \end{bmatrix} = \begin{bmatrix} a & b & c & d \\ b & a & c & d \end{bmatrix},$$

because the left side maps a to a and then to b; b maps to c and then to a; and so on.

If the ordering of X is fixed and known from context, we may omit the top line of the two-line form. This leads to one-line notation for a function defined on X.

7.13. Definition: One-Line Form of a Function. Let $X = \{x_1 < x_2 < \cdots < x_n\}$ be a finite totally ordered set. The *one-line form* of a function $f : X \to X$ is the array $[f(x_1)\ f(x_2) \cdots f(x_n)]$.

We use square brackets to avoid a conflict with the cycle notation to be introduced below. Sometimes we omit the brackets, identifying f with the word $f(x_1)f(x_2)\cdots f(x_n)$.

7.14. Example. The functions f and g in the preceding example are given in one-line form by writing $f = [b\ c\ e\ a\ b]$ and $g = [2\ 4\ 5\ 1\ 3]$. In word notation, $f = bceab$ and $g = 24513$. Note that the one-line form of an element of $\mathrm{Sym}(X)$ is literally a permutation (rearrangement) of the elements of X, as defined in §1.3. This explains why elements of this group are called permutations.

7.15. Cycle Notation for Permutations. Assume X is a finite set. Recall from §3.6 that any function $f : X \to X$ can be represented by a digraph with vertex set X and a directed edge $(i, f(i))$ for each i in X. An arbitrary digraph on X is the digraph of some function $f : X \to X$ iff every vertex in X has outdegree 1. In Theorem 3.45 we proved that the digraph of a permutation is a disjoint union of directed cycles. For example, Figure 7.1 displays the digraph of the permutation

$$h = \begin{bmatrix} 1 & 2 & 3 & 4 & 5 & 6 & 7 & 8 & 9 & 10 \\ 7 & 8 & 4 & 3 & 10 & 2 & 5 & 6 & 9 & 1 \end{bmatrix}.$$

We can describe a directed cycle in a digraph by traversing the edges in the cycle and listing the elements we encounter in the order of visitation, enclosing the whole list in parentheses. For example, the cycle containing 1 in Figure 7.1 can be described by writing

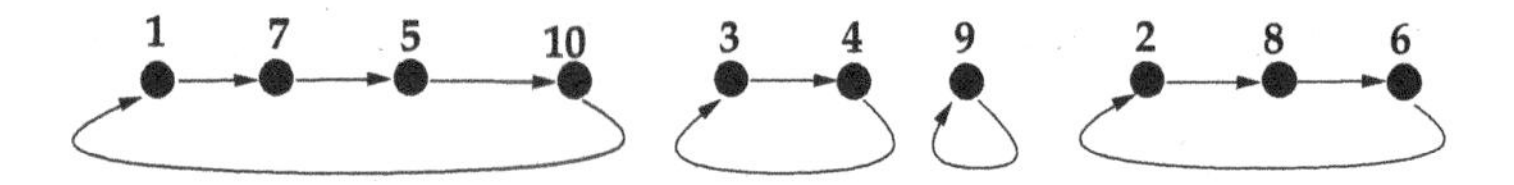

FIGURE 7.1
Digraph associated to the permutation h.

$(1, 7, 5, 10)$. The cycle containing 9 is denoted by (9). To describe the entire digraph of a permutation, we list all the cycles in the digraph, one after the other. For example, h can be written in cycle notation as

$$h = (1, 7, 5, 10)(2, 8, 6)(3, 4)(9).$$

This cycle notation is not unique. We are free to begin our description of each cycle at any vertex in the cycle, and we may also rearrange the order of the cycles. Furthermore, by convention it is permissible to omit some or all cycles of length 1. For example, some other cycle notations for h are

$$h = (5, 10, 1, 7)(3, 4)(9)(6, 2, 8) = (2, 8, 6)(4, 3)(7, 5, 10, 1).$$

To compute the inverse of a permutation written in cycle notation, we reverse the orientation of each cycle. For example,

$$h^{-1} = (10, 5, 7, 1)(6, 8, 2)(4, 3)(9).$$

7.16. Example. Using word notation, the group S_3 consists of these six elements:

$$S_3 = \{123,\ 213,\ 321,\ 132,\ 231,\ 312\}.$$

Using cycle notation, we can describe the elements of S_3 as follows:

$$S_3 = \{(1)(2)(3),\ (1, 2),\ (1, 3),\ (2, 3),\ (1, 2, 3),\ (1, 3, 2)\}.$$

7.17. Example. To compose permutations written in cycle notation, we must see how the composite function acts on each element. For instance, consider the product $(3, 5)(1, 2, 4) \circ (3, 5, 2, 1)$ in S_5. This composite function sends 1 to 3 and then 3 to 5, so 1 maps to 5. Next, 2 maps first to 1 and then to 2, so 2 maps to 2. Continuing similarly, we find that

$$(3, 5)(1, 2, 4) \circ (3, 5, 2, 1) = \left[\begin{array}{ccccc} 1 & 2 & 3 & 4 & 5 \\ 5 & 2 & 3 & 1 & 4 \end{array}\right] = (1, 5, 4)(2)(3).$$

With enough practice, one can proceed immediately to the cycle form of the answer without writing the two-line form or doing other scratch work.

7.18. Definition: k-cycles. For $k > 1$, a *k-cycle* is a permutation $f \in \mathrm{Sym}(X)$ whose digraph consists of one cycle of length k and all other cycles of length 1.

7.19. Remark. We can view the cycle notation for a permutation f as a way of factorizing *f in the group S_n* into a product of cycles. For example,

$$(1, 7, 5, 10)(2, 8, 6)(3, 4)(9) = (1, 7, 5, 10) \circ (2, 8, 6) \circ (3, 4) \circ (9).$$

Here we have expressed the single permutation on the left side as a product of four other permutations in S_{10}. The stated equality may be verified by checking that both sides have the same effect on each $x \in \{1, 2, \ldots, 10\}$.

7.20. Definition: cyc(f) and type(f). Given a permutation $f \in \mathrm{Sym}(X)$, let cyc(f) be the number of components (cycles) in the digraph for f. Let type(f) be the list of sizes of these components, including repetitions, and written in weakly decreasing order.

Note that type(f) is an integer partition of $n = |X|$.

7.21. Example. The permutation h in Figure 7.1 has $\mathrm{cyc}(h) = 4$ and $\mathrm{type}(h) = (4,3,2,1)$. The identity element of S_n, namely $\mathrm{id} = (1)(2)\cdots(n)$, has $\mathrm{cyc}(\mathrm{id}) = n$ and $\mathrm{type}(\mathrm{id}) = (1,\ldots,1) = (1^n)$. Table 7.1 displays the 24 elements of S_4 in cycle notation, collecting together all permutations with the same type and counting the number of permutations of each type. In Theorem 7.115, we give a general formula for the number of permutations of n objects having a given type.

Type	Permutations	Count
$(1,1,1,1)$	$(1)(2)(3)(4)$	1
$(2,1,1)$	$(1,2)$, $(1,3)$, $(1,4)$, $(2,3)$, $(2,4)$, $(3,4)$	6
$(2,2)$	$(1,2)(3,4)$, $(1,3)(2,4)$, $(1,4)(2,3)$	3
$(3,1)$	$(1,2,3)$, $(1,2,4)$, $(1,3,4)$, $(1,3,2)$, $(1,4,2)$, $(1,4,3)$, $(2,3,4)$, $(2,4,3)$	8
(4)	$(1,2,3,4)$, $(1,2,4,3)$, $(1,3,2,4)$, $(1,3,4,2)$, $(1,4,2,3)$, $(1,4,3,2)$	6

TABLE 7.1
Elements of S_4.

7.4 Inversions and Sign of a Permutation

In this section, we use inversions of permutations to define the *sign* function $\mathrm{sgn} : S_n \to \{+1,-1\}$. We then study factorizations of permutations into products of transpositions to derive facts about the sgn function.

7.22. Definition: Inversions and Sign of a Permutation. Let $w = w_1w_2\cdots w_n \in S_n$ be a permutation written in one-line form. An *inversion* of w is a pair of positions (i,j) such that $i < j$ and $w_i > w_j$. The number of inversions of w is denoted $\mathrm{inv}(w)$. The *sign* of w is defined to be $\mathrm{sgn}(w) = (-1)^{\mathrm{inv}(w)}$.

7.23. Example. Given $w = 42531$, we have $\mathrm{inv}(w) = 7$ and $\mathrm{sgn}(w) = -1$. The seven inversions of w are $(1,2)$, $(1,4)$, $(1,5)$, $(2,5)$, $(3,4)$, $(3,5)$, and $(4,5)$. For instance, $(1,4)$ is an inversion because $w_1 > w_4$ $(4 > 3)$. Table 7.2 displays $\mathrm{inv}(f)$ and $\mathrm{sgn}(f)$ for all $f \in S_3$.

Our next goal is to understand how the group operation in S_n (composition of permutations) is related to inversions and sign. For this purpose, we introduce special permutations called transpositions.

7.24. Definition: Transpositions. A *transposition* in S_n is a permutation f of the form (i,j), for some $i \neq j$ in $\{1,2,\ldots,n\}$. A *basic transposition* in S_n is a transposition $(i,i+1)$, for some $i \in \{1,2,\ldots,n-1\}$.

$f \in S_3$	inv(f)	sgn(f)
123	0	+1
132	1	−1
213	1	−1
231	2	+1
312	2	+1
321	3	−1

TABLE 7.2
Computing inv(f) and sgn(f) for $f \in S_3$.

Note that the transposition $f = (i, j)$ satisfies $f(i) = j$, $f(j) = i$, and $f(k) = k$ for all $k \neq i, j$. The following lemmas illuminate the connection between basic transpositions and the process of sorting the one-line form of a permutation into increasing order.

7.25. Lemma: Basic Transpositions and Sorting. Let $w = w_1 \cdots w_i w_{i+1} \cdots w_n \in S_n$ be a permutation in one-line form. For each $i \in \{1, 2, \ldots, n-1\}$,

$$w \circ (i, i+1) = w_1 \cdots w_{i+1} w_i \cdots w_n.$$

So *right-multiplication by the basic transposition* $(i, i+1)$ *interchanges the elements in positions* i *and* $i+1$ *of* w.

Proof. Let us evaluate the function $f = w \circ (i, i+1)$ at each k between 1 and n. When $k = i$, $f(i) = w(i+1)$. When $k = i+1$, $f(i+1) = w(i)$. When $k \neq i$ and $k \neq i+1$, $f(k) = w(k)$. So the one-line form of f is $w_1 \cdots w_{i+1} w_i \cdots w_n$, as needed. □

7.26. Lemma: Basic Transpositions and Inversions. Let $w = w_1 \cdots w_n \in S_n$ be a permutation in one-line form. For each $i \in \{1, 2, \ldots, n-1\}$,

$$\operatorname{inv}(w \circ (i, i+1)) = \begin{cases} \operatorname{inv}(w) + 1 & \text{if } w_i < w_{i+1}; \\ \operatorname{inv}(w) - 1 & \text{if } w_i > w_{i+1}. \end{cases}$$

Consequently, in all cases, we have

$$\operatorname{sgn}(w \circ (i, i+1)) = -\operatorname{sgn}(w).$$

Proof. We use the result of the previous lemma to compare the inversions of w and $w' = w \circ (i, i+1)$. Let $j < k$ be two indices between 1 and n, and consider various cases. First, if $j \neq i, i+1$ and $k \neq i, i+1$, then (j, k) is an inversion of w iff (j, k) is an inversion of w', since $w_j = w'_j$ and $w_k = w'_k$. Second, if $j = i$ and $k > i+1$, then (i, k) is an inversion of w iff $(i+1, k)$ is an inversion of w', since $w_i = w'_{i+1}$ and $w_k = w'_k$. Similar results hold in the cases $(j = i+1 < k)$, $(j < k = i)$, and $(j < i, k = i+1)$. The critical case is when $j = i$ and $k = i+1$. If $w_i < w_{i+1}$, then (j, k) is an inversion of w' but not of w. If $w_i > w_{i+1}$, then (j, k) is an inversion of w but not of w'. This establishes the first formula in the lemma. The remaining formula follows since $\operatorname{sgn}(w') = (-1)^{\operatorname{inv}(w) \pm 1} = (-1)^{\operatorname{inv}(w)}(-1) = -\operatorname{sgn}(w)$. □

The next lemma follows immediately from the definitions.

7.27. Lemma. For all $n \geq 1$, the identity permutation id $= [1\,2 \ldots n]$ is the unique element of S_n satisfying inv(id) $= 0$. Also, sgn(id) $= +1$.

If $f = (i, i+1)$ is a basic transposition, then the ordered pair $(i, i+1)$ is the only inversion of f, so inv(f) $= 1$ and sgn(f) $= -1$. More generally, we now show that any transposition has sign -1.

7.28. Lemma. If $f = (i, j)$ is any transposition in S_n, then $\text{sgn}(f) = -1$.

Proof. Since $(i, j) = (j, i)$, we may assume that $i < j$. Let us write f in two-line form:

$$f = \begin{bmatrix} 1 & \cdots & i & \cdots & j & \cdots & n \\ 1 & \cdots & j & \cdots & i & \cdots & n \end{bmatrix}.$$

We can find the inversions of f by inspecting the two-line form. The inversions are: all (i, k) with $i < k \leq j$; and all (k, j) with $i < k < j$. There are $j - i$ inversions of the first type and $j - i - 1$ inversions of the second type, hence $2(j - i) - 1$ inversions total. Since this number is odd, we conclude that $\text{sgn}(f) = -1$. □

7.29. Theorem: Inversions and Sorting. Let $w = w_1 w_2 \cdots w_n \in S_n$ be a permutation in one-line form. The number $\text{inv}(w)$ is the minimum number of steps required to sort the word w into increasing order by repeatedly interchanging two adjacent elements. Furthermore, w can be factored in S_n into the product of $\text{inv}(w)$ basic transpositions.

Proof. Given $w \in S_n$, it is certainly possible to sort w into increasing order in finitely many steps by repeatedly swapping adjacent elements. For instance, we can move 1 to the far left position in at most $n - 1$ moves, then move 2 to its proper position in at most $n - 2$ moves, and so on. Let m be the *minimum* number of moves of this kind that are needed to sort w. By Lemma 7.25, we can accomplish each sorting move by starting with w and repeatedly multiplying on the right by an appropriately chosen basic transposition. Each such multiplication either increases or decreases the inversion count by 1, according to Lemma 7.26. At the end, we have transformed w into the identity permutation. Combining these observations, we see that $0 = \text{inv}(\text{id}) \geq \text{inv}(w) - m$, so that $m \geq \text{inv}(w)$. On the other hand, consider the following particular sequence of sorting moves starting from w. If the current permutation w^* is not the identity, there exists a smallest index i with $w^*_i > w^*_{i+1}$. Apply the basic transposition $(i, i + 1)$, which reduces $\text{inv}(w^*)$ by 1, and continue. This sorting method will end in exactly $\text{inv}(w)$ steps, since id is the unique permutation with zero inversions. This proves it is possible to sort w in $\text{inv}(w)$ steps, so that $m \leq \text{inv}(w)$.

To prove the last part of the theorem, recall that the sorting process just described can be implemented by right-multiplying by basic transpositions. We therefore have an equation in S_n of the form

$$w \circ (i_1, i_1 + 1) \circ (i_2, i_2 + 1) \circ \cdots \circ (i_m, i_m + 1) = \text{id}.$$

Solving for w, and using the fact that $(i, j)^{-1} = (j, i) = (i, j)$, we get

$$w = (i_m, i_m + 1) \circ \cdots \circ (i_2, i_2 + 1) \circ (i_1, i_1 + 1),$$

which expresses w as a product of m basic transpositions. □

7.30. Example. Let us trace through the sorting algorithm in the preceding proof to write $w = 42531$ as a product of $\text{inv}(w) = 7$ basic transpositions. Since $4 > 2$, we first multiply w on the right by $(1, 2)$ to obtain

$$w \circ (1, 2) = 24531.$$

Observe that $\text{inv}(24531) = 6 = \text{inv}(w) - 1$. Next, since $5 > 3$, we multiply on the right by $(3, 4)$ to get

$$w \circ (1, 2) \circ (3, 4) = 24351.$$

The computation continues as follows:

$$\begin{aligned}
w \circ (1,2) \circ (3,4) \circ (2,3) &= 23451; \\
w \circ (1,2) \circ (3,4) \circ (2,3) \circ (4,5) &= 23415; \\
w \circ (1,2) \circ (3,4) \circ (2,3) \circ (4,5) \circ (3,4) &= 23145; \\
w \circ (1,2) \circ (3,4) \circ (2,3) \circ (4,5) \circ (3,4) \circ (2,3) &= 21345; \\
w \circ (1,2) \circ (3,4) \circ (2,3) \circ (4,5) \circ (3,4) \circ (2,3) \circ (1,2) &= 12345 = \mathrm{id}.
\end{aligned}$$

We now solve for w, which has the effect of reversing the order of the basic transpositions used to reach the identity:

$$w = (1,2) \circ (2,3) \circ (3,4) \circ (4,5) \circ (2,3) \circ (3,4) \circ (1,2).$$

It is also possible to find such a factorization by starting with the identity word and unsorting to reach w. Here it will not be necessary to reverse the order of the transpositions at the end. We illustrate this idea with the following computation:

$$\begin{aligned}
\mathrm{id} &= 12345; \\
\mathrm{id} \circ (3,4) &= 12435; \\
\mathrm{id} \circ (3,4) \circ (2,3) &= 14235; \\
\mathrm{id} \circ (3,4) \circ (2,3) \circ (1,2) &= 41235; \\
\mathrm{id} \circ (3,4) \circ (2,3) \circ (1,2) \circ (2,3) &= 42135; \\
\mathrm{id} \circ (3,4) \circ (2,3) \circ (1,2) \circ (2,3) \circ (4,5) &= 42153; \\
\mathrm{id} \circ (3,4) \circ (2,3) \circ (1,2) \circ (2,3) \circ (4,5) \circ (3,4) &= 42513; \\
\mathrm{id} \circ (3,4) \circ (2,3) \circ (1,2) \circ (2,3) \circ (4,5) \circ (3,4) \circ (4,5) &= 42531 = w.
\end{aligned}$$

So $w = (3,4) \circ (2,3) \circ (1,2) \circ (2,3) \circ (4,5) \circ (3,4) \circ (4,5)$. Observe that this is a different factorization of w from the one obtained earlier, although both involve seven basic transpositions. This shows that *factorizations of permutations into products of basic transpositions are not unique.* It is also possible to find factorizations involving more than seven factors, by interchanging two entries that are already in the correct order during the sorting of w into id. So the number of factors in such factorizations is not unique either; but we will see shortly that the *parity* of the number of factors (odd or even) *is* uniquely determined by w. In fact, the parity is odd when $\mathrm{sgn}(w) = -1$ and even when $\mathrm{sgn}(w) = +1$.

We now have enough machinery to prove the fundamental properties of the sign function.

7.31. Theorem: Properties of Sign. (a) For all $f, g \in S_n$, $\mathrm{sgn}(f \circ g) = \mathrm{sgn}(f) \cdot \mathrm{sgn}(g)$. (b) For all $f \in S_n$, $\mathrm{sgn}(f^{-1}) = \mathrm{sgn}(f)$.

Proof. (a) If $g = \mathrm{id}$, then the result is true since $f \circ g = f$ and $\mathrm{sgn}(g) = 1$ in this case. If $t = (i, i+1)$ is a basic transposition, then Lemma 7.26 shows that $\mathrm{sgn}(f \circ t) = -\mathrm{sgn}(f)$. Given a non-identity permutation g, use Theorem 7.29 to write g as a nonempty product of basic transpositions, say $g = t_1 \circ t_2 \circ \cdots \circ t_k$. Then, for every $f \in S_n$, repeated use of Lemma 7.26 gives

$$\begin{aligned}
\mathrm{sgn}(f \circ g) &= \mathrm{sgn}(ft_1 \cdots t_{k-1}t_k) = -\mathrm{sgn}(ft_1 \cdots t_{k-1}) \\
&= (-1)^2 \mathrm{sgn}(ft_1 \cdots t_{k-2}) = \cdots = \mathrm{sgn}(f)(-1)^k.
\end{aligned}$$

In particular, this equation is true when $f = \mathrm{id}$; in that case, we obtain $\mathrm{sgn}(g) = (-1)^k$.

Using this fact in the preceding equation (for arbitrary $f \in S_n$) produces $\text{sgn}(f \circ g) = \text{sgn}(f)\,\text{sgn}(g)$.

(b) By part (a), $\text{sgn}(f) \cdot \text{sgn}(f^{-1}) = \text{sgn}(f \circ f^{-1}) = \text{sgn}(\text{id}) = +1$. If $\text{sgn}(f) = +1$, it follows that $\text{sgn}(f^{-1}) = +1$. If instead $\text{sgn}(f) = -1$, then it follows that $\text{sgn}(f^{-1}) = -1$. So $\text{sgn}(f^{-1}) = \text{sgn}(f)$ in both cases. □

Iteration of Theorem 7.31 shows that

$$\text{sgn}(f_1 \circ \cdots \circ f_k) = \prod_{i=1}^{k} \text{sgn}(f_i). \tag{7.2}$$

7.32. Theorem: Factorizations into Transpositions. Let $f = t_1 \circ t_2 \circ \cdots \circ t_k$ be *any* factorization of $f \in S_n$ into a product of transpositions (not necessarily basic ones). Then $\text{sgn}(f) = (-1)^k$. In particular, the parity of k (odd or even) is uniquely determined by f.

Proof. By Lemma 7.28, $\text{sgn}(t_i) = -1$ for all i. The conclusion now follows by setting $f_i = t_i$ in (7.2). □

7.33. Theorem: Sign of a k-cycle. The sign of any k-cycle $(i_1, i_2, \ldots, i_k)$ is $(-1)^{k-1}$.

Proof. The result is already known for $k = 1$ and $k = 2$. For $k > 2$, one may check that the given k-cycle can be written as the following product of $k - 1$ transpositions:

$$(i_1, i_2, \ldots, i_k) = (i_1, i_2) \circ (i_2, i_3) \circ (i_3, i_4) \circ \cdots (i_{k-1}, i_k).$$

So the result follows from Theorem 7.32. □

We now show that the sign of a permutation f can be computed from $\text{type}(f)$ or $\text{cyc}(f)$.

7.34. Theorem: Cycle Type and Sign. Suppose $f \in S_n$ has $\text{type}(f) = \mu$. Then

$$\text{sgn}(f) = \prod_{i=1}^{\ell(\mu)} (-1)^{\mu_i - 1} = (-1)^{n-\ell(\mu)} = (-1)^{n-\text{cyc}(f)}.$$

Proof. Let the cycle decomposition of f be $f = C_1 \circ \cdots \circ C_{\ell(\mu)}$, where C_i is a μ_i-cycle. The result follows from the relations $\text{sgn}(f) = \prod_{i=1}^{\ell(\mu)} \text{sgn}(C_i)$ and $\text{sgn}(C_i) = (-1)^{\mu_i - 1}$. □

7.35. Example. The permutation $f = (4, 6, 2, 8)(3, 9, 1)(5, 10, 7)$ in S_{10} has $\text{type}(f) = (4, 3, 3)$, $\text{cyc}(f) = 3$, and $\text{sgn}(f) = (-1)^{10-3} = -1$.

7.5 Subgroups

Suppose G is a group with operation $\star : G \times G \to G$, and H is a subset of G. One might wonder if H is also a group with operation obtained by restricting $\star$ to the domain $H \times H$. This construction succeeds when H is a subgroup, as defined below.

7.36. Definition: Subgroups. Let G be a group with operation $\star$, and let H be a subset of G. H is called a *subgroup* of G, written $H \leq G$, iff H satisfies the following three *closure conditions*:

$e_G \in H$	(closure under identity);
$\forall a, b \in H,\ a \star b \in H$	(closure under the operation);
$\forall a \in H,\ a^{-1} \in H$	(closure under inverses).

A subgroup H is called *normal in* G, written $H \trianglelefteq G$, iff

$$\forall g \in G, \forall h \in H, ghg^{-1} \in H \qquad \text{(closure under conjugation)}.$$

Let us verify that when H is a subgroup of G, H really is a group using the restriction $\star'$ of $\star$ to $H \times H$. Since H is closed under the operation, $\star'$ does map $H \times H$ *into the codomain* H (not just into G), so the closure axiom holds for H. Since H is a subset of G, associativity holds in H because it is known to hold in G. The identity e of G lies in H by assumption. Since $e \star g = g = g \star e$ holds for all $g \in G$, the relation $e \star' h = h = h \star' e$ certainly holds for all $h \in H \subseteq G$. Finally, every element x of H has an inverse y (relative to $\star$) that lies *in* H, by assumption. Now y is still an inverse of x relative to $\star'$, so the inverse axiom holds. This completes the proof of the group axioms for H. It can be checked that if G is commutative then H is commutative, but the converse statement is not always true. Henceforth, we use the same symbol $\star$ (instead of $\star'$) to denote the operation in the subgroup H.

7.37. Example. We have the following chain of subgroups of the additive group $\mathbb{C}$:

$$\{0\} \leq \mathbb{Z} \leq \mathbb{Q} \leq \mathbb{R} \leq \mathbb{C}.$$

Likewise, $\{-1, 1\}$ and $\mathbb{Q}_{>0}$ are both subgroups of $\mathbb{Q}-\{0\}$ under multiplication. The set $\{0, 3, 6, 9\}$ is a subgroup of the additive group $\mathbb{Z}_{12}$; one can prove closure under addition and inverses by a finite case analysis, or by inspection of the relevant portion of the addition table for $\mathbb{Z}_{12}$. On the other hand, $\{0, 3, 6, 9\}$ is not a subgroup of $\mathbb{Z}$, since this set is not closed under addition or inverses.

7.38. Example. The sets $H = \{(1)(2)(3), (1, 2, 3), (1, 3, 2)\}$ and $K = \{(1)(2)(3), (1, 3)\}$ are subgroups of S_3, as one readily verifies. Moreover, H is normal in S_3, but K is not. The set $J = \{(1)(2)(3), (1, 3), (2, 3), (1, 3)\}$ is not a subgroup of S_3, since closure under the operation fails: $(1, 3)$ and $(2, 3)$ are in J, but $(1, 3) \circ (2, 3) = (1, 3, 2)$ is not in J. Here is a four-element normal subgroup of S_4:

$$V = \{(1)(2)(3)(4),\ (1, 2)(3, 4),\ (1, 3)(2, 4),\ (1, 4)(2, 3)\}.$$

Each element of V is its own inverse, and one confirms closure of V under the operation by checking all possible products. To prove the normality of V in S_4, it is helpful to use Theorem 7.112 later in this chapter.

7.39. Example. The set of even integers is a subgroup of the additive group $\mathbb{Z}$. This follows because the identity element zero is even; the sum of two even integers is even; and if x is even then $-x$ is even. More generally, let k be any fixed integer, and let $H = \{kn : n \in \mathbb{Z}\}$ consist of all integer multiples of k. A routine verification shows that H is a subgroup of $\mathbb{Z}$. We write $H = k\mathbb{Z}$ for brevity. The next theorem shows that we have found *all* the subgroups of $\mathbb{Z}$.

7.40. Theorem: Subgroups of $\mathbb{Z}$. Every subgroup H of the additive group $\mathbb{Z}$ has the form $k\mathbb{Z}$ for a unique integer $k \geq 0$.

Proof. We have noted that all the subsets $k\mathbb{Z}$ are indeed subgroups. Given an arbitrary subgroup H, consider two cases. If $H = \{0\}$, then $H = 0\mathbb{Z}$. Otherwise, H contains at least one nonzero integer m. If m is negative, then $-m \in H$ since H is closed under inverses. So, H contains strictly positive integers. Take k to be the least positive integer in H. We claim that $H = k\mathbb{Z}$. Let us prove that $kn \in H$ for all $n \in \mathbb{Z}$, so that $k\mathbb{Z} \subseteq H$. For $n \geq 0$, we use induction on n. When $n = 0$, we must prove $k0 = 0 \in H$, which holds since H contains the identity of $\mathbb{Z}$. When $n = 1$, we must prove $k1 = k \in H$, which is true by choice of

k. Assume $n \geq 1$ and $kn \in H$. Then $k(n+1) = kn + k \in H$ since $kn \in H$, $k \in H$, and H is closed under addition. Finally, for negative n, write $n = -m$ for some $m > 0$. Then $kn = -(km) \in H$ since $km \in H$ and H is closed under inverses.

The key step is to prove the reverse inclusion $H \subseteq k\mathbb{Z}$. Fix $z \in H$. Dividing z by k, we obtain $z = kq + r$ for some integers q, r with $0 \leq r < k$. By what we proved in the last paragraph, $k(-q) \in H$. So, $r = z - kq = z + k(-q) \in H$ since H is closed under addition. Now, since k is the *least* positive integer in H, we cannot have $0 < r < k$. The only possibility left is $r = 0$, so $z = kq \in k\mathbb{Z}$, as needed.

To prove that k is uniquely determined by H, suppose $k\mathbb{Z} = m\mathbb{Z}$ for $k, m \geq 0$. Now $k = 0$ iff $m = 0$, so we may assume $k, m > 0$. Since $k \in k\mathbb{Z} = m\mathbb{Z}$, k is a multiple of m. Similarly, m is a multiple of k. As both k and m are positive, this forces $k = m$, completing the proof. □

How can we find subgroups of a given group G? As we see next, each element $x \in G$ can be used to produce a subgroup of G.

7.41. Definition: Cyclic Subgroups and Cyclic Groups. Let G be a multiplicative group, and let $x \in G$. The *cyclic subgroup of G generated by x* is $\langle x \rangle = \{x^n : n \in \mathbb{Z}\}$. G is called a *cyclic* group iff there exists $x \in G$ with $G = \langle x \rangle$. When G is an additive group, $\langle x \rangle = \{nx : n \in \mathbb{Z}\}$.

One checks, using the Laws of Exponents, that the subset $\langle x \rangle$ of G really is a subgroup.

7.42. Example. The additive group $\mathbb{Z}$ is cyclic, since $\mathbb{Z} = \langle 1 \rangle = \langle -1 \rangle$. The subgroups $k\mathbb{Z} = \langle k \rangle$ considered above are cyclic subgroups of $\mathbb{Z}$. Theorem 7.40 shows that *every subgroup of $\mathbb{Z}$ is cyclic.* The additive groups $\mathbb{Z}_n$ are also cyclic; each of these groups (for $n > 1$) is generated by 1. The group $\{a, b, c, d\}$ with multiplication table given in Example 7.7 is *not* cyclic. To prove this, we compute all the cyclic subgroups of this group:

$$\langle a \rangle = \{a\}, \quad \langle b \rangle = \{a, b\}, \quad \langle c \rangle = \{a, c\}, \quad \langle d \rangle = \{a, d\}.$$

None of the cyclic subgroups equals the whole group, so the group is not cyclic. For a bigger example of a non-cyclic group, consider $\mathbb{Q}$ under addition. Any nonzero cyclic subgroup has the form $\langle a/b \rangle$ for some positive rational number a/b. One may check that $a/(2b)$ does not lie in this subgroup, so $\mathbb{Q} \neq \langle a/b \rangle$. Non-commutative groups furnish additional examples of non-cyclic groups, as the next result shows.

7.43. Theorem. Every cyclic group is commutative.

Proof. Let $G = \langle x \rangle$ be cyclic. Given $y, z \in G$, we can write $y = x^n$ and $z = x^m$ for some $n, m \in \mathbb{Z}$. Since integer addition is commutative, the Laws of Exponents give

$$yz = x^n x^m = x^{n+m} = x^{m+n} = x^m x^n = zy.$$ □

By adapting the proof of Theorem 7.40, one can show that *every subgroup of a cyclic group is cyclic.*

7.44. Example. The cyclic group $\mathbb{Z}_6$ has the following cyclic subgroups (which are *all* the subgroups of this group):

$$\langle 0 \rangle = \{0\}; \quad \langle 1 \rangle = \{0, 1, 2, 3, 4, 5\} = \langle 5 \rangle; \quad \langle 2 \rangle = \{0, 2, 4\} = \langle 4 \rangle; \quad \langle 3 \rangle = \{0, 3\}.$$

In the group S_4, $(1, 3, 4, 2)$ generates the cyclic subgroup

$$\langle (1, 3, 4, 2) \rangle = \{(1, 3, 4, 2),\ (1, 4)(3, 2),\ (1, 2, 4, 3),\ (1)(2)(3)(4)\}.$$

7.6 Automorphism Groups of Graphs

This section uses graphs to construct examples of subgroups of symmetric groups. These subgroups are needed later when we discuss applications of group theory to counting problems.

7.45. Definition: Automorphism Group of a Graph. Let K be a simple graph with vertex set X and edge set E. A *graph automorphism* of K is a bijection $f : X \to X$ such that, for all u, v in X, $\{u, v\} \in E$ iff $\{f(u), f(v)\} \in E$. Let $\text{Aut}(K)$ denote the set of all graph automorphisms of K. We make an analogous definition for directed simple graphs; here, the bijection f must satisfy $(u, v) \in E$ iff $(f(u), f(v)) \in E$ for all $u, v \in X$.

Using this definition, it can be checked that the automorphism group $\text{Aut}(K)$ is a subgroup of the symmetric group $\text{Sym}(X)$ under the operation $\circ$ (function composition). Informally, we can visualize a bijection $f : X \to X$ as a relabeling of the vertices that replaces each vertex label $u \in X$ by a new label $f(u)$. Such an f is an automorphism of K iff the relabeled graph has exactly the same edges as the original graph.

7.46. Example. Consider the graphs shown in Figure 7.2. The undirected cycle C_5 has

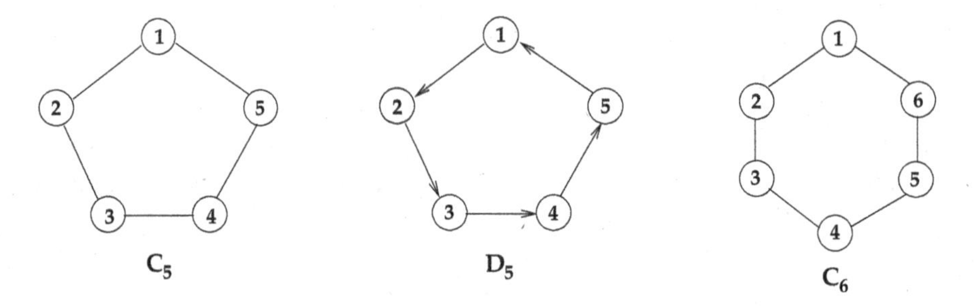

FIGURE 7.2
Graphs used to illustrate automorphism groups.

exactly ten automorphisms. They are given in one-line form in the following list:

$$\begin{array}{ccccc} [1\,2\,3\,4\,5], & [2\,3\,4\,5\,1], & [3\,4\,5\,1\,2], & [4\,5\,1\,2\,3], & [5\,1\,2\,3\,4], \\ [5\,4\,3\,2\,1], & [4\,3\,2\,1\,5], & [3\,2\,1\,5\,4], & [2\,1\,5\,4\,3], & [1\,5\,4\,3\,2]. \end{array}$$

The same automorphisms, written in cycle notation, look like this:

$$\begin{array}{lllll} (1)(2)(3)(4)(5), & (1,2,3,4,5), & (1,3,5,2,4), & (1,4,2,5,3), & (1,5,4,3,2), \\ (1,5)(2,4)(3), & (1,4)(2,3)(5), & (1,3)(4,5)(2), & (1,2)(3,5)(4), & (2,5)(3,4)(1). \end{array}$$

Geometrically, we can think of C_5 as a *necklace* with five beads. The first five automorphisms on each list can be implemented by rotating the necklace through various angles (rotation by zero is the identity map). The next five automorphisms arise by reflecting the necklace in five possible axes of symmetry.

Now consider the automorphism group of the *directed* cycle D_5. Every automorphism of the directed graph D_5 is automatically an automorphism of the associated undirected graph C_5, so $\text{Aut}(D_5) \leq \text{Aut}(C_5)$. However, not every automorphism of C_5 is an automorphism of D_5. In this example, the five rotations preserve the direction of the edges, hence are automorphisms of D_5. But the five reflections reverse the direction of the edges, so these are not elements of $\text{Aut}(D_5)$. We can write $\text{Aut}(D_5) = \langle(1,2,3,4,5)\rangle$, so that this automorphism group is cyclic of size 5.

The 6-cycle C_6 can be analyzed in a similar way. The automorphism group consists of six rotations and six reflections, which are given in cycle form below:

$$\begin{array}{llll} (1)(2)(3)(4)(5)(6), & (1,2,3,4,5,6), & (1,3,5)(2,4,6), & (1,4)(2,5)(3,6), \\ (1,5,3)(2,6,4), & (1,6,5,4,3,2), & (2,6)(3,5)(1)(4), & (1,2)(3,6)(4,5), \\ (1,3)(4,6)(2)(5), & (2,3)(5,6)(1,4), & (1,5)(2,4)(3)(6), & (1,6)(2,5)(3,4). \end{array}$$

The observations in the previous example generalize as follows.

7.47. Theorem: Automorphism Group of a Cycle. For $n \geq 3$, let C_n be the graph with vertex set $X = \{1, 2, \ldots, n\}$ and edges $\{1,2\}, \{2,3\}, \ldots, \{n-1,n\}, \{n,1\}$. Then $\text{Aut}(C_n)$ is a subgroup of S_n of size $2n$, called the *dihedral group of order* $2n$. The elements of this group (in one-line form) are the n permutations

$$[i,\ i+1,\ i+2,\ \ldots\ n,\ 1,\ 2,\ \ldots\ i-1] \qquad \text{for } 1 \leq i \leq n, \tag{7.3}$$

together with the reversals of these n words.

Proof. It is routine to check that the $2n$ permutations mentioned in the theorem preserve the edges of C_n and hence are automorphisms of this graph. We must show that these are the *only* automorphisms of C_n. Let g be any automorphism of C_n, and let $i = g(1)$. Now, since 1 and 2 are adjacent in C_n, $g(1)$ and $g(2)$ must also be adjacent in C_n. There are two cases: $g(2) = i+1$ or $g(2) = i-1$ (we read all outputs mod n, interpreting $n+1$ as 1 and $1-1$ as n). Suppose the first case occurs. Since 2 is adjacent to 3, we must have $g(3) = i+2$ or $g(3) = i$. But $i = g(1)$ and g is injective, so it must be that $g(3) = i+2$. Continuing around the cycle in this way, we see that g must be one of the permutations displayed in (7.3). Similarly, in the case where $g(2) = i-1$, we are forced to have $g(3) = i-2$, $g(4) = i-3$, and so on, so that g must be the reversal of one of the permutations in (7.3). □

The reasoning used in the preceding proof can be adapted to determine the automorphism groups of more complicated graphs. We stress that a graph automorphism does not necessarily arise from a rigid geometric motion such as a rotation or a reflection; the only requirement is that the vertex relabeling map preserves all edges.

7.48. Example. Consider the graph B displayed in Figure 7.3, which models a 5×5 chessboard. What are the automorphisms of B? We note that B has four vertices of degree 2: a, e, v, and z. An automorphism ϕ of B must restrict to give a permutation of these four vertices, since automorphisms preserve degree. Suppose, for example, that $\phi(a) = v$. What can $\phi(b)$ be in this situation? Answer: $\phi(b)$ must be a vertex adjacent to $\phi(a)$, namely q or w. In the case $\phi(b) = q$, $\phi(c) = k$ is forced by degree considerations, whereas $\phi(b) = w$ forces $\phi(c) = x$. Continuing this reasoning as we work around the outer border of the figure, we see that the action of ϕ on all vertices on the border is completely determined by where a and b go. A tedious but routine argument then shows that the images of the remaining vertices are also forced. Since a can map to one of the four corners, and then b can map to one of the two neighbors of $\phi(a)$, there are at most $4 \times 2 = 8$ automorphisms of B. Here are

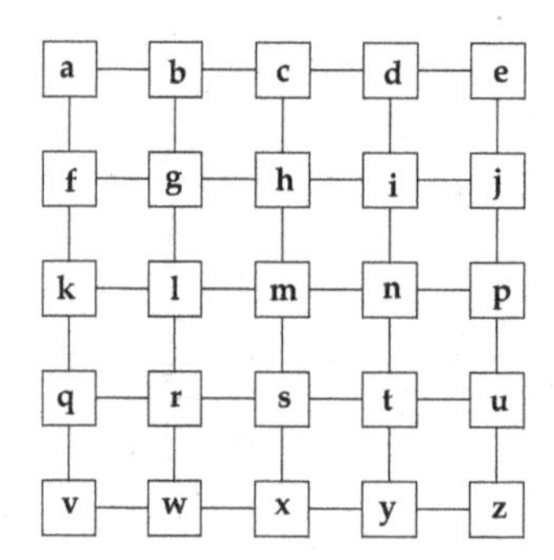

FIGURE 7.3
Graph representing a 5×5 chessboard.

the eight possibilities in cycle form:

$$\begin{aligned}
r_0 &= (a)(b)(c)\cdots(x)(y)(z) = \text{id};\\
r_1 &= (a,e,z,v)(b,j,y,q)(c,p,x,k)(d,u,w,f)(g,i,t,r)(h,n,s,l)(m);\\
r_2 &= (a,z)(b,y)(c,x)(d,w)(e,v)(j,q)(p,k)(u,f)(g,t)(h,s)(i,r)(n,l)(m);\\
r_3 &= (a,v,z,e)(b,q,y,j)(c,k,x,p)(d,f,w,u)(g,r,t,i)(h,l,s,n)(m);\\
s_{y=0} &= (a,v)(b,w)(c,x)(d,y)(e,z)(f,q)(g,r)(h,s)(i,t)(j,u)(k)(l)(m)(n)(p);\\
s_{x=0} &= (a,e)(b,d)(f,j)(g,i)(k,p)(l,n)(q,u)(r,t)(v,z)(w,y)(c)(h)(m)(s)(x);\\
s_{y=x} &= (a,z)(b,u)(c,p)(d,j)(f,y)(g,t)(h,n)(k,x)(l,s)(q,w)(e)(i)(m)(r)(v);\\
s_{y=-x} &= (b,f)(c,k)(d,q)(e,v)(h,l)(i,r)(j,w)(n,s)(p,x)(u,y)(a)(g)(m)(t)(z).
\end{aligned}$$

One may check that all of these maps really are automorphisms of B, so $|\operatorname{Aut}(B)| = 8$. This graph has essentially the same symmetries as the cycle C_4: four rotations and four reflections. (The subscripts of the reflections indicate the axis of reflection, taking m to be located at the origin.) By directing edges appropriately, we could produce a graph with only four automorphisms (the rotations). These graphs and groups play a crucial role in solving the chessboard-coloring problem mentioned in the introduction to this chapter.

7.49. Example. As a final illustration of the calculation of an automorphism group, consider the graph C shown in Figure 7.4, which models a three-dimensional cube. We have taken the vertex set of C to be $\{0,1\}^3$, the set of binary words of length 3. Which bijections $f : \{0,1\}^3 \to \{0,1\}^3$ might be automorphisms of C? First, $f(000)$ can be any of the eight vertices. Next, the three neighbors of 000 (namely 001, 010, and 100) can be mapped bijectively onto the three neighbors of $f(000)$ in any of $3! = 6$ ways. The images of the remaining four vertices are now uniquely determined, as one may check. By the Product Rule, there are at most $8 \times 6 = 48$ automorphisms of C. A routine but tedious verification shows that all of these potential automorphisms really are automorphisms, so $|\operatorname{Aut}(C)| = 48$. By chance, all of these automorphisms can be implemented geometrically using appropriate rotations and reflections in three-dimensional space. Here are the six automorphisms of C that send 000 to 110:

$$f_1 = \begin{bmatrix} 000 & 001 & 010 & 011 & 100 & 101 & 110 & 111 \\ 110 & 100 & 111 & 101 & 010 & 000 & 011 & 001 \end{bmatrix},$$

$$f_2 = \begin{bmatrix} 000 & 001 & 010 & 011 & 100 & 101 & 110 & 111 \\ 110 & 100 & 010 & 000 & 111 & 101 & 011 & 001 \end{bmatrix},$$

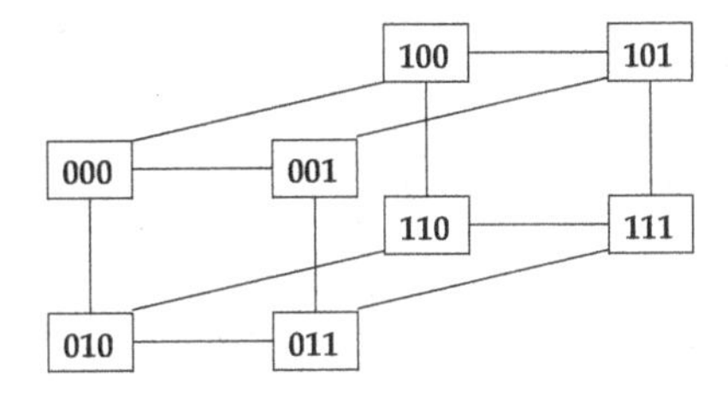

FIGURE 7.4
The cube graph.

$$f_3 = \begin{bmatrix} 000 & 001 & 010 & 011 & 100 & 101 & 110 & 111 \\ 110 & 111 & 100 & 101 & 010 & 011 & 000 & 001 \end{bmatrix},$$

$$f_4 = \begin{bmatrix} 000 & 001 & 010 & 011 & 100 & 101 & 110 & 111 \\ 110 & 111 & 010 & 011 & 100 & 101 & 000 & 001 \end{bmatrix},$$

$$f_5 = \begin{bmatrix} 000 & 001 & 010 & 011 & 100 & 101 & 110 & 111 \\ 110 & 010 & 100 & 000 & 111 & 011 & 101 & 001 \end{bmatrix},$$

$$f_6 = \begin{bmatrix} 000 & 001 & 010 & 011 & 100 & 101 & 110 & 111 \\ 110 & 010 & 111 & 011 & 100 & 000 & 101 & 001 \end{bmatrix}.$$

7.7 Group Homomorphisms

We turn next to group homomorphisms, which are functions from one group to another group that preserve the algebraic structure.

7.50. Definition: Group Homomorphisms. Let G and H be groups with operations $\star$ and $\bullet$, respectively. A function $f : G \to H$ is called a *group homomorphism* iff

$$f(x \star y) = f(x) \bullet f(y) \qquad \text{for all } x, y \in G.$$

A *group isomorphism* is a bijective group homomorphism.

7.51. Example. Define $f : \mathbb{R} \to \mathbb{R}_{>0}$ by $f(x) = e^x$ for all $x \in \mathbb{R}$. This function is a group homomorphism from the additive group $\mathbb{R}$ to the multiplicative group $\mathbb{R}_{>0}$, since $f(x+y) = e^{x+y} = e^x \cdot e^y = f(x) \cdot f(y)$ for all $x, y \in \mathbb{R}$. In fact, f is a group isomorphism since $g : \mathbb{R}_{>0} \to \mathbb{R}$, given by $g(x) = \ln x$ for $x > 0$, is a two-sided inverse for f.

7.52. Example. Define $h : \mathbb{C} \to \mathbb{R}$ by $h(x + iy) = x$ for all $x, y \in \mathbb{R}$. It can be checked that h is an additive group homomorphism that is surjective but not injective. Next, define $r : \mathbb{C}-\{0\} \to \mathbb{R}-\{0\}$ by setting $r(x+iy) = |x+iy| = \sqrt{x^2+y^2}$. Given nonzero $w = x+iy$ and $z = u+iv$ with $x, y, u, v \in \mathbb{R}$, we calculate

$$\begin{aligned} r(wz) &= r((xu - yv) + i(yu + xv)) = \sqrt{(xu-yv)^2 + (yu+xv)^2} \\ &= \sqrt{(x^2+y^2)(u^2+v^2)} = r(w)r(z). \end{aligned}$$

So r is a homomorphism of multiplicative groups.

7.53. Example. For any group G, the identity map $\mathrm{id}_G : G \to G$ is a group isomorphism. More generally, if $H \leq G$, then the inclusion map $j : H \to G$ (defined by $j(h) = h$ for all $h \in H$) is a group homomorphism. If $f : G \to K$ and $g : K \to P$ are group homomorphisms, then $g \circ f : G \to P$ is a group homomorphism, since for all $x, y \in G$,

$$(g \circ f)(xy) = g(f(xy)) = g(f(x)f(y)) = g(f(x))g(f(y)) = (g \circ f)(x)(g \circ f)(y).$$

Moreover, $g \circ f$ is an isomorphism if f and g are isomorphisms, since the composition of bijections is a bijection. If $f : G \to K$ is an isomorphism, then $f^{-1} : K \to G$ is also an isomorphism. For suppose $u, v \in K$. Write $x = f^{-1}(u)$ and $y = f^{-1}(v)$, so $u = f(x)$ and $v = f(y)$. Since f is a group homomorphism, it follows that $uv = f(xy)$. Applying f^{-1} to this relation, we get $f^{-1}(uv) = xy = f^{-1}(u)f^{-1}(v)$.

7.54. Definition: Automorphisms of a Group. An *automorphism* of a group G is a group isomorphism $f : G \to G$. Let $\mathrm{Aut}(G)$ denote the set of all such automorphisms.

The remarks in the preceding example (taking $K = P = G$) show that $\mathrm{Aut}(G)$ *is a subgroup of the symmetric group* $\mathrm{Sym}(G)$, where the operation is composition of functions.

7.55. Example: Inner Automorphisms. Let G be any group, and fix an element $g \in G$. Define a map $C_g : G \to G$ (called *conjugation by* g) by setting $C_g(x) = gxg^{-1}$ for $x \in G$. The map C_g is a group homomorphism, since $C_g(xy) = g(xy)g^{-1} = (gxg^{-1})(gyg^{-1}) = C_g(x)C_g(y)$ for all $x, y \in G$. Furthermore, C_g is a group isomorphism, since a calculation shows that $C_{g^{-1}}$ is the two-sided inverse of C_g. It follows that $C_g \in \mathrm{Aut}(G)$ for every $g \in G$. We call automorphisms of the form C_g *inner automorphisms of* G. It is possible for different group elements to induce the same inner automorphism of G. For example, if G is commutative, then $C_g(x) = gxg^{-1} = gg^{-1}x = x$ for all $g, x \in G$, so that all of the conjugation maps C_g reduce to id_G.

The next theorem clarifies our initial comment that group homomorphisms "preserve the algebraic structure."

7.56. Theorem: Properties of Group Homomorphisms. Let $f : G \to H$ be a group homomorphism. For all $n \in \mathbb{Z}$ and all $x \in G$, $f(x^n) = f(x)^n$. In particular, $f(e_G) = e_H$ and $f(x^{-1}) = f(x)^{-1}$. We say that f *preserves powers, identities, and inverses.*

Proof. First we prove the result for all $n \geq 0$ by induction on n. When $n = 0$, we must prove that $f(e_G) = e_H$. We know $e_G e_G = e_G$. Applying f to both sides of this equation gives

$$f(e_G)f(e_G) = f(e_G e_G) = f(e_G) = f(e_G)e_H.$$

By left cancellation of $f(e_G)$ in H, we conclude that $f(e_G) = e_H$. For the induction step, assume $n \geq 0$ and $f(x^n) = f(x)^n$; we will prove $f(x^{n+1}) = f(x)^{n+1}$. Using the definition of exponent notation, we calculate

$$f(x^{n+1}) = f(x^n x) = f(x^n)f(x) = f(x)^n f(x) = f(x)^{n+1}.$$

Next, let us prove the result when $n = -1$. Given $x \in G$, apply f to the equation $xx^{-1} = e_G$ to obtain

$$f(x)f(x^{-1}) = f(xx^{-1}) = f(e_G) = e_H = f(x)f(x)^{-1}.$$

Left cancellation of $f(x)$ gives $f(x^{-1}) = f(x)^{-1}$. Finally, consider an arbitrary negative integer $n = -m$, where $m > 0$. We have

$$f(x^n) = f((x^{-1})^m) = f(x^{-1})^m = (f(x)^{-1})^m = f(x)^{-m} = f(x)^n. \qquad \square$$

We can use group homomorphisms to construct more examples of subgroups.

7.57. Definition: Kernel and Image of a Homomorphism. Let $f : G \to H$ be a group homomorphism. The *kernel* of f, denoted $\ker(f)$, is the set of all $x \in G$ such that $f(x) = e_H$. The *image* of f, denoted $\text{img}(f)$ or $f[G]$, is the set of all $y \in H$ such that $y = f(z)$ for some $z \in G$.

One readily verifies that $\ker(f) \trianglelefteq G$ and $\text{img}(f) \leq H$; i.e., *the kernel of f is a normal subgroup of the domain of f*, and *the image of f is a subgroup of the codomain of f*.

7.58. Example. Consider the group homomorphisms h and r from Example 7.52, given by $h(x + iy) = x$ and $r(z) = |z|$ for $x, y \in \mathbb{R}$ and nonzero $z \in \mathbb{C}$. The kernel of h is the set of pure imaginary numbers $\{iy : y \in \mathbb{R}\}$. The kernel of r is the unit circle $\{z \in \mathbb{C} : |z| = 1\}$. The image of h is all of $\mathbb{R}$, while the image of r is $\mathbb{R}_{>0}$.

7.59. Example: Even Permutations. By Theorem 7.31, the function $\text{sgn} : S_n \to \{+1, -1\}$ is a group homomorphism. The kernel of this homomorphism, which is denoted A_n, consists of all $f \in S_n$ such that $\text{sgn}(f) = +1$. Such f are called *even permutations* since they can be written as products of an even number of transpositions. A_n is called the *alternating group on n symbols.* We show later (Corollary 7.102) that $|A_n| = |S_n|/2 = n!/2$ for all $n \geq 2$.

The next example illustrates how group homomorphisms can reveal structural information about groups.

7.60. Example: Analysis of Cyclic Subgroups. Let G be any group, written multiplicatively, and fix an element $x \in G$. Define $f : \mathbb{Z} \to G$ by setting $f(n) = x^n$ for all $n \in \mathbb{Z}$. By the Laws of Exponents, f is a group homomorphism. The image of f is precisely $\langle x \rangle$, the cyclic subgroup of G generated by x. The kernel of f is some subgroup of $\mathbb{Z}$, which by Theorem 7.40 has the form $m\mathbb{Z}$ for some integer $m \geq 0$. Consider the case where $m = 0$. For all $i \in \mathbb{Z}$, $x^i = e_G$ iff $f(i) = e_G$ iff $i \in \ker(f)$ iff $i = 0$, so x^0 is the only power of x that equals the identity of G. We say that x has *infinite order* when $m = 0$. In this case, for all $i, j \in \mathbb{Z}$, $i \neq j$ implies $x^i \neq x^j$, since

$$x^i = x^j \Rightarrow x^{i-j} = e_G \Rightarrow i - j = 0 \Rightarrow i = j.$$

So the sequence of group elements $(\ldots, x^{-2}, x^{-1}, x^0, x^1, x^2, \ldots)$ contains no repetitions, and the function $f : \mathbb{Z} \to G$ is injective. By shrinking the codomain of f, we obtain a group isomorphism $f' : \mathbb{Z} \to \langle x \rangle$.

Now consider the case where $m > 0$. For all $i \in \mathbb{Z}$, $x^i = e_G$ iff $f(i) = e_G$ iff $i \in \ker(f)$ iff i is a multiple of m. We say that x *has order m* in this case; thus, the order of x is the least positive exponent n such that $x^n = e_G$. We claim that the cyclic group $\langle x \rangle$ consists of the m *distinct* elements $x^0, x^1, x^2, \ldots, x^{m-1}$. For, given an arbitrary element $x^n \in \langle x \rangle$, we can divide n by m to get $n = mq + r$ for some $q, r \in \mathbb{Z}$ with $0 \leq r < m$. Then $x^n = x^{mq+r} = (x^m)^q x^r = e_G^q x^r = x^r$, so x^n is equal to one of the elements in our list. Furthermore, the listed elements are distinct. For suppose $0 \leq i \leq j < m$ and $x^i = x^j$. Then $x^{j-i} = e_G$, forcing m to divide $j - i$. But $0 \leq j - i < m$, so the only possibility is $j - i = 0$, hence $i = j$. Consider the function $g : \mathbb{Z}_m \to \langle x \rangle$ given by $g(i) = x^i$ for $0 \leq i < m$. This function is a well-defined bijection by the preceding remarks. Furthermore, g is a group homomorphism. To check this, let $i, j \in \mathbb{Z}_m$. If $i + j < m$, then

$$g(i \oplus j) = g(i + j) = x^{i+j} = x^i x^j = g(i)g(j).$$

If $i + j \geq m$, then

$$g(i \oplus j) = g(i + j - m) = x^{i+j-m} = x^i x^j (x^m)^{-1} = x^i x^j = g(i)g(j).$$

So g is an isomorphism from the additive group $\mathbb{Z}_m$ to the cyclic subgroup generated by x.

To summarize, we have shown that *every cyclic group $\langle x \rangle$ is isomorphic to one of the additive groups $\mathbb{Z}$ or $\mathbb{Z}_m$ for some $m > 0$.* The first case occurs when x has infinite order, and the second case occurs when x has order m.

7.8 Group Actions

The fundamental tool needed to solve counting problems involving symmetry is the notion of a *group action.*

7.61. Definition: Group Actions. Suppose G is a group and X is a set. An *action* of G on X is a function $* : G \times X \to X$ satisfying the following axioms:

1. For all $g \in G$ and all $x \in X$, $g * x \in X$ (closure).
2. For all $x \in X$, $e_G * x = x$ (identity).
3. For all $g, h \in G$ and all $x \in X$, $g * (h * x) = (gh) * x$ (associativity).

The set X with the given action $*$ is called a *G-set.*

7.62. Example. For any set X, the symmetric group $G = \mathrm{Sym}(X)$ acts on X via the rule $g*x = g(x)$ for $g \in G$ and $x \in X$. Axiom 1 holds because each $g \in G$ is a function from X to X, hence $g*x = g(x) \in X$ for all $x \in X$. Axiom 2 holds since $e_G * x = \mathrm{id}_X * x = \mathrm{id}_X(x) = x$ for all $x \in X$. Axiom 3 holds because

$$g * (h * x) = g(h(x)) = (g \circ h)(x) = (gh) * x \quad \text{for all } g, h \in \mathrm{Sym}(X) \text{ and all } x \in X.$$

7.63. Example. Let G be any group, written multiplicatively, and let X be the set G. Define $* : G \times X \to X$ by $g * x = gx$ for all $g, x \in G$. We say that "G acts on itself by left multiplication." In this example, the action axioms reduce to the corresponding group axioms for G.

We can define another action of G on the set $X = G$ by letting $g \bullet x = xg^{-1}$ for all $g, x \in G$. The first two group action axioms are immediately verified; the third axiom follows by calculating (for $g, h, x \in G$)

$$g \bullet (h \bullet x) = g \bullet (xh^{-1}) = (xh^{-1})g^{-1} = x(h^{-1}g^{-1}) = x(gh)^{-1} = (gh) \bullet x.$$

We say that "G acts on itself by inverted right multiplication." It can be checked that the rule $g \cdot x = xg$ (for $g, x \in G$) does *not* define a group action for non-commutative groups G, because Axiom 3 fails. But see the discussion of right group actions below.

7.64. Example. Let the group G act on the set $X = G$ as follows:

$$g * x = gxg^{-1} \quad \text{for all } g \in G \text{ and } x \in X.$$

We say that "G acts on itself by conjugation." The reader may verify that the axioms for an action are satisfied.

7.65. Example. Suppose we are given a group action $* : G \times X \to X$. Let H be any subgroup of G. By restricting the action function to the domain $H \times X$, we obtain an action of H on X, as one immediately verifies. Combining this construction with previous examples, we obtain many additional instances of group actions. For example, any subgroup

H of a group G acts on G by left multiplication, and by inverted right multiplication, and by conjugation. Any subgroup H of $\mathrm{Sym}(X)$ acts on X via $f \star x = f(x)$ for $f \in H$ and $x \in X$. In particular, the automorphism group $\mathrm{Aut}(G)$ of a group G is a subgroup of $\mathrm{Sym}(G)$, so $\mathrm{Aut}(G)$ acts on G via $f \star x = f(x)$ for $f \in \mathrm{Aut}(G)$ and $x \in G$. Similarly, if K is a graph with vertex set X, then $\mathrm{Aut}(K)$ is a subgroup of $\mathrm{Sym}(X)$, and therefore $\mathrm{Aut}(K)$ acts on X via $f \star x = f(x)$ for $f \in \mathrm{Aut}(K)$ and $x \in X$.

7.66. Example. Suppose X is a G-set with action $*$. Let $\mathcal{P}(X)$ be the power set of X, which is the set of all subsets of X. It is routine to check that $\mathcal{P}(X)$ is a G-set under the action

$$g \bullet S = \{g * s : s \in S\} \quad \text{for all } g \in G \text{ and } S \in \mathcal{P}(X).$$

7.67. Example. Consider the set $X = \mathbb{R}[x_1, x_2, \ldots, x_n]$ consisting of polynomials in $x_1, \ldots, x_n$ with real coefficients. The symmetric group S_n acts on $\{1, 2, \ldots, n\}$ via $f * i = f(i)$ for $f \in S_n$ and $1 \leq i \leq n$. We can transfer this to an action of S_n on $\{x_1, \ldots, x_n\}$ by defining

$$f * x_i = x_{f(i)} \quad \text{for all } f \in S_n \text{ and } i \text{ between 1 and } n.$$

Each function on $\{x_1, \ldots, x_n\}$ sending x_i to $x_{f(i)}$ extends to a ring homomorphism $E_f : X \to X$ that sends a polynomial $p = p(x_1, \ldots, x_n) \in X$ to $E_f(p) = p(x_{f(1)}, \ldots, x_{f(n)})$ (see the Appendix). One may check that the rule $f * p = E_f(p)$ (for $f \in S_n$ and $p \in X$) defines an action of S_n on X. In particular, $g * (h * p) = (g \circ h) * p$ follows since both sides are the image of p under the unique ring homomorphism sending x_i to $x_{g(h(i))}$ for all i.

7.68. Example. By imitating ideas in the previous example, we can define certain group actions on vector spaces. Suppose V is a real vector space and $X = (x_1, \ldots, x_n)$ is an ordered basis of V. For $f \in S_n$, the map sending each x_i to $x_{f(i)}$ uniquely extends by linearity to a linear map $T_f : V \to V$, given explicitly by

$$T_f(a_1x_1 + \cdots + a_nx_n) = a_1x_{f(1)} + \cdots + a_nx_{f(n)} \quad \text{for all } a_1, \ldots, a_n \in \mathbb{R}.$$

One may check that $f * v = T_f(v)$ (for $f \in S_n$ and $v \in V$) defines an action of the group S_n on the set V.

7.69. Example. Suppose G is a group, X is a G-set with action $*$, and W and Z are any sets. Let $\mathrm{Fun}(W, X)$ denote the set of all functions $F : W \to X$. This set of functions can be turned into a G-set by defining

$$(g \bullet F)(w) = g * (F(w)) \quad \text{for all } g \in G,\ F \in \mathrm{Fun}(W, X), \text{ and } w \in W,$$

as is readily verified.

Now consider the set $\mathrm{Fun}(X, Z)$ of all functions $F : X \to Z$. We claim this set of functions becomes a G-set if we define

$$(g \bullet F)(x) = F(g^{-1} * x) \quad \text{for all } g \in G,\ F \in \mathrm{Fun}(X, Z), \text{ and } x \in X.$$

Let us carefully prove this claim. First, given $g \in G$ and $F \in \mathrm{Fun}(X, Z)$, the map $g \bullet F$ is a well-defined function from X to Z because $g^{-1} * x \in X$ and F maps X into Z. So, $g \bullet F \in \mathrm{Fun}(X, Z)$, verifying closure. Second, to prove that $e \bullet F = F$ for all $F \in \mathrm{Fun}(X, Z)$, we fix $x \in X$ and check that $(e \bullet F)(x) = F(x)$. We calculate

$$(e \bullet F)(x) = F(e^{-1} * x) = F(e * x) = F(x).$$

Third, we verify the associativity axiom for $\bullet$. Fix $g, h \in G$ and $F \in \mathrm{Fun}(X, Z)$. The two functions $g \bullet (h \bullet F)$ and $(gh) \bullet F$ both have domain X and codomain Z. Fix $x \in X$. On one hand,

$$[(gh) \bullet F](x) = F((gh)^{-1} * x) = F((h^{-1}g^{-1}) * x).$$

On the other hand, using the definition of $\bullet$ twice,

$$[g \bullet (h \bullet F)](x) = [h \bullet F](g^{-1} * x) = F(h^{-1} * (g^{-1} * x)).$$

Since $*$ is known to be an action, we see that $g \bullet (h \bullet F)$ and $(gh) \bullet F$ have the same value at x. So the third action axiom is proved. One may check that this axiom would fail, in general, if we omitted the inverse in the definition of $\bullet$.

7.70. Example. Let n be a fixed integer, let Y be a set, and let

$$U = Y^n = \{(y_1, \ldots, y_n) : y_i \in Y\}$$

be the set of all sequences of n elements of Y. The group S_n acts on U via the rule

$$f \cdot (y_1, y_2, \ldots, y_n) = (y_{f^{-1}(1)}, y_{f^{-1}(2)}, \ldots, y_{f^{-1}(n)}) \quad \text{for all } f \in S_n \text{ and all } y_1, \ldots, y_n \in Y.$$

The inverses in this formula are *essential.* To see why, we observe that the action here is in fact a special case of the previous example. This is true because a sequence in U is formally defined to be a function $y : \{1, 2, \ldots, n\} \to Y$ where $y(i) = y_i$. Using this function notation for sequences, we have (for all $f \in S_n$ and all i between 1 and n)

$$(f \cdot y)(i) = y(f^{-1}(i)) = (f \bullet y)(i),$$

in agreement with the previous example. We can also say that acting by f moves the object z originally in *position* i to *position* $f(i)$ in the new sequence. This is true because $(f \cdot y)(f(i)) = y(f^{-1}(f(i))) = y(i) = z$.

The reader may now be disturbed by the *lack* of inverses in the formula $f * x_i = x_{f(i)}$ from Example 7.68. However, that situation is different since $x_1, \ldots, x_n$ in that example are fixed basis elements in a vector space V, not the entries in a sequence. Indeed, recall that the action on V is given by $f * v = T_f(v)$ where T_f is the linear extension of the map sending x_i to $x_{f(i)}$. Writing $v = \sum_{i=1}^{n} a_i x_i$, the *coordinates* of v relative to this basis are the entries in the sequence $(a_1, a_2, \ldots, a_n) \in \mathbb{R}^n$. Applying f to v gives

$$f * v = \sum_{i=1}^{n} a_i x_{f(i)} = \sum_{j=1}^{n} a_{f^{-1}(j)} x_j,$$

after changing variables by letting $j = f(i)$, $i = f^{-1}(j)$. We now see that the coordinates of $f * v$ relative to the ordered basis $(x_1, \ldots, x_n)$ are $(a_{f^{-1}(1)}, \ldots, a_{f^{-1}(n)})$. For example,

$$(1, 2, 3) * (a_1x_1 + a_2x_2 + a_3x_3) = a_1x_2 + a_2x_3 + a_3x_1 = a_3x_1 + a_1x_2 + a_2x_3,$$

or equivalently, in coordinate notation,

$$(1, 2, 3) * (a_1, a_2, a_3) = (a_3, a_1, a_2).$$

To summarize, when f acts directly on the *objects* x_i, no inverse is needed; but when f permutes the *positions* in a list, one must apply f^{-1} to each subscript.

7.71. Remark: Right Actions. A *right action* of a group G on a set X is a function $* : X \times G \to X$ such that $x * e = x$ and $x * (gh) = (x * g) * h$ for all $x \in X$ and all $g, h \in G$. For example, $x * g = xg$ (with no inverse) defines a right action of a group G on the set $X = G$. Similarly, we get a *right* action of S_n on the set of sequences in the previous example by writing

$$(y_1, \ldots, y_n) * f = (y_{f(1)}, \ldots, y_{f(n)}).$$

Group actions (as defined at the beginning of this section) are sometimes called *left* actions to avoid confusion with right actions. We shall mostly consider left group actions, but right actions are occasionally more convenient to use (see, for example, Definition 7.90).

7.9 Permutation Representations

Group actions are closely related to symmetric groups. To understand the precise nature of this relationship, we need the following definition.

7.72. Definition: Permutation Representations. A *permutation representation* of a group G on a set X is a group homomorphism $\phi : G \to \mathrm{Sym}(X)$.

This definition seems quite different from the definition of a group action given in the last section. But we show in this section that group actions and permutation representations are essentially the same thing. Both viewpoints turn out to be pertinent in the application of group actions to problems in combinatorics and algebra.

We first show that any group action of G on X can be used to construct an associated permutation representation of G on X. The key idea appears in the next definition.

7.73. Definition: Left Multiplication Maps. Let $* : G \times X \to X$ be an action of the group G on the set X. For each $g \in G$, *left multiplication by* G (relative to this action) is the function $L_g : X \to X$ defined by

$$L_g(x) = g * x \quad \text{for all } x \in X.$$

Note that the outputs of L_g are members of the codomain X, by the closure axiom for group actions.

7.74. Theorem: Properties of Left Multiplication Maps. Let a group G act on a set X. (a) $L_e = \mathrm{id}_X$. (b) For all $g, h \in G$, $L_{gh} = L_g \circ L_h$. (c) For all $g \in G$, $L_g \in \mathrm{Sym}(X)$, and $L_g^{-1} = L_{g^{-1}}$.

Proof. All functions appearing here have domain X and codomain X. So it suffices to check that the relevant functions take the same value at each $x \in X$. For (a), $L_e(x) = e * x = x = \mathrm{id}_X(x)$ by the identity axiom for group actions. For (b), $L_{gh}(x) = (gh) * x = g * (h * x) = L_g(L_h(x)) = (L_g \circ L_h)(x)$ by the associativity axiom for group actions. Finally, using (a) and (b) with $h = g^{-1}$ shows that $\mathrm{id}_X = L_g \circ L_{g^{-1}}$. Similarly, $\mathrm{id}_X = L_{g^{-1}} \circ L_g$. This means that $L_{g^{-1}}$ is the two-sided inverse of L_g; in particular, both of these maps must be bijections. □

Using the theorem, we can pass from a group action $*$ to a permutation representation ϕ as follows. Define $\phi : G \to \mathrm{Sym}(X)$ by setting $\phi(g) = L_g$ for all $g \in G$. We just saw that ϕ does map into the codomain $\mathrm{Sym}(X)$, and

$$\phi(gh) = L_{gh} = L_g \circ L_h = \phi(g) \circ \phi(h) \quad \text{for all } g, h \in G.$$

So ϕ is a group homomorphism.

7.75. Example: Cayley's Theorem. We know that any group G acts on the set $X = G$ by left multiplication. The preceding construction produces a group homomorphism $\phi : G \to \text{Sym}(G)$ such that $\phi(g) = L_g$ for all $g \in G$, and $L_g(x) = gx$ for all $x \in G$. We claim that ϕ is injective in this situation. For, suppose $g, h \in G$ and $L_g = L_h$. Applying these two functions to e (the identity of G) gives $L_g(e) = L_h(e)$, so $ge = he$, so $g = h$. It follows that G is isomorphic (via ϕ) to the image of ϕ, which is a subgroup of the symmetric group $\text{Sym}(G)$. This proves *Cayley's Theorem*, which says that *any group is isomorphic to a subgroup of some symmetric group.* If G has n elements, one can check that $\text{Sym}(G)$ is isomorphic to S_n. So *every n-element group is isomorphic to a subgroup of the symmetric group S_n.*

7.76. Example. For any set X, we know $\text{Sym}(X)$ acts on X via $f * x = f(x)$ for $f \in \text{Sym}(X)$ and $x \in X$. What is the associated permutation representation $\phi : \text{Sym}(X) \to \text{Sym}(X)$? First note that for $f \in \text{Sym}(X)$, left multiplication by f is the map $L_f : X \to X$ such that $L_f(x) = f * x = f(x)$. In other words, $L_f = f$, so that $\phi(f) = L_f = f$. This means that ϕ is the identity homomorphism. More generally, whenever a subgroup H of $\text{Sym}(X)$ acts on X via $h * x = h(x)$, the corresponding permutation representation is the inclusion map of H into $\text{Sym}(X)$.

So far, we have seen that every group action of G on X has an associated permutation representation. We can reverse this process by starting with an arbitrary permutation representation $\phi : G \to \text{Sym}(X)$ and building a group action, as follows. Given ϕ, define $* : G \times X \to X$ by setting $g * x = \phi(g)(x)$ for all $g \in G$ and $x \in X$. Note that $\phi(g)$ is a function with domain X, so the expression $\phi(g)(x)$ denotes a well-defined element of X. In particular, $*$ satisfies the closure axiom for an action. Since group homomorphisms preserve identities, $\phi(e) = \text{id}_X$, and so $e * x = \phi(e)(x) = \text{id}_X(x) = x$ for all $x \in X$. So the identity axiom holds. Finally, using the fact that ϕ is a group homomorphism, we calculate

$$\begin{aligned} (gh) * x &= \phi(gh)(x) = (\phi(g) \circ \phi(h))(x) \\ &= \phi(g)(\phi(h)(x)) = g * (h * x) \quad \text{for all } g, h \in G \text{ and all } x \in X. \end{aligned}$$

So the associativity axiom holds, completing the proof that $*$ is a group action.

The following theorem is the formal enunciation of our earlier claim that group actions and permutation representations are essentially the same thing.

7.77. Theorem: Equivalence of Group Actions and Permutation Representations. Fix a group G and a set X. Let A be the set of all group actions of G on X, and let P be the set of all permutation representations of G on X. There are mutually inverse bijections $F : A \to P$ and $H : P \to A$, given by

$$F(*) = \phi : G \to \text{Sym}(X) \text{ where } \phi(g) = L_g \text{ and } L_g(x) = g * x \text{ for all } x \in X \text{ and } g \in G;$$

$$H(\phi) = * : G \times X \to X \text{ where } g * x = \phi(g)(x) \text{ for all } g \in G \text{ and } x \in X.$$

Proof. The discussion preceding the theorem shows that F maps the set A into the codomain P, and that H maps the set P into the codomain A. We must verify that $F \circ H = \text{id}_P$ and $H \circ F = \text{id}_A$.

To show $F \circ H = \text{id}_P$, fix $\phi \in P$, and write $* = H(\phi)$ and $\psi = F(*)$. We must confirm the equality of ψ and ϕ, which are both functions from G into $\text{Sym}(X)$. To do this, fix $g \in G$, and ask whether the two functions $\psi(g), \phi(g) : X \to X$ are equal. For each $x \in X$,

$$\psi(g)(x) = L_g(x) = g * x = \phi(g)(x).$$

So $\psi(g) = \phi(g)$ for all g, hence $\psi = \phi$ as needed.

To show $H \circ F = \mathrm{id}_A$, fix $* \in A$, and write $\phi = F(*)$ and $\bullet = H(\phi)$. We must confirm the equality of $\bullet$ and $*$, which are both functions from $G \times X$ into X. For this, fix $g \in G$ and $x \in X$. Now compute

$$g \bullet x = \phi(g)(x) = L_g(x) = g * x.$$ □

7.78. Example. We can use permutation representations to generate new constructions of group actions. For instance, suppose X is a G-set with action $*$ and associated permutation representation $\phi : G \to \mathrm{Sym}(X)$. Now suppose we are given a group homomorphism $u : K \to G$. Composing with ϕ gives a homomorphism $\phi \circ u : K \to \mathrm{Sym}(X)$. This is a permutation representation of K on X, which leads to an action of K on X. Specifically, by applying the map H from the theorem, we see that the new action is given by

$$k \bullet x = u(k) * x \quad \text{for all } k \in K \text{ and } x \in X.$$

7.10 Stable Subsets and Orbits

One way to gain information about a group is to study its subgroups. The analogous concept for G-sets appears in the next definition.

7.79. Definition: G-Stable Subsets. Let a group G act on a set X. A subset Y of X is called a *G-stable subset* iff $g * y \in Y$ for all $g \in G$ and all $y \in Y$.

When Y is a G-stable subset, the restriction of $*$ to $G \times Y$ maps into the codomain Y, by definition. Since the identity axiom and associativity axiom still hold for the restricted action, we see that Y is a G-set.

Recall that every element of a group generates a cyclic subgroup. Similarly, we can pass from an element of a G-set to a G-stable subset as follows.

7.80. Definition: Orbits. Suppose a group G acts on a set X, and $x \in X$. The *orbit* of x under this action is the set

$$Gx = G * x = \{g * x : g \in G\} \subseteq X.$$

Every orbit is a G-stable subset: for, given $h \in G$ and $g*x \in Gx$, the associativity axiom gives $h * (g * x) = (hg) * x \in Gx$. Furthermore, by the identity axiom, $x = e * x \in Gx$ for each $x \in X$. The orbit Gx is the smallest G-stable subset of X containing x.

7.81. Example. Let S_5 act on the set $X = \{1, 2, 3, 4, 5\}$ via $f * x = f(x)$ for $f \in S_5$ and $x \in X$. For each $i \in X$, the orbit $S_5 * i = \{f(i) : f \in S_5\}$ is all of X. The reason is that for any given j in X, we can find an $f \in S_5$ such that $f(i) = j$; for instance, take $f = (i, j)$. On the other hand, consider the subgroup $H = \langle (1, 3)(2, 4, 5) \rangle$ of S_5. If we let H act on X via $f * x = f(x)$ for $f \in H$ and $x \in X$, we get different orbits. It can be checked that

$$H * 1 = H * 3 = \{1, 3\}, \qquad H * 2 = H * 4 = H * 5 = \{2, 4, 5\}.$$

Note that the H-orbits are precisely the connected components of the digraph representing the generator $(1, 3)(2, 4, 5)$ of H. This observation holds more generally whenever a cyclic subgroup of S_n acts on $\{1, 2, \ldots, n\}$.

Now consider the action of A_5 on X. As in the case of S_5, we have $A_5 * i = X$ for all $i \in X$, but for a different reason. Given $j \in X$, we must now find an *even* permutation sending i to j. We can use the identity permutation if $i = j$. Otherwise, choose two distinct elements k, l that are different from i and j, and use the permutation $(i, j)(k, l)$.

7.82. Example. Let S_4 act on the set X of all 4-tuples of integers by permuting positions:

$$f * (x_1, x_2, x_3, x_4) = (x_{f^{-1}(1)}, x_{f^{-1}(2)}, x_{f^{-1}(3)}, x_{f^{-1}(4)}) \quad \text{for all } f \in S_4 \text{ and } x_i \in \mathbb{Z}.$$

The S_4-orbit of a sequence $x = (x_1, x_2, x_3, x_4)$ consists of all possible sequences reachable from x by permuting the entries. For example,

$$S_4 * (5,1,5,1) = \{(1,1,5,5), (1,5,1,5), (1,5,5,1), (5,1,1,5), (5,1,5,1), (5,5,1,1)\}.$$

As another example, $S_4 * (3,3,3,3) = \{(3,3,3,3)\}$ and $S_4 * (1,3,5,7)$ is the set of all 24 permutations of $1,3,5,7$. Now consider the cyclic subgroup $H = \langle(1,2,3,4)\rangle$ of S_4. Restricting the action turns X into an H-set. When computing orbits relative to the action of H, we are only allowed to cyclically shift the elements in each 4-tuple. So, for instance,

$$\begin{aligned} H * (5,1,5,1) &= \{(5,1,5,1),\ (1,5,1,5)\}; \\ H * (1,3,5,7) &= \{(1,3,5,7),\ (3,5,7,1),\ (5,7,1,3),\ (7,1,3,5)\}. \end{aligned}$$

We see that the orbit of a given $x \in X$ depends heavily on which group is acting on X.

7.83. Example. Let a group G act on itself by left multiplication: $g * x = gx$ for $g, x \in G$. For every $x \in G$, the orbit Gx is all of G. For, given any $y \in G$, we have $(yx^{-1}) * x = y$. In the next section, we study what happens when a subgroup H acts on G by left (or right) multiplication.

7.84. Example: Conjugacy Classes. Let G be a group. We have seen that G acts on itself by conjugation: $g * x = gxg^{-1}$ for $g, x \in G$. The orbit of $x \in G$ under this action is the set

$$G * x = \{gxg^{-1} : g \in G\}.$$

This set is called the *conjugacy class of* x *in* G. For example, when $G = S_3$, the conjugacy classes are:

$$\begin{aligned} G * \text{id} &= \{\text{id}\}; \\ G * (1,2) = G * (1,3) = G * (2,3) &= \{(1,2), (1,3), (2,3)\}; \\ G * (1,2,3) = G * (1,3,2) &= \{(1,2,3), (1,3,2)\}. \end{aligned}$$

One can confirm this with the aid of the identities

$$f \circ (i,j) \circ f^{-1} = (f(i), f(j)); \qquad f \circ (i,j,k) \circ f^{-1} = (f(i), f(j), f(k)) \qquad \text{for all } f \in S_3.$$

The generalization of this example to any S_n is discussed in §7.13. We observe in passing that $G * x = \{x\}$ iff $gxg^{-1} = x$ for all $g \in G$ iff $gx = xg$ for all $g \in G$ iff x commutes with every element of G. In particular, for G commutative, every conjugacy class consists of a single element.

7.85. Example. Let $B = (X, E)$ be the graph representing a 5×5 chessboard shown in Figure 7.3. Let the graph automorphism group $G = \text{Aut}(B)$ act on X via $f * x = f(x)$ for $f \in G$ and $x \in X$. We explicitly determined the elements of G in Example 7.48. We can use this calculation to find all the distinct orbits of the action. They are:

$$\begin{aligned} Ga &= \{a, e, z, v\} = Ge = Gz = Gv; \\ Gb &= \{b, d, j, u, y, w, q, f\} = Gd = Gj = \cdots; \\ Gc &= \{c, p, x, k\}; \\ Gg &= \{g, i, t, r\}; \\ Gh &= \{h, n, s, l\}; \\ Gm &= \{m\}. \end{aligned}$$

The reader may have noticed in these examples that distinct orbits of the group action are always pairwise disjoint subsets of X. We now prove that this always happens.

7.86. Theorem: Orbit Decomposition of a G-set. Let X be a G-set. Every element $x \in X$ belongs to exactly one orbit, namely Gx. In other words, the distinct orbits of the action of G on X form a set partition of X.

Proof. Define a relation on X by setting, for all $x, y \in X$, $x \sim y$ iff $y = g * x$ for some $g \in G$. The relation $\sim$ is reflexive on X: given $x \in X$, we have $x = e * x$, so $x \sim x$. The relation $\sim$ is symmetric: given $x, y \in X$ with $x \sim y$, we know $y = g * x$ for some $g \in G$. A routine calculation shows that $x = g^{-1} * y$, so $y \sim x$. The relation $\sim$ is transitive: given $x, y, z \in X$ with $x \sim y$ and $y \sim z$, we know $y = g * x$ and $z = h * y$ for some $g, h \in G$. So $z = h*(g*x) = (hg)*x$, and $x \sim z$. Thus $\sim$ is an equivalence relation on X. Recall from the proof of Theorem 2.51 that the equivalence classes of any equivalence relation on X form a set partition of X. In this situation, the equivalence classes are precisely the G-orbits, since the equivalence class of x is

$$\{y \in X : x \sim y\} = \{y \in X : y = g * x \text{ for some } g \in G\} = Gx. \qquad \square$$

7.87. Corollary. Every group G is the disjoint union of its conjugacy classes.

Everything we have said can be adapted to give results on right actions. In particular, if G acts on X on the right, then X is partitioned into a disjoint union of the *right orbits*

$$xG = \{x * g : g \in G\} \qquad \text{where } x \in X.$$

7.11 Cosets

The concept of *cosets* plays a central role in group theory. Cosets appear as the orbits of a certain group action.

7.88. Definition: Right Cosets. Let G be a group, and let H be any subgroup of G. Let H act on G by left multiplication: $h * x = hx$ for $h \in H$ and $x \in G$. The orbit of x under this action, namely

$$Hx = \{hx : h \in H\}$$

is called the *right coset of H determined by x*.

By the general theory of group actions, we know that G is the disjoint union of its right cosets.

7.89. Example. Let $G = S_3$ and $H = \{\text{id}, (1,2)\}$. The right cosets of H in G are

$$\begin{aligned} H\,\text{id} = H(1,2) &= \{\text{id}, (1,2)\} = H; \\ H(1,3) = H(1,3,2) &= \{(1,3), (1,3,2)\}; \\ H(2,3) = H(1,2,3) &= \{(2,3), (1,2,3)\}. \end{aligned}$$

For the subgroup $K = \{\text{id}, (1,2,3), (1,3,2)\}$, the distinct right cosets are

$$K\,\text{id} = K \text{ and } K(1,2) = \{(1,2), (2,3), (1,3)\}.$$

Note that the subgroup itself is always a right coset, but the other right cosets are not subgroups (they do not contain the identity of G).

By letting H act on the right, we obtain the notion of a *left* coset, which will be used frequently below.

7.90. Definition: Left Cosets. Let G be a group, and let H be any subgroup of G. Let H act on G by right multiplication: $x * h = xh$ for $h \in H$ and $x \in G$. The orbit of x under this action, namely

$$xH = \{xh : h \in H\}$$

is called the *left coset of H determined by x*.

By Theorem 7.86, G is the disjoint union of its left cosets.

7.91. Example. Let $G = S_3$ and $H = \{\text{id}, (1,2)\}$ as above. The left cosets of H in G are

$$\begin{aligned} \text{id}\, H = (1,2)H &= \{\text{id}, (1,2)\} = H; \\ (1,3)H = (1,2,3)H &= \{(1,3), (1,2,3)\}; \\ (2,3)H = (1,3,2)H &= \{(2,3), (1,3,2)\}. \end{aligned}$$

In this example, $xH \neq Hx$ except when $x \in H$. This shows that left cosets and right cosets do not coincide in general. On the other hand, for the subgroup $K = \{\text{id}, (1,2,3), (1,3,2)\}$, the left cosets are K and $(1,2)K = \{(1,2), (1,3), (2,3)\}$. One checks that $xK = Kx$ for all $x \in S_3$, so that left cosets and right cosets do coincide for some subgroups.

Although $x = y$ certainly implies $xH = yH$, one must remember that the converse is almost always false. The next result gives criteria for deciding when two cosets xH and yH are equal; it is used constantly in arguments involving cosets.

7.92. The Coset Equality Theorem. Let H be a subgroup of G. For all $x, y \in G$, the following conditions are logically equivalent:

(a) $xH = yH$	(a′) $yH = xH$
(b) $x \in yH$	(b′) $y \in xH$
(c) $\exists h \in H, x = yh$	(c′) $\exists h' \in H, y = xh'$
(d) $y^{-1}x \in H$	(d′) $x^{-1}y \in H$

Proof. We first prove (a)⇒(b)⇒(c)⇒(d)⇒(a). If $xH = yH$, then $x = xe \in xH = yH$, so $x \in yH$. If $x \in yH$, then $x = yh$ for some $h \in H$ by definition of yH. If $x = yh$ for some $h \in H$, then multiplying by y^{-1} on the left gives $y^{-1}x \in H$. Finally, assume that $y^{-1}x \in H$. Then $y(y^{-1}x) = x$ lies in the orbit yH. We also have $x = xe \in xH$. As orbits are either disjoint or equal, we must have $xH = yH$.

Interchanging x and y in the last paragraph proves the equivalence of (a′), (b′), (c′), and (d′). Since (a) and (a′) are visibly equivalent, the proof is complete. □

7.93. Remark. The equivalence of (a) and (d) in the last theorem is used quite frequently. Note too that the subgroup H is a coset (namely eH), and $xH = H$ iff $xH = eH$ iff $e^{-1}x \in H$ iff $x \in H$. Finally, one can prove an analogous theorem for right cosets. The key difference is that $Hx = Hy$ iff $xy^{-1} \in H$ iff $yx^{-1} \in H$ (so that inverses occur on the right for right cosets).

We can use cosets to construct more examples of G-sets.

7.94. Example: The G-set G/H. Let G be a group, and let H be any subgroup of G. Let G/H be the set of all distinct left cosets of H in G. Every element of G/H is a subset of G of the form $xH = \{xh : h \in H\}$ for some $x \in G$ (x is not unique when $|H| > 1$). So,

G/H is a subset of the power set $\mathcal{P}(G)$. Let the group G act on the set $X = G/H$ by left multiplication:

$$g * S = \{gs : s \in S\} \quad \text{for all } g \in G \text{ and } S \in G/H.$$

This action is the restriction of the action from Example 7.66 to the domain $G \times X$. To see that the action makes sense, we must check that X is a *G-stable subset* of $\mathcal{P}(G)$. Let xH be an element of X and let $g \in G$; then

$$g * (xH) = \{g(xh) : h \in H\} = \{(gx)h : h \in H\} = (gx)H \in X.$$

Let $[G : H] = |G/H|$, which may be infinite; this cardinal number is called the *index of H in G*. Lagrange's Theorem (proved below) shows that $|G/H| = |G|/|H|$ when G is finite.

7.95. Remark. Using the Coset Equality Theorem, it can be shown that the map sending each left coset xH to the right coset Hx^{-1} gives a well-defined bijection between G/H and the set of right cosets of H in G. So, we would obtain the same number $[G : H]$ if we had used right cosets in the definition of G/H. It is more convenient to use left cosets here, so that G can act on G/H on the left.

7.96. Example. If $G = S_3$ and $H = \{\text{id}, (1,2)\}$, then

$$G/H = \{\{\text{id}, (1,2)\}, \{(1,3), (1,2,3)\}, \{(2,3), (1,3,2)\}\} = \{\text{id}\, H, (1,3)H, (2,3)H\}.$$

We have $[G : H] = |G/H| = 3$. Note that $|G|/|H| = 6/2 = 3 = |G/H|$. This is a special case of Lagrange's Theorem, proved below.

To prepare for Lagrange's Theorem, we first show that every left coset of H in G has the same cardinality as H.

7.97. The Coset Size Theorem. Let H be a subgroup of G. For all $x \in G$, $|xH| = |H|$.

Proof. We have seen that the left multiplication map $L_x : G \to G$, given by $L_x(g) = xg$ for $g \in G$, is a bijection with inverse $L_{x^{-1}}$. Restricting the domain of L_x to H gives an injective map $L'_x : H \to G$. The image of this map is $\{xh : h \in H\} = xH$. So, restricting the codomain gives a bijection from H to xH. Thus, the sets H and xH have the same cardinality. □

7.98. Lagrange's Theorem. Let H be any subgroup of a finite group G. Then

$$[G : H] \cdot |H| = |G|.$$

So $|H|$ and $[G : H]$ are divisors of $|G|$, and $|G/H| = [G : H] = |G|/|H|$.

Proof. We know that G is the disjoint union of its distinct left cosets: $G = \bigcup_{S \in G/H} S$. By the Coset Size Theorem, $|S| = |H|$ for every $S \in G/H$. So, by the Sum Rule,

$$|G| = \sum_{S \in G/H} |S| = \sum_{S \in G/H} |H| = |G/H| \cdot |H| = [G : H] \cdot |H|. \qquad \square$$

7.99. Remark. The equality of cardinal numbers $|H| \cdot [G : H] = |G|$ holds even when G is infinite, with the same proof.

7.100. Theorem: Order of Group Elements. If G is a finite group of size n and $x \in G$, then the order of x is a divisor of n, and $x^n = e_G$.

Proof. Consider the subgroup $H = \langle x \rangle$ generated by x. The order d of x is $|H|$, which divides $|G| = n$ by Lagrange's Theorem. Writing $n = cd$, we see that $x^n = (x^d)^c = e^c = e$. □

The next result gives an interpretation for cosets xK in the case where K is the kernel of a group homomorphism.

7.101. Theorem: Cosets of the Kernel of a Homomorphism. Let $f : G \to L$ be a group homomorphism with kernel K. For every $x \in G$,

$$xK = \{y \in G : f(y) = f(x)\} = Kx.$$

If G is finite and I is the image of f, it follows that $|G| = |K| \cdot |I|$.

Proof. Fix $x \in G$, and set $S = \{y \in G : f(y) = f(x)\}$. We will prove that $xK = S$. First suppose $y \in xK$, so $y = xk$ for some $k \in K$. Applying f, we find that $f(y) = f(xk) = f(x)f(k) = f(x)e_L = f(x)$, so $y \in S$. Next suppose $y \in S$, so $f(y) = f(x)$. Note that $f(x^{-1}y) = f(x)^{-1}f(y) = e$, so $x^{-1}y \in \ker(f) = K$. So $y = x(x^{-1}y) \in xK$. The proof that $S = Kx$ is analogous. To obtain the formula for $|G|$, note that G is the disjoint union

$$G = \bigcup_{z \in I} \{y \in G : f(y) = z\}.$$

Every $z \in I$ has the form $z = f(x)$ for some $x \in G$. So, by what we have just proved, each set appearing in the union is a coset of K, which has the same cardinality as K. So the Sum Rule gives $|G| = \sum_{z \in I} |K| = |K| \cdot |I|$. □

7.102. Corollary: Size of A_n. For $n \geq 2$, $|A_n| = n!/2$.

Proof. We know that $\text{sgn} : S_n \to \{1, -1\}$ is a surjective group homomorphism with kernel A_n. So $n! = |S_n| = |A_n| \cdot 2$. □

7.12 The Size of an Orbit

In Theorem 7.86, we saw that every G-set X breaks up into a disjoint union of orbits. This result leads to two combinatorial questions. First, given $x \in X$, what is the size of the orbit Gx? Second, how many orbits are there? We answer the first question here; the second question will be solved in §7.15.

The key to computing the orbit size $|Gx|$ is to relate the G-set Gx to one of the special G-sets G/H defined in Example 7.94. For this purpose, we need to associate a subgroup H of G to the given orbit Gx.

7.103. Definition: Stabilizers. Let X be a G-set. For each $x \in X$, the *stabilizer of x in G* is

$$\text{Stab}(x) = \{g \in G : g * x = x\}.$$

Sometimes the notation G_x is used to denote $\text{Stab}(x)$.

We now show that $\text{Stab}(x)$ *is a subgroup of G for each $x \in X$*. First, since $e * x = x$, $e \in \text{Stab}(x)$. Second, given $g, h \in \text{Stab}(x)$, we know $g * x = x = h * x$, so $(gh) * x = g * (h * x) = x$, so $gh \in \text{Stab}(x)$. Third, given $g \in \text{Stab}(x)$, we know $g * x = x$, so $x = g^{-1} * x$, so $g^{-1} \in \text{Stab}(x)$.

7.104. Example. Let S_n act on $X = \{1, 2, \ldots, n\}$ via $f * x = f(x)$ for $f \in S_n$ and $x \in X$. The stabilizer of a point $i \in X$ consists of all permutations of X for which i is a fixed point. In particular, $\text{Stab}(n)$ consists of all bijections $f : X \to X$ with $f(n) = n$. Restricting the domain and codomain to $\{1, 2, \ldots, n-1\}$ defines a group isomorphism between $\text{Stab}(n)$ and S_{n-1}.

7.105. Example. Let a group G act on itself by left multiplication. Right cancellation of x shows that $gx = x$ iff $g = e$. Therefore, $\text{Stab}(x) = \{e\}$ for all $x \in G$. At the other extreme, we can let G act on any set X by declaring $g * x = x$ for all $g \in G$ and all $x \in X$. Relative to this action, $\text{Stab}(x) = G$ for all $x \in X$.

7.106. Example: Centralizers. Let G act on itself by conjugation: $g * x = gxg^{-1}$ for all $g, x \in G$. For a given $x \in G$, $g \in \text{Stab}(x)$ iff $gxg^{-1} = x$ iff $gx = xg$ iff g commutes with x. This stabilizer subgroup is often denoted $C_G(x)$ and called the *centralizer of x in G*. The intersection $\bigcap_{x \in G} C_G(x)$ consists of all $g \in G$ that commute with *every* $x \in G$. This is a subgroup called the *center* of G and denoted $Z(G)$.

7.107. Example: Normalizers. Let G be a group, and let X be the set of all subgroups of G. G acts on X by conjugation: $g * H = gHg^{-1} = \{ghg^{-1} : h \in H\}$. (Note that $g * H$ *is* a subgroup, since it is the image of a subgroup under the inner automorphism of conjugation by g; see Example 7.55.) For this action, $g \in \text{Stab}(H)$ iff $gHg^{-1} = H$. This stabilizer subgroup is denoted $N_G(H)$ and called the *normalizer of H in G*. One may check that $N_G(H)$ always contains H.

7.108. Example. Let S_4 act on 4-tuples of integers by permuting the positions. Then $\text{Stab}((5,1,5,1)) = \{\text{id}, (1,3), (2,4), (1,3)(2,4)\}$; $\text{Stab}((2,2,2,2)) = S_4$; $\text{Stab}((1,2,3,4)) = \{\text{id}\}$; and $\text{Stab}((2,5,2,2))$ is a subgroup of S_4 isomorphic to $\text{Sym}(\{1,3,4\})$, which is in turn isomorphic to S_3.

The following fundamental theorem calculates the size of an orbit of a group action.

7.109. The Orbit Size Theorem. Assume G is a group and X is a G-set. For each $x \in X$, there is a bijection $f : G/\text{Stab}(x) \to Gx$ given by $f(g\,\text{Stab}(x)) = g * x$ for all $g \in G$. So, when G is finite, *the size of the orbit of x is the index of the stabilizer of x, which is a divisor of $|G|$:*

$$|Gx| = [G : \text{Stab}(x)] = |G|/|\text{Stab}(x)|.$$

Proof. Fix $x \in X$, and write $H = \text{Stab}(x)$ for brevity. We first check that the function $f : G/H \to Gx$ given by $f(gH) = g * x$ is well-defined. Assume $g, k \in G$ satisfy $gH = kH$; we must check that $g * x = k * x$. By the Coset Equality Theorem, $gH = kH$ means $k^{-1}g \in H = \text{Stab}(x)$, and hence $(k^{-1}g) * x = x$. Acting on both sides by k and simplifying, we obtain $g * x = k * x$. Second, is f one-to-one? Fix $g, k \in G$ with $f(gH) = f(kH)$; we must prove $gH = kH$. Now, $f(gH) = f(kH)$ means $g * x = k * x$. Acting on both sides by k^{-1}, we find that $(k^{-1}g) * x = x$, so $k^{-1}g \in H$, so $gH = kH$. Third, is f surjective? Given $y \in Gx$, the definition of Gx says that $y = g * x$ for some $g \in G$, so $y = f(gH)$. In summary, f is a well-defined bijection. □

7.110. Remark. One can prove a stronger version of the theorem, analogous to the Fundamental Homomorphism Theorem for Groups (see Exercise 7-57) by introducing the following definition. Given G-sets X and Y with respective actions $*$ and $\bullet$, a *G-map* is a function $p : X \to Y$ such that $p(g * x) = g \bullet p(x)$ for all $g \in G$ and all $x \in X$. A *G-isomorphism* is a bijective G-map. The theorem above gives us a bijection $p = f^{-1}$ from the G-set Gx to the G-set $G/\text{Stab}(x)$ such that $p(g_0 * x) = g_0\,\text{Stab}(x)$. This bijection is in fact a G-isomorphism, because

$$p(g * (g_0 * x)) = p((gg_0) * x) = (gg_0)\,\text{Stab}(x) = g \bullet (g_0\,\text{Stab}(x)) = g \bullet p(g_0 * x).$$

Since every G-set is a disjoint union of orbits, this result shows that the special G-sets of the form G/H are the building blocks from which all G-sets are constructed.

Applying Theorem 7.109 to some of the preceding examples gives the following corollary.

7.111. Corollary: Counting Conjugates of Group Elements and Subgroups. The size of the conjugacy class of x in a finite group G is $[G : \text{Stab}(x)] = [G : C_G(x)] = |G|/|C_G(x)|$. If H is a subgroup of G, the number of distinct conjugates of H (i.e., subgroups of the form gHg^{-1}) is $[G : \text{Stab}(H)] = [G : N_G(H)] = |G|/|N_G(H)|$.

7.13 Conjugacy Classes in S_n

The conjugacy classes in the symmetric groups S_n can be described explicitly. We shall prove that the conjugacy class of $f \in S_n$ consists of all $g \in S_n$ with the same cycle type as f (see Definition 7.20). The proof employs the following result showing how to compute conjugates of permutations written in cycle notation.

7.112. Theorem: Conjugation in S_n. For all $f, g \in S_n$, the permutation gfg^{-1} is obtained by applying g to each entry in the disjoint cycle decomposition of f. More specifically, if f is written in cycle notation as

$$f = (i_1, i_2, i_3, \ldots)(j_1, j_2, \ldots)(k_1, k_2, \ldots)\cdots,$$

then

$$gfg^{-1} = (g(i_1), g(i_2), g(i_3), \ldots)(g(j_1), g(j_2), \ldots)(g(k_1), g(k_2), \ldots)\cdots.$$

In particular, $\text{type}(gfg^{-1}) = \text{type}(f)$.

Proof. First assume f is a k-cycle, say $f = (i_1, i_2, \ldots, i_k)$. We prove that the functions gfg^{-1} and $h = (g(i_1), g(i_2), \ldots, g(i_k))$ are equal by showing that both have the same effect on every $x \in \{1, 2, \ldots, n\}$. We consider various cases. First, if $x = g(i_s)$ for some $s < k$, then $gfg^{-1}(x) = gfg^{-1}(g(i_s)) = g(f(i_s)) = g(i_{s+1}) = h(x)$. Second, if $x = g(i_k)$, then $gfg^{-1}(x) = g(f(i_k)) = g(i_1) = h(x)$. Finally, if x does not equal any $g(i_s)$, then $g^{-1}(x)$ does not equal any i_s. So f fixes $g^{-1}(x)$, and $gfg^{-1}(x) = g(g^{-1}(x)) = x = h(x)$.

In the general case, write $f = C_1 \circ C_2 \circ \cdots \circ C_t$ where each C_i is a cycle. Since conjugation by g is a homomorphism,

$$gfg^{-1} = (gC_1g^{-1}) \circ (gC_2g^{-1}) \circ \cdots \circ (gC_tg^{-1}).$$

By the previous paragraph, we can compute gC_ig^{-1} by applying g to each element appearing in the cycle notation for C_i. Since this holds for each i, gfg^{-1} can be computed by the same process. □

7.113. Theorem: Conjugacy Classes of S_n. The conjugacy class of $f \in S_n$ consists of all $h \in S_n$ with $\text{type}(h) = \text{type}(f)$. The number of conjugacy classes is $p(n)$, the number of integer partitions of n.

Proof. Fix $f \in S_n$; let $T = \{gfg^{-1} : g \in S_n\}$ be the conjugacy class of f, and let $U = \{h \in S_n : \text{type}(h) = \text{type}(f)\}$. Using Theorem 7.112, we see that $T \subseteq U$. For the reverse inclusion, let $h \in S_n$ have the same cycle type as f. We give an algorithm for finding a $g \in S_n$ such that $h = gfg^{-1}$. First write any complete cycle notation for f (including 1-cycles), writing longer cycles before shorter cycles. Immediately below this, write a complete cycle notation for h. Now erase all the parentheses and regard the resulting array as the two-line form of a permutation g. Theorem 7.112 now shows that $gfg^{-1} = h$. For example, suppose

$$\begin{aligned} f &= (1,7,3)(2,8,9)(4,5)(6); \\ h &= (4,9,2)(6,3,5)(1,8)(7). \end{aligned}$$

Then

$$g=\begin{bmatrix}1&7&3&2&8&9&4&5&6\\4&9&2&6&3&5&1&8&7\end{bmatrix}=\begin{bmatrix}1&2&3&4&5&6&7&8&9\\4&6&2&1&8&7&9&3&5\end{bmatrix}.$$

By the very definition of g, applying g to each symbol in the chosen cycle notation for f produces the chosen cycle notation for h. The g constructed here is not unique; we could obtain other permutations g satisfying $gfg^{-1}=h$ by starting with different complete cycle notations for f and h.

The last statement of the theorem follows since the possible cycle types of permutations of n objects are exactly the integer partitions of n (weakly decreasing sequences of positive integers that sum to n). □

We now apply Corollary 7.111 to determine the sizes of the conjugacy classes of S_n.

7.114. Definition: z_μ**.** Let μ be an integer partition of n consisting of a_1 ones, a_2 twos, etc. Define

$$z_\mu = 1^{a_1}2^{a_2}\cdots n^{a_n}a_1!a_2!\cdots a_n!.$$

For example, for $\mu=(3,3,2,2,2,2,1,1,1,1,1)$, we have $a_1=5$, $a_2=4$, $a_3=2$, and $z_\mu=1^5 2^4 3^2 5!4!2!=829{,}440$.

7.115. Theorem: Size of Conjugacy Classes of S_n**.** Given an integer partition μ of n, the number of permutations $f\in S_n$ with $\text{type}(f)=\mu$ is $n!/z_\mu$.

Proof. Fix a particular $f\in S_n$ with $\text{type}(f)=\mu$. By Corollary 7.111 and the fact that $|S_n|=n!$, it is enough to show that $|C_{S_n}(f)|=z_\mu$. We illustrate the reasoning through a specific example. Let $\mu=(3,3,2,2,2,2,1,1,1,1,1)$, and take

$$f=(1,2,3)(4,5,6)(7,8)(9,10)(11,12)(13,14)(15)(16)(17)(18)(19).$$

A permutation $g\in S_n$ lies in $C_{S_n}(f)$ iff $gfg^{-1}=f$ iff applying g to each symbol in the given cycle notation for f produces another cycle notation for f. So we need only count the number of ways of writing a complete cycle notation for f such that longer cycles come before shorter cycles. Note that we have freedom to rearrange the order of all cycles of a given length, and we also have freedom to cyclically permute the entries in any given cycle of f. For example, we could permute the five 1-cycles of f in any of 5! ways; we could replace $(4,5,6)$ by one of the three cyclic shifts $(4,5,6)$ or $(5,6,4)$ or $(6,4,5)$; and so on. For this particular f, the Product Rule gives $2!4!5!3^2 2^4 1^5=z_\mu$ different possible complete cycle notations for f. The proof of the general case is the same: the term $a_i!$ in z_μ arises when we choose a permutation of the a_i cycles of length i, while the term i^{a_i} arises when we choose one of i possible cyclic shifts for each of the a_i cycles of length i. Multiplying these contributions gives z_μ, as needed. □

7.14 Applications of the Orbit Size Formula

When a finite group G acts on a finite set X, the Orbit Size Theorem 7.109 asserts that the size of the orbit Gx is $|G|/|\text{Stab}(x)|$, which is a divisor of $|G|$. We now use this fact to establish several famous theorems from algebra, number theory, and combinatorics. Recall that for $a,b,p\in\mathbb{Z}$, $a\equiv b \pmod p$ means that $a-b$ is an integer multiple of p. For fixed p, $\equiv$ is an equivalence relation on $\mathbb{Z}$.

7.116. Fermat's Little Theorem. For every integer $a > 0$ and every prime p, $a^p \equiv a \pmod{p}$.

Proof. Let $Y = \{1, 2, \ldots, a\}$, and let $X = Y^p$ be the set of all p-tuples $(y_1, \ldots, y_p)$ of elements of Y. By the Product Rule, $|X| = a^p$. We know that S_p acts on X by permuting positions (see Example 7.70). Let $H = \langle(1, 2, \ldots, p)\rangle$, which is a cyclic subgroup of S_p of size p. Restricting the action to H, we see that H acts on X by cyclically shifting positions. The only divisors of the prime p are 1 and p, so all orbits of X under the action of H have size 1 or p. Now, $w = (y_1, y_2, \ldots, y_p)$ is in an orbit of size 1 iff all cyclic shifts of w are equal to w iff $y_1 = y_2 = \cdots = y_p$. So there are precisely a orbits of size 1, corresponding to the a possible choices for y_1 in Y. Let k be the number of orbits of size p. Since X is the disjoint union of the orbits, $a^p = |X| = kp + a$. So $a^p - a = kp$ is a multiple of p, as needed. □

7.117. Cauchy's Theorem. Suppose G is a finite group and p is a prime divisor of $|G|$. Then there exists an element $x \in G$ of order p.

Proof. As in the previous proof, the group $H = \langle(1, 2, \ldots, p)\rangle$ acts on the set G^p by cyclically permuting positions. Let X consist of all p-tuples $(g_1, \ldots, g_p) \in G^p$ such that $g_1 g_2 \cdots g_p = e$. We can build a typical element of X by choosing $g_1, \ldots, g_{p-1}$ arbitrarily from G; then we are forced to choose $g_p = (g_1 \cdots g_{p-1})^{-1}$ to achieve the condition $g_1 g_2 \cdots g_{p-1} g_p = e$. The Product Rule therefore gives $|X| = |G|^{p-1}$, which is a multiple of p.

We next claim that X is an *H-stable subset* of G^p. This means that for $1 \le i \le p$, $g_1 g_2 \cdots g_p = e$ implies $g_i g_{i+1} \cdots g_p g_1 \cdots g_{i-1} = e$. To prove this, multiply both sides of the equation $g_1 g_2 \cdots g_p = e$ by $(g_1 g_2 \cdots g_{i-1})^{-1}$ on the left and by $(g_1 g_2 \cdots g_{i-1})$ on the right. We now know that X is an H-set, so it is a union of orbits of size 1 and size p. Since $|X|$ is a multiple of p, the number of orbits of size 1 must be a multiple of p as well. Now, $(e, e, \ldots, e)$ is one orbit of size 1; so there must exist at least $p - 1 > 0$ additional orbits of size 1. By definition of the action of H, such an orbit looks like $(x, x, \ldots, x)$ where $x \neq e$. By definition of X, we must have $x^p = e$. Since p is prime, we have proved the existence of an element x of order p (in fact, the proof shows there are at least $p - 1$ such elements). □

7.118. Lucas's Congruence for Binomial Coefficients. Suppose $p, k, n \in \mathbb{Z}_{\ge 0}$, p is prime, and $0 \le k \le n$. Let n and k have base-p expansions $n = \sum_{i \ge 0} n_i p^i$, $k = \sum_{i \ge 0} k_i p^i$, where $0 \le n_i, k_i < p$. Then

$$\binom{n}{k} \equiv \prod_{i \ge 0} \binom{n_i}{k_i} \pmod{p}, \tag{7.4}$$

where $\binom{0}{0} = 1$ and $\binom{a}{b} = 0$ whenever $b > a$.

Proof. Step 1. For all $p, m, j \in \mathbb{Z}_{\ge 0}$ with p prime, we show that

$$\binom{m+p}{j} \equiv \binom{m}{j} + \binom{m}{j-p} \pmod{p}. \tag{7.5}$$

To prove this identity, let $X = \{1, 2, \ldots, m+p\}$, and let Y be the set of all j-element subsets of X. By the Subset Rule, $|Y| = \binom{m+p}{j}$. Consider the subgroup $G = \langle(1, 2, \ldots, p)\rangle$ of S_{m+p}, which is cyclic of size p. G acts on Y via $g \star S = \{g(s) : s \in S\}$ for $g \in G$ and $S \in Y$.

Y is a disjoint union of orbits under this action. Since every orbit has size 1 or p, $|Y|$ is congruent modulo p to the number M of orbits of size 1. We show that $M = \binom{m}{j} + \binom{m}{j-p}$. The orbits of size 1 correspond to the j-element subsets S of X such that $g \star S = S$ for all $g \in G$. It is equivalent to require that $f \star S = S$ for the generator $f = (1, 2, \ldots, p)$ of G. Suppose S satisfies this condition, and consider two cases. Case 1: $S \cap \{1, 2, \ldots, p\} = \emptyset$. Since $f(x) = x$ for $x > p$, we have $f \star S = S$ for all such subsets S. Since S can be an arbitrary subset of the m-element set $\{p+1,\ldots,p+m\}$, there are $\binom{m}{j}$ subsets of this form. Case 2: $S\cap\{1,2,\ldots,p\}\neq\emptyset$. Say $i\in S$ where $1\leq i\leq p$. Applying f repeatedly and noting that $f\star S=S$, we see that $\{1,2,\ldots,p\}\subseteq S$. The remaining $j-p$ elements of S can be chosen arbitrarily from the m-element set $\{p+1,\ldots,p+m\}$. So there are $\binom{m}{j-p}$ subsets of this form. Combining the two cases, we see that $M=\binom{m}{j}+\binom{m}{j-p}$.

Step 2. Assume p is prime, $a,c\geq 0$, and $0\leq b,d<p$; we show that $\binom{ap+b}{cp+d}\equiv\binom{a}{c}\binom{b}{d}$ (mod p). The idea is to use Step 1 and Pascal's Identity $\binom{n+1}{k}=\binom{n}{k}+\binom{n}{k-1}$. We proceed by induction on a. The base step is $a=0$. If $a=0$ and $c>0$, both sides of the congruence are zero; if $a=0=c$, then both sides of the congruence are $\binom{b}{d}$. Assuming that the result holds for a given integer a (and all b,c,d), the following computation shows that it holds for $a+1$:

$$\binom{(a+1)p+b}{cp+d}=\binom{(ap+b)+p}{cp+d}\equiv\binom{ap+b}{cp+d}+\binom{ap+b}{(c-1)p+d}\equiv\binom{a}{c}\binom{b}{d}+\binom{a}{c-1}\binom{b}{d}$$
$$=\left[\binom{a}{c}+\binom{a}{c-1}\right]\binom{b}{d}=\binom{a+1}{c}\binom{b}{d}\pmod{p}.$$

Step 3. We prove Lucas's Congruence (7.4) by induction on n. If $k>n$, then $k_i>n_i$ for some i, so that both sides of the congruence are zero. From now on, assume $k\leq n$. The result holds in the base cases $0\leq n<p$, since $n=n_0$, $k=k_0$, and all higher digits of the base p expansions of n and k are zero. For the induction step, note that $n=ap+n_0$, $k=cp+k_0$, where $a=\sum_{i\geq 0}n_{i+1}p^i$ and $c=\sum_{i\geq 0}k_{i+1}p^i$ in base p. (We obtain a and c from n and k, respectively, by chopping off the final base p digits n_0 and k_0.) By Step 2 and induction, we have

$$\binom{n}{p}\equiv\binom{a}{c}\binom{n_0}{k_0}\equiv\binom{n_0}{k_0}\prod_{i\geq 1}\binom{n_i}{k_i}=\prod_{i\geq 0}\binom{n_i}{k_i}\pmod{p}.\qquad\square$$

7.119. Corollary. Given $a,b,p\in\mathbb{Z}_{\geq 0}$ with p prime and p not dividing b,

$$p\text{ does not divide }\binom{p^a b}{p^a}.$$

Proof. Write $b=\sum_{i\geq 0}b_ip^i$ in base p. The base-p expansions of p^ab and p^a are $p^ab=\cdots b_3b_2b_1b_000\cdots 0$ and $p^a=100\cdots 0$, respectively, where each expansion ends in a zeroes. Since $b_0\neq 0$ by hypothesis, Lucas's Congruence gives

$$\binom{p^ab}{p^a}\equiv\binom{b_0}{1}=b_0\not\equiv 0\pmod{p}.\qquad\square$$

This corollary can also be proved directly, by writing out the fraction defining $\binom{p^ab}{p^a}$ and counting powers of p in the numerator and denominator (see Exercise 7-90).

7.120. Sylow's Theorem. Let G be a finite group of size p^ab, where p is prime, $a>0$, and p does not divide b. There exists a subgroup H of G of size p^a.

Proof. Let X be the collection of all subsets of G of size p^a. By the Subset Rule, $|X|=\binom{p^ab}{p^a}$. Corollary 7.119 says that p does not divide $|X|$. G acts on X by left multiplication: $g*S=\{gs:s\in S\}$ for $g\in G$ and $S\in X$. (The set $g*S$ still has size p^a, since left multiplication by g is injective.) Not every orbit of X has size divisible by p, since $|X|$ itself is not divisible by p. Choose $T\in X$ such that $|GT|\not\equiv 0$ (mod p). Let $H=\mathrm{Stab}(T)=\{g\in G:g*T=T\}$, which is a subgroup of G. The size of the orbit of T is $|G|/|H| = p^a b/|H|$. This integer is not divisible by p, forcing $|H|$ to be a multiple of p^a. So $|H| \geq p^a$. To obtain the reverse inequality, let t_0 be any fixed element of T. Given any $h \in H$, $h * T = T$ implies $ht_0 \in T$. So the right coset $Ht_0 = \{ht_0 : h \in H\}$ is contained in T. We conclude that $|H| = |Ht_0| \leq |T| = p^a$. Thus H is a subgroup of size p^a (and T is in fact one of the right cosets of H). □

7.15 The Number of Orbits

The following theorem, which is sometimes called *Burnside's Lemma*, allows us to count the number of orbits in a given G-set.

7.121. The Orbit-Counting Theorem. Let a finite group G act on a finite set X. For each $g \in G$, let $\text{Fix}(g) = \{x \in X : gx = x\}$ be the set of *fixed points* of g, and let N be the number of distinct orbits. Then

$$N = \frac{1}{|G|} \sum_{g \in G} |\text{Fix}(g)|.$$

So *the number of orbits is the average number of fixed points of elements of* G.

Proof. Define $f : X \to \mathbb{R}$ by setting $f(x) = 1/|Gx|$ for each $x \in X$. We will compute $\sum_{x \in X} f(x)$ in two ways. Let $\{O_1, \ldots, O_N\}$ be the distinct orbits of the group action. On one hand, by grouping summands based on which orbit they are in, we get

$$\sum_{x \in X} f(x) = \sum_{i=1}^{N} \sum_{x \in O_i} f(x) = \sum_{i=1}^{N} \sum_{x \in O_i} \frac{1}{|O_i|} = \sum_{i=1}^{N} 1 = N.$$

On the other hand, the Orbit Size Theorem 7.109 says that $|Gx| = |G|/|\text{Stab}(x)|$. Recall the notation $\chi(gx = x) = 1$ if $gx = x$, and $\chi(gx = x) = 0$ otherwise (see Definition 4.23). We compute:

$$\begin{aligned} \sum_{x \in X} f(x) &= \sum_{x \in X} \frac{|\text{Stab}(x)|}{|G|} = \frac{1}{|G|} \sum_{x \in X} \sum_{g \in G} \chi(gx = x) \\ &= \frac{1}{|G|} \sum_{g \in G} \sum_{x \in X} \chi(gx = x) = \frac{1}{|G|} \sum_{g \in G} |\text{Fix}(g)|. \quad \square \end{aligned}$$

We are finally ready to solve the counting problems involving symmetry that were mentioned in the introduction to this chapter. The strategy is to introduce a set of objects X on which a certain group of symmetries acts. Each orbit of the group action consists of a set of objects in X that are identified with one another when symmetries are taken into account. So the solution to the counting problem is the number of orbits, and this number may be calculated by the formula of the previous theorem.

7.122. Example: Counting Necklaces. How many ways can we build a five-bead circular necklace if there are seven available types of gemstones (repeats allowed) and all rotations of a given necklace are considered equivalent? We can model the set of necklaces (before accounting for symmetries) by the set of words $X = \{(y_1, y_2, y_3, y_4, y_5) : 1 \leq y_i \leq 7\}$. Now let $G = \langle (1, 2, 3, 4, 5) \rangle$ act on X by cyclically permuting positions (see Example 7.70). Every orbit of G consists of a set of necklaces that are identified with one another when symmetry is taken into account. To count the orbits, let us compute $|\operatorname{Fix}(g)|$ for each $g \in G$. First, $\mathrm{id} = (1)(2)(3)(4)(5)$ fixes every object in X, so $|\operatorname{Fix}(\mathrm{id})| = |X| = 7^5$ by the Product Rule. Second, the generator $g = (1,2,3,4,5)$ fixes $(y_1, y_2, y_3, y_4, y_5)$ iff

$$(y_1, y_2, y_3, y_4, y_5) = (y_5, y_1, y_2, y_3, y_4).$$

Comparing coordinates, this holds iff $y_1 = y_2 = y_3 = y_4 = y_5$. So $|\operatorname{Fix}((1,2,3,4,5))| = 7$ since there are seven choices for y_1, and then y_2, y_3, y_4, and y_5 are determined. Next, what is $|\operatorname{Fix}(g^2)|$? We have $g^2 = (1,3,5,2,4)$, so that g^2 fixes $(y_1, y_2, y_3, y_4, y_5)$ iff

$$(y_1, y_2, y_3, y_4, y_5) = (y_4, y_5, y_1, y_2, y_3),$$

which holds iff $y_1 = y_3 = y_5 = y_2 = y_4$. So $|\operatorname{Fix}(g^2)| = 7$. Similarly, $|\operatorname{Fix}(g^3)| = |\operatorname{Fix}(g^4)| = 7$, so the answer is

$$\frac{7^5 + 7 + 7 + 7 + 7}{5} = 3367.$$

Now suppose we are counting six-bead necklaces, identifying all rotations of a given necklace. Here, the group of symmetries is

$$G = \{\mathrm{id},\ (1,2,3,4,5,6),\ (1,3,5)(2,4,6),\ (1,4)(2,5)(3,6),\ (1,5,3)(2,6,4),\ (1,6,5,4,3,2)\}.$$

As before, id has 7^6 fixed points, and each of the two 6-cycles has 7 fixed points. What is $\operatorname{Fix}((1,3,5)(2,4,6))$? We have

$$(1,3,5)(2,4,6) * (y_1, y_2, y_3, y_4, y_5, y_6) = (y_5, y_6, y_1, y_2, y_3, y_4),$$

and this equals $(y_1, \ldots, y_6)$ iff $y_1 = y_3 = y_5$ and $y_2 = y_4 = y_6$. Here there are 7 choices for y_1, 7 choices for y_2, and the remaining y_i's are then forced. So $|\operatorname{Fix}((1,3,5)(2,4,6))| = 7^2$. Likewise, $|\operatorname{Fix}((1,5,3)(2,6,4))| = 7^2$. Similarly, we find that $(y_1, \ldots, y_6)$ is fixed by $(1,4)(2,5)(3,6)$ iff $y_1 = y_4$ and $y_2 = y_5$ and $y_3 = y_6$, so that there are 7^3 such fixed points. In each case, $\operatorname{Fix}(f)$ turned out to be $7^{\operatorname{cyc}(f)}$ where $\operatorname{cyc}(f)$ is the number of cycles in a complete cycle notation for f (including 1-cycles). The number of necklaces is

$$\frac{7^6 + 7 + 7^2 + 7^3 + 7^2 + 7}{6} = 19{,}684.$$

Now consider the question of counting five-bead necklaces using q types of beads, where rotations and reflections of a given necklace are considered equivalent. For this problem, the appropriate group of symmetries is the automorphism group of the cycle graph C_5 (see Example 7.46). In addition to the five powers of $(1,2,3,4,5)$, this group contains the following five permutations corresponding to reflections of the necklace:

$$(1,5)(2,4)(3),\ (1,4)(2,3)(5),\ (1,3)(4,5)(2),\ (1,2)(3,5)(4),\ (2,5)(3,4)(1).$$

It can be checked that each of the five new permutations has $q^3 = q^{\operatorname{cyc}(f)}$ fixed points. For example, a necklace $(y_1, \ldots, y_5)$ is fixed by $(1,5)(2,4)(3)$ iff $y_1 = y_5$ and $y_2 = y_4$ and y_3 is arbitrary. Thus, we may build such a fixed point by choosing y_1 and y_2 and y_3 (q choices each), and then setting $y_4 = y_2$ and $y_5 = y_1$. Using the Orbit-Counting Theorem, the number of necklaces is

$$\frac{q^5 + 5q^3 + 4q^1}{10}.$$

The next example can be used to solve many counting problems involving symmetry.

7.123. Example: Counting Colorings under Symmetries. Suppose V is a finite set of objects, Q is a finite set of q colors, and $G \leq \mathrm{Sym}(V)$ is a group of symmetries of the objects V. (For example, if V is the vertex set of a graph, we could take G to be the automorphism group of the graph.) G acts on V via $g \cdot x = g(x)$ for $g \in G$ and $x \in V$. Now let X be the set of all functions $f : V \to Q$. We think of a function f as a *coloring* of V such that x receives color $f(x)$ for all $x \in V$. As we saw in Example 7.69, G acts on X via $g * f = f \circ g^{-1}$ for $g \in G$ and $f \in X$. Informally, if f assigns color c to object x, then $g * f$ assigns color c to object $g(x)$. The G-orbits consist of colorings that are identified when we take into account the symmetries in G. So the number of colorings up to symmetry is $\frac{1}{|G|} \sum_{g \in G} |\mathrm{Fix}(g)|$.

In the previous example, we observed that $|\mathrm{Fix}(g)| = q^{\mathrm{cyc}(g)}$. To see why this holds in general, let $g \in G$ have a complete cycle notation $g = C_1 C_2 \cdots C_k$, where $k = \mathrm{cyc}(g)$. Let V_i be the elements appearing in cycle C_i, so V is the disjoint union of the sets V_i. Consider C_1, for example. Say $C_1 = (x_1, x_2, \ldots, x_s)$, so that $V_1 = \{x_1, \ldots, x_s\}$. Suppose $f \in X$ is fixed by g, so $f = g * f$. Then

$$f(x_2) = (g * f)(x_2) = f(g^{-1}(x_2)) = f(x_1).$$

Similarly, $f(x_3) = f(x_2)$, and in general $f(x_{j+1}) = f(x_j)$ for all j with $1 \leq j < s$. It follows that f is constant on V_1. Similarly, f is constant on every V_i in the sense that f assigns the same color to every $x \in V_i$. This argument is reversible, so $\mathrm{Fix}(g)$ consists precisely of the colorings $f \in X$ that are constant on each V_i. To build such an f, choose a common color for all the vertices in V_i (for $1 \leq i \leq k$). By the Product Rule, $|\mathrm{Fix}(g)| = q^k = q^{\mathrm{cyc}(g)}$ as claimed. Therefore, the answer to the counting problem is

$$\frac{1}{|G|} \sum_{g \in G} q^{\mathrm{cyc}(g)}. \tag{7.6}$$

7.124. Example: Counting Chessboards. We now answer the question asked at the beginning of this chapter: how many ways can we color a 5×5 chessboard with seven colors, if all rotations and reflections of a given colored board are considered the same? We apply the method of the preceding example. Let $B = (V, E)$ be the graph that models the chessboard (see Figure 7.3). Let $Q = \{1, 2, \ldots, 7\}$ be the set of colors, and let X be the set of colorings before accounting for symmetry. The symmetry group $G = \mathrm{Aut}(B)$ was computed in Example 7.48. By inspecting the cycle notation for the eight elements $g \in G$, the answer follows from (7.6):

$$\frac{7^{25} + 7^7 + 7^{13} + 7^7 + 4 \cdot 7^{15}}{8} = 167{,}633{,}579{,}843{,}887{,}699{,}759.$$

7.16 Pólya's Formula

Consider the following variation of the chessboard coloring example: how many ways can we color a 5×5 chessboard so that 10 squares are red, 12 are blue, and 3 are green, if all rotations and reflections of a colored board are equivalent? We can answer questions like this with the aid of Pólya's Formula, which extends the Orbit-Counting Theorem to weighted sets.

Let a finite group G act on a finite set X. Let $\{O_1, \ldots, O_N\}$ be the orbits of this action. Suppose each $x \in X$ has a weight $\mathrm{wt}(x)$ that is a monomial in variables $z_1, \ldots, z_q$. Also assume that the weights are *G-invariant:* $\mathrm{wt}(g * x) = \mathrm{wt}(x)$ for all $g \in G$ and all $x \in X$.

This condition implies that every object in a given orbit has the same weight. So we can assign a well-defined weight to each orbit by letting $\mathrm{wt}(O_i) = \mathrm{wt}(x_i)$ for any $x_i \in O_i$. The next result lets us compute the generating function for the set of weighted orbits.

7.125. The Orbit-Counting Theorem for Weighted Sets. With the notation of the preceding paragraph,

$$\sum_{i=1}^{N} \mathrm{wt}(O_i) = \frac{1}{|G|} \sum_{g \in G} \sum_{x \in \mathrm{Fix}(g)} \mathrm{wt}(x).$$

So, *the weighted sum of the orbits is the average over G of the weighted fixed point sets of elements of G.*

Proof. We adapt the proof of the original Orbit-Counting Theorem 7.121 to include weights. Define $f : X \to \mathbb{R}[z_1, \ldots, z_q]$ by setting $f(x) = \mathrm{wt}(x)/|Gx|$ for each $x \in X$. On one hand,

$$\sum_{x \in X} f(x) = \sum_{i=1}^{N} \sum_{x \in O_i} f(x) = \sum_{i=1}^{N} \sum_{x \in O_i} \frac{\mathrm{wt}(x)}{|O_i|} = \sum_{i=1}^{N} \sum_{x \in O_i} \frac{\mathrm{wt}(O_i)}{|O_i|} = \sum_{i=1}^{N} \mathrm{wt}(O_i).$$

On the other hand, using $|Gx| = |G|/|\operatorname{Stab}(x)|$, we get

$$\begin{aligned} \sum_{x \in X} f(x) &= \sum_{x \in X} \frac{|\operatorname{Stab}(x)|\,\mathrm{wt}(x)}{|G|} = \frac{1}{|G|} \sum_{x \in X} \sum_{g \in G} \chi(gx = x)\,\mathrm{wt}(x) \\ &= \frac{1}{|G|} \sum_{g \in G} \sum_{x \in X} \chi(gx = x)\,\mathrm{wt}(x) = \frac{1}{|G|} \sum_{g \in G} \sum_{x \in \mathrm{Fix}(g)} \mathrm{wt}(x). \quad \square \end{aligned}$$

We now extend the setup of Example 7.123 to count weighted colorings. We are given finite sets V and $Q = \{1, \ldots, q\}$, a subgroup G of $\mathrm{Sym}(V)$, and the set of colorings X consisting of all functions $f : V \to Q$. G acts on X by permuting the domain: $g * f = f \circ g^{-1}$ for $g \in G$ and $f \in X$. We define a *weight* for a given coloring by setting

$$\mathrm{wt}(f) = \prod_{x \in V} z_{f(x)} \in \mathbb{R}[z_1, z_2, \ldots, z_q].$$

In other words, $\mathrm{wt}(f) = z_1^{e_1} \cdots z_q^{e_q}$ where e_i is the number of objects in V that f colors i. By making the change of variables $v = g^{-1}(x)$, we see that

$$\mathrm{wt}(g * f) = \prod_{x \in V} z_{f(g^{-1}(x))} = \prod_{v \in V} z_{f(v)} = \mathrm{wt}(f) \quad \text{for all } g \in G \text{ and } f \in X.$$

So the Orbit-Counting Theorem for Weighted Sets is applicable. In the unweighted case (see Example 7.123), we found that $|\mathrm{Fix}(g)| = q^{\mathrm{cyc}(g)}$ by observing that $f \in \mathrm{Fix}(g)$ must be constant on each connected component $V_1, \ldots, V_k$ of the digraph of the permutation g.

To take weights into account, let us construct colorings $f \in \mathrm{Fix}(g)$ using the Product Rule for Weighted Sets. Suppose the components $V_1, \ldots, V_k$ in the digraph of g have sizes $n_1 \geq n_2 \geq \cdots \geq n_k$, so that $\mathrm{type}(g) = (n_1, n_2, \ldots, n_k)$. First choose a common color for the n_1 vertices in V_1. The generating function for this choice is $z_1^{n_1} + z_2^{n_1} + \cdots + z_q^{n_1}$; the term $z_i^{n_1}$ arises by coloring all n_1 vertices in V_1 with color i. Second, choose a common color for the n_2 vertices in V_2. The generating function for this choice is $z_1^{n_2} + \cdots + z_q^{n_2}$. Continuing similarly, we arrive at the formula

$$\sum_{x \in \mathrm{Fix}(g)} \mathrm{wt}(x) = \prod_{i=1}^{k} (z_1^{n_i} + z_2^{n_i} + \cdots + z_q^{n_i}).$$

We can abbreviate this formula by introducing the power-sum polynomials (which are studied in more detail in Chapter 9). For each integer $k \geq 1$, set $p_k(z_1, \ldots, z_q) = z_1^k + z_2^k + \cdots + z_q^k$. For each integer partition $\mu = (\mu_1, \mu_2, \ldots, \mu_k)$, set $p_\mu(z_1, \ldots, z_q) = \prod_{i=1}^k p_{\mu_i}(z_1, \ldots, z_q)$. Then the weighted orbit-counting formula assumes the following form.

7.126. Pólya's Formula. With the above notation, the generating function for weighted colorings with q colors relative to the symmetry group G is

$$\sum_{i=1}^{N} \mathrm{wt}(O_i) = \frac{1}{|G|} \sum_{g \in G} p_{\mathrm{type}(g)}(z_1, z_2, \ldots, z_q).$$

The coefficient of $z_1^{e_1} \cdots z_q^{e_q}$ in this polynomial is the number of colorings (taking the symmetries in G into account) in which color i is used e_i times, for $1 \leq i \leq q$.

7.127. Example. The generating function for five-bead necklaces using q types of beads (identifying all rotations and reflections of a given necklace) is

$$(p_{(1,1,1,1,1)} + 4p_{(5)} + 5p_{(2,2,1)})/10,$$

where all power-sum polynomials involve the variables $z_1, \ldots, z_q$.

7.128. Example. Let us use Pólya's Formula to count 5×5 chessboards with 10 red squares, 12 blue squares, and 3 green squares. We may as well take $q = 3$ here. Consulting the cycle decompositions in Example 7.48, we find that the group G has one element of type $(1^{25}) = (1, 1, \cdots, 1)$, two elements of type $(4^6, 1)$, one element of type $(2^{12}, 1)$, and four elements of type $(2^{10}, 1^5)$. Therefore, $\sum_{i=1}^{N} \mathrm{wt}(O_i)$ is

$$\frac{p_{(1^{25})}(z_1, z_2, z_3) + 2p_{(4^6,1)}(z_1, z_2, z_3) + p_{(2^{12},1)}(z_1, z_2, z_3) + 4p_{(2^{10},1^5)}(z_1, z_2, z_3)}{8}.$$

Using a computer algebra system, we can compute this polynomial and extract the coefficient of $z_1^{10} z_2^{12} z_3^3$. The final answer is 185,937,878.

Summary

Table 7.3 summarizes some definitions from group theory used in this chapter. Table 7.4 contains definitions pertinent to the theory of group actions.

- *Examples of Groups.* (i) Additive commutative groups: $\mathbb{Z}$, $\mathbb{Q}$, $\mathbb{R}$, $\mathbb{C}$, and $\mathbb{Z}_n$ (the integers mod n); (ii) Multiplicative commutative groups: invertible elements in $\mathbb{Z}$, $\mathbb{Q}$, $\mathbb{R}$, $\mathbb{C}$, and $\mathbb{Z}_n$; (iii) Non-commutative groups: invertible matrices in $M_n(\mathbb{R})$, the group $\mathrm{Sym}(X)$ of bijections on X under composition, dihedral groups (automorphism groups of cycle graphs); (iv) Constructions of groups: product groups (Exercise 7-6), subgroups, quotient groups (Exercise 7-55), cyclic subgroup generated by a group element, automorphism group of a graph, automorphism group of a group.

- *Basic Properties of Groups.* The identity of a group is unique, as is the inverse of each group element. In a group, there are left and right cancellation laws: $(ax = ay) \Rightarrow (x = y)$ and $(xa = ya) \Rightarrow (x = y)$; inverse rules: $(x^{-1})^{-1} = x$ and $(x_1 \cdots x_n)^{-1} = x_n^{-1} \cdots x_1^{-1}$; and the Laws of Exponents: $x^{m+n} = x^m x^n$; $(x^m)^n = x^{mn}$; and, when $xy = yx$, $(xy)^n = x^n y^n$.

TABLE 7.3
Definitions in group theory.

Concept	Definition
group axioms	$\begin{cases} \forall x,y \in G, xy \in G \text{ (closure)} \\ \forall x,y,z \in G, x(yz) = (xy)z \text{ (associativity)} \\ \exists e \in G, \forall x \in G, xe = x = ex \text{ (identity)} \\ \forall x \in G, \exists y \in G, xy = e = yx \text{ (inverses)} \end{cases}$
commutative group	group G with $xy = yx$ for all $x, y \in G$
H is a subgroup of G	$\begin{cases} e_G \in H \text{ (closure under identity)} \\ \forall a,b \in H, ab \in H \text{ (closure under operation)} \\ \forall a \in H, a^{-1} \in H \text{ (closure under inverses)} \end{cases}$
H is *normal* in G ($H \trianglelefteq G$)	$\forall g \in G, \forall h \in H, ghg^{-1} \in H$ (closure under conjugation)
exponent notation	$x^0 = 1_G$, $x^{n+1} = x^n x$, $x^{-n} = (x^{-1})^n$ $(n > 0)$
multiple notation	$0x = 0_G$, $(n+1)x = nx + x$, $(-n)x = n(-x)$ $(n > 0)$
k-cycle	permutation of the form $(i_1, i_2, \cdots, i_k)$ (cycle notation)
transposition	2-cycle (i, j)
basic transposition	2-cycle $(i, i+1)$ in S_n
$\text{cyc}(f)$	number of components in digraph of $f \in \text{Sym}(X)$
$\text{type}(f)$	list of cycle lengths of $f \in \text{Sym}(X)$ in decreasing order
$\text{inv}(w_1 \cdots w_n)$	number of $i < j$ with $w_i > w_j$
$\text{sgn}(w)$ for $w \in S_n$	$(-1)^{\text{inv}(w)}$
cyclic subgroup $\langle x \rangle$	$\{x^n : n \in \mathbb{Z}\}$ or (in additive notation) $\{nx : n \in \mathbb{Z}\}$
cyclic group	group G such that $G = \langle x \rangle$ for some $x \in G$
order of $x \in G$	least $n > 0$ with $x^n = e_G$, or ∞ if no such n
graph automorphism of K	bijection on vertex set of K preserving edges of K
group homomorphism	map $f : G \to H$ with $f(xy) = f(x)f(y)$ for all $x, y \in G$
kernel of hom. $f : G \to H$	$\ker(f) = \{x \in G : f(x) = e_H\}$
image of hom. $f : G \to H$	$\text{img}(f) = \{y \in H : y = f(x) \text{ for some } x \in G\}$
group isomorphism	bijective group homomorphism
group automorphism	group isomorphism from G to itself
inner automorphism C_g	automorphism sending $x \in G$ to gxg^{-1}

- *Notation for Permutations.* A bijection $f \in S_n$ can be described in two-line form $\begin{bmatrix} 1 & 2 & \cdots & n \\ f(1) & f(2) & \cdots & f(n) \end{bmatrix}$, in one-line form $[f(1), f(2), \ldots, f(n)]$, or in cycle notation. The cycle notation is obtained by listing the elements going around each directed cycle in the digraph of f, enclosing each cycle in parentheses, and optionally omitting cycles of length 1. The cycle notation for f is not unique.

- *Sorting, Inversions, and Sign.* A permutation $w = w_1 w_2 \cdots w_n \in S_n$ can be sorted to the identity permutation $\text{id} = 12 \cdots n$ by applying $\text{inv}(w)$ basic transpositions to switch adjacent elements that are out of order. It follows that w can be written as the composition of $\text{inv}(w)$ basic transpositions. Any factorization of w into a product of transpositions must involve an even number of terms when $\text{sgn}(w) = +1$, or an odd number when $\text{sgn}(w) = -1$. Sign is a group homomorphism: $\text{sgn}(f \circ g) = \text{sgn}(f) \cdot \text{sgn}(g)$ for $f, g \in S_n$. The sign of a k-cycle is $(-1)^{k-1}$. For all $f \in S_n$, $\text{sgn}(f) = (-1)^{n - \text{cyc}(f)}$.

- *Properties of Cyclic Groups.* Every cyclic group is commutative and isomorphic to $\mathbb{Z}$ or $\mathbb{Z}_n$ (under addition) for some $n \geq 1$. More precisely, if $G = \langle x \rangle$ is an infinite multiplicative cyclic group, then $f : \mathbb{Z} \to G$ given by $f(i) = x^i$ for $i \in \mathbb{Z}$ is a group isomorphism. If

TABLE 7.4
Definitions involving group actions.

Concept	Definition
action axioms for G-set X	$\begin{cases} \forall g \in G, \forall x \in X, g * x \in X \text{ (closure)} \\ \forall x \in X, e_G * x = x \text{ (identity)} \\ \forall g, h \in G, \forall x \in X, g * (h * x) = (gh) * x \text{ (assoc.)} \end{cases}$
perm. representation of G on X	group homomorphism $R : G \to \text{Sym}(X)$
G-stable subset Y of X	$\forall g \in G, \forall y \in Y, g * y \in Y$ (closure under action)
orbit of x in G-set X	$Gx = G * x = \{g * x : g \in G\}$
stabilizer of x rel. to G-set X	$\text{Stab}(x) = \{g \in G : g * x = x\} \leq G$
fixed points of g in G-set X	$\text{Fix}(g) = \{x \in X : g * x = x\}$
conjugacy class of x in G	$\{gxg^{-1} : g \in G\}$
centralizer of x in G	$C_G(x) = \{g \in G : gx = xg\} \leq G$
center of G	$Z(G) = \{g \in G : gx = xg \text{ for all } x \in G\} \trianglelefteq G$
normalizer of H in G	$N_G(H) = \{g \in G : gHg^{-1} = H\} \leq G$
left coset of H	$xH = \{xh : h \in H\}$
right coset of H	$Hx = \{hx : h \in H\}$
set of left cosets G/H	for $H \leq G$, $G/H = \{xH : x \in G\}$
index $[G : H]$	$[G : H] = \|G/H\|$ =number of left cosets of H in G

$G = \langle x \rangle$ has size n, then $g : \mathbb{Z}_n \to G$ given by $g(i) = x^i$ for $i \in \mathbb{Z}_n$ is a group isomorphism; moreover, $x^m = e$ iff n divides m. Every subgroup of the additive group $\mathbb{Z}$ has the form $k\mathbb{Z}$ for a unique $k \geq 0$. Every subgroup of a cyclic group is cyclic.

- *Properties of Group Homomorphisms.* If $f : G \to H$ is a group homomorphism, then $\ker(f) \trianglelefteq G$ and $\text{img}(f) \leq H$. Moreover, $f(x^n) = f(x)^n$ for all $x \in G$ and $n \in \mathbb{Z}$. The composition of group homomorphisms (respectively isomorphisms) is a group homomorphism (respectively isomorphism), and the inverse of a group isomorphism is a group isomorphism.

- *Main Results on Group Actions.* Actions $*$ of a group G on a set X correspond bijectively to permutation representations $R : G \to \text{Sym}(X)$, via the formula $R(g)(x) = g * x$ for $g \in G$ and $x \in X$. Every G-set X is the disjoint union of orbits; more precisely, each $x \in X$ lies in a unique orbit Gx. The size of the orbit Gx is the index (number of cosets) of the stabilizer $\text{Stab}(x)$ in G, which (for finite G) is a divisor of $|G|$. The number of orbits is the average number of fixed points of elements of G (for G finite); this extends to weighted sets where the weight is constant on each orbit.

- *Examples of Group Actions.* A subgroup H of a group G can act on G by left multiplication ($h * x = hx$), by inverted right multiplication ($h * x = xh^{-1}$), and by conjugation ($h * x = hxh^{-1}$). The orbits of x under these respective actions are the right coset Hx, the left coset xH, and (when $H = G$) the conjugacy class of x in G. Similarly, G and its subgroups act on the set of all subsets of G by left multiplication, and G acts by conjugation on the set of subgroups of G. The set of subsets of a fixed size k is also a G-set under these actions. Centralizers of elements and normalizers of subgroups are stabilizers of group actions, hence they are subgroups of G. Any subgroup G of $\text{Sym}(X)$ acts on X by $g * x = g(x)$ for $g \in G$ and $x \in X$. For any set X, S_n (or its subgroups) acts on X^n via $f \cdot (x_1, \ldots, x_n) = (x_{f^{-1}(1)}, \ldots, x_{f^{-1}(n)})$. For any subgroup H of G, G acts on G/H via $g * (xH) = (gx)H$ for $g, x \in G$.

- *Facts about Cosets.* Given a subgroup H of a group G, G is the disjoint union of its left cosets, which all have the same cardinality as H (similarly for right cosets). This implies Lagrange's Theorem: $|G| = |H| \cdot [G : H]$, so that (for finite G) the order and index of any subgroup of G are both divisors of $|G|$. To test equality of left cosets, one may check any of the following equivalent conditions: $xH = yH$; $x \in yH$; $x = yh$ for some $h \in H$; $y^{-1}x \in H$; $x^{-1}y \in H$. Similarly, $Hx = Hy$ iff $xy^{-1} \in H$ iff $yx^{-1} \in H$. Left and right cosets coincide (i.e., $xH = Hx$ for all $x \in G$) iff H is normal in G iff all conjugates xHx^{-1} equal H iff H is a union of conjugacy classes of G. Given a group homomorphism $f : G \to L$ with kernel K, $Kx = xK = \{y \in G : f(y) = f(x)\}$ for all $x \in G$.

- *Conjugacy Classes.* Every group G is the disjoint union of its conjugacy classes, where the conjugacy class of x is $\{gxg^{-1} : g \in G\}$. Conjugacy classes need not all have the same size. The size of the conjugacy class of x is the index $[G : C_G(x)]$, where $C_G(x)$ is the subgroup $\{y \in G : xy = yx\}$; this index is a divisor of $|G|$ for G finite. For $x \in G$, the conjugacy class of x has size 1 iff x is in the center $Z(G)$. So (Exercise 7-92) groups G of size p^n (where p is prime and $n \geq 1$) have $|Z(G)| > 1$. Each conjugacy class of S_n consists of those $f \in S_n$ with a fixed cycle type μ. This follows from the fact that a cycle notation for gfg^{-1} can be found from a cycle notation for f by replacing each value x by $g(x)$. The size of the conjugacy class indexed by μ is $n!/z_\mu$.

- *Cayley's Theorem on Permutation Representations.* Every group G is isomorphic to a subgroup of $\text{Sym}(G)$, via the homomorphism sending $g \in G$ to the left multiplication map $L_g : G \to G$ given by $L_g(x) = gx$ for $x \in G$. Every n-element group is isomorphic to a subgroup of S_n.

- *Theorems Provable by Group Actions.* (i) Fermat's Little Theorem: $a^p \equiv a \pmod{p}$ for $a \in \mathbb{Z}_{>0}$ and p prime. (ii) Cauchy's Theorem: If G is a group and p is a prime divisor of $|G|$, then there exists $x \in G$ of order p. (iii) Lucas's Congruence: For $0 \leq k \leq n$ and prime p, $\binom{n}{k} \equiv \prod_{i \geq 0} \binom{n_i}{k_i} \pmod{p}$, where the n_i and k_i are the base-p digits of n and k. (iv) Sylow's Theorem: If G is a group and $|G|$ has prime factorization $|G| = p_1^{n_1} \cdots p_k^{n_k}$, then G has a subgroup of size $p_i^{n_i}$ for $1 \leq i \leq k$.

- *Counting Colorings under Symmetries.* Given a finite set V, a group of symmetries $G \leq \text{Sym}(V)$, and a set Q of q colors, the number of colorings $f : V \to Q$ taking symmetries into account is $|G|^{-1} \sum_{g \in G} q^{\text{cyc}(g)}$. If the colors are weighted using $z_1, \ldots, z_q$, the generating function for weighted colorings is given by Pólya's Formula

$$\frac{1}{|G|} \sum_{g \in G} p_{\text{type}(g)}(z_1, \ldots, z_q),$$

where $p_\mu = \prod_i (\sum_{j=1}^{q} z_j^{\mu_i})$ is a power-sum symmetric polynomial. The coefficient of $z_1^{e_1} \cdots z_q^{e_q}$ gives the number of colorings (taking the symmetries in G into account) where color i is used e_i times.

Exercises

7-1. Let X be a set with more than one element. Define $a \star b = b$ for all $a, b \in X$. (a) Prove that X satisfies the closure axiom and associativity axiom in Definition 7.1. (b) Does there exist $e \in X$ such that $e \star x = x$ for all $x \in X$? If so, is this e unique? (c) Does there exist $e \in X$ such that $x \star e = x$ for all $x \in X$? If so, is this e unique? (d) Is X a group?

7-2. Let G be the set of odd integers. For all $x, y \in G$, define $x \star y = x + y + 5$. Prove that G is a commutative group.

7-3. Let G be the set of real numbers unequal to 1. For each $a, b \in G$, define $a \star b = a + b - ab$. Prove that G is a commutative group.

7-4. Assume G is a group such that $x \star x = e$ for *all* $x \in G$, where e is the identity element of G. Prove that G is commutative.

7-5. Let G be a group with operation $\star$. Define a new operation $\bullet : G \times G \to G$ by setting $a \bullet b = b \star a$ for all $a, b \in G$. Prove that G is a group using the operation $\bullet$.

7-6. Product Groups. Let G and H be groups with operations $\star$ and $\bullet$, respectively. (a) Show that $G \times H$ is a group using the operation $(g_1, h_1) * (g_2, h_2) = (g_1 \star g_2, h_1 \bullet h_2)$ for $g_1, g_2 \in G$, $h_1, h_2 \in H$. (b) Show $G \times H$ is commutative iff G and H are commutative.

7-7. Prove that the operation $\oplus$ on $\mathbb{Z}_n$ satisfies the associative axiom by verifying the relations (7.1). [*Hint:* One approach is to use the fact that for all $u, v \in \mathbb{Z}_n$, there exists $k \in \mathbb{Z}$ with $u \oplus v = u + v - kn$.]

7-8. Suppose G is a set, $\star : G \times G \to G$ is associative, and there exists $e \in G$ such that for all $x \in G$, $e \star x = x$ and there is $y \in G$ with $y \star x = e$. Prove G is a group.

7-9. For x, y in a group G, define the *commutator* $[x, y] = xyx^{-1}y^{-1}$, and let $C_x(y) = xyx^{-1}$. Verify that the following identities hold for all $x, y, z \in G$.
(a) $[x, y]^{-1} = [y, x]$
(b) $[x, yz] = [x, y]C_y([x, z])$
(c) $[x, yz][y, zx][z, xy] = e_G$
(d) $[[x, y], C_y(z)][[y, z], C_z(x)][[z, x], C_x(y)] = e_G$

7-10. Give complete proofs of the three Laws of Exponents in Theorem 7.10.

7-11. Let G be a group. For each $g \in G$, define a function $R_g : G \to G$ by setting $R_g(x) = xg$ for each $x \in G$. R_g is called *right multiplication by g*. (a) Prove that R_g is a bijection. (b) Prove that $R_e = \mathrm{id}_G$ (where e is the identity of G) and $R_g \circ R_h = R_{hg}$ for all $g, h \in G$. (c) Point out why R_g is an element of $\mathrm{Sym}(G)$. Give two answers, one based on (a) and one based on (b). (d) Define $\phi : G \to \mathrm{Sym}(G)$ by setting $\phi(g) = R_g$ for $g \in G$. Prove that ϕ is one-to-one. (e) Prove that for all $g, h \in G$, $L_g \circ R_h = R_h \circ L_g$ (where L_g is left multiplication by g).

7-12. Let G be a group. (a) Prove that for all $a, b \in G$, there exists a unique $x \in G$ with $ax = b$. (b) Prove the *Sudoku Theorem*: in the multiplication table for a group G, every group element appears exactly once in each row and column.

7-13. A certain group G has a multiplication table that is partially given here:

$\star$	1	2	3	4
1	4			
2		1		
3			1	
4				

Use properties of groups to fill in the rest of the table.

7-14. Suppose $G = \{u, v, w, x, y\}$ is a group such that $e_G = w$, $y \star y = x$, and $x \star v = y$. Use this information to find (with explanation) the complete multiplication table for G.

7-15. Suppose $G = \{e, f, g, h, p, r\}$ is a group such that $e \star p = p$, $f = f^{-1}$, $g = g^{-1}$, $p = r^{-1}$, $h \star f = p$, and $g \star p = h$. Find (with explanation) the complete multiplication table for G.

7-16. Let $f, g \in S_8$ be given in one-line form by $f = [3, 2, 7, 5, 1, 4, 8, 6]$ and $g =$

$[4,5,1,3,2,6,8,7]$. (a) Write f and g in cycle notation. (b) Compute $f \circ g$, $g \circ f$, $g \circ g$, and f^{-1}, giving final answers in one-line form.

7-17. Let $f, g \in S_8$ be given in cycle notation by $f = (3,2,7,5)(4,8,6)$ and $g = (4,5,1,3,2,6,8,7)$. (a) Write f and g in one-line form. (b) Compute $f \circ g$, $g \circ f$, $g \circ g$, and f^{-1}, giving final answers in cycle notation.

7-18. Find all cases where $[f_1\ f_2\ \cdots\ f_n] \in S_n$ (given in one-line form) is equal to $(f_1, f_2, \ldots, f_n)$ (given in cycle notation).

7-19. Let $h = [4,1,3,6,5,2]$ in one-line form. Compute $\operatorname{inv}(h)$ and $\operatorname{sgn}(h)$. Write h as a product of $\operatorname{inv}(h)$ basic transpositions.

7-20. Let $f = (1,3,6)(2,8)(4)(5,7)$ and $g = (5,4,3,2,1)(7,8)$. (a) Compute fg, gf, fgf^{-1}, and gfg^{-1}, giving all answers in cycle notation. (b) Compute $\operatorname{sgn}(f)$ and $\operatorname{sgn}(g)$ without counting inversions. (c) Find an $h \in S_8$ such that $hfh^{-1} = (1,2,3)(4,5)(6)(7,8)$; give the answer in two-line form.

7-21. Suppose that $f \in S_n$ has cycle type $\mu = (\mu_1, \ldots, \mu_k)$. What is the order of f?

7-22. The *support* of a bijection $f \in \operatorname{Sym}(X)$ is the set $\operatorname{supp}(f) = \{x \in X : f(x) \neq x\}$. Two permutations $f, g \in \operatorname{Sym}(X)$ are called *disjoint* iff $\operatorname{supp}(f) \cap \operatorname{supp}(g) = \emptyset$. (a) Prove that for all $x \in X$ and $f \in \operatorname{Sym}(X)$, $x \in \operatorname{supp}(f)$ implies $f(x) \in \operatorname{supp}(f)$. (b) Prove that *disjoint permutations commute*, i.e., for all disjoint $f, g \in \operatorname{Sym}(X)$, $f \circ g = g \circ f$. (c) Suppose $f \in \operatorname{Sym}(X)$ is given in cycle notation by $f = C_1 C_2 \cdots C_k$, where the C_i are cycles involving pairwise disjoint subsets of X. Show that the C_i's commute with one another, and prove carefully that $f = C_1 \circ C_2 \circ \cdots \circ C_k$ (see Remark 7.19).

7-23. Prove Lemma 7.27.

7-24. (a) Verify the formula $(i_1, i_2, \ldots, i_k) = (i_1, i_2) \circ (i_2, i_3) \circ (i_3, i_4) \circ \cdots (i_{k-1}, i_k)$ used in the proof of Theorem 7.33. (b) Prove that every transposition has sign -1 by finding an explicit formula for (i, j) as a product of an odd number of basic transpositions (which have sign -1 by Lemma 7.26 with $w = \mathrm{id}$).

7-25. Given $f \in S_n$, how are the one-line forms of f and $f \circ (i, j)$ related? How are the one-line forms of f and $(i, j) \circ f$ related?

7-26. Given $f \in S_n$, how are the one-line forms of f and $f \circ (1, 2, \ldots, n)$ related? How are the one-line forms of f and $(1, 2, \ldots, n) \circ f$ related?

7-27. Let $f \in S_n$ and $h = (i, i+1) \circ f$. (a) Prove an analogue of Lemma 7.26 relating $\operatorname{inv}(f)$ to $\operatorname{inv}(h)$ and $\operatorname{sgn}(f)$ to $\operatorname{sgn}(h)$. (b) Use (a) to give another proof of the formula $\operatorname{sgn}(f \circ g) = \operatorname{sgn}(f)\operatorname{sgn}(g)$ that proceeds by induction on $\operatorname{inv}(f)$.

7-28. Prove that for all $n \geq 3$, every f in the alternating group A_n (see Example 7.59) can be written as a product of 3-cycles.

7-29. Prove that for all $n \geq 2$, every $f \in S_n$ can be written as a composition of factors, where each factor is $(1, 2)$ or $(1, 2, \ldots, n)$. Give an algorithm, based on the one-line form of f, for finding such a factorization. Illustrate this algorithm for $f = 36241875$.

7-30. For which choices of n and k can every $f \in S_n$ be written as a product of factors, where each factor is $(1, k)$ or $(1, 2, \ldots, n)$?

7-31. Verify all the assertions in Example 7.38.

7-32. Let x be an element of a group G, written multiplicatively. Use the Laws of Exponents to verify that $\langle x \rangle = \{x^n : n \in \mathbb{Z}\}$ is a subgroup of G.

7-33. The Subgroup Generated by a Set. Let S be a nonempty subset of a group G. Let $\langle S \rangle$ be the set of elements of G of the form $x_1 x_2 \cdots x_n$, where $n \in \mathbb{Z}_{>0}$ and, for $1 \leq i \leq n$, either $x_i \in S$ or $x_i^{-1} \in S$. Prove that $\langle S \rangle \leq G$, and for all T with $S \subseteq T \leq G$, $\langle S \rangle \leq T$.

7-34. Prove that every subgroup of a cyclic group is cyclic.

7-35. For subsets S and T of a multiplicative group G, define $ST = \{st : s \in S, t \in T\}$. (a) Show that if $S \trianglelefteq G$ and $T \leq G$, then $ST = TS$ and $ST \leq G$. Give an example to show ST may not be normal in G. (b) Show that if $S \trianglelefteq G$ and $T \trianglelefteq G$, then $ST \trianglelefteq G$. (c) Give an example of a group G and subgroups S and T such that ST is not a subgroup of G.

7-36. Let S and T be finite subgroups of a group G. Prove that $|S| \cdot |T| = |ST| \cdot |S \cap T|$.

7-37. Assume that G is a group and $H \leq G$. Let $H^{-1} = \{h^{-1} : h \in H\}$. (a) Show that $HH = H^{-1} = HH^{-1} = H$. (b) Prove that $H \trianglelefteq G$ iff $gHg^{-1} = H$ for all $g \in G$.

7-38. Show that a subgroup H of a group G is normal in G iff H is a union of conjugacy classes of G.

7-39. Find all the subgroups of S_4. Which subgroups are normal? Confirm that Sylow's Theorem 7.120 is true for this group.

7-40. Find all normal subgroups of S_5, and prove that you have found them all (Lagrange's Theorem and Exercise 7-38 can be helpful here).

7-41. Suppose H is a finite, nonempty subset of a group G such that $xy \in H$ for all $x, y \in H$. Prove that $H \leq G$. Give an example to show this result may not be true if H is not finite.

7-42. Given any simple graph or digraph K with vertex set X, show that $\text{Aut}(K)$ is a subgroup of $\text{Sym}(X)$.

7-43. Determine the automorphism groups of the following graphs and digraphs: (a) the path graph P_n with vertex set $\{1, 2, \ldots, n\}$ and edges $\{1, 2\}, \{2, 3\}, \ldots, \{n-1, n\}$; (b) the complete graph K_n with vertex set $\{1, 2, \ldots, n\}$ and an edge between every pair of distinct vertices; (c) the empty graph on $\{1, 2, \ldots, n\}$ with no edges; (d) the directed cycle with vertex set $\{1, 2, \ldots, n\}$ and edges $(1, 2), (2, 3), \ldots, (n-1, n), (n, 1)$; (e) the graph with vertex set $\{\pm 1, \pm 2, \ldots, \pm n\}$ and edge set $\{\{i, -i\} : 1 \leq i \leq n\}$.

7-44. Let K be the Petersen graph defined in Exercise 3-101. (a) Given two paths $P = (y_0, y_1, y_2, y_3)$ and $Q = (z_0, z_1, z_2, z_3)$ in K, prove that there exists a unique automorphism of K that maps y_i to z_i for $0 \leq i \leq 3$. (b) Prove that K has exactly $5! = 120$ automorphisms. (c) Is $\text{Aut}(K)$ isomorphic to S_5?

7-45. Let Q_k be the simple graph with vertex set $V = \{0, 1\}^k$ and edge set $E = \{(v, w) \in V : v, w \text{ differ in exactly one position}\}$. Q_k is called a *k-dimensional hypercube.* (a) Compute $|V(Q_k)|$, $|E(Q_k)|$, and $\deg(Q_k)$. (b) Show that Q_k has exactly $\binom{k}{i} 2^{k-i}$ induced subgraphs isomorphic to Q_i. (c) Find all the automorphisms of Q_k. How many are there?

7-46. (a) Construct an undirected graph whose automorphism group has size three. What is the minimum number of vertices in such a graph? (b) For each $n \geq 1$, construct an undirected graph whose automorphism group is cyclic of size n.

7-47. Let G be a simple graph with connected components $C_1, \ldots, C_k$. Assume that C_i is not isomorphic to C_j for all $i \neq j$. Show that $\text{Aut}(G)$ is isomorphic to the product group $\text{Aut}(C_1) \times \cdots \times \text{Aut}(C_k)$.

7-48. Let $f : G \to H$ be a group homomorphism. (a) Show that if $K \leq G$, then $f[K] = \{f(x) : x \in K\}$ is a subgroup of H. If $K \trianglelefteq G$, must $f[K]$ be normal in H? (b) Show that if $L \leq H$, then $f^{-1}[L] = \{x \in G : f(x) \in L\}$ is a subgroup of G. If $L \trianglelefteq H$, must $f^{-1}[L]$ be normal in G? (c) Deduce from (a) and (b) that the kernel and image of a group homomorphism are subgroups.

7-49. Show that the group of nonzero complex numbers under multiplication is isomorphic to the product of the subgroups $\mathbb{R}_{>0}$ and $\{z \in \mathbb{C} : |z| = 1\}$.

7-50. Give examples of four non-isomorphic groups of size 12.

7-51. Suppose G is a commutative group with subgroups H and K, such that $G = HK$

and $H \cap K = \{e_G\}$. (a) Prove that the map $f : H \times K \to G$, given by $f(h, k) = hk$ for $h \in H$ and $k \in K$, is a group isomorphism. (b) Does any analogous result hold if G is not commutative? What if H and K are normal in G?

7-52. (a) Let G be a group and $x \in G$. Show there exists a unique group homomorphism $f : \mathbb{Z} \to G$ with $f(1) = x$. (b) Use (a) to determine the group $\mathrm{Aut}(\mathbb{Z})$.

7-53. (a) Suppose G is a group, $x \in G$, and $x^n = e_G$ for some $n \geq 2$. Show there exists a unique group homomorphism $f : \mathbb{Z}_n \to G$ with $f(1) = x$. (b) Use (a) to prove that $\mathrm{Aut}(\mathbb{Z}_n)$ is isomorphic to the group $\mathbb{Z}_n^*$ of invertible elements of $\mathbb{Z}_n$ under multiplication modulo n.

7-54. Properties of Order. Let G be a group and $x \in G$. (a) Prove x and x^{-1} have the same order. (b) Show that if x has infinite order, then so does x^i for all nonzero integers i. (c) Suppose x has finite order n. Show that the order of x^k is $n/\gcd(k, n)$ for all $k \in \mathbb{Z}$. (d) Show that if $f : G \to H$ is a group isomorphism, then x and $f(x)$ have the same order. (e) What can be said in part (d) if f is only a group homomorphism?

7-55. Quotient Groups. (a) Suppose H is a *normal* subgroup of G. Show that the set G/H of left cosets of H in G becomes a group of size $[G : H]$ if we define $(xH) \star (yH) = (xy)H$ for all $x, y \in G$. (One must first show that this operation is *well-defined*: i.e., for all $x_1, x_2, y_1, y_2 \in G$, $x_1H = x_2H$ and $y_1H = y_2H$ imply $x_1y_1H = x_2y_2H$. For this, use the Coset Equality Theorem.) (b) With the notation in (a), define $\pi : G \to G/H$ by $\pi(x) = xH$ for $x \in G$. Show that π is a surjective group homomorphism with kernel H.
(c) Let $H = \{\mathrm{id}, (1, 2)\} \leq S_3$. Find $x_1, x_2, y_1, y_2 \in S_3$ with $x_1H = x_2H$ and $y_1H = y_2H$, but $x_1y_1H \neq x_2y_2H$. This shows that normality of H is needed for the product in (a) to be well-defined.

7-56. Let H be a normal subgroup of a group G. (a) Prove that G/H is commutative if G is commutative. (b) Prove that G/H is cyclic if G is cyclic. (c) Does the converse of (a) or (b) hold? Explain.

7-57. The Fundamental Homomorphism Theorem for Groups. Suppose G and H are groups and $f : G \to H$ is a group homomorphism. Let $K = \{x \in G : f(x) = e_H\}$ be the kernel of f, and let $I = \{y \in H : \exists x \in G, y = f(x)\}$ be the image of f. Show that $K \trianglelefteq G$, $I \leq H$, and there exists a unique group isomorphism $\overline{f} : G/K \to I$ given by $\overline{f}(xK) = f(x)$ for $x \in G$.

7-58. The Universal Mapping Property for Quotient Groups. Let G be a group with normal subgroup N, let $\pi : G \to G/N$ be the homomorphism $\pi(x) = xN$ for $x \in G$, and let H be any group. (a) Show that if $h : G/N \to H$ is a group homomorphism, then $h \circ \pi$ is a group homomorphism from G to H sending each $n \in N$ to e_H. (b) Conversely, given any group homomorphism $f : G \to H$ such that $f(n) = e_H$ for all $n \in N$, show that there exists a unique group homomorphism $h : G/N \to H$ such that $f = h \circ \pi$. (c) Conclude that the map sending h to $h \circ \pi$ is a *bijection* from the set of all group homomorphisms from G/N to H to the set of all group homomorphisms from G to H that map everything in N to e_H.

7-59. The Diamond Isomorphism Theorem for Groups. Suppose G is a group, $S \trianglelefteq G$, and $T \leq G$. Show that $TS = ST \leq G$, $S \trianglelefteq TS$, $(S \cap T) \trianglelefteq T$, and there is a well-defined group isomorphism $f : T/(S \cap T) \to (TS)/S$ given by $f(x(S \cap T)) = xS$ for all $x \in T$. Use this to give another solution to Exercise 7-36 in the case where S is *normal* in G.

7-60. The Double-Quotient Isomorphism Theorem for Groups. Assume $A \leq B \leq C$ are groups with A and B both normal in C. Show that $A \trianglelefteq B$, $B/A \trianglelefteq C/A$, and $(C/A)/(B/A)$ is isomorphic to C/B via the map sending $(xA)B/A$ to xB for $x \in C$.

7-61. The Correspondence Theorem for Quotient Groups. Let H be a normal subgroup of a group G. Let X be the set of subgroups of G containing H, and let Y be the set of subgroups of G/H. Show that the map sending $L \in X$ to $L/H = \{xH : x \in L\}$

is an inclusion-preserving bijection from X onto Y with inverse map sending $M \in Y$ to $\{x \in G : xH \in M\}$. If L maps to M under this correspondence, show $[G : L] = [G/H : M]$, $[L : H] = |M|$, $L \trianglelefteq G$ iff $M \trianglelefteq G/H$, and G/L is isomorphic to $(G/H)/M$ whenever $L \trianglelefteq G$.

7-62. Let G be a non-commutative group. Show that the rule $g \cdot x = xg$ (for $g, x \in G$) does not define a (left) action of G on the set G.

7-63. Let G act on itself by conjugation: $g * x = gxg^{-1}$ for $g, x \in G$. Verify that the axioms for a group action are satisfied.

7-64. Let X be a G-set with action $*$, and let $f : K \to G$ be a group homomorphism. Verify the K-set axioms for the action given by $k \bullet x = f(k) * x$ for all $k \in K$ and $x \in X$.

7-65. Suppose $* : G \times X \to X$ is a group action. (a) Show that $\mathcal{P}(X)$ is a G-set via the action $g \bullet S = \{g * s : s \in S\}$ for $g \in G$ and $S \in \mathcal{P}(X)$. (b) For fixed k, show that the set of all k-element subsets of X is a G-stable subset of $\mathcal{P}(X)$.

7-66. Verify the action axioms for the action of S_n on V in Example 7.68.

7-67. Suppose X is a G-set with action $*$, and W is a set. Show that the set of functions $F : W \to X$ is a G-set via the action $(g \bullet F)(w) = g * (F(w))$ for all $g \in G$, $F \in \text{Fun}(W, X)$, and $w \in W$.

7-68. Let a subgroup H of a group G act on G via $h * x = xh^{-1}$ for $h \in H$ and $x \in G$. Show that the orbit $H * x$ is the left coset xH, for all $x \in G$.

7-69. (a) Suppose $f : X \to Y$ is a bijection. Show that the map $T : \text{Sym}(X) \to \text{Sym}(Y)$ given by $T(g) = f \circ g \circ f^{-1}$ for $g \in \text{Sym}(X)$ is a group isomorphism. (b) Use (a) and Cayley's Theorem to conclude that every n-element group is isomorphic to a subgroup of S_n.

7-70. Let a group G act on a set X. Show that $\bigcap_{x \in X} \text{Stab}(x)$ is a *normal* subgroup of G. Give an example to show that a stabilizer subgroup $\text{Stab}(x)$ may not be normal in G.

7-71. Let G act on itself by conjugation. (a) By considering the associated permutation representation and using the Fundamental Homomorphism Theorem, deduce that $G/Z(G)$ is isomorphic to the subgroup of inner automorphisms in $\text{Aut}(G)$. (b) Show that the subgroup of inner automorphisms is normal in $\text{Aut}(G)$.

7-72. Let the additive group $\mathbb{R}$ act on the set of column vectors $\mathbb{R}^2$ by the rule

$$\theta * \begin{bmatrix} x \\ y \end{bmatrix} = \begin{bmatrix} \cos\theta & -\sin\theta \\ \sin\theta & \cos\theta \end{bmatrix} \begin{bmatrix} x \\ y \end{bmatrix} \quad \text{for all } \theta, x, y \in \mathbb{R}.$$

Verify that this is a group action, and describe the orbit and stabilizer of each point in $\mathbb{R}^2$.

7-73. Let $f \in S_n$, and let $\langle f \rangle$ act on $\{1, 2, \ldots, n\}$ via $g \cdot x = g(x)$ for all $g \in \langle f \rangle$ and $x \in \{1, 2, \ldots, n\}$. Prove that the orbits of this action are the connected components of the digraph of f.

7-74. Suppose X is a G-set and $x, y \in X$. Without appealing to equivalence relations, give a direct proof that $Gx \cap Gy \neq \emptyset$ implies $Gx = Gy$.

7-75. Let $*$ be a right action of a group G on a set X. (a) Prove that X is the disjoint union of orbits $x * G$. (b) Prove that $|x * G| = [G : \text{Stab}(x)]$, where $\text{Stab}(x) = \{g \in G : x * g = x\}$.

7-76. State and prove a version of the Coset Equality Theorem 7.92 for right cosets.

7-77. Let G be a group with subgroup H. Prove that the map $T(xH) = Hx^{-1}$ for $x \in G$ is a well-defined bijection from the set of left cosets of H in G onto the set of right cosets of H in G.

7-78. Let X be a G-set. For $x \in X$ and $g \in G$, prove that $g\,\text{Stab}(x) = \{h \in G : h*x = g*x\}$. (This shows that each left coset of the stabilizer of x consists of those group elements sending x to a particular element in its orbit Gx. Compare to Theorem 7.101.)

7-79. Let G be a group with subgroup H. Prove the following facts about the normalizer of H in G (see Example 7.107). (a) $N_G(H)$ contains H; (b) $H \trianglelefteq N_G(H)$; (c) for any $L \leq G$ such that $H \trianglelefteq L$, $L \leq N_G(H)$; (d) $H \trianglelefteq G$ iff $N_G(H) = G$.

7-80. Let X be a G-set. Prove: for all $g \in G$ and $x \in X$, $\text{Stab}(gx) = g\,\text{Stab}(x)g^{-1}$.

7-81. Let H and K be subgroups of a group G. Prove that the G-sets G/H and G/K are isomorphic (as defined in Remark 7.110) iff H and K are conjugate subgroups of G (i.e., $K = gHg^{-1}$ for some $g \in G$).

7-82. Calculate z_μ for all integer partitions μ with $|\mu| = 6$.

7-83. List all elements in the centralizer of $g = (2,4,7)(1,6)(3,8)(5) \in S_8$. How large is this centralizer? How large is the conjugacy class of g?

7-84. Suppose $f = (2,4,7)(8,10,15)(1,9)(11,12)(17,20)(18,19)$ and $g = (7,8,9)(1,4,5)(11,20)(2,6)(3,18)(13,19)$. How many $h \in S_{20}$ satisfy $h \circ f = g \circ h$?

7-85. Find all integer partitions μ of n for which $z_\mu = n!$. Use your answer to calculate $Z(S_n)$ for all $n \geq 1$.

7-86. Prove that for all $n \geq 1$, $p(n) = \frac{1}{n!}\sum_{f \in S_n} z_{\text{type}(f)}$.

7-87. Conjugacy Classes of A_n. For $f \in A_n$, write $[f]_{A_n}$ to denote the conjugacy class of f in A_n, and write $[f]_{S_n}$ to denote the conjugacy class of f in S_n. (a) Prove: for all $f \in A_n$, $[f]_{A_n} \subseteq [f]_{S_n}$. (b) Prove: for all $f \in A_n$, if there exists $g \in S_n - A_n$ with $fg = gf$, then $[f]_{A_n} = [f]_{S_n}$; but if no such g exists, then $[f]_{S_n}$ is the disjoint union of $[f]_{A_n}$ and $[(1,2) \circ f \circ (1,2)]_{A_n}$, and the latter two conjugacy classes are equal in size. (c) What are the conjugacy classes of A_5? How large are they?

7-88. Prove that the only normal subgroups of A_5 are $\{\text{id}\}$ and A_5 (use part (c) of the previous exercise).

7-89. Suppose G is a finite group and p is a prime divisor of $|G|$. Show that the number of elements in G of order p is congruent to $-1 \pmod{p}$.

7-90. (a) Compute $\binom{8936}{5833} \bmod 7$. (b) Compute $\binom{843}{212} \bmod 10$.

7-91. Prove Corollary 7.119 without using Lucas's Congruence, by counting powers of p in the numerator and denominator of $\binom{p^a b}{p^a} = (p^a b)(p^a b - 1)\cdots(p^a b - p^a + 1)/(p^a)!$.

7-92. The Class Equation. Let G be a finite group with center $Z(G)$ (see Example 7.106), and let $x_1, \ldots, x_k \in G$ be such that each conjugacy class of G of size greater than 1 contains exactly one x_i. Prove that $|G| = |Z(G)| + \sum_{i=1}^{k}[G : C_G(x_i)]$, where each term in the sum is a divisor of $|G|$ greater than 1.

7-93. A *p-group* is a finite group of size p^e for some $e \geq 1$. Prove that every p-group G has $|Z(G)| > 1$.

7-94. Wilson's Theorem. Use group actions to prove that if an integer $p > 1$ is prime, then $(p-1)! \equiv -1 \pmod{p}$. Is the converse true?

7-95. How many ways are there to color an $n \times n$ chessboard with q possible colors if: (a) no symmetries are allowed; (b) rotations of a given board are considered equivalent; (c) rotations and reflections of a given board are considered equivalent?

7-96. Consider an $m \times n$ chessboard where $m \neq n$. (a) Describe all symmetries of this board (rotations and reflections). (b) How many ways can we color such a board with q possible colors taking symmetries into account?

7-97. How many n-letter words can be made using a k-letter alphabet if we identify each word with its reversal?

7-98. Consider necklaces that can use q kinds of gemstones, where rotations and reflections of a given necklace are considered equivalent. How many such necklaces are there with: (a) eight gems; (b) nine gems; (c) n gems?

7-99. Taking rotational symmetries into account, how many ways can we color the vertices of a regular tetrahedron with 7 available colors?

7-100. Taking rotational symmetries into account, how many ways can we color the vertices of a cube with 8 available colors?

7-101. Taking rotational symmetries into account, how many ways can we color the faces of a cube with q available colors?

7-102. Taking rotational symmetries into account, how many ways can we color the edges of a cube with q available colors?

7-103. Taking all symmetries into account, how many ways are there to color the vertices of the cycle C_3 with three *distinct* colors chosen from a set of five colors?

7-104. Taking all symmetries into account, how many ways are there to color the vertices of the cycle C_6 so that three vertices are blue, two are red, and one is yellow?

7-105. Taking rotational symmetries into account, how many ways are there to color the vertices of a regular tetrahedron so that: (a) two are blue and two are red; (b) one is red, one is blue, one is green, and one is yellow?

7-106. Taking rotational symmetries into account, how many ways are there to color the vertices of a cube so that four are blue, two are red, and two are green?

7-107. Taking rotational symmetries into account, how many ways are there to color the faces of a cube so that: (a) three are red, two are blue, and one is green; (b) two are red, two are blue, one is green, and one is yellow?

7-108. Taking rotational symmetries into account, how many ways are there to color the edges of a cube so that four are red, four are blue, and four are yellow?

7-109. How many ways can we color a 4×4 chessboard with five colors (identifying rotations of a given board) if each color must be used at least once?

7-110. How many ways can we build an eight-gem necklace using five kinds of gems (identifying rotations and reflections of a given necklace) if each type of gem must be used at least once?

Notes

For a more detailed development of group theory, we recommend the excellent book by Rotman [114]. More information on groups, rings, and fields may be found in textbooks on abstract algebra such as [26, 65, 66]. The proof of Cauchy's Theorem in 7.117 is due to James McKay [86]. The proof of Lucas's Congruence in 7.118 is due to Sagan [116]. The proof of Sylow's Theorem in 7.120 is often attributed to Wielandt [131], although G. A. Miller [87] gave a proof along similar lines over 40 years earlier. Proofs of Fermat's Little Theorem and Wilson's Theorem using group actions were given by Peterson [98].

8

Permutation Statistics and q-Analogues

In Chapter 1, we used the Sum Rule and the Product Rule to count permutations, lattice paths, anagrams, and many other collections of combinatorial objects. We found that the factorial $n!$ counts permutations of an n-element set; the binomial coefficient $\binom{n}{k}$ counts the number of lattice paths from $(0,0)$ to $(k, n-k)$; and the multinomial coefficient $\binom{n_1+\cdots+n_k}{n_1,\ldots,n_k}$ counts anagrams that contain n_i copies of letter a_i for $1 \leq i \leq k$.

In this chapter, we generalize these results by introducing various weight functions (called *statistics*) on permutations, lattice paths, anagrams, and other combinatorial objects. For example, we could weight all permutations $w \in S_n$ by the inversion statistic $\text{inv}(w)$, which was defined in §7.4. Letting q denote a formal variable, we then study the polynomial $\sum_{w\in S_n} q^{\text{inv}(w)}$, which is a sum of $n!$ monomials in the variable q. This polynomial is called a *q-analogue* of $n!$, since setting $q = 1$ in the polynomial produces $n!$. This construction is a special case of the generating function of a weighted set, which we studied in Chapter 5. However, all weighted sets considered here will be finite, so that the associated generating functions are polynomials rather than formal power series.

This chapter studies several permutation statistics (including inversions, descents, and major index) and their generalizations to statistics on words and lattice paths. We develop algebraic and combinatorial formulas for q-factorials, q-binomial coefficients, q-multinomial coefficients, q-Catalan numbers, and q-Stirling numbers. We also encounter some remarkable bijections proving that two different statistics on a given set of objects have the same generating function.

8.1 Statistics on Finite Sets

Generating functions for arbitrary weighted sets have already been introduced in Chapter 5. To keep this chapter self-contained, we begin with a short summary of the relevant concepts in the special case of finite weighted sets.

8.1. Definition: Generating Function of a Finite Weighted Set. Given a finite set S, a *statistic on* S is any function $\text{wt} : S \to \mathbb{Z}_{\geq 0}$. We call the pair (S, wt) a *weighted set.* The *generating function* for such a set is the polynomial

$$G_{S,\text{wt}}(q) = \sum_{z\in S} q^{\text{wt}(z)},$$

where q is a formal variable.

8.2. Example. Suppose $S = \{a, b, c, d, e, f\}$, and $\text{wt} : S \to \mathbb{Z}_{\geq 0}$ is given by

$$\text{wt}(a) = 4,\ \text{wt}(b) = 1,\ \text{wt}(c) = 0,\ \text{wt}(d) = 4,\ \text{wt}(e) = 4,\ \text{wt}(f) = 1.$$

The generating function for (S, wt) is

$$G_{S,\text{wt}}(q) = q^{\text{wt}(a)} + q^{\text{wt}(b)} + \cdots + q^{\text{wt}(f)} = q^4 + q^1 + q^0 + q^4 + q^4 + q^1 = 1 + 2q + 3q^4.$$

Define another statistic $w : S \to \mathbb{Z}_{\geq 0}$ by setting $w(a) = 0$, $w(b) = 1$, $w(c) = 2$, $w(d) = 3$, $w(e) = 4$, and $w(f) = 5$. This new statistic leads to a different generating function, namely

$$G_{S,w}(q) = 1 + q + q^2 + q^3 + q^4 + q^5.$$

8.3. Example. Suppose S is the set of all subsets of $\{1, 2, 3\}$, so

$$S = \{\emptyset, \{1\}, \{2\}, \{3\}, \{1,2\}, \{1,3\}, \{2,3\}, \{1,2,3\}\}.$$

Define three statistics on subsets $A \in S$ as follows:

$$w_1(A) = |A|; \qquad w_2(A) = \sum_{i \in A} i; \qquad w_3(A) = \min\{i : i \in A\} \text{ for } A \neq \emptyset; \quad w_3(\emptyset) = 0.$$

Each statistic produces its own generating function:

$$\begin{aligned}
G_{S,w_1}(q) &= q^0 + q^1 + q^1 + q^1 + q^2 + q^2 + q^2 + q^3 = 1 + 3q + 3q^2 + q^3 = (1+q)^3; \\
G_{S,w_2}(q) &= q^0 + q^1 + q^2 + q^3 + q^3 + q^4 + q^5 + q^6 = 1 + q + q^2 + 2q^3 + q^4 + q^5 + q^6; \\
G_{S,w_3}(q) &= q^0 + q^1 + q^2 + q^3 + q^1 + q^1 + q^2 + q^1 = 1 + 4q + 2q^2 + q^3.
\end{aligned}$$

8.4. Example. For each integer $n \geq 0$, we have introduced the notation $[\![n]\!]$ for the set $\{0, 1, 2, \ldots, n-1\}$. Define a weight function on this set by letting $\mathrm{wt}(i) = i$ for all $i \in [\![n]\!]$. The associated generating function is

$$G_{[\![n]\!],\mathrm{wt}}(q) = q^0 + q^1 + q^2 + \cdots + q^{n-1} = \frac{q^n - 1}{q - 1}.$$

The last equality can be verified by using the distributive law to calculate

$$(q-1)(1 + q + q^2 + \cdots + q^{n-1}) = q^n - 1.$$

The generating function in this example will be a recurring building block in our later work, so we give it a special name.

8.5. Definition: q-Integers. If n is a positive integer and q is any variable, define the *q-integer*

$$[n]_q = 1 + q + q^2 + \cdots + q^{n-1} = \frac{q^n - 1}{q - 1},$$

which is a polynomial in q. Also set $[0]_q = 0$.

8.6. Example. Let S be the set of all lattice paths from $(0,0)$ to $(2,3)$. For $P \in S$, let $w(P)$ be the number of unit squares in the region bounded by P, the x-axis, and the line $x = 2$. Let $w'(P)$ be the number of unit squares in the region bounded by P, the y-axis, and the line $y = 3$. By examining the paths in Figure 1.2, we compute

$$\begin{aligned}
G_{S,w}(q) &= q^6 + q^5 + q^4 + q^4 + q^3 + q^2 + q^3 + q^2 + q^1 + q^0 \\
&= 1 + q + 2q^2 + 2q^3 + 2q^4 + q^5 + q^6; \\
G_{S,w'}(q) &= q^0 + q^1 + q^2 + q^2 + q^3 + q^4 + q^3 + q^4 + q^5 + q^6 \\
&= 1 + q + 2q^2 + 2q^3 + 2q^4 + q^5 + q^6.
\end{aligned}$$

Although the two weight functions are not equal (since there are paths P with $w(P) \neq w'(P)$), it happens that $G_{S,w}(q) = G_{S,w'}(q)$ in this example.

Next, consider the set T of Dyck paths from $(0,0)$ to $(3,3)$. For $P \in T$, let $\mathrm{wt}(P)$ be the number of complete unit squares located between P and the diagonal line $y = x$. Using Figure 1.3, we find that

$$G_{T,\mathrm{wt}}(q) = q^3 + q^2 + q^1 + q^1 + q^0 = 1 + 2q + q^2 + q^3.$$

For any finite weighted set (S, wt), we know $G_{S,\text{wt}}(q) = \sum_{z \in S} q^{\text{wt}(z)}$. By collecting together equal powers of q (as done in the calculations above), we can write this polynomial in the standard form $G_{S,\text{wt}}(q) = a_0q^0 + a_1q^1 + a_2q^2 + \cdots + a_mq^m$, where each $a_i \in \mathbb{Z}_{\geq 0}$. Comparing the two formulas for the generating function, we see that *the coefficient a_i of q^i in $G_{S,\text{wt}}(q)$ is the number of objects z in S such that* $\text{wt}(z) = i$. The next examples illustrate this observation.

8.7. Example. Suppose T is the set of all set partitions of an n-element set, and the weight of a partition is the number of blocks in the partition. By definition of the Stirling number of the second kind (see §2.12), we have $G_T(q) = \sum_{k=0}^{n} S(n,k)q^k$. Similarly, if U is the set of all permutations of n elements, weighted by the number of cycles in the disjoint cycle decomposition, then $G_U(q) = \sum_{k=0}^{n} s'(n,k)q^k$, where $s'(n,k)$ is a signless Stirling number of the first kind (see §3.6). Finally, if V is the set of all integer partitions of n, weighted by number of parts, then $G_V(q) = \sum_{k=0}^{n} p(n,k)q^k$. This is also the generating function for V if we weight a partition by the length of its largest part (see §2.11).

8.8. Remark. Suppose we replace the variable q in $G_S(q)$ by the value 1. We obtain $G_S(1) = \sum_{z \in S} 1^{\text{wt}(z)} = \sum_{z \in S} 1 = |S|$. For instance, in Example 8.6, $G_{T,\text{wt}}(1) = 5 = C_3$. In Example 8.7, $G_T(1) = B(n)$ (the Bell number), $G_U(1) = n!$, and $G_V(1) = p(n)$. Thus, the generating function $G_S(q)$ can be viewed as a weighted analogue of the number of elements in S. This is the origin of the term "q-analogue." On the other hand, using the convention that $0^0 = 1$, $G_S(0)$ is the number of objects in S having weight zero.

We also note that the polynomial $G_S(q)$ can sometimes be factored or otherwise simplified, as illustrated by the first weight function in Example 8.3. Different statistics on S often lead to different generating functions, but this is not always true (see Examples 8.3 and 8.6).

8.2 Counting Rules for Finite Weighted Sets

This section reviews the counting rules for weighted sets, which were given earlier in §5.8. We also give proofs of the Sum Rule and Product Rule for finite weighted sets that are a bit simpler than the proofs given earlier for the general case.

8.9. The Sum Rule for Finite Weighted Sets. Suppose S is a finite weighted set that is the disjoint union of k weighted sets $S_1, S_2, \ldots, S_k$. Assume $\text{wt}_{S_i}(u) = \text{wt}_S(u)$ whenever $1 \leq i \leq k$ and $u \in S_i$. Then

$$G_S(q) = G_{S_1}(q) + G_{S_2}(q) + \cdots + G_{S_k}(q).$$

8.10. The Product Rule for Finite Weighted Sets. Suppose k is a fixed positive integer and $S_1, \ldots, S_k$ are finite weighted sets. Suppose S is a weighted set such that every $u \in S$ can be constructed in exactly one way by choosing $u_1 \in S_1$, choosing $u_2 \in S_2$, and so on, and then assembling the chosen objects $u_1, \ldots, u_k$ in a prescribed manner. Assume that whenever u is constructed from $u_1, u_2, \ldots, u_k$, the *weight-additivity condition* $\text{wt}_S(u) = \text{wt}_{S_1}(u_1) + \text{wt}_{S_2}(u_2) + \cdots + \text{wt}_{S_k}(u_k)$ holds. Then

$$G_S(q) = G_{S_1}(q) \cdot G_{S_2}(q) \cdot \ldots \cdot G_{S_k}(q).$$

8.11. The Bijection Rule for Finite Weighted Sets. Suppose S and T are finite weighted sets and $f : S \to T$ is a *weight-preserving* bijection, meaning that $\text{wt}_T(f(u)) = \text{wt}_S(u)$ for all $u \in S$. Then $G_S(q) = G_T(q)$.

We prove the Sum Rule for Finite Weighted Sets as follows. Assume the setup in that rule. By definition, $G_{S,\mathrm{wt}}(q) = \sum_{z\in S} q^{\mathrm{wt}(z)}$. Because addition of polynomials is commutative and associative, we can order the terms of this sum so that the objects in S_1 come first, followed by the objects in S_2, and so on, ending with the objects in S_k. No term is duplicated in the resulting list, since the sets S_j are pairwise disjoint. We now compute

$$\begin{aligned} G_{S,\mathrm{wt}}(q) &= \sum_{z\in S_1} q^{\mathrm{wt}(z)} + \sum_{z\in S_2} q^{\mathrm{wt}(z)} + \cdots + \sum_{z\in S_k} q^{\mathrm{wt}(z)} \\ &= \sum_{z\in S_1} q^{\mathrm{wt}_1(z)} + \sum_{z\in S_2} q^{\mathrm{wt}_2(z)} + \cdots + \sum_{z\in S_k} q^{\mathrm{wt}_k(z)} \\ &= G_{S_1,\mathrm{wt}_1}(q) + G_{S_2,\mathrm{wt}_2}(q) + \cdots + G_{S_k,\mathrm{wt}_k}(q). \end{aligned}$$

Turning to the proof of the Product Rule for Finite Weighted Sets, first consider the special case where $S = A \times B$ and $\mathrm{wt}_S(a,b) = \mathrm{wt}_A(a) + \mathrm{wt}_B(b)$ for all $a \in A$ and $b \in B$. Using the distributive law, associative law, and commutative law to simplify finite sums, we compute:

$$\begin{aligned} G_S(q) &= \sum_{z\in S} q^{\mathrm{wt}_S(z)} = \sum_{(a,b)\in A\times B} q^{\mathrm{wt}_S(a,b)} = \sum_{a\in A}\sum_{b\in B} q^{\mathrm{wt}_A(a)+\mathrm{wt}_B(b)} = \sum_{a\in A}\sum_{b\in B} q^{\mathrm{wt}_A(a)} q^{\mathrm{wt}_B(b)} \\ &= \sum_{a\in A}\left(q^{\mathrm{wt}_A(a)}\sum_{b\in B} q^{\mathrm{wt}_B(b)}\right) = \left(\sum_{a\in A} q^{\mathrm{wt}_A(a)}\right)\cdot\left(\sum_{b\in B} q^{\mathrm{wt}_B(b)}\right) = G_A(q)\cdot G_B(q). \end{aligned}$$

It follows by induction on k that the set $S' = S_1 \times S_2 \times \cdots \times S_k$ with weight function $\mathrm{wt}_{S'}(s_1,\ldots,s_k) = \sum_{i=1}^k \mathrm{wt}_{S_i}(s_i)$ has generating function $G_{S'}(q) = \prod_{i=1}^k G_{S_i}(q)$. Finally, the hypothesis of the Product Rule 8.10 says that there is a weight-preserving bijection from S' to S. Thus the Product Rule is a consequence of the preceding remarks and the Bijection Rule.

Later, we need the following generalization of the Bijection Rule for Weighted Sets; we leave the proof as an exercise.

8.12. The Weight-Shifting Rule. Suppose (S, w_S) and (T, w_T) are finite weighted sets, $f : S \to T$ is a bijection, and b is a fixed integer such that for all $z \in S$, $w_S(z) = w_T(f(z))+b$. Then $G_{S,w_S}(q) = q^b G_{T,w_T}(q)$.

8.3 Inversions

The inversion statistic for permutations was introduced in §7.4. We now define inversions for general words.

8.13. Definition: Inversions. Suppose $w = w_1w_2\cdots w_n$ is a word, where each letter w_i is an integer. An *inversion* of w is a pair of indices $i < j$ such that $w_i > w_j$. We write $\mathrm{inv}(w)$ for the number of inversions of w. Also let $\mathrm{Inv}(w)$ be the set of all inversion pairs (i, j).

Thus, $\mathrm{inv}(w) = |\mathrm{Inv}(w)|$ counts pairs of letters in w (not necessarily adjacent) that are out of numerical order. If S is any finite set of words using the alphabet $\mathbb{Z}$, then

$$G_{S,\mathrm{inv}}(q) = \sum_{w\in S} q^{\mathrm{inv}(w)}$$

is the *inversion generating function for* S. These definitions extend to words using any totally ordered alphabet.

8.14. Example. Consider the word $w = 414253$; here $w_1 = 4$, $w_2 = 1$, $w_3 = 4$, etc. The pair $(1, 2)$ is an inversion of w since $w_1 = 4 > 1 = w_2$. The pair $(2, 3)$ is not an inversion, since $w_2 = 1 \leq 4 = w_3$. Similarly, $(1, 3)$ is not an inversion. Continuing in this way, we find that $\mathrm{Inv}(w) = \{(1,2), (1,4), (1,6), (3,4), (3,6), (5,6)\}$, so $\mathrm{inv}(w) = 6$.

8.15. Example. Let S be the set of all permutations of $\{1, 2, 3\}$. We know that

$$S = \{123,\ 132,\ 213,\ 231,\ 312,\ 321\}.$$

Counting inversions, we conclude that

$$G_{S,\mathrm{inv}}(q) = q^0 + q^1 + q^1 + q^2 + q^2 + q^3 = 1 + 2q + 2q^2 + q^3 = 1(1+q)(1+q+q^2).$$

Note that $G_S(1) = 6 = 3! = |S|$. Similarly, if T is the set of all permutations of $\{1, 2, 3, 4\}$, a longer calculation leads to

$$G_{T,\mathrm{inv}}(q) = 1 + 3q + 5q^2 + 6q^3 + 5q^4 + 3q^5 + q^6 = 1(1+q)(1+q+q^2)(1+q+q^2+q^3).$$

The factorization patterns in this example are explained and generalized in §8.4.

8.16. Example. Let $S = \mathcal{R}(0^2 1^3)$ be the set of all rearrangements of two 0's and three 1's. We know that

$$S = \{00111,\ 01011,\ 01101,\ 01110,\ 10011,\ 10101,\ 10110,\ 11001,\ 11010,\ 11100\}.$$

Counting inversions, we conclude that

$$G_{S,\mathrm{inv}}(q) = q^0+q^1+q^2+q^3+q^2+q^3+q^4+q^4+q^5+q^6 = 1+q+2q^2+2q^3+2q^4+q^5+q^6.$$

8.17. Example. Let $S = \mathcal{R}(\mathrm{a}^1\mathrm{b}^1\mathrm{c}^2)$, where we use $\mathrm{a} < \mathrm{b} < \mathrm{c}$ as the ordering of the alphabet. We know that

$$S = \{\mathrm{abcc, acbc, accb, bacc, bcac, bcca, cabc, cacb, cbac, cbca, ccab, ccba}\}.$$

Counting inversions leads to

$$G_{S,\mathrm{inv}}(q) = 1 + 2q + 3q^2 + 3q^3 + 2q^4 + q^5.$$

Now let $T = \mathcal{R}(\mathrm{a}^1\mathrm{b}^2\mathrm{c}^1)$ and $U = \mathcal{R}(\mathrm{a}^2\mathrm{b}^1\mathrm{c}^1)$ with the same ordering of the alphabet. One may check that $G_{S,\mathrm{inv}}(q) = G_{T,\mathrm{inv}}(q) = G_{U,\mathrm{inv}}(q)$, although the sets of words in question are all different. This phenomenon will be explained in §8.9.

8.18. Remark. It can be shown that for any word w, $\mathrm{inv}(w)$ is the minimum number of transpositions of adjacent letters required to sort the letters of w into weakly increasing order (Theorem 7.29 proves a special case of this result).

8.4 q-Factorials and Inversions

This section studies the generating functions for sets of permutations weighted by inversions. The answer turns out to be the following q-analogue of $n!$.

8.19. Definition: q-Factorials. Given $n \in \mathbb{Z}_{>0}$ and a formal variable q, define the *q-analogue of $n!$* to be the polynomial

$$[n]!_q = \prod_{i=1}^{n} [i]_q = \prod_{i=1}^{n} (1 + q + q^2 + \cdots + q^{i-1}) = \prod_{i=1}^{n} \frac{q^i - 1}{q - 1}.$$

Also define $[0]!_q = 1$.

The q-factorials satisfy the recursion $[n]!_q = [n-1]!_q [n]_q$ for all $n \geq 1$.

8.20. Example. We have $[0]!_q = 1$, $[1]!_q = 1$, $[2]!_q = [2]_q = 1 + q$,

$$\begin{aligned}
[3]!_q &= (1+q)(1+q+q^2) = 1 + 2q + 2q^2 + q^3, \\
[4]!_q &= (1+q)(1+q+q^2)(1+q+q^2+q^3) = 1 + 3q + 5q^2 + 6q^3 + 5q^4 + 3q^5 + q^6, \\
[5]!_q &= 1 + 4q + 9q^2 + 15q^3 + 20q^4 + 22q^5 + 20q^6 + 15q^7 + 9q^8 + 4q^9 + q^{10}.
\end{aligned}$$

We can use other variables besides q; for instance, $[3]!_t = 1 + 2t + 2t^2 + t^3$. Occasionally we replace the variable by a specific integer or real number; then the q-factorial evaluates to some specific number. For example, when $q = 4$, $[3]!_q = 1 + 8 + 32 + 64 = 105$. As another example, when $t = 1$, $[n]!_t = n!$.

Now we prove that $[n]!_q$ is the generating function for permutations of n symbols weighted by inversions.

8.21. Theorem: q-Factorials and Inversions. For every $n \in \mathbb{Z}_{\geq 0}$, let S_n be the set of all permutations of $\{1, 2, \ldots, n\}$, weighted by inversions. Then

$$G_{S_n,\text{inv}}(q) = \sum_{w \in S_n} q^{\text{inv}(w)} = [n]!_q. \tag{8.1}$$

Proof. We use induction on n. When $n = 0$ or $n = 1$, both sides of (8.1) are 1. Now, fix $n \geq 2$, and assume we already know that $\sum_{w' \in S_{n-1}} q^{\text{inv}(w')} = [n-1]!_q$. To prove the corresponding result for n, we define a weight-preserving bijection $F : S_n \to [\![n]\!] \times S_{n-1}$, where the weight of $(k, w') \in [\![n]\!] \times S_{n-1}$ is $k + \text{inv}(w')$. By Example 8.4, the induction hypothesis, and the Product Rule for Weighted Sets, the codomain of F has generating function $[n]_q \cdot [n-1]!_q = [n]!_q$. So the needed result will follow from the Bijection Rule for Weighted Sets.

Given a permutation $w = w_1 w_2 \cdots w_n \in S_n$, we need to define $F(w) = (k, w') \in [\![n]\!] \times S_{n-1}$ in such a way that $\text{inv}(w) = k + \text{inv}(w')$. One way to pass from $w \in S_n$ to $w' \in S_{n-1}$ is to erase the unique occurrence of the symbol n in the word w. To preserve weights, we are forced to define $k = \text{inv}(w) - \text{inv}(w')$. To see that the map sending w to (k, w') is invertible, we describe k in another way. Suppose $w_i = n$. Since n is larger than every other symbol in w, we see that (i, j) is an inversion pair of w for $i < j \leq n$, whereas (r, i) is not an inversion pair of w for $1 \leq r < i$. All other inversion pairs of w do not involve the symbol n; these inversion pairs correspond bijectively to the inversion pairs of w'. Thus, $\text{inv}(w) = \text{inv}(w') + n - i$, where $k = n - i$ is the number of symbols to the right of n in w. It follows that the two-sided inverse of F sends $(k, w') \in [\![n]\!] \times S_{n-1}$ to the permutation obtained from the word w' by inserting the symbol n so that n is followed by k symbols. The formula $k = n - i$ also shows that $k \in \{0, 1, 2, \ldots, n-1\} = [\![n]\!]$, so that F does map into the required codomain. □

8.22. Example. For $n = 6$, we compute $F(351642) = (2, 35142)$. Observe that

$$\begin{aligned}
\text{Inv}(351642) &= \{(1,3), (1,6), (2,3), (2,5), (2,6), (4,5), (4,6), (5,6)\}, \\
\text{Inv}(35142) &= \{(1,3), (1,5), (2,3), (2,4), (2,5), (4,5)\},
\end{aligned}$$

and $\text{inv}(351642) = 8 = 2 + 6 = 2 + \text{inv}(35142)$. To pass from $\text{Inv}(351642)$ to $\text{Inv}(35142)$, we delete the two inversion pairs $(4, 5)$ and $(4, 6)$ involving the symbol $w_4 = 6 = n$, and then renumber positions in the remaining inversion pairs to account for the erasure of the symbol in position 4.

Next we compute $F^{-1}(k, 25143)$ for $k = 0, 1, 2, 3, 4, 5$. The answers are

$$25143\underline{6}, \quad 2514\underline{6}3, \quad 251\underline{6}43, \quad 25\underline{6}143, \quad 2\underline{6}5143, \quad \underline{6}25143.$$

The inversion counts for these permutations are 5, 6, 7, 8, 9, and 10, respectively. As the new largest symbol moves through the permutation w' from right to left, one symbol at a time, the total inversion count increases by 1 with each step.

We can iterate the construction in the proof of Theorem 8.21, removing $n-1$ from w' to go from the set $[\![n]\!] \times S_{n-1}$ to $[\![n]\!] \times [\![n-1]\!] \times S_{n-2}$, and so on. Ultimately, we arrive at a weight-preserving bijection $G : S_n \to [\![n]\!] \times [\![n-1]\!] \times \cdots \times [\![2]\!] \times [\![1]\!]$ that maps $w \in S_n$ to an n-tuple $G(w) = (k_n, k_{n-1}, \ldots, k_1) \in [\![n]\!] \times [\![n-1]\!] \times \cdots \times [\![1]\!]$. Here, k_n is the number of symbols to the right of n in w; k_{n-1} is the number of symbols to the right of $n-1$ in w with n erased; k_{n-2} is the number of symbols to the right of $n-2$ in w with n and $n-1$ erased; and so on. More succinctly, for $1 \le r \le n$, *k_r is the number of symbols to the right of r in w that are less than r.* In other words,

$$k_r = \{(i, j) : 1 \le i < j \le n \text{ and } r = w_i > w_j\} \quad \text{for } 1 \le r \le n.$$

Thus, k_r counts the number of inversion pairs in $\text{Inv}(w)$ such that the left symbol in the pair is r. The sequence $G(w) = (k_n, k_{n-1}, \ldots, k_1)$ is called an *inversion table* for w. We can reconstruct the permutation w from its inversion table by repeatedly applying the map F^{-1} from the proof of Theorem 8.21. Starting with the empty word, we insert symbols $r = 1, 2, \ldots, n$ in this order. We insert symbol r to the left of k_r symbols in the current word.

8.23. Example. The inversion table for $w = 42851673 \in S_8$ is $G(w) = (5, 1, 1, 2, 3, 0, 1, 0)$. We have $k_5 = 2$, for instance, because of the two inversion pairs $(4, 5)$ and $(4, 8)$ caused by the symbols $5 > 1$ and $5 > 3$ in w. Now let us compute $G^{-1}(5, 5, 1, 2, 3, 1, 0, 0)$. We begin with the empty word, insert 1, then insert 2 at the right end to obtain 12, since $k_0 = k_1 = 0$. Since $k_3 = 1$, we insert 3 to the left of the 2 to get 132. Since $k_4 = 3$, we insert 4 to the left of all 3 existing symbols to get 4132. The process continues, leading to 41532, then 415362, then 4715362, and finally to 47815362. Having found this answer, one may quickly check that $G(47815362) = (5, 5, 1, 2, 3, 1, 0, 0)$, as needed.

Other types of inversion tables for permutations can be constructed by classifying the inversions of w in different ways. Our discussion above classified inversions by the value of the leftmost symbol in the inversion pair. One can also classify inversions using the value of the rightmost symbol, the position of the leftmost symbol, or the position of the rightmost symbol. These possibilities are explored in the exercises.

8.5 Descents and Major Index

This section introduces more statistics on words, which leads to another combinatorial interpretation for the q-factorial $[n]!_q$.

8.24. Definition: Descents and Major Index. Let $w = w_1w_2\cdots w_n$ be a word where each w_i comes from some totally ordered alphabet. The *descent set* of w, denoted $\mathrm{Des}(w)$, is the set of all $i < n$ such that $w_i > w_{i+1}$. The *descent statistic* of w is $\mathrm{des}(w) = |\,\mathrm{Des}(w)|$. The *major index* of w, denoted $\mathrm{maj}(w)$, is the sum of the elements of the set $\mathrm{Des}(w)$.

Thus, $\mathrm{Des}(w)$ is the set of *positions* in w where a letter is immediately followed by a smaller letter; $\mathrm{des}(w)$ is the number of such positions; and $\mathrm{maj}(w)$ is the sum of these positions.

8.25. Example. If $w = 47815362$, then $\mathrm{Des}(w) = \{3,5,7\}$, $\mathrm{des}(w) = 3$, and $\mathrm{maj}(w) = 3+5+7 = 15$. If $w = 101100101$, then $\mathrm{Des}(w) = \{1,4,7\}$, $\mathrm{des}(w) = 3$, and $\mathrm{maj}(w) = 12$. If $w = 33555789$, then $\mathrm{Des}(w) = \emptyset$, $\mathrm{des}(w) = 0$, and $\mathrm{maj}(w) = 0$.

8.26. Theorem: q-Factorials and Major Index. For all $n \in \mathbb{Z}_{\geq 0}$, let S_n be the set of all permutations of $\{1,2,\ldots,n\}$, weighted by major index. Then

$$G_{S_n,\mathrm{maj}}(q) = \sum_{w\in S_n} q^{\mathrm{maj}(w)} = [n]!_q. \tag{8.2}$$

Proof. We imitate the proof of Theorem 8.21. Both sides of (8.2) are 1 when $n = 0$ or $n = 1$. Proceeding by induction, fix $n \geq 2$, and assume $\sum_{w'\in S_{n-1}} q^{\mathrm{maj}(w')} = [n-1]!_q$. To prove the corresponding result for n, it suffices to define a weight-preserving bijection $F : S_n \to [\![n]\!] \times S_{n-1}$, since $[n]_q \cdot [n-1]!_q = [n]!_q$. If $F(w) = (k, w')$, we must have $\mathrm{maj}(w) = k + \mathrm{maj}(w')$ to preserve weights.

As before, we can map $w \in S_n$ to $w' \in S_{n-1}$ by erasing the unique occurrence of n in the word $w = w_1w_2\cdots w_n$. To preserve weights, we are forced to define $F(w) = (\mathrm{maj}(w) - \mathrm{maj}(w'), w')$. However, it is not obvious that the map F defined in this way is invertible. To check this, we need to know that w can be recovered from w' and $k = \mathrm{maj}(w) - \mathrm{maj}(w')$. Given any $w' \in S_{n-1}$, there are n positions between the $n-1$ symbols of w' (including the far left and far right positions) where we might insert n to get a permutation in S_n. Imagine inserting n into each of these n positions and seeing how $\mathrm{maj}(w')$ changes. If we can show that the changes in major index are the numbers $0, 1, 2, \ldots, n-1$ (in some order), then the invertibility of F will follow. This argument also proves that $k = \mathrm{maj}(w) - \mathrm{maj}(w') \in \{0, 1, \ldots, n-1\} = [\![n]\!]$, so that F does map into the required codomain. We prove the required fact in Lemma 8.28 below, after considering an example. □

8.27. Example. Let $n = 8$ and $w' = 4251673 \in S_7$, so that $\mathrm{maj}(w') = 1+3+6 = 10$. There are eight gaps in w' where the new symbol 8 might be placed. We compute the major index of each of the resulting permutations $w \in S_8$:

$$\begin{aligned}
&\mathrm{maj}(8>4>2<5>1<6<7>3) &&= 1+2+4+7 &&= 14 &&= \mathrm{maj}(w')+4;\\
&\mathrm{maj}(4<8>2<5>1<6<7>3) &&= 2+4+7 &&= 13 &&= \mathrm{maj}(w')+3;\\
&\mathrm{maj}(4>2<8>5>1<6<7>3) &&= 1+3+4+7 &&= 15 &&= \mathrm{maj}(w')+5;\\
&\mathrm{maj}(4>2<5<8>1<6<7>3) &&= 1+4+7 &&= 12 &&= \mathrm{maj}(w')+2;\\
&\mathrm{maj}(4>2<5>1<8>6<7>3) &&= 1+3+5+7 &&= 16 &&= \mathrm{maj}(w')+6;\\
&\mathrm{maj}(4>2<5>1<6<8>7>3) &&= 1+3+6+7 &&= 17 &&= \mathrm{maj}(w')+7;\\
&\mathrm{maj}(4>2<5>1<6<7<8>3) &&= 1+3+7 &&= 11 &&= \mathrm{maj}(w')+1;\\
&\mathrm{maj}(4>2<5>1<6<7>3<8) &&= 1+3+6 &&= 10 &&= \mathrm{maj}(w')+0.
\end{aligned}$$

We see that the possible values of $k = \mathrm{maj}(w) - \mathrm{maj}(w')$ are $4,3,5,2,6,7,1,0$, which form a rearrangement of $0,1,2,3,4,5,6,7$.

The next lemma explains precisely how the major index changes when we insert the symbol n into a permutation of $\{1,2,\ldots,n-1\}$.

8.28. Lemma. Suppose $v = v_1v_2 \cdots v_{n-1} \in S_{n-1}$ has descents at positions $i_1 > i_2 > \cdots > i_d$. Number the n gaps between the $n-1$ symbols in v as follows. The gap to the right of v_{n-1} is numbered 0. For $1 \leq j \leq d$, the gap between v_{i_j} and v_{i_j+1} is numbered j. The remaining gaps are numbered $d+1, d+2, \ldots, n-1$, starting at the gap left of v_1 and working left to right. For all j in the range $0 \leq j < n$, if w is obtained from v by inserting n in the gap numbered j, then $\text{maj}(w) = \text{maj}(v) + j$.

Proof. As an example of how gaps are numbered, consider $v = 4251673$ from the example above. The gaps in v are numbered as follows:

$$\begin{array}{ccccccccccccccc} . & 4 & > & 2 & < & 5 & > & 1 & < & 6 & < & 7 & > & 3 & . \\ 4 & & 3 & & 5 & & 2 & & 6 & & 7 & & 1 & & 0 \end{array}$$

The calculations in the example show that inserting the symbol 8 into the gap numbered j causes the major index to increase by j, as predicted by the lemma.

Now we analyze the general case. If we insert n into the far right gap of v (which is numbered 0), there are no new descents, so $\text{maj}(w) = \text{maj}(v) + 0$ as needed. Next suppose we insert n into the gap numbered j, where $1 \leq j \leq d$. We had $v_{i_j} > v_{i_j+1}$ in v, but the insertion of n changes this configuration to $w_{i_j} < w_{i_j+1} = n > w_{i_j+2}$. This pushes the descent in v at position i_j one position to the right. Furthermore, the descents that occur in v at positions $i_{j-1}, \ldots, i_1$ (which are to the right of i_j) also get pushed one position to the right in w because of the new symbol n. It follows that the major index increases by exactly j, as needed.

Finally, suppose $d < j \leq n-1$. Let the gap numbered j occur at position u in w, and let t be the number of descents in v preceding this gap. By definition of the gap labeling, we must have $j = (u-t)+d$. On the other hand, inserting n in this gap produces a new descent in w at position u, and pushes the $(d-t)$ descents in v located to the right of position u one position further right. The net change in the major index is therefore $u + (d-t) = j$, as needed. □

As in the case of inversions, we can iterate the proof of Theorem 8.26 to obtain a weight-preserving bijection $G : (S_n, \text{maj}) \to [\![n]\!] \times [\![n-1]\!] \times \cdots \times [\![1]\!]$. For $w \in S_n$, define $w^{(r)}$ to be w with all symbols greater than r erased. We have $G(w) = (k_n, k_{n-1}, \ldots, k_1)$, where $k_r = \text{maj}(w^{(r)}) - \text{maj}(w^{(r-1)})$ for $1 \leq r \leq n$. The list $(k_n, \ldots, k_1)$ is a *major index table* for w. We can recover w from its major index table by inserting symbols $1, 2, \ldots, n$, in this order, into an initially empty word. We insert symbol r into the unique position that increases the major index of the current word by k_r. This position can be found by numbering gaps as described in Lemma 8.28.

8.29. Example. Given $w = 42851673 \in S_8$, we compute $\text{maj}(w) = 15$, $\text{maj}(4251673) = 10$, $\text{maj}(425163) = 9$, $\text{maj}(42513) = 4$, $\text{maj}(4213) = 3$, $\text{maj}(213) = 1$, $\text{maj}(21) = 1$, $\text{maj}(1) = 0$, and $\text{maj}(\epsilon) = 0$, where ϵ denotes the empty word. So the major index table of w is $(5, 1, 5, 1, 2, 0, 1, 0)$.

Next, taking $n = 6$, we find $G^{-1}(3, 4, 0, 1, 1, 0)$. Using the insertion procedure from Lemma 8.28, we generate the following sequence of permutations:

$$1, \quad 21, \quad 231, \quad 2314, \quad 23154, \quad 623154.$$

8.6 q-Binomial Coefficients

The formula $[n]!_q = \prod_{i=1}^n [i]_q$ for the q-factorial is analogous to the formula $n! = \prod_{i=1}^n i$ for the ordinary factorial. We can extend this analogy to binomial coefficients and multinomial coefficients. This leads to the following definitions.

8.30. Definition: q-Binomial Coefficients. Given $k, n \in \mathbb{Z}_{\geq 0}$ with $0 \leq k \leq n$ and a variable q, define the *q-binomial coefficient*

$$\begin{bmatrix} n \\ k \end{bmatrix}_q = \frac{[n]!_q}{[k]!_q [n-k]!_q} = \frac{(q^n-1)(q^{n-1}-1)\cdots(q-1)}{(q^k-1)(q^{k-1}-1)\cdots(q-1)(q^{n-k}-1)(q^{n-k-1}-1)\cdots(q-1)}.$$

8.31. Definition: q-Multinomial Coefficients. Given $n_1, \ldots, n_k \in \mathbb{Z}_{\geq 0}$ and a variable q, define the *q-multinomial coefficient*

$$\begin{bmatrix} n_1 + \cdots + n_k \\ n_1, \ldots, n_k \end{bmatrix}_q = \frac{[n_1 + \cdots + n_k]!_q}{[n_1]!_q [n_2]!_q \cdots [n_k]!_q} = \frac{(q^{n_1+\cdots+n_k}-1)\cdots(q^2-1)(q-1)}{\prod_{i=1}^k [(q^{n_i}-1)(q^{n_i-1}-1)\cdots(q-1)]}.$$

It is not immediately evident from the defining formulas that q-binomial coefficients and q-multinomial coefficients really are polynomials in q (rather than ratios of polynomials). Below, we prove the stronger fact that these objects are polynomials with nonnegative integer coefficients. We also give several combinatorial interpretations for q-binomial coefficients and q-multinomial coefficients as generating functions for weighted sets. Before doing so, we need to develop a few more tools.

By consulting the definitions, we see at once that $\begin{bmatrix} n \\ k \end{bmatrix}_q = \begin{bmatrix} n \\ n-k \end{bmatrix}_q = \begin{bmatrix} n \\ k, n-k \end{bmatrix}_q$. In particular, q-binomial coefficients are special cases of q-multinomial coefficients. We often prefer to use multinomial coefficients, writing $\begin{bmatrix} a+b \\ a,b \end{bmatrix}_q$ rather than $\begin{bmatrix} a+b \\ a \end{bmatrix}_q$ or $\begin{bmatrix} a+b \\ b \end{bmatrix}_q$, because in most combinatorial settings the parameters a and b are more natural than a and $a+b$.

Before describing the combinatorics of q-binomial coefficients, we give an algebraic proof of two recursions satisfied by the q-binomial coefficients, which are q-analogues of Pascal's Recursion 2.3 for ordinary binomial coefficients.

8.32. Theorem: Recursions for q-Binomial Coefficients. For all $a, b \in \mathbb{Z}_{>0}$,

$$\begin{aligned} \begin{bmatrix} a+b \\ a,b \end{bmatrix}_q &= q^b \begin{bmatrix} a+b-1 \\ a-1,b \end{bmatrix}_q + \begin{bmatrix} a+b-1 \\ a,b-1 \end{bmatrix}_q; \\ \begin{bmatrix} a+b \\ a,b \end{bmatrix}_q &= \begin{bmatrix} a+b-1 \\ a-1,b \end{bmatrix}_q + q^a \begin{bmatrix} a+b-1 \\ a,b-1 \end{bmatrix}_q. \end{aligned}$$

The initial conditions are $\begin{bmatrix} a \\ a,0 \end{bmatrix}_q = \begin{bmatrix} b \\ 0,b \end{bmatrix}_q = 1$ for all $a, b \in \mathbb{Z}_{\geq 0}$.

Proof. We prove the first equality, leaving the second one as an exercise. Writing out the definitions, the right side of the first recursion is

$$q^b \begin{bmatrix} a+b-1 \\ a-1,b \end{bmatrix}_q + \begin{bmatrix} a+b-1 \\ a,b-1 \end{bmatrix}_q = \frac{q^b [a+b-1]!_q}{[a-1]!_q [b]!_q} + \frac{[a+b-1]!_q}{[a]!_q [b-1]!_q}.$$

Multiply the first fraction by $[a]_q/[a]_q$ and the second fraction by $[b]_q/[b]_q$ to create a common denominator $[a]!_q[b]!_q$. Bringing out common factors, we obtain

$$\left(\frac{[a+b-1]!_q}{[a]!_q [b]!_q}\right) \cdot (q^b [a]_q + [b]_q).$$

By definition of q-integers,

$$\begin{aligned}[b]_q + q^b[a]_q &= (1 + q + q^2 + \cdots + q^{b-1}) + q^b(1 + q + \cdots + q^{a-1}) \\ &= (1 + \cdots + q^{b-1}) + (q^b + q^{b+1} + \cdots + q^{a+b-1}) = [a + b]_q.\end{aligned}$$

Putting this into the previous formula, we get

$$\frac{[a+b-1]!_q[a+b]_q}{[a]!_q[b]!_q} = \begin{bmatrix} a+b \\ a,b \end{bmatrix}_q.$$

The initial conditions follow immediately from the definitions. □

8.33. Corollary: Polynomiality of q-Binomial Coefficients. For all $k, n \in \mathbb{Z}_{\geq 0}$ with $0 \leq k \leq n$, $\begin{bmatrix} n \\ k \end{bmatrix}_q$ is a polynomial in q with nonnegative integer coefficients.

Proof. Use induction on $n \geq 0$. The base case ($n = 0$) holds because $\begin{bmatrix} 0 \\ 0 \end{bmatrix}_q = 1$. Fix $n > 0$, and assume that $\begin{bmatrix} n-1 \\ j \end{bmatrix}_q$ is already known to be a polynomial with coefficients in $\mathbb{Z}_{\geq 0}$ for all j in the range $0 \leq j \leq n - 1$. Then, by the first recursion above (with a replaced by k and b replaced by $n - k$),

$$\begin{bmatrix} n \\ k \end{bmatrix}_q = q^{n-k} \begin{bmatrix} n-1 \\ k-1 \end{bmatrix}_q + \begin{bmatrix} n-1 \\ k \end{bmatrix}_q.$$

By the induction hypothesis, each q-binomial coefficient on the right side is a polynomial with coefficients in $\mathbb{Z}_{\geq 0}$, so $\begin{bmatrix} n \\ k \end{bmatrix}_q$ is also such a polynomial. □

8.7 Combinatorial Interpretations of q-Binomial Coefficients

We now show that q-binomial coefficients count various weighted sets of anagrams, lattice paths, and integer partitions.

8.34. Theorem: Combinatorial Interpretations of q-Binomial Coefficients. Fix integers $a, b \geq 0$. Let $\mathcal{R}(0^a 1^b)$ be the set of anagrams consisting of a zeroes and b ones. Let $L(a, b)$ be the set of all lattice paths from $(0, 0)$ to (a, b). For $\pi \in L(a, b)$, let $\text{area}(\pi)$ be the area of the region between π and the x-axis, and let $\text{area}'(\pi)$ be the area of the region between π and the y-axis. Let $P(a, b)$ be the set of integer partitions μ with largest part at most a and with at most b parts. Then

$$\begin{bmatrix} a+b \\ a,b \end{bmatrix}_q = \sum_{w \in \mathcal{R}(0^a 1^b)} q^{\text{inv}(w)} = \sum_{\pi \in L(a,b)} q^{\text{area}(\pi)} = \sum_{\pi \in L(a,b)} q^{\text{area}'(\pi)} = \sum_{\mu \in P(a,b)} q^{|\mu|}.$$

Proof. Step 1. For $a, b \in \mathbb{Z}_{\geq 0}$, define $g(a, b) = \sum_{\pi \in L(a,b)} q^{\text{area}(\pi)} = G_{L(a,b),\text{area}}(q)$. We show that this function satisfies the same recursion and initial conditions as the q-binomial coefficients, namely

$$g(a, b) = q^b g(a - 1, b) + g(a, b - 1) \text{ for } a, b > 0; \quad g(a, 0) = g(0, b) = 1 \text{ for } a, b \geq 0.$$

It then follows by a routine induction on $a + b$ that $\begin{bmatrix} a+b \\ a,b \end{bmatrix}_q = g(a, b)$ for all $a, b \in \mathbb{Z}_{\geq 0}$.

To check the initial conditions, note that there is only one lattice path from $(0, 0)$ to $(a, 0)$, and the area underneath this path is zero. So $g(a, 0) = q^0 = 1$. Similarly, $g(0, b) = 1$.

Now we prove the recursion for $g(a,b)$, for fixed $a,b>0$. The set $L(a,b)$ is the disjoint union of sets L_1 and L_2, where L_1 consists of all paths from $(0,0)$ to (a,b) ending in an east step, and L_2 consists of all paths from $(0,0)$ to (a,b) ending in a north step. See Figure 8.1. Deleting the final north step from a path in L_2 defines a bijection from L_2 to $L(a,b-1)$, which is weight-preserving since the area below the path is not affected by the deletion of the north step. By the Bijection Rule for Weighted Sets, $\sum_{\pi\in L_2} q^{\text{area}(\pi)} = \sum_{\pi\in L(a,b-1)} q^{\text{area}(\pi)} = g(a,b-1)$. On the other hand, deleting the final east step from a path in L_1 defines a bijection from L_1 to $L(a-1,b)$ that is *not* weight-preserving. The reason is that the b area cells below the final east step in a path in L_1 no longer contribute to the area of the path in $L(a-1,b)$. However, since the area drops by b for all objects in L_1, we conclude from the Weight-Shifting Rule that $\sum_{\pi\in L_1} q^{\text{area}(\pi)} = q^b \sum_{\pi\in L(a-1,b)} q^{\text{area}(\pi)} = q^b g(a-1,b)$. By the Sum Rule for Weighted Sets,

$$g(a,b) = G_{L(a,b)}(q) = G_{L_1}(q) + G_{L_2}(q) = q^b g(a-1,b) + g(a,b-1).$$

We remark that a similar argument involving deletion of the initial step of a path in $L(a,b)$ establishes the dual recursion $g(a,b) = g(a-1,b) + q^a g(a,b-1)$.

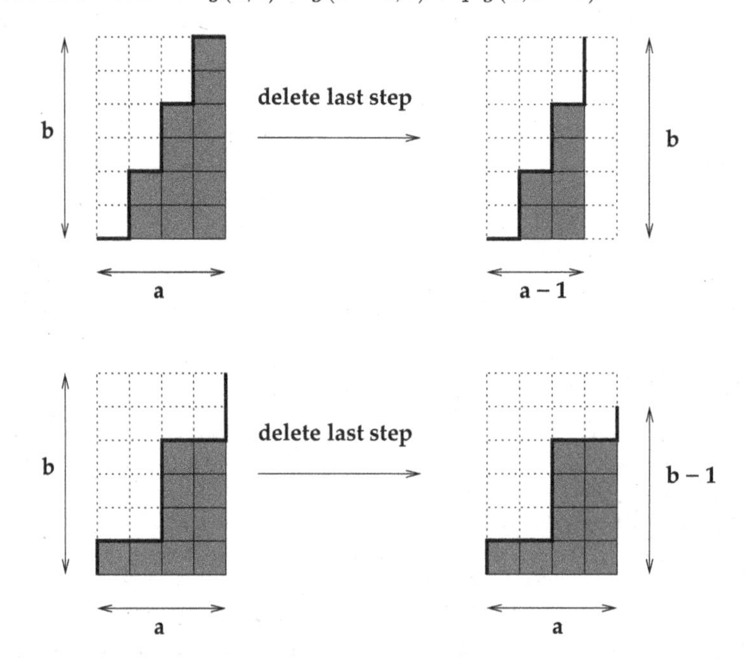

FIGURE 8.1
Deleting the final step of a lattice path.

Step 2. We define a weight-preserving bijection $g:\mathcal{R}(0^a1^b)\to L(a,b)$, where $\mathcal{R}(0^a1^b)$ is weighted by inv and $L(a,b)$ is weighted by area. For $w\in\mathcal{R}(0^a1^b)$, $g(w)$ is the path obtained by converting each 0 in w to an east step and each 1 in w to a north step. By examining a picture, one sees that $\text{inv}(w) = \text{area}(g(w))$ for all $w\in\mathcal{R}(0^a1^b)$. For example, given $w = 1001010$, $g(w)$ is the lattice path shown in Figure 8.2. The four area cells in the lowest row correspond to the inversions between the first symbol 1 in w and the four zeroes occurring later. The two area cells in the next lowest row come from the inversions between the second 1 in w and the two zeroes occurring later. And so on. By the Bijection Rule for Weighted Sets, $G_{\mathcal{R}(0^a1^b),\text{inv}}(q) = G_{L(a,b),\text{area}}(q) = \begin{bmatrix} a+b \\ a,b \end{bmatrix}_q$.

Step 3. We define a weight-preserving bijection $I: L(a,b)\to L(a,b)$, where the domain

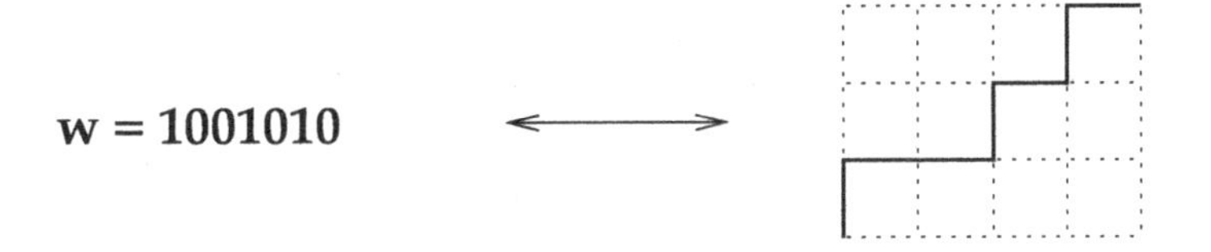

FIGURE 8.2
A weight-preserving bijection from words to lattice paths.

is weighted by area and the codomain is weighted by area′. Given $\pi \in L(a,b)$, $I(\pi)$ is the path obtained by reversing the sequence of north and east steps in π. Geometrically, $I(\pi)$ is obtained by rotating the path 180 degrees through $(a/2, b/2)$. One readily checks from a picture that $\text{area}'(I(\pi)) = \text{area}(\pi)$. So $G_{L(a,b),\text{area}'}(q) = G_{L(a,b),\text{area}}(q) = \begin{bmatrix} a+b \\ a,b \end{bmatrix}_q$.

Step 4. Inspection of Figure 2.17 shows that the set $P(a,b)$, weighted by $\text{wt}(\mu) = |\mu|$ for $\mu \in P(a,b)$, can be identified with the set $L(a,b)$, weighted by area′. So $\sum_{\mu \in P(a,b)} q^{|\mu|} = G_{L(a,b),\text{area}'}(q) = \begin{bmatrix} a+b \\ a,b \end{bmatrix}_q$. □

Step 1 of the preceding proof used the recursion for q-binomial coefficients to connect the algebraic and combinatorial interpretations for these coefficients. Here is another approach to proving the theorem, based on clearing denominators in $\begin{bmatrix} a+b \\ a,b \end{bmatrix}_q$. We show that

$$[a+b]!_q = [a]!_q [b]!_q \sum_{w \in \mathcal{R}(0^a 1^b)} q^{\text{inv}(w)}, \tag{8.3}$$

which reproves the first equality in Theorem 8.34.

We know from Theorem 8.21 that the left side of (8.3) is the generating function for the set S_{a+b} of permutations of $\{1, 2, \ldots, a+b\}$, weighted by inversions. By the Product Rule for Weighted Sets, the right side is the generating function for the Cartesian product $S_a \times S_b \times \mathcal{R}(0^a 1^b)$, with weight $\text{wt}(u,v,w) = \text{inv}(u) + \text{inv}(v) + \text{inv}(w)$ for $u \in S_a$, $v \in S_b$, and $w \in \mathcal{R}(0^a 1^b)$. By the Bijection Rule for Weighted Sets, it suffices to define a bijection

$$f : S_a \times S_b \times \mathcal{R}(0^a 1^b) \to S_{a+b}$$

such that $\text{inv}(f(u,v,w)) = \text{inv}(u) + \text{inv}(v) + \text{inv}(w)$ for $u \in S_a$, $v \in S_b$, and $w \in \mathcal{R}(0^a 1^b)$.

Given (u,v,w) in the domain of f, note that u is a permutation of the a symbols $1, 2, \ldots, a$. Replace the a zeroes in w with these a symbols, in the same order that they occur in u. Next, add a to each of the values in the permutation v. Then replace the b ones from the original word w by these new values in the same order that they occur in v. The resulting word z is evidently a permutation of $\{1, 2, \ldots, a+b\}$. For example, if $a = 3$ and $b = 5$, then

$$f(132, 24531, 01100111) = 15732864.$$

Since a and b are fixed and known, we can invert the action of f. Starting with a permutation z of $\{1, 2, \ldots, a+b\}$, we first recover the word $w \in \mathcal{R}(0^a 1^b)$ by replacing the numbers $1, 2, \ldots, a$ in z by zeroes and replacing the numbers $a+1, \ldots, a+b$ in z by ones. Next, we take the subword of z consisting of the numbers $1, 2, \ldots, a$ to recover u. Similarly, let v' be the subword of z consisting of the numbers $a+1, \ldots, a+b$. We recover v by subtracting a from each of these numbers. This algorithm defines a two-sided inverse map to f. For example, still taking $a = 3$ and $b = 5$, we have

$$f^{-1}(35162847) = (312, 23514, 01010111).$$

To finish, we check that f is weight-preserving. Fix u, v, w, z with $z = f(u, v, w)$. Let A be the set of positions in z occupied by letters in u, and let B be the remaining positions (occupied by shifted letters of v). Equivalently, by definition of f, $A = \{i : w_i = 0\}$ and $B = \{i : w_i = 1\}$. The inversions of z can be classified into three kinds. First, there are inversions (i, j) such that $i, j \in A$. These inversions correspond bijectively to the inversions of u. Second, there are inversions (i, j) such that $i, j \in B$. These inversions correspond bijectively to the inversions of v. Third, there are inversions (i, j) such that $i \in A$ and $j \in B$, or $i \in B$ and $j \in A$. The first case ($i \in A, j \in B$) cannot occur, because every position in A is filled with a lower number than every position in B. The second case ($i \in B, j \in A$) occurs iff $i < j$ and $w_i = 1$ and $w_j = 0$. This means that the inversions of the third kind in z correspond bijectively to the inversions of the binary word w. Adding the three kinds of inversions, we conclude that $\text{inv}(z) = \text{inv}(u) + \text{inv}(v) + \text{inv}(w)$, as needed.

8.35. Remark. We have discussed several combinatorial interpretations of the q-binomial coefficients. These coefficients are also relevant to the study of linear algebra over finite fields. Specifically, let F be a finite field with q elements, where (by a theorem of abstract algebra) q is necessarily a prime power. Then the q-binomial coefficient $\begin{bmatrix} n \\ k \end{bmatrix}_q$ is an integer that counts the number of k-dimensional subspaces of the n-dimensional vector space F^n. We prove this fact in §12.7. Incidentally, this is why the letter q appears so frequently as the variable in q-analogues: in algebra, p is often used to denote a prime integer, and $q = p^e$ is used to denote a prime power.

8.8 q-Binomial Coefficient Identities

Like ordinary binomial coefficients, the q-binomial coefficients satisfy many algebraic identities, which often have combinatorial proofs. This section gives two examples of such identities: a q-analogue of the Chu–Vandermonde identity 2.17, and a q-analogue of the Binomial Theorem 2.8.

8.36. Theorem: q-Chu–Vandermonde Identity. For all integers $a, b, c \geq 0$,

$$\begin{bmatrix} a+b+c+1 \\ a, b+c+1 \end{bmatrix}_q = \sum_{k=0}^{a} q^{(b+1)(a-k)} \begin{bmatrix} k+b \\ k, b \end{bmatrix}_q \begin{bmatrix} a-k+c \\ a-k, c \end{bmatrix}_q.$$

Proof. Recall the diagram we used to prove the original version of the identity, which is redrawn here as Figure 8.3. The path dissection in this diagram defines a bijection

$$f : L(a, b+c+1) \to \bigcup_{k=0}^{a} L(k, b) \times L(a-k, c).$$

Here, k is the x-coordinate where the given path in $L(a, b+c+1)$ crosses the line $y = b+(1/2)$. If a path $P \in L(a, b+c+1)$ maps to $(Q, R) \in L(k, b) \times L(a-k, c)$ under f, then we see from the diagram that

$$\text{area}(P) = \text{area}(Q) + \text{area}(R) + (b+1)(a-k).$$

The term $(b+1)(a-k)$ is the area of the lower-right rectangle of width $a-k$ and height $b+1$. It now follows from the Weight-Shifting Rule, the Sum Rule, and the Product Rule for Weighted Sets that

$$G_{L(a,b+c+1),\text{area}}(q) = \sum_{k=0}^{a} q^{(b+1)(a-k)} G_{L(k,b),\text{area}}(q) \cdot G_{L(a-k,c),\text{area}}(q).$$

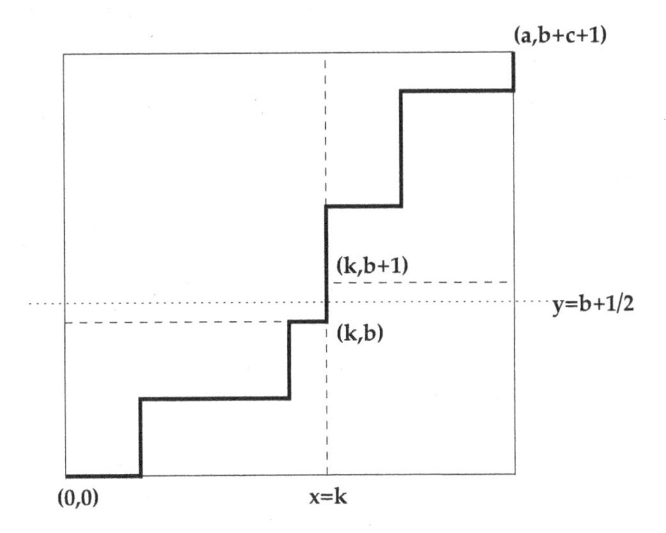

FIGURE 8.3
Diagram used to prove the q-Chu Vandermonde identity.

We complete the proof by using Theorem 8.34 to replace each area generating function by the appropriate q-binomial coefficient. □

Next we prove a q-analogue of the Binomial Theorem.

8.37. The q-Binomial Theorem. For all variables x, y, q and $n \in \mathbb{Z}_{>0}$,

$$(x+qy)(x+q^2y)(x+q^3y)\cdots(x+q^ny) = \sum_{k=0}^{n} q^{k(k+1)/2} \begin{bmatrix} n \\ k \end{bmatrix}_q x^{n-k}y^k. \tag{8.4}$$

Proof. We first prove the special case of the theorem where $x = 1$. Fix $n > 0$, and let P be the set of integer partitions consisting of distinct parts chosen from $\{1, 2, \ldots, n\}$. On one hand, the two-variable version of the Product Rule for Weighted Sets shows that

$$\sum_{\mu \in P} q^{|\mu|} y^{\ell(\mu)} = \prod_{i=1}^{n} (1 + q^i y) \tag{8.5}$$

(this is a finite version of Theorem 5.48). Specifically, we build $\mu \in P$ by choosing to exclude or include a part of size i, for each i between 1 and n. The generating function for the ith choice is $q^0y^0 + q^iy^1 = 1 + q^iy$, since the area of μ increases by i and the length of μ increases by 1 if the part i is included, whereas area and length increase by 0 if part i is excluded.

On the other hand, we can write P as the disjoint union of sets $P_0, P_1, \ldots, P_n$, where $P_k = \{\mu \in P : \ell(\mu) = k\}$. Recall from Theorem 8.34 that $P(a, b) = \{\nu \in \text{Par} : \nu_1 \leq a \text{ and } \ell(\nu) \leq b\}$. Fix k, and define a map $F : P_k \to P(n-k, k)$ by $F(\mu_1, \mu_2, \ldots, \mu_k) = (\mu_1 - k, \mu_2 - (k-1), \ldots, \mu_k - 1)$. One readily checks that F does map into the stated codomain, and F is bijective with two-sided inverse $F^{-1}(\nu_1, \ldots, \nu_k) = (\nu_1 + k, \nu_2 + (k-1), \ldots, \nu_k + 1)$. Here we use the convention $\nu_i = 0$ for $\ell(\nu) < i \leq k$. Looking only at area, applying the bijection F decreases the area of μ by $k + (k-1) + \cdots + 1 = k(k+1)/2$. On the other hand, every object in P_k has length k. By the Weight-Shifting Rule and Theorem 8.34, we conclude that

$$\sum_{\mu \in P_k} q^{|\mu|} y^{\ell(\mu)} = q^{k(k+1)/2} \begin{bmatrix} n \\ k \end{bmatrix}_q y^k.$$

Adding these equations for k ranging from 0 to n and comparing to (8.5), we obtain the special case of the theorem where $x = 1$.

To prove the general case, replace y by y/x in what we just proved to get

$$\prod_{i=1}^{n}[1+q^i(y/x)] = \sum_{k=0}^{n} q^{k(k+1)/2}\begin{bmatrix} n \\ k \end{bmatrix}_q (y/x)^k.$$

Multiplying both sides by x^n and simplifying, we obtain the needed result. $\square$

8.9 q-Multinomial Coefficients

Let $n, n_1, \ldots, n_s \in \mathbb{Z}_{\geq 0}$ satisfy $n = n_1 + \cdots + n_s$. Recall from Recursion 2.23 that the ordinary multinomial coefficients $C(n; n_1, \ldots, n_s) = \binom{n}{n_1,\ldots,n_s} = \frac{n!}{n_1!\cdots n_s!}$ satisfy

$$C(n; n_1, \ldots, n_s) = \sum_{k=1}^{s} C(n-1; n_1, \ldots, n_k - 1, \ldots, n_s),$$

with initial conditions $C(0; 0, \ldots, 0) = 1$ and $C(m; m_1, \ldots, m_s) = 0$ whenever some m_i is negative. The q-multinomial coefficients satisfy the following analogous recursion.

8.38. Theorem: Recursion for q-Multinomial Coefficients. Let $n_1, \ldots, n_s$ be nonnegative integers, and set $n = \sum_{k=1}^{s} n_k$. Then

$$\begin{bmatrix} n \\ n_1, \ldots, n_s \end{bmatrix}_q = \sum_{k=1}^{s} q^{n_1+n_2+\cdots+n_{k-1}} \begin{bmatrix} n-1 \\ n_1, \ldots, n_k - 1, \ldots, n_s \end{bmatrix}_q,$$

where we interpret the kth summand on the right side to be zero if $n_k = 0$. The initial condition is $\begin{bmatrix} 0 \\ 0,\ldots,0 \end{bmatrix}_q = 1$. Moreover, $\begin{bmatrix} n \\ n_1,\ldots,n_s \end{bmatrix}_q$ is a polynomial in q with coefficients in $\mathbb{Z}_{\geq 0}$.

Proof. Neither side of the claimed recursion changes if we delete all n_i's that are equal to zero; so, without loss of generality, assume every n_i is positive. We can create a common factor of $[n-1]!_q / \prod_{j=1}^{s}[n_j]!_q$ on the right side by multiplying the kth summand by $[n_k]_q/[n_k]_q$, for $1 \leq k \leq s$. Pulling out this common factor, we are left with

$$\sum_{k=1}^{s} q^{n_1+n_2+\cdots+n_{k-1}}[n_k]_q = \sum_{k=1}^{s} q^{n_1+\cdots+n_{k-1}}(1+q+q^2+\cdots+q^{n_k-1}).$$

The kth summand consists of the sum of consecutive powers of q starting at $q^{n_1+\cdots+n_{k-1}}$ and ending at $q^{n_1+\cdots+n_k-1}$. Chaining these together, we see that the sum evaluates to $q^0 + q^1 + \cdots + q^{n-1} = [n]_q$. Multiplying by the common factor mentioned above, we obtain

$$\frac{[n]!_q}{\prod_{j=1}^{s}[n_j]!_q} = \begin{bmatrix} n \\ n_1, \ldots, n_s \end{bmatrix}_q,$$

as needed. The initial condition follows from $[0]!_q = 1$. Finally, we deduce polynomiality of the q-multinomial coefficients using induction on n and the recursion just proved, as in the proof of Corollary 8.33. $\square$

8.39. Theorem: q-Multinomial Coefficients and Anagrams. For any totally ordered alphabet $A = \{a_1 < a_2 < \cdots < a_s\}$ and all integers $n_1, \ldots, n_s \geq 0$,

$$\begin{bmatrix} n_1 + \cdots + n_s \\ n_1, n_2, \ldots, n_s \end{bmatrix}_q = \sum_{w \in \mathcal{R}(a_1^{n_1} \cdots a_s^{n_s})} q^{\text{inv}(w)}.$$

Proof. For all integers $n_1, \ldots, n_s$, define

$$g(n_1, \ldots, n_s) = \sum_{w \in \mathcal{R}(a_1^{n_1} \cdots a_s^{n_s})} q^{\text{inv}(w)}.$$

(This is zero by convention if any n_i is negative.) By induction on $\sum_k n_k$, it suffices to show that g satisfies the recursion in Theorem 8.38. Now $g(0, 0, \ldots, 0) = q^0 = 1$, so the initial condition is correct. Next, fix $n_1, \ldots, n_s \geq 0$, and let W be the set of words appearing in the definition of $g(n_1, \ldots, n_s)$. Write W as the disjoint union of sets $W_1, \ldots, W_s$, where W_k consists of the words in W with first letter a_k. By the Sum Rule for Weighted Sets,

$$g(n_1, \ldots, n_s) = G_W(q) = \sum_{k=1}^{s} G_{W_k}(q).$$

Fix a value of k in the range $1 \leq k \leq s$ such that W_k is nonempty. Erasing the first letter of a word w in W_k defines a bijection from W_k to the set $\mathcal{R}(a_1^{n_1} \cdots a_k^{n_k - 1} \cdots a_s^{n_s})$. The generating function for the latter set is $g(n_1, \ldots, n_k - 1, \ldots, n_s)$. The bijection in question does not preserve weights, because inversions involving the first letter of $w \in W_k$ disappear when this letter is erased. However, no matter what word w we pick in W_k, the number of inversions that involve the first letter in w will always be the same. Specifically, this first letter (namely a_k) will cause inversions with all of the occurrences of $a_1, a_2, \ldots, a_{k-1}$ that follow it in w. The number of such letters is $n_1 + \cdots + n_{k-1}$. Therefore, by the Weight-Shifting Rule,

$$G_{W_k, \text{inv}}(q) = q^{n_1 + \cdots + n_{k-1}} g(n_1, \ldots, n_k - 1, \ldots, n_s).$$

This equation is also correct if $W_k = \emptyset$ (which occurs iff $n_k = 0$). Using these results in the formula above, we conclude that

$$g(n_1, \ldots, n_s) = \sum_{k=1}^{s} q^{n_1 + \cdots + n_{k-1}} g(n_1, \ldots, n_k - 1, \ldots, n_s),$$

which is precisely the recursion occurring in Theorem 8.38. □

8.40. Remark. Theorem 8.39 can also be proved by generalizing (8.3) in the second proof of Theorem 8.34. Specifically, one can prove

$$[n_1 + \cdots + n_s]!_q = [n_1]!_q \cdots [n_s]!_q \sum_{w \in \mathcal{R}(1^{n_1} \cdots s^{n_s})} q^{\text{inv}(w)}$$

by defining a weight-preserving bijection

$$f : S_{n_1 + \cdots + n_s} \to S_{n_1} \times \cdots \times S_{n_s} \times \mathcal{R}(1^{n_1} \cdots s^{n_s})$$

where S_{n_i} is the set of all permutations of $\{1, 2, \ldots, n_i\}$, and all sets in the Cartesian product are weighted by inversions. We leave the details as an exercise.

8.10 Foata's Bijection

We know from Theorems 8.21 and 8.26 that $\sum_{w\in S_n} q^{\mathrm{inv}(w)} = [n]!_q = \sum_{w\in S_n} q^{\mathrm{maj}(w)}$, where S_n is the set of permutations of $\{1,2,\ldots,n\}$. We can express this result by saying that the statistics inv and maj are *equidistributed* on S_n. We have just derived a formula for the distribution of inv on more general sets of words, namely

$$\sum_{w\in\mathcal{R}(1^{n_1}\cdots s^{n_s})} q^{\mathrm{inv}(w)} = \begin{bmatrix} n_1+\cdots+n_s \\ n_1,\ldots,n_s \end{bmatrix}_q .$$

Could it be true that inv and maj are still equidistributed on these more general sets? MacMahon proved that this is indeed the case. We present a combinatorial proof of this result based on a bijection due to Dominique Foata. For each set $S = \mathcal{R}(1^{n_1}\cdots s^{n_s})$, our goal is to define a weight-preserving bijection $f : (S,\mathrm{maj}) \to (S,\mathrm{inv})$.

To achieve our goal, let W be the set of *all* words in the alphabet $\{1,2,\ldots,s\}$. We shall define a function $g : W \to W$ with the following properties: (a) g is a bijection; (b) for all $w \in W$, w and $g(w)$ are anagrams (see §1.5); (c) if w is not the empty word, then w and $g(w)$ have the same last letter; (d) for all $w \in W$, $\mathrm{inv}(g(w)) = \mathrm{maj}(w)$. We can then obtain the required weight-preserving bijections f by restricting g to the various anagram classes $\mathcal{R}(1^{n_1}\cdots s^{n_s})$.

We define g by recursion on the length of $w \in W$. If this length is 0 or 1, set $g(w) = w$. Then conditions (b), (c), and (d) hold in this case. Now suppose w has length $n \geq 2$. Write $w = w'yz$, where $w' \in W$ and y, z are the last two letters of w. We can assume by induction that $u = g(w'y)$ has already been defined, and that u is an anagram of $w'y$ ending in y such that $\mathrm{inv}(u) = \mathrm{maj}(w'y)$. We define $g(w) = h_z(u)z$, where $h_z : W \to W$ is a certain map (to be described momentarily) that satisfies conditions (a) and (b) above. No matter what the details of the definition of h_z, it already follows that g satisfies conditions (b) and (c) for words of length n.

To motivate the definition of h_z, we first give a lemma that analyzes the effect on inv and maj of appending a letter to the end of a word. The lemma uses the following notation. If u is any word and z is any letter, let $n_{\leq z}(u)$ be the number of letters in u (counting repetitions) that are $\leq z$; define $n_{<z}(u)$, $n_{>z}(u)$, and $n_{\geq z}(u)$ similarly.

8.41. Lemma. Suppose u is a word of length m with last letter y, and z is any letter. (a) If $y \leq z$, then $\mathrm{maj}(uz) = \mathrm{maj}(u)$. (b) If $y > z$, then $\mathrm{maj}(uz) = \mathrm{maj}(u) + m$. (c) $\mathrm{inv}(uz) = \mathrm{inv}(u) + n_{>z}(u)$.

Proof. All statements follow routinely from the definitions of inv and maj. □

We now describe the map $h_z : W \to W$. First, h_z sends the empty word to itself. Suppose u is a nonempty word ending in y. There are two cases.

Case 1: $y \leq z$. In this case, we break the word u into *runs* of consecutive letters such that the last letter in each run is $\leq z$, while all preceding letters in the run are $> z$. For example, if $u = 1342434453552$ and $z = 3$, then the decomposition of u into runs is

$$u = 1/3/4,2/4,3/4,4,5,3/5,5,2/$$

where we use slashes to delimit consecutive runs. Now, h_z operates on u by cyclically shifting the letters in each run one step to the right. Continuing the preceding example,

$$h_3(u) = 1/3/2,4/3,4/3,4,4,5/2,5,5/.$$

What effect does this process have on $\mathrm{inv}(u)$? In u, the last element in each run (which is $\leq z$) is strictly less than all elements before it in its run (which are $> z$). So, moving the last element to the front of its run causes the inversion number to drop by the number of elements $> z$ in the run. Adding up these changes over all the runs, we see that

$$\mathrm{inv}(h_z(u)) = \mathrm{inv}(u) - n_{>z}(u) \quad \text{in Case 1.} \tag{8.6}$$

Furthermore, note that the first letter of $h_z(u)$ is always $\leq z$ in this case.

Case 2: $y > z$. Again we break the word u into runs, but here the last letter of each run must be $> z$, while all preceding letters in the run are $\leq z$. For example, if $z = 3$ and $u = 134243445355$, we decompose u as

$$u = 1, 3, 4/2, 4/3, 4/4/5/3, 5/5/.$$

As before, we cyclically shift the letters in each run one step right, which gives

$$h_3(u) = 4, 1, 3/4, 2/4, 3/4/5/5, 3/5/$$

in our example. This time, the last element in each run of u is $> z$ and is strictly greater than the elements $\leq z$ that precede it in its run. So, the cyclic shift of each run will increase the inversion count by the number of elements $\leq z$ in the run. Adding over all runs, we see that

$$\mathrm{inv}(h_z(u)) = \mathrm{inv}(u) + n_{\leq z}(u) \quad \text{in Case 2.} \tag{8.7}$$

Furthermore, note that the first letter of $h_z(u)$ is always $> z$ in this case.

In both cases, $h_z(u)$ is an anagram of u. Moreover, we can invert the action of h_z as follows. Examination of the first letter of $h_z(u)$ tells us whether we were in Case 1 or Case 2 above. To invert in Case 1, break the word into runs whose first letter is $\leq z$ and whose other letters are $> z$, and cyclically shift each run one step left. To invert in Case 2, break the word into runs whose first letter is $> z$ and whose other letters are $\leq z$, and cyclically shift each run one step left. We now see that h_z is a bijection. For example, to compute $h_3^{-1}(1342434453552)$, first write

$$1/3, 4/2, 4/3, 4, 4, 5/3, 5, 5/2/$$

and then cyclically shift to get the answer $1/4, 3/4, 2/4, 4, 5, 3/5, 5, 3/2/$.

Now we return to the discussion of g. Recall that we have set $g(w) = g(w'yz) = h_z(u)z$, where $u = g(w'y)$ is an anagram of $w'y$ ending in y and satisfying $\mathrm{inv}(u) = \mathrm{maj}(w'y)$. To check condition (d) for this w, we must show that $\mathrm{inv}(h_z(u)z) = \mathrm{maj}(w)$. Again consider two cases. If $y \leq z$, then on one hand,

$$\mathrm{maj}(w) = \mathrm{maj}(w'yz) = \mathrm{maj}(w'y) = \mathrm{inv}(u).$$

On the other hand, by Lemma 8.41 and (8.6), we have

$$\mathrm{inv}(h_z(u)z) = \mathrm{inv}(h_z(u)) + n_{>z}(h_z(u)) = \mathrm{inv}(u) - n_{>z}(u) + n_{>z}(u) = \mathrm{inv}(u),$$

where $n_{>z}(h_z(u)) = n_{>z}(u)$ since $h_z(u)$ and u are anagrams. If $y > z$, then on one hand,

$$\mathrm{maj}(w) = \mathrm{maj}(w'yz) = \mathrm{maj}(w'y) + n - 1 = \mathrm{inv}(u) + n - 1.$$

On the other hand, Lemma 8.41 and (8.7) give

$$\mathrm{inv}(h_z(u)z) = \mathrm{inv}(h_z(u)) + n_{>z}(h_z(u)) = \mathrm{inv}(u) + n_{\leq z}(u) + n_{>z}(u) = \mathrm{inv}(u) + n - 1,$$

since u has $n-1$ letters, each of which is either $\leq z$ or $> z$.

To prove that g is a bijection, we describe the two-sided inverse map g^{-1}. This is the identity map on words of length at most 1. To compute $g^{-1}(uz)$, first compute $u' = h_z^{-1}(u)$. Then return the answer $g^{-1}(uz) = (g^{-1}(u'))z$. Here is a non-recursive description of the maps g and g^{-1}, obtained by unrolling the recursive applications of g and g^{-1} in the preceding definitions.

> *To compute $g(w_1 w_2 \cdots w_n)$:* for $i = 2, \ldots, n$ in this order, apply h_{w_i} to the first $i-1$ letters of the current word.
>
> *To compute $g^{-1}(z_1 z_2 \cdots z_n)$:* for $i = n, n-1, \ldots, 2$ in this order, let z_i' be the ith letter of the current word, and apply $h_{z_i'}^{-1}$ to the first $i-1$ letters of the current word.

8.42. Example. Figure 8.4 illustrates the computation of $g(w)$ for $w = 21331322$. We find that $g(w) = 23131322$. Observe that $\text{maj}(w) = 1+4+6 = 11 = \text{inv}(g(w))$. Next, Figure 8.5 illustrates the calculation of $g^{-1}(w)$. We have $g^{-1}(w) = 33213122$, and $\text{inv}(w) = 10 = \text{maj}(g^{-1}(w))$.

$$\begin{array}{rcll} & & & \text{current word:} \\ & & & 2,1,3,3,1,3,2,2 \\ h_1(2) & = & 2; & 2,1,3,3,1,3,2,2 \\ h_3(2,1) & = & 2,1; & 2,1,3,3,1,3,2,2 \\ h_3(2,1,3) & = & 2,1,3; & 2,1,3,3,1,3,2,2 \\ h_1(2,1,3,3) & = & 2,3,1,3; & 2,3,1,3,1,3,2,2 \\ h_3(2,3,1,3,1) & = & 2,3,1,3,1; & 2,3,1,3,1,3,2,2 \\ h_2(2,3,1,3,1,3) & = & 3,2,3,1,3,1; & 3,2,3,1,3,1,2,2 \\ h_2(3,2,3,1,3,1,2) & = & 2,3,1,3,1,3,2; & 2,3,1,3,1,3,2,2 \end{array}$$

FIGURE 8.4
Computation of $g(w)$.

$$\begin{array}{rcll} & & & \text{current word:} \\ & & & 2,1,3,3,1,3,2,2 \\ h_2^{-1}(2,1,3,3,1,3,2) & = & 2,3,3,1,3,1,2; & 2,3,3,1,3,1,2,2 \\ h_2^{-1}(2,3,3,1,3,1) & = & 3,3,2,3,1,1; & 3,3,2,3,1,1,2,2 \\ h_1^{-1}(3,3,2,3,1) & = & 3,3,2,1,3; & 3,3,2,1,3,1,2,2 \\ h_3^{-1}(3,3,2,1) & = & 3,3,2,1; & 3,3,2,1,3,1,2,2 \\ h_1^{-1}(3,3,2) & = & 3,3,2; & 3,3,2,1,3,1,2,2 \\ h_2^{-1}(3,3) & = & 3,3; & 3,3,2,1,3,1,2,2 \\ h_3^{-1}(3) & = & 3; & 3,3,2,1,3,1,2,2 \end{array}$$

FIGURE 8.5
Computation of $g^{-1}(w)$.

We summarize the results of this section in the following theorem.

8.43. Theorem. For all $n_1, \ldots, n_s \geq 0$,

$$\sum_{w \in \mathcal{R}(1^{n_1} \cdots s^{n_s})} q^{\text{maj}(w)} = \sum_{w \in \mathcal{R}(1^{n_1} \cdots s^{n_s})} q^{\text{inv}(w)} = \begin{bmatrix} n_1 + \cdots + n_s \\ n_1, \ldots, n_s \end{bmatrix}_q.$$

More precisely, there is a bijection on $\mathcal{R}(1^{n_1} \cdots s^{n_s})$ sending maj to inv and preserving the last letter of each word.

8.11 q-Catalan Numbers

In this section, we investigate two weighted analogues of the Catalan numbers. Recall that the Catalan number $C_n = \frac{1}{n+1}\binom{2n}{n} = \binom{2n}{n,n} - \binom{2n}{n-1,n+1}$ counts the collection of all lattice paths from $(0,0)$ to (n,n) that never go below the line $y = x$ (see §1.14). Let D_n be the collection of these paths, which are called *Dyck paths.* Also, let W_n be the set of words that encode the Dyck paths, where we use 0 to encode a north step and 1 to encode an east step. Elements of W_n are called *Dyck words.*

8.44. Definition: Statistics on Dyck Paths. For every Dyck path $P \in D_n$, let area(P) be the number of complete unit squares located between P and the line $y = x$. If P is encoded by the Dyck word $w \in W_n$, let $\text{inv}(P) = \text{inv}(w)$ and $\text{maj}(P) = \text{maj}(w)$.

For example, the path P shown in Figure 8.6 has area$(P) = 23$. One sees that inv(P) is the number of unit squares in the region bounded by P, the y-axis, and the line $y = n$. We also have $\text{inv}(P) + \text{area}(P) = \binom{n}{2}$ since $\binom{n}{2}$ is the total number of area squares in the bounding triangle. The statistic maj(P) is the sum of the number of steps in the path that precede each *left-turn* where an east step (1) is immediately followed by a north step (0). For the path in Figure 8.6, we have $\text{inv}(P) = 97$ and $\text{maj}(P) = 4+6+10+16+18+22+24+28 = 128$.

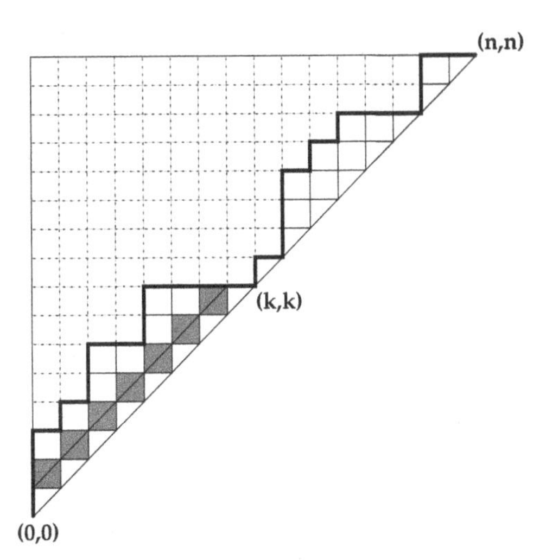

FIGURE 8.6
First-return analysis for weighted Dyck paths.

8.45. Example. When $n = 3$, examination of Figure 1.3 shows that

$$\begin{aligned} G_{D_3,\text{area}}(q) &= 1 + 2q + q^2 + q^3; \\ G_{D_3,\text{inv}}(q) &= 1 + q + 2q^2 + q^3; \\ G_{D_3,\text{maj}}(q) &= 1 + q^2 + q^3 + q^4 + q^6. \end{aligned}$$

When $n = 4$, a longer calculation gives

$$\begin{aligned} G_{D_4,\text{area}}(q) &= 1+3q+3q^2+3q^3+2q^4+q^5+q^6; \\ G_{D_4,\text{maj}}(q) &= 1+q^2+q^3+2q^4+q^5+2q^6+q^7+2q^8+q^9+q^{10}+q^{12}. \end{aligned}$$

There is no particularly simple closed formula for $G_{D_n,\text{area}}(q)$, although determinant formulas do exist for this polynomial. However, these generating functions satisfy a recursion, which is a q-analogue of the first-return recursion used in the unweighted case (see §2.10).

8.46. Theorem: Recursion for Dyck Paths Weighted by Area. For all $n \geq 0$, set $C_n(q) = G_{D_n,\text{area}}(q) = \sum_{P \in D_n} q^{\text{area}(P)}$. Then $C_0(q) = 1$, and for all $n \geq 1$,

$$C_n(q) = \sum_{k=1}^{n} q^{k-1} C_{k-1}(q) C_{n-k}(q).$$

Proof. We imitate the proof of Theorem 2.28, but now we must take weights into account. For $1 \leq k \leq n$, write $D_{n,k}$ for the set of Dyck paths ending at (n,n) whose first return to the line $y = x$ occurs at (k,k). Since D_n is the disjoint union of the $D_{n,k}$ as k ranges from 1 to n, the Sum Rule for Weighted Sets gives

$$C_n(q) = \sum_{k=1}^{n} G_{D_{n,k},\text{area}}(q). \tag{8.8}$$

For fixed k, we have a bijection from $D_{n,k}$ to $D_{k-1} \times D_{n-k}$ defined by sending $P = \text{N}, P_1, \text{E}, P_2$ to (P_1, P_2), where the displayed E is the east step that arrives at (k,k). See Figure 8.6. Examination of the figure shows that

$$\text{area}(P) = \text{area}(P_1) + \text{area}(P_2) + (k-1),$$

where the $k-1$ counts the shaded cells in the figure that are not included in the calculation of area(P_1). By the Product Rule and Weight-Shifting Rule, we see that

$$G_{D_{n,k},\text{area}}(q) = C_{k-1}(q) C_{n-k}(q) q^{k-1}.$$

Inserting this expression into (8.8) proves the recursion. □

Now let us consider the generating function $G_{D_n,\text{maj}}(q)$. This polynomial does have a nice closed formula, as we see in the next theorem.

8.47. Theorem: Dyck Paths Weighted by Major Index. For all $n \geq 0$,

$$G_{D_n,\text{maj}}(q) = \sum_{P \in D_n} q^{\text{maj}(P)} = \begin{bmatrix} 2n \\ n, n \end{bmatrix}_q - q \begin{bmatrix} 2n \\ n-1, n+1 \end{bmatrix}_q = \frac{1}{[n+1]_q} \begin{bmatrix} 2n \\ n, n \end{bmatrix}_q.$$

Proof. The last equality follows from the manipulation

$$\begin{aligned} &\begin{bmatrix} 2n \\ n, n \end{bmatrix}_q - q \begin{bmatrix} 2n \\ n-1, n+1 \end{bmatrix}_q = \begin{bmatrix} 2n \\ n, n \end{bmatrix}_q \cdot \left(1 - \frac{q[n]_q}{[n+1]_q}\right) \\ &= \begin{bmatrix} 2n \\ n, n \end{bmatrix}_q \cdot \left(\frac{(1+q+q^2+\cdots+q^n) - q(1+q+\cdots+q^{n-1})}{[n+1]_q}\right) = \begin{bmatrix} 2n \\ n, n \end{bmatrix}_q \cdot \frac{1}{[n+1]_q}. \end{aligned}$$

The second equality in the theorem statement can be rewritten

$$q \begin{bmatrix} 2n \\ n-1, n+1 \end{bmatrix}_q + \sum_{P \in D_n} q^{\text{maj}(P)} = \begin{bmatrix} 2n \\ n, n \end{bmatrix}_q.$$

We now give a bijective proof of this result reminiscent of André's Reflection Principle (see the proof of the Dyck Path Rule 1.101). Consider the set of words $S = \mathcal{R}(0^n1^n)$, weighted by major index. By Theorem 8.43, the generating function for this set is $\left[{2n \atop n,n}\right]_q$. On the other hand, we can write S as the disjoint union of W_n and T, where W_n is the set of Dyck words and T consists of all other words in $\mathcal{R}(0^n1^n)$. We define a bijection $g : T \to \mathcal{R}(0^{n+1}1^{n-1})$ such that $\text{maj}(w) = 1 + \text{maj}(g(w))$ for all $w \in T$. By the Bijection Rule and the Weight-Shifting Rule, it follows that

$$\left[\begin{matrix} 2n \\ n,n \end{matrix}\right]_q = G_{S,\text{maj}}(q) = G_{T,\text{maj}}(q) + G_{W_n,\text{maj}}(q) = q\left[\begin{matrix} 2n \\ n-1,n+1 \end{matrix}\right]_q + G_{D_n,\text{maj}}(q),$$

as needed.

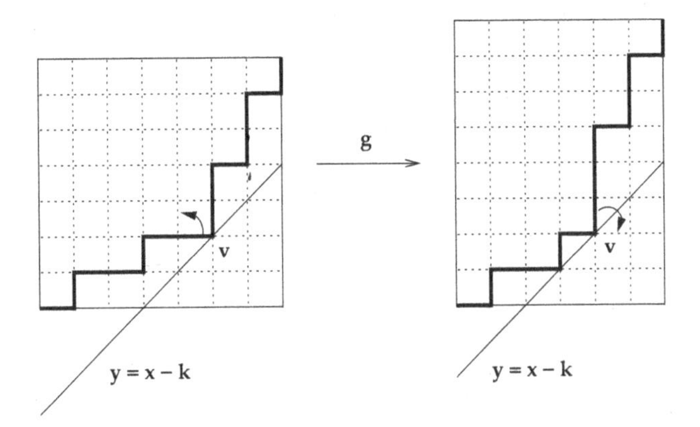

FIGURE 8.7
The tipping bijection.

To define $g(w)$ for $w \in T$, regard w as a lattice path in an $n \times n$ rectangle by interpreting 0's as north steps and 1's as east steps. Find the largest $k > 0$ such that the path w touches the line $y = x - k$. Such a k must exist, because $w \in T$ is not a Dyck path. Consider the first vertex v on w that touches the line in question. (See Figure 8.7 for an example.) The path w must arrive at v by taking an east step, and w must leave v by taking a north step. These steps correspond to certain adjacent letters $w_i = 1$ and $w_{i+1} = 0$ in the word w. Furthermore, since v is the first arrival at the line $y = x - k$, we must have either $i = 1$ or $w_{i-1} = 1$ (i.e., the step before w_i must be an east step if it exists). Let $g(w)$ be the word obtained by changing w_i from 1 to 0. Pictorially, we tip the east step arriving at v upwards, changing it to a north step (which causes the following steps to shift to the northwest). The word $w = \cdots 1, 1, 0 \cdots$ turns into $g(w) = \cdots 1, 0, 0 \cdots$, so the major index drops by exactly 1 when we pass from w to $g(w)$. This result also holds if $i = 1$. The new word $g(w)$ has $n-1$ east steps and $n+1$ north steps, so $g(w) \in \mathcal{R}(0^{n+1}1^{n-1})$.

Finally, we show g is invertible. Given a word (or path) $P \in \mathcal{R}(0^{n+1}1^{n-1})$, take the largest $k \geq 0$ such that P touches the line $y = x - k$, and let v be the *last* time P touches this line. Here, v is preceded by an east step and followed by two north steps (or v is the origin and is followed by two north steps). Changing the first north step following v into an east step produces a path $g'(P) \in \mathcal{R}(0^n1^n)$. It can be checked that $g'(P)$ cannot be a Dyck path (so g' maps into T), and that g' is the two-sided inverse of g. The key point to verify is that the selection rules for v ensure that the same step is tipped when we apply g followed by g', and similarly in the other order. □

8.12 Set Partitions and q-Stirling Numbers

Recall from §2.12 that the Stirling number of the second kind, denoted $S(n,k)$, counts set partitions of $\{1,2,\ldots,n\}$ into k blocks. This section develops a q-analogue of $S(n,k)$ that counts weighted set partitions.

To define an appropriate statistic on set partitions, we use the following method of encoding a set partition P by a permutation $f(P)$. Recall that a set partition of $\{1,2,\ldots,n\}$ into k blocks is a set $P = \{B_1, B_2, \ldots, B_k\}$ of pairwise disjoint, nonempty blocks B_j such that the union of $B_1, \ldots, B_k$ is $\{1,2,\ldots,n\}$. We choose the indexing of the blocks so that $\min(B_1) > \min(B_2) > \cdots > \min(B_k)$, where $\min(B_j)$ denotes the least element of block B_j. Furthermore, we present the elements within each block B_j in increasing order. Finally, we erase all the set braces to obtain a permutation of $\{1,2,\ldots,n\}$. For example, given the set partition $P = \{\{8,4,5\},\{7,2\},\{1,3,9\},\{6\}\}$, we first present the partition in standard form $P = \{\{6\},\{4,5,8\},\{2,7\},\{1,3,9\}\}$ and then erase the braces to get the permutation $f(P) = 645827139$.

Note that each block B_j in P becomes an *ascending run* in $f(P)$ consisting of symbols that increase from left to right. On the other hand, for all $j < k$, the largest symbol in block B_j must be greater than the smallest symbol in block B_{j+1}, because $\max(B_j) \geq \min(B_j) > \min(B_{j+1})$ by the choice of indexing. Thus there is a descent in $f(P)$ every time we go from one block to the next block. It follows that we can recover the set partition P from the permutation $f(P) = w_1 \cdots w_n$ by finding the descent set $\mathrm{Des}(f(P)) = \{i_1 < i_2 < \cdots < i_{k-1}\}$ and setting

$$P = \{\{w_1, w_2, \ldots, w_{i_1}\}, \{w_{i_1+1}, \ldots, w_{i_2}\}, \{w_{i_2+1}, \ldots, w_{i_3}\}, \ldots, \{w_{i_{k-1}+1}, \ldots, w_n\}\}.$$

For example, $f^{-1}(794582613) = \{\{7,9\},\{4,5,8\},\{2,6\},\{1,3\}\}$.

Let $SP(n,k)$ be the set of set partitions of $\{1,2,\ldots,n\}$ into k blocks. The preceding discussion defines a one-to-one map $f : SP(n,k) \to S_n$. We call permutations in the image of f *Stirling permutations of type* (n,k). A permutation w is in the image of f iff $\mathrm{des}(w) = k-1$ and the list of first elements in the ascending runs of w form a decreasing subsequence of w. We define a *major index statistic* on $SP(n,k)$ by letting $\mathrm{maj}(P) = \mathrm{maj}(f(P))$ for each $P \in SP(n,k)$. The *q-Stirling numbers of the second kind* are defined to be

$$S_q(n,k) = G_{SP(n,k),\mathrm{maj}}(q) = \sum_{P \in SP(n,k)} q^{\mathrm{maj}(P)}.$$

These q-Stirling numbers are characterized by the following recursion (compare to Recursion 2.45).

8.48. Theorem: Recursion for q-Stirling Numbers of the Second Kind. For all integers k and n with $1 < k < n$,

$$S_q(n,k) = q^{k-1} S_q(n-1,k-1) + [k]_q S_q(n-1,k).$$

The initial conditions are $S_q(0,0) = 1$, $S_q(n,1) = 1$ for all $n \geq 1$, and $S_q(n,n) = q^{n(n-1)/2}$ for all $n \geq 1$.

Proof. Fix integers k and n with $1 < k < n$. The set $SP(n,k)$ is the disjoint union of sets A and B, where A consists of set partitions of $\{1,2,\ldots,n\}$ into k blocks where n is in a block by itself, and B consists of set partitions where n is in a block with other elements. There is a bijection from A to $SP(n-1,k-1)$ that sends $P \in A$ to the set partition P'

obtained from P by deleting the block $\{n\}$. Consider the Stirling permutations $w = f(P)$ and $w' = f(P')$ that encode P and P'. From the definition of the encoding, we see that $w = n, w'$ where w' is some permutation of $\{1, 2, \ldots, n-1\}$ with $k-2$ descents. Adding n as the new first symbol pushes each of these descents to the right one position, and we also get a new descent at position 1 of w, since $w_1 = n > w_2 = w'_1$. This means that $\mathrm{maj}(w) = \mathrm{maj}(w') + k - 1$. By the Weight-Shifting Rule, we conclude that

$$\sum_{P \in A} q^{\mathrm{maj}(P)} = q^{k-1} \sum_{P' \in SP(n-1,k-1)} q^{\mathrm{maj}(P')} = q^{k-1} S_q(n-1, k-1). \tag{8.9}$$

We can build each set partition $P \in B$ by first choosing a set partition $P^* = \{B_1, B_2, \ldots, B_k\} \in SP(n-1, k)$, then choosing an index $j \in \{1, 2, \ldots, k\}$, then inserting n as a new element of block B_j. Here, as above, we index the blocks of P^* so that $\min(B_1) > \min(B_2) > \cdots > \min(B_k)$. Let us compare the permutations $w = f(P)$ and $w^* = f(P^*)$. We obtain w from w^* by inserting the symbol n at the end of the jth ascending run of w^*. This causes the $k-j$ descents following this run in w^* to shift right one position in w, whereas the $j-1$ descents preceding this run in w^* maintain their current positions. Also, since n is the largest symbol, no new descent is created by the insertion of n. It follows that $\mathrm{maj}(w) = \mathrm{maj}(w^*) + k - j$. The generating function for the choice of P^* is $S_q(n-1, k)$, whereas the generating function for the choice of $j \in \{1, 2, \ldots, k\}$ is $q^{k-1} + q^{k-2} + \cdots + q^0 = [k]_q$. Thus, by the Product Rule and Bijection Rule for Weighted Sets, we get

$$\sum_{P \in B} q^{\mathrm{maj}(P)} = [k]_q S_q(n-1, k). \tag{8.10}$$

The recursion in the theorem follows by adding (8.9) and (8.10).

The initial condition $S_q(0,0) = q^0 = 1$ holds since the major index of the empty word is 0. The initial condition $S_q(n,1) = q^0 = 1$ holds since the unique set partition P of $\{1, 2, \ldots, n\}$ with one block has $w(P) = 1, 2, \ldots, n$ and $\mathrm{maj}(w(P)) = 0$. The initial condition $S_q(n,n) = q^{n(n-1)/2}$ holds since the unique set partition $P \in SP(n,n)$ has $w(P) = n, n-1, \ldots, 3, 2, 1$ and $\mathrm{maj}(w(P)) = 1 + 2 + \cdots + (n-1) = n(n-1)/2$. □

Summary

- *Generating Functions for Weighted Sets.* A weighted set is a pair (S, wt) where S is a set and $\mathrm{wt} : S \to \mathbb{Z}_{\geq 0}$ is a function called a *statistic* on S. The generating function for this weighted set is $G_{S,\mathrm{wt}}(q) = G_S(q) = \sum_{z \in S} q^{\mathrm{wt}(z)}$. Writing $G_S(q) = \sum_{k \geq 0} a_k q^k$, a_k is the number of objects in S having weight k.

- *The Bijection Rule and the Weight-Shifting Rule.* A weight-preserving bijection from (S, wt) to (T, wt') is a bijection $f : S \to T$ with $\mathrm{wt}'(f(z)) = \mathrm{wt}(z)$ for all $z \in S$. When such an f exists, $G_{S,\mathrm{wt}}(q) = G_{T,\mathrm{wt}'}(q)$. More generally, if there is $b \in \mathbb{Z}$ with $\mathrm{wt}'(f(z)) = b + \mathrm{wt}(z)$ for all $z \in S$, then $G_{T,\mathrm{wt}'}(q) = q^b G_{S,\mathrm{wt}}(q)$.

- *The Sum Rule for Weighted Sets.* Suppose (S_i, w_i) are weighted sets for $1 \leq i \leq k$, S is the disjoint union of the S_i, and we define $w : S \to \mathbb{Z}_{\geq 0}$ by $w(z) = w_i(z)$ for $z \in S_i$. Then $G_S(q) = \sum_{i=1}^{k} G_{S_i}(q)$.

- *The Product Rule for Weighted Sets.* Suppose S is a weighted set such that every $u \in S$ can be constructed in exactly one way by choosing $u_1 \in S_1$, choosing $u_2 \in S_2$, and so on, finally choosing $u_k \in S_k$, and then assembling the chosen objects $u_1, \ldots, u_k$ in a prescribed manner. Assume that whenever u is constructed from $u_1, u_2, \ldots, u_k$, $\mathrm{wt}_S(u) = \mathrm{wt}_{S_1}(u_1) + \mathrm{wt}_{S_2}(u_2) + \cdots + \mathrm{wt}_{S_k}(u_k)$. Then $G_S(q) = \prod_{i=1}^k G_{S_i}(q)$.

- *q-Integers, q-Factorials, q-Binomial Coefficients, and q-Multinomial Coefficients.* Suppose q is a variable, $n, k, n_i \in \mathbb{Z}_{\geq 0}$, $0 \leq k \leq n$, and $\sum_i n_i = n$. We define $[n]_q = \sum_{i=0}^{n-1} q^i = \frac{q^n - 1}{q-1}$, $[n]!_q = \prod_{i=1}^n [i]_q$, $\begin{bmatrix} n \\ k \end{bmatrix}_q = \frac{[n]!_q}{[k]!_q [n-k]!_q}$, and $\begin{bmatrix} n \\ n_1, \ldots, n_s \end{bmatrix}_q = \frac{[n]!_q}{\prod_{i=1}^s [n_i]!_q}$. These are all polynomials in q with coefficients in $\mathbb{Z}_{\geq 0}$.

- *Recursions for q-Binomial Coefficients, etc.* The following recursions hold:

$$[n]!_q = [n-1]!_q \cdot [n]_q.$$

$$\begin{bmatrix} a+b \\ a, b \end{bmatrix}_q = q^b \begin{bmatrix} a+b-1 \\ a-1, b \end{bmatrix}_q + \begin{bmatrix} a+b-1 \\ a, b-1 \end{bmatrix}_q = \begin{bmatrix} a+b-1 \\ a-1, b \end{bmatrix}_q + q^a \begin{bmatrix} a+b-1 \\ a, b-1 \end{bmatrix}_q.$$

$$\begin{bmatrix} n_1 + \cdots + n_s \\ n_1, \ldots, n_s \end{bmatrix}_q = \sum_{k=1}^{s} q^{n_1 + \cdots + n_{k-1}} \begin{bmatrix} n_1 + \cdots + n_s - 1 \\ n_1, \ldots, n_k - 1, \ldots, n_s \end{bmatrix}_q.$$

- *Statistics on Words.* Given a word $w = w_1 w_2 \cdots w_n$ using a totally ordered alphabet, $\mathrm{Inv}(w) = \{(i,j) : i < j \text{ and } w_i > w_j\}$, $\mathrm{inv}(w) = |\,\mathrm{Inv}(w)|$, $\mathrm{Des}(w) = \{i < n : w_i > w_{i+1}\}$, $\mathrm{des}(w) = |\,\mathrm{Des}(w)|$, and $\mathrm{maj}(w) = \sum_{i \in \mathrm{Des}(w)} i$. We have

$$\begin{bmatrix} n_1 + \cdots + n_s \\ n_1, \ldots, n_s \end{bmatrix}_q = \sum_{w \in \mathcal{R}(a_1^{n_1} \cdots a_s^{n_s})} q^{\mathrm{inv}(w)} = \sum_{w \in \mathcal{R}(a_1^{n_1} \cdots a_s^{n_s})} q^{\mathrm{maj}(w)}.$$

The second equality follows from a bijection due to Foata, which maps maj to inv while preserving the last letter of the word. In particular, letting $S_n = \mathcal{R}(1^1 2^1 \cdots n^1)$ be the set of permutations of $\{1, 2, \ldots, n\}$,

$$[n]!_q = \sum_{w \in S_n} q^{\mathrm{inv}(w)} = \sum_{w \in S_n} q^{\mathrm{maj}(w)}.$$

These formulas can be proved bijectively by mapping $w \in S_n$ to its inversion table (or major index table) $(t_n, \ldots, t_1)$, where t_i records the change in inversions (or major index) caused by inserting the symbol i into the subword of w consisting of symbols $1, 2, \ldots, i-1$.

- *The q-Binomial Theorem.* For all variables x, y, q and all $n \in \mathbb{Z}_{>0}$,

$$(x + qy)(x + q^2 y)(x + q^3 y) \cdots (x + q^n y) = \sum_{k=0}^{n} q^{k(k+1)/2} \begin{bmatrix} n \\ k \end{bmatrix}_q x^{n-k} y^k.$$

- *Weighted Lattice Paths.* The q-binomial coefficient $\begin{bmatrix} a+b \\ a,b \end{bmatrix}_q = \begin{bmatrix} b+a \\ b,a \end{bmatrix}_q$ counts lattice paths in an $a \times b$ (or $b \times a$) rectangle, weighted either by area above the path or area below the path. This coefficient also counts integer partitions with first part at most a and length at most b, weighted by area, as well as anagrams in $\mathcal{R}(0^a 1^b)$, weighted by inversions or major index.

- *Weighted Dyck Paths.* Let $C_n(q)$ be the generating function for Dyck paths of order n, weighted by area between the path and $y = x$. Then $C_0(q) = 1$ and $C_n(q) = \sum_{k=1}^n q^{k-1} C_{k-1}(q) C_{n-k}(q)$. The generating function for Dyck paths (viewed as words in $\mathcal{R}(0^n 1^n)$) weighted by major index is $\frac{1}{[n+1]_q} \begin{bmatrix} 2n \\ n,n \end{bmatrix}_q = \begin{bmatrix} 2n \\ n,n \end{bmatrix}_q - q \begin{bmatrix} 2n \\ n-1,n+1 \end{bmatrix}_q$.

- *Weighted Set Partitions.* We encode a set partition P as a Stirling permutation $f(P)$ by listing the blocks of P in decreasing order of their minimum elements, with entries in each block written in increasing order. Define $\text{maj}(P) = \text{maj}(f(P))$ and $S_q(n,k) = \sum q^{\text{maj}(P)}$ summed over all set partitions of $\{1,2,\ldots,n\}$ with k blocks. These q-analogues of the Stirling numbers of the second kind satisfy the recursion

$$S_q(n,k) = q^{k-1}S_q(n-1,k-1) + [k]_q S_q(n-1,k) \quad \text{for } 1 < k < n$$

with initial conditions $S_q(0,0) = 1$, $S_q(n,1) = 1$ for all $n \geq 1$, and $S_q(n,n) = q^{n(n-1)/2}$ for all $n \geq 1$.

Exercises

In the exercises below, S_n denotes the set of permutations of $\{1,2,\ldots,n\}$, unless otherwise specified.

8-1. Let $S = \mathcal{R}(\text{a}^1\text{b}^1\text{c}^2)$, $T = \mathcal{R}(\text{a}^1\text{b}^2\text{c}^1)$, and $U = \mathcal{R}(\text{a}^2\text{b}^1\text{c}^1)$. Confirm that $G_{S,\text{inv}}(q) = G_{T,\text{inv}}(q) = G_{U,\text{inv}}(q)$ (as asserted in Example 8.17) by listing all weighted objects in T and U.

8-2. (a) Compute $\text{inv}(w)$, $\text{des}(w)$, and $\text{maj}(w)$ for each $w \in S_4$. (b) Use (a) to find the generating functions $G_{S_4,\text{inv}}(q)$, $G_{S_4,\text{des}}(q)$, and $G_{S_4,\text{maj}}(q)$. (c) Compute $[4]!_q$ by polynomial multiplication, and compare to the answers in (b).

8-3. (a) Compute $\text{inv}(w)$ for the following words w: 4251673, 101101110001, 314423313, 55233514425331. (b) Compute $\text{Des}(w)$, $\text{des}(w)$, and $\text{maj}(w)$ for each word w in (a).

8-4. Confirm the formulas for $G_{D_4,\text{area}}(q)$ and $G_{D_4,\text{maj}}(q)$ stated in Example 8.45 by listing all weighted Dyck paths of order 4.

8-5. (a) Find the maximum value of $\text{inv}(w)$, $\text{des}(w)$, and $\text{maj}(w)$ as w ranges over S_n. (b) Repeat (a) for w ranging over $\mathcal{R}(1^{n_1}2^{n_2}\cdots s^{n_s})$.

8-6. Let S be the set of k-letter words over the alphabet $[\![n]\!]$. For $w \in S$, let $\text{wt}(w)$ be the sum of all letters in w. Compute $G_{S,\text{wt}}(q)$.

8-7. Let S be the set of 5-letter words using the 26-letter English alphabet. For $w \in S$, let $\text{wt}(w)$ be the number of vowels in w. Compute $G_{S,\text{wt}}(q)$.

8-8. Let S be the set of all subsets of $\{1,2,\ldots,n\}$. For $A \in S$, let $\text{wt}(A) = |A|$. Use the Product Rule for Weighted Sets to compute $G_{S,\text{wt}}(q)$.

8-9. Let S be the set of all k-element multisets using the alphabet $[\![n]\!]$. For $M \in S$, let $\text{wt}(M)$ be the sum of the elements in M, counting multiplicities. Express $G_{S,\text{wt}}(q)$ in terms of q-binomial coefficients.

8-10. (a) How many permutations of $\{1,2,\ldots,8\}$ have exactly 17 inversions? (b) How many permutations of $\{1,2,\ldots,9\}$ have major index 29?

8-11. (a) How many lattice paths from $(0,0)$ to $(8,6)$ have area 21? (b) How many words in $\mathcal{R}(0^5 1^6 2^7)$ have ten inversions? (c) How many Dyck paths of order 7 have major index 30?

8-12. Use an involution to prove $\sum_{k=0}^{n}(-1)^k q^{k(k-1)/2}\begin{bmatrix} n \\ k \end{bmatrix}_q = 0$ for all integers $n > 0$.

8-13. Compute each of the following polynomials by any method, expressing the answer in the form $\sum_{k\geq 0} a_k q^k$. (a) $[7]_q$ (b) $[6]!_q$ (c) $\begin{bmatrix} 8 \\ 5 \end{bmatrix}_q$ (d) $\begin{bmatrix} 7 \\ 2,3,2 \end{bmatrix}_q$ (e) $\frac{1}{[8]_q}\begin{bmatrix} 8 \\ 5,3 \end{bmatrix}_q$ (f) $\frac{1}{[11]_q}\begin{bmatrix} 11 \\ 5,6 \end{bmatrix}_q$.

8-14. (a) Factor the polynomials $[4]_q$, $[5]_q$, $[6]_q$, and $[12]_q$ in $\mathbb{Z}[q]$. (b) How do these polynomials factor in $\mathbb{C}[q]$?

8-15. Compute $\left[\begin{smallmatrix}4\\2\end{smallmatrix}\right]_q$ in six ways, by: (a) simplifying the formula in Definition 8.30; (b) using the first recursion in Theorem 8.32; (c) using the second recursion in Theorem 8.32; (d) enumerating words in $\mathcal{R}(0011)$ by inversions; (e) enumerating words in $\mathcal{R}(0011)$ by major index; (f) enumerating partitions contained in a 2×2 box by area.

8-16. (a) Prove the identity $\begin{bmatrix} n_1 + \cdots + n_k \\ n_1, \ldots, n_k \end{bmatrix}_q = \prod_{i=1}^{k} \begin{bmatrix} n_i + \cdots + n_k \\ n_i, n_{i+1} + \cdots + n_k \end{bmatrix}_q$ algebraically. (b) Give a combinatorial proof of the identity in (a).

8-17. For $1 \leq i \leq 3$, let (T_i, w_i) be a set of weighted objects. (a) Prove that $\mathrm{id}_{T_1} : T_1 \to T_1$ is a weight-preserving bijection. (b) Prove that if $f : T_1 \to T_2$ is a weight-preserving bijection, then $f^{-1} : T_2 \to T_1$ is weight-preserving. (c) Prove that if $f : T_1 \to T_2$ and $g : T_2 \to T_3$ are weight-preserving bijections, so is $g \circ f$.

8-18. Prove the second recursion in Theorem 8.32: (a) by an algebraic manipulation; (b) by removing the first step from a lattice path in an $a \times b$ rectangle.

8-19. Let f be the map used to prove (8.3), with $a = b = 4$. Compute each of the following, and verify that weights are preserved.
(a) $f(2413, 1423, 10011010)$
(b) $f(4321, 4321, 11110000)$
(c) $f(2134, 3214, 01010101)$

8-20. Let f be the map used to prove (8.3), with $a = 5$ and $b = 4$. For each w given here, compute $f^{-1}(w)$ and verify that weights are preserved: (a) $w = 123456789$; (b) $w = 371945826$; (c) $w = 987456321$.

8-21. Repeat the previous exercise, taking $a = 2$ and $b = 7$.

8-22. Prove that $[n_1 + \cdots + n_s]!_q = [n_1]!_q \cdots [n_s]!_q \sum_{w \in \mathcal{R}(1^{n_1} \cdots s^{n_s})} q^{\mathrm{inv}(w)}$ by defining a weight-preserving bijection $f : S_{n_1+\cdots+n_s} \to S_{n_1} \times \cdots \times S_{n_s} \times \mathcal{R}(1^{n_1} \cdots s^{n_s})$.

8-23. (a) Find and prove a q-analogue of the identity $\sum_{k=0}^{n} \binom{n}{k}^2 = \binom{2n}{n}$ involving q-binomial coefficients (compare to Theorem 2.4 and Figure 2.1). (b) Similarly, derive a q-analogue of the identity $\sum_{k=0}^{a} \binom{k+b-1}{k, b-1} = \binom{a+b}{a, b}$.

8-24. Let S be the set of two-card poker hands. For $H \in S$, let $\mathrm{wt}(H)$ be the sum of the values of the two cards in H, where aces count as 11 and jacks, queens, and kings count as 10. Find $G_{S,\mathrm{wt}}(q)$.

8-25. Define the weight of a five-card poker hand to be the number of face cards in the hand (the face cards are aces, jacks, queens, and kings). Compute the generating functions for the following sets of poker hands relative to this weight: (a) full house hands; (b) three-of-a-kind hands; (c) flush hands; (d) straight hands.

8-26. Define the weight of a five-card poker hand to be the number of diamond cards in the hand. Compute the generating functions for the following sets of poker hands relative to this weight: (a) full house hands; (b) three-of-a-kind hands; (c) flush hands; (d) straight hands.

8-27. Let T_n be the set of connected simple graphs with vertex set $\{1, 2, \ldots, n\}$. Let the weight of a graph in T_n be the number of edges. Compute $G_{T_n}(q)$ for $1 \leq n \leq 5$.

8-28. Let G be the bijection in §8.4. Compute $G(341265)$ and $G^{-1}(3, 2, 3, 0, 0, 1)$, and verify that weights are preserved for these two objects.

8-29. Let G be the bijection in §8.4. Compute $G(35261784)$ and $G^{-1}(5, 6, 4, 2, 3, 0, 1, 0)$, and verify that weights are preserved for these two objects.

8-30. In §8.4, we constructed an inversion table for $w \in S_n$ by classifying inversions $(i,j) \in \mathrm{Inv}(w)$ based on the left value w_i. Define a new map $F : S_n \to [\![n]\!] \times [\![n-1]\!] \times \cdots \times [\![1]\!]$ by classifying inversions $(i,j) \in \mathrm{Inv}(w)$ based on the right value w_j. Show that F is a bijection, and compute $F(35261784)$ and $F^{-1}(5,6,4,3,2,0,1,0)$.

8-31. Define a map $F : S_n \to [\![n]\!] \times [\![n-1]\!] \times \cdots \times [\![1]\!]$ by setting $f(w) = (t_1, \ldots, t_n)$, where $t_i = |\{j : (i,j) \in \mathrm{Inv}(w)\}|$. Show that F is a bijection. (Informally, F classifies inversions of w based on the left position of the inversion pair.) Compute $F(35261784)$ and $F^{-1}(5,6,4,3,2,0,1,0)$.

8-32. Define a map $F : S_n \to [\![n]\!] \times [\![n-1]\!] \times \cdots \times [\![1]\!]$ that classifies inversions of w based on the right position of the inversion pair (compare to the previous exercise). Show that F is a bijection, and compute $F(35261784)$ and $F^{-1}(5,6,4,3,2,0,1,0)$.

8-33. Let G be the bijection in §8.5. Compute $G(341265)$ and $G^{-1}(3,2,3,1,0,0)$, and verify that weights are preserved for these two objects.

8-34. Let G be the bijection in §8.5. Compute $G(35261784)$ and $G^{-1}(5,6,4,2,3,0,1,0)$, and verify that weights are preserved for these two objects.

8-35. (a) Define a bijection $H : S_n \to S_n$ such that $\mathrm{maj}(H(w)) = \mathrm{inv}(w)$ for all $w \in S_n$ by combining the bijections in §8.4 and §8.5. (b) Compute $H(41627853)$ and $H^{-1}(41627853)$. (c) Compute $H(13576428)$ and $H^{-1}(13576428)$.

8-36. Coinversions. Define the *coinversions* of a word $w = w_1 w_2 \cdots w_n$, denoted $\mathrm{coinv}(w)$, to be the number of pairs (i,j) with $1 \le i < j \le n$ and $w_i < w_j$. Prove $\sum_{w \in \mathcal{R}(1^{n_1} 2^{n_2} \cdots s^{n_s})} q^{\mathrm{coinv}(w)} = \begin{bmatrix} n_1+\cdots+n_s \\ n_1,\ldots,n_s \end{bmatrix}_q$: (a) by using a bijection to reduce to the corresponding result for inv; (b) by verifying an appropriate recursion.

8-37. Given a word $w = w_1 \cdots w_n$, let $\mathrm{comaj}(w)$ be the sum of all $i < n$ with $w_i < w_{i+1}$, and let $\mathrm{rlmaj}(w)$ be the sum of $n-i$ for all $i < n$ with $w_i > w_{i+1}$. Calculate $\sum_{w \in S_n} q^{\mathrm{comaj}(w)}$ and $\sum_{w \in S_n} q^{\mathrm{rlmaj}(w)}$.

8-38. For $w \in S_n$, let $\mathrm{wt}(w)$ be the sum of all $i < n$ such that $i+1$ appears to the left of i in w. Compute $G_{S_n,\mathrm{wt}}(q)$.

8-39. (a) Suppose $w = w_1 w_2 \cdots w_{n-1}$ is a fixed permutation of $n-1$ distinct letters. Let a be a new letter less than all letters appearing in w. Let S be the set of n words that can be obtained from w by inserting a in some gap. Prove that $\sum_{z \in S} q^{\mathrm{maj}(z)} = q^{\mathrm{maj}(w)} [n]_q$. (b) Use (a) to obtain another proof that $\sum_{w \in S_n} q^{\mathrm{maj}(w)} = [n]!_q$.

8-40. Suppose k is fixed in $\{1,2,\ldots,n\}$, and $w = w_1 w_2 \cdots w_{n-1}$ is a fixed permutation of $\{1,2,\ldots,k-1,k+1,\ldots,n\}$. Let S be the set of n words that can be obtained from w by inserting k in some gap. Prove or disprove: $\sum_{z \in S} q^{\mathrm{maj}(z)} = q^{\mathrm{maj}(w)} [n]_q$.

8-41. Define a *cyclic shift* function $c : \{1,2,\ldots,n\} \to \{1,2,\ldots,n\}$ by $c(i) = i+1$ for $i < n$, and $c(n) = 1$. Define a map $C : S_n \to S_n$ by setting $C(w_1 w_2 \cdots w_n) = c(w_1)c(w_2)\cdots c(w_n)$. (a) Prove: for all $w \in S_n$, $\mathrm{maj}(C(w)) = \mathrm{maj}(w) - 1$ if $w_n \ne n$, and $\mathrm{maj}(C(w)) = \mathrm{maj}(w) + n - 1$ if $w_n = n$. (b) Use (a) to show combinatorially that, for $1 \le k \le n$, $\sum_{w \in S_n:\ w_n = k} q^{\mathrm{maj}(w)} = q^{n-k} \sum_{v \in S_{n-1}} q^{\mathrm{maj}(v)}$. (c) Use (b), the Sum Rule, and induction to obtain another proof of Theorem 8.26.

8-42. Prove the following q-analogue of the Negative Binomial Theorem: for all $n \in \mathbb{Z}_{>0}$,

$$\frac{1}{(1-t)(1-tq)\cdots(1-tq^{n-1})} = \sum_{k=0}^{\infty} \begin{bmatrix} k+n-1 \\ k, n-1 \end{bmatrix}_q t^k.$$

8-43. For all $n \ge 1$, all $T \subseteq \{1,2,\ldots,n-1\}$, and $1 \le k \le n$, let $G(n,T,k)$ be the number of permutations w of $\{1,2,\ldots,n\}$ with $\mathrm{Des}(w) = T$ and $w_n = k$. (a) Find a recursion for the quantities $G(n,T,k)$. (b) Count the number of permutations of 10 objects with descent set $\{2,3,5,7\}$.

8-44. Let w be the word 4523351452511332, and let h_z be the map from §8.10. Compute $h_z(w)$ for $z = 1,2,3,4,5,6$. Verify that (8.6) or (8.7) holds in each case.

8-45. Let w be the word 4523351452511332, and let h_z be the map from §8.10. Compute $h_z^{-1}(w)$ for $z = 1,2,3,4,5,6$.

8-46. Compute the image of each $w \in S_4$ under the map g from §8.10.

8-47. Let g be the map in §8.10. Compute $g(w)$ for each of these words, and verify that $\text{inv}(g(w)) = \text{maj}(w)$. (a) 4251673 (b) 27418563 (c) 101101110001 (d) 314423313

8-48. Let g be the map in §8.10. Compute $g^{-1}(w)$ for each word w in the preceding exercise.

8-49. Let g be the bijection in the proof of Theorem 8.47. Compute $g(w)$ for each non-Dyck word $w \in \mathcal{R}(0^3 1^3)$.

8-50. **q-Fibonacci Numbers.**(a) Let W_n be the set of words in $\{0,1\}^n$ with no two consecutive 0's, and let the weight of a word be the number of 0's in it. Find a recursion for the generating functions $G_{W_n}(q)$, and use this to compute $G_{W_6}(q)$. (b) Repeat part (a), taking the weight to be the number of 1's in the word.

8-51. Let T_n be the set of trees with vertex set $\{1,2,\ldots,n\}$. Can you find a statistic on trees such that the associated generating function satisfies $G_{T_n}(q) = [n]_q^{n-2}$?

8-52. **Multivariable Generating Functions.** Suppose S is a finite set, and $w_1,\ldots,w_n : S \to \mathbb{Z}_{\geq 0}$ are n statistics on S. The *generating function for S relative to the n weights* $w_1,\ldots,w_n$ is the polynomial $G_{S,w_1,\ldots,w_n}(q_1,\ldots,q_n) = \sum_{z\in S}\prod_{i=1}^n q_i^{w_i(z)}$. Formulate and prove versions of the Sum Rule, Product Rule, Bijection Rule, and Weight-Shifting Rule for such generating functions.

8-53. Recall that we can view permutations $w \in S_n$ as bijective maps of $\{1,2,\ldots,n\}$ into itself. Define $I : S_n \to S_n$ by $I(w) = w^{-1}$ for $w \in S_n$. (a) Show that $I \circ I = \text{id}_{S_n}$. (b) Show that $\text{inv}(I(w)) = \text{inv}(w)$ for all $w \in S_n$. (c) Define $\text{imaj}(w) = \text{maj}(I(w))$ for all $w \in S_n$. Compute the two-variable generating function $G_n(q,t) = \sum_{w\in S_n} q^{\text{maj}(w)}t^{\text{imaj}(w)}$ for $1 \leq n \leq 4$. Prove that $G_n(q,t) = G_n(t,q)$.

8-54. Let g be the map in §8.10, and let $\text{IDes}(w) = \text{Des}(w^{-1})$ for $w \in S_n$. (a) Show that for all $w \in S_n$, $\text{IDes}(g(w)) = \text{IDes}(w)$. (b) Construct a bijection $h : S_n \to S_n$ such that, for all $w \in S_n$, $\text{inv}(h(w)) = \text{maj}(w)$ and $\text{maj}(h(w)) = \text{inv}(w)$.

8-55. Let P_n be the set of integer partitions whose diagrams fit in the diagram of $(n-1, n-2, \ldots, 2, 1, 0)$, i.e., $\mu \in P_n$ iff $\ell(\mu) < n$ and $\mu_i \leq n-i$ for $1 \leq i < n$. Let $G_n(q) = \sum_{\mu\in P_n} q^{|\mu|}$. Find a recursion satisfied by $G_n(q)$ and use this to calculate $G_5(q)$. What is the relation between $G_n(q)$ and the q-Catalan number $C_n(q)$ from §8.11?

8-56. For each set partition P, find the associated Stirling permutation $f(P)$ and compute $\text{maj}(P)$. (a) $\{\{3,7,5\},\{8,2\},\{4,1,6\}\}$ (b) $\{\{1,2n\},\{2,2n-1\},\{3,2n-2\},\ldots,\{n,n+1\}\}$ (c) $\{\{1,2\},\{3,4\},\ldots,\{2n-1,2n\}\}$

8-57. Find the q-Stirling numbers $S_q(4,k)$ for $1 \leq k \leq 4$ by enumerating all weighted set partitions of $\{1,2,3,4\}$ and the associated Stirling permutations.

8-58. Use the recursion in Theorem 8.48 to compute the q-Stirling numbers $S_q(n,k)$ for $1 \leq k \leq n \leq 6$.

8-59. For a set partition $P = \{B_1,\ldots,B_k\}$ with $\min(B_1) > \cdots > \min(B_k)$, show that $\text{maj}(P) = \sum_{i=1}^k (k-i)|B_i|$.

8-60. Let $R(n,k)$ be the set of non-attacking placements of $n-k$ rooks on the board Δ_n (see 2.55). Define the weight of such a placement as follows. Each rook in the placement *cancels* all squares above it in its column. The weight of the placement is the total number of uncanceled squares located due west of rooks in the placement. Find a recursion for the generating functions $g(n,k) = G_{R(n,k),\mathrm{wt}}(q)$, which are q-analogues of the Stirling numbers of the second kind. Compute these generating functions for $1 \leq k \leq n \leq 5$. (Compare to the q-Stirling numbers $S_q(n,k)$ from §8.12.)

8-61. A *left-to-right minimum* of a permutation $w = w_1 w_2 \cdots w_n$ is a value w_i such that $w_i = \min\{w_1, w_2, \ldots, w_i\}$. For $w \in S_n$, let $\mathrm{lrmin}(w)$ be the number of left-to-right minima of w. (a) Define a bijection from the set of permutations in S_n with k cycles onto the set $\{w \in S_n : \mathrm{lrmin}(w) = k\}$. (b) Define $c_q(n,k) = \sum_{w \in S_n : \mathrm{lrmin}(w)=k} q^{\mathrm{inv}(w)}$, which is a q-analogue of the signless Stirling number of the first kind. Prove the recursion

$$c_q(n,k) = q^{n-1} c_q(n-1,k-1) + [n-1]_q c_q(n-1,k).$$

For which k and n is the recursion valid? What are the initial conditions?

8-62. Calculate the polynomials $c_q(n,k)$ in the preceding exercise for $1 \leq k \leq n \leq 5$.

8-63. Find and prove a matrix identity relating the q-Stirling numbers $S_q(n,k)$ and $(-1)^{n+k} c_q(n,k)$.

8-64. Prove that

$$S_q(n,k) = \sum_{j=0}^{k} (-1)^j q^{\binom{j}{2}} \begin{bmatrix} k \\ j \end{bmatrix}_q \frac{[k-j]_q^n}{[k]!_q}.$$

8-65. For $n \geq 1$, write $\sum_{w \in S_n} q^{\mathrm{des}(w)} = \sum_{k=0}^{n-1} a_{n,k} q^k$. Prove that

$$a_{n,k} = \sum_{i=0}^{k+1} (-1)^i \binom{n+1}{i} (k+1-i)^n.$$

8-66. Bounce Statistic on Dyck Paths. Given a Dyck path $P \in D_n$, define a new weight $\mathrm{bounce}(P)$ as follows. A ball starts at $(0,0)$ and moves north and east to (n,n) according to the following rules. The ball moves north v_0 steps until blocked by an east step of P, then moves east v_0 steps to the line $y = x$. The ball then moves north v_1 steps until blocked by the east step of P starting on the line $x = v_0$, then moves east v_1 steps to the line $y = x$. This bouncing process continues, generating a sequence $(v_0, v_1, \ldots, v_s)$ of vertical moves adding to n. We define $\mathrm{bounce}(P) = \sum_{i=0}^{s} i v_i$ and $C_n(q,t) = \sum_{P \in D_n} q^{\mathrm{area}(P)} t^{\mathrm{bounce}(P)}$. (a) Calculate $C_n(q,t)$ for $1 \leq n \leq 4$ by enumerating Dyck paths. (b) Let $C_{n,k}(q,t) = \sum_{P \in D_n : v_0(P)=k} q^{\mathrm{area}(P)} t^{\mathrm{bounce}(P)}$ be the generating function for Dyck paths that start with exactly k north steps. Establish the recursion

$$C_{n,k}(q,t) = \sum_{r=1}^{n-k} t^{n-k} q^{k(k-1)/2} \begin{bmatrix} r+k-1 \\ r, k-1 \end{bmatrix}_q C_{n-k,r}(q,t)$$

by removing the first bounce. Show also that $C_n(q,t) = t^{-n} C_{n+1,1}(q,t)$. (c) Use the recursion in (b) to calculate $C_n(q,t)$ for $n = 5, 6$. (d) Prove that $q^{n(n-1)/2} C_n(q, 1/q) = \sum_{P \in D_n} q^{\mathrm{maj}(P)}$. (e) Can you prove bijectively that $C_n(q,t) = C_n(t,q)$ for all $n \geq 1$? (If so, please contact the author.)

8-67. Let G_n be the set of sequences $g = (g_0, g_1, \ldots, g_{n-1})$ of nonnegative integers with $g_0 = 0$ and $g_{i+1} \leq g_i + 1$ for all $i < n-1$. For $g \in G_n$, define $\mathrm{area}(g) = \sum_{i=0}^{n-1} g_i$, and let $\mathrm{dinv}(g)$ be the number of pairs $i < j$ with $g_i - g_j \in \{0, 1\}$. (a) Find a bijection $k : G_n \to D_n$ such that $\mathrm{area}(k(g)) = \mathrm{area}(g)$ for all $g \in G_n$. (b) Find a bijection $h : G_n \to D_n$ such that $\mathrm{area}(h(g)) = \mathrm{dinv}(g)$ and $\mathrm{bounce}(h(g)) = \mathrm{area}(g)$ for all $g \in G_n$ (see the previous exercise). Conclude that the statistics dinv, bounce, area (on G_n), and area (on D_n) all have the same distribution.

Notes

The idea used to prove Theorem 8.26 seems to have first appeared in [56]. The bijection in §8.10 is due to Foata [33]. For related material, see [35]. Much of the early work on permutation statistics, including proofs of Theorems 8.43 and 8.47, is due to Major Percy MacMahon [85]. The bijective proof of Theorem 8.47, along with other material on q-Catalan numbers, may be found in [42]. The bounce statistic in Exercise 8-66 was introduced by Haglund [57]; for more on this topic, see Haglund's book [58].

9

Tableaux and Symmetric Polynomials

In this chapter, we study combinatorial objects called *tableaux*. Informally, a tableau is a filling of the cells in the diagram of an integer partition with values that weakly increase reading across rows and strictly increase reading down columns. We use tableaux to give a combinatorial definition of Schur polynomials, which are examples of symmetric polynomials. The theory of symmetric polynomials nicely demonstrates the interplay between combinatorics and algebra. We give an introduction to this vast subject in this chapter, stressing bijective proofs throughout.

9.1 Fillings and Tableaux

Before defining tableaux, we review the definitions of integer partitions and their diagrams from §2.11. A *partition*[1] *of* n is a weakly decreasing sequence of positive integers with sum n. Given a partition $\mu = (\mu_1, \mu_2, \ldots, \mu_k)$, we call μ_i the ith *part* of μ. We write $|\mu| = \mu_1 + \cdots + \mu_k$ for the sum of the parts of μ, and we write $\ell(\mu) = k$ for the *length* (number of positive parts) of μ. It is convenient to define $\mu_i = 0$ for all $i > \ell(\mu)$. The *diagram* of the partition μ is the set

$$\mathrm{dg}(\mu) = \{(i,j) \in \mathbb{Z}_{>0} \times \mathbb{Z}_{>0} : 1 \leq i \leq \ell(\mu),\ 1 \leq j \leq \mu_i\}.$$

We represent the diagram of μ as an array of unit squares with μ_i left-justified squares in the ith row from the top. This is called the *English notation* for partition diagrams. Some authors use *French notation* for partition diagrams, in which there are μ_i squares in the ith row from the bottom.

We obtain new combinatorial objects called *fillings* by putting a number in each box of a partition diagram. For example, taking $\mu = (5, 5, 2)$, here is a filling of shape μ:

4	3	3	7	2
1	1	3	9	1
5	6			

We can define fillings formally as follows.

9.1. Definition: Fillings. Given a partition μ and a set X, a *filling* of shape μ with values in X is a function $T : \mathrm{dg}(\mu) \to X$.

For each cell $(i,j) \in \mathrm{dg}(\mu)$, $T(i,j)$ is the value appearing in that cell of the partition diagram. In the filling depicted above, we have $T(1,1) = 4$, $T(1,2) = 3$, $T(1,3) = 3$, $T(1,4) = 7$, $T(1,5) = 2$, $T(2,1) = 1$, and so on. In most cases, the alphabet X is the set $[N] = \{1, 2, \ldots, N\}$ for some fixed $N > 0$, or $\mathbb{Z}_{>0}$, or $\mathbb{Z}$.

Next we define special fillings called *tableaux* (the terms *semistandard tableaux*, *Young*

[1] In this chaper, the unqualified term "partition" will always refer to an integer partition rather than a set partition.

tableaux, and *column-strict tableaux* are also used). A tableau is a filling where the values weakly increase as we read each row from left to right, and the values strictly increase as we read each column from top to bottom. A *standard tableau* is a tableau with n boxes containing each value in $\{1, 2, \ldots, n\}$ exactly once. We can state these definitions more formally as follows.

9.2. Definition: Tableaux and Standard Tableaux. Given a filling $T : \mathrm{dg}(\mu) \to \mathbb{Z}$, T is called a *(semistandard) tableau* of shape μ iff $T(i, j) \leq T(i, j + 1)$ for all i, j such that (i, j) and $(i, j+1)$ both belong to $\mathrm{dg}(\mu)$, and $T(i, j) < T(i+1, j)$ for all i, j such that (i, j) and $(i + 1, j)$ both belong to $\mathrm{dg}(\mu)$. A tableau T is called a *standard tableau* of shape μ iff T is a bijection from $\mathrm{dg}(\mu)$ to $\{1, 2, \ldots, n\}$, where $n = |\mu|$.

Note that the plural form of "tableau" is "tableaux;" both words are pronounced *tab–loh.*

9.3. Example. Consider the three fillings of shape $(3, 2, 2)$ shown here:

$$T_1 = \begin{array}{|c|c|c|}\hline 1&1&3\\ \hline 3&4 \\ \cline{1-2} 5&5 \\ \cline{1-2}\end{array} \qquad T_2 = \begin{array}{|c|c|c|}\hline 1&2&6\\ \hline 3&5 \\ \cline{1-2} 4&7 \\ \cline{1-2}\end{array} \qquad T_3 = \begin{array}{|c|c|c|}\hline 1&2&5\\ \hline 3&2 \\ \cline{1-2} 4&5 \\ \cline{1-2}\end{array}$$

T_1 is a tableau that is not standard, and T_2 is a standard tableau. T_3 is not a tableau, because of the strict decrease $3 > 2$ in row 2, and also because of the weak increase $2 \leq 2$ in column 2. Here is a tableau of shape $(6, 5, 3, 3)$:

$$S = \begin{array}{|c|c|c|c|c|c|}\hline 2&2&4&4&4&5\\ \hline 4&4&5&5&7 \\ \cline{1-5} 5&5&7 \\ \cline{1-3} 7&8&9 \\ \cline{1-3}\end{array}$$

9.4. Example. There are five standard tableaux of shape $(3, 2)$, as shown here:

$$S_1 = \begin{array}{|c|c|c|}\hline 1&2&3\\ \hline 4&5 \\ \cline{1-2}\end{array} \qquad S_2 = \begin{array}{|c|c|c|}\hline 1&2&4\\ \hline 3&5 \\ \cline{1-2}\end{array} \qquad S_3 = \begin{array}{|c|c|c|}\hline 1&2&5\\ \hline 3&4 \\ \cline{1-2}\end{array} \qquad S_4 = \begin{array}{|c|c|c|}\hline 1&3&4\\ \hline 2&5 \\ \cline{1-2}\end{array} \qquad S_5 = \begin{array}{|c|c|c|}\hline 1&3&5\\ \hline 2&4 \\ \cline{1-2}\end{array}.$$

As mentioned in the Introduction, there is a surprising formula for counting the number of standard tableaux of a given shape μ. We discuss this formula in §12.12.

We introduce the following notation for certain sets of tableaux.

9.5. Definition: $\mathrm{SSYT}_N(\mu)$, $\mathrm{SSYT}(\mu)$, and $\mathrm{SYT}(\mu)$. Given a partition μ and a positive integer N, let $\mathrm{SSYT}_N(\mu)$ be the set of all (semistandard Young) tableaux of shape μ with values in $\{1, 2, \ldots, N\}$. Let $\mathrm{SSYT}(\mu)$ be the set of all tableaux of shape μ with values in $\mathbb{Z}_{>0}$. Let $\mathrm{SYT}(\mu)$ be the set of all standard tableaux of shape μ.

For each fixed μ and N, $\mathrm{SYT}(\mu)$ and $\mathrm{SSYT}_N(\mu)$ are finite sets, but $\mathrm{SSYT}(\mu)$ is infinite.

9.2 Schur Polynomials

We now define a weight function on fillings that keeps track of the number of times each value appears.

9.6. Definition: Content of a Filling. Given a filling $T : \mathrm{dg}(\mu) \to \mathbb{Z}_{>0}$, the *content* of T is the infinite sequence $c(T) = (c_1, c_2, \ldots, c_k, \ldots)$, where c_k is the number of times the value k appears in T. Formally, $c_k = |\{(i,j) \in \mathrm{dg}(\mu) : T(i,j) = k\}|$. Given formal variables $x_1, x_2, \ldots, x_k, \ldots$, the *content monomial* of T is

$$\mathbf{x}^T = \mathbf{x}^{c(T)} = x_1^{c_1} x_2^{c_2} \cdots x_k^{c_k} \cdots = \prod_{(i,j)\in \mathrm{dg}(\mu)} x_{T(i,j)}.$$

9.7. Example. The filling T_1 shown in Example 9.3 has content $c(T_1) = (2,0,2,1,2,0,0,\ldots)$ and content monomial $\mathbf{x}^{T_1} = x_1^2 x_3^2 x_4 x_5^2$. For the other fillings in that example,

$$\mathbf{x}^{T_2} = x_1 x_2 x_3 x_4 x_5 x_6 x_7, \qquad \mathbf{x}^{T_3} = x_1 x_2^2 x_3 x_4 x_5^2, \qquad \mathbf{x}^S = x_2^2 x_4^5 x_5^5 x_7^3 x_8 x_9.$$

All five standard tableaux in Example 9.4 have content monomial $x_1x_2x_3x_4x_5$. More generally, if μ is a partition of n, then the content monomial of any $S \in \mathrm{SYT}(\mu)$ is $x_1 x_2 \cdots x_n$.

We now define the Schur polynomials, which can be viewed as generating functions for semistandard tableaux weighted by content.

9.8. Definition: Schur Polynomials. Given a partition μ and an integer $N \geq 1$, the *Schur polynomial in N variables indexed by μ* is

$$s_\mu(x_1, \ldots, x_N) = \sum_{T \in \mathrm{SSYT}_N(\mu)} \mathbf{x}^T.$$

9.9. Example. Let us compute the Schur polynomials $s_\mu(x_1, x_2, x_3)$ for all μ with $|\mu| = 3$. First, when $\mu = (3)$, we have the following semistandard tableaux of shape (3) using the alphabet $\{1,2,3\}$:

$$\boxed{1|1|1}\quad \boxed{1|1|2}\quad \boxed{1|1|3}\quad \boxed{1|2|2}\quad \boxed{1|2|3}$$

$$\boxed{1|3|3}\quad \boxed{2|2|2}\quad \boxed{2|2|3}\quad \boxed{2|3|3}\quad \boxed{3|3|3}$$

It follows that

$$s_{(3)}(x_1, x_2, x_3) = x_1^3 + x_1^2 x_2 + x_1^2 x_3 + x_1 x_2^2 + x_1 x_2 x_3 + x_1 x_3^2 + x_2^3 + x_2^2 x_3 + x_2 x_3^2 + x_3^3.$$

Second, when $\mu = (2,1)$, we obtain the following semistandard tableaux:

$$\begin{array}{|c|c|}\hline 1&1\\\hline 2\\\cline{1-1}\end{array}\quad \begin{array}{|c|c|}\hline 1&1\\\hline 3\\\cline{1-1}\end{array}\quad \begin{array}{|c|c|}\hline 2&2\\\hline 3\\\cline{1-1}\end{array}\quad \begin{array}{|c|c|}\hline 1&2\\\hline 2\\\cline{1-1}\end{array}\quad \begin{array}{|c|c|}\hline 1&2\\\hline 3\\\cline{1-1}\end{array}\quad \begin{array}{|c|c|}\hline 1&3\\\hline 2\\\cline{1-1}\end{array}\quad \begin{array}{|c|c|}\hline 1&3\\\hline 3\\\cline{1-1}\end{array}\quad \begin{array}{|c|c|}\hline 2&3\\\hline 3\\\cline{1-1}\end{array}$$

So $s_{(2,1)}(x_1, x_2, x_3) = x_1^2 x_2 + x_1^2 x_3 + x_2^2 x_3 + x_1 x_2^2 + 2x_1 x_2 x_3 + x_1 x_3^2 + x_2 x_3^2$. Third, when $\mu = (1,1,1)$, we see that $s_{(1,1,1)}(x_1, x_2, x_3) = x_1 x_2 x_3$, since there is only one semistandard tableau in this case.

Now consider what happens when we change N, the number of variables. Suppose first that we use $N = 2$ instead of $N = 3$. This means that the allowed alphabet for the tableaux has changed to $\{1,2\}$. Consulting the tableaux just computed, but disregarding those that use the letter 3, we conclude that

$$s_{(3)}(x_1, x_2) = x_1^3 + x_1^2 x_2 + x_1 x_2^2 + x_2^3; \qquad s_{(2,1)}(x_1, x_2) = x_1^2 x_2 + x_1 x_2^2; \qquad s_{(1,1,1)}(x_1, x_2) = 0.$$

In these examples, note that we can obtain the polynomial $s_\mu(x_1, x_2)$ from $s_\mu(x_1, x_2, x_3)$ by setting $x_3 = 0$. More generally, we claim that for any μ and any $N > M$, we can obtain

$s_\mu(x_1, \ldots, x_M)$ from $s_\mu(x_1, \ldots, x_M, \ldots, x_N)$ by setting the last $N - M$ variables equal to zero. To verify this, consider the defining formula

$$s_\mu(x_1, x_2, \ldots, x_N) = \sum_{T \in \mathrm{SSYT}_N(\mu)} \mathbf{x}^T.$$

Upon setting $x_{M+1} = \cdots = x_N = 0$ in this formula, the terms coming from tableaux T that use values larger than M become zero. We are left with the sum over $T \in \mathrm{SSYT}_M(\mu)$, which is precisely $s_\mu(x_1, x_2, \ldots, x_M)$.

Suppose instead that we increase the number of variables from $N = 3$ to $N = 5$. Here we must draw new tableaux to find the new Schur polynomial. For instance, the tableaux for $\mu = (1, 1, 1)$ are:

$$\begin{array}{|c|}\hline 1\\\hline 2\\\hline 3\\\hline\end{array}\quad \begin{array}{|c|}\hline 1\\\hline 2\\\hline 4\\\hline\end{array}\quad \begin{array}{|c|}\hline 1\\\hline 2\\\hline 5\\\hline\end{array}\quad \begin{array}{|c|}\hline 1\\\hline 3\\\hline 4\\\hline\end{array}\quad \begin{array}{|c|}\hline 1\\\hline 3\\\hline 5\\\hline\end{array}\quad \begin{array}{|c|}\hline 1\\\hline 4\\\hline 5\\\hline\end{array}\quad \begin{array}{|c|}\hline 2\\\hline 3\\\hline 4\\\hline\end{array}\quad \begin{array}{|c|}\hline 2\\\hline 3\\\hline 5\\\hline\end{array}\quad \begin{array}{|c|}\hline 2\\\hline 4\\\hline 5\\\hline\end{array}\quad \begin{array}{|c|}\hline 3\\\hline 4\\\hline 5\\\hline\end{array}$$

Accordingly,

$$s_{(1,1,1)}(x_1, x_2, x_3, x_4, x_5) = x_1x_2x_3 + x_1x_2x_4 + x_1x_2x_5 + \cdots + x_3x_4x_5.$$

9.10. Example. A semistandard tableau of shape[2] (1^k) using the alphabet $[N] = \{1, 2, \ldots, N\}$ is essentially a strictly increasing sequence of k elements of $[N]$, which can be identified with a k-element subset of $[N]$. Combining this remark with the definition of Schur polynomials, we conclude that

$$s_{(1^k)}(x_1, \ldots, x_N) = \sum_{1 \leq i_1 < i_2 < \cdots < i_k \leq N} x_{i_1}x_{i_2}\cdots x_{i_k}. \tag{9.1}$$

Similarly, a semistandard tableau of shape (k) is a weakly increasing sequence of k elements of $[N]$, which can be identified with a k-element multiset using values in $[N]$. So

$$s_{(k)}(x_1, \ldots, x_N) = \sum_{1 \leq i_1 \leq i_2 \leq \cdots \leq i_k \leq N} x_{i_1}x_{i_2}\cdots x_{i_k}. \tag{9.2}$$

9.11. Example. Given any integer $N \geq 4$, what is the coefficient of $x_1^2x_2^2x_3^2x_4$ in the Schur polynomial $s_{(4,3)}(x_1, \ldots, x_N)$? The answer is the number of semistandard tableaux of shape $(4, 3)$ where 1, 2, and 3 each appear twice, and 4 appears once. Equivalently, we seek all tableaux with shape $(4, 3)$ and content $(2, 2, 2, 1)$. (It is customary to omit trailing zeroes from the content vector.) The tableaux satisfying these conditions are shown here:

$$\begin{array}{|c|c|c|c|}\hline 1&1&2&2\\\hline 3&3&4\\\cline{1-3}\end{array}\qquad \begin{array}{|c|c|c|c|}\hline 1&1&2&3\\\hline 2&3&4\\\cline{1-3}\end{array}\qquad \begin{array}{|c|c|c|c|}\hline 1&1&2&4\\\hline 2&3&3\\\cline{1-3}\end{array}\qquad \begin{array}{|c|c|c|c|}\hline 1&1&3&3\\\hline 2&2&4\\\cline{1-3}\end{array}$$

So the requested coefficient is 4. Next, what is the coefficient of $x_1x_2^2x_3^2x_4^2$? Now we must find the tableaux of shape $(4, 3)$ and content $(1, 2, 2, 2)$, which are the following:

$$\begin{array}{|c|c|c|c|}\hline 1&2&2&4\\\hline 3&3&4\\\cline{1-3}\end{array}\qquad \begin{array}{|c|c|c|c|}\hline 1&2&2&3\\\hline 3&4&4\\\cline{1-3}\end{array}\qquad \begin{array}{|c|c|c|c|}\hline 1&2&3&3\\\hline 2&4&4\\\cline{1-3}\end{array}\qquad \begin{array}{|c|c|c|c|}\hline 1&2&3&4\\\hline 2&3&4\\\cline{1-3}\end{array}$$

Again there are four tableaux, so the coefficient of $x_1x_2^2x_3^2x_4^2$ is 4. Next, what is the coefficient

[2]Recall that when listing the parts of integer partitions, 1^k abbreviates a sequence of k ones.

of $x_1^2 x_2 x_3^2 x_4^2$? Drawing the tableaux of shape $(4,3)$ and content $(2,1,2,2)$ produces these objects:

1	1	2	3
3	4	4	

1	1	2	4
3	3	4	

1	1	3	3
2	4	4	

1	1	3	4
2	3	4	

The coefficient is 4 once again. One may check that for the content $(2,2,1,2)$, and indeed for any rearrangement of $(2,2,2,1,0,0,\ldots)$ with all zeroes after position N, the number of semistandard tableaux of shape $(4,3)$ having this content is always 4. This is not a coincidence; it is a consequence of the fact that Schur polynomials are *symmetric*, which we prove in §9.5.

9.12. Remark. We have presented a combinatorial definition of Schur polynomials as a weighted sum of semistandard tableaux. We can also define Schur polynomials algebraically as a quotient of two determinants; see Theorem 10.45. Alternatively, we can define Schur polynomials via determinants of matrices whose entries are the symmetric polynomials e_n and h_n defined below; see Theorems 10.60 and 10.61. Many properties of Schur polynomials can be established either combinatorially or algebraically. In this text, we focus on the combinatorial proofs. Macdonald's comprehensive monograph [84] approaches the subject from a much more algebraic perspective.

9.3 Symmetric Polynomials

The examples of Schur polynomials computed in the last section were all *symmetric*; in other words, permuting the subscripts of the x-variables in any fashion did not change the answer. This section begins our examination of the general theory of symmetric polynomials. Throughout the discussion, we consider polynomials with real coefficients in variables $x_1, x_2, \ldots, x_N$, where N is a fixed positive integer. Everything said below applies more generally to polynomials with coefficients in any field K containing $\mathbb{Q}$ (the rational numbers). The symbol $K[x_1, x_2, \ldots, x_N]$ denotes the set of all polynomials in $x_1, \ldots, x_N$ with coefficients in K. Given a polynomial f in this set, the expression $f(a_1, a_2, \ldots, a_N)$ is the object we get by replacing each formal variable x_i by a_i for $1 \leq i \leq N$. (This is a special case of an *evaluation homomorphism*, which is discussed in the Appendix.) Next, recall that S_N is the set of all permutations of the alphabet $[N] = \{1, 2, \ldots, N\}$, which are bijections $w : [N] \to [N]$. With this notation in hand, we can now formally define symmetric polynomials.

9.13. Definition: Symmetric Polynomials. A polynomial $f \in \mathbb{R}[x_1, \ldots, x_N]$ is *symmetric* iff

$$f(x_{w(1)}, x_{w(2)}, \ldots, x_{w(N)}) = f(x_1, \ldots, x_N) \qquad \text{for all } w \in S_N.$$

This means that any permutation of the variables x_i leaves f unchanged. Since any permutation can be achieved by a finite sequence of basic transpositions (by Theorem 7.29), f is symmetric iff for all i with $1 \leq i < N$, interchanging x_i and x_{i+1} in f leaves f unchanged.

We now define some particular symmetric polynomials.

9.14. Definition: Power Sums. For every $k \geq 1$, the kth *power-sum polynomial* in N variables is

$$p_k(x_1, x_2, \ldots, x_N) = x_1^k + x_2^k + \cdots + x_N^k.$$

For example, $p_3(x_1, x_2, x_3, x_4, x_5) = x_1^3 + x_2^3 + x_3^3 + x_4^3 + x_5^3$. It is immediate that each power-sum polynomial $p_k(x_1, \ldots, x_N)$ is indeed symmetric.

9.15. Definition: Elementary Symmetric Polynomials. For fixed k with $1 \leq k \leq N$, the kth *elementary symmetric polynomial* in N variables is

$$e_k(x_1, x_2, \ldots, x_N) = \sum_{1 \leq i_1 < i_2 < \cdots < i_k \leq N} x_{i_1} x_{i_2} \cdots x_{i_k}.$$

We also set $e_0(x_1, \ldots, x_N) = 1$ and $e_k(x_1, \ldots, x_N) = 0$ for all $k > N$.

For example, $e_2(x_1, x_2, x_3, x_4) = x_1x_2 + x_1x_3 + x_1x_4 + x_2x_3 + x_2x_4 + x_3x_4$. One readily verifies that each e_k really is symmetric. By formula (9.1), we see that $e_k(x_1, \ldots, x_N) = s_{(1^k)}(x_1, \ldots, x_N)$, so that elementary symmetric polynomials are special cases of Schur polynomials.

9.16. Definition: Complete Symmetric Polynomials. For fixed $k \geq 1$, the kth *complete homogeneous symmetric polynomial* in N variables is

$$h_k(x_1, x_2, \ldots, x_N) = \sum_{1 \leq i_1 \leq i_2 \leq \cdots \leq i_k \leq N} x_{i_1} x_{i_2} \cdots x_{i_k}.$$

We also set $h_0(x_1, \ldots, x_N) = 1$.

One may verify that each h_k really is symmetric. We call h_k "complete" because it is the sum of *all* monomials of degree k in the given variables. For example, $h_2(x_1, x_2, x_3) = x_1^2 + x_2^2 + x_3^2 + x_1x_2 + x_1x_3 + x_2x_3$. By formula (9.2), we see that $h_k(x_1, \ldots, x_N) = s_{(k)}(x_1, \ldots, x_N)$, so that complete symmetric polynomials are also special cases of Schur polynomials.

We now show how a partition μ of length at most N can be used to create a symmetric polynomial in N variables. First we need some notation.

9.17. Definition: Sets of Partitions. Given integers $k \geq 0$ and $N > 0$, let $\mathrm{Par}_N(k)$ be the set of partitions μ with $|\mu| = k$ and $\ell(\mu) \leq N$. (Recall that $\mu_i = 0$ for $\ell(\mu) < i \leq N$.) Let Par_N be the set of partitions of length at most N. Let Par be the set of all partitions, and let $\mathrm{Par}(k)$ be the set of all partitions of k.

For $\alpha = (\alpha_1, \ldots, \alpha_N) \in \mathbb{Z}_{\geq 0}^N$, let $\mathbf{x}^\alpha = x_1^{\alpha_1} x_2^{\alpha_2} \cdots x_N^{\alpha_N}$. Also, let $\mathrm{sort}(\alpha) \in \mathrm{Par}_N$ be the unique partition obtained by sorting the entries of α into weakly decreasing order.

9.18. Definition: Monomial Symmetric Polynomials. Given a partition $\mu \in \mathrm{Par}_N$, the *monomial symmetric polynomial* in N variables indexed by μ is

$$m_\mu(x_1, \ldots, x_N) = \sum_{\substack{\alpha \in \mathbb{Z}_{\geq 0}^N: \\ \mathrm{sort}(\alpha) = \mu}} \mathbf{x}^\alpha.$$

Informally, $m_\mu(x_1, \ldots, x_N)$ is the sum of all distinct monomials $x_1^{\alpha_1} \cdots x_N^{\alpha_N}$ whose exponent vector can be rearranged to give μ. For example,

$$m_{(3,3,2)}(x_1, x_2, x_3) = x_1^3 x_2^3 x_3^2 + x_1^3 x_2^2 x_3^3 + x_1^2 x_2^3 x_3^3.$$

Some of our previous examples are instances of monomial symmetric polynomials. Namely, we have $p_k(x_1, \ldots, x_N) = m_{(k)}(x_1, \ldots, x_N)$ and $e_k(x_1, \ldots, x_N) = m_{(1^k)}(x_1, \ldots, x_N)$.

Let us check that m_μ really is symmetric. Given $w \in S_N$ with inverse w', we have

$$m_\mu(x_{w(1)}, \ldots, x_{w(N)}) = \sum_{\alpha:\ \mathrm{sort}(\alpha) = \mu} x_{w(1)}^{\alpha_1} \cdots x_{w(N)}^{\alpha_N} = \sum_{\alpha:\ \mathrm{sort}(\alpha) = \mu} x_1^{\alpha_{w'(1)}} \cdots x_N^{\alpha_{w'(N)}}.$$

The last step follows by noting that $w(j) = i$ iff $j = w'(i)$, so that $x_{w(j)}^{\alpha_j} = x_i^{\alpha_{w'(i)}}$ holds for

$i = w(j)$. To continue, introduce a new summation variable $\beta = (\alpha_{w'(1)}, \ldots, \alpha_{w'(N)})$. The entries of β are obtained by rearranging the entries of α, so $\text{sort}(\beta) = \text{sort}(\alpha) = \mu$. As α ranges over all vectors in $\mathbb{Z}_{\geq 0}^N$ that sort to μ, so does β. The calculation therefore continues:

$$\sum_{\alpha:\ \text{sort}(\alpha)=\mu} x_1^{\alpha_{w'(1)}} \cdots x_N^{\alpha_{w'(N)}} = \sum_{\beta:\ \text{sort}(\beta)=\mu} x_1^{\beta_1} \cdots x_N^{\beta_N} = m_\mu(x_1, \ldots, x_N).$$

9.19. Definition: The Space Λ_N. Let Λ_N be the set of all symmetric polynomials in $\mathbb{R}[x_1, \ldots, x_N]$.

If two polynomials f and g are symmetric, so are $f+g$, $-f$, and fg. For example, given $f, g \in \Lambda_N$ and $w \in S_N$,

$$\begin{aligned}(fg)(x_{w(1)}, \ldots, x_{w(N)}) &= f(x_{w(1)}, \ldots, x_{w(N)})g(x_{w(1)}, \ldots, x_{w(N)}) \\ &= f(x_1, \ldots, x_N)g(x_1, \ldots, x_N) \\ &= (fg)(x_1, \ldots, x_N).\end{aligned}$$

Also, any constant polynomial c is certainly symmetric, and hence any scalar multiple cf of a symmetric polynomial f is symmetric. These comments imply that Λ_N is a subring and vector subspace of $\mathbb{R}[x_1, \ldots, x_N]$, so Λ_N is a subalgebra of the real algebra of polynomials in N variables (see the Appendix for the definitions of these terms).

We have just seen that Λ_N is closed under products. So, we can multiply together polynomials of the form e_k, h_k, or p_k to obtain even more examples of symmetric polynomials. This leads to the following definition.

9.20. Definition: The Symmetric Polynomials e_α, h_α, and p_α. Let $\alpha = (\alpha_1, \ldots, \alpha_s)$ be any sequence of positive integers. Define

$$\begin{aligned}e_\alpha(x_1, \ldots, x_N) &= \prod_{i=1}^{s} e_{\alpha_i}(x_1, \ldots, x_N); \\ h_\alpha(x_1, \ldots, x_N) &= \prod_{i=1}^{s} h_{\alpha_i}(x_1, \ldots, x_N); \\ p_\alpha(x_1, \ldots, x_N) &= \prod_{i=1}^{s} p_{\alpha_i}(x_1, \ldots, x_N).\end{aligned}$$

We call e_α the *elementary symmetric polynomial* indexed by α; h_α the *complete homogeneous symmetric polynomial* indexed by α; and p_α the *power-sum symmetric polynomial* indexed by α (in N variables).

These definitions are most frequently used when α is a partition. Suppose the sequence α can be sorted to give the partition μ. Then $e_\alpha = e_\mu$, $h_\alpha = h_\mu$, and $p_\alpha = p_\mu$, because multiplication of polynomials is commutative. More generally, if α and β are rearrangements of each other, then $e_\alpha = e_\beta$, $h_\alpha = h_\beta$, and $p_\alpha = p_\beta$.

9.21. Remark. The power-sum polynomials p_α have already appeared in our discussion of Pólya's Formula (§7.16), where they were used to count weighted colorings with symmetries taken into account.

9.4 Vector Spaces of Symmetric Polynomials

When studying symmetric polynomials, it is often helpful to focus attention on those polynomials that are *homogeneous* of a given degree. We recall that the *degree* of an individual monomial $\mathbf{x}^\alpha = x_1^{\alpha_1} \cdots x_N^{\alpha_N}$ is $\alpha_1 + \cdots + \alpha_N$. A polynomial p is called *homogeneous of degree* k iff every monomial $\mathbf{x}^\alpha$ appearing in p with nonzero coefficient has degree k. In particular, the zero polynomial is homogeneous of every degree.

9.22. Definition: The Space Λ_N^k. For all $k \geq 0$ and $N > 0$, let Λ_N^k be the set of symmetric polynomials $p \in \Lambda_N$ such that p is homogeneous of degree k.

It can be checked that for all $f, g \in \Lambda_N^k$ and $c \in \mathbb{R}$, $f + g \in \Lambda_N^k$ and $cf \in \Lambda_N^k$. This means that Λ_N^k *is a subspace of* Λ_N. Furthermore, each symmetric polynomial can be written uniquely as a finite sum of its nonzero homogeneous components; this means that the vector space Λ_N is the direct sum of these subspaces: $\Lambda_N = \bigoplus_{k=0}^{\infty} \Lambda_N^k$. Finally, for all $p \in \Lambda_N^k$ and $q \in \Lambda_N^j$, we have $pq \in \Lambda_N^{k+j}$, which means that this direct sum decomposition turns Λ_N into a *graded algebra*.

The vector space Λ_N is infinite-dimensional, but each homogeneous subspace Λ_N^k is finite-dimensional. A recurring theme in the theory of symmetric polynomials is the problem of finding different bases of the vector space Λ_N^k and understanding the relations between these bases. We begin in this section by considering the most straightforward basis for this vector space, which consists of certain monomial symmetric polynomials.

9.23. Theorem: Monomial Basis of Λ_N^k. For every $k \geq 0$ and $N > 0$,

$$\{m_\mu(x_1, \ldots, x_N) : \mu \in \mathrm{Par}_N(k)\}$$

is a basis for the vector space Λ_N^k.

Proof. For $\mu \in \mathrm{Par}_N(k)$, recall that m_μ is the sum of all distinct monomials $\mathbf{x}^\alpha$ such that $\alpha \in \mathbb{Z}_{\geq 0}^N$ can be rearranged to give μ. Each of these monomials has degree $|\mu| = k$, so that each m_μ in the given set is homogeneous of degree k and thus belongs to Λ_N^k. Next, let us prove the linear independence of the given monomial symmetric polynomials. Suppose some linear combination of these polynomials is the zero polynomial, say

$$\sum_{\mu \in \mathrm{Par}_N(k)} c_\mu m_\mu(x_1, \ldots, x_N) = 0 \qquad \text{where } c_\mu \in \mathbb{R}. \tag{9.3}$$

Consider a fixed $\nu \in \mathrm{Par}_N(k)$. Given any partition $\mu \neq \nu$, we cannot rearrange the parts of ν to obtain μ. It follows that m_ν is the only monomial symmetric polynomial in the sum in which $\mathbf{x}^\nu$ appears with nonzero coefficient. The coefficient of $\mathbf{x}^\nu$ in m_ν is 1. Extracting the coefficient of $\mathbf{x}^\nu$ on both sides of (9.3) therefore gives $c_\nu \cdot 1 = 0$. Since ν was arbitrary, every coefficient c_ν is zero, completing the proof of linear independence.

Next, let us prove that the given monomial symmetric polynomials span Λ_N^k. Let $f \in \Lambda_N^k$ be any homogeneous symmetric polynomial of degree k. For each $\mu \in \mathrm{Par}_N(k)$, define $d_\mu \in \mathbb{R}$ to be the coefficient of $\mathbf{x}^\mu$ in f. We claim that

$$\sum_{\mu \in \mathrm{Par}_N(k)} d_\mu m_\mu(x_1, \ldots, x_N) = f(x_1, \ldots, x_N). \tag{9.4}$$

Since both sides are homogeneous of degree k, it suffices to check that the coefficients of $\mathbf{x}^\alpha$ on both sides of (9.4) are equal, for all $\alpha \in \mathbb{Z}_{\geq 0}^N$ with $\alpha_1 + \cdots + \alpha_N = k$. Fix such an α, and note that there is a unique partition $\nu \in \mathrm{Par}_N(k)$ such that $\mathrm{sort}(\alpha) = \nu$. As in the previous paragraph, the coefficient of $\mathbf{x}^\alpha$ is 1 in m_ν and is 0 in m_μ for all $\mu \neq \nu$. Thus, the coefficient of $\mathbf{x}^\alpha$ in $\sum_\mu d_\mu m_\mu$ is d_ν. On the other hand, since f is symmetric, the coefficient of $\mathbf{x}^\alpha$ in f must be the same as the coefficient of $\mathbf{x}^\nu$ in f, since some permutation of the variables changes $\mathbf{x}^\alpha$ into $\mathbf{x}^\nu$. Since the coefficient of $\mathbf{x}^\nu$ in f is d_ν by definition, we are done. □

9.24. Remark. In the previous theorem, we need to restrict attention to those partitions μ of length at most N, since $m_\mu(x_1, \ldots, x_N)$ is not defined when $\ell(\mu) > N$. On the other hand, if the number of variables N is at least k, then $\mathrm{Par}_N(k) = \mathrm{Par}(k)$ since the length of any partition of k is at most k. Therefore, *when* $N \geq k$, Λ_N^k has basis $\{m_\mu(x_1, \ldots, x_N) : \mu \in \mathrm{Par}(k)\}$, and the dimension of Λ_N^k is $p(k)$, the number of integer partitions of k.

9.5 Symmetry of Schur Polynomials

Recall the formula for Schur polynomials from Definition 9.8:

$$s_\mu(x_1, \ldots, x_N) = \sum_{T \in \mathrm{SSYT}_N(\mu)} \mathbf{x}^T.$$

This section gives a bijective proof that all Schur polynomials are symmetric. First, we give names to the coefficients in these polynomials.

9.25. Definition: Kostka Numbers. For each partition $\mu \in \mathrm{Par}_N$ and each $\alpha \in \mathbb{Z}_{\geq 0}^N$, define the *Kostka number* $K_{\mu,\alpha}$ to be the coefficient of $\mathbf{x}^\alpha$ in $s_\mu(x_1, \ldots, x_N)$. So, $K_{\mu,\alpha}$ is the number of semistandard tableaux of shape μ and content α.

9.26. Example. The calculations in Example 9.11 show that

$$K_{(4,3),(2,2,2,1)} = K_{(4,3),(1,2,2,2)} = K_{(4,3),(2,1,2,2)} = 4.$$

Similarly, we see from Example 9.4 that $K_{(3,2),(1,1,1,1,1)} = 5$.

The following result is the key to proving the symmetry of Schur polynomials.

9.27. Theorem: Symmetry of Kostka Numbers. For all partitions $\mu \in \mathrm{Par}_N$ and all $\alpha, \beta \in \mathbb{Z}_{\geq 0}^N$ such that $\mathrm{sort}(\alpha) = \mathrm{sort}(\beta)$, we have $K_{\mu,\alpha} = K_{\mu,\beta}$.

Proof. Fix μ, α, β as in the theorem statement. Since $\mathrm{sort}(\alpha) = \mathrm{sort}(\beta)$, we can pass from α to β by an appropriate permutation of the entries of α. This permutation can be achieved in finitely many steps by repeatedly interchanging two consecutive entries of α (compare to Theorem 7.29).By induction, it is enough to prove the result when β is obtained from α by switching α_i and α_{i+1} for some i.

Let Y be the set of all tableaux $T \in \mathrm{SSYT}_N(\mu)$ such that $c(T) = \alpha$, and let Z be the set of all tableaux $T \in \mathrm{SSYT}_N(\mu)$ such that $c(T) = \beta$. Since $|Y| = K_{\mu,\alpha}$ and $|Z| = K_{\mu,\beta}$, it suffices to define a bijection $f_i : Y \to Z$. An input to the map f_i is a semistandard tableau T of shape μ and content α. The output $f_i(T)$ must be a new semistandard tableau of shape μ in which the number of i's and $(i+1)$'s are switched, while the number of k's (for all $k \neq i, i+1$) is unchanged. We illustrate the action of f_3 on the following tableau:

1	1	1	1	1	1	1	2	3
2	2	2	3	3	3	4	4	4
3	3	3	4	4	5	5	6	
4	5	6	6	6	7	9		
5	6	7	7	8				

Observe that certain occurrences of 3 are matched with an occurrence of 4 in the cell directly below. Let us underline the 3's and 4's that are *not* part of these matched pairs:

1	1	1	1	1	1	1	2	3
2	2	2	3	3	$\underline{3}$	$\underline{4}$	$\underline{4}$	4
3	$\underline{3}$	$\underline{3}$	4	4	5	5	6	
4	5	6	6	6	7	9		
5	6	7	7	8				

Notice that each row of the tableau contains a (possibly empty) run of consecutive cells consisting of underlined 3's and 4's. The entries directly above these cells (when they exist) are less than 3, while any entries directly below are greater than 4. So we are free to change the frequencies of 3's and 4's within each run without altering the fact that the filling is a semistandard tableau. If the run of underlined entries in a given row consists of j threes followed by k fours (where $j, k \geq 0$), we change this to a run consisting of k threes followed by j fours. Doing this in every row switches the overall frequency of 3's and 4's in the entire filling. Note in particular that the matched pairs are not altered, and these pairs contribute equally to the frequency counts for 3 and 4. Our example tableau is mapped by f_3 to the following tableau:

1	1	1	1	1	1	1	2	3
2	2	2	3	3	$\underline{3}$	$\underline{3}$	$\underline{4}$	4
3	$\underline{4}$	$\underline{4}$	4	4	5	5	6	
4	5	6	6	6	7	9		
5	6	7	7	8				

Applying the same run-modification process to this new tableau restores the original tableau; this means that f_3 is a bijection. As another example of the action of f_3, we have:

$$f_3\left(\begin{array}{|c|c|c|c|c|c|c|c|c|}\hline 3&3&3&4&4&4&4&4&6\\\hline 4&\multicolumn{8}{l}{}\\\cline{1-1}\end{array}\right) = \begin{array}{|c|c|c|c|c|c|c|c|c|}\hline 3&3&3&3&3&3&4&4&6\\\hline 4&\multicolumn{8}{l}{}\\\cline{1-1}\end{array}$$

The definition of f_i for general i is analogous. We locate and ignore matched pairs consisting of an i directly atop an $i+1$, then underline the remaining i's and $(i+1)$'s, then switch the relative frequencies of the underlined i's and $(i+1)$'s in each row. This action maintains the property of being a tableau, switches the overall frequency of i's and $(i+1)$'s, and preserves the frequency of all other letters. Applying the algorithm twice restores the original tableau, so we have found the required bijection. □

9.28. Example. Let us trace through the preceding proof to construct bijections between the sets of tableaux in Example 9.11. We must chain together appropriate maps f_i, where the values of i are chosen to rearrange the starting content vector α into the target content vector β. For example, we can go from content $(2, 2, 2, 1)$ to content $(2, 1, 2, 2)$ by applying f_3 and then f_2. The first tableau of content $(2, 2, 2, 1)$ displayed in Example 9.11 maps to a tableau of content $(2, 1, 2, 2)$ via these steps:

$$\begin{array}{|c|c|c|c|}\hline 1&1&2&2\\\hline 3&3&4&\multicolumn{1}{l}{}\\\cline{1-3}\end{array} \xrightarrow{f_3} \begin{array}{|c|c|c|c|}\hline 1&1&2&2\\\hline 3&4&4&\multicolumn{1}{l}{}\\\cline{1-3}\end{array} \xrightarrow{f_2} \begin{array}{|c|c|c|c|}\hline 1&1&3&3\\\hline 2&4&4&\multicolumn{1}{l}{}\\\cline{1-3}\end{array}$$

If we continue by applying the map f_1, we reach a tableau with content $(1, 2, 2, 2)$:

$$\begin{array}{|c|c|c|c|}\hline 1&1&3&3\\\hline 2&4&4&\multicolumn{1}{l}{}\\\cline{1-3}\end{array} \xrightarrow{f_1} \begin{array}{|c|c|c|c|}\hline 1&2&3&3\\\hline 2&4&4&\multicolumn{1}{l}{}\\\cline{1-3}\end{array}$$

The inverse bijection is computed by applying the maps in the reverse order. For example, the first tableau of content $(1, 2, 2, 2)$ in Example 9.11 is mapped to a tableau of content $(2, 2, 2, 1)$ via the following steps.

$$\begin{array}{|c|c|c|c|}\hline 1&2&2&4\\\hline 3&3&4\\\cline{1-3}\end{array} \xrightarrow{f_1} \begin{array}{|c|c|c|c|}\hline 1&1&2&4\\\hline 3&3&4\\\cline{1-3}\end{array} \xrightarrow{f_2} \begin{array}{|c|c|c|c|}\hline 1&1&3&4\\\hline 2&2&4\\\cline{1-3}\end{array} \xrightarrow{f_3} \begin{array}{|c|c|c|c|}\hline 1&1&3&3\\\hline 2&2&4\\\cline{1-3}\end{array}$$

We can now deduce the symmetry of Schur polynomials. In fact, we can even expand these polynomials as specific linear combinations of monomial symmetric polynomials using the Kostka numbers.

9.29. Theorem: Monomial Expansion of Schur Polynomials. For all partitions $\mu \in \mathrm{Par}_N(k)$ and all $N \geq 1$,

$$s_\mu(x_1, \ldots, x_N) = \sum_{\rho \in \mathrm{Par}_N(k)} K_{\mu,\rho} m_\rho(x_1, \ldots, x_N). \tag{9.5}$$

So, $s_\mu(x_1, \ldots, x_N)$ is a *symmetric* polynomial belonging to the space Λ_N^k.

Proof. We calculate as follows:

$$\begin{aligned} s_\mu(x_1, \ldots, x_N) &= \sum_{\alpha \in \mathbb{Z}_{\geq 0}^N} K_{\mu,\alpha} \mathbf{x}^\alpha = \sum_{\rho \in \mathrm{Par}_N(k)} \sum_{\substack{\alpha \in \mathbb{Z}_{\geq 0}^N: \\ \mathrm{sort}(\alpha) = \rho}} K_{\mu,\alpha} \mathbf{x}^\alpha \\ &= \sum_{\rho \in \mathrm{Par}_N(k)} \sum_{\substack{\alpha \in \mathbb{Z}_{\geq 0}^N: \\ \mathrm{sort}(\alpha) = \rho}} K_{\mu,\rho} \mathbf{x}^\alpha = \sum_{\rho \in \mathrm{Par}_N(k)} K_{\mu,\rho} \sum_{\substack{\alpha \in \mathbb{Z}_{\geq 0}^N: \\ \mathrm{sort}(\alpha) = \rho}} \mathbf{x}^\alpha \\ &= \sum_{\rho \in \mathrm{Par}_N(k)} K_{\mu,\rho} m_\rho(x_1, \ldots, x_N). \end{aligned}$$

The first equality follows from the definition of Kostka numbers. In the second equality, we reorganize the sum by grouping together those $\alpha \in \mathbb{Z}_{\geq 0}^N$ with $\mathrm{sort}(\alpha) = \rho$. We only need to sum over partitions $\rho \in \mathrm{Par}_N(k)$, since $\mathbf{x}^T$ is a monomial of degree k for every tableau T of shape μ. The third equality follows from Theorem 9.27. The fourth equality uses the fact that $K_{\mu,\rho}$ does not depend on the inner summation index α. The final equality follows by definition of m_ρ. Since s_μ is a linear combination of basis elements of the subspace Λ_N^k, we see that s_μ belongs to this subspace. In particular, s_μ is symmetric. □

9.6 Orderings on Partitions

We intend to use Theorem 9.29 to find bases for the vector spaces Λ_N^k consisting of certain Schur polynomials. For this purpose, we must first introduce some order relations on partitions.

9.30. Definition: Lexicographic Ordering of Partitions. Given $\mu, \nu \in \mathrm{Par}(k)$, we say that μ is *lexicographically smaller* than ν, written $\mu \leq_{\mathrm{lex}} \nu$, iff either $\mu = \nu$ or the first nonzero entry in the sequence $\nu - \mu$ is positive.

In other words, $\mu \leq_{\mathrm{lex}} \nu$ iff $\mu = \nu$ or there exists j such that $\mu_1 = \nu_1$, $\mu_2 = \nu_2$, $\ldots$, $\mu_{j-1} = \nu_{j-1}$, and $\mu_j < \nu_j$. This definition uses the convention that $\mu_i = 0$ for $i > \ell(\mu)$ and $\nu_i = 0$ for $i > \ell(\nu)$. It is routine to check that $\leq_{\mathrm{lex}}$ is a total ordering of the set $\mathrm{Par}(k)$ for each $k \geq 0$.

9.31. Example. Here is a list of all integer partitions of 6, written in lexicographic order from smallest to largest:

$$(1,1,1,1,1,1) \leq_{\text{lex}} (2,1,1,1,1) \leq_{\text{lex}} (2,2,1,1) \leq_{\text{lex}} (2,2,2) \leq_{\text{lex}} (3,1,1,1)$$
$$\leq_{\text{lex}} (3,2,1) \leq_{\text{lex}} (3,3) \leq_{\text{lex}} (4,1,1) \leq_{\text{lex}} (4,2) \leq_{\text{lex}} (5,1) \leq_{\text{lex}} (6).$$

In more detail, $(3,1,1,1) \leq_{\text{lex}} (3,2,1)$ since

$$(3,2,1,0,0,\ldots) - (3,1,1,1,0,\ldots) = (0,1,0,-1,0,\ldots)$$

and the first nonzero entry in this sequence is positive.

In later sections, we often need matrices and vectors whose rows and columns are indexed by integer partitions. Unless otherwise specified, we always use the lexicographic ordering of partitions to determine which partition labels each row and column of the matrix. For instance, a matrix $A = [c_{\mu,\nu} : \mu, \nu \in \text{Par}(3)]$ is displayed as follows:

$$A = \begin{bmatrix} c_{(1,1,1),(1,1,1)} & c_{(1,1,1),(2,1)} & c_{(1,1,1),(3)} \\ c_{(2,1),(1,1,1)} & c_{(2,1),(2,1)} & c_{(2,1),(3)} \\ c_{(3),(1,1,1)} & c_{(3),(2,1)} & c_{(3),(3)} \end{bmatrix}.$$

We now define a partial ordering on partitions that occurs frequently in the theory of symmetric polynomials.

9.32. Definition: Dominance Ordering on Partitions. Given partitions $\mu, \nu \in \text{Par}(k)$, we say that μ *is dominated by* ν, written $\mu \trianglelefteq \nu$, iff

$$\mu_1 + \mu_2 + \cdots + \mu_i \leq \nu_1 + \nu_2 + \cdots + \nu_i \quad \text{for all } i \geq 1.$$

Note that $\mu \ntrianglelefteq \nu$ iff there exists $i \geq 1$ with $\mu_1 + \cdots + \mu_i > \nu_1 + \cdots + \nu_i$.

9.33. Example. We have $(2,2,1,1) \trianglelefteq (4,2)$ since $2 \leq 4$, $2+2 \leq 4+2$, $2+2+1 \leq 4+2+0$, and $2+2+1+1 \leq 4+2+0+0$. On the other hand, $(3,1,1,1) \ntrianglelefteq (2,2,2)$ since $3 > 2$, and $(2,2,2) \ntrianglelefteq (3,1,1,1)$ since $2+2+2 > 3+1+1$. This example shows that not every pair of partitions is comparable under the dominance relation.

9.34. Theorem: Dominance Partial Order. For all $k \geq 0$, the dominance relation is a partial ordering on $\text{Par}(k)$.

Proof. We show that $\trianglelefteq$ is reflexive, antisymmetric, and transitive on $\text{Par}(k)$.
Reflexivity: Given $\mu \in \text{Par}(k)$, we have $\mu_1 + \cdots + \mu_i \leq \mu_1 + \cdots + \mu_i$ for all $i \geq 1$. So $\mu \trianglelefteq \mu$.
Antisymmetry: Suppose $\mu, \nu \in \text{Par}(k)$, $\mu \trianglelefteq \nu$, and $\nu \trianglelefteq \mu$. We know $\mu_1 + \cdots + \mu_i \leq \nu_1 + \cdots + \nu_i$ and also $\nu_1 + \cdots + \nu_i \leq \mu_1 + \cdots + \mu_i$ for all i, hence $\mu_1 + \cdots + \mu_i = \nu_1 + \cdots + \nu_i$ for all $i \geq 1$. In particular, taking $i = 1$ gives $\mu_1 = \nu_1$. For each $i > 1$, subtracting the $(i-1)$th equation from the ith equation shows that $\mu_i = \nu_i$. So $\mu = \nu$.
Transitivity: Fix $\mu, \nu, \rho \in \text{Par}(k)$, and assume $\mu \trianglelefteq \nu$ and $\nu \trianglelefteq \rho$; we must prove $\mu \trianglelefteq \rho$. We know $\mu_1 + \cdots + \mu_i \leq \nu_1 + \cdots + \nu_i$ for all i, and also $\nu_1 + \cdots + \nu_i \leq \rho_1 + \cdots + \rho_i$ for all i. Combining these inequalities yields $\mu_1 + \cdots + \mu_i \leq \rho_1 + \cdots + \rho_i$ for all i, so $\mu \trianglelefteq \rho$. □

One may check that $\trianglelefteq$ is a *total* ordering of $\text{Par}(k)$ iff $k \leq 5$.

9.35. Theorem: Lexicographic and Dominance Ordering. For all $\mu, \nu \in \text{Par}(k)$, if $\mu \trianglelefteq \nu$ then $\mu \leq_{\text{lex}} \nu$.

Proof. Fix $\mu, \nu \in \text{Par}(k)$ such that $\mu \not\leq_{\text{lex}} \nu$; we will prove that $\mu \ntrianglelefteq \nu$. By definition of the lexicographic order, there must exist an index $j \geq 1$ such that $\mu_i = \nu_i$ for all $i < j$, but $\mu_j > \nu_j$. Adding these relations together, we see that $\mu_1 + \cdots + \mu_j > \nu_1 + \cdots + \nu_j$, and so $\mu \ntrianglelefteq \nu$. □

The next definition and theorem allow us to visualize the dominance relation in terms of partition diagrams.

9.36. Definition: Raising Operation. Let μ and ν be two partitions of k. We say that ν is related to μ by a *raising operation*, denoted $\mu R\nu$, iff there exist $i < j$ such that $\nu_i = \mu_i+1$, $\nu_j = \mu_j - 1$, and $\nu_s = \mu_s$ for all $s \neq i, j$.

Intuitively, $\mu R\nu$ means that we can go from the diagram for μ to the diagram for ν by taking the last square from some row of $\text{dg}(\mu)$ and moving it to the end of a higher row.

9.37. Example. The following pictures illustrate a sequence of raising operations.

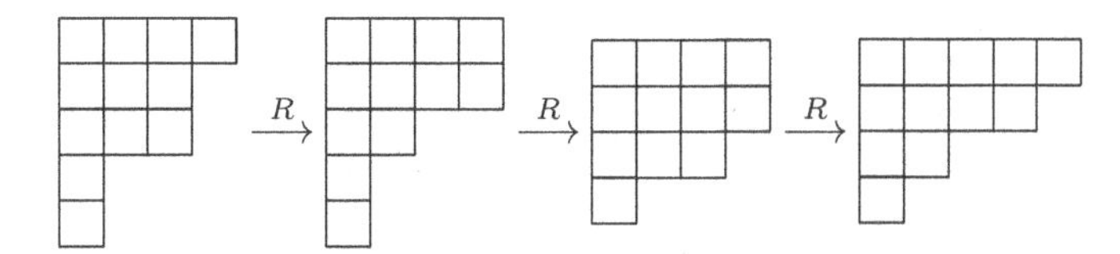

Observe that $(4,3,3,1,1) \trianglelefteq (5,4,2,1)$, so that the last partition in the sequence dominates the first one. The next result shows that this always happens.

9.38. Theorem: Dominance and Raising Operations. Given $\mu, \nu \in \text{Par}(k)$, $\mu \trianglelefteq \nu$ iff there exist $m \geq 0$ and partitions $\mu^0, \ldots, \mu^m$ such that $\mu = \mu^0$, $\mu^{i-1} R \mu^i$ for $1 \leq i \leq m$, and $\mu^m = \nu$.

Proof. We first show that $\mu R\nu$ implies $\mu \trianglelefteq \nu$. Suppose $\nu = (\mu_1, \ldots, \mu_i+1, \ldots, \mu_j-1, \ldots)$ as in the definition of raising operations. Let us check that $\nu_1 + \cdots + \nu_k \geq \mu_1 + \cdots + \mu_k$ holds for all $k \geq 1$. This is true for $k < i$, since equality holds for these values of k. If $k = i$, then $\nu_1 + \cdots + \nu_k = \mu_1 + \cdots + \mu_{i-1} + (\mu_i + 1) > \mu_1 + \cdots + \mu_i$. Similarly, for all k with $i \leq k < j$, we have $\nu_1 + \cdots + \nu_k = \mu_1 + \cdots + \mu_k + 1 > \mu_1 + \cdots + \mu_k$. Finally, for all $k \geq j$, we have $\mu_1 + \cdots + \mu_k = \nu_1 + \cdots + \nu_k$ since the $+1$ and -1 adjustments to parts i and j cancel out.

Next, suppose μ and ν are linked by a chain of raising operations as in the theorem statement. By the previous paragraph, we know $\mu = \mu^0$, $\mu^{i-1} \trianglelefteq \mu^i$ for $1 \leq i \leq m$, and $\mu^m = \nu$. Since $\trianglelefteq$ is transitive, we conclude that $\mu \trianglelefteq \nu$, as needed.

Conversely, suppose that $\mu \trianglelefteq \nu$. Consider the vector $(d_1, d_2, \ldots)$ such that $d_s = (\nu_1 + \cdots + \nu_s) - (\mu_1 + \cdots + \mu_s)$. Since $\mu \trianglelefteq \nu$, we have $d_s \geq 0$ for all s. Also, $d_s = 0$ for all large enough s, since μ and ν are both partitions of k. We use strong induction on $n = \sum_s d_s$ to show that we can go from μ to ν by a sequence of raising operations. If $n = 0$, then $\mu = \nu$, and we can take $m = 0$ and $\mu = \mu^0 = \nu$. Otherwise, let i be the least index such that $d_i > 0$, and let j be the least index after i such that $d_j = 0$. The choice of i shows that $\mu_s = \nu_s$ for all $s < i$, but $\mu_i < \nu_i$. If $i > 1$, the inequality $\mu_i < \nu_i \leq \nu_{i-1} = \mu_{i-1}$ shows that it is possible to add one box to the end of row i in $\text{dg}(\mu)$ and still get a partition diagram. If $i = 1$, the addition of this box certainly gives a partition diagram. On the other hand, the relations $d_{j-1} > 0$, $d_j = 0$ mean that $\mu_1 + \cdots + \mu_{j-1} < \nu_1 + \cdots + \nu_{j-1}$ but $\mu_1 + \cdots + \mu_j = \nu_1 + \cdots + \nu_j$, so that $\mu_j > \nu_j$. Furthermore, from $d_j = 0$ and $d_{j+1} \geq 0$ we deduce that $\mu_{j+1} \leq \nu_{j+1}$. So, $\mu_{j+1} \leq \nu_{j+1} \leq \nu_j < \mu_j$, which shows that we can remove a box from row j of $\text{dg}(\mu)$ and still get a partition diagram.

We have just shown that it is permissible to modify μ by a raising operator that moves the box at the end of row j to the end of row i. Let μ^1 be the new partition obtained in this way, so that $\mu R \mu^1$. Consider how the partial sums $\mu_1 + \cdots + \mu_s$ change when we replace μ by μ^1. For $s < i$ or $s \geq j$, the partial sums are the same for μ and μ^1. For $i \leq s < j$, the partial sums increase by 1. Since $d_s > 0$ in the range $i \leq s < j$, it follows that the new differences $d'_s = (\nu_1 + \cdots + \nu_s) - (\mu^1_1 + \cdots + \mu^1_s)$ are all nonnegative; in other words, $\mu^1 \trianglelefteq \nu$. We have $d'_s = d_s - 1$ for $i \leq s < j$, and $d'_s = d_s$ for all other s; so $\sum_s d'_s < \sum_s d_s$.

By the induction hypothesis, we can find a chain of raising operations linking μ^1 to ν. This completes the induction step. $\square$

As an application of the previous result, we prove a theorem relating the dominance ordering to the conjugation operation on partitions. We recall that for a partition μ, the conjugate partition μ' is the partition whose parts are the column heights in the diagram of μ. We obtain $\text{dg}(\mu')$ from $\text{dg}(\mu)$ by interchanging rows and columns in the diagram.

9.39. Theorem: Dominance and Conjugation. For all $\mu, \nu \in \text{Par}(k)$, $\mu \trianglelefteq \nu$ iff $\nu' \trianglelefteq \mu'$.

Proof. Fix $\mu, \nu \in \text{Par}(k)$. We first claim that $\mu R \nu$ implies $\nu' R \mu'$. This assertion follows from the pictorial description of the raising operation, since the box that moves from a lower row in $\text{dg}(\mu)$ to a higher row in $\text{dg}(\nu)$ necessarily moves from an earlier column to a later column (scanning columns from left to right). If we perform this move backward on the transposed partition diagrams, we see that we can pass from ν' to μ' by moving a box in $\text{dg}(\nu')$ from a lower row to a higher row.

Now, assume $\mu \trianglelefteq \nu$. By Theorem 9.38, there exist partitions $\mu^0, \ldots, \mu^m$ with $\mu = \mu^0$, $\mu^{i-1} R \mu^i$ for $1 \leq i \leq m$, and $\mu^m = \nu$. Applying the claim in the previous paragraph, we see that $(\mu^i)' R (\mu^{i-1})'$ for $1 \leq i \leq m$, so we can go from ν' to μ' by a chain of raising operations. Invoking Theorem 9.38 again, we conclude that $\nu' \trianglelefteq \mu'$.

Conversely, assume that $\nu' \trianglelefteq \mu'$. Applying the result just proved, we get $\mu'' \trianglelefteq \nu''$. Since $\mu'' = \mu$ and $\nu'' = \nu$, we have $\mu \trianglelefteq \nu$. $\square$

9.7 Schur Bases

We now have all the necessary tools to find bases for the vector spaces Λ_N^k consisting of Schur polynomials. First we illustrate the key ideas with an example.

9.40. Example. In Example 9.9, we computed the Schur polynomials $s_\mu(x_1, x_2, x_3)$ for all partitions $\mu \in \text{Par}(3)$. We can use Theorem 9.29 to write these Schur polynomials as linear combinations of monomial symmetric polynomials $m_\nu(x_1, x_2, x_3)$, where the coefficients are Kostka numbers:

$$\begin{aligned} s_{(1,1,1)} &= 1m_{(1,1,1)} \\ s_{(2,1)} &= 2m_{(1,1,1)} + 1m_{(2,1)} \\ s_{(3)} &= 1m_{(1,1,1)} + 1m_{(2,1)} + 1m_{(3)}. \end{aligned}$$

These equations can be combined to give the following matrix identity:

$$\begin{bmatrix} s_{(1,1,1)} \\ s_{(2,1)} \\ s_{(3)} \end{bmatrix} = \begin{bmatrix} 1 & 0 & 0 \\ 2 & 1 & 0 \\ 1 & 1 & 1 \end{bmatrix} \begin{bmatrix} m_{(1,1,1)} \\ m_{(2,1)} \\ m_{(3)} \end{bmatrix}.$$

The 3×3 matrix appearing here is lower-triangular with 1's on the main diagonal, so this matrix is invertible. Multiplying by the inverse matrix, we find that

$$\begin{bmatrix} m_{(1,1,1)} \\ m_{(2,1)} \\ m_{(3)} \end{bmatrix} = \begin{bmatrix} 1 & 0 & 0 \\ -2 & 1 & 0 \\ 1 & -1 & 1 \end{bmatrix} \begin{bmatrix} s_{(1,1,1)} \\ s_{(2,1)} \\ s_{(3)} \end{bmatrix}.$$

This says that each monomial symmetric polynomial $m_\nu(x_1, x_2, x_3)$ is expressible as a linear combination of the Schur polynomials $s_\mu(x_1, x_2, x_3)$. Since $\{m_\nu : \nu \in \text{Par}(3)\}$ is a basis of the vector space Λ_3^3, the Schur polynomials must span this space. Since $\dim(\Lambda_3^3) = p(3) = 3$, the three-element set $\{s_\mu(x_1, x_2, x_3) : \mu \in \text{Par}(3)\}$ is in fact a basis of Λ_3^3.

The reasoning in this example extends to the general case. The key fact is that the transition matrix from Schur polynomials to monomial symmetric polynomials is always lower-triangular with 1's on the main diagonal, as shown next. We refer to this matrix as a *Kostka matrix*, since its entries are Kostka numbers.

9.41. Theorem: Lower Unitriangularity of the Kostka Matrix. For all partitions λ, $K_{\lambda,\lambda} = 1$. For all partitions λ and μ, $K_{\lambda,\mu} \neq 0$ implies $\mu \trianglelefteq \lambda$ (and also $\mu \leq_{\text{lex}} \lambda$, by Theorem 9.35).

Proof. The Kostka number $K_{\lambda,\lambda}$ is the number of semistandard tableaux T of shape λ and content λ. Such a tableau must contain λ_i copies of i for each $i \geq 1$. In particular, T contains λ_1 1's. Since T is semistandard, all these 1's must occur in the top row, which has λ_1 boxes. So the top row of T contains all 1's. For the same reason, the λ_2 2's in T must all occur in the second row, which has λ_2 boxes. Continuing similarly, we see that T must be the tableau whose ith row contains all i's, for all $i \geq 1$. Thus, there is exactly one semistandard tableau of shape λ and content λ. For example, when $\lambda = (4, 2, 2, 1)$, T is this tableau:

$$T = \begin{array}{|c|c|c|c|} \hline 1 & 1 & 1 & 1 \\ \hline 2 & 2 \\ \cline{1-2} 3 & 3 \\ \cline{1-2} 4 \\ \cline{1-1} \end{array}$$

For the second part of the theorem, assume $\lambda, \mu \in \text{Par}(k)$ and $K_{\lambda,\mu} \neq 0$; we prove $\mu \trianglelefteq \lambda$. Since the Kostka number is nonzero, there exists a semistandard tableau T of shape λ and content μ. Every value in T is a positive integer. Because the columns of T must strictly increase, all 1's in T must occur in row 1; all 2's in T must occur in row 1 or row 2; and, in general, all j's in T must occur in the top j rows of $\text{dg}(\lambda)$. Now fix $i \geq 1$. Since T has content μ, the total number of occurrences of the symbols $1, 2, \ldots, i$ in T is $\mu_1 + \cdots + \mu_i$. All of these symbols must appear in the top i rows of $\text{dg}(\lambda)$; these rows contain $\lambda_1 + \cdots + \lambda_i$ boxes. We conclude that $\mu_1 + \cdots + \mu_i \leq \lambda_1 + \cdots + \lambda_i$. This holds for every i, so $\mu \trianglelefteq \lambda$. $\square$

9.42. Example. Let $\lambda = (3, 2, 2)$ and $\mu = (2, 2, 2, 1)$. The Kostka number $K_{\lambda,\mu}$ is 3, as we see by listing the semistandard tableaux of shape λ and content μ:

$$\begin{array}{|c|c|c|} \hline 1 & 1 & 2 \\ \hline 2 & 3 \\ \cline{1-2} 3 & 4 \\ \cline{1-2} \end{array} \qquad \begin{array}{|c|c|c|} \hline 1 & 1 & 3 \\ \hline 2 & 2 \\ \cline{1-2} 3 & 4 \\ \cline{1-2} \end{array} \qquad \begin{array}{|c|c|c|} \hline 1 & 1 & 4 \\ \hline 2 & 2 \\ \cline{1-2} 3 & 3 \\ \cline{1-2} \end{array}$$

In each tableau, all occurrences of i appear in the top i rows, and hence $\mu \trianglelefteq \lambda$.

9.43. Theorem: Schur Basis of Λ_N^k. For all $k \geq 0$ and $N > 0$,

$$\{s_\lambda(x_1, \ldots, x_N) : \lambda \in \text{Par}_N(k)\}$$

is a basis for the vector space Λ_N^k.

Proof. Let $p = |\text{Par}_N(k)|$, and let $\mathbf{S}$ be the $p \times 1$ column vector consisting of the Schur polynomials $\{s_\lambda(x_1, \ldots, x_N) : \lambda \in \text{Par}_N(k)\}$, arranged in lexicographic order. Let $\mathbf{M}$ be the $p \times 1$ column vector consisting of the monomial symmetric polynomials $\{m_\mu(x_1, \ldots, x_N) : \mu \in \text{Par}_N(k)\}$, also arranged in lexicographic order. Finally, let $\mathbf{K}$ be the $p \times p$ matrix, with rows and columns indexed by elements of $\text{Par}_N(k)$ in lexicographic order, such that the entry in row λ and column μ is the Kostka number $K_{\lambda,\mu}$. Theorem 9.29 says that, for every $\lambda \in \text{Par}_N(k)$,

$$s_\lambda(x_1, \ldots, x_N) = \sum_{\mu \in \text{Par}_N(k)} K_{\lambda,\mu} m_\mu(x_1, \ldots, x_N).$$

These scalar equations are equivalent to the matrix-vector equation $\mathbf{S} = \mathbf{KM}$. Moreover, Theorem 9.41 asserts that $\mathbf{K}$ is a lower-triangular matrix of integers with 1's on the main diagonal. So $\mathbf{K}$ has an inverse matrix (whose entries are also integers, since $\det(\mathbf{K}) = 1$). Multiplying on the left by this inverse matrix, we get $\mathbf{M} = \mathbf{K}^{-1}\mathbf{S}$. This equation means that every m_μ is a linear combination of Schur polynomials. Since the m_μ generate Λ_N^k, the Schur polynomials must also generate this space. Linear independence follows automatically since the number of Schur polynomials in the proposed basis (namely p) equals the dimension of the vector space, by Theorem 9.23. □

9.44. Remark. The entries of the *inverse Kostka matrix* $\mathbf{K}^{-1}$ tell us how to expand monomial symmetric polynomials in terms of Schur polynomials. As seen in the 3×3 example, these entries are integers that can be negative. We give a combinatorial interpretation for these *inverse Kostka numbers* in §10.16, using signed objects called *special rim-hook tableaux*.

9.45. Remark. If $\lambda \in \text{Par}(k)$ has more than N parts, then $s_\lambda(x_1, \ldots, x_N) = 0$. This follows since there are not enough values available in the alphabet to fill the first column of $\text{dg}(\lambda)$ with a strictly increasing sequence. So there are no semistandard tableaux of this shape using this alphabet.

9.8 Tableau Insertion

We have seen that the Kostka numbers give the coefficients in the monomial expansion of Schur polynomials. Surprisingly, the Kostka numbers also relate Schur polynomials to the elementary and complete homogeneous symmetric polynomials. This fact is a consequence of the *Pieri Rules*, which tell us how to rewrite products of the form $s_\mu e_k$ and $s_\mu h_k$ as linear combinations of Schur polynomials.

To develop these results, we need a fundamental combinatorial construction on tableaux called *tableau insertion*. Given a semistandard tableau T of shape μ and a value x, we will build a new semistandard tableau by inserting x into T, as follows.

9.46. Definition: Tableau Insertion Algorithm. Given a semistandard tableau T of shape μ and a value $x \in \mathbb{Z}$, define the *insertion of x into T*, denoted $T \leftarrow x$, by the following procedure.

1. If $\mu = (0)$, so that T is the empty tableau, then $T \leftarrow x$ is the tableau of shape (1) whose sole entry is x.
2. Otherwise, let $y_1 \leq y_2 \leq \cdots \leq y_m$ be the entries in the top row of T.
   - 2a. If $y_m \leq x$, then $T \leftarrow x$ is the tableau of shape $(\mu_1 + 1, \mu_2, \ldots)$ obtained by placing a new box containing x at the right end of the top row of T.
   - 2b. Otherwise, choose the least $i \in \{1, 2, \ldots, m\}$ such that $x < y_i$. Let T' be the semistandard tableau consisting of all rows of T below the top row. To form $T \leftarrow x$, first replace y_i by x in the top row of T. Then replace T' by $T' \leftarrow y_i$, which is computed recursively by the same algorithm.

If step 2b occurs, we say that x has *bumped* y_i out of row 1. In turn, y_i may bump an element from row 2 to row 3, and so on.

This recursive insertion algorithm always terminates, since the number of times we execute step 2b is at most $\ell(\mu)$, which is finite. We must also prove that the algorithm always produces a *semistandard tableau of partition shape*. We prove these facts after considering some examples.

9.47. Example. Let us compute $T \leftarrow 3$ for the following tableau T:

$$T = \begin{array}{lllllll} 1 & 1 & 2 & 3 & 4 & 4 & 6 \\ 2 & 4 & 5 & 6 & 6 & & \\ 3 & 5 & 7 & 8 & & & \\ 4 & 6 & & & & & \end{array}$$

We scan the top row of T from left to right, looking for the first entry *strictly larger* than 3. This entry is the 4 in the fifth box. In step 2b, the 3 bumps the 4 into the second row. The current situation looks like this:

$$\begin{array}{lllllll} 1 & 1 & 2 & 3 & \underline{3} & 4 & 6 \\ 2 & 4 & 5 & 6 & 6 & & \\ 3 & 5 & 7 & 8 & & & \\ 4 & 6 & & & & & \end{array} \qquad \begin{array}{l} \\ \leftarrow 4 \\ \\ \\ \end{array}$$

Now we scan the second row from left to right, looking for the first entry strictly larger than 4. It is the 5, so the 4 bumps the 5 into the third row:

$$\begin{array}{lllllll} 1 & 1 & 2 & 3 & \underline{3} & 4 & 6 \\ 2 & 4 & \underline{4} & 6 & 6 & & \\ 3 & 5 & 7 & 8 & & & \\ 4 & 6 & & & & & \end{array} \qquad \begin{array}{l} \\ \\ \leftarrow 5 \\ \\ \end{array}$$

Next, the 5 bumps the 7 into the fourth row:

$$\begin{array}{lllllll} 1 & 1 & 2 & 3 & \underline{3} & 4 & 6 \\ 2 & 4 & \underline{4} & 6 & 6 & & \\ 3 & 5 & \underline{5} & 8 & & & \\ 4 & 6 & & & & & \end{array} \qquad \begin{array}{l} \\ \\ \\ \leftarrow 7 \end{array}$$

Now, everything in the fourth row is weakly smaller than 7. So, as directed by step 2a, we insert 7 at the end of this row. The final tableau $T \leftarrow 3$ is therefore

$$\begin{array}{lllllll} 1 & 1 & 2 & 3 & \underline{3} & 4 & 6 \\ 2 & 4 & \underline{4} & 6 & 6 & & \\ 3 & 5 & \underline{5} & 8 & & & \\ 4 & 6 & \underline{7} & & & & \end{array}$$

We have underlined the entries of $T \leftarrow 3$ that were affected by the insertion process. These entries are the starting value $x = 3$ together with those entries that got bumped during the insertion. Call these entries the *bumping sequence*; in this example, the bumping sequence is $(3, 4, 5, 7)$. The sequence of boxes occupied by the bumping sequence is called the *bumping path*. The lowest box in the bumping path is called the *new box*. It is the only box in $T \leftarrow 3$ that was not present in the original diagram for T.

Here is a simpler exampler of tableau insertion:

$$T \leftarrow 6 \quad = \quad \begin{array}{llllllll} 1 & 1 & 2 & 3 & 4 & 4 & 6 & \underline{6} \\ 2 & 4 & 5 & 6 & 6 & & & \\ 3 & 5 & 7 & 8 & & & & \\ 4 & 6 & & & & & & \end{array}$$

The reader may check that:

$$T \leftarrow 1 \quad = \quad \begin{array}{|c|c|c|c|c|c|c|} \hline 1 & 1 & \underline{1} & 3 & 4 & 4 & 6 \\ \hline 2 & \underline{2} & 5 & 6 & 6 \\ \cline{1-5} 3 & \underline{4} & 7 & 8 \\ \cline{1-4} 4 & \underline{5} \\ \cline{1-2} \underline{6} \\ \cline{1-1} \end{array} \qquad T \leftarrow 0 \quad = \quad \begin{array}{|c|c|c|c|c|c|c|} \hline \underline{0} & 1 & 2 & 3 & 4 & 4 & 6 \\ \hline \underline{1} & 4 & 5 & 6 & 6 \\ \cline{1-5} \underline{2} & 5 & 7 & 8 \\ \cline{1-4} \underline{3} & 6 \\ \cline{1-2} \underline{4} \\ \cline{1-1} \end{array}$$

To prove that $T \leftarrow x$ is always a semistandard tableau of partition shape, we need the following result.

9.48. Theorem: Bumping Sequence and Bumping Path. Given a semistandard tableau T and element x, let $(x_1, x_2, \ldots, x_k)$ be the bumping sequence and let $((1, j_1), (2, j_2), \ldots, (k, j_k))$ be the bumping path arising in the computation of $T \leftarrow x$. Then $x = x_1 < x_2 < \cdots < x_k$ and $j_1 \geq j_2 \geq \cdots \geq j_k > 0$. So the bumping sequence *strictly increases* and the bumping path *moves weakly left* as it goes down.

Proof. By definition of the bumping sequence, $x = x_1$ and x_i bumps x_{i+1} from row i into row $i + 1$, for $1 \leq i < k$. By definition of bumping, x_i bumps an entry strictly larger than itself, so $x_i < x_{i+1}$ for all $i < k$. Next, consider what happens to x_{i+1} when it is bumped out of row i. Before being bumped, x_{i+1} occupied the cell (i, j_i). After being bumped, x_{i+1} will occupy the cell $(i + 1, j_{i+1})$, which is either an existing cell in row $i + 1$ of T, or a new cell at the end of this row. Consider the cell $(i + 1, j_i)$ directly below (i, j_i). If this cell is outside the shape of T, the previous observation shows that $(i + 1, j_{i+1})$ must be located weakly left of this cell, so that $j_{i+1} \leq j_i$. On the other hand, if $(i + 1, j_i)$ is part of T and contains some value z, then $x_{i+1} < z$ because T is semistandard. Now, x_{i+1} bumps the *leftmost* entry in row $i + 1$ that is strictly larger than x_{i+1}. Since z is such an entry, x_{i+1} bumps z or some entry to the left of z. In either case, we again have $j_{i+1} \leq j_i$. $\square$

9.49. Theorem: Output of a Tableau Insertion. If T is a semistandard tableau of shape μ, then $T \leftarrow x$ is a semistandard tableau whose shape is a partition diagram obtained by adding one new box to $\mathrm{dg}(\mu)$.

Proof. Let us first show that the shape of $T \leftarrow x$ is a partition diagram. This shape is obtained from $\mathrm{dg}(\mu)$ by adding one new box (the last box in the bumping path). If this new box is in the top row, then the resulting shape is a partition diagram, being the diagram of $(\mu_1 + 1, \mu_2, \mu_3, \ldots)$. Suppose the new box is in row $i > 1$. In this case, Theorem 9.48 shows that the new box is located weakly left of a box in the previous row that belongs to $\mathrm{dg}(\mu)$. This implies that $\mu_i < \mu_{i-1}$, so adding the new box to row i still gives a partition diagram.

Next we prove that each time an entry of T is bumped during the insertion of x, the resulting filling is still a semistandard tableau. Suppose, at some stage in the insertion process, that an element y bumps z out of the following configuration:

$$\begin{array}{|c|c|c|} \cline{2-2} \multicolumn{1}{c|}{} & a & \multicolumn{1}{c}{} \\ \hline b & z & c \\ \hline \multicolumn{1}{c|}{} & d & \multicolumn{1}{c}{} \\ \cline{2-2} \end{array}$$

(Some of the boxes containing a, b, c, d may be absent, in which case the following proof must be modified appropriately.) The original configuration is part of a semistandard tableau, so $b \leq z \leq c$ and $a < z < d$. Because y bumps z, z must be the first entry strictly larger than y in its row. This means that $b \leq y < z \leq c$, so replacing z by y still leaves a weakly increasing row. Does the column containing z still strictly increase after the bumping? On one hand, $y < d$, since $y < z < d$. On the other hand, if the box containing a exists (i.e., if z is below the top row), then y was the element bumped out of a's row. Since the bumping path moves weakly left, the original location of y must have been weakly right of z in the row above z. If y was directly above z, then a must have bumped y, and so $a < y$ by definition of bumping. Otherwise, y was located strictly to the right of a before y was bumped, so $a \leq y$. We cannot have $a = y$ in this situation, since otherwise a (or something to its left) would have been bumped instead of y. Thus, $a < y$ in all cases.

Finally, consider what happens at the end of the insertion process, when an element w is inserted in a new box at the end of a (possibly empty) row. This only happens when w weakly exceeds all entries in its row, so the row containing w is weakly increasing. There is no cell below w in this case. Repeating the argument at the end of the last paragraph, we see that w is strictly greater than the entry directly above it (if any). This completes the proof that $T \leftarrow x$ is a semistandard tableau. □

9.9 Reverse Insertion

Given the output $T \leftarrow x$ of a tableau insertion operation, it is generally not possible to determine what T and x were. However, if we also know the location of the new box created by this insertion, then we can recover T and x. More generally, we can start with any semistandard tableau S and any corner box of S, and *uninsert* the value in this box to obtain a semistandard tableau T and value x such that $S = T \leftarrow x$. (Here we do not assume in advance that S has the form $T \leftarrow x$.) This process is called *reverse tableau insertion.* Before giving the general definition, we look at some examples.

9.50. Example. Consider the following semistandard tableau S:

$$\begin{array}{|c|c|c|c|c|}\hline 1&1&2&2&4\\\hline 2&2&3&5\\\cline{1-4} 3&4&4&6\\\cline{1-4} 4&5\\\cline{1-2} 6&6\\\cline{1-2} 7&8\\\cline{1-2}\end{array}$$

There are three corner boxes whose removal from S still leaves a partition diagram; they are the boxes at the end of the first, third, and sixth rows. Removing the corner box in the top row, we have $S = T_1 \leftarrow 4$, where T_1 is this tableau:

$$\begin{array}{|c|c|c|c|}\hline 1&1&2&2\\\hline 2&2&3&5\\\hline 3&4&4&6\\\hline 4&5\\\cline{1-2} 6&6\\\cline{1-2} 7&8\\\cline{1-2}\end{array}$$

Suppose instead that we remove the 6 at the end of the third row of S. Reversing the bumping process, we see that 6 must have been bumped into the third row from the second row. What element bumped it? In this case, it is the 5 in the second row. In turn, the 5 must have originally resided in the first row, before being bumped into the second row by the 4. In summary, we have $S = T_2 \leftarrow \underline{4}$, where T_2 is this tableau:

1	1	2	2	$\underline{5}$
2	2	3	$\underline{6}$	
3	4	4		
4	5			
6	6			
7	8			

Here we have underlined the entries in the *reverse bumping sequence*, which occupy boxes of S in the *reverse bumping path.* Finally, consider what happens when we uninsert the 8 at the end of the last row of S. The 8 was bumped to its current location by one of the 6's in the previous row; it must have been bumped by the rightmost 6, since the 8 could not appear to the left of the 6 in the previous semistandard tableau. Next, the rightmost 6 was bumped by the 5 in row 4; the 5 was bumped by the rightmost 4 in row 3; and so on. In general, to determine which element in row i bumped some value z into row $i+1$, we look for the *rightmost* entry in row i that is *strictly less* than z. Continuing in this way, we discover that $S = T_3 \leftarrow \underline{2}$, where T_3 is the tableau shown here:

1	1	2	$\underline{3}$	4
2	2	$\underline{4}$	5	
3	4	$\underline{5}$	6	
4	$\underline{6}$			
6	$\underline{8}$			
7				

With these examples in hand, we are ready to give the general definition of reverse tableau insertion.

9.51. Definition: Reverse Tableau Insertion. Suppose S is a semistandard tableau of shape ν. A *corner box* of ν is a box $(i,j) \in \mathrm{dg}(\nu)$ such that $\mathrm{dg}(\nu) - \{(i,j)\}$ is still the diagram of some partition μ. Given S and a corner box (i,j) of ν, we define a tableau T and a value x as follows. We construct a *reverse bumping sequence* $(x_i, x_{i-1}, \ldots, x_1)$ and a *reverse bumping path* $((i,j_i),(i-1,j_{i-1}),\ldots,(1,j_1))$ as follows.

1. Set $j_i = j$ and $x_i = S(i,j)$, which is the value of S in the given corner box.
2. Once x_k and j_k have been found, for some $i \geq k > 1$, scan row $k-1$ of S for the rightmost entry that is strictly less than x_k. Define x_{k-1} to be this entry, and let j_{k-1} be the column in which this entry occurs.
3. At the end, let $x = x_1$, and let T be the tableau obtained by erasing box (i, j_i) from S and replacing the contents of box $(k-1, j_{k-1})$ by x_k for $i \geq k > 1$.

The next results show that reverse insertion really is the two-sided inverse of ordinary insertion (given knowledge of the location of the new box).

9.52. Theorem: Properties of Reverse Insertion. Suppose we perform reverse tableau insertion on S and (i,j) to obtain a filling T and value x. (a) The reverse bumping sequence satisfies $x_i > x_{i-1} > \cdots > x_1 = x$. (b) The reverse bumping path satisfies $j_i \leq j_{i-1} \leq \cdots \leq j_1$. (c) T is a semistandard tableau of partition shape. (d) $(T \leftarrow x) = S$.

Proof. Part (a) follows from the definition of x_{k-1} in the reverse insertion algorithm. Note that there does exist an entry in row $k-1$ strictly less than x_k, since the entry directly above x_k (in cell $(k-1, j_k)$ of S) is such an entry. This observation also shows that the rightmost entry strictly less than x_k in row $k-1$ occurs in column j_k or later, proving (b).

Part (c) follows from (a) and (b) by an argument similar to that given in Theorem 9.49; we ask the reader to fill in the details.

For part (d), consider the bumping sequence $(x'_1, x'_2, \ldots)$ and bumping path $((1, j'_1), (2, j'_2), \ldots)$ for the forward insertion $T \leftarrow x$. We have $x'_1 = x = x_1$ by definition. Recall that $x_1 = S(1, j_1)$ is the rightmost entry in row 1 of S that is strictly less than x_2, and $T(1, j_1) = x_2$ by definition of T. All other entries in row 1 are the same in S and T. Then $T(1, j_1) = x_2$ is the leftmost entry of row 1 of T strictly larger than x_1. So, in the insertion $T \leftarrow x$, x_1 bumps x_2 out of cell $(1, j_1)$. In particular, $j'_1 = j_1$ and $x'_2 = x_2$. Repeating this argument in each successive row, we see by induction that $x'_k = x_k$ and $j'_k = j_k$ for all k. At the end of the insertion, we have recovered the original tableau S. □

9.53. Theorem: Reversing Insertion. Suppose $S = (T \leftarrow x)$ for some semistandard tableau T and value x. Let (i, j) be the new box created by this insertion. If we perform reverse insertion on S starting with box (i, j), then we obtain the original T and x.

Proof. This can be proved by induction, showing step by step that the forward and reverse bumping paths and bumping sequences are the same. The reasoning is similar to part (d) of Theorem 9.52, so we ask the reader to supply the proof. □

The next theorem summarizes the results of the last two sections.

9.54. Theorem: Invertibility of Tableau Insertion. For a fixed partition μ and positive integer N, let $P(\mu)$ be the set of all partitions that can be obtained from μ by adding a single box at the end of some row. There exist mutually inverse bijections

$$I : \mathrm{SSYT}_N(\mu) \times [N] \to \bigcup_{\nu \in P(\mu)} \mathrm{SSYT}_N(\nu), \qquad R : \bigcup_{\nu \in P(\mu)} \mathrm{SSYT}_N(\nu) \to \mathrm{SSYT}_N(\mu) \times [N],$$

where $I(T, x)$ is $T \leftarrow x$, and $R(S)$ is the result of applying reverse tableau insertion to S starting at the unique box of S not in μ. A similar statement holds for tableaux using the alphabet $\mathbb{Z}_{>0}$.

Proof. We have seen that I and R are well-defined functions mapping into the stated codomains. Theorem 9.52(d) says that $I \circ R$ is the identity map on $\bigcup_{\nu \in P(\mu)} \mathrm{SSYT}_N(\nu)$, while Theorem 9.53 says that $R \circ I$ is the identity map on $\mathrm{SSYT}_N(\mu) \times [N]$. Hence I and R are bijections. □

We can regard the set $[N] = \{1, 2, \ldots, N\}$ as a weighted set with $\mathrm{wt}(i) = x_i$ for $1 \leq i \leq N$. The generating function for this weighted set is $x_1 + x_2 + \cdots + x_N = h_1(x_1, \ldots, x_N) = s_{(1)}(x_1, \ldots, x_N) = e_1(x_1, \ldots, x_N) = p_1(x_1, \ldots, x_N)$. The content monomial $\mathbf{x}^{T \leftarrow j}$ is $\mathbf{x}^T x_j$, since $T \leftarrow j$ contains all the entries of T together with one new entry equal to j. This means that $\mathrm{wt}(I(T, j)) = \mathrm{wt}(T)\,\mathrm{wt}(j)$, so that the bijection I in Theorem 9.54 is *weight-preserving*. Using the Product Rule for Weighted Sets and the definition of Schur polynomials, the generating function for the domain of I is $s_\mu(x_1, \ldots, x_N)h_1(x_1, \ldots, x_N)$. Using the Sum Rule for Weighted Sets, the generating function for the codomain of I is $\sum_{\nu \in P(\mu)} s_\nu(x_1, \ldots, x_N)$. In conclusion, the tableau insertion algorithms have furnished a combinatorial proof of the following multiplication rules:

$$s_\mu h_1 = s_\mu e_1 = s_\mu s_{(1)} = s_\mu p_1 = \sum_{\nu \in P(\mu)} s_\nu,$$

where we sum over all partitions ν obtained by adding *one* corner box to μ. We have discovered the simplest instance of the *Pieri Rules* mentioned at the beginning of §9.8.

9.10 The Bumping Comparison Theorem

We now extend the analysis of the previous section to prove the general Pieri Rules for expanding $s_\mu h_k$ and $s_\mu e_k$ in terms of Schur polynomials. The key idea is to see what happens when we successively insert k weakly increasing numbers (or k strictly decreasing numbers) into a semistandard tableau by repeated tableau insertion. We begin with some examples to build intuition.

9.55. Example. Let T be the semistandard tableau shown here:

$$\begin{array}{ccccc} 1&1&2&3&4\\ 2&3&3&4\\ 3&4&5&6\\ 5&5&6&7\\ 6 \end{array}$$

Let us compute the tableaux that result by successively inserting the weakly increasing sequence of values $2, 3, 3, 5$ into T:

$$T_1 = T \leftarrow 2 \quad = \quad \begin{array}{ccccc} 1&1&2&\underline{2}&4\\ 2&3&3&\underline{3}\\ 3&4&\underline{4}&6\\ 5&5&\underline{5}&7\\ 6&\underline{6} \end{array} \qquad T_2 = T_1 \leftarrow 3 \quad = \quad \begin{array}{ccccc} 1&1&2&2&\underline{3}\\ 2&3&3&3&\underline{4}\\ 3&4&4&6\\ 5&5&5&7\\ 6&6 \end{array}$$

$$T_3 = T_2 \leftarrow 3 \quad = \quad \begin{array}{cccccc} 1&1&2&2&3&\underline{3}\\ 2&3&3&3&4\\ 3&4&4&6\\ 5&5&5&7\\ 6&6 \end{array} \qquad T_4 = T_3 \leftarrow 5 \quad = \quad \begin{array}{ccccccc} 1&1&2&2&3&3&\underline{5}\\ 2&3&3&3&4\\ 3&4&4&6\\ 5&5&5&7\\ 6&6 \end{array}$$

Consider the four new boxes in the diagram of T_4 that are not in the diagram of T, which are marked by asterisks in the following picture:

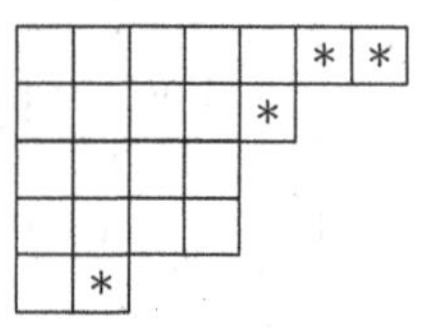

These boxes form what is called a *horizontal strip* of size 4, since no two new boxes are in the same column. Next, compare the bumping paths in the successive insertions of $2, 3, 3, 5$. We see that each path lies *strictly right* of the previous bumping path and ends with a new box in a *weakly higher* row.

Now return to the original tableau T, and consider the insertion of a strictly decreasing sequence $5, 4, 2, 1$. We obtain the following tableaux:

$$S_1 = T \leftarrow 5 \quad = \quad \begin{array}{cccccc} 1&1&2&3&4&\underline{5}\\ 2&3&3&4\\ 3&4&5&6\\ 5&5&6&7\\ 6 \end{array} \qquad S_2 = S_1 \leftarrow 4 \quad = \quad \begin{array}{cccccc} 1&1&2&3&4&\underline{4}\\ 2&3&3&4&\underline{5}\\ 3&4&5&6\\ 5&5&6&7\\ 6 \end{array}$$

$$S_3 = S_2 \leftarrow 2 \quad = \quad \begin{array}{cccccc} 1&1&2&\underline{2}&4&4\\ 2&3&3&\underline{3}&5\\ 3&4&\underline{4}&6\\ 5&5&\underline{5}&7\\ 6&\underline{6} \end{array} \qquad S_4 = S_3 \leftarrow 1 \quad = \quad \begin{array}{cccccc} 1&1&\underline{1}&2&4&4\\ 2&\underline{2}&3&3&5\\ 3&\underline{3}&4&6\\ \underline{4}&5&5&7\\ \underline{5}&6\\ \underline{6} \end{array}$$

This time, each successive bumping path is *weakly left* of the previous one and ends in a *strictly lower* row. Accordingly, the new boxes in S_4 form a *vertical strip*, where no two boxes occupy the same row:

					*
				*	
	*				
*					

We now show that the observations in this example hold in general.

9.56. The Bumping Comparison Theorem. Given a semistandard tableau T and values $x, y \in \mathbb{Z}$, let the new box in $T \leftarrow x$ be (i, j), and let the new box in $(T \leftarrow x) \leftarrow y$ be (r, s). (a) $x \leq y$ iff $i \geq r$ and $j < s$; (b) $x > y$ iff $i < r$ and $j \geq s$.

Proof. It suffices to prove the forward implications, since exactly one of $x \leq y$ or $x > y$ is true. Let the bumping path for the insertion of x be $((1, j_1), (2, j_2), \ldots, (i, j_i))$, where $j_i = j$, and let the bumping sequence be $(x = x_1, x_2, \ldots, x_i)$. Let the bumping path for the insertion of y be $((1, s_1), (2, s_2), \ldots, (r, s_r))$, where $s_r = s$, and let the bumping sequence be $(y = y_1, y_2, \ldots, y_r)$.

Assume $x \leq y$. We prove the following statement by induction: for all k with $1 \leq k \leq r$, we have $i \geq k$ and $x_k \leq y_k$ and $j_k < s_k$. When $k = 1$, we have $i \geq 1$ and $x_1 \leq y_1$ (by assumption). Note that x_1 appears in box $(1, j_1)$ of $T \leftarrow x$. We cannot have $s_1 \leq j_1$, for this would mean that y_1 bumps an entry weakly left of $(1, j_1)$, and this entry is at most $x_1 \leq y_1$, contrary to the definition of bumping. So $j_1 < s_1$. Now consider the induction step. Suppose $k < r$ and the induction hypothesis holds for k; does it hold for $k+1$? Since $k < r$, y_k must have bumped something from position (k, s_k) into the next row. Since $j_k < s_k$, x_k must also have bumped something out of row k, proving that $i \geq k + 1$. The object bumped by x_k, namely x_{k+1}, appears to the left of the object bumped by y_k, namely y_{k+1}, in the same row of a semistandard tableau. Therefore, $x_{k+1} \leq y_{k+1}$. Now we can repeat the argument used for the first row to see that $j_{k+1} < s_{k+1}$. Now that the induction is complete, take $k = r$ to see that $i \geq r$ and $j = j_i \leq j_r < s_r = s$ (the first inequality holding since the bumping path for $T \leftarrow x$ moves weakly left as we go down).

Next, assume $x > y$. This time we prove the following by induction: for all k with $1 \leq k \leq i$, we have $r > k$ and $x_k > y_k$ and $j_k \geq s_k$. When $k = 1$, we have $x_1 = x > y = y_1$. Since x appears somewhere in the first row of $T \leftarrow x$, y will necessarily bump something into the second row, so $r > 1$. In fact, the value bumped by y occurs weakly left of the position $(1, j_1)$ occupied by x, so $s_1 \leq j_1$. For the induction step, assume the induction hypothesis is known for some $k < i$, and try to prove it for $k + 1$. Since $k < i$ and $k < r$, both x_k and y_k must bump elements out of row k into row $k+1$. The element y_{k+1} bumped by y_k occurs in column s_k, which is weakly left of the cell (k, j_k) occupied by x_k in $T \leftarrow x$. Therefore, $y_{k+1} \leq x_k$, and x_k is *strictly* less than x_{k+1}, the original occupant of cell (k, j_k) in T. So $x_{k+1} > y_{k+1}$. Repeating the argument used in the first row for row $k + 1$, we now see that y_{k+1} must bump something in row $k + 1$ into row $k + 2$ (so that $r > k + 1$), and $s_{k+1} \leq j_{k+1}$. This completes the induction. Taking $k = i$, we finally conclude that $r > i$ and $j = j_i \geq s_i \geq s_r = s$. □

9.11 The Pieri Rules

To state the Pieri Rules, we first formally define horizontal and vertical strips.

9.57. Definition: Horizontal and Vertical Strips. A *horizontal strip* is a set of cells in $\mathbb{Z}^2_{>0}$, no two in the same column. A *vertical strip* is a set of cells in $\mathbb{Z}^2_{>0}$, no two in the same row.

9.58. Theorem: Inserting a Monotone Sequence into a Tableau. Let T be a semistandard tableau of shape μ, and let S be the semistandard tableau obtained from T by insertion of $z_1, z_2, \ldots, z_k$ in this order; we write $S = (T \leftarrow z_1 z_2 \cdots z_k)$ in this situation. Let ν be the shape of S. (a) If $z_1 \leq z_2 \leq \cdots \leq z_k$, then $\mathrm{dg}(\nu)$ is obtained from $\mathrm{dg}(\mu)$ by adding a horizontal strip of size k. (b) If $z_1 > z_2 > \cdots > z_k$, then $\mathrm{dg}(\nu)$ is obtained from $\mathrm{dg}(\mu)$ by adding a vertical strip of size k.

Proof. For (a), assume $z_1 \leq z_2 \leq \cdots \leq z_k$. By the Bumping Comparison Theorem, the new boxes $(i_1, j_1), \ldots, (i_k, j_k)$ created by the insertion of $z_1, \ldots, z_k$ satisfy $j_1 < j_2 < \cdots < j_k$. Thus, these boxes form a horizontal strip of size k. For (b), assume $z_1 > z_2 > \cdots > z_k$. In this case, the new boxes satisfy $i_1 < i_2 < \cdots < i_k$, so these boxes form a vertical strip of size k. □

Since tableau insertion is reversible given the location of the new box, we can also reverse the insertion of a monotone sequence, in the following sense.

9.59. Theorem: Reverse Insertion of a Monotone Sequence. Suppose μ and ν are given partitions, and S is any semistandard tableau of shape ν. (a) If $\mathrm{dg}(\nu) - \mathrm{dg}(\mu)$ is a horizontal strip of size k, then there exists a unique sequence $z_1 \leq z_2 \leq \cdots \leq z_k$ and a unique semistandard tableau T of shape μ such that $S = (T \leftarrow z_1 z_2 \cdots z_k)$. (b) If $\mathrm{dg}(\nu) - \mathrm{dg}(\mu)$ is a vertical strip of size k, then there exists a unique sequence $z_1 > z_2 > \cdots > z_k$ and a unique semistandard tableau T of shape μ such that $S = (T \leftarrow z_1 z_2 \cdots z_k)$.

Proof. To prove the existence of T and $z_1, \ldots, z_k$ in part (a), we repeatedly perform reverse tableau insertion, erasing each cell in the horizontal strip $\mathrm{dg}(\nu) - \mathrm{dg}(\mu)$ *from right to left.* This produces a sequence of elements $z_k, \ldots, z_2, z_1$ and a semistandard tableau T of shape μ such that $(T \leftarrow z_1 z_2 \cdots z_k) = S$. By comparing the relative locations of the new boxes created by z_i and z_{i+1}, we see from the Bumping Comparison Theorem that $z_i \leq z_{i+1}$ for all i.

As for uniqueness, suppose T' and $z'_1 \leq z'_2 \leq \cdots \leq z'_k$ also satisfy $S = (T' \leftarrow z'_1 z'_2 \cdots z'_k)$. Since $z'_1 \leq z'_2 \leq \cdots \leq z'_k$, the Bumping Comparison Theorem shows that the insertion of $z'_1, \ldots, z'_k$ creates the new boxes in ν in order from left to right, just as the insertion of $z_1, \ldots, z_k$ does. Write $T_0 = T$, $T_i = (T \leftarrow z_1 z_2 \cdots z_i)$, $T'_0 = T'$, and $T'_i = (T' \leftarrow z'_1 z'_2 \cdots z'_i)$. Since reverse tableau insertion produces a unique answer given the location of the new box, we see by reverse induction on i that $T_i = T'_i$ and $z_i = z'_i$ for $k \geq i \geq 0$.

Part (b) is proved similarly; here we erase cells in $\mathrm{dg}(\nu) - \mathrm{dg}(\mu)$ *from bottom to top.* □

9.60. Theorem: Pieri Rules. Given an integer partition μ and a positive integer k, let $H_k(\mu)$ consist of all partitions ν such that $\mathrm{dg}(\nu) - \mathrm{dg}(\mu)$ is a horizontal strip of size k, and let $V_k(\mu)$ consist of all partitions ν such that $\mathrm{dg}(\nu) - \mathrm{dg}(\mu)$ is a vertical strip of size k. For every $N > 0$, there are weight-preserving bijections

$$F : \mathrm{SSYT}_N(\mu) \times \mathrm{SSYT}_N((k)) \to \bigcup_{\nu \in H_k(\mu)} \mathrm{SSYT}_N(\nu);$$

$$G : \mathrm{SSYT}_N(\mu) \times \mathrm{SSYT}_N((1^k)) \to \bigcup_{\nu \in V_k(\mu)} \mathrm{SSYT}_N(\nu).$$

Consequently, we have the *Pieri Rules* in Λ_N:

$$s_\mu h_k = \sum_{\nu \in H_k(\mu)} s_\nu; \qquad s_\mu e_k = \sum_{\nu \in V_k(\mu)} s_\nu.$$

Proof. Recall that a semistandard tableau of shape (k) can be identified with a weakly increasing sequence $z_1 \leq z_2 \leq \cdots \leq z_k$ of elements of $[N]$. So, we can define $F(T, z_1 z_2 \ldots z_k) = (T \leftarrow z_1 z_2 \cdots z_k)$. By Theorem 9.58, F does map into the stated codomain. Theorem 9.59 shows that F is a bijection. Moreover, F is weight-preserving, since the content monomial of $(T \leftarrow z_1 z_2 \cdots z_k)$ is $\mathbf{x}^T x_{z_1} x_{z_2} \cdots x_{z_k}$.

Similarly, a semistandard tableau of shape (1^k) can be identified with a strictly increasing sequence $y_1 < y_2 < \cdots < y_k$. Reversing this gives a strictly decreasing sequence. So we define $G(T, y_1 y_2 \ldots y_k) = (T \leftarrow y_k \cdots y_2 y_1)$. As above, Theorems 9.58 and 9.59 show that G is a well-defined bijection.

Finally, the Pieri Rules follow by passing from weighted sets to generating functions, applying the Sum and Product Rules for Weighted Sets, and using $h_k = s_{(k)}$ and $e_k = s_{(1^k)}$. □

9.61. Example. We have

$$s_{(4,3,1)} h_2 = s_{(6,3,1)} + s_{(5,4,1)} + s_{(5,3,2)} + s_{(5,3,1,1)} + s_{(4,4,2)} + s_{(4,4,1,1)} + s_{(4,3,3)} + s_{(4,3,2,1)},$$

as we see by drawing the following diagrams:

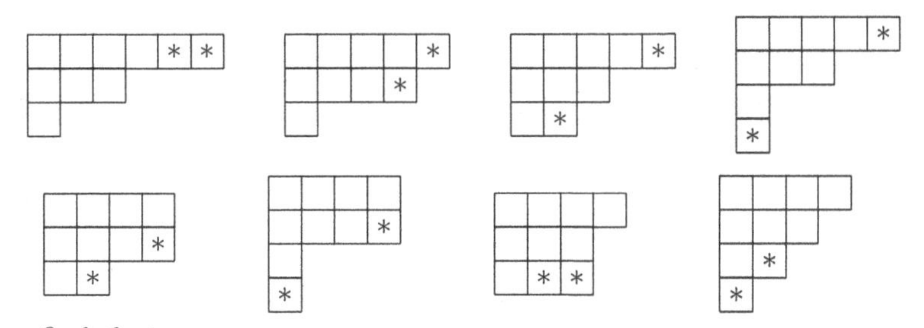

Similarly, we find that

$$s_{(2,2)} e_3 = s_{(2,2,1,1,1)} + s_{(3,2,1,1)} + s_{(3,3,1)}$$

by adding vertical strips to $\mathrm{dg}(2,2)$ as shown here:

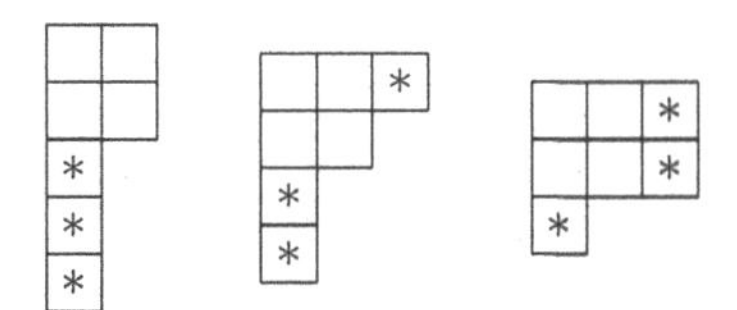

9.12 Schur Expansion of h_α

Iteration of the Pieri Rules lets us compute the Schur expansions of products of the form $s_\mu h_{\alpha_1} h_{\alpha_2} \cdots h_{\alpha_s}$, or $s_\mu e_{\alpha_1} e_{\alpha_2} \cdots e_{\alpha_s}$, or even mixed products involving both h's and e's. Taking $\mu = 0$, so that $s_\mu = 1$, we obtain in particular the expansions of h_α and e_α into sums of Schur polynomials. As we will see, examination of these expansions leads to another appearance of the Kostka matrix.

9.62. Example. Let us use the Pieri Rule to find the Schur expansion of $h_{(2,1,3)} = h_2h_1h_3$. To start, recall that $h_2 = s_{(2)}$. Adding one box to dg(2) in all possible ways gives

$$h_2h_1 = s_{(3)} + s_{(2,1)}.$$

Now we add a horizontal strip of size 3 in all possible ways to get

$$\begin{aligned} h_2h_1h_3 &= s_{(3)}h_3 + s_{(2,1)}h_3 \\ &= [s_{(6)} + s_{(5,1)} + s_{(4,2)} + s_{(3,3)}] + [s_{(5,1)} + s_{(4,2)} + s_{(4,1,1)} + s_{(3,2,1)}] \\ &= s_{(6)} + 2s_{(5,1)} + 2s_{(4,2)} + s_{(4,1,1)} + s_{(3,3)} + s_{(3,2,1)}. \end{aligned}$$

Observe that each of the Schur polynomials $s_{(5,1)}$ and $s_{(4,2)}$ occurs twice in the final expansion. Now, consider the computation of $h_{(2,3,1)} = h_2h_3h_1$. Since multiplication of polynomials is commutative, this symmetric polynomial must be the same as $h_{(2,1,3)}$. But the computations with the Pieri Rule involve different intermediate objects. We initially calculate

$$h_2h_3 = s_{(2)}h_3 = s_{(5)} + s_{(4,1)} + s_{(3,2)}.$$

Multiplying this expression by h_1 gives

$$\begin{aligned} h_2h_3h_1 &= s_{(5)}h_1 + s_{(4,1)}h_1 + s_{(3,2)}h_1 \\ &= [s_{(6)} + s_{(5,1)}] + [s_{(5,1)} + s_{(4,2)} + s_{(4,1,1)}] + [s_{(4,2)} + s_{(3,3)} + s_{(3,2,1)}], \end{aligned}$$

which is the same as the previous answer after collecting terms. As an exercise, the reader may compute $h_{(3,2,1)} = h_3h_2h_1$ and verify that the final answer is again the same.

9.63. Example. We have seen that a given Schur polynomial may appear several times in the Schur expansion of h_α. Is there some way to find the coefficient of a particular Schur polynomial in this expansion, without listing all the shapes generated by iteration of the Pieri Rule? To answer this question, consider the problem of finding the coefficient of $s_{(5,4,3)}$ when $h_{(4,3,3,2)}$ is expanded into a sum of Schur polynomials. Consider the shapes that appear when we repeatedly use the Pieri Rule on the product $h_4h_3h_3h_2$. Initially, we have a single shape (4) corresponding to h_4. Next, we add a horizontal strip of size 3 in all possible ways. Then we add another horizontal strip of size 3 in all possible ways. Finally, we add a horizontal strip of size 2 in all possible ways. The coefficient we seek is *the number of ways that the shape* $(5,4,3)$ *can be built by making the ordered sequence of choices just described.* For example, here is one choice sequence that leads to the shape $(5,4,3)$:

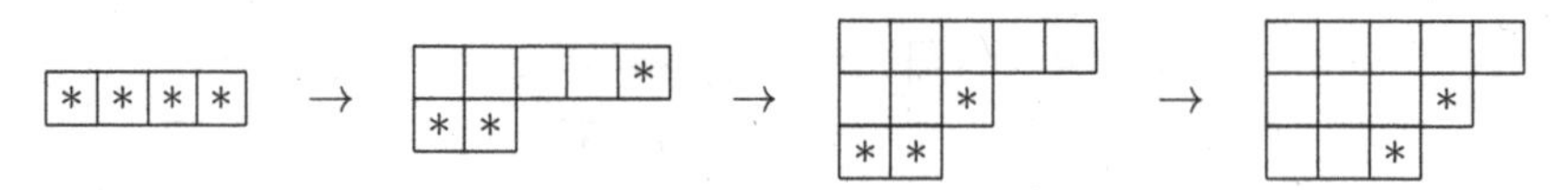

Here is a second choice sequence that leads to the same shape:

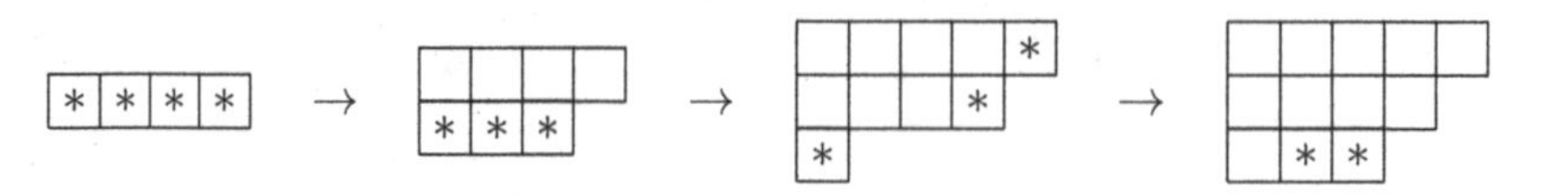

Here is a third choice sequence that leads to the same shape:

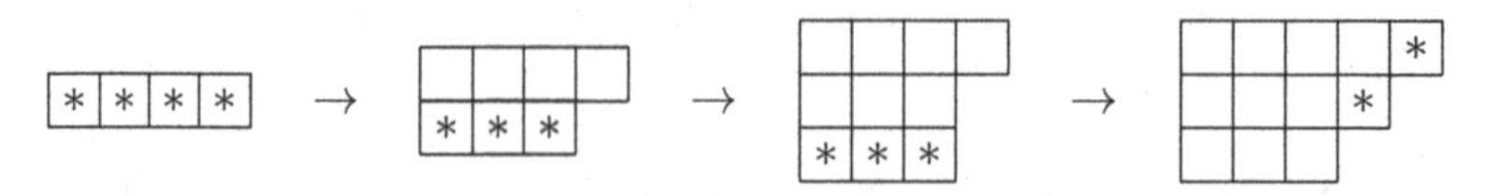

Now comes the key observation. We have exhibited each choice sequence by drawing a succession of shapes showing the addition of each new horizontal strip. The same information can be encoded by drawing *one* copy of the final shape $(5,4,3)$ and putting a label in each box to show which horizontal strip caused that box to first appear in the shape. For example, the three choice sequences displayed above are encoded (in order) by the following three objects:

1	1	1	1	2
2	2	3	4	
3	3	4		

1	1	1	1	3
2	2	2	3	
3	4	4		

1	1	1	1	4
2	2	2	4	
3	3	3		

These three objects are semistandard tableaux of shape $(5,4,3)$ and content $(4,3,3,2)$. By definition of the encoding just described, we see that every choice sequence under consideration is encoded by some filling of content $(4,3,3,2)$. Since we build the filling by adding horizontal strips one at a time using increasing labels, it follows that the filling we get is always a *semistandard tableau.* Finally, we can go backward in the sense that *any* semistandard tableau of content $(4,3,3,2)$ can be built uniquely by choosing a succession of horizontal strips telling us where the 1's, 2's, 3's, and 4's appear in the tableau. To summarize these remarks, our encoding scheme proves that *the coefficient of* $s_{(5,4,3)}$ *in the Schur expansion of* $h_{(4,3,3,2)}$ *is the number of semistandard tableaux of shape* $(5,4,3)$ *and content* $(4,3,3,2)$. In addition to the three semistandard tableaux already drawn, we have the following tableaux of this shape and content:

1	1	1	1	2
2	2	3	3	
3	4	4		

1	1	1	1	4
2	2	2	3	
3	3	4		

1	1	1	1	3
2	2	2	4	
3	3	4		

So the requested coefficient in this example is 6.

The reasoning in the last example generalizes to prove the following result.

9.64. Theorem: Schur Expansion of h_α. Let $\alpha = (\alpha_1, \alpha_2, \ldots, \alpha_s)$ be any sequence of nonnegative integers with sum k. Then

$$h_\alpha(x_1, \ldots, x_N) = \sum_{\lambda \in \operatorname{Par}_N(k)} K_{\lambda,\alpha} s_\lambda(x_1, \ldots, x_N).$$

(It is also permissible to sum over $\operatorname{Par}(k)$ here.)

Proof. By the Pieri Rule, the coefficient of s_λ in h_α is the number of sequences of partitions

$$0 = \mu^0 \subseteq \mu^1 \subseteq \mu^2 \subseteq \cdots \subseteq \mu^s = \lambda \tag{9.6}$$

such that $\operatorname{dg}(\mu^i) - \operatorname{dg}(\mu^{i-1})$ is a horizontal strip of size α_i, for $1 \leq i \leq s$. (This is a formal way of describing which horizontal strips we choose at each application of the Pieri Rule to the product h_α.) On the other hand, $K_{\lambda,\alpha}$ is the number of semistandard tableaux of shape λ and content α. There is a bijection between the sequences (9.6) and these tableaux, defined by filling each horizontal strip $\operatorname{dg}(\mu^i) - \operatorname{dg}(\mu^{i-1})$ with α_i copies of the letter i. The resulting filling has content α and is a semistandard tableau. The inverse map sends a semistandard tableau T to the sequence $(\mu^i : 0 \leq i \leq s)$, where $\operatorname{dg}(\mu^i)$ consists of the cells of T containing symbols in $\{1, 2, \ldots, i\}$. □

9.65. Remark. Suppose α, β are sequences such that $\operatorname{sort}(\alpha) = \operatorname{sort}(\beta)$. Note that $h_\alpha = h_\beta$ since multiplication of polynomials is commutative. Expanding each side into Schur polynomials gives

$$\sum_{\lambda \in \operatorname{Par}(k)} K_{\lambda,\alpha} s_\lambda(x_1, \ldots, x_N) = \sum_{\lambda \in \operatorname{Par}(k)} K_{\lambda,\beta} s_\lambda(x_1, \ldots, x_N).$$

For $N \geq k$, the Schur polynomials appearing here are linearly independent by Theorem 9.43. So $K_{\lambda,\alpha} = K_{\lambda,\beta}$ for all λ, in agreement with Theorem 9.27. (This remark leads to an algebraic proof of Theorem 9.27, provided one first gives an algebraic proof of the linear independence of Schur polynomials.)

9.66. Theorem: Complete Homogeneous Basis of Λ_N^k. For all $k \geq 0$ and $N > 0$,

$$\{h_\mu(x_1, \ldots, x_N) : \mu \in \text{Par}_N(k)\}$$

is a basis of the vector space Λ_N^k.

Proof. Consider the column vectors $\mathbf{S} = (s_\lambda(x_1, \ldots, x_N) : \lambda \in \text{Par}_N(k))$ and $\mathbf{H} = (h_\mu(x_1, \ldots, x_N) : \mu \in \text{Par}_N(k))$, where the entries are listed in lexicographic order. As in the proof of Theorem 9.43, let $\mathbf{K} = [K_{\lambda,\mu}]$ be the Kostka matrix with rows and columns indexed by partitions in $\text{Par}_N(k)$ in lexicographic order. Recall from Theorem 9.41 that $\mathbf{K}$ is a lower-triangular matrix with 1's on the main diagonal. In matrix notation, Theorem 9.64 asserts that $\mathbf{H} = \mathbf{K}^{\text{tr}}\mathbf{S}$, where $\mathbf{K}^{\text{tr}}$ is the transpose of the Kostka matrix. This transpose is upper-triangular with 1's on the main diagonal, hence is invertible. Since $\mathbf{H}$ is obtained from $\mathbf{S}$ by multiplying by an invertible matrix of scalars, we see that the elements of $\mathbf{H}$ form a basis by the same reasoning used in the proof of Theorem 9.43. □

9.67. Remark. Combining Theorems 9.66 and 9.43, we can write $\mathbf{H} = (\mathbf{K}^{\text{tr}}\mathbf{K})\mathbf{M}$, where $\mathbf{M}$ is the vector of monomial symmetric polynomials indexed by $\text{Par}_N(k)$. This matrix equation gives the monomial expansion of the complete homogeneous symmetric polynomials h_μ.

9.13 Schur Expansion of e_α

Now we turn to the elementary symmetric polynomials e_α. We can iterate the Pieri Rule as we did for h_α, but here we must add vertical strips at each stage.

9.68. Example. Let us compute the Schur expansion of $e_{(2,2,2)} = e_2e_2e_2$. First, $e_2e_2 = s_{(1,1)}e_2 = s_{(2,2)} + s_{(2,1,1)} + s_{(1,1,1,1)}$. Next,

$$\begin{aligned} e_2e_2e_2 &= [s_{(3,3)} + s_{(3,2,1)} + s_{(2,2,1,1)}] \\ &\quad + [s_{(3,2,1)} + s_{(3,1,1,1)} + s_{(2,2,2)} + s_{(2,2,1,1)} + s_{(2,1,1,1,1)}] \\ &\quad + [s_{(2,2,1,1)} + s_{(2,1,1,1,1)} + s_{(1,1,1,1,1,1)}] \\ &= s_{(3,3)} + 2s_{(3,2,1)} + s_{(3,1,1,1)} + s_{(2,2,2)} + 3s_{(2,2,1,1)} + 2s_{(2,1^4)} + s_{(1^6)}. \end{aligned}$$

As in the case of h_α, we can use fillings to encode the sequence of vertical strips chosen in the repeated application of the Pieri Rule. For example, the following fillings encode the three choice sequences that lead to the shape $(2,2,1,1)$ in the expansion of $e_{(2,2,2)}$:

$$\begin{array}{|c|c|}\hline 1&2\\\hline 1&2\\\hline 3\\\cline{1-1} 3\\\cline{1-1}\end{array} \quad \begin{array}{|c|c|}\hline 1&2\\\hline 1&3\\\hline 2\\\cline{1-1} 3\\\cline{1-1}\end{array} \quad \begin{array}{|c|c|}\hline 1&3\\\hline 1&3\\\hline 2\\\cline{1-1} 2\\\cline{1-1}\end{array}$$

We see at once that these fillings are not semistandard tableaux. However, transposing the diagrams will produce semistandard tableaux of shape $(2,2,1,1)' = (4,2)$ and content $(2,2,2)$, as shown here:

$$\begin{array}{|c|c|c|c|}\hline 1&1&3&3\\\hline 2&2\\\cline{1-2}\end{array} \quad \begin{array}{|c|c|c|c|}\hline 1&1&2&3\\\hline 2&3\\\cline{1-2}\end{array} \quad \begin{array}{|c|c|c|c|}\hline 1&1&2&2\\\hline 3&3\\\cline{1-2}\end{array}$$

This encoding gives a bijection from the relevant choice sequences to the collection of semistandard tableaux of this shape and content. So the coefficient of $s_{(2,2,1,1)}$ in the Schur expansion of $e_{(2,2,2)}$ is the Kostka number $K_{(4,2),(2,2,2)} = 3$. This reasoning generalizes to prove the following theorem.

9.69. Theorem: Schur Expansion of e_α. Let $\alpha = (\alpha_1, \alpha_2, \ldots, \alpha_s)$ be any sequence of nonnegative integers with sum k. Then

$$e_\alpha(x_1, \ldots, x_N) = \sum_{\lambda \in \mathrm{Par}_N(k)} K_{\lambda', \alpha} s_\lambda(x_1, \ldots, x_N) = \sum_{\nu \in \mathrm{Par}_N(k)'} K_{\nu, \alpha} s_{\nu'}(x_1, \ldots, x_N).$$

9.70. Remark. We have written $\mathrm{Par}_N(k)'$ for the set $\{\lambda' : \lambda \in \mathrm{Par}_N(k)\}$. Since conjugation of a partition interchanges the number of parts with the length of the largest part, we have

$$\mathrm{Par}_N(k)' = \{\nu \in \mathrm{Par}(k) : \nu_1 \leq N\} = \{\nu \in \mathrm{Par}(k) : \nu_i \leq N \text{ for all } i \geq 1\}.$$

It is also permissible to sum over all partitions of k in the theorem, since this only adds zero terms to the sum. If the number of variables is large enough ($N \geq k$), then we are already summing over all partitions of k.

9.71. Theorem: Elementary Basis of Λ_N^k. For all $k \geq 0$ and $N > 0$,

$$\{e_\mu(x_1, \ldots, x_N) : \mu \in \mathrm{Par}_N(k)'\} = \{e_{\mu'}(x_1, \ldots, x_N) : \mu \in \mathrm{Par}_N(k)\}$$

is a basis of the vector space Λ_N^k. Consequently, the set of all polynomials $e_1^{i_1} \cdots e_N^{i_N}$, where the i_j are arbitrary nonnegative integers, is a basis of Λ_N.

Proof. We use the same matrix argument employed earlier, adjusted to account for the shape conjugation. As in the past, let us index the rows and columns of matrices and vectors by the partitions in $\mathrm{Par}_N(k)$, listed in lexicographic order. Introduce column vectors $\mathbf{S} = (s_\lambda(x_1, \ldots, x_N) : \lambda \in \mathrm{Par}_N(k))$ and $\mathbf{E} = (e_{\mu'}(x_1, \ldots, x_N) : \mu \in \mathrm{Par}_N(k))$. Next, consider the modified Kostka matrix $\hat{\mathbf{K}}$ whose entry in row μ and column λ is $K_{\lambda', \mu'}$. Theorem 9.69 asserts that

$$e_{\mu'}(x_1, \ldots, x_N) = \sum_{\lambda \in \mathrm{Par}_N(k)} K_{\lambda', \mu'} s_\lambda(x_1, \ldots, x_N).$$

By definition of matrix-vector multiplication, the equations just written are equivalent to $\mathbf{E} = \hat{\mathbf{K}}\mathbf{S}$. Since the entries of $\mathbf{S}$ are known to be a basis, it suffices (as in the proofs of Theorems 9.43 and 9.66) to show that $\hat{\mathbf{K}}$ is a triangular matrix with 1's on the diagonal. There are 1's on the diagonal, since $K_{\mu', \mu'} = 1$. On the other hand, using Theorem 9.41, Theorem 9.39, and Theorem 9.35, we have the implications

$$\hat{\mathbf{K}}(\mu, \lambda) \neq 0 \Rightarrow K_{\lambda', \mu'} \neq 0 \Rightarrow \mu' \trianglelefteq \lambda' \Rightarrow \lambda \trianglelefteq \mu \Rightarrow \lambda \leq_{\mathrm{lex}} \mu.$$

So $\hat{\mathbf{K}}$ is lower-triangular.

Since the vector space Λ_N is the direct sum of its subspaces Λ_N^k for $k \geq 0$, we get a basis for Λ_N by combining bases for these subspaces. The resulting basis of Λ_N consists of all e_ν with $\nu \in \mathrm{Par}'_N$, meaning that $\nu_i \leq N$ for all i. If ν has i_j parts equal to j for $1 \leq j \leq N$, the definition of e_ν shows that $e_\nu = e_1^{i_1} e_2^{i_2} \cdots e_N^{i_N}$. Thus we obtain the basis of Λ_N in the theorem statement. □

9.72. Remark. Combining this theorem with Theorem 9.43, we can write $\mathbf{E} = (\hat{\mathbf{K}}\mathbf{K})\mathbf{M}$, where $\mathbf{M}$ is the vector of monomial symmetric polynomials indexed by $\mathrm{Par}_N(k)$. This matrix equation gives the monomial expansion of the elementary symmetric polynomials e_μ.

9.14 Algebraic Independence

We now use Theorem 9.71 to obtain structural information about the ring Λ_N of symmetric polynomials in N variables. First we need the following definition.

9.73. Definition: Algebraic Independence. Let $z_1, \ldots, z_m$ be a list of polynomials in $\mathbb{R}[x_1, \ldots, x_N]$. We say that this list is *algebraically independent* iff the collection of all monomials

$$\{\mathbf{z}^\alpha = z_1^{\alpha_1} z_2^{\alpha_2} \cdots z_N^{\alpha_N} : \alpha \in \mathbb{Z}_{\geq 0}^N\}$$

is linearly independent. This means that whenever a finite linear combination of the monomials $\mathbf{z}^\alpha$ is zero, say

$$\sum_{\alpha \in F} c_\alpha \mathbf{z}^\alpha = 0 \quad (F \text{ finite, each } c_\alpha \in \mathbb{R}),$$

then $c_\alpha = 0$ for all $\alpha \in F$.

9.74. Example. Consider the list $x_1, \ldots, x_N$ of all formal variables in the polynomial ring $\mathbb{R}[x_1, \ldots, x_N]$. By the very definition of (formal) polynomials, if $\sum_{\alpha \in F} c_\alpha \mathbf{x}^\alpha = 0$, then every c_α is zero. So the formal variables $x_1, \ldots, x_N$ are algebraically independent. On the other hand, consider the three polynomials $z_1 = x_1 + x_2$, $z_2 = x_1^2 + x_2^2$, and $z_3 = x_1^3 + x_2^3$ in $\mathbb{R}[x_1, x_2]$. The polynomials z_1, z_2, z_3 are *linearly* independent, as one may check. However, they are not *algebraically* independent, because of the relation

$$1z_1^3 - 3z_1z_2 + 2z_3 = 0.$$

Later, we show that z_1, z_2 is an algebraically independent list.

Here is a more sophisticated way of looking at algebraic independence. Suppose $z_1, \ldots, z_m$ is any list of polynomials in $B = \mathbb{R}[x_1, \ldots, x_N]$, and $A = \mathbb{R}[y_1, \ldots, y_m]$ is a polynomial ring in new formal variables $y_1, \ldots, y_m$. There is an algebra homomorphism $E : A \to B$ that sends $f(y_1, \ldots, y_m) \in A$ to $f(z_1, \ldots, z_m) \in B$. (This means that E preserves addition, ring multiplication, and multiplication by real scalars; see the Appendix for more discussion.) The map E is called an *evaluation homomorphism* because it acts on an input f by evaluating each formal variable y_i at the value z_i. On one hand, the *image* of the homomorphism E is the subalgebra of B generated by $z_1, \ldots, z_m$. On the other hand, the *kernel* of E consists of all polynomials $f \in A$ such that $E(f) = 0$. Writing $f = \sum_\alpha c_\alpha \mathbf{y}^\alpha$, we see that $E(f) = 0$ iff $\sum_\alpha c_\alpha \mathbf{z}^\alpha = 0$.

Now suppose the given list $z_1, \ldots, z_m$ is algebraically independent. This means that $\sum_\alpha c_\alpha \mathbf{z}^\alpha = 0$ implies that all c_α are 0, and hence $f = 0$. Thus, the kernel of E is zero, and E is therefore one-to-one. The reasoning is reversible, so we conclude that *algebraic independence of the list $z_1, \ldots, z_m$ is equivalent to injectivity of the evaluation homomorphism determined by this list.* In this case, E is an algebra *isomorphism* from the formal polynomial ring $\mathbb{R}[y_1, \ldots, y_m]$ onto the subalgebra of $\mathbb{R}[x_1, \ldots, x_N]$ generated by $z_1, \ldots, z_m$.

We can use these comments to rephrase Theorem 9.71. That theorem states that the set of all monomials $e_1^{i_1} \cdots e_N^{i_N}$ forms a *basis* for Λ_N, where e_j denotes the jth elementary symmetric polynomial in variables $x_1, \ldots, x_N$. On one hand, the fact that these monomials span Λ_N means that Λ_N is the subalgebra of $\mathbb{R}[x_1, \ldots, x_N]$ generated by $e_1, \ldots, e_N$. On the other hand, the linear independence of the monomials means that $e_1, \ldots, e_N$ is an algebraically independent list, so that the evaluation homomorphism E sending y_j to e_j is one-to-one. Thus, $E : \mathbb{R}[y_1, \ldots, y_N] \to \Lambda_N$ is an algebra isomorphism. The following theorem summarizes these structural results.

9.75. The Fundamental Theorem of Symmetric Polynomials. The elementary symmetric polynomials $\{e_j(x_1,\ldots,x_N) : 1 \leq j \leq N\}$ are algebraically independent. The evaluation map $E : \mathbb{R}[y_1,\ldots,y_N] \to \Lambda_N$ sending each y_j to e_j is an algebra isomorphism. So Λ_N is isomorphic to a polynomial ring in N variables. Moreover, for every symmetric polynomial $f(x_1,\ldots,x_N)$, there *exists* a *unique* polynomial $g(y_1,\ldots,y_N)$ such that $f = E(g) = g(e_1,\ldots,e_N)$.

9.15 Power-Sum Symmetric Polynomials

Recall that the *power-sum* symmetric polynomials in N variables are defined by $p_k(x_1,\ldots,x_N) = x_1^k + x_2^k + \cdots + x_N^k$ for all $k \geq 1$ and $p_\alpha(x_1,\ldots,x_N) = \prod_{j\geq 1} p_{\alpha_j}(x_1,\ldots,x_N)$. It turns out that the polynomials $p_1,\ldots,p_N$ are algebraically independent. One way to prove this is to invoke the following determinant criterion for algebraic independence.

9.76. Theorem: Determinant Test for Algebraic Independence. Given a list $g_1,\ldots,g_N$ of N polynomials in $\mathbb{R}[x_1,\ldots,x_N]$, let $\mathbf{A}$ be the $N \times N$ matrix whose j,k-entry is the partial derivative $\partial g_k/\partial x_j$, and let $J \in \mathbb{R}[x_1,\ldots,x_N]$ be the determinant of $\mathbf{A}$. If $J \neq 0$, then the list $g_1,\ldots,g_N$ is algebraically independent.

Proof. We prove the contrapositive. Assume the list $g_1,\ldots,g_N$ is algebraically dependent. Then there exist nonzero polynomials $h \in \mathbb{R}[y_1,\ldots,y_N]$ such that $h(g_1,\ldots,g_N) = 0$. Choose such an h whose degree (in the y-variables) is as small as possible. We can find the partial derivative of $h(g_1,\ldots,g_N)$ with respect to each variable x_j by applying the multivariable chain rule. We obtain the N equations

$$\sum_{k=1}^{N} \frac{\partial h}{\partial y_k}(g_1,\ldots,g_N)\frac{\partial g_k}{\partial x_j} = 0 \quad \text{for } 1 \leq j \leq N. \tag{9.7}$$

Let $\mathbf{v}$ be the column vector whose kth entry is $\frac{\partial h}{\partial y_k}(g_1,\ldots,g_N)$. The equations in (9.7) are equivalent to the matrix identity $\mathbf{Av} = \mathbf{0}$. We now show that $\mathbf{v}$ is not the zero vector.

Note that $h \in \mathbb{R}[y_1,\ldots,y_N]$ cannot be a constant polynomial, since $h \neq 0$ but $h(g_1,\ldots,g_N) = 0$. So at least one partial derivative of h must be a nonzero polynomial in $\mathbb{R}[y_1,\ldots,y_N]$. For any k with $\frac{\partial h}{\partial y_k} \neq 0$, the degree of $\frac{\partial h}{\partial y_k}$ in the y-variables is lower than the degree of h. By choice of h, it follows that $\frac{\partial h}{\partial y_k}(g_1,\ldots,g_N)$ is nonzero in $\mathbb{R}[y_1,\ldots,y_N]$. This polynomial is the kth entry of $\mathbf{v}$, so $\mathbf{v} \neq \mathbf{0}$. Now, $\mathbf{Av} = \mathbf{0}$ forces $\mathbf{A}$ to be non-invertible, so that $J = \det(\mathbf{A}) = 0$ by a theorem of linear algebra. □

9.77. Remark. The converse of Theorem 9.76 is also true: if the list $g_1,\ldots,g_N$ is algebraically independent, then the *Jacobian* J is nonzero. We do not need this fact, so we omit the proof; see [64, §3.10].

9.78. Theorem: Algebraic Independence of Power-Sums. The power-sum polynomials $\{p_k(x_1,\ldots,x_N) : 1 \leq k \leq N\}$ are algebraically independent.

Proof. We apply the determinant criterion from Theorem 9.76. The j,k-entry of the matrix $\mathbf{A}$ is

$$\frac{\partial p_k}{\partial x_j} = \frac{\partial}{\partial x_j}(x_1^k + x_2^k + \cdots + x_j^k + \cdots + x_N^k) = kx_j^{k-1}.$$

Therefore, $J = \det[kx_j^{k-1}]_{1\leq j,k\leq N}$. For each column k, we may factor out the scalar k to see that $J = N! \det[x_j^{k-1}]$. The resulting determinant (after reversing the order of the columns) is called a Vandermonde determinant. This determinant evaluates to $\pm \prod_{1 \leq r < s \leq N}(x_r - x_s)$, which is a nonzero polynomial (see §12.11 for a combinatorial proof of this formula). We conclude that $J \neq 0$, which proves the result. □

Now that we know that the list $p_1, \ldots, p_N$ is algebraically independent, we can obtain power-sum bases for the vector spaces Λ_N^k and Λ_N.

9.79. Theorem: Power-Sum Basis. For all $k \geq 0$ and $N > 0$,

$$\{p_\mu(x_1, \ldots, x_N) : \mu \in \text{Par}_N(k)'\}$$

is a basis of the vector space Λ_N^k. The collection $\{p_1^{i_1} \cdots p_N^{i_N} : i_1, \ldots, i_N \geq 0\}$ is a basis of the vector space Λ_N. The evaluation map $E : \mathbb{R}[y_1, \ldots, y_N] \to \Lambda_N$ sending each y_j to p_j is an algebra isomorphism. So, for every symmetric polynomial $f(x_1, \ldots, x_N)$, there *exists* a *unique* polynomial $g(y_1, \ldots, y_N)$ such that $f = E(g) = g(p_1, \ldots, p_N)$.

9.80. Remark. As mentioned earlier, everything said here is valid if we replace $\mathbb{R}$ by any field K containing $\mathbb{Q}$. But, the results in this section *fail* if we use coefficients from a field K of characteristic $p > 0$. For example, if $\text{char}(K) = 3$, then p_1, p_2, p_3 is an algebraically dependent list in $K[x_1, x_2, x_3]$. To see why, note that $g(y_1, y_2, y_3) = y_1^3 - y_3$ is nonzero in $K[y_1, y_2, y_3]$, but $g(p_1, p_2, p_3) = (x_1 + x_2 + x_3)^3 - (x_1^3 + x_2^3 + x_3^3)$ is zero in $K[x_1, x_2, x_3]$. This follows by expanding $(x_1 + x_2 + x_3)^3$ using the Multinomial Theorem and noting that all terms other than $x_1^3 + x_2^3 + x_3^3$ are multiples of $3 = 1_K + 1_K + 1_K = 0_K$.

9.16 Relations between e's and h's

We have seen that the lists $e_1, \ldots, e_N$ and $p_1, \ldots, p_N$ are algebraically independent in $\mathbb{R}[x_1, \ldots, x_N]$. The reader may wonder whether the list $h_1, \ldots, h_N$ is also algebraically independent. This fact would follow (as it did for $e_1, \ldots, e_N$) if we knew that $\{h_\mu : \mu \in \text{Par}_N(k)'\}$ was a basis of Λ_N^k for all $k \geq 0$. However, the basis we found in Theorem 9.66 was $\{h_\mu : \mu \in \text{Par}_N(k)\}$, which is indexed by partitions of k with at most N parts, instead of partitions of k with each part at most N. The next result allows us to overcome this difficulty by providing equations relating $e_1, \ldots, e_N$ to $h_1, \ldots, h_N$.

9.81. Theorem: Recursion involving e_i and h_j. For all $m \geq 0$ and $N > 0$,

$$\sum_{i=0}^{m} (-1)^i e_i(x_1, \ldots, x_N) h_{m-i}(x_1, \ldots, x_N) = \chi(m = 0) = \begin{cases} 1 & \text{if } m = 0; \\ 0 & \text{if } m > 0. \end{cases} \tag{9.8}$$

Proof. If $m = 0$, the identity becomes $1 = 1$, so let us assume $m > 0$. We can model the left side of the identity using a collection Z of signed weighted objects. A typical object in Z is a triple $z = (i, S, T)$, where $0 \leq i \leq m$, $S \in \text{SSYT}_N((1^i))$, and $T \in \text{SSYT}_N((m - i))$. The *weight* of (i, S, T) is $\mathbf{x}^S \mathbf{x}^T$, and the *sign* of (i, S, T) is $(-1)^i$. For example, taking $N = 9$ and $m = 7$, a typical object in Z is

$$z = \left(3, \begin{array}{|c|}\hline 2 \\ \hline 4 \\ \hline 7 \\ \hline\end{array}, \begin{array}{|c|c|c|c|}\hline 3 & 3 & 4 & 6 \\ \hline\end{array}\right).$$

The signed weight of this object is $(-1)^3(x_2x_4x_7)(x_3^2x_4x_6) = -x_2x_3^2x_4^2x_6x_7$. Recalling that $e_i = s_{(1^i)}$ and $h_{m-i} = s_{(m-i)}$, we see that the left side of (9.8) is precisely

$$\sum_{z\in Z} \operatorname{sgn}(z)\operatorname{wt}(z).$$

To prove this expression is zero, we define a sign-reversing, weight-preserving involution $I : Z \to Z$ with no fixed points. Given $z = (i, S, T) \in Z$, we compute $I(z)$ as follows. Let $j = S(1,1)$ be the smallest entry in S, and let $k = T(1,1)$ be the leftmost entry in T. If $i = 0$, then S is empty and j is undefined; if $i = m$, then T is empty and k is undefined. Since $m > 0$, at least one of j or k is defined. If $j \leq k$ or k is not defined, move the box containing j from S to T, so that this box is the new leftmost entry in T, and decrement i by 1. Otherwise, if $k < j$ or j is not defined, move the box containing k from T to S, so that this box is the new topmost box in S, and increment i by 1. For example, if z is the object shown above, then

$$I(z) = \left(2,\ \begin{array}{|c|}\hline 4\\\hline 7\\\hline\end{array},\ \begin{array}{|c|c|c|c|c|}\hline 2&3&3&4&6\\\hline\end{array}\right).$$

As another example,

$$I\left(0,\ \emptyset,\ \begin{array}{|c|c|c|c|c|c|c|}\hline 2&2&3&5&5&7&9\\\hline\end{array}\right) = \left(1,\ \begin{array}{|c|}\hline 2\\\hline\end{array},\ \begin{array}{|c|c|c|c|c|c|}\hline 2&3&5&5&7&9\\\hline\end{array}\right).$$

From the definition of I, we can check that I does map Z into Z, that $I \circ I = \mathrm{id}_Z$, that I is weight-preserving and sign-reversing, and that I has no fixed points. □

9.82. Theorem: Complete Homogeneous Basis. For all $k \geq 0$ and $N > 0$,

$$\{h_\mu(x_1, \ldots, x_N) : \mu \in \operatorname{Par}_N(k)'\}$$

is a basis of the vector space Λ_N^k. The collection $\{h_1^{i_1} \cdots h_N^{i_N} : i_1, \ldots, i_N \geq 0\}$ is a basis of the vector space Λ_N, and $h_1, \ldots, h_N$ is an algebraically independent list. The evaluation map $E : \mathbb{R}[y_1, \ldots, y_N] \to \Lambda_N$ sending each y_j to h_j is an algebra isomorphism. So, for every symmetric polynomial $f(x_1, \ldots, x_N)$, there *exists* a *unique* polynomial $g(y_1, \ldots, y_N)$ such that $f = E(g) = g(h_1, \ldots, h_N)$.

Proof. It suffices to prove the statement about the basis of Λ_N^k, from which the other assertions follow. Since $|\operatorname{Par}_N(k)'| = |\operatorname{Par}_N(k)|$, we have

$$|\{h_\mu(x_1, \ldots, x_N) : \mu \in \operatorname{Par}_N(k)'\}| \leq |\{h_\mu(x_1, \ldots, x_N) : \mu \in \operatorname{Par}_N(k)\}|,$$

where the right side is the dimension of Λ_N^k by Theorem 9.66. (Strict inequality could occur if $h_\mu = h_\nu$ for some $\mu \neq \nu$ in $\operatorname{Par}_N(k)'$.) By a theorem of linear algebra, it is enough to prove that $\{h_\mu : \mu \in \operatorname{Par}_N(k)'\}$ spans the entire subspace Λ_N^k. For each $k \geq 0$, let W_N^k be the vector subspace of Λ_N^k spanned by h_μ with $\mu \in \operatorname{Par}_N(k)'$. We must prove $W_N^k = \Lambda_N^k$ for all k. It suffices to show that $e_1^{i_1} \cdots e_N^{i_N} \in W_N^k$ for all $i_1, \ldots, i_N$ that sum to k, since these elementary symmetric polynomials are known to be a basis of Λ_N^k. Now, it is routine to check that $f \in W_N^k$ and $g \in W_N^m$ imply $fg \in W_N^{k+m}$. (This holds when f and g are products of $h_1, \ldots, h_N$, and the general case follows by linearity and the Distributive Law.) Using this fact, we can further reduce to proving that $e_j(x_1, \ldots, x_N) \in W_N^j$ for $1 \leq j \leq N$.

We prove this by induction on j. The result is true for $j = 1$, since $e_1 = \sum_{k=1}^N x_k = h_1 \in W_N^1$. Assume $1 < j \leq N$ and the result is known to hold for all smaller values of j. Taking $m = j$ in the recursion (9.8), we have

$$e_j = e_{j-1}h_1 - e_{j-2}h_2 + e_{j-3}h_3 - \cdots \pm e_1h_{j-1} \mp h_j.$$

Since $e_{j-s} \in W_N^{j-s}$ (by induction) and $h_s \in W_N^s$ (by definition) for $1 \leq s \leq j$, each term on the right side lies in W_N^j. Since W_N^j is a subspace, it follows that $e_j \in W_N^j$, completing the induction. □

9.17 Generating Functions for e's and h's

Another approach to the identity (9.8) involves generating functions.

9.83. Definition: $E_N(t)$ and $H_N(t)$. For each $N \geq 1$, define

$$E_N(t) = \prod_{i=1}^{N}(1 + x_i t), \qquad H_N(t) = \prod_{i=1}^{N} \frac{1}{1 - x_i t}.$$

9.84. Remark. E_N is an element of the polynomial ring $K[t]$, where $K = \mathbb{R}(x_1, \ldots, x_N)$ is the field of rational functions (formal ratios of polynomials) in $x_1, \ldots, x_N$. Similarly, H_N is an element of the formal power series ring $K[[t]]$.

9.85. Theorem: Expansion of $E_N(t)$. For all $N \geq 1$,

$$E_N(t) = \sum_{k=0}^{N} e_k(x_1, \ldots, x_N) t^k.$$

Proof. We can use the Generalized Distributive Law (see Exercises 2-16 and 4-68) to expand the product in the definition of $E_N(t)$. For each i between 1 and N, we choose either 1 or $x_i t$ from the factor $1 + x_i t$, and then multiply all these choices together. We can encode each choice sequence by a subset S of $\{1, \ldots, N\}$, where $i \in S$ iff $x_i t$ is chosen. Therefore,

$$E_N(t) = \prod_{i=1}^{N}(1 + x_i t) = \sum_{S \subseteq \{1,2,\ldots,N\}} \prod_{i \in S}(x_i t) \prod_{i \notin S} 1.$$

To get terms involving t^k, we must restrict the sum to subsets S of size k. Such subsets can be identified with increasing sequences $1 \leq i_1 < i_2 < \cdots < i_k \leq N$. Therefore, for all $k \geq 0$, the coefficient of t^k in $E_N(t)$ is

$$\sum_{1 \leq i_1 < i_2 < \cdots < i_k \leq N} x_{i_1} x_{i_2} \cdots x_{i_k} = e_k(x_1, \ldots, x_N).$$

Note that this coefficient is zero when $k > N$. □

Here is a famous algebraic application of elementary symmetric functions, which explains the relation between the roots and the coefficients of a polynomial in one variable.

9.86. Theorem: Roots and Coefficients of a Polynomial. Suppose a polynomial $p(X) = X^N + a_1 X^{N-1} + \cdots + a_i X^{N-i} + \cdots + a_{N-1} X + a_N \in \mathbb{R}[X]$ can be factored as $p(X) = (X - r_1)(X - r_2) \cdots (X - r_N)$ for some $r_1, \ldots, r_N \in \mathbb{R}$. For all i in the range $1 \leq i \leq N$,

$$a_i = (-1)^i e_i(r_1, \ldots, r_N).$$

Proof. This can be proved by expanding $\prod_{i=1}^{N}(X - r_i)$ using the Generalized Distributive Law, as in the preceding proof. Alternatively, we can deduce the result from Theorem 9.85 as follows. Replacing t by $1/X$ and x_i by $-r_i$ in $E_N(t)$ gives $\prod_{i=1}^{N}(1 - r_i/X) = X^{-N}p(X)$. Using Theorem 9.85, we conclude that

$$p(X) = X^N \sum_{k=0}^{n} e_k(-r_1, \ldots, -r_N)X^{-k} = \sum_{k=0}^{N}(-1)^k e_k(r_1, \ldots, r_N)X^{N-k}.$$

Taking the coefficient of X^{N-i} gives the result. □

9.87. Theorem: Expansion of $H_N(t)$. For all $N \geq 1$,

$$H_N(t) = \sum_{k=0}^{\infty} h_k(x_1, \ldots, x_N)t^k.$$

Proof. Using a formal version of the Geometric Series Formula (see §11.3 for a rigorous development), we have

$$H_N(t) = \prod_{i=1}^{N} \frac{1}{1 - x_i t} = \prod_{i=1}^{N} \sum_{j_i=0}^{\infty} (x_i t)^{j_i}.$$

Next, using a generalization of the distributive law to finite products of formal power series (see Theorem 5.35), we get

$$H_N(t) = \sum_{(j_1,\ldots,j_N)\in\mathbb{Z}_{\geq 0}^N} \prod_{i=1}^{N} x_i^{j_i} t^{j_i} = \sum_{(j_1,\ldots,j_N)\in\mathbb{Z}_{\geq 0}^N} t^{j_1+\cdots+j_N} x_1^{j_1} x_2^{j_2} \cdots x_N^{j_N}.$$

The coefficient of t^k consists of the sum of all possible monomials in $x_1, \ldots, x_N$ of degree k, which is precisely $h_k(x_1, \ldots, x_N)$. □

Now we can give an algebraic proof of Theorem 9.81. For each $N > 0$,

$$H_N(t)E_N(-t) = \prod_{i=1}^{N} \frac{1}{(1 - x_i t)} \prod_{i=1}^{N}(1 - x_i t) = 1. \tag{9.9}$$

Equating the coefficients of t^m on both sides gives (9.8).

9.18 Relations between p's, e's, and h's

In this section, we study recursions similar to (9.8) that relate the complete and elementary symmetric polynomials to the power-sum symmetric polynomials. These recursions can be used to deduce the algebraic independence of $p_1, \ldots, p_N$ from the algebraic independence of $h_1, \ldots, h_N$ (or $e_1, \ldots, e_N$) by adapting the proof of Theorem 9.82 to the new recursions.

9.88. Theorem: Recursion involving h_i and p_j. For all $n, N \geq 1$, the following identity is valid in Λ_N:

$$h_0 p_n + h_1 p_{n-1} + h_2 p_{n-2} + \cdots + h_{n-1} p_1 = n h_n. \tag{9.10}$$

Proof. Let us interpret each side of the equation as the generating function for a collection of weighted objects. For the left side, let X be the set of all triples (k, T, U), where: $0 \leq k < n$; $T \in \mathrm{SSYT}_N((k))$; and U consists of a row of $n - k$ boxes all filled with the same integer $i \in [N]$. The *weight* of such a triple is $\mathbf{x}^T\mathbf{x}^U = \mathbf{x}^T x_i^{n-k}$. For example, letting $n = 8$ and $N = 9$, here is a typical object in X of weight $x_1^2 x_2 x_3^3 x_4^2$:

$$z_0 = (5, \boxed{1|1|2|4|4}, \boxed{3|3|3}).$$

For a fixed value of k, the generating function for the possible T's is $h_k(x_1, \ldots, x_N)$, and the generating function for the possible U's is $p_{n-k}(x_1, \ldots, x_N)$. By the Sum and Product Rules for Weighted Sets, the left side of (9.10) is the generating function for X.

Now let Y be the set of all pairs (V, j), where $V \in \mathrm{SSYT}_N((n))$ and $1 \leq j \leq n$. We can visualize an object in Y as a semistandard tableau of shape (n) in which the jth cell has been *marked*. For example, here is a typical object in Y of weight $x_1^2 x_3^5 x_4$:

$$y_0 = \boxed{1|1|3|3^*|3|3|3|4}.$$

The generating function for the weighted set Y is $nh_n(x_1, \ldots, x_N)$.

To prove (9.10), it suffices to define a weight-preserving bijection $f : X \to Y$. Given $(k, T, U) \in X$, note that U consists of a run of $n - k$ copies of some value i. To compute $f(k, T, U)$, mark the first box in U and splice the boxes of U into T in the appropriate position to get a weakly increasing sequence. If T already contains one or more i's, the first box of U is inserted immediately after these i's. For example, the triple z_0 above maps to

$$f(z_0) = \boxed{1|1|2|3^*|3|3|4|4}.$$

This insertion process is reversible, thanks to the marker. More precisely, define $g : Y \to X$ as follows. Given $(V, j) \in Y$, let i be the entry in the jth cell of V. Starting at cell j and scanning right, remove each cell equal to i from V to get a pair of tableaux T and U as in the definition of X. Define $g(V, j) = (k, T, U)$, where k is the number of boxes in T. For example, the object y_0 above maps to

$$g(y_0) = (4, \boxed{1|1|3|4}, \boxed{3|3|3|3}).$$

One may check that f and g are weight-preserving functions that are two-sided inverses of each other. □

9.89. Theorem: Recursion involving e_i and p_j. For all $n, N \geq 1$, the following identity is valid in Λ_N:

$$e_0 p_n - e_1 p_{n-1} + e_2 p_{n-2} - \cdots + (-1)^{n-1} e_{n-1} p_1 = (-1)^{n-1} n e_n. \tag{9.11}$$

Proof. This time we interpret each side of the equation using signed weighted objects. For the left side, let X be the set of all triples (k, T, U), where: $0 \leq k < n$; $T \in \mathrm{SSYT}_N((1^k))$; and U consists of a row of $n - k$ boxes all filled with the same integer $j \in [N]$. The *weight* of this triple is $\mathbf{x}^T\mathbf{x}^U$, and the *sign* of this triple is $(-1)^k$. For example, here is a typical object in X whose signed weight is $(-1)^4 x_2 x_4^5 x_5 x_7$:

$$z_0 = \left(4, \begin{array}{|c|}\hline 2\\\hline 4\\\hline 5\\\hline 7\\\hline\end{array}, \boxed{4|4|4|4}\right).$$

Using the Sum and Product Rules for Weighted Sets, one sees that $\sum_{z \in X} \mathrm{sgn}(z)\,\mathrm{wt}(z)$ is the left side of (9.11).

Now let $Y = \{(T, j) : T \in \mathrm{SSYT}_N((1^n)), 1 \leq j \leq n\}$. We can think of each element of Y as a strictly increasing sequence of n elements of $\{1, 2, \ldots, N\}$ in which the jth element has been *marked*. The generating function for the weighted set Y is $ne_n(x_1, \ldots, x_N)$.

Let us define a weight-preserving, sign-reversing involution $I : X \to X$. Fix $(k, T, U) \in X$. Since $k < n$, U is not empty; let j be the integer appearing in each box of U. The map I acts as follows. On one hand, if $k < n - 1$ and j does not appear in T, then increase k by 1, remove one copy of j from U, and insert this number in the proper position in T to get a sorted sequence. On the other hand, if j does appear in T, then decrease k by 1, remove the unique copy of j from T, and place another copy of j in U. If neither of the two preceding cases occurs, (k, T, U) is a fixed point of I. For example,

$$I(z_0) = \left(3, \begin{array}{|c|}\hline 2 \\ \hline 5 \\ \hline 7 \\ \hline \end{array}, \begin{array}{|c|c|c|c|c|}\hline 4 & 4 & 4 & 4 & 4 \\ \hline \end{array}\right).$$

It can be checked that I is a well-defined, weight-preserving, sign-reversing involution on X.

Let Z be the set of fixed points of I. We see from the description of I that Z consists of all triples $(n-1, T, \boxed{j})$ where j does not appear in T. All of these triples have sign $(-1)^{n-1}$. The proof will be complete if we can find a weight-preserving bijection $g : Z \to Y$. We define g by inserting a marked copy of j into its proper position in the increasing sequence T. The inverse map takes an increasing sequence of size n with one marked element and removes the marked element. For example,

$$g\left(4, \begin{array}{|c|}\hline 2 \\ \hline 5 \\ \hline 7 \\ \hline 8 \\ \hline \end{array}, \boxed{3}\right) = \begin{array}{|c|}\hline 2 \\ \hline 3^* \\ \hline 5 \\ \hline 7 \\ \hline 8 \\ \hline \end{array} \quad \text{and} \quad g^{-1}\left(\begin{array}{|c|}\hline 1^* \\ \hline 4 \\ \hline 5 \\ \hline 6 \\ \hline 9 \\ \hline \end{array}\right) = \left(4, \begin{array}{|c|}\hline 4 \\ \hline 5 \\ \hline 6 \\ \hline 9 \\ \hline \end{array}, \boxed{1}\right). \qquad \square$$

9.19 Power-Sum Expansions of h_n and e_n

We can use the recursions in Theorems 9.88 and 9.89 to compute expansions for h_n and e_n in terms of the power-sum symmetric polynomials p_μ.

9.90. Example. We know that $h_0 = 1$ and $h_1 = p_1$. Next, since $h_0 p_2 + h_1 p_1 = 2h_2$, we find that $h_2 = (p_{(2)} + p_{(1,1)})/2$. For $n = 3$, we have

$$h_0 p_3 + h_1 p_2 + h_2 p_1 = 3h_3,$$

so that

$$h_3 = \frac{1}{3}\left(p_3 + p_1 p_2 + \left[\frac{p_2 + p_1^2}{2}\right] p_1\right) = (1/3)p_{(3)} + (1/2)p_{(2,1)} + (1/6)p_{(1,1,1)}.$$

For $n = 4$, we use the relation

$$h_0 p_4 + h_1 p_3 + h_2 p_2 + h_3 p_1 = 4h_4$$

to find, after some calculations,

$$h_4 = (1/4)p_{(4)} + (1/3)p_{(3,1)} + (1/8)p_{(2,2)} + (1/4)p_{(2,1,1)} + (1/24)p_{(1,1,1,1)}.$$

We can eliminate the fractions in the formula for h_n by multiplying both sides by $n!$. For instance,

$$\begin{aligned} 3!h_3 &= 2p_{(3)} + 3p_{(2,1)} + 1p_{(1,1,1)}; \\ 4!h_4 &= 6p_{(4)} + 8p_{(3,1)} + 3p_{(2,2)} + 6p_{(2,1,1)} + 1p_{(1,1,1,1)}. \end{aligned}$$

Similar formulas can be derived for $n!e_n$, but here some signs occur. For instance, calculations with (9.11) lead to the identities

$$\begin{aligned} 3!e_3 &= 2p_{(3)} - 3p_{(2,1)} + 1p_{(1,1,1)}; \\ 4!e_4 &= -6p_{(4)} + 8p_{(3,1)} + 3p_{(2,2)} - 6p_{(2,1,1)} + 1p_{(1,1,1,1)}. \end{aligned}$$

Notice that the coefficients in the power-sum expansion of $4!h_4$ match the entries in Table 7.1. This suggests the following result.

9.91. Theorem: Power-Sum Expansion of h_n. For all positive integers n and N, the following identity is valid in Λ_N:

$$n!h_n = \sum_{\mu \in \mathrm{Par}(n)} (n!/z_\mu) p_\mu. \tag{9.12}$$

Proof. Recall from Theorem 7.115 that $n!/z_\mu$ is the number of permutations $w \in S_n$ with cycle type μ. This leads to the following combinatorial interpretations for the two sides of (9.12). The left side counts all pairs (w, T), where $w = w_1 w_2 \cdots w_n \in S_n$ is a permutation written in *one-line form* and $T = (i_1 \leq i_2 \leq \cdots \leq i_n)$ is an element of $\mathrm{SSYT}_N((n))$. Let X be the set of all such pairs, with $\mathrm{wt}(w, T) = \mathbf{x}^T$. For example, here is a typical element of X when $n = 8$, written as a two-rowed array:

$$z_0 = \begin{bmatrix} w: & 4 & 2 & 5 & 8 & 3 & 7 & 1 & 6 \\ T: & 1 & 1 & 1 & 2 & 2 & 2 & 2 & 3 \end{bmatrix}.$$

The right side of (9.12) counts all triples (μ, σ, C), where $\mu \in \mathrm{Par}(n)$, $\sigma \in S_n$ is a permutation with cycle type μ, and $C : \{1, 2, \ldots, n\} \to \{1, 2, \ldots, N\}$ is a *coloring* of the numbers $1, \ldots, n$ using N available colors such that all elements in the same cycle of σ are assigned the same color (cf. §7.16). Let the *weight* of (μ, σ, C) be $\prod_{k=1}^{n} x_{C(k)}$, and let Y be the set of all such weighted triples. For example, a typical element of Y is shown here:

$$y_0 = \left((3,2,2,1),\ (1,6,3)(2,5)(7,4)(8),\ \begin{bmatrix} 1 & 2 & 3 & 4 & 5 & 6 & 7 & 8 \\ 3 & 2 & 3 & 3 & 2 & 3 & 3 & 3 \end{bmatrix} \right).$$

To see why the factor $p_\mu(x_1, \ldots, x_N)$ arises, consider how we may choose the coloring function C once μ and σ have been selected. We know σ is a product of cycles of lengths $\mu_1, \mu_2, \ldots, \mu_l$. Choose the common color of the elements in the first cycle in any of N ways. Since μ_1 elements all receive the same color, the generating function for this choice is $x_1^{\mu_1} + x_2^{\mu_1} + \cdots + x_N^{\mu_1} = p_{\mu_1}(x_1, \ldots, x_N)$. Next, choose the common color of the elements in the second cycle, which gives a factor of p_{μ_2}, and so on. Multiplying the generating functions for these choices gives $p_\mu(x_1, \ldots, x_N)$.

To complete the proof, we define weight-preserving maps $f : Y \to X$ and $g : X \to Y$ that are inverses of each other. To understand the definition of f, recall that a given $\sigma \in S_n$ can be written in cycle notation in several different ways, since the cycles can be presented in any order, and elements within each cycle can be cyclically permuted. Given (μ, σ, C), we specify one particular cycle notation for σ that depends on C, as follows. First, cycles colored with smaller colors are written before cycles colored with larger colors. Second, elements within each cycle are cyclically shifted so that the first element in each cycle is the smallest element appearing in that cycle. Third, if several cycles have the same color, then these cycles are ordered so that their minimum elements decrease from left to right. For example, starting with the object y_0 above, we obtain the following cycle notation for σ: $(2,5)(8)(4,7)(1,6,3)$. Note that $(2,5)$ is written first because this cycle has color 2. The other cycles, which are all colored 3, are presented in the given order because $8 > 4 > 1$. Finally, to compute $f(\mu,\sigma,C)$, we erase the parentheses from the chosen cycle notation for σ and write the color $C(i)$ directly beneath each i in the resulting word. For example,

$$f(y_0) = \begin{bmatrix} w: & 2 & 5 & 8 & 4 & 7 & 1 & 6 & 3 \\ T: & 2 & 2 & 3 & 3 & 3 & 3 & 3 & 3 \end{bmatrix}.$$

It can be checked that f is well-defined, maps into X, and preserves weights.

Now consider how to define the inverse map $g : X \to Y$. Given $(w,T) \in X$ with $w = w_1 \cdots w_n$ and $T = i_1 \leq \cdots \leq i_n$, the coloring map C is defined by setting $C(w_j) = i_j$ for all j. To recover σ from w and T, we need to add parentheses to w to recreate the cycle notation satisfying the rules above. For each color i in turn, look at the substring of w consisting of the symbols located above the i's in T. Scan this substring from left to right, and begin a new cycle each time a number smaller than all preceding numbers in this substring is encountered. (The numbers that begin new cycles are called *left-to-right minima relative to color* i.) This procedure defines σ, and finally we set $\mu = \text{type}(\sigma)$. For example, for the object z_0 above,

$$g(z_0) = \left((2,2,1,1,1,1),(4)(2,5)(8)(3,7)(1)(6), \begin{bmatrix} 1 & 2 & 3 & 4 & 5 & 6 & 7 & 8 \\ 2 & 1 & 2 & 1 & 1 & 3 & 2 & 2 \end{bmatrix}\right).$$

We find that $g(f(y_0)) = y_0$ and $f(g(z_0)) = z_0$. The reader may similarly verify that $g \circ f = \text{id}_Y$ and $f \circ g = \text{id}_X$, so the proof is complete. □

Before considering the analogous theorem for e_n, we introduce the following notation.

9.92. Definition: The Sign Factor ϵ_μ. For every partition $\mu \in \text{Par}(n)$, let

$$\epsilon_\mu = (-1)^{n-\ell(\mu)} = \prod_{i=1}^{\ell(\mu)} (-1)^{\mu_i - 1}.$$

We proved in Theorem 7.34 that $\epsilon_\mu = \text{sgn}(\sigma)$ for all $\sigma \in S_n$ such that $\text{type}(\sigma) = \mu$.

9.93. Theorem: Power-Sum Expansion of e_n. For all positive integers n and N, the following identity is valid in Λ_N:

$$n!e_n = \sum_{\mu \in \text{Par}(n)} \epsilon_\mu (n!/z_\mu) p_\mu. \tag{9.13}$$

Proof. We use the notation X, Y, f, g, z_0, and y_0 from the proof of Theorem 9.91. We saw in that proof that $\sum_{\mu \in \text{Par}(n)} (n!/z_\mu) p_\mu$ is the generating function for the weighted set Y. To model the right side of (9.13), we need to get the sign factors ϵ_μ into this sum. We accomplish this by assigning *signs* to objects in Y as follows. Given $(\mu,\sigma,C) \in Y$, write σ in cycle notation as described previously. Attach a $+$ to the first (minimum) element of each cycle, and attach a $-$ to the remaining elements in each cycle. The overall sign of (μ,σ,C) is the product of these signs, which is $\prod_i (-1)^{\mu_i - 1} = \epsilon_\mu$. For example, the object y_0 considered previously is now written

$$y_0 = \left((3,2,2,1),\ (2^+,5^-)(8^+)(4^+,7^-)(1^+,6^-,3^-),\ \begin{bmatrix} 1 & 2 & 3 & 4 & 5 & 6 & 7 & 8 \\ 3 & 2 & 3 & 3 & 2 & 3 & 3 & 3 \end{bmatrix}\right);$$

the sign of this object is $(-1)^4 = +1$.

The next step is to transfer these signs to the objects in X using the weight-preserving bijection $f : Y \to X$. Given $(w,T) \in X$, find the left-to-right minima relative to each color i (as discussed in the definition of g in the proof of Theorem 9.91). Attach a $+$ to these numbers and a $-$ to all other numbers in w. For example, $f(y_0)$ is now written

$$f(y_0) = \begin{bmatrix} w: & 2^+ & 5^- & 8^+ & 4^+ & 7^- & 1^+ & 6^- & 3^- \\ T: & 2 & 2 & 3 & 3 & 3 & 3 & 3 & 3 \end{bmatrix}.$$

As another example,

$$z_0 = \begin{bmatrix} w: & 4^+ & 2^+ & 5^- & 8^+ & 3^+ & 7^- & 1^+ & 6^+ \\ T: & 1 & 1 & 1 & 2 & 2 & 2 & 2 & 3 \end{bmatrix}.$$

The bijections f and g now preserve both signs and weights, by the way we defined signs of objects in X. It follows that $\sum_{z\in X} \operatorname{sgn}(z)\operatorname{wt}(z) = \sum_{y\in Y} \operatorname{sgn}(y)\operatorname{wt}(y)$, and the sum over Y is the right side of (9.13).

Now we define a sign-reversing, weight-preserving involution $I : X \to X$. Fix $(w,T) \in X$. If all the entries of T are distinct, then (w,T) is a fixed point of I. We observe at once that such a fixed point is necessarily positive, and the generating function for such objects is $n!e_n(x_1,\ldots,x_N)$. On the other hand, suppose some color i appears more than one time in T. Choose the smallest color i with this property, and let w_k, w_{k+1} be the first two symbols in the substring of w located above this color. Define $I(w,T)$ by switching w_k and w_{k+1}; one checks that this is a weight-preserving involution. Furthermore, it can be verified that switching these two symbols changes the number of left-to-right minima (relative to color i) by exactly 1. For example,

$$I(f(y_0)) = \begin{bmatrix} w: & 5^+ & 2^+ & 8^+ & 4^+ & 7^- & 1^+ & 6^- & 3^- \\ T: & 2 & 2 & 3 & 3 & 3 & 3 & 3 & 3 \end{bmatrix}.$$

As another example,

$$I(z_0) = \begin{bmatrix} w: & 2^+ & 4^- & 5^- & 8^+ & 3^+ & 7^- & 1^+ & 6^+ \\ T: & 1 & 1 & 1 & 2 & 2 & 2 & 2 & 3 \end{bmatrix}.$$

In general, note that w_k is always labeled by $+$; w_{k+1} is labeled $+$ iff $w_k > w_{k+1}$; and the signs attached to numbers following w_{k+1} do not depend on the order of the two symbols w_k, w_{k+1}. We have now shown that I is sign-reversing, so the proof is complete. □

9.20 The Involution ω

Recall from §9.15 that the algebra Λ_N is generated by the algebraically independent list of power-sums $p_1,\ldots,p_N$. So the polynomial ring $\mathbb{R}[y_1,\ldots,y_N]$ is isomorphic to Λ_N via the evaluation map sending y_j to $p_j(x_1,\ldots,x_N)$. Because of this isomorphism, we can forget about the original variables $x_1,\ldots,x_N$ and regard the symbols $p_1,\ldots,p_N$ as formal variables, writing $\Lambda_N = \mathbb{R}[p_1,\ldots,p_N]$. We can then define evaluation homomorphisms with domain Λ_N by sending each p_j to an arbitrarily chosen element a_j in a given algebra over $\mathbb{R}$. It follows that for all partitions μ and polynomials f, p_μ maps to $\prod_i a_{\mu_i}$ and $f(p_1,\ldots,p_N)$ maps to $f(a_1,\ldots,a_N)$ under this homomorphism. The next definition uses this technique to define a homomorphism from Λ_N to itself.

9.94. Definition: The Map ω. Let $\omega : \Lambda_N \to \Lambda_N$ be the unique algebra homomorphism such that $\omega(p_j) = (-1)^{j-1}p_j$ for all j between 1 and n.

9.95. Theorem: Properties of ω. Let ν be a partition with $\nu_1 \leq N$. (a) $\omega \circ \omega = \mathrm{id}_{\Lambda_N}$, so ω is an isomorphism. (b) $\omega(p_\nu) = \epsilon_\nu p_\nu$. (c) $\omega(h_\nu) = e_\nu$. (d) $\omega(e_\nu) = h_\nu$.

Proof. (a) Observe that $\omega \circ \omega(p_j) = \omega((-1)^{j-1}p_j) = (-1)^{j-1}(-1)^{j-1}p_j = p_j = \mathrm{id}(p_j)$ for $1 \leq j \leq N$. Since $\omega \circ \omega$ and id are algebra homomorphisms with domain Λ_N that have the same effect on every generator p_j, these two maps are equal. Thus, ω has a two-sided inverse (namely, $\omega^{-1} = \omega$), so ω is an isomorphism.

(b) The homomorphism ω preserves multiplication, so

$$\omega(p_\nu) = \omega\left(\prod_{i=1}^{\ell(\nu)} p_{\nu_i}\right) = \prod_{i=1}^{\ell(\nu)} \omega(p_{\nu_i}) = \prod_{i=1}^{\ell(\nu)} (-1)^{\nu_i - 1} p_{\nu_i} = \epsilon_\nu p_\nu.$$

(c) For $1 \leq n \leq N$, we use Theorems 9.91 and 9.93 and part (b) to compute

$$\omega(h_n) = \omega\left(\sum_{\mu \in \mathrm{Par}(n)} z_\mu^{-1} p_\mu\right) = \sum_{\mu \in \mathrm{Par}(n)} z_\mu^{-1}\omega(p_\mu) = \sum_{\mu \in \mathrm{Par}(n)} \epsilon_\mu z_\mu^{-1} p_\mu = e_n.$$

Since ω preserves multiplication, $\omega(h_\nu) = e_\nu$ follows.

(d) Part (d) follows by applying ω to both sides of (c), since $\omega \circ \omega = \mathrm{id}$. □

The next theorem shows how ω acts on the Schur basis.

9.96. Theorem: Action of ω on s_λ. If $\lambda \in \mathrm{Par}(n)$ and $n \leq N$, then $\omega(s_\lambda) = s_{\lambda'}$ in Λ_N.

Proof. From Theorem 9.64, we know that for each $\mu \in \mathrm{Par}(n)$,

$$h_\mu(x_1, \ldots, x_N) = \sum_{\lambda \in \mathrm{Par}(n)} K_{\lambda,\mu} s_\lambda(x_1, \ldots, x_N).$$

We can combine these equations into a single vector equation $\mathbf{H} = \mathbf{K}^{\mathrm{tr}}\mathbf{S}$ using column vectors $\mathbf{H} = (h_\mu : \mu \in \mathrm{Par}(n))$ and $\mathbf{S} = (s_\lambda : \lambda \in \mathrm{Par}(n))$. Since $\mathbf{K}^{\mathrm{tr}}$ (the transpose of the Kostka matrix) is unitriangular and hence invertible, $\mathbf{S} = (\mathbf{K}^{\mathrm{tr}})^{-1}\mathbf{H}$ is the *unique* vector $\mathbf{v}$ satisfying $\mathbf{H} = \mathbf{K}^{\mathrm{tr}}\mathbf{v}$.

From Theorem 9.69, we know that for each $\mu \in \mathrm{Par}(n)$,

$$e_\mu(x_1, \ldots, x_N) = \sum_{\lambda \in \mathrm{Par}(n)} K_{\lambda,\mu} s_{\lambda'}(x_1, \ldots, x_N).$$

Applying the linear map ω to these equations produces the equations

$$h_\mu = \sum_{\lambda \in \mathrm{Par}(n)} K_{\lambda,\mu}\omega(s_{\lambda'}).$$

This says that the vector $\mathbf{v} = (\omega(s_{\lambda'}) : \lambda \in \mathrm{Par}(n))$ satisfies $\mathbf{H} = \mathbf{K}^{\mathrm{tr}}\mathbf{v}$. By the uniqueness property mentioned above, $\mathbf{v} = \mathbf{S}$. So, for all $\lambda \in \mathrm{Par}(n)$, $s_\lambda = \omega(s_{\lambda'})$. Replacing λ by λ' (or applying ω to both sides) gives the result. □

What happens if we apply ω to the monomial basis of Λ_N? Since ω is a linear map, we get another basis of Λ_N that turns out to be different from those discussed so far. This basis is hard to describe directly, so it is given the following name.

9.97. Definition: Forgotten Basis for Λ_N. For each $\lambda \in \mathrm{Par}_N$, define the *forgotten symmetric polynomial* $\mathrm{fgt}_\lambda = \omega(m_\lambda)$. The set $\{\mathrm{fgt}_\lambda : \lambda \in \mathrm{Par}_N(k)\}$ is a basis of Λ_N^k.

9.21 Permutations and Tableaux

Iteration of the tableau insertion algorithm (§9.8) leads to some remarkable bijections that map permutations, words, and matrices to certain pairs of tableaux. These bijections were studied by Robinson, Schensted, and Knuth, and are therefore called *RSK correspondences.* We begin in this section by showing how permutations can be encoded using pairs of standard tableaux of the same shape.

9.98. Theorem: RSK Correspondence for Permutations. For all $n > 0$, there is a bijection $\text{RSK} : S_n \to \bigcup_{\lambda\in\text{Par}(n)} \text{SYT}(\lambda) \times \text{SYT}(\lambda)$. Given $\text{RSK}(w) = (P(w), Q(w))$, we call $P(w)$ the *insertion tableau for* w and $Q(w)$ the *recording tableau for* w.

Proof. Let $w \in S_n$ have one-line form $w = w_1w_2\cdots w_n$. We construct a sequence of tableaux $P_0, P_1, \ldots, P_n = P(w)$ and a sequence of tableaux $Q_0, Q_1, \ldots, Q_n = Q(w)$ as follows. Initially, let P_0 and Q_0 be empty tableaux of shape (0). Suppose $1 \leq i \leq n$ and P_{i-1}, Q_{i-1} have already been constructed. Define $P_i = P_{i-1} \leftarrow w_i$ (the semistandard tableau obtained by insertion of w_i into P_i). Let (a, b) be the new cell in P_i created by this insertion. Define Q_i to be the filling obtained from Q_{i-1} by placing the value i in the new cell (a, b). Informally, we build $P(w)$ by inserting $w_1, \ldots, w_n$ (in this order) into an initially empty tableau. We build $Q(w)$ by placing the numbers $1, 2, \ldots, n$ (in this order) in the new boxes created by each insertion. By construction, $Q(w)$ has the same shape as $P(w)$. Furthermore, since the new box at each stage is a corner box, one sees that $Q(w)$ is a standard tableau. Since w is a permutation, the semistandard tableau $P(w)$ contains the values $1, 2, \ldots, n$ once each, so $P(w)$ is also a standard tableau. We define $\text{RSK}(w) = (P(w), Q(w))$.

To see that RSK is a bijection, we present an algorithm for computing the inverse map. Let (P, Q) be any pair of standard tableaux of the same shape $\lambda \in \text{Par}(n)$. The idea is to recover the one-line form $w_1 \cdots w_n$ in reverse by uninserting entries from P, using the entries in Q to decide which box to remove at each stage (cf. §9.9). To begin, note that n occurs in some corner box (a, b) of Q (since Q is standard). Apply reverse insertion to P starting at (a, b) to obtain the unique tableau P_{n-1} and value w_n such that $P_{n-1} \leftarrow w_n$ is P with new box (a, b) (see Theorem 9.54). Let Q_{n-1} be the tableau obtained by erasing n from Q. Continue similarly: having computed P_i and Q_i such that Q_i is a standard tableau with i cells, let (a, b) be the corner box of Q_i containing i. Apply reverse insertion to P_i starting at (a, b) to obtain P_{i-1} and w_i. Then delete i from Q_i to obtain a standard tableau Q_{i-1} with $i-1$ cells. The resulting word $w = w_1w_2\cdots w_n$ is a permutation of $\{1, 2, \ldots, n\}$ (since P contains each of these values exactly once), and our argument has shown that w is the *unique* object satisfying $\text{RSK}(w) = (P, Q)$. So RSK is a bijection. □

9.99. Example. Let $w = 35164872 \in S_8$. Figure 9.1 illustrates the computation of $\text{RSK}(w) = (P(w), Q(w))$. As an example of the inverse computation, let us determine the permutation $v = \text{RSK}^{-1}(Q(w), P(w))$ (note that we have switched the order of the insertion and recording tableaux). Figure 9.2 displays the reverse insertions used to find $v_n, v_{n-1}, \ldots, v_1$. We see that $v = 38152476$.

Let us compare the two-line forms of w and v:

$$w = \begin{bmatrix} 1 & 2 & 3 & 4 & 5 & 6 & 7 & 8 \\ 3 & 5 & 1 & 6 & 4 & 8 & 7 & 2 \end{bmatrix}; \qquad v = \begin{bmatrix} 1 & 2 & 3 & 4 & 5 & 6 & 7 & 8 \\ 3 & 8 & 1 & 5 & 2 & 4 & 7 & 6 \end{bmatrix}.$$

We notice that v and w are inverse permutations.

	Insertion Tableau	Recording Tableau
insert 3:	3	1
insert 5:	3 5	1 2
insert 1:	1 5 3	1 2 3
insert 6:	1 5 6 3	1 2 4 3
insert 4:	1 4 6 3 5	1 2 4 3 5
insert 8:	1 4 6 8 3 5	1 2 4 6 3 5
insert 7:	1 4 6 7 3 5 8	1 2 4 6 3 5 7
insert 2:	1 2 6 7 3 4 8 5	1 2 4 6 3 5 7 8

FIGURE 9.1
Computation of RSK(35164872).

	Insertion Tableau	Recording Tableau	Output Value
initial tableau:	1 2 4 6 3 5 $\underline{7}$ 8	1 2 6 7 3 4 $\underline{8}$ 5	
uninsert 7:	1 2 4 $\underline{7}$ 3 5 8	1 2 6 $\underline{7}$ 3 4 5	6
uninsert 7:	1 2 $\underline{4}$ 3 5 8	1 2 $\underline{6}$ 3 4 5	7
uninsert 4:	1 2 3 5 $\underline{8}$	1 2 3 4 $\underline{5}$	4
uninsert 8:	1 5 3 $\underline{8}$	1 2 3 $\underline{4}$	2
uninsert 8:	1 8 $\underline{3}$	1 2 $\underline{3}$	5
uninsert 3:	3 $\underline{8}$	1 $\underline{2}$	1
uninsert 8:	$\underline{3}$	$\underline{1}$	8
uninsert 3:	empty	empty	3

FIGURE 9.2
Mapping pairs of standard tableaux to permutations.

9.22 Inversion Property of RSK

The phenomenon observed in the last example holds in general: if w maps to (P,Q) under the RSK correspondence, then w^{-1} maps to (Q,P). To prove this fact, we must introduce a new way of visualizing the construction of the insertion and recording tableaux for w.

9.100. Definition: Cartesian Graph of a Permutation. Given a permutation $w = w_1w_2\cdots w_n \in S_n$, the graph of w (in the xy-plane) is the set $G(w) = \{(i,w_i) : 1 \le i \le n\}$.

For example, the graph of $w = 35164872$ is drawn in Figure 9.3.

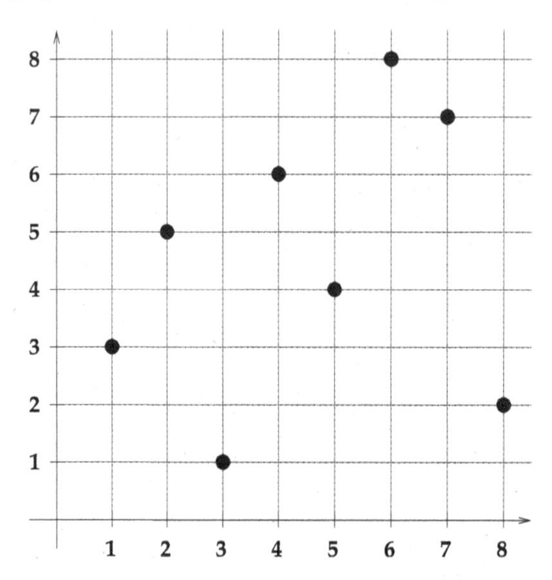

FIGURE 9.3
Cartesian graph of a permutation.

To analyze the creation of the insertion and recording tableaux for RSK(w), we annotate the graph of w by drawing lines as described in the following definitions.

9.101. Definition: Shadow Lines. Let $S = \{(x_1,y_1),\ldots,(x_k,y_k)\}$ be a finite set of points in the first quadrant. The *shadow* of S is

$$\text{Shd}(S) = \{(u,v) \in \mathbb{R}^2 : \text{for some } i, u \ge x_i \text{ and } v \ge y_i\}.$$

Informally, the shadow consists of all points northeast of some point in S. The *first shadow line* $L_1(S)$ is the boundary of Shd(S). This boundary consists of an infinite vertical ray (part of the line $x = a_1$, say), followed by zero or more alternating horizontal and vertical line segments, followed by an infinite horizontal ray (part of the line $y = b_1$, say). Call a_1 and b_1 the *x-coordinate* and *y-coordinate* associated to this shadow line. Next, let S_1 be the set of points in S that lie on the first shadow line of S. The *second shadow line* $L_2(S)$ is the boundary of Shd($S-S_1$), which has associated coordinates (a_2,b_2). Letting S_2 be the points in S that lie on the second shadow line, the *third shadow line* $L_3(S)$ is the boundary of Shd($S-(S_1\cup S_2)$). We continue to generate shadow lines in this way until all points of S lie on some shadow line. Finally, the *first-order shadow diagram of $w \in S_n$* consists of all shadow lines associated to the graph $G(w)$.

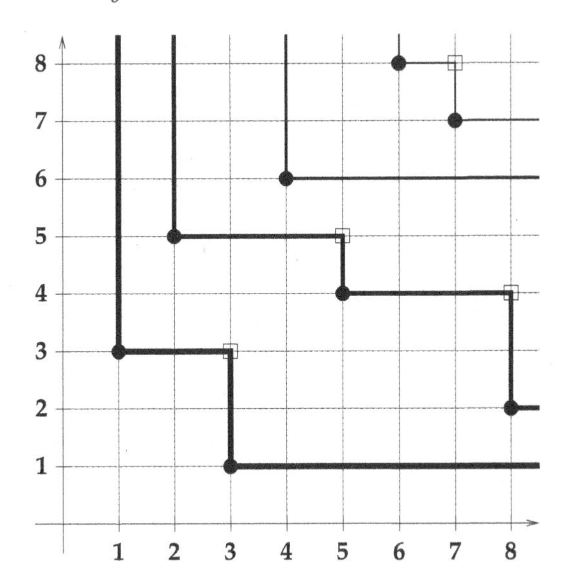

FIGURE 9.4
Shadow lines for a permutation graph.

9.102. Example. The first-order shadow diagram of $w = 35164872$ is drawn in Figure 9.4. The x-coordinates associated to the shadow lines of w are $1, 2, 4, 6$. These x-coordinates agree with the entries in the first row of the recording tableau $Q(w)$, which we computed in Example 9.21. Similarly, the y-coordinates of the shadow lines are $1, 2, 6, 7$, which are precisely the entries in the first row of the insertion tableau $P(w)$. The next result explains why this happens, and shows that the shadow diagram contains complete information about the evolution of the first rows of $P(w)$ and $Q(w)$.

9.103. Theorem: Shadow Lines and RSK. Let $w \in S_n$ have first-order shadow lines $L_1, \ldots, L_k$ with associated coordinates $(x_1, y_1), \ldots, (x_k, y_k)$. Let $P_0, P_1, \ldots, P_n = P(w)$ and $Q_0, Q_1, \ldots, Q_n = Q(w)$ be the sequences of tableaux generated in the computation of $\mathrm{RSK}(w)$. For $0 \leq i \leq n$, the y-coordinates of the intersections of the shadow lines with the line $x = i + (1/2)$ are the entries in the first row of P_i, whereas the entries in the first row of Q_i consist of all $x_j \leq i$. Whenever some shadow line L_r has a vertical segment from (i, a) down to (i, b), then $b = w_i$ and the insertion $P_i = P_{i-1} \leftarrow w_i$ bumps the value a out of the rth cell in the first row of P_{i-1}.

Proof. We proceed by induction on $i \geq 0$. The theorem holds when $i = 0$, since P_0 and Q_0 are empty, and no shadow lines intersect the line $x = 1/2$. Fix i between 1 and n, and assume the result holds for $i - 1$. Then the first row of P_{i-1} is $a_1 < a_2 < \cdots < a_j$, which are the y-coordinates where the shadow lines hit the line $x = i - 1/2$. Consider the point (i, w_i), which is the unique point in $G(w)$ on the line $x = i$. First consider the case $w_i > a_j$. In this case, the first j shadow lines all pass underneath (i, w_i). It follows that (i, w_i) is the first point of $G(w)$ on shadow line $L_{j+1}(G(w))$, so $x_{j+1} = i$. When we insert w_i into P_{i-1}, w_i goes at the end of the first row of P_{i-1} (since it exceeds the last entry a_j), and we place i in the corresponding cell in the first row of Q_i. The statements in the theorem regarding P_i and Q_i are true in this case. Now consider the case $w_i < a_j$. Suppose a_r is the smallest value in the first row of P_{i-1} exceeding w_i. Then insertion of w_i into P_{i-1} bumps a_r out of the first row. On the other hand, the point (i, w_i) lies between the points (i, a_{r-1}) and (i, a_r) in the shadow diagram (taking $a_0 = 0$). It follows from the way the shadow lines are

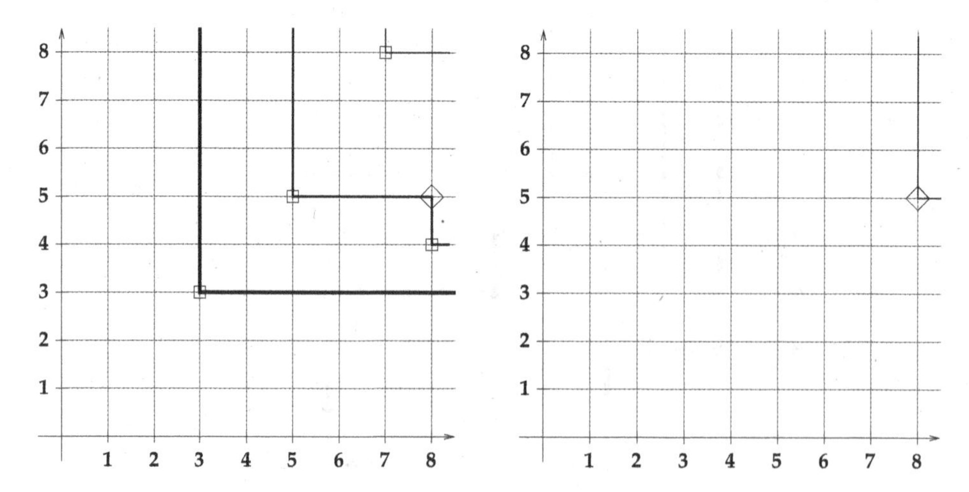

FIGURE 9.5
Higher-order shadow diagrams.

drawn that shadow line L_r must drop from (i, a_r) to (i, w_i) when it reaches the line $x = i$. The statements of the theorem therefore hold for i in this case as well. □

To analyze the rows of $P(w)$ and $Q(w)$ below the first row, we iterate the shadow diagram construction as follows.

9.104. Definition: Iterated Shadow Diagrams. Let $L_1, \ldots, L_k$ be the shadow lines associated to a given subset S of $\mathbb{R}^2$. An *inner corner* is a point (a, b) at the top of one of the vertical segments of some shadow line. Let S' be the set of inner corners associated to S. The *second-order shadow diagram* of S is the shadow diagram associated to S'. We iterate this process to define all higher-order shadow diagrams of S.

For example, taking $w = 35164872$, Figure 9.5 displays the second-order and third-order shadow diagrams for $G(w)$.

9.105. Theorem: Higher-Order Shadows and RSK. For $w \in S_n$, let $L_1, \ldots, L_k$ be the shadow lines in the rth-order shadow diagram for $G(w)$, with associated coordinates $(x_1, y_1), \ldots, (x_k, y_k)$. Let $P_0, P_1, \ldots, P_n = P(w)$ and $Q_0, Q_1, \ldots, Q_n = Q(w)$ be the sequences of tableaux generated in the computation of RSK(w). For $0 \leq i \leq n$, the y-coordinates of the intersections of the shadow lines with the line $x = i + (1/2)$ are the entries in the rth row of P_i, whereas the entries in the rth row of Q_i consist of all $x_j \leq i$. Whenever some shadow line L_c has a vertical segment from (i, a) down to (i, b), then b is the value bumped out of row $r - 1$ by the insertion $P_i = P_{i-1} \leftarrow w_i$, and b bumps the value a out of the cth cell in row r of P_{i-1}. (Take $b = w_i$ when $r = 1$.)

Proof. We use induction on $r \geq 1$. The base case $r = 1$ was proved in Theorem 9.103. Consider $r = 2$ next. The proof of Theorem 9.103 shows that the inner corners of the first-order shadow diagram of w are precisely those points (i, b) such that b is bumped out of the first row of P_{i-1} and inserted into the second row of P_{i-1} when forming P_i. The reasoning used in the previous proof can now be applied to this set of points. Whenever a point (i, b) lies above all second-order shadow lines approaching the line $x = i$ from the left, b gets inserted in a new cell at the end of the second row of P_i, and the corresponding cell in Q_i receives the label i. Otherwise, if (i, b) lies between shadow lines L_{c-1} and L_c in the second-order diagram, then b bumps the value in the cth cell of the second row of P_{i-1} into the third row, and shadow line L_c moves down to level b when it reaches $x = i$. The statements in the theorem (for $r = 2$) follow exactly as before by induction on i. Iterating this argument establishes the analogous results for each $r > 2$. □

9.106. Theorem: RSK and Inversion. For all $w \in S_n$, if $\mathrm{RSK}(w) = (P, Q)$, then $\mathrm{RSK}(w^{-1}) = (Q, P)$.

Proof. Consider the picture consisting of $G(w)$ and its first-order shadow diagram. Suppose the shadow lines have associated x-coordinates $(a_1, \ldots, a_k)$ and y-coordinates $(b_1, \ldots, b_k)$. Let us reflect the picture through the line $y = x$ (which interchanges x-coordinates and y-coordinates). This reflection changes $G(w)$ into $G(w^{-1})$, since $(x, y) \in G(w)$ iff $y = w(x)$ iff $x = w^{-1}(y)$ iff $(y, x) \in G(w^{-1})$. We see from the geometric definition that the shadow lines for w get reflected into the shadow lines for w^{-1}. It follows from Theorem 9.103 that the first row of both $Q(w)$ and $P(w^{-1})$ is $a_1, \ldots, a_k$, whereas the first row of both $P(w)$ and $Q(w^{-1})$ is $b_1, \ldots, b_k$. The inner corners for w^{-1} are the reflections of the inner corners for w. So, we can apply the same argument to the higher-order shadow diagrams of w and w^{-1}. It follows that each row of $P(w^{-1})$ matches the corresponding row of $Q(w)$, and similarly for $Q(w^{-1})$ and $P(w)$. □

9.23 Words and Tableaux

The RSK algorithm in the previous section sends permutations to pairs of standard tableaux of the same shape. We now extend this algorithm to operate on words w. Now the output $\mathrm{RSK}(w)$ is a pair of tableaux of the same shape, where the insertion tableau $P(w)$ is semistandard and the recording tableau $Q(w)$ is standard.

9.107. Theorem: RSK Correspondence for Words. Let $W = [N]^n$ be the set of n-letter words using the alphabet $[N]$. There is a bijection

$$\mathrm{RSK} : W \to \bigcup_{\lambda \in \mathrm{Par}(n)} \mathrm{SSYT}_N(\lambda) \times \mathrm{SYT}(\lambda).$$

For all $i \in [N]$, i occurs the same number of times in w and in $P(w)$.

Proof. Given $w = w_1 w_2 \cdots w_n \in W$, we define sequences of tableaux $P_0, P_1, \ldots, P_n$ and $Q_0, Q_1, \ldots, Q_n$ as follows. P_0 and Q_0 are the empty tableau. If P_{i-1} and Q_{i-1} have been computed for some i with $1 \leq i \leq n$, let $P_i = P_{i-1} \leftarrow w_i$. Suppose this insertion creates a new box (c, d); then we form Q_i from Q_{i-1} by placing the value i in the box (c, d). By induction on i, we see that every P_i is a semistandard tableau, every Q_i is a standard tableau, and P_i and Q_i have the same shape. We set $\mathrm{RSK}(w) = (P_n, Q_n)$. The letters in P_n (counting repetitions) are exactly the letters in w, so the last statement of the theorem holds.

Next we describe the inverse algorithm. Given (P, Q) with P semistandard and Q standard of the same shape, we construct semistandard tableaux $P_n, P_{n-1}, \ldots, P_0$, standard tableaux $Q_n, Q_{n-1}, \ldots, Q_0$, and letters $w_n, w_{n-1}, \ldots, w_1$ as follows. Initially, $P_n = P$ and $Q_n = Q$. Suppose, for some i with $1 \leq i \leq n$, that we have already constructed tableaux P_i and Q_i such that these tableaux have the same shape and consist of i boxes, P_i is semistandard, and Q_i is standard. The value i lies in a corner cell of Q_i; perform uninsertion starting from the same cell in P_i to get a smaller semistandard tableau P_{i-1} and a letter

w_i. Let Q_{i-1} be Q_i with the i erased. At the end, output the word $w_1w_2\cdots w_n$. Using Theorem 9.54 and induction, it can be checked that $w = w_1\cdots w_n$ is the unique word w with $\mathrm{RSK}(w) = (P, Q)$. So the RSK algorithm is a bijection. □

9.108. Example. Let $w = 21132131$. We compute $\mathrm{RSK}(w)$ in Figure 9.6.

	Insertion Tableau	Recording Tableau
insert 2:	2	1
insert 1:	1 2	1 2
insert 1:	1 1 2	1 3 2
insert 3:	1 1 3 2	1 3 4 2
insert 2:	1 1 2 2 3	1 3 4 2 5
insert 1:	1 1 1 2 2 3	1 3 4 2 5 6
insert 3:	1 1 1 3 2 2 3	1 3 4 7 2 5 6
insert 1:	1 1 1 1 2 2 3 3	1 3 4 7 2 5 8 6

FIGURE 9.6
Computation of RSK(21132131).

Next we investigate how the RSK algorithm is related to certain statistics on words and tableaux.

9.109. Definition: Descents and Major Index for Standard Tableaux. Let Q be a standard tableau with n cells. The *descent set of* Q, denoted $\mathrm{Des}(Q)$, is the set of all $k < n$ such that $k+1$ appears in a lower row of Q than k. The *descent count of* Q, denoted $\mathrm{des}(Q)$, is $|\mathrm{Des}(Q)|$. The *major index of* Q, denoted $\mathrm{maj}(Q)$, is $\sum_{k\in\mathrm{Des}(Q)} k$. (Compare to Definition 8.24, which gives the analogous definitions for words.)

9.110. Example. For the standard tableau $Q = Q(w)$ shown at the bottom of Figure 9.6, we have $\mathrm{Des}(Q) = \{1, 4, 5, 7\}$, $\mathrm{des}(Q) = 4$, and $\mathrm{maj}(Q) = 17$. Here, $w = 21132131$. Note that $\mathrm{Des}(w) = \{1, 4, 5, 7\}$, $\mathrm{des}(w) = 4$, and $\mathrm{maj}(w) = 17$. This is not a coincidence.

9.111. Theorem: RSK Preserves Descents and Major Index. For every word $w \in [N]^n$ with recording tableau $Q = Q(w)$, we have $\mathrm{Des}(w) = \mathrm{Des}(Q)$, $\mathrm{des}(w) = \mathrm{des}(Q)$, and $\mathrm{maj}(w) = \mathrm{maj}(Q)$.

Proof. It suffices to prove $\mathrm{Des}(w) = \mathrm{Des}(Q)$. Let $P_0, P_1, \ldots, P_n$ and $Q_0, Q_1, \ldots, Q_n = Q$

be the sequences of tableaux computed when we apply the RSK algorithm to w. For each $k < n$, note that $k \in \text{Des}(w)$ iff $w_k > w_{k+1}$, whereas $k \in \text{Des}(Q)$ iff $k+1$ appears in a row below k in Q. So, for each $k < n$, we must prove $w_k > w_{k+1}$ iff $k+1$ is in a row lower than k in Q. For this, we use the Bumping Comparison Theorem 9.56. Consider the double insertion $(P_{k-1} \leftarrow w_k) \leftarrow w_{k+1}$. Let the new box in $P_{k-1} \leftarrow w_k$ be (i,j), and let the new box in $(P_{k-1} \leftarrow w_k) \leftarrow w_{k+1}$ be (r,s). By definition of the recording tableau, $Q(i,j) = k$ and $Q(r,s) = k+1$. Now, if $w_k > w_{k+1}$, part 2 of Theorem 9.56 says that $i < r$ (and $j \geq s$). So $k+1$ appears in a lower row than k in Q. If instead $w_k \leq w_{k+1}$, part 1 of Theorem 9.56 says that $i \geq r$ (and $j < s$). So $k+1$ does not appear in a lower row than k in Q. □

Define the *weight* of a value $j \in [N]$ to be x_j, and define the *weight* of a word $w = w_1 w_2 \dots w_n \in [N]^n$ to be $\mathbf{x}^w = x_{w_1} x_{w_2} \cdots x_{w_n}$. Using the Product Rule for Weighted Sets, we find that

$$\sum_{w \in [N]^n} \text{wt}(w) = (x_1 + \cdots + x_N)^n = p_{(1^n)}(x_1, \dots, x_N) = h_{(1^n)}(x_1, \dots, x_N). \tag{9.14}$$

Recall that $h_{(1^n)} = \sum_{\lambda \in \text{Par}(n)} K_{\lambda,(1^n)} s_\lambda$, where $K_{\lambda,(1^n)} = |\,\text{SYT}(\lambda)|$. It follows that

$$p_{(1^n)} = (x_1 + \cdots + x_N)^n = \sum_{\lambda \in \text{Par}(n)} |\,\text{SYT}(\lambda)| s_\lambda(x_1, \dots, x_N).$$

Using the RSK algorithm and Theorem 9.111, we obtain the following t-analogue of this identity.

9.112. Theorem: Schur Expansion of Words Weighted by Major Index. For all positive integers n and N,

$$\sum_{w \in [N]^n} t^{\text{maj}(w)} \mathbf{x}^w = \sum_{\lambda \in \text{Par}(n)} \left(\sum_{Q \in \text{SYT}(\lambda)} t^{\text{maj}(Q)} \right) s_\lambda(x_1, \dots, x_N). \tag{9.15}$$

Proof. The left side of (9.15) is the generating function for the weighted set $[N]^n$, where the weight of a word w is $t^{\text{maj}(w)} \mathbf{x}^w$. On the other side, $\text{SSYT}_N(\lambda)$ is a weighted set with $\text{wt}(P) = \mathbf{x}^P$ for each semistandard tableau P. The generating function for this weighted set is precisely the Schur polynomial $s_\lambda(x_1, \dots, x_N)$. Next, define weights on the set $\text{SYT}(\lambda)$ by taking $\text{wt}(Q) = t^{\text{maj}(Q)}$ for $Q \in \text{SYT}(\lambda)$. By the Sum and Product Rules for Weighted Sets, the generating function for the weighted set $Z = \bigcup_{\lambda \in \text{Par}(n)} \text{SSYT}_N(\lambda) \times \text{SYT}(\lambda)$ is the right side of (9.15). To complete the proof, note that the RSK map is a weight-preserving bijection between $[N]^n$ and Z, because of Theorems 9.107 and 9.111. □

9.113. Remark. The RSK correspondence can also be used to find the length of the longest weakly increasing or strictly decreasing subsequence of a given word. For details, see §12.13.

9.24 Matrices and Tableaux

Performing the RSK map on a word produces a pair consisting of one semistandard tableau and one standard tableau. We now define an RSK operation on matrices that maps each matrix to a pair of semistandard tableaux of the same shape. The first step is to encode the matrix using an object called a biword.

9.114. Definition: Biword of a Matrix. Let $A = [a_{ij}]$ be an $M \times N$ matrix with entries in $\mathbb{Z}_{\geq 0}$. The *biword* of A is a two-row array

$$\text{bw}(A) = \begin{bmatrix} i_1 & i_2 & \cdots & i_k \\ j_1 & j_2 & \cdots & j_k \end{bmatrix}$$

constructed as follows. Start with an empty array, and scan the rows of A from top to bottom, reading each row from left to right. Whenever a nonzero integer a_{ij} is encountered in the scan, write down a_{ij} copies of the column $\begin{bmatrix} i \\ j \end{bmatrix}$ at the end of the current biword. The top row of bw(A) is called the *row word of* A and denoted $r(A)$. The bottom row of bw(A) is called the *column word of* A and denoted $c(A)$.

9.115. Example. Suppose A is the matrix

$$\begin{bmatrix} 2 & 0 & 1 & 0 \\ 0 & 1 & 0 & 3 \\ 0 & 0 & 1 & 0 \end{bmatrix}.$$

The biword of A is

$$\text{bw}(A) = \begin{bmatrix} 1 & 1 & 1 & 2 & 2 & 2 & 2 & 3 \\ 1 & 1 & 3 & 2 & 4 & 4 & 4 & 3 \end{bmatrix}.$$

9.116. Theorem: Matrices and Biwords. Let X be the set of all $M \times N$ matrices with entries in $\mathbb{Z}_{\geq 0}$. Let Y be the set of all biwords $w = \begin{bmatrix} i_1 & i_2 & \cdots & i_k \\ j_1 & j_2 & \cdots & j_k \end{bmatrix}$ satisfying the following conditions: (a) $i_1 \leq i_2 \leq \cdots \leq i_k$; (b) if $i_s = i_{s+1}$, then $j_s \leq j_{s+1}$; (c) $1 \leq i_s \leq M$ for all s; (d) $1 \leq j_s \leq N$ for all s. The map bw : $X \to Y$ is a bijection. For all $A \in X$, i appears $\sum_j a_{ij}$ times in $r(A)$, j appears $\sum_i a_{ij}$ times in $c(A)$, and bw(A) has length $k = \sum_{i,j} a_{ij}$.

Proof. To show that bw maps X *into* Y, we must show that bw(A) satisfies conditions (a) through (d). Condition (a) holds since we scan the rows of A from top to bottom. Condition (b) holds since each row is scanned from left to right. Condition (c) holds since A has M rows. Condition (d) holds since A has N columns. We can invert the map bw as follows. Given a biword $w \in Y$, let A be the $M \times N$ matrix such that, for all i, j satisfying $1 \leq i \leq M$ and $1 \leq j \leq N$, a_{ij} is the number of indices s with $i_s = i$ and $j_s = j$. The last statements in the theorem follow from the way we constructed $r(A)$ and $c(A)$. □

9.117. Theorem: RSK Correspondence for Biwords. Let Y be the set of biwords defined in Theorem 9.116. Let $Z = \bigcup_{\lambda \in \text{Par}} \text{SSYT}_N(\lambda) \times \text{SSYT}_M(\lambda)$. There is a bijection RSK : $Y \to Z$. If $\begin{bmatrix} v \\ w \end{bmatrix} \in Y$ maps to $(P, Q) \in Z$, then v and Q contain the same number of i's for all i, and w and P contain the same number of j's for all j.

Proof. Given a biword $\begin{bmatrix} v \\ w \end{bmatrix} \in Y$, write $v = i_1 \leq i_2 \leq \cdots \leq i_k$ and $w = j_1, j_2, \ldots, j_k$, where $i_s = i_{s+1}$ implies $j_s \leq j_{s+1}$. As in the previous RSK maps, we build sequences of insertion tableaux $P_0, P_1, \ldots, P_k$ and recording tableaux $Q_0, Q_1, \ldots, Q_k$. Initially, P_0 and Q_0 are empty. Having constructed P_s and Q_s, let $P_{s+1} = P_s \leftarrow j_{s+1}$. If the new box created by this insertion is (a, b), obtain Q_{s+1} from Q_s by setting $Q_{s+1}(a, b) = i_{s+1}$. The final output is the pair (P_k, Q_k).

By construction, P_k is a semistandard tableau with entries consisting of the letters in w, and the entries of Q_k are the letters in v. But, is $Q = Q_k$ a semistandard tableau? To see that it is, note that we obtain Q by successively placing a weakly increasing sequence of numbers $i_1 \leq i_2 \leq \cdots \leq i_k$ into new corner boxes of an initially empty tableau. It follows that the rows and columns of Q weakly increase. To see that columns of Q *strictly* increase, consider what happens during the placement of a run of equal numbers into Q, say $i = i_s = i_{s+1} = \cdots = i_t$. By definition of Y, we have $j_s \leq j_{s+1} \leq \cdots \leq j_t$. When we insert this weakly increasing sequence into the P-tableau, the resulting sequence of new boxes forms a horizontal strip by Theorem 9.58. So, the corresponding boxes in Q (which consist of all the boxes labeled i in Q) also form a horizontal strip. This means that there are never two equal numbers in a given column of Q.

The inverse algorithm reconstructs the words v and w in reverse, starting with i_k and j_k. Given (P, Q), look for the rightmost occurrence of the largest letter in Q, which must reside in a corner box. Let i_k be this letter. Erase this cell from Q, and perform reverse insertion on P starting at the same cell to recover j_k. Iterate this process on the resulting smaller tableaux. We have $i_k \geq \cdots \geq i_1$ since we remove the largest letter in Q at each stage. When we remove a string of equal letters from Q, say $i = i_t = i_{t-1} = \cdots = i_s$, the associated letters removed from P must satisfy $j_t \geq j_{t-1} \geq \cdots \geq j_s$. This follows from the Bumping Comparison Theorem 9.56. For instance, if $j_{t-1} > j_t$, then the new box created at stage t would be weakly left of the new box created at stage $t-1$, which contradicts the requirement of choosing the *rightmost* i in Q when recovering i_t and j_t. It follows that the inverse algorithm does produce a biword in Y, as required. □

Composing the preceding bijections between the sets X, Y, and Z gives the following result.

9.118. Theorem: RSK Correspondence for Matrices. For every $M, N \geq 1$, there is a bijection between the set of $M \times N$ matrices with entries in $\mathbb{Z}_{\geq 0}$ and the set

$$\bigcup_{\lambda \in \mathrm{Par}} \mathrm{SSYT}_N(\lambda) \times \mathrm{SSYT}_M(\lambda),$$

which sends the matrix A to $\mathrm{RSK}(\mathrm{bw}(A))$. If $[a_{ij}]$ maps to (P, Q) under this bijection, then the number of j's in P is $\sum_i a_{ij}$, and the number of i's in Q is $\sum_j a_{ij}$.

9.119. Example. Let us compute the pair of tableaux associated to the matrix A from Example 9.115. Looking at the biword of A, we must insert the sequence $c(A) = (1, 1, 3, 2, 4, 4, 4, 3)$ into the P-tableau, recording the entries in $r(A) = (1, 1, 1, 2, 2, 2, 2, 3)$ in the Q-tableau. This computation appears in Figure 9.7.

9.120. Theorem: Cauchy Identity for Schur Polynomials. For all $M, N \geq 1$,

$$\prod_{i=1}^{M} \prod_{j=1}^{N} \frac{1}{1 - x_i y_j} = \sum_{\lambda \in \mathrm{Par}} s_\lambda(y_1, \ldots, y_N) s_\lambda(x_1, \ldots, x_M). \tag{9.16}$$

Proof. We interpret each side as the generating function for a certain set of weighted objects. For the left side, consider $M \times N$ matrices with entries in $\mathbb{Z}_{\geq 0}$. Let the weight of a matrix $A = [a_{ij}]$ be

$$\mathrm{wt}(A) = \prod_{i=1}^{M} \prod_{j=1}^{N} (x_i y_j)^{a_{ij}}.$$

We can build such a matrix by choosing the entries $a_{ij} \in \mathbb{Z}_{\geq 0}$ one at a time. For fixed i and j, the generating function for the choice of a_{ij} is

$$1 + x_i y_j + (x_i y_j)^2 + \cdots + (x_i y_j)^k + \cdots = \frac{1}{1 - x_i y_j}.$$

	Insertion Tableau	Recording Tableau
insert 1, record 1:	1	1
insert 1, record 1:	1 1	1 1
insert 3, record 1:	1 1 3	1 1 1
insert 2, record 2:	1 1 2 3	1 1 1 2
insert 4, record 2:	1 1 2 4 3	1 1 1 2 2
insert 4, record 2:	1 1 2 4 4 3	1 1 1 2 2 2
insert 4, record 2:	1 1 2 4 4 4 3	1 1 1 2 2 2 2
insert 3, record 3:	1 1 2 3 4 4 3 4	1 1 1 2 2 2 2 3

FIGURE 9.7
Applying the RSK map to a biword.

By the Product Rule for Weighted Sets, we see that the left side of (9.16) is the generating function for this set of matrices. On the other hand, the RSK bijection converts each matrix A in this set to a pair (P,Q) of semistandard tableaux of the same shape. This bijection is weight-preserving provided that we weight each occurrence of j in P by y_j and each occurrence of i in Q by x_i. With these weights, the generating function for $\mathrm{SSYT}_N(\lambda)$ is $s_\lambda(y_1,\ldots,y_N)$, and the generating function for $\mathrm{SSYT}_M(\lambda)$ is $s_\lambda(x_1,\ldots,x_M)$. It now follows from the Sum and Product Rules for Weighted Sets that the right side of (9.16) is the generating function for the weighted set $\bigcup_{\lambda\in\mathrm{Par}}\mathrm{SSYT}_N(\lambda)\times\mathrm{SSYT}_M(\lambda)$. Since RSK is a weight-preserving bijection, the proof is complete. □

9.25 Cauchy's Identities

In the last section, we found a formula expressing the product $\prod_{i,j}(1-x_iy_j)^{-1}$ as a sum of products of Schur polynomials. Next we derive other formulas for this product that involve other kinds of symmetric polynomials. Throughout, we operate in the formal power series ring $\mathbb{R}[[x_1,\ldots,x_M,y_1,\ldots,y_N]]$.

9.121. Theorem: Cauchy's Identities. For all $M,N\geq 1$,

$$\begin{aligned}\prod_{i=1}^{M}\prod_{j=1}^{N}\frac{1}{1-x_iy_j} &= \sum_{\lambda\in\mathrm{Par}_N} h_\lambda(x_1,\ldots,x_M)m_\lambda(y_1,\ldots,y_N)\\ &= \sum_{\lambda\in\mathrm{Par}_M} m_\lambda(x_1,\ldots,x_M)h_\lambda(y_1,\ldots,y_N)\\ &= \sum_{\lambda\in\mathrm{Par}} \frac{p_\lambda(x_1,\ldots,x_M)p_\lambda(y_1,\ldots,y_N)}{z_\lambda}.\end{aligned}$$

Proof. Recall from Theorem 9.87 the product expansion

$$\prod_{i=1}^{M} \frac{1}{1-x_i t} = \sum_{k=0}^{\infty} h_k(x_1,\ldots,x_M)t^k.$$

Replacing t by y_j, where j is a fixed index between 1 and n, we obtain

$$\prod_{i=1}^{M} \frac{1}{1-x_i y_j} = \sum_{k=0}^{\infty} h_k(x_1,\ldots,x_M)y_j^k.$$

Taking the product over j gives

$$\prod_{i=1}^{M}\prod_{j=1}^{N} \frac{1}{1-x_i y_j} = \prod_{j=1}^{N}\sum_{k_j=0}^{\infty} h_{k_j}(x_1,\ldots,x_M)y_j^{k_j}.$$

We can expand the product on the right side using the Generalized Distributive Law for formal power series (cf. Exercise 2-16). We obtain

$$\prod_{i=1}^{M}\prod_{j=1}^{N} \frac{1}{1-x_i y_j} = \sum_{k_1=0}^{\infty}\cdots\sum_{k_N=0}^{\infty}\prod_{j=1}^{N} h_{k_j}(x_1,\ldots,x_M)y_j^{k_j}.$$

Let us reorganize the sum on the right side by grouping together summands indexed by sequences $(k_1,\ldots,k_N)$ that can be sorted to give the same partition λ. Since $h_{k_1}h_{k_2}\cdots h_{k_N} = h_\lambda$ for all such sequences, the right side becomes

$$\sum_{\lambda\in\mathrm{Par}_N} h_\lambda(x_1,\ldots,x_M) \sum_{\substack{(k_1,\ldots,k_N)\in\mathbb{Z}_{\geq 0}^N:\\ \mathrm{sort}(k_1,\ldots,k_N)=\lambda}} y_1^{k_1}y_2^{k_2}\cdots y_N^{k_N}.$$

The inner sum is precisely the definition of $m_\lambda(y_1,\ldots,y_N)$. So the first formula of the theorem is proved. The second formula follows from the same reasoning, interchanging the roles of the x-variables and the y-variables.

To derive the formula involving power sums, we again start with Theorem 9.87, which can be written

$$\prod_{k=1}^{MN} \frac{1}{1-z_k t} = \sum_{n=0}^{\infty} h_n(z_1,\ldots,z_{MN})t^n.$$

Replace the MN variables z_k by the MN quantities $x_i y_j$, where $1\leq i\leq M$ and $1\leq j\leq N$. We obtain

$$\prod_{i=1}^{M}\prod_{j=1}^{N} \frac{1}{1-x_i y_j t} = \sum_{n=0}^{\infty} h_n(x_1y_1, x_1y_2,\ldots,x_My_N)t^n.$$

Now use Theorem 9.91 to rewrite the right side in terms of power sums:

$$\prod_{i=1}^{M}\prod_{j=1}^{N} \frac{1}{1-x_i y_j t} = \sum_{n=0}^{\infty} t^n \sum_{\lambda\in\mathrm{Par}(n)} p_\lambda(x_1y_1, x_1y_2,\ldots,x_My_N)/z_\lambda.$$

Observe next that, for all $k\geq 1$,

$$\begin{aligned} p_k(x_1y_1,\ldots,x_My_N) &= \sum_{i=1}^{M}\sum_{j=1}^{N}(x_iy_j)^k = \sum_{i=1}^{M}\sum_{j=1}^{N} x_i^k y_j^k \\ &= \left(\sum_{i=1}^{M} x_i^k\right)\cdot\left(\sum_{j=1}^{N} y_j^k\right) = p_k(x_1,x_2,\ldots,x_M)p_k(y_1,y_2,\ldots,y_N). \end{aligned}$$

Therefore, for any partition λ,

$$p_\lambda(x_1y_1,\ldots,x_My_N) = p_\lambda(x_1,x_2,\ldots,x_M)p_\lambda(y_1,y_2,\ldots,y_N).$$

It follows that

$$\prod_{i=1}^{M}\prod_{j=1}^{N}\frac{1}{1-x_iy_jt} = \sum_{n=0}^{\infty}t^n\sum_{\lambda\in\text{Par}(n)}\frac{p_\lambda(x_1,x_2,\ldots,x_M)p_\lambda(y_1,y_2,\ldots,y_N)}{z_\lambda}. \tag{9.17}$$

Setting $t=1$ gives the final formula of the theorem. □

9.26 Dual Bases

Now we introduce a scalar product on the vector spaces Λ_N^k. We only consider the case $N \geq k$, so that the various bases of Λ_N^k are indexed by *all* the integer partitions of k.

9.122. Definition: Hall Scalar Product on Λ_N^k. For $N \geq k$, define the *Hall scalar product* on the vector space Λ_N^k by setting (for all $\mu,\nu \in \text{Par}(k)$)

$$\langle p_\mu, p_\nu\rangle = 0 \text{ if } \mu \neq \nu, \qquad \langle p_\mu, p_\mu\rangle = z_\mu,$$

and extending by bilinearity. In more detail, given $f,g \in \Lambda_N^k$, choose scalars $a_\mu, b_\mu \in \mathbb{R}$ such that $f = \sum_\mu a_\mu p_\mu$ and $g = \sum_\nu b_\nu p_\nu$. Then $\langle f,g\rangle = \sum_\mu a_\mu b_\mu z_\mu \in \mathbb{R}$.

In the next definition, recall that $\chi(\mu=\nu)$ is 1 if $\mu=\nu$, and 0 otherwise.

9.123. Definition: Orthonormal Bases and Dual Bases. Suppose $N \geq k$ and $B_1 = \{f_\mu : \mu \in \text{Par}(k)\}$ and $B_2 = \{g_\mu : \mu \in \text{Par}(k)\}$ are two bases of Λ_N^k. B_1 is called an *orthonormal basis* iff $\langle f_\mu, f_\nu\rangle = \chi(\mu=\nu)$ for all $\mu,\nu \in \text{Par}(k)$. B_1 and B_2 are called *dual bases* iff $\langle f_\mu, g_\nu\rangle = \chi(\mu=\nu)$ for all $\mu,\nu\in\text{Par}(k)$.

For example, it follows from Definition 9.122 that $\{p_\mu/\sqrt{z_\mu} : \mu \in \text{Par}(k)\}$ is an orthonormal basis of Λ_N^k. The next theorem allows us to detect dual bases by looking at expansions of the product $\prod_{i,j}(1-x_iy_j)^{-1}$.

9.124. Theorem: Characterization of Dual Bases. Suppose $N \geq k$ and $B_1 = \{f_\mu : \mu \in \text{Par}(k)\}$ and $B_2 = \{g_\mu : \mu\in\text{Par}(k)\}$ are two bases of Λ_N^k. B_1 and B_2 are dual bases iff

$$\left.\left(\prod_{i=1}^{N}\prod_{j=1}^{N}\frac{1}{1-x_iy_jt}\right)\right|_{t^k} = \sum_{\mu\in\text{Par}(k)} f_\mu(x_1,\ldots,x_N)g_\mu(y_1,\ldots,y_N),$$

where the left side denotes the coefficient of t^k in the indicated product.

Proof. Comparing the displayed equation to (9.17), we must prove that B_1 and B_2 are dual bases iff

$$\sum_{\mu\in\text{Par}(k)} p_\mu(x_1,\ldots,x_N)p_\mu(y_1,\ldots,y_N)/z_\mu = \sum_{\mu\in\text{Par}(k)} f_\mu(x_1,\ldots,x_N)g_\mu(y_1,\ldots,y_N).$$

The idea of the proof is to convert each condition into a statement about matrices. Since $\{p_\mu\}$ and $\{p_\mu/z_\mu\}$ are bases of Λ_N^k, there exist scalars $a_{\mu,\nu}, b_{\mu,\nu} \in \mathbb{R}$ satisfying

$$f_\nu = \sum_{\mu\in\text{Par}(k)} a_{\mu,\nu}p_\mu, \qquad g_\nu = \sum_{\mu\in\text{Par}(k)} b_{\mu,\nu}(p_\mu/z_\mu).$$

Define matrices $\mathbf{A} = [a_{\mu,\nu}]$ and $\mathbf{B} = [b_{\mu,\nu}]$. For all $\lambda, \nu \in \operatorname{Par}(k)$, we use bilinearity to compute

$$\begin{aligned}
\langle f_\lambda, g_\nu \rangle &= \left\langle \sum_{\mu \in \operatorname{Par}(k)} a_{\mu,\lambda} p_\mu, \sum_{\rho \in \operatorname{Par}(k)} b_{\rho,\nu} p_\rho / z_\rho \right\rangle \\
&= \sum_{\mu,\rho \in \operatorname{Par}(k)} a_{\mu,\lambda} b_{\rho,\nu} \langle p_\mu, p_\rho / z_\rho \rangle \\
&= \sum_{\mu \in \operatorname{Par}(k)} a_{\mu,\lambda} b_{\mu,\nu} = (\mathbf{A}^{\mathrm{tr}} \mathbf{B})_{\lambda,\nu}.
\end{aligned}$$

It follows that $\{f_\lambda\}$ and $\{g_\nu\}$ are dual bases iff $\mathbf{A}^{\mathrm{tr}}\mathbf{B} = \mathbf{I}$, where $\mathbf{I}$ is the identity matrix of size $|\operatorname{Par}(k)|$.

On the other hand, writing $\mathbf{x} = (x_1, \ldots, x_N)$ and $\mathbf{y} = (y_1, \ldots, y_N)$, we have

$$\sum_{\mu \in \operatorname{Par}(k)} f_\mu(\mathbf{x}) g_\mu(\mathbf{y}) = \sum_{\mu,\alpha,\beta \in \operatorname{Par}(k)} a_{\alpha,\mu} b_{\beta,\mu} p_\alpha(\mathbf{x}) p_\beta(\mathbf{y}) / z_\beta.$$

Now, one may check that the indexed set of polynomials

$$\{p_\alpha(\mathbf{x}) p_\beta(\mathbf{y}) / z_\beta : (\alpha, \beta) \in \operatorname{Par}(k) \times \operatorname{Par}(k)\}$$

is linearly independent, using the fact that the power-sum polynomials in one set of variables are linearly independent. It follows that the expression given above for $\sum_{\mu \in \operatorname{Par}(k)} f_\mu(\mathbf{x}) g_\mu(\mathbf{y})$ is equal to $\sum_{\alpha \in \operatorname{Par}(k)} p_\alpha(\mathbf{x}) p_\alpha(\mathbf{y}) / z_\alpha$ iff $\sum_\mu a_{\alpha,\mu} b_{\beta,\mu} = \chi(\alpha = \beta)$ for all α, β. In matrix form, these equations say that $\mathbf{A}\mathbf{B}^{\mathrm{tr}} = \mathbf{I}$. This matrix equation is equivalent to $\mathbf{B}^{\mathrm{tr}}\mathbf{A} = \mathbf{I}$ (since all the matrices are square), which is equivalent in turn to $\mathbf{A}^{\mathrm{tr}}\mathbf{B} = \mathbf{I}$. We saw above that this last condition holds iff B_1 and B_2 are dual bases, so the proof is complete. □

9.125. Theorem: Dual Bases of Λ_N^k. For $N \geq k$, $\{s_\mu(x_1, \ldots, x_N) : \mu \in \operatorname{Par}(k)\}$ is an orthonormal basis of Λ_N^k. Also, $\{m_\mu(x_1, \ldots, x_N) : \mu \in \operatorname{Par}(k)\}$ and $\{h_\mu(x_1, \ldots, x_N) : \mu \in \operatorname{Par}(k)\}$ are dual bases of Λ_N^k.

Proof. In Theorem 9.120, replace every x_i by tx_i. Since s_λ is homogeneous of degree $|\lambda|$, we obtain

$$\prod_{i=1}^N \prod_{j=1}^N \frac{1}{1 - tx_i y_j} = \sum_{\lambda \in \operatorname{Par}} s_\lambda(y_1, \ldots, y_N) s_\lambda(x_1, \ldots, x_N) t^{|\lambda|}.$$

Extracting the coefficient of t^k gives

$$\left. \left(\prod_{i=1}^N \prod_{j=1}^N \frac{1}{1 - x_i y_j t} \right) \right|_{t^k} = \sum_{\lambda \in \operatorname{Par}(k)} s_\lambda(y_1, \ldots, y_N) s_\lambda(x_1, \ldots, x_N).$$

Theorem 9.124 now applies to show that $\{s_\lambda : \lambda \in \operatorname{Par}(k)\}$ is an orthonormal basis. We proceed similarly to see that $\{m_\mu\}$ and $\{h_\mu\}$ are dual bases, starting with Theorem 9.121. □

9.126. Theorem: ω is an Isometry. For $N \geq k$, the map $\omega : \Lambda_N^k \to \Lambda_N^k$ is an isometry relative to the Hall scalar product. In other words, for all $f, g \in \Lambda_N^k$, $\langle \omega(f), \omega(g) \rangle = \langle f, g \rangle$. Therefore, ω maps orthonormal bases to orthonormal bases and dual bases to dual bases.

Proof. Given $f, g \in \Lambda_N^k$, write $f = \sum_\mu a_\mu p_\mu$ and $g = \sum_\nu b_\nu p_\nu$ for certain scalars $a_\mu, b_\nu \in \mathbb{R}$. By linearity of ω and bilinearity of the Hall scalar product, we compute

$$\begin{aligned}\langle \omega(f), \omega(g)\rangle &= \left\langle \omega\left(\sum_{\mu\in\mathrm{Par}(k)} a_\mu p_\mu\right), \omega\left(\sum_{\nu\in\mathrm{Par}(k)} b_\nu p_\nu\right)\right\rangle \\ &= \sum_{\mu\in\mathrm{Par}(k)}\sum_{\nu\in\mathrm{Par}(k)} a_\mu b_\nu \langle \omega(p_\mu), \omega(p_\nu)\rangle \\ &= \sum_{\mu\in\mathrm{Par}(k)}\sum_{\nu\in\mathrm{Par}(k)} a_\mu b_\nu \epsilon_\mu \epsilon_\nu \langle p_\mu, p_\nu\rangle \\ &= \sum_{\mu\in\mathrm{Par}(k)} a_\mu b_\mu \epsilon_\mu^2 z_\mu.\end{aligned}$$

The last step follows since we only get a nonzero scalar product when $\nu = \mu$. Now, the last expression is

$$\sum_{\mu\in\mathrm{Par}(k)} a_\mu b_\mu z_\mu = \sum_{\mu\in\mathrm{Par}(k)}\sum_{\nu\in\mathrm{Par}(k)} a_\mu b_\nu \langle p_\mu, p_\nu\rangle = \langle f, g\rangle. \qquad \square$$

9.127. Theorem: Duality of e_μ and fgt_ν. For $N \geq k$, the bases $\{e_\mu : \mu \in \mathrm{Par}(k)\}$ and $\{\mathrm{fgt}_\mu : \mu \in \mathrm{Par}(k)\}$ (the forgotten basis) are dual. Moreover,

$$\left.\left(\prod_{i=1}^N \prod_{j=1}^N \frac{1}{1 - x_i y_j t}\right)\right|_{t^k} = \sum_{\lambda\in\mathrm{Par}(k)} e_\lambda(x_1, \ldots, x_N)\mathrm{fgt}_\lambda(y_1, \ldots, y_N).$$

Proof. We know that $\{m_\mu\}$ and $\{h_\mu\}$ are dual bases. Since $\mathrm{fgt}_\mu = \omega(m_\mu)$ and $e_\mu = \omega(h_\mu)$, $\{\mathrm{fgt}_\mu\}$ and $\{e_\mu\}$ are dual bases. The product formula now follows from Theorem 9.124. $\square$

9.27 Skew Schur Polynomials

In this section we study skew Schur polynomials, which are the generating functions for generalized tableaux that can have missing squares in the upper-left corner. First we describe the diagrams we can use for these new tableaux.

9.128. Definition: Skew Shapes. Let μ and ν be integer partitions such that $\mathrm{dg}(\nu) \subseteq \mathrm{dg}(\mu)$, or equivalently $\nu_i \leq \mu_i$ for all $i \geq 1$. Define the *skew shape*

$$\mu/\nu = \mathrm{dg}(\mu) - \mathrm{dg}(\nu) = \{(i,j) \in \mathbb{Z}_{\geq 0}^2 : 1 \leq i \leq \ell(\mu), \nu_i < j \leq \mu_i\}.$$

We visualize μ/ν as the collection of unit squares obtained by starting with the diagram of μ and erasing the squares in the diagram of ν. If $\nu = (0)$, then $\mu/(0) = \mathrm{dg}(\mu)$. A skew shape of the form $\mu/(0)$ is sometimes called a *straight shape.*

9.129. Example. Let $\mu = (7,7,3,3,2,1)$, $\nu = (5,2,2,2,1)$, $\rho = (6,5,4,3,2,2)$, and $\tau = (3,3,3)$. The skew shapes μ/ν and ρ/τ are shown here:

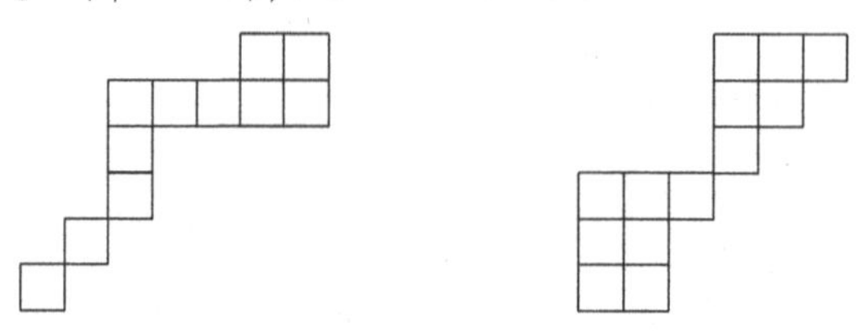

Skew shapes need not be connected; for instance, $(5,2,2,1)/(3,2)$ looks like this:

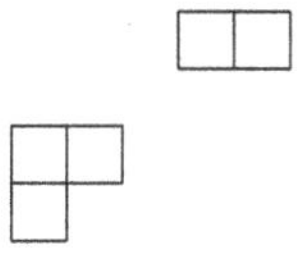

The skew shape μ/ν does not always determine μ and ν uniquely; for example, $(5,2,2,1)/(3,2) = (5,3,2,1)/(3,3)$.

9.130. Definition: Skew Tableaux. Given a skew shape μ/ν, a *filling* of this shape is a function $T : \mu/\nu \to \mathbb{Z}$. Such a filling T is a *(semistandard) tableau* of shape μ/ν iff $T(i,j) \le T(i,j+1)$ and $T(i,j) < T(i+1,j)$ whenever both sides are defined. A tableau T is *standard* iff T is a bijection from μ/ν to $\{1,2,\ldots,n\}$, where $n = |\mu/\nu| = |\mu| - |\nu|$. Let $\mathrm{SSYT}_N(\mu/\nu)$ be the set of semistandard tableaux of shape μ/ν with values in $\{1,2,\ldots,N\}$, and let $\mathrm{SYT}(\mu/\nu)$ be the set of standard tableaux of shape μ/ν. For any filling T of μ/ν, the *content monomial* of T is $\mathbf{x}^T = \prod_{(i,j)\in\mu/\nu} x_{T(i,j)}$.

9.131. Example. Here is a semistandard tableau of shape $(6,5,5,3)/(3,2,2)$ with content monomial $x_1^2x_2x_3^2x_4^4x_5x_7^2$:

			1	1	4
		2	3	4	
		4	4	5	
3	7	7			

9.132. Definition: Skew Schur Polynomials. Given a skew shape μ/ν and a positive integer N, define the *skew Schur polynomial* in N variables by

$$s_{\mu/\nu}(x_1,\ldots,x_N) = \sum_{T\in\mathrm{SSYT}_N(\mu/\nu)} \mathbf{x}^T.$$

9.133. Example. For $\mu = (2,2)$, $\nu = (1)$, and $N = 3$, $\mathrm{SSYT}_N(\mu/\nu)$ is the following set of tableaux:

	1		1		1		1		1		2		2		2
1	2	1	3	2	2	2	3	3	3	1	3	2	3	3	3

So, $s_{\mu/\nu}(x_1,x_2,x_3) = x_1^2x_2 + x_1^2x_3 + x_1x_2^2 + 2x_1x_2x_3 + x_1x_3^2 + x_2^2x_3 + x_2x_3^2$. By chance, $s_{(2,2)/(1)} = s_{(2,1)}$.

9.134. Remark. The polynomials e_α and h_α are special cases of skew Schur polynomials. For example, consider $h_\alpha = h_{\alpha_1}h_{\alpha_2}\cdots h_{\alpha_s}$. We have seen that each factor h_{α_i} is the generating function for semistandard tableaux of shape (α_i). There exists a skew shape μ/ν consisting of disconnected horizontal rows of lengths $\alpha_1,\ldots,\alpha_s$. When building a semistandard tableau of this shape, each row can be filled with labels independently of the others. So the Product Rule for Weighted Sets shows that $h_\alpha(x_1,\ldots,x_N) = s_{\mu/\nu}(x_1,\ldots,x_N)$. For example, given $h_{(2,4,3,2)} = h_2h_4h_3h_2 = h_{(4,3,2,2)}$, we draw the skew shape:

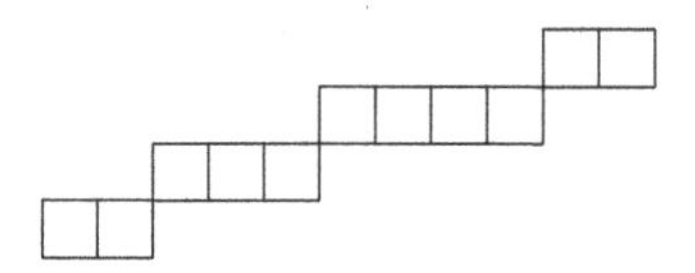

Then $h_{(2,4,3,2)} = s_{(11,9,5,2)/(9,5,2)}$. An analogous procedure works for e_α, but now we use disconnected vertical columns with lengths $\alpha_1, \ldots, \alpha_s$. For example, $e_{(3,3,1)} = s_{\mu/\nu}$ if we take

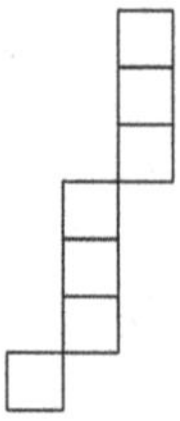

$$\mu/\nu = \quad = (3,3,3,2,2,2,1)/(2,2,2,1,1,1).$$

For any skew shape μ/ν and $\alpha \in \mathbb{Z}_{\geq 0}^N$, define the *skew Kostka number* $K_{\mu/\nu,\alpha}$ to be the coefficient of $\mathbf{x}^\alpha$ in $s_{\mu/\nu}(x_1, \ldots, x_N)$, which is the number of semistandard tableaux with shape μ/ν and content α. The next result is proved exactly like the theorems in §9.5.

9.135. Theorem: Symmetry and Monomial Expansion of Skew Schur Polynomials. For all skew shapes μ/ν and all $\alpha, \beta \in \mathbb{Z}_{\geq 0}^N$ with $\text{sort}(\alpha) = \text{sort}(\beta)$, $K_{\mu/\nu,\alpha} = K_{\mu/\nu,\beta}$. Hence, letting $k = |\mu/\nu|$, we have $s_{\mu/\nu} \in \Lambda_N^k$ and

$$s_{\mu/\nu}(x_1, \ldots, x_N) = \sum_{\lambda \in \text{Par}_N(k)} K_{\mu/\nu,\lambda} m_\lambda(x_1, \ldots, x_N).$$

Similarly, the proofs of Theorems 9.64 and 9.69 extend to prove the following result.

9.136. Theorem: Skew Pieri Rules. For all $\mu \in \text{Par}_N$ and all $\alpha \in \mathbb{Z}_{\geq 0}^s$,

$$s_\mu h_\alpha = \sum_{\lambda \in \text{Par}_N} K_{\lambda/\mu,\alpha} s_\lambda; \qquad s_\mu e_\alpha = \sum_{\lambda \in \text{Par}_N} K_{\lambda'/\mu',\alpha} s_\lambda.$$

In the next chapter, we prove that skew Schur polynomials are related to the Hall scalar product as follows: for any skew shape μ/ν with k cells and any $f \in \Lambda_N^k$ with $N \geq k$, $\langle s_{\mu/\nu}, f \rangle = \langle s_\mu, s_\nu f \rangle$. This identity can be used to prove that $\omega(s_{\mu/\nu}) = s_{\mu'/\nu'}$.

9.28 Abstract Symmetric Functions

So far, we have discussed symmetric *polynomials*, which involve only finitely many variables $x_1, \ldots, x_N$. For most purposes, N (the number of variables) is not important as long as we restrict attention to symmetric polynomials of degree $k \leq N$. This section formally defines *symmetric functions*, which can be regarded intuitively as symmetric polynomials in infinitely many variables.

Let K be a field containing $\mathbb{Q}$. Recall that Λ_N is the algebra of symmetric polynomials in N variables $x_1, \ldots, x_N$ with coefficients in K. The key initial idea is to view Λ_N as the polynomial ring $K[p_1, \ldots, p_N]$, thinking of the power-sums p_j as algebraically independent formal variables, *not* as polynomials in the x-variables (see §9.15). We can define a new K-algebra Λ by taking the union of these rings:

$$\Lambda = \bigcup_{N=1}^{\infty} K[p_1, \ldots, p_N] = K[p_n : n \in \mathbb{Z}_{>0}].$$

Given $f, g \in \Lambda$, f and g both lie in $K[p_1, \ldots, p_N]$ for some N, so $f + g$ and $f \cdot g$ are defined. We call Λ the algebra of *symmetric functions* with coefficients in K. By definition, every

$f \in \Lambda$ is a finite linear combination of monomials, where each monomial is a product of a finite sequence of p_n's (repeats allowed). As before, given a sequence of positive integers $\alpha = (\alpha_1, \ldots, \alpha_s)$, we define $p_\alpha = p_{\alpha_1} \cdots p_{\alpha_s}$. Since multiplication in Λ is commutative, it follows that every $f \in \Lambda$ can be written uniquely as a sum $f = \sum_{\mu \in \mathrm{Par}} c_\mu p_\mu$ where $c_\mu \in K$ and only finitely many c_μ are nonzero.

In an ordinary polynomial ring $K[x_1, \ldots, x_N]$, each formal variable x_i has degree 1. However, in the polynomial rings Λ_N and Λ, we define the degree of the formal power-sum p_j to be j. The degree of a monomial cp_α is then $|\alpha| = \alpha_1 + \cdots + \alpha_s$. Let Λ^k be the set of $f \in \Lambda$ that are homogeneous of degree k. It is immediate that for all $k \geq 0$, $\{p_\mu : \mu \in \mathrm{Par}(k)\}$ is a basis for Λ^k, and that Λ is the direct sum of its subspaces Λ^k. Moreover, for $f \in \Lambda^k$ and $g \in \Lambda^m$, $fg \in \Lambda^{k+m}$. So Λ is a graded commutative K-algebra.

We can use power-sum expansions developed earlier to define symmetric function versions of h_α and e_α. Specifically, for all $n > 0$, let

$$h_n = \sum_{\mu \in \mathrm{Par}(n)} z_\mu^{-1} p_\mu, \qquad e_n = \sum_{\mu \in \mathrm{Par}(n)} \epsilon_\mu z_\mu^{-1} p_\mu.$$

Then set $h_0 = e_0 = 1$, $h_\alpha = \prod_{i \geq 1} h_{\alpha_i}$, and $e_\alpha = \prod_{i \geq 1} e_{\alpha_i}$. Schur symmetric functions can be defined either by the power-sum expansion given later in Theorem 10.53, or by inverting the linear system

$$h_\mu = \sum_{\lambda \in \mathrm{Par}} K_{\lambda,\mu} s_\lambda.$$

As before, we define the Hall scalar product on Λ by letting $\langle p_\lambda, p_\mu \rangle = \chi(\lambda = \mu) z_\mu$ for all partitions λ and μ. The symmetric functions m_μ can be defined as the unique basis dual to $\{h_\mu : \mu \in \mathrm{Par}\}$ with respect to this scalar product. Similarly, the forgotten basis is dual to the elementary basis. The Schur basis is orthonormal relative to the Hall scalar product. We define the algebra isomorphism ω on Λ by specifying its effect on the generators: $\omega(p_n) = (-1)^{n-1} p_n$ for all $n \geq 1$. It follows that $\omega(e_\mu) = h_\mu$, $\omega(h_\mu) = e_\mu$, $\omega(p_\mu) = \epsilon_\mu p_\mu$, and $\omega(s_\mu) = s_{\mu'}$ for all partitions μ.

We can define evaluation homomorphisms with domain Λ as follows. Given any K-algebra B and arbitrary elements $b_j \in B$, there exists a unique algebra homomorphism $T : \Lambda \to B$ such that $T(p_j) = b_j$ for all $j \geq 1$. The map T sends the symmetric function $f = \sum_{\mu \in \mathrm{Par}} c_\mu p_\mu$ to $T(f) = \sum_{\mu \in \mathrm{Par}} c_\mu \prod_{i \geq 1} b_{\mu_i}$. These homomorphisms enable us to connect abstract symmetric functions to concrete symmetric polynomials. Given $N > 0$, consider the evaluation homomorphism $E_N : \Lambda \to K[x_1, \ldots, x_N]$ sending the abstract power-sum p_j to the power-sum polynomial $p_j(x_1, \ldots, x_N) = x_1^j + x_2^j + \cdots + x_N^j$ for all $j \geq 1$. The image of E_N is Λ_N, viewed as a subring of $K[x_1, \ldots, x_N]$. By Theorem 9.79, E_N restricts to a vector space isomorphism from Λ^k onto Λ_N^k as long as $N \geq k$.

We can use these homomorphisms to transfer information about Λ_N to information about Λ. For example, let us show that the infinite list $e_1, e_2, \ldots, e_n, \ldots$ is algebraically independent in Λ. This means that whenever a linear combination $\sum_{\mu \in F} c_\mu e_\mu$ is zero (where each $c_\mu \in K$ and F is a *finite* set of partitions), then every c_μ must be zero. Given such a linear combination, let N be the maximum of all parts μ_i of all partitions $\mu \in F$. Applying E_N to the given linear combination, we find that $\sum_{\mu \in F} c_\mu e_\mu(x_1, \ldots, x_N) = 0$ in $K[x_1, \ldots, x_N]$. Since we know that the symmetric *polynomials* $e_1(x_1, \ldots, x_N), \ldots, e_N(x_1, \ldots, x_N)$ are algebraically independent, we can conclude that every c_μ is zero. A similar argument shows that $h_1, h_2, \ldots, h_n, \ldots$ is an algebraically independent list in Λ. So, if we prefer, we could have defined Λ to be the polynomial ring $K[h_n : n \in \mathbb{Z}_{>0}]$ or $K[e_n : n \in \mathbb{Z}_{>0}]$, viewing all h_n (or e_n) as formal variables.

It may be tempting to think of the abstract symmetric function p_j as the formal infinite sum $\sum_{n=1}^{\infty} x_n^j$. To justify this rigorously, we need the evaluation homomorphism from Λ into the formal power series ring $K[[x_n : n \in \mathbb{Z}_{>0}]]$ such that p_j maps to $p_j(\mathbf{x}) = \sum_{n=1}^{\infty} x_n^j$. It can be shown that this homomorphism sends e_k to $e_k(\mathbf{x}) = \sum_{i_1<i_2<\cdots<i_k : i_j \in \mathbb{Z}_{>0}} x_{i_1} x_{i_2} \cdots x_{i_k}$ and s_λ to $s_\lambda(\mathbf{x}) = \sum_{T \in \mathrm{SSYT}(\lambda)} \mathbf{x}^T$. However, it can be dangerous to think of Λ in this way. For example, we could try to extend Theorem 9.85 to the following identity involving infinitely many variables x_i:

$$\prod_{i=1}^{\infty}(1+x_i) = \sum_{n=0}^{\infty} e_n(\mathbf{x}).$$

But the right side of this formula is *not* in the image of Λ under the evaluation homomorphism sending $f \in \Lambda$ to $f(\mathbf{x})$. The reason is that $\sum_{n=0}^{\infty} e_n$ is not an element of Λ, since only *finite* linear combinations of the p_j are allowed.

One way around this difficulty is to introduce a new formal variable t and work in the formal power series ring $K[[t, x_1, x_2, \ldots]]$. In this ring,

$$\prod_{i=1}^{\infty}(1+tx_i) = \sum_{n=0}^{\infty} e_n(\mathbf{x})t^n \quad \text{and} \quad \prod_{i=1}^{\infty} \frac{1}{1-tx_i} = \sum_{n=0}^{\infty} h_n(\mathbf{x})t^n,$$

although it must still be proved that the infinite products converge. For the Cauchy identities, we work in the ring $K[[t, x_1, x_2, \ldots, y_1, y_2, \ldots]]$, where (for example)

$$\prod_{i=1}^{\infty}\prod_{j=1}^{\infty} \frac{1}{1-tx_iy_j} = \sum_{k=0}^{\infty} t^k \sum_{\mu \in \mathrm{Par}(k)} h_\mu(\mathbf{x}) m_\mu(\mathbf{y}) = \sum_{k=0}^{\infty} t^k \sum_{\mu \in \mathrm{Par}(k)} s_\mu(\mathbf{x}) s_\mu(\mathbf{y}).$$

Summary

Table 9.1 summarizes information about five bases for the vector space Λ_N^k of symmetric polynomials in N variables that are homogeneous of degree k. The statements about dual bases assume $N \geq k$. Recall that $\mathrm{Par}_N(k)$ is the set of integer partitions of k into at most N parts, while $\mathrm{Par}_N(k)'$ is the set of partitions of k where every part is at most N. Table 9.2 gives formulas and recursions for expressing certain symmetric polynomials as linear combinations of other symmetric polynomials. More identities of this type appear in the summary of Chapter 10.

- *Tableaux and Schur Polynomials.* Given a partition μ, a semistandard tableau of shape μ is a filling of the cells in $\mathrm{dg}(\mu)$ so that rows weakly increase and columns strictly increase. The Schur polynomial in N variables indexed by μ is

$$s_\mu(x_1, \ldots, x_N) = \sum_{T \in \mathrm{SSYT}_N(\mu)} \mathbf{x}^T,$$

  where the power of x_i is the number of i's in T. Schur polynomials are symmetric, since an involution exists that switches the frequencies of i's and $(i+1)$'s in semistandard tableaux of shape μ. Similar remarks hold for skew Schur polynomials $s_{\mu/\nu}$, which enumerate semistandard tableaux using the shape μ/ν obtained by removing the diagram of ν from the diagram of μ.

- *Orderings on Partitions.* For $\mu, \nu \in \mathrm{Par}(k)$, $\mu \leq_{\mathrm{lex}} \nu$ means that $\mu = \nu$ or the first nonzero entry of $\nu - \mu$ is positive; $\leq_{\mathrm{lex}}$ is a total ordering on $\mathrm{Par}(k)$. We say $\mu \trianglelefteq \nu$ (μ is dominated by ν) iff $\mu_1 + \cdots + \mu_i \leq \nu_1 + \cdots + \nu_i$ for all $i \geq 1$; $\trianglelefteq$ is a partial ordering on $\mathrm{Par}(k)$. We have $\mu \trianglelefteq \nu$ iff $\nu' \trianglelefteq \mu'$ iff μ can be transformed into ν by a sequence of raising operators (moving one box to a higher row). Also, $\mu \trianglelefteq \nu$ implies $\mu \leq_{\mathrm{lex}} \nu$.

TABLE 9.1
Bases for Λ_N^k.

Basis of Λ_N^k	Definition	Dual Basis	Action of ω
Monomial $\{m_\mu : \mu \in \text{Par}_N(k)\}$	$m_\mu = \sum_{\substack{\alpha\in\mathbb{Z}_{\geq 0}^N:\\ \text{sort}(\alpha)=\mu}} x_1^{\alpha_1}\cdots x_N^{\alpha_N}$	$\{h_\mu\}$	$\omega(m_\mu) = \text{fgt}_\mu$
Elementary $\{e_\mu : \mu \in \text{Par}_N(k)'\}$	$e_k = \sum_{1\leq i_1<i_2<\cdots<i_k\leq N} x_{i_1}x_{i_2}\cdots x_{i_k}$ $e_\mu = e_{\mu_1}e_{\mu_2}\cdots e_{\mu_s}$	$\{\text{fgt}_\mu\}$	$\omega(e_\mu) = h_\mu$
Complete $\{h_\mu : \mu \in \text{Par}_N(k)\}$ or $\{h_\mu : \mu \in \text{Par}_N(k)'\}$	$h_k = \sum_{1\leq i_1\leq i_2\leq\cdots\leq i_k\leq N} x_{i_1}x_{i_2}\cdots x_{i_k}$ $h_\mu = h_{\mu_1}h_{\mu_2}\cdots h_{\mu_s}$	$\{m_\mu\}$	$\omega(h_\mu) = e_\mu$
Power-sum $\{p_\mu : \mu \in \text{Par}_N(k)'\}$	$p_k = \sum_{i=1}^N x_i^k$, $p_\mu = \prod_j p_{\mu_j}$	$\{p_\mu/z_\mu\}$	$\omega(p_\mu) = \epsilon_\mu p_\mu$
Schur $\{s_\mu : \mu \in \text{Par}_N(k)\}$	$s_\mu = \sum_{T\in\text{SSYT}_N(\mu)} \mathbf{x}^T$	$\{s_\mu\}$	$\omega(s_\mu) = s_{\mu'}$

- *Kostka Numbers.* For $\mu, \nu \in \text{Par}$ and $\alpha \in \mathbb{Z}_{\geq 0}^N$, the Kostka number $K_{\mu/\nu,\alpha}$ is the number of semistandard tableaux of shape μ/ν and content α. We have $K_{\lambda,\lambda} = 1$ for all $\lambda \in \text{Par}$, and $K_{\lambda,\mu} \neq 0$ implies $\mu \trianglelefteq \lambda$ and $\mu \leq_{\text{lex}} \lambda$.

- *Tableau Insertion.* Given a semistandard tableau T and value x, we obtain a new semistandard tableau $T \leftarrow x$ as follows. The element x bumps the leftmost value $y > x$ in the top row into the second row, and this bumping continues recursively until a value is placed in a new box at the end of some row. The bumping path moves weakly left as it goes down. Insertion is invertible if we know which corner box is the new one. If we insert a weakly increasing sequence into T, the new boxes move strictly right and weakly higher, producing a horizontal strip. If we insert a strictly decreasing sequence into T, the new boxes move weakly left and strictly lower, producing a vertical strip.

- *The Pieri Rules.* (a) $s_\mu h_k = \sum_\nu s_\nu$ where we sum over all ν such that ν/μ is a horizontal strip of size k. (b) $s_\mu e_k = \sum_\nu s_\nu$ where we sum over all ν such that ν/μ is a vertical strip of size k. If there are N variables, only shapes ν with at most N parts contribute nonzero terms to the sum.

- *Algebraic Independence.* A list of polynomials $f_1, \ldots, f_k$ is algebraically independent iff the set of monomials $\{f_1^{i_1} \cdots f_k^{i_k} : i_j \in \mathbb{Z}_{\geq 0}\}$ is linearly independent. Equivalently, the evaluation homomorphism with domain $\mathbb{R}[y_1, \ldots, y_k]$ sending y_j to f_j is one-to-one. The list $f_1, \ldots, f_k \in \mathbb{R}[x_1, \ldots, x_k]$ is algebraically independent if and only if $\det[\frac{\partial f_j}{\partial x_i}]_{1\leq i,j\leq k} \neq 0$.

- *Algebraically Independent Symmetric Polynomials.* In the ring $\mathbb{R}[x_1, \ldots, x_N]$, the lists $p_1, \ldots, p_N$ and $h_1, \ldots, h_N$ and $e_1, \ldots, e_N$ are algebraically independent. So there are three isomorphisms from the polynomial ring $\mathbb{R}[z_1, \ldots, z_N]$ onto the algebra Λ_N (we can send each z_j to p_j, to h_j, or to e_j).

TABLE 9.2
Expansions and recursions for symmetric polynomials.

Monomial expansion of Schur basis	$s_\lambda = \sum_{\mu\in\text{Par}(\|\lambda\|)} K_{\lambda,\mu} m_\mu$
Schur expansion of complete basis	$h_\alpha = \sum_{\lambda\in\text{Par}(\|\alpha\|)} K_{\lambda,\alpha} s_\lambda$
Schur expansion of elementary basis	$e_\alpha = \sum_{\lambda\in\text{Par}(\|\alpha\|)} K_{\lambda',\alpha} s_\lambda$
Power-sum expansion of h_n	$h_n = \sum_{\mu\in\text{Par}(n)} z_\mu^{-1} p_\mu$
Power-sum expansion of e_n	$e_n = \sum_{\mu\in\text{Par}(n)} \epsilon_\mu z_\mu^{-1} p_\mu$
Schur expansion of $p_{(1^n)}$	$p_{(1^n)} = \sum_{\lambda\in\text{Par}(n)} \|\text{SYT}(\lambda)\| s_\lambda$
Monomial expansion of skew Schur	$s_{\mu/\nu} = \sum_{\rho\in\text{Par}(\|\mu/\nu\|)} K_{\mu/\nu,\rho} m_\rho$
Schur expansion of $s_\mu h_\alpha$	$s_\mu h_\alpha = \sum_\lambda K_{\lambda/\mu,\alpha} s_\lambda$
Schur expansion of $s_\mu e_\alpha$	$s_\mu e_\alpha = \sum_\lambda K_{\lambda'/\mu',\alpha} s_\lambda$
Recursion linking e's and h's	$\sum_{i=0}^{m} (-1)^i e_i h_{m-i} = \chi(m=0)$
Recursion linking h's and p's	$\sum_{i=0}^{n-1} h_i p_{n-i} = n h_n$
Recursion linking e's and p's	$\sum_{i=0}^{n-1} (-1)^i e_i p_{n-i} = (-1)^{n-1} n e_n$

- *Generating Functions for e's and h's.* We have

$$E_N(t) = \prod_{i=1}^{N}(1+x_i t) = \sum_{k=0}^{N} e_k(x_1,\ldots,x_N)t^k;$$

$$H_N(t) = \prod_{i=1}^{N}(1-x_i t)^{-1} = \sum_{k=0}^{N} h_k(x_1,\ldots,x_N)t^k;$$

so $H_N(t)E_N(-t) = 1$.

- *Dual Bases and Cauchy Identities.* Assume $N \geq k$. The Hall scalar product on Λ_N^k is defined by setting $\langle p_\mu, p_\nu\rangle = z_\mu \chi(\mu = \nu)$ and extending by bilinearity. Two bases $\{f_\mu : \mu \in \text{Par}(k)\}$ and $\{g_\mu : \mu \in \text{Par}(k)\}$ of Λ_N^k are dual relative to this inner product iff they satisfy the Cauchy identity

$$\text{coefficient of } t^k \text{ in } \prod_{i=1}^{N}\prod_{j=1}^{N} \frac{1}{1-x_i y_j t} = \sum_{\mu\in\text{Par}(k)} f_\mu(x_1,\ldots,x_N) g_\mu(y_1,\ldots,y_N).$$

In particular, this identity holds with $f_\mu = g_\mu = s_\mu$; $f_\mu = m_\mu$ and $g_\mu = h_\mu$; and $f_\mu = p_\mu$ and $g_\mu = p_\mu / z_\mu$.

- *The Map ω.* There is a unique algebra isomorphism $\omega : \Lambda_N \to \Lambda_N$ such that $\omega(p_i) = (-1)^{i-1} p_i$ for all i between 1 and N. We have $\omega(p_\mu) = \epsilon_\mu p_\mu$, where $\epsilon_\mu = (-1)^{|\mu|-\ell(\mu)}$; $\omega(e_\mu) = h_\mu$, $\omega(h_\mu) = e_\mu$, and $\omega(s_\mu) = s_{\mu'}$. The map ω is an involution ($\omega \circ \omega = \text{id}$). For $k \leq N$, ω is an isometry of Λ_N^k, which means $\langle \omega(f), \omega(g)\rangle = \langle f, g\rangle$ for all $f, g \in \Lambda_N^k$.

- *RSK Correspondences.* There are bijections between: (a) permutations in S_n and pairs (P,Q) of standard tableaux of the same shape $\lambda \in \text{Par}(n)$; (b) words in $[N]^n$ and pairs (P,Q) where $P \in \text{SSYT}_N(\lambda)$ and $Q \in \text{SYT}(\lambda)$ for some $\lambda \in \text{Par}(n)$; (c) $M \times N$ matrices with values in $\mathbb{Z}_{\geq 0}$ and pairs (P,Q) where $P \in \text{SSYT}_N(\lambda)$ and $Q \in \text{SSYT}_M(\lambda)$. In each case, one inserts successive entries into P, using Q to record the locations of new boxes. For (c), one must first encode the matrix as a biword. If $w \in S_n$ maps to (P,Q), then w^{-1} maps to (Q,P). If $w \in [N]^n$ maps to (P,Q), then $\text{Des}(w) = \text{Des}(Q)$, $\text{des}(w) = \text{des}(Q)$, and $\text{maj}(w) = \text{maj}(Q)$, where $\text{Des}(Q)$ is the set of $k < n$ such that $k+1$ is in a lower row of Q than k, $\text{des}(Q) = |\text{Des}(Q)|$, and $\text{maj}(Q) = \sum_{k \in \text{Des}(Q)} k$.

- *Symmetric Functions.* Viewing each power-sum p_n as a formal variable, we can define the algebra of symmetric functions $\Lambda = K[p_n : n \in \mathbb{Z}_{>0}]$. Symmetric functions h_μ, e_μ, and s_μ can be defined via their expansions in the power-sum basis. We can define evaluation homomorphisms with domain Λ by sending each p_n to any element of a given K-algebra. Many identities for symmetric polynomials in finitely many variables extend to the setting of symmetric functions. In particular, $\{e_n : n \in \mathbb{Z}_{>0}\}$ and $\{h_n : n \in \mathbb{Z}_{>0}\}$ are algebraically independent subsets of Λ.

Exercises

9-1. Consider fillings of shape $\mu \in \text{Par}(k)$ using values in $[N]$. (a) How many fillings are there? (b) How many fillings have weakly increasing rows? (c) How many fillings have strictly increasing columns?

9-2. List all the tableaux in: (a) $\text{SSYT}_5((3,2))$; (b) $\text{SSYT}_2((3,2))$; (c) $\text{SYT}((3,2,1))$.

9-3. Give a direct counting argument to find $|\text{SYT}(\mu)|$ when $\mu = (a, 1^b)$ is a hook shape.

9-4. Compute $s_{(2,2)}(x_1, \ldots, x_N)$ for $N = 3, 4, 5$ by enumerating tableaux.

9-5. Find the coefficients of the following monomials in $s_{(3,2,1)}(x_1, \ldots, x_6)$ by enumerating tableaux: (a) $x_1x_2x_3x_4x_5x_6$; (b) $x_1^2x_2^2x_3^2$; (c) $x_1^3x_2^3$; (d) $x_1^2x_2x_3x_4x_5$; (e) $x_1x_2x_3^2x_4x_5$; (f) $x_1x_2x_3x_4x_5^2$.

9-6. List all terms in: (a) $p_4(x_1, x_2, x_3)$; (b) $e_3(x_1, x_2, x_3, x_4, x_5)$; (c) $h_3(x_1, x_2, x_3)$; (d) $m_{(3,2,2)}(x_1, x_2, x_3, x_4)$.

9-7. For $\mu = (2,1)$, compute: (a) $p_\mu(x_1, x_2, x_3)$; (b) $e_\mu(x_1, x_2, x_3)$; (c) $h_\mu(x_1, x_2, x_3)$.

9-8. Find how many monomials appear with nonzero coefficient in: (a) $e_k(x_1, \ldots, x_N)$; (b) $h_k(x_1, \ldots, x_N)$; (c) $m_\mu(x_1, \ldots, x_N)$.

9-9. Give a direct proof that the polynomials $e_k(x_1, \ldots, x_N)$ and $h_k(x_1, \ldots, x_N)$ (as defined in Definitions 9.15 and 9.16) are symmetric.

9-10. Check that the set of homogeneous polynomials of degree k is a vector subspace of $\mathbb{R}[x_1, \ldots, x_N]$ for all $k \geq 0$. Conclude that Λ_N^k is a subspace of Λ_N for each $k \geq 0$.

9-11. List five bases for the real vector space Λ_3^7.

9-12. Compute the dimension of Λ_N^k for all k, N in the range $1 \leq k \leq 6$ and $1 \leq N \leq 6$.

9-13. Suppose $\{f_i : i \in I\}$ is a collection of nonzero polynomials in $\mathbb{R}[x_1, \ldots, x_N]$ such that, whenever some $\mathbf{x}^\alpha$ appears in some f_i with nonzero coefficient, the coefficient of $\mathbf{x}^\alpha$ in every other f_j is zero. Prove that $\{f_i : i \in I\}$ is linearly independent.

9-14. Compute the following Kostka numbers: (a) $K_{(3,3,2),(2,1,2,1,1,1)}$; (b) $K_{(3,2,2,1),(2,2,1,1,1,1)}$; (c) $K_{(5,5),(1^{10})}$; (d) $K_{(3,3,3)/(2,1),(2,2,1,1)}$.

9-15. Compute the image of the first tableau in the proof of Theorem 9.27 under the maps f_i, for $i = 1, 2, 4, 5, 6, 7, 8$.

9-16. Express the Schur polynomials $s_\mu(x_1, x_2, x_3, x_4, x_5)$ as explicit linear combinations of monomial symmetric polynomials, for all μ in $\text{Par}(4)$ and $\text{Par}(5)$.

9-17. (a) Find a recursion characterizing the Kostka numbers $K_{\mu/\nu,\alpha}$. (b) Use (a) to write a computer program for computing Kostka numbers.

9-18. Check that $\leq_{\text{lex}}$ is a total ordering of the set $\text{Par}(k)$, for each $k \geq 0$.

9-19. Prove that $\trianglelefteq$ is a total ordering of $\text{Par}(k)$ iff $k \leq 5$.

9-20. (a) List the integer partitions of 7 in lexicographic order. (b) Find all pairs $\mu, \nu \in \text{Par}(7)$ such that $\mu \leq_{\text{lex}} \nu$ but $\mu \not\trianglelefteq \nu$.

9-21. (a) Find an ordered sequence of raising operators that changes $\mu = (5,4,2,1,1)$ to $\nu = (7,3,2,1)$. (b) How many such sequences are there?

9-22. Prove or disprove: for all partitions $\mu, \nu \in \text{Par}(k)$, $\mu \leq_{\text{lex}} \nu$ iff $\nu' \leq_{\text{lex}} \mu'$.

9-23. Let $\mu, \nu \in \text{Par}(k)$. Can you prove that $\mu \trianglelefteq \nu$ implies $\nu' \trianglelefteq \mu'$ directly from the definitions, without using raising operators?

9-24. Define an ordering $\leq_{\text{lex}}$ on the set $\mathbb{Z}_{\geq 0}^N$ as in Definition 9.30. Show that $\leq_{\text{lex}}$ is a total ordering of $\mathbb{Z}_{\geq 0}^N$ satisfying the following *well-ordering* property: every nonempty subset of $\mathbb{Z}_{\geq 0}^N$ has a least element relative to $\leq_{\text{lex}}$.

9-25. Define the *lex degree* of a nonzero polynomial $f(x_1, \ldots, x_N) \in \mathbb{R}[x_1, \ldots, x_N]$, denoted $\text{degl}(f)$, to be the largest $\alpha \in \mathbb{Z}_{\geq 0}^N$ (relative to the lexicographic ordering defined in the previous exercise) such that $\mathbf{x}^\alpha$ occurs with nonzero coefficient in f. Prove that $\text{degl}(gh) = \text{degl}(g) + \text{degl}(h)$ for all nonzero g, h, and $\text{degl}(g+h) \leq \max(\text{degl}(g), \text{degl}(h))$ whenever both sides are defined.

9-26. (a) Find the Kostka matrix indexed by all partitions of 4. (b) Invert this matrix, and thereby express the monomial symmetric polynomials $m_\mu(x_1, x_2, x_3, x_4)$ (for $\mu \in \text{Par}(4)$) as linear combinations of Schur polynomials.

9-27. Find the Kostka matrix indexed by partitions in $\text{Par}_3(7)$. Invert this matrix.

9-28. Let $\mathbf{K}$ be the Kostka matrix indexed by all partitions of 8. How many nonzero entries does this matrix have?

9-29. Suppose A is an $n \times n$ matrix with integer entries such that $\det(A) = \pm 1$. Prove that A^{-1} has all integer entries. In particular, this problem applies when A is a Kostka matrix. [*Hint:* See Corollary 12.54.]

9-30. Suppose $\{v_i : i \in I\}$ is a basis for a finite-dimensional real vector space V, $\{w_i : i \in I\}$ is an indexed family of vectors in V, and for some total ordering $\leq$ of I and some scalars $a_{ij} \in \mathbb{R}$ with $a_{ii} \neq 0$, we have $w_i = \sum_{j \leq i} a_{ij} v_j$ for all $i \in I$. Prove that $\{w_i : i \in I\}$ is a basis of V.

9-31. Fix $N \geq k$, and define column vectors $\mathbf{s}$, $\mathbf{m}$, $\mathbf{e}$, $\mathbf{h}$ whose entries are s_μ, m_μ, e_μ, and h_μ (respectively) with μ ranging over partitions of k. For each ordered pair of column vectors $\mathbf{v}$, $\mathbf{w}$ in this list, find the matrix $\mathbf{A}$ such that $\mathbf{v} = \mathbf{A}\mathbf{w}$. Express each answer using the Kostka matrix and its variations.

9-32. Let T be the tableau in Example 9.47. Confirm that $T \leftarrow 1$ and $T \leftarrow 0$ are as stated in that example. Also, compute $T \leftarrow i$ for $i = 2, 4, 5, 7$, and verify that Theorem 9.48 holds.

9-33. Let T be the semistandard tableau shown here:

$$\begin{array}{|c|c|c|c|c|c|c|}
\hline
2 & 2 & 3 & 5 & 5 & 7 & 7 \\ \hline
3 & 3 & 4 & 6 & 7 & 8 \\ \cline{1-6}
4 & 5 & 5 & 8 & 8 \\ \cline{1-5}
6 & 6 & 6 & 9 \\ \cline{1-4}
7 & 8 & 8 \\ \cline{1-3}
8 \\ \cline{1-1}
\end{array}$$

Compute $T \leftarrow i$ for $1 \leq i \leq 9$.

9-34. Give a non-recursive description of $T \leftarrow x$ in the case where: (a) x is larger than every entry of T; (b) x is smaller than every entry of T.

9-35. Let T be the tableau in Example 9.47. Perform reverse insertion starting at each corner box of T to obtain smaller tableaux T_i and values x_i. Verify that $T_i \leftarrow x_i = T$ for each answer.

9-36. Let T be the tableau in Exercise 9-33. Perform reverse insertion starting at each corner box of T, and verify that Theorem 9.52(a) and (b) hold in each case.

9-37. Prove Theorem 9.52(c).

9-38. Prove Theorem 9.53.

9-39. Express $s_{(4,4,3,1,1)}h_1$ as a sum of Schur polynomials.

9-40. Let T be the tableau in Example 9.47. Successively insert $1, 2, 2, 3, 5, 5$ into T, and verify that the assertions of the Bumping Comparison Theorem hold.

9-41. Let T be the tableau in Example 9.47. Successively insert $7, 5, 3, 2, 1$ into T, and verify that the assertions of the Bumping Comparison Theorem hold.

9-42. Let T be the tableau in Exercise 9-33. Successively insert $1, 1, 3, 3, 3, 4$ into T, and verify that the assertions of the Bumping Comparison Theorem hold.

9-43. Let T be the tableau in Exercise 9-33. Successively insert $7, 6, 5, 3, 2, 1$ into T, and verify that the assertions of the Bumping Comparison Theorem hold.

9-44. Let T be the tableau in Example 9.55 of shape $\mu = (5, 4, 4, 4, 1)$. For each shape ν such that ν/μ is a horizontal strip of size 3, find a weakly increasing sequence $x_1 \leq x_2 \leq x_3$ such that $(((T \leftarrow x_1) \leftarrow x_2) \leftarrow x_3)$ has shape ν, or prove that no such sequence exists.

9-45. Repeat the previous exercise, replacing horizontal strips by vertical strips and weakly increasing sequences by strictly decreasing sequences.

9-46. Prove Theorem 9.59(b).

9-47. Let T be the tableau in Exercise 9-33. Find the unique semistandard tableau S of shape $(7, 5, 4, 4, 1, 1)$ and $z_1 \leq z_2 \leq z_3 \leq z_4$ such that $T = S \leftarrow z_1 z_2 z_3 z_4$.

9-48. Let T be the tableau in Exercise 9-33. Find the unique semistandard tableau S of shape $(6, 6, 5, 3, 2)$ and $z_1 > z_2 > z_3 > z_4$ such that $T = S \leftarrow z_1 z_2 z_3 z_4$.

9-49. Expand each symmetric polynomial into sums of Schur polynomials: (a) $s_{(4,3,1)}e_2$; (b) $s_{(2,2)}h_3$; (c) $s_{(2,2,1,1,1)}h_4$; (d) $s_{(3,3,2)}e_3$.

9-50. Use the Pieri Rule to find the Schur expansions of $h_{(3,2,1)}$, $h_{(3,1,2)}$, $h_{(1,2,3)}$, and $h_{(1,3,2)}$, and verify that the answers agree with those found in Example 9.62.

9-51. Expand each symmetric polynomial into sums of Schur polynomials: (a) $h_{(2,2,2)}$; (b) $h_{(5,3)}$; (c) $s_{(3,2)}h_{(2,1)}$; (d) $s_{(6,3,2,2)/(3,2)}$.

9-52. Find the coefficients of the following Schur polynomials in the Schur expansion of $h_{(3,2,2,1,1)}$: (a) $s_{(9)}$; (b) $s_{(5,4)}$; (c) $s_{(4,4,1)}$; (d) $s_{(2,2,2,2,1)}$; (e) $s_{(3,3,3)}$; (f) $s_{(3,2,2,1,1)}$.

9-53. Use Remark 9.67 to compute the monomial expansions of $h_\mu(x_1, x_2, x_3, x_4)$ for all partitions μ of size at most four.

9-54. Let $\alpha = (\alpha_1, \ldots, \alpha_s)$. Prove that the coefficient of $m_\lambda(x_1, \ldots, x_N)$ in the monomial expansion of $h_\alpha(x_1, \ldots, x_N)$ is the number of $s \times N$ matrices A with entries in $\mathbb{Z}_{\geq 0}$ such that $\sum_{j=1}^{N} A(i,j) = \alpha_i$ for $1 \leq i \leq s$ and $\sum_{i=1}^{s} A(i,j) = \lambda_j$ for $1 \leq j \leq N$.

9-55. Find and prove a combinatorial interpretation (similar to the one in the previous exercise) for the coefficient of $m_\lambda(x_1, \ldots, x_n)$ in the monomial expansion of $e_\alpha(x_1, \ldots, x_N)$.

9-56. Use the Pieri Rules to compute the Schur expansions of: (a) $e_{(3,3,1)}$; (b) $e_{(5,3)}$; (c) $s_{(3,2)}e_{(2,1)}$; (d) $s_{(4,3,3,3,1,1)/(3,1,1,1)}$.

9-57. Find the coefficients of the following Schur polynomials in the Schur expansion of $e_{(4,3,2,1)}$: (a) $s_{(4,3,2,1)}$; (b) $s_{(5,5)}$; (c) $s_{(2,2,2,2,2)}$; (d) $s_{(2,2,2,1^4)}$; (e) $s_{(1^{10})}$.

9-58. Use Remark 9.72 to express $e_{(2,2,1)}$ and $e_{(3,2)}$ as linear combinations of monomial symmetric polynomials.

9-59. Prove: for all $N \geq k$, $p_k(x_1, \ldots, x_N) = \sum_{a=0}^{k-1} (-1)^a s_{(k-a,1^a)}(x_1, \ldots, x_N)$.

9-60. Express $s_{(2,2)} p_3$ as a sum of Schur polynomials.

9-61. Conjecture a formula for expressing $s_\lambda p_k$ as a linear combination of Schur polynomials. Can you prove your formula?

9-62. Prove that the following lists of polynomials are algebraically dependent by exhibiting a non-trivial linear combination of monomials equal to zero: (a) $h_i(x_1, x_2)$ for $1 \leq i \leq 3$; (b) $e_i(x_1, x_2, x_3)$ for $1 \leq i \leq 4$; (c) $p_i(x_1, x_2, x_3)$ for $1 \leq i \leq 4$.

9-63. Prove that any sublist of an algebraically independent list is algebraically independent.

9-64. Suppose $\alpha = (\alpha_1, \ldots, \alpha_N) \in \mathbb{Z}_{\geq 0}^N$ is a partition. Show that

$$\operatorname{degl}(e_1^{\alpha_1-\alpha_2} e_2^{\alpha_2-\alpha_3} \cdots e_{N-1}^{\alpha_{N-1}-\alpha_N} e_N^{\alpha_N}) = \alpha$$

(see Exercise 9-25 for the definition of lex degree).

9-65. Algorithmic Proof of the Fundamental Theorem of Symmetric Polynomials. Prove that the following algorithm will express any $f \in \Lambda_N$ as a polynomial in the elementary symmetric polynomials $e_i(x_1, \ldots, x_N)$ (where $1 \leq i \leq N$) in finitely many steps. If $f = 0$, use the zero polynomial. Otherwise, let the term of largest lex degree in f be $c\mathbf{x}^\alpha$ where $c \in \mathbb{R}$ is nonzero. Use symmetry of f to show that $\alpha_1 \geq \alpha_2 \geq \cdots \geq \alpha_N$, and that $f - ce_1^{\alpha_1-\alpha_2} e_2^{\alpha_2-\alpha_3} \cdots e_{N-1}^{\alpha_{N-1}-\alpha_N} e_N^{\alpha_N}$ is either 0 or has lex degree $\beta <_{\text{lex}} \alpha$. Continue similarly to express this new polynomial (and hence f) as a polynomial in the e_i.

9-66. Use the algorithm in the preceding exercise to express $m_{(2,1)}(x_1, x_2, x_3, x_4)$ and $p_3(x_1, x_2, x_3, x_4)$ as polynomials in $\{e_i(x_1, x_2, x_3, x_4) : 1 \leq i \leq 4\}$.

9-67. Use the test in Theorem 9.76 to verify that the polynomials $h_i(x_1, x_2, x_3)$ for $1 \leq i \leq 3$ are algebraically independent. Can you generalize this computation to more than three variables?

9-68. Use the test in Theorem 9.76 to verify that the polynomials $e_i(x_1, x_2, x_3, x_4)$ for $1 \leq i \leq 4$ are algebraically independent. Can you generalize this computation to more than four variables?

9-69. Compute the images of

$$\left(4, \begin{array}{|c|}\hline 1\\\hline 3\\\hline 4\\\hline 5\\\hline\end{array}, \begin{array}{|c|c|c|c|}\hline 1&2&2&3\\\hline\end{array}\right) \text{ and } \left(4, \begin{array}{|c|}\hline 2\\\hline 3\\\hline 4\\\hline 5\\\hline\end{array}, \begin{array}{|c|c|c|c|}\hline 1&1&2&3\\\hline\end{array}\right)$$

under the involution I in the proof of Theorem 9.81.

9-70. List all the matched pairs $(z, I(z))$ in the proof of Theorem 9.81 when: (a) $N = 2$ and $m = 3$; (b) $N = 3$ and $m = 2$.

9-71. Imitate the proof of Theorem 9.82 to show that algebraic independence of $h_1, \ldots, h_N$ in $\mathbb{R}[x_1, \ldots, x_N]$ implies algebraic independence of $e_1, \ldots, e_N$.

9-72. (a) Prove the recursion $e_k(x_1, \ldots, x_N) = e_k(x_1, \ldots, x_{N-1}) + e_{k-1}(x_1, \ldots, x_{N-1})x_N$ for $k, N \geq 1$. What are the initial conditions? (b) Find a similar recursion for $h_k(x_1, \ldots, x_N)$.

9-73. (a) Prove $s'(n, k) = e_{n-k}(1, 2, \ldots, n-1)$. (b) Prove $S(n, k) = h_{n-k}(1, 2, \ldots, k)$.

9-74. Prove Theorem 9.86 by expanding $\prod_{i=1}^N (X - r_i)$ using the Generalized Distributive Law.

9-75. Consider the polynomial $p = x^5 - 2x^4 + 5x^3 + 7x^2 - x - 4$, which has five roots $r_1, \ldots, r_5 \in \mathbb{C}$. Compute: (a) the sum of the roots; (b) the product of the roots; (c) $e_3(r_1, \ldots, r_5)$; (d) the sum of the squares of the roots; (e) $\sum_{i \neq j} r_i^2 r_j$.

9-76. Use Theorem 9.87 to calculate the coefficient of x^4 in the multiplicative inverse of $(1-2x)(1-3x)(1-5x)$.

9-77. Let A be an $n \times n$ complex matrix. What is the relationship between the coefficients of the characteristic polynomial $\det(tI - A)$ and the eigenvalues $r_1, \ldots, r_n$ of A?

9-78. Use (9.10) to show that $p_i(x_1, \ldots, x_N)$ (for $1 \leq i \leq N$) are algebraically independent iff $h_i(x_1, \ldots, x_N)$ (for $1 \leq i \leq N$) are algebraically independent.

9-79. Use (9.11) to show that $p_i(x_1, \ldots, x_N)$ (for $1 \leq i \leq N$) are algebraically independent iff $e_i(x_1, \ldots, x_N)$ (for $1 \leq i \leq N$) are algebraically independent.

9-80. Consider the maps f and g from the proof of Theorem 9.88. Compute

$$f\left(5, \begin{array}{|c|c|c|c|c|}\hline 2&4&4&5&5\\\hline\end{array}, \begin{array}{|c|c|}\hline 4&4\\\hline\end{array}\right) \text{ and } g\left(\begin{array}{|c|c|c|c|c|c|c|c|}\hline 1&1^*&1&2&2&4&6&6\\\hline\end{array}\right).$$

9-81. Consider the maps I and g from the proof of Theorem 9.89. Compute

$$I\left(3, \begin{array}{|c|}\hline 1\\\hline 3\\\hline 4\\\hline\end{array}, \begin{array}{|c|c|c|}\hline 3&3&3\\\hline\end{array}\right), \quad I\left(3, \begin{array}{|c|}\hline 1\\\hline 3\\\hline 4\\\hline\end{array}, \begin{array}{|c|c|c|}\hline 2&2&2\\\hline\end{array}\right), \quad I\left(3, \begin{array}{|c|}\hline 1\\\hline 3\\\hline 4\\\hline\end{array}, \begin{array}{|c|}\hline 5\\\hline\end{array}\right), \quad I\left(3, \begin{array}{|c|}\hline 1\\\hline 3\\\hline 4\\\hline\end{array}, \begin{array}{|c|}\hline 4\\\hline\end{array}\right).$$

For any objects that are fixed points of I, compute the images of those objects under g.

9-82. Write $\sum_{i=1}^{N} \frac{x_i}{1-x_i t}$ in terms of symmetric polynomials.

9-83. Obtain Theorems 9.88 and 9.89 algebraically by differentiating the generating functions $H_N(t)$ and $E_N(-t)$.

9-84. Use the recursions in Theorems 9.88 and 9.89 to verify the formulas for h_4, $3!e_3$, and $4!e_4$ stated in Example 9.90.

9-85. Complete the proof of Theorem 9.91 by checking that $g(f(y_0)) = y_0$ and $f(g(z_0)) = z_0$, and, in general, $g \circ f = \mathrm{id}_Y$ and $f \circ g = \mathrm{id}_X$.

9-86. Let g be the map in the proof of Theorem 9.91. Compute $g(z_1)$ and $g(z_2)$, where

$$z_1 = \begin{bmatrix} w: & 3 & 7 & 2 & 5 & 4 & 6 & 8 & 1 \\ T: & 2 & 4 & 4 & 4 & 4 & 6 & 6 & 6 \end{bmatrix}; \quad z_2 = \begin{bmatrix} w: & 2 & 1 & 4 & 3 & 7 & 5 & 6 & 8 \\ T: & 1 & 1 & 1 & 2 & 3 & 3 & 3 & 3 \end{bmatrix}.$$

9-87. Let f be the map in the proof of Theorem 9.91. Compute $f(y)$, where

$$y = \left((2,2,2,2), (2,5)(3,8)(4,6)(1,7), \begin{bmatrix} 1 & 2 & 3 & 4 & 5 & 6 & 7 & 8 \\ 3 & 2 & 3 & 2 & 2 & 2 & 3 & 3 \end{bmatrix}\right).$$

9-88. Let I be the involution in the proof of Theorem 9.93. Compute $I(z_1)$, $I(z_2)$, and $I(f(y))$, where z_1, z_2, and y are the objects given in the preceding two exercises.

9-89. Let A be an $n \times n$ complex matrix with eigenvalues $r_1, \ldots, r_n$. (a) Show that the trace of A, defined by $\mathrm{tr}(A) = \sum_{i=1}^{n} A(i,i)$, is $p_1(r_1, \ldots, r_n)$. (b) For $k \geq 1$, express $\mathrm{tr}(A^k)$ as a function of $r_1, \ldots, r_n$. (c) Suppose $n = 5$ and $(\mathrm{tr}(A^k) : k = 1, 2, \ldots, 5) = (3, 41, -93, 693, -2957)$. Find the characteristic polynomial of A.

9-90. Compute: (a) $\omega(h_3)$; (b) $\omega(p_{(3,2,1,1)})$; (c) $\omega(e_{(4,4)})$; (d) $\omega(s_{(5,3,3,1,1,1)})$.

9-91. Use facts about ω to deduce Theorem 9.69 from Theorem 9.64.

9-92. Use facts about ω to deduce Theorem 9.89 from Theorem 9.88.

9-93. Use ω to deduce Theorem 9.93 from Theorem 9.91.

9-94. Show that there exists an algebra isomorphism of Λ_N sending each p_i to $-p_i$. Compute the images of h_n and e_n under this isomorphism.

9-95. In the proof of Theorem 9.95(b), where is the assumption $n \leq N$ needed?

9-96. Compute the polynomials $\text{fgt}_\lambda(x_1, x_2, x_3)$ for all partitions of size at most 3.

9-97. Compute $\text{RSK}(w)$ for all $w \in S_3$.

9-98. Compute $\text{RSK}^{-1}(P, Q)$ for all pairs P, Q of standard tableaux of shape $(2, 2)$.

9-99. Let $w = 41572863 \in S_8$. Compute $\text{RSK}(w)$ and $\text{RSK}(w^{-1})$. Verify that Theorem 9.106 holds in this case.

9-100. Given the pair of standard tableaux:

$$P = \begin{array}{|c|c|c|}\hline 1 & 3 & 6 \\ \hline 2 & 4 & 8 \\ \hline 5 & 7 \\ \cline{1-2}\end{array}, \qquad Q = \begin{array}{|c|c|c|}\hline 1 & 2 & 5 \\ \hline 3 & 6 & 7 \\ \hline 4 & 8 \\ \cline{1-2}\end{array}$$

Compute $w = \text{RSK}^{-1}(P, Q)$ and $v = \text{RSK}^{-1}(Q, P)$, and verify that Theorem 9.106 holds in this case.

9-101. (a) Verify that Theorem 9.103 holds for the example $w = 35164872$ by comparing the first rows of the tableaux in Figure 9.1 with the shadow diagram in Figure 9.4. (b) Verify the assertions in Theorem 9.105 using Figure 9.5.

9-102. Draw all the shadow diagrams for the permutations w and w^{-1} in Exercise 9-99, and use them to verify the assertions in Theorem 9.105 for this example.

9-103. Draw all the shadow diagrams for the permutations w and v in Exercise 9-100, and use them to verify the assertions in Theorem 9.105 for this example.

9-104. (a) Prove: for all $n \geq 1$, $n! = \sum_{\lambda \in \text{Par}(n)} |\text{SYT}(\lambda)|^2$. (b) Verify this identity directly for $n = 5$.

9-105. Show that the number of $w \in S_n$ such that $w^2 = \text{id}$ is given by $\sum_{\lambda \in \text{Par}(n)} |\text{SYT}(\lambda)|$.

9-106. Suppose $w' \in S_{n-1}$ is obtained from $w \in S_n$ by deleting n from the one-line form of w. How is $P(w')$ related to $P(w)$?

9-107. Compute $\text{RSK}(w)$ for all words $w \in \{1, 2\}^3$.

9-108. Compute RSK(313211231), and verify that Theorem 9.111 holds in this case.

9-109. Compute the word w such that

$$\text{RSK}(w) = \left(\begin{array}{|c|c|c|c|}\hline 1 & 1 & 2 & 2 \\ \hline 2 & 3 & 4 & 4 \\ \hline 4 & 5 & 5 \\ \cline{1-3}\end{array}, \ \begin{array}{|c|c|c|c|}\hline 1 & 2 & 4 & 6 \\ \hline 3 & 5 & 8 & 10 \\ \hline 7 & 9 & 11 \\ \cline{1-3}\end{array} \right).$$

Verify that Theorem 9.111 holds in this case.

9-110. (a) Compute $\sum_{T \in \text{SYT}((4,1))} q^{\text{maj}(T)}$. (b) Compute $\sum_{T \in \text{SYT}((3,2,1))} q^{\text{maj}(T)}$.

9-111. The *reading word* of a semistandard tableau T is the word $\text{rw}(T)$ obtained by listing the values in each row of T from left to right, working from the bottom row to the top row. Prove that $P(\text{rw}(T)) = T$. What is $Q(\text{rw}(T))$?

9-112. Given $w = w_1 \cdots w_n \in S_n$, let $\text{rev}(w) = w_n \cdots w_1$ be the reversal of w. By looking at examples, conjecture a relationship between $P(\text{rev}(w))$ and $P(w)$. Can you prove your conjecture? Does the conjecture extend to arbitrary words w?

9-113. Express $p_{(1^4)}$ as a linear combination of Schur polynomials.

9-114. (a) Compute the biword and the pair of tableaux associated to the matrix $\begin{bmatrix} 1 & 0 & 2 \\ 0 & 1 & 1 \\ 3 & 2 & 0 \end{bmatrix}$. (b) Do the same for the transpose of this matrix.

9-115. (a) Compute the matrix and pair of tableaux associated to the biword

$$\begin{bmatrix} 1 & 1 & 2 & 2 & 2 & 3 & 3 & 5 \\ 2 & 4 & 1 & 1 & 3 & 3 & 3 & 2 \end{bmatrix}.$$

(b) Do the same for the biword obtained by switching the two rows and sorting the new top row into increasing order (using the values in the bottom row to break ties).

9-116. (a) Compute the biword and matrix associated to the pair of tableaux:

$$P = \begin{array}{|c|c|c|} \hline 1 & 1 & 3 \\ \hline 2 & 3 \\ \cline{1-2} 4 & 4 \\ \cline{1-2} 5 & 5 \\ \cline{1-2} \end{array} \qquad Q = \begin{array}{|c|c|c|} \hline 1 & 2 & 2 \\ \hline 3 & 3 \\ \cline{1-2} 4 & 5 \\ \cline{1-2} 6 & 6 \\ \cline{1-2} \end{array}$$

(b) Do the same for the pair of tableaux (Q, P).

9-117. Show that if a matrix A maps to (P, Q) under the RSK correspondence, then the transpose matrix A^{tr} maps to (Q, P) under RSK. Do this by generalizing the shadow constructions in §9.21, allowing more than one dot to occupy a given point (i, j) in the graph.

9-118. Verify the fact (used in the proof of Theorem 9.124) that the polynomials

$$\{p_\alpha(\mathbf{x})p_\beta(\mathbf{y})/z_\beta : (\alpha, \beta) \in \text{Par}(k) \times \text{Par}(k)\}$$

are linearly independent.

9-119. Suppose A and B are $n \times n$ matrices such that $AB = I$. Prove that $BA = I$.

9-120. Suppose $\{f_\mu : \mu \in \text{Par}(k)\}$ is an orthonormal basis of Λ_N^k, and $g \in \Lambda_N^k$. Prove that $g = \sum_{\mu \in \text{Par}(k)} \langle g, f_\mu \rangle f_\mu$.

9-121. Prove: $\prod_{i=1}^M \prod_{j=1}^N (1 + x_i y_j) = \sum_{\mu \in \text{Par}} e_\mu(x_1, \ldots, x_M) m_\mu(y_1, \ldots, y_N)$.

9-122. Prove: $\prod_{i=1}^M \prod_{j=1}^N (1 + x_i y_j) = \sum_{\mu \in \text{Par}} \epsilon_\mu p_\mu(x_1, \ldots, x_M) p_\mu(y_1, \ldots, y_N)/z_\mu$.

9-123. Prove: $\prod_{i=1}^M \prod_{j=1}^N (1+x_i y_j) = \sum_{\mu \in \text{Par}} s_\mu(x_1, \ldots, x_M) s_{\mu'}(y_1, \ldots, y_N)$. (Hint: Use ω.)

9-124. Given a skew shape $S \subseteq \mathbb{Z}_{\geq 0}^2$, describe how to calculate the number of different pairs of partitions (μ, ν) such that $S = \mu/\nu$.

9-125. Find necessary and sufficient algebraic conditions on the parts of μ and ν to ensure that the skew shape μ/ν is: (a) a horizontal strip; (b) a vertical strip.

9-126. How many horizontal strips are contained in $\{1, 2, \ldots, a\} \times \{1, 2, \ldots, b\}$?

9-127. Prove that $|\,\text{SYT}(\mu/\nu)| = |\,\text{SYT}(\mu'/\nu')|$ for all skew shapes μ/ν.

9-128. Compute $s_{(3,2)/(1)}(x_1, \ldots, x_N)$ for $N = 2, 3, 4$ by enumerating tableaux.

9-129. Let $N \geq 4$. Enumerate tableaux to confirm that the coefficients of $x_1^2 x_2 x_3^2 x_4$, $x_1 x_2^2 x_3 x_4^2$, and $x_1 x_2^2 x_3^2 x_4$ in $s_{(4,3)/(1)}(x_1, \ldots, x_N)$ are all equal to 6. What happens to these coefficients if $N < 4$?

9-130. For which values of N is $s_{\mu/\nu}(x_1, \ldots, x_N) = 0$?

9-131. Find a skew shape μ/ν such that $e_3 h_4 h_2 e_5 h_1 = s_{\mu/\nu}$.

9-132. Prove that any finite product of skew Schur polynomials is a skew Schur polynomial.

9-133. Express the skew Schur polynomial $s_{(3,3,2)/(1)}(x_1, \ldots, x_8)$ as a linear combination of monomial symmetric polynomials.

9-134. For all partitions $\mu \in \text{Par}(3)$, express h_μ and e_μ in terms of monomial symmetric polynomials by viewing h_μ and e_μ as instances of skew Schur polynomials.

9-135. Suppose we apply the Tableau Insertion Algorithm 9.46 to a tableau T of skew shape. Are Theorems 9.48 and 9.49 still true?

9-136. Prove Theorem 9.135.

9-137. Prove Theorem 9.136.

Notes

Ian Macdonald's book [84] contains a comprehensive treatment of symmetric polynomials, with a heavy emphasis on algebraic methods. A more combinatorial development is given by Stanley in Chapter 7 of [121]; see the references to that chapter for an extensive bibliography of the literature in this area. Two other relevant references are [41], which treats tableaux and their connections to representation theory and geometry, and [117], which explains the role of symmetric polynomials in the representation theory of symmetric groups.

The bijective proof of Theorem 9.27 is due to Bender and Knuth [8]. The algorithmic proof of the existence part of the fundamental theorem of symmetric polynomials (outlined in Exercise 9-65) is usually attributed to Waring [124]. Some of the seminal papers by Robinson, Schensted, and Knuth on the RSK correspondence are [74, 111, 118]. The symmetry property in Theorem 9.106 was first proved by Schützenberger [120]; the combinatorial proof using shadow lines is due to Viennot [129].

10

Abaci and Antisymmetric Polynomials

In Chapter 9, we used combinatorial operations on tableaux to establish algebraic properties of Schur polynomials and other symmetric polynomials. This chapter investigates the interplay between the combinatorics of abaci and the algebraic properties of antisymmetric polynomials. These concepts are used to prove additional facts about integer partitions and symmetric polynomials. In particular, we derive some formulas for expanding skew Schur polynomials in terms of various bases. Key results include the Jacobi–Trudi Formulas and the Littlewood–Richardson Rule for the Schur expansion of the product of two Schur polynomials.

10.1 Abaci and Integer Partitions

An *abacus* is an instrument used in ancient times for performing arithmetical calculations. The abacus consists of one or more runners that contain sliding beads. The following combinatorial object gives a mathematical model of an abacus.

10.1. Definition: One-Runner Abacus. An *abacus with one runner* is a function $w : \mathbb{Z} \to \{0,1\}$ such that for some m and n, $w_i = 1$ for all $i \leq m$ and $w_i = 0$ for all $i \geq n$. We think of w as an infinite word $\cdots w_{-2}w_{-1}\underline{w_0}w_1w_2w_3\cdots$ that begins with an infinite string of 1's and ends with an infinite string of 0's. Each 1 is called a *bead*, and each 0 is called a *gap*. Let Abc denote the set of all 1-runner abaci. An abacus w is called *justified at position* m iff $w_i = 1$ for all $i \leq m$ and $w_i = 0$ for all $i > m$. Intuitively, an abacus is justified iff all the beads have been pushed to the left as far as they will go. The *weight* of an abacus w, denoted $\mathrm{wt}(w)$, is the number of pairs $i < j$ with $w_i < w_j$ (or equivalently, $w_i = 0$ and $w_j = 1$).

10.2. Example. Here is a picture of a 1-runner abacus:

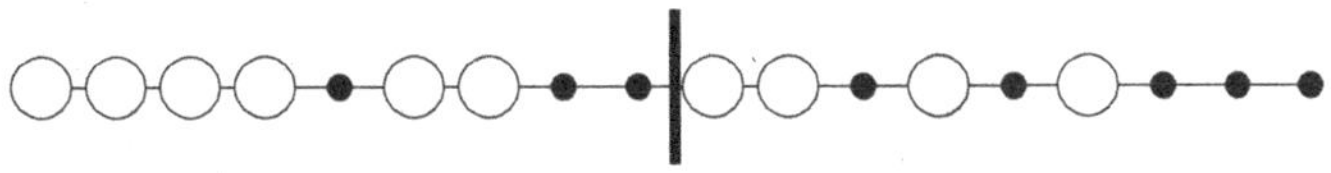

This picture corresponds to the mathematical abacus

$$w = \cdots 111101100\underline{1}10101000\cdots,$$

where the underlined 1 is w_0. All positions to the left of the displayed region contain beads, and all positions to the right contain gaps.

Consider the actions required to transform w into a justified abacus. We begin with the bead following the leftmost gap, which slides one position to the left, producing

$$w' = \cdots 111110100\underline{1}10101000\cdots.$$

The next bead now slides into the position vacated by the previous bead, producing

$$w'' = \cdots 111111000\underline{1}10101000 \cdots .$$

The next bead moves three positions to the left to give the abacus

$$w^{(3)} = \cdots 111111100\underline{0}10101000 \cdots .$$

In the next three steps, the remaining beads move left by three, four, and five positions respectively, leading to the abacus

$$w^{*} = \cdots 111111111\underline{1}00000000 \cdots ,$$

which is justified at position 0. If we list the number of positions that each bead moved, we obtain a weakly increasing sequence: $1 \leq 1 \leq 3 \leq 3 \leq 4 \leq 5$. This sequence can be identified with the integer partition $\lambda = (5, 4, 3, 3, 1, 1)$. Observe that $\mathrm{wt}(w) = 17 = |\lambda|$. This example generalizes as follows.

10.3. Theorem: Partitions and Abaci. Justification of abaci defines a bijection $J : \mathrm{Abc} \to \mathbb{Z} \times \mathrm{Par}$ with inverse $U : \mathbb{Z} \times \mathrm{Par} \to \mathrm{Abc}$. If $J(w) = (m, \lambda)$, then $\mathrm{wt}(w) = |\lambda|$.

Proof. Given an abacus w, let n be the least integer with $w_n = 0$ (the position of the leftmost gap), which exists since w begins with an infinite string of 1's. Since w ends with an infinite string of 0's, there are only finitely many $j > n$ with $w_j = 1$; let these indices be $j_1 < j_2 < \cdots < j_t$, where $n < j_1$. We justify the abacus by moving the bead at position j_1 left $\lambda_t = j_1 - n$ places. Then we move the bead at position j_2 left $\lambda_{t-1} = j_2 - (n + 1)$ places. (We subtract $n + 1$ since the leftmost gap is now at position $n + 1$.) In general, at stage k we move the bead at position j_k left $\lambda_{t+1-k} = j_k - (n + k - 1)$ places. After moving all t beads, we have a justified abacus with the leftmost gap located at position $n + t$. Since $n < j_1 < j_2 < \cdots < j_t$, it follows that $0 < \lambda_t \leq \lambda_{t-1} \leq \cdots \leq \lambda_1$. We define $J(w) = (n + t - 1, \lambda)$ where $\lambda = (\lambda_1, \ldots, \lambda_t)$. For all k, moving the bead at position j_k left λ_{t+1-k} places decreases the weight of the abacus by λ_{t+1-k}. Since a justified abacus has weight zero, it follows that the weight of the original abacus is precisely $\lambda_t + \cdots + \lambda_1 = |\lambda|$.

J is a bijection because *unjustification* is a two-sided inverse for J. More precisely, given $(m, \mu) \in \mathbb{Z} \times \mathrm{Par}$, we create an abacus $U(m, \mu)$ as follows. Start with an abacus justified at position m. Move the rightmost bead to the right μ_1 places, then move the next bead to the right μ_2 places, and so on. This process reverses the action of J. □

10.4. Remark: Computing U. The unjustification map U can also be computed using partition diagrams. We can reconstruct the bead-gap sequence in the abacus $U(m, \mu)$ by traversing the *frontier* of the diagram of μ (traveling northeast) and recording a gap (0) for each horizontal step and a bead (1) for each vertical step. For example, if $\mu = (5, 4, 3, 3, 1, 1)$, the diagram of μ is

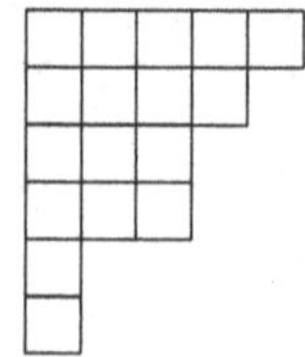

and the bead-gap sequence is 01100110101. To obtain the abacus w, we prepend an infinite string of 1's, append an infinite string of 0's, and finally use m to determine which symbol in the resulting string is considered to be w_0. It can be checked that this procedure produces the same abacus as the map U in the previous proof. We can also confirm that the map U

is weight-preserving via the following bijection between the set of cells of the diagram of μ and the set of pairs $i < j$ with $w_i = 0$ and $w_j = 1$. Starting at a cell c, travel south to reach a horizontal edge on the frontier (encoded by some $w_i = 0$). Travel east from c to reach a vertical edge on the frontier (encoded by some $w_j = 1$ with $j > i$). For example, the cell in the second row and third column of the diagram above corresponds to the marked gap-bead pair in the associated abacus:

$$\cdots 01100\hat{0}110\hat{1}01 \cdots .$$

10.2 The Jacobi Triple Product Identity

The *Jacobi Triple Product Identity* is a partition identity that has several applications in combinatorics and number theory. Here we give a bijective proof (due to Borcherds) of this formal power series identity by using cleverly chosen weights on abaci.

10.5. Theorem: Jacobi Triple Product Identity.

$$\sum_{m \in \mathbb{Z}} q^{m(m+1)/2} u^m = \prod_{n=1}^{\infty}(1 + uq^n) \prod_{n=0}^{\infty}(1 + u^{-1}q^n) \prod_{n=1}^{\infty}(1 - q^n).$$

Proof. Since the formal power series $\prod_{n=1}^{\infty}(1 - q^n)$ is invertible, it suffices to prove the equivalent identity

$$\sum_{m \in \mathbb{Z}} q^{m(m+1)/2} u^m \prod_{n=1}^{\infty} \frac{1}{1 - q^n} = \prod_{n=1}^{\infty}(1 + uq^n) \prod_{n=0}^{\infty}(1 + u^{-1}q^n). \tag{10.1}$$

Let the weight of an integer m be $\text{wt}(m) = q^{m(m+1)/2} u^m$, and let the weight of a partition μ be $q^{|\mu|}$. Since $\prod_{n=1}^{\infty} 1/(1 - q^n) = \sum_{\mu \in \text{Par}} q^{|\mu|}$ by Theorem 5.45, the left side of (10.1) is

$$\sum_{(m,\mu) \in \mathbb{Z} \times \text{Par}} \text{wt}(m)\,\text{wt}(\mu),$$

which is the generating function for the weighted set $\mathbb{Z} \times \text{Par}$.

On the other hand, let us define new weights on the set Abc as follows. Given an abacus w, let $N(w) = \{i \leq 0 : w_i = 0\}$ be the set of nonpositive positions in w not containing a bead, and let $P(w) = \{i > 0 : w_i = 1\}$ be the set of positive positions in w containing a bead. Both $N(w)$ and $P(w)$ are finite sets. Define

$$\text{wt}(w) = \prod_{i \in N(w)} (u^{-1} q^{|i|}) \prod_{i \in P(w)} (u^1 q^i).$$

We can build an abacus by choosing a bead or a gap in each nonpositive position (choosing a bead all but finitely many times), and then choosing a bead or a gap in each positive position (choosing a gap all but finitely many times). The generating function for the choice at position $i \leq 0$ is $1 + u^{-1} q^{|i|}$, while the generating function for the choice at position $i > 0$ is $1 + u^1 q^i$. By the Product Rule for Weighted Sets, the right side of (10.1) is $\sum_{w \in \text{Abc}} \text{wt}(w)$.

To complete the proof, it suffices to verify that the justification bijection $J : \text{Abc} \to \mathbb{Z} \times \text{Par}$ is weight-preserving. Suppose $J(w) = (m, \mu)$ for some abacus w. The map J converts w to an abacus w^*, justified at position m, by $|\mu|$ steps in which some bead moves one position to the left. *Claim 1:* The weight of the justified abacus w^* is $\text{wt}(m) =$

$u^m q^{m(m+1)/2}$. We prove this by considering three cases. When $m = 0$, $N(w^*) = \emptyset = P(w^*)$, so $\mathrm{wt}(w^*) = 1 = \mathrm{wt}(0)$. When $m > 0$, $N(w^*) = \emptyset$ and $P(w^*) = \{1, 2, \ldots, m\}$, so

$$\mathrm{wt}(w^*) = u^m q^{1+2+\cdots+m} = u^m q^{m(m+1)/2} = \mathrm{wt}(m).$$

When $m < 0$, $N(w^*) = \{0, -1, -2, \ldots, -(|m| - 1)\}$ and $P(w^*) = \emptyset$, so

$$\mathrm{wt}(w^*) = u^{-|m|} q^{0+1+2+\cdots+(|m|-1)} = u^m q^{|m|(|m|-1)/2} = u^m q^{m(m+1)/2} = \mathrm{wt}(m).$$

Claim 2: If we move one bead one step left in a given abacus y, the u-weight stays the same and the q-weight drops by 1. Let i be the initial position of the moved bead, and let y' be the abacus obtained by moving the bead to position $i - 1$. If $i > 1$, then $N(y') = N(y)$ and $P(y') = (P(y) - \{i\}) \cup \{i - 1\}$, so $\mathrm{wt}(y') = \mathrm{wt}(y)/q$ as needed. If $i \leq 0$, then $P(y') = P(y)$ and $N(y') = (N(y) - \{i - 1\}) \cup \{i\}$ (since the N-set records positions of gaps), and so $\mathrm{wt}(y') = \mathrm{wt}(y) q^{|i|}/q^{|i-1|} = \mathrm{wt}(y)/q$. If $i = 1$, then $P(y') = P(y) - \{1\}$ and $N(y') = N(y) - \{0\}$, so the total u-weight is preserved and the q-weight still drops by 1.

To finish, use Claim 2 $|\mu|$ times to conclude that

$$\mathrm{wt}(w) = \mathrm{wt}(w^*) q^{|\mu|} = \mathrm{wt}(m)\,\mathrm{wt}(\mu) = \mathrm{wt}(J(w)).$$ □

Variations of the preceding proof can be used to establish other partition identities. As an example, we now sketch a bijective proof of Euler's Pentagonal Number Theorem. Unlike our earlier proof in §5.17, the current proof does not use an involution to cancel oppositely signed objects. We remark that Euler's identity also follows by appropriately specializing the Jacobi Triple Product Identity.

10.6. Euler's Pentagonal Number Theorem.

$$\prod_{n=1}^{\infty} (1 - q^n) = \sum_{k \in \mathbb{Z}} (-1)^k q^{\frac{3}{2}k^2 - \frac{1}{2}k}.$$

Proof. Note first that

$$\prod_{n=1}^{\infty} (1 - q^n) = \prod_{i=1}^{\infty} (1 - q^{3i}) \prod_{i=1}^{\infty} (1 - q^{3i-1}) \prod_{i=1}^{\infty} (1 - q^{3i-2}).$$

It therefore suffices to prove the identity

$$\prod_{i=1}^{\infty} (1 - q^{3i-1}) \prod_{i=1}^{\infty} (1 - q^{3i-2}) = \sum_{k \in \mathbb{Z}} (-1)^k q^{(3k^2-k)/2} \prod_{i=1}^{\infty} \frac{1}{1 - q^{3i}} = \sum_{(k,\mu) \in \mathbb{Z} \times \mathrm{Par}} (-1)^k q^{3|\mu| + (3k^2-k)/2}. \tag{10.2}$$

Consider abaci $w = \{w_{3k+1} : k \in \mathbb{Z}\}$ whose positions are indexed by integers congruent to 1 mod 3. Define $N(w) = \{i \leq 0 : i \equiv 1 \pmod 3, w_i = 0\}$ and $P(w) = \{i > 0 : i \equiv 1 \pmod 3, w_i = 1\}$. Let $\mathrm{sgn}(w) = (-1)^{|N(w)| + |P(w)|}$ and $\mathrm{wt}(w) = \sum_{i \in N(w) \cup P(w)} |i|$. We can compute the generating function $\sum_w \mathrm{sgn}(w) q^{\mathrm{wt}(w)}$ in two ways. On one hand, placing a bead or a gap in each negative position and each positive position leads to the double product on the left side of (10.2). On the other hand, justifying the abacus transforms w into a pair $(3k - 2, \mu)$ for some $k \in \mathbb{Z}$. As in the proof of the Jacobi Triple Product Identity, one checks that the justified abacus associated to a given integer k has signed weight $(-1)^k q^{(3k^2-k)/2}$, while each of the $|\mu|$ bead moves in the justification process reduces the q-weight by 3 and preserves the sign. So the right side of (10.2) is also the generating function for these abaci, completing the proof. □

10.3 Ribbons and k-Cores

Recall the following fact about division of integers: given integers $a \geq 0$ and $k > 0$, there exist a unique quotient q and remainder r satisfying $a = kq + r$ and $0 \leq r < k$. Our next goal is to develop an analogous operation for dividing an integer partition μ by a positive integer k. The result of this operation consists of k *quotient partitions* together with a *remainder partition* with special properties. We begin by describing the calculation of the remainder, which is called a k-core. Abaci can then be used to establish the uniqueness of the remainder, and this leads us to the definition of the k quotient partitions.

To motivate our construction, consider the following pictorial method for performing integer division. Suppose we are dividing $a = 17$ by $k = 5$, obtaining quotient $q = 3$ and remainder $r = 2$. To find these answers geometrically, first draw a row of 17 boxes:

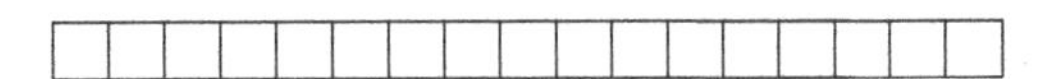

Now, starting at the right end, repeatedly remove strings of five consecutive cells until this is no longer possible. We depict this process by placing an i in every cell removed at stage i, and writing a star in every leftover cell:

*	*	3	3	3	3	3	2	2	2	2	2	1	1	1	1	1

The quotient q is the number of 5-cell blocks we removed (here 3), and the remainder r is the number of leftover cells (here 2). This geometric procedure corresponds to the algebraic process of subtracting k from a repeatedly until a remainder less than k is reached. For the purposes of partition division, we now introduce a two-dimensional version of this strip-removal process.

10.7. Definition: Ribbons. A *ribbon* is a skew shape that can be formed by starting at a given square, repeatedly moving left or down one step at a time, and including all squares visited in this way. A ribbon consisting of k cells is called a *k-ribbon*. A *border ribbon* of a partition μ is a ribbon R contained in $\text{dg}(\mu)$ such that $\text{dg}(\mu) - R$ is also a partition diagram.

10.8. Example. Here are two examples of ribbons:

$(6, 6, 4, 3)/(5, 3, 2) =$ $\qquad$ $(7, 4, 4, 4)/(3, 3, 3, 2) =$

The first ribbon is a 9-ribbon and a border ribbon of $(6, 6, 4, 3)$. The partition $(4, 3, 1)$ with diagram

has exactly eight border ribbons, four of which begin at the cell $(1, 4)$.

10.9. Definition: k-cores. Let k be a positive integer. An integer partition ν is called a *k-core* iff no border ribbon of ν is a k-ribbon.

For example, $(4, 3, 1)$ is a 5-core, but not a k-core for any $k < 5$.

Suppose μ is any partition and k is a positive integer. If μ has no border ribbons of size k, then μ is a k-core. Otherwise, we can pick one such ribbon and remove it from the diagram of μ to obtain a smaller partition diagram. We can iterate this process, repeatedly removing a border k-ribbon from the current partition diagram until this is no longer possible. Since the number of cells decreases at each step, the process must eventually terminate. The final partition ν (which may be empty) must be a k-core. This partition is the *remainder* when μ is divided by k.

10.10. Example. Consider the partition $\mu = (5, 5, 4, 3)$ with diagram

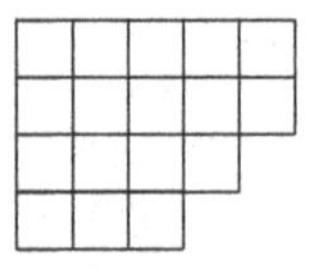

Let us divide μ by $k = 4$. We record the removal of border 4-ribbons by entering an i in each square that is removed at stage i. Any leftover squares at the end are marked by a star. One possible removal sequence is the following:

*	*	*	*	1
*	3	3	1	1
3	3	2	1	
2	2	2		

Another possible sequence is:

*	*	*	*	2
*	3	2	2	2
3	3	1	1	
3	1	1		

Notice that the three 4-ribbons removed are different, but the final 4-core is the same, namely $\nu = (4, 1)$.

We can use abaci to show that the k-core obtained when dividing μ by k depends only on μ and k, not on the choice of which border k-ribbon is removed at each stage.

10.11. Definition: Abacus with k Runners. A *k-runner abacus* is an ordered k-tuple of abaci. The set of all such objects is denoted Abc^k.

10.12. Theorem: Decimation of Abaci. For each $k \geq 1$, there are mutually inverse bijections $D_k : \text{Abc} \to \text{Abc}^k$ (decimation) and $I_k : \text{Abc}^k \to \text{Abc}$ (interleaving).

Proof. Given $w = (w_i : i \in \mathbb{Z}) \in \text{Abc}$, set $D_k(w) = (w^0, w^1, \ldots, w^{k-1})$, where

$$w^r = (w_{qk+r} : q \in \mathbb{Z}) \qquad \text{for all } r \text{ with } 0 \leq r < k.$$

Thus, the abacus w^r is obtained by reading every kth symbol in the original abacus (in both directions), starting at position r. It is routine to check that each w^r is an abacus. The inverse map interleaves these abaci to reconstruct the original 1-runner abacus. More precisely, given $v = (v^0, v^1, \ldots, v^{k-1}) \in \text{Abc}^k$, let $I_k(v) = z$ where $z_{qk+r} = v^r_q$ for all $q, r \in \mathbb{Z}$ with $0 \leq r < k$. One readily checks that $I_k(v)$ is an abacus and that D_k and I_k are two-sided inverses. $\square$

By computing $D_k(U(-1, \mu))$, we can convert any partition into a k-runner abacus. We now show that moving one bead left one step on a k-runner abacus corresponds to removing a border k-ribbon from the associated partition diagram.

10.13. Theorem: Bead Motions Encode Ribbon Removals. Suppose a partition μ is encoded by a k-runner abacus $w = (w^0, w^1, \ldots, w^{k-1})$. Suppose that v is a k-runner abacus obtained from w by changing one substring $\ldots 01 \ldots$ to $\ldots 10 \ldots$ in some w^i. Then the partition ν associated to v can be obtained by removing one border k-ribbon from μ. Moreover, there is a bijection between the set of removable border k-ribbons in μ and the set of occurrences of the substring 01 in the components of w.

Proof. Recall from Remark 10.4 that we can encode the frontier of a partition μ by writing a 0 (gap) for each horizontal step and writing a 1 (bead) for each vertical step. The word so obtained (when preceded by 1's and followed by 0's) is a 1-runner abacus associated to this partition, and w is the k-decimation of this abacus.

Let R be a border k-ribbon of μ. The southeast border of R, which is part of the frontier of μ, gets encoded as a string of $k+1$ symbols $r_0, r_1, \ldots, r_k$, where $r_0 = 0$ and $r_k = 1$. For instance, the first ribbon in Example 10.8 has southeast border 0001010011. Note that the *northwest* border of this ribbon is encoded by 1001010010, which is the string obtained by interchanging the initial 0 and the terminal 1 in the original string. The following picture indicates why this property holds for general k-ribbons.

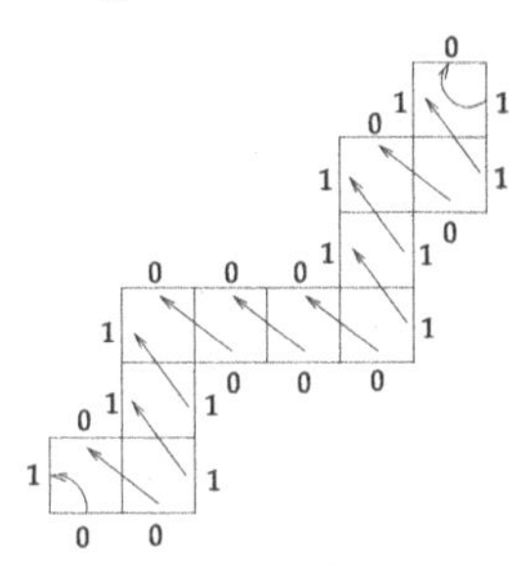

Since $r_0 = 0$ and $r_k = 1$ are separated by k positions in the 1-runner abacus, these two symbols map to two *consecutive* symbols 01 on one of the runners in the k-runner abacus for μ. Changing these symbols to 10 will interchange r_0 and r_k in the original word. Hence, the portion of the frontier of μ consisting of the southeast border of R gets replaced by the northwest border of R. So, this bead motion transforms μ into the partition ν obtained by removing the ribbon R.

Conversely, each substring 01 in the k-runner abacus for μ corresponds to a unique pair of symbols $0 \cdots 1$ in the 1-runner abacus that are k positions apart. This pair corresponds to a unique pair of steps $\mathrm{H} \cdots \mathrm{V}$ on the frontier that are k steps apart. Finally, this pair of steps corresponds to a unique removable border k-ribbon of μ. So the map from these ribbons to occurrences of 01 on the runners of w is a bijection. □

10.14. Example. Let us convert the partition $\mu = (5, 5, 4, 3)$ from Example 10.10 to a 4-runner abacus. First, the 1-runner abacus $U(-1, \mu)$ is

$$\cdots 1110001\underline{0}1011000 \cdots .$$

Decimating by 4 produces the following 4-runner abacus:

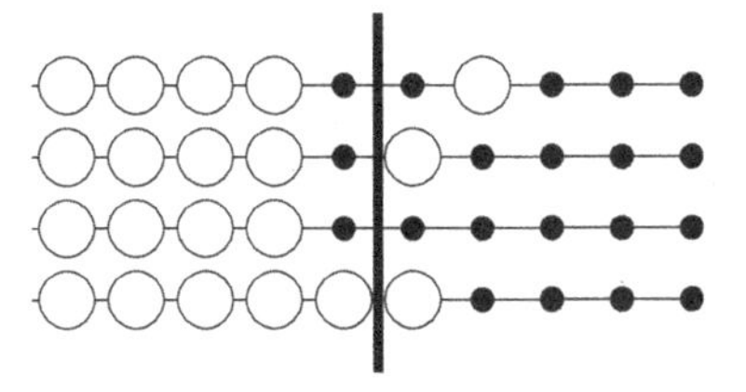

Note that the bead-gap pattern in this abacus can be read directly from the frontier of μ by filling in the runners one column at a time, working from left to right. For the purposes of ribbon removal, we may decide arbitrarily where to place the gap corresponding to the first step of the frontier; this decision determines the integer m in the expression $U(m, \mu)$.

Now let us start removing ribbons. Suppose we push the rightmost bead on the top runner left one position, producing the following abacus:

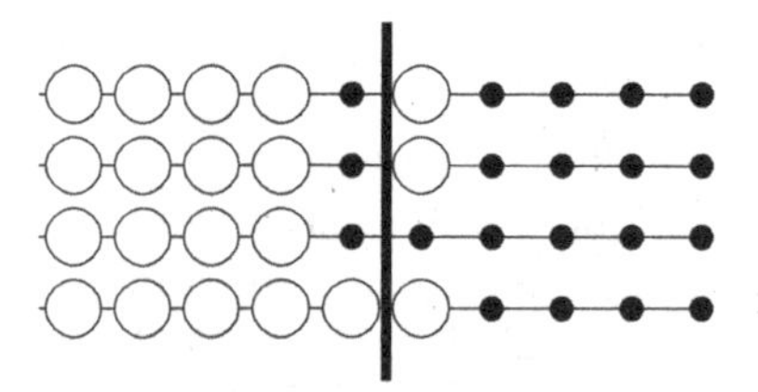

Reading down columns to recover the frontier of the new partition, we obtain the partition $\nu = (4, 3, 3, 3)$. We get ν from μ by removing one border 4-ribbon, as shown here:

				1
			1	1
			1	

Pushing the same bead one more step on its runner produces the following abacus:

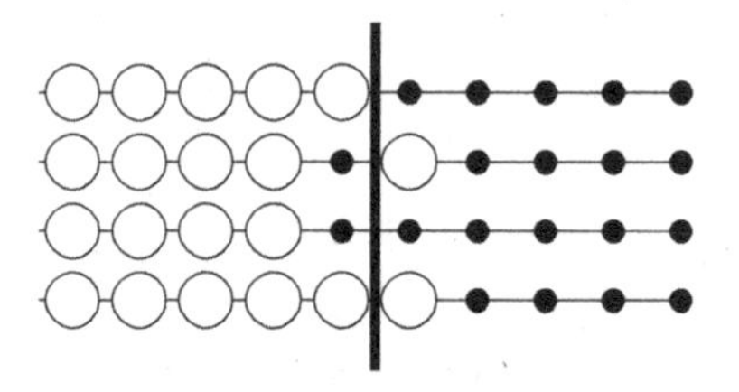

The new partition is $(4, 3, 2)$, which arises by removing one border 4-ribbon from ν, as shown here:

				1
			1	1
		2	1	
2	2	2		

Finally, we push the rightmost bead on the second runner left one position to get the following abacus:

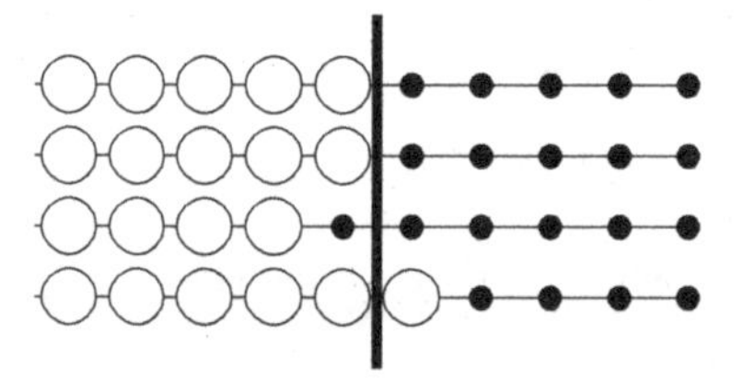

The associated partition is $(4, 1)$, as shown here:

				1
	3	3	1	1
3	3	2	1	
2	2	2		

At this point, all runners on the abacus are justified, so no further bead motion is possible. This reflects the fact that we can remove no further border 4-ribbons from the 4-core $(4, 1)$.

Now return to the original partition μ and the associated 4-runner abacus. Suppose we start by moving the bead on the second runner left one position, producing the following abacus:

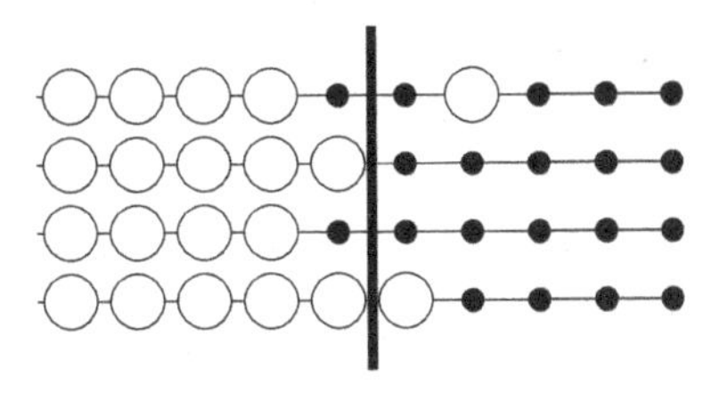

This corresponds to removing a different border 4-ribbon from μ:

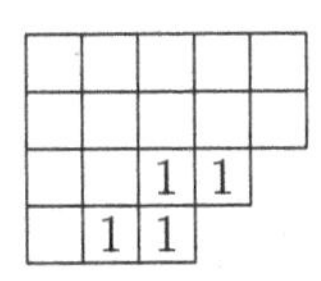

Observe that μ has exactly two removable border 4-ribbons, whereas the 4-runner abacus for μ has exactly two movable beads, in accordance with the last assertion of Theorem 10.13.

10.15. Example. Consider the following 3-runner abacus:

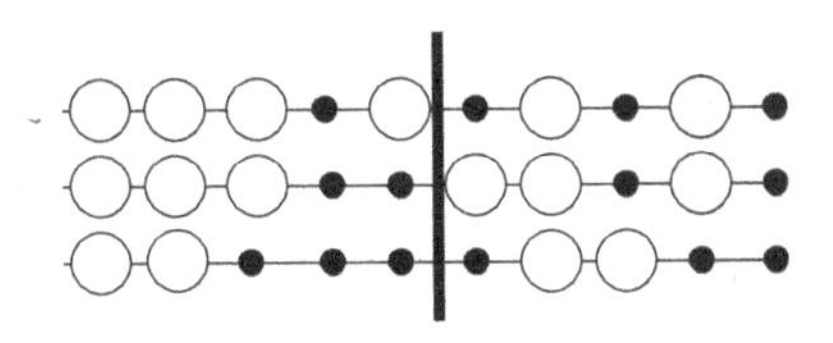

We count six beads on this abacus that can be moved one position left without bumping into another bead. Accordingly, we expect the associated partition to have exactly six removable border 3-ribbons. This is indeed the case, as shown below (we have marked the southwestmost cell of each removable 3-ribbon with an asterisk):

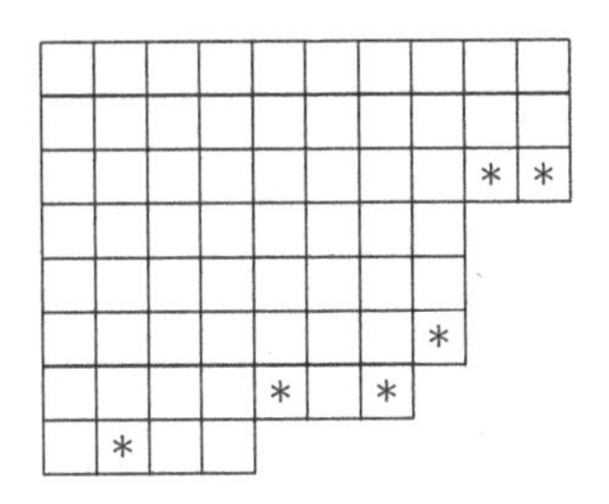

Now we can prove that the k-core obtained from a partition μ by repeated removal of border k-ribbons is uniquely determined by μ and k.

10.16. Theorem: Uniqueness of k-cores. Suppose μ is an integer partition and $k \geq 1$ is an integer. There is exactly one k-core ρ obtainable from μ by repeatedly removing border k-ribbons. We call ρ *the k-core of μ.*

Proof. Let w be a fixed k-runner abacus associated to μ, say $w = D_k(U(-1,\mu))$ for definiteness. As we have seen, a particular sequence of ribbon-removal operations on μ corresponds to a particular sequence of bead motions on w. The operations on μ terminate when we reach a k-core, whereas the corresponding operations on w terminate when the beads on all runners of w have been justified. Now ρ is uniquely determined by the justified k-runner abacus by applying I_k and then J. The key observation is that the justified abacus obtained from w does not depend on the order in which individual bead moves are made. Thus, the k-core ρ does not depend on the order in which border ribbons are removed from μ. □

10.17. Example. The theorem shows that we can calculate the k-core of μ by justifying any k-runner abacus associated to μ. For example, consider the partition $\mu = (10, 10, 10, 8, 8, 8, 7, 4)$ from Example 10.15. Justifying the 3-runner abacus in that example produces the following abacus:

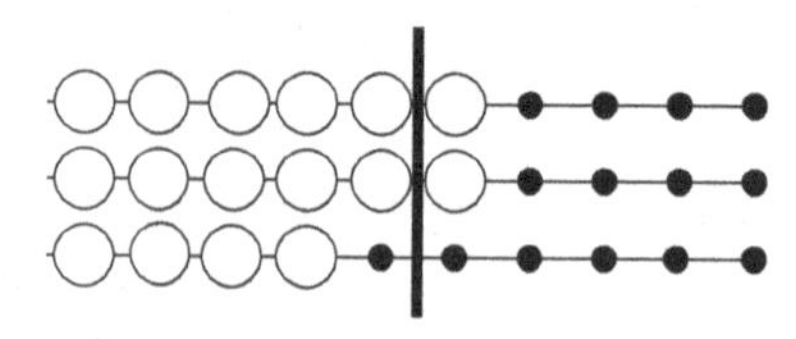

We find that the 3-core of μ is $(1, 1)$.

10.4 k-Quotients of a Partition

Each runner of a k-runner abacus can be regarded as a 1-runner abacus, which corresponds (under the justification bijection J) to an element of $\mathbb{Z} \times \text{Par}$. This observation leads to the definition of the k-quotients of a partition.

10.18. Definition: k-Quotients of a Partition. Let μ be a partition and $k \geq 1$ an integer. Consider the k-runner abacus $(w^0, w^1, \ldots, w^{k-1}) = D_k(U(-1, \mu))$. Write $J(w^i) = (m_i, \nu^i)$ for $0 \leq i < k$. The partitions appearing in the k-tuple $(\nu^0, \nu^1, \ldots, \nu^{k-1})$ are called the *k-quotients* of μ.

10.19. Example. Let $\mu = (5, 5, 4, 3)$. In Example 10.14, we computed the 4-runner abacus $D_k(U(-1, \mu))$:

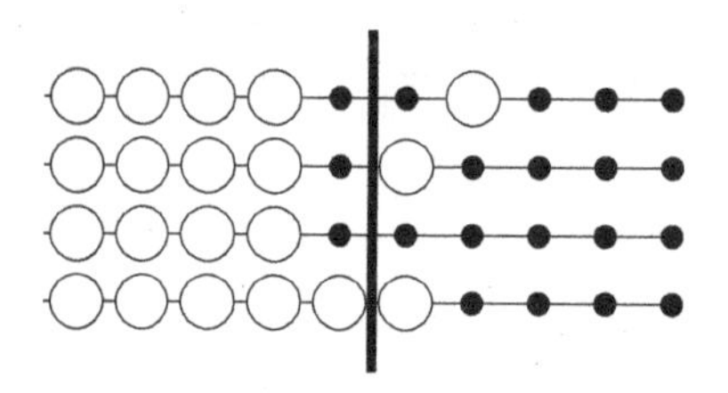

Justifying each runner and converting the resulting 4-runner abacus back to a partition produces the 4-core of μ, namely $(4, 1)$. On the other hand, converting each runner to a separate partition produces the 4-tuple of 4-quotients of μ, namely

$$(\nu^0, \nu^1, \nu^2, \nu^3) = ((2), (1), (0), (0)).$$

10.20. Example. Consider the partition $\mu = (10, 10, 10, 8, 8, 8, 7, 4)$ from Example 10.15. We compute

$$U(-1, \mu) = \cdots 1111000010001\underline{0}1110011100 \cdots.$$

Decimation by 3 produces the 3-runner abacus shown here:

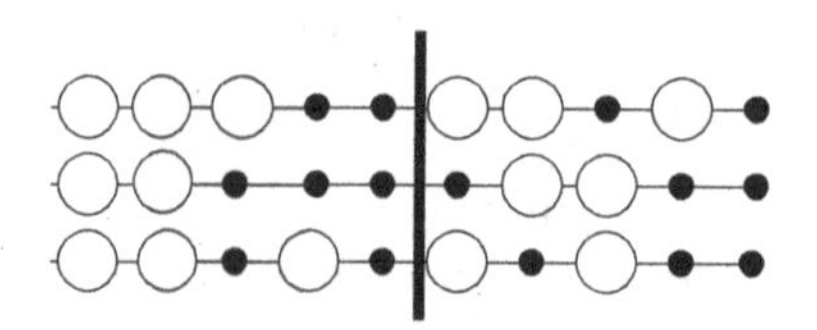

Justifying each runner shows that the 3-core of μ is $\rho = (1,1)$. On the other hand, by regarding each runner separately as a partition, we obtain the 3-tuple of 3-quotients of μ:

$$(\nu^0, \nu^1, \nu^2) = ((3,2,2), (4,4), (3,2,1)).$$

Observe that $|\mu| = 65 = 2 + 3 \cdot (7 + 8 + 6) = |\rho| + 3|\nu^0| + 3|\nu^1| + 3|\nu^2|$.

Now consider what would have happened if we had performed similar computations on the 3-runner abacus for μ displayed in Example 10.15, which is $D_3(U(0,\mu))$. The 3-core coming from this abacus is still $(1,1)$, but converting each runner to a partition produces the following 3-tuple:

$$((3,2,1), (3,2,2), (4,4)).$$

This 3-tuple arises by cyclically shifting the previous 3-tuple one step to the right. One can check that this holds in general: if the k-quotients for μ are $(\nu^0, \ldots, \nu^{k-1})$, then the k-quotients computed using $D_k(U(m,\mu))$ are $(\nu^{k-m'}, \ldots, \nu^{k-1}, \nu^0, \nu^1, \ldots)$, where m' is the integer remainder when $m+1$ is divided by k.

10.21. Remark. Here is a way to compute $(w^0, w^1, \ldots, w^{k-1}) = D_k(U(-1,\mu))$ from the frontier of μ without writing the intermediate abacus $U(-1,\mu)$. Draw a line of slope -1 starting at the northwest corner of the diagram of μ. The first step on the frontier of μ lying northeast of this line corresponds to position 0 of the zeroth runner w^0. The next step is position 0 on w^1, and so on. The step just southwest of the diagonal line is position -1 on w^{k-1}, the previous step is position -1 on w^{k-2}, and so on. To see that this works, it must be checked that the first step northeast of the diagonal line gets mapped to position 0 on the 1-runner abacus $U(-1,\mu)$.

10.22. Theorem: Partition Division. Let Core(k) be the set of all k-cores. There is a bijection

$$\Delta_k : \text{Par} \to \text{Core}(k) \times \text{Par}^k$$

such that $\Delta_k(\mu) = (\rho, \nu^0, \ldots, \nu^{k-1})$, where ρ is the k-core of μ and $\nu^0, \ldots, \nu^{k-1}$ are the k-quotients of μ. We have $|\mu| = |\rho| + k\sum_{i=0}^{k-1} |\nu^i|$.

Proof. We have already seen that the function Δ_k is well-defined and maps into the stated codomain. To see that this function is a bijection, we describe its inverse. Given $(\rho, \nu^0, \ldots, \nu^{k-1}) \in \text{Core}(k) \times \text{Par}^k$, first compute the k-runner abacus $(w^0, \ldots, w^{k-1}) = D_k(U(-1,\rho))$. Each w^i is a *justified* 1-runner abacus because ρ is a k-core; say w^i is justified at position m_i. Now replace each w^i by $v^i = U(m_i, \nu^i)$. Finally, let μ be the unique partition satisfying $J(I_k(v^0, \ldots, v^{k-1})) = (-1, \mu)$. This construction reverses the one used to produce k-cores and k-quotients, so μ is the unique partition mapped to $(\rho, \nu^0, \ldots, \nu^{k-1})$ by Δ_k.

To prove the formula for $|\mu|$, consider the bead movements used to justify the runners of the k-runner abacus $D_k(U(-1,\mu))$. On one hand, every time we move a bead one step left on this abacus, the area of μ drops by k since the bead motion removes one border k-ribbon. When we finish moving all the beads, we are left with the k-core ρ. It follows that $|\mu| = |\rho| + km$ where m is the total number of bead motions on all k runners. On the other hand, for $0 \leq i < k$, let m_i be the number of times we move a bead one step left on runner i. Then $m = m_0 + m_1 + \cdots + m_{k-1}$, whereas $m_i = |\rho^i|$ by Theorem 10.3. Substituting these expressions into $|\mu| = |\rho| + km$ gives the stated formula for $|\mu|$. □

10.5 k-Quotients and Hooks

We close our discussion of partition division by describing a way to compute the k-quotients of μ directly from the diagram of μ, without recourse to abaci. We need the following device for labeling cells of $\text{dg}(\mu)$ and steps on the frontier of μ by integers in $\{0, 1, \ldots, k-1\}$.

10.23. Definition: Content and k-Content. Consider a partition diagram for μ, drawn with the longest row on top. Introduce a coordinate system so that the northwest corner of the diagram is $(0,0)$ and (i,j) is located i steps south and j steps east of the origin. The *content* of the point (i,j) is $c(i,j) = j - i$. The content of a cell in the diagram of μ is the content of its southeast corner. The content of a frontier step from (i,j) to $(i,j+1)$ is $j-i$. The content of a frontier step from (i,j) to $(i-1,j)$ is $j-i$. If z is a lattice point, cell, or step in the diagram, then the k-content $c_k(z)$ is the unique value $r \in \{0, 1, \ldots, k-1\}$ such that $c_k(z) \equiv r \pmod{k}$.

10.24. Example. The left side of Figure 10.1 shows the diagram of the partition $\mu = (10, 10, 10, 8, 8, 8, 7, 4)$ with each cell and frontier step labeled by its content. On the right side of the figure, each cell and step is labeled by its 3-content.

content (cells; frontier step labels at right of each row, and below the last row)

										frontier
0	1	2	3	4	5	6	7	8	9	9
−1	0	1	2	3	4	5	6	7	8	8
−2	−1	0	1	2	3	4	5	6	7	7
−3	−2	−1	0	1	2	3	4			4, 5, 6
−4	−3	−2	−1	0	1	2	3			3
−5	−4	−3	−2	−1	0	1	2			2
−6	−5	−4	−3	−2	−1	0				0, 1
−7	−6	−5	−4							−4, −3, −2, −1
−8	−7	−6	−5							

3−content

										frontier
0	1	2	0	1	2	0	1	2	0	0
2	0	1	2	0	1	2	0	1	2	2
1	2	0	1	2	0	1	2	0	1	1
0	1	2	0	1	2	0	1			1, 2, 0
2	0	1	2	0	1	2	0			0
1	2	0	1	2	0	1	2			2
0	1	2	0	1	2	0				0, 1
2	0	1	2							2, 0, 1, 2
1	2	0	1							

FIGURE 10.1
Content and 3-content of cells and steps.

Given a cell in the diagram of μ, we obtain an associated pair of steps on the frontier of μ by traveling due south or due east from the cell in question. Suppose we mark all cells whose associated steps both have content zero. Then erase all other cells and shift the marked cells up and left as far as possible. The following diagram results:

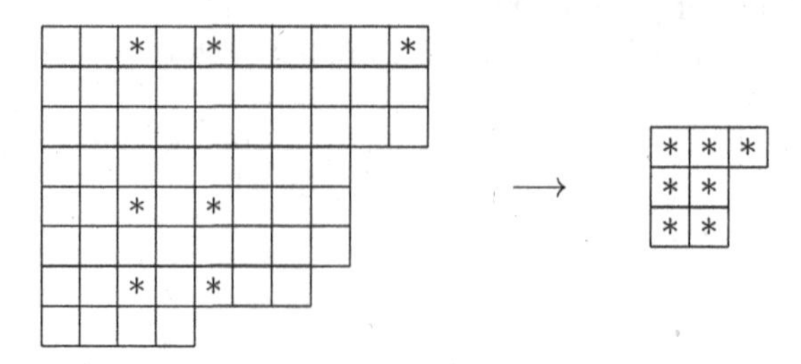

This partition $(3,2,2)$ is precisely the zeroth 3-quotient of μ. Similarly, marking the cells whose associated steps both have 3-content equal to 1 produces the next 3-quotient of μ:

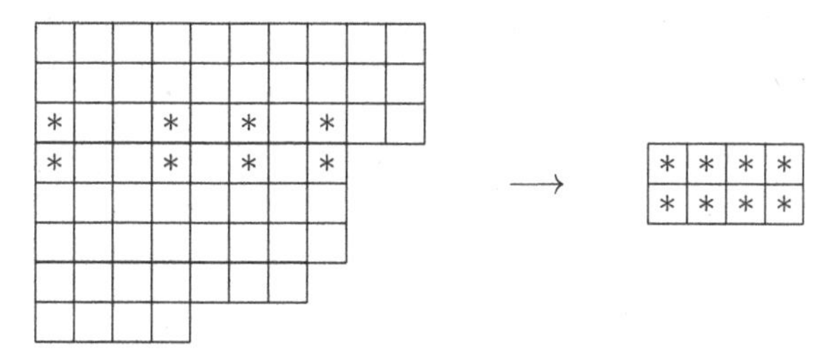

Finally, marking the cells whose associated steps both have 3-content equal to 2 produces the last 3-quotient of μ:

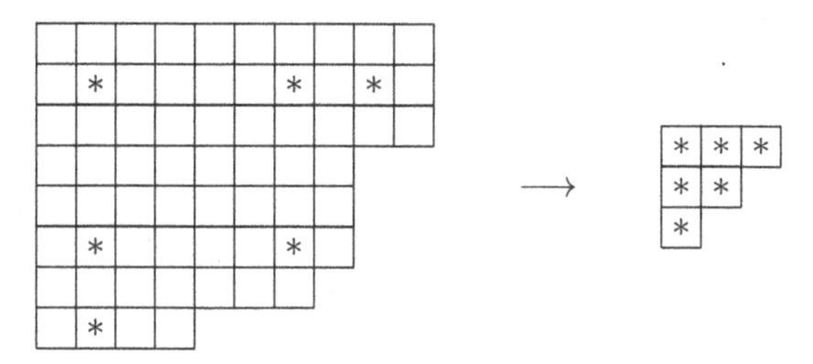

In general, to obtain the ith k-quotient ν^i of μ from the diagram of μ, label each row (respectively column) of the diagram with the k-content of the frontier step located in that row (respectively column). Erase all rows and columns not labeled i. The number of cells remaining in the jth unerased row is the jth part of ν^i. To see why this works, recall that the cells of ν^i correspond bijectively to the pairs of symbols $0 \cdots 1$ on the ith runner of the k-runner abacus for μ. In turn, these pairs correspond to pairs of symbols $w_s = 0$, $w_t = 1$ on the 1-runner abacus for μ where $s < t$ and $s \equiv i \equiv t \pmod{k}$. The symbols in positions congruent to i mod k come from the steps on the frontier of μ whose k-content is i. Finally, the relevant pairs of steps on the frontier correspond to the unerased cells in the construction described above. Composing all these bijections, we see that the cells of ν^i are in one-to-one correspondence with the unerased cells of the construction. Furthermore, cells in row j of ν^i are mapped onto the unerased cells in the jth unerased row of μ. It follows that the construction at the beginning of this paragraph does indeed produce the k-quotient ν^i.

10.6 Antisymmetric Polynomials

We now define antisymmetric polynomials, which form a vector space similar to the vector space of symmetric polynomials studied in Chapter 9.

10.25. Definition: Antisymmetric Polynomials. A polynomial $f \in \mathbb{R}[x_1, \ldots, x_N]$ is *antisymmetric* iff for all $w \in S_N$,

$$f(x_{w(1)}, x_{w(2)}, \ldots, x_{w(N)}) = \text{sgn}(w) f(x_1, x_2, \ldots, x_N).$$

More generally, everything said below holds for antisymmetric polynomials with coefficients in any field K containing $\mathbb{Q}$.

10.26. Remark. The group S_N acts on the set $\{x_1, \ldots, x_N\}$ via $w \bullet x_i = x_{w(i)}$ for all

$w \in S_N$ and i between 1 and N. This action extends to an action of S_N on $\mathbb{R}[x_1, \ldots, x_N]$ given by $w \bullet f = f(x_{w(1)}, \ldots, x_{w(N)})$ for all $w \in S_N$ and $f \in \mathbb{R}[x_1, \ldots, x_N]$. The polynomial f is antisymmetric iff $w \bullet f = \text{sgn}(w) f$ for all $w \in S_N$. It suffices to check this condition when w is a basic transposition $(i, i+1)$. For, any $w \in S_N$ can be written as a product of basic transpositions, say $w = t_1 t_2 \cdots t_k$. If we have checked that $t_i \bullet f = \text{sgn}(t_i) f$ for all i between 1 and $N-1$, then

$$w \bullet f = t_1 \bullet \cdots \bullet (t_k \bullet f) = (-1)^k f = \text{sgn}(w) f.$$

We can restate this result by saying that $f \in \mathbb{R}[x_1, \ldots, x_N]$ is antisymmetric iff for all i in the range $1 \leq i < N$,

$$f(x_1, \ldots, x_{i+1}, x_i, \ldots, x_N) = -f(x_1, \ldots, x_i, x_{i+1}, \ldots, x_N).$$

10.27. Example. The polynomial $f(x_1, \ldots, x_N) = \prod_{1 \leq j < k \leq N} (x_j - x_k)$ is antisymmetric. To check this, consider what happens to the factors in the product when we interchange x_i and x_{i+1}. Factors not involving x_i or x_{i+1} are unchanged; factors of the form $(x_i - x_k)$ with $k > i+1$ get interchanged with factors of the form $(x_{i+1} - x_k)$; and factors of the form $(x_j - x_i)$ with $j < i$ get interchanged with factors of the form $(x_j - x_{i+1})$. Finally, the factor $(x_i - x_{i+1})$ becomes $(x_{i+1} - x_i) = -(x_i - x_{i+1})$. Thus, $(i, i+1) \bullet f = -f$ for $1 \leq i < N$, proving antisymmetry of f.

The polynomial $f = \prod_{j<k} (x_j - x_k)$ in this example is the *Vandermonde determinant*

$$\det[x_j^{N-i}]_{1 \leq i,j \leq N} = \sum_{w \in S_N} \text{sgn}(w) \prod_{i=1}^{N} x_{w(i)}^{N-i}.$$

(see §12.11 for a combinatorial proof of this assertion).

We can use determinants similar to the Vandermonde determinant to manufacture additional examples of antisymmetric polynomials. Here we need the combinatorial definition of determinants, which is covered in §12.9.

10.28. Definition: Monomial Antisymmetric Polynomials. Let $\mu = (\mu_1 > \mu_2 > \cdots > \mu_N)$ be a strictly decreasing sequence of N nonnegative integers. Define a polynomial $a_\mu(x_1, \ldots, x_N)$ by the formula

$$a_\mu(x_1, \ldots, x_N) = \det[x_j^{\mu_i}]_{1 \leq i,j \leq N} = \sum_{w \in S_N} \text{sgn}(w) \prod_{i=1}^{N} x_{w(i)}^{\mu_i}.$$

We call a_μ a *monomial antisymmetric polynomial indexed by* μ.

To see that a_μ really is antisymmetric, note that interchanging x_k and x_{k+1} has the effect of interchanging columns k and $k+1$ in the determinant defining a_μ. This column switch changes the sign of a_μ (for a proof of this determinant fact, see Theorem 12.50).

10.29. Example. Let $N = 3$ and $\mu = (5, 4, 2)$. Then

$$a_\mu(x_1, x_2, x_3) = +x_1^5 x_2^4 x_3^2 + x_1^4 x_2^2 x_3^5 + x_1^2 x_2^5 x_3^4 - x_1^4 x_2^5 x_3^2 - x_1^5 x_2^2 x_3^4 - x_1^2 x_2^4 x_3^5.$$

As the previous example shows, $a_\mu(x_1, \ldots, x_N)$ is a sum of $N!$ distinct monomials obtained by rearranging the subscripts (or equivalently, the exponents) in the monomial $x_1^{\mu_1} x_2^{\mu_2} \cdots x_N^{\mu_N}$. Each monomial appears in the sum with sign $+1$ or -1, where the sign of $x_1^{e_1} \cdots x_N^{e_N}$ depends on the parity of the number of basic transpositions needed to transform the sequence $(e_1, \ldots, e_N)$ to the sorted sequence $(\mu_1, \ldots, \mu_N)$. It follows from these remarks that a_μ is a nonzero homogeneous polynomial of degree $|\mu| = \mu_1 + \cdots + \mu_N$.

10.30. Definition: $\delta(N)$. For each $N \geq 1$, let $\delta(N) = (N-1, N-2, \ldots, 2, 1, 0)$.

The *strictly* decreasing sequences $\mu = (\mu_1 > \mu_2 > \cdots > \mu_N)$ correspond bijectively to the *weakly* decreasing sequences $\lambda = (\lambda_1 \geq \lambda_2 \geq \cdots \geq \lambda_N)$ via the maps sending μ to $\mu - \delta(N)$ and λ to $\lambda + \delta(N)$. It follows that each polynomial a_μ can be written $a_{\lambda+\delta(N)}$ for a unique partition $\lambda \in \mathrm{Par}_N$. This indexing scheme is used frequently below. Note that when $\lambda = (0, \ldots, 0)$, we have $\mu = \delta(N)$ and $a_{\delta(N)} = \prod_{1 \leq j < k \leq N}(x_j - x_k)$ by Example 10.27. Observe that $a_{\delta(N)}$ is a homogeneous polynomial of degree $N(N-1)/2 = \binom{N}{2}$.

10.31. Definition: Spaces of Antisymmetric Polynomials. Let A_N be the set of all antisymmetric polynomials in $\mathbb{R}[x_1, \ldots, x_N]$. Let A_N^n consist of those polynomials in A_N that are homogeneous of degree n, together with the zero polynomial.

It can be verified that A_N is a vector subspace of $\mathbb{R}[x_1, \ldots, x_N]$, and each A_N^n is a subspace of A_N. We now exhibit bases for these vector spaces involving monomial antisymmetric polynomials. We use the notation $\mathrm{Par}_N^d(n)$ to denote the set of all partitions of n into N distinct nonnegative parts, and let $\mathrm{Par}_N^d = \bigcup_{n \geq \binom{N}{2}} \mathrm{Par}_N^d(n)$.

10.32. Theorem: Monomial Basis for A_N^n. If $n < \binom{N}{2}$, then $A_N^n = \{0\}$. If $n \geq \binom{N}{2}$, then

$$\{a_\mu : \mu \in \mathrm{Par}_N^d(n)\} = \{a_{\lambda+\delta(N)} : \lambda \in \mathrm{Par}_N(n - \binom{N}{2})\}$$

is a basis of the real vector space A_N^n. Hence, the collection

$$\{a_\mu : \mu \in \mathrm{Par}_N^d\} = \{a_{\lambda+\delta(N)} : \lambda \in \mathrm{Par}_N\}$$

is a basis of A_N.

Proof. Suppose $\mathbf{e} = (e_1, \ldots, e_N)$ is any exponent sequence, $f \in A_N$ is an arbitrary antisymmetric polynomial, and $w \in S_N$. Let c be the coefficient of $\mathbf{x}^\mathbf{e} = x_1^{e_1} \cdots x_N^{e_N}$ in f, so

$$f = c x_1^{e_1} \cdots x_N^{e_N} + \text{other terms}.$$

Acting by w, we see that

$$\begin{aligned} \mathrm{sgn}(w) f = w \bullet f &= c x_{w(1)}^{e_1} \cdots x_{w(N)}^{e_N} + \text{other terms} \\ &= c x_1^{e_{w^{-1}(1)}} \cdots x_N^{e_{w^{-1}(N)}} + \text{other terms} \\ &= c \mathbf{x}^{w * \mathbf{e}} + \text{other terms}, \end{aligned}$$

where $w * \mathbf{e} = (e_{w^{-1}(1)}, \ldots, e_{w^{-1}(N)})$. In other words, writing $f|_{\mathbf{x}^\alpha}$ for the coefficient of $\mathbf{x}^\alpha$ in f, we have $f|_{\mathbf{x}^{w*\mathbf{e}}} = \mathrm{sgn}(w)(f|_{\mathbf{x}^\mathbf{e}})$.

Let us apply this fact to an exponent sequence $\mathbf{e}$ such that $e_i = e_j$ for some $i \neq j$. Let $w = (i, j)$, so that $w * \mathbf{e} = \mathbf{e}$ and $\mathrm{sgn}(w) = -1$. It follows that $c = -c$, so $c = 0$. This means that no antisymmetric polynomial contains any monomial with a repeated value in its exponent vector. In particular, the smallest possible degree of a monomial that can appear with nonzero coefficient in any antisymmetric polynomial in N variables is $0 + 1 + 2 + \cdots + (N-1) = \binom{N}{2}$. This proves that $A_N^n = \{0\}$ for $n < \binom{N}{2}$.

For the second assertion, recall that the map sending λ to $\lambda + \delta(N)$ is a bijection from $\mathrm{Par}_N(n - \binom{N}{2})$ to $\mathrm{Par}_N^d(n)$. So we need only show that $\{a_\mu : \mu \in \mathrm{Par}_N^d(n)\}$ is a basis for A_N^n. To show that this set spans A_N^n, fix $f \in A_N^n$. By the previous paragraph, we can write $f = \sum_\alpha c_\alpha \mathbf{x}^\alpha$ where we sum over all sequences $(\alpha_1, \ldots, \alpha_N) \in \mathbb{Z}_{\geq 0}^N$ with *distinct* entries summing to n, and each c_α is a real scalar. We claim

$$f = \sum_{\nu \in \mathrm{Par}_N^d(n)} c_\nu a_\nu.$$

To prove this, we compare the coefficient of $\mathbf{x}^\alpha$ on each side. Choose $\mu \in \mathrm{Par}_N^d(n)$ and $w \in S_N$ such that $w * \mu = \alpha$ (μ consists of the entries of α sorted into decreasing order). By the first paragraph of this proof,

$$f|_{\mathbf{x}^\alpha} = f|_{\mathbf{x}^{w*\mu}} = \mathrm{sgn}(w)(f|_{\mathbf{x}^\mu}) = \mathrm{sgn}(w)c_\mu.$$

On the other side, $a_\nu|_{\mathbf{x}^\alpha} = 0$ for all $\nu \neq \mu$ (since no rearrangement of ν equals α). For $\nu = \mu$, antisymmetry gives $a_\mu|_{\mathbf{x}^\alpha} = \mathrm{sgn}(w)(a_\mu|_{\mathbf{x}^\mu}) = \mathrm{sgn}(w)$. Multiplying by c_ν and summing over all ν, the coefficient of $\mathbf{x}^\alpha$ in $\sum_\nu c_\nu a_\nu$ is $\mathrm{sgn}(w)c_\mu$, as needed.

To prove linear independence, suppose

$$0 = \sum_{\nu \in \mathrm{Par}_N^d(n)} d_\nu a_\nu \text{ with all } d_\nu \in \mathbb{R}.$$

For a fixed $\mu \in \mathrm{Par}_N^d(n)$, a_μ is the only polynomial among the a_ν's that involves the monomial $\mathbf{x}^\mu$. Extracting this coefficient on both sides of the given equation, we find that $0 = d_\mu \cdot 1 = d_\mu$. Since μ was arbitrary, all d_μ are zero. □

The next result explains the relationship between the various vector spaces Λ_N^k and A_N^n.

10.33. Theorem: Symmetric and Antisymmetric Polynomials. For each $k \geq 0$, the vector spaces Λ_N^k and $A_N^{k+\binom{N}{2}}$ are isomorphic, as are the vector spaces Λ_N and A_N. In each of these cases, an isomorphism is given by the formula $M(f) = f \cdot a_{\delta(N)}$ for $f \in \Lambda_N$, and the inverse isomorphism sends $g \in A_N$ to $g/a_{\delta(N)}$. In particular, every antisymmetric polynomial in N variables is divisible by the polynomial $a_{\delta(N)}$.

Proof. Fix $k \geq 0$, and consider the map $M_k : \Lambda_N^k \to \mathbb{R}[x_1, \ldots, x_N]$ defined by $M_k(f) = f \cdot a_{\delta(N)}$ for $f \in \Lambda_N^k$. First, f is homogeneous of degree k and $a_{\delta(N)}$ is homogeneous of degree $\binom{N}{2}$, so $M_k(f)$ is homogeneous of degree $k + \binom{N}{2}$. Second, $M_k(f)$ is antisymmetric, since for any $w \in S_N$,

$$w \bullet (f a_{\delta(N)}) = (w \bullet f) \cdot (w \bullet a_{\delta(N)}) = f \cdot (\mathrm{sgn}(w) a_{\delta(N)}) = \mathrm{sgn}(w)(f a_{\delta(N)}).$$

So M_k takes values in the codomain $A_N^{k+\binom{N}{2}}$. Third, one immediately verifies that M_k is a linear map. Fourth, the kernel of this linear map is zero: $M_k(f) = 0$ implies $f \cdot a_{\delta(N)} = 0$, which implies $f = 0$ since $a_{\delta(N)}$ is a nonzero element of the integral domain $\mathbb{R}[x_1, \ldots, x_N]$. So M is injective. Fifth, M must also be surjective since its domain and codomain are vector spaces having the same finite dimension $|\mathrm{Par}_N(k)|$. So each M_k is an isomorphism. Since Λ_N (respectively A_N) is the direct sum of subspaces Λ_N^k (respectively $A_N^{k+\binom{N}{2}}$), it follows that Λ_N and A_N are isomorphic as well. Finally, surjectivity of the map sending f to $f a_{\delta(N)}$ means that every antisymmetric polynomial g has the form $f a_{\delta(N)}$ for some symmetric polynomial f. So g is divisible by $a_{\delta(N)}$ in $\mathbb{R}[x_1, \ldots, x_N]$. □

10.34. Remark. Suppose we apply the inverse of the isomorphism M_k to the basis $\{a_{\lambda+\delta(N)} : \lambda \in \mathrm{Par}_N(k)\}$ of $A_N^{k+\binom{N}{2}}$. We obtain a basis $\{a_{\lambda+\delta(N)}/a_{\delta(N)} : \lambda \in \mathrm{Par}_N(k)\}$ of Λ_N^k. It turns out that $a_{\lambda+\delta(N)}/a_{\delta(N)}$ is none other than the Schur symmetric polynomial $s_\lambda(x_1, \ldots, x_N)$. To prove this fact and other properties of antisymmetric polynomials, we use the *labeled abaci* introduced below.

10.7 Labeled Abaci

Given a sequence of distinct nonnegative integers $\mu = (\mu_1 > \mu_2 > \cdots > \mu_N)$, recall that the monomial antisymmetric polynomial indexed by μ is defined by

$$a_\mu(x_1, \ldots, x_N) = \sum_{w \in S_N} \mathrm{sgn}(w) \prod_{i=1}^{N} x_{w(i)}^{\mu_i}.$$

The next definition introduces a set of signed weighted combinatorial objects to model this formula.

10.35. Definition: Labeled Abaci. A *labeled abacus with N beads* is a word $v = (v_i : i \geq 0)$ such that each symbol $1, \ldots, N$ appears exactly once in v, and all other symbols in v are zero. We think of the indices i as positions on an abacus containing one runner that extends to infinity in the positive direction. When $v_i = 0$, there is a gap at position i on the abacus; when $v_i = j > 0$, there is a bead labeled j at position i. The *weight* of the abacus v is

$$\mathrm{wt}(v) = \prod_{i:\ v_i > 0} x_{v_i}^{i}.$$

So if bead j is located at position i, this bead contributes a factor of x_j^i to the weight.

We can encode a labeled abacus by specifying the positions occupied by the beads and the ordering of the bead labels. Formally, define $\mathrm{pos}(v) = (\mu_1 > \mu_2 > \cdots > \mu_N)$ to be the list of indices i such that $v_i > 0$, written in decreasing order. Then define $w(v) = (v_{\mu_1}, v_{\mu_2}, \ldots, v_{\mu_N}) \in S_N$. We define the *sign* of v to be the sign of the permutation $w(v)$, which is $(-1)^{\mathrm{inv}(w(v))}$. Let LAbc be the set of all labeled abaci, and for each $\mu \in \mathrm{Par}_N^d$, let

$$\mathrm{LAbc}(\mu) = \{v \in \mathrm{LAbc} : \mathrm{pos}(v) = \mu\}.$$

For each fixed $\mu \in \mathrm{Par}_N^d$, there is a bijection between $\mathrm{LAbc}(\mu)$ and S_N sending $v \in \mathrm{LAbc}(\mu)$ to $w(v) \in S_N$. Furthermore, an abacus $v \in \mathrm{LAbc}(\mu)$ has sign $\mathrm{sgn}(w(v))$ and weight $\prod_{i=1}^{N} x_{w(v)_i}^{\mu_i}$. So

$$\sum_{v \in \mathrm{LAbc}(\mu)} \mathrm{sgn}(v)\,\mathrm{wt}(v) = \sum_{w \in S_N} \mathrm{sgn}(w) \prod_{i=1}^{N} x_{w(i)}^{\mu_i} = a_\mu(x_1, \ldots, x_N).$$

10.36. Example. Let $N = 3$ and $\nu = (5, 4, 2)$. Earlier, we computed

$$a_\nu(x_1, x_2, x_3) = +x_1^5x_2^4x_3^2 + x_1^4x_2^2x_3^5 + x_1^2x_2^5x_3^4 - x_1^4x_2^5x_3^2 - x_1^5x_2^2x_3^4 - x_1^2x_2^4x_3^5.$$

The six terms in this polynomial come from the six labeled abaci in $\mathrm{LAbc}(\nu)$ shown in Figure 10.2. Observe that we read labels from right to left in v to obtain the permutation $w(v)$. This is necessary so that the leading term $x_1^{\nu_1} \cdots x_N^{\nu_N}$ will correspond to the identity permutation and have positive sign.

Informally, we *justify* a labeled abacus $v \in \mathrm{LAbc}(\mu)$ by moving all beads to the left as far as they will go. This produces a justified labeled abacus $J(v) = (w_N, \ldots, w_2, w_1, 0, 0, \ldots) \in \mathrm{LAbc}(\delta(N))$, where $(w_N, \ldots, w_1) = w(v)$. To recover v from $J(v)$, first write $\mu = \lambda + \delta(N)$ for some $\lambda \in \mathrm{Par}_N$. Move the rightmost bead (labeled w_1) to the right λ_1 positions from position $N-1$ to position $N - 1 + \lambda_1 = \mu_1$. Then move the next bead (labeled w_2) to the right λ_2 positions from position $N-2$ to position $N - 2 + \lambda_2 = \mu_2$, and so on.

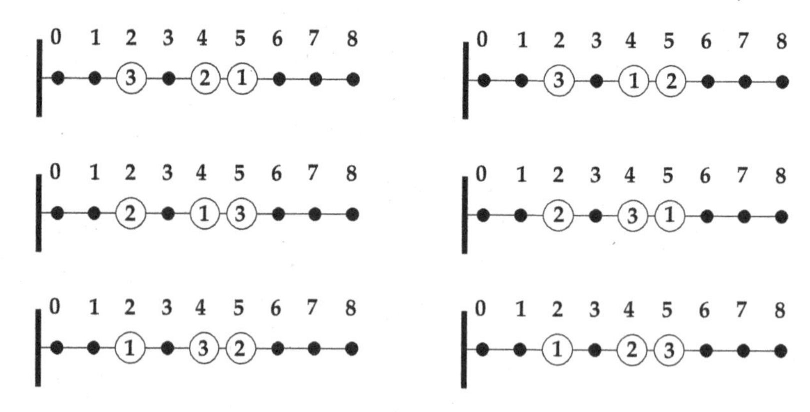

FIGURE 10.2
Labeled abaci.

10.8 The Pieri Rule for p_k

It is routine to check that the product of an antisymmetric polynomial and a symmetric polynomial is an antisymmetric polynomial, so such a product can be written as a linear combination of the monomial antisymmetric polynomials. In the next few sections, we derive several Pieri-type rules for expressing a product $a_{\lambda+\delta(N)}g$ (where g is symmetric) in terms of the a_μ. We begin by considering the case where $g = p_k(x_1, \ldots, x_N) = \sum_{i=1}^{N} x_i^k$ is a power-sum symmetric polynomial.

We know $a_{\lambda+\delta(N)}(x_1, \ldots, x_N)$ is a sum of signed terms, each of which represents a labeled abacus with beads in positions given by $\mu = \lambda + \delta(N)$. If we multiply some term in this sum by x_i^k, what happens to the associated abacus? Recalling that the power of x_i tells us where bead i is located, we see that this multiplication should move bead i to the right k positions. This bead motion occurs all at once, not one step at a time, so bead i is allowed to jump over any beads between its original position and its destination. However, there is a problem if the new position for bead i already contains a bead. In the proofs below, we will see that two objects of opposite sign cancel whenever a *bead collision* like this occurs. If there is no collision, the motion of bead i produces a new labeled abacus whose x_i-weight has increased by k. However, the sign of the new abacus (compared to the original) depends on the parity of the number of beads that bead i jumps over when it moves to its new position.

To visualize these ideas more conveniently, we *decimate* our labeled abacus to obtain a labeled abacus with k runners. Formally, the k-decimation of the labeled abacus $v = (v_j : j \geq 0) \in \mathrm{LAbc}(\lambda+\delta(N))$ is the k-tuple $(v^0, v^1, \ldots, v^{k-1})$, where $v_q^r = v_{qk+r}$. Moving a bead from position j to position $j+k$ on the original abacus corresponds to moving a bead one position along its runner on the k-runner abacus. If there is already a bead in position $j+k$, we say that this bead move causes a *bead collision*. Otherwise, the bead motion produces a new labeled abacus in $\mathrm{LAbc}(\nu + \delta(N))$, for some $\nu \in \mathrm{Par}_N$. By ignoring the labels in the decimated abacus, we see that ν arises from λ by adding one k-ribbon at the border. The shape of this ribbon determines the sign change caused by the bead move, as illustrated in the following example.

10.37. Example. Take $N = 6$, $k = 4$, $\lambda = (3, 3, 2, 0, 0, 0)$, and $\mu = \lambda + \delta(6) = (8, 7, 5, 2, 1, 0)$. Consider the following labeled abacus v in $\mathrm{LAbc}(\mu)$:

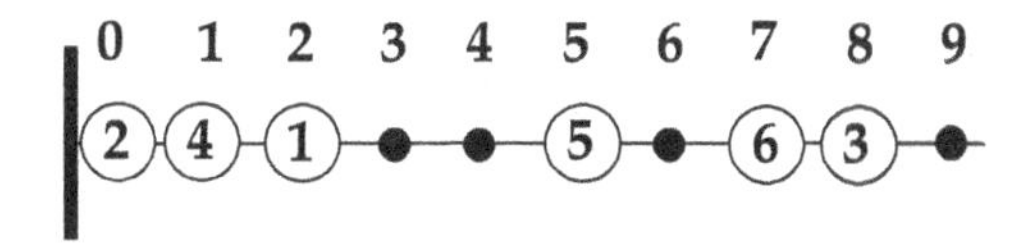

This abacus has weight $x_1^2x_2^0x_3^8x_4^1x_5^5x_6^7$ and sign $\text{sgn}(3,6,5,1,4,2) = (-1)^{10} = +1$. Decimation by 4 produces the following 4-runner abacus:

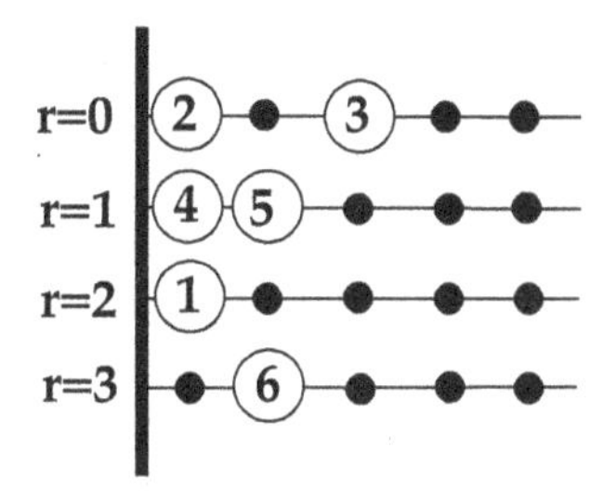

Suppose we move bead 1 four positions to the right in the original abacus, from position 2 to position 6:

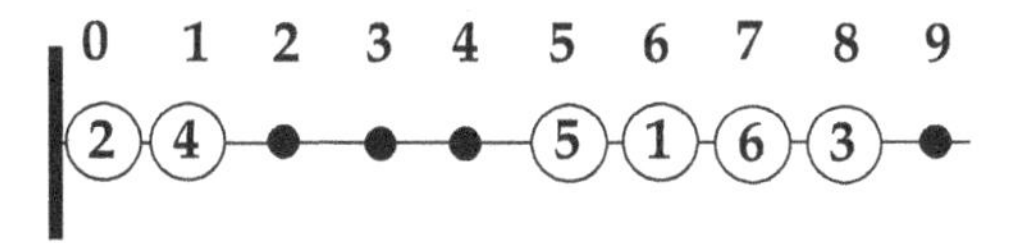

The new abacus has weight $x_1^6x_2^0x_3^8x_4^1x_5^5x_6^7 = \text{wt}(v)x_1^4$ and sign $\text{sgn}(3,6,1,5,4,2) = (-1)^9 = -1$. The change in weight occurs since bead 1 moved four positions to the right. The change in sign occurs since bead 1 passed one other bead (bead 5) to reach its new position, and one basic transposition is needed to transform the permutation $3,6,5,1,4,2$ into $3,6,1,5,4,2$. The decimation of the new abacus looks like:

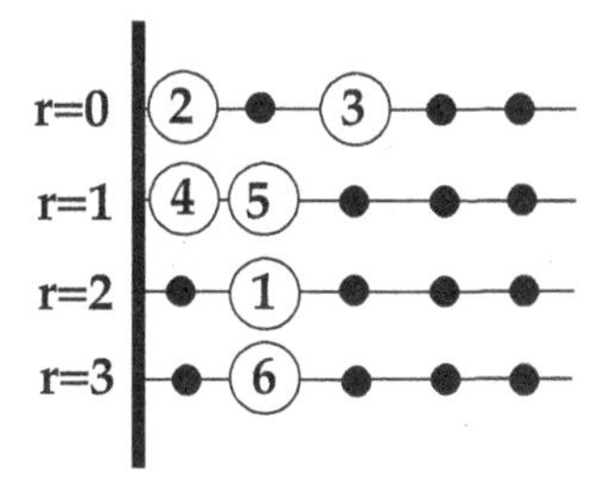

This abacus is in $\text{LAbc}(\nu) = \text{LAbc}(\alpha + \delta(6))$, where $\nu = (8,7,6,5,1,0)$ and $\alpha = (3,3,3,3,0,0)$. Compare the diagrams of the partitions λ and α:

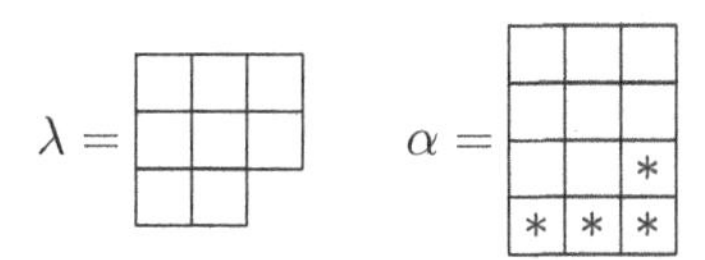

We obtain α from λ by adding a new border 4-ribbon. To go from λ to α, we change part of the frontier of λ from NEENE (where the first N step corresponds to bead 1) to EEENN (where the last N step corresponds to bead 1). There is one other N in this string, corresponding to the one bead (labeled 5) that bead 1 passes when it moves to position 6. Thus the number of passed beads (1 in this example) is one less than the number of rows occupied by the new border ribbon (2 in this example).

Let us return to the original abacus v and move bead 5 four positions, from position 5 to position 9:

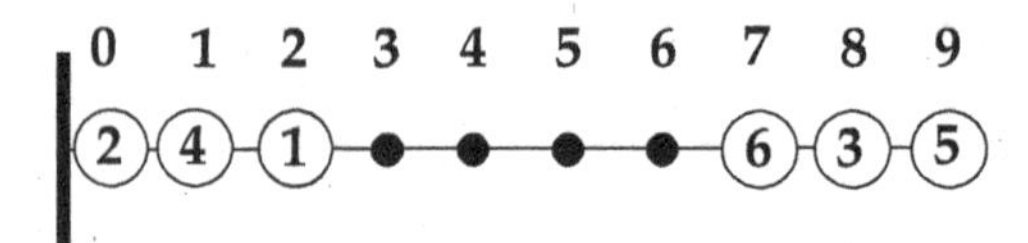

This abacus has weight $x_1^2 x_2^0 x_3^8 x_4^1 x_5^9 x_6^7 = \mathrm{wt}(v)x_5^4$ and sign $\mathrm{sgn}(5,3,6,1,4,2) = (-1)^{10} = +1$. Note that the sign is unchanged since two basic transpositions are required to change the permutation $3,6,5,1,4,2$ into $5,3,6,1,4,2$. The decimation of the new abacus is:

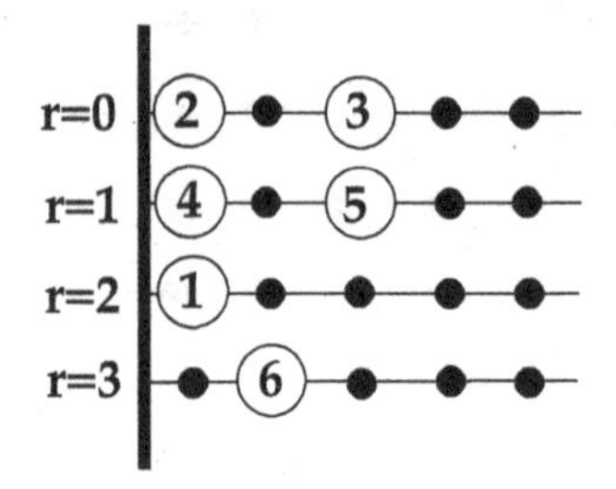

This abacus lies in $\mathrm{LAbc}(\beta + \delta(6))$ where $\beta = (4,4,4,0,0,0)$. The diagram of β arises by adding a border 4-ribbon to the diagram of λ:

$\lambda =$

$\beta =$

			*
			*
		*	*

This time the frontier changed from ...NENNE... (where the first N is bead 5) to ...EENNN... (where the last N is bead 5). The moved bead passed two other beads (beads 3 and 6), which is one less than the number of rows in the new ribbon (three). In general, the number of passed beads is one less than the number of N's in the frontier substring associated to the added ribbon. So the number of passed beads is one less than the number of rows in the added ribbon.

Finally, consider what would happen if we tried to move bead 4 (in the original abacus) four positions to the right. A collision occurs with bead 5, so this move is impossible. Now consider the labeled abacus v' obtained by interchanging the labels 4 and 5 in v:

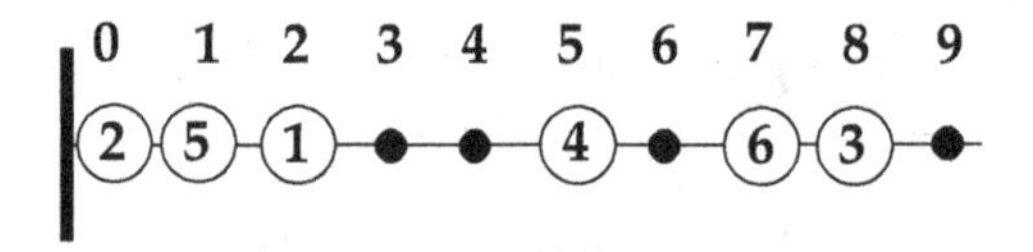

Moving bead 5 four positions to the right in v' causes a bead collision with bead 4. Notice that $\mathrm{sgn}(v') = -\mathrm{sgn}(v)$ since $[3,6,4,1,5,2] = (4,5)\circ[3,6,5,1,4,2]$. Also note that $\mathrm{wt}(v)x_4^4 = \mathrm{wt}(v')x_5^4$; this equality is valid precisely because of the bead collisions. The abaci v and v' are examples of a matched pair of oppositely signed objects that cancel in the proof of the Pieri Rule given below.

The observations in the last example motivate the following definition.

10.38. Definition: Spin and Sign of Ribbons. The *spin* of a ribbon R, denoted $\mathrm{spin}(R)$, is one less than the number of rows occupied by the ribbon. The *sign* of R is $\mathrm{sgn}(R) = (-1)^{\mathrm{spin}(R)}$.

We now have all the combinatorial ingredients needed to prove the Pieri Rule for multiplication by a power-sum polynomial.

10.39. The Antisymmetric Pieri Rule for p_k. For all $\lambda \in \mathrm{Par}_N$ and all $k \geq 1$,

$$a_{\lambda+\delta(N)}(x_1,\ldots,x_N)p_k(x_1,\ldots,x_N) = \sum_{\substack{\beta\in\mathrm{Par}_N: \\ \beta/\lambda \text{ is a } k\text{-ribbon } R}} \mathrm{sgn}(R)a_{\beta+\delta(N)}(x_1,\ldots,x_N).$$

Proof. Let X be the set of pairs (v,i), where $v \in \mathrm{LAbc}(\lambda+\delta(N))$ and $1 \leq i \leq N$. For $(v,i) \in X$, set $\mathrm{sgn}(v,i) = \mathrm{sgn}(v)$ and $\mathrm{wt}(v,i) = \mathrm{wt}(v)x_i^k$. Then $a_{\lambda+\delta(N)}p_k = \sum_{z\in X}\mathrm{sgn}(z)\,\mathrm{wt}(z)$. We introduce a weight-preserving, sign-reversing involution I on X. Given (v,i) in X, try to move bead i to the right k positions in v. If this move causes a bead collision with bead j, let v' be v with beads i and j switched, and set $I(v,i) = (v',j)$. Otherwise, set $I(v,i) = (v,i)$. It can be verified that I is an involution.

Consider the case where $I(v,i) = (v',j) \neq (v,i)$. Since the label permutation $w(v')$ is obtained from $w(v)$ by multiplying by the basic transposition (i,j), $\mathrm{sgn}(v',j) = \mathrm{sgn}(v') = -\mathrm{sgn}(v) = -\mathrm{sgn}(v,i)$. The weight of v must have the form $x_i^a x_j^{a+k}\cdots$ because of the bead collision, so $\mathrm{wt}(v') = x_j^a x_i^{a+k}\cdots$. It follows that $\mathrm{wt}(v,i) = \mathrm{wt}(v)x_i^k = x_i^{a+k}x_j^{a+k} = \mathrm{wt}(v')x_j^k = \mathrm{wt}(v',j)$. Thus, I is a weight-preserving, sign-reversing map.

Now consider a fixed point (v,i) of I. Let v^* be the abacus obtained from v by moving bead i to the right k positions, so $\mathrm{wt}(v^*) = \mathrm{wt}(v)x_i^k = \mathrm{wt}(v,i)$. Since the unlabeled k-runner abacus for v^* arises from the unlabeled k-runner abacus for v by moving one bead one step along its runner, it follows that $v^* \in \mathrm{LAbc}(\beta+\delta(N))$ for a unique $\beta \in \mathrm{Par}_N$ such that $R = \beta/\lambda$ is a k-ribbon. As explained earlier, $\mathrm{sgn}(v^*)$ differs from $\mathrm{sgn}(v)$ by $\mathrm{sgn}(R) = (-1)^{\mathrm{spin}(R)}$, which is the number of beads that bead i passes over when it moves. Conversely, any abacus y counted by $a_{\beta+\delta(N)}$ (for some shape β as above) arises from a unique fixed point $(v,i) \in X$, since the moved bead i is uniquely determined by the shapes λ and β, and v is determined from y by moving the bead i back k positions. These remarks show that the sum appearing on the right side of the theorem is the generating function for the fixed point set of I, which completes the proof. □

10.40. Example. When $N = 6$, we calculate

$$a_{(3,3,2)+\delta(6)}p_4 = -a_{(3,3,3,3)+\delta(6)} + a_{(4,4,4)+\delta(6)} - a_{(6,4,2)+\delta(6)} + a_{(7,3,2)+\delta(6)} + a_{(3,3,2,2,1,1)+\delta(6)}$$

by adding border 4-ribbons to the shape $(3,3,2)$, as shown here:

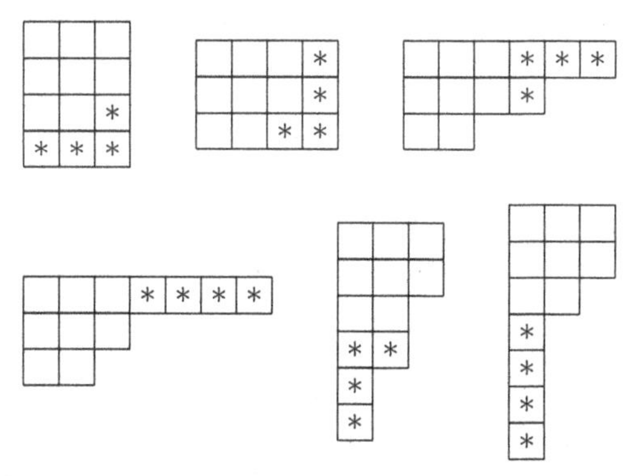

Observe that the last shape pictured does *not* contribute to the sum because it has more than N parts. An antisymmetric polynomial indexed by this shape would appear for $N \geq 7$.

10.9 The Pieri Rule for e_k

Next we derive Pieri Rules for calculating $a_{\lambda+\delta(N)}e_k$ and $a_{\lambda+\delta(N)}h_k$. Our starting point is the following expression for the elementary symmetric polynomial e_k:

$$e_k(x_1,\ldots,x_N) = \sum_{\substack{S\subseteq\{1,2,\ldots,N\}\\ |S|=k}} \prod_{j\in S} x_j.$$

Let $S = \{j_1,\ldots,j_k\}$ be a fixed k-element subset of $\{1,2,\ldots,N\}$. Then $\prod_{j\in S} x_j = x_{j_1}x_{j_2}\cdots x_{j_k}$ is a typical term in the polynomial e_k. On the other hand, a typical term in $a_{\lambda+\delta(N)}$ corresponds to a signed weighted abacus v. Let us investigate what happens to the abacus when we multiply such a term by $x_{j_1}\cdots x_{j_k}$.

Since the power of x_j indicates which position bead j occupies, multiplication by $x_{j_1}\cdots x_{j_k}$ should cause each of the beads labeled $j_1,\ldots,j_k$ to move one position to the right. We execute this action by scanning the positions of v *from right to left*. Whenever we see a bead labeled j for some $j\in S$, we move this bead one step to the right, thus multiplying the weight by x_j. Bead collisions may occur, which will lead to object cancellations in the proof below. In the case where no bead collisions happen, we obtain a new abacus $v^*\in a_{\nu+\delta(N)}$. The beads on this abacus occur in the same order as on v, so $w(v^*) = w(v)$ and $\operatorname{sgn}(v^*) = \operatorname{sgn}(v)$. Recalling that the parts of λ (respectively ν) count the number of bead moves needed to justify the beads in v (respectively v^*), it follows that $\nu\in\operatorname{Par}_N$ is a partition obtained from $\lambda\in\operatorname{Par}_N$ by *adding 1 to k distinct parts of* λ. This means that the skew shape ν/λ is a vertical strip of size k (see Definition 9.57).

10.41. Example. Let $N = 6$ and $\lambda = (3,3,2,2)$. Let v be the abacus in $\operatorname{LAbc}(\lambda+\delta(6))$ shown here:

0 1 2 3 4 5 6 7 8 9
5 1 • • 3 2 • 4 6 •

Suppose $k = 3$ and $S = \{1,2,3\}$. We move bead 2, then bead 3, then bead 1 one step right on the abacus. No bead collision occurs, and we get the following abacus:

0 1 2 3 4 5 6 7 8 9
5 • 1 • • 3 2 4 6 •

This abacus lies in $\operatorname{LAbc}(\nu+\delta(6))$, where $\nu = (3,3,3,3,1)$. Drawing the diagrams, we see that ν arises from λ by adding a vertical 3-strip:

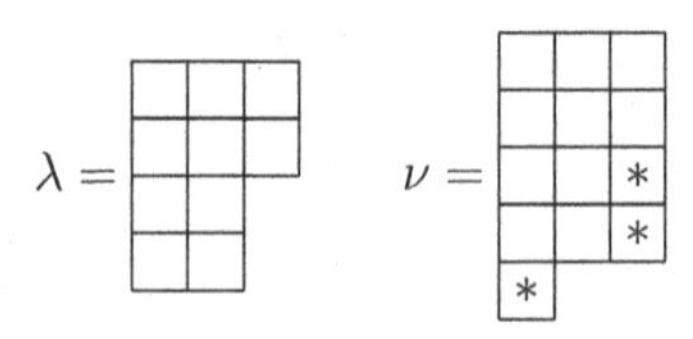

Suppose instead that $S = \{1,2,6\}$. This time we obtain the abacus

0 1 2 3 4 5 6 7 8 9
5 • 1 • 3 • 2 4 • 6

which is in $\text{LAbc}((4,3,3,2,1)+\delta(6))$. Now the partition diagrams look like this:

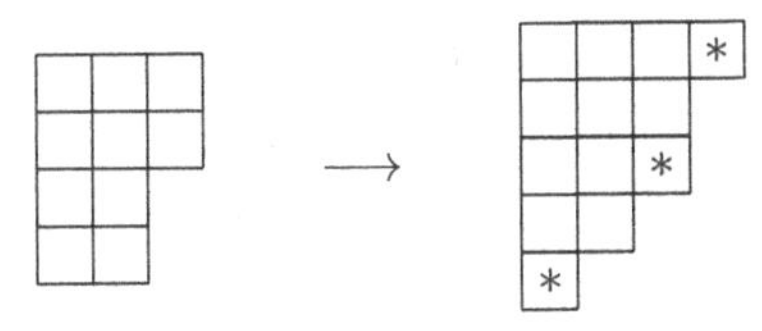

However, suppose we start with the subset $S = \{3,5,6\}$. When we move bead 6, then bead 3, then bead 5 on the abacus v, bead 3 collides with bead 2. We can match the pair (v,S) to (v',S'), where $S' = \{2,5,6\}$ and v' is this abacus:

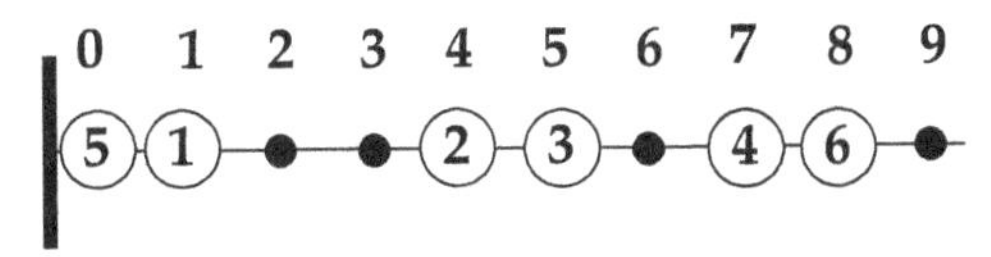

Observe that $\text{sgn}(v') = -\text{sgn}(v)$ and $\text{wt}(v)x_3x_5x_6 = \text{wt}(v')x_2x_5x_6$. This example illustrates the cancellation idea used in the proof below.

10.42. The Antisymmetric Pieri Rule for e_k. For all $\lambda \in \text{Par}_N$ and all $k \geq 1$,

$$a_{\lambda+\delta(N)}(x_1,\ldots,x_N)e_k(x_1,\ldots,x_N) = \sum_{\substack{\beta\in\text{Par}_N:\\ \beta/\lambda \text{ is a vertical } k\text{-strip}}} a_{\beta+\delta(N)}(x_1,\ldots,x_N).$$

Proof. Let X be the set of pairs (v,S) where $v \in \text{LAbc}(\lambda+\delta(N))$ and S is a k-element subset of $\{1,2,\ldots,N\}$. Letting $\text{sgn}(v,S) = \text{sgn}(v)$ and $\text{wt}(v,S) = \text{wt}(v)\prod_{j\in S} x_j$, the remarks at the start of this section show that

$$a_{\lambda+\delta(N)}e_k = \sum_{z\in X} \text{sgn}(z)\,\text{wt}(z).$$

Define an involution $I : X \to X$ as follows. Given $(v,S) \in X$, scan the abacus v from right to left and move each bead in S one step to the right. If this can be done with no bead collisions, we obtain an abacus v^* counted by the sum on the right side of the theorem, such that $\text{sgn}(v) = \text{sgn}(v^*)$ and $\text{wt}(v,S) = \text{wt}(v^*)$. In this case, (v,S) is a fixed point of I, and the bead motion rule defines a sign-preserving, weight-preserving bijection between these fixed points and the abaci counted by the right side of the theorem.

Now suppose a bead collision does occur. Then for some $j \in S$ and some $k \notin S$, bead k lies one step to the right of bead j in v. Take j to be the rightmost bead in v for which this is true. Let $I(v,S) = (v',S')$ where v' is v with beads j and k interchanged, and S' is S with j removed and k added. One immediately verifies that $\text{sgn}(v',S') = -\text{sgn}(v,S)$, $\text{wt}(v,S) = \text{wt}(v',S')$, and $I(v',S') = (v,S)$. So I cancels all objects in which a bead collision occurs. □

10.10 The Pieri Rule for h_k

In the last section, we computed $a_{\lambda+\delta(N)}e_k$ by using a k-element *subset* of $\{1,2,\ldots,N\}$ to move beads on a labeled abacus. Now we compute $a_{\lambda+\delta(N)}h_k$ by moving beads based on a k-element *multiset*. This approach is motivated by the formula

$$h_k(x_1,\ldots,x_N) = \sum_{\substack{k\text{-element multisets}\\ M \text{ of } \{1,\ldots,N\}}} \prod_{j\in M} x_j,$$

where the factor x_j is repeated as many times as j appears in M.

Suppose v is an abacus counted by $a_{\lambda+\delta(N)}$, and $x_1^{m_1}\cdots x_N^{m_N}$ is a typical term in h_k (so each $m_j \geq 0$ and $m_1+\cdots+m_N=k$). Scan the beads in v *from left to right*. Whenever we encounter a bead labeled j, we move it right, one step at a time, for a total of m_j positions. Bead collisions may occur and will lead to object cancellations later. If no collision occurs, we have a new abacus $v^* \in \mathrm{LAbc}(\nu+\delta(N))$ with the same sign as v and weight $\mathrm{wt}(v^*)=\mathrm{wt}(v)x_1^{m_1}\cdots x_N^{m_N}$. It follows from the bead motion rule that the shape ν arises from λ by adding a *horizontal k-strip* to λ (see Definition 9.57). Conversely, any abacus indexed by such a shape can be constructed from an abacus indexed by λ by an appropriate choice of the bead multiset. These ideas are illustrated in the following example, which should be compared to the example in the preceding section.

10.43. Example. Let $N=6$ and $\lambda=(3,3,2,2)$. Let v be the abacus in $\mathrm{LAbc}(\lambda+\delta(6))$ shown here:

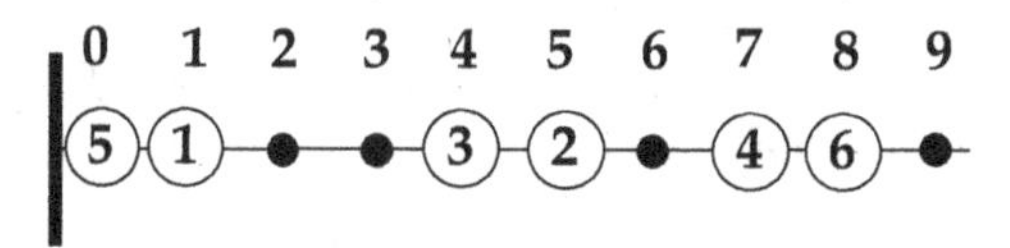

Let M be the multiset $[1,1,2]$. It is possible to move bead 1 to the right twice in a row, and then move bead 2 once, without causing any collisions. This produces the following abacus:

This abacus lies in $\mathrm{LAbc}(\nu+\delta(6))$, where $\nu=(3,3,3,2,2)$ arises from λ by adding a horizontal 3-strip:

If instead we take $M=[1,2,6]$, we move bead 1, then bead 2, then bead 6, leading to this abacus in $\mathrm{LAbc}((4,3,3,2,1)+\delta(6))$:

On the other hand, suppose we try to modify v using the multiset $M=[1,2,3]$. When scanning v from left to right, bead 3 moves before bead 2 and collides with bead 2. We match the pair (v,M) with the pair (v',M'), where $M'=[1,2,2]$ and v' is the following abacus:

Observe that $\mathrm{sgn}(v')=-\mathrm{sgn}(v)$ and $\mathrm{wt}(v)x_1x_2x_3=x_1^2x_2^6x_3^5x_4^7x_5^0x_6^8=\mathrm{wt}(v')x_1x_2^2$. This example illustrates the cancellation idea used in the proof below.

10.44. The Antisymmetric Pieri Rule for h_k. For all $\lambda \in \text{Par}_N$ and all $k \geq 1$,

$$a_{\lambda+\delta(N)}(x_1,\ldots,x_N)h_k(x_1,\ldots,x_N) = \sum_{\substack{\beta\in\text{Par}_N:\\ \beta/\lambda \text{ is a horizontal } k\text{-strip}}} a_{\beta+\delta(N)}(x_1,\ldots,x_N).$$

Proof. Let X be the set of pairs (v, M) where $v \in \text{LAbc}(\lambda + \delta(N))$ and $M = [1^{m_1}2^{m_2}\cdots N^{m_N}]$ is a k-element multiset. Defining $\text{sgn}(v, M) = \text{sgn}(v)$ and $\text{wt}(v, M) = \text{wt}(v)\prod_{j=1}^{N} x_j^{m_j}$, we have

$$a_{\lambda+\delta(N)}h_k = \sum_{z\in X} \text{sgn}(z)\,\text{wt}(z).$$

Define an involution $I : X \to X$ as follows. Given $(v, M) \in X$, scan the abacus v from left to right. When bead j is encountered in the scan, move it m_j steps right, one step at a time. If all bead motions are completed with no bead collisions, we obtain an abacus v^* counted by the sum on the right side of the theorem, such that $\text{sgn}(v) = \text{sgn}(v^*)$ and $\text{wt}(v, M) = \text{wt}(v^*)$. In this case, (v, M) is a fixed point of I, and the bead motion rule defines a sign-preserving, weight-preserving bijection between these fixed points and the abaci counted by the right side of the theorem.

Now consider the case where a bead collision does occur. Suppose the first collision occurs when bead j hits a bead k that is located $p \leq m_j$ positions to the right of bead j's initial position. Define $I(v, M) = (v', M')$, where v' is v with beads j and k interchanged, and M' is obtained from M by letting j occur $m_j - p \geq 0$ times in M', letting k occur $m_k + p$ times in M', and leaving all other multiplicities the same. One may check that $\text{sgn}(v', M') = -\text{sgn}(v, M)$, $\text{wt}(v, M) = \text{wt}(v', M')$, and $I(v', M') = (v, M)$. So I cancels all objects in which a bead collision occurs. □

10.11 Antisymmetric Polynomials and Schur Polynomials

The Pieri Rule for computing $a_{\lambda+\delta(N)}h_k$ closely resembles the rule for computing $s_\lambda h_k$ from §9.11. This resemblance leads to an algebraic proof of a formula expressing Schur polynomials as quotients of antisymmetric polynomials.

10.45. Theorem: Schur Polynomials and Antisymmetric Polynomials. For all $\lambda \in \text{Par}_N$,

$$s_\lambda(x_1,\ldots,x_N) = \frac{a_{\lambda+\delta(N)}(x_1,\ldots,x_N)}{a_{\delta(N)}(x_1,\ldots,x_N)} = \frac{\det[x_j^{\lambda_i+N-i}]_{1\leq i,j\leq N}}{\det[x_j^{N-i}]_{1\leq i,j\leq N}}.$$

Proof. In Theorem 9.64, we iterated the Pieri Rule

$$s_\nu(x_1,\ldots,x_N)h_k(x_1,\ldots,x_N) = \sum_{\substack{\beta\in\text{Par}_N:\\ \beta/\nu \text{ is a horizontal } k\text{-strip}}} s_\beta(x_1,\ldots,x_N)$$

to deduce the formula

$$h_\mu(x_1,\ldots,x_N) = \sum_{\lambda\in\text{Par}_N} K_{\lambda,\mu}s_\lambda(x_1,\ldots,x_N) \quad \text{for all } \mu \in \text{Par}_N. \tag{10.3}$$

Recall that this derivation used semistandard tableaux to encode the sequence of horizontal strips that were added to go from the empty shape to the shape λ. Now, precisely the same idea can be applied to iterate the antisymmetric Pieri Rule

$$a_{\nu+\delta(N)}(x_1,\ldots,x_N)h_k(x_1,\ldots,x_N) = \sum_{\substack{\beta\in\mathrm{Par}_N:\\ \beta/\nu \text{ is a horizontal } k\text{-strip}}} a_{\beta+\delta(N)}(x_1,\ldots,x_N).$$

If we start with $\nu = (0)$ and multiply successively by $h_{\mu_1}, h_{\mu_2}, \ldots$, we obtain the formula

$$a_{0+\delta(N)}(x_1,\ldots,x_N)h_\mu(x_1,\ldots,x_N) = \sum_{\lambda\in\mathrm{Par}_N} K_{\lambda,\mu}a_{\lambda+\delta(N)}(x_1,\ldots,x_N) \text{ for all } \mu\in\mathrm{Par}_N. \tag{10.4}$$

Now restrict attention to partitions $\lambda,\mu \in \mathrm{Par}_N(m)$. As in Theorem 9.66, we can write the equations in (10.3) in the form $\mathbf{H} = \mathbf{K}^{\mathrm{tr}}\mathbf{S}$, where $\mathbf{H} = (h_\mu : \mu \in \mathrm{Par}_N(m))$ and $\mathbf{S} = (s_\lambda : \lambda \in \mathrm{Par}_N(m))$ are column vectors, and $\mathbf{K}^{\mathrm{tr}}$ is the transpose of the Kostka matrix. Letting $\mathbf{A} = (a_{\lambda+\delta(N)}/a_{\delta(N)} : \lambda \in \mathrm{Par}_N(m))$, we can similarly write the equations in (10.4) in the form $\mathbf{H} = \mathbf{K}^{\mathrm{tr}}\mathbf{A}$. Finally, since the Kostka matrix is invertible (being unitriangular), we can conclude that

$$\mathbf{A} = (\mathbf{K}^{\mathrm{tr}})^{-1}\mathbf{H} = \mathbf{S}.$$

Equating entries of these vectors gives the result. □

A combinatorial proof of the identity $a_{\lambda+\delta(N)} = s_\lambda a_{\delta(N)}$ is given in §10.13.

10.12 Rim-Hook Tableaux

The connection between Schur polynomials and antisymmetric polynomials lets us deduce the following Pieri Rule for calculating the product $s_\lambda p_k$.

10.46. The Symmetric Pieri Rule for p_k. For all $\lambda \in \mathrm{Par}_N$ and all $k \geq 1$,

$$s_\lambda(x_1,\ldots,x_N)p_k(x_1,\ldots,x_N) = \sum_{\substack{\beta\in\mathrm{Par}_N:\\ \beta/\lambda \text{ is a } k\text{-ribbon } R}} \mathrm{sgn}(R)s_\beta(x_1,\ldots,x_N).$$

Proof. Start with the identity

$$a_{\lambda+\delta(N)}(x_1,\ldots,x_N)p_k(x_1,\ldots,x_N) = \sum_{\substack{\beta\in\mathrm{Par}_N:\\ \beta/\lambda \text{ is a } k\text{-ribbon } R}} \mathrm{sgn}(R)a_{\beta+\delta(N)}(x_1,\ldots,x_N)$$

(proved in Theorem 10.39), divide both sides by $a_{\delta(N)}$, and use Theorem 10.45. □

10.47. Example. Suppose we multiply $s_{(0)} = 1$ by p_4 using the Pieri Rule. The result is a signed sum of Schur polynomials indexed by 4-ribbons:

$$p_4 = s_{(0)}p_4 = s_{(4)} - s_{(3,1)} + s_{(2,1,1)} - s_{(1,1,1,1)}.$$

To expand $p_{(4,3)}$ into Schur polynomials, first multiply both sides of the previous equation by p_3:

$$p_{(4,3)} = p_4p_3 = s_{(4)}p_3 - s_{(3,1)}p_3 + s_{(2,1,1)}p_3 - s_{(1,1,1,1)}p_3.$$

Now use the Pieri Rule on each term on the right side. This leads to the diagrams shown in Figure 10.3. Taking signs into account, we obtain

$$\begin{aligned} p_{(4,3)} &= s_{(7)} + s_{(4,3)} - s_{(4,2,1)} + s_{(4,1,1,1)} \\ &\quad - s_{(6,1)} + s_{(3,2,2)} - s_{(3,1,1,1,1)} \\ &\quad + s_{(5,1,1)} - s_{(3,3,1)} + s_{(2,1,1,1,1,1)} \\ &\quad - s_{(4,1,1,1)} + s_{(3,2,1,1)} - s_{(2,2,2,1)} - s_{(1,1,1,1,1,1,1)}. \end{aligned}$$

Here we are assuming N (the number of variables) is at least 7.

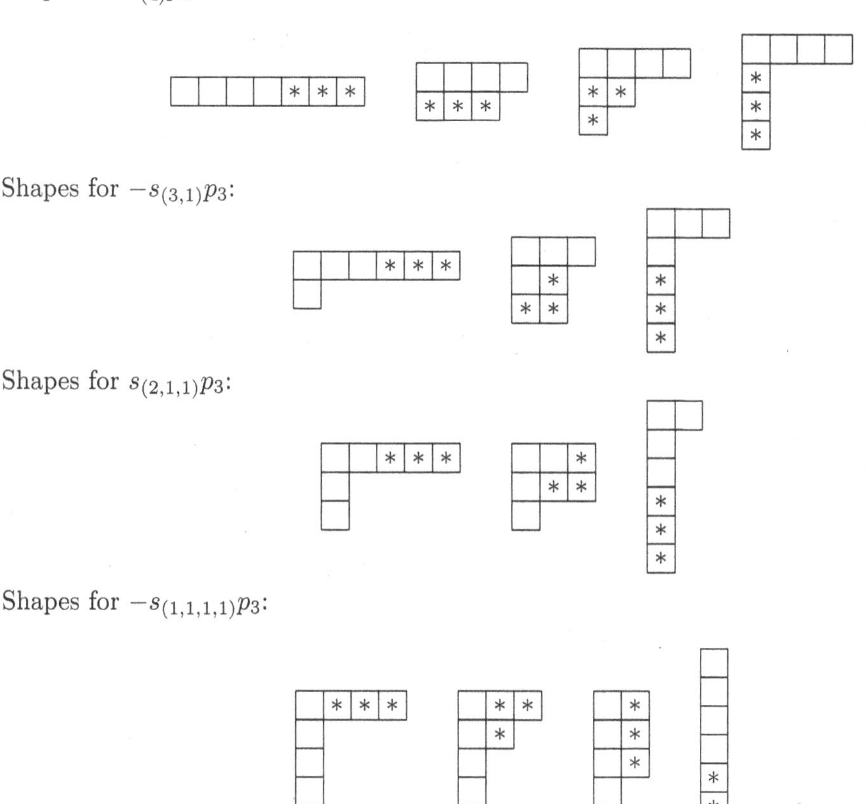

FIGURE 10.3
Adding k-ribbons to compute $s_\lambda p_k$.

Just as we used semistandard tableaux to encode successive additions of horizontal strips, we can use the following notion of a *rim-hook tableau* to encode successive additions of signed ribbons.

10.48. Definition: Rim-Hook Tableaux. Given a partition λ and a sequence $\alpha \in \mathbb{Z}^s_{\geq 0}$, a *rim-hook tableau of shape* λ *and content* α is a sequence T of partitions

$$(0) = \nu^0 \subseteq \nu^1 \subseteq \nu^2 \subseteq \cdots \subseteq \nu^s = \lambda$$

such that ν^i/ν^{i-1} is an α_i-ribbon for all i between 1 and s. We represent this tableau pictorially by drawing the diagram of λ and entering the number i in each cell of the ribbon ν^i/ν^{i-1}. The *sign* of the rim-hook tableau T is the product of the signs of the ribbons ν^i/ν^{i-1}. (Recall that the sign of a ribbon occupying r rows is $(-1)^{r-1}$.) Let $\mathrm{RHT}(\lambda,\alpha)$ be the set of all rim-hook tableaux of shape λ and content α. Finally, define the integer

$$\chi^{\lambda}_{\alpha} = \sum_{T\in\mathrm{RHT}(\lambda,\alpha)} \mathrm{sgn}(T).$$

Rim-hook tableaux of skew shape λ/μ are defined analogously; now we require that $\nu^0 = \mu$, so that the cells of μ do not get filled with ribbons. The set $\mathrm{RHT}(\lambda/\mu,\alpha)$ and the integer $\chi^{\lambda/\mu}_{\alpha}$ are defined as above.

10.49. Example. Suppose we expand the product $p_4p_2p_3p_1$ into a sum of Schur polynomials. We can do this by applying the Pieri Rule four times, starting with the empty shape. Each application of the Pieri Rule adds a new border ribbon to the shape. The lengths of the ribbons are given by the content vector $\alpha = (4,2,3,1)$. Here is one possible sequence of ribbon additions:

$$0 \longrightarrow \begin{array}{|c|c|}\hline * & * \\ \hline * \\ \cline{1-1} * \\ \cline{1-1}\end{array} \longrightarrow \begin{array}{|c|c|}\hline & \\ \hline & * \\ \hline & * \\ \hline\end{array} \longrightarrow \begin{array}{|c|c|c|c|}\hline & & * & * \\ \hline & & * \\ \cline{1-3} & \\ \cline{1-2}\end{array} \longrightarrow \begin{array}{|c|c|c|c|}\hline & & & \\ \hline & & \\ \cline{1-3} & & * \\ \cline{1-3}\end{array}$$

This sequence of shapes defines a rim-hook tableau

$$T = ((0),(2,1,1),(2,2,2),(4,3,2),(4,3,3)),$$

which can be visualized using the following diagram:

$$T = \begin{array}{|c|c|c|c|}\hline 1 & 1 & 3 & 3 \\ \hline 1 & 2 & 3 \\ \cline{1-3} 1 & 2 & 4 \\ \cline{1-3}\end{array}$$

Note that the ribbons we added have signs $+1$, -1, -1, and $+1$, so $\mathrm{sgn}(T) = +1$. This particular choice of ribbon additions therefore produces a term $+s_{(4,3,3)}$ in the Schur expansion of $p_{(4,2,3,1)}$.

Now suppose we want to know the coefficient of $s_{(4,3,3)}$ in the Schur expansion of $p_4p_2p_3p_1$. The preceding discussion shows that we obtain a term $\pm s_{(4,3,3)}$ for every rim-hook tableau of shape $(4,3,3)$ and content $(4,2,3,1)$, where the sign of the term is the sign of the tableau. To find the required coefficient, we must enumerate all the objects in $\mathrm{RHT}((4,3,3),(4,2,3,1))$. In addition to the rim-hook tableau T displayed above, we find the following rim-hook tableaux:

$$\begin{array}{|c|c|c|c|}\hline 1 & 1 & 3 & 4 \\ \hline 1 & 2 & 3 \\ \cline{1-3} 1 & 2 & 3 \\ \cline{1-3}\end{array} \qquad \begin{array}{|c|c|c|c|}\hline 1 & 1 & 2 & 2 \\ \hline 1 & 3 & 3 \\ \cline{1-3} 1 & 3 & 4 \\ \cline{1-3}\end{array} \qquad \begin{array}{|c|c|c|c|}\hline 1 & 1 & 1 & 4 \\ \hline 1 & 2 & 2 \\ \cline{1-3} 3 & 3 & 3 \\ \cline{1-3}\end{array} \qquad \begin{array}{|c|c|c|c|}\hline 1 & 1 & 1 & 1 \\ \hline 2 & 3 & 3 \\ \cline{1-3} 2 & 3 & 4 \\ \cline{1-3}\end{array}$$

The signs of the new tableaux are -1, -1, -1, and $+1$, so the coefficient is $+1-1-1-1+1 = -1$.

The calculations in the preceding example generalize to give the following rule for expanding power-sum polynomials into sums of Schur polynomials.

10.50. Theorem: Schur Expansion of Power-Sum Polynomials. For all $\alpha \in \mathbb{Z}^t_{\geq 0}$ and all $N \geq 1$,

$$p_\alpha(x_1,\ldots,x_N) = \sum_{\lambda\in\mathrm{Par}_N} \chi^{\lambda}_{\alpha} s_\lambda(x_1,\ldots,x_N).$$

Proof. By iteration of the Pieri Rule, the coefficient of s_λ in $p_\alpha = s_{(0)}p_{\alpha_1}\cdots p_{\alpha_t}$ is the signed sum of all sequences of partitions

$$0 = \nu^0 \subseteq \nu^1 \subseteq \nu^2 \subseteq \cdots \subseteq \nu^t = \lambda$$

such that the skew shape ν^i/ν^{i-1} is an α_i-ribbon for all i. By the definition of rim-hook tableaux, this sum is precisely χ_α^λ. □

10.51. Theorem: Symmetry of χ_α^λ. If α and β are compositions with $\mathrm{sort}(\alpha) = \mathrm{sort}(\beta)$, then $\chi_\alpha^\lambda = \chi_\beta^\lambda$ for all partitions λ.

Proof. The hypothesis implies that the sequence $\alpha = (\alpha_1, \alpha_2, \ldots)$ can be rearranged to the sequence $\beta = (\beta_1, \beta_2, \ldots)$. It follows from this that $p_\alpha = \prod_i p_{\alpha_i} = \prod_i p_{\beta_i} = p_\beta$, since multiplication of polynomials is commutative. Let $k = \sum_i \alpha_i$ and take $N \geq k$. Two applications of the previous theorem give

$$\sum_{\lambda \in \mathrm{Par}(k)} \chi_\alpha^\lambda s_\lambda = p_\alpha = p_\beta = \sum_{\lambda \in \mathrm{Par}(k)} \chi_\beta^\lambda s_\lambda.$$

By linear independence of the Schur polynomials $\{s_\lambda(x_1, \ldots, x_N) : \lambda \in \mathrm{Par}(k)\}$, we conclude that $\chi_\alpha^\lambda = \chi_\beta^\lambda$ for all λ. □

10.52. Remark. These results extend to skew shapes as follows. If μ is a partition, then

$$s_\mu(x_1, \ldots, x_N)p_\alpha(x_1, \ldots, x_N) = \sum_{\substack{\lambda \in \mathrm{Par}_N: \\ \mu \subseteq \lambda}} \chi_\alpha^{\lambda/\mu} s_\lambda(x_1, \ldots, x_N).$$

Furthermore, if $\mathrm{sort}(\alpha) = \mathrm{sort}(\beta)$ then $\chi_\alpha^{\lambda/\mu} = \chi_\beta^{\lambda/\mu}$. The proof is the same as before, replacing (0) by μ throughout.

We have just seen how to expand power-sum symmetric polynomials into sums of Schur polynomials. Conversely, it is possible to express Schur polynomials in terms of the p_μ. We can use the Hall scalar product from §9.26 to derive this expansion from the previous one.

10.53. Theorem: Power-Sum Expansion of Schur Polynomials. For all $N \geq k$ and $\lambda \in \mathrm{Par}(k)$,

$$s_\lambda(x_1, \ldots, x_N) = \sum_{\mu \in \mathrm{Par}(k)} \frac{\chi_\mu^\lambda}{z_\mu} p_\mu(x_1, \ldots, x_N).$$

Proof. For all $\mu \in \mathrm{Par}(k)$, we know that $p_\mu = \sum_{\nu \in \mathrm{Par}(k)} \chi_\mu^\nu s_\nu$. Therefore, for a given partition $\lambda \in \mathrm{Par}(k)$,

$$\langle p_\mu, s_\lambda \rangle = \sum_{\nu \in \mathrm{Par}(k)} \chi_\mu^\nu \langle s_\nu, s_\lambda \rangle = \chi_\mu^\lambda$$

since the Schur polynomials are orthonormal relative to the Hall scalar product. Now, since the p_ν form a basis of Λ_N^k, we know there exist scalars $c_\nu \in \mathbb{R}$ with $s_\lambda = \sum_\nu c_\nu p_\nu$. To find a given coefficient c_μ, we compute

$$\chi_\mu^\lambda = \langle p_\mu, s_\lambda \rangle = \sum_\nu c_\nu \langle p_\mu, p_\nu \rangle = c_\mu z_\mu,$$

where the last equality follows by definition of the Hall scalar product. We see that $c_\mu = \chi_\mu^\lambda / z_\mu$, as needed. □

10.13 Abaci and Tableaux

This section contains a combinatorial proof of the identity

$$a_{\delta(N)}(x_1,\ldots,x_N)s_\lambda(x_1,\ldots,x_N) = a_{\lambda+\delta(N)}(x_1,\ldots,x_N),$$

which we proved algebraically in §10.11.

Let X be the set of pairs (v,T), where v is a justified labeled abacus with N beads and T is a semistandard tableau using letters in $\{1,2,\ldots,N\}$. We need to use the following non-standard total ordering on this alphabet that depends on v: define $i <_v j$ iff bead i is to the right of bead j on the abacus v. Equivalently, we can describe the total order by writing

$$v_{N-1} <_v v_{N-2} <_v \cdots <_v v_1 <_v v_0.$$

Here are two examples of objects in X when $N=7$ and $\lambda=(7,7,5,3,2)$:

$$(v,T) = \left(7654321000\cdots,\ \begin{array}{|c|c|c|c|c|c|c|}\hline 1&1&1&1&1&1&1\\\hline 2&2&2&2&2&2&2\\\hline 3&3&5&5&5\\\cline{1-5} 6&6&6\\\cline{1-3} 7&7\\\cline{1-2}\end{array}\right)$$

$$(v',T') = \left(2451763000\cdots,\ \begin{array}{|c|c|c|c|c|c|c|}\hline 3&3&3&3&3&3&3\\\hline 6&6&6&6&6&6&6\\\hline 7&7&5&5&5\\\cline{1-5} 4&4&4\\\cline{1-3} 2&2\\\cline{1-2}\end{array}\right)$$

Note that we can pass from the first tableau (which is semistandard under the usual ordering of integers) to the second tableau (which is semistandard relative to one of the non-standard orderings) by applying the permutation sending 7 to 2, 6 to 4, etc., to each entry in the first tableau. It follows that the generating function for the set $\mathrm{SSYT}_N(\lambda)$ relative to one of the non-standard orderings $<_v$ can be obtained from the generating function for semistandard tableaux relative to $<$ (namely $s_\lambda(x_1,\ldots,x_N)$) by applying the permutation sending x_7 to x_2, x_6 to x_4, etc. Since Schur polynomials are symmetric, the answer is still $s_\lambda(x_1,\ldots,x_N)$. By the Product Rule for Weighted Sets, we conclude that

$$\sum_{(v,T)\in X} \mathrm{sgn}(v)\,\mathrm{wt}(v)\mathbf{x}^T = a_{\delta(N)}(x_1,\ldots,x_N)s_\lambda(x_1,\ldots,x_N).$$

On the other hand, let $Y=\mathrm{LAbc}(\lambda+\delta(N))$ be the set of N-bead labeled abaci with beads in positions $\lambda+\delta(N)$. The generating function for the signed weighted set Y is $a_{\lambda+\delta(N)}(x_1,\ldots,x_N)$. So it suffices to define a sign-reversing, weight-preserving involution $I: X\to X$ where the fixed point set of I corresponds bijectively to Y. The main idea is that the tableau T encodes a sequence of bead motions on the abacus v. If performing these movements causes a bead collision, then (v,T) will cancel with some other object in X. Otherwise, the abacus obtained from v by the bead motions is one of the objects in Y.

A tableau T specifies bead motions as follows. Define the *reading word of* T to be the word $w(T)=w_1w_2\cdots w_n$ (where $n=|\lambda|$) obtained by concatenating the rows of T from bottom to top. For example, the object (v',T') shown above has

$$w(T') = 224447755566666663333333.$$

Given $(v, T) \in X$, scan the symbols in $w(T)$ from right to left. When a symbol j is encountered, move the bead labeled j in v one step to the right.

Let us first determine which objects (v, T) have no bead collisions. Suppose $v = v_0 \dots v_{N-1}00\dots$. Let i be the last entry in the top row of T, which is the rightmost letter in $w(T)$. We must first move bead i one step to the right. This move already causes a collision (since v is justified) unless $i = v_{N-1}$. Since v_{N-1} is the smallest letter relative to $<_v$ and T is semistandard, $i = v_{N-1}$ iff all entries in the top row of T are equal to v_{N-1}. In this situation, we move the rightmost bead v_{N-1} to the right λ_1 positions with no collisions.

Now we repeat the argument on the second row of T. The rightmost entry j in this row cannot be v_{N-1} (otherwise we would not have a strict increase in every column). The only way to avoid an immediate bead collision is when $j = v_{N-2}$, in which case all entries in the second row must equal v_{N-2}. In this situation, bead v_{N-2} moves to the right λ_2 positions with no collisions.

Continuing similarly, we see that (v, T) has no collisions iff for all k, the kth row of T consists of λ_k copies of the kth smallest letter v_{N-k}. Moving the beads on v according to T has the effect of unjustifying v to an abacus $v^* \in Y = \mathrm{LAbc}(\lambda + \delta(N))$. Defining $I(v, T) = (v, T)$ in this case, we have specified a bijection between the set of fixed points of I and Y. For example,

$$(v,T) = \left(2451763000\cdots, \begin{array}{|c|c|c|c|c|c|c|} \hline 3&3&3&3&3&3&3 \\ \hline 6&6&6&6&6&6&6 \\ \hline 7&7&7&7&7 \\ \cline{1-5} 1&1&1 \\ \cline{1-3} 5&5 \\ \cline{1-2} \end{array} \right) \quad \text{maps to } v^* = 24005010070063000\cdots.$$

The map sending (v, T) to v^* preserves signs and weights.

To complete the proof, we describe a cancellation mechanism to pair off objects (v, T) in which bead collisions do occur. Suppose the first bead collision for (v, T) occurs when some bead i moves to the right one step and bumps into bead j. Note that $i >_v j$, and i and j must be two adjacent letters in the total ordering $>_v$. Define $(v', T') = I(v, T)$ as follows. We obtain v' from v by interchanging the adjacent beads i and j, so that $\mathrm{sgn}(v') = -\mathrm{sgn}(v)$, $\mathrm{wt}(v')x_j = \mathrm{wt}(v)x_i$, and $<_{v'}$ agrees with $<_v$ except that now $i <_{v'} j$.

We obtain T' from T by modifying the occurrences of i and j in $w(T)$ by a procedure similar to the one used in §9.5. By the same reasoning used to determine the fixed points of I, we know that the occurrence of i in T that caused the bead collision is the rightmost entry in some row of T, say the kth row; furthermore, for $1 \leq l < k$, row l consists of λ_l copies of v_{N-l}. Now $i >_v v_{N-k}$ (or this entry of T would not cause a collision), and so $j \geq_v v_{N-k}$. This means that no entry in the first $k-1$ rows of T equals i or j, so these rows can be ignored in the following discussion.

We now describe how to change T into T'. Whenever j occurs directly above i in T (call these occurrences *matched pairs*), interchange these two symbols. Some rows of T may contain unmatched i's and j's, in which $a \geq 0$ copies of j are followed by $b \geq 0$ copies of i. In particular, row k has $a \geq 0$ and $b > 0$, since the i at the end of the row cannot be matched with a j above it. In row k, replace the unmatched symbols $j^a i^b$ by $j^{a+1} i^{b-1}$. Then, in all rows containing unmatched i's and j's (including the new row k), replace the unmatched symbols $j^a i^b$ by $i^b j^a$. The following assertions can now be checked: T' is a semistandard tableau relative to $<_{v'}$; T' has one fewer i and one more j than T does; $\mathbf{x}^{T'} x_i = \mathbf{x}^T x_j$; $\mathrm{wt}(v', T') = \mathrm{wt}(v, T)$; $\mathrm{sgn}(v', T') = -\mathrm{sgn}(v)$; the last symbol in row k of T' is an unmatched j; this unmatched j causes the first bead collision when T' is used to move the beads on v'; and $I(v', T') = (v, T)$.

10.54. Example. Consider the object

$$(v,T)=\left(2451763000\cdots,\ \begin{array}{|c|c|c|c|c|c|c|}\hline 3&3&3&3&3&3&3\\\hline 6&6&6&6&6&6&6\\\hline 7&7&5&5&5\\\cline{1-5} 4&4&4\\\cline{1-3} 2&2\\\cline{1-2}\end{array}\right).$$

Processing the first two rows of T, we move bead 3 right seven positions, then move bead 6 right seven positions with no collisions. But in row 3, the rightmost symbol $i=5$ causes a collision with bead $j=1$. There are no matched pairs of 5's and 1's in this tableau, so we first change the 555 in row 3 to 155, and then change this string to 551 to preserve semistandardness under the new ordering. We have

$$I(v,T)=\left(2415763000\cdots,\ \begin{array}{|c|c|c|c|c|c|c|}\hline 3&3&3&3&3&3&3\\\hline 6&6&6&6&6&6&6\\\hline 7&7&5&5&1\\\cline{1-5} 4&4&4\\\cline{1-3} 2&2\\\cline{1-2}\end{array}\right).$$

If we apply I to this object, bead 1 bumps into bead 5, and we find that $I(I(v,T))=(v,T)$.

10.55. Example. Consider the object

$$(v,T)=\left(2451763000\cdots,\ \begin{array}{|c|c|c|c|c|c|c|}\hline 3&3&3&3&3&3&3\\\hline 6&6&6&7&7&7&7\\\hline 7&7&5&5&5\\\cline{1-5} 4&4&4\\\cline{1-3} 2&2\\\cline{1-2}\end{array}\right).$$

Now the first collision occurs when bead $i=7$ bumps into bead $j=6$ because of the 7 at the end of the second row of T. The first two 6's in that row are matched with 7's below, so the unmatched is and js in row 2 form the word 67777. We replace this string first by 66777, then by 77766. Interchanging the matched 6's and 7's leads to

$$I(v,T)=\left(2451673000\cdots,\ \begin{array}{|c|c|c|c|c|c|c|}\hline 3&3&3&3&3&3&3\\\hline 7&7&7&7&7&6&6\\\hline 6&6&5&5&5\\\cline{1-5} 4&4&4\\\cline{1-3} 2&2\\\cline{1-2}\end{array}\right).$$

10.56. Example. The reader may check that

$$I\left(76543210\cdots,\ \begin{array}{|c|c|c|c|c|c|c|}\hline 1&1&1&2&2&3&4\\\hline 2&3&3&3&4&4&7\\\hline 3&4&5&5&5\\\cline{1-5}\end{array}\right)=\left(76534210\cdots,\ \begin{array}{|c|c|c|c|c|c|c|}\hline 1&1&1&2&2&4&3\\\hline 2&4&4&3&3&3&7\\\hline 3&3&5&5&5\\\cline{1-5}\end{array}\right).$$

10.14 Skew Schur Polynomials

In the remainder of this chapter, we develop further combinatorial properties of skew Schur polynomials. Recall Definition 9.132: for every skew shape λ/μ,

$$s_{\lambda/\mu}(x_1,\ldots,x_N)=\sum_{T\in\mathrm{SSYT}_N(\lambda/\mu)}\mathbf{x}^T.$$

Theorem 9.135 states that skew Schur polynomials are symmetric. More precisely, we have the following expansion in the monomial basis:

$$s_{\lambda/\mu}(x_1,\ldots,x_N) = \sum_{\nu\in\mathrm{Par}_N} K_{\lambda/\mu,\nu} m_\nu(x_1,\ldots,x_N),$$

where $K_{\lambda/\mu,\nu}$ is the number of semistandard tableaux of shape λ/μ and content ν. Our current goal is to find combinatorial formulas for the expansion of skew Schur polynomials relative to some other bases for the vector space Λ. We begin by proving an algebraic fact involving the Hall scalar product.

10.57. Theorem: Skew Schur Polynomials and the Hall Scalar Product. Suppose $\lambda,\mu\in\mathrm{Par}$, $k=|\lambda|-|\mu|$, $N\geq|\lambda|$, and $f\in\Lambda_N^k$. Then $\langle s_{\lambda/\mu},f\rangle=\langle s_\lambda,s_\mu f\rangle$.

Proof. We first prove the result for $f=h_\nu$, where $\nu\in\mathrm{Par}(k)$. On one hand, we have the expansion

$$s_{\lambda/\mu} = \sum_{\rho\in\mathrm{Par}(k)} K_{\lambda/\mu,\rho} m_\rho.$$

Taking the scalar product of both sides with h_ν gives $\langle s_{\lambda/\mu},h_\nu\rangle = K_{\lambda/\mu,\nu}$, by Theorem 9.125.

On the other hand, the Pieri Rule shows that

$$s_\mu h_\nu = \sum_\rho K_{\rho/\mu,\nu} s_\rho$$

(see Theorem 9.136). Taking the scalar product of both sides with s_λ gives $\langle s_\lambda,s_\mu h_\nu\rangle = K_{\lambda/\mu,\nu}$. Thus the result holds for every f in the complete homogeneous basis.

The general case now follows by linearity: given any $f\in\Lambda_N^k$, write $f=\sum_\nu c_\nu h_\nu$ for certain real scalars c_ν. Then compute

$$\begin{aligned}\langle s_{\lambda/\mu},f\rangle &= \left\langle s_{\lambda/\mu},\sum_\nu c_\nu h_\nu\right\rangle = \sum_\nu c_\nu\langle s_{\lambda/\mu},h_\nu\rangle\\ &= \sum_\nu c_\nu\langle s_\lambda,s_\mu h_\nu\rangle = \left\langle s_\lambda,\sum_\nu c_\nu s_\mu h_\nu\right\rangle = \langle s_\lambda,s_\mu f\rangle. \quad\square\end{aligned}$$

We can use Theorem 10.57 to expand skew Schur polynomials in terms of power-sum symmetric polynomials.

10.58. Theorem: Power-Sum Expansion of Skew Schur Polynomials. Suppose $\mu\subseteq\lambda$ are partitions with $k=|\lambda|-|\mu|$. For all $N\geq|\lambda|$,

$$s_{\lambda/\mu}(x_1,\ldots,x_N) = \sum_{\nu\in\mathrm{Par}(k)} \frac{\chi_\nu^{\lambda/\mu}}{z_\nu} p_\nu(x_1,\ldots,x_N).$$

Proof. We imitate the proof of Theorem 10.53. Start with the expansion

$$s_\mu p_\nu = \sum_\lambda \chi_\nu^{\lambda/\mu} s_\lambda.$$

Now take the scalar product of both sides with a given partition λ:

$$\langle s_\lambda,s_\mu p_\nu\rangle = \chi_\nu^{\lambda/\mu}.$$

We know the symmetric polynomial $s_{\lambda/\mu}$ has some expansion in the power-sum basis, say $s_{\lambda/\mu} = \sum_{\nu} a_\nu p_\nu$ for some $a_\nu \in \mathbb{R}$. To find a particular a_ν, take the scalar product with p_ν/z_ν to get

$$a_\nu = \langle s_{\lambda/\mu}, p_\nu/z_\nu \rangle = \langle s_\lambda, s_\mu p_\nu/z_\nu \rangle = \langle s_\lambda, s_\mu p_\nu \rangle /z_\nu = \chi_\nu^{\lambda/\mu}/z_\nu. \qquad \square$$

We also deduce the effect of the involution ω on skew Schur polynomials.

10.59. Theorem: Action of ω on Skew Schur Polynomials. For all partitions $\mu \subseteq \lambda$ and all $N \geq |\lambda|$,

$$\omega(s_{\lambda/\mu}(x_1, \ldots, x_N)) = s_{\lambda'/\mu'}(x_1, \ldots, x_N).$$

Proof. We already know that the involution ω is a ring homomorphism and isometry sending every s_α to $s_{\alpha'}$. For each partition ν of size $|\lambda| - |\mu|$, we can therefore write:

$$\begin{aligned} \langle \omega(s_{\lambda/\mu}), s_\nu \rangle &= \langle \omega(\omega(s_{\lambda/\mu})), \omega(s_\nu) \rangle = \langle s_{\lambda/\mu}, s_{\nu'} \rangle = \langle s_\lambda, s_\mu s_{\nu'} \rangle \\ &= \langle \omega(s_\lambda), \omega(s_\mu s_{\nu'}) \rangle = \langle s_{\lambda'}, s_{\mu'} s_\nu \rangle \rangle = \langle s_{\lambda'/\mu'}, s_\nu \rangle. \end{aligned}$$

Thus $\omega(s_{\lambda/\mu})$ and $s_{\lambda'/\mu'}$ have the same expansion in the Schur basis and are therefore equal. $\square$

10.15 The Jacobi–Trudi Formulas

Our next goal is to obtain formulas expressing skew Schur polynomials as determinants involving the complete symmetric polynomials h_k or the elementary symmetric polynomials e_k. To derive these results, we need a new combinatorial construction relating tableaux to collections of non-intersecting lattice paths.

We begin by interpreting $h_k(x_1, \ldots, x_N)$ in terms of lattice paths. Fix an integer a and consider the set S of lattice paths from $(a, 1)$ to $(a + k, N)$ that take unit steps up (u) and east (e). We can encode a path p in this set by listing the y-coordinates of the successive east steps of p. For example, the path eeuueuee corresponds to the sequence $1, 1, 3, 4, 4$. This gives a bijection from S to the set of weakly increasing sequences $1 \leq i_1 \leq i_2 \leq \cdots \leq i_k \leq N$. Let the weight of the path corresponding to this sequence be $x_{i_1} x_{i_2} \cdots x_{i_k}$. Comparing to the definition of h_k, we see that

$$h_k(x_1, \ldots, x_N) = \sum_{p \in S} \mathrm{wt}(p).$$

This formula holds for all integers k (possibly negative), if we use the convention $h_0 = 1$ and $h_k = 0$ for negative k.

Now let λ be a partition with $n \leq N$ parts, and let $\mu \subseteq \lambda$. Let X be the set of fillings of the skew shape λ/μ using letters in $\{1, 2, \ldots, N\}$ such that each row weakly increases. Let Y be the set of sequences $P = (p_1, \ldots, p_n)$ where p_i is a lattice path from $(n - i + \mu_i, 1)$ to $(n - i + \lambda_i, N)$. Let $\mathrm{wt}(P) = \mathrm{wt}(p_1) \cdots \mathrm{wt}(p_n)$, so $\mathrm{wt}(P)$ keeps track of the y-coordinates of all the east steps of the paths in P. As explained above, we can encode each row i of a filling $U \in X$ as a lattice path p_i from $(a, 1)$ to $(a + \lambda_i - \mu_i, N)$, where $a = n - i + \mu_i$. The function sending U to $(p_1, \ldots, p_n)$ is a weight-preserving bijection $f : X \to Y$. Some examples are shown in Figure 10.4.

We say that two lattice paths *intersect* iff they share a common edge or vertex. Let Y' be the set of $P \in Y$ such that no two paths in P intersect. Inspection of Figure 10.4 suggests

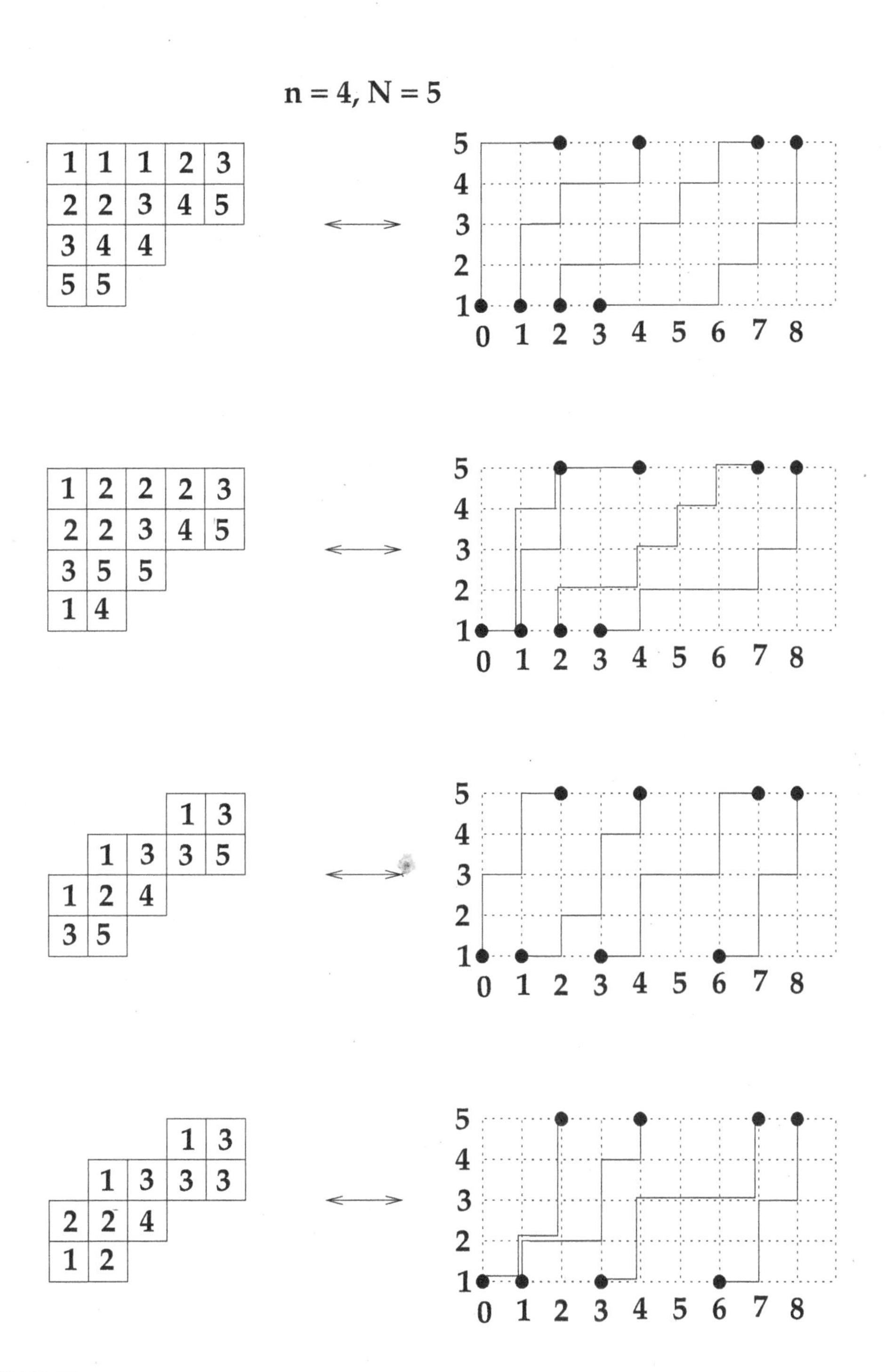

FIGURE 10.4
Encoding fillings of a skew shape by sequences of lattice paths.

that f restricts to a weight-preserving bijection from $\mathrm{SSYT}_N(\lambda/\mu)$ to Y'. To see why this holds, consider consecutive entries $U(i,j) = a$ and $U(i+1,j) = b$ in column j of a filling $U \in X$. In $f(U)$, path p_i has an east step from $(n-i+\mu_i+(j-\mu_i)-1, a) = (n+j-i-1, a)$ to $(n+j-i, a)$, whereas p_{i+1} has an east step from $(n+j-i-2, b)$ to $(n+j-i-1, b)$. Suppose $a \geq b$. Since the beginning of p_i goes from $(n-i+\mu_i, 1)$ to $(n+j-i-1, a)$, there is no way for p_{i+1} (which starts to the left of p_i) to reach the point $(n+j-i-1, b)$ without intersecting p_i. Conversely, suppose two paths intersect. Then there must exist i such that p_i and p_{i+1} intersect. The earliest intersection of these paths must occur when p_{i+1} reaches p_i by taking an east step ending at some point $(n+j-i-1, b)$. One may now check that there must exist an east step in p_i starting at $(n+j-i-1, a)$ for some $a \geq b$, which shows that $U(i,j) \geq U(i+1,j)$ in the filling U.

Now we are ready to prove the Jacobi–Trudi Formulas. The idea is to introduce a large collection of signed weighted sequences of paths that model the terms of a determinant. Cancellations will remove all sequences of intersecting paths, leaving only the objects in Y', which correspond to semistandard skew tableaux.

10.60. The First Jacobi–Trudi Formula. Suppose λ is a partition with $n \leq N$ parts, and $\mu \subseteq \lambda$. Then

$$s_{\lambda/\mu}(x_1, \ldots, x_N) = \det[h_{\lambda_i-\mu_j+j-i}(x_1, \ldots, x_N)]_{1\leq i,j\leq n}.$$

Proof. By the definition of a determinant (see Definition 12.40), the right side of the formula to be proved can be written

$$\sum_{w\in S_n} \mathrm{sgn}(w) \prod_{i=1}^{n} h_{\lambda_i-\mu_{w(i)}+w(i)-i}(x_1, \ldots, x_N).$$

This is the generating function for the following signed weighted set. Let Z be the set of sequences $(w, p_1, \ldots, p_n)$ such that $w \in S_n$ and p_i is a path from $(n - w(i) + \mu_{w(i)}, 1)$ to $(\lambda_i + n - i, N)$. The weight of such a sequence is $\prod_{i=1}^{n} \mathrm{wt}(p_i)$, and the sign is $\mathrm{sgn}(w)$.

The following involution cancels all objects $(w, p_1, \ldots, p_n)$ in which two or more paths intersect. Among all lattice points (u, v) where two paths intersect, choose the one for which u is minimized; if there are ties, choose the point that minimizes v. Let $i < j$ be the two least indices such that p_i and p_j pass through (u, v). Write $p_i = qr$ where q (respectively r) is the part of p_i before (respectively after) the point (u, v). Similarly write $p_j = st$. Now, pair the given object with the object $(w', p'_1, \ldots, p'_n)$ where $w' = w \circ (i, j)$, $p'_i = sr$, $p'_j = qt$, and $p'_k = p_k$ for all $k \neq i, j$. (Thus we have switched the initial segments of the two intersecting paths.) It can be checked that the new object lies in Z and has the same weight and opposite sign as the original object. Moreover, applying the map a second time restores the original object, so we have an involution. Some examples are shown in Figure 10.5. (Note that path p_i goes from the $w(i)$th point from the right on the line $y = 1$ to the ith point from the right on the line $y = N$.)

Let us consider an object $(w, p_1, \ldots, p_n)$ in Z that is not canceled by the involution. No two paths in this object can intersect. We claim that this forces $w = \mathrm{id}$. For otherwise, there would exist $i < j$ with $w(i) > w(j)$. But then p_i would start to the left of p_j on the line $y = 1$ and end to the right of p_j on the line $y = N$, which would force p_i and p_j to intersect. So $w = \mathrm{id}$. Erasing w maps the fixed points in Z bijectively to the set Y', which in turn maps bijectively to $\mathrm{SSYT}_N(\lambda/\mu)$, as shown in the discussion preceding the theorem. □

10.61. The Second Jacobi–Trudi Formula. Suppose λ is a partition with $\lambda_1 = n \leq N$, and $\mu \subseteq \lambda$. Then

$$s_{\lambda/\mu}(x_1, \ldots, x_N) = \det[e_{\lambda'_i-\mu'_j+j-i}(x_1, \ldots, x_N)]_{1\leq i,j\leq n}.$$

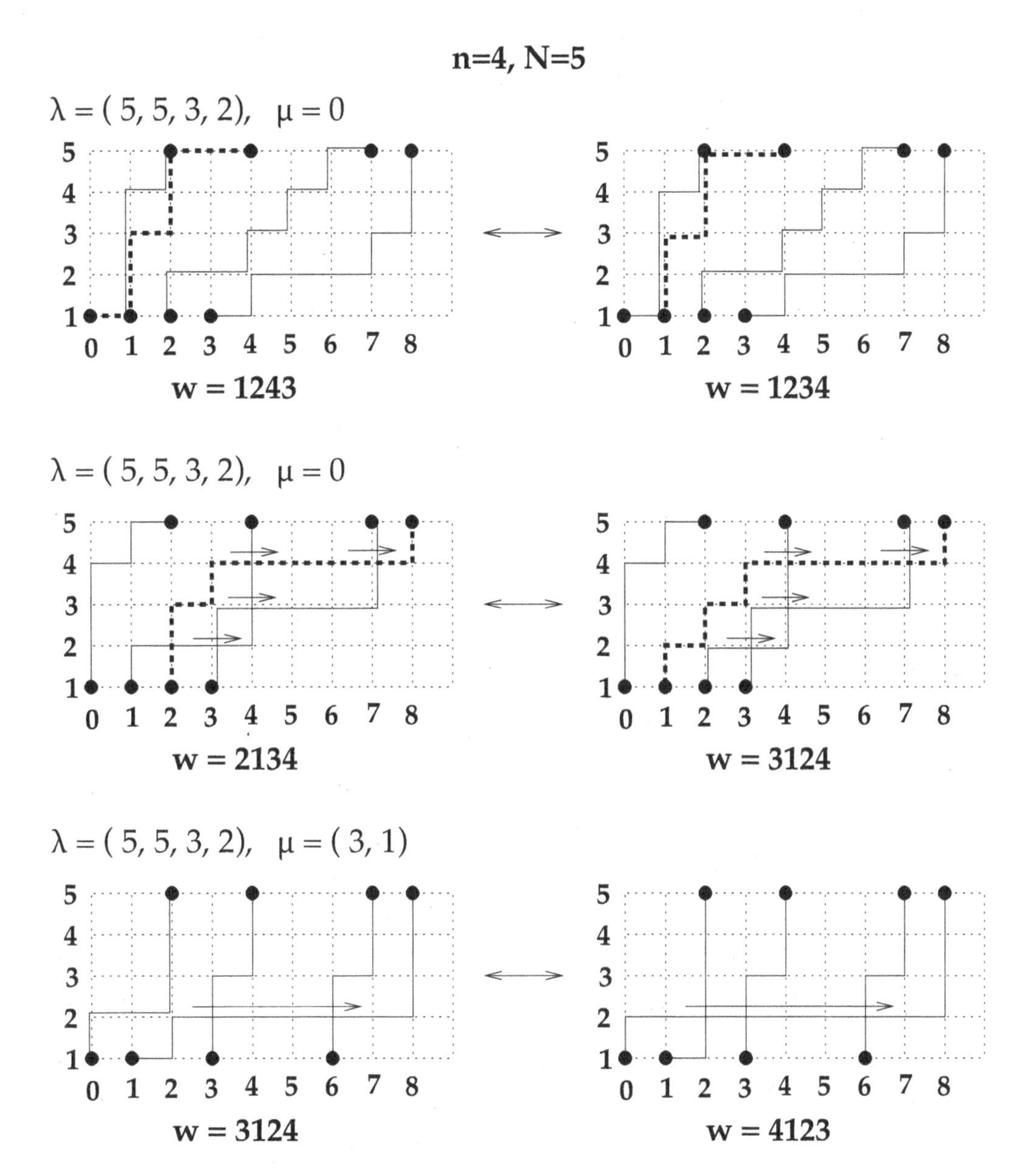

FIGURE 10.5
Cancellation mechanism for intersecting paths.

Proof. For all $f_{ij} \in \Lambda_N$, we have $\omega(\det[f_{ij}]) = \det[\omega(f_{ij})]$. This follows from the defining formula for determinants and the fact that ω is a ring homomorphism. So, we obtain the second Jacobi–Trudi Formula by applying ω to both sides of the first Jacobi–Trudi Formula

$$s_{\lambda'/\mu'} = \det[h_{\lambda'_i-\mu'_j+j-i}].$$ □

10.62. Example. According to the first Jacobi–Trudi Formula,

$$s_{(3,3,1)} = \det \begin{bmatrix} h_3 & h_4 & h_5 \\ h_2 & h_3 & h_4 \\ 0 & 1 & h_1 \end{bmatrix} = h_{(3,3,1)} + h_{(5,2)} - h_{(4,3)} - h_{(4,2,1)}.$$

Note that the main diagonal entries in the formula for s_λ are $h_{\lambda_1}, h_{\lambda_2}, \ldots, h_{\lambda_n}$, and the subscripts increase by 1 (respectively decrease by 1) as we read to the right (respectively left) along each row. Similarly,

$$s_{(3,3,1)} = \det \begin{bmatrix} e_3 & e_4 & e_5 \\ e_1 & e_2 & e_3 \\ 1 & e_1 & e_2 \end{bmatrix} = e_{(3,2,2)} + e_{(4,3)} + e_{(5,1,1)} - e_{(5,2)} - e_{(3,3,1)} - e_{(4,2,1)}.$$

Here is a typical expansion of a skew Schur polynomial:

$$s_{(5,5,3)/(3,2,0)} = \det \begin{bmatrix} h_2 & h_4 & h_7 \\ h_1 & h_3 & h_6 \\ 0 & 1 & h_3 \end{bmatrix} = h_{(3,3,2)} + h_{(7,1)} - h_{(4,3,1)} - h_{(6,2)}.$$

10.16 The Inverse Kostka Matrix

In Chapter 9, the Kostka matrix played a prominent role in relating the Schur basis of Λ_N to several other bases. More specifically, we proved the formulas

$$s_\lambda = \sum_\mu K_{\lambda,\mu} m_\mu, \qquad h_\mu = \sum_\lambda K_{\lambda,\mu} s_\lambda, \qquad e_\mu = \sum_\lambda K_{\lambda,\mu} s_{\lambda'},$$

where all symmetric polynomials have N variables and all summations extend over Par_N. Letting $\mathbf{K} = \mathbf{K}_N$ be the matrix of Kostka numbers with rows and columns indexed by elements of Par_N, and letting $\mathbf{s}$, $\mathbf{m}$, and $\mathbf{e}$ be column vectors with entries s_μ, m_μ, and e_μ, these relations can also be written

$$\mathbf{s} = \mathbf{K}\mathbf{m}, \qquad \mathbf{h} = \mathbf{K}^{\text{tr}}\mathbf{s}, \qquad \mathbf{e} = \mathbf{K}^{\text{tr}}\omega(\mathbf{s}).$$

We know that the Kostka matrix is invertible (being unitriangular). Let $K'_{\lambda,\mu}$ be the entry in row λ and column μ of the inverse of the Kostka matrix. Inverting the relations above, we see that

$$m_\lambda = \sum_\mu K'_{\lambda,\mu} s_\mu, \qquad s_\mu = \sum_\lambda K'_{\lambda,\mu} h_\lambda, \qquad s_{\mu'} = \sum_\lambda K'_{\lambda,\mu} e_\lambda.$$

Observe that the determinant formulas in the previous section, which express Schur polynomials in terms of complete homogeneous symmetric polynomials, give algebraic interpretations for the coefficients $K'_{\lambda,\mu}$. In this section, we derive combinatorial interpretations for these coefficients. To do this, we need the concept of a special rim-hook tableau.

10.63. Definition: Special Rim-hook Tableaux. For $\lambda, \mu \in \text{Par}_N$, a *special rim-hook tableau of shape* μ *and type* λ is a rim-hook tableau S of shape μ and content α such that $\text{sort}(\alpha) = \lambda$ and every nonzero rim-hook in S contains a cell in the leftmost column of the diagram of μ. The *sign* of such a tableau is defined as in Definition 10.48. Let $\text{SRHT}(\mu, \lambda)$ be the set of special rim-hook tableaux of shape μ and type λ.

10.64. Theorem: Combinatorial Interpretation of the Inverse Kostka Matrix. For all $\lambda, \mu \in \text{Par}_N$,

$$K'_{\lambda,\mu} = \sum_{S \in \text{SRHT}(\mu,\lambda)} \text{sgn}(S).$$

Proof. We give a combinatorial proof of the identity

$$a_{\delta(N)}(x_1, \ldots, x_N) m_\lambda(x_1, \ldots, x_N) = \sum_{\mu \in \text{Par}_N} \sum_{S \in \text{SRHT}(\mu,\lambda)} \text{sgn}(S) a_{\mu+\delta(N)}(x_1, \ldots, x_N).$$

Once this is done, the theorem follows by dividing both sides by $a_{\delta(N)}$ and comparing the resulting identity to the known expansion $m_\lambda = \sum_\mu K'_{\lambda,\mu} s_\mu$, which is the unique way of writing m_λ as a linear combination of Schur symmetric polynomials.

To prove the identity, we study a combinatorial interpretation of the product $a_{\delta(N)} m_\lambda$ involving abaci. Each term in the polynomial $a_{\delta(N)}$ is modeled by a justified abacus containing N beads labeled $w(N), \ldots, w(1)$ in positions $0, \ldots, N-1$ (respectively). Given such an abacus, we can view $m_\lambda(x_1, \ldots, x_N)$ as the sum of all distinct monomials $\prod_{i=1}^N x_{w(i)}^{e(i)}$ such that the exponent sequence $(e(1), \ldots, e(N))$ is a rearrangement of $(\lambda_1, \ldots, \lambda_N)$. Here and below, we view elements of Par_N as partitions with *exactly* N parts, some of which may be zero. The multiplication of $a_{\delta(N)}$ by one of these monomials can be implemented on the abacus as follows. Imagine moving the N justified beads from their current runner to a new, initially empty runner, by moving each bead $w(i)$ from position $N - i$ on the old runner to position $N - i + e(i)$ on the new runner. Call such a transformation of the justified abacus a λ-*move.* A given λ-move either causes a bead collision on the new runner, or else produces a new abacus, which is enumerated by a monomial in $a_{\mu+\delta(N)}(x_1, \ldots, x_N)$ for some $\mu \in \text{Par}_N$.

Consider the situation where a bead collision occurs. Choose i minimal such that bead $w(i)$ collides with some other bead on the new runner, and then choose j minimal such that bead $w(i)$ collides with $w(j)$. Create a new object counted by $a_{\delta(N)} m_\lambda$ by switching beads $w(i)$ and $w(j)$ on the old abacus, and switching $e(i)$ and $e(j)$ in the exponent vector. This defines a sign-reversing, weight-preserving involution that cancels all objects in which bead collisions occur.

To complete the proof, we must find a sign-preserving, weight-preserving bijection ϕ from the set X of uncanceled objects counted by $a_{\delta(N)} m_\lambda$ to the signed weighted set

$$\bigcup_{\mu \in \text{Par}_N} \text{SRHT}(\mu, \lambda) \times \text{LAbc}(\mu + \delta(N)).$$

For this purpose, we fix $\mu \in \text{Par}_N$ and consider the ways in which a justified abacus with N beads can be transformed into an abacus in $\text{LAbc}(\mu + \delta(N))$ by means of a λ-move. Let us temporarily ignore bead labels and signs, concentrating at first only on the positions of the N beads. The positions of the N beads on the old runner are the entries in the sequence $\delta(N) = (N-1, N-2, \ldots, 2, 1, 0)$. A λ-move adds some rearrangement of the sequence $\lambda = (\lambda_1, \ldots, \lambda_N)$ to the sequence $\delta(N)$. We obtain an abacus in $\text{LAbc}(\mu+\delta(N))$ iff the sum of these sequences is some rearrangement of the sequence

$$\mu + \delta(N) = (\mu_1 + N - 1, \mu_2 + N - 2, \ldots, \mu_N + N - N).$$

$$S = \begin{array}{|c|c|c|c|c|c|c|}\hline 1&2&2&2&2&2&2\\ \hline 2&2&4&4&4 \\ \cline{1-5} 4&4&4&5 \\ \cline{1-4} 4&5&5&5 \\ \cline{1-4} 5&5 \\ \cline{1-2}\end{array}$$

FIGURE 10.6
A special rim-hook tableau.

We now show that the rearrangements of λ that produce abaci in $\text{LAbc}(\mu+\delta(N))$ can be encoded by special rim-hook tableaux of shape μ and type λ. The proof uses induction on N. Let us first illustrate the idea of the proof by considering an example. Take $N = 5$, $\mu = (7,5,4,4,2)$, and $\lambda = (8,7,6,1,0)$. We seek rearrangements of the vector $(8,7,6,1,0)$ which, when added to the vector $(4,3,2,1,0)$, produce a rearrangement of $\mu+\delta(N) = (11,8,6,5,2)$. In this example, the only solution turns out to be $(1,8,0,7,6)+(4,3,2,1,0) = (5,11,2,8,6)$. We can visualize this solution using the special rim-hook tableau in Figure 10.6, in which the rim-hooks (from top to bottom) have lengths $(1,8,0,7,6)$. If we start with a labeled justified abacus $54321000\cdots$ and perform a λ-move using the rearrangement $(1,8,0,7,6)$, we obtain the abacus $003001504002000\cdots \in \text{LAbc}(11,8,6,5,2)$. The sign of this abacus, namely $\text{sgn}(24513) = -1$, differs from the sign of the original abacus, namely $\text{sgn}(12345) = +1$, by a factor of $(-1)^5 = \text{sgn}(S)$. A similar remark holds if the original abacus had involved some other permutation of the five labels.

With this example in mind, we return to the general proof. We are seeking permutations $j_1\cdots j_N$ and $k_1\cdots k_N$ satisfying the system of equations

$$\begin{array}{rcl} 0+\lambda_{j_N} & = & \mu_{k_N}+N-k_N \\ 1+\lambda_{j_{N-1}} & = & \mu_{k_{N-1}}+N-k_{N-1} \\ \cdots & & \cdots \\ N-1+\lambda_{j_1} & = & \mu_{k_1}+N-k_1. \end{array} \tag{10.5}$$

In particular, to satisfy the first equation, we need an index $j = j_N$ and an index $k = k_N$ such that $\lambda_j = \mu_k + N - k$. If such an index exists, we encode it by drawing the unique border ribbon of length λ_j starting in the leftmost cell of row N of μ. By choice of j and k, this border ribbon must end in the rightmost cell of row k of μ. In terms of the abaci, the λ-move encoded by $j_1\cdots j_N$ moves the bead in position 0 on the old runner (the Nth bead from the right) to position $\mu_k + N - k$ on the new runner (which becomes the kth bead from the right). Thus this bead moves past $N-k$ other beads during the λ-move, which causes a sign change of $(-1)^{N-k}$ for any choice of labels. But $N-k$ is precisely the spin of the border ribbon we just drew.

To finish solving system (10.5), let λ^* be the partition obtained by dropping one part λ_j from λ, and let μ^* be the partition in Par_{N-1} obtained by erasing the cells of μ occupied by the ribbon that starts in row N. Suppose we ignore the first equation in the system (10.5) and subtract 1 from both sides of the remaining $N-1$ equations. One may check that the resulting system of $N-1$ equations is precisely the system we must solve to change a justified abacus to an abacus in $\text{LAbc}(\mu^*+\delta(N-1))$ by means of a λ^*-move.

For instance, in the example considered earlier, after we move a bead from position 0 to position 6 (accounting for the lowest rim-hook in the displayed tableau), we have $\lambda^* = (8,7,1,0)$ and $\mu^* = (7,5,3,1)$. Having moved one bead, we are left with the task of moving beads from positions $(4,3,2,1) = (1,1,1,1)+\delta(4)$ to positions $(11,8,5,2) = (1,1,1,1)+\mu^*+\delta(4)$ using the moves in $\lambda^* = (8,7,1,0)$.

By induction on N, the solutions of the reduced system are encoded by special rim-hook tableaux S^* of shape μ^* and type λ^*; and furthermore, the net sign change going from the old abacus to the new abacus (disregarding the bead originally in position 0) is $\text{sgn}(S^*)$. It follows that all solutions of the original system are encoded by special rim-hook tableaux S of shape μ and type λ; and furthermore, the net sign change going from the old abacus to the new abacus (taking all beads into account) is $\text{sgn}(S)$.

The preceding discussion contains an implicit recursive definition of the required bijection ϕ. More explicitly, suppose $z = (w(N)\cdots w(1)000\cdots, e(N)\cdots e(1)) \in X$ is an uncanceled object counted by $a_{\delta(N)}m_\lambda$. Then $\phi(z) = (S, v)$ where $v \in \text{LAbc}(\mu + \delta(N))$ is obtained from the first component of z by moving bead $w(i)$ right $e(i)$ positions for all i, and S is the unique special rim-hook tableau (of shape μ determined by v) that has a rim-hook of length $e(i)$ starting in the leftmost cell of row i of the diagram. The preceding arguments show that ϕ preserves signs and weights. To compute $\phi^{-1}(S, v)$, it suffices to note that the sequence $(e(1), \ldots, e(N))$ is the content of the rim-hook tableau S. Knowledge of this sequence allows us to reverse the λ-move and recover $w(N)\cdots w(1)$. Thus, ϕ is a bijection. □

10.65. Remark. An alternate approach to the theorem is to *define*

$$K'_{\lambda,\mu} = \sum_{S \in \text{SRHT}(\mu,\lambda)} \text{sgn}(S)$$

and then give a combinatorial proof of the matrix identity $\mathbf{KK}' = \mathbf{I}$ (see Exercise 10-55). Since $\mathbf{K}$ is known to be invertible, it follows that $\mathbf{K}'$ must be the (two-sided) matrix inverse of $\mathbf{K}$.

10.17 Schur Expansion of Skew Schur Polynomials

We now consider the expansion of skew Schur polynomials as linear combinations of ordinary Schur polynomials. Since the ordinary Schur polynomials are a basis of Λ_N and the skew Schur polynomials are in this vector space, we know there exist unique scalars $c^\lambda_{\nu,\mu} \in \mathbb{R}$ such that

$$s_{\lambda/\nu}(x_1, \ldots, x_N) = \sum_\mu c^\lambda_{\nu,\mu} s_\mu(x_1, \ldots, x_N), \tag{10.6}$$

where it suffices to sum over partitions μ of size $|\lambda/\nu|$. The scalars $c^\lambda_{\nu,\mu}$ are called *Littlewood–Richardson coefficients.* The following result shows that these coefficients are all nonnegative integers. Recall that, for a semistandard tableau T of any shape, the *word of* T is obtained by concatenating the rows of T from bottom to top.

10.66. The Littlewood–Richardson Rule for Skew Schur Polynomials. For all partitions λ, μ, ν, $c^\lambda_{\nu,\mu}$ is the number of semistandard tableaux T of shape λ/ν and content μ such that every suffix of the word of T has partition content. In other words, writing $w(T) = w_1 w_2 \cdots w_n$, we require that for all k between 1 and n and all $i \geq 1$, the number of i's in the suffix $w_k w_{k+1} \cdots w_n$ equals or exceeds the number of $i + 1$'s in this suffix.

Proof. Multiplying both sides of (10.6) by $a_{\delta(N)}$, it suffices to prove the identity

$$a_{\delta(N)}(x_1, \ldots, x_N) s_{\lambda/\nu}(x_1, \ldots, x_N) = \sum_\mu c^\lambda_{\nu,\mu} a_{\mu+\delta(N)}(x_1, \ldots, x_N).$$

The idea is to generalize the proof of the special case $\nu = (0)$ given in §10.13. We model the left side of the identity by the set X of pairs (v, T), where v is a justified N-bead labeled abacus and T is a semistandard tableau of shape λ/ν using the alphabet $\{1, 2, \ldots, N\}$ ordered by $<_v$. Since skew Schur polynomials are symmetric, the generating function for the signed weighted set X is $a_{\delta(N)} s_{\lambda/\nu}$.

We now define an involution $I : X \to X$. Given $(v, T) \in X$, T determines a sequence of bead motions on v by reading $w(T)$ from right to left and moving bead k one step to the right each time the symbol k is seen. If these bead motions cause a collision, define $I(v, T) = (v', T')$ by the following rules. Suppose the first collision occurs when bead i bumps into bead j, where $i >_v j$ are adjacent beads in v. Let v' be v with beads i and j switched, so $\text{sgn}(v') = -\text{sgn}(v)$ and $\text{wt}(v')x_j = \text{wt}(v)x_i$.

Next, we calculate T' from T as follows. Starting with the word of T, replace each i by a left parenthesis, each j by a right parenthesis, and ignore all other symbols. Match left and right parentheses in the resulting string of parentheses, and ignore these matched pairs of parentheses hereafter. The remaining unmatched parentheses must consist of a string of $a \geq 0$ right parentheses followed by a string of $b \geq 0$ left parentheses, since if a left parenthesis appeared somewhere to the left of a right parenthesis we could find another matched pair of parentheses.

Note that $b > 0$, since otherwise bead i would never bump into bead j. Indeed, the first bead collision occurs when we reach the rightmost unmatched left parenthesis (occurrence of i) in the word of T. Now, change the subword of unmatched parentheses from "$)^a(^b$" to "$)^{b-1}(^{a+1}$", and then convert all left parentheses to j's and all right parentheses to i's. One may verify that the new word is the word of a tableau $T' \in \text{SSYT}_N(\lambda/\nu)$, relative to the ordering $<_{v'}$, because i and j are adjacent relative to the orderings $<_v$ and $<_{v'}$, and the status of a given parenthesis symbol in T' (matched or unmatched) is the same as its status in T. See the example following the proof for more discussion of this point.

Because T' has one less i than T and T' has one more j than T, we have $\text{wt}(T')x_i = \text{wt}(T)x_j$. Since we also had $\text{wt}(v')x_j = \text{wt}(v)x_i$, we see that $\text{wt}(v', T') = \text{wt}(v, T)$. Thus I is sign-reversing and weight-preserving. Finally, to check that I is an involution, consider what happens when we use T' to move the beads on v'. Bead j on v' moves the same way as bead i did on v (and vice versa) until we reach the rightmost unmatched parenthesis (relative to i and j) in $w(T')$. When this symbol is reached, bead j bumps into bead i on v', just as bead i bumped into bead j on v. To compute $I(v', T')$, we therefore apply the parenthesis modification rule to the i's and j's appearing in $w(T')$. This rule changes the unmatched parentheses from "$)^{b-1}(^{a+1}$" back to "$)^a(^b$", which shows that $I(v', T') = (v, T)$. So I is an involution.

All that remains is to analyze the fixed points of I, which are (by definition) the pairs (v, T) for which no bead collision occurs. Recall that we are starting with a justified abacus v, scanning the symbols in $w(T) = w_1 \cdots w_n$ from right to left, and moving the corresponding beads on v. Suppose all suffixes of T have partition content relative to the ordering $<_v$ (which means the rightmost bead label occurs at least as often in each suffix as the next bead label, and so on). We see from the description of the bead motion that no collision occurs. Conversely, if the condition is first violated by some suffix $w_k w_{k+1} \cdots w_n$, then a collision occurs at this point in the scan. Thus the fixed points of I are the pairs (v, T) such that each suffix of T has partition content relative to $<_v$. We map each such fixed point to the abacus v^* obtained from v by performing the bead motions specified by T. The abacus v^* lies in the set $\text{LAbc}(\mu + \delta(N))$, where μ is the content of T (calculating content relative to $<_v$, so μ_1 is the number of times the rightmost bead moves, etc.).

We can obtain all the fixed points of I from fixed points of the form (v^0, T), where $v^0 = (N, N-1, \ldots, 1, 0, 0, \ldots)$, $<_{v^0}$ is the usual ordering on integers, and T is a semistandard tableau satisfying the conditions in the theorem statement. We need only permute the bead labels in v^0 by any $w \in S_N$, and permute the entries of T in the same way. The object (v^0, T) thereby generates $N!$ fixed points, which together contribute one copy of $a_{\mu+\delta(N)}(x_1, \ldots, x_N)$ to the generating function for the fixed points of I. The total number of times this term appears in the generating function is the total number of semistandard tableaux T of content μ satisfying the conditions in the theorem. Since the generating function for X must equal the generating function for the fixed point set of I, the proof is complete. $\square$

10.67. Example. To illustrate the parenthesis construction, we compute $I(5432100\cdots, T)$, where

$T =$

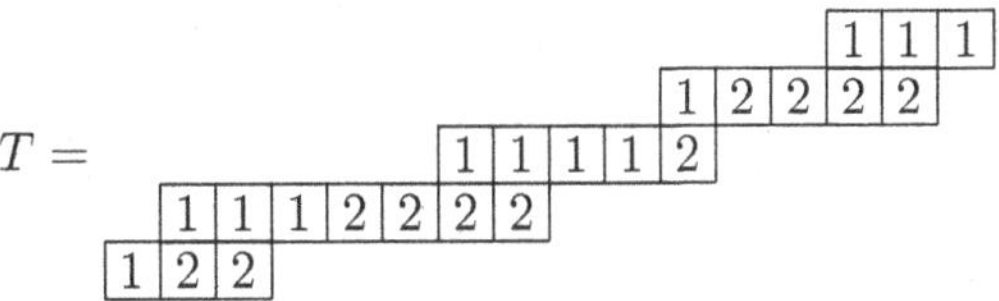

The word of T is 12211122221111212222111. The suffix 2222111 of $w(T)$ does not have partition content, so this object cancels with some object $(5431200\cdots, T')$. To find T', first convert 1's to right parentheses and 2's to left parentheses in $w(T)$:

```
12211122221111212222111
)(()))((((()))()(((()))
```

Now we balance parentheses and mark the remaining unmatched symbols:

```
)(()))((((()))()(((()))
*    *         *
```

The substring of unmatched parentheses is `))(`. Observe that the rightmost symbol in this substring is a left parenthesis corresponding to the first 2 in the offending suffix 2222111, and this 2 is the symbol in $w(T)$ causing the first bead collision. As directed by the proof, we convert the unmatched parentheis string to `(((` and then replace left parentheses by 1's and right parentheses by 2's:

```
*    *         *
((()((((((()))()(((()))
11122111112222121111222
```

This new word is $w(T')$, so finally

$T' =$

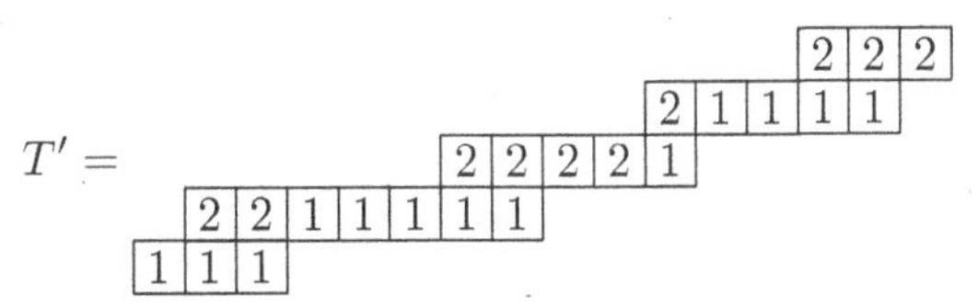

Observe that T' is a semistandard tableau relative to the ordering $5 > 4 > 3 > 1 > 2$. In particular, columns of T' strictly increase because whenever 1 appears above 2 in T, these occurrences of 1 and 2 become matched parentheses. Rearranging the unmatched parentheses does not affect these symbols, so in the end we get a 2 above a 1 in T'. Also, rows of T' weakly increase since a strict decrease in some row would be encoded as a matched parenthesis pair in $w(T')$, which would have also been matched in $w(T)$, implying that T had a strict decrease in some row. But T is a semistandard tableau so this cannot happen. Finally, note that the shortest suffix of T' that does not have partition content (relative to the new ordering) is 1111222, where the leftmost 1 corresponds to the rightmost unbalanced parenthesis in $w(T')$. Consequently, $I(5431200\cdots, T') = (5432100\cdots, T)$. Observe that these two objects have opposite sign, but both have weight $x_1^{16}x_2^{14}x_3^2x_4^1x_5^0$.

10.68. Example. Let us compute $I(v,T)$, where

$$v = 5432100\cdots, \quad T = \begin{array}{ccccccc} & & & & 1 & 1 & 1 \\ & & & 2 & 3 & 3 & \\ 1 & 1 & 2 & 4 & 4 & 5 & \\ 2 & 2 & 3 & 5 & & & \\ 3 & 5 & & & & & \end{array}$$

Moving beads on v according to the word $w(T) = 352235112445233111$, bead 3 bumps into bead 2 when we have scanned the suffix 3111 (which is the shortest suffix without partition content). We therefore modify the 2's and 3's in the word as follows:

```
352235112445233111
3 223   2   233
( ))(   )   )((
   *         ***
( ))(   )   (((
2 332   3   222
253325113445222111
```

Therefore $I(v,T) = (v',T')$, where

$$v' = 5423100\cdots, \quad T' = \begin{array}{ccccccc} & & & & 1 & 1 & 1 \\ & & & 2 & 2 & 2 & \\ 1 & 1 & 3 & 4 & 4 & 5 & \\ 3 & 3 & 2 & 5 & & & \\ 2 & 5 & & & & & \end{array}$$

Observe that $\text{wt}(v,T) = \text{wt}(v',T') = x_1^9 x_2^7 x_3^6 x_4^3 x_5^3$, $\text{sgn}(v',T') = -\,\text{sgn}(v,T)$, and $I(v',T') = (v,T)$.

10.69. Example. Let us compute $c^{\lambda}_{\nu,\mu}$ when $\lambda = (5,4,4,1)$, $\nu = (3,1)$, and $\mu = (4,4,2)$. We draw the semistandard tableaux of shape λ/ν whose words have the required suffix property. The following two tableaux are the only ones, so $c^{\lambda}_{\nu,\mu} = 2$:

$$T_1 = \begin{array}{ccccc} & & & 1 & 1 \\ & 1 & 1 & 2 & \\ 2 & 2 & 2 & 3 & \\ 3 & & & & \end{array} \qquad T_2 = \begin{array}{ccccc} & & & 1 & 1 \\ & 1 & 2 & 2 & \\ 1 & 2 & 3 & 3 & \\ 2 & & & & \end{array}$$

Let us see how these tableaux correspond to fixed points of I when $N = 5$. The first tableau changes the standard abacus $v^0 = (5432100\cdots)$ to the abacus $(54003002100\cdots)$ counted by $\text{LAbc}(\mu + \delta(5))$ by moving bead 1 twice, then bead 2 once, then bead 1 twice, and so on. Permuting the labels gives the other 119 signed objects that make up one copy of $a_{\mu+\delta(5)}(x_1,\ldots,x_5)$; for instance,

$$\left(3425100\cdots, \begin{array}{ccccc} & & & 1 & 1 \\ & 1 & 1 & 5 & \\ 5 & 5 & 5 & 2 & \\ 2 & & & & \end{array}\right) \quad \text{maps to} \quad (34002005100\cdots).$$

On the other hand, the second tableau changes the standard abacus $(5432100\cdots)$ to the abacus $(54003002100\cdots)$ via a different sequence of collision-free bead moves: move bead 1 twice, then bead 2 twice, then bead 1 once, and so on. This pair and its permutations produce another copy of the generating function $a_{\mu+\delta(5)}(x_1,\ldots,x_5)$. Dividing by $a_{\delta(5)}$, we conclude that

$$s_{\lambda/\nu} = 2s_{\mu} + \cdots.$$

Now let us compute $c^{\lambda}_{\mu,\nu}$. The required skew tableaux, which have shape $(5,4,4,1)/(4,4,2)$ and content $(3,1)$, are:

$$\begin{array}{ccccc} & & & & 1 \\ \\ & & 1 & 1 & \\ 2 & & & & \end{array} \qquad \begin{array}{ccccc} & & & & 1 \\ \\ & & 1 & 2 & \\ 1 & & & & \end{array}$$

So $c^{\lambda}_{\mu,\nu} = 2$. This illustrates the general symmetry property $c^{\lambda}_{\nu,\mu} = c^{\lambda}_{\mu,\nu}$, which is true but not immediately evident from our combinatorial description of these coefficients. We prove this property in §10.18 when we discuss products of Schur polynomials.

10.70. Example. For $N \geq 7$, let us find the Schur expansion of the skew Schur polynomial $s_{(3,3,2,2)/(2,1)}$ in N variables. This expansion is found by enumerating all semistandard skew tableaux of shape $(3,3,2,2)/(2,1)$ satisfying the required suffix property. Each such tableau of content μ contributes one term s_{μ} to the expansion. The relevant tableaux are shown here:

$$\begin{array}{ccc} & & 1 \\ & 1 & 2 \\ 1 & 2 & \\ 2 & 3 & \end{array} \qquad \begin{array}{ccc} & & 1 \\ & 1 & 2 \\ 1 & 2 & \\ 3 & 3 & \end{array} \qquad \begin{array}{ccc} & & 1 \\ & 1 & 2 \\ 1 & 3 & \\ 2 & 4 & \end{array} \qquad \begin{array}{ccc} & & 1 \\ & 1 & 2 \\ 2 & 3 & \\ 3 & 4 & \end{array}$$

We conclude that

$$s_{(3,3,2,2)/(2,1)} = 1s_{(3,3,1)} + 1s_{(3,2,2)} + 1s_{(3,2,1,1)} + 1s_{(2,2,2,1)}.$$

10.18 Products of Schur Polynomials

Given partitions $\mu \in \text{Par}_N(m)$ and $\nu \in \text{Par}_N(n)$, the product $s_{\mu}(x_1,\ldots,x_N)s_{\nu}(x_1,\ldots,x_N)$ is a symmetric polynomial, so it can be expressed uniquely in terms of Schur polynomials $s_{\lambda}(x_1,\ldots,x_N)$ indexed by $\lambda \in \text{Par}_N(m+n)$:

$$s_{\mu}s_{\nu} = \sum_{\lambda} a(\lambda,\mu,\nu)s_{\lambda} \quad \text{for some } a(\lambda,\mu,\nu) \in \mathbb{R}. \tag{10.7}$$

By Theorem 10.57, the coefficients here are precisely the Littlewood–Richardson numbers:

$$a(\lambda,\mu,\nu) = \langle s_{\mu}s_{\nu}, s_{\lambda}\rangle = \langle s_{\mu}, s_{\lambda/\nu}\rangle = c^{\lambda}_{\nu,\mu}.$$

Since $s_{\mu}s_{\nu} = s_{\nu}s_{\mu}$ (because multiplication of polynomials is commutative), we deduce the symmetry property:

$$c^{\lambda}_{\nu,\mu} = c^{\lambda}_{\mu,\nu}.$$

We now derive another combinatorial expression for these integers by viewing the product $s_{\mu}s_{\nu}$ as a skew Schur polynomial. We claim that $s_{\mu}s_{\nu} = s_{\alpha/\beta}$, where

$$\alpha = (\mu_1+\nu_1, \mu_1+\nu_2, \ldots, \mu_1+\nu_N, \mu_1, \ldots, \mu_N), \text{ and } \beta = (\mu_1^N).$$

This follows since the skew shape α/β consists of two disconnected pieces, one of shape ν and one of shape μ. A semistandard skew tableau of this shape can be formed by choosing a semistandard tableau of shape ν and independently choosing a semistandard tableau of shape μ; thus the result follows from the Product Rule for Weighted Sets. We conclude that

$$c^{\lambda}_{\mu,\nu} = c^{\alpha}_{\beta,\lambda} \qquad \text{(with } \alpha, \beta \text{ as above).}$$

This formula is illustrated in the next example.

10.71. Example. Let us compute the Schur expansion of $s_{(2,1)}s_{(2,1)}$ using the observation $s_{(2,1)}s_{(2,1)} = s_{(4,3,2,1)/(2,2)}$. The following skew tableaux have words such that all suffixes have partition content:

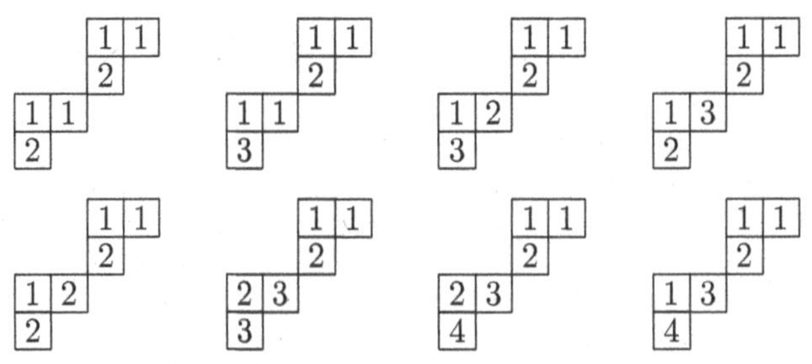

Looking at contents, we conclude that

$$s_{(2,1)}s_{(2,1)} = s_{(4,2)} + s_{(4,1,1)} + 2s_{(3,2,1)} + s_{(3,3)} + s_{(2,2,2)} + s_{(2,2,1,1)} + s_{(3,1,1,1)}.$$

Observe that the upper-right portion of the skew tableau could be filled in only one way. So we could ignore this part of the tableau and just consider allowable fillings of the lower shape. Generalizing this remark leads to the following prescription for the Littlewood–Richardson coefficients.

10.72. Theorem: Alternate Formula for Littlewood–Richardson Coefficients. For all partitions $\lambda, \mu, \nu \in \text{Par}_N$, the coefficent $c^{\lambda}_{\mu,\nu} = c^{\lambda}_{\nu,\mu}$ is the number of semistandard tableaux T of shape μ and content $\lambda - \nu = (\lambda_i - \nu_i : 1 \leq i \leq N)$ such that $w(T) = w_1 \cdots w_n$ satisfies the following condition: for all k between 1 and n, the exponent vector of the monomial $\prod_{j=1}^{N} x_j^{\nu_j} \prod_{i=k}^{n} x_{w_i}$ is a partition. (This condition means that for all j, k with $1 \leq j < N$ and $1 \leq k \leq n$, if there are a copies of j and b copies of $j+1$ in the suffix $w_k w_{k+1} \cdots w_n$, then $\nu_j + a \geq \nu_{j+1} + b$.)

Proof. We already know that $c^{\lambda}_{\mu,\nu} = c^{\alpha}_{\beta,\lambda}$ where the skew shape α/β consists of an upper part of shape ν and a lower part of shape μ. We also know that $c^{\alpha}_{\beta,\lambda}$ is the number of skew tableaux U of shape α/β and content λ such that every suffix of $w(U)$ has partition content. Consider the last $|\nu|$ symbols in $w(U)$. The last symbol is the label in the rightmost cell of the first row of the skew shape α/β. The partition content condition forces this symbol to be 1, and then all symbols in the first row of the skew tableau U must be 1. The symbol at the end of the second row must be strictly greater than 1, so it is 2 (by the partition content condition), and then every entry in the second row must be 2. Proceeding in this way, we see that for k between 1 and N, every entry in row k of U must equal k. Equivalently, the last $|\nu|$ symbols of $w(U)$ must be $N^{\nu_N} \cdots 2^{\nu_2} 1^{\nu_1}$. Call this suffix z.

Now we must fill the lower part of the shape α/β by choosing a semistandard tableau T of shape μ. Because the upper part of U has content ν, the content of the entire skew tableau U is λ iff the content of the lower part T is $\lambda - \nu$. The other condition imposed on T is that for every suffix y of $w(T)$, the suffix yz of $w(U)$ has partition content. Given the formula for z above, this condition is equivalent to the condition on $w(T)$ in the theorem statement. □

Summary

Table 10.1 contains formulas derived in this chapter for computing with antisymmetric and symmetric polynomials.

TABLE 10.1
Formulas for manipulating antisymmetric and symmetric polynomials.

The Pieri Rules:	$a_{\lambda+\delta(N)}p_k = \sum_{\beta:\ \beta/\lambda \text{ is a } k\text{-ribbon } R} \operatorname{sgn}(R) a_{\beta+\delta(N)}$
	$a_{\lambda+\delta(N)}e_k = \sum_{\beta:\ \beta/\lambda \text{ is a vertical } k\text{-strip}} a_{\beta+\delta(N)}$
	$a_{\lambda+\delta(N)}h_k = \sum_{\beta:\ \beta/\lambda \text{ is a horizontal } k\text{-strip}} a_{\beta+\delta(N)}$
	$s_\lambda p_k = \sum_{\beta:\ \beta/\lambda \text{ is a } k\text{-ribbon } R} \operatorname{sgn}(R) s_\beta$
	$s_\mu p_\alpha = \sum_\lambda \chi_\alpha^{\lambda/\mu} s_\lambda$
Determinant formula for s_λ:	$s_\lambda = \frac{a_{\lambda+\delta(N)}}{a_{\delta(N)}} = \frac{\det[x_j^{\lambda_i+N-i}]_{1\le i,j\le N}}{\det[x_j^{N-i}]_{1\le i,j\le N}}$
Schur expansion of power-sums:	$p_\alpha = \sum_\lambda \chi_\alpha^\lambda s_\lambda$
Power-sum expansion of Schur polynomials:	$s_\lambda = \sum_\mu \frac{\chi_\mu^\lambda}{z_\mu} p_\mu$
Formulas for skew Schur polynomials:	$s_{\lambda/\mu} = \sum_\nu \frac{\chi_\nu^{\lambda/\mu}}{z_\nu} p_\nu$
	$\langle s_{\lambda/\mu}, f\rangle = \langle s_\lambda, s_\mu f\rangle$ for $f \in \Lambda_N$
	$\omega(s_{\lambda/\mu}) = s_{\lambda'/\mu'}$
	$s_{\lambda/\mu} = \det[h_{\lambda_i-\mu_j+j-i}]_{1\le i,j\le \ell(\lambda)}$
	$s_{\lambda/\mu} = \det[e_{\lambda'_i-\mu'_j+j-i}]_{1\le i,j\le \lambda_1}$
	$s_{\lambda/\mu} = \sum_\nu c_{\mu,\nu}^\lambda s_\nu$
	$s_\mu s_\nu = \sum_\lambda c_{\mu,\nu}^\lambda s_\lambda$
Inverse Kostka formulas:	$K'_{\lambda,\mu} = \sum_{S\in\mathrm{SRHT}(\mu,\lambda)} \operatorname{sgn}(S)$
	$m_\lambda = \sum_\mu K'_{\lambda,\mu} s_\mu$
	$s_\mu = \sum_\lambda K'_{\lambda,\mu} h_\lambda$
	$s_{\mu'} = \sum_\lambda K'_{\lambda,\mu} e_\lambda$

- *Unlabeled Abaci.* An abacus is a function $w : \mathbb{Z} \to \{0,1\}$ with $w(i) = 1$ for all small enough i and $w(j) = 0$ for all large enough j. Justification of abaci gives a bijection to pairs $(m, \lambda) \in \mathbb{Z} \times \text{Par}$. The inverse bijection can be computed by traversing the frontier of $\text{dg}(\lambda)$, converting north steps to beads (1's) and east steps to gaps (0's), and using m to decide which step on the frontier corresponds to position 0 of w.

- *The Jacobi Triple Product Identity.* Abaci can be used to prove

$$\sum_{m \in \mathbb{Z}} q^{m(m+1)/2} u^m = \prod_{n=1}^{\infty}(1 + uq^n) \prod_{n=0}^{\infty}(1 + u^{-1}q^n) \prod_{n=1}^{\infty}(1 - q^n).$$

One consequence is the formula $\prod_{n=1}^{\infty}(1 - q^n) = \sum_{k \in \mathbb{Z}}(-1)^k q^{(3k^2-k)/2}$.

- *k-Cores and k-Quotients.* Given a partition μ, repeated removal of border ribbons of size k (in any order) leads to a unique partition from which no further ribbons of this kind can be removed. This partition is called the k-core of μ. We can also find the k-core by converting μ to an abacus, decimating the abacus to give a k-runner abacus, justifying all runners, and converting back to a partition. Each ribbon removal corresponds to moving one bead one step to the left on the k-runner abacus. Justifying each separate runner on the k-runner abacus for μ produces the k-quotients $(\nu^0, \ldots, \nu^{k-1})$ of μ. Alternatively, $\text{dg}(\nu^i)$ can be found by taking the cells of $\text{dg}(\mu)$ lying due north and due west of steps of k-content i on the frontier of μ. We get a bijection $\Delta_k : \text{Par} \to \text{Core}(k) \times \text{Par}^k$ by mapping μ to its k-core and k-quotients.

- *Labeled Abaci and Antisymmetric Polynomials.* A polynomial f in N variables is antisymmetric iff interchanging any two adjacent variables changes the sign of f. For each $\mu = (\mu_1 > \mu_2 > \cdots > \mu_N \geq 0)$, the polynomial $a_\mu(x_1, \ldots, x_N) = \det[x_j^{\mu_i}]_{1 \leq i,j \leq N}$ is antisymmetric. Writing $\delta(N) = (N-1, N-2, \ldots, 2, 1, 0)$, the set $\{a_{\lambda+\delta(N)} : \lambda \in \text{Par}_N\}$ is a basis for the vector space A_N of antisymmetric polynomials. Division by $a_{\delta(N)} = \prod_{1 \leq i < j \leq N}(x_i - x_j)$ gives a vector space isomorphism from A_N to Λ_N sending $a_{\lambda+\delta(N)}$ to the Schur polynomial $s_\lambda = a_{\lambda+\delta(N)}/a_{\delta(N)}$. To model the terms in $a_{\lambda+\delta(N)}$, we use the $N!$ labeled abaci consisting of beads $1, 2, \ldots, N$ (in any order) at positions given by $\lambda + \delta(N)$.

- *Rim-Hook Tableaux.* A rim-hook tableau of shape λ/μ and content α is obtained by enlarging the diagram of μ using border ribbons of lengths $\alpha_1, \alpha_2, \ldots$ (in this order) until the diagram of λ is obtained. The set of such tableaux is denoted $\text{RHT}(\lambda/\mu, \alpha)$. A ribbon occupying r rows has sign $(-1)^{r-1}$, and the sign of a rim-hook tableau is the product of the signs of its ribbons. We write $\chi_\alpha^{\lambda/\mu} = \sum_{T \in \text{RHT}(\lambda/\mu,\alpha)} \text{sgn}(T)$. We have $\chi_\alpha^{\lambda/\mu} = \chi_\beta^{\lambda/\mu}$ whenever $\text{sort}(\alpha) = \text{sort}(\beta)$.

- *Interactions between Abaci and Tableaux.* One can give combinatorial proofs of several identities in Table 10.1 by using the word of a tableau to encode bead motions on abaci. When these motions lead to bead collisions, one obtains two objects of opposite sign and equal weight that cancel terms on one side of the formula to be proved. Objects with no collisions are fixed points that can be reorganized to give the other side of the formula.

- *The Inverse Kostka Matrix.* A rim-hook tableau is called special iff each ribbon in the tableau begins in the leftmost column; $\text{SRHT}(\mu, \lambda)$ is the set of such tableaux of shape μ and content α with $\text{sort}(\alpha) = \lambda$. Letting $K'_{\lambda,\mu} = \sum_{S \in \text{SRHT}(\mu,\lambda)} \text{sgn}(S)$, we have $\mathbf{KK'} = \mathbf{I}$.

- *Littlewood–Richardson Coefficients.* The scalars $c_{\nu,\mu}^\lambda = c_{\mu,\nu}^\lambda$ appearing in the Schur expansions of $s_{\lambda/\nu}$ and $s_\mu s_\nu$ count semistandard tableaux T of shape λ/ν and content μ such that every suffix of the word of T has partition content. The scalars $c_{\nu,\mu}^\lambda$ also count semistandard tableaux T of shape μ and content $\lambda-\nu$ such that $w(T) = w_1 \cdots w_n$ satisfies the following condition: for all k between 1 and n, the exponent vector of $\prod_j x_j^{\nu_j} \prod_{i=k}^{n} x_{w_i}$ is a partition.

Exercises

10-1. Let $w = \cdots 11011011\underline{1}0101001100 \cdots$. Compute $\text{wt}(w)$ and $J(w)$.

10-2. Compute $U(-1,\mu)$ for each $\mu \in \text{Par}(5)$.

10-3. In the computation of $U(m,\mu)$ in Remark 10.4, describe in detail how to use m to decide which symbol in the bead-gap sequence is w_0.

10-4. Given $\mu \in \text{Par}$, what is the relationship between the abaci $U(-1,\mu)$ and $U(-1,\mu')$?

10-5. Show that the abacus w in Example 10.2 and its justification $J(w)$ have the same weight if we use the weights defined in the proof of Theorem 10.5.

10-6. Show how to deduce Euler's Pentagonal Number Theorem as an algebraic consequence of the Jacobi Triple Product Identity.

10-7. Fill in the details in the proof of Theorem 10.6.

10-8. Use Theorem 10.5 to simplify the product $\prod_{n=0}^{\infty}(1-x^{5n+1})^{-1}\prod_{n=0}^{\infty}(1-x^{5n+4})^{-1}$ appearing in one of the Rogers–Ramanujan Identities. Can you give a direct proof of the resulting identity using abaci?

10-9. Use Remark 10.4 to find a bijective proof of Theorem 10.5 that makes no reference to abaci, instead using combinatorial operations on partition diagrams and their frontiers.

10-10. Complete the proof of Theorem 10.12 by verifying that $D_k(w) \in \text{Abc}^k$, $I_k(v) \in \text{Abc}$, and D_k and I_k are two-sided inverses.

10-11. (a) Verify that the 3-core of $\mu = (10,10,10,8,8,8,7,4)$ is $(1,1)$ by removing border 3-ribbons from μ in several different orders. (b) Use the 3-runner abacus encoding μ to determine exactly how many ways there are to change μ into $(1,1)$ by removing an ordered sequence of border 3-ribbons.

10-12. Let $\mu = (8,7,6,4,4,4,3,1,1,1)$. Use abaci to compute the k-core and k-quotients of μ for $1 \leq k \leq 6$.

10-13. Find all integer partitions that are 2-cores, and draw some of their diagrams.

10-14. Find all 3-cores with at most 8 cells.

10-15. Verify the assertion in the last sentence of Example 10.20.

10-16. Let $\mu = (8,8,8,8,8,8,8,8)$. (a) Use abaci to compute the k-core and k-quotients of μ for $3 \leq k \leq 8$. (b) Use the construction at the end of §10.5 to compute the k-quotients of μ (for $3 \leq k \leq 8$) directly from the diagram of μ.

10-17. For $k = 3,4,5$, compute the k-quotients of $\mu = (6,6,6,3,3,2,2,2,1,1)$ without using abaci.

10-18. Consider the construction at the end of §10.5 for computing k-quotients of μ. Show that the hook-length of each unerased cell is divisible by k, and these are the only cells in the diagram of μ whose hook-lengths are divisible by k.

10-19. For each $k \geq 1$, find a formula for the generating function $\sum_{\mu \in \text{Core}(k)} q^{|\mu|}$.

10-20. Given that μ has k-core ρ and k-quotients $\nu^0, \ldots, \nu^{k-1}$, find a formula for the number of ways we can go from μ to ρ by removing an ordered sequence of border k-ribbons.

10-21. Compute $a_\mu(x_1,x_2,x_3)$ and $a_{\lambda+\delta(3)}(x_1,x_2,x_3)$ for $\mu=(6,3,1)$ and $\lambda=(2,2,1)$.

10-22. Verify by direct calculation that, for $N=3$ and $\lambda=(2,1,0)$, $a_{\lambda+\delta(N)}$ is divisible by $a_{\delta(N)}$ and $a_{\lambda+\delta(N)}/a_{\delta(N)}=s_\lambda(x_1,\ldots,x_N)$.

10-23. Verify that A_N and A_N^n are subspaces of $\mathbb{R}[x_1,\ldots,x_N]$, and that the map sending $f\in\Lambda_N$ to $fa_{\delta(N)}$ is linear.

10-24. (a) Show that the product of two antisymmetric polynomials is symmetric. (b) Show that the product of a symmetric polynomial and an antisymmetric polynomial is antisymmetric.

10-25. Define a map $T:\mathbb{R}[x_1,\ldots,x_N]\to\mathbb{R}[x_1,\ldots,x_N]$ by setting

$$T(f)=\frac{1}{N!}\sum_{w\in S_N}\operatorname{sgn}(w)f(x_{w(1)},\ldots,x_{w(N)}).$$

Show that T is a linear map with image A_N whose restriction to A_N is the identity map. Can you describe the kernel of T?

10-26. Let v be the labeled abacus $v=0041000300502600\cdots$. Compute $\operatorname{wt}(v)$, $w(v)$, $\operatorname{pos}(v)$, and $\operatorname{sgn}(v)$. For which λ is v in $\operatorname{LAbc}(\lambda+\delta(6))$?

10-27. Draw all the labeled abaci in $\operatorname{LAbc}(6,5,1)$, and compute the sign of each abacus.

10-28. Using $N=6$ variables, compute all terms in: (a) $a_{(4,2,1)+\delta(6)}p_4$; (b) $a_{(3,3,3)+\delta(6)}p_3$; (c) $a_{(1,1,1,1,1)+\delta(6)}p_2$. How would the answers change if we changed N?

10-29. Let $v=0310040206500\cdots\in\operatorname{LAbc}(\lambda+\delta(6))$ and $k=4$. For $1\le i\le 6$, compute $I(v,i)$ where I is the involution in the proof of Theorem 10.39. For any fixed points that arise, compute v^* and indicate which border 4-ribbon is added to $\operatorname{dg}(\lambda)$ in the passage from v to v^*.

10-30. Using $N=6$ variables, compute all terms in: (a) $a_{(4,2,1)+\delta(6)}e_3$; (b) $a_{(3,3,3)+\delta(6)}e_2$; (c) $a_{(5,4,3,1,1)+\delta(6)}e_4$. How would the answers change if we changed N?

10-31. Let $v=0310040206500\cdots\in\operatorname{LAbc}(\lambda+\delta(6))$. Compute $I(v,S)$ for $S=\{2,5,6\}$, $S=\{1,4,5\}$, $S=\{1,3,4\}$, and $S=\{3,4,6\}$, where I is the involution in the proof of Theorem 10.42. For any fixed points that arise, compute v^* and indicate which vertical strip is added to $\operatorname{dg}(\lambda)$ in the passage from v to v^*.

10-32. Using $N=6$ variables, compute all terms in: (a) $a_{(4,2,1)+\delta(6)}h_3$; (b) $a_{(3,3,3)+\delta(6)}h_3$; (c) $a_{(5,4,3,1,1)+\delta(6)}h_4$. How would the answers change if we changed N?

10-33. Let $v=0310040206500\cdots\in\operatorname{LAbc}(\lambda+\delta(6))$. Compute $I(v,M)$ for $M=[1,1,4,5]$, $M=[2,2,5,6]$, $M=[2,4,5,5]$, and $M=[1,2,3,4]$, where I is the involution in the proof of Theorem 10.44. For any fixed points that arise, compute v^* and indicate which horizontal strip is added to $\operatorname{dg}(\lambda)$ in the passage from v to v^*.

10-34. Explain in detail why the bead motion rule in §10.10 leads to the addition of a horizontal k-strip to the shape λ, assuming no bead collision occurs.

10-35. In the proof of Theorem 10.44, check in detail that I reverses signs, preserves weights, and is an involution.

10-36. Reprove Theorem 10.45 by comparing the symmetric and antisymmetric Pieri Rules for multiplication by e_k.

10-37. Expand the following symmetric polynomials into linear combinations of Schur polynomials: (a) $s_{(3,3,2)}p_3$; (b) $p_{(3,1,3)}$; (c) $s_{(2,2)}p_{(2,1)}$.

10-38. Compute the coefficients of the following Schur polynomials in the Schur expansion of $p_{(3,3,2,1)}$: (a) $s_{(9)}$; (b) $s_{(3,3,3)}$; (c) $s_{(4,4,1)}$; (d) $s_{(1^9)}$.

10-39. Show that, for $\lambda\in\operatorname{Par}(n)$, $\chi^\lambda_{(1^n)}=|\operatorname{SYT}(\lambda)|$.

10-40. Write $s_{(3,2,1)}$ as a linear combination of power-sum polynomials.

10-41. For each $\mu \in \text{Par}(4)$, write p_μ in terms of Schur polynomials, and write s_μ in terms of power-sum polynomials.

10-42. Let I be the involution in §10.13. For each $(v, T) \in X$ given below, compute $I(v, T)$. If (v, T) is a fixed point, compute $v^* \in Y$.

(a) $v = 5432100\cdots$, $T = \begin{array}{|c|c|c|c|c|} \hline 1 & 1 & 1 & 2 & 2 \\ \hline 2 & 3 & 4 \\ \cline{1-3} 3 & 5 \\ \cline{1-2} \end{array}$

(b) $v = 2431500\cdots$, $T = \begin{array}{|c|c|c|c|c|} \hline 5 & 5 & 5 & 5 & 5 \\ \hline 1 & 1 & 1 \\ \cline{1-3} 3 & 3 \\ \cline{1-2} \end{array}$

(c) $v = 3452100\cdots$, $T = \begin{array}{|c|c|c|c|c|} \hline 1 & 1 & 1 & 1 & 1 \\ \hline 2 & 2 & 4 \\ \cline{1-3} 5 & 3 \\ \cline{1-2} \end{array}$

10-43. Let I, X, and Y be defined as in §10.13. Take $N = 3$ and $\lambda = (2, 1, 0)$. List all the elements of X and Y, compute the action of I on X, and show how the fixed points of I map bijectively to Y.

10-44. Verify all the assertions stated before Example 10.54.

10-45. Express $s_{(4,3,1)/(2,1)}$ as a linear combination of power-sums.

10-46. Explain why the formulas $\omega(h_\mu) = e_\mu$ and $\omega(e_\mu) = h_\mu$ are special cases of Theorem 10.59.

10-47. For $N \geq k$, two linear operators S and T on Λ_N^k are called *adjoint* iff $\langle S(f), g\rangle = \langle f, T(g)\rangle$ for all $f, g \in \Lambda_N^k$. Prove that this condition holds for all such f, g iff it holds for all f in some basis of Λ_N^k and all g in some (possibly different) basis of Λ_N^k.

10-48. Write the following Schur polynomials in terms of the complete symmetric polynomials h_μ: (a) $s_{(5,3)}$; (b) $s_{(4,1,1)}$; (c) $s_{(5,5,2,2)}$.

10-49. Write the following Schur polynomials in terms of the elementary symmetric polynomials e_μ: (a) $s_{(2,2,2,2)}$; (b) $s_{(3,2,1)}$; (c) $s_{(4,2)}$.

10-50. Write the skew Schur polynomial $s_{(4,4,3)/(2,1,1)}$ in terms of the following bases: (a) h_μ; (b) e_μ; (c) p_μ; (d) m_μ.

10-51. Modify the definition of the involution used in the proof of Theorem 10.60 as follows. If two or more paths in $(w, p_1, \ldots, p_n)$ intersect, choose i minimal and then j minimal such that p_i and p_j intersect. Let (u, v) be the earliest vertex on p_i that is also a vertex of p_j, and switch the initial segments of these two paths as in the original proof. Show that the map just defined is *not* always an involution.

10-52. Can you find a way to rephrase the combinatorial proof of Theorem 10.60 in terms of abaci?

10-53. Enumerate special rim-hook tableaux to compute $K'_{\mu,\lambda}$ for all partitions λ, μ with at most four cells. Use this to confirm by direct calculation that $\mathbf{KK'} = \mathbf{I}$.

10-54. Find and prove a Pieri-type rule giving the Schur expansion of a product $s_\nu m_\lambda$.

10-55. Let $\mathbf{K'}$ be the matrix defined combinatorially by $K'_{\lambda,\mu} = \sum_{S \in \text{SRHT}(\mu,\lambda)} \text{sgn}(S)$. Find involutions that prove $\mathbf{KK'} = \mathbf{I}$.

10-56. Let $\mathbf{K'}$ be the inverse Kostka matrix, defined using special rim-hook tableaux. Can you prove the identity $\mathbf{K'K} = \mathbf{I}$ combinatorially?

10-57. Let I be the involution in the proof of Theorem 10.66. (a) Compute $I(v^0, T)$, where

$$v_0 = 5432100\cdots, \qquad T = \begin{array}{|c|c|c|c|c|} \cline{3-5} \multicolumn{1}{c}{} & \multicolumn{1}{c|}{} & 1 & 1 & 1 \\ \cline{3-5} \multicolumn{1}{c}{} & \multicolumn{1}{c|}{} & 2 & 2 & 3 \\ \hline 1 & 2 & 3 & 4 & 4 \\ \hline 2 & 3 & 5 \\ \cline{1-3} \end{array}$$

(b) Answer (a) if the last 1 in the top row of T is changed to a 2. (c) Answer (a) if the last 3 in row 2 of T is changed to a 2.

10-58. Compute $c^{\lambda}_{\nu,\mu}$ and $c^{\lambda}_{\mu,\nu}$ using Theorem 10.66, where: (a) $\lambda = (5,3,1,1)$, $\mu = (3,1)$, $\nu = (3,2,1)$; (b) $\lambda = (5,4,4,3,1)$, $\mu = (4,3,3,1)$, $\nu = (3,1,1,1)$.

10-59. Repeat the previous exercise, but use Theorem 10.72 to compute the Littlewood–Richardson coefficients.

10-60. Continuing Example 10.69, find the expansion of $s_{(5,4,4,1)/(3,1)}$ into a sum of Schur polynomials.

10-61. Expand the following skew Schur polynomials into sums of Schur polynomials: (a) $s_{(3,3,3)/(2,1)}$; (b) $s_{(5,4)/(2)}$; (c) $s_{(4,3,2,1)/(1,1,1)}$.

10-62. Expand $s_{(3,2)}s_{(2,2)}$ into a sum of Schur polynomials.

10-63. In the Schur expansion of $s^2_{(3,2,1,1)}$, find the coefficients of: (a) $s_{(5,4,2,2,1)}$; (b) $s_{(5,3,3,1,1,1)}$; (c) $s_{(4,3,3,2,1,1)}$.

10-64. Give a combinatorial proof of Theorem 10.72 based on abaci.

Notes

The proof of the Jacobi Triple Product Identity in §10.2 is adapted from a lecture of Richard Borcherds. One source for material on unlabeled abaci, k-cores, and k-quotients is the book by James and Kerber [67]; for labeled abaci, see [79]. Gessel and Viennot have used intersecting lattice path models to prove many enumeration results [46]. The combinatorial interpretation of the inverse Kostka matrix is due to Eğecioğlu and Remmel [29]. The proof of the Littlewood–Richardson rule given in §10.17 may be viewed as a combinatorialization of the algebraic proof in [106]. Many other proofs of this rule may be found in the literature; see, e.g., the bibliographic notes in [41] and [121, Ch. 7].

11

Algebraic Aspects of Generating Functions

In Chapter 5, we gave an introduction to generating functions emphasizing their applications to combinatorial problems. This chapter takes a closer look at some algebraic aspects of formal power series. We study some new operations on formal power series such as infinite sums, infinite products, formal exponentials, formal logarithms, and formal composition. To define these operations, we need to use the analytic concepts of limits and continuity for formal power series. A major goal of this chapter is to develop algebraic and combinatorial formulas for the coefficients in the multiplicative inverse and the compositional inverse of a formal power series, when these inverses exist. We also prove some theorems regarding partial fraction decompositions and recursions with constant coefficients, along with infinite versions of the Sum Rule and Product Rule for Weighted Sets. These results were used informally in Chapter 5.

Throughout this chapter, let K denote a field of characteristic zero (as defined in the Appendix). Assuming that the characteristic is zero ensures that n^{-1} exists in K for each positive integer n. This enables us to define power series such as $e^z = \sum_{n=0}^{\infty} z^n/n!$.

11.1 Limit Concepts for Formal Power Series

This section studies various limit concepts for formal power series, including the ideas of infinite sums and products of formal series. We begin by reviewing the definitions of the algebraic operations on formal power series from Chapter 5. First recall that an element F of a formal power series ring $K[[z]]$ is defined to be an infinite sequence $F = (a_n : n \in \mathbb{Z}_{\geq 0})$ where each a_n belongs to the field K. We often write $F = F(z) = \sum_{n=0}^{\infty} a_n z^n$ and call a_n the *coefficient of* z^n *in* F. But, at the moment, the summation symbol appearing here is merely notation designed to suggest the analogy between formal power series and polynomials; it does not mean that we are adding up infinitely many terms. Similarly, the formal power series F is *not* a function of the variable z. The powers of z are notational placeholders used to display the coefficients a_n. To emphasize this point, we may refer to z as a *formal indeterminate* or *formal variable*.

Given $F = \sum_{n=0}^{\infty} a_n z^n$ and $G = \sum_{n=0}^{\infty} b_n z^n$ in $K[[z]]$, we have $F = G$ (*equality* of formal series) iff $a_n = b_n$ for all $n \in \mathbb{Z}_{\geq 0}$. We define the *sum* $F + G = \sum_{n=0}^{\infty}(a_n + b_n)z^n$ and the *product* $FG = \sum_{n=0}^{\infty} c_n z^n$, where $c_n = \sum_{k=0}^{n} a_k b_{n-k}$. Note that each particular coefficient in $F + G$ or FG can be computed using only finitely many operations in the field K. We use the notation $F|_{z^k}$ to denote a_k, the coefficient of z^k in F. We can identify each scalar $c \in K$ with the *constant* power series $(c, 0, 0, \ldots) = c + 0z + 0z^2 + \cdots$. Then scalar multiplication satisfies $cF = \sum_{n=0}^{\infty} ca_n z^n$, which is a special case of the formula for multiplying two series. The constant 0 is the additive identity in $K[[z]]$, and the constant 1 is the multiplicative identity. Similarly, for each $m \in \mathbb{Z}_{\geq 0}$, z^m is the formal power series $(0, 0, \ldots, 1, 0, \ldots)$, where the 1 is in the position indexed by m. We have $z^m F = \sum_{n=0}^{\infty} d_n z^n$, where $d_n = 0$ for $0 \leq n < m$, and $d_n = a_{n-m}$ for $n \geq m$. One may verify that $K[[z]]$ is an integral domain (a commutative ring with $1 \neq 0$ having no zero divisors). $K[[z]]$ is also a vector space over K and a K-algebra. (See the Appendix for the definitions of these algebraic structures.)

We define *formal polynomials* to be formal power series $F = \sum_{n=0}^{\infty} a_n z^n$ such that all but finitely many coefficients a_n are zero. Let $K[z]$ denote the set of all formal polynomials with coefficients in K. It can be checked that $K[z]$ is a subring, subspace, and subalgebra of $K[[z]]$.

11.1. Definition: Degree and Order. Given a nonzero $F \in K[z]$, the *degree* of F (denoted $\deg(F)$) is the largest $n \in \mathbb{Z}_{\geq 0}$ such that $F|_{z^n} \neq 0$. Given a nonzero $G \in K[[z]]$, the *order* of G (denoted $\operatorname{ord}(G)$) is the smallest $n \in \mathbb{Z}_{\geq 0}$ such that $G|_{z^n} \neq 0$. Note $\deg(0)$ and $\operatorname{ord}(0)$ are not defined, and $\deg(G)$ is not defined if G is not a polynomial.

One readily proves the following properties of degree and order.

11.2. Theorem: Degree and Order. Let $P, Q \in K[z]$ and $F, G \in K[[z]]$ be nonzero.
(a) $\deg(PQ) = \deg(P) + \deg(Q)$. If $P + Q \neq 0$, then $\deg(P + Q) \leq \max(\deg(P), \deg(Q))$.
(b) $\operatorname{ord}(FG) = \operatorname{ord}(F) + \operatorname{ord}(G)$. If $F + G \neq 0$, then $\operatorname{ord}(F + G) \geq \min(\operatorname{ord}(F), \operatorname{ord}(G))$.
(c) For all $m \in \mathbb{Z}_{\geq 0}$, $\deg(P^m) = m \deg(P)$ and $\operatorname{ord}(F^m) = m \operatorname{ord}(F)$.

Now we define limit concepts for formal power series.

11.3. Definition: Limit of a Sequence of Formal Power Series. For each $m \in \mathbb{Z}_{\geq 0}$, let $F_m \in K[[z]]$ be a formal power series. We say that $F \in K[[z]]$ is the *limit* of the sequence (F_m), denoted $\lim_{m\to\infty} F_m = F$ or $(F_m) \to F$, iff for each $k \in \mathbb{Z}_{\geq 0}$, there exists M (depending on k) such that for all $m \geq M$, $F_m|_{z^k} = F|_{z^k}$.

Intuitively, $(F_m) \to F$ means that as m increases, the coefficient of any fixed z^k in F_m eventually stabilizes at the value $F|_{z^k}$. For example, take $F_m = z^m$ for each m. Then $\lim_{m\to\infty} z^m = 0$. To prove this, fix $k \in \mathbb{Z}_{\geq 0}$ and choose $M = k + 1$. For all $m \geq M$, the coefficient of z^k in z^m is 0, which equals the coefficient of z^k in the zero series. (Remember that z is still a formal indeterminate here. Compare this result to the analytic theorem that for all $z \in \mathbb{C}$ with $|z| < 1$, $\lim_{m\to\infty} z^m = 0$.) More generally, for any formal series G with $\operatorname{ord}(G) > 0$, one may check that $(G^m) \to 0$.

In calculus, a function $f : \mathbb{R} \to \mathbb{R}$ is continuous iff whenever (x_k) is a real sequence converging to $x \in \mathbb{R}$, the sequence $(f(x_k))$ converges to $f(x)$. This suggests the following formal version of continuity.

11.4. Definition: Formal Continuity. Suppose $p : D \to K[[z]]$ is a function with domain D. When $D \subseteq K[[z]]$, we say p is *continuous* iff $(F_k) \to F$ implies $(p(F_k)) \to p(F)$ for all $F_k, F \in D$. When $D \subseteq K[[z]] \times K[[z]]$, continuity of p means that for all $(F_k, G_k), (F, G) \in D$, if $(F_k) \to F$ and $(G_k) \to G$ then $p(F_k, G_k) \to p(F, G)$.

The next theorem lists some properties of formal continuity.

11.5. Theorem: Formal Continuity.
(a) The addition operation on $K[[z]]$, sending (F, G) to $F + G$, is continuous.
(b) The multiplication operation on $K[[z]]$, sending (F, G) to $F \cdot G$, is continuous.
(c) For fixed $G \in K[[z]]$, the maps $A_G(F) = F + G$ and $M_G(F) = F \cdot G$ are continuous.
(d) For each $n \geq 0$, the coefficient extraction operation (sending F to $F|_{z^n}$) is continuous.
(e) The composition of two continuous functions is continuous when defined.

Proof. We sketch the proof of (b), leaving the other parts of the theorem as exercises. Suppose $F_k, G_k, F, G \in K[[z]]$, $(F_k) \to F$, and $(G_k) \to G$; we must show $(F_k \cdot G_k) \to F \cdot G$. Fix $n \geq 0$. Since $(F_k) \to F$, there exists K_1 so that for all $k \geq K_1$ and all $i \in \{0, 1, \ldots, n\}$,

$F_k|_{z^i} = F_{z^i}$ (Exercise 11-9). Since $(G_k) \to G$, there exists K_2 so that for all $k \geq K_2$ and all $i \in \{0, 1, \ldots, n\}$, $G_k|_{z^i} = G_{z^i}$. It follows that for all $k \geq \max\{K_1, K_2\}$,

$$(F_k G_k)|_{z^n} = \sum_{i=0}^{n} F_k|_{z^i} G_k|_{z^{n-i}} = \sum_{i=0}^{n} F|_{z^i} G|_{z^{n-i}} = (FG)|_{z^n}.$$

This proves that $(F_k G_k) \to FG$. □

Now let us consider infinite sums and products. We can define the sum or product of finitely many formal power series by recursion. For instance, $\prod_{k=0}^{0} F_k = F_0$ and $\prod_{k=0}^{m+1} F_k = (\prod_{k=0}^{m} F_k) \cdot F_{m+1}$. We use limits to define the sum or product of infinitely many formal power series.

11.6. Definition: Infinite Sums and Products of Formal Series. For each $m \in \mathbb{Z}_{\geq 0}$, let $F_m \in K[[z]]$ be a formal power series. (a) We say that $\sum_{m=0}^{\infty} F_m$ *converges* to the *sum* $F \in K[[z]]$ iff $\lim_{N\to\infty} \sum_{m=0}^{N} F_m = F$. (b) We say that $\prod_{m=0}^{\infty} F_m$ *converges* to the *product* $F \in K[[z]]$ iff $\lim_{N\to\infty} \prod_{m=0}^{N} F_m = F$. If these limits do not exist, we say that the corresponding infinite sum or product *diverges*.

For example, fix $G = (b_n : n \geq 0) = \sum_{n=0}^{\infty} b_n z^n \in K[[z]]$. Defining $F_m = b_m z^m$ for each $m \geq 0$, one readily checks that $\sum_{m=0}^{\infty} F_m = G$. This means that our original *notation* $\sum_{n=0}^{\infty} b_n z^n$ for G agrees with the infinite summation process defined above.

In contrast to the situation for real or complex series, there is an easily tested criterion for the convergence of an infinite sum or product of formal power series. This criterion uses the following limit notation: given a sequence $(k_n : n \geq 0)$ of integers, we write $(k_n) \to \infty$ or $\lim_{n\to\infty} k_n = \infty$ to mean that for all $M \in \mathbb{Z}_{\geq 0}$, there exists $N \in \mathbb{Z}_{\geq 0}$ such that for all $n \geq N$, $k_n > M$.

11.7. Theorem: Convergence Criterion For Infinite Sums and Products in $K[[z]]$. For each $n \in \mathbb{Z}_{\geq 0}$, let $F_n \in K[[z]]$ be a nonzero formal series.

(a) $\sum_{n=0}^{\infty} F_n$ converges iff $(\mathrm{ord}(F_n)) \to \infty$.

(b) $\prod_{n=0}^{\infty}(1 + F_n)$ converges iff some $F_n = -1$ or $(\mathrm{ord}(F_n)) \to \infty$.

In this theorem, we assume that every F_n is nonzero. This can always be arranged by dropping summands equal to 0 in $\sum_{n=0}^{\infty} F_n$, or dropping factors equal to 1 in $\prod_{n=0}^{\infty}(1+F_n)$; this adjustment does not affect convergence of the sum or product (Exercise 11-13).

We prove the backward direction in (b), leaving the rest of the proof as an exercise. If some $F_n = -1$, then all partial products from some point on are zero, so $\prod_{n=0}^{\infty} F_n$ converges to zero. Now assume $F_n \neq -1$ for all n, and $(\mathrm{ord}(F_n)) \to \infty$. For each $m \geq 0$, let $P_m = \prod_{n=0}^{m} F_n$ be the mth partial product of the factors F_n. Fix $k \in \mathbb{Z}_{\geq 0}$; we must show that the coefficient $P_m|_{z^k}$ eventually stabilizes as m increases. Because $(\mathrm{ord}(F_n)) \to \infty$, there exists N so that for all $n \geq N$, $\mathrm{ord}(F_n) > k$. We show by induction on m that for all $m \geq N$, $P_m|_{z^k} = P_N|_{z^k}$. This certainly holds for $m = N$. For the induction step, fix $m \geq N$, assume $P_m|_{z^k} = P_N|_{z^k}$, and prove $P_{m+1}|_{z^k} = P_N|_{z^k}$. By the recursive definition of finite products, $P_{m+1} = P_m(1 + F_{m+1})$. Taking the coefficient of z^k, we get

$$P_{m+1}|_{z^k} = \sum_{i=0}^{k} P_m|_{z^{k-i}}(1 + F_{m+1})|_{z^i}.$$

Because $m + 1 \geq N$, $\mathrm{ord}(F_{m+1}) > k$, which means $F_{m+1}|_{z^i} = 0$ for all i between 0 and k. So $(1 + F_{m+1})|_{z^i}$ is 1 for $i = 0$ and is 0 for $0 < i \leq k$. Putting these values into the sum above, we find that $P_{m+1}|_{z^k} = P_m|_{z^k} = P_N|_{z^k}$, as needed.

11.8. Example: Formal Geometric Series. Suppose $G \in K[[z]]$ is any nonzero formal series with zero constant term, and let $\operatorname{ord}(G) = d > 0$. We claim $\sum_{m=0}^{\infty} G^m$ converges in $K[[z]]$. For, $\operatorname{ord}(G^m) = m \operatorname{ord}(G) = md$, and $\lim_{m\to\infty} md = \infty$ since $d > 0$. The significance of the power series $\sum_{m=0}^{\infty} G^m$ is revealed in §11.3.

11.2 The Infinite Sum and Product Rules

The Sum and Product Rules for Weighted Sets (discussed in §5.8) provide combinatorial interpretations for finite sums and products of formal power series. This section extends these rules to the case of infinite sums and products of formal series. These more general rules were already used informally in our discussion of generating functions for trees (§5.10) and integer partitions (§5.15).

11.9. The Infinite Sum Rule for Weighted Sets. Suppose S_k is a nonempty weighted set for each $k \in \mathbb{Z}_{\geq 0}$, $S_k \cap S_j = \emptyset$ for all $j \neq k$, and $S = \bigcup_{k=0}^{\infty} S_k$. Assume that for all k and all $u \in S_k$, $\operatorname{wt}_S(u) = \operatorname{wt}_{S_k}(u)$. For each k, let $\operatorname{minwt}(S_k) = \min\{\operatorname{wt}_{S_k}(u) : u \in S_k\}$. If $\lim_{k\to\infty} \operatorname{minwt}(S_k) = \infty$, then $\operatorname{GF}(S; z) = \sum_{k=0}^{\infty} \operatorname{GF}(S_k; z)$.

Proof. We have $\operatorname{minwt}(S_k) = \operatorname{ord}(\operatorname{GF}(S_k; z))$ for each k, so the hypothesis of the theorem means that $(\operatorname{ord}(\operatorname{GF}(S_k; z))) \to \infty$. By Theorem 11.7(a), it follows that $\sum_{k=0}^{\infty} \operatorname{GF}(S_k; z)$ converges to some formal power series H. We need only show that for each $n \in \mathbb{Z}_{\geq 0}$, $\operatorname{GF}(S; z)|_{z^n} = H|_{z^n}$. Fix n, and then choose K so that for all $k > K$, $\operatorname{minwt}(S_k) > n$. Now define $H^* = \sum_{k=0}^{K} \operatorname{GF}(S_k; z)$ and $S^* = \bigcup_{k=0}^{K} S_k$. Since all objects in $S_{k+1}, S_{k+2}, \ldots$ have weight exceeding n, we see that $\{u \in S : \operatorname{wt}(u) = n\} = \{u \in S^* : \operatorname{wt}(u) = n\}$. Using this fact and the Weighted Sum Rule for finitely many summands, we conclude

$$\operatorname{GF}(S; z)|_{z^n} = \operatorname{GF}(S^*; z)|_{z^n} = H^*|_{z^n} = H|_{z^n}.$$

The final equality follows since $\operatorname{GF}(S_k; z)|_{z^n} = 0$ for all $k > K$. □

11.10. The Infinite Product Rule for Weighted Sets. For each $k \in \mathbb{Z}_{\geq 1}$, let S_k be a weighted set that contains a unique object o_k of weight zero, which we call the *default value for choice* k. Assume $\lim_{k\to\infty} \operatorname{minwt}(S_k - \{o_k\}) = \infty$. Suppose S is a weighted set such that every $u \in S$ can be constructed in exactly one way as follows. For each $k \geq 1$, choose $u_k \in S_k$, subject to the restriction that we must choose the default value o_k for all but finitely many k's. Then assemble the chosen objects in a prescribed manner. Assume that whenever u is constructed from $(u_k : k \geq 0)$, the *weight-additivity condition* $\operatorname{wt}_S(u) = \sum_{k=1}^{\infty} \operatorname{wt}_{S_k}(u_k)$ holds. Then $\operatorname{GF}(S; z) = \prod_{k=1}^{\infty} \operatorname{GF}(S_k; z)$.

Proof. The assumed condition on $\operatorname{minwt}(S_k - \{o_k\})$ means that $(\operatorname{ord}(\operatorname{GF}(S_k; z) - 1)) \to \infty$, so that $\prod_{k=1}^{\infty} \operatorname{GF}(S_k; z)$ converges to some formal series H by Theorem 11.7(b). Now fix $n \in \mathbb{Z}_{\geq 0}$, and choose K so that for all $k > K$, $\operatorname{minwt}(S_k - \{o_k\}) > n$. Define $H^* = \prod_{k=1}^{K} \operatorname{GF}(S_k; z)$, and let S^* be the subset of $u \in S$ that can be built by choosing $u_1 \in S_1, \ldots, u_K \in S_K$ arbitrarily, but then choosing $u_k = o_k$ for all $k > K$. The weight-additivity condition implies that $\{u \in S : \operatorname{wt}(u) = n\} = \{u \in S^* : \operatorname{wt}(u) = n\}$. Using the Weighted Product Rule for finitely many factors, one may now check that

$$\operatorname{GF}(S; z)|_{z^n} = \operatorname{GF}(S^*; z)|_{z^n} = H^*|_{z^n} = H|_{z^n}.$$ □

11.3 Multiplicative Inverses of Formal Power Series

In a field K, every nonzero element x has a *multiplicative inverse*, which is an element $y \in K$ satisfying $xy = 1_K = yx$. In commutative rings that are not fields, some elements have multiplicative inverses and others do not. One can ask for characterizations of the invertible elements in such a ring. For example, it can be shown that for the ring $\mathbb{Z}_n$ of integers modulo n, $k \in \mathbb{Z}_n$ is invertible iff $\gcd(k, n) = 1$. The next theorem gives characterizations of the invertible elements in the rings $K[z]$ and $K[[z]]$. Roughly speaking, the theorem says that almost all formal power series can be inverted, whereas almost all formal polynomials cannot be inverted.

11.11. Theorem: Invertible Polynomials and Formal Power Series. (a) A polynomial $P \in K[z]$ is invertible in $K[z]$ iff $\deg(P) = 0$ (i.e., P is a nonzero constant). (b) A formal power series $F \in K[[z]]$ is invertible in $K[[z]]$ iff $\operatorname{ord}(F) = 0$ (i.e., F has nonzero constant term).

Proof. (a) First assume P is a nonzero constant $c \in K$. Since K is a field, c^{-1} exists in K, and c^{-1} is also a multiplicative inverse of c in $K[z]$. Conversely, assume $P \in K[z]$ has an inverse $Q \in K[z]$, so $PQ = 1$. We must have $P \neq 0 \neq Q$, since $1_K \neq 0_K$. Taking degrees, we find that $0 = \deg(1) = \deg(PQ) = \deg(P) + \deg(Q)$. Since $\deg(P)$ and $\deg(Q)$ are nonnegative integers, this forces $\deg(P) = 0 = \deg(Q)$. This means that P (and Q) are nonzero constants.

(b) First assume F is invertible in $K[[z]]$ with inverse $G \in K[[z]]$. Taking orders in the equation $FG = 1$, we get $0 = \operatorname{ord}(1) = \operatorname{ord}(FG) = \operatorname{ord}(F) + \operatorname{ord}(G)$, forcing $\operatorname{ord}(F) = 0 = \operatorname{ord}(G)$. In turn, this means that F (and G) have nonzero constant terms.

Conversely, assume $F = \sum_{n=0}^{\infty} a_n z^n$ with $a_0 \neq 0$. We construct a multiplicative inverse $G = \sum_{n=0}^{\infty} u_n z^n$ for F recursively. For any $G \in K[[z]]$, we have $FG = 1$ iff $(FG)|_{z^0} = 1$ and $(FG)|_{z^n} = 0$ for all $n > 0$. So $FG = 1$ iff $(u_n : n \geq 0)$ solves the infinite system of equations

$$a_0 u_0 = 1; \quad a_0 u_n + a_1 u_{n-1} + a_2 u_{n-2} + \cdots + a_n u_0 = 0 \text{ for all } n > 0. \tag{11.1}$$

Since $a_0 \neq 0$, a_0^{-1} exists in the field K. So we can recursively define $u_0 = a_0^{-1}$ and (assuming $u_0, \ldots, u_{n-1}$ have already been found) $u_n = -a_0^{-1} \sum_{k=1}^{n} a_k u_{n-k} \in K$. It can be checked by induction that the sequence $(u_n : n \geq 0)$ does solve (11.1), so $FG = 1$ does hold for this choice of the u_n's. □

The preceding proof provides a recursive algorithm for calculating the coefficients of F^{-1} from the coefficients of F. Note that $u_n = F^{-1}|_{z^n}$ only depends on the values of $a_k = F|_{z^k}$ for $0 \leq k \leq n$. Next we develop a closed formula for the coefficients of F^{-1} based on a formal version of the geometric series.

11.12. Theorem: Formal Geometric Series. For all $G \in K[[z]]$ satisfying $G|_{z^0} = 0$,

$$(1 - G)^{-1} = \sum_{m=0}^{\infty} G^m.$$

Proof. Fix $G \in K[[z]]$ with $G|_{z^0} = 0$. For each $m \geq 0$, the distributive law shows that

$$(1-G)(1+G+\cdots+G^m) = (1+G+G^2+\cdots+G^m) - (G+G^2+\cdots+G^m+G^{m+1}) = 1-G^{m+1},$$

since all intermediate terms cancel. (One can prove this equation more rigorously by induction on m.) Let $H_M = \sum_{m=0}^{M} G^m$ and $H = \sum_{m=0}^{\infty} G^m$; Example 11.8 shows that this infinite series converges. By continuity of the map $p(F) = (1-G)F$, we now compute

$$(1-G)H = (1-G)\lim_{M\to\infty} H_M = \lim_{M\to\infty}[(1-G)H_M] = \lim_{M\to\infty}[1-G^{M+1}] = 1-0 = 1;$$

note $(G^{M+1}) \to 0$ as $M \to \infty$ since $\mathrm{ord}(G) > 0$ or $G = 0$. Thus H is the multiplicative inverse of $1-G$, as needed. □

We can use this theorem to invert any formal series $F = \sum_{n=0}^{\infty} a_n z^n$ with $a_0 \neq 0$. To do so, write $F = a_0(1-G)$ where $G = \sum_{n=1}^{\infty}(-a_n/a_0)z^n$. So $1/F = a_0^{-1}\sum_{m=0}^{\infty} G^m$. Next, we can use Theorem 5.35 to give an explicit formula for each coefficient $G^m|_{z^n}$. Defining $b_n = G|_{z^n} = -a_n/a_0$ for each $n > 0$, the formula is

$$G^m|_{z^n} = \sum_{\substack{(i_1,i_2,\ldots,i_m)\in\mathbb{Z}_{>0}^m:\\ i_1+i_2+\cdots+i_m=n}} b_{i_1}b_{i_2}\cdots b_{i_m}. \tag{11.2}$$

Then the coefficient of z^n in $1/F$ is a_0^{-1} times the sum of these expressions over all m. For fixed n, it suffices to let m range from 0 to n. To summarize, we have proved the following.

11.13. Theorem: Coefficients of Multiplicative Inverses. For all $F = \sum_{n=0}^{\infty} a_n z^n$ in $K[[z]]$ with $a_0 \neq 0$,

$$F^{-1}|_{z^n} = \sum_{m=0}^{n}\frac{(-1)^m}{a_0^{m+1}}\sum_{\substack{(i_1,i_2,\ldots,i_m)\in\mathbb{Z}_{>0}^m:\\ i_1+i_2+\cdots+i_m=n}} a_{i_1}a_{i_2}\cdots a_{i_m}. \tag{11.3}$$

There is another formula for the multiplicative inverse based on symmetric functions. Let Λ be the ring of abstract symmetric functions with coefficients in K (see §9.28). For $n \in \mathbb{Z}_{>0}$, let e_n and h_n denote the elementary and complete symmetric functions of degree n, and let $e_0 = h_0 = 1$. Define $E(z) = \sum_{n=0}^{\infty} e_n z^n$ and $H(z) = \sum_{n=0}^{\infty}(-1)^n h_n z^n$, which are formal power series with coefficients in Λ. The key observation is that $E(z)H(z) = 1$. This holds because the quantities $a_n = e_n$ and $u_n = (-1)^n h_n$ solve the system of equations (11.1), by Theorem 9.81 (also compare to the finite version (9.9)).

Suppose we are trying to invert a formal series of the form $F(z) = 1 + \sum_{n=1}^{\infty} a_n z^n$ where each a_n is in K. Recall that $(e_n : n \in \mathbb{Z}_{>0})$ is algebraically independent over K. So there exists a unique evaluation homomorphism $\phi : \Lambda \to K$ such that $\phi(e_n) = a_n$ for all $n > 0$. It can be checked that this map induces a K-algebra homomorphism $\phi^* : \Lambda[[z]] \to K[[z]]$ given by

$$\phi^*\left(\sum_{n=0}^{\infty} f_n z^n\right) = \sum_{n=0}^{\infty}\phi(f_n)z^n \quad \text{for all } f_n \in \Lambda. \tag{11.4}$$

Now define $b_n = \phi((-1)^n h_n)$ for each $n > 0$, and set $G(z) = 1 + \sum_{n=1}^{\infty} b_n z^n$. Applying ϕ^* to the equation $E(z)H(z) = 1$ in $\Lambda[[z]]$ produces the equation $F(z)G(z) = 1$ in $K[[z]]$, so $G = F^{-1}$. In summary, we can find the coefficients in the multiplicative inverse of F if we can figure out how the homomorphism ϕ acts on the symmetric functions h_n. It is possible to express each h_n as a specific linear combination of the symmetric functions e_λ, which are products of e_k's (see Exercise 11-40). Then each $b_n = (-1)^n\phi(h_n)$ is determined by the values $a_k = \phi(e_k)$ and the fact that ϕ is a K-algebra homomorphism. Some algebraic manipulation eventually reproduces Formula (11.3).

For yet another approach to multiplicative inversion based on the formal exponential function, see Theorem 11.27(b) below.

11.4 Partial Fraction Expansions

This section proves the existence and uniqueness of partial fraction decompositions for ratios of complex polynomials. Suppose f and g are polynomials with complex coefficients such that g has nonzero constant term. We saw in §11.3 that g (viewed as a formal power series) has a multiplicative inverse, so we can write $f/g = \sum_{n=0}^{\infty} b_n z^n$ for some $b_n \in \mathbb{C}$. The partial fraction decomposition of f/g leads to explicit formulas for the coefficients b_n. Our starting point is the Fundamental Theorem of Algebra, which we state here without proof.

11.14. The Fundamental Theorem of Algebra. Let $p \in \mathbb{C}[z]$ be a monic polynomial of degree $n \geq 1$. There exist pairwise distinct complex numbers $r_1, \ldots, r_k$ (unique up to reordering) and unique positive integers $n_1, \ldots, n_k$ such that

$$p = (z - r_1)^{n_1}(z - r_2)^{n_2} \cdots (z - r_k)^{n_k}.$$

The number r_i is called a *root* of p of *multiplicity* n_i.

The following variant of the Fundamental Theorem is more convenient for partial fraction problems because it allows us to use the Negative Binomial Theorem (see §5.3).

11.15. Theorem: Factorization of Polynomials in $\mathbb{C}[z]$. Let $p \in \mathbb{C}[z]$ be a polynomial of degree $n \geq 1$ with $p(0) = 1$. There exist pairwise distinct, nonzero complex numbers $r_1, \ldots, r_k$ and positive integers $n_1, \ldots, n_k$ such that

$$p(z) = (1 - r_1 z)^{n_1}(1 - r_2 z)^{n_2} \cdots (1 - r_k z)^{n_k}.$$

Proof. Write $p = \sum_{i=0}^{n} p_i z^i$ with $p_0 = 1$, and consider the polynomial $q = z^n p(1/z) = \sum_{i=0}^{n} p_{n-i} z^i$. Intuitively, q is obtained from p by reversing the coefficient sequence. Since $p_0 = 1$, q is a monic polynomial of degree n. Using the Fundamental Theorem of Algebra, we write

$$z^n p(1/z) = q(z) = \prod_{i=1}^{k} (z - r_i)^{n_i},$$

where $\sum_{i=1}^{k} n_i = n$. Since the constant term of q is nonzero, no r_i is equal to zero. Reversing the coefficient sequence again, it follows that

$$p(z) = z^n q(1/z) = z^n \prod_{i=1}^{k} ((1/z) - r_i)^{n_i} = \prod_{i=1}^{k} z^{n_i} \left(\frac{1 - r_i z}{z}\right)^{n_i} = \prod_{i=1}^{k} (1 - r_i z)^{n_i}. \qquad \square$$

The next step is to rewrite a general fraction f/g as a sum of fractions whose denominators have the form $(1 - rz)^m$. Note that we can always arrange that g has constant coefficient 1 (assuming $g(0) \neq 0$ initially) by multiplying the numerator and denominator of f/g by $1/g(0)$.

11.16. Theorem: Splitting a Denominator. Suppose $f, g \in \mathbb{C}[z]$ are polynomials such that $g(0) = 1$, and let g have factorization $g(z) = \prod_{i=1}^{k} (1 - r_i z)^{n_i}$, where $r_1, \ldots, r_k \in \mathbb{C}$ are distinct and nonzero. There exist polynomials $p_0, p_1, \ldots, p_k$ with $\deg(p_i) < n_i$ or $p_i = 0$ for i between 1 and k, such that

$$\frac{f}{g} = p_0 + \sum_{i=1}^{k} \frac{p_i}{(1 - r_i z)^{n_i}}.$$

Proof. For i between 1 and k, define a polynomial $h_i = g/(1-r_iz)^{n_i} = \prod_{j:j\neq i}(1-r_jz)^{n_j}$. Since $r_1, \ldots, r_k$ are distinct, $\gcd(h_1, \ldots, h_k) = 1$. By a known result from polynomial algebra, it follows that there exist polynomials $a_1, \ldots, a_k \in \mathbb{C}[z]$ with $a_1h_1 + \cdots + a_kh_k = 1$. Therefore,

$$\frac{f}{g} = \frac{f\cdot 1}{g} = \frac{fa_1h_1 + \cdots + fa_kh_k}{g} = \sum_{i=1}^{k} \frac{fa_i}{(1-r_iz)^{n_i}}.$$

This is almost the answer we seek, but the degrees of the numerators may be too high. Using polynomial division we can write $fa_i = q_i(1-r_iz)^{n_i} + p_i$ where $q_i, p_i \in \mathbb{C}[z]$, and either $p_i = 0$ or $\deg(p_i) < n_i$. Dividing by $(1-r_iz)^{n_i}$, we see that

$$\frac{f}{g} = p_0 + \sum_{i=1}^{k} \frac{p_i}{(1-r_iz)^{n_i}}$$

holds if we take $p_0 = \sum_{i=1}^{k} q_i \in \mathbb{C}[z]$. □

The fractions $p_i/(1-r_iz)^{n_i}$ (with $\deg(p_i) < n_i$ or $p_i = 0$) can be further reduced into sums of fractions where the numerators are complex constants.

11.17. Theorem: Division by $(1-rz)^n$. Given a fraction $p/(1-rz)^n$ where $p \in \mathbb{C}[z]$, $\deg(p) < n$ or $p = 0$, and $0 \neq r \in \mathbb{C}$, there exist complex numbers $a_1, \ldots, a_n$ such that

$$\frac{p}{(1-rz)^n} = \sum_{j=1}^{n} \frac{a_j}{(1-rz)^j}.$$

Proof. This proof uses some facts about evaluation homomorphisms for polynomial rings from Exercise 11-42. Consider the evaluation homomorphism $E : \mathbb{C}[z] \to \mathbb{C}[z]$ such that $E(z) = 1 - rz$. The evaluation homomorphism $E^* : \mathbb{C}[z] \to \mathbb{C}[z]$ such that $E^*(z) = (1-z)/r$ is a two-sided inverse to E (since $\mathrm{id}_{\mathbb{C}[z]}$, $E \circ E^*$, and $E^* \circ E$ are all $\mathbb{C}$-algebra homomorphisms sending z to z, forcing $\mathrm{id} = E \circ E^* = E^* \circ E$ by uniqueness), so E is a bijection. In particular, E is surjective, so $p = E(q)$ for some $q \in \mathbb{C}[z]$. Now, one may check that E and E^* each map polynomials of degree less than n to polynomials of degree less than n, and it follows that $\deg(q) < n$ or $q = 0$. Write $q = c_0 + c_1z + c_2z^2 + \cdots + c_{n-1}z^{n-1}$, with $c_i \in \mathbb{C}$. Then

$$p = E(q) = c_0 + c_1(1-rz) + c_2(1-rz)^2 + \cdots + c_{n-1}(1-rz)^{n-1}.$$

Dividing by $(1-rz)^n$, we see that we may take $a_1 = c_{n-1}, \ldots, a_{n-1} = c_1, a_n = c_0$. □

The next result summarizes the partial fraction manipulations in the last two theorems. The uniqueness proof given below also provides an algorithm for finding the coefficients in the partial fraction decomposition.

11.18. Theorem: Partial Fraction Decompositions. Suppose $f, g \in \mathbb{C}[z]$ are polynomials with $g(0) = 1$; let $g = \prod_{i=1}^{k}(1-r_iz)^{n_i}$ where the r_i are distinct nonzero complex numbers. There exist a unique polynomial $h \in \mathbb{C}[z]$ and unique complex numbers a_{ij} (for i, j in the range $1 \leq i \leq k$ and $1 \leq j \leq n_i$) with

$$\frac{f}{g} = h + \sum_{i=1}^{k}\sum_{j=1}^{n_i} \frac{a_{ij}}{(1-r_iz)^j}. \tag{11.5}$$

The coefficient of z^n in the power series expansion of f/g is

$$\left.\frac{f}{g}\right|_{z^n} = h|_{z^n} + \sum_{i=1}^{k}\sum_{j=1}^{n_i} a_{ij}\binom{n+j-1}{n,j-1} r_i^n.$$

If we view z as a complex variable (rather than a formal indeterminate), the series for f/g in (11.5) converges for all $z \in \mathbb{C}$ such that $|z| < \min\{1/|r_i| : 1 \leq i \leq k\}$.

Proof. Existence of the decomposition (11.5) follows by combining Theorems 11.16 and 11.17. The formula for the coefficient of z^n follows from the Negative Binomial Theorem 5.13. We must still prove uniqueness of h and the a_{ij}. Note first that the numbers r_i and n_i appearing in the factorization of g are unique because of the uniqueness assertion in the Fundamental Theorem of Algebra. Now consider any expression of the form (11.5). Multiplying both sides by g produces an equation

$$f = gh + \sum_{i=1}^{k}\sum_{j=1}^{n_i} a_{ij}(1-r_iz)^{n_i-j}\prod_{s\neq i}(1-r_sz)^{n_s}, \tag{11.6}$$

where both sides are polynomials. Furthermore, the terms in the double sum add up to a polynomial that is either zero or has degree less than $\deg(g)$. Thus h must be the quotient when f is divided by g using the polynomial division algorithm, and this quotient is known to be unique. Next, we show how to recover the top coefficients a_{i,n_i} for $1 \leq i \leq k$. Fix i, and evaluate the polynomials on each side of (11.6) at $z = 1/r_i$. Since any positive power of $(1 - r_iz)$ becomes zero for this choice of z, all but one term on the right side becomes zero. We are left with

$$f(1/r_i) = a_{i,n_i}\prod_{s\neq i}(1-r_s/r_i)^{n_s}.$$

Since $r_s \neq r_i$ for $s \neq i$, the product is nonzero. Thus there is a unique $a_{i,n_i} \in \mathbb{C}$ for which this equation holds. We can use the displayed formula to calculate each a_{i,n_i} given f and g.

To find the remaining a_{ij}, subtract the recovered summands $a_{i,n_i}/(1-r_iz)^{n_i}$ from both sides of (11.5) (thus replacing f/g by a new fraction f_1/g_1) to obtain a new problem in which all n_i have been reduced by 1. We now repeat the procedure of the previous paragraph to find a_{i,n_i-1} for all i such that $n_i > 1$. Continuing similarly, we eventually recover all the a_{ij}. This process is illustrated in the examples below. □

11.19. Example. Let us find the partial fraction expansion of

$$\frac{f}{g} = \frac{z^2-2}{1-2z-z^2+2z^3}.$$

To find the required factorization of the denominator, we first reverse the coefficient sequence to obtain $z^3 - 2z^2 - z + 2$. This polynomial factors as $(z-2)(z-1)(z+1)$, so the original denominator can be rewritten as

$$1 - 2z - z^2 + 2z^3 = (1-2z)(1-z)(1+z)$$

(see the proof of Theorem 11.15). We know that

$$\frac{z^2-2}{1-2z-z^2+2z^3} = \frac{A}{1-2z} + \frac{B}{1-z} + \frac{C}{1+z} \tag{11.7}$$

for certain complex constants A, B, C. To find A, multiply both sides by $1-2z$ to get

$$\frac{z^2-2}{(1-z)(1+z)} = A + \frac{B(1-2z)}{1-z} + \frac{C(1-2z)}{1+z}.$$

Now set $z = 1/2$ to see that $A = (-7/4)/(3/4) = -7/3$. Similarly,

$$\begin{aligned} B &= \left.\frac{z^2-2}{(1-2z)(1+z)}\right|_{z=1} = 1/2; \\ C &= \left.\frac{z^2-2}{(1-2z)(1-z)}\right|_{z=-1} = -1/6. \end{aligned}$$

Expanding (11.7) into a power series, we obtain

$$\frac{z^2-2}{1-2z-z^2+2z^3} = \sum_{n=0}^{\infty}\left[-\frac{7}{3}\cdot 2^n + \frac{1}{2} - \frac{1}{6}\cdot(-1)^n\right] z^n.$$

11.20. Example. We find the partial fraction expansion of

$$\frac{f}{g} = \frac{1}{1-9z+30z^2-46z^3+33z^4-9z^5}.$$

Factoring the denominator as in the last example, we find that $g(z) = (1-z)^3(1-3z)^2$. We can therefore write

$$\frac{f}{g} = \frac{A}{(1-z)^3} + \frac{B}{(1-z)^2} + \frac{C}{1-z} + \frac{D}{(1-3z)^2} + \frac{E}{1-3z}. \tag{11.8}$$

To find A, multiply both sides by $(1-z)^3$ and then substitute $z = 1$ to get $A = 1/(-2)^2 = 1/4$. Similarly, multiplying by $(1-3z)^2$ and setting $z = 1/3$ reveals that $D = 1/(2/3)^3 = 27/8$. Having found A and D, we subtract $A/(1-z)^3$ and $D/(1-3z)^2$ from both sides of (11.8). After simplifying, we are left with

$$\frac{(3/8)(3z-7)}{(1-z)^2(1-3z)} = \frac{B}{(1-z)^2} + \frac{C}{1-z} + \frac{E}{1-3z}.$$

Now we repeat the process. Multiplying by $(1-z)^2$ and setting $z = 1$ shows that $B = 3/4$. Similarly, $E = -81/16$. Subtracting these terms from both sides leaves $(27/16)/(1-z)$, so $C = 27/16$. Using (11.8) and the Negative Binomial Theorem, we conclude that

$$\frac{f}{g} = \sum_{n=0}^{\infty}\left[\frac{1}{4}\binom{n+2}{2} + \frac{3}{4}\binom{n+1}{1} + \frac{27}{16} + \frac{27}{8}\binom{n+1}{1}3^n - \frac{81}{16}\cdot 3^n\right] z^n.$$

11.5 Generating Functions for Recursively Defined Sequences

In §5.5, we gave examples of the generating function method for solving recursions. This section proves a general theorem about the generating function for a recursively defined sequence.

11.21. Theorem: Recursions with Constant Coefficients. Suppose we are given the following data: a positive integer k, complex constants $c_1, c_2, \ldots, c_k, d_0, \ldots, d_{k-1}$, and a function $g : \mathbb{Z}_{\geq k} \to \mathbb{C}$. The recursion

$$a_n = c_1 a_{n-1} + c_2 a_{n-2} + \cdots + c_k a_{n-k} + g(n) \quad \text{for all } n \geq k \tag{11.9}$$

with initial conditions $a_i = d_i$ for $0 \leq i < k$ has a unique solution. Defining $F = \sum_{n=0}^{\infty} a_n z^n$, $d_i' = d_i - c_1 d_{i-1} - c_2 d_{i-2} - \cdots - c_i d_0$, $G = \sum_{i=0}^{k-1} d_i' z^i + \sum_{n=k}^{\infty} g(n) z^n$, and $P = 1 - c_1 z - c_2 z^2 - \cdots - c_k z^k$, we have $F = G/P$.

Proof. The existence and uniqueness of the sequence $(a_n : n \geq 0)$ satisfying the given recursion and initial conditions is intuitively plausible and can be informally established by an induction argument. (A formal proof requires the *Recursion Theorem* from set theory; see Section 12 of [59] for a discussion of this theorem.) It follows that the formal power series F in the theorem statement is well-defined. Consider next the formal power series

$$H = (1 - c_1 z - c_2 z^2 - \cdots - c_k z^k)F = PF.$$

For each $n \geq k$, the recursion shows that G and H have the same coefficient of z^n, namely

$$g(n) = a_n - c_1 a_{n-1} - c_2 a_{n-2} - \cdots - c_k a_{n-k}.$$

On the other hand, for $0 \leq n < k$, the initial conditions show that the coefficient of z^n in both G and H is d'_n. So $G = H$, and the formula for F follows by dividing the equation $G = PF$ by the formal power series P, which has a multiplicative inverse since $P(0) \neq 0$. □

We can now justify Method 2.70 for solving a homogeneous recursion with constant coefficients. In this case, the function $g(n)$ in (11.9) is identically zero. So the generating function $F(z) = \sum_{n=0}^{\infty} a_n z^n$ for the solution to the recursion has the form

$$F(z) = \frac{G(z)}{1 - c_1 z - c_2 z^2 - \cdots - c_k z^k},$$

where $G(z)$ is a polynomial of degree less than k. Recall that the *characteristic polynomial* of the given recursion is $\chi(z) = z^k - c_1 z^{k-1} - c_2 z^{k-2} - \cdots - c_k$. Factoring $\chi(z)$ as $\prod_{i=1}^{k}(z - r_i)^{m_i}$, we know from the proof of Theorem 11.15 that the denominator P in the formula for $F(z)$ factors as $\prod_{i=1}^{k}(1 - r_i z)^{m_i}$. By Theorem 11.18 on Partial Fraction Decompositions, we conclude that

$$a_n = \sum_{i=1}^{k} \sum_{j=1}^{m_i} b_{i,j} \binom{n+j-1}{n, j-1} r_i^n$$

for some complex constants $b_{i,j}$. This means that the sequence $(a_n : n \geq 0)$ solving the recursion is some linear combination of the *basic sequences* $(r_i^n (n+s)\!\downarrow_s / s! : n \geq 0)$, where $1 \leq i \leq k$ and $0 \leq s < m_i$.

Conversely, since $(1 - r_i z)^{m_i}$ is a factor of P, one may check that each of these basic sequences does solve the recursion in (11.9) for an appropriate choice of initial conditions. To obtain the basic solutions mentioned in 2.70, note that $(n+s)\!\downarrow_s / s!$ is a polynomial in n of degree s. The collection of all such polynomials for $0 \leq s < m_i$ is thus a basis for the vector space of polynomials in n of degree less than m_i. We know that $(n^s : 0 \leq s < m_i)$ is also a basis for this vector space. Therefore, each sequence of the form $(r_i^n (n+s)\!\downarrow_s / s! : n \geq 0)$ can be expressed as a linear combination of sequences of the form $(r_i^n n^t : n \geq 0)$ and vice versa, where $0 \leq s, t < m_i$. In conclusion, when $g = 0$, every solution to the recursion in (11.9) is a linear combination of the basic solutions $(r_i^n n^t : n \geq 0)$, where r_i is a root of the recursion's characteristic polynomial of multiplicity m_i and $0 \leq t < m_i$.

11.6 Formal Composition and Derivative Rules

In analysis, the composition of two functions $f : X \to Y$ and $g : Y \to Z$ is a new function $g \circ f : X \to Z$ given by $(g \circ f)(x) = g(f(x))$ for all $x \in X$. This section studies a formal version of composition involving formal power series.

11.22. Definition: Composition of Formal Series. Suppose $F, G \in K[[z]]$ are formal power series such that G has constant term zero. Writing $F = \sum_{n=0}^{\infty} a_n z^n$, we define the *composition* $F \circ G$ [also denoted $F(G)$ or $F(G(z))$] by $F \circ G = \sum_{n=0}^{\infty} a_n G^n$.

Since G has constant term zero, the infinite sum of formal power series $\sum_{n=0}^{\infty} a_n G^n$ does converge, since the order of the nonzero summands $a_n G^n$ is tending to infinity (see Theorem 11.7). Intuitively, we obtain $F(G(z))$ from $F(z)$ by substituting the series $G(z)$ for the formal variable z. We point out once again that F is not a function of z, so this operation is not a special case of ordinary function composition, although the same symbol $\circ$ is used for both operations.

11.23. Example. Let $F = \sum_{n=0}^{\infty} a_n z^n$. Taking G to be $z = 0+1z+0z^2+\cdots$, the definition of composition gives $F \circ G = F$. So our earlier notational convention $F(z) = F$ is consistent with the current notation $F(G)$ for formal composition. Similarly, taking G to be the zero series, we get $F(0) = F \circ 0 = a_0 = F|_{z^0}$. Thus, $F(0)$ denotes the constant coefficient of F. If $F(0)$ is zero, we see at once that $z \circ F = F = F \circ z$. Thus, the formal power series z acts as a two-sided identity element for the operation of formal composition (applied to formal series with constant term zero).

Suppose $F = \sum_{n=0}^{\infty} a_n z^n$ and $G = \sum_{n=0}^{\infty} b_n z^n$ are formal power series with $G(0) = 0$. Given $m \in \mathbb{Z}_{\geq 0}$, let $F^* = \sum_{n=0}^{m} a_n z^n$ and $G^* = \sum_{n=0}^{m} b_n z^n$. One readily checks that $(F \circ G)|_{z^m} = (F^* \circ G^*)|_{z^m}$, since the discarded powers of z cannot contribute to the coefficient of z^m in the composition. This shows that any particular coefficient of $F \circ G$ can be computed from a finite amount of data with finitely many algebraic operations.

The next theorem lists some algebraic properties of formal composition.

11.24. Theorem: Formal Composition. Let $F, G, H, F_k, G_k \in K[[z]]$ (for $k \in \mathbb{Z}_{>0}$) satisfy $G(0) = G_k(0) = 0$, and let $c \in K$.

(a) Continuity of Composition: If $(F_k) \to F$ and $(G_k) \to G$, then $(F_k \circ G_k) \to F \circ G$.

(b) Homomorphism Properties: $(F + H) \circ G = (F \circ G) + (H \circ G)$, $(F \cdot H) \circ G = (F \circ G) \cdot (H \circ G)$, and $c \circ G = c$.

(c) Associativity: If $H(0) = 0$, then $(F \circ G) \circ H = F \circ (G \circ H)$.

Proof. To prove (a), assume $(F_k) \to F$ and $(G_k) \to G$. Fix $m \in \mathbb{Z}_{\geq 0}$, and choose k_0 large enough so that for all $k \geq k_0$ and all $n \in \{0, 1, \ldots, m\}$, $F_k|_{z^n} = F|_{z^n}$ and $G_k|_{z^n} = G|_{z^n}$. Let F_k^*, F^*, G_k^*, G^* be the series obtained from F_k, F, G_k, and G (respectively) by discarding all powers of z beyond z^m. Our choice of k_0 shows that $F_k^* = F^*$ and $G_k^* = G^*$ for all $k \geq k_0$. By the remarks preceding the theorem statement, we see that

$$(F_k \circ G_k)|_{z^m} = (F_k^* \circ G_k^*)|_{z^m} = (F^* \circ G^*)|_{z^m} = (F \circ G)|_{z^m} \quad \text{for all } k \geq k_0.$$

Since m was arbitrary, it follows that $(F_k \circ G_k) \to F \circ G$, as needed.

We ask the reader to prove (b) as an exercise. Parts (a) and (b) can be restated as follows. For fixed G with $G(0) = 0$, define a map $R_G : K[[z]] \to K[[z]]$ (called *right composition by G* or *evaluation at G*) by $R_G(F) = F \circ G$ for $F \in K[[z]]$. Part (b) says that R_G is a K-algebra homomorphism. Part (a) says, among other things, that this homorphism is *continuous*: whenever $(F_k) \to F$ in $K[[z]]$, $R_G(F_k) \to R_G(F)$. Note that $R_G(z) = z \circ G = G$. In fact, one may check that R_G is the *unique* continuous K-algebra homomorphism of $K[[z]]$ sending z to G (Exercise 11-48).

We use these remarks to prove (c). Assuming $G(0) = H(0) = 0$, we also have $(G \circ H)(0) = 0$, so that $F \circ (G \circ H)$ is defined. Now $F \circ (G \circ H) = R_{G \circ H}(F)$, whereas $(F \circ G) \circ H = R_H(R_G(F)) = (R_H \circ R_G)(F)$. Here $R_H \circ R_G$ denotes ordinary function composition of the functions R_H and R_G. It is routine to check that the composition of two continuous

K-algebra homomorphisms on $K[[z]]$ is also a continuous homomorphism. So, on one hand, $R_H \circ R_G$ is a continuous homomorphism sending z to $R_H \circ R_G(z) = (z \circ G) \circ H = G \circ H$. On the other hand, $R_{G \circ H}$ is also a continuous homomorphism sending z to $G \circ H$. By the uniqueness assertion in the last paragraph, the functions $R_H \circ R_G$ and $R_{G \circ H}$ must be equal. So applying these functions to F must give equal results, which proves (c). □

Recall that the formal derivative of a formal power series $F = \sum_{n=0}^{\infty} a_n z^n$ is the formal power series $F' = \frac{d}{dz}F(z) = \sum_{n=1}^{\infty} n a_n z^{n-1} = \sum_{m=0}^{\infty}(m+1)a_{m+1}z^m$. The next theorem summarizes some rules for computing formal derivatives, including a formal version of the Chain Rule.

11.25. Theorem: Formal Derivatives. Let $F, G \in K[[z]]$ be formal power series.

(a) The Sum Rule: $(F+G)' = F' + G'$.
(b) The Scalar Rule: For all $c \in K$, $(cF)' = c(F')$.
(c) The Product Rule: $(F \cdot G)' = (F') \cdot G + F \cdot (G')$.
(d) The Power Rule: For all $n \in \mathbb{Z}_{\geq 0}$, $(F^n)' = nF^{n-1} \cdot F'$.
(e) The Chain Rule: If $G(0) = 0$, then $(F \circ G)' = (F' \circ G) \cdot G'$. In other words,

$$\frac{d}{dz}[F(G(z))] = F'(G(z)) \cdot G'(z).$$

(f) Continuity of Differentiation: If $(F_k) \to F$, then $(F_k') \to F'$.

Proof. We sketch the proofs of (a) through (e). Write $F = \sum_{n=0}^{\infty} a_n z^n$ and $G = \sum_{n=0}^{\infty} b_n z^n$. We verify (a), (b), and (c) by comparing the coefficients of z^n on both sides of each rule. For (a), we get $(n+1)(a_{n+1} + b_{n+1})$; for (b), we get $c(n+1)a_{n+1}$; and for (c), the coefficient is

$$\begin{aligned}(n+1)\sum_{k=0}^{n+1} a_k b_{n+1-k} &= \sum_{k=0}^{n+1}(k + (n+1-k))a_k b_{n+1-k} \\ &= \sum_{k=1}^{n+1} k a_k b_{n+1-k} + \sum_{k=0}^{n}(n+1-k)a_k b_{n+1-k} \\ &= \sum_{j=0}^{n}(j+1)a_{j+1}b_{n-j} + \sum_{k=0}^{n} a_k (n-k+1)b_{n-k+1}.\end{aligned}$$

Now (d) follows from (c) by induction on n.

For (e), fix G with $G(0) = 0$, and define functions $p, q : K[[z]] \to K[[z]]$ by $p(F) = (F \circ G)'$ and $q(F) = (F' \circ G) \cdot G'$ for $F \in K[[z]]$. Using earlier results on composition, products, and derivatives, one readily verifies that both p and q are K-linear and continuous. Furthermore, for any $n \in \mathbb{Z}_{\geq 0}$, we compute

$$p(z^n) = (z^n \circ G)' = (G^n)' = nG^{n-1} \cdot G' \text{ and } q(z^n) = (nz^{n-1} \circ G) \cdot G' = nG^{n-1} \cdot G'.$$

Since $p(z^n) = q(z^n)$ for all n, it follows from Exercise 11-51 that $p = q$ as functions, so that $p(F) = q(F)$ for all $F \in K[[z]]$. This proves (e). □

11.7 Formal Exponentials and Logarithms

The exponential and logarithm functions play a central role in calculus. This section introduces formal versions of these functions that satisfy many of the same properties as their analytic counterparts.

11.26. Definition: Formal Exponentials. Let $\exp(z) = e^z = \sum_{n=0}^{\infty} z^n/n!$. For any formal power series $G \in K[[z]]$ with constant term zero, define $\exp(G) = e^{G(z)} = \sum_{n=0}^{\infty} G^n/n!$, which is the composition of e^z with $G(z)$.

The next theorem lists some fundamental properties of the formal exponential function.

11.27. Theorem: Formal Exponentials. Let F, G, and G_k (for $k \in \mathbb{Z}_{>0}$) be formal power series with constant term zero.

(a) $\frac{d}{dz}\exp(z) = \exp(z)$, $\exp(0) = 1$, and $\frac{d}{dz}\exp(F(z)) = \exp(F(z))F'(z)$.
(b) $\exp(F) \neq 0$, and $\exp(-F) = 1/\exp(F)$.
(c) $\exp(F+G) = \exp(F)\exp(G)$.
(d) exp is continuous: if $(G_k) \to G$ then $(\exp(G_k)) \to \exp(G)$.
(e) For all $N \in \mathbb{Z}_{>0}$, $\exp(\sum_{k=1}^{N} G_k) = \prod_{k=1}^{N} \exp(G_k)$.
(f) If $\sum_{k=1}^{\infty} G_k$ converges, then $\prod_{k=1}^{\infty} \exp(G_k)$ converges to $\exp(\sum_{k=1}^{\infty} G_k)$.

Proof. (a) By definition of formal derivatives, $\frac{d}{dz}\exp(z) = \sum_{n=1}^{\infty} \frac{nz^{n-1}}{n!} = \sum_{n=1}^{\infty} \frac{z^{n-1}}{(n-1)!} = \sum_{m=0}^{\infty} \frac{z^m}{m!} = \exp(z)$. Next, $\exp(0) = 1 + \sum_{n=1}^{\infty} 0^n/n! = 1$. The formula for the derivative of $\exp(F(z))$ now follows from the Formal Chain Rule.

(b) Consider the formal series $H(z) = \exp(F(z))\exp(-F(z))$. By the Formal Product Rule, the Formal Chain Rule, and (a), we compute

$$H'(z) = \exp(F(z))[\exp(-F(z)) \cdot (-F'(z))] + [\exp(F(z))F'(z)]\exp(-F(z)) = 0.$$

So $H(z)$ must be a constant formal power series. Since $H(0) = \exp(F(0))\exp(-F(0)) = \exp(0)\exp(-0) = 1 \cdot 1 = 1$, H is the constant 1. This means that $1 = \exp(F)\exp(-F)$, so that $\exp(-F)$ is the multiplicative inverse of F in $K[[z]]$. We also see that $\exp(F)$ cannot be zero, since $1 \neq 0$ in K. In fact, one readily checks that $\exp(F)$ has constant term 1.

(c) Consider the formal series $P = \exp(F+G)\exp(-F)\exp(-G)$. The formal derivative of P is

$$P' = (F'+G')e^{F+G}e^{-F}e^{-G} - F'e^{F+G}e^{-F}e^{-G} - G'e^{F+G}e^{-F}e^{-G} = 0,$$

so P is constant. Evaluating at $z = 0$, we find that the constant is $P(0) = e^{0+0}e^{-0}e^{-0} = 1$, so $P = 1$. Using (b), it follows that $\exp(F+G) = P\exp(F)\exp(G) = \exp(F)\exp(G)$, as needed.

Part (d) follows from the continuity of formal composition. Part (e) follows from (c) by induction on N. Part (f) follows from (d) and (e) by a formal limiting argument. We ask the reader to give the details of these proofs in an exercise. □

Next we define a formal version of the natural logarithm function.

11.28. Definition: Formal Logarithms. Define a formal power series

$$L(z) = \log(1+z) = \sum_{n=1}^{\infty} \frac{(-1)^{n-1}}{n} z^n = z - z^2/2 + z^3/3 - z^4/4 + \cdots.$$

For any formal power series $G \in K[[z]]$ with constant term 0, define $\log(1+G) = \sum_{n=1}^{\infty}(-1)^{n-1}G^n/n$, which is the composition $L \circ G$. For any formal power series H with constant term 1, define $\log(H) = \sum_{n=1}^{\infty}(-1)^{n-1}(H-1)^n/n$, which is the composition $L \circ (H-1)$.

11.29. Theorem: Formal Logarithms. Let F, G, H, and H_k (for $k \in \mathbb{Z}_{>0}$) be formal power series with $F(0) = 0$ and $G(0) = H(0) = H_k(0) = 1$.

(a) $\frac{d}{dz}\log(1+z) = (1+z)^{-1}$, $\log(1) = 0$, $\frac{d}{dz}\log(1+F(z)) = F'(z)/(1+F(z))$, and $\frac{d}{dz}\log(G(z)) = G'(z)/G(z)$.
(b) $\log(\exp(F)) = F$ and $\exp(\log(H)) = H$.
(c) $\log(GH) = \log(G) + \log(H)$.
(d) log is continuous: if $(H_k) \to H$, then $(\log(H_k)) \to \log(H)$.
(e) For all $N \in \mathbb{Z}_{>0}$, $\log(\prod_{k=1}^{N} H_k) = \sum_{k=1}^{N} \log(H_k)$.
(f) If $\prod_{k=1}^{\infty} H_k$ converges, then $\sum_{k=1}^{\infty} \log(H_k)$ converges to $\log(\prod_{k=1}^{\infty} H_k)$.

Proof. (a) By the definition of formal derivatives and the Formal Geometric Series formula,

$$\frac{d}{dz}\log(1+z) = \sum_{n=1}^{\infty} \frac{(-1)^{n-1} n z^{n-1}}{n} = \sum_{m=0}^{\infty} (-1)^m z^m = (1-(-z))^{-1} = (1+z)^{-1}.$$

Taking $G = 0$ in the definition gives $\log(1) = \log(1+0) = \sum_{n=1}^{\infty}(-1)^{n-1}0^n/n = 0$. The formula for the derivative of $\log(1+F(z))$ follows from the Formal Chain Rule, and the formula for $\frac{d}{dz}\log(G(z))$ follows by taking F to be $G-1$.

(b) Note that $\exp(F)$ has constant term 1, so $\log(\exp(F))$ is defined. Let $Q(z) = \log(\exp(F(z))) - F(z)$. Taking the formal derivative of Q, we get $Q'(z) = F'(z)\exp(F(z))/\exp(F(z)) - F'(z) = 0$, so Q is a constant. Since the constant term of Q is zero, Q is the zero power series. Thus, $\log(\exp(F)) = F$ as needed. A similar proof shows that $\exp(\log(H)) = H$.

(c) Consider the formal series $P = \log(GH) - \log(G) - \log(H)$. Computing derivatives,

$$P'(z) = \frac{G'(z)H(z) + G(z)H'(z)}{G(z)H(z)} - \frac{G'(z)}{G(z)} - \frac{H'(z)}{H(z)} = 0.$$

So P is constant. As P has constant term zero, $P = 0$, proving (c).

Now (d) follows from the continuity of formal composition, (e) follows from (c) by induction on N, and (f) follows from (d) and (e) by a formal limiting argument (Exercise 11-62). □

11.8 The Exponential Formula

Many combinatorial structures can be decomposed into disjoint unions of smaller structures that are *connected* in some sense. For example, set partitions consist of a collection of disjoint blocks; permutations can be regarded as a collection of disjoint cycles; and graphs are disjoint unions of connected graphs. The Exponential Formula allows us to compute the generating function for such structures from the generating functions for their connected components. This formula reveals the combinatorial significance of the exponential of a formal power series.

First we need to review set partitions and ordered set partitions. Recall from §2.12 that a *set partition* of a set X is a set $P = \{B_1, B_2, \ldots, B_m\}$ of nonempty subsets of X such that every $a \in X$ belongs to exactly one block B_i of P. Since P is a set, the blocks in a set partition can be presented in any order; for example, $\{\{1,2,4\},\{3,5\}\}$ and $\{\{3,5\},\{1,2,4\}\}$ are equal set partitions of the set $\{1,2,3,4,5\}$. In contrast, an *ordered set partition* of X is a sequence $Q = (B_1, B_2, \ldots, B_m)$ of distinct sets such that $\{B_1, B_2, \ldots, B_m\}$ is a set partition of X. Here the order of the blocks in the list Q is important, but we can still list the elements within each block in any order. For example, $(\{1,2,4\},\{3,5\})$ and $(\{3,5\},\{1,2,4\})$ are two different ordered set partitions of $\{1,2,3,4,5\}$.

Let SetPar(n,m) be the set of all set partitions of $\{1,2,\ldots,n\}$ consisting of m blocks, and let OrdPar(n,m) be the set of all ordered set partitions of $\{1,2,\ldots,n\}$ consisting of m blocks. We know that $|\text{SetPar}(n,m)| = S(n,m)$, the Stirling number of the second kind. Define a map $f : \text{OrdPar}(n,m) \to \text{SetPar}(n,m)$ by sending $Q = (B_1,\ldots,B_m) \in \text{OrdPar}(n,m)$ to $f(Q) = \{B_1,\ldots,B_m\}$. This map forgets the ordering of the blocks in an ordered set partition to produce an ordinary set partition. We see that f is onto but not one-to-one. More precisely, since we can order a set of m distinct blocks in $m!$ ways, we see that for each $P \in \text{SetPar}(n,m)$, there are exactly $m!$ objects $Q \in \text{OrdPar}(n,m)$ with $f(Q) = P$. Consequently, $|\text{OrdPar}(n,m)| = m!S(n,m)$.

We also need the following encoding of ordered set partitions. Let $W(n,m)$ be the set of all lists $(k_1,\ldots,k_m,w)$, where $k_1,\ldots,k_m$ are positive integers with $k_1+\cdots+k_m = n$, and w is an anagram in $\mathcal{R}(1^{k_1}\cdots m^{k_m})$ (so for j between 1 and m, the letter j appears k_j times in w). We define a bijection $g : W(n,m) \to \text{OrdPar}(n,m)$ as follows. Given $(k_1,\ldots,k_m,w) \in W(n,m)$, g maps this object to the ordered set partition $(B_1,\ldots,B_m)$ such that $B_j = \{i : w_i = j\}$ for $1 \le j \le m$. Note that $|B_j| = k_j$ for $1 \le j \le m$. The inverse of g sends $(B_1,\ldots,B_m) \in \text{OrdPar}(n,m)$ to the list $(|B_1|,\ldots,|B_m|,w)$, where $w_i = j$ iff $i \in B_j$ for $1 \le i \le n$ and $1 \le j \le m$.

We are now ready to state the Exponential Formula. Compare the next theorem to the EGF Product Rule in §5.12.

11.30. The Exponential Formula. Let C and S be sets such that each object x in C or S has a *weight* wt(x) and a *size* sz(x), where sz$(c) > 0$ for all $c \in C$. (Intuitively, C is a set of *connected structures* of various positive sizes and S is a set of objects that are *labeled disjoint unions* of these structures.) Suppose that every object $s \in S$ of size $n \ge 0$ can be constructed uniquely by the following process. First, choose a set partition P of $\{1,2,\ldots,n\}$. Next, for each block B in the set partition P, choose a connected structure $c(B)$ from C such that $\text{sz}(c(B)) = |B|$. Finally, assemble these choices in a prescribed manner to produce s. Assume that whenever s is built from P and $(c(B) : B \in P)$, the *weight-additivity condition* $\text{wt}(s) = \sum_{B\in P}\text{wt}(c(B))$ holds. Define generating functions

$$G_C = \sum_{c\in C} t^{\text{wt}(c)}\frac{z^{\text{sz}(c)}}{\text{sz}(c)!} \quad\text{and}\quad G_S = \sum_{s\in S} t^{\text{wt}(s)}\frac{z^{\text{sz}(s)}}{\text{sz}(s)!}.$$

Then $G_S = \exp(G_C)$.

Proof. For each $n \ge 0$, define $S_n = \{s \in S : \text{sz}(s) = n\}$, $C_n = \{c \in C : \text{sz}(c) = n\}$, $\text{GF}(S_n) = \sum_{s\in S_n} t^{\text{wt}(s)}$, and $\text{GF}(C_n) = \sum_{c\in C_n} t^{\text{wt}(c)}$. We show that $G_S|_{z^n} = [\exp(G_C)]|_{z^n}$ by computing both sides. The constant term on each side is 1, so fix $n > 0$ from now on.

First, $G_S|_{z^n} = \text{GF}(S_n)/n!$. To compute this more explicitly, we use the description of objects in S_n in the theorem statement. For each $P \in \text{SetPar}(n,m)$, let S_P be the set of $s \in S_n$ that are built by choosing P at the first stage. By the Sum Rule for Weighted Sets,

$$\text{GF}(S_n) = \sum_{m=1}^{n}\sum_{P\in\text{SetPar}(n,m)}\text{GF}(S_P).$$

Now fix $m \in \{1,2,\ldots,n\}$, and fix a set partition $P \in \text{SetPar}(n,m)$. Write $P = \{B_1,B_2,\ldots,B_m\}$, where we choose the indexing so that $\min(B_1) < \min(B_2) < \cdots < \min(B_m)$, and let $k_i = |B_i|$ for $1 \le i \le m$. By assumption, we can build each object $s \in S_P$ by choosing $c(B_1) \in C_{k_1}$, $c(B_2) \in C_{k_2}$, ..., $c(B_m) \in C_{k_m}$, and assembling these choices in a prescribed manner. By the Product Rule for Weighted Sets,

$\text{GF}(S_P) = \text{GF}(C_{k_1})\,\text{GF}(C_{k_2})\cdots\text{GF}(C_{k_m}) = \prod_{B\in P}\text{GF}(C_{|B|})$. Thus,

$$G_S|_{z^n} = \frac{1}{n!}\text{GF}(S_n) = \frac{1}{n!}\sum_{m=1}^{n}\sum_{P\in\text{SetPar}(n,m)}\prod_{B\in P}\text{GF}(C_{|B|}). \tag{11.10}$$

Next, we find the coefficient of z^n in $\exp(G_C)$. By Definition 11.26,

$$\exp(G_C) = \sum_{m=0}^{\infty}\frac{(G_C)^m}{m!} = \sum_{m=0}^{\infty}\frac{1}{m!}\left(\sum_{k=1}^{\infty}\frac{\text{GF}(C_k)}{k!}z^k\right)^m.$$

Using Theorem 5.35, we find that

$$[\exp(G_C)]|_{z^n} = \sum_{m=1}^{n}\frac{1}{m!}\sum_{\substack{(k_1,k_2,\ldots,k_m)\in\mathbb{Z}_{>0}^m:\\ k_1+k_2+\cdots+k_m=n}}\frac{\text{GF}(C_{k_1})\,\text{GF}(C_{k_2})\cdots\text{GF}(C_{k_m})}{k_1!k_2!\cdots k_m!}$$

(compare to (11.2)). Now multiply and divide each term by $n!$ to make a multinomial coefficient appear. We find that $[\exp(G_C)]|_{z^n}$ equals

$$\frac{1}{n!}\sum_{m=1}^{n}\frac{1}{m!}\sum_{\substack{(k_1,k_2,\ldots,k_m)\in\mathbb{Z}_{>0}^m:\\ k_1+k_2+\cdots+k_m=n}}\binom{n}{k_1,k_2,\ldots,k_m}\text{GF}(C_{k_1})\,\text{GF}(C_{k_2})\cdots\text{GF}(C_{k_m}).$$

By the Anagram Rule,

$$\binom{n}{k_1,k_2,\ldots,k_m} = \sum_{w\in\mathcal{R}(1^{k_1}2^{k_2}\cdots m^{k_m})}1.$$

This observation turns the inner sum into a sum indexed by objects $(k_1,\ldots,k_m,w)$ in $W(n,m)$:

$$[\exp(G_C)]|_{z^n} = \frac{1}{n!}\sum_{m=1}^{n}\frac{1}{m!}\sum_{(k_1,\ldots,k_m,w)\in W(n,m)}\text{GF}(C_{k_1})\,\text{GF}(C_{k_2})\cdots\text{GF}(C_{k_m}).$$

We use the bijection g to convert to a sum indexed by ordered set partitions. Recall that for $g(k_1,\ldots,k_m,w) = (B_1,\ldots,B_m)$, we have $|B_j| = k_j$ for all j. So we get

$$[\exp(G_C)]|_{z^n} = \frac{1}{n!}\sum_{m=1}^{n}\frac{1}{m!}\sum_{(B_1,\ldots,B_m)\in\text{OrdPar}(n,m)}\prod_{j=1}^{m}\text{GF}(C_{|B_j|}).$$

The final step is to use the function f to change the sum over ordered set partitions to a sum over ordinary set partitions. Note that multiplication of generating functions is commutative. So for each of the $m!$ ordered set partitions $(B_1,\ldots,B_m)$ that map to a given set partition $P\in\text{SetPar}(n,m)$, we have $\prod_{j=1}^{m}\text{GF}(C_{|B_j|}) = \prod_{B\in P}\text{GF}(C_{|B|})$. This produces an extra factor of $m!$ that cancels the division by $m!$ in the previous formula. In conclusion,

$$[\exp(G_C)]|_{z^n} = \frac{1}{n!}\sum_{m=1}^{n}\sum_{P\in\text{SetPar}(n,m)}\prod_{B\in P}\text{GF}(C_{|B|}).$$

This agrees with (11.10), so the proof is complete. □

11.9 Examples of the Exponential Formula

This section gives examples illustrating the Exponential Formula. We begin by showing how the generating functions for Stirling numbers (derived in §5.13 and §5.14) are consequences of this formula.

11.31. Example: Stirling Numbers of the Second Kind. Let us find the generating function for the collection S of all set partitions of one of the sets $\{1, 2, \ldots, n\}$ for some $n \geq 0$. For a set partition $P \in \mathrm{SetPar}(n, m) \subseteq S$, we define $\mathrm{sz}(P) = n$ and $\mathrm{wt}(P) = m$, the number of blocks of P. In this case, the connected pieces used to build a set partition P are the individual blocks of P. Each connected piece has no additional structure beyond the set of labels appearing in the block. We model this by letting $C = \mathbb{Z}_{>0} = \{1, 2, 3, \ldots\}$, and setting $\mathrm{sz}(c) = c$ and $\mathrm{wt}_C(c) = 1$ for all $c \in C$. To build a typical object in S of size n, we first choose a set partition $P = \{B_1, \ldots, B_m\}$ of $\{1, 2, \ldots, n\}$. For $1 \leq i \leq m$, we then choose $c(B_i) = |B_i|$, which is the unique element in C of size $|B_i|$. The final object constructed from these choices is P itself. The weight-additivity condition holds because $\mathrm{wt}(P) = m = \sum_{i=1}^{m} 1 = \sum_{i=1}^{m} \mathrm{wt}_C(c(B_i))$. By the definition of C,

$$G_C = \sum_{c \in C} t^{\mathrm{wt}(c)} \frac{z^{\mathrm{sz}(c)}}{\mathrm{sz}(c)!} = \sum_{j=1}^{\infty} t \frac{z^j}{j!} = t(e^z - 1).$$

Applying the Exponential Formula, we get

$$1 + \sum_{n=1}^{\infty} \sum_{m=1}^{n} S(n, m) t^m \frac{z^n}{n!} = G_S = \exp(G_C) = e^{t(e^z - 1)},$$

in agreement with Theorem 5.43. Recall that the Bell number $B(n)$ is the number of set partitions of $\{1, 2, \ldots, n\}$ with any number of blocks. Setting $t = 1$ in the previous generating function, we get the EGF for Bell numbers:

$$\sum_{n=0}^{\infty} B(n) \frac{z^n}{n!} = e^{e^z - 1}.$$

This calculation can be adjusted to count set partitions with restrictions on the allowable block sizes. For example, suppose we are counting set partitions that contain no blocks of size 1 or 3. We modify G_C by making the coefficients of z^1 and z^3 zero, giving $G_C = t(e^z - 1 - z - z^3/3!)$. Then $G_S = \exp(G_C) = \exp(t(e^z - 1 - z - z^3/3!))$. Setting $t = 1$ and extracting the coefficient of z^{12}, we find (using a computer algebra system) that the number of such set partitions of $\{1, 2, \ldots, 12\}$ is 159,457.

11.32. Example: Stirling Numbers of the First Kind. Let S be the set of all permutations of one of the sets $\{1, 2, \ldots, n\}$ for some $n \geq 0$. For $w \in S$ permuting $\{1, 2, \ldots, n\}$, let $\mathrm{sz}(w) = n$, and let $\mathrm{wt}(w)$ be the number of cycles in the digraph of w (see §3.6). Recall that the signless Stirling number of the first kind, denoted $s'(n, k)$, is the number of objects in S of size n and weight k. We use the Exponential Formula to find the generating function for these numbers.

In this case, we assemble permutations $w \in S$ from connected pieces that are the individual directed cycles in the digraph of w. To model this, let C be the set of all k-cycles on $\{1, 2, \ldots, k\}$, as k ranges through positive integers. Given a k-cycle $c \in C$, define $\mathrm{wt}(c) = 1$ and $\mathrm{sz}(c) = k$. For each $k > 0$, C contains $(k-1)!$ objects of size k. Therefore

$$G_C = \sum_{c \in C} t^{\mathrm{wt}(c)} \frac{z^{\mathrm{sz}(c)}}{\mathrm{sz}(c)!} = \sum_{k=1}^{\infty} t(k-1)! \frac{z^k}{k!} = t \sum_{k=1}^{\infty} \frac{z^k}{k} = -t \log(1 - z) = \log[(1 - z)^{-t}],$$

where the last step uses Exercise 11-64(e). To check the hypothesis of the Exponential Formula, note that each permutation $s \in S$ of size n can be built uniquely as follows. Choose a set partition $P = \{B_1, \ldots, B_m\}$ of $\{1, 2, \ldots, n\}$. For each k-element block B of P, choose a k-cycle $c(B) \in C$. Suppose $B = \{i_1 < i_2 < \cdots < i_k\}$. Replace the numbers $1, 2, \ldots, k$ appearing in the k-cycle $c(B)$ by the numbers $i_1, i_2, \ldots, i_k$, respectively. We obtain one of the cycles in the digraph of s. After all choices have been made, we have built the full digraph of s, which uniquely determines s. For example, suppose $n = 8$, $P = \{\{2, 4, 7, 8\}, \{1, 3, 5\}, \{6\}\}$, $c(\{2, 4, 7, 8\}) = (1, 3, 4, 2)$, $c(\{1, 3, 5\}) = (1, 3, 2)$, and $c(\{6\}) = (1)$. These choices create the permutation $s = (2, 7, 8, 4)(1, 5, 3)(6)$. The Exponential Formula tells us that

$$1 + \sum_{n=1}^{\infty} \sum_{m=1}^{n} s'(n, m) t^m \frac{z^n}{n!} = G_S = \exp(G_C) = \exp(\log[(1-z)^{-t}]) = (1-z)^{-t},$$

in agreement with (5.11).

To get a generating function for derangements (permutations with no 1-cycles), we modify G_C by making the coefficient of z^1 be zero. This gives $G_C = \log[(1-z)^{-t}] - tz$, so $G_S = \exp(G_C) = e^{-tz}(1-z)^{-t}$. (Compare to the generating function found in §5.7.) Extracting the coefficient of $t^5 z^{12}/12!$, we find that there are 866,250 derangements of $\{1, 2, \ldots, 12\}$ consisting of five cycles.

The Exponential Formula tells us that $G_S = \exp(G_C)$ when structures in S are built from labeled disjoint unions of connected structures in C. Sometimes we know the generating function G_S and need to know the generating function G_C. Solving for G_C, we get $G_C = \log(G_S)$. The next example uses this formula to obtain information about connected graphs.

11.33. Example: Connected Components of Graphs. Let C be the set of all connected graphs on one of the vertex sets $\{1, 2, \ldots, k\}$, with $k > 0$. Given a k-vertex graph $c \in C$, let $\mathrm{wt}(c) = 0$ and $\mathrm{sz}(c) = k$. Direct computation of the generating function G_C is difficult. On the other hand, consider the set S of objects we can build from C by the procedure in the Exponential Formula. Suppose we choose a set partition $P = \{B_1, B_2, \ldots, B_m\}$ of $\{1, 2, \ldots, n\}$ for some $n \geq 0$, then choose a connected graph $c(B_i) \in C$ of size $|B_i|$ for $1 \leq i \leq m$. We can assemble these choices to get an arbitrary (simple, undirected) graph with vertex set $\{1, 2, \ldots, n\}$ by relabeling the vertices $1, 2, \ldots, |B_i|$ in each graph $c(B_i)$ with the labels in B_i in increasing order. Thus S consists of all graphs on one of the vertex sets $\{1, 2, \ldots, n\}$, where the size of the graph is the number of vertices and the weight of the graph is zero. By the Product Rule, there are $2^{\binom{n}{2}}$ graphs with vertex set $\{1, 2, \ldots, n\}$, since we can either include or exclude each of the $\binom{n}{2}$ possible edges. Accordingly,

$$G_S = \sum_{s \in S} t^{\mathrm{wt}(s)} \frac{z^{\mathrm{sz}(s)}}{\mathrm{sz}(s)!} = \sum_{n=0}^{\infty} \frac{2^{\binom{n}{2}} z^n}{n!}.$$

By the Exponential Formula, $G_C = \log(G_S)$. Extracting the coefficient of $z^n/n!$ on both sides leads to the exact formula

$$\sum_{m=1}^{n} \sum_{\substack{(k_1, \ldots, k_m) \in \mathbb{Z}_{>0}^m: \\ k_1 + \cdots + k_m = n}} \frac{(-1)^{m-1}}{m} \binom{n}{k_1, \ldots, k_m} 2^{\binom{k_1}{2} + \cdots + \binom{k_m}{2}} \tag{11.11}$$

for the number of connected simple graphs on n vertices. When $n = 7$, we find there are 1,866,256 such graphs.

11.10 Ordered Trees and Terms

The last main topic in this chapter is compositional inverses of formal power series. Our goal is to develop algebraic and combinatorial formulas for the coefficients in these inverses. To prepare for this, we must first study combinatorial structures called ordered trees and ordered forests. Ordered trees are defined recursively as follows.

11.34. Definition: Ordered Trees. The symbol 0 is an ordered tree. If $n \in \mathbb{Z}_{>0}$ and $T_1, \ldots, T_n$ is a sequence of ordered trees, then the $(n+1)$-tuple $(n, T_1, T_2, \ldots, T_n)$ is an ordered tree. All ordered trees arise by applying these two rules a finite number of times.

We can visualize ordered trees as follows. The ordered tree 0 is depicted as a single node. The ordered tree $(n, T_1, T_2, \ldots, T_n)$ is drawn by putting a single *root node* at the top of the picture with n edges leading down. At the ends of these edges, reading from left to right, we recursively draw pictures of the trees $T_1, T_2, \ldots, T_n$ in this order. The term *ordered tree* emphasizes the fact that the left-to-right order of the children of each node is significant. Note that an ordered tree is *not* a tree in the graph-theoretic sense, and ordered trees are not the same as rooted trees.

11.35. Example. Figure 11.1 illustrates the ordered tree

$$T = (4, (2, 0, (1, 0)), 0, (3, 0, (3, 0, 0, 0), 0), 0).$$

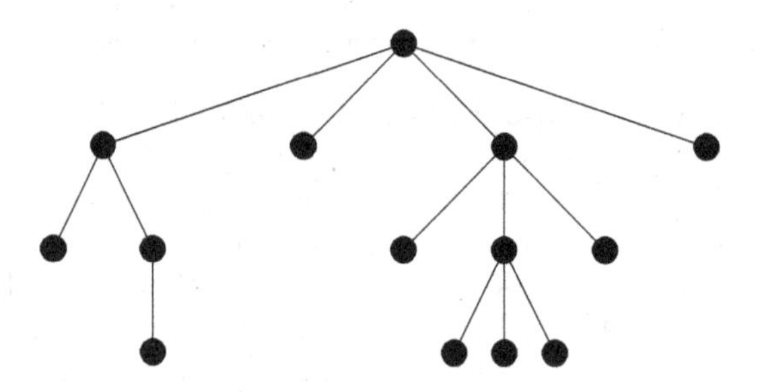

FIGURE 11.1
Diagram of an ordered tree.

Ordered trees can be used to model algebraic expressions that are built up by applying functions to lists of inputs. For example, the tree T in the previous example represents the syntactic structure of the following algebraic expression:

$$f(g(x_1, h(x_2)), x_3, k(x_4, j(x_5, x_6, x_7), x_8), x_9).$$

More specifically, if we replace each function symbol f, g, h, k, j by its *arity* (number of inputs) and replace each variable x_i by zero, we obtain

$$4(2(0, 1(0)), 0, 3(0, 3(0, 0, 0), 0), 0).$$

This string becomes T if we move each left parenthesis to the left of the positive integer immediately preceding it and put a comma in its original location.

Surprisingly, the syntactic structure of such an algebraic expression is uniquely determined even if we erase all the parentheses. To prove this statement, we introduce a combinatorial object called a term that is like an ordered tree, but contains no parentheses. For example, the algebraic expression above will be modeled by the term 4201003030000.

11.36. Definition: Words and Terms. A *word* is a finite sequence of symbols in $\mathbb{Z}_{\geq 0}$. We define *terms* recursively as follows. The word 0 is a term. If $n > 0$ and $T_1, T_2, \ldots, T_n$ are terms, then the word $nT_1T_2\cdots T_n$ is a term. All terms arise by applying these two rules a finite number of times.

We see from this definition that every term is a *nonempty* word.

11.37. Definition: Weight of a Word. Given a word $w = w_1w_2\cdots w_s$, the *weight* of w is $\text{wt}(w) = w_1 + w_2 + \cdots + w_s - s$.

For example, $\text{wt}(42010030300000) = 13 - 14 = -1$. Note that $\text{wt}(vw) = \text{wt}(v) + \text{wt}(w)$ for all words v, w. The next result uses weights to characterize terms.

11.38. Theorem: Characterization of Terms. A word $w = w_1w_2\cdots w_s$ is a term iff $\text{wt}(w) = -1$ and $\text{wt}(w_1w_2\cdots w_k) \geq 0$ for all $k \in \{1, 2, \ldots, s-1\}$.

Proof. We use strong induction on the length s of the word. First suppose $w = w_1w_2\cdots w_s$ is a term of length s. If $w = 0$, then the weight condition holds. Otherwise, we must have $w = nT_1T_2\cdots T_n$ where $n > 0$ and $T_1, T_2, \ldots, T_n$ are terms. Since each T_i has length less than s, the induction hypothesis shows that $\text{wt}(T_i) = -1$ and every proper prefix of T_i has nonnegative weight. So, first of all, $\text{wt}(w) = \text{wt}(n) + \text{wt}(T_1) + \cdots + \text{wt}(T_n) = (n-1) - n = -1$. On the other hand, consider a proper prefix $w_1w_2\cdots w_k$ of w. If $k = 1$, the weight of this prefix is $n - 1$, which is nonnegative since $n > 0$. If $k > 1$, we must have $w_1w_2\cdots w_k = nT_1\cdots T_iz$ where $0 \leq i < n$ and z is a proper prefix of T_{i+1}. Using the induction hypothesis, the weight of $w_1w_2\cdots w_k$ is therefore $(n-1) - i + \text{wt}(z) \geq (n-i) - 1 \geq 0$.

For the converse, we also use strong induction on the length of the word. Let $w = w_1w_2\cdots w_s$ satisfy the weight conditions. The empty word has weight zero, so $s > 0$. If $s = 1$, then $\text{wt}(w_1) = -1$ forces $w = 0$, so that w is a term in this case. Now suppose $s > 1$. The first symbol w_1 must be an integer $n > 0$, lest the proper prefix w_1 of w have negative weight. Observe that appending one more letter to any word decreases the weight by at most 1. Since $\text{wt}(w_1) = n - 1$ and $\text{wt}(w_1w_2\cdots w_s) = -1$, there exists a least integer k_1 with $\text{wt}(w_1w_2\cdots w_{k_1}) = n - 2$. Now if $n \geq 2$, there exists a least integer $k_2 > k_1$ with $\text{wt}(w_1w_2\cdots w_{k_2}) = n - 3$. We continue similarly, obtaining integers $k_1 < k_2 < \cdots < k_n$ such that k_i is the least index following k_{i-1} such that $\text{wt}(w_1w_2\cdots w_{k_i}) = n - 1 - i$. Because w satisfies the weight conditions, we must have $k_n = s$. Now define n subwords $T_1 = w_2w_3\cdots w_{k_1}$, $T_2 = w_{k_1+1}w_{k_1+2}\cdots w_{k_2}$, ..., $T_n = w_{k_{n-1}+1}\cdots w_s$. Evidently $w = nT_1T_2\cdots T_n$. For $1 \leq i \leq n$, $nT_1T_2\cdots T_{i-1}$ has weight $n-i$, $nT_1T_2\cdots T_i$ has weight $n-i-1$, and (by minimality of k_i) no proper prefix of $nT_1T_2\cdots T_i$ of length at least k_{i-1} has weight less than $n-i$. It follows that T_i has weight -1 but every proper prefix of T_i has nonnegative weight. Thus each T_i satisfies the weight conditions and has length less than w. By induction, every T_i is a term. Then $w = nT_1T_2\cdots T_n$ is also a term, completing the induction. □

11.39. Corollary. No proper prefix of a term is a term.

11.40. Theorem: Unique Readability of Terms. For every term w, there exists a unique integer $n \geq 0$ and unique terms $T_1, \ldots, T_n$ such that $w = nT_1\cdots T_n$.

Proof. Existence follows from the recursive definition of terms. We prove uniqueness by induction on the length of w. Suppose $w = nT_1\cdots T_n = mT'_1\cdots T'_m$ where $n, m \geq 0$ and every T_i and T'_j is a term. We must prove $n = m$ and $T_i = T'_i$ for $1 \leq i \leq n$. First, $n = w_1 = m$. If $T_1 \neq T'_1$, then one of T_1 and T'_1 must be a proper prefix of the other, in violation of the preceding corollary. So $T_1 = T'_1$. Then if $T_2 \neq T'_2$, one of T_2 and T'_2 must be a proper prefix of the other, in violation of the corollary. Continuing similarly, we see that $T_i = T'_i$ for $i = 1, 2, \ldots, n$. □

Using the previous theorem and induction, it can be checked that erasing all parentheses defines a bijection from ordered trees to terms. Therefore, to count various collections of ordered trees, it suffices to count the corresponding collections of terms. We give examples of this technique in the next section.

11.11 Ordered Forests and Lists of Terms

We continue our study of ordered trees and terms by introducing two more general concepts: ordered forests and lists of terms.

11.41. Definition: Ordered Forests. For $n \in \mathbb{Z}_{\geq 0}$, an *ordered forest of n trees* is a list $(T_1, T_2, \ldots, T_n)$, where each T_i is an ordered tree.

11.42. Definition: Lists of Terms. For $n \in \mathbb{Z}_{\geq 0}$, a *list of n terms* is a word w of the form $w = T_1 T_2 \cdots T_n$, where each T_i is a term.

11.43. Theorem: Weight Characterization of Lists of Terms. A word $w = w_1 w_2 \cdots w_s$ is a list of n terms iff $\text{wt}(w) = -n$ and $\text{wt}(w_1 w_2 \cdots w_k) > -n$ for all $k \in \{1, 2, \ldots, s-1\}$.

Proof. First suppose w is a list of n terms, say $w = T_1 T_2 \cdots T_n$. Then $nw = nT_1 T_2 \cdots T_n$ is a single term. This term has weight -1, by Theorem 11.38, so w has weight $-1 - \text{wt}(n) = -n$. If $\text{wt}(w_1 \cdots w_k) \leq -n$ for some $k < s$, then the proper prefix $nw_1 \cdots w_k$ of the term nw would have negative weight, contradicting Theorem 11.38.

Conversely, suppose w satisfies the weight conditions in Theorem 11.43. Then the word nw satisfies the weight conditions in Theorem 11.38, as one may verify. So nw is a term, which must have the form $nT_1 T_2 \cdots T_n$ for certain terms $T_1, \ldots, T_n$. Then $w = T_1 T_2 \cdots T_n$ is a list of n terms. □

11.44. Theorem: Unique Readability of Lists of Terms. If $w = T_1 T_2 \cdots T_n$ is a list of n terms, then n and the terms T_i are uniquely determined by w.

Proof. First, $n = -\text{wt}(w)$ is uniquely determined by w. To see that the T_i are unique, add an n to the beginning of w and then appeal to Theorem 11.40. □

We deduce that erasing parentheses gives a bijection between ordered forests of n trees and lists of n terms.

The next lemma reveals a key property that will allow us to count lists of terms.

11.45. The Cycle Lemma for Lists of Terms. Suppose $w = w_1 w_2 \cdots w_s$ is a word of weight $-n < 0$. There exist exactly n indices $i \in \{1, 2, \ldots, s\}$ such that the cyclic rotation

$$R_i(w) = w_i w_{i+1} \cdots w_s w_1 w_2 \cdots w_{i-1}$$

is a list of n terms.

Proof. Step 1. We prove the result when w itself is a list of n terms. Say $w = T_1 T_2 \cdots T_n$ where T_j is a term of length k_j. Then $R_i(w)$ is a list of n terms for the n indices $i \in \{1, k_1 + 1, k_1 + k_2 + 1, \ldots, k_1 + k_2 + \cdots + k_{n-1} + 1\}$. Suppose i is another index (different from those just listed) such that $R_i(w)$ is a list of n terms. For some j in the range $1 \leq j \leq n$, we must have

$$R_i(w) = yT_{j+1} \cdots T_n T_1 \cdots T_{j-1} z$$

where $T_j = zy$ and z, y are nonempty words. Since $\text{wt}(z) \geq 0$ but $\text{wt}(T_j) = -1$, we must have $\text{wt}(y) < 0$. So

$$\text{wt}(yT_{j+1} \cdots T_{j-1}) = \text{wt}(y) + \text{wt}(T_{j+1}) + \cdots + \text{wt}(T_{j-1}) < -(n-1).$$

Then $yT_{j+1} \cdots T_{j-1}$ is a proper prefix of $R_i(w)$ with weight $\leq -n$, in violation of Theorem 11.43.

Step 2. We prove the result for a general word w. It suffices to show that there exists at least one i such that $R_i(w)$ is a list of n terms. For if this holds, since we obtain the same collection of words by cyclically shifting w and $R_i(w)$, the result follows from Step 1.

First note that all cyclic rotations of w have weight $\sum_{j=1}^{s} \text{wt}(w_j) = \text{wt}(w) = -n$. Let m be the minimum weight of any prefix $w_1 w_2 \cdots w_k$ of w, where $1 \leq k \leq s$. Choose k minimal such that $\text{wt}(w_1 w_2 \cdots w_k) = m$. If $k = s$, then $m = -n$, and by minimality of k and Theorem 11.43, w itself is already a list of n terms. Otherwise, let $i = k+1$. We claim $R_i(w)$ is a list of n terms. It suffices to check that each proper prefix of $R_i(w)$ has weight $> -n$. On one hand, for all j with $i \leq j \leq s$, the prefix $w_i \cdots w_j$ of $R_i(w)$ cannot have negative weight; otherwise, $\text{wt}(w_1 \cdots w_k w_i \cdots w_j) < m$ violates the minimality of m. So $\text{wt}(w_i \cdots w_j) \geq 0 > -n$. Note that when $j = s$, we have $\text{wt}(w_i \cdots w_s) = \text{wt}(w) - \text{wt}(w_1 \cdots w_k) = -n - m$. Now consider j in the range $1 \leq j < k$. If $\text{wt}(w_i \cdots w_s w_1 \cdots w_j) \leq -n$, then

$$\text{wt}(w_1 \cdots w_j) = \text{wt}(w_i \cdots w_s w_1 \cdots w_j) - \text{wt}(w_i \cdots w_s) \leq -n - (-n - m) = m.$$

But this violates the choice of k as the *least* index such that the prefix ending at k has minimum weight. So $\text{wt}(w_i \cdots w_s w_1 \cdots w_j) > -n$. It now follows from Theorem 11.43 that $R_i(w)$ is indeed a list of n terms. □

Suppose w is a list of n terms containing exactly k_i occurrences of i for each $i \geq 0$. We have

$$-n = \text{wt}(w) = \sum_{i \geq 0} k_i \,\text{wt}(i) = \sum_{i \geq 0} (i-1)k_i = -k_0 + \sum_{i \geq 1} (i-1)k_i.$$

It follows that $k_0 = n + \sum_{i \geq 1} (i-1)k_i$ in this situation. Conversely, if k_0 satisfies this relation, then $\text{wt}(w) = -n$ for all $w \in \mathcal{R}(0^{k_0} 1^{k_1} 2^{k_2} \cdots)$. We now have all the ingredients needed for our main counting result.

11.46. Theorem: Counting Lists of Terms. Let $n > 0$ and $k_0, k_1, \ldots, k_t \geq 0$ be given integers such that $k_0 = n + \sum_{i=1}^{t} (i-1)k_i$. The number of words w such that w is a list of n terms containing k_i copies of i for $0 \leq i \leq t$ is

$$\frac{n}{s} \binom{s}{k_0, k_1, \ldots, k_t} = \frac{n(s-1)!}{k_0! k_1! \cdots k_t!},$$

where $s = \sum_{i=0}^{t} k_i = n + \sum_{i=1}^{t} i k_i$ is the common length of all such words.

Proof. Let A be the set of all pairs (w, j), where $w \in \mathcal{R}(0^{k_0} 1^{k_1} \cdots t^{k_t})$ is a word and $j \in \{1, 2, \ldots, s\}$ is an index such that the cyclic rotation $R_j(w)$ is a list of n terms. Combining Lemma 11.45 and the Anagram Rule, we see that $|A| = n\binom{s}{k_0, k_1, \ldots, k_t}$.

Let B be the set of all words $w \in \mathcal{R}(0^{k_0} 1^{k_1} \cdots t^{k_t})$ such that w is a list of n terms (necessarily of length s). To complete the proof, we show $|A| = s|B|$ by exhibiting mutually inverse bijections $f : A \to B \times \{1, 2, \ldots, s\}$ and $g : B \times \{1, 2, \ldots, s\} \to A$. We define $f(w, j) = (R_j(w), j)$ for all $(w, j) \in A$, and $g(w, i) = (R_i^{-1}(w), i)$ for all $(w, i) \in B \times \{1, 2, \ldots, s\}$. □

11.12 Compositional Inversion

Earlier in the chapter, we studied multiplicative inverses of formal power series $F \in K[[z]]$ such that $F(0) \neq 0$. In this section, we study compositional inverses of formal power series $G \in K[[z]]$ such that $G(0) = 0$. We find algebraic and combinatorial formulas for the coefficients in these inverses. We begin by showing that a certain subset of $K[[z]]$ is a group under the operation of formal composition.

11.47. Theorem: Group Axioms for Formal Composition. The set

$$\mathcal{G} = \{F \in K[[z]] : F|_{z^0} = 0 \text{ and } F|_{z^1} \neq 0\}$$

is a group under the operation $\circ$ (composition of formal power series).

Proof. We check the group axioms in Definition 7.1. For closure, fix $F, G \in \mathcal{G}$. Write $F = \sum_{n=0}^{\infty} a_n z^n$ and $G = \sum_{n=0}^{\infty} b_n z^n$ where $a_0 = 0 = b_0$ and $a_1 \neq 0 \neq b_1$. By Definition 11.22, $F \circ G = \sum_{n=1}^{\infty} a_n G^n$. Since $G(0) = 0$, each summand $a_n G^n$ has constant term zero, so $(F \circ G)(0) = 0$. On the other hand, the only summand $a_n G^n$ that has a nonzero coefficient of z^1 is $a_1 G^1$. We see that $(F \circ G)|_{z^1} = (a_1 G)|_{z^1} = a_1 b_1$, which is nonzero since a_1 and b_1 are nonzero. Thus $F \circ G$ is in $\mathcal{G}$, so the closure axiom holds. We have seen in Example 11.23 that $z \in \mathcal{G}$ is a two-sided identity element relative to composition: $z \circ F = F = F \circ z$ for all $F \in \mathcal{G}$. Theorem 11.24(c) shows that $\circ$ is associative: $F \circ (G \circ H) = (F \circ G) \circ H$ for all $F, G, H \in \mathcal{G}$.

We must also verify the inverse axiom: for all $F \in \mathcal{G}$, there exists $G \in \mathcal{G}$ with $G \circ F = z = F \circ G$. Fix $F = \sum_{n=1}^{\infty} a_n z^n \in \mathcal{G}$; we first prove there exists a unique $G = \sum_{m=1}^{\infty} b_m z^m \in \mathcal{G}$ solving the equation $G \circ F = z$. For each $n \in \mathbb{Z}_{\geq 0}$, the coefficient of z^n in $G \circ F$ is

$$(G \circ F)|_{z^n} = \left.\left(\sum_{m=1}^{n} b_m F^m\right)\right|_{z^n}.$$

We show there is a unique choice of the sequence of scalars $(b_m : m \geq 1)$ that makes this coefficient equal 1 for $n = 1$ and 0 otherwise. When $n = 1$, we need $1 = b_1 F^1|_{z^1} = b_1 a_1$. We know $a_1 \neq 0$ in K, so this equation is satisfied iff $b_1 = 1/a_1$, which is a nonzero element of the field K.

Now fix an integer $n > 1$, and assume we have already found unique $b_1, \ldots, b_{n-1} \in K$ making the coefficients of z^k in $G \circ F$ and z agree for all $k < n$. Since $(b_n F^n)|_{z^n} = b_n a_1^n$, we need to choose b_n so that

$$0 = \left.\left(\sum_{m=1}^{n} b_m F^m\right)\right|_{z^n} = b_n a_1^n + \left.\left(\sum_{m=1}^{n-1} b_m F^m\right)\right|_{z^n}.$$

There is a unique $b_n \in K$ that works, namely

$$b_n = -\frac{1}{a_1^n} \left.\left(\sum_{m=1}^{n-1} b_m F^m\right)\right|_{z^n}. \tag{11.12}$$

We have now found a unique $G = \sum_{m=1}^{\infty} b_m z^m$ in $\mathcal{G}$ such that $G \circ F = z$. We call G the *left inverse* of F in $\mathcal{G}$.

To finish, we show that $G \circ F = z$ automatically implies $F \circ G = z$. We have shown that every element of $\mathcal{G}$ has a left inverse. So let $H \in \mathcal{G}$ be the left inverse of G, which satsifies $H \circ G = z$. Using the associativity and identity axioms, we compute

$$H = H \circ z = H \circ (G \circ F) = (H \circ G) \circ F = z \circ F = F.$$

Since $H = F$ and $H \circ G = z$, we conclude $F \circ G = z$, as needed. □

Our proof of the inverse axiom contains a recursive formula for finding the coefficients in the compositional inverse of $F \in \mathcal{G}$. Our next goal is to find other combinatorial and algebraic formulas for these coefficients that can sometimes be more convenient to use. Given $F \in \mathcal{G}$, write $F = zF^*$, where $F^* = f_1 + f_2 z + f_3 z^2 + \cdots$ is a formal series with nonzero constant term. By Theorem 11.11, the series F^* has a multiplicative inverse $R = \sum_{n=0}^{\infty} r_n z^n$ with $r_0 \neq 0$. Writing $F(z) = z/R(z)$, we have $F \circ G = z$ iff $F(G(z)) = z$ iff $G(z)/R(G(z)) = z$ iff $G(z) = zR(G(z))$ iff $G = z \cdot (R \circ G)$.

It turns out that we can solve the equation $G = z(R \circ G)$ by taking G to be the generating function for the set of ordered trees (or equivalently, terms) relative to a certain weight function. This idea is the essence of the following combinatorial formula for G.

11.48. Theorem: Combinatorial Compositional Inversion Formula. Let $F(z) = z/R(z)$ where $R(z) = \sum_{n=0}^{\infty} r_n z^n$ is a given series in $K[[z]]$ with $r_0 \neq 0$. Let T be the set of all terms, and let the *weight* of a term $w = w_1 w_2 \cdots w_s \in T$ be $\text{wt}(w) = r_{w_1} r_{w_2} \cdots r_{w_s} z^s$. Then $G(z) = \text{GF}(T) = \sum_{w \in T} \text{wt}(w)$ is the compositional inverse of $F(z)$.

Proof. For any two words v and w, we have $\text{wt}(vw) = \text{wt}(v)\,\text{wt}(w)$. Also $G(0) = 0$, since every term has positive length. By Theorem 11.40, we know that for every term $w \in T$, there exist a unique integer $n \geq 0$ and unique terms $t_1, \ldots, t_n \in T$ such that $w = nt_1 t_2 \cdots t_n$. For fixed n, we build such a term by choosing the symbol n (which has weight zr_n), then choosing terms $t_1 \in T$, $t_2 \in T$, ..., $t_n \in T$. By the Product Rule for Weighted Sets (adapted to weights satisfying the multiplicative condition $\text{wt}(vw) = \text{wt}(v)\,\text{wt}(w)$), the generating function for terms starting with n is therefore $zr_n G(z)^n$. By the Infinite Sum Rule for Weighted Sets (§11.2), we conclude that

$$G(z) = \sum_{n=0}^{\infty} zr_n G(z)^n = zR(G(z)),$$

so $G = z(R \circ G)$. By the remarks preceding the theorem, this shows that $F \circ G = z$, as needed. Recall that $G \circ F = z$ automatically follows, as in the proof of Theorem 11.47. □

Theorem 11.46 provides a formula counting all terms in a given anagram class $\mathcal{R}(0^{k_0} 1^{k_1} 2^{k_2} \cdots)$. Combining this formula with the previous result, we deduce the following algebraic recipe for the coefficients of G.

11.49. The Lagrange Inversion Formula. Let $F(z) = z/R(z)$ where $R(z) = \sum_{n=0}^{\infty} r_n z^n$ is a given series in $K[[z]]$ with $r_0 \neq 0$. Let G be the compositional inverse of F. For all $n \in \mathbb{Z}_{\geq 1}$,

$$G(z)|_{z^n} = \frac{1}{n} R(z)^n|_{z^{n-1}} = \frac{1}{n!} \left[\left(\frac{d}{dz} \right)^{n-1} R(z)^n \right] \Bigg|_{z^0}.$$

Proof. The second equality follows routinely from the definition of formal differentiation. To prove the first equality, let T_n be the set of terms of length n. By Theorem 11.48, we know that

$$G(z)|_{z^n} = \sum_{w \in T_n} r_{w_1} r_{w_2} \cdots r_{w_n}.$$

Let us group together summands on the right side corresponding to terms of length n that contain k_0 zeroes, k_1 ones, etc., where $\sum_{i\geq 0} k_i = n$. Each such term has weight $z^n r_0^{k_0} r_1^{k_1} \cdots$, and the number of such terms is $\frac{1}{n}\binom{n}{k_0,k_1,\ldots}$, provided that $k_0 = 1 + \sum_{i\geq 1}(i-1)k_i$ (see Theorem 11.46). Summing over all possible choices of the k_i, we get

$$G(z)|_{z^n} = \sum_{\substack{k_0+k_1+k_2+\cdots=n,\\ k_0=1+0k_1+1k_2+2k_3+\cdots}} \frac{1}{n}\binom{n}{k_0,k_1,\ldots}\prod_{i\geq 0} r_i^{k_i}.$$

In the presence of the condition $\sum_{i\geq 0} k_i = n$, the equation $k_0 = 1+\sum_{i\geq 1}(i-1)k_i$ holds iff $\sum_{i\geq 0}(i-1)k_i = -1$ iff $\sum_{i\geq 0} ik_i = n-1$. So

$$G(z)|_{z^n} = \sum_{\substack{k_0+k_1+k_2+\cdots=n,\\ 0k_0+1k_1+2k_2+\cdots=n-1}} \frac{1}{n}\binom{n}{k_0,k_1,\ldots}\prod_{i\geq 0} r_i^{k_i}.$$

On the other hand, Theorem 5.35 shows that

$$\frac{1}{n}R(z)^n|_{z^{n-1}} = \frac{1}{n}\sum_{i_1+i_2+\cdots+i_n=n-1} r_{i_1}r_{i_2}\cdots r_{i_n}.$$

Each summand containing k_0 copies of r_0, k_1 copies of r_1, etc., can be rearranged to $\prod_{i\geq 0} r_i^{k_i}$, and there are $\binom{n}{k_0,k_1,\ldots}$ such summands by the Anagram Rule. So this formula for $\frac{1}{n}R(z)^n|_{z^{n-1}}$ reduces to the previous formula for $G(z)|_{z^n}$. □

11.50. Example. Given $F(z) = z/e^z$, let us use Theorem 11.49 to find the compositional inverse G of F. Here $R(z) = e^z = \sum_{k=0}^{\infty} z^k/k!$, and $R(z)^n = e^{nz} = \sum_{k=0}^{\infty}(n^k/k!)z^k$. So

$$G(z)|_{z^n} = \frac{1}{n}R(z)^n|_{z^{n-1}} = \frac{n^{n-1}}{n\cdot(n-1)!} = \frac{n^{n-1}}{n!},$$

and $G(z) = \sum_{n=1}^{\infty}(n^{n-1}/n!)z^n$.

Summary

- **Algebraic Operations on Formal Power Series.** $K[[z]]$ is the set of sequences $(a_n : n \in \mathbb{Z}_{\geq 0})$ with all $a_n \in K$. Given $F = \sum_{n=0}^{\infty} a_n z^n$ and $G = \sum_{n=0}^{\infty} b_n z^n$ in $K[[z]]$, we define:
  (a) *Equality.* $F = G$ iff $a_n = b_n$ for all $n \in \mathbb{Z}_{\geq 0}$.
  (b) *Addition.* $F + G = \sum_{n=0}^{\infty}(a_n + b_n)z^n$.
  (c) *Multiplication.* $F \cdot G = \sum_{n=0}^{\infty}\left(\sum_{k=0}^{n} a_k b_{n-k}\right) z^n$.
  (d) *Coefficient Extraction.* For $n \geq 0$, $F|_{z^n} = a_n$ and $F(0) = a_0$.
  (e) *Order.* If $F \neq 0$, $\operatorname{ord}(F)$ is the least n with $a_n \neq 0$.
  (f) *Formal Differentiation.* $F' = \sum_{n=1}^{\infty} na_n z^{n-1} = \sum_{m=0}^{\infty}(m+1)a_{m+1}z^m$.
  (g) *Formal Composition.* If $G(0) = 0$, $F \circ G = \sum_{n=0}^{\infty} a_n G^n$.
  (h) *Formal Exponentiation.* If $G(0) = 0$, $\exp(G) = e^G = \sum_{n=0}^{\infty} G^n/n!$.
  (i) *Formal Logarithm.* if $F(0) = 1$, $\log(F) = \sum_{n=1}^{\infty}(-1)^{n-1}(F-1)^n/n$.
- **Limit Operations on Formal Power Series.** Let F and F_m (for $m \in \mathbb{Z}_{\geq 0}$) be formal

power series in $K[[z]]$.
(a) *Limits.* $(F_m) \to F$ iff $\lim_{m\to\infty} F_m = F$ iff

$$\forall k \in \mathbb{Z}_{\geq 0}, \exists M \in \mathbb{Z}_{\geq 0}, \forall m \geq M, F_m|_{z^k} = F|_{z^k}.$$

(b) *Infinite Sums.* $\sum_{m=0}^{\infty} F_m = F$ iff $\lim_{N\to\infty} \sum_{m=0}^{N} F_m = F$.
If all $F_m \neq 0$, the infinite sum converges iff $\lim_{m\to\infty} \operatorname{ord}(F_m) = \infty$.
(c) *Infinite Products.* $\prod_{m=0}^{\infty} F_m = F$ iff $\lim_{N\to\infty} \prod_{m=0}^{N} F_m = F$.
If all $F_m \neq 0$, $\prod_{m=0}^{\infty}(1 + F_m)$ converges iff some $F_m = -1$ or $\lim_{m\to\infty} \operatorname{ord}(F_m) = \infty$.

- **Formal Continuity.** A function p mapping a subset of $K[[z]]$ into $K[[z]]$ is *continuous* iff for all F_m, F in the domain of p, $(F_m) \to F$ implies $(p(F_m)) \to p(F)$. If the domain of p is a subset of $K[[z]] \times K[[z]]$, continuity means that whenever $(F_m) \to F$ and $(G_m) \to G$, $(p(F_m, G_m)) \to p(F, G)$. Formal addition, multiplication, coefficient extraction, differentiation, composition, exponentiation, and logarithm are all continuous. The composition of continuous functions is continuous.

- **The Infinite Sum Rule for Weighted Sets.** Suppose S_k is a nonempty weighted set for each $k \in \mathbb{Z}_{\geq 0}$, $S_k \cap S_j = \emptyset$ for all $j \neq k$, and $S = \bigcup_{k=0}^{\infty} S_k$. Assume that for all k and all $u \in S_k$, $\operatorname{wt}_S(u) = \operatorname{wt}_{S_k}(u)$. For each k, let $\operatorname{minwt}(S_k) = \min\{\operatorname{wt}_{S_k}(u) : u \in S_k\}$. If $\lim_{k\to\infty} \operatorname{minwt}(S_k) = \infty$, then $\operatorname{GF}(S; z) = \sum_{k=0}^{\infty} \operatorname{GF}(S_k; z)$.

- **The Infinite Product Rule for Weighted Sets.** For each $k \in \mathbb{Z}_{\geq 1}$, let S_k be a weighted set that contains a unique object o_k of weight zero. Assume $\lim_{k\to\infty} \operatorname{minwt}(S_k - \{o_k\}) = \infty$. Suppose S is a weighted set such that every $u \in S$ can be constructed in exactly one way as follows. For each $k \geq 1$, choose $u_k \in S_k$, subject to the restriction that we must choose o_k for all but finitely many k's. Then assemble the chosen objects in a prescribed manner. Assume that whenever u is constructed from $(u_k : k \geq 0)$, $\operatorname{wt}_S(u) = \sum_{k=1}^{\infty} \operatorname{wt}_{S_k}(u_k)$. Then $\operatorname{GF}(S; z) = \prod_{k=1}^{\infty} \operatorname{GF}(S_k; z)$.

- **Multiplicative Inverses of Formal Series.** $F = \sum_{n=0}^{\infty} a_n z^n \in K[[z]]$ is invertible iff there exists $G = \sum_{n=0}^{\infty} u_n z^n \in K[[z]]$ with $FG = 1$ iff $a_0 \neq 0$.
(a) *Recursion for Coefficients of* $1/F$. When $a_0 \neq 0$, the coefficients of G are determined recursively by $u_0 = a_0^{-1}$ and $u_n = -a_0^{-1} \sum_{k=1}^{n} a_k u_{n-k}$ for all $n > 0$.
(b) *Closed Formula for Coefficients of* $1/F$. When $a_0 \neq 0$,

$$u_n = F^{-1}|_{z^n} = \sum_{m=0}^{n} \frac{(-1)^m}{a_0^{m+1}} \sum_{\substack{(i_1,i_2,\ldots,i_m)\in\mathbb{Z}_{>0}^m:\\ i_1+i_2+\cdots+i_m=n}} a_{i_1} a_{i_2} \cdots a_{i_m}.$$

(c) *Formal Geometric Series.* When $a_0 = 0$, $(1 - F)^{-1} = \sum_{m=0}^{\infty} F^m$.
(d) *Symmetric Function Formula for* $1/F$. $H(z) = \sum_{n=0}^{\infty}(-1)^n h_n z^n$ is the multiplicative inverse of $E(z) = \sum_{n=0}^{\infty} e_n z^n$, where e_n and h_n denote elementary and complete symmetric functions. If $\phi : \Lambda \to K$ is the homomorphism sending e_n to a_n, then $u_n = \phi((-1)^n h_n)$.
(e) *Exponential Formula for* $1/F$. When $a_0 = 1$, $1/F = \exp(-\log(F))$.

- **Partial Fractions.** Given polynomials $f, g \in \mathbb{C}[z]$ with $g(0) = 1$, let $g = \prod_{i=1}^{k}(1 - r_i z)^{n_i}$ where the r_i are distinct nonzero complex numbers. There exist a unique polynomial

$h \in \mathbb{C}[z]$ and unique complex numbers a_{ij} (for $1 \le i \le k$ and $1 \le j \le n_i$) with $f/g = h + \sum_{i=1}^{k}\sum_{j=1}^{n_i} a_{ij}/(1-r_i z)^j$. The coefficient of z^n in the power series expansion of f/g is $(f/g)|_{z^n} = h|_{z^n} + \sum_{i=1}^{k}\sum_{j=1}^{n_i} a_{ij}\binom{n+j-1}{n,j-1} r_i^n$.

- **Recursions with Constant Coefficients.** Given constants $c_j, d_i \in \mathbb{C}$ and $g : \mathbb{Z}_{\ge k} \to \mathbb{C}$, let $(a_n : n \ge 0)$ be defined recursively by $a_n = c_1 a_{n-1} + c_2 a_{n-2} + \cdots + c_k a_{n-k} + g(n)$ for $n \ge k$, with initial conditions $a_i = d_i$ for $0 \le i < k$. The generating function $F = \sum_{n=0}^{\infty} a_n z^n$ is given by $F = G/P$, where $P = 1 - c_1 z - c_2 z^2 - \cdots - c_k z^k$, $G = \sum_{i=0}^{k-1} d_i' z^i + \sum_{n=k}^{\infty} g(n) z^n$, and $d_i' = d_i - c_1 d_{i-1} - c_2 d_{i-2} - \cdots - c_i d_0$.

- **Properties of Formal Composition.** For fixed $G \in K[[z]]$ with $G(0) = 0$, the map $R_G : K[[z]] \to K[[z]]$ given by $R_G(F) = F \circ G$ is the unique continuous K-algebra homomorphism sending z to G. The set $\mathcal{G} = \{F \in K[[z]] : F|_{z^0} = 0 \text{ and } F|_{z^1} \neq 0\}$ is a group under formal composition. In particular, formal composition is associative when defined, with identity element z.

- **Properties of Formal Derivatives.** For all $F, G \in K[[z]]$:
  (a) *The Sum Rule.* $(F+G)' = F' + G'$.
  (b) *The Scalar Rule.* For all $c \in K$, $(cF)' = c(F')$.
  (c) *The Product Rule.* $(F \cdot G)' = (F') \cdot G + F \cdot (G')$.
  (d) *The Power Rule.* For all $n \in \mathbb{Z}_{\ge 0}$, $(F^n)' = nF^{n-1} \cdot F'$.
  (e) *The Chain Rule.* If $G(0) = 0$, then $\frac{d}{dz}[F(G(z))] = F'(G(z)) \cdot G'(z)$.

- **Properties of Formal Exponentials.** Let F, G, and G_k (for $k \in \mathbb{Z}_{>0}$) be formal power series with constant term zero.
  (a) $\frac{d}{dz}\exp(z) = \exp(z)$, $\exp(0) = 1$, and $\frac{d}{dz}\exp(F(z)) = \exp(F(z))F'(z)$.
  (b) $\exp(F) \neq 0$, and $\exp(-F) = 1/\exp(F)$.
  (c) $\exp(F+G) = \exp(F)\exp(G)$.
  (d) For all $N \in \mathbb{Z}_{>0}$, $\exp(\sum_{k=1}^{N} G_k) = \prod_{k=1}^{N} \exp(G_k)$.
  (e) If $\sum_{k=1}^{\infty} G_k$ converges, then $\prod_{k=1}^{\infty} \exp(G_k)$ converges to $\exp(\sum_{k=1}^{\infty} G_k)$.

- **Properties of Formal Logarithms.** Let F, G, H, and H_k (for $k \in \mathbb{Z}_{>0}$) be formal power series with $F(0) = 0$ and $G(0) = H(0) = H_k(0) = 1$.
  (a) $\frac{d}{dz}\log(1+z) = (1+z)^{-1}$, $\log(1) = 0$, $\frac{d}{dz}\log(1+F(z)) = F'(z)/(1+F(z))$, and $\frac{d}{dz}\log(G(z)) = G'(z)/G(z)$.
  (b) $\log(\exp(F)) = F$ and $\exp(\log(H)) = H$.
  (c) $\log(GH) = \log(G) + \log(H)$.
  (d) For all $N \in \mathbb{Z}_{>0}$, $\log(\prod_{k=1}^{N} H_k) = \sum_{k=1}^{N} \log(H_k)$.
  (e) If $\prod_{k=1}^{\infty} H_k$ converges, then $\sum_{k=1}^{\infty} \log(H_k)$ converges to $\log(\prod_{k=1}^{\infty} H_k)$.

- **The Exponential Formula.** Let C and S be sets such that each object x in C or S has a *weight* $\text{wt}(x)$ and a *size* $\text{sz}(x)$, where $\text{sz}(c) > 0$ for all $c \in C$. Suppose that every object $s \in S$ of size $n \ge 0$ can be constructed uniquely by the following process. First, choose a set partition P of $\{1, 2, \ldots, n\}$. Next, for each block B in the set partition P, choose a connected structure $c(B)$ from C such that $\text{sz}(c(B)) = |B|$. Finally, assemble these choices in a prescribed manner to produce s. Assume that whenever s is built from P and $(c(B) : B \in P)$, $\text{wt}(s) = \sum_{B \in P} \text{wt}(c(B))$. Define generating functions $G_C = \sum_{c \in C} t^{\text{wt}(c)} z^{\text{sz}(c)}/\text{sz}(c)!$ and $G_S = \sum_{s \in S} t^{\text{wt}(s)} z^{\text{sz}(s)}/\text{sz}(s)!$. Then $G_S = \exp(G_C)$.

- **Terms and Ordered Trees.** For every term T, there exist a unique integer $n \ge 0$ and unique terms $T_1, \ldots, T_n$ such that $T = nT_1T_2\cdots T_n$. A word $w_1 \cdots w_s$ is a term iff $w_1 + \cdots + w_i - i \ge 0$ for all $i < s$ and $w_1 + \cdots + w_s - s = -1$. No proper prefix of a term is a term. Terms correspond bijectively to ordered trees.

- **Lists of Terms and Ordered Forests.** Every list of terms has the form $T_1 \cdots T_n$ for some unique integer $n \geq 0$ and unique terms $T_1, \ldots, T_n$. A word $w_1 \cdots w_s$ is a list of n terms iff $w_1 + \cdots + w_i - i > -n$ for all $i < s$ and $w_1 + \cdots + w_s - s = -n$. Lists of terms correspond bijectively to ordered forests.

- **The Cycle Lemma for Counting Lists of Terms.** For a word $w = w_1 \cdots w_s$ with $w_1 + \cdots + w_s - s = -n$, there exist exactly n cyclic shifts of w that are lists of n terms. Consequently, the number of lists of n terms using k_i copies of i (for $0 \leq i \leq t$) is $\frac{n}{s}\binom{s}{k_0, k_1, \ldots, k_s}$, where $s = \sum_{i=0}^{t} k_i$ and $k_0 = n + \sum_{i=1}^{t}(i-1)k_i$.

- **Compositional Inverses of Formal Series.** Given $F = \sum_{n=0}^{\infty} a_n z^n$ with $a_0 = 0 \neq a_1$, F has a compositional inverse $G = \sum_{n=0}^{\infty} b_n z^n$ with coefficients given recursively by $b_0 = 0$, $b_1 = 1/a_1$, and $b_n = (-1/a_1^n)\left[\sum_{m=1}^{n-1} b_m F^m\right]\Big|_{z^n}$. Writing $F = x/R$ where $R = \sum_{n=0}^{\infty} r_n z^n \in K[[z]]$ and $r_0 \neq 0$, G is also the generating function for the set of terms, where the weight of a term $w_1 \cdots w_n$ is $z^n r_{w_1} \cdots r_{w_n}$. For all $n \geq 1$, $b_n = \frac{1}{n}(R(z)^n)|_{z^{n-1}} = \frac{1}{n!}\left[(d/dz)^{n-1} R(z)^n\right]\Big|_{z^0}$.

Exercises

11-1. Let $f = z - z^2 + 3z^4$ and $g = 1 - 2z - 3z^4$. Compute $f + g$, fg, and the degrees and orders of f, g, $f + g$, and fg.

11-2. Let $F = (1, 0, 1, 0, 1, 0, \ldots)$ and $G = \sum_{n=0}^{\infty} nz^n$. Compute $F + G$, FG, $F(1 + z)$, $F(1 - z^2)$, $G(1 + z)$, F', G', and the orders of these formal power series.

11-3. Prove that $K[[z]]$ is a commutative ring, a vector space over K, and a K-algebra by verifying the axioms.

11-4. Prove that $K[z]$ is a subring, subspace, and subalgebra of $K[[z]]$.

11-5. (a) Prove Theorem 11.2. (b) Deduce that $K[z]$ and $K[[z]]$ have no zero divisors.

11-6. When does equality hold in the formulas $\deg(P + Q) \leq \max(\deg(P), \deg(Q))$ and $\text{ord}(F + G) \geq \min(\text{ord}(F), \text{ord}(G))$ from Theorem 11.2?

11-7. Given nonzero $F_k \in K[[z]]$, prove $(F_k) \to 0$ in $K[[z]]$ iff $(\text{ord}(F_k)) \to \infty$.

11-8. Fix $G \in K[[z]]$. Prove $(G^m) \to 0$ iff $G = 0$ or $\text{ord}(G) > 0$.

11-9. Suppose $(F_m) \to F$ with $F_m, F \in K[[z]]$. Prove: for each $k \geq 0$, there exists $M \geq 0$ such that for all $m \geq M$ and all i in the range $0 \leq i \leq k$, $F_m|_{z^i} = F|_{z^i}$.

11-10. Finish the proof of Theorem 11.5.

11-11. Let $G = (b_n : n \geq 0) \in K[[z]]$ and define $F_m = b_m z^m \in K[[z]]$ for each $m \geq 0$. Prove $\sum_{m=0}^{\infty} F_m = G$.

11-12. Finish the proof of Theorem 11.7.

11-13. Given a sequence (F_n) of formal series, let (F_{k_n}) be the subsequence of nonzero terms of (F_n). (a) Prove $\sum_{n=0}^{\infty} F_n$ converges to the sum G iff $\sum_{n=0}^{\infty} F_{k_n}$ converges to the sum G. (b) Prove $\prod_{n=0}^{\infty}(1 + F_n)$ converges to the product G iff $\prod_{n=0}^{\infty}(1 + F_{k_n})$ converges to the product G.

11-14. Density of $K[z]$ in $K[[z]]$. Show that for all $F \in K[[z]]$, there exists a sequence of polynomials $P_n \in K[z]$ with $(P_n) \to F$.

11-15. For fixed $m, n \in \mathbb{Z}_{>0}$, evaluate $\sum \frac{m!}{k_0!0!^{k_0} k_1!1!^{k_1} \cdots k_n!n!^{k_n}}$, where the sum extends over all lists $(k_0, k_1, \ldots, k_n) \in \mathbb{Z}_{\geq 0}^{n+1}$ such that $k_0+k_1+\cdots+k_n = m$ and $0k_0+1k_1+\cdots+nk_n = n$.

11-16. Carefully justify the following calculation:

$$\prod_{n=1}^{\infty}(1-z^{2n-1})^{-1} = \prod_{i=1}^{\infty}(1-z^{2i})\prod_{j=1}^{\infty}(1-z^j)^{-1} = \prod_{k=1}^{\infty}(1+z^k).$$

In particular, explain why all the infinite products appearing here converge.

11-17. Carefully check that the infinite products at the end of §9.28 converge to the indicated formal series.

11-18. Find a necessary and sufficient condition on series $F_k \in K[[z]]$ so that the infinite product $\prod_{k=1}^{\infty}(1+F_k)^{-1}$ exists.

11-19. Evaluate $\prod_{n=0}^{\infty}(1+z^{2^n})$.

11-20. Ideal Structure of $K[z]$. Show that every nonzero ideal I in the ring $K[z]$ has the form $\langle P \rangle = \{PQ : Q \in K[z]\}$ for some monic polynomial $P \in K[z]$. [*Hint:* Let P be a monic polynomial in I of least degree. Use polynomial division with remainder to show $I = \langle P \rangle$.]

11-21. Ideal Structure of $K[[z]]$. Show that every nonzero ideal I in the ring $K[[z]]$ has the form $\langle z^m \rangle = \{z^m G : G \in K[[z]]\} = \{\sum_{n=m}^{\infty} a_n z^n : a_n \in K\}$ for some $m \in \mathbb{Z}_{\geq 0}$. Draw a diagram of the poset of all ideals of $K[[z]]$ ordered by set inclusion.

11-22. Formal Laurent Series. A *formal Laurent series* is a sequence $F = (a_n : n \in \mathbb{Z})$, denoted $F(z) = \sum_{n=-\infty}^{\infty} a_n z^n$, such that all a_n are in K, and for some $d \in \mathbb{Z}$, $a_n = 0_K$ for all $n < d$. Let $K((z))$ be the set of all such formal Laurent series. Define a K-algebra structure on $K((z))$ by analogy with $K[[z]]$, and verify the algebra axioms.

11-23. Prove that the Laurent series ring $K((z))$ is a field containing $K[[z]]$. Prove that every $F \in K((z))$ has the form $F = GH^{-1}$ for some $G, H \in K[[z]]$; in fact, H can be chosen to be a power of z.

11-24. Compute the multiplicative inverse of $\sum_{n=0}^{\infty} n^2 z^n$ in $K((z))$.

11-25. Convert the following expressions to formal Laurent series: (a) $(z^2+3)/(z^3-z^2)$; (b) $z/(z^3-5z^2+6z)$.

11-26. Differentiation of Laurent Series. Define a version of the formal derivative operator for the ring $K((z))$ of formal Laurent series. Extend the derivative rules (in particular, the Quotient Rule) to this ring.

11-27. Carefully check the final equality in the proof of the Infinite Sum Rule for Weighted Sets.

11-28. Generalize the Infinite Sum Rule 11.9 to the case where finitely many of the sets S_k have multiple objects of weight zero.

11-29. Prove: for all $n \in \mathbb{Z}_{>1}$ and all $k \in \mathbb{Z}_n$, k is invertible in the ring $\mathbb{Z}_n$ iff $\gcd(k,n) = 1$.

11-30. Check that the sequence (u_n) defined below (11.1) is the unique solution to the system (11.1).

11-31. Solve the system (11.1) to find the first five terms in the multiplicative inverse of each of the following series: (a) e^z; (b) $1-2z+z^3+3z^4$; (c) $1+\log(1+z)$.

11-32. Use the Formal Geometric Series Formula to find the first five terms in $(1-z+z^3)^{-1}$.

11-33. Use (11.1) to find the first nine coefficients in the multiplicative inverse of the formal series $\cos z = 1 - z^2/2! + z^4/4! - \cdots$.

11-34. Suppose $F(z) = 2 - 6z + 3z^2 + 5z^3 - z^4 + \cdots$. Say as much as you can about the coefficients of the power series $1/F$.

11-35. Suppose $F \in K[[z]]$ is a formal series such that $\operatorname{ord}(F) = 0$ and there is $k \in \mathbb{Z}_{>0}$ such that $F|_{z^n} = 0$ whenever n is not divisible by k. Show that $1/F$ also has this property.

11-36. Use geometric series to find the inverses of these formal series: (a) $1-3z$; (b) $1+z^3$; (c) $a - bz$ where $a, b \neq 0$.

11-37. Prove by induction on m: for all $G \in K[[z]]$ and all $m \in \mathbb{Z}_{\geq 0}$, $(1-G)\sum_{k=0}^{m} G^k = 1 - G^{m+1}$.

11-38. Derive (11.3) from (11.1).

11-39. Check that ϕ^* defined in (11.4) is a K-algebra homomorphism.

11-40. (a) Prove: for all $n \in \mathbb{Z}_{\geq 0}$,

$$(-1)^n h_n = \sum_{m=0}^{n} (-1)^m \sum_{\substack{(i_1, i_2, \ldots, i_m) \in \mathbb{Z}_{>0}^m: \\ i_1 + i_2 + \cdots + i_m = n}} e_{i_1} e_{i_2} \cdots e_{i_m}.$$

(b) Use part (a) to show that the symmetric function formula for the multiplicative inverse of a formal power series agrees with Formula (11.3).

11-41. Find the partial fraction decomposition and the power series expansion of each ratio of polynomials.

(a) $F(z) = (10 + 2z)/(1 - 2z - 8z^2)$.
(b) $F(z) = (1 - 7z)/(15z^2 - 8z + 1)$.
(c) $F(z) = (2z^3 - 4z^2 - z - 3)/(2z^2 - 4z + 2)$.
(d) $F(z) = (15z^6 + 30z^5 - 15z^4 - 35z^3 - 15z^2 - 12z - 8)/(15(z^4 + 2z^3 - 2z - 1))$.

11-42. Evaluation Homomorphisms for Polynomial Rings. Let A be any K-algebra. Prove that for all $a \in A$, there exists a unique K-algebra homomorphism $E_a : K[z] \to A$ such that $E_a(z) = a$. This map is called the *evaluation homomorphism determined by a*.

11-43. Prove Theorem 11.24(b). [*Hint:* One approach is to use the previous exercise and 11.24(a).]

11-44. Even and Odd Formal Series. A series $F \in K[[z]]$ is *even* iff $F(-z) = F$; F is *odd* iff $F(-z) = -F$. (a) Show that F is even iff $F|_{z^n} = 0$ for all odd n, and F is odd iff $F|_{z^n} = 0$ for all even n. (b) Give rules for determining the parity (even or odd) of $F + G$, FG, and (when defined) F^{-1}, given the parity of F and G.

11-45. (a) Show that the differentiation map $D : K[[z]] \to K[[z]]$, given by $D(F) = F'$ for $F \in K[[z]]$, is continuous and K-linear. (b) For fixed $H \in K[[z]]$, show that the map $M_H : K[[z]] \to K[[z]]$, given by $M(F) = F \cdot H$ for $F \in K[[z]]$, is continuous and K-linear.

11-46. Prove: if $F = \sum_{k=0}^{\infty} F_k$ in $K[[z]]$, then $F' = \sum_{k=0}^{\infty} F_k'$.

11-47. Given K-algebras A, B, C and functions $p : A \to B$ and $g : B \to C$. (a) Prove that if p and q are K-linear, then $q \circ p : A \to C$ is K-linear. (b) Prove that if p and q are K-algebra homomorphisms, then $q \circ p$ is a K-algebra homomorphism.

11-48. Uniqueness of Evaluation Homomorphisms. Prove: Given $G \in K[[z]]$ with $G(0) = 0$, if $h : K[[z]] \to K[[z]]$ is any continuous K-algebra homomorphism such that $h(z) = G$, then $h(F) = F \circ G$ for all $F \in K[[z]]$ (so that $h = R_G$).

11-49. Complete the following outline to give a new proof of the Formal Product Rule $(F \cdot G)' = F' \cdot G + F \cdot G'$ for $F, G \in K[[z]]$. (a) Show that the result holds when $F = z^i$ and $G = z^j$, for all $i, j \in \mathbb{Z}_{\geq 0}$. (b) Deduce from (a) that the result holds for all $F, G \in K[z]$. (c) Use a continuity argument to obtain the result for all $F, G \in K[[z]]$.

11-50. Verify that the maps p and q used in the proof of the Formal Chain Rule (Theorem 11.25(e)) are K-linear and continuous.

11-51. Suppose $p, q : K[[z]] \to K[[z]]$ are continuous K-linear maps such that $p(z^n) = q(z^n)$ for all $n \in \mathbb{Z}_{\geq 0}$. Prove that $p = q$.

11-52. The Formal Quotient Rule. Suppose $F, G \in K[[x]]$ where $G(0) \neq 0$. Prove the derivative rule $(F/G)' = (G \cdot F' - F \cdot G')/G^2$.

11-53. Formal Integrals. The *formal integral* or *antiderivative* of a formal series $F = \sum_{n=0}^{\infty} a_n z^n \in K[[z]]$ is the formal series

$$\int F\,dz = \sum_{n=1}^{\infty} \frac{a_{n-1}}{n} z^n \in K[[z]],$$

which has constant term zero. Compute the formal integrals of the following formal power series: (a) $3+2z-7z^2+12z^5$; (b) $\sum_{n=0}^{\infty} n^2 z^n$; (c) $\sum_{n=0}^{\infty}(n+1)!z^n$; (d) e^z; (e) $\sin z$; (f) $\cos z$; (g) $(1+z)^{-1}$; (h) $\frac{3+2z}{1-3z+2z^2}$.

11-54. Prove the following facts about formal integrals.

(a) The Sum Rule: $\int F + G\,dz = \int F\,dz + \int G\,dz$ for all $F, G \in K[[z]]$.

(b) The Scalar Rule: $\int cF\,dz = c\int F\,dz$ for all $c \in K$ and $F \in K[[z]]$.

(c) The Linearity Rule: $\int \sum_{i=1}^{n} c_i H_i\,dz = \sum_{i=1}^{n} c_i \int H_i\,dz$ for $c_i \in K$ and $H_i \in K[[z]]$. Can you formulate a similar statement for infinite sums?

(d) The Power Rule: $\int z^k\,dz = \frac{1}{k+1} z^{k+1}$ for all $k \geq 0$.

(e) General Antiderivatives: For all $F, G \in K[[z]]$, $G' = F$ iff there exists $c \in K$ with $G = \int F\,dz + c$.

(f) The Formal Fundamental Theorems of Calculus: For all $F \in K[[z]]$, $F = \frac{d}{dz}\int F\,dz$ and $\int F'\,dz = F - F(0)$.

(g) Continuity of Integration: For all $F_k, H \in K[[z]]$, if $(F_k) \to H$ then $(\int F_k\,dz) \to \int H\,dz$.

11-55. Formulate and prove an Integration by Parts Rule and a Substitution Rule for formal integrals.

11-56. (a) Prove that $(\sin z)^2 + (\cos z)^2 = 1$ in $K[[z]]$ by computing the coefficient of z^n on each side. (b) Prove that $(\sin z)^2 + (\cos z)^2 = 1$ in $K[[z]]$ by showing the derivative of the left side is zero.

11-57. The Product Rule for Multiple Factors. Let $F_1, \ldots, F_k \in K[[z]]$. Prove that

$$\frac{d}{dz}(F_1 F_2 \cdots F_k) = \sum_{j=1}^{k} F_1 \cdots F_{j-1}\left(\frac{d}{dz}F_j\right) F_{j+1} \cdots F_k.$$

Does a version of this rule hold for infinite products?

11-58. Use evaluation homomorphisms to show that when $F \in K[[z]]$ is a polynomial, we can define $F \circ G$ for all $G \in K[[z]]$, not just those G with $G(0) = 0$. Extend the results of §11.6 to this setting.

11-59. Compute the first four nonzero terms in: (a) $\exp(\sin x)$; (b) $\log(\cos x)$.

11-60. Suppose $F, G \in K[[z]]$ satisfy $F(0) = 0 = G(0)$. Prove $\exp(F+G) = \exp(F)\exp(G)$ by computing the coefficient of z^n on both sides.

11-61. Prove Theorem 11.27 parts (d), (e), and (f).

11-62. Prove Theorem 11.29 parts (d), (e), and (f).

11-63. Let $X = \{F \in K[[z]] : F(0) = 0\}$ and $Y = \{G \in K[[z]] : G(0) = 1\}$. Prove that $\exp : X \to Y$ is a bijection with inverse $\log : Y \to X$.

11-64. Formal Powers. Given $F \in K[[z]]$ with $F(0) = 1$ and $r \in K[[z]]$, define the formal power $F^r = \exp(r\log(F))$. Prove the following facts. (a) F^r is well-defined, and $F^r(0) = 1$.

(b) For $F, r, s \in K[[z]]$ with $F(0) = 1$, $F^{r+s} = F^r F^s$. (c) For $F, r \in K[[z]]$ with $F(0) = 1$, $F^{-r} = 1/F^r$. (d) For $F, G, r \in K[[z]]$ with $F(0) = G(0) = 1$, $(FG)^r = F^r G^r$. (e) For $F, r \in K[[z]]$ with $F(0) = 1$, $\log(F^r) = r\log(F)$. (f) $F^0 = 1$; if n is a positive integer, F^n (as defined here) equals the product of n copies of F; and if n is a negative integer, F^n (as defined here) equals the product of $|n|$ copies of $1/F$. (g) For $F, r \in K[[z]]$ with $F(0) = 1$, $\frac{d}{dz}F^r = rF^{r-1} \cdot F'$. (h) The operation sending (F, r) to F^r is continuous.

11-65. Prove a formal version of the Extended Binomial Theorem: for all $r \in K[[z]]$, $(1+z)^r = \sum_{n=0}^{\infty} \frac{(r)\downarrow n}{n!} z^n$, where $(1+z)^r$ is defined in the previous exercise.

11-66. Formal nth Roots. Given $F \in K[[z]]$ with $F(0) = 1$ and a positive integer n, show there exists a unique $G \in K[[z]]$ with $G(0) = 1$ and $G^n = F$, where G^n is the product of n copies of G.

11-67. State and prove a formal version of the Quadratic Formula for solving $AF^2 + BF + C = 0$, where $A, B, C \in K[[z]]$ are known series and $F \in K[[z]]$ is unknown. What hypotheses must you impose on A, B, C? Is the solution F unique?

11-68. (a) Find the generating function for the set of set partitions of one of the sets $\{1, 2, \ldots, n\}$ where every block has more than one element, weighted by number of blocks. (b) Find the number of such set partitions of $\{1, 2, \ldots, 9\}$. (c) Find the number of such set partitions of $\{1, 2, \ldots, 12\}$ with four blocks.

11-69. (a) Find the generating function for the set of set partitions of one of the sets $\{1, 2, \ldots, n\}$ where all blocks have even size, weighted by number of blocks. (b) Find the number of such set partitions of $\{1, 2, \ldots, 12\}$. (c) Find the number of such set partitions of $\{1, 2, \ldots, 16\}$ consisting of four blocks.

11-70. (a) Find the generating function for the set of permutations of one of the sets $\{1, 2, \ldots, n\}$ where no cycle has length more than five, weighted by number of cycles. (b) Find the number of such permutations of $\{1, 2, \ldots, 10\}$. (c) Find the number of such permutations of $\{1, 2, \ldots, 12\}$ having four cycles.

11-71. For $1 \leq n \leq 6$, find the number of connected simple graphs with n vertices.

11-72. (a) Find the generating function for connected simple digraphs with vertex set $\{1, 2, \ldots, n\}$ for some $n \geq 1$ (see Definition 3.50). (b) Use (a) to find a summation formula counting connected simple digraphs with n vertices. (c) For $1 \leq n \leq 7$, find the number of connected simple digraphs with n vertices.

11-73. Carry out the computations showing how the equation $G_C = \log(G_S)$ leads to formula (11.11) in Example 11.33.

11-74. Rewrite (11.11) as a sum over partitions of n.

11-75. (a) Modify (11.11) to include a power of t that keeps track of the number of edges in the connected graph. (b) How many connected simple graphs with vertex set $\{1, 2, 3, 4, 5, 6\}$ have exactly seven edges?

11-76. Let S be the set of simple graphs with vertex set $\{1, 2, \ldots, n\}$ for some $n \geq 0$ such that every component of the graph is an undirected cycle, weighted by number of components. (a) Find G_S. (b) Find the number of such graphs with 12 vertices and 5 components.

11-77. A *star* is a tree with at most one vertex of degree greater than 1. Let S be the set of simple graphs with vertex set $\{1, 2, \ldots, n\}$ for some $n \geq 0$ such that every component of the graph is a star, weighted by number of components. (a) Find G_S. (b) Find the number of such graphs with 14 vertices and 4 components.

11-78. Verify that $\frac{1}{n}(R(z)^n)|_{z^{n-1}} = \frac{1}{n!}(d/dz)^{n-1}R(z)^n|_{z^0}$ for all $n \in \mathbb{Z}_{\geq 1}$ and $R \in K[[z]]$.

11-79. Use (11.12) to find the first several coefficients in the compositional inverses of each of the following power series: (a) $\sin z$; (b) $\tan z$; (c) $z/(1-z)$.

11-80. (a) List all terms of length at most 5. (b) Use (a) and Theorem 11.48 to find explicit formulas for the first five coefficients of the compositional inverse of $z/R(z)$ as combinations of the coefficients of $R(z)$. (c) Use (b) to find the first five terms in the compositional inverse of $z/(1-3z+2z^2+5z^4)$.

11-81. Use Theorem 11.49 to compute the compositional inverse of the following formal series: (a) ze^{2z}; (b) $z-z^2$; (c) $z/(1+az)$; (d) $z-4z^4+4z^7$.

11-82. (a) Find a bijection from the set of terms of length n to the set of binary trees with $n-1$ nodes. (b) Use (a) to formulate a version of Theorem 11.48 that expresses the coefficients in the compositional inverse of $z/R(z)$ as sums of weighted binary trees.

11-83. (a) Find a bijection from the set of terms of length n to the set of Dyck paths ending at $(n-1, n-1)$. (b) Use (a) to formulate a version of Theorem 11.48 that expresses the coefficients in the compositional inverse of $z/R(z)$ as sums of weighted Dyck paths.

11-84. Let $A(n,k)$ be the number of ways to assign n people to k committees in such a way that each person belongs to exactly one committee, and each committee has one member designated as chairman. Find a formula for $\sum_{n=0}^{\infty}\sum_{k=0}^{n}\frac{A(n,k)}{n!}t^k z^n$.

11-85. Formal Linear Ordinary Differential Equations (ODEs). Suppose $P, Q \in K[[z]]$ are given formal series, and we are trying to find a formal series $F \in K[[z]]$ satisfying the linear ODE $F' + PF = Q$ and initial condition $F(0) = c \in K$. Solve this ODE by multiplying by the integrating factor $\exp(\int P\,dz)$ and using the Product Rule to simplify the left side.

11-86. Formal ODEs with Constant Coefficients. Let V be the set of all formal series $F \in \mathbb{C}[[z]]$ satisfying the ODE

$$F^{(k)} + c_1F^{(k-1)} + c_2F^{(k-2)} + \cdots + c_kF = 0. \tag{11.13}$$

The *characteristic polynomial* for this ODE is $q(z) = z^k + c_1z^{k-1} + c_2z^{k-2} + \cdots + c_k \in \mathbb{C}[z]$. Suppose $q(z)$ factors as $(z-r_1)^{k_1}\cdots(z-r_s)^{k_s}$ for certain $k_i > 0$ and distinct $r_i \in \mathbb{C}$. (a) Show that the k series $z^j\exp(r_iz)$ (for $1 \le i \le s$ and $0 \le j < k_i$) are in V. (b) Show that V is a complex vector space, and the k series in (a) form a basis for V. (c) Describe a procedure for expressing a given sequence $F \in V$ as a linear combination of the sequences in the basis from part (a), given the *initial conditions* $F(0), F'(0), \ldots, F^{(k-1)}(0)$. (d) Let W be the set of formal series $G \in \mathbb{C}[[z]]$ satisfying the non-homogeneous ODE

$$G^{(k)} + c_1G^{(k-1)} + c_2G^{(k-2)} + \cdots + c_kG = H,$$

where $H \in \mathbb{C}[[z]]$ is a given series. If G^* is one particular series in W, show that $W = \{F + G^* : F \in V\}$.

Notes

A detailed but very technical treatment of the algebraic theory of polynomials and formal power series is given in [18, Ch. IV]. Discussions of formal power series from a more combinatorial perspective may be found in [121, Ch. 1] and [132]. Our treatment of compositional inversion closely follows the presentation in [102].

12

Additional Topics

This chapter covers a variety of topics illustrating different aspects of enumerative combinatorics and probability. The treatment of each topic is essentially self-contained.

12.1 Cyclic Shifting of Paths

This section illustrates a technique for enumerating certain collections of lattice paths. The basic idea is to introduce an equivalence relation on paths by cyclically shifting the steps of a path. A similar idea was used in §11.11 to enumerate lists of terms.

12.1. Theorem: Rational-Slope Dyck Paths. Let r and s be positive integers such that $\gcd(r,s)=1$. The number of lattice paths from $(0,0)$ to (r,s) that never go below the diagonal line $sx=ry$ is

$$\frac{1}{r+s}\binom{r+s}{r,s}.$$

We call the paths counted by this theorem *Dyck paths of slope s/r*.

Proof. Step 1. Let $X=\mathcal{R}(\mathrm{E}^r\mathrm{N}^s)$, which is the set of all rearrangements of r copies of E and s copies of N. Thinking of E as an east step and N as a north step, we see that X can be identified with the set of all lattice paths from $(0,0)$ to (r,s). Given $v=v_1v_2\cdots v_{r+s}\in X$, we define an associated *label vector* $L(v)=(m_0,m_1,\ldots,m_{r+s})$ as follows. We set $m_0=0$. Then we recursively calculate $m_i=m_{i-1}+r$ if $v_i=\mathrm{N}$, and $m_i=m_{i-1}-s$ if $v_i=\mathrm{E}$. For example, if $r=5$, $s=3$, and $v=\mathrm{NEEENENE}$, then $L(v)=(0,5,2,-1,-4,1,-2,3,0)$. We can also describe this construction in terms of the lattice path encoded by v. If we label each lattice point (x,y) on this path by the integer $ry-sx$, then $L(v)$ is the sequence of labels encountered as we traverse the path from $(0,0)$ to (r,s). This construction is illustrated by the lattice paths in Figure 12.1. Note that v is recoverable from $L(v)$, since $v_i=\mathrm{N}$ iff $m_i-m_{i-1}=r$, and $v_i=\mathrm{E}$ iff $m_i-m_{i-1}=-s$.

Step 2. We prove that for all $v\in X$, if $L(v)=(m_0,m_1,\ldots,m_{r+s})$ then m_0, m_1, …, m_{r+s-1} are *distinct*, whereas $m_{r+s}=0=m_0$. To see this, suppose there exist x,y,a,b with $0<a\leq r$ and $0<b\leq s$, such that (x,y) and $(x+a,y+b)$ are two points on the lattice path for v that have the same label. This means that $ry-sx=r(y+b)-s(x+a)$, which simplifies to $rb=sa$. Thus the number $rb=sa$ is a common multiple of r and s. Since $\gcd(r,s)=1$, we have $\mathrm{lcm}(r,s)=rs$, so that $rb\geq rs$ and $sa\geq rs$. Thus $b\geq s$ and $a\geq r$, forcing $b=s$ and $a=r$. But then (x,y) must be $(0,0)$ and $(x+a,y+b)$ must be (r,s). So the only two points on the path with equal labels are $(0,0)$ and (r,s), which correspond to m_0 and m_{r+s}.

Step 3. Introduce an equivalence relation $\sim$ on X (see Definition 2.50) by setting $v\sim w$ iff v is a cyclic shift of w. More precisely, defining $C(w_1w_2\cdots w_{r+s})=w_2w_3\cdots w_{r+s}w_1$, we have $v\sim w$ iff $v=C^i(w)$ for some integer i (which can be chosen in the range $0\leq i<r+s$).

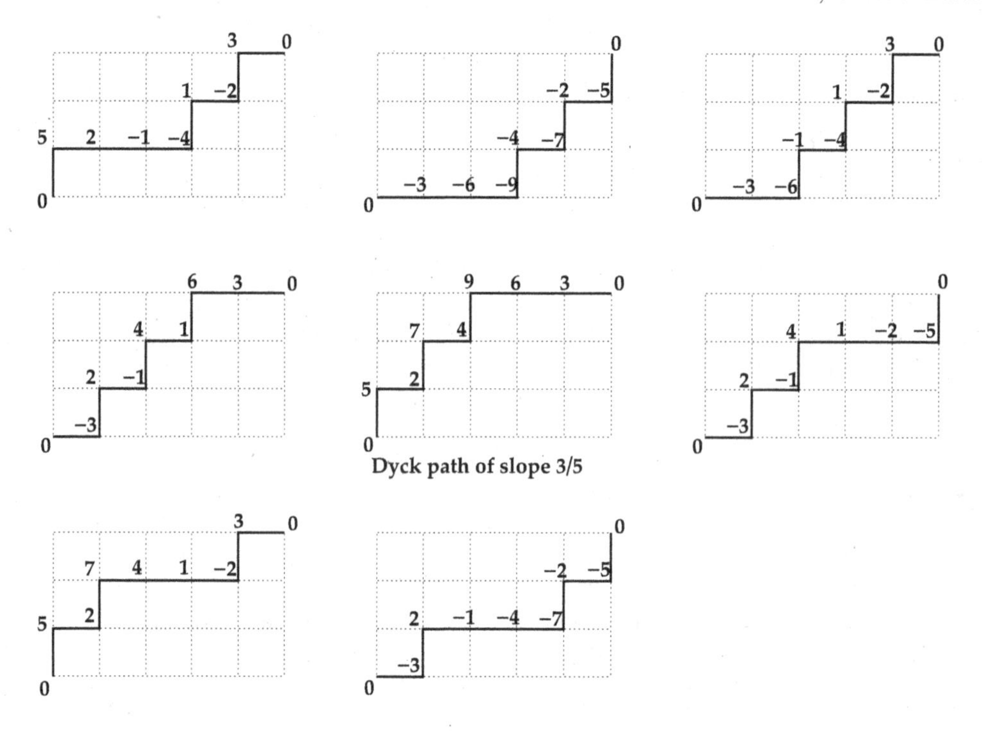

FIGURE 12.1
Cyclic shifts of a lattice path.

For each $v \in X$, let $[v] = \{w \in X : w \sim v\}$ be the equivalence class of v relative to this equivalence relation. Figure 12.1 shows the paths in the class [NEEENENE].

Step 4. We show that for all $v \in X$, the equivalence class $[v]$ has size $r + s$, which means that all $r + s$ cyclic shifts of v are *distinct*. Suppose $v = v_1 v_2 \cdots v_{r+s}$ has $L(v) = (m_0, m_1, \ldots, m_{r+s})$. By definition of L, for each i with $0 \le i < r+s$, the label vector of the cyclic shift $C^i(v) = v_{i+1} \cdots v_{r+s} v_1 \cdots v_i$ is

$$L(C^i(v)) = (0, m_{i+1} - m_i, m_{i+2} - m_i, \ldots, m_{r+s} - m_i, m_1 - m_i, \ldots, m_i - m_i)$$

(see Figure 12.1 for examples). The set of integers appearing in the label vector $L(C^i(v))$ is therefore obtained from the set of integers in $L(v)$ by subtracting m_i from each integer in the latter set. In particular, if μ is the smallest integer in $L(v)$, then the smallest integer in $L(C^i(v))$ is $\mu - m_i$. Since the numbers $m_0, m_1, \ldots, m_{r+s-1}$ are distinct (by Step 2), we see that the minimum elements in the sequences $L(C^i(v))$ are distinct, as i ranges from 0 to $r+s-1$. This implies that the sequences $L(C^i(v))$, and hence the words $C^i(v)$, are pairwise distinct.

Step 5. We show that, for all $v \in X$, there exists a unique word $w \in [v]$ such that w encodes a Dyck path of slope s/r. By the way we defined the labels, w is a Dyck path of slope s/r iff $L(w)$ has no negative entries. We know from Step 4 that the set of labels in $L(C^i(v))$ is obtained from the set of labels in $L(v)$ by subtracting m_i from each label in the latter set. By Step 2, there is a unique i in the range $0 \le i < r+s$ such that $m_i = \mu$, the minimum value in $L(v)$. For this choice of i, we have $m_j \ge \mu = m_i$ for every j, so that $m_j - m_i \ge 0$ and $L(C^i(v))$ has no negative labels. For any other choice of i, $m_i > \mu$ by Step 2, so that $L(C^i(v))$ contains the negative label $\mu - m_i$.

Step 6. Suppose $\sim$ has n equivalence classes in X. By Step 5, n is also the number of

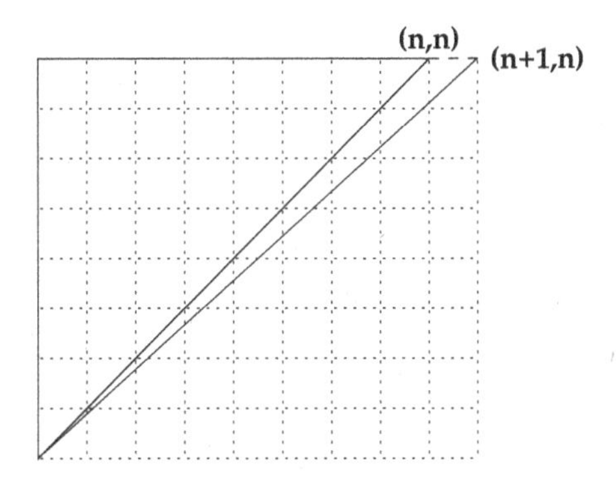

FIGURE 12.2
Comparing Dyck paths to Dyck paths of slope $n/(n+1)$.

Dyck paths of slope s/r. By Step 4, each equivalence class has size $r+s$. Since X is the disjoint union of its equivalence classes, we see from the Sum Rule and the Anagram Rule that

$$\binom{r+s}{r,s} = |X| = n(r+s).$$

Dividing by $r+s$ gives the formula stated in the theorem. □

We can use the previous theorem to enumerate Dyck paths and certain m-ballot numbers, as follows.

12.2. Corollary. For all $n \in \mathbb{Z}_{\geq 1}$, the number of Dyck paths ending at (n,n) is

$$\frac{1}{2n+1}\binom{2n+1}{n+1,n}.$$

For all $m,n \in \mathbb{Z}_{\geq 1}$, the number of lattice paths from $(0,0)$ to (mn,n) that stay weakly above $x = my$ is

$$\frac{1}{(m+1)n+1}\binom{(m+1)n+1}{mn+1,n}.$$

Proof. Let X be the set of Dyck paths ending at (n,n), and let X' be the set of Dyck paths of slope $n/(n+1)$ ending at $(n+1,n)$. Since $\gcd(n+1,n) = 1$, we know that $|X'| = \frac{1}{2n+1}\binom{2n+1}{n+1,n}$. On the other hand, tilting the line $y = x$ to the line $(n+1)y = nx$ does not introduce any new lattice points in the region visited by paths in X and X', except for $(n+1,n)$, as shown in Figure 12.2. It follows that appending a final east step gives a bijection from X onto X', proving the first formula in the corollary. The second formula is proved in the same way: appending a final east step gives a bijection from the set of lattice paths described in the corollary to the set of Dyck paths of slope $n/(mn+1)$. □

12.2 The Chung–Feller Theorem

In §1.14, we defined Dyck paths and proved that the number of Dyck paths of order n is the Catalan number $C_n = \frac{1}{n+1}\binom{2n}{n,n}$. This section discusses a remarkable generalization of this result called the *Chung–Feller Theorem*.

12.3. Definition: Flawed Paths. Suppose π is a lattice path from $(0,0)$ to (n,n), viewed as a sequence of lattice points: $\pi = ((x_0, y_0), \ldots, (x_{2n}, y_{2n}))$. For $1 \leq j \leq n$, we say that π has a *flaw in row* j iff there exists a point (x_i, y_i) visited by π such that $y_i = j-1$, $y_i < x_i$, and $(x_{i+1}, y_{i+1}) = (x_i, y_i + 1)$. This means that the jth north step of π occurs in the region southeast of the diagonal line $y = x$. For $1 \leq j \leq n$, define

$$X_j(\pi) = \begin{cases} 1 & \text{if } \pi \text{ has a flaw in row } j; \\ 0 & \text{otherwise.} \end{cases}$$

Also define the *number of flaws of* π by setting $\text{flaw}(\pi) = X_1(\pi) + X_2(\pi) + \cdots + X_n(\pi)$.

For example, the paths shown in Figure 12.3 have zero and six flaws, respectively. The paths shown in Figure 12.4 have five and zero flaws, respectively. Observe that π is a Dyck path iff $\text{flaw}(\pi) = 0$.

12.4. The Chung–Feller Theorem. Fix $n \in \mathbb{Z}_{\geq 0}$, and let A be the set of lattice paths from $(0,0)$ to (n,n). For $0 \leq k \leq n$, let $A_k = \{\pi \in A : \text{flaw}(\pi) = k\}$. Then $|A_k| = |A_0|$ for all k. In particular, for $0 \leq k \leq n$,

$$|A_k| = \frac{|A|}{n+1} = \frac{1}{n+1}\binom{2n}{n,n} = C_n.$$

Proof. Fix k with $0 < k \leq n$. To prove that $|A_0| = |A_k|$, we define a bijection $\phi_k : A_0 \to A_k$. See Figure 12.3 for an example where $n = 10$ and $k = 6$. Given a Dyck path $\pi \in A_0$, we begin by drawing the line $y = k$ superimposed on the Dyck path. There is a unique point (x_i, y_i) on π such that $y_i = k$ and π arrives at (x_i, y_i) by taking a vertical step. Call this step the *special vertical step*. Let $(a_1, a_2, \ldots) \in \{\text{H}, \text{V}\}^{2n-x_i-y_i}$ be the ordered sequence of steps of π reading northeast from (x_i, y_i), where H means horizontal step and V means vertical step. Let $(b_0, b_1, b_2, \ldots) \in \{\text{H}, \text{V}\}^{x_i+y_i}$ be the ordered sequence of steps of π reading southwest from (x_i, y_i), where $b_0 = \text{V}$ is the special vertical step. For the Dyck path shown on the left in Figure 12.3, we have

$$a_1 a_2 \cdots a_{11} = \text{VHVHHHVHVHH}; \qquad b_0 b_1 \cdots b_9 = \text{VVHVHVVHV}.$$

We compute the lattice path $\phi_k(\pi)$, viewed as a sequence of steps $c_1 c_2 \cdots c_{2n} \in \{\text{V}, \text{H}\}^{2n}$, as follows. Let $c_1 = a_1$, $c_2 = a_2$, etc., until we reach a horizontal step $c_k = a_k$ that ends strictly below the diagonal $y = x$. Then set $c_{k+1} = b_1$, $c_{k+2} = b_2$, etc., until we reach a vertical step $c_{k+m} = b_m$ that ends on the line $y = x$. Then set $c_{k+m+1} = a_{k+1}$, $c_{k+m+2} = a_{k+2}$, etc., until we take a horizontal step that ends strictly below $y = x$. Then switch back to using the steps b_{m+1}, b_{m+2}, etc., until we return to $y = x$. Continue in this way until all steps are used. By convention, the special vertical step $b_0 = \text{V}$ is the last step from the b-sequence to be consumed.

For example, given the path π in Figure 12.3, we have labeled the steps of the path A through T for reference purposes. The special vertical step is step I. We begin by transferring steps J, K, L, M, N to the image path (starting at the origin). Step N goes below the diagonal, so we jump to the section of π prior to the special vertical step and work southwest. After taking only one step (the vertical step labeled H), we have returned to the diagonal. Now we jump back to our previous location in the top part of π and take step O. This again takes us below the diagonal, so we jump back to the bottom part of π and transfer steps G, F, E, D, C. Now we return to the top part and transfer steps P, Q, R, S, T. Then we return to the bottom part of π and transfer steps B, A, and finally the special vertical step I.

This construction has the following crucial property. Vertical steps above the line $y = k$ in π get transferred to vertical steps above the line $y = x$ in $\phi_k(\pi)$, while vertical steps

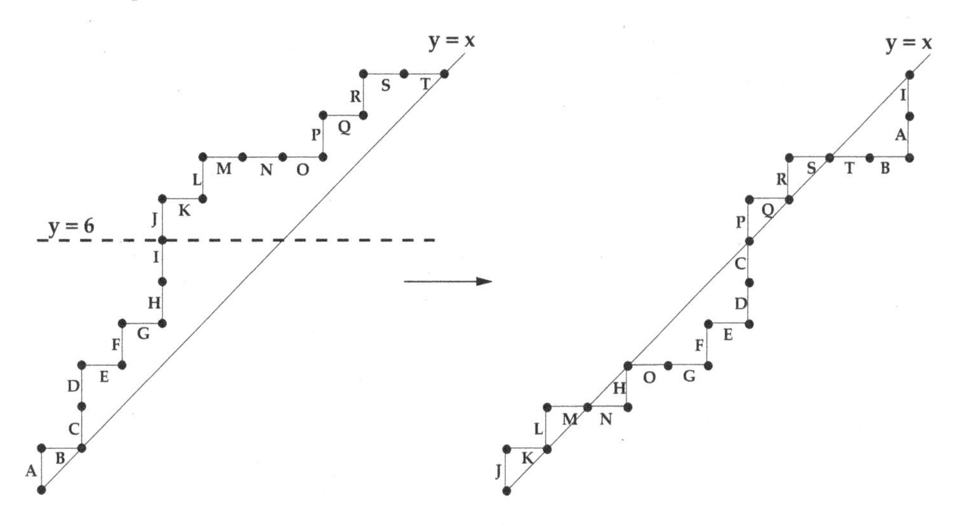

FIGURE 12.3
Mapping Dyck paths to flawed paths.

below the line $y = k$ in π get transferred to vertical steps below the line $y = x$ in $\phi_k(\pi)$. Thus, $\phi_k(\pi)$ has exactly k flaws, so that $\phi_k(\pi)$ is an element of A_k.

Moreover, consider the coordinates of the special point (x_i, y_i). By definition, $y_i = k = \text{flaw}(\phi_k(\pi))$. On the other hand, we claim that $y_i - x_i$ equals the number of horizontal steps in $\phi_k(\pi)$ that start on $y = x$ and end to the right of $y = x$. Each such horizontal step came from a step northeast of (x_i, y_i) in π that brings the path π closer to the main diagonal $y = x$. For instance, these steps are N, O, and T in Figure 12.3. The definition of ϕ_k shows that the steps in question (in π) are the earliest east steps after (x_i, y_i) that arrive on the lines $y = x + d$ for $d = y_i - x_i - 1, \ldots, 2, 1, 0$. The number of such steps is therefore $y_i - x_i$ as claimed.

The observations in the last paragraph allow us to compute the inverse map $\phi'_k : A_k \to A_0$. For, suppose $\pi^* \in A_k$ is a path with k flaws. We can recover (x_i, y_i) since $y_i = k$ and $y_i - x_i$ is the number of east steps of π^* departing from $y = x$. Next, we transfer the steps of π^* to the top and bottom portions of $\phi'_k(\pi^*)$ by reversing the process described earlier. Figure 12.4 gives an example where $n = 10$ and $k = 5$. First we find the special point $(x_i, y_i) = (2, 5)$. We start by transferring the initial steps A, B, C of π^* to the part of the image path starting at $(2, 5)$ and moving northeast. Since C goes below the diagonal in π^*, we now switch to the bottom part of the image path. The special vertical step must be skipped, so we work southwest from $(2, 4)$. We transfer steps D, E, F, G, H. Since H returns to $y = x$ in π^*, we then switch back to the top part of the image path. We only get to transfer one step (step I) before returning to the bottom part of the image path. We transfer step J, then move back to the top part and transfer steps K through S. Finally, step T is transferred to become the special vertical step from $(2, 4)$ to $(2, 5)$. It can be checked that ϕ'_k is the two-sided inverse of ϕ_k, so $\phi_k : A_0 \to A_k$ is a bijection.

We now know that $|A_k| = |A_0|$ for all k between 0 and n. Note that A is the disjoint union of the $n+1$ sets $A_0, A_1, \ldots, A_n$, all of which have cardinality $|A_0|$. By the Sum Rule,

$$|A| = |A_0| + |A_1| + \cdots + |A_n| = (n+1)|A_0|.$$

For each k between 0 and n, the Anagram Rule gives

$$|A_k| = |A_0| = \frac{|A|}{n+1} = \frac{1}{n+1}\binom{2n}{n, n} = C_n. \qquad \square$$

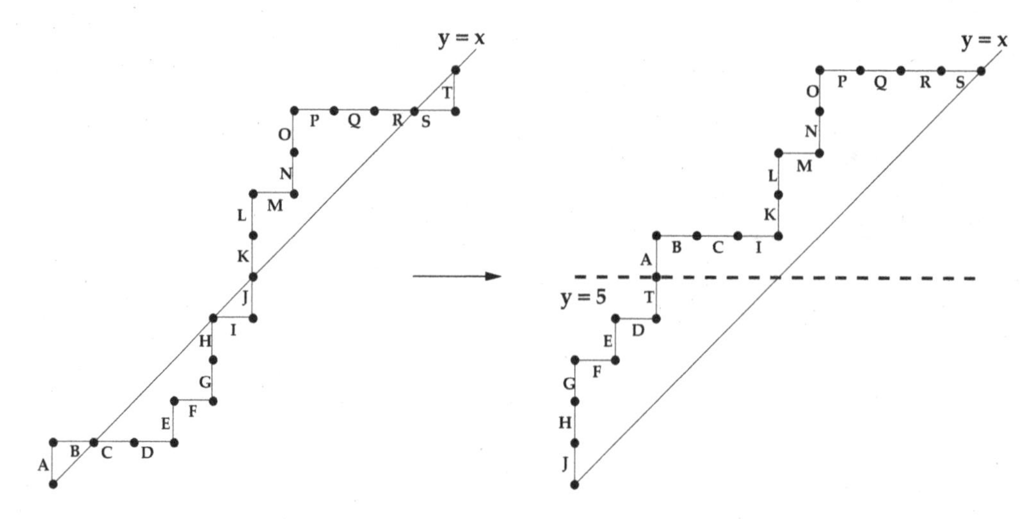

FIGURE 12.4
Mapping flawed paths to Dyck paths.

In probabilistic language, the Chung–Feller Theorem can be stated as follows.

12.5. Corollary. Suppose we pick a random lattice path π from the origin to (n, n). The number of flaws in this path is uniformly distributed on $\{0, 1, 2, \ldots, n\}$. In other words,

$$P(\text{flaw}(\pi) = k) = \frac{1}{n+1} \qquad \text{for all } k \text{ between } 0 \text{ and } n.$$

Proof. Given k with $0 \leq k \leq n$, we compute

$$P(\text{flaw}(\pi) = k) = \frac{|A_k|}{|A|} = \frac{\frac{1}{n+1}\binom{2n}{n,n}}{\binom{2n}{n,n}} = \frac{1}{n+1}. \qquad \square$$

12.6. Remark. The Chung–Feller Theorem is significant in probability theory for the following reason. One of the major theorems of probability is the *Central Limit Theorem.* Roughly speaking, this theorem says that the sum of a large number of independent, identically distributed random variables (appropriately normalized) converges to a *normal distribution.* The normal distribution is described by the bell curve that appears ubiquitously in probability and statistics. One often deals with situations involving random variables that are *not* identically distributed and are *not* independent of one another. One might hope that a generalization of the Central Limit Theorem would still hold in such situations.

Chung and Feller used the example of flawed lattice paths to show that such a generalization is not always possible. Fix $n > 0$, and let the sample space S consist of all lattice paths from the origin to (n, n). Given a lattice path $\pi \in S$, recall that

$$\text{flaw}(\pi) = X_1(\pi) + X_2(\pi) + \cdots + X_n(\pi),$$

where $X_j(\pi)$ is 1 if π has a flaw in row j and 0 otherwise. The random variables $X_1, X_2, \ldots, X_n$ are identically distributed; in fact, $P(X_j = 0) = 1/2 = P(X_j = 1)$ for all j (see Exercise 12-7). But we have seen that the sum of these random variables, namely the flaw statistic $X_1 + X_2 + \cdots + X_n$, is uniformly distributed on $\{0, 1, 2, \ldots, n\}$ for every n. A uniform distribution is about as far as we can get from a normal distribution! The trouble is that the random variables $X_1, \ldots, X_n$ are not independent.

12.3 Rook-Equivalence of Ferrers Boards

This section continues the investigation of rook theory begun in §2.14. We first define the concepts of Ferrers boards and rook polynomials. Then we derive a characterization of when two Ferrers boards have the same rook polynomial.

12.7. Definition: Ferrers Boards and Rook Polynomials. Let $\mu = (\mu_1 \geq \mu_2 \geq \cdots \geq \mu_s > 0)$ be an integer partition of n. The *Ferrers board* F_μ is a diagram consisting of s left-justified rows of squares with μ_i squares in row i. A *non-attacking placement* of k rooks on F_μ is a subset of k squares in F_μ such that no two squares lie in the same row or column. For all $k \in \mathbb{Z}_{\geq 0}$, let $r_k(\mu)$ be the number of non-attacking placements of k rooks on F_μ. The *rook polynomial* of μ is

$$R_\mu(x) = \sum_{k=0}^{n} r_k(\mu)x^k.$$

12.8. Example. If $\mu = (4,1,1,1)$, then $R_\mu(x) = 9x^2+7x+1$. To see this, note that there is one empty subset of F_μ, which is a non-attacking placement of zero rooks. We can place one rook on any of the seven squares in F_μ, so the coefficient of x^1 in $R_\mu(x)$ is 7. To place two non-attacking rooks, we place one rook in the first column but not in the first row (three ways), and we place the second rook in the first row but not in the first column (three ways). The Product Rule gives 9 as the coefficient of x^2 in $R_\mu(x)$. It is impossible to place three or more non-attacking rooks on F_μ, so all higher coefficients in $R_\mu(x)$ are zero.

As seen in the previous example, the constant term in any rook polynomial is 1, whereas the linear coefficient of a rook polynomial is the number $|\mu|$ of squares on the board F_μ. Furthermore, $R_\mu(x)$ has degree at most $\min(\mu_1, \ell(\mu))$, since all rooks must be placed in distinct rows and columns of the board.

It is possible for two different partitions to have the same rook polynomial. For example, it can be checked that

$$R_{(2,2)}(x) = 2x^2 + 4x + 1 = R_{(3,1)}(x) = R_{(2,1,1)}(x).$$

More generally, $R_\mu(x) = R_{\mu'}(x)$ for any partition μ.

12.9. Definition: Rook-Equivalence. We say that two integer partitions μ and ν are *rook-equivalent* iff they have the same rook polynomial, which means $r_k(\mu) = r_k(\nu)$ for all integers $k \geq 0$.

A necessary condition for μ and ν to be rook equivalent is that $|\mu| = |\nu|$. The next theorem gives an easily tested necessary and sufficient criterion for deciding whether two partitions are rook-equivalent.

12.10. Theorem: Rook-Equivalence of Ferrers Boards. Suppose μ and ν are partitions of n. Write $\mu = (\mu_1 \geq \cdots \geq \mu_n)$ and $\nu = (\nu_1 \geq \ldots \geq \nu_n)$ by adding zero parts if necessary. The rook polynomials $R_\mu(x)$ and $R_\nu(x)$ are equal iff the multisets

$$[\mu_1 + 1, \mu_2 + 2, \ldots, \mu_n + n] \quad \text{and} \quad [\nu_1 + 1, \nu_2 + 2, \ldots, \nu_n + n]$$

are equal.

Proof. The idea of the proof is to use modified versions of the rook polynomials that involve linear combinations of the falling factorial polynomials $(x){\downarrow_n}$ instead of the monomials x^n (see Definition 2.63). For any partition λ, define the *falling rook polynomial*

$$R^*_\lambda(x) = \sum_{k=0}^{n} r_{n-k}(\lambda)(x){\downarrow_k} = \sum_{k=0}^{n} r_{n-k}(\lambda)x(x-1)(x-2)\cdots(x-k+1).$$

Note that $R_\mu(x) = R_\nu(x)$ iff $r_k(\mu) = r_k(\nu)$ for all k between 0 and n (by linear independence of the monomial basis) iff $R^*_\mu(x) = R^*_\nu(x)$ (by linear independence of the falling factorial basis). We now prove that $R^*_\mu(x) = R^*_\nu(x)$ iff the multisets mentioned in the theorem are equal.

First we use rook combinatorics to prove a factorization formula for $R^*_\mu(x)$. Fix a positive integer x. Consider the extended board $F_\mu(x)$, which has $\mu_i + x$ squares in row i, for $1 \leq i \leq n$. We obtain $F_\mu(x)$ from the board F_μ by adding x new squares on the left end of each of the n rows. Let us count the number of placements of n non-attacking rooks on $F_\mu(x)$. On one hand, we can build such a placement by working up the rows from bottom to top, placing a rook in a valid column of each successive row. For $j \geq 0$, when we place the rook in the $(j+1)$th row from the bottom, there are $x + \mu_{n-j}$ columns in this row, but we must avoid the j distinct columns that already have rooks in lower rows. By the Product Rule, the number of valid placements is

$$(x+\mu_n)(x+\mu_{n-1}-1)\cdots(x+\mu_1-(n-1)) = \prod_{i=1}^{n}(x+[\mu_i-(n-i)]).$$

On the other hand, let us count the number of placements of n non-attacking rooks on $F_\mu(x)$ that have exactly k rooks on the original board F_μ. We can place these rooks first in $r_k(\mu)$ ways. The remaining $n-k$ rooks must go in the remaining $n-k$ unused rows in one of the leftmost x squares. Placing these rooks one at a time, we obtain $r_k(\mu)x(x-1)(x-2)\cdots(x-(n-k-1))$ valid placements. Adding over k gives the identity

$$\sum_{k=0}^{n} r_k(\mu)(x){\downarrow_{n-k}} = \prod_{i=1}^{n}(x+[\mu_i-(n-i)]). \tag{12.1}$$

Replacing k by $n-k$ in the summation, we find that

$$R^*_\mu(x) = \prod_{i=1}^{n}(x+[\mu_i-(n-i)]).$$

This polynomial identity holds for infinitely many values of x (namely, for each positive integer x), so the identity must hold in the polynomial ring $\mathbb{R}[x]$. Similarly,

$$R^*_\nu(x) = \prod_{i=1}^{n}(x+[\nu_i-(n-i)]).$$

The proof is now completed by invoking the uniqueness of prime factorizations for one-variable polynomials with real coefficients. More precisely, note that we have exhibited factorizations of $R^*_\mu(x)$ and $R^*_\nu(x)$ into products of linear factors. These two monic polynomials are equal iff their linear factors (counting multiplicities) are the same, which holds iff the multisets

$$[\mu_i-(n-i) : 1 \leq i \leq n] \quad \text{and} \quad [\nu_i-(n-i) : 1 \leq i \leq n]$$

are the same. Adding n to everything, this is equivalent to the multiset equality in the theorem statement. □

12.11. Example. The partitions $(2, 2, 0, 0)$ and $(3, 1, 0, 0)$ are rook-equivalent, because $[3, 4, 3, 4] = [4, 3, 3, 4]$. The partitions $(4, 2, 1)$ and $(5, 2)$ are not rook-equivalent, since $[5, 4, 4, 4, 5, 6, 7] \neq [6, 4, 3, 4, 5, 6, 7]$.

12.4 Parking Functions

This section defines combinatorial objects called parking functions. We then count parking functions using a probabilistic argument.

12.12. Definition: Parking Functions. For $n \in \mathbb{Z}_{>0}$, a *parking function of order n* is a function $f : \{1, 2, \ldots, n\} \to \{1, 2, \ldots, n\}$ such that for each i in the range $1 \leq i \leq n$, there are at least i inputs x satisfying $f(x) \leq i$.

12.13. Example. For $n = 8$, the function f defined by

$$f(1) = 2,\ f(2) = 6,\ f(3) = 3,\ f(4) = 2,\ f(5) = 6,\ f(6) = 2,\ f(7) = 2,\ f(8) = 1$$

is a parking function. The function g defined by

$$g(1) = 5,\ g(2) = 6,\ g(3) = 1,\ g(4) = 5,\ g(5) = 6,\ g(6) = 1,\ g(7) = 7,\ g(8) = 1$$

is not a parking function because $g(x) \leq 4$ is true for only three values of x, namely $x = 3, 6, 8$.

Here is the reason these functions are called "parking functions." Consider a one-way street with n parking spaces numbered $1, 2, \ldots, n$. Cars numbered $1, 2, \ldots, n$ arrive at the beginning of this street in numerical order. Each car wants to park in its own preferred spot on the street. We encode these parking preferences by a function $h : \{1, 2, \ldots, n\} \to \{1, 2, \ldots, n\}$, by letting $h(x)$ be the parking spot preferred by car x. Given h, the cars park in the following way. For $x = 1, 2, \ldots, n$, car x arrives and drives forward along the street to the spot $h(x)$. If that spot is empty, car x parks there. Otherwise, the car continues to drive forward on the one-way street and parks in the first available spot after $h(x)$, if any. The cars cannot return to the beginning of the street, so it is possible that not every car will be able to park.

For example, suppose the parking preferences are given by the parking function f defined in Example 12.13. Car 1 arrives first and parks in spot 2. Cars 2 and 3 arrive and park in spots 6 and 3, respectively. When car 4 arrives, spots 2 and 3 are full, so car 4 parks in spot 4. This process continues. At the end, every car has parked successfully, and the parking order is $8, 1, 3, 4, 6, 2, 5, 7$. Now suppose the parking preferences are given by the non-parking function g from Example 12.13. After the first six cars have arrived, the parking spots on the street are filled as follows:

$$3, 6, -, -, 1, 2, 4, 5.$$

Car 7 arrives and drives to spot $g(7) = 7$. Since spots 7 and 8 are both full at this point, car 7 cannot park.

12.14. Lemma. A function $h : \{1, 2, \ldots, n\} \to \{1, 2, \ldots, n\}$ is a parking function iff every car is able to park using the parking preferences determined by h.

Proof. We prove the contrapositive in each direction. Suppose first that h is not a parking function. Then there exists $i \in \{1, 2, \ldots, n\}$ such that $h(x) \leq i$ holds for fewer than i choices of x. This means that fewer than i cars prefer to park in the first i spots. But then the first i spots cannot all be used, since a car never parks in a spot prior to the spot it prefers. Since there are n cars and n spots, the existence of an unused spot implies that not every car can park.

Conversely, assume not every car can park. Let i be the earliest spot that is not taken after every car has attempted to park. Then no car preferred spot i. Suppose i or more cars preferred the first $i-1$ spots. Not all of these cars can park in the first $i-1$ spots. But then one of these cars would have parked in spot i, a contradiction. We conclude that fewer than i cars preferred one of the first i spots, so that $h(x) \leq i$ is true for fewer than i choices of x. This means that h is not a parking function. □

12.15. The Parking Function Rule. For all $n \in \mathbb{Z}_{>0}$, there are $(n+1)^{n-1}$ parking functions of order n.

Proof. Fix $n \in \mathbb{Z}_{>0}$. Define a *circular parking function* of order n to be any function $f : \{1, 2, \ldots, n\} \to \{1, 2, \ldots, n+1\}$. Let Z be the set of all such functions; by the Function Rule, $|Z| = (n+1)^n$. We interpret circular parking functions as follows. Imagine a roundabout (circular street) with $n+1$ parking spots numbered $1, 2, \ldots, n+1$. See Figure 12.5. As before, f encodes the parking preferences of n cars that wish to park on the roundabout. Thus, for $1 \leq x \leq n$ and $1 \leq y \leq n+1$, $y = f(x)$ iff car x prefers to park in spot y. Cars $1, 2, \ldots, n$ arrive at the roundabout in increasing order. Each car x enters just before spot 1, then drives around to spot $f(x)$ and parks there if possible. If spot $f(x)$ is full, car x keeps driving around the roundabout and parks in the first empty spot that it encounters.

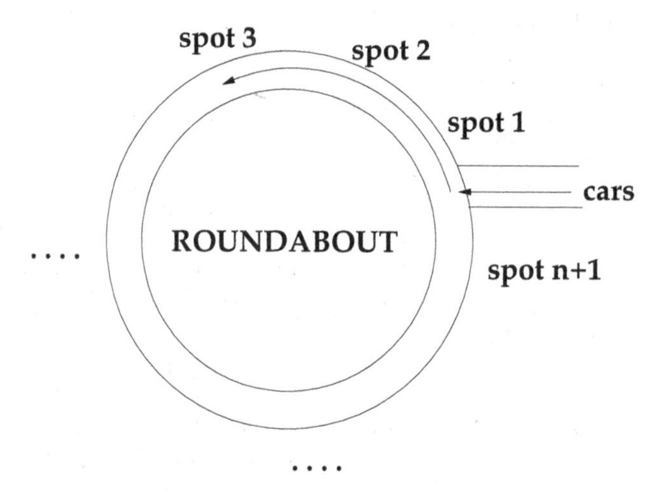

FIGURE 12.5
Parking on a roundabout.

No matter what f is, every car succeeds in parking in the circular situation. Moreover, since there are now $n+1$ spots and only n cars, there is always one empty spot at the end. Suppose we randomly select a circular parking function. Because of the symmetry of the roundabout, each of the $n+1$ parking spaces is equally likely to be the empty one. The fact that the entrance to the roundabout is at spot 1 is irrelevant here, since for parking purposes we may as well assume that car x enters the roundabout at its preferred spot $f(x)$. Thus, the probability that spot k is empty is $\frac{1}{n+1}$, for $1 \leq k \leq n+1$.

On the other hand, spot $n+1$ is the empty spot iff f is a parking function of order n. For, if spot $n+1$ is empty, then no car preferred spot $n+1$, and no car passed spot $n+1$

during the parking process. In this case, the circular parking process on the roundabout coincides with the original parking process on the one-way street. The converse is established similarly. Since spot $n+1$ is empty with probability $1/(n+1)$ and the sample space Z has size $(n+1)^n$, we conclude that the number of ordinary parking functions must be $|Z|/(n+1) = (n+1)^{n-1}$. □

12.16. Remark. Let $A_{n,k}$ be the set of circular parking functions of order n with empty spot k. The preceding proof shows that $|A_{n,k}| = (n+1)^{n-1}$ for k between 1 and $n+1$. We established this counting result by a probabilistic argument, using symmetry to deduce that $P(A_{n,k}) = 1/(n+1)$ for all k. Readers bothered by the vague appeal to symmetry may prefer the following more rigorous argument. Suppose $f \in A_{n,k_1}$ and k_2 are given. Let $\phi(f)$ be the function sending i to $(f(i)+k_2-k_1)$ mod $(n+1)$ for $1 \leq i \leq n$, taking the remainder to lie in the range $\{1,2,\ldots,n+1\}$. Informally, $\phi(f)$ rotates all of the parking preferences in f by k_2-k_1. One may check that ϕ is a bijection from A_{n,k_1} onto A_{n,k_2}. These bijections prove that all the sets $A_{n,k}$ (for k between 1 and $n+1$) have the same cardinality.

12.17. Remark. One of the original motivations for studying parking functions was their connection to hashing protocols. Computer programs often store information in a data structure called a *hash table*. We consider a simplified model where n items are to be stored in a linear array of n cells. A *hash function* $h : \{1,2,\ldots,n\} \to \{1,2,\ldots,n\}$ is used to determine where each item is stored. We store item i in position $h(i)$, unless that position has already been taken by a previous item—this circumstance is called a *collision*. We handle collisions via the following *collision resolution policy*: if $h(i)$ is full, we store item i in the earliest position after position i that is not yet full (if any). If there is no such position, the collision resolution fails (we do not allow wraparound). This scenario is exactly like that of the cars parking on a one-way street according to the preferences encoded by h. Thus, we can store all n items in the hash table iff h is a parking function.

12.5 Parking Functions and Trees

This section uses parking functions (defined in §12.4) to give a bijective proof of the Tree Rule 3.71 for the number of n-vertex trees. The proof involves labeled lattice paths, which we now define.

12.18. Definition: Labeled Lattice Paths. A *labeled lattice path* consists of a lattice path P from $(0,0)$ to (a,b), together with a labeling of the b north steps of P with labels $1,2,\ldots,b$ (each used exactly once) such that the labels for the north steps in a given column increase from bottom to top.

We illustrate a labeled lattice path by drawing the path inside an $(a+1) \times b$ grid of unit squares and placing the label of each north step in the unit square to the right of that north step. For example, Figure 12.6 displays a labeled lattice path ending at $(5,7)$.

12.19. Theorem: Labeled Paths. There are $(a+1)^b$ labeled lattice paths from $(0,0)$ to (a,b).

Proof. It suffices to construct a bijection between the set of labeled lattice paths ending at (a,b) and the set of all functions $f : \{1,2,\ldots,b\} \to \{1,2,\ldots,a+1\}$. Given a labeled lattice path P, define the associated function by setting $f(i) = j$ for all labels i in column j of P.

The inverse map acts as follows. Given a function $f : \{1,2,\ldots,b\} \to \{1,2,\ldots,a+1\}$, let

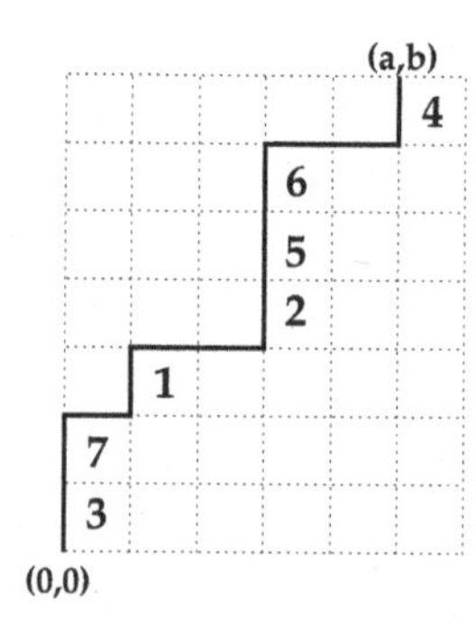

FIGURE 12.6
A labeled lattice path.

$S_j = \{x : f(x) = j\}$ and $s_j = |S_j|$ for j between 1 and $a+1$. The labeled path associated to f is the lattice path $\mathrm{N}^{s_1}\mathrm{E}\mathrm{N}^{s_2}\mathrm{E}\cdots\mathrm{N}^{s_{a+1}}$ where the jth string of consecutive north steps is labeled by the elements of S_j in increasing order. □

12.20. Example. The function f associated to the labeled path P in Figure 12.6 is given by

$$f(1) = 2,\ f(2) = 4,\ f(3) = 1,\ f(4) = 6,\ f(5) = 4,\ f(6) = 4,\ f(7) = 1.$$

Going the other way, the function $g : \{1, 2, \ldots, 7\} \to \{1, 2, \ldots, 6\}$ defined by

$$g(1) = 2,\ g(2) = 5,\ g(3) = 4,\ g(4) = 2,\ g(5) = 4,\ g(6) = 2,\ g(7) = 2$$

is mapped to the labeled lattice path shown in Figure 12.7.

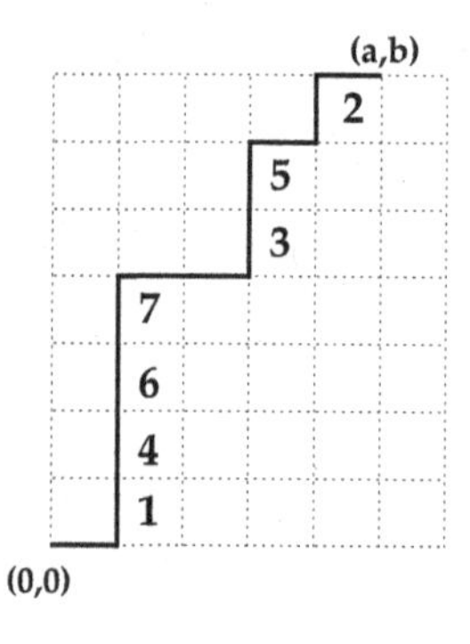

FIGURE 12.7
Converting a function to a labeled path.

A *labeled Dyck path of order* n is a Dyck path ending at (n, n) that is labeled according to the rules in Definition 12.18. For example, Figure 12.8 displays the 16 labeled Dyck paths of order 3.

12.21. Theorem: Labeled Dyck Paths. For all $n \in \mathbb{Z}_{\geq 1}$, there are $(n + 1)^{n-1}$ labeled Dyck paths of order n.

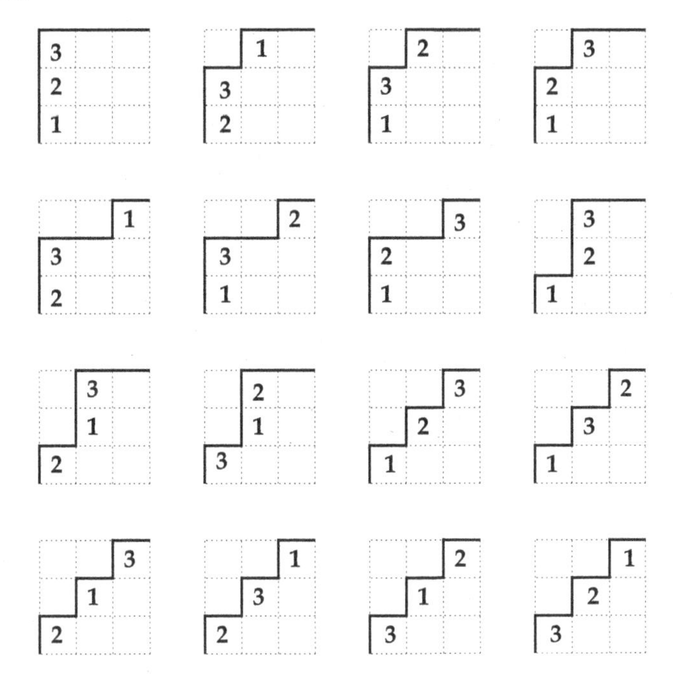

FIGURE 12.8
Labeled Dyck paths.

Proof. Using the bijection from the proof of Theorem 12.19, we can regard labeled lattice paths from $(0,0)$ to (n,n) as functions $f : \{1,2,\ldots,n\} \to \{1,2,\ldots,n+1\}$. We first show that labeled paths that are not Dyck paths correspond to non-parking functions under this bijection. A labeled path P is not a Dyck path iff some east step of P goes from $(i-1,j)$ to (i,j) for some $i > j$. This condition holds for P iff the function f associated to P satisfies $|\{x : f(x) \leq i\}| = j$ for some $i > j$. In turn, this condition on f is equivalent to the existence of i such that $|\{x : f(x) \leq i\}| < i$. But this means that f is not a parking function (see Definition 12.12). We now see that labeled Dyck paths are in bijective correspondence with parking functions. So the result follows from Theorem 12.15. □

12.22. Parking Function Proof of the Tree Rule. There are $(n+1)^{n-1}$ trees with vertex set $\{0,1,2,\ldots,n\}$.

Proof. Because of the previous result, it suffices to define bijections between the set B of labeled Dyck paths of order n and the set C of all trees with vertex set $\{0,1,2,\ldots,n\}$. To define $f : B \to C$, let P be a labeled Dyck path of order n. Let $(a_1,a_2,\ldots,a_n)$ be the sequence of labels in the diagram of P, reading from the bottom row to the top row, and set $a_0 = 0$. Define a graph $T = f(P)$ as follows. For $0 \leq j \leq n$, there is an edge in T from vertex a_j to each vertex whose label appears in column $j+1$ of the diagram of P. These are all the edges of T. Using the fact that P is a labeled Dyck path, one proves by induction on j that every a_j is either 0 or appears to the left of column $j+1$ in the diagram, so that every vertex in column $j+1$ of P is reachable from vertex 0 in T. Thus, $T = f(P)$ is a connected graph with n edges and $n+1$ vertices, so T is a tree by Theorem 3.70.

12.23. Example. Figure 12.9 shows a parking function f, the labeled Dyck path P corresponding to f, and the tree $T = f(P)$. We can use the figure to compute the edges of T by writing a_j underneath column $j+1$, for $0 \le j \le n$. If we regard zero as the ancestor of all other vertices, then the labels in column $j+1$ are the children of vertex a_j.

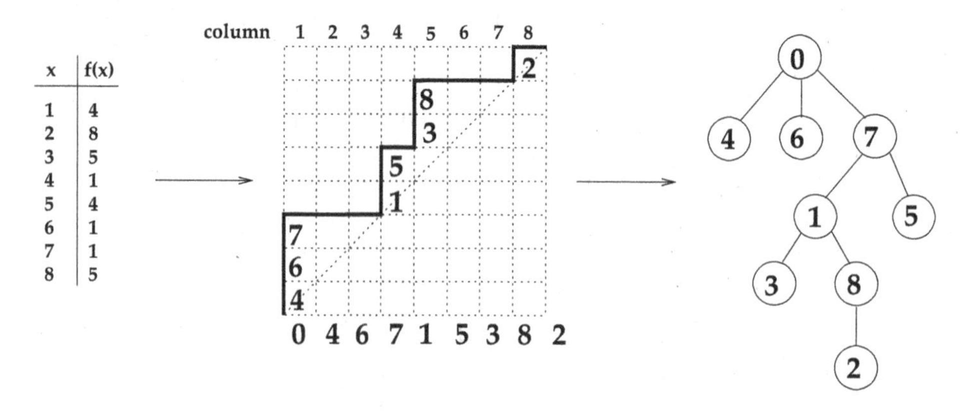

FIGURE 12.9
Mapping parking functions to labeled Dyck paths to trees.

Continuing the proof, we define the inverse map $f' : C \to B$. Let $T \in C$ be a tree with vertex set $\{0, 1, 2, \ldots, n\}$. We generate the diagram for $f'(T)$ by inserting labels into an $n \times n$ grid from bottom to top. Denote these labels by $(a_1, \ldots, a_n)$, and set $a_0 = 0$. The labels $a_1, a_2, \ldots$ in column 1 are the vertices of T adjacent to vertex $a_0 = 0$, written in increasing order from bottom to top. The labels in the second column are the neighbors of a_1 other than vertex 0. The labels in the third column are the neighbors of a_2 not in the set $\{a_0, a_1\}$. In general, the labels in column $j+1$ are the neighbors of a_j not in the set $\{a_0, a_1, \ldots, a_{j-1}\}$, written in increasing order from bottom to top. Observe that we do not know the full sequence $(a_1, \ldots, a_n)$ in advance, but we reconstruct this sequence as we go along. We show momentarily that a_j is always known by the time we reach column $j+1$.

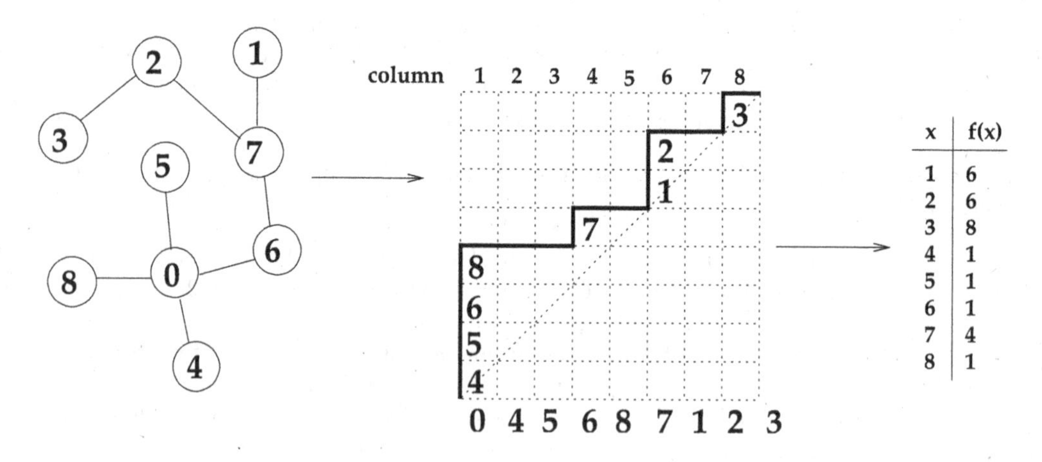

FIGURE 12.10
Mapping trees to labeled Dyck paths to parking functions.

12.24. Example. Figure 12.10 shows a tree T, the labeled Dyck path $P = f'(T)$, and the parking function associated to P.

Let us check that f' is well-defined. We break up the computation of $f'(T)$ into stages, where stage j consists of choosing the increasing sequence of labels $a_i < \cdots < a_k$ that occur in column j. We claim that at each stage j with $1 \leq j \leq n+1$, a_{j-1} has already been computed, so that the labels entered in column j occur in rows j or higher. This will show that the algorithm for computing f' is well-defined and produces a labeled Dyck path. We proceed by induction on j. The claim holds for $j = 1$, since $a_0 = 0$ by definition. Assume that $1 < j \leq n+1$ and that the claim holds for all $j' < j$. To get a contradiction, assume that a_{j-1} is not known when we reach column j. Since the claim holds for $j-1$, we must have already recovered the labels in the set $W = \{a_0 = 0, a_1, \ldots, a_{j-2}\}$, which are precisely the labels that occur in the first $j-1$ columns. Let z be a vertex of T not in W. Since T is a tree, there is a path from 0 to z in T. Let y be the earliest vertex on this path not in W, and let x be the vertex just before y on the path. By choice of y, we have $x \in W$, so that $x = 0$ or $x = a_k$ for some $k \leq j-2$. But if $x = 0$, then y occurs in column 1 and hence $y \in W$. And if $x = a_k$, then the algorithm for f' would have placed y in column $k+1 \leq j-1$, and again $y \in W$. These contradictions show that the claim holds for j. It is now routine to check that f' is the two-sided inverse of f. □

12.6 Möbius Inversion and Field Theory

This section gives two applications of the material in §4.9 to field theory. We show that every finite subgroup of the multiplicative group of any field must be cyclic; and we count the number of irreducible polynomials of a given degree with coefficients in a given finite field. The starting point for proving the first result is the relation $n = \sum_{d|n} \phi(d)$, proved in Theorem 4.37. We begin by giving a combinatorial interpretation of this identity in terms of the orders of elements in a cyclic group of size n.

12.25. Theorem: Order of Elements in a Cyclic Group. Suppose G is a cyclic group of size $d < \infty$, written multiplicatively. If $x \in G$ generates G and $c \in \mathbb{Z}_{\geq 1}$, then x^c generates a cyclic subgroup of G of size $d/\gcd(c,d) = \operatorname{lcm}(c,d)/c$.

Proof. Since the cyclic subgroup generated by x^c is a subset of the finite group G, the order of x^c must be finite. Let k be the order of x^c. We have seen in Example 7.60 that k is the smallest positive integer such that $(x^c)^k = 1_G$, and that the k elements x^c, $(x^c)^2 = x^{2c}, \ldots,$ $(x^c)^k = x^{kc}$ are distinct and constitute the cyclic subgroup of G generated by x^c. Since x has order d, we know from Example 7.60 that $x^m = 1_G$ iff $d|m$. It follows from this and the definition of k that kc is the least positive multiple of c that is also a multiple of d. In other words, $kc = \operatorname{lcm}(c,d)$. It follows that the order of x^c is $k = \operatorname{lcm}(c,d)/c$. Since $cd = \operatorname{lcm}(c,d)\gcd(c,d)$, we also have $k = d/\gcd(c,d)$. □

12.26. Theorem: Number of Generators of a Cyclic Group. If G is a cyclic group of size $d < \infty$, then there are exactly $\phi(d)$ elements in G that generate G.

Proof. Let x be a fixed generator of G. By Example 7.60, the d distinct elements of G are $x^1, x^2, \ldots, x^d = 1_G$. By Theorem 12.25, the element x^c generates all of G iff $\gcd(c,d) = 1$. By Definition 4.12, the number of such integers c between 1 and d is precisely $\phi(d)$. □

12.27. Theorem: Subgroup Structure of Cyclic Groups. Let G be a cyclic group of size $n < \infty$. For each d dividing n, there exists exactly one subgroup of G of size d, and this subgroup is cyclic.

Proof. We only sketch the proof, which uses some results about group homomorphisms that were stated as exercises in Chapter 7. We know from Theorem 7.40 that every subgroup of the cyclic group $\mathbb{Z}$ has the form $k\mathbb{Z}$ for some unique $k \geq 0$, and is therefore cyclic. Next, any finite cyclic group G can be viewed as the quotient group $\mathbb{Z}/n\mathbb{Z}$ for some $n \geq 1$. This follows by applying the Fundamental Homomorphism Theorem (Exercise 7-57) to the homomorphism from $\mathbb{Z}$ to G sending 1 to a generator of G. By the Correspondence Theorem (Exercise 7-61), each subgroup H of G has the form $H = m\mathbb{Z}/n\mathbb{Z}$ for some subgroup $m\mathbb{Z}$ of $\mathbb{Z}$ containing $n\mathbb{Z}$. Now, $m\mathbb{Z}$ contains $n\mathbb{Z}$ iff $m|n$, and in this case $|m\mathbb{Z}/n\mathbb{Z}| = n/m$. It follows that there is a bijection between the positive divisors of n and the subgroups of G. Each such subgroup is the homomorphic image of a cyclic group $m\mathbb{Z}$, so each subgroup of G is cyclic. □

Suppose G is cyclic of size n. For each d dividing n, let G_d be the unique (cyclic) subgroup of G of size d. On one hand, each of the n elements y of G generates exactly one of the subgroups G_d (namely, y generates the group G_d such that d is the order of y). On the other hand, we have shown that G_d has exactly $\phi(d)$ generators. Invoking the Sum Rule, we obtain a new group-theoretic proof of the fact that $n = \sum_{d|n} \phi(d)$.

12.28. Theorem: Detecting Cyclic Groups. If G is a group of size $n < \infty$ such that for each d dividing n, G has at most one subgroup of size d, then G is cyclic.

Proof. For each d dividing n, let T_d be the set of elements in G of order d. G is the disjoint union of the sets T_d by Theorem 7.100. Consider a fixed choice of d such that T_d is nonempty. Then G has an element of order d, hence has a cyclic subgroup of size d. By assumption, this is the only subgroup of G of size d, and we know this subgroup has $\phi(d)$ generators. Therefore, $|T_d| = \phi(d)$ whenever $|T_d| \neq 0$. We conclude that

$$n = |G| = \sum_{d|n} |T_d| \leq \sum_{d|n} \phi(d) = n.$$

Since the extreme ends of this calculation both equal n, the middle inequality here must in fact be an equality. This is only possible if every T_d is nonempty. In particular, T_n is nonempty. Therefore, G is cyclic, since it is generated by each of the elements in T_n. □

12.29. Theorem: Multiplicative Subgroups of Fields. Let F be any field, possibly infinite. If G is a finite subgroup of the multiplicative group of F, then G is cyclic.

Proof. Suppose G is an n-element subgroup of the multiplicative group of nonzero elements of F, where $n < \infty$. By Theorem 12.28, it suffices to show that G has at most one subgroup of size d, for each d dividing n. If not, let H and K be two distinct subgroups of G of size d. Then $H \cup K$ is a set with at least $d+1$ elements; and for each $z \in H \cup K$, it follows from Theorem 7.100 that z is a root of the polynomial $x^d - 1$ in F. But any polynomial of degree d over F has at most d distinct roots in the field F, by Exercise 12-72. This contradiction completes the proof. □

Our next goal is to count irreducible polynomials of a given degree over a finite field. We shall assume a number of results from field theory, whose proofs may be found in Chapter V of the algebra text by Hungerford [65]. Let F be a finite field with q elements. It is known that q must be a prime power, say $q = p^e$, and F is uniquely determined (up to isomorphism) by its cardinality q. Every finite field F with $q = p^e$ elements is a splitting field for the polynomial $x^q - x$ over $\mathbb{Z}/p\mathbb{Z}$.

12.30. Theorem: Counting Irreducible Polynomials. Let F be a field with $q = p^e$ elements. For each $n \in \mathbb{Z}_{\geq 1}$, let $I(n, q)$ be the number of monic irreducible polynomials of degree n in the polynomial ring $F[x]$. Then

$$q^n = \sum_{d|n} dI(d, q), \qquad \text{and hence} \qquad I(n, q) = \frac{1}{n} \sum_{d|n} q^d \mu(n/d).$$

Proof. The strategy of the proof is to classify the elements in a finite field K of size q^n based on their minimal polynomials. From field theory, we know that each element $u \in K$ is the root of a uniquely determined monic, irreducible polynomial in $F[x]$ (called the *minimal polynomial of u over F*). The degree d of this minimal polynomial is $d = [F(u) : F]$, where for any field extension $E \subseteq H$, $[H : E]$ denotes the dimension of H viewed as a vector space over E. It is known that $n = [K : F] = [K : F(u)] \cdot [F(u) : F]$, so that $d|n$. Conversely, given any divisor d of n, we claim that every irreducible polynomial of degree d in $F[x]$ has d distinct roots in K. Sketch of proof: Suppose g is such a polynomial and $z \neq 0$ is a root of g in a splitting field of g over K. Since z lies in $F(z)$, which is a field with q^d elements, it follows from Theorem 7.100 (applied to the multiplicative group of the field $F(z)$) that $z^{q^d-1} = 1$. It can be checked that $q^d - 1$ divides $q^n - 1$ (since $d|n$), so that $z^{q^n-1} = 1$, and hence z is a root of $x^{q^n} - x$. It follows that every root z of g must lie in K, which is a splitting field for $x^{q^n} - x$. Furthermore, since z is a root of $x^{q^n} - x$, it follows that the minimal polynomial for z over F (namely g) divides $x^{q^n} - x$ in $F[x]$. We conclude that g divides $x^{q^n} - x$ in $K[x]$ also. The polynomial $x^{q^n} - x$ is known to split into a product of q^n *distinct* linear factors over K; in fact, $x^{q^n} - x = \prod_{x_0 \in K}(x - x_0)$. By unique factorization in the polynomial ring $K[x]$, g must also be a product of d *distinct* linear factors. This completes the proof of the claim.

We can now write K as the disjoint union of sets R_g indexed by all monic irreducible polynomials g in $F[x]$ whose degrees divide n, where R_g consists of the $\deg(g)$ distinct roots of g in K. Invoking the Sum Rule and then grouping together terms indexed by polynomials of the same degree d, we obtain

$$q^n = |K| = \sum_{\substack{\text{monic irred. } g \in F[x]: \\ \deg(g)|n}} |R_g| = \sum_{\substack{\text{monic irred. } g \in F[x]: \\ \deg(g)|n}} \deg(g) = \sum_{d|n} dI(d, q).$$

We can now apply the Möbius Inversion Formula 4.33 to the functions $f(n) = q^n$ and $g(n) = nI(n, q)$ to obtain

$$nI(n, q) = \sum_{d|n} q^d \mu(n/d).$$

□

12.7 q-Binomial Coefficients and Subspaces

Recall from Definition 8.30 that the *q-binomial coefficients* are polynomials in a formal variable q defined by the formula

$$\begin{bmatrix} n \\ k \end{bmatrix}_q = \frac{[n]!_q}{[k]!_q [n-k]!_q} = \frac{\prod_{i=1}^n (q^i - 1)}{\prod_{a=1}^k (q^a - 1) \prod_{b=1}^{n-k} (q^b - 1)}.$$

We gave several combinatorial interpretations of these polynomials in §8.6. In this section, we discuss a linear-algebraic interpretation of the integers $\left[\begin{smallmatrix} n \\ k \end{smallmatrix}\right]_q$, where the variable q is set equal to a prime power. To read this section, the reader should have some previous experience with fields and vector spaces. We begin by using bases to determine the possible sizes of vector spaces over finite fields.

12.31. Theorem: Size of a Finite Vector Space. Suppose V is a d-dimensional vector space over a finite field F with q elements. Then $|V| = q^d$.

Proof. Let $(v_1, \ldots, v_d)$ be an ordered basis for V. By definition of a basis, for each $v \in V$, there exists exactly one d-tuple of scalars $(c_1, \ldots, c_d) \in F^d$ such that $v = c_1v_1 + c_2v_2 + \cdots + c_dv_d$. In other words, the function from F^d to V sending $(c_1, \ldots, c_d)$ to $\sum_{i=1}^{d} c_iv_i$ is a bijection. Because $|F| = q$, the Product Rule gives $|V| = |F^d| = |F|^d = q^d$. □

12.32. Theorem: Size of a Finite Field. If K is a finite field, then $|K| = p^e$ for some prime p and some $e \in \mathbb{Z}_{\geq 1}$.

Proof. Given a finite field K, let F be the cyclic subgroup of the additive group of K generated by 1_K. The size of F is some finite number p (since K is finite), and $p > 1$ since $1_K \neq 0_K$. We know that p is the smallest positive integer such that $p1_K = 0_K$. One checks (using the distributive laws) that F is not only an additive subgroup of K, but also a subring of K. If p were not prime, say $p = ab$ with $1 < a, b < p$, then $(a1_K) \cdot (b1_K) = ab1_K = p1_K = 0_K$, and yet $a1_K, b1_K \neq 0$. This contradicts the fact that fields have no zero divisors. Thus, p must be prime. It now follows that F is a *field* isomorphic to the field of integers modulo p. K can be regarded as a vector space over its subfield F, by defining scalar multiplication (which is a map from $F \times K$ into K) to be the restriction of the field multiplication $m : K \times K \to K$. Since K is finite, it must be a finite-dimensional vector space over F. Thus the required result follows from Theorem 12.31. □

12.33. Remark. One can show that, for every prime power p^e, there exists a finite field of size p^e, which is unique up to isomorphism. The existence proof is sketched in Exercise 12-36.

We now give the promised linear-algebraic interpretation of q-binomial coefficients.

12.34. Theorem: Subspaces of Finite Vector Spaces. Let K be a finite field with q elements. For all integers k, n with $0 \leq k \leq n$ and all n-dimensional vector spaces V over K, the integer $\begin{bmatrix} n \\ k \end{bmatrix}_q$ is the number of k-dimensional subspaces of V.

Proof. Let $f(n, k, q)$ be the number of k-dimensional subspaces of V. One may check that this number depends only on k, q, and $n = \dim(V)$. Recall from Theorem 12.31 that $|V| = q^n$ and each d-dimensional subspace of V has size q^d. By rearranging factors in the defining formula for $\begin{bmatrix} n \\ k \end{bmatrix}_q$, we see that $\begin{bmatrix} n \\ k \end{bmatrix}_q = f(n, k, q)$ holds iff

$$f(n, k, q)(q^k - 1)(q^{k-1} - 1) \cdots (q^1 - 1) = (q^n - 1)(q^{n-1} - 1) \cdots (q^{n-k+1} - 1).$$

We establish this equality by the following counting argument. Let S be the set of all ordered lists $(v_1, \ldots, v_k)$ of k linearly independent vectors in V. Here is one way to build such a list. First, choose a nonzero vector $v_1 \in V$ in any of $q^n - 1$ ways. This vector spans a one-dimensional subspace W_1 of V of size $q = q^1$. Second, choose a vector $v_2 \in V - W_1$ in any of $q^n - q$ ways. The list (v_1, v_2) must be linearly independent since v_2 is not in the space W_1 spanned by v_1. The vectors v_1 and v_2 span a two-dimensional subspace W_2 of V of size q^2. Third, choose $v_3 \in V - W_2$ in $q^n - q^2$ ways. Continue similarly. When choosing v_i, we have already found $i - 1$ linearly independent vectors $v_1, \ldots, v_{i-1}$ that span a subspace of V of size q^{i-1}. Consequently, $(v_1, \ldots, v_i)$ is linearly independent iff we choose $v_i \in V - W_i$, which is a set of size $q^n - q^{i-1}$. By the Product Rule, we conclude that

$$|S| = \prod_{i=1}^{k}(q^n - q^{i-1}) = \prod_{i=1}^{k} q^{i-1}(q^{n+1-i} - 1) = q^{k(k-1)/2}(q^n - 1)(q^{n-1} - 1) \cdots (q^{n-k+1} - 1).$$

Now we count S in a different way. Observe that the vectors in each list $(v_1, \ldots, v_k) \in S$ span some k-dimensional subspace of V. So we begin by choosing such a subspace W in any of $f(n,k,q)$ ways. Next we choose $v_1, \ldots, v_k \in W$ one at a time, following the same process used in the first part of the proof. We can choose v_1 in $|W| - 1 = q^k - 1$ ways, then v_2 in $q^k - q$ ways, and so on. By the Product Rule,

$$|S| = f(n,k,q) \prod_{i=1}^{k} (q^k - q^{i-1}) = f(n,k,q) q^{k(k-1)/2} (q^k - 1)(q^{k-1} - 1) \cdots (q^1 - 1).$$

Equating the two formulas for $|S|$ and cancelling $q^{k(k-1)/2}$ gives the required result. □

In Theorem 8.34, we saw that

$$\begin{bmatrix} n \\ k \end{bmatrix}_q = \sum_{\mu \in P(k,n-k)} q^{|\mu|},$$

where $P(k, n-k)$ is the set of all integer partitions μ that fit in a $k \times (n-k)$ rectangle. In the rest of this section, we give a second proof of Theorem 12.34 by showing that

$$f(n,k,q) = \sum_{\mu \in P(k,n-k)} q^{|\mu|}.$$

This proof is longer than the one already given, but it reveals a close connection between the enumeration of subspaces on one hand, and the enumeration of partitions in a box (or, equivalently, lattice paths) on the other hand.

For definiteness, we work with the vector space $V = K^n$ whose elements are n-tuples of elements of K. We regard elements of V as row vectors of length n. The key linear-algebraic fact we need is that every k-dimensional subspace of $V = K^n$ has a unique reduced row-echelon form basis, as defined below.

12.35. Definition: Reduced Row-Echelon Form. Let A be a $k \times n$ matrix with entries in K. Let $A_1, \ldots, A_k \in K^n$ be the k rows of A. We say A is a *reduced row-echelon form* (RREF) matrix iff the following conditions hold: (i) $A_i \neq 0$ for all i, and the leftmost nonzero entry of A_i is 1_K (call these entries *leading 1's*); (ii) if the leading 1 of A_i occurs in column $j(i)$, then $j(1) < j(2) < \cdots < j(k)$; (iii) every leading 1 is the only nonzero entry in its column. An ordered basis $B = (v_1, \ldots, v_k)$ for a k-dimensional subspace of K^n is called a *RREF basis* iff the matrix whose rows are $v_1, \ldots, v_k$ is a RREF matrix.

12.36. Theorem: RREF Bases. Let K be any field. Every k-dimensional subspace of K^n has a unique RREF basis. Conversely, the rows of every $k \times n$ RREF matrix comprise an ordered basis for a k-dimensional subspace of K^n. Consequently, there is a bijection between the set of such subspaces and the set of $k \times n$ RREF matrices with entries in K.

Proof. We sketch the proof, asking the reader to supply the missing linear algebra details.

Step 1: We use row-reduction to show that any given k-dimensional subspace W of K^n has *at least one* RREF basis. Start with any ordered basis $v_1, \ldots, v_k$ of W, and let A be the matrix with rows $v_1, \ldots, v_k$. There are three *elementary row operations* we can use to simplify A: interchange two rows; multiply one row by a nonzero scalar (element of K); add any scalar multiple of one row to a different row. A routine verification shows that performing any one of these operations has no effect on the subspace spanned by the rows of A. Therefore, we can create new ordered bases for W by performing sequences of row operations on A. Using the well-known Gaussian elimination algorithm (also called "row reduction"), we can bring the matrix A into reduced row-echelon form. The rows of the new matrix give the required RREF basis of W.

Step 2: We show that a given subspace W has *at most one* RREF basis. Use induction on k, the base case $k = 0$ being immediate. For the induction step, assume $n \geq 1$ and $k \geq 1$ are fixed, and the uniqueness result is known for smaller values of k. Let A and B be two RREF matrices whose rows form bases of W; we must prove $A = B$. Let $j(1) < j(2) < \cdots < j(k)$ be the positions of the leading 1's in A, and let $r(1) < r(2) < \cdots < r(k)$ be the positions of the leading 1's in B. If $j(1) < r(1)$, then the first row of A (which is a vector in W) has a 1 in position $j(1)$. This vector cannot possibly be a linear combination of the rows of B, all of whose nonzero entries occur in columns after $j(1)$. Thus, $j(1) < r(1)$ is impossible. Similar reasoning rules out $r(1) < j(1)$, so we must have $j(1) = r(1)$. Let W' be the subspace of W consisting of vectors with zeroes in positions $1, 2, \ldots, j(1)$. Consideration of leading 1's shows that rows 2 through k of A must form a basis for W', and rows 2 through k of B also form a basis for W'. Since $\dim(W') = k - 1$, the induction hypothesis implies that rows 2 through k of A equal the corresponding rows of B. In particular, we now know that $r(i) = j(i)$ for all i between 1 and k. To finish, we must still check that row 1 of A equals row 1 of B. Let the rows of B be $w_1, \ldots, w_k$, and write v_1 for the first row of A. Since $v_1 \in W$, we have $v_1 = a_1w_1 + \cdots + a_kw_k$ for some unique scalars a_k. Consideration of column $j(1)$ shows that $a_1 = 1$. On the other hand, if $a_i \neq 0$ for some $i > 1$, then $a_1w_1 + \cdots + a_kw_k$ would have a nonzero entry in position $j(i)$, whereas v_1 has a zero entry in this position (since the leading 1's occur in the same columns in A and B). This is a contradiction, so $a_2 = \cdots = a_k = 0$. Thus $v_1 = w_1$, as needed, and we have now proved that $A = B$.

Step 3: We show that the k rows $v_1, \ldots, v_k$ of a given RREF matrix form an ordered basis for some k-dimensional subspace of K^n. It suffices to show that the rows in question are linearly independent vectors. Suppose $c_1v_1 + \cdots + c_kv_k = 0$, where $c_i \in K$. Recall that the leading 1 in position $(i, j(i))$ is the only nonzero entry in its column. Therefore, taking the $j(i)$th component of the preceding equation, we get $c_i = 0$ for all i between 1 and k. □

Because of the preceding theorem, the problem of counting k-dimensional subspaces of K^n (where $|K| = q$) reduces to the problem of counting $k \times n$ RREF matrices with entries in K. Our second proof of Theorem 12.34 is therefore complete once we prove the following result.

12.37. Theorem: RREF Matrices. Let K be a finite field with q elements. The number of $k \times n$ RREF matrices with entries in K is

$$\sum_{\mu \in P(k, n-k)} q^{|\mu|} = \begin{bmatrix} n \\ k \end{bmatrix}_q.$$

Proof. Let us classify the $k \times n$ RREF matrices based on the columns $j(1) < j(2) < \cdots < j(k)$ where the leading 1's occur. To build a RREF matrix with the leading 1's in these positions, we must put 0's in all matrix positions (i, p) such that $p < j(i)$; we must also put 0's in all matrix positions $(r, j(i))$ such that $r < i$. However, in all the other positions to the right of the leading 1's, there is no restriction on the elements that occur except that they must come from the field K of size q. How many such free positions are there? The first row contains $n - j(1)$ entries after the leading 1, but $k - 1$ of these entries are in columns above other leading 1's. So there are $\mu_1 = n - j(1) - (k - 1)$ free positions in this row. The next row contains $n - j(2)$ entries after the leading 1, but $k - 2$ of these occur in columns above other leading 1's. So there are $\mu_2 = n - j(2) - (k - 2)$ free positions in row 2. Similarly, there are $\mu_i = n - j(i) - (k - i) = n - k + i - j(i)$ free positions in row i for i between 1 and k. The condition $1 \leq j(1) < j(2) < \cdots < j(k) \leq n$ is logically equivalent to $0 \leq j(1) - 1 \leq j(2) - 2 \leq \cdots \leq j(k) - k \leq n - k$, which is in turn equivalent to $n - k \geq \mu_1 \geq \mu_2 \geq \cdots \geq \mu_k \geq 0$. Thus, *there is a bijection between the set of valid positions $j(1) < j(2) < \ldots < j(k)$ for the leading 1's, and the set of integer partitions $\mu = (\mu_1, \ldots, \mu_k)$ whose diagrams fit in a $k \times (n-k)$ box; this bijection is given by $\mu_i = n - k + i - j(i)$.* Furthermore, $|\mu| = \mu_1 + \cdots + \mu_k$ is the total number of free positions in each RREF matrix with leading 1's in the positions $j(i)$. Using the Product Rule to fill these free positions one at a time with elements of K, we see that there are $q^{|\mu|}$ RREF matrices with leading 1's in the given positions. The theorem now follows from the Sum Rule, keeping in mind the bijection just constructed between the set of possible j-sequences and the set $P(k, n-k)$. □

12.38. Example. To illustrate the preceding proof, take $n = 10$ and $k = 4$, and consider RREF matrices of the form

$$\begin{bmatrix} 0 & 1 & * & * & 0 & 0 & * & * & 0 & * \\ 0 & 0 & 0 & 0 & 1 & 0 & * & * & 0 & * \\ 0 & 0 & 0 & 0 & 0 & 1 & * & * & 0 & * \\ 0 & 0 & 0 & 0 & 0 & 0 & 0 & 0 & 1 & * \end{bmatrix}.$$

The stars mark the free positions in the matrix, and $(j(1), j(2), j(3), j(4)) = (2, 5, 6, 9)$. The associated partition is $\mu = (5, 3, 3, 1)$, which does fit in a 4×6 box. We can see a reflected version of the diagram of μ in the matrix by erasing the columns without stars and right-justifying the remaining columns. We see that there are $q^{12} = q^{|\mu|}$ ways of filling in this template to get an RREF matrix with the leading 1's in the indicated positions.

Going the other way, consider another partition $\mu = (6, 2, 2, 0)$ that fits in a 4×6 box. Using the formula $j(i) = n - k + i - \mu_i$, we recover $(j(1), j(2), j(3), j(4)) = (1, 6, 7, 10)$, which tells us the locations of the leading 1's. So this particular partition corresponds to RREF matrices that match the following template:

$$\begin{bmatrix} 1 & * & * & * & * & 0 & 0 & * & * & 0 \\ 0 & 0 & 0 & 0 & 0 & 1 & 0 & * & * & 0 \\ 0 & 0 & 0 & 0 & 0 & 0 & 1 & * & * & 0 \\ 0 & 0 & 0 & 0 & 0 & 0 & 0 & 0 & 0 & 1 \end{bmatrix}.$$

12.8 Tangent and Secant Numbers

In calculus, one learns the following power series expansions for the trigonometric functions sine, cosine, and arctangent:

$$\sin x = x - \frac{x^3}{3!} + \frac{x^5}{5!} - \frac{x^7}{7!} + \cdots = \sum_{k=0}^{\infty} (-1)^k \frac{x^{2k+1}}{(2k+1)!};$$

$$\cos x = 1 - \frac{x^2}{2!} + \frac{x^4}{4!} - \frac{x^6}{6!} + \cdots = \sum_{k=0}^{\infty} (-1)^k \frac{x^{2k}}{(2k)!};$$

$$\arctan x = x - \frac{x^3}{3} + \frac{x^5}{5} - \frac{x^7}{7} + \cdots = \sum_{k=0}^{\infty} (-1)^k \frac{x^{2k+1}}{2k+1}.$$

These expansions are all special cases of Taylor's formula $f(x) = \sum_{k=0}^{\infty} \frac{f^{(k)}(0)}{k!} x^k$. Using Taylor's formula, one can also find power series expansions for the tangent and secant functions:

$$\tan x = x + \frac{1}{3}x^3 + \frac{2}{15}x^5 + \frac{17}{315}x^7 + \frac{62}{2835}x^9 + \frac{1382}{155925}x^{11} + \cdots;$$

$$\sec x = 1 + \frac{1}{2}x^2 + \frac{5}{24}x^4 + \frac{61}{720}x^6 + \frac{277}{8064}x^8 + \frac{50521}{3628800}x^{10} + \cdots.$$

The coefficients of these series seem quite irregular and unpredictable compared to the preceding three series. Remarkably, as we shall see in this section, these coefficients encode the solution to a counting problem involving permutations.

It can be shown that the power series expansions for $\tan x$ and $\sec x$ given by Taylor's formula do converge for all x in a neighborhood of 0. Furthermore, it is permissible to compute coefficients in these power series by algebraic manipulation of the identities $\tan x = \sin x/\cos x$ and $\sec x = 1/\cos x$, using the series expansions for $\sin x$ and $\cos x$ given above. We could avoid worrying about these technical points by working with formal power series throughout.

For each $n \in \mathbb{Z}_{\geq 0}$, let a_n be the nth derivative of $\tan x$ evaluated at $x = 0$, and let b_n be the nth derivative of $\sec x$ evaluated at $x = 0$. By Taylor's formula, we know

$$\tan x = \sum_{n=0}^{\infty} \frac{a_n}{n!}x^n; \qquad \sec x = \sum_{n=0}^{\infty} \frac{b_n}{n!}x^n. \tag{12.2}$$

The first several values of a_n and b_n are

$$\begin{aligned}(a_0, a_1, a_2, \ldots) &= (0, 1, 0, 2, 0, 16, 0, 272, 0, 7936, 0, 353792, \ldots); \\ (b_0, b_1, b_2, \ldots) &= (1, 0, 1, 0, 5, 0, 61, 0, 1385, 0, 50521, \ldots).\end{aligned} \tag{12.3}$$

Since $\tan x$ is an odd function and $\sec x$ is an even function, it readily follows that $a_n = 0$ for all even n and $b_n = 0$ for all odd n.

Next, for each integer $n \geq 1$, let c_n be the number of permutations $w = w_1 w_2 \cdots w_n$ of $\{1, 2, \ldots, n\}$ (or any fixed n-letter ordered alphabet) such that

$$w_1 < w_2 > w_3 < w_4 > \cdots < w_{n-1} > w_n. \tag{12.4}$$

Note that $c_n = 0$ for all even $n > 0$; we also define $c_0 = 0$. For each integer $n \geq 0$, let d_n be the number of permutations w of $\{1, 2, \ldots, n\}$ (or any fixed n-letter ordered alphabet) such that

$$w_1 < w_2 > w_3 < w_4 > \cdots > w_{n-1} < w_n; \tag{12.5}$$

note that $d_0 = 1$ and $d_n = 0$ for all odd n. By reversing the ordering of the letters, one sees that d_n also counts permutations w of n letters such that

$$w_1 > w_2 < w_3 > w_4 < \cdots < w_{n-1} > w_n. \tag{12.6}$$

Permutations of the form (12.4) or (12.5) are called *up-down permutations.* We now prove that $a_n = c_n$ *and* $b_n = d_n$ *for all integers* $n \geq 0$. The proof consists of five steps.

Step 1. Differentiating $\tan x = \sin x/\cos x$, one readily calculates that $\frac{d}{dx}(\tan x) = \sec^2 x$. Differentiating the first series in (12.2), squaring the second series using Definition 5.15(e), and equating the coefficients of x^n, we obtain

$$\frac{a_{n+1}}{n!} = \sum_{k=0}^{n} \frac{b_k}{k!} \cdot \frac{b_{n-k}}{(n-k)!}$$

or equivalently,

$$a_{n+1} = \sum_{k=0}^{n} \binom{n}{k} b_k b_{n-k} \qquad \text{for all } n \geq 0. \tag{12.7}$$

Step 2. Differentiating $\sec x = 1/\cos x$, we find that $\frac{d}{dx}(\sec x) = \tan x \sec x$. Differentiating the second series in (12.2), multiplying the two series together using Definition 5.15(e), and equating the coefficients of x^n, we obtain

$$\frac{b_{n+1}}{n!} = \sum_{k=0}^{n} \frac{a_k}{k!} \frac{b_{n-k}}{(n-k)!}$$

or equivalently,

$$b_{n+1} = \sum_{k=0}^{n} \binom{n}{k} a_k b_{n-k} \qquad \text{for all } n \geq 0. \tag{12.8}$$

Step 3. We give a counting argument to prove the equation

$$c_{n+1} = \sum_{k=0}^{n} \binom{n}{k} d_k d_{n-k} \qquad \text{for all } n \geq 0. \tag{12.9}$$

If n is odd, then both sides of this equation are zero, since at least one of k or $n-k$ is odd for each k. Now suppose n is even. How can we build a typical permutation

$$w = w_1 < w_2 > w_3 < \cdots > w_{n+1}$$

counted by c_{n+1}? Let us first choose the position of 1 in w; say $w_{k+1} = 1$ for some k between 0 and n. The required inequalities at position $k+1$ are satisfied if and only if k is even. In the case where k is odd, we know $d_k d_{n-k} = 0$, so this term contributes nothing to the right side of (12.9). Given that k is even, choose a k-element subset A of the n remaining letters in $\binom{n}{k}$ ways. Use these k letters to fill in the first k positions of w, subject to the required inequalities (12.5), in any of d_k ways. Use the remaining letters to fill in the last $(n+1)-(k+1) = n-k$ positions of w (subject to the inequalities (12.6), reindexed to begin at index $k+2$), in any of d_{n-k} ways. Equation (12.9) now follows from the Sum Rule and Product Rule. See Figure 12.11, in which w is visualized as a sequence of line segments connecting the points (i, w_i) for i between 1 and $n+1$.

Step 4. We give a counting argument to prove the equation

$$d_{n+1} = \sum_{k=0}^{n} \binom{n}{k} c_k d_{n-k} \qquad \text{for all } n \geq 0. \tag{12.10}$$

Both sides are zero if n is even. If n is odd, we must build a permutation

$$w = w_1 < w_2 > w_3 < \cdots < w_{n+1}.$$

First choose an index k with $0 \leq k \leq n$, and define $w_{k+1} = n+1$. This time, to get a nonzero contribution from this value of k, we need k to be odd. Now pick a k-element subset A of the n remaining letters in $\binom{n}{k}$ ways. Use the letters in A to fill in $w_1 w_2 \cdots w_k$ (c_k ways), and use the remaining letters to fill in $w_{k+2} \cdots w_{n+1}$ (d_{n-k} ways). See Figure 12.12.

Step 5: A routine induction argument now shows that $a_n = c_n$ and $b_n = d_n$ for all $n \in \mathbb{Z}_{\geq 0}$, since the pair of sequences (a_n), (b_n) satisfy the same system of recursions and initial conditions as the pair of sequences (c_n), (d_n). This completes the proof.

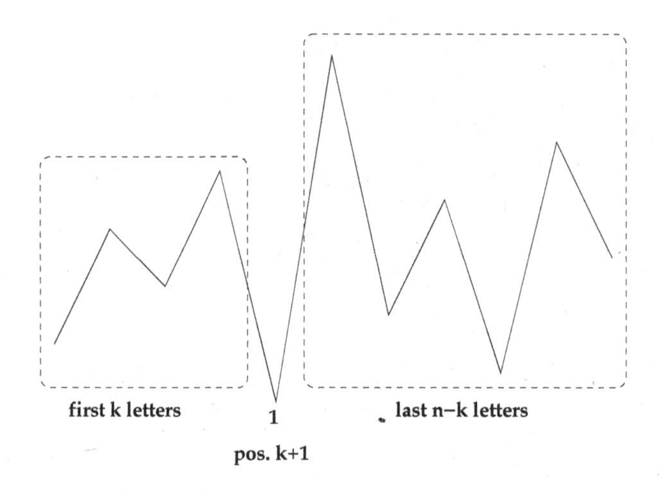

FIGURE 12.11
Counting up-down permutations of odd length.

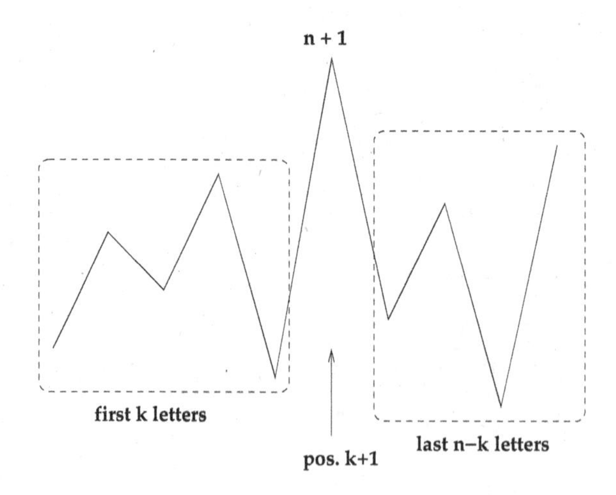

FIGURE 12.12
Counting up-down permutations of even length.

12.9 Combinatorial Definition of Determinants

This section applies material from Chapter 7 to the theory of determinants. Our goal is to explain how the combinatorial properties of permutations underlie many basic facts about determinants. First we recall the definitions of operations on matrices.

12.39. Definition: Matrix Operations. For each positive integer n, let $M_n(\mathbb{R})$ be the set of $n \times n$ matrices with entries in $\mathbb{R}$. Given $A \in M_n(\mathbb{R})$, we denote the entry in row i, column j of A by $A(i,j)$. For $A, B \in M_n(\mathbb{R})$ and $c \in \mathbb{R}$, define $A+B$, AB, and cA by setting

$$\begin{aligned}(A+B)(i,j) &= A(i,j)+B(i,j);\\ (AB)(i,j) &= \sum_{k=1}^{n} A(i,k)B(k,j);\\ (cA)(i,j) &= c(A(i,j)) \qquad \text{for all } i,j \in \{1,2,\ldots,n\}.\end{aligned}$$

Routine verifications show that $M_n(\mathbb{R})$ with these operations is a ring; the multiplicative identity element I_n in $M_n(\mathbb{R})$ is given by $I_n(i,j) = 1$ if $i = j$, and $I_n(i,j) = 0$ if $i \neq j$. This ring is non-commutative for all $n > 1$. More generally, we could replace $\mathbb{R}$ by any commutative ring R throughout the following discussion, considering matrices with entries in R.

12.40. Definition: Determinants. For a matrix $A \in M_n(\mathbb{R})$, the *determinant* of A is

$$\det(A) = \sum_{w \in S_n} \operatorname{sgn}(w) \prod_{i=1}^{n} A(i, w(i)).$$

In this definition, S_n is the set of all permutations of $\{1,2,\ldots,n\}$, and $\operatorname{sgn}(w) = (-1)^{\operatorname{inv}(w)}$ for $w \in S_n$ (see §7.4). Note that $\det(A)$ is an element of $\mathbb{R}$.

12.41. Example. When $n = 1$, $\det(A) = A(1,1)$. When $n = 2$, the possible permutations w (in one-line form) are $w = 12$ with $\operatorname{sgn}(w) = +1$, and $w = 21$ with $\operatorname{sgn}(w) = -1$. Therefore, the definition gives

$$\det(A) = \det\begin{bmatrix} A(1,1) & A(1,2) \\ A(2,1) & A(2,2)\end{bmatrix} = +A(1,1)A(2,2) - A(1,2)A(2,1).$$

When $n = 3$, we find (using the table in Example 7.23) that

$$\begin{aligned}\det(A) &= \det\begin{bmatrix} A(1,1) & A(1,2) & A(1,3)\\ A(2,1) & A(2,2) & A(2,3)\\ A(3,1) & A(3,2) & A(3,3)\end{bmatrix}\\ &= +A(1,1)A(2,2)A(3,3) - A(1,1)A(2,3)A(3,2) - A(1,2)A(2,1)A(3,3)\\ &\quad +A(1,2)A(2,3)A(3,1) + A(1,3)A(2,1)A(3,2) - A(1,3)A(2,2)A(3,1).\end{aligned}$$

For general n, we see that $\det(A)$ is a sum of $n!$ signed terms. A given term arises by choosing one factor $A(i, w(i))$ from each row of A; since w is a permutation, each of the chosen factors must come from a different column of A. The term in question is the product of the n chosen factors, times $\operatorname{sgn}(w)$. Since $\operatorname{sgn}(w) = (-1)^{\operatorname{inv}(w)}$, the sign attached to this term depends on the parity of the number of basic transpositions needed to sort the column indices $w(1), w(2), \ldots, w(n)$ into increasing order (see Theorem 7.29).

The next result shows that we can replace $A(i, w(i))$ by $A(w(i), i)$ in the defining formula for $\det(A)$. This corresponds to interchanging the roles of rows and columns in the description above.

12.42. Definition: Transpose of a Matrix. Given $A \in M_n(\mathbb{R})$, the *transpose* of A is the matrix $A^{\text{tr}} \in M_n(\mathbb{R})$ such that $A^{\text{tr}}(i,j) = A(j,i)$ for all $i, j \in \{1, 2, \ldots, n\}$.

12.43. Theorem: Determinant of a Transpose. For all $A \in M_n(\mathbb{R})$, $\det(A^{\text{tr}}) = \det(A)$.

Proof. By definition,

$$\det(A^{\text{tr}}) = \sum_{w \in S_n} \text{sgn}(w) \prod_{k=1}^{n} A^{\text{tr}}(k, w(k)) = \sum_{w \in S_n} \text{sgn}(w) \prod_{k=1}^{n} A(w(k), k).$$

For a fixed $w \in S_n$, we make a change of variables in the product indexed by w by letting $j = w(k)$, so $k = w^{-1}(j)$. Since w is a permutation and multiplication in $\mathbb{R}$ is commutative,

$$\prod_{k=1}^{n} A(w(k), k) = \prod_{j=1}^{n} A(j, w^{-1}(j))$$

because the second product contains the same factors as the first product in a different order. Using Theorem 7.31, we now calculate

$$\det(A^{\text{tr}}) = \sum_{w \in S_n} \text{sgn}(w) \prod_{j=1}^{n} A(j, w^{-1}(j)) = \sum_{w \in S_n} \text{sgn}(w^{-1}) \prod_{j=1}^{n} A(j, w^{-1}(j)).$$

Now consider the change of variable $v = w^{-1}$. As w ranges over S_n, so does v, since the map sending w to w^{-1} is a bijection on S_n. Furthermore, we can reorder the terms of the sum since addition in $\mathbb{R}$ is commutative. We conclude that

$$\det(A^{\text{tr}}) = \sum_{v \in S_n} \text{sgn}(v) \prod_{j=1}^{n} A(j, v(j)) = \det(A). \qquad \square$$

Next we obtain a formula for the determinant of an upper-triangular matrix.

12.44. Theorem: Determinants of Triangular and Diagonal Matrices. Suppose $A \in M_n(\mathbb{R})$ is upper-triangular, which means $A(i,j) = 0$ whenever $i > j$. Then $\det(A) = \prod_{i=1}^{n} A(i,i)$. Consequently, if A is either upper-triangular, lower-triangular, or diagonal, then $\det(A)$ is the product of the diagonal entries of A.

Proof. By definition, $\det(A) = \sum_{w \in S_n} \text{sgn}(w) \prod_{i=1}^{n} A(i, w(i))$. In order for a given summand to be nonzero, we must have $i \leq w(i)$ for all $i \in \{1, 2, \ldots, n\}$. Since w is a permutation, we successively deduce that $w(n) = n$, $w(n-1) = n-1$, $\ldots$, $w(1) = 1$. Thus, the only summand that could be nonzero is the one indexed by $w = \text{id}$. Since $\text{sgn}(\text{id}) = +1$ and $\text{id}(i) = i$ for all i, the stated formula for $\det(A)$ follows when A is upper-triangular. The result for lower-triangular A follows by considering A^{tr}. Since diagonal matrices are upper-triangular, the proof is complete. $\square$

12.45. Corollary: Determinant of Identity Matrix. For all $n > 0$, $\det(I_n) = 1$.

Next we discuss the multilinearity property of determinants and some of its consequences.

12.46. Definition: Linear Maps. For $n \in \mathbb{Z}_{>0}$, a function $T : \mathbb{R}^n \to \mathbb{R}$ is called a *linear map* iff $T(\mathbf{v} + \mathbf{z}) = T(\mathbf{v}) + T(\mathbf{z})$ and $T(c\mathbf{v}) = cT(\mathbf{v})$ for all $\mathbf{v}, \mathbf{z} \in \mathbb{R}^n$ and all $c \in \mathbb{R}$.

12.47. Example. Suppose $b_1, \ldots, b_n \in \mathbb{R}$ are fixed constants, and $T : \mathbb{R}^n \to \mathbb{R}$ is defined by

$$T(v_1, \ldots, v_n) = b_1 v_1 + b_2 v_2 + \cdots + b_n v_n \qquad \text{for all } (v_1, \ldots, v_n) \in \mathbb{R}^n.$$

It is routine to check that T is a linear map. Conversely, one can show that every linear map from $\mathbb{R}^n$ to $\mathbb{R}$ must be of this form for some (unique) choice of $b_1, \ldots, b_n \in \mathbb{R}$.

12.48. Theorem: Multilinearity of Determinants. Let $A \in M_n(\mathbb{R})$, and let k be a fixed row index in $\{1, \ldots, n\}$. Given a row vector $\mathbf{v} \in \mathbb{R}^n$, let $A[\mathbf{v}]$ denote the matrix A with row k replaced by $\mathbf{v}$. Then $T : \mathbb{R}^n \to \mathbb{R}$ given by $T(\mathbf{v}) = \det(A[\mathbf{v}])$ is a linear map. The same result holds with "row" replaced by "column" everywhere.

Proof. By Example 12.47, it suffices to show that there exist constants $b_1, \ldots, b_n \in \mathbb{R}$ such that for all $\mathbf{v} = (v_1, v_2, \ldots, v_n) \in \mathbb{R}^n$,

$$T(\mathbf{v}) = b_1 v_1 + b_2 v_2 + \cdots + b_n v_n. \tag{12.11}$$

To prove this, consider the defining formula for $\det(A[\mathbf{v}])$:

$$T(\mathbf{v}) = \det(A[\mathbf{v}]) = \sum_{w \in S_n} \operatorname{sgn}(w) \prod_{i=1}^{n} A[\mathbf{v}](i, w(i)) = \sum_{w \in S_n} \operatorname{sgn}(w) \left[\prod_{\substack{i=1 \\ i \neq k}}^{n} A(i, w(i)) \right] v_{w(k)}.$$

The bracketed expressions depend only on the fixed matrix A, not on $\mathbf{v}$. So (12.11) holds with

$$b_j = \sum_{\substack{w \in S_n: \\ w(k)=j}} \operatorname{sgn}(w) \prod_{\substack{i=1 \\ i \neq k}}^{n} A(i, w(i)) \qquad \text{for } j \text{ between 1 and } n. \tag{12.12}$$

To obtain the analogous fact for the columns of A, apply the result just proved to A^{tr}. □

We sometimes use the following notation when invoking the multilinearity of determinants. For $A \in M_n(\mathbb{R})$, let $A_1, A_2, \ldots, A_n$ denote the n rows of A; thus each A_i lies in $\mathbb{R}^n$. We write $\det(A) = \det(A_1, \ldots, A_n)$, viewing the determinant as a function of n inputs (each of which is a row vector). The previous result says if we let the kth input vary while holding the other $n-1$ inputs fixed, the resulting function sending $\mathbf{v} \in \mathbb{R}^n$ to $\det(A_1, \ldots, \mathbf{v}, \ldots, A_n)$ is a linear map from $\mathbb{R}^n$ to $\mathbb{R}$.

12.49. Theorem: Alternating Property of Determinants. If $A \in M_n(\mathbb{R})$ has two equal rows or two equal columns, then $\det(A) = 0$.

Proof. Recall $\det(A)$ is a sum of $n!$ signed terms of the form $T(w) = \operatorname{sgn}(w) \prod_{i=1}^{n} A(i, w(i))$, where w ranges over S_n. Suppose rows r and s of A are equal, so $A(r,k) = A(s,k)$ for all k. We define an involution I on S_n with no fixed points such that $T(I(w)) = -T(w)$ for all $w \in S_n$. It follows that the $n!$ terms cancel in pairs, so that $\det(A) = 0$. Define $I(w) = w \circ (r,s)$ for $w \in S_n$. We see at once that $I \circ I = \mathrm{id}_{S_n}$ and I has no fixed points. On one hand, $\operatorname{sgn}(I(w)) = \operatorname{sgn}(w) \cdot \operatorname{sgn}((r,s)) = -\operatorname{sgn}(w)$ by Theorem 7.31. On the other hand,

$$\begin{aligned} \prod_{i=1}^{n} A(i, [w \circ (r,s)](i)) &= A(r, w(s)) A(s, w(r)) \prod_{i \neq r,s} A(i, w(i)) \\ &= A(s, w(s)) A(r, w(r)) \prod_{i \neq r,s} A(i, w(i)) = \prod_{i=1}^{n} A(i, w(i)). \end{aligned}$$

Combining these facts, we see that $T(I(w)) = -T(w)$, as needed. Now if B has two equal columns, then B^{tr} has two equal rows, so $\det(B) = \det(B^{\text{tr}}) = 0$. □

The next theorem shows how elementary row operations (used in the Gaussian elimination algorithm) affect the determinant of a square matrix.

12.50. Theorem: Elementary Row Operations and Determinants. Suppose $A, B \in M_n(\mathbb{R})$, $j,k \in \{1,2,\ldots,n\}$, $j \neq k$, and $c \in \mathbb{R}$.

(a) If B is obtained from A by multiplying row j by c, then $\det(B) = c\det(A)$.

(b) If B is obtained from A by interchanging rows j and k, then $\det(B) = -\det(A)$.

(c) If B is obtained from A by adding c times row j to row k, then $\det(B) = \det(A)$.

Analogous results hold with "row" replaced by "column" everywhere.

Proof. Part (a) is a special case of the multilinearity of determinants (see Theorem 12.48). Part (b) is a consequence of multilinearity and the alternating property. Specifically, define $T : \mathbb{R}^n \times \mathbb{R}^n \to \mathbb{R}$ by letting $T(\mathbf{v}, \mathbf{w}) = \det(A_1, \ldots, \mathbf{v}, \ldots, \mathbf{w}, \ldots, A_n)$, where the row vectors $\mathbf{v}$ and $\mathbf{w}$ occur in positions j and k. Since det is multilinear and alternating, we have for each $\mathbf{v}, \mathbf{w} \in \mathbb{R}^n$:

$$0 = T(\mathbf{v}+\mathbf{w}, \mathbf{v}+\mathbf{w}) = T(\mathbf{v},\mathbf{v}) + T(\mathbf{w},\mathbf{v}) + T(\mathbf{v},\mathbf{w}) + T(\mathbf{w},\mathbf{w}) = T(\mathbf{w},\mathbf{v}) + T(\mathbf{v},\mathbf{w}).$$

Thus, $T(\mathbf{w},\mathbf{v}) = -T(\mathbf{v},\mathbf{w})$, which reduces to statement (b) upon choosing $\mathbf{v} = A_j$ and $\mathbf{w} = A_k$. Part (c) follows for similar reasons, since

$$T(\mathbf{v}, c\mathbf{v}+\mathbf{w}) = cT(\mathbf{v},\mathbf{v}) + T(\mathbf{v},\mathbf{w}) = T(\mathbf{v},\mathbf{w}). \qquad \square$$

Our next result is a recursive formula for computing determinants.

12.51. Theorem: Laplace Expansions of Determinants. For $A \in M_n(\mathbb{R})$ and all $i, j \in \{1,2,\ldots,n\}$, let $A[i|j]$ be the matrix in $M_{n-1}(\mathbb{R})$ obtained by deleting row i and column j of A. For each fixed k in the range $1 \leq k \leq n$,

$$\begin{aligned} \det(A) &= \sum_{i=1}^{n} (-1)^{i+k} A(i,k) \det(A[i|k]) \quad \text{(expansion along column } k) \\ &= \sum_{j=1}^{n} (-1)^{j+k} A(k,j) \det(A[k|j]) \quad \text{(expansion along row } k). \end{aligned}$$

Proof. We first prove the expansion formula along row $k = n$. By the proof of multilinearity (see Equations (12.11) and (12.12)), we know that

$$\det(A) = b_1 A(n,1) + b_2 A(n,2) + \cdots + b_n A(n,n)$$

where

$$b_j = \sum_{\substack{w \in S_n: \\ w(n)=j}} \operatorname{sgn}(w) \prod_{i=1}^{n-1} A(i, w(i)) \quad \text{for } j \text{ between 1 and } n.$$

To prove the formula in the theorem, it is enough to show that $b_j = (-1)^{j+n} \det(A[n|j])$ for all j between 1 and n.

Fix an index j. Let $S_{n,j} = \{w \in S_n : w(n) = j\}$. We define a bijection $f : S_{n,j} \to S_{n-1}$ as follows. Every $w \in S_{n,j}$ can be written in one-line form as $w = w_1 w_2 \cdots w_{n-1} w_n$ where $w_n = j$. Define $f(w) = w'_1 w'_2 \cdots w'_{n-1}$ where $w'_t = w_t$ if $w_t < j$, and $w'_t = w_t - 1$ if $w_t > j$. In other words, we drop the j at the end of w and subtract 1 from all letters larger than j. The inverse map acts on $w' \in S_{n-1}$ by adding 1 to all letters in w' weakly exceeding j and then putting a j at the end of the one-line form. Observe that the deletion of j decreases $\operatorname{inv}(w)$ by $n - j$ (which is the number of letters to the left of j that are greater than j), and the subtraction operation has no further effect on the inversion count. So, $\text{inv}(f(w)) = \text{inv}(w) - (n-j)$ and $\text{sgn}(f(w)) = (-1)^{j+n}\,\text{sgn}(w)$. We also note that for $w' = f(w)$, we have $A(i, w(i)) = A[n|j](i, w'(i))$ for all $i < n$, since all columns in A after column j get shifted left one column when column j is deleted. Now we use the bijection f to change the summation variable in the formula for b_j. Writing $w' = f(w)$, we obtain

$$\begin{aligned} b_j &= \sum_{w\in S_{n,j}} \text{sgn}(w) \prod_{i=1}^{n-1} A(i, w(i)) \\ &= \sum_{w'\in S_{n-1}} (-1)^{j+n}\,\text{sgn}(w') \prod_{i=1}^{n-1} A[n|j](i, w'(i)) = (-1)^{j+n} \det(A[n|j]). \end{aligned}$$

The expansion along an arbitrary row k follows from the special case $k = n$. Given $k < n$, let B be the matrix obtained from A by interchanging rows k and $k+1$, then rows $k+1$ and $k+2$, and so on, until the original row k has reached the last row of the matrix. The procedure converting A to B involves $n-k$ row interchanges, so $\det(B) = (-1)^{n-k}\det(A)$. Moreover, $B(n, j) = A(k, j)$ and $B[n|j] = A[k|j]$ for j between 1 and n. So

$$\begin{aligned} \det(A) &= (-1)^{k-n}\det(B) = (-1)^{k-n}\sum_{j=1}^{n}(-1)^{j+n}B(n,j)\det(B[n|j]) \\ &= \sum_{j=1}^{n}(-1)^{j+k}A(k,j)\det(A[k|j]). \end{aligned}$$

Finally, to derive the expansion along column k, we transpose the matrix. We see from the definitions that $A^{\text{tr}}[k|j] = A[j|k]^{\text{tr}}$, and this matrix has the same determinant as $A[j|k]$. Therefore,

$$\begin{aligned} \det(A) &= \det(A^{\text{tr}}) = \sum_{j=1}^{n}(-1)^{j+k}A^{\text{tr}}(k,j)\det(A^{\text{tr}}[k|j]) \\ &= \sum_{j=1}^{n}(-1)^{j+k}A(j,k)\det(A[j|k]). \quad \square \end{aligned}$$

Next we show how the Laplace expansions for $\det(A)$ lead to explicit formulas for the entries in the inverse of a matrix.

12.52. Definition: Classical Adjoint of a Matrix. Given $A \in M_n(\mathbb{R})$, let $\text{adj}\,A \in M_n(\mathbb{R})$ be the matrix with i,j-entry $(-1)^{i+j}\det(A[j|i])$ for $i, j \in \{1, 2, \ldots, n\}$.

The next result explains why $A[j|i]$ appears instead of $A[i|j]$ in the preceding definition.

12.53. Theorem: Adjoint Formula. For all $A \in M_n(\mathbb{R})$, we have

$$A(\text{adj}\,A) = \det(A)I_n = (\text{adj}\,A)A.$$

Proof. For i between 1 and n, the i,i-entry of the product $A(\text{adj}\,A)$ is

$$\sum_{k=1}^{n} A(i,k)[\text{adj}\,A](k,i) = \sum_{k=1}^{n}(-1)^{i+k}A(i,k)\det(A[i|k]) = \det(A),$$

by Laplace expansion along row i of A. Now suppose $i \neq j$. The i,j-entry of $A(\text{adj}\,A)$ is

$$\sum_{k=1}^{n} A(i,k)[\text{adj}\,A](k,j) = \sum_{k=1}^{n}(-1)^{j+k}A(i,k)\det(A[j|k]).$$

Let C be the matrix obtained from A by replacing row j of A by row i of A. Then $C(j,k) = A(i,k)$ and $C[j|k] = A[j|k]$ for all k. So the preceding expression is the Laplace expansion for $\det(C)$ along row j. On the other hand, $\det(C) = 0$ because C has two equal rows. So $[A(\operatorname{adj} A)](i,j) = 0$. We have proved that $A(\operatorname{adj} A)$ is a diagonal matrix with all diagonal entries equal to $\det(A)$, so that $A(\operatorname{adj} A) = \det(A)I_n$. The analogous result for $(\operatorname{adj} A)A$ is proved similarly, using column expansions. □

12.54. Corollary: Formula for the Inverse of a Matrix. If $A \in M_n(\mathbb{R})$ and $\det(A)$ is a nonzero element of $\mathbb{R}$, then the matrix A is invertible in $M_n(\mathbb{R})$ with inverse

$$A^{-1} = \frac{1}{\det(A)} \operatorname{adj} A.$$

12.55. Remark. Conversely, if A is invertible in $M_n(\mathbb{R})$, then $\det(A)$ is a nonzero element of $\mathbb{R}$. The proof uses the following *Product Formula for Determinants*:

$$\det(AB) = \det(A)\det(B) = \det(B)\det(A) \qquad \text{for all } A, B \in M_n(\mathbb{R}).$$

Taking $B = A^{-1}$, the left side becomes $\det(I_n) = 1$, so $\det(B)$ is a two-sided inverse of $\det(A)$ in $\mathbb{R}$. In particular, $\det(A)$ and $\det(B)$ are nonzero. We deduce the Product Formula as a consequence of the Cauchy–Binet Theorem, which is proved in the next section.

More generally, when $\mathbb{R}$ is replaced by a commutative ring R, the same proof shows that a matrix A is an invertible element of the ring $M_n(R)$ iff $\det(A)$ is an invertible element of the ring R.

12.10 The Cauchy–Binet Theorem

We continue our study of determinants by giving a combinatorial proof of the Cauchy–Binet Theorem, which expresses the determinant of a product of rectangular matrices as the sum of products of determinants of certain submatrices. This proof is a nice application of the properties of inversions and determinants.

To state the Cauchy–Binet Theorem, we need the following notation. Given a $c \times d$ matrix M, write M_i for the ith row of M and M^j for the jth column of M. Given indices $j_1, \ldots, j_c \in \{1, 2, \ldots, d\}$, let $(M^{j_1}, \ldots, M^{j_c})$ denote the $c \times c$ matrix whose columns are $M^{j_1}, \ldots, M^{j_c}$ in this order. Similarly, given $i_1, \ldots, i_d \in \{1, 2, \ldots, c\}$, let $(M_{i_1}, \ldots, M_{i_d})$ be the $d \times d$ matrix whose rows are $M_{i_1}, \ldots, M_{i_d}$ in this order.

12.56. The Cauchy–Binet Theorem. Suppose $m \le n$, A is an $m \times n$ matrix, and B is an $n \times m$ matrix. Let J be the set of all lists $\mathbf{j} = (j_1, j_2, \ldots, j_m)$ such that $1 \le j_1 < j_2 < \cdots < j_m \le n$. Then

$$\det(AB) = \sum_{\mathbf{j} \in J} \det(A^{j_1}, A^{j_2}, \ldots, A^{j_m}) \det(B_{j_1}, B_{j_2}, \ldots, B_{j_m}).$$

Proof. All matrices appearing in the displayed formula are $m \times m$, so all the determinants are defined. We begin by using the definitions of matrix products and determinants (§12.9) to write

$$\det(AB) = \sum_{w \in S_m} \operatorname{sgn}(w) \prod_{i=1}^{m} (AB)(i, w(i)) = \sum_{w \in S_m} \operatorname{sgn}(w) \prod_{i=1}^{m} \left[\sum_{k_i=1}^{n} A(i, k_i) B(k_i, w(i)) \right].$$

The Generalized Distributive Law changes the product of sums into a sum of products:

$$\det(AB) = \sum_{w\in S_m} \sum_{k_1=1}^{n} \cdots \sum_{k_m=1}^{n} \operatorname{sgn}(w) \prod_{i=1}^{m} A(i,k_i) \prod_{i=1}^{m} B(k_i, w(i)).$$

Let K be the set of all lists $\mathbf{k} = (k_1, \ldots, k_m)$ with every $k_i \in \{1, 2, \ldots, n\}$, and let K' be the set of lists in K whose entries k_i are distinct. We can combine the m separate sums over $k_1, \ldots, k_m$ into a single sum over lists $\mathbf{k} \in K$. We can also reorder the summations to get

$$\det(AB) = \sum_{\mathbf{k}\in K} \sum_{w\in S_m} \operatorname{sgn}(w) \prod_{i=1}^{m} A(i,k_i) \prod_{i=1}^{m} B(k_i, w(i)).$$

Next, factor out quantities that do not depend on w:

$$\det(AB) = \sum_{k\in K} \prod_{i=1}^{m} A(i,k_i) \left[\sum_{w\in S_m} \operatorname{sgn}(w) \prod_{i=1}^{m} B(k_i, w(i)) \right].$$

The term in brackets is the defining formula for $\det(B_{k_1}, \ldots, B_{k_m})$. If any two entries in $(k_1, \ldots, k_m)$ are equal, this matrix has two equal rows, so its determinant is zero. Discarding these terms, we are reduced to summing over lists $\mathbf{k} \in K'$. So now we have

$$\det(AB) = \sum_{\mathbf{k}\in K'} \prod_{i=1}^{m} A(i,k_i) \det(B_{k_1}, \ldots, B_{k_m}).$$

To continue, observe that for every list $\mathbf{k} \in K'$ there exists a unique list $\mathbf{j} \in J$, denoted $\mathbf{j} = \operatorname{sort}(\mathbf{k})$, obtained by sorting the entries of $\mathbf{k}$ into increasing order. Grouping summands gives

$$\det(AB) = \sum_{\mathbf{j}\in J} \sum_{\substack{\mathbf{k}\in K': \\ \operatorname{sort}(\mathbf{k})=\mathbf{j}}} \prod_{i=1}^{m} A(i,k_i) \det(B_{k_1}, \ldots, B_{k_m}).$$

Given that $\operatorname{sort}(\mathbf{k}) = \mathbf{j}$, we can change the matrix $(B_{k_1}, \ldots, B_{k_m})$ into the matrix $(B_{j_1}, \ldots, B_{j_m})$ by repeatedly switching adjacent rows. Each such switch flips the sign of the determinant, and one checks that the number of row switches required is $\operatorname{inv}(k_1 k_2 \cdots k_m)$. (To see this, adapt the proof of Theorem 7.29 to the case where the objects being sorted are $j_1 < j_2 < \cdots < j_m$ instead of $1 < 2 < \cdots < m$.) Letting $\operatorname{sgn}(\mathbf{k}) = (-1)^{\operatorname{inv}(\mathbf{k})}$, we can therefore write

$$\det(AB) = \sum_{\mathbf{j}\in J} \sum_{\substack{\mathbf{k}\in K': \\ \operatorname{sort}(\mathbf{k})=\mathbf{j}}} \operatorname{sgn}(\mathbf{k}) \prod_{i=1}^{m} A(i,k_i) \det(B_{j_1}, \ldots, B_{j_m}).$$

The determinant in this formula depends only on $\mathbf{j}$, not on $\mathbf{k}$, so it can be brought out of the inner summation:

$$\det(AB) = \sum_{\mathbf{j}\in J} \det(B_{j_1}, \ldots, B_{j_m}) \sum_{\substack{\mathbf{k}\in K': \\ \operatorname{sort}(\mathbf{k})=\mathbf{j}}} \operatorname{sgn}(\mathbf{k}) \prod_{i=1}^{m} A(i,k_i).$$

To finish, note that every $\mathbf{k} \in K'$ that sorts to $\mathbf{j}$ can be written as $(k_1, \ldots, k_m) =$

$(j_{v(1)},\dots,j_{v(m)})$ for a uniquely determined permutation $v \in S_m$. Since j is an increasing sequence, it follows that $\text{inv}(\mathbf{k}) = \text{inv}(v)$ and $\text{sgn}(\mathbf{k}) = \text{sgn}(v)$. Changing variables in the inner summation, we get

$$\det(AB) = \sum_{\mathbf{j}\in J} \det(B_{j_1},\dots,B_{j_m}) \left[\sum_{v\in S_m} \text{sgn}(v) \prod_{i=1}^{m} A(i, j_{v(i)}) \right].$$

The term in brackets is none other than $\det(A^{j_1},\dots,A^{j_m})$, so the proof is complete. □

12.57. The Product Formula for Determinants. If A and B are $m \times m$ matrices, then $\det(AB) = \det(A)\det(B)$.

Proof. Take $n = m$ in the Cauchy–Binet Theorem. The index set J consists of the single list $(1, 2, \dots, m)$, and the summand corresponding to this list reduces to $\det(A)\det(B)$. □

Other examples of combinatorial proofs of determinant formulas appear in §10.15 and §12.11.

12.11 Tournaments and the Vandermonde Determinant

This section uses the combinatorics of tournaments to prove Vandermonde's determinant formula.

12.58. Definition: Tournaments. An *n-player tournament* is a digraph τ with vertex set $\{1, 2, \dots, n\}$ such that for all vertices $i \neq j$, exactly one of the directed edges (i, j) or (j, i) is an edge of τ. Also, no loop edge (i, i) is an edge of τ. Let T_n be the set of all n-player tournaments.

Intuitively, the n vertices represent n players who compete in a series of one-on-one matches. Each player plays every other player exactly once, and there are no ties. If player i beats player j, the edge (i, j) is part of the tournament; otherwise, the edge (j, i) is included.

12.59. Definition: Weights, Inversions, and Sign for Tournaments. Given a tournament $\tau \in T_n$, the *weight* of τ is $\text{wt}(\tau) = \prod_{i=1}^{n} x_i^{\text{outdeg}_\tau(i)}$. The *inversion statistic* for τ, denoted $\text{inv}(\tau)$, is the number of $i < j$ such that (j, i) is an edge of τ. The *sign* of τ is $\text{sgn}(\tau) = (-1)^{\text{inv}(\tau)}$.

Informally, $\text{wt}(\tau) = x_1^{e_1} \cdots x_n^{e_n}$ means that player i beats e_i other players for all i between 1 and n. If we think of the players' numbers $1, 2, \dots, n$ as giving the initial rankings of each player, with 1 being the highest rank, then $\text{inv}(\tau)$ counts the number of times a lower-ranked player beats a higher-ranked player in the tournament τ.

12.60. Example. Consider the tournament $\tau \in T_5$ with edge set

$$\{(1,3),(1,4),(1,5),(2,1),(2,4),(3,2),(3,4),(3,5),(5,2),(5,4)\}.$$

We have $\text{wt}(\tau) = x_1^3 x_2^2 x_3^3 x_5^2$, $\text{inv}(\tau) = 4$, and $\text{sgn}(\tau) = +1$.

12.61. Theorem: Tournament Generating Function. For all integers $n \geq 1$,

$$\sum_{\tau\in T_n} \text{sgn}(\tau)\,\text{wt}(\tau) = \prod_{1\leq i<j\leq n} (x_i - x_j).$$

Proof. We can build a tournament $\tau \in T_n$ by making a sequence of binary choices, indexed by the pairs (i,j) with $1 \leq i < j \leq n$. For each $i < j$, we choose one of the edges (i,j) or (j,i) and add it to the tournament's edge set. Let us examine the effect of this choice on $\mathrm{wt}(\tau)$, $\mathrm{inv}(\tau)$, and $\mathrm{sgn}(\tau)$. If we pick the edge (i,j) (so i beats j), then the exponent of x_i goes up by 1, inversions go up by 0, and the sign is unchanged. If we pick edge (j,i) instead (so j beats i), then the exponent of x_j goes up by 1, inversions go up by 1, and the sign is multiplied by -1. The generating function $(+x_i - x_j)$ records the effect of this choice. The proof is completed by invoking the Product Rule for Generating Functions. □

Given a tournament τ, there may exist three players u, v, w where u beats v, v beats w, and w beats u. This situation occurs whenever the digraph τ contains a directed 3-cycle. Tournaments where this circularity condition does *not* occur are given a special name.

12.62. Definition: Transitive Tournaments. A tournament $\tau \in T_n$ is *transitive* iff for all vertices u, v, w, if (u,v) and (v,w) are edges of τ, then (u,w) is an edge of τ.

Note that if (u,v) and (v,w) are edges in τ, we must have $w \neq u$, and then (w,u) is an edge of τ iff (u,w) is not an edge of τ. It follows that a tournament τ is not transitive iff there exist vertices u, v, w such that (u,v), (v,w), and (w,u) are all edges of τ.

12.63. Theorem: Generating Function for Transitive Tournaments. Let T'_n be the set of transitive tournaments in T_n. Then

$$\sum_{\tau \in T'_n} \mathrm{sgn}(\tau)\,\mathrm{wt}(\tau) = \sum_{w \in S_n} \mathrm{sgn}(w) \prod_{k=1}^{n} x_{w(k)}^{n-k}.$$

Proof. We define a bijection $f : T'_n \to S_n$ that will be used to transfer signs and weights from T'_n to S_n. Given $\tau \in T'_n$, define an associated relation $\preceq$ on $\{1, 2, \ldots, n\}$ by setting $u \preceq v$ iff $u = v$ or (u,v) is an edge of τ. This relation is reflexive, antisymmetric (since τ is a tournament), and transitive (since τ is transitive). Furthermore, for all $u, v \in \{1, 2, \ldots, n\}$, $u \preceq v$ or $v \preceq u$ since τ is a tournament. So $\preceq$ is a total ordering of $\{1, 2, \ldots, n\}$. This ordering determines a unique permutation $w = f(\tau) \in S_n$ that satisfies $w_1 \prec w_2 \prec \cdots \prec w_n$. For each k, player w_k beats players w_m for all $m > k$ and loses to players w_m for all $m < k$. One readily checks that f is a bijection; the inverse map sends $w \in S_n$ to the transitive tournament with edge set $\{(w_i, w_j) : 1 \leq i < j \leq n\}$.

Given $\tau \in T'_n$ and $w = f(\tau) \in S_n$, let us compare $\mathrm{inv}(\tau)$ to $\mathrm{inv}(w)$. On one hand, $\mathrm{inv}(w)$ is defined as the number of $i < j$ with $w_i > w_j$. On the other hand, the description of the edge set of $\tau = f^{-1}(w)$ at the end of the last paragraph shows that the number of $i < j$ with $w_i > w_j$ is also equal to $\mathrm{inv}(\tau)$. So $\mathrm{inv}(\tau) = \mathrm{inv}(w)$, and hence $\mathrm{sgn}(\tau) = \mathrm{sgn}(w)$.

Next, let us express $\mathrm{wt}(\tau)$ in terms of w. Since player w_k beats exactly those players w_m such that $k < m \leq n$, we see that $\mathrm{wt}(\tau) = \prod_{k=1}^{n} x_{w_k}^{n-k}$. Define $\mathrm{wt}(w)$ by the right side of this formula. The theorem now follows because f is a weight-preserving, sign-preserving bijection. □

We can use the bijection f to characterize transitive tournaments.

12.64. Theorem: Criterion for Transitive Tournaments. A tournament $\tau \in T_n$ is transitive iff no two vertices of τ have the same outdegree.

Proof. Given a transitive $\tau \in T_n$, let $w = f(\tau)$ be the permutation constructed in the preceding proof. We have shown that $\mathrm{wt}(\tau) = \prod_{k=1}^{n} x_{w_k}^{n-k}$. The exponents $n-k$ are all distinct, so every vertex of τ has a different outdegree.

Conversely, suppose $\tau \in T_n$ is such that every vertex has a different outdegree. There are n vertices and n possible outdegrees (namely $0, 1, \ldots, n-1$), so each possible outdegree occurs at exactly one vertex. Let w_1 be the unique vertex with outdegree $n-1$. Then w_1 beats all other players. Next, let w_2 be the unique vertex with outdegree $n-2$. Then w_2 must beat all players except w_1. Continuing similarly, we obtain a permutation $w = w_1 w_2 \cdots w_n$ of $\{1, 2, \ldots, n\}$ such that w_j beats w_k iff $j < k$. To confirm that τ is transitive, consider three players w_i, w_j, w_k such that (w_i, w_j) and (w_j, w_k) are edges of τ. Then $i < j$ and $j < k$, so $i < k$, so (w_i, w_k) is an edge of τ. (In fact, $\tau = f^{-1}(w)$.) □

12.65. Theorem: Vandermonde Determinant Formula. Let $x_1, \ldots, x_n$ be real numbers or formal variables. Define an $n \times n$ matrix V by setting $V(i,j) = x_j^{n-i}$ for $1 \leq i, j \leq n$. Then

$$\det(V) = \prod_{1 \leq i < j \leq n} (x_i - x_j).$$

Proof. According to Definition 12.40,

$$\det(V) = \sum_{w \in S_n} \operatorname{sgn}(w) \prod_{k=1}^{n} V(k, w(k)) = \sum_{w \in S_n} \operatorname{sgn}(w) \prod_{k=1}^{n} x_{w(k)}^{n-k}. \tag{12.13}$$

This is the generating function for transitive tournaments, whereas $\prod_{i<j}(x_i - x_j)$ is the generating function for all tournaments with n players. So, it suffices to define a sign-reversing, weight-preserving involution $I : T_n \to T_n$ with fixed point set T'_n. Define $I(\tau) = \tau$ for $\tau \in T'_n$. Now consider a non-transitive tournament $\tau \in T_n - T'_n$. By Theorem 12.64, there exist two vertices $i < j$ with the same outdegree in τ. If several pairs of vertices have the same outdegree, then choose the pair such that i and then j is minimized. Define $I(\tau)$ by switching the roles of i and j in τ; more precisely, replace every directed edge (u, v) in τ by $(s_{i,j}(u), s_{i,j}(v))$, where $s_{i,j}$ is the transposition $(i, j) \in S_n$. The resulting tournament is non-transitive (since i and j still have the same outdegree in $I(\tau)$) and has the same weight as τ. Furthermore, $I(I(\tau)) = \tau$.

To finish, we show that $\operatorname{sgn}(I(\tau)) = -\operatorname{sgn}(\tau)$. Consider the factorization of $(i, j) \in S_n$ into $2(j - i) - 1$ basic transpositions:

$$(i, j) = (j-1, j)(j-2, j-1) \cdots (i+1, i+2)(i, i+1)(i+1, i+2) \cdots (j-2, j-1)(j-1, j).$$

We can pass from τ to $I(\tau)$ in stages, by applying these basic transpositions one at a time to the endpoints of the directed edges in τ. We claim that each such step changes the sign of the tournament. For, consider what happens to the inversion count when we pass from a tournament σ to σ' by switching labels k and $k+1$. The inversion $(k+1, k)$ is present in exactly one of the tournaments σ and σ', and the other inversions are unaffected by the label switch. So $\operatorname{inv}(\sigma')$ differs from $\operatorname{inv}(\sigma)$ by ± 1, and hence $\operatorname{sgn}(\sigma') = -\operatorname{sgn}(\sigma)$. Since we pass from τ to $I(\tau)$ by an odd number of moves of this type (namely $2(j - i) - 1$ moves), we see that $\operatorname{sgn}(I(\tau)) = -\operatorname{sgn}(\tau)$, as needed. □

12.12 The Hook-Length Formula

This section presents a probabilistic proof of the Hook-Length Formula for the number of standard tableaux of a given shape. This formula was first stated in the Introduction. For the reader's convenience, we begin by recalling the relevant definitions.

12.66. Definitions. An *integer partition of* n is a weakly decreasing sequence $\lambda = (\lambda_1 \geq \lambda_2 \geq \cdots \geq \lambda_l)$ of positive integers with $\lambda_1 + \cdots + \lambda_l = n$. The *diagram* of λ is

$$\mathrm{dg}(\lambda) = \{(i,j) \in \mathbb{Z}_{>0} \times \mathbb{Z}_{>0} : 1 \leq i \leq l, 1 \leq j \leq \lambda_i\}.$$

Each $(i,j) \in \mathrm{dg}(\lambda)$ is called a *box* or a *cell*. We take i as the row index and j as the column index, where the highest row is row 1. Given any cell $c = (i,j) \in \mathrm{dg}(\lambda)$, the *hook* of c in λ is

$$H(c) = \{(i,k) \in \mathrm{dg}(\lambda) : k \geq j\} \cup \{(k,j) \in \mathrm{dg}(\lambda) : k \geq i\}.$$

The *hook length* of c in λ is $h(c) = |H(c)|$. A *corner box* of λ is a cell $c \in \mathrm{dg}(\lambda)$ with $h(c) = 1$. A *standard tableau of shape* λ is a bijection $S : \mathrm{dg}(\lambda) \to \{1, 2, \ldots, n\}$ such that $S(i,j) < S(i,j+1)$ for all i,j such that (i,j) and $(i,j+1)$ are in $\mathrm{dg}(\lambda)$, and $S(i,j) < S(i+1,j)$ for all i,j such that (i,j) and $(i+1,j)$ are in $\mathrm{dg}(\lambda)$. Let $\mathrm{SYT}(\lambda)$ be the set of standard tableaux of shape λ, and let $f^\lambda = |\,\mathrm{SYT}(\lambda)|$.

12.67. Example. If $\lambda = (7,5,5,4,2,1)$ and $c = (3,2)$, then

$$H(c) = \{(3,2), (3,3), (3,4), (3,5), (4,2), (5,2)\},$$

and $h(c) = 6$. We can visualize $\mathrm{dg}(\lambda)$ and $H(c)$ using the following picture.

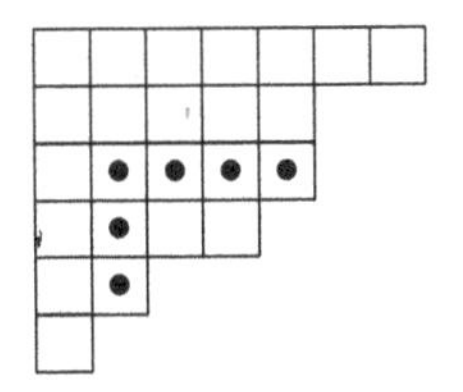

Let λ'_j be the number of boxes in column j of $\mathrm{dg}(\lambda)$. Then $h(i,j) = (\lambda_i - j) + (\lambda'_j - i) + 1$. We use this formula to establish the following lemma.

12.68. Lemma. Suppose λ is a partition of n, (r,s) is a corner box of λ, and $(i,j) \in \mathrm{dg}(\lambda)$ satisfies $i < r$ and $j < s$. Then $h(i,j) = h(r,j) + h(i,s) - 1$.

Proof. Since (r,s) is a corner box, $\lambda_r = s$ and $\lambda'_s = r$. So

$$\begin{aligned} h(r,j) + h(i,s) - 1 &= [(\lambda_r - j) + (\lambda'_j - r) + 1] + [(\lambda_i - s) + (\lambda'_s - i) + 1] - 1 \\ &= s - j + \lambda'_j - r + \lambda_i - s + r - i + 1 \\ &= (\lambda_i - j) + (\lambda'_j - i) + 1 = h(i,j). \quad \square \end{aligned}$$

12.69. The Hook-Length Formula. For any partition λ of n,

$$f^\lambda = \frac{n!}{\prod_{c \in \mathrm{dg}(\lambda)} h(c)}.$$

The idea of the proof is to define a *random algorithm* that takes a partition λ of n as input and produces a standard tableau $S \in \mathrm{SYT}(\lambda)$ as output. We prove in Theorem 12.74 that this algorithm outputs any given standard tableau S with probability

$$p = \frac{\prod_{c \in \mathrm{dg}(\lambda)} h(c)}{n!}.$$

This probability depends only on λ, not on S, so we obtain a uniform probability distribution on the sample space $\mathrm{SYT}(\lambda)$. So, on one hand, each standard tableau is produced with probability p; and on the other hand, each standard tableau is produced with probability $1/|\operatorname{SYT}(\lambda)| = 1/f^\lambda$. Thus $f^\lambda = 1/p$, and we obtain the Hook-Length Formula.

Here is an informal description of the the algorithm for generating a random standard tableau of shape λ. Start at a random cell in the shape λ. As long as we are not at a corner box, we jump from our current box c to some other cell in $H(c)$; each cell in the hook is chosen with equal probability. This jumping process eventually takes us to a corner cell. We place the entry n in this box, and then pretend this cell is no longer there. We are left with a partition μ of size $n-1$. Proceed recursively to select a random standard tableau of shape μ. Adding back the corner cell containing n gives the standard tableau of shape λ produced by the algorithm.

Now we give a formal description of the algorithm. Every random choice below is to be independent of all other choices.

12.70. Tableau Generation Algorithm. The input to the algorithm is a partition λ of n. The output is a tableau $S \in \operatorname{SYT}(\lambda)$, constructed according to the following random procedure. As a base case, if $n = 0$, return the empty tableau of shape 0.

1. Choose a random cell $c \in \operatorname{dg}(\lambda)$. Each cell in $\operatorname{dg}(\lambda)$ is chosen with probability $1/n$.
2. While $h(c) > 1$, do the following.
   2a. Choose a random cell $c' \in H(c) - \{c\}$. Each cell in $H(c) - \{c\}$ is chosen with probability $1/(h(c) - 1)$.
   2b. Replace c by c' and go back to Step 2.
3. Now c is a corner box of $\operatorname{dg}(\lambda)$, so $\operatorname{dg}(\lambda) - \{c\}$ is the diagram of some partition μ of $n-1$. Recursively use the same algorithm to generate a random standard tableau $S' \in \operatorname{SYT}(\mu)$. Extend this to a standard tableau $S \in \operatorname{SYT}(\lambda)$ by setting $S(c) = n$, and output S as the answer.

Let $(c_1, c_2, c_3, \ldots, c_k)$ be the sequence of cells chosen in Steps 1 and 2. Call this sequence the *hook walk for n*. Note that the hook walk must be finite, since $h(c_1) > h(c_2) > h(c_3) > \cdots$. Writing $c_s = (i_s, j_s)$ for each s, define $I = \{i_1, \ldots, i_{k-1}\} - \{i_k\}$ and $J = \{j_1, \ldots, j_{k-1}\} - \{j_k\}$. We call I and J the *row set* and *column set* for this hook walk.

12.71. Example. Given $n = 24$ and $\lambda = (7, 5, 5, 4, 2, 1)$, the first iteration of the algorithm might proceed as follows.

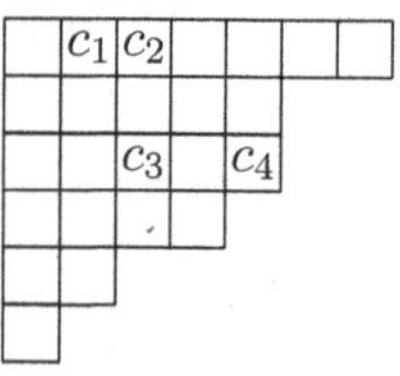

In this situation, we place $n = 24$ in corner box c_4 and proceed recursively to fill in the rest of the tableau. The probability that the algorithm chooses this particular hook walk for n is

$$\frac{1}{n} \cdot \frac{1}{h(c_1) - 1} \cdot \frac{1}{h(c_2) - 1} \cdot \frac{1}{h(c_3) - 1} = \frac{1}{24} \cdot \frac{1}{9} \cdot \frac{1}{7} \cdot \frac{1}{3}.$$

The row set and column set for this hook walk are $I = \{1\}$ and $J = \{2, 3\}$.

The next lemma is the key technical fact needed to analyze the behavior of the tableau generation algorithm.

12.72. Lemma. Given a partition λ of n, a corner box $c=(r,s)$, and sets $I\subseteq\{1,2,\ldots,r-1\}$ and $J\subseteq\{1,2,\ldots,s-1\}$, the probability that the hook walk for n ends at c with row set I and column set J is

$$p(\lambda,c,I,J)=\frac{1}{n}\prod_{i\in I}\frac{1}{h(i,s)-1}\prod_{j\in J}\frac{1}{h(r,j)-1}.$$

Proof. Write $I=\{i_1<i_2<\cdots<i_\ell\}$ and $J=\{j_1<j_2<\cdots<j_m\}$, where $\ell,m\geq 0$. First we consider some degenerate cases. Say $I=J=\emptyset$. Then the hook walk for n consists of the single cell c. This happens with probability $1/n$, in agreement with the formula in the lemma (interpreting the empty products as 1). Next, suppose I is empty but J is not. The hook walk for n in this case must be $c_1=(r,j_1)$, $c_2=(r,j_2)$, …, $c_m=(r,j_m)$, $c_{m+1}=(r,s)$. The probability of this hook walk is

$$\frac{1}{n}\cdot\frac{1}{h(c_1)-1}\cdot\frac{1}{h(c_2)-1}\cdot\cdots\cdot\frac{1}{h(c_m)-1}=\frac{1}{n}\prod_{j\in J}\frac{1}{h(r,j)-1}.$$

Similarly, the result holds when J is empty and I is nonempty.

Now consider the case where both I and J are nonempty. We use induction on $|I|+|J|$. A hook walk with row set I and column set J ending at c must begin with the cell $c_1=(i_1,j_1)$; this cell is chosen in Step 1 of the algorithm with probability $1/n$. Now, there are two possibilities for cell c_2: either $c_2=(i_1,j_2)$ or $c_2=(i_2,j_1)$. Each possibility for c_2 is chosen with probability $1/(h(c)-1)=1/(h(i_1,j_1)-1)$. When $c_2=(i_1,j_2)$, the sequence $(c_2,\ldots,c_k)$ is a hook walk ending at c with row set I and column set $J'=J-\{j_1\}$. By induction, such a hook walk occurs with probability

$$\frac{1}{n}\prod_{i\in I}\frac{1}{h(i,s)-1}\prod_{j\in J'}\frac{1}{h(r,j)-1}.$$

However, since the walk really started at c_1 and proceeded to c_2, we replace the first factor $1/n$ by $\frac{1}{n}\cdot\frac{1}{h(c_1)-1}$. Similarly, when $c_2=(i_2,j_1)$, the sequence $(c_2,\ldots,c_k)$ is a hook walk ending at c with row set $I'=I-\{i_1\}$ and column set J. So the probability that the hook walk starts at c_1 and proceeds through $c_2=(i_2,j_1)$ is

$$\frac{1}{n}\cdot\frac{1}{h(c_1)-1}\prod_{i\in I'}\frac{1}{h(i,s)-1}\prod_{j\in J}\frac{1}{h(r,j)-1}.$$

Adding these two terms, we see that

$$p(\lambda,c,I,J)=\frac{1}{n}\cdot\frac{1}{h(c_1)-1}\prod_{i\in I'}\frac{1}{h(i,s)-1}\prod_{j\in J'}\frac{1}{h(r,j)-1}\cdot\left(\frac{1}{h(i_1,s)-1}+\frac{1}{h(r,j_1)-1}\right).$$

The factor in parentheses is

$$\frac{h(r,j_1)+h(i_1,s)-2}{(h(i_1,s)-1)(h(r,j_1)-1)}.$$

Using Lemma 12.68, the numerator simplifies to $h(i_1,j_1)-1=h(c_1)-1$. This factor cancels and leaves us with

$$p(\lambda,c,I,J)=\frac{1}{n}\prod_{i\in I}\frac{1}{h(i,s)-1}\prod_{j\in J}\frac{1}{h(r,j)-1}.$$

This completes the induction proof. □

12.73. Theorem: Probability that a Hook Walk ends at c. Given a partition λ of n and a corner box $c = (r, s)$ of $\text{dg}(\lambda)$, the probability that the hook walk for n ends at c is

$$p(\lambda, c) = \frac{1}{n} \prod_{i=1}^{r-1} \frac{h(i,s)}{h(i,s)-1} \prod_{j=1}^{s-1} \frac{h(r,j)}{h(r,j)-1}.$$

Proof. Write $[r-1] = \{1, 2, \ldots, r-1\}$ and $[s-1] = \{1, 2, \ldots, s-1\}$. By the Sum Rule for probabilities,

$$\begin{aligned} p(\lambda, c) &= \sum_{I \subseteq [r-1]} \sum_{J \subseteq [s-1]} p(\lambda, c, I, J) \\ &= \frac{1}{n} \sum_{I \subseteq [r-1]} \sum_{J \subseteq [s-1]} \prod_{i \in I} \frac{1}{h(i,s)-1} \prod_{j \in J} \frac{1}{h(r,j)-1} \\ &= \frac{1}{n} \left(\sum_{I \subseteq [r-1]} \prod_{i \in I} \frac{1}{h(i,s)-1} \right) \cdot \left(\sum_{J \subseteq [s-1]} \prod_{j \in J} \frac{1}{h(r,j)-1} \right). \end{aligned}$$

By induction on r, or by the Generalized Distributive Law, it can be checked that

$$\sum_{I \subseteq [r-1]} \prod_{i \in I} \frac{1}{h(i,s)-1} = \prod_{i=1}^{r-1} \left(1 + \frac{1}{h(i,s)-1} \right) = \prod_{i=1}^{r-1} \frac{h(i,s)}{h(i,s)-1}$$

(cf. the proof of (4.3)). The sum over J can be simplified in a similar way, giving the formula in the theorem. □

The next theorem is the final step in the proof of the Hook-Length Formula.

12.74. Theorem: Probability of Generating a Given Tableau. If λ is a fixed partition of n and $S \in \text{SYT}(\lambda)$, the tableau generation algorithm for λ outputs S with probability

$$\frac{\prod_{c \in \text{dg}(\lambda)} h(c)}{n!}.$$

Proof. We prove the theorem by induction on n. Note first that the result does hold for $n = 0$ and $n = 1$. For the induction step, assume the result is known for partitions and tableaux with fewer than n boxes. Let $c^* = (r, s)$ be the cell such that $S(c^*) = n$, let μ be the partition obtained by removing c^* from $\text{dg}(\lambda)$, and let $S' \in \text{SYT}(\mu)$ be the tableau obtained by erasing n from S. First, the probability that the hook walk for n (in Steps 1 and 2 of the algorithm) ends at c^* is $p(\lambda, c^*)$. Given that this event has occurred, induction tells us that the probability of generating S' in Step 3 is

$$\frac{\prod_{c \in \text{dg}(\mu)} h_\mu(c)}{(n-1)!},$$

where $h_\mu(c)$ refers to the hook length of c relative to $\text{dg}(\mu)$. Multiplying these probabilities, the probability of generating S is therefore

$$\frac{1}{n!} \prod_{c \in \text{dg}(\mu)} h_\mu(c) \prod_{i=1}^{r-1} \frac{h_\lambda(i,s)}{h_\lambda(i,s)-1} \prod_{j=1}^{s-1} \frac{h_\lambda(r,j)}{h_\lambda(r,j)-1}.$$

Now, consider what happens to the hook lengths of cells when we pass from μ to λ by restoring the box $c^* = (r,s)$. For every cell $c = (i,j) \in \mathrm{dg}(\mu)$ with $i \neq r$ and $j \neq s$, we have $h_\mu(c) = h_\lambda(c)$. If $c = (i,s) \in \mathrm{dg}(\mu)$ with $i < r$, then $h_\mu(c) = h_\lambda(c) - 1 = h_\lambda(i,s) - 1$. Thus, the fractions in the second product convert $h_\mu(c)$ to $h_\lambda(c)$ for each such c. Similarly, if $c = (r,j) \in \mathrm{dg}(\mu)$ with $j < s$, then $h_\mu(c) = h_\lambda(c) - 1 = h_\lambda(r,j) - 1$. So the fractions in the third product convert $h_\mu(c)$ to $h_\lambda(c)$ for each such c. So we are left with

$$\frac{1}{n!} \prod_{c \in \mathrm{dg}(\mu)} h_\lambda(c) = \frac{\prod_{c \in \mathrm{dg}(\lambda)} h_\lambda(c)}{n!},$$

where the last equality follows since $h_\lambda(c^*) = 1$. This completes the proof by induction. □

12.13 Knuth Equivalence

Let $[N]$ denote the set $\{1, 2, \ldots, N\}$, and let $[N]^* = \bigcup_{k=0}^{\infty} [N]^k$ be the set of all words using the alphabet $[N]$. Given a word $\mathbf{w} \in [N]^*$, we can use the RSK algorithm to construct the insertion tableau $P(\mathbf{w})$, which is a semistandard tableau using the same multiset of letters as $\mathbf{w}$ (see §9.23). This section studies some of the relationships between $\mathbf{w}$ and $P(\mathbf{w})$. In particular, we show that the shape of $P(\mathbf{w})$ contains information about increasing and decreasing subsequences of $\mathbf{w}$. First we show how to encode semistandard tableaux using words.

12.75. Definition: Reading Word of a Tableau. Let $\lambda = (\lambda_1, \ldots, \lambda_k)$ be an integer partition, and let $T \in \mathrm{SSYT}_N(\lambda)$. The *reading word of T* is

$$\begin{aligned} \mathrm{rw}(T) \quad = \quad & T(k,1), T(k,2), \ldots, T(k,\lambda_k), \\ & T(k-1,1), T(k-1,2), \ldots, T(k-1,\lambda_{k-1}), \quad \ldots, \\ & T(1,1), T(1,2), \ldots, T(1,\lambda_1). \end{aligned}$$

Thus, $\mathrm{rw}(T)$ is the concatenation of the weakly increasing words appearing in each row of T, reading the rows from bottom to top. Note that $T(j,\lambda_j) \geq T(j,1) > T(j-1,1)$ for all $j > 1$. This implies that we can recover the tableau T from $\mathrm{rw}(T)$ by starting a new row whenever we see a strict descent in $\mathrm{rw}(T)$.

12.76. Example. Given the tableau

$$T = \begin{array}{|c|c|c|c|c|c|c|} \hline 1 & 1 & 2 & 3 & 4 & 4 & 6 \\ \hline 2 & 4 & 5 & 6 & 6 \\ \cline{1-5} 3 & 5 & 7 & 8 \\ \cline{1-4} 4 & 6 \\ \cline{1-2} \end{array}$$

the reading word of T is $\mathrm{rw}(T) = 463578245661123446$. Given that the word $\mathbf{w} = 7866453446223511224$ is the reading word of some tableau S, we deduce that S must be

$$\begin{array}{|c|c|c|c|c|} \hline 1 & 1 & 2 & 2 & 4 \\ \hline 2 & 2 & 3 & 5 \\ \cline{1-4} 3 & 4 & 4 & 6 \\ \cline{1-4} 4 & 5 \\ \cline{1-2} 6 & 6 \\ \cline{1-2} 7 & 8 \\ \cline{1-2} \end{array}$$

by looking at the descents in $\mathbf{w}$.

Next we introduce two equivalence relations on $[N]^*$ that are related to the map sending $\mathbf{w} \in [N]^*$ to $P(\mathbf{w})$.

12.77. Definition: P-Equivalence. Two words $\mathbf{v}, \mathbf{w} \in [N]^*$ are called *P-equivalent*, denoted $\mathbf{v} \equiv_P \mathbf{w}$, iff $P(\mathbf{v}) = P(\mathbf{w})$.

12.78. Definition: Knuth Equivalence. The *elementary Knuth relation of the first kind on* $[N]^*$ is

$$K_1 = \{(\mathbf{u}yxz\mathbf{v}, \mathbf{u}yzx\mathbf{v}) : \mathbf{u}, \mathbf{v} \in [N]^*, x, y, z \in [N], \text{ and } x < y \le z\}.$$

The *elementary Knuth relation of the second kind on* $[N]^*$ is

$$K_2 = \{(\mathbf{u}xzy\mathbf{v}, \mathbf{u}zxy\mathbf{v}) : \mathbf{u}, \mathbf{v} \in [N]^*, x, y, z \in [N], \text{ and } x \le y < z\}.$$

Two words $\mathbf{v}, \mathbf{w} \in [N]^*$ are *Knuth equivalent*, denoted $\mathbf{v} \equiv_K \mathbf{w}$, iff there is a finite sequence of words $\mathbf{v} = \mathbf{v}^0, \mathbf{v}^1, \mathbf{v}^2, \ldots, \mathbf{v}^k = \mathbf{w}$ such that, for all i in the range $1 \le i \le k$, either $(\mathbf{v}^{i-1}, \mathbf{v}^i) \in K_1 \cup K_2$ or $(\mathbf{v}^i, \mathbf{v}^{i-1}) \in K_1 \cup K_2$.

12.79. Remark. Informally, Knuth equivalence allows us to modify words by repeatedly changing subsequences of three consecutive letters according to certain rules. Specifically, if the middle *value* among the three letters does not occupy the middle *position*, then the other two values can switch positions. To determine which value is the middle value in the case of repeated letters, use the rule that the letter to the right is larger. These comments should aid the reader in remembering the inequalities in the definitions of K_1 and K_2.

It is routine to check that $\equiv_P$ and $\equiv_K$ are equivalence relations on $[N]^*$. Our current goal is to prove that these equivalence relations are equal. First we show that we can simulate each step in the Tableau Insertion Algorithm 9.46 using the elementary Knuth relations.

12.80. Theorem: Reading Words and Knuth Equivalence. For all $\mathbf{w} \in [N]^*$, $\mathbf{w} \equiv_K \mathrm{rw}(P(\mathbf{w}))$.

Proof. First note that, for any words $\mathbf{u}, \mathbf{z}, \mathbf{v}, \mathbf{v}' \in [N]^*$, if $\mathbf{v} \equiv_K \mathbf{v}'$ then $\mathbf{uvz} \equiv_K \mathbf{uv'z}$. Now, fix $\mathbf{w} = w_1 w_2 \cdots w_k \in [N]^*$ and use induction on k. The theorem holds if $k \le 1$, since $\mathrm{rw}(P(\mathbf{w})) = \mathbf{w}$ in this case. For the induction step, assume $k > 1$ and write $T' = P(w_1 w_2 \cdots w_{k-1})$ and $T = P(\mathbf{w})$. By the induction hypothesis, $w_1 \cdots w_{k-1} \equiv_K \mathrm{rw}(T')$, so $\mathbf{w} = (w_1 \cdots w_{k-1})w_k \equiv_K \mathrm{rw}(T')w_k$. Therefore, it suffices to prove that $\mathrm{rw}(T')w_k$ is Knuth equivalent to $\mathrm{rw}(T) = \mathrm{rw}(T' \leftarrow w_k)$. We prove this by induction on ℓ, the number of rows in the tableau T'.

For the base case, let $\ell = 1$. Then $\mathrm{rw}(T')$ is a weakly increasing sequence $u_1 u_2 \cdots u_{k-1}$. If $u_{k-1} \le w_k$, then T is obtained from T' by appending w_k at the end of the first row. In this situation, $\mathrm{rw}(T')w_k = u_1 \cdots u_{k-1} w_k = \mathrm{rw}(T)$, so the required result holds. On the other hand, if $w_k < u_{k-1}$, let j be the least index with $w_k < u_j$. When inserting w_k into T', w_k will bump u_j into the second row, so that

$$\mathrm{rw}(T) = u_j u_1 u_2 \cdots u_{j-1} w_k u_{j+1} \cdots u_{k-1}.$$

We now show that this word can be obtained from $\mathrm{rw}(T')w_k = u_1 \cdots u_{k-1} w_k$ by a sequence of elementary Knuth equivalences. If $j \le k-2$, then $w_k < u_{k-2} \le u_{k-1}$ implies

$$(u_1 \cdots u_{k-3} u_{k-2} w_k u_{k-1}, u_1 \cdots u_{k-3} u_{k-2} u_{k-1} w_k) \in K_1.$$

So $\mathrm{rw}(T')w_k$ is Knuth-equivalent to the word obtained by interchanging w_k with the letter u_{k-1} to its immediate left. Similarly, if $j \le k-3$, the inequality $w_k < u_{k-3} \le u_{k-2}$

lets us interchange w_k with u_{k-2}. We can continue in this way, using elementary Knuth equivalences of the first kind, to see that

$$\mathrm{rw}(T')w_k \equiv_K u_1 \cdots u_{j-1} u_j w_k u_{j+1} \cdots u_{k-1}.$$

Now, we have $u_{j-1} \leq w_k < u_j$, so an elementary Knuth equivalence of the second kind transforms this word into

$$u_1 \cdots u_{j-2} u_j u_{j-1} w_k u_{j+1} \cdots u_{k-1}.$$

If $j > 2$, we now have $u_{j-2} \leq u_{j-1} < u_j$, so we can interchange u_j with u_{j-2}. We can continue in this way until u_j reaches the left end of the word. We have now transformed $\mathrm{rw}(T')w_k$ into $\mathrm{rw}(T)$ by elementary Knuth equivalences, so $\mathrm{rw}(T')w_k \equiv_K \mathrm{rw}(T)$.

For the induction step, assume $\ell > 1$. Let T'' be the tableau T' with its first (longest) row erased. Then $\mathrm{rw}(T') = \mathrm{rw}(T'')u_1 \cdots u_p$ where $u_1 \leq \cdots \leq u_p$ is the weakly increasing sequence in the first row of T'. If $u_p \leq w_k$, then $\mathrm{rw}(T')w_k = \mathrm{rw}(T)$. Otherwise, assume w_k bumps u_j in the insertion $T' \leftarrow w_k$. By the result in the last paragraph,

$$\mathrm{rw}(T')w_k \equiv_K \mathrm{rw}(T'')u_j u_1 \cdots u_{j-1} w_k u_{j+1} \cdots u_p.$$

Now, by the induction hypothesis, $\mathrm{rw}(T'')u_j \equiv_K \mathrm{rw}(T'' \leftarrow u_j)$. Thus,

$$\mathrm{rw}(T')w_k \equiv_K \mathrm{rw}(T'' \leftarrow u_j)\mathbf{u}'$$

where $\mathbf{u}'$ is $u_1 \cdots u_p$ with u_j replaced by w_k. But, by definition of tableau insertion, $\mathrm{rw}(T) = \mathrm{rw}(T'' \leftarrow u_j)\mathbf{u}'$. This completes the induction step. $\square$

12.81. Example. Let us illustrate how elementary Knuth equivalences implement the steps in the insertion $T \leftarrow 3$, where

$$T = \begin{array}{|c|c|c|c|c|c|}\hline 1&1&3&4&4&6\\ \hline 2&2&4&5 \\ \cline{1-4} 3&4 \\ \cline{1-2}\end{array}$$

Appending a 3 at the right end of $\mathrm{rw}(T)$, we first compute

$$34\ 2245\ 113446\ 3 \equiv_K 34\ 2245\ 1134436 \equiv_K 34\ 2245\ 1134346 \equiv_K$$

$$34\ 2245\ 1143346 \equiv_K 34\ 2245\ 1413346 \equiv_K 34\ 2245\ 4\ 113346.$$

The steps so far correspond to the insertion of 3 into the first row of T, which bumps the leftmost 4 into the second row. Continuing,

$$34\ 22454\ 113346 \equiv_K 34\ 22544\ 113346 \equiv_K 34\ 25244\ 113346 \equiv_K 34\ 5\ 2244\ 113346,$$

and now the incoming 4 has bumped the 5 into the third row. The process stops here with the word

$$3452244113346 = \mathrm{rw}\left(\begin{array}{|c|c|c|c|c|c|}\hline 1&1&3&3&4&6\\ \hline 2&2&4&4 \\ \cline{1-4} 3&4&5 \\ \cline{1-3}\end{array}\right) = \mathrm{rw}(T \leftarrow 3).$$

We see that $\mathrm{rw}(T)3 \equiv_K \mathrm{rw}(T \leftarrow 3)$, in agreement with the proof above.

12.82. Definition: Increasing and Decreasing Subsequences. Let $\mathbf{w} = w_1 w_2 \cdots w_n \in [N]^*$. An *increasing subsequence of* $\mathbf{w}$ *of length* ℓ is a subset $I = \{i_1 < i_2 < \cdots < i_\ell\}$ of $\{1, 2, \ldots, n\}$ such that $w_{i_1} \leq w_{i_2} \leq \cdots \leq w_{i_\ell}$. A *decreasing subsequence of* $\mathbf{w}$ *of length* ℓ is a subset $I = \{i_1 < i_2 < \cdots < i_\ell\}$ such that $w_{i_1} > w_{i_2} > \cdots > w_{i_\ell}$. A *set of* k *disjoint increasing subsequences of* $\mathbf{w}$ is a set $\{I_1, \ldots, I_k\}$ of pairwise disjoint increasing subsequences of $\mathbf{w}$. For each integer $k \geq 1$, let $\mathrm{inc}_k(\mathbf{w})$ be the maximum value of $|I_1| + \cdots + |I_k|$ over all such sets. Similarly, let $\mathrm{dec}_k(\mathbf{w})$ be the maximum total length of a set of k disjoint decreasing subsequences of $\mathbf{w}$.

12.83. Theorem: Knuth Equivalence and Monotone Subsequences. For all $\mathbf{v}, \mathbf{w}$ in $[N]^*$ and all $k \in \mathbb{Z}_{\geq 1}$, $\mathbf{v} \equiv_K \mathbf{w}$ implies $\text{inc}_k(\mathbf{v}) = \text{inc}_k(\mathbf{w})$ and $\text{dec}_k(\mathbf{v}) = \text{dec}_k(\mathbf{w})$.

Proof. It suffices to consider the case where $\mathbf{v}$ and $\mathbf{w}$ differ by a single elementary Knuth equivalence. First suppose $x, y, z \in [N]$,

$$\mathbf{v} = \mathbf{a}yxz\mathbf{b}, \quad \mathbf{w} = \mathbf{a}yzx\mathbf{b}, \quad x < y \leq z,$$

and the y occurs at position i. If I is an increasing subsequence of $\mathbf{w}$, then $i+1$ and $i+2$ do not both belong to I (since $z > x$). Therefore, if $\{I_1, \ldots, I_k\}$ is any set of k disjoint increasing subsequences of $\mathbf{w}$, we can obtain a set $\{I_1', \ldots, I_k'\}$ of disjoint increasing subsequences of $\mathbf{v}$ by replacing $i+1$ by $i+2$ and $i+2$ by $i+1$ in any I_j in which one of these indices appears. Since $|I_1'| + \cdots + |I_k'| = |I_1| + \cdots + |I_k|$, we deduce that $\text{inc}_k(\mathbf{w}) \leq \text{inc}_k(\mathbf{v})$.

To establish the opposite inequality, let $\mathbf{I} = \{I_1, I_2, \ldots, I_k\}$ be any set of k disjoint increasing subsequences of $\mathbf{v}$. We construct a set of k disjoint increasing subsequences of $\mathbf{w}$ having the same total size as $\mathbf{I}$. The device used in the previous paragraph works here, unless some member of $\mathbf{I}$ (say I_1) contains both $i+1$ and $i+2$. In this case, we cannot have $i \in I_1$, since $y > x$. If no other member of $\mathbf{I}$ contains i, we replace I_1 by $(I_1 - \{i+2\}) \cup \{i\}$, which is an increasing subsequence of $\mathbf{w}$. On the other hand, suppose $i+1, i+2 \in I_1$, and some other member of $\mathbf{I}$ (say I_2) contains i. Write

$$\begin{aligned} I_1 &= \{j_1 < j_2 < \cdots < j_r < i+1 < i+2 < j_{r+1} < \cdots < j_p\}, \\ I_2 &= \{k_1 < k_2 < \cdots < k_s < i < k_{s+1} < \cdots < k_q\}, \end{aligned}$$

and note that $v_{j_r} \leq x < z \leq v_{j_{r+1}}$ and $v_{k_s} \leq y \leq v_{k_{s+1}}$. Replace these two disjoint increasing subsequences of $\mathbf{v}$ by

$$\begin{aligned} I_1' &= \{j_1 < j_2 < \cdots < j_r < i+2 < k_{s+1} < \cdots < k_q\}, \\ I_2' &= \{k_1 < k_2 < \cdots < k_s < i < i+1 < j_{r+1} < \cdots < j_p\}. \end{aligned}$$

Since $w_{j_r} \leq x \leq w_{k_{s+1}}$ and $w_{k_s} \leq y \leq z \leq w_{j_{r+1}}$, I_1' and I_2' are two disjoint increasing subsequences of $\mathbf{w}$ having the same total length as I_1 and I_2. This completes the proof that $\text{inc}_k(\mathbf{w}) \geq \text{inc}_k(\mathbf{v})$.

Similar reasoning proves the result in the case where

$$\mathbf{v} = \mathbf{a}xzy\mathbf{b}, \quad \mathbf{w} = \mathbf{a}zxy\mathbf{b}, \quad \text{and } x \leq y < z.$$

We also ask the reader to prove the statement about decreasing subsequences. □

12.84. Theorem: Subsequences and the Shape of Insertion Tableaux. Assume $\mathbf{w}$ is in $[N]^*$ and $P(\mathbf{w})$ has shape λ. For all integers $k \geq 1$,

$$\text{inc}_k(\mathbf{w}) = \lambda_1 + \cdots + \lambda_k, \qquad \text{dec}_k(\mathbf{w}) = \lambda_1' + \cdots + \lambda_k'.$$

In particular, λ_1 is the length of the longest increasing subsequence of $\mathbf{w}$, whereas $\ell(\lambda)$ is the length of the longest decreasing subsequence of $\mathbf{w}$.

Proof. Let $\mathbf{w}' = \text{rw}(P(\mathbf{w}))$. We know $\mathbf{w} \equiv_K \mathbf{w}'$ by Theorem 12.80, so $\text{inc}_k(\mathbf{w}) = \text{inc}_k(\mathbf{w}')$ and $\text{dec}_k(\mathbf{w}) = \text{dec}_k(\mathbf{w}')$ by Theorem 12.83. So we need only prove

$$\text{inc}_k(\mathbf{w}') = \lambda_1 + \cdots + \lambda_k, \qquad \text{dec}_k(\mathbf{w}') = \lambda_1' + \cdots + \lambda_k'.$$

Now, $\mathbf{w}'$ consists of increasing sequences of letters of successive lengths $\lambda_l, \ldots, \lambda_2, \lambda_1$, where $l = \ell(\lambda)$. By taking $I_1, I_2, \ldots, I_k$ to be the set of positions of the last k of these sequences, we obtain k disjoint increasing subsequences of $\mathbf{w}'$ of length $\lambda_1 + \cdots + \lambda_k$. Therefore, $\text{inc}_k(\mathbf{w}') \geq \lambda_1 + \cdots + \lambda_k$.

On the other hand, let $\{I_1, \ldots, I_k\}$ be any k disjoint increasing subsequences of $\mathbf{w}'$. Each position i in $\mathbf{w}'$ is associated to a particular box in the diagram of λ, via Definition 12.75. For example, position 1 corresponds to the first box in the last row, while the last position corresponds to the last box in the first row. For each position i that belongs to some I_j, place an X in the corresponding box in the diagram of λ. Since entries in a given column of $P(\mathbf{w})$ strictly decrease reading from bottom to top, the X's coming from a given increasing subsequence I_j must all lie in different columns of the diagram. It follows that every column of the diagram contains k or fewer X's. Suppose we push all these X's up their columns as far as possible. Then all the X's in the resulting figure must lie in the top k rows of λ. It follows that the number of X's, which is $|I_1| + \cdots + |I_k|$, cannot exceed $\lambda_1 + \cdots + \lambda_k$. This gives $\text{inc}_k(\mathbf{w}') \leq \lambda_1 + \cdots + \lambda_k$. The proof for $\text{dec}_k(\mathbf{w})$ is similar. □

12.85. Theorem: Knuth Equivalence and Tableau Shape. For all $\mathbf{v}, \mathbf{w} \in [N]^*$, $\mathbf{v} \equiv_K \mathbf{w}$ implies that $P(\mathbf{v})$ and $P(\mathbf{w})$ have the same shape.

Proof. Fix $\mathbf{v}, \mathbf{w} \in [N]^*$ with $\mathbf{v} \equiv_K \mathbf{w}$. Let λ and μ be the shapes of $P(\mathbf{v})$ and $P(\mathbf{w})$, respectively. Using Theorems 12.83 and 12.84, we see that for all $k \geq 1$,

$$\lambda_k = \text{inc}_k(\mathbf{v}) - \text{inc}_{k-1}(\mathbf{v}) = \text{inc}_k(\mathbf{w}) - \text{inc}_{k-1}(\mathbf{w}) = \mu_k.$$ □

12.86. Example. Consider the word $\mathbf{w} = 35164872$. As shown in Figure 9.1, we have

$$P(\mathbf{w}) = \begin{array}{|c|c|c|c|} \hline 1 & 2 & 6 & 7 \\ \hline 3 & 4 & 8 & \\ \cline{1-3} 5 & & & \\ \cline{1-1} \end{array}$$

Since the shape is $\lambda = (4, 3, 1)$, the longest increasing subsequence of $\mathbf{w}$ has length 4. Two such subsequences are $I_1 = \{1, 2, 4, 7\}$ (corresponding to the subword 3567) and $I_2 = \{1, 2, 4, 6\}$. Note that the first row of $P(\mathbf{w})$, namely 1267, does *not* appear as a subword of $\mathbf{w}$. Since the column lengths of λ are $(3, 2, 2, 1)$, the longest length of two disjoint decreasing subsequences of $\mathbf{w}$ is $3 + 2 = 5$. For example, we could take $I_1 = \{6, 7, 8\}$ and $I_2 = \{4, 5\}$ to achieve this. Note that $\mathbf{w}' = \text{rw}(P(\mathbf{w})) = 5\ 348\ 1267$. To illustrate the end of the previous proof, consider the two disjoint increasing subsequences $I_1 = \{1, 4\}$ and $I_2 = \{2, 3, 7, 8\}$ of $\mathbf{w}'$ (this pair does not achieve the maximum length for such subsequences). Drawing X's in the boxes of the diagram associated to the positions in I_1 (respectively I_2) produces

$$\begin{array}{|c|c|c|c|} \hline \ & \ & \ & \ \\ \hline & & X & \\ \cline{1-3} X & & & \\ \cline{1-1} \end{array} \quad \left(\text{respectively } \begin{array}{|c|c|c|c|} \hline \ & \ & X & X \\ \hline X & X & & \\ \cline{1-3} \ & & & \\ \cline{1-1} \end{array} \right).$$

Combining these diagrams and pushing the X's up as far as they can go, we get

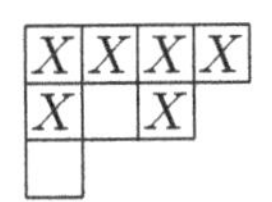

So, indeed, the combined length of I_1 and I_2 does not exceed $\lambda_1 + \lambda_2$.

The next lemma provides the remaining ingredients needed to establish that P-equivalence and Knuth equivalence are the same.

12.87. Lemma. Suppose $\mathbf{v}, \mathbf{w} \in [N]^*$ and z is the largest symbol appearing in both $\mathbf{v}$ and $\mathbf{w}$. Let $\mathbf{v}'$ (respectively $\mathbf{w}'$) be the word obtained by erasing the rightmost z from $\mathbf{v}$ (respectively $\mathbf{w}$). If $\mathbf{v} \equiv_K \mathbf{w}$, then $\mathbf{v}' \equiv_K \mathbf{w}'$. Furthermore, if $T = P(\mathbf{v})$ and $T' = P(\mathbf{v}')$, then T' can be obtained from T by erasing the rightmost box containing z.

Proof. Write $\mathbf{v} = \mathbf{a}z\mathbf{b}$ and $\mathbf{w} = \mathbf{c}z\mathbf{d}$ where $\mathbf{a}, \mathbf{b}, \mathbf{c}, \mathbf{d} \in [N]^*$ and z does not appear in $\mathbf{b}$ or $\mathbf{d}$. First assume that $\mathbf{v}$ and $\mathbf{w}$ differ by a single elementary Knuth relation. If the triple of letters affected by this relation are part of the subword $\mathbf{a}$, then $\mathbf{a} \equiv_K \mathbf{c}$ and $\mathbf{b} = \mathbf{d}$, so $\mathbf{v}' = \mathbf{ab} \equiv_K \mathbf{cd} = \mathbf{w}'$. Similarly, the result holds if the triple of letters is part of the subword $\mathbf{b}$. The next possibility is that for some x, y with $x < y \leq z$,

$$\mathbf{v} = \mathbf{a}'yxz\mathbf{b} \text{ and } \mathbf{w} = \mathbf{a}'yzx\mathbf{b}$$

or vice versa. Then $\mathbf{v}' = \mathbf{a}'yx\mathbf{b} = \mathbf{w}'$, so certainly $\mathbf{v}' \equiv_K \mathbf{w}'$. Another possibility is that for some x, y with $x \leq y < z$,

$$\mathbf{v} = \mathbf{c}xzy\mathbf{b}' \text{ and } \mathbf{w} = \mathbf{c}zxy\mathbf{b}'$$

or vice versa, and we again have $\mathbf{v}' = \mathbf{c}xy\mathbf{b}' = \mathbf{w}'$. Since the z under consideration is the rightmost occurrence of the largest symbol in both $\mathbf{v}$ and $\mathbf{w}$, the possibilities already considered are the only elementary Knuth relations that involve this symbol. So the result holds when $\mathbf{v}$ and $\mathbf{w}$ differ by one elementary Knuth relation. Now, if $\mathbf{v} = \mathbf{v}^0, \mathbf{v}^1, \mathbf{v}^2, \ldots, \mathbf{v}^k = \mathbf{w}$ is a sequence of words as in Definition 12.78, we can write each $\mathbf{v}^i = \mathbf{a}^i z \mathbf{b}^i$ where z does not appear in $\mathbf{b}^i$. Letting $(\mathbf{v}^i)' = \mathbf{a}^i\mathbf{b}^i$ for each i, the chain $\mathbf{v}' = (\mathbf{v}^0)', (\mathbf{v}^1)', \ldots, (\mathbf{v}^k)' = \mathbf{w}'$ proves that $\mathbf{v}' \equiv_K \mathbf{w}'$.

Now consider the actions of the Tableau Insertion Algorithm 9.46 applied to $\mathbf{v} = \mathbf{a}z\mathbf{b}$ and to $\mathbf{v}' = \mathbf{ab}$. We prove the statement about T and T' by induction on the length of $\mathbf{b}$. The statement holds if $\mathbf{b}$ is empty. Assume $\mathbf{b}$ has length $k > 0$ and the statement is known for smaller values of k. Write $\mathbf{b} = \mathbf{b}'x$ where $x \in [N]$. Define $T_1' = P(\mathbf{ab}')$ and $T_1 = P(\mathbf{a}z\mathbf{b}')$. By induction hypothesis, T_1' is T_1 with the rightmost z erased. By definition, $T' = (T_1' \leftarrow x)$ and $T = (T_1 \leftarrow x)$. When we insert the x into these two tableaux, the bumping paths are the same (and hence the required result holds), unless x bumps the rightmost z in T_1. If this happens, the rightmost z (which must have been the only z in its row) gets bumped into the next lower row. It comes to rest there without bumping anything else, and it is still the rightmost z in the tableau. Thus it is still true that erasing this z in T produces T'. The induction is therefore complete. □

12.88. Theorem: P-Equivalence and Knuth Equivalence. For all $\mathbf{v}, \mathbf{w} \in [N]^*$, $\mathbf{v} \equiv_P \mathbf{w}$ iff $\mathbf{v} \equiv_K \mathbf{w}$.

Proof. First, if $\mathbf{v} \equiv_P \mathbf{w}$, then Theorem 12.80 shows that $\mathbf{v} \equiv_K \operatorname{rw}(P(\mathbf{v})) = \operatorname{rw}(P(\mathbf{w})) \equiv_K \mathbf{w}$, so $\mathbf{v} \equiv_K \mathbf{w}$ by transitivity of $\equiv_K$. Conversely, assume $\mathbf{v} \equiv_K \mathbf{w}$. We prove $\mathbf{v} \equiv_P \mathbf{w}$ by induction on the length k of $\mathbf{v}$. For $k \leq 1$, we have $\mathbf{v} = \mathbf{w}$ and so $\mathbf{v} \equiv_P \mathbf{w}$. Now assume $k > 1$ and the result is known for words of length $k - 1$. Write $\mathbf{v} = \mathbf{a}z\mathbf{b}$ and $\mathbf{w} = \mathbf{c}z\mathbf{d}$ where z is the largest symbol in $\mathbf{v}$ and $\mathbf{w}$, and z does not occur in $\mathbf{b}$ or $\mathbf{d}$. Write $\mathbf{v}' = \mathbf{ab}$ and $\mathbf{w}' = \mathbf{cd}$. By Theorem 12.87, $\mathbf{v}' \equiv_K \mathbf{w}'$, $P(\mathbf{v}')$ is $P(\mathbf{v})$ with the rightmost z erased, and $P(\mathbf{w}')$ is $P(\mathbf{w})$ with the rightmost z erased. By induction, $P(\mathbf{v}') = P(\mathbf{w}')$. If we knew that $P(\mathbf{v})$ and $P(\mathbf{w})$ had the same shape, it would follow that $P(\mathbf{v}) = P(\mathbf{w})$. But $P(\mathbf{v})$ and $P(\mathbf{w})$ do have the same shape, because of Theorem 12.85. So $\mathbf{v} \equiv_P \mathbf{w}$. □

We conclude with an application of Theorem 12.84.

12.89. The Erdös–Szekeres Subsequence Theorem. Every word of length exceeding mn either has a weakly increasing subsequence of length $m + 1$ or a strictly decreasing subsequence of length $n + 1$.

Proof. Suppose $\mathbf{w}$ is a word with no increasing subsequence of length $m+1$ and no decreasing subsequence of length $n+1$. Let λ be the shape of $P(\mathbf{w})$. Then Theorem 12.84 implies that $\lambda_1 \leq m$ and $\ell(\lambda) \leq n$. Therefore the length of $\mathbf{w}$, which is $|\lambda|$, can be no greater than $\lambda_1 \ell(\lambda) \leq mn$. □

12.14 Quasisymmetric Polynomials

This section introduces generalizations of symmetric polynomials called quasisymmetric polynomials. Our main goal is to prove a combinatorial formula expanding Schur polynomials as linear combinations of fundamental quasisymmetric polynomials. The proof of this formula illuminates the relationship between semistandard tableaux and standard tableaux.

12.90. Definition: Quasisymmetric Polynomials. Given $\alpha \in \mathbb{Z}_{\geq 0}^N$, let $\text{del}_0(\alpha)$ be the sequence obtained by deleting all zeroes in α. We say that two exponent sequences $\alpha, \beta \in \mathbb{Z}_{\geq 0}^N$ are *shifts* of each other iff $\text{del}_0(\alpha) = \text{del}_0(\beta)$. A polynomial $f \in \mathbb{R}[x_1, \ldots, x_N]$ is called *quasisymmetric* iff the coefficients of $\mathbf{x}^\alpha$ and $\mathbf{x}^\beta$ in f are equal whenever α and β are shifts of each other. Let Q_N be the set of all quasisymmetric polynomials in N variables. For $k \geq 0$, let Q_N^k be the set of all $f \in Q_N$ that are homogeneous of degree k.

12.91. Example. For $N = 4$, the sequences $(2,1,2,0)$, $(2,1,0,2)$, $(2,0,1,2)$, and $(0,2,1,2)$ are all shifts of each other, but are not shifts of $(2,2,1,0)$. The polynomial

$$f = 3x_1^2x_2x_3^2 + 3x_1^2x_2x_4^2 + 3x_1^2x_3x_4^2 + 3x_2^2x_3x_4^2 + 2x_1^2x_2^2x_3 + 2x_1^2x_2^2x_4 + 2x_1^2x_3^2x_4 + 2x_2^2x_3^2x_4$$

is quasisymmetric but not symmetric. All monomials in f have degree 5, so $f \in Q_4^5$.

One readily checks that Q_N and each Q_N^k is a real vector space, and Q_N is the direct sum of its subspaces Q_N^k. Q_N is also closed under polynomial multiplication, so Q_N is a subring of $\mathbb{R}[x_1, \ldots, x_N]$ and a graded algebra over $\mathbb{R}$. Moreover, every symmetric polynomial is a quasisymmetric polynomial, so $\Lambda_N \subseteq Q_N$ and $\Lambda_N^k \subseteq Q_N^k$ for all N and k. The next step is to find bases for the vector spaces Q_N^k. We know that for $N \geq k$, Λ_N^k has many bases indexed by integer partitions of k. On the other hand, we are about to see that Q_N^k has bases indexed by compositions of k or by subsets of $[k-1] = \{1, 2, \ldots, k-1\}$.

Recall that a *composition* of k is a sequence of positive integers, say $\alpha = (\alpha_1, \alpha_2, \ldots, \alpha_s)$, with $\sum_{i=1}^s \alpha_i = k$. We write $|\alpha| = k$ and $\ell(\alpha) = s$. Integer partitions are compositions where the parts occur in weakly decreasing order. Let $\text{Comp}(k)$ be the set of all compositions of k, and let $\text{Comp}_N(k)$ be the set of compositions of k with at most N parts. For $\alpha \in \text{Comp}_N(k)$, we define $\alpha_i = 0$ for $\ell(\alpha) < i \leq N$. The Composition Rule (proved in §1.12) says that $|\text{Comp}(k)| = 2^{k-1}$ for all $k \geq 1$. In the proof of that rule, we defined a bijection from $\text{Comp}(k)$ onto the set $\{0,1\}^{k-1}$. There is also a bijection from $\{0,1\}^{k-1}$ onto the set of all subsets of $[k-1]$, sending a bit string $b_1b_2\cdots b_{k-1}$ to the subset $\{i \in [k-1] : b_i = 1\}$. When we compose these bijections, $\alpha = (\alpha_1, \ldots, \alpha_s) \in \text{Comp}(k)$ maps to the subset

$$\text{sub}(\alpha) = \{\alpha_1, \alpha_1 + \alpha_2, \ldots, \alpha_1 + \cdots + \alpha_{s-1}\} \subseteq [k-1].$$

The inverse bijection sends a subset $S = \{i_1 < i_2 < \cdots < i_t\} \subseteq [k-1]$ to the composition

$$\text{comp}(S) = (i_1, i_2 - i_1, i_3 - i_2, \ldots, i_t - i_{t-1}, k - i_t) \in \text{Comp}(k).$$

Now we can define our first basis for quasisymmetric polynomials, which is analogous to the monomial basis for symmetric polynomials.

12.92. Definition: Monomial Quasisymmetric Polynomials. Given $N > 0$, $k \geq 0$, and $\alpha \in \text{Comp}_N(k)$, the *monomial quasisymmetric polynomial* in N variables indexed by α is

$$M_\alpha(x_1, \ldots, x_N) = \sum_{\substack{\beta \in \mathbb{Z}_{\geq 0}^N: \\ \text{del}_0(\beta) = \text{del}_0(\alpha)}} \mathbf{x}^\beta.$$

12.93. Example. For $N=5$ and $\alpha=(3,2,3)$, $M_\alpha(x_1,\ldots,x_5)$ is

$$x_1^3x_2^2x_3^3+x_1^3x_2^2x_4^3+x_1^3x_2^2x_5^3+x_1^3x_3^2x_4^3+x_1^3x_3^2x_5^3+x_1^3x_4^2x_5^3+x_2^3x_3^2x_4^3+x_2^3x_3^2x_5^3+x_2^3x_4^2x_5^3+x_3^3x_4^2x_5^3.$$

In Example 12.91, $f=3M_{(2,1,2)}+2M_{(2,2,1)}$.

More generally, every quasisymmetric polynomial $f\in Q_N^k$ can be expanded uniquely as a linear combination of M_α where $\alpha\in\mathrm{Comp}_N(k)$. This fact can be proved by adapting the proof of Theorem 9.23. The intuition for the proof is that M_α groups together all monomials $\mathbf{x}^\beta$ that *must* have the same coefficient as $\mathbf{x}^\alpha$ in any quasisymmetric polynomial. Hence, we have the following result.

12.94. Theorem: Monomial Basis of Q_N^k. For all $k\geq 0$ and $N>0$,

$$\{M_\alpha(x_1,\ldots,x_N):\alpha\in\mathrm{Comp}_N(k)\}$$

is a basis for the vector space Q_N^k. So for all $N\geq k$, the dimension of Q_N^k is $|\mathrm{Comp}(k)|=2^{k-1}$.

Recall the definitions of elementary and complete symmetric polynomials:

$$e_k=\sum_{1\leq i_1<i_2<\cdots<i_k\leq N}x_{i_1}x_{i_2}\cdots x_{i_k}\quad\text{and}\quad h_k=\sum_{1\leq i_1\leq i_2\leq\cdots\leq i_k\leq N}x_{i_1}x_{i_2}\cdots x_{i_k}.$$

The next definition generalizes these formulas by allowing mixtures of strict and weak inequalities among the subscripts i_j.

12.95. Definition: Fundamental Quasisymmetric Polynomials. Given $k\geq 0$, $N>0$, and $S\subseteq[k-1]$, the *fundamental quasisymmetric polynomial* in N variables indexed by k and S is

$$\mathrm{FQ}_{k,S}(x_1,\ldots,x_N)=\sum_{\substack{1\leq i_1\leq i_2\leq\cdots\leq i_k\leq N:\\ j\in S\Rightarrow i_j<i_{j+1}}}x_{i_1}x_{i_2}\cdots x_{i_k}.$$

Intuitively, the set S used to index $\mathrm{FQ}_{k,S}$ consists of the positions j where we are forced to have a *strict* increase $i_j<i_{j+1}$ in the subscript sequence for the x-variables. We prove in Theorem 12.97 that each $\mathrm{FQ}_{k,S}$ is quasisymmetric, as the name suggests.

12.96. Example. For $N=3$ and $k=3$, we have

$$\begin{aligned}
\mathrm{FQ}_{3,\emptyset} &= x_1^3+x_2^3+x_3^3+x_1^2x_2+x_1^2x_3+x_2^2x_3+x_1x_2^2+x_1x_3^2+x_2x_3^2+x_1x_2x_3\\
&= M_{(3)}+M_{(2,1)}+M_{(1,2)}+M_{(1,1,1)}=h_3;\\
\mathrm{FQ}_{3,\{1\}} &= x_1x_2^2+x_1x_3^2+x_2x_3^2+x_1x_2x_3=M_{(1,2)}+M_{(1,1,1)};\\
\mathrm{FQ}_{3,\{2\}} &= x_1^2x_2+x_1^2x_3+x_2^2x_3+x_1x_2x_3=M_{(2,1)}+M_{(1,1,1)};\\
\mathrm{FQ}_{3,\{1,2\}} &= x_1x_2x_3=M_{(1,1,1)}=e_3.
\end{aligned}$$

More generally, $\mathrm{FQ}_{k,\emptyset}=h_k$ and $\mathrm{FQ}_{k,[k-1]}=e_k$ for all k and N, so that the fundamental quasisymmetric polynomials *interpolate* between the symmetric polynomials h_k and e_k. The patterns in the preceding example also suggest how to write $\mathrm{FQ}_{k,S}$ as a linear combination of the M_α. The next theorem gives the general formula, which also proves that the polynomials $\mathrm{FQ}_{k,S}$ really are quasisymmetric.

12.97. Theorem: Monomial Expansion of Fundamental Quasisymmetric Polynomials. For all $k \geq 0$, $N > 0$, and $S \subseteq [k-1]$,

$$\mathrm{FQ}_{k,S}(x_1,\ldots,x_N) = \sum_{\substack{\alpha\in\mathrm{Comp}_N(k):\\ S\subseteq\mathrm{sub}(\alpha)}} M_\alpha(x_1,\ldots,x_N).$$

Therefore, $\mathrm{FQ}_{k,S} \in Q_N^k$.

Proof. The key observation is the following reformulation of the definition of M_α: for all $\alpha \in \mathrm{Comp}_N(k)$,

$$M_\alpha(x_1,\ldots,x_N) = \sum_{\substack{1\leq i_1\leq i_2\leq\cdots\leq i_k\leq N:\\ j\in\mathrm{sub}(\alpha)\Leftrightarrow i_j<i_{j+1}}} x_{i_1}x_{i_2}\cdots x_{i_k}. \tag{12.14}$$

We explain this formula through an example where $\alpha = (3,2,3)$ and $\mathrm{sub}(\alpha) = \{3,5\}$. In this case, the right side of (12.14) is

$$\sum_{1\leq i_1=i_2=i_3<i_4=i_5<i_6=i_7=i_8\leq N} x_{i_1}x_{i_2}\cdots x_{i_8} = \sum_{1\leq i_1<i_4<i_6\leq N} x_{i_1}^3x_{i_4}^2x_{i_6}^3 = M_{(3,2,3)}.$$

In general, the condition $j \in \mathrm{sub}(\alpha) \Leftrightarrow i_j < i_{j+1}$ ensures that the right side of (12.14) is the sum of all monomials whose exponent sequences are shifts of $\mathbf{x}^\alpha$, and this sum is M_α.

Let $\mathcal{I}$ be the set of weakly increasing sequences $I = (i_1 \leq i_2 \leq \cdots \leq i_k)$ with each i_j in $[N]$. For $I \in \mathcal{I}$, define the *ascent set* $\mathrm{Asc}(I) = \{j \in [k-1] : i_j < i_{j+1}\}$, and let $\mathbf{x}_I = x_{i_1}x_{i_2}\cdots x_{i_k}$. So far, we know that

$$\mathrm{FQ}_{k,S} = \sum_{I\in\mathcal{I}:\ S\subseteq\mathrm{Asc}(I)} \mathbf{x}_I \quad\text{and}\quad M_\alpha = \sum_{I\in\mathcal{I}:\ \mathrm{sub}(\alpha)=\mathrm{Asc}(I)} \mathbf{x}_I.$$

In the sum for $\mathrm{FQ}_{k,S}$, let us group together all terms indexed by subscript sequences I that have the same ascent set. We get

$$\mathrm{FQ}_{k,S} = \sum_{T:\ S\subseteq T\subseteq[k-1]} \left(\sum_{I\in\mathcal{I}:\ T=\mathrm{Asc}(I)} \mathbf{x}_I\right).$$

Replacing each subset T by the associated composition $\alpha = \mathrm{comp}(T)$, this becomes

$$\mathrm{FQ}_{k,S} = \sum_{\substack{\alpha\in\mathrm{Comp}_N(k):\\ S\subseteq\mathrm{sub}(\alpha)}} \left(\sum_{I\in\mathcal{I}:\ \mathrm{sub}(\alpha)=\mathrm{Asc}(I)} \mathbf{x}_I\right) = \sum_{\substack{\alpha\in\mathrm{Comp}_N(k):\\ S\subseteq\mathrm{sub}(\alpha)}} M_\alpha.$$

Finally, $\mathrm{FQ}_{k,S} \in Q_N^k$ follows because we have written $\mathrm{FQ}_{k,S}$ as a linear combination of the M_α, which form a basis for Q_N^k. □

For the rest of this section, we assume that $N \geq k$, so that bases of Q_N^k are indexed by all compositions of k (or all subsets of $[k-1]$).

12.98. Theorem: Fundamental Quasisymmetric Basis of Q_N^k. For all $N \geq k$,

$$\{\mathrm{FQ}_{k,S}(x_1,\ldots,x_N) : S \subseteq [k-1]\}$$

is a basis for the vector space Q_N^k.

Proof. Define column vectors $\mathbf{F} = (\mathrm{FQ}_{k,S} : S \subseteq [k-1])$ and $\mathbf{M} = (M_{\mathrm{comp}(S)} : S \subseteq [k-1])$. Define a matrix $\mathbf{A}$, with rows and columns indexed by subsets of $[k-1]$, such that the entry of $\mathbf{A}$ in row S and column T is 1 if $S \subseteq T$ and 0 otherwise. Order the rows and columns of $\mathbf{A}$, $\mathbf{F}$, and $\mathbf{M}$ using a fixed total ordering on the set of subsets of $[k-1]$ such that smaller subsets precede larger subsets in the ordering. Theorem 12.97 says that for all $S \subseteq [k-1]$,

$$\mathrm{FQ}_{k,S} = \sum_{T \subseteq [k-1]} \mathbf{A}(S,T) M_{\mathrm{comp}(T)}.$$

In matrix notation, this becomes $\mathbf{F} = \mathbf{AM}$.

It now suffices to show that $\mathbf{A}$ is an upper-triangular matrix with 1's on the diagonal, as in the proof of Theorem 9.43. For each S, the diagonal entry of $\mathbf{A}$ in row S, column S is 1, since $S \subseteq S$. Next, if the S,T-entry of $\mathbf{A}$ is nonzero for some $S \neq T$, then $S \subsetneq T$, so $|S| < |T|$. This means that S precedes T in the chosen ordering on subsets, so this entry of $\mathbf{A}$ appears above the diagonal, as needed. □

An inclusion-exclusion calculation yields an explicit formula for the inverse of the transition matrix $\mathbf{A}$ in the preceding proof. This leads to a formula expanding M_α as a linear combination of fundamental quasisymmetric polynomials (see Exercise 12-98). Our next result describes the fundamental quasisymmetric expansion of Schur symmetric polynomials. Recall that for a standard tableau U with n cells, $\mathrm{Des}(U)$ is the set of all $j \in \{1,2,\ldots,n-1\}$ such that $j+1$ appears in a lower row than j in U.

12.99. Theorem: Fundamental Quasisymmetric Expansion of Schur Polynomials. For all $N \geq n$ and all $\lambda \in \mathrm{Par}(n)$,

$$s_\lambda(x_1,\ldots,x_N) = \sum_{U \in \mathrm{SYT}(\lambda)} \mathrm{FQ}_{n,\mathrm{Des}(U)}(x_1,\ldots,x_N).$$

Proof. Expanding the definitions of s_λ and $\mathrm{FQ}_{n,\mathrm{Des}(U)}$, we must prove

$$\sum_{T \in \mathrm{SSYT}_N(\lambda)} \mathbf{x}^T = \sum_{U \in \mathrm{SYT}(\lambda)} \sum_{\substack{I \in \mathcal{I}: \\ \mathrm{Des}(U) \subseteq \mathrm{Asc}(I)}} \mathbf{x}_I,$$

where $\mathcal{I}$ is the set of weakly increasing subscript sequences $I = (i_1 \leq i_2 \leq \cdots \leq i_n)$ with each $i_j \in [N]$, $\mathrm{Asc}(I) = \{j : i_j < i_{j+1}\}$, and $\mathbf{x}_I = x_{i_1} \cdots x_{i_n}$. Let $X = \mathrm{SSYT}_N(\lambda)$, and let Y be the set of pairs (U,I), where $U \in \mathrm{SYT}(\lambda)$, $I \in \mathcal{I}$, and $\mathrm{Des}(U) \subseteq \mathrm{Asc}(I)$. It suffices to define a weight-preserving bijection $F : X \to Y$.

Given a semistandard tableau $T \in X$, we compute $F(T) = (U,I)$ as follows. Suppose T has k_1 1's, k_2 2's, and so on. Because T is semistandard, the k_j cells containing j in T form a horizontal strip for every j. To create the standard tableau U, we use the following *standardization algorithm.* Replace the k_1 1's in T, from left to right, with the integers $1,2,\ldots,k_1$. Then replace the k_2 2's in T, from left to right, with the integers $k_1+1, k_1+2,\ldots,k_1+k_2$. In general, replace the k_j copies of j in T, from left to right, with the integers $(\sum_{i<j} k_i)+1, (\sum_{i<j} k_i)+2, \ldots, (\sum_{i<j} k_i)+k_j$. Furthermore, let I be the weakly increasing sequence consisting of k_1 1's, k_2 2's, and so on. For example,

$$F\left(\begin{array}{|c|c|c|c|c|}\hline 1&1&1&2&3\\\hline 2&2&3&3\\\cline{1-4} 3&4&5&5\\\cline{1-4}\end{array}\right) = \left(\begin{array}{|c|c|c|c|c|}\hline 1&2&3&6&10\\\hline 4&5&8&9\\\cline{1-4} 7&11&12&13\\\cline{1-4}\end{array}\;,\quad 1112223333455\right).$$

We must check that (U,I) does belong to the claimed codomain of F. First, is U really a standard tableau? On one hand, U is a filling containing the integers $1, 2, \ldots, n = |\lambda|$ once each. Note that if $i < j$ are two values somewhere in T, the standardization process always relabels i with a lower integer than j. If there are multiple copies of i in T, these copies get relabeled with an increasing sequence of consecutive integers moving from left to right. We see from these comments that since the rows of T weakly increase, the rows of U strictly increase, and similarly for columns. Thus U is standard, as needed.

Next, is $\mathrm{Des}(U) \subseteq \mathrm{Asc}(I)$? We prove the contrapositive: fix $k \in \{1, 2, \ldots, n-1\}$ with $k \notin \mathrm{Asc}(I)$, and show $k \notin \mathrm{Des}(U)$. We have assumed $i_k = i_{k+1} = i$. By definition of standardization, the unique copies of k and $k+1$ in U were used to relabel two occurrences of i in T. Now, the cells containing i in T form a horizontal strip that is relabeled from left to right. So $k+1$ must appear strictly right and weakly above k in U, which means $k \notin \mathrm{Des}(U)$. We have now proved that $F(T) = (U, I)$ is in Y. By definition of I, $\mathbf{x}^T = \mathbf{x}_I$, so F is weight-preserving.

To finish, we construct a map $G : Y \to X$ that is the two-sided inverse of F. Given $(U, I) \in Y$, let $T = G(U, I)$ be the filling of shape λ obtained by replacing each k in U by i_k. For example,

$$G\left(\begin{array}{ccccc} 1 & 3 & 4 & 5 & 6 \\ 2 & 7 & 8 & 11 & \\ 9 & 10 & 12 & 13 & \end{array}, \quad 1333445566788\right) \quad = \quad \begin{array}{ccccc} 1 & 3 & 3 & 4 & 4 \\ 3 & 5 & 5 & 7 & \\ 6 & 6 & 8 & 8 & \end{array}.$$

Since U is standard and I is weakly increasing, the new filling T has weakly increasing rows and columns. Since $\mathrm{Des}(U) \subseteq \mathrm{Asc}(I)$, we see (as in the previous paragraph) that every run of equal values in I is used to relabel a horizontal strip of cells in $\mathrm{dg}(\lambda)$. Thus T has strictly increasing columns, so $T \in \mathrm{SSYT}_N(\lambda)$. Finally, it is routine to check that $F \circ G = \mathrm{id}_Y$ and $G \circ F = \mathrm{id}_X$. □

12.15 Pfaffians and Perfect Matchings

Given a square matrix A with N rows and N columns, we have defined the determinant of A by the formula

$$\det(A) = \sum_{w \in S_N} \mathrm{sgn}(w) \prod_{i=1}^{N} A(i, w(i))$$

(see §12.9). This section studies the *Pfaffian*, which is a number associated to a *triangular* array $(a_{i,j} : 1 \leq i < j \leq N)$ where N is even. Pfaffians arise in the theory of skew-symmetric matrices.

12.100. Definition: Skew-Symmetric Matrices. An $N \times N$ matrix A is called *skew-symmetric* iff $A^{\mathrm{tr}} = -A$, which means $A(i,j) = -A(j,i)$ for all $i, j \in \{1, 2, \ldots, N\}$.

If A is a real or complex skew-symmetric matrix, then $A(i,i) = 0$ for all i. Moreover, A is completely determined by the triangular array of numbers $(A(i,j) : 1 \leq i < j \leq N)$ lying strictly above the main diagonal. The starting point for the theory of Pfaffians is the observation that, for all even N and all skew-symmetric A, $\det(A)$ is a perfect square. (For odd N, the condition $A^{\mathrm{tr}} = -A$ can be used to show that $\det(A) = 0$.)

12.101. Example. A general skew-symmetric 2×2 matrix has the form $A = \begin{bmatrix} 0 & a \\ -a & 0 \end{bmatrix}$.

In this case, $\det(A) = a^2$ is a square. A skew-symmetric 4×4 matrix looks like

$$A = \begin{bmatrix} 0 & a & b & c \\ -a & 0 & d & e \\ -b & -d & 0 & f \\ -c & -e & -f & 0 \end{bmatrix}.$$

A somewhat tedious calculation reveals that

$$\begin{aligned} \det(A) &= a^2f^2 + b^2e^2 + c^2d^2 - 2abef + 2acdf - 2bcde \\ &= (af + cd - be)^2. \end{aligned}$$

The remainder of this section develops the theory needed to explain the phenomenon observed in the last example.

12.102. Definition: Pfaffians. Suppose N is even and A is a skew-symmetric $N \times N$ matrix. Let SPf_N be the set of all permutations $w \in S_N$ such that

$$w_1 < w_3 < w_5 < \cdots < w_{N-1}, \quad w_1 < w_2,\ w_3 < w_4,\ w_5 < w_6,\ \ldots,\ \text{and } w_{N-1} < w_N.$$

The *Pfaffian* of A, denoted $\mathrm{Pf}(A)$, is the number

$$\mathrm{Pf}(A) = \sum_{w \in \mathrm{SPf}_N} \mathrm{sgn}(w) A(w_1, w_2) A(w_3, w_4) A(w_5, w_6) \cdots A(w_{N-1}, w_N).$$

12.103. Example. If $N = 2$, $\mathrm{SPf}_2 = \{12\}$ and $\mathrm{Pf}(A) = A(1, 2)$ (we write permutations in one-line form here). If $N = 4$, $\mathrm{SPf}_4 = \{1234, 1423, 1324\}$ and

$$\mathrm{Pf}(A) = A(1, 2)A(3, 4) + A(1, 4)A(2, 3) - A(1, 3)A(2, 4).$$

For a general $N \times N$ matrix A, $\det(A)$ is a sum of $|S_N| = N!$ terms. Similarly, for a skew-symmetric matrix A, $\mathrm{Pf}(A)$ is a sum of $|\,\mathrm{SPf}_N\,|$ terms.

12.104. Theorem: Size of SPf_N. For even $N > 0$, $|\,\mathrm{SPf}_N\,| = 1 \cdot 3 \cdot 5 \cdot \ldots \cdot (N - 1)$.

Proof. We can construct each permutation $w \in \mathrm{SPf}_N$ as follows. First, w_1 must be 1. There are $N - 1$ choices for w_2, which can be anything other than 1. To finish building w, choose an arbitrary permutation $v = v_1 v_2 \cdots v_{N-2} \in \mathrm{SPf}_{N-2}$. For i between 1 and $N - 2$, set

$$w_{i+2} = \begin{cases} v_i + 1 & \text{if } v_i < w_2 - 1; \\ v_i + 2 & \text{otherwise.} \end{cases}$$

Informally, we are renumbering the v's to use symbols in $\{1, 2, \ldots, N\} - \{w_1, w_2\} = [N] - \{1, w_2\}$ and then appending this word to $w_1 w_2$. By the Product Rule, $|\,\mathrm{SPf}_N\,| = (N - 1) \cdot |\,\mathrm{SPf}_{N-2}\,|$. Since $|\,\mathrm{SPf}_2\,| = 1$, the formula in the theorem follows by induction. □

Recall that the Laplace expansions in Theorem 12.51 provide recursive formulas for evaluating determinants. Similar recursive formulas exist for evaluating Pfaffians. The key difference is that two rows and columns are erased at each stage, whereas in Laplace expansions only one row and column are erased at a time.

12.105. Theorem: Pfaffian Expansion along Row 1. Suppose N is even and A is an $N \times N$ skew-symmetric matrix. For each $i < j$, let $A[[i, j]]$ be the matrix obtained from A by deleting row i, row j, column i, and column j; this is a skew-symmetric matrix of size $(N - 2) \times (N - 2)$. We have

$$\mathrm{Pf}(A) = \sum_{j=2}^{N} (-1)^j A(1, j)\, \mathrm{Pf}(A[[1, j]]).$$

Proof. By definition,

$$\mathrm{Pf}(A) = \sum_{w \in \mathrm{SPf}_N} \mathrm{sgn}(w) \prod_{\substack{i=1 \\ i \text{ odd}}}^{N} A(w_i, w_{i+1}).$$

By the proof of Theorem 12.104, there is a bijection $\mathrm{SPf}_N \to \{2, 3, \ldots, N\} \times \mathrm{SPf}_{N-2}$ that maps $w \in \mathrm{SPf}_N$ to (j, v), where $j = w_2$ and v is obtained from $w_3 w_4 \cdots w_N$ by renumbering the symbols to be $1, 2, \ldots, N-2$. We use this bijection to change the indexing set for the summation from SPf_N to $\{2, \ldots, N\} \times \mathrm{SPf}_{N-2}$. Counting inversions, we see that $\mathrm{inv}(w) = \mathrm{inv}(v) + j - 2$ since $w_2 = j$ exceeds $j - 2$ symbols to its right. So $\mathrm{sgn}(w) = (-1)^j \mathrm{sgn}(v)$. Next, $A(w_1, w_2) = A(1, j)$. For odd $i > 1$, it follows from the definitions that $A(w_i, w_{i+1}) = A[[1, j]](v_{i-2}, v_{i-1})$. Putting all this information into the formula, and replacing i by $i + 2$ in the product over odd i from 3 to N, we see that

$$\mathrm{Pf}(A) = \sum_{j=2}^{N} (-1)^j A(1, j) \sum_{v \in \mathrm{SPf}_{N-2}} \mathrm{sgn}(v) \prod_{\substack{i=1 \\ i \text{ odd}}}^{N-2} A[[1, j]](v_i, v_{i+1}).$$

The inner sum is precisely $\mathrm{Pf}(A[[1, j]])$, so the proof is complete. □

12.106. Example. Let us compute the Pfaffian of the matrix

$$A = \begin{bmatrix} 0 & x & -y & 0 & 0 & 0 \\ -x & 0 & 0 & y & 0 & 0 \\ y & 0 & 0 & x & -y & 0 \\ 0 & -y & -x & 0 & 0 & y \\ 0 & 0 & y & 0 & 0 & x \\ 0 & 0 & 0 & -y & -x & 0 \end{bmatrix}.$$

Expanding along row 1 gives

$$\mathrm{Pf}(A) = x \,\mathrm{Pf} \begin{bmatrix} 0 & x & -y & 0 \\ -x & 0 & 0 & y \\ y & 0 & 0 & x \\ 0 & -y & -x & 0 \end{bmatrix} - (-y) \,\mathrm{Pf} \begin{bmatrix} 0 & y & 0 & 0 \\ -y & 0 & 0 & y \\ 0 & 0 & 0 & x \\ 0 & -y & -x & 0 \end{bmatrix}.$$

By expanding these 4×4 Pfaffians in the same way, or by using the formula in Example 12.103, we obtain

$$\mathrm{Pf}(A) = x(x^2 + y^2) + y(xy) = x^3 + 2xy^2.$$

The combinatorial significance of this Pfaffian evaluation is revealed in §12.16.

Pfaffians are closely related to perfect matchings of graphs, which we now discuss.

12.107. Definition: Perfect Matchings. Let G be a simple graph with vertex set V and edge set E. A *perfect matching of* G is a subset M of E such that each $v \in V$ is the endpoint of exactly one edge in M. Let $\mathrm{PM}(G)$ be the set of perfect matchings of G.

12.108. Example. For the graph shown in Figure 12.13, one perfect matching is

$$M_1 = \{\{1, 6\}, \{2, 10\}, \{3, 9\}, \{4, 8\}, \{5, 7\}\}.$$

Another perfect matching is

$$M_2 = \{\{1, 2\}, \{3, 4\}, \{5, 7\}, \{6, 9\}, \{8, 10\}\}.$$

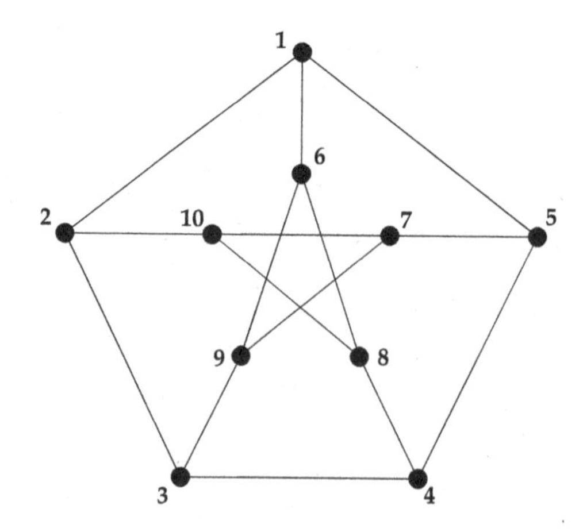

FIGURE 12.13
Graph used to illustrate perfect matchings.

A perfect matching on a graph G is a set partition of the vertex set of G into blocks of size 2 where each such block is an edge of G. Therefore, if G has N vertices and a perfect matching exists for G, then N must be even. The next result shows that perfect matchings on a *complete* graph can be encoded by permutations in SPf_N.

12.109. Theorem: Perfect Matchings on a Complete Graph. Suppose N is even and K_N is the simple graph with vertex set $\{1, 2, \ldots, N\}$ and edge set $\{\{i, j\} : 1 \leq i < j \leq N\}$. The map $f : \mathrm{SPf}_N \to \mathrm{PM}(K_N)$ defined by

$$f(w_1 w_2 \cdots w_N) = \{\{w_1, w_2\}, \{w_3, w_4\}, \ldots, \{w_{N-1}, w_N\}\}$$

is a bijection. Consequently,

$$|\,\mathrm{PM}(K_N)| = 1 \cdot 3 \cdot 5 \cdot \ldots \cdot (N-1).$$

Proof. Note first that f does map into the set $\mathrm{PM}(K_N)$. Next, a matching $M \in \mathrm{PM}(K_N)$ is a set of $N/2$ edges $M = \{\{i_1, i_2\}, \{i_3, i_4\}, \ldots, \{i_{N-1}, i_N\}\}$. Since $\{i, j\} = \{j, i\}$, we can choose the notation so that $i_1 < i_2$, $i_3 < i_4$, ..., and $i_{N-1} < i_N$. Similarly, since the $N/2$ edges of M can be presented in any order, we can change notation again (if needed) to arrange that $i_1 < i_3 < i_5 < \cdots < i_{N-1}$. Then the permutation $w = i_1 i_2 i_3 \cdots i_N$ is in SPf_N and satisfies $f(w) = M$. Thus f maps *onto* $\mathrm{PM}(K_N)$. To see that f is one-to-one, suppose $v = j_1 j_2 j_3 \cdots j_N$ is another element of SPf_N such that $f(v) = M = f(w)$. We must have $j_1 = 1 = i_1$. Since M has only one edge incident to vertex 1, and since $\{i_1, i_2\} \in M$ and $\{j_1, j_2\} \in M$ by definition of f, we conclude that $i_2 = j_2$. Now i_3 and j_3 must both be the smallest vertex in the set $\{1, 2, \ldots, N\} - \{i_1, i_2\}$, so $i_3 = j_3$. Then $i_4 = j_4$ follows, as above, since M is a perfect matching. Continuing similarly, we see that $i_k = j_k$ for all k, so $v = w$ and f is one-to-one. Since f is a bijection, the formula for $|\,\mathrm{PM}(K_N)|$ follows from Theorem 12.104. □

The preceding theorem leads to the following combinatorial interpretation for Pfaffians. Given a perfect matching $M \in \mathrm{PM}(K_N)$, use Theorem 12.109 to write $M = f(w)$ for a unique $w \in \mathrm{SPf}_N$. Define the *sign* of M to be $\mathrm{sgn}(w)$, and define the *weight* of M to be

$$\mathrm{wt}(M) = \prod_{\{i,j\} \in M} x_{i,j} = \prod_{\substack{i=1 \\ i \text{ odd}}}^{N} x_{w_i, w_{i+1}},$$

where the $x_{i,j}$ (for $1 \leq i < j \leq N$) are formal variables. Let X be the skew-symmetric matrix with entries $x_{i,j}$ above the main diagonal. It follows from Theorem 12.109 and the definition of a Pfaffian that

$$\sum_{M \in \mathrm{PM}(K_N)} \mathrm{sgn}(M)\,\mathrm{wt}(M) = \mathrm{Pf}(X).$$

More generally, we have the following result.

12.110. Theorem: Pfaffians and Perfect Matchings. Let N be even, and let G be a simple graph with vertex set $V = \{1, 2, \ldots, N\}$ and edge set $E(G)$. Let $x_{i,j}$ be formal variables, and let $X = X(G)$ be the skew-symmetric matrix with entries

$$X(i,j) = \begin{cases} x_{i,j} & \text{if } i < j \text{ and } \{i,j\} \in E(G); \\ -x_{i,j} & \text{if } i > j \text{ and } \{i,j\} \in E(G); \\ 0 & \text{otherwise.} \end{cases}$$

Then $\sum_{M \in \mathrm{PM}(G)} \mathrm{sgn}(M)\,\mathrm{wt}(M) = \mathrm{Pf}(X(G))$.

Proof. We have already observed that

$$\sum_{M \in \mathrm{PM}(K_N)} \mathrm{sgn}(M)\,\mathrm{wt}(M) = \mathrm{Pf}(X(K_N)). \tag{12.15}$$

Given the graph G, let ϵ be the unique algebra homomorphism on the polynomial ring $\mathbb{R}[x_{i,j} : 1 \leq i < j \leq N]$ that sends $x_{i,j}$ to $x_{i,j}$ if $\{i,j\} \in E(G)$ and sends $x_{i,j}$ to 0 if $\{i,j\} \notin E(G)$. (See the Appendix for more discussion of evaluation homomorphisms.) Applying ϵ to the left side of (12.15) produces

$$\sum_{M \in \mathrm{PM}(G)} \mathrm{sgn}(M)\,\mathrm{wt}(M),$$

since all matchings of K_N that use an edge not in $E(G)$ are mapped to zero. On the other hand, since ϵ is an algebra homomorphism and the Pfaffian of a matrix is a polynomial in the entries of the matrix, we can compute $\epsilon(\mathrm{Pf}(X(K_N)))$ by applying ϵ to each entry of $X(K_N)$ and taking the Pfaffian of the resulting matrix. So, applying ϵ to the right side of (12.15) gives

$$\epsilon(\mathrm{Pf}(X(K_N))) = \mathrm{Pf}(\epsilon(X(K_N))) = \mathrm{Pf}(X(G)).$$ □

12.111. Remark. The last result shows that $\mathrm{Pf}(X(G))$ is a *signed* sum of distinct monomials, where there is one monomial for each perfect matching of G. Because of the signs, one cannot compute $|\mathrm{PM}(G)|$ by setting $x_{i,j} = 1$ for each $\{i,j\} \in E(G)$. However, for certain graphs G, one can introduce extra signs into the upper part of the matrix $X(G)$ to counteract the sign arising from $\mathrm{sgn}(M)$. This process is illustrated in the next section.

We can now give a combinatorial proof of the main result linking Pfaffians and determinants.

12.112. Theorem: Pfaffians and Determinants. For every even $N > 0$ and every $N \times N$ skew-symmetric matrix A, $\det(A) = \mathrm{Pf}(A)^2$.

Proof. First we use the skew-symmetry of A to cancel some terms in the sum

$$\det(A) = \sum_{w \in S_N} \mathrm{sgn}(w) \prod_{i=1}^{N} A(i, w(i)).$$

We can cancel every term indexed by a permutation w whose functional digraph contains at least one cycle of odd length (see §3.6). If w has a cycle of length 1, then $w(i) = i$ for some i. So $A(i, w(i)) = A(i, i) = 0$ by skew-symmetry, and the term indexed by this w is zero. On the other hand, suppose w has no fixed points, but w does have at least one cycle of odd length. Among all the odd-length cycles of w, choose the cycle $(i_1, i_2, \ldots, i_k)$ whose minimum element is as small as possible. Reverse the orientation of this cycle to get a permutation $w' \neq w$. For example, if $w = (3,8,4)(2,5,7)(1,6)(9,10)$, then $w' = (3,8,4)(7,5,2)(1,6)(9,10)$. In general, $\text{sgn}(w') = \text{sgn}(w)$ since w and w' have the same cycle structure (see Theorem 7.34). However, since k is odd and A is skew-symmetric,

$$A(i_1,i_2)A(i_2,i_3)\cdots A(i_{k-1},i_k)A(i_k,i_1) = -A(i_2,i_1)A(i_3,i_2)\cdots A(i_k,i_{k-1})A(i_1,i_k).$$

It follows that the term in $\det(A)$ indexed by w' is the negative of the term in $\det(A)$ indexed by w, so these two terms cancel. Since the map sending w to w' is an involution, we conclude that

$$\det(A) = \sum_{w \in S_N^{ev}} \text{sgn}(w) \prod_{i=1}^{N} A(i, w(i)),$$

where S_N^{ev} denotes the set of permutations of $\{1, 2, \ldots, N\}$ with only even-length cycles.

The next step is to compare the terms in this sum to the terms in $\text{Pf}(A)^2$. Using the distributive law to square the defining formula for $\text{Pf}(A)$, we see that

$$\text{Pf}(A)^2 = \sum_{u \in \text{SPf}_N} \sum_{v \in \text{SPf}_N} \text{sgn}(u)\,\text{sgn}(v) \prod_{i \text{ odd}} [A(u_i, u_{i+1})A(v_i, v_{i+1})].$$

For each $w \in S_N^{ev}$ indexing an uncanceled term in $\det(A)$, we associate a pair $g(w) = (u, v) \in \text{SPf}_N^2$ indexing a summand in $\text{Pf}(A)^2$ as follows. Consider the functional digraph $G(w)$ with vertex set $\{1, 2, \ldots, N\}$ and edge set $\{(i, w(i)) : 1 \leq i \leq N\}$, which is a disjoint union of cycles. Define a perfect matching M_1 on $G(w)$ (viewed as an undirected graph) by starting at the minimum element in each cycle and including every other edge as one travels around the cycle. Define another perfect matching M_2 on $G(w)$ by taking all the edges not used in M_1. Finally, let u and v be the permutations in SPf_N that encode M_1 and M_2 via the bijection in Theorem 12.109. For example, if $w = (1,5,2,8,6,3)(4,7)$, then $M_1 = \{\{1,5\},\{2,8\},\{6,3\},\{4,7\}\}$ and $M_2 = \{\{5,2\},\{8,6\},\{3,1\},\{7,4\}\}$, so $u = 15283647$ and $v = 13254768$.

The function g sending w to (u, v) is a bijection from S_N^{ev} to SPf_N^2. Given $(u, v) \in \text{SPf}_N^2$, we find $g^{-1}(u, v)$ as follows. First take the union of the perfect matchings encoded by u and v. This produces a graph that is a disjoint union of cycles of even length, as is readily checked. One can restore the directions on each cycle by recalling that the outgoing edge from the minimum element in each cycle belongs to the matching encoded by u. For example, the pair $(u, v) = (15234867, 12374856)$ maps to $g^{-1}(u, v) = (1,5,6,7,3,2)(4,8)$.

Throughout the following discussion, fix $w \in S_N^{ev}$ and $(u, v) \in \text{SPf}_N^2$ with $(u, v) = g(w)$. To complete the proof, it suffices to show that the term in $\det(A)$ indexed by w equals the term in $\text{Pf}(A)^2$ indexed by (u, v). Write w in cycle form as

$$w = (m_1, n_1, \ldots, z_1)(m_2, n_2, \ldots, z_2)\cdots(m_k, n_k, \ldots, z_k)$$

where $m_1 < m_2 < \cdots < m_k$ are the minimum elements in their cycles. Define two words (permutations in one-line form)

$$\begin{aligned} u^* &= m_1 n_1 \cdots z_1\; m_2 n_2 \cdots z_2\; \cdots\; m_k n_k \cdots z_k; \\ v^* &= n_1 \cdots z_1 m_1\; n_2 \cdots z_2 m_2\; \cdots\; n_k \cdots z_k m_k. \end{aligned}$$

Thus u^* is obtained by erasing the parentheses in the particular cycle notation for w just mentioned, and v^* is obtained similarly after first cycling the values in each cycle one step to the left. Since each m_i is the smallest value in its cycle, it follows that

$$\operatorname{inv}(v^*) = N - k + \operatorname{inv}(u^*),$$

where $k = \operatorname{cyc}(w)$ is the number of cycles in w. Using Theorem 7.34, we get $\operatorname{sgn}(u^*)\operatorname{sgn}(v^*) = (-1)^{N-\operatorname{cyc}(w)} = \operatorname{sgn}(w)$. Since all the edges $(i, w(i))$ in $G(w)$ arise by pairing off consecutive letters in u^* and v^*, we have

$$\operatorname{sgn}(w)\prod_{i=1}^{N} A(i, w(i)) = \operatorname{sgn}(u^*)\operatorname{sgn}(v^*)\prod_{i \text{ odd}} [A(u_i^*, u_{i+1}^*)A(v_i^*, v_{i+1}^*)].$$

We now transform the right side to the term indexed by (u, v) in $\operatorname{Pf}(A)^2$, as follows. Note that the words u^* and v^* provide *non-standard* encodings of the perfect matchings M_1 and M_2 encoded by u and v (where u^* encodes the matching $\{\{u_1^*, u_2^*\}, \{u_3^*, u_4^*\}, \ldots\}$, and similarly for v^*). To convert these encodings to the standard encodings, first reverse each pair of consecutive letters u_i^*, u_{i+1}^* in u^* such that $u_i^* > u_{i+1}^*$ and i is odd. Each such reversal causes $\operatorname{sgn}(u^*)$ to change, but this change is balanced by the fact that $A(u_{i+1}^*, u_i^*) = -A(u_i^*, u_{i+1}^*)$. Similarly, we can reverse pairs of consecutive letters in v^* that are out of order. The next step is to sort the pairs in u^* to force $u_1 < u_3 < u_5 < \cdots < u_{N-1}$. This sorting can be achieved by repeatedly swapping adjacent pairs $a < b; c < d$ in the word, where $a > c$ and a is in an odd position. The swap sending a, b, c, d to c, d, a, b can be achieved by applying the two transpositions (a, c) and (b, d) on the left. So this modification of u^* does not change $\operatorname{sgn}(u^*)$, nor does it affect the product of the factors $A(u_i^*, u_{i+1}^*)$ (since multiplication is commutative). Similarly, we can sort the pairs in v^* to obtain v without changing the formula. We conclude finally that

$$\begin{aligned}\operatorname{sgn}(w)\prod_{i=1}^{N} A(i, w(i)) &= \operatorname{sgn}(u^*)\operatorname{sgn}(v^*)\prod_{i \text{ odd}} [A(u_i^*, u_{i+1}^*)A(v_i^*, v_{i+1}^*)] \\ &= \operatorname{sgn}(u)\operatorname{sgn}(v)\prod_{i \text{ odd}} [A(u_i, u_{i+1})A(v_i, v_{i+1})]. \quad \square\end{aligned}$$

The following example illustrates the calculations at the end of the preceding proof.

12.113. Example. Suppose $w = (3, 8)(11, 4, 2, 9)(1, 10, 6, 7, 5, 12) \in S_{12}^{ev}$, so $k = \operatorname{cyc}(w) = 3$. We begin by writing the standard cycle notation for w:

$$w = (1, 10, 6, 7, 5, 12)(2, 9, 11, 4)(3, 8).$$

Next we set

$$u^* = 1, 10; 6, 7; 5, 12; 2, 9; 11, 4; 3, 8; \qquad v^* = 10, 6; 7, 5; 12, 1; 9, 11; 4, 2; 8, 3.$$

Observe that $\operatorname{inv}(v^*) = \operatorname{inv}(u^*) + (12 - 3)$ due to the cyclic shifting of $1, 2, 3$, so that $\operatorname{sgn}(u^*)\operatorname{sgn}(v^*) = (-1)^{12-3} = \operatorname{sgn}(w)$. Now we modify u^* and v^* so that the elements in each pair increase:

$$u' = 1, 10; 6, 7; 5, 12; 2, 9; 4, 11; 3, 8; \qquad v' = 6, 10; 5, 7; 1, 12; 9, 11; 2, 4; 3, 8.$$

Note that $\operatorname{sgn}(u') = -\operatorname{sgn}(u^*)$ since we switched 11 and 4, but this is offset by the fact that $A(11, 4) = -A(4, 11)$. So $\operatorname{sgn}(u^*)\prod_i A(u_i^*, u_{i+1}^*) = \operatorname{sgn}(u')\prod_i A(u_i', u_{i+1}')$, and similarly for v^* and v'. Finally, we sort the pairs so that the minimum elements increase, obtaining

$$u = 1, 10; 2, 9; 3, 8; 4, 11; 5, 12; 6, 7; \qquad v = 1, 12; 2, 4; 3, 8; 5, 7; 6, 10; 9, 11.$$

This sorting does not introduce any further sign changes, so we have successfully transformed the term indexed by w in $\det(A)$ to the term indexed by (u, v) in $\operatorname{Pf}(A)^2$.

12.16 Domino Tilings of Rectangles

This section presents P. W. Kasteleyn's proof of a formula for the number of ways to tile a rectangle with dominos. Let $\mathrm{Dom}(m,n)$ be the set of domino tilings of a rectangle of width m and height n. This set is empty if m and n are both odd, so we assume throughout that m is even. Given a tiling $T \in \mathrm{Dom}(m,n)$, let $N_h(T)$ and $N_v(T)$ be the number of horizontal and vertical dominos (respectively) appearing in T. Define the *weight* of the tiling T to be $\mathrm{wt}(T) = x^{N_h(T)}y^{N_v(T)}$.

12.114. Theorem: Domino Tiling Formula. For all even $m \geq 1$ and all $n \geq 1$,

$$\sum_{T\in\mathrm{Dom}(m,n)} \mathrm{wt}(T) = 2^{mn/2}\prod_{j=1}^{m/2}\prod_{k=1}^{n}\sqrt{x^2\cos^2\left(\frac{j\pi}{m+1}\right)+y^2\cos^2\left(\frac{k\pi}{n+1}\right)}. \tag{12.16}$$

Setting $x = y = 1$ gives the expression for $|\mathrm{Dom}(m,n)|$ stated in the Introduction.

Step 1: Conversion to a Perfect Matching Problem. Introduce a simple graph $G(m,n)$ with vertex set $V = \{1,2,\ldots,mn\}$ and edge set $E = E_x \cup E_y$, where

$$E_x = \{\{k,k+1\} : k \not\equiv 0 \pmod{m}\}, \quad E_y = \{\{k,k+m\} : 1 \leq k \leq m(n-1)\}.$$

This graph models an $m \times n$ rectangle R, as follows. The unit square in the ith row from the bottom and the jth column from the left in R corresponds to the vertex $(i-1)m+j$, for $1 \leq i \leq n$ and $1 \leq j \leq m$. There is an edge in E_x for each pair of two horizontally adjacent squares in R, and there is an edge in E_y for each pair of two vertically adjacent squares in R. There is a bijection between the set $\mathrm{Dom}(m,n)$ of domino tilings of R and the set $\mathrm{PM}(G(m,n))$ of perfect matchings of $G(m,n)$. Given a domino tiling, we need only replace each domino covering two adjacent squares by the edge corresponding to these two squares. This does give a perfect matching, since each square is covered by exactly one domino. If a tiling T corresponds to a matching M under this bijection, we have $N_h(T) = |M \cap E_x|$ and $N_v(T) = |M \cap E_y|$. So, defining $\mathrm{wt}(M) = x^{|M\cap E_x|}y^{|M\cap E_y|}$, we have

$$\sum_{T\in\mathrm{Dom}(m,n)} \mathrm{wt}(T) = \sum_{M\in\mathrm{PM}(G(m,n))} \mathrm{wt}(M).$$

12.115. Example. Figure 12.14 shows the rectangle R and associated graph $G(m,n)$ when $m = 4$ and $n = 5$.

Figure 12.15 shows a domino tiling of R and the associated perfect matching. The tiling and matching shown both have weight x^4y^6.

Step 2: Enumeration via Pfaffians. Let X_1 be the skew-symmetric matrix defined in Theorem 12.110, taking G there to be $G(m,n)$. We know that

$$\sum_{M\in\mathrm{PM}(G(m,n))} \mathrm{sgn}(M) \prod_{i<j:\{i,j\}\in M} x_{i,j} = \mathrm{Pf}(X_1). \tag{12.17}$$

We introduce the terms *horizontal edge*, *odd vertical edge*, and *even vertical edge* to refer (respectively) to edges in E_x, edges $\{k,k+m\}$ in E_y with k odd, and edges $\{k,k+m\}$ in E_y with k even. Consider the algebra homomorphism $\epsilon : \mathbb{R}[\{x_{i,j}\}] \to \mathbb{R}[x,y]$ that sends $x_{i,j}$ to x if $\{i,j\}$ is a horizontal edge, sends $x_{i,j}$ to y if $\{i,j\}$ is an even vertical edge, and sends

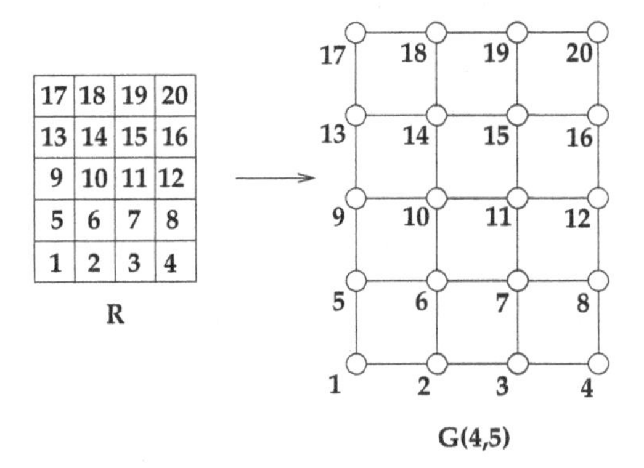

FIGURE 12.14
Graph used to model domino tilings.

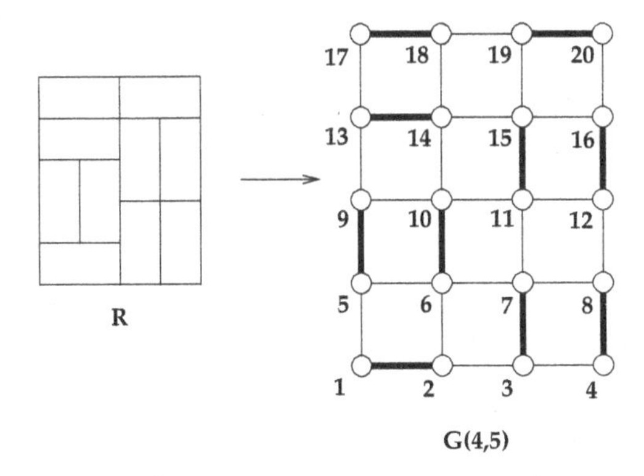

FIGURE 12.15
A domino tiling and a perfect matching.

$x_{i,j}$ to $-y$ if $\{i,j\}$ is an odd vertical edge. Let X be the matrix obtained by applying this homomorphism to each entry of the matrix X_1. Explicitly, X is the $mn \times mn$ matrix with entries

$$X(i,j) = \begin{cases} x & \text{if } j = i+1 \text{ and } i \not\equiv 0 \pmod{m}; \\ y & \text{if } j = i+m \text{ and } i \equiv 0 \pmod{2}; \\ -y & \text{if } j = i+m \text{ and } i \equiv 1 \pmod{2}; \\ -x & \text{if } i = j+1 \text{ and } j \not\equiv 0 \pmod{m}; \\ -y & \text{if } i = j+m \text{ and } j \equiv 0 \pmod{2}; \\ y & \text{if } i = j+m \text{ and } j \equiv 1 \pmod{2}; \\ 0 & \text{otherwise.} \end{cases} \tag{12.18}$$

For example, the matrix X when $m = 4$ and $n = 3$ appears in Figure 12.16. Let $\text{sgn}^*(M) = \text{sgn}(M)(-1)^t$, where t is the number of odd vertical edges in M. Applying the ring homomorphism ϵ to each side of (12.17) gives

$$\sum_{M \in \text{PM}(G(m,n))} \text{sgn}^*(M)\,\text{wt}(M) = \text{Pf}(X).$$

Step 3: Sign Analysis. The crucial fact to be verified is that $\text{sgn}^*(M) = +1$ *for every perfect matching* M. Before proving this fact, we consider an example.

$$\begin{bmatrix}
0 & x & 0 & 0 & -y & 0 & 0 & 0 & 0 & 0 & 0 & 0 \\
-x & 0 & x & 0 & 0 & y & 0 & 0 & 0 & 0 & 0 & 0 \\
0 & -x & 0 & x & 0 & 0 & -y & 0 & 0 & 0 & 0 & 0 \\
0 & 0 & -x & 0 & 0 & 0 & 0 & y & 0 & 0 & 0 & 0 \\
y & 0 & 0 & 0 & 0 & x & 0 & 0 & -y & 0 & 0 & 0 \\
0 & -y & 0 & 0 & -x & 0 & x & 0 & 0 & y & 0 & 0 \\
0 & 0 & y & 0 & 0 & -x & 0 & x & 0 & 0 & -y & 0 \\
0 & 0 & 0 & -y & 0 & 0 & -x & 0 & 0 & 0 & 0 & y \\
0 & 0 & 0 & 0 & y & 0 & 0 & 0 & 0 & x & 0 & 0 \\
0 & 0 & 0 & 0 & 0 & -y & 0 & 0 & -x & 0 & x & 0 \\
0 & 0 & 0 & 0 & 0 & 0 & y & 0 & 0 & -x & 0 & x \\
0 & 0 & 0 & 0 & 0 & 0 & 0 & -y & 0 & 0 & -x & 0
\end{bmatrix}$$

FIGURE 12.16
Matrix used to enumerate domino tilings ($m = 4, n = 3$).

12.116. Example. Consider the following domino tiling of a 16×4 rectangle:

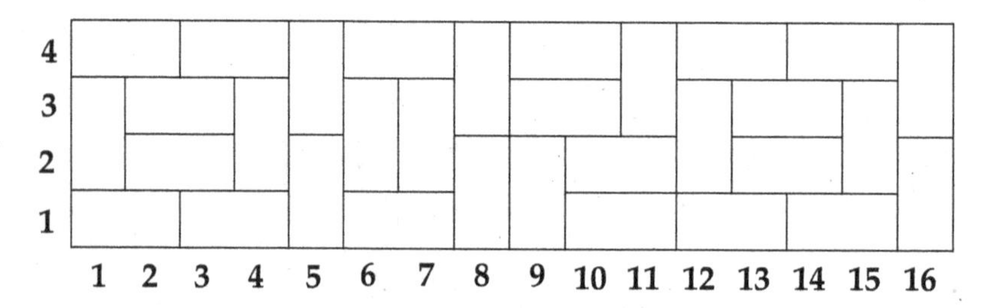

This tiling corresponds to a perfect matching M of $G(16, 4)$, which is encoded (as in Theorem 12.109) by a word $w \in \mathrm{SPf}_{64}$. By definition, $\mathrm{sgn}(M) = (-1)^{\mathrm{inv}(w)}$. In our example, the word of M is

$$\begin{aligned} w \;=\; & 1,2;3,4;5,21;6,7;8,24;9,25;10,11;12,13;14,15;16,32; \\ & 17,33;18,19;20,36;22,38;23,39;26,27;28,44;29,30;31,47;\ldots;60,61;62,63. \end{aligned}$$

Note that w consists of pairs of letters indicating the two squares occupied by each domino in the tiling. We imagine placing dominos on the board one at a time, in the order specified by w, and updating $\mathrm{sgn}(M)$ and $\mathrm{sgn}^*(M)$ as we go along. When computing $\mathrm{inv}(w)$, the second symbol in each pair sometimes causes inversions with symbols following it in w. Pairs corresponding to horizontal dominos never cause any inversions. Consider the inversions caused by a vertical domino (i.e., a vertical edge in M). The first vertical edge appearing in w is $\{5, 21\}$. The 21 is greater than the fifteen symbols $6, 7, \ldots, 20$ corresponding to squares to the right of column 5 in row 1 and squares to the left of column 5 in row 2, which have not been covered by a domino yet. So this edge increases $\mathrm{inv}(w)$ by $15 = m - 1$, which causes a sign change in $\mathrm{sgn}(M)$. However, since this edge is an odd vertical edge, that sign change is counteracted in $\mathrm{sgn}^*(M)$.

The next vertical edge in w is $\{8, 24\}$. The symbol 24 causes $14 = m - 2$ new inversions, corresponding to squares to the right of column 8 in row 1 and squares to the left of column 8 in row 2, excluding column 5. These inversions do not change $\mathrm{sgn}(M)$, and $\mathrm{sgn}^*(M)$ is also unchanged since $\{8, 24\}$ is an even vertical edge.

Continuing similarly, we eventually come to the odd vertical edge $\{23, 39\}$ in w. Recalling the order of domino placement, we see that the 39 causes inversions with the following nine symbols to its right in w: 37, 35, 34, 31, 30, 29, 28, 27, 26. Since nine is odd, we get a sign change in $\text{sgn}(M)$, but this is counteracted in $\text{sgn}^*(M)$ since we have just added an odd vertical edge. After accounting for all the dominos, we find (Exercise 12-110) that indeed $\text{sgn}^*(M) = +1$, since the insertion of each vertical domino never leads to a net sign change.

Now we are ready to prove that $\text{sgn}^*(M) = +1$ for a general $M \in \text{PM}(G(m,n))$. Let $w \in \text{SPf}_{mn}$ be the word encoding M. As in the example, we calculate $\text{sgn}^*(M) = (-1)^{\text{inv}(w)}(-1)^t$ incrementally by scanning w from left to right. Initially, before scanning any edges, $\text{sgn}^*(M)$ is $+1$. Suppose the next edge in the scan is the horizontal edge $\{k, k+1\}$. By definition of w (see Theorem 12.109), k is the smallest symbol that has not appeared previously in w. So k and $k+1$ cannot cause any new inversions with symbols following them. Similarly, t (the number of odd vertical edges) does not increase when we scan this edge. So $\text{sgn}^*(M)$ is still $+1$ after scanning this edge.

Before continuing, we need the following observation: for every row $i \geq 1$, the number of vertical dominos that start in row i and end in row $i+1$ is even (possibly zero). This is proved by induction on i. To prove the case $i = 1$, suppose there are a horizontal dominos in row 1. Then there must be $m - 2a$ vertical dominos starting in row 1. This number is even, since m is even. Now assume the result holds in row $i-1$. In row i, suppose there are a horizontal dominos, b vertical dominos coming up from row $i-1$, and c vertical dominos leading up into row $i+1$. Then $c = m - 2a - b$. Since m is even and (by hypothesis) b is even, c must also be even.

Now suppose the next edge in the scan is a vertical edge $\{k, k+m\}$ in column j that covers rows i and $i+1$ (so $k = (i-1)m+j$). As before, the symbol k causes no new inversions. Let us count the inversions in w between $k+m$ and symbols to its right. There are $m-1$ symbols that might cause inversions with $k+m$, namely $k+1, k+2, \ldots, k+(m-1)$, but some of these symbols may have already appeared in w. Specifically, if there are a vertical dominos covering rows i and $i+1$ to the left of column j, and b vertical dominos covering rows $i-1$ and i to the right of column j, then $a+b$ of the symbols just mentioned have already appeared in w. So, the inclusion of the new edge increases $\text{inv}(w)$ by $(m-1)-(a+b)$. Now, let there be b' vertical dominos covering rows $i-1$ and i to the left of column j, and c horizontal dominos in row i to the left of column j. Since $m-1 \equiv 1 \pmod{2}$, $-a \equiv a \pmod{2}$, $-b \equiv b' \pmod{2}$ (by the observation in the last paragraph), and $2c \equiv 0 \pmod{2}$, we see that

$$(m-1) - a - b \equiv 1 + a + b' + 2c \pmod{2}.$$

But $1 + a + b' + 2c = j$ since $a + b' + 2c$ counts all the columns left of column j in row i. We conclude, finally, that the increase in $\text{inv}(w)$ caused by the insertion of the edge $\{k, k+m\}$ has the same parity as the column index j. Since j and k have the same parity, the number of new inversions is odd iff the new vertical edge is an odd vertical edge. So there is no net change in $\text{sgn}^*(M) = (-1)^{\text{inv}(w)}(-1)^t$ when we add this edge. This completes the proof that $\text{sgn}^*(M) = +1$.

Step 4: Evaluation of the Pfaffian. Combining Steps 1 through 3 and Theorem 12.112, we have

$$\sum_{T \in \text{Dom}(m,n)} \text{wt}(T) = \sum_{M \in \text{PM}(G(m,n))} \text{wt}(M) = \text{Pf}(X) = \sqrt{\det(X)},$$

where X is the $mn \times mn$ matrix defined by (12.18). So we are reduced to evaluating the determinant of X. The idea is to replace X by a similar matrix $U^{-1}XU$ whose determinant is easier to evaluate. For this purpose, we pause to introduce tensor products of matrices.

12.117. Definition: Tensor Product of Matrices. If $A = [a_{i,j}]$ is any $n \times n$ matrix and

B is any $m \times m$ matrix, let $A \otimes B$ be the $mn \times mn$ matrix given in block form by

$$A \otimes B = \begin{bmatrix} a_{1,1}B & a_{1,2}B & \cdots & a_{1,n}B \\ a_{2,1}B & a_{2,2}B & \cdots & a_{2,n}B \\ \cdots & \cdots & \cdots & \cdots \\ a_{n,1}B & a_{n,2}B & \cdots & a_{n,n}B \end{bmatrix}.$$

Formally, $(A \otimes B)(m(i_1 - 1) + i_2, m(j_1 - 1) + j_2) = A(i_1, j_1)B(i_2, j_2)$ for all i_1, j_1, i_2, j_2 satisfying $1 \leq i_1, j_1 \leq n$ and $1 \leq i_2, j_2 \leq m$.

The following properties of tensor products may be routinely verified:
(a) $(A_1 + A_2) \otimes B = (A_1 \otimes B) + (A_2 \otimes B)$ and $A \otimes (B_1 + B_2) = (A \otimes B_1) + (A \otimes B_2)$.
(b) For any scalar c, $(cA) \otimes B = c(A \otimes B) = A \otimes (cB)$.
(c) $(A_1 \otimes B_1)(A_2 \otimes B_2) = (A_1 A_2) \otimes (B_1 B_2)$.
(d) If A and B are invertible, then $(A \otimes B)^{-1} = A^{-1} \otimes B^{-1}$.

For every $k \geq 1$, let I_k denote the $k \times k$ identity matrix, let F_k denote the $k \times k$ diagonal matrix with diagonal entries $-1, 1, -1, 1, \ldots, (-1)^k$, let I'_k denote the $k \times k$ matrix with 1's on the antidiagonal, and let Q_k denote the $k \times k$ matrix with 1's on the diagonal above the main diagonal, -1's on the diagonal below the main diagonal, and 0's elsewhere. For example,

$$I_5 = \begin{bmatrix} 1 & 0 & 0 & 0 & 0 \\ 0 & 1 & 0 & 0 & 0 \\ 0 & 0 & 1 & 0 & 0 \\ 0 & 0 & 0 & 1 & 0 \\ 0 & 0 & 0 & 0 & 1 \end{bmatrix}, \quad F_5 = \begin{bmatrix} -1 & 0 & 0 & 0 & 0 \\ 0 & 1 & 0 & 0 & 0 \\ 0 & 0 & -1 & 0 & 0 \\ 0 & 0 & 0 & 1 & 0 \\ 0 & 0 & 0 & 0 & -1 \end{bmatrix},$$

$$I'_5 = \begin{bmatrix} 0 & 0 & 0 & 0 & 1 \\ 0 & 0 & 0 & 1 & 0 \\ 0 & 0 & 1 & 0 & 0 \\ 0 & 1 & 0 & 0 & 0 \\ 1 & 0 & 0 & 0 & 0 \end{bmatrix}, \quad Q_5 = \begin{bmatrix} 0 & 1 & 0 & 0 & 0 \\ -1 & 0 & 1 & 0 & 0 \\ 0 & -1 & 0 & 1 & 0 \\ 0 & 0 & -1 & 0 & 1 \\ 0 & 0 & 0 & -1 & 0 \end{bmatrix}.$$

The definition of X in (12.18) can now be written

$$X = x(I_n \otimes Q_m) + y(Q_n \otimes F_m).$$

(Compare to Figure 12.16.) The following lemma can be established by routine calculations.

12.118. Lemma: Eigenvectors of Q_k. For $0 \leq a \leq k+1$ and $1 \leq b \leq k$, define complex numbers

$$U_k(a, b) = i^a \sin\left(\frac{\pi ab}{k+1}\right), \quad \lambda_k(b) = 2i \cos\left(\frac{b\pi}{k+1}\right).$$

For $1 \leq a, b \leq k$, we have

$$U_k(a+1, b) - U_k(a-1, b) = \lambda_k(b) U_k(a, b).$$

Therefore, the column vector $[U_k(1, b), U_k(2, b), \ldots, U_k(a, b)]^{\text{tr}}$ is an eigenvector of Q_k associated to the eigenvalue $\lambda_k(b)$. Let $U_k = [U_k(a, b)]_{1 \leq a,b \leq k}$, and let D_k be the $k \times k$ diagonal matrix with diagonal entries $\lambda_k(b)$. Then $Q_k U_k = U_k D_k$, $(-1)^a U_k(a, b) = -U_k(a, k+1-b)$ for all a, b between 1 and k, and so $F_k U_k = -U_k I'_k$.

The columns of U_k are linearly independent, because they are eigenvectors of Q_k associated to *distinct* eigenvalues. Therefore, U_k is invertible, so the lemma gives $U_k^{-1} Q_k U_k = D_k$

$$2i\begin{bmatrix} xr_1 & 0 & 0 & -ys_1 & 0 & 0 & 0 & 0 & 0 & 0 & 0 & 0 \\ 0 & xr_2 & -ys_1 & 0 & 0 & 0 & 0 & 0 & 0 & 0 & 0 & 0 \\ 0 & -ys_1 & xr_3 & 0 & 0 & 0 & 0 & 0 & 0 & 0 & 0 & 0 \\ -ys_1 & 0 & 0 & xr_4 & 0 & 0 & 0 & 0 & 0 & 0 & 0 & 0 \\ 0 & 0 & 0 & 0 & xr_1 & 0 & 0 & -ys_2 & 0 & 0 & 0 & 0 \\ 0 & 0 & 0 & 0 & 0 & xr_2 & -ys_2 & 0 & 0 & 0 & 0 & 0 \\ 0 & 0 & 0 & 0 & 0 & -ys_2 & xr_3 & 0 & 0 & 0 & 0 & 0 \\ 0 & 0 & 0 & 0 & -ys_2 & 0 & 0 & xr_4 & 0 & 0 & 0 & 0 \\ 0 & 0 & 0 & 0 & 0 & 0 & 0 & 0 & xr_1 & 0 & 0 & -ys_3 \\ 0 & 0 & 0 & 0 & 0 & 0 & 0 & 0 & 0 & xr_2 & -ys_3 & 0 \\ 0 & 0 & 0 & 0 & 0 & 0 & 0 & 0 & 0 & -ys_3 & xr_3 & 0 \\ 0 & 0 & 0 & 0 & 0 & 0 & 0 & 0 & -ys_3 & 0 & 0 & xr_4 \end{bmatrix}$$

FIGURE 12.17
The matrix $U^{-1}XU$ for $m = 4$, $n = 3$; here $r_a = 2i\cos(\pi a/5)$ and $s_b = 2i\cos(\pi b/4)$.

and $U_k^{-1}F_kU_k = -I_k'$. Let $U = U_n \otimes U_m$, so $U^{-1} = U_n^{-1} \otimes U_m^{-1}$. Using properties of tensor products, we calculate

$$\begin{aligned} U^{-1}XU &= x(U_n^{-1} \otimes U_m^{-1})(I_n \otimes Q_m)(U_n \otimes U_m) + y(U_n^{-1} \otimes U_m^{-1})(Q_n \otimes F_m)(U_n \otimes U_m) \\ &= x(U_n^{-1}I_nU_n) \otimes (U_m^{-1}Q_mU_m) + y(U_n^{-1}Q_nU_n) \otimes (U_m^{-1}F_mU_m) \\ &= x(I_n \otimes D_m) - y(D_n \otimes I_m'). \end{aligned}$$

For example, if X is the matrix shown in Figure 12.16, then $U^{-1}XU$ is the matrix shown in Figure 12.17. In general, $U^{-1}XU$ is a block-diagonal matrix consisting of n $m \times m$ blocks. The bth block has entries $-y\lambda_n(b)$ on the anti-diagonal and entries $x\lambda_m(a)$ (for $1 \le a \le m$) on the diagonal. Now, since m is even, we can reorder the rows and columns of each block into this order: $1, m, 2, m-1, 3, m-2, \ldots, m/2, m/2+1$. This reordering can be accomplished by performing an even number of row and column switches on $U^{-1}XU$, so the determinant does not change. The new matrix is also block-diagonal, consisting of $(mn/2)$ 2×2 blocks that look like

$$\begin{bmatrix} x\lambda_m(a) & -y\lambda_n(b) \\ -y\lambda_n(b) & x\lambda_m(m+1-a) \end{bmatrix} \qquad \text{for } 1 \le a \le m/2 \text{ and } 1 \le b \le n.$$

Now, $\lambda_m(m+1-a) = 2i\cos(\pi(m+1-a)/(m+1)) = -2i\cos(\pi a/(m+1)) = -\lambda_m(a)$. It follows that the determinant of the 2×2 block just mentioned is

$$-x^2\lambda_m(a)^2 - y^2\lambda_n(b)^2 = 4\left[x^2\cos^2\left(\frac{\pi a}{m+1}\right) + y^2\cos^2\left(\frac{\pi b}{n+1}\right)\right].$$

Finally, $\det(X) = \det(U^{-1}XU)$ is the product of these determinants as a ranges from 1 to $m/2$ and b ranges from 1 to n. Taking the square root of $\det(X)$ and factoring out powers of 2 produces formula (12.16). Remarkable!

Summary

- *Rational-Slope Dyck Paths.* If $\gcd(r, s) = 1$, then the number of lattice paths from $(0, 0)$ to (r, s) that never go below the line $sx = ry$ is $\frac{1}{r+s}\binom{r+s}{r,s}$. For any lattice path ending at (r, s), the $r + s$ cyclic shifts of this path are all distinct, and exactly one of them is a Dyck path of slope s/r.

- *The Chung–Feller Theorem.* A lattice path from $(0, 0)$ to (n, n) has k flaws iff the path has k north steps starting below $y = x$. For k between 0 and n, there are $C_n = \frac{1}{n+1}\binom{2n}{n,n}$ paths ending at (n, n) with k flaws. So the number of flaws in a random lattice path from $(0, 0)$ to (n, n) is uniformly distributed on $\{0, 1, 2, \ldots, n\}$.

- *Rook-Equivalence of Ferrers Boards.* For each integer partition μ, $r_k(\mu)$ is the number of ways to place k non-attacking rooks on $F_\mu = \mathrm{dg}(\mu)$, and $R_\mu(x) = \sum_{k\geq 0} r_k(\mu)x^k$. For all partitions $\mu = (\mu_1 \geq \mu_2 \geq \cdots \geq \mu_n \geq 0)$ and $\nu = (\nu_1 \geq \nu_2 \geq \cdots \geq \nu_n \geq 0)$ with $|\mu| = n = |\nu|$, we have $R_\mu(x) = R_\nu(x)$ iff the multisets $[\mu_i + i : 1 \leq i \leq n]$ and $[\nu_i + i : 1 \leq i \leq n]$ are equal.

- *Parking Functions.* A function $f : \{1, 2, \ldots, n\} \to \{1, 2, \ldots, n\}$ is a *parking function* iff $|\{x : f(x) \leq i\}| \geq i$ for all i between 1 and n. There are $(n + 1)^{n-1}$ parking functions of order n. A bijection from parking functions to labeled Dyck paths is given by listing the labels $\{x : f(x) = i\}$ in increasing order in column i for $i = 1, 2, \ldots, n$ in turn, putting one label in each row from bottom to top. A bijection from labeled Dyck paths to trees is given by letting the children of a_i be the labels in column $i + 1$, for all $i \geq 0$ (where $a_0 = 0$ and $a_1, \ldots, a_n$ are the labels from bottom to top).

- *Facts about Cyclic Groups.* If G is a cyclic group of size $n < \infty$, then G has a unique cyclic subgroup of size d for each divisor d of n, and these are all the subgroups of G. Any cyclic group of size d has $\phi(d)$ generators, and hence $n = \sum_{d|n} \phi(d)$. If G is a group of size n with at most one subgroup of size d for each divisor d of n, then G must be cyclic. Hence, any finite subgroup of the multiplicative group of a field is cyclic.

- *Counting Irreducible Polynomials.* The size of a finite field must be a prime power. For each prime power q, there exists a field F with q elements, which is unique up to isomorphism. For such a field F, let $I(n, q)$ be the number of monic irreducible polynomials of degree n in $F[x]$. Classifying elements in the field of size q^n by their minimal polynomials in $F[x]$ gives $q^n = \sum_{d|n} dI(d, q)$. Hence, by Möbius inversion, $I(n, q) = \frac{1}{n}\sum_{d|n} q^d\mu(n/d)$, where μ is the Möbius function from Definition 4.31.

- *Subspaces of Vector Spaces over Finite Fields.* A d-dimensional vector space over a q-element field has size q^d. The number of k-dimensional subspaces of an n-dimensional vector space over a q-element field is the integer $\begin{bmatrix} n \\ k \end{bmatrix}_q$. Each such subspace has a unique basis in reduced row-echelon form (RREF). The number of $k \times n$ RREF matrices with entries in a q-element field is thus $\begin{bmatrix} n \\ k \end{bmatrix}_q$.

- *Combinatorial Meaning of Tangent and Secant Power Series.* $\tan x = \sum_{n=1}^{\infty}(a_n/n!)x^n$, where a_n counts permutations w satisfying $w_1 < w_2 > w_3 < w_4 > \cdots > w_n$; and $\sec x = \sum_{n=0}^{\infty}(b_n/n!)x^n$, where b_n counts permutations w satisfying $w_1 < w_2 > w_3 < \cdots < w_n$.

- *Properties of Determinants.* The determinant of a matrix $A \in M_n(\mathbb{R})$ is a multilinear, alternating function of the rows (or the columns) of A such that $\det(I_n) = 1$. This means that $\det(A)$ is a linear function of any given row when the other rows are fixed, and the determinant is zero if A has two equal rows; similarly for columns. We have $\det(A^{\mathrm{tr}}) = \det(A)$. For triangular or diagonal A, $\det(A) = \prod_{i=1}^{n} A(i,i)$. The Laplace expansions for $\det(A)$ along row k and column k are

$$\det(A) = \sum_{j=1}^{n} (-1)^{j+k} A(k,j) \det(A[k|j]) = \sum_{i=1}^{n} (-1)^{i+k} A(i,k) \det(A[i|k]),$$

where $A[k|j]$ is A with row k and column j deleted. We have $A(\operatorname{adj} A) = (\det(A))I_n = (\operatorname{adj} A)A$, so that $A^{-1} = (\det(A))^{-1} \operatorname{adj}(A)$ when $\det(A)$ is invertible in $\mathbb{R}$. Similar results hold with $\mathbb{R}$ replaced by any commutative ring R.

- *The Cauchy–Binet Theorem.* Given an $m \times n$ matrix A and an $n \times m$ matrix B with $m \leq n$,
$$\det(AB) = \sum_{1 \leq j_1 < j_2 < \cdots < j_m \leq n} \det(A^{j_1}, \ldots, A^{j_m}) \det(B_{j_1}, \ldots, B_{j_m}),$$
where A^j is the jth column of A, and B_j is the jth row of B. In particular, $\det(AB) = \det(A)\det(B)$ for all $n \times n$ matrices A and B.

- *Tournaments.* A tournament is a digraph with no loop edges and exactly one directed edge between each pair of distinct vertices. A tournament τ is transitive iff for all vertices u, v, w, if (u,v) and (v,w) are edges of τ, then (u,w) is an edge of τ. Moreover, τ is transitive iff τ has no directed 3-cycle iff the list of outdegrees of the vertices of τ has no repetitions. A sign-reversing involution exists that cancels all non-transitive tournaments, leading to the following formula for the Vandermonde determinant:
$$\det[x_j^{n-i}]_{1 \leq i,j \leq n} = \sum_{w \in S_n} \operatorname{sgn}(w) \prod_{k=1}^{n} x_{w(k)}^{n-k} = \prod_{1 \leq i < j \leq n} (x_i - x_j).$$

- *The Hook-Length Formula.* For a partition λ with n boxes, the number of standard tableaux of shape λ is $n!/\prod_{c \in \mathrm{dg}(\lambda)} h(c)$, where $h(c)$ is the hook length of cell c. This can be proved probabilistically by defining a random algorithm that generates each $S \in \mathrm{SYT}(\lambda)$ with probability $\prod_{c \in \mathrm{dg}(\lambda)} h(c)/n!$. To build S, start at a random cell in $\mathrm{dg}(\lambda)$, then repeatedly jump to a random new cell in the hook of the current cell until reaching a corner. Place n in this corner and proceed recursively to fill the other cells in $\mathrm{dg}(\lambda)$.

- *Knuth Equivalence and Monotone Subsequences of Words.* Two words $\mathbf{v}$ and $\mathbf{w}$ are Knuth equivalent iff $\mathbf{v}$ can be changed into $\mathbf{w}$ by a sequence of moves of the form $\cdots yxz \cdots \leftrightarrow \cdots yzx \cdots$ (where $x < y \leq z$) or $\cdots xzy \cdots \leftrightarrow \cdots zxy \cdots$ (where $x \leq y < z$). These moves simulate tableau insertion (when applied to reading words), so every $\mathbf{w}$ is Knuth equivalent to the reading word of its insertion tableau $P(\mathbf{w})$. Words $\mathbf{v}$ and $\mathbf{w}$ are Knuth equivalent iff $P(\mathbf{v}) = P(\mathbf{w})$. If $P(\mathbf{w})$ has shape λ, then $\lambda_1 + \cdots + \lambda_k$ is the maximum total length of a set of k disjoint weakly increasing subsequences of $\mathbf{w}$, and $\lambda'_1 + \cdots + \lambda'_k$ is the maximum total length of a set of k disjoint strictly decreasing subsequences of $\mathbf{w}$.

- *Quasisymmetric Polynomials.* A polynomial $f \in \mathbb{R}[x_1, \ldots, x_N]$ is quasisymmetric iff the coefficients of $\mathbf{x}^\alpha$ and $\mathbf{x}^\beta$ in f are equal whenever the exponent sequences α and β are shifts of each other. For a composition α with at most N parts, the monomial quasisymmetric polynomial M_α is the sum of all monomials $\mathbf{x}^\beta$ where β is a shift of α. The set $\{M_\alpha : \alpha \in \mathrm{Comp}_N(k)\}$ is a basis of the vector space Q_N^k of quasisymmetric polynomials that are homogeneous of degree k. For $S \subseteq [k-1]$, the fundamental quasisymmetric polynomial $\mathrm{FQ}_{k,S}$ is the sum of all products $x_{i_1} x_{i_2} \cdots x_{i_k}$ where $i_1, \ldots, i_k$ is a weakly increasing sequence such that $j \in S$ implies $i_j < i_{j+1}$. For $N \geq k$, $\{\mathrm{FQ}_{k,S} : S \subseteq [k-1]\}$ is a basis of Q_N^k. We have $\mathrm{FQ}_{k,S} = \sum_{\alpha:S\subseteq \mathrm{sub}(\alpha)} M_\alpha$ and $s_\lambda = \sum_{U\in\mathrm{SYT}(\lambda)} \mathrm{FQ}_{|\lambda|,\mathrm{Des}(U)}$. The proof of the last formula uses a standardization bijection to convert semistandard tableaux to standard tableaux. This bijection relabels equal entries in a tableau T by a run of consecutive integers.

- *Pfaffians.* Let N be even. Given an $N \times N$ skew-symmetric matrix A (meaning $A^{\mathrm{tr}} = -A$), the Pfaffian of A is
$$\mathrm{Pf}(A) = \sum_{w\in \mathrm{SPf}_N} \mathrm{sgn}(w) \prod_{i \text{ odd}} A(w_i, w_{i+1}),$$
where $w \in \mathrm{SPf}_N$ iff $w \in S_N$, $w_i < w_{i+1}$, and $w_i < w_{i+2}$ for all odd i. We have $\det(A) = \mathrm{Pf}(A)^2$. Each term of $\mathrm{Pf}(A)$ counts a signed, weighted perfect matching of a graph with vertex set $\{1, 2, \ldots, N\}$, where an edge from i to j (for $i < j$) is weighted by $A(i, j)$. There is a recursion $\mathrm{Pf}(A) = \sum_{j=2}^{N} (-1)^j A(1, j)\,\mathrm{Pf}(A[[1, j]])$, where $A[[1, j]]$ is the matrix obtained by deleting rows 1 and j and columns 1 and j of A.

- *Domino Tilings.* For all $m, n \in \mathbb{Z}_{>0}$ with m even, the coefficient of $x^a y^b$ in
$$2^{mn/2} \prod_{j=1}^{m/2} \prod_{k=1}^{n} \sqrt{x^2 \cos^2\left(\frac{j\pi}{m+1}\right) + y^2 \cos^2\left(\frac{k\pi}{n+1}\right)}$$
is the number of ways to tile an $m \times n$ board with a horizontal dominos and b vertical dominos. The steps in the proof are: (a) model domino tilings by perfect matchings of a grid-shaped graph; (b) use a Pfaffian to enumerate these signed perfect matchings; (c) adjust signs in the matrix so every perfect matching has sign $+1$; (d) rewrite the Pfaffian as the square root of the determinant of the matrix; (e) evaluate the determinant by performing a similarity transformation that nearly diagonalizes the matrix, creating 2×2 blocks running down the diagonal. Each 2×2 block contributes one of the factors in the product formula above.

Exercises

12-1. Let $\sim$ be the cyclic shift relation from §12.1. Find all the equivalence classes of $\sim$ for: (a) the set of lattice paths ending at $(3, 4)$; (b) the set of lattice paths ending at $(3, 3)$.

12-2. In this problem, we do not assume $\gcd(r, s) = 1$. For $v, w \in \mathcal{R}(\mathrm{N}^s\mathrm{E}^r)$, write $v \sim w$ iff w can be obtained from v by a cyclic shift. Which of the following statements must be true for all $r, s \geq 1$? Explain. (a) Every equivalence class of $\sim$ has size $r + s$. (b) Every equivalence class of $\sim$ contains at least one path staying weakly above $sx = ry$. (c) Every equivalence class of $\sim$ contains at most one path staying weakly above $sx = ry$.

12-3. Fix $h, k \in \mathbb{Z}_{\geq 0}$ and $m \in \mathbb{Z}_{\geq 1}$. Show that the number of lattice paths from $(0, 0)$ to $(k + mh, h)$ that never go below the line $x = k + my$ is
$$\binom{k + (m+1)h}{k + mh, h} - m\binom{k + (m+1)h}{k + mh + 1, h - 1}$$
via a bijective proof analogous to the proof of the Dyck Path Rule 1.101 in §1.14. (*Hint:* Label each point (x, y) by the integer $x - k - my$ and thereby divide the bad paths into m classes. Reflections do not work for $m > 1$, so look for other symmetries.)

12-4. Verify the Chung–Feller theorem directly for $n = 3$ by drawing all lattice paths from $(0,0)$ to $(3,3)$ with: (a) 0 flaws; (b) 1 flaw; (c) 2 flaws; (d) 3 flaws.

12-5. Let π be the Dyck path NNENEENNNENNENNEEENENEEE. Use the bijections from the proof of Theorem 12.4 to compute the associated lattice path with: (a) 5 flaws; (b) 8 flaws; (c) 10 flaws.

12-6. For each flawed path π, find the Dyck path associated to π via the bijections used to prove Theorem 12.4.

(a) NENNEEEENENNNEENEEENENNNE

(b) NEEENNENEEENNNNNEENE

12-7. Let π be a random lattice path from $(0,0)$ to (n,n), and for $1 \leq j \leq n$, let $X_j(\pi)$ be 1 if π has a flaw in row j and 0 otherwise. Prove bijectively that $P(X_j = 0) = 1/2 = P(X_j = 1)$.

12-8. Let $X_1, X_2, \ldots, X_n$ be *independent* random variables such that $P(X_i = 1) = 1/2 = P(X_i = 0)$ for all i. (This means that, for all $v_1, \ldots, v_n \in \{0,1\}$, the events $X_1 = v_1, X_2 = v_2, \ldots, X_n = v_n$ are independent as in Definition 1.66.) Compute $P(X_1+X_2+\cdots+X_n = k)$ for k between 0 and n. Contrast your answer with the Chung–Feller theorem.

12-9. Find a formula for the number of lattice paths from $(0,0)$ to (n,n) with k flaws and j east steps departing from the line $y = x$.

12-10. Compute the rook polynomial for each of the following partitions. (a) $(3,2,1)$ (b) $(8,8,8,8,8,8,8,8)$ (c) (n) (d) $(n,n,1^k)$.

12-11. Draw the diagrams of all integer partitions of 8 and determine which pairs of partitions are rook-equivalent.

12-12. Prove that for any integer partition μ, $R_\mu(x) = R_{\mu'}(x)$.

12-13. (a) For any $n \geq 1$, prove that the partition μ consisting of n copies of n is rook-equivalent to the partition $\nu = (2n-1, 2n-3, \ldots, 5, 3, 1)$. (b) Define a bijection between the set of non-attacking placements of k rooks on μ and the set of non-attacking placements of k rooks on ν.

12-14. Let μ be an integer partition such that $\mathrm{dg}(\mu) \subseteq \mathrm{dg}(\Delta_N)$, where $\Delta_N = (N-1, N-2, \ldots, 3, 2, 1, 0)$. Let the sequence $(N-1-\mu_1, N-2-\mu_2, \ldots, 0-\mu_N)$ contain a_k copies of k for each $k \geq 0$. (This sequence gives the row lengths of the skew shape Δ_N/μ.) Prove that the number of partitions that are rook-equivalent to μ is

$$\prod_{k\geq 1} \binom{a_{k-1}+a_k-1}{a_{k-1}-1, a_k}.$$

12-15. Let $\Delta_N = (N-1, N-2, \ldots, 3, 2, 1, 0)$. (a) Show that for all rook-equivalent partitions μ and ν, $\mathrm{dg}(\mu) \subseteq \mathrm{dg}(\Delta_N)$ iff $\mathrm{dg}(\nu) \subseteq \mathrm{dg}(\Delta_N)$. (b) Part (a) says that the set $\{\mu : \mathrm{dg}(\mu) \subseteq \mathrm{dg}(\Delta_N)\}$ is a union of equivalence classes of the rook-equivalence relation. Determine the number of equivalence classes in this union.

12-16. Show that for each integer partition μ, there is a unique integer partition ν with distinct parts that is rook-equivalent to μ.

12-17. Given a non-attacking rook placement π on a Ferrers board F_μ, we define a q-weight $\mathrm{wt}(\pi)$ as follows. Each rook in π *cancels* its own square, all squares above the rook in its column, and all squares left of the rook in its row. Let $\mathrm{wt}(\pi)$ be the number of uncanceled squares in F_μ. Define $r_k(\mu; q) = \sum_\pi q^{\mathrm{wt}(\pi)}$, where we sum over all non-attacking placements of k rooks on F_μ. Find and prove a q-analogue of the factorization formula (12.1) where $r_k(\mu)$ is replaced by $r_k(\mu; q)$.

12-18. Suppose μ is an integer partition with $\text{dg}(\mu) \subseteq \text{dg}(\Delta_N)$, where $\Delta_N = (N-1, N-2, \ldots, 2, 1, 0)$. (a) Use an involution to prove

$$r_k(\mu) = \sum_{i=0}^{k} S(N-i, N-k)(-1)^i e_i(N-1-\mu_1, N-2-\mu_2, \ldots, N-N-\mu_N),$$

where $S(u,v)$ is a Stirling number of the second kind and e_i is an elementary symmetric polynomial. (b) Use (a) to deduce a combinatorial proof of Theorem 2.64(d). (c) Use (a) to deduce that the multiset condition in Theorem 12.10 is sufficient for $R_\mu(x) = R_\nu(x)$. (d) Assume μ and ν are rook-equivalent partitions. Use (a) and the Garsia–Milne Involution Principle (Exercise 4-88) to construct a bijection from the set of non-attacking placements of k rooks on F_μ to the set of non-attacking placements of k rooks on F_ν.

12-19. For each labeled Dyck path in Figure 12.8, compute the associated parking function and tree (see Theorems 12.21 and 12.22).

12-20. (a) Convert the parking function f in Example 12.13 to a labeled Dyck path and a tree. (b) Convert the labeled Dyck path NNENNEENEENENNEE with labels $5, 8, 2, 4, 1, 6, 3, 7$ (from bottom to top) to a parking function and a tree. (c) Convert the tree

$$T = (\{0, 1, \ldots 10\}, \{\{0,9\}, \{5,7\}, \{5,8\}, \{9,4\}, \{7,6\}, \{6,9\}, \{7,10\}, \{10,1\}, \{3,9\}, \{2,9\}\})$$

to a labeled Dyck path and a parking function.

12-21. Suppose we represent a function $f : \{1, 2, \ldots, b\} \to \{1, 2, \ldots, a+1\}$ as a labeled lattice path ending at (a, b). Find conditions on the labeled path that are equivalent to f being (a) surjective; (b) injective.

12-22. (a) Given nonnegative integers $c_1, \ldots, c_{a+1}$ adding to b, how many labeled lattice paths from $(0,0)$ to (a,b) have c_i labels in column i for all i? (b) Use the bijections in §12.5 to translate (a) into enumeration results for parking functions and trees.

12-23. (a) Let p_n be the number of parking functions of order n. Give a combinatorial proof of the recursion

$$p_n = \sum_{m=1}^{n} m \binom{n-1}{m-1} p_{m-1} p_{n-m}.$$

(b) Use (a) and Exercise 3-74 to define a bijection between parking functions and trees.

12-24. Let $\mathcal{P}_n$ be the set of parking functions of order n. For $f \in \mathcal{P}_n$, define $\text{wt}(f) = n(n+1)/2 - \sum_{i=1}^{n} f(i)$. Let $P_n(q) = \sum_{f \in \mathcal{P}_n} q^{\text{wt}(f)}$. Prove the recursion

$$P_n(q) = \sum_{m=1}^{n} [m]_q \binom{n-1}{m-1} P_{m-1}(q) P_{n-m}(q).$$

12-25. Let S be a k-element subset of $\{1, 2, \ldots, n\}$. Prove that there are kn^{n-k-1} parking functions f such that $S = \{x : f(x) = 1\}$.

12-26. For each $n, k, m \in \mathbb{Z}_{\geq 0}$, let $\mathcal{P}_{n,k,m}$ be the set of labeled lattice paths ending at $(k + mn, n)$ that never go below the line $x = k + my$. Find a recursion satisfied by the quantities $|\mathcal{P}_{n,k,m}|$.

12-27. Find a bijection between the set of parking functions of order n and the quotient group $(\mathbb{Z}_{n+1})^n / H$, where H is the subgroup generated by $(1, 1, \ldots, 1)$.

12-28. How many generators does an infinite cyclic group have?

12-29. Prove or disprove: if every proper subgroup of a finite group G is cyclic, then G itself must be cyclic.

12-30. Suppose G is a group such that, for all $d \in \mathbb{Z}_{\geq 1}$, G has at most d elements x such that $x^d = 1$. Prove that every finite subgroup of G is cyclic.

12-31. Describe all the finite subgroups of the field $\mathbb{C}$.

12-32. Quaternions. Let $\mathbb{H}$ be a four-dimensional real vector space with basis $\mathbf{1}, \mathbf{i}, \mathbf{j}, \mathbf{k}$. Define multiplication on $\mathbb{H}$ by letting $\mathbf{1}$ act as the identity, setting $\mathbf{i}^2 = \mathbf{j}^2 = \mathbf{k}^2 = -\mathbf{1}$, $\mathbf{ij} = \mathbf{k} = -\mathbf{ji}$, $\mathbf{jk} = \mathbf{i} = -\mathbf{kj}$, $\mathbf{ki} = \mathbf{j} = -\mathbf{ik}$, and extending by linearity. (a) Show that $\mathbb{H}$ with this multiplication is a division ring (i.e., $\mathbb{H}$ satisfies all the axioms in the definition of a field except commutativity of multiplication). (b) Find a non-cyclic finite subgroup of the multiplicative group $\mathbb{H} - \{0\}$ (cf. Theorem 12.29). (c) Show that the equation $x^2 = -\mathbf{1}$ has infinitely many solutions in $\mathbb{H}$.

12-33. Prove that the product of all the nonzero elements in a finite field F is -1_F. Deduce Wilson's Theorem: for p prime, $(p-1)! \equiv -1 \pmod{p}$.

12-34. Compute the number of monic irreducible polynomials of degree 12 over a 9-element field.

12-35. (a) Enumerate all the irreducible polynomials in $\mathbb{Z}_2[x]$ of degree at most 5. (b) Use the formula in Theorem 12.30 to compute $I(n, 2)$ for $1 \leq n \leq 8$ (compare with the results in (a) for $1 \leq n \leq 5$).

12-36. Construction of Finite Fields. Let F be a field with q elements, let $h \in F[x]$ be a fixed monic irreducible polynomial of degree n, and let

$$K = \{f \in F[x] : f = 0 \text{ or } \deg(f) < n\}.$$

For $f, g \in K$, define $f + g$ to be the sum of these polynomials in $F[x]$, and define $f \otimes g$ to be the remainder when fg is divided by h. Show that K, with these operations, is a field of size q^n. The field K is denoted $F[x]/(h)$.

12-37. Let $K = \mathbb{Z}_2[x]/(x^3 + x + 1)$ (see the previous exercise). Construct addition and multiplication tables for K. Explicitly confirm that the multiplicative group $K - \{0\}$ is generated by x by computing x^i for $1 \leq i \leq 7$.

12-38. Let $h = x^4 + x + 1 \in \mathbb{Z}_2[x]$, and let $K = \mathbb{Z}_2[x]/(h)$, which is a 16-element field. (a) Explain why every element $y \in K$ satisfies $y^{16} = y$. (b) List all the elements of K and their minimal polynomials over $\mathbb{Z}_2$. (c) Factor the polynomial $x^{16} - x \in \mathbb{Z}_2[x]$ into a product of irreducible polynomials. (d) Explain the relation between part (b), part (c), and the formulas in Theorem 12.30. (e) Find all generators of the cyclic group of nonzero elements of K.

12-39. (a) Use Theorem 12.30 to show that $I(n, q) > 0$ for all prime powers q and all $n \geq 1$. (b) Prove that for every prime power p^n, there exists a field of size p^n.

12-40. Let F be a finite field of size q. A polynomial $h \in F[x]$ is called *primitive* iff h is a monic irreducible polynomial such that x is a generator of the multiplicative group of the field $K = F[x]/(h)$. (a) Count the primitive polynomials of degree n in $F[x]$. (b) Give an example of an irreducible polynomial in $\mathbb{Z}_2[x]$ that is not primitive.

12-41. Let K be a q-element field. How many $n \times n$ matrices with entries in K are: (a) upper-triangular; (b) strictly upper-triangular (zeroes on the main diagonal); (c) unitriangular (ones on the main diagonal); (d) upper-triangular and invertible?

12-42. How many 2×2 matrices with entries in a q-element field have determinant 1?

12-43. Count the number of invertible $n \times n$ matrices with entries in a q-element field F. How is the answer related to $[n]!_q$?

12-44. How many three-dimensional subspaces does the vector space $\mathbb{Z}_7^5$ have?

12-45. For each integer partition μ that fits in a box with two rows and three columns, draw a picture of the RREF matrix associated to μ in the proof of Theorem 12.37.

12-46. Find the RREF basis for the subspace of $\mathbb{Z}_5^5$ spanned by $v_1 = (1,4,2,3,4)$, $v_2 = (2,3,1,0,0)$, and $v_3 = (0,0,3,1,1)$.

12-47. Find the RREF basis for the subspace of $\mathbb{Z}_2^6$ spanned by $v_1 = (0,1,1,1,1,0)$, $v_2 = (1,1,1,0,1,1)$, and $v_3 = (1,0,1,0,0,0)$.

12-48. Let V be an n-dimensional vector space over a field K. A *flag of subspaces of* V is a chain of subspaces $V = V_0 \supseteq V_1 \supseteq V_2 \supseteq \cdots \supseteq V_s = \{0\}$. Suppose $|K| = q$. Given $n_1, \ldots, n_s$ and $n = n_1 + \cdots + n_s$, count the number of such flags in V such that $\dim_K(V_{i-1}) - \dim_K(V_i) = n_i$ for i between 1 and s.

12-49. (a) Give a linear-algebraic proof of the symmetry property $\left[{n \atop k}\right]_q = \left[{n \atop n-k}\right]_q$ when q is a prime power. (b) Explain how the equality of formal polynomials $\left[{n \atop k}\right]_q = \left[{n \atop n-k}\right]_q$ can be deduced from (a).

12-50. Let V be an n-dimensional vector space over a q-element field, and let X be the poset of all subspaces of V, ordered by inclusion. Show that the Möbius function of X is given by $\mu_X(W,Y) = (-1)^d q^{d(d-1)/2}$ if $W \subseteq Y$ and $d = \dim(Y) - \dim(W)$, and $\mu_X(W,Y) = 0$ if $W \not\subseteq Y$. (Use Exercise 8-12.)

12-51. Use the recursions for a_n and b_n in §12.8 to verify the values in (12.3).

12-52. Give probabilistic interpretations for the rational numbers appearing as coefficients in the Maclaurin series for $\tan x$ and $\sec x$.

12-53. Fill in the details of Step 5 of the proof in §12.8.

12-54. (a) List the permutations satisfying (12.4) for $n = 1,3,5$. (b) List the permutations satisfying (12.5) for $n = 0,2,4$.

12-55. (a) Develop ranking and unranking algorithms for up-down permutations. (b) Unrank 147 to get an up-down permutation in S_7. (c) Find the rank of 25364817 among up-down permutations of length 8.

12-56. (a) Develop successor algorithms for up-down permutations. (b) Find the successor of 25364817 among up-down permutations of length 8. (c) Find the successor of 385927164 among up-down permutation of length 9.

12-57. Let $(q;q)_0 = 1$ and $(q;q)_n = (1-q)(1-q^2)\cdots(1-q^n)$ for $n \in \mathbb{Z}_{\geq 1}$. Consider the following formal q-analogues of trigonometric functions:

$$\sin_q x = \sum_{k=0}^{\infty} (-1)^k \frac{x^{2k+1}}{(q;q)_{2k+1}}; \qquad \cos_q x = \sum_{k=0}^{\infty} (-1)^k \frac{x^{2k}}{(q;q)_{2k}};$$

$$\tan_q x = \sin_q x / \cos_q x; \qquad \sec_q x = 1/\cos_q x.$$

Define *q-tangent numbers* t_n and *q-secant numbers* s_n by

$$\tan_q x = \sum_{k=0}^{\infty} \frac{t_n}{(q;q)_n} x^n; \qquad \sec_q x = \sum_{k=0}^{\infty} \frac{s_n}{(q;q)_n} x^n.$$

(a) Show that for each $n \in \mathbb{Z}_{\geq 0}$,

$$t_n = \sum_{w \text{ satisfying } (12.4)} q^{\text{inv}(w)}; \qquad s_n = \sum_{w \text{ satisfying } (12.5)} q^{\text{inv}(w)}.$$

(b) Use (a) to conclude that t_n and s_n are polynomials in q with nonnegative integer coefficients. Compute t_n for $n = 1, 3, 5$ and s_n for $n = 0, 2, 4$.

12-58. Suppose an $n \times n$ matrix A is given in block form as $A = \begin{bmatrix} B & \mathbf{0} \\ C & D \end{bmatrix}$, where B is $k \times k$, C is $(n-k) \times k$, D is $(n-k) \times (n-k)$, and $\mathbf{0}$ denotes a $k \times (n-k)$ block of zeroes. Prove that $\det(A) = \det(B)\det(D)$.

12-59. Algorithmic Complexity of Determinant Evaluation. Let $A \in M_n(\mathbb{R})$. (a) How many additions and multiplications in $\mathbb{R}$ are needed to compute $\det(A)$ directly from Definition 12.40? (b) How many additions and multiplications in $\mathbb{R}$ are needed to compute $\det(A)$ recursively, using Theorem 12.51? (c) Explain how to use Theorems 12.44 and 12.50 to compute $\det(A)$ efficiently (using about cn^3 operations in $\mathbb{R}$ for some constant c).

12-60. Permanents. The *permanent* of an $n \times n$ matrix $A \in M_n(\mathbb{R})$ is $\operatorname{per}(A) = \sum_{w \in S_n} \prod_{i=1}^n A(i, w(i))$. Prove the following facts about permanents: (a) $\operatorname{per}(A^{\text{tr}}) = \operatorname{per}(A)$; (b) if A is diagonal, then $\operatorname{per}(A) = \prod_{i=1}^n A(i,i)$; (c) $\operatorname{per}(I_n) = 1$; (d) $\operatorname{per}(A)$ is an $\mathbb{R}$-multilinear function of the rows and columns of A (cf. Theorem 12.48); (e) if B is obtained from A by permuting the rows in any fashion, then $\operatorname{per}(B) = \operatorname{per}(A)$.

12-61. State and prove versions of the expansions in Theorem 12.51 for computing permanents.

12-62. Verify the characterization of linear maps stated in Example 12.47.

12-63. Complete the proof of Theorem 12.53 by showing that $(\operatorname{adj} A)A = \det(A)I_n$.

12-64. Cramer's Rule. Let $A \in M_n(\mathbb{R})$ where $\det(A)$ is invertible in $\mathbb{R}$, let $\mathbf{b}$ be a given $n \times 1$ vector, and let $\mathbf{x} = [x_1 \ \cdots \ x_n]^{\text{tr}}$. Show that the unique solution of the linear system $A\mathbf{x} = \mathbf{b}$ is given by $x_i = \det(A_i)/\det(A)$, where A_i is the matrix obtained from A by replacing the ith column by $\mathbf{b}$.

12-65. Verify the Cauchy–Binet Theorem by hand computation for the matrices

$$A = \begin{bmatrix} 2 & 1 & 0 & 3 \\ 1 & -1 & 1 & 2 \\ 4 & 0 & 2 & 1 \end{bmatrix}, \qquad B = \begin{bmatrix} -1 & 1 & -1 \\ 0 & 2 & 5 \\ 1 & 1 & 4 \\ -2 & 0 & -1 \end{bmatrix}.$$

12-66. Consider a function $w : \{1, 2, \ldots, k\} \to \{1, 2, \ldots, n\}$, which we regard as a word $w = w_1 w_2 \cdots w_k$. Show that there exist basic transpositions $t_1, \ldots, t_m \in S_k$ such that $w \circ (t_1 t_2 \cdots t_m)$ is a weakly increasing word, and the minimum possible value of m is $\operatorname{inv}(w)$.

12-67. Let A and B be $n \times n$ matrices. Prove that $\det(AB) = \det(A)\det(B)$ by imitating (and simplifying) the proof of the Cauchy–Binet Theorem 12.56.

12-68. Given an $m \times n$ matrix A and an $n \times m$ matrix B with $m > n$, what is $\det(AB)$? Explain.

12-69. Let τ be the tournament with edge set $\{(2,1), (1,3), (4,1), (1,5), (6,1), (2,3), (4,2), (5,2), (2,6), (3,4), (3,5), (6,3), (4,5), (6,4), (6,5)\}$. Compute $\operatorname{wt}(\tau)$, $\operatorname{inv}(\tau)$, and $\operatorname{sgn}(\tau)$. Is τ transitive?

12-70. Let τ be the tournament in Example 12.60, and let I be the involution used to prove Theorem 12.65. Compute $\tau' = I(\tau)$, and verify directly that $\operatorname{wt}(\tau') = \operatorname{wt}(\tau)$, $\operatorname{sgn}(\tau') = -\operatorname{sgn}(\tau)$, and $I(\tau') = \tau$.

12-71. Use induction and Theorem 12.50 to give an algebraic proof of Theorem 12.65.

12-72. Suppose $x_0, x_1, \ldots, x_N$ are *distinct* elements of a field F. State why the Vandermonde matrix $[x_j^{N-i}]_{0\leq i,j\leq N}$ is invertible. Use this to prove that if $p \in F[x]$ has degree at most N and satisfies $p(x_i) = 0$ for $0 \leq i \leq N$, then p must be the zero polynomial.

12-73. A *king* in a tournament τ is a vertex v from which every other vertex can be reached by following at most two directed edges. Show that every vertex of maximum outdegree in a tournament is a king; in particular, every tournament has a king.

12-74. Use the Hook-Length Formula to compute f^λ for the following shapes λ: (a) $(3,2,1)$; (b) $(4,4,4)$; (c) $(6,3,2,2,1,1,1)$; (d) $(n,n-1)$; (e) $(a,1^b)$.

12-75. Show that $f^{(0)} = 1$ and, for all nonzero partitions λ, $f^\lambda = \sum_\mu f^\mu$ where we sum over all μ that can be obtained from λ by removing some corner square. Use this recursion to calculate f^λ for all λ with $|\lambda| \leq 6$.

12-76. (a) Develop ranking and unranking algorithms for standard tableaux of shape λ based on the recursion in Exercise 12-75. (b) Unrank 46 to get a standard tableau of shape $(4,3,1)$. (c) Rank the standard tableau

$$\begin{array}{|c|c|c|}\hline 1 & 3 & 4 \\ \hline 2 & 5 & 8 \\ \hline 6 & 7 \\ \cline{1-2}\end{array}.$$

12-77. (a) Develop successor algorithms for standard tableaux of shape λ based on the recursion in Exercise 12-75. (b) Find the successors of the three standard tableaux of shape $(4,2,2,1)$ shown in the Introduction of this book. (c) Find the successor of the standard tableau in part (c) of the previous exercise.

12-78. Enumerate all the hook walks for the shape $\lambda = (4,3,2,1)$ that end in the corner cell $(2,3)$, and compute the probability of each walk. Use this computation to verify Theorem 12.73 in this case.

12-79. Suppose $\lambda \in \text{Par}(p)$ where p is prime. (a) Show that p divides f^λ if λ is not a hook shape (a hook shape is a partition of the form $(a,1^{p-a})$). (b) Compute $f^\lambda \bmod p$ if λ is a hook shape.

12-80. Does the Hook-Length Formula extend to enumerate standard tableaux of skew shape? Either adapt the probabilistic proof to this situation, or find the steps in the proof that cannot be generalized.

12-81. Confirm that $\equiv_P$ and $\equiv_K$ are equivalence relations on $[N]^*$, as asserted in §12.13.

12-82. Let T be the tableau in Example 12.81. Find an explicit chain of elementary Knuth equivalences demonstrating that $\text{rw}(T)1 \equiv_K \text{rw}(T \leftarrow 1)$.

12-83. Find the length of the longest increasing and decreasing subsequences of the word $w = 41353214627311324231 42$.

12-84. Complete the proofs of Theorems 12.83 and 12.84.

12-85. For any semistandard tableau T, prove that $P(\text{rw}(T)) = T$. Show that the set of reading words of semistandard tableaux intersects every Knuth equivalence class in exactly one point.

12-86. Prove Theorem 12.89 without using the RSK algorithm.

12-87. (a) Show that each Q_N^k is a real vector space. (b) Prove: for all $f \in Q_N^k$ and $g \in Q_N^m$, $fg \in Q_N^{k+m}$. Conclude that Q_N is a subalgebra of $\mathbb{R}[x_1,\ldots,x_N]$.

12-88. List all compositions of 4, all bit strings in $\{0,1\}^3$, and all subsets of $[3]$. Show how these objects correspond under the bijections described in the text.

12-89. Assume $\alpha \in \text{Comp}_N(k)$ has s parts. (a) How many monomials appear in $M_\alpha(x_1,\ldots,x_N)$? (b) How many monomials appear in $\text{FQ}_{k,\text{sub}(\alpha)}(x_1,\ldots,x_N)$?

12-90. Prove Theorem 12.94.

12-91. Prove: for all $\lambda \in \text{Par}_N(k)$,

$$m_\lambda(x_1, \ldots, x_N) = \sum_{\substack{\alpha \in \text{Comp}_N(k): \\ \text{sort}(\alpha) = \lambda}} M_\alpha(x_1, \ldots, x_N).$$

Conclude that $\Lambda_N^k \subseteq Q_N^k$.

12-92. Let $N = k = 4$. For all $S \subseteq \{1, 2, 3\}$, write $\text{FQ}_{4,S}$ as a sum of monomials in the x-variables and as a linear combination of the M_α.

12-93. For all partitions $\lambda \in \text{Par}(5)$, write s_λ as an explicit linear combination of fundamental quasisymmetric polynomials.

12-94. For each semistandard tableau T of shape $(4, 3)$ and content $(2, 2, 2, 1)$, find the image of T under the standardization map F from the proof of Theorem 12.99.

12-95. Describe all semistandard tableaux that standardize to this standard tableau:

1	2	6
3	5	7
4	8	9

12-96. Does Theorem 12.99 extend to skew Schur polynomials $s_{\lambda/\nu}$?

12-97. For $k \geq 1$, let A_k be the matrix with rows and columns indexed by subsets of $[k-1]$, such that the entry in row S and column T is 1 if $S \subseteq T$ and 0 otherwise. Order the rows and columns of A_k by identifying a subset S with a bit string of length $k - 1$ and regarding this string as an integer written in binary. For example, when $k = 5$ and $S = \{1, 2, 4\}$, the associated bit string is 1101, which is the integer 13. (a) Find the entries in A_k for $1 \leq k \leq 4$. (b) How is A_{k+1} related to A_k? (Describe A_{k+1} in block form.) (c) Find A_k^{-1} for $1 \leq k \leq 4$. (d) How is A_{k+1}^{-1} related to A_k^{-1}?

12-98. Assume $N \geq k$ and $\alpha \in \text{Comp}(k)$. Find a formula expressing M_α as an explicit linear combination of the basis $\{\text{FQ}_{n,S} : S \subseteq [k-1]\}$.

12-99. Suppose Y is a set and $G : \mathbb{Z}_{\geq 0}^N \to Y$ is a surjective function. For each $y \in Y$, define $f_y \in \mathbb{R}[x_1, \ldots, x_N]$ to be the sum of all monomials $\mathbf{x}^\alpha$ such that $G(\alpha) = y$. (a) Prove that $\{f_y : y \in Y\}$ is linearly independent. (b) Define $f \in \mathbb{R}[x_1, \ldots, x_N]$ to be *G-symmetric* iff for all $\alpha, \beta \in \mathbb{Z}_{\geq 0}^N$ such that $G(\alpha) = G(\beta)$, the coefficients of $\mathbf{x}^\alpha$ and $\mathbf{x}^\beta$ in f are the same. Show that $\{f_y : y \in Y\}$ is a basis for the vector space of G-symmetric polynomials in N variables. (c) Explain how Theorems 9.23 and 12.94 are special cases of this exercise.

12-100. Multiplication Rule for Fundamental Quasisymmetric Polynomials. Given a word $w = w_1 w_2 \cdots w_n$ containing the symbols $1, 2, \ldots, n$ once each, and given a word $v = v_1 v_2 \cdots v_m$ containing the symbols $n+1, n+2, \ldots, n+m$ once each, let $\text{Shuf}(w, v)$ be the set of words $u = u_1 u_2 \cdots u_{n+m}$ that contain w and v as subsequences; u is called a *shuffle* of w and v. Show that

$$\text{FQ}_{n,\text{Des}(w)} \, \text{FQ}_{m,\text{Des}(v)} = \sum_{u \in \text{Shuf}(w,v)} \text{FQ}_{n+m,\text{Des}(u)} \, .$$

12-101. Show that if A is an $N \times N$ skew-symmetric matrix with N odd, then $\det(A) = 0$.

12-102. Verify by direct calculation that $\det(A) = (af + cd - be)^2$ for the 4×4 matrix A in Example 12.101.

12-103. Find the Pfaffian of a general 6×6 skew-symmetric matrix.

12-104. Count the number of perfect matchings for the graph shown in Figure 12.13.

12-105. Let G be the simple graph with $V(G) = \{1, 2, 3, 4, 5, 6\}$ and

$$E(G) = \{\{2, 3\}, \{3, 4\}, \{4, 5\}, \{2, 5\}, \{1, 2\}, \{1, 5\}, \{3, 6\}, \{4, 6\}, \{2, 4\}\}.$$

Find all perfect matchings of G. Use this to compute $\sum_{M \in \mathrm{PM}(G)} \mathrm{sgn}(M)\,\mathrm{wt}(M)$, and verify your answer by evaluating a Pfaffian.

12-106. Compute the images of the following permutations in S_N^{ev} under the bijection $g : S_N^{ev} \to \mathrm{SPf}_N^2$ used in the proof of Theorem 12.112: (a) $(3, 1, 5, 7)(2, 4, 8, 6)$; (b) $(1, 4)(2, 3)(5, 7)(6, 8)$; (c) $(2, 5, 1, 6, 8, 4, 7, 3)$; (d) $(3, 2, 1, 5, 6, 7)(4, 8)$.

12-107. Compute the images w of the following pairs $(u, v) \in \mathrm{SPf}_N^2$ under the bijection $g^{-1} : \mathrm{SPf}_N^2 \to S_N^{ev}$ used in the proof of Theorem 12.112: (a) $u = 13254768$, $v = 15283647$; (b) $u = 13254768$, $v = 12374856$; (c) $u = 15243867 = v$. In each case, confirm that the term indexed by w in $\det(A)$ equals the term indexed by (u, v) in $\mathrm{Pf}(A)^2$.

12-108. Compute the exact number of domino tilings of a 10×10 board and a 6×9 board.

12-109. How many domino tilings of an 8×8 board use: (a) 24 horizontal dominos and 8 vertical dominos; (b) 4 horizontal dominos and 28 vertical dominos?

12-110. Complete Example 12.116 by writing the full word w and showing that the placement of every new domino never causes $\mathrm{sgn}^*(M)$ to become negative.

12-111. Verify properties (a) through (d) of tensor products of matrices stated after Definition 12.117.

12-112. Prove Lemma 12.118.

12-113. Let U_k be the matrix defined in Lemma 12.118. Show that $\sqrt{2/(k+1)}U_k$ is a unitary matrix (i.e., $U^{-1} = U^*$, where U^* is the conjugate-transpose of U).

12-114. (a) Prove that, for all even $m > 0$,

$$\prod_{j=1}^{m/2} 4(u^2 + \cos^2(j\pi/(m+1))) = \frac{[u + \sqrt{1+u^2}]^{m+1} - [u - \sqrt{1+u^2}]^{m+1}}{2\sqrt{1+u^2}}.$$

(b) Deduce that $\prod_{j=1}^{m/2} 2\cos(j\pi/(m+1)) = 1$.

12-115. Show that formula (12.16) simplifies to

$$\begin{cases} 2^{mn/2} \displaystyle\prod_{j=1}^{m/2} \prod_{k=1}^{n/2} \left[x^2 \cos^2\left(\frac{j\pi}{m+1}\right) + y^2 \cos^2\left(\frac{k\pi}{n+1}\right) \right] & \text{for } n \text{ even;} \\ 2^{m(n-1)/2} x^{m/2} \displaystyle\prod_{j=1}^{m/2} \prod_{k=1}^{(n-1)/2} \left[x^2 \cos^2\left(\frac{j\pi}{m+1}\right) + y^2 \cos^2\left(\frac{k\pi}{n+1}\right) \right] & \text{for } n \text{ odd.} \end{cases}$$

Notes

§12.1. Detailed treatments of the theory of lattice paths may be found in [89, 93]. **§12.2.** The Chung–Feller Theorem was originally proved in [22]; the bijective proof given here is due to Eu, Fu, and Yeh [30]. **§12.3.** There is a growing literature on rook theory; some of the early papers on this subject are [36, 48, 49, 69]. **§12.4.** More information about parking functions may be found in [34, 38, 76, 119]. **§12.6.** Expositions of field theory may be found in [65] or Chapter 5 of [18]. An encyclopedic reference for finite fields is [78]. **§12.7.** For more on Gaussian elimination and RREF matrices, see linear algebra texts such as [62]. **§12.8.** The combinatorial interpretation of the coefficients of the tangent and secant power series is due to André [2, 4]. For more information on q-analogues of the tangent and secant series, see [6, 7, 32]. **§12.11.** Many facts about matrices and determinants, including the Cauchy–Binet Formula, appear in the matrix theory text by Lancaster [77]. The text [92] gives a thorough account of tournaments. The combinatorial derivation of the Vandermonde Determinant is due to Gessel [45]. **§12.12.** The probabilistic proof of the Hook-Length Formula is due to Greene, Nijenhuis, and Wilf [55]. **§12.13.** A discussion of Knuth equivalence and its connection to the RSK correspondence appears in [72]. Theorem 12.84 on disjoint monotone subsequences was proved by Curtis Greene [54]; this generalizes Schensted's original result [118] on the size of the longest increasing subsequence of a word. **§12.16.** Our treatment of the domino tiling formula closely follows the presentation in Kasteleyn's original paper [70].

Appendix: Definitions from Algebra

This appendix reviews some definitions from abstract algebra and linear algebra that are used in certain parts of the main text.

Rings and Fields

This section defines abstract algebraic structures called rings and fields. (Groups are studied in detail in Chapter 7.) Our main purpose for introducing these concepts is to specify the most general setting in which certain algebraic identities are true.

Definition of a Ring. A *ring* consists of a set R and two binary operations $+$ (addition) and $\cdot$ (multiplication) with domain $R \times R$, subject to the following axioms.

$\forall x, y \in R,\ x + y \in R$	(closure under addition)
$\forall x, y, z \in R,\ x + (y + z) = (x + y) + z$	(associativity of addition)
$\forall x, y \in R,\ x + y = y + x$	(commutativity of addition)
$\exists 0_R \in R, \forall x \in R, x + 0_R = x = 0_R + x$	(existence of additive identity)
$\forall x \in R, \exists {-x} \in R, x + (-x) = 0_R = (-x) + x$	(existence of additive inverses)
$\forall x, y \in R,\ x \cdot y \in R$	(closure under multiplication)
$\forall x, y, z \in R,\ x \cdot (y \cdot z) = (x \cdot y) \cdot z$	(associativity of multiplication)
$\exists 1_R \in R, \forall x \in R, x \cdot 1_R = x = 1_R \cdot x$	(existence of multiplicative identity)
$\forall x, y, z \in R,\ x \cdot (y + z) = x \cdot y + x \cdot z$	(left distributive law)
$\forall x, y, z \in R,\ (x + y) \cdot z = x \cdot z + y \cdot z$	(right distributive law)

We often write xy instead of $x \cdot y$. R is a *commutative ring* iff R satisfies the additional axiom

$$\forall x, y \in R,\ xy = yx \qquad \text{(commutativity of multiplication).}$$

For example, $\mathbb{Z}$ (the set of integers), $\mathbb{Q}$ (the set of rational numbers), $\mathbb{R}$ (the set of real numbers), and $\mathbb{C}$ (the set of complex numbers) are all commutative rings under the usual operations of addition and multiplication. For $n > 0$, the set $\mathbb{Z}_n = \{0, 1, 2, \ldots, n-1\}$ of *integers modulo* n is a commutative ring using the operations of addition and multiplication mod n. For $n > 1$, the set $M_n(\mathbb{R})$ of $n \times n$ matrices with real entries is a non-commutative ring.

Definition of an Integral Domain. An *integral domain* is a commutative ring R such that $1_R \neq 0_R$ and R has no zero divisors:

$$\forall x, y \in R, xy = 0_R \Rightarrow x = 0_R \text{ or } y = 0_R.$$

For example, $\mathbb{Z}$ is an integral domain. $\mathbb{Z}_6$ is not an integral domain, since 2 and 3 are nonzero elements of $\mathbb{Z}_6$ whose product in this ring is 0.

Definition of a Field. A *field* is a commutative ring F such that $1_F \neq 0_F$ and every nonzero element of F has a multiplicative inverse:

$$\forall x \in F, x \neq 0_F \Rightarrow \exists y \in F, xy = 1_F = yx.$$

For example, $\mathbb{Q}$, $\mathbb{R}$, and $\mathbb{C}$ are fields, but $\mathbb{Z}$ is not a field. One can show that $\mathbb{Z}_n$ is a field iff n is prime.

Let R be a ring, and suppose $x_1, x_2, \ldots, x_n \in R$. Because addition is associative, we can unambiguously write a sum like $x_1 + x_2 + x_3 + \cdots + x_n$ without parentheses. Similarly, associativity of multiplication implies that we can write the product $x_1 x_2 \cdots x_n$ without parentheses. Because addition in the ring is commutative, we can permute the summands in a sum like $x_1 + x_2 + \cdots + x_n$ without changing the answer. More formally, for any bijection $f : \{1, 2, \ldots, n\} \to \{1, 2, \ldots, n\}$ and all $x_1, \ldots, x_n \in R$, we have

$$x_{f(1)} + x_{f(2)} + \cdots + x_{f(n)} = x_1 + x_2 + \cdots + x_n.$$

It follows that if $\{x_i : i \in I\}$ is a finite indexed family of ring elements, then the sum of all these elements (denoted $\sum_{i \in I} x_i$) is well-defined. Similarly, if A is a finite subset of R, then $\sum_{x \in A} x$ is well-defined. On the other hand, the products $\prod_{i \in I} x_i$ and $\prod_{x \in A} x$ are *not* well-defined (when R is non-commutative) unless we specify in advance a total ordering on I and A.

Let F be a field with additive identity 0_F and multiplicative identity 1_F. It sometimes happens that there exist positive integers n such that $n.1_F$ (the sum of n copies of 1_F) is equal to 0_F. The *characteristic* of F is defined to be the least $n > 0$ such that $n.1_F = 0_F$; if no such n exists, the characteristic of F is zero. For example, $\mathbb{Q}$, $\mathbb{R}$, and $\mathbb{C}$ are fields of characteristic zero, whereas $\mathbb{Z}_p$ (the integers modulo p for a prime p) is a field of characteristic p. It can be shown that the characteristic of every field is either zero or a prime positive integer. When F has characteristic zero, all the field elements $n.1_F$ (for $n \in \mathbb{Z}_{>0}$) are nonzero and hence invertible in F.

Vector Spaces and Algebras

Next we recall the definitions of the fundamental algebraic structures of linear algebra.

Definition of a Vector Space. Given a field F, a *vector space over* F consists of a set V, an addition operation $+$ with domain $V \times V$, and a scalar multiplication $\cdot$ with domain $F \times V$, which must satisfy the following axioms.

$\forall x, y \in V,\ x + y \in V$	(closure under addition)
$\forall x, y, z \in V,\ x + (y + z) = (x + y) + z$	(associativity of addition)
$\forall x, y \in V,\ x + y = y + x$	(commutativity of addition)
$\exists 0_V \in V, \forall x \in V, x + 0_V = x = 0_V + x$	(existence of additive identity)
$\forall x \in V, \exists -x \in V, x + (-x) = 0_V = (-x) + x$	(existence of additive inverses)
$\forall c \in F, \forall v \in V,\ c \cdot v \in V$	(closure under scalar multiplication)
$\forall c \in F, \forall v, w \in V,\ c \cdot (v + w) = (c \cdot v) + (c \cdot w)$	(left distributive law)
$\forall c, d \in F, \forall v \in V,\ (c + d) \cdot v = (c \cdot v) + (d \cdot v)$	(right distributive law)
$\forall c, d \in F, \forall v \in V,\ (cd) \cdot v = c \cdot (d \cdot v)$	(associativity of scalar multiplication)
$\forall v \in V,\ 1 \cdot v = v$	(identity property)

When discussing vector spaces, elements of V are often called *vectors*, while elements of F

are called *scalars*. For any field F, the set $F^n = \{(x_1, \ldots, x_n) : x_i \in F\}$ is a vector space over F with operations

$$(x_1, \ldots, x_n) + (y_1, \ldots, y_n) = (x_1 + y_1, \ldots, x_n + y_n); \qquad c(x_1, \ldots, x_n) = (cx_1, \ldots, cx_n).$$

Definition of an Algebra. Given a field F, an (associative) *algebra over* F is a set A that is both a ring and a vector space over F such that the ring multiplication (denoted $v \bullet w$) and the scalar multiplication (denoted $c \cdot v$) satisfy this axiom:

$$\forall c \in F, \forall v, w \in A, c \cdot (v \bullet w) = (c \cdot v) \bullet w = v \bullet (c \cdot w).$$

For example, given any field F, let $F[x]$ be the set of all formal polynomials $a_0 + a_1 x + \cdots + a_k x^k$ where all coefficients $a_0, a_1, \ldots, a_k$ come from F. This is a commutative algebra over F (called the *one-variable polynomial ring with coefficients in* F) using the standard operations of polynomial addition, polynomial multiplication, and multiplication of a polynomial by a scalar. Some relatives of this algebra appear in the main text when we study formal power series and Laurent series. An example of a non-commutative algebra is the set $M_n(\mathbb{R})$ of $n \times n$ real-valued matrices, where $n > 1$. More generally, for any field F, the set $M_n(F)$ of $n \times n$ matrices with entries in F is an algebra over F using the usual formulas for the matrix operations.

We can also consider algebras where the field F of scalars is a replaced by any commutative ring R. For example, the set of polynomials $R[x_1, \ldots, x_N]$ is a commutative algebra over R, and the set of matrices $M_n(R)$ is a non-commutative algebra over R when $n > 1$ and $|R| > 1$.

Subgroups of groups are defined and studied in §7.5. Here we define the analogous concepts for other algebraic structures: subrings, ideals, subspaces, and subalgebras. In each case, we are looking at subsets that are *closed* under the relevant algebraic operations.

Definition of Subrings and Ideals. Let R be a ring. A *subring* of R is a subset S of R such that $0_R \in S$, $1_R \in S$, and for all $x, y \in S$, $x + y$ and $-x$ and xy are in S. An *ideal* of R is a subset I of R such that $0_R \in I$ and for all $x, y \in I$ and all $r \in R$, $x + y$ and $-x$ and rx and xr are in I.

Definition of Subspaces. Let V be a vector space over a field F. A *subspace* of V is a subset W of V such that $0_V \in W$ and for all $x, y \in W$ and all $c \in F$, $x + y$ and cx are in W.

Definition of Subalgebras and Ideals. Let A be an algebra over a field (or commutative ring) F. A *subalgebra* of A is a subset B of A such that $0_A \in B$, $1_A \in B$, and for all $x, y \in B$ and all $c \in F$, $x + y$ and $x \bullet y$ and $c \cdot x$ are in B. An *ideal* of A is a subset I of A such that $0_A \in I$ and for all $x, y \in I$, all $z \in A$, and all $c \in F$, $x + y$ and $c \cdot x$ and $z \bullet x$ and $x \bullet z$ are in I.

A *graded vector space* is a vector space V together with subspaces V_n for each integer $n \geq 0$, such that V is the direct sum $\bigoplus_{n \geq 0} V_n$. This means that for every $v \in V$, there exist unique vectors $v_n \in V_n$ such that all but finitely many v_n are zero, and $v = \sum_{n \geq 0} v_n$. A *graded algebra* is an algebra A and subalgebras A_n such that $A = \bigoplus_{n \geq 0} A_n$ as vector spaces, and whenever $n, m \geq 0$, $v \in A_n$, and $w \in A_m$, we have $v \bullet w \in A_{n+m}$. For example, the polynomial algebra $A = \mathbb{R}[x_1, \ldots, x_N]$ is a graded algebra where A_n is the subspace of all homogeneous polynomials of degree n (including zero). In §9.28, we study the algebra of symmetric functions $\Lambda = K[p_m : m \geq 0]$, where Λ_n is the subspace spanned by the power-sum symmetric functions p_μ as μ ranges over integer partitions of n.

Homomorphisms

This section defines homomorphisms for various kinds of algebraic structures. Group homomorphisms are studied in the main text (see §7.7).

Definition of a Ring Homomorphism. Given rings R and S, a *ring homomorphism* is a function $f : R \to S$ such that $f(1_R) = 1_S$ and for all $x, y \in R$, $f(x+y) = f(x) + f(y)$ and $f(x \cdot y) = f(x) \cdot f(y)$.

It automatically follows from these conditions that $f(0_R) = 0_S$, $f(nx) = nf(x)$ for all $x \in R$ and $n \in \mathbb{Z}$ (see Theorem 7.56) and $f(x^n) = f(x)^n$ for all $x \in R$ and $n \in \mathbb{Z}_{>0}$. Not all texts require that ring homomorphisms preserve the multiplicative identity.

Definition of a Vector Space Homomorphism. Suppose V and W are vector spaces over a field F. A *vector space homomorphism* (also called a *linear map* or *linear transformation*) is a function $T : V \to W$ such that for all $v, w \in V$ and all $c \in F$, $T(v + w) = T(v) + T(w)$ and $T(cv) = cT(v)$.

It automatically follows from these conditions that T preserves linear combinations: for all $v_1, \ldots, v_n \in V$ and all $c_1, \ldots, c_n \in F$, $T(c_1v_1 + \cdots + c_nv_n) = c_1T(v_1) + \cdots + c_nT(v_n)$.

Definition of an Algebra Homomorphism. Suppose A and B are algebras over a field F. An *algebra homomorphism* is a function $T : A \to B$ that is both a ring homomorphism and a vector space homomorphism.

Here is an important example of an algebra homomorphism. Let F be any field, and let $A = F[z]$ be the algebra of polynomials in the formal variable z with coefficients in F. Suppose B is any algebra over F and c is any element of B. One may check that there *exists* a *unique* algebra homomorphism $E_c : F[z] \to B$ such that $E_c(z) = c$, given explicitly by

$$E_c(a_0 + a_1z + \cdots + a_nz^n) = a_0 \cdot 1_B + a_1c + \cdots + a_nc^n \quad \text{for all } a_0, \ldots, a_n \in F.$$

The map E_c is called the *evaluation homomorphism* sending z to c. More generally, given any ordered list $c_1, \ldots, c_k$ of elements of B, there exists a unique algebra homomorphism $E : F[z_1, \ldots, z_k] \to B$ such that $E(z_i) = c_i$ for all i between 1 and k. These evaluation homomorphisms often arise in algebraic combinatorics, especially in the theory of symmetric functions.

Let A and B be algebraic structures of the same type (e.g., two rings, two vector spaces, or two algebras), and let $g : A \to B$ be a homomorphism for that type of algebraic structure. The *kernel* of g, denoted $\ker(g)$, is the set $\{x \in A : g(x) = 0_B\}$, where 0_B is the additive identity element of B. The *image* of g, denoted $\operatorname{img}(g)$, is the set $\{y \in B : \exists x \in A, y = g(x)\}$. One can show that g is one-to-one iff $\ker(g) = \{0_A\}$, and g is onto iff $\operatorname{img}(g) = B$.

If A and B are rings, then $\ker(g)$ is an ideal of A and $\operatorname{img}(g)$ is a subring of B. If A and B are vector spaces, then $\ker(g)$ is a subspace of A and $\operatorname{img}(g)$ is a subspace of B. If A and B are algebras, then $\ker(g)$ is an ideal of A and $\operatorname{img}(g)$ is a subalgebra of B. Continuing the example above, consider the evaluation homomorphism $E : F[z_1, \ldots, z_k] \to B$ such that $E(z_i) = c_i$ for i between 1 and k. The image of E is the subalgebra of B *generated* by $c_1, \ldots, c_k$, which is the set of all finite F-linear combinations of finite products of $c_1, \ldots, c_k$. The kernel of E is the set of all formal polynomials $p(z_1, \ldots, z_k)$ such that $p(c_1, \ldots, c_k) = 0$. Each $p \in \ker(E)$ represents an algebraic relation between the elements $c_1, \ldots, c_k$. If E consists of zero alone, we say that $c_1, \ldots, c_k$ is an *algebraically independent* list. (See §9.14 for more discussion.)

Linear Algebra Concepts

Here we review some concepts from linear algebra that lead to the idea of the dimension of a vector space. Throughout, let V be a fixed vector space over a field F.

Definition of Spanning Sets. A subset S of the vector space V *spans* V iff for every $v \in V$, there exists a *finite* list of vectors $v_1, \ldots, v_k \in S$ and scalars $c_1, \ldots, c_k \in F$ with $v = c_1 v_1 + \cdots + c_k v_k$. Any expression of the form $c_1 v_1 + \cdots + c_k v_k$ is called a *linear combination* of $v_1, \ldots, v_k$. A linear combination must be a *finite* sum of vectors.

Definition of Linear Independence. A finite list $(v_1, \ldots, v_k)$ of vectors in V is called *linearly dependent* iff there exist scalars $c_1, \ldots, c_k \in F$ such that $c_1 v_1 + \cdots + c_k v_k = 0_V$ and at least one c_i is not zero. Otherwise, the list $(v_1, \ldots, v_k)$ is called *linearly independent.* A set $S \subseteq V$ (possibly infinite) is *linearly dependent* iff there is a finite list of *distinct* elements of S that is linearly dependent; otherwise, S is *linearly independent.*

Definition of a Basis. A *basis* of a vector space V is a set $S \subseteq V$ that is linearly independent and spans V.

For example, define $\vec{e_i} \in F^n$ to be the vector with 1_F in position i and 0_F in all other positions. Then $\{\vec{e_1}, \ldots, \vec{e_n}\}$ is a basis for F^n. Similarly, one may check that the infinite set $S = \{1, x, x^2, x^3, \ldots, x^n, \ldots\}$ is a basis for the vector space $V = F[x]$ of polynomials in x with coefficients in F. S spans V since every polynomial must be a *finite* linear combination of powers of x. The linear independence of S follows from the definition of equality of formal polynomials: the only linear combination $c_0 1 + c_1 x + c_2 x^2 + \cdots$ that can equal the zero polynomial is the one where $c_0 = c_1 = c_2 = \cdots = 0_F$.

We now state without proof some of the fundamental facts about spanning sets, linear independence, and bases.

Linear Algebra Facts. Every vector space V over a field F has a basis (possibly infinite). Any two bases of V have the same cardinality, which is called the *dimension* of V and denoted $\dim(V)$. Given a basis of V, every $v \in V$ can be expressed in exactly one way as a linear combination of the basis elements. Any linearly independent set in V can be enlarged to a basis of V. Any spanning set for V contains a basis of V. A set $S \subseteq V$ with $|S| > \dim(V)$ must be linearly dependent. A set $T \subseteq V$ with $|T| < \dim(V)$ cannot span V.

For example, $\dim(F^n) = n$ for all $n \geq 1$, whereas the polynomial ring $F[x]$ is an infinite-dimensional vector space.

Bibliography

[1] Martin Aigner, *Combinatorial Theory*, Springer Science & Business Media, New York (2012).

[2] D. André, "Développement de $\sec x$ et $\operatorname{tg} x$," *C. R. Acad. Sci. Paris* **88** (1879), 965–967.

[3] D. André, "Solution directe du problème résolu par M. Bertrand," *C. R. Acad. Sci. Paris* **105** (1887), 436–437.

[4] D. André, "Sur les permutations alternées," *J. Math. Pures et Appl.* **7** (1881), 167–184.

[5] George Andrews, *The Theory of Partitions*, Encyclopedia of Mathematics and its Applications Series, Cambridge University Press, New York (1998).

[6] George Andrews and Dominique Foata, "Congruences for the q-secant numbers," *Europ. J. Combin.* **1** (1980), 283–297.

[7] George Andrews and Ira Gessel, "Divisibility properties of the q-tangent numbers," *Proc. Amer. Math. Soc.* **68** (1978), 380–384.

[8] Edward Bender and Donald Knuth, "Enumeration of plane partitions," *J. Combin. Theory Ser. A* **13** (1972), 40–54.

[9] Edward Bender and S. Gill Williamson, *Foundations of Combinatorics with Applications*, Dover, New York (2006).

[10] Patrick Billingsley, *Probability and Measure* (3rd ed.), John Wiley & Sons, New York (1995).

[11] Garrett Birkhoff, *Lattice Theory* (rev. ed.), American Mathematical Society, New York (1949).

[12] Kenneth Bogart, *Introductory Combinatorics*, Harcourt/Academic Press, San Diego (2000).

[13] Béla Bollobás, *Modern Graph Theory*, Graduate Texts in Mathematics **184**, Springer, New York (2013).

[14] Miklòs Bòna, *Combinatorics of Permutations*, Chapman & Hall/CRC Press (2004).

[15] Miklòs Bòna, *Introduction to Enumerative Combinatorics*, McGraw-Hill Higher Education, Boston (2007).

[16] J. A. Bondy and U. S. R. Murty, *Graph Theory*, Springer Science & Business Media, New York (2007).

[17] J. A. Bondy and U. S. R. Murty, *Graph Theory with Applications*, North Holland, New York (1976).

[18] Nicolas Bourbaki, *Elements of Mathematics: Algebra II*, Springer Science & Business Media, New York (2013).

[19] Richard Brualdi, *Introductory Combinatorics*, North-Holland, New York (1977).

[20] Peter Cameron, *Combinatorics: Topics, Techniques, Algorithms*, Cambridge University Press, Cambridge (1994).

[21] A. Cayley, "A theorem on trees," *Quart. J. Math.* **23** (1889), 376–378.

[22] K. L. Chung and K. Feller, "On fluctuations in coin-tossing," *Proc. Nat. Acad. Sci. USA* **35** (1949), 605–608.

[23] James Brown and Ruel Churchill, *Complex Variables and Applications* (9th ed.), McGraw Hill, New York (2014).

[24] Louis Comtet, *Advanced Combinatorics*, D. Reidel, Dordrecht (1974).

[25] Reinhard Diestel, *Graph Theory*, Graduate Texts in Mathematics **173**, Springer, New York (2000).

[26] David Dummit and Richard Foote, *Abstract Algebra* (3rd ed.), John Wiley & Sons, New York (2004).

[27] Richard Durrett, *Probability: Theory and Examples* (3rd ed.), Duxbury Press, Pacific Grove, CA (2004).

[28] Ö. Eğecioğlu and J. Remmel, "Bijections for Cayley trees, spanning trees, and their q-analogues," *J. Combin. Theory Ser. A* **42** (1986), 15–30.

[29] Ö. Eğecioğlu and J. Remmel, "A combinatorial interpretation of the inverse Kostka matrix," *Linear Multilinear Algebra* **26** (1990), 59–84.

[30] Sen-Peng Eu, Tung-Shan Fu, and Yeong-Nan Yeh, "Refined Chung-Feller theorems for lattice paths," *J. Combin. Theory Ser. A* **112** (2005), 143–162.

[31] Michael Fisher and H. N. V. Temperley, "Dimer problem in statistical mechanics—an exact result," *Philos. Mag.* **6** (1961), 1061–1063.

[32] Dominique Foata, "Further divisibility properties of the q-tangent numbers," *Proc. Amer. Math. Soc.* **81** (1981), 143–148.

[33] Dominique Foata, "On the Netto inversion number of a sequence," *Proc. Amer. Math. Soc.* **19** (1968), 236–240.

[34] Dominique Foata and John Riordan, "Mappings of acyclic and parking functions," *Aequationes Math.* **10** (1974), 10–22.

[35] Dominique Foata and M.-P. Schützenberger, "Major index and inversion number of permutations," *Math. Nachr.* **83** (1978), 143–159.

[36] Dominique Foata and M.-P. Schützenberger, "On the rook polynomials of Ferrers relations," in *Combinatorial Theory and its Applications II*, North-Holland, Amsterdam (1970), pp. 413–436.

[37] J. Frame, G. Robinson, and R. Thrall, "The hook graphs of the symmetric groups," *Canad. J. Math.* **6** (1954), 316–324.

[38] J. Françon, "Acyclic and parking functions," *J. Combin. Theory Ser. A* **18** (1975), 27–35.

[39] F. Franklin, "Sur le développement du produit infini $(1-x)(1-x^2)(1-x^3)\cdots$," *C. R. Acad. Sci. Paris Ser. A* **92** (1881), 448–450.

[40] D. Franzblau and Doron Zeilberger, "A bijective proof of the hook-length formula," *J. Algorithms* **3** (1982), 317–343.

[41] William Fulton, *Young Tableaux*, Cambridge University Press, Cambridge (1997).

[42] J. Fürlinger and J. Hofbauer, "q-Catalan numbers," *J. Combin. Theory Ser. A* **40** (1985), 248–264.

[43] Adriano Garsia and Stephen Milne, "Method for constructing bijections for classical partition identities," *Proc. Natl. Acad. Sci. USA* **78** (1981), 2026–2028.

[44] Adriano Garsia and Stephen Milne, "A Rogers–Ramanujan bijection," *J. Comb. Theory Ser. A* **31** (1981), 289–339.

[45] Ira Gessel, "Tournaments and Vandermonde's determinant," *J. Graph Theory* **3** (1979), 305–7.

[46] Ira Gessel and Xavier G. Viennot, "Determinants, paths, and plane partitions," preprint (1989), 36 pages. Available online at Professor Gessel's website.

[47] J. W. L. Glaisher, "A theorem in partitions," *Messenger of Math.* N.S. **12** (1883), 158–170.

[48] Jay Goldman, J. Joichi, David Reiner, and Dennis White, "Rook theory II. Boards of binomial type," *SIAM J. Appl. Math.* **31** (1976), 618–633.

[49] Jay Goldman, J. Joichi, and Dennis White, "Rook theory I. Rook equivalence of Ferrers boards," *Proc. Amer. Math. Soc.* **52** (1975), 485–492.

[50] B. Gordon, "Sieve-equivalence and explicit bijections," *J. Combin. Theory Ser. A* **34** (1983), 90–93.

[51] Henry Gould, *Combinatorial Identities; a standardized set of tables listing 500 binomial coefficient summations*, Morgantown, WV (1972).

[52] R. Gould, *Graph Theory*, Benjamin/Cummings, London (1988).

[53] Ronald Graham, Donald Knuth, and Orem Patashnik, *Concrete Mathematics: a Foundation for Computer Science*, Addison-Wesley, Reading, MA (1989).

[54] Curtis Greene, "An extension of Schensted's theorem," *Adv. in Math.* **14** (1974), 254–265.

[55] Curtis Greene, Albert Nijenhuis, and Herbert Wilf, "A probabilistic proof of a formula for the number of Young tableaux of a given shape," *Adv. in Math.* **31** (1979), 104–109.

[56] H. Gupta, "A new look at the permutation of the first n natural numbers," *Indian J. Pure Appl. Math.* **9** (1978), 600–631.

[57] James Haglund, "Conjectured statistics for the q,t-Catalan numbers," *Adv. in Math.* **175** (2003), 319–334.

[58] James Haglund, *The q, t-Catalan Numbers and the Space of Diagonal Harmonics, with an Appendix on the Combinatorics of Macdonald Polynomials*, AMS University Lecture Series (2008).

[59] Paul Halmos, *Naive Set Theory*, Springer Science & Business Media, New York (1998).

[60] Frank Harary, *Graph Theory*, Addison-Wesley, Reading, MA (1969).

[61] Paul Hoel, Sidney Port, and Charles Stone, *Introduction to Probability Theory*, Houghton Mifflin, Boston (1971).

[62] Kenneth Hoffman and Ray Kunze, *Linear Algebra* (2nd ed.), Prentice Hall, Upper Saddle River, NJ (1971).

[63] Karel Hrbacek and Thomas Jech, *Introduction to Set Theory* (3rd ed.), Chapman & Hall/CRC Press, Boca Raton, FL (1999).

[64] James Humphreys, *Reflection Groups and Coxeter Groups*, Cambridge University Press, Cambridge (1990).

[65] Thomas Hungerford, *Algebra*, Graduate Texts in Mathematics **73**, Springer, New York (1980).

[66] Nathan Jacobson, *Basic Algebra I*, Dover paperback reprint of 1985 edition (2009).

[67] G. James and A. Kerber, *The Representation Theory of the Symmetric Group*, Addison-Wesley, Reading, MA (1981).

[68] J. Joichi and D. Stanton, "Bijective proofs of basic hypergeometric series identities," *Pacific J. Math.* **127** (1987), 103–120.

[69] Irving Kaplansky and John Riordan, "The problem of the rooks and its applications," *Duke Math. J.* **13** (1946), 259–268.

[70] P. W. Kasteleyn, "The statistics of dimers on a lattice I. The number of dimer arrangements on a quadratic lattice," *Physica* **27** (1961), 1209–1225.

[71] G. Kirchhoff, "Über die Auflösung der Gleichungen, auf welche man bei der Untersuchung der linearen Verteilung galvanischer Ströme geführt wird," *Ann. Phys. Chem.* **72** (1847), 497–508.

[72] Donald Knuth, *The Art of Computer Programming*, Volume 3 (multiple fascicles), Addison-Wesley, Reading, MA (1973).

[73] Donald Knuth, *The Art of Computer Programming*, Volume 4 (multiple fascicles), Addison-Wesley, Reading, MA (2005).

[74] Donald Knuth, "Permutations, matrices, and generalized Young tableaux," *Pacific J. Math.* **34** (1970), 709–727.

[75] Wolfram Koepf, *Hypergeometric Summation*, Braunschweig/Wiesbaden: Vieweg, Heidelberg (1998).

[76] A. G. Konheim and B. Weiss, "An occupancy discipline and applications," *SIAM J. Applied Math.* **14** (1966), 17–76.

[77] Peter Lancaster, *Theory of Matrices*, Academic Press, New York (1969).

[78] Rudolf Lidl and Harald Niederreiter, *Finite fields* (2nd ed.), Cambridge University Press, New York (1997).

[79] Nicholas Loehr, "Abacus proofs of Schur function identities," *SIAM J. Discrete Math.* **24** (2010), 1356–1370.

[80] Nicholas Loehr, "The major index specialization of the q,t-Catalan," *Ars Combinatoria* **83** (2007), 145–160.

[81] Nicholas Loehr, "Successor algorithms via counting arguments," preprint (2017).

[82] Nicholas Loehr and Anthony Mendes, "Bijective matrix algebra," *Linear Algebra Appl.* **416** (2006), 917–944.

[83] Nicholas Loehr and Jeffrey Remmel, "Rook-by-rook rook theory: bijective proofs of rook and hit equivalences," *Adv. in Appl. Math.* **42** (2009), 483–503.

[84] Ian Macdonald, *Symmetric Functions and Hall Polynomials* (2nd edition), Oxford University Press, Oxford (1995).

[85] P. MacMahon, *Combinatory Analysis*, AMS Chelsea Publishing, Providence (1983).

[86] James McKay, "Another proof of Cauchy's group theorem," *Amer. Math Monthly* **66** (1959), 119.

[87] G. A. Miller, "A new proof of Sylow's theorem," *Annals of Mathematics* **16** (1915), 169–171.

[88] J. Susan Milton and Jesse Arnold, *Introduction to Probability and Statistics* (4th ed.), McGraw-Hill, Boston (2003).

[89] Sri Gopal Mohanty, *Lattice Path Counting and Applications*, Academic Press, New York (1979).

[90] J. Donald Monk, *Introduction to Set Theory*, McGraw-Hill, New York (1969).

[91] John Moon, *Counting Labeled Trees*, Canadian Mathematical Congress (1970).

[92] John Moon, *Topics on Tournaments*, Holt, Rinehart, and Winston, New York (1968).

[93] T. V. Narayana, *Lattice Path Combinatorics, with Statistical Applications*, University of Toronto Press, Toronto (1979).

[94] C. Nash-Williams, "Decomposition of finite graphs into forests," *J. London Math. Soc.* **39** (1964), 12.

[95] Albert Nijenhuis and Herbert Wilf, *Combinatorial Algorithms*, Academic Press, New York (1975).

[96] Jean-Christophe Novelli, Igor Pak, and Alexander Stoyanovskii, "A direct bijective proof of the hook-length formula," *Discrete Math. Theor. Comput. Sci.* **1** (1997), 53–67.

[97] Igor Pak, "Partition bijections, a survey," *Ramanujan J.* **12** (2006), 5–75.

[98] J. Peterson, "Beviser for Wilsons og Fermats Theoremer," *Tidsskrift for Mathematik* **2** (1872), 64–65.

[99] Marko Petkovsek, Herb Wilf, and Doron Zeilberger, $A = B$, AK Peters Ltd., Wellesley MA (1996).

[100] H. Prüfer, "Neuer Beweis eines Satzes über Permutationen," *Arch. Math. Phys.* **27** (1918), 742–744.

[101] S. Ramanujan, "Proof of certain identities in combinatory analysis," *Proc. Cambridge Philos. Soc.* **19** (1919), 214–216.

[102] G. N. Raney, "Functional composition patterns and power series reversion," *Trans. Amer. Math. Soc.* **94** (1960), 441–451.

[103] Jeffrey Remmel, "Bijective proofs of formulae for the number of standard Young tableaux," *Linear and Multilinear Algebra* **11** (1982), 45–100.

[104] Jeffrey Remmel, "Bijective proofs of some classical partition identities," *J. Combin. Theory Ser. A* **33** (1982), 273–286.

[105] Jeffrey Remmel, "A note on a recursion for the number of derangements," *European J. Combin.* **4** (1983), 371–374.

[106] Jeffrey Remmel and Mark Shimozono, "A simple proof of the Littlewood–Richardson rule and applications," *Discrete Math.* **193** (1998), 257–266.

[107] John Riordan, *An Introduction to Combinatorial Analysis*, John Wiley & Sons, New York (1958).

[108] Romeo Rizzi, "A short proof of König's Theorem," *J. Graph Th.* **33** (2000), 138–139.

[109] Fred Roberts, *Graph Theory and its Applications to Problems of Society*, SIAM, Philadelphia (1978).

[110] Fred Roberts and Barry Tesman, *Applied Combinatorics*, Prentice Hall, Upper Saddle River, NJ (2005).

[111] G. de B. Robinson, "On the representations of the symmetric group," *Amer. J. Math.* **60** (1938), 745–760.

[112] L. J. Rogers, "Second memoir on the expansion of certain infinite products," *Proceedings London Math. Soc.* **25** (1894), 318–343.

[113] Gian-Carlo Rota, "On the foundations of combinatorial theory I. Theory of Möbius functions," *Zeitschrift für Wahrscheinlichkeitstheorie* **2** (1964), 340–368.

[114] Joseph Rotman, *An Introduction to the Theory of Groups*, Graduate Texts in Mathematics **148**, Springer Science & Business Media, New York (2012).

[115] Walter Rudin, *Principles of Mathematical Analysis* (3rd ed.), McGraw-Hill, New York (1976).

[116] Bruce Sagan, "Congruences via abelian groups," *J. Number Theory* **20** (1985), 210–237.

[117] Bruce Sagan, *The Symmetric Group: Representations, Combinatorial Algorithms, and Symmetric Functions*, Graduate Texts in Mathematics **203**, Springer Science & Business Media, New York (2013).

[118] C. Schensted, "Longest increasing and decreasing subsequences," *Canad. J. Math.* **13** (1961), 179–191.

[119] M.-P. Schützenberger, "On an enumeration problem," *J. Combin. Theory* **4** (1968), 219–221.

[120] M.-P. Schützenberger, "Quelques remarques sur une construction de Schensted," *Math. Scand.* **12** (1963), 117–128.

[121] Richard Stanley, *Enumerative Combinatorics* (2 volumes), Cambridge University Press, Cambridge (1997 and 1999).

[122] Dennis Stanton and Dennis White, *Constructive Combinatorics*, Springer Science & Business Media, New York (2012).

[123] James Sylvester, "A constructive theory of partitions, arranged in three acts, an interact, and an exodion," *Amer. J. Math.* **5** (1882), 251–330.

[124] J.-P. Tignol, *Galois' Theory of Algebraic Equations*, World Scientific Publishing, Singapore (2001).

[125] Alan Tucker, *Applied Combinatorics*, John Wiley & Sons, New York (2002).

[126] W. T. Tutte, "The dissection of equilateral triangles into equilateral triangles," *Proc. Cambridge Philos. Soc.* **44** (1948), 463–482.

[127] T. van Aardenne-Ehrenfest and N. G. de Bruijn, "Circuits and trees in oriented linear graphs," *Simon Stevin* **28** (1951), 203–217.

[128] J. H. van Lint and R. M. Wilson, *A Course in Combinatorics*, Cambridge University Press, Cambridge (1992).

[129] Xavier G. Viennot, "Une forme géométrique de la correspondance de Robinson-Schensted," in *Combinatoire et Représentation du Groupe Symétrique*, Lecture Notes in Math. **579**, Springer, New York (1977), 29–58.

[130] Douglas West, *Introduction to Graph Theory* (2nd edition), Prentice Hall, Upper Saddle River, NJ (2001).

[131] H. Wielandt, "Ein Beweis für die Existenz der Sylowgruppen," *Archiv der Mathematik* **10** (1959), 401–402.

[132] Herbert Wilf, *Generatingfunctionology*, Elsevier, Amsterdam (2013).

[133] Herbert Wilf, "Sieve-equivalence in generalized partition theory," *J. Combin. Theory Ser. A* **34** (1983), 80–89.

[134] Herbert Wilf, "A unified setting for selection algorithms II. Algorithmic aspects of combinatorics," *Ann. Discrete Math.* **2** (1978), 135–148.

[135] Herbert Wilf, "A unified setting for sequencing, ranking, and selection algorithms for combinatorial objects," *Adv. in Math.* **24** (1977), 281–291.

[136] R. Wilson, *Introduction to Graph Theory*, Longman, London (1985).

Index